LINEAR SYSTEMS AND SIGNALS

THE OXFORD SERIES IN ELECTRICAL AND COMPUTER ENGINEERING

Adel S. Sedra, Series Editor

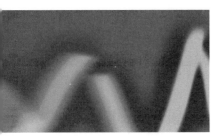

LINEAR SYSTEMS AND SIGNALS

SECOND EDITION

B. P. Lathi

New York   Oxford
OXFORD UNIVERSITY PRESS
2005

Oxford University Press

Oxford   New York
Auckland   Bangkok   Buenos Aires   Cape Town   Chennai
Dar es Salaam   Delhi   Hong Kong   Istanbul   Karachi   Kolkata
Kuala Lumpur   Madrid   Melbourne   Mexico City   Mumbai   Nairobi
São Paulo   Shanghai   Taipei   Tokyo   Toronto

Published by Oxford University Press, Inc.
198 Madison Avenue, New York, New York 10016
www.oup.com

Oxford is a registered trademark of Oxford University Press

Library of Congress Cataloging-in-Publication Data

Lathi, B. P. (Bhagwandas Pannalal)
  Linear systems and signals / B. P. Lathi.—2nd ed.
    p. cm.
  ISBN-13 978-0-19-515833-5

  1. Signal processing—Mathematics. 2. System analysis. 3. Linear time invariant systems.
  4. Digital filters (Mathematics) I. Title.

TK5102.5L298 2004
621.382'2—dc22   2003064978

Printing number: 9 8 7

Printed in the United States of America
on acid-free paper

CONTENTS

1 SIGNALS AND SYSTEMS

2 TIME-DOMAIN ANALYSIS OF CONTINUOUS-TIME SYSTEMS

3 TIME-DOMAIN ANALYSIS OF DISCRETE-TIME SYSTEMS

4 CONTINUOUS-TIME SYSTEM ANALYSIS USING THE LAPLACE TRANSFORM

5   DISCRETE-TIME SYSTEM ANALYSIS USING THE z-TRANSFORM

6 CONTINUOUS-TIME SIGNAL ANALYSIS: THE FOURIER SERIES

7 CONTINUOUS-TIME SIGNAL ANALYSIS: THE FOURIER TRANSFORM

8 SAMPLING: THE BRIDGE FROM CONTINUOUS TO DISCRETE

9 FOURIER ANALYSIS OF DISCRETE-TIME SIGNALS

10 STATE-SPACE ANALYSIS

PREFACE

This book, *Linear Systems and Signals,* presents a comprehensive treatment of signals and linear systems at an introductory level. Like all my other books, it emphasizes physical appreciation of concepts through heuristic reasoning, and the use of metaphors, analogies, and creative explanations. Such an approach is much different from a purely deductive technique that uses mere mathematical manipulation of symbols. There is a temptation to treat engineering subjects as a branch of applied mathematics. Such an approach is a perfect match to the public image of engineering as a dry and dull discipline. It ignores the physical meaning behind various derivations and deprives a student of intuitive grasp and the enjoyable experience of logical uncovering of the subject matter. Here I have used mathematics not so much to prove axiomatic theory as to support and enhance physical and intuitive understanding. Wherever possible, theoretical results are interpreted heuristically and are enhanced by carefully chosen examples and analogies.

This second edition, which closely follows the organization of the first edition, has been refined by incorporating suggestions and changes provided by various reviewers. The added topics include Bode plots, use of digital filters in an impulse-invariance method of designing analog systems, convergence of infinite series, bandpass systems, group and phase delay, and Fourier applications to communication systems. A significant and sizable addition in the area of MATLAB® (a registered trademark of The MathWorks, Inc.) has been provided by Dr. Roger Green of North Dakota State University. Dr. Green discusses his contribution at the conclusion of this preface.

ORGANIZATION

The book may be conceived as divided into five parts:

1. Introduction (Background and Chapter 1).
2. Time-domain analysis of linear time-invariant (LTI) systems (Chapters 2 and 3).
3. Frequency-domain (transform) analysis of LTI systems (Chapters 4 and 5).
4. Signal analysis (Chapters 6, 7, 8, and 9).
5. State-space analysis of LTI systems (Chapter 10).

The organization of the book permits much flexibility in teaching the continuous-time and discrete-time concepts. The natural sequence of chapters is meant to integrate continuous-time and discrete-time analysis. It is also possible to use a sequential approach in which all the continuous-time analysis is covered first (Chapters 1, 2, 4, 6, 7, and 8), followed by discrete-time analysis (Chapters 3, 5, and 9).

SUGGESTIONS FOR USING THIS BOOK

The book can be readily tailored for a variety of courses spanning 30 to 45 lecture hours. Most of the material in the first eight chapters can be covered at a brisk pace in about 45 hours. The book can also be used for a 30-lecture-hour course by covering only analog material (Chapters 1, 2, 4, 6, 7, and possibly selected topics in Chapter 8). Alternately, one can also select Chapters 1 to 5 for courses purely devoted to systems analysis or transform techniques. To treat continuous- and discrete-time systems by using an integrated (or parallel) approach, the appropriate sequence of Chapters is 1, 2, 3, 4, 5, 6, 7, and 8. For a sequential approach, where the continuous-time analysis is followed by discrete-time analysis, the proper chapter sequence is 1, 2, 4, 6, 7, 8, 3, 5, and possibly 9 (depending on the time availability).

Logically, the Fourier transform should precede the Laplace transform. I have used such an approach in the companion volume, *Signal Processing and Linear Systems* (Oxford, 1998). However, a sizable number of instructors feel that it is easier for students to learn Fourier after Laplace. Such an approach has an appeal because of the gradual progression of difficulty, in the sense that the relatively more difficult concepts of Fourier are treated after the simpler area of Laplace. This book is written to accommodate that viewpoint. For those who wish to see Fourier before Laplace, there is *Signal Processing and Linear Systems*.

NOTABLE FEATURES

The notable features of this book include the following:

1. Intuitive and heuristic understanding of the concepts and physical meaning of mathematical results are emphasized throughout. Such an approach not only leads to deeper appreciation and easier comprehension of the concepts, but also makes learning enjoyable for students.

2. Many students are handicapped by an inadequate background in basic material such as complex numbers, sinusoids, quick sketching of functions, Cramer's rule, partial fraction expansion, and matrix algebra. I have added a chapter that addresses these basic and pervasive topics in electrical engineering. Response by students has been unanimously enthusiastic.

3. There are more than 200 worked examples in addition to exercises (usually with answers) for students to test their understanding. There is also a large number of selected problems of varying difficulty at the end of each chapter.

4. For instructors who like to get students involved with computers, several examples are worked out by means of MATLAB, which is becoming a standard software package in electrical engineering curricula. There is also a MATLAB session at the end of each chapter. The problem set contains several computer problems. Working computer examples or problems, though not essential for the use of this book, is highly recommended.

5. The discrete-time and continuous-time systems may be treated in sequence, or they may be integrated by using a parallel approach.

6. The summary at the end of each chapter proves helpful to students in summing up essential developments in the chapter.

7. There are several historical notes to enhance student's interest in the subject. This information introduces students to the historical background that influenced the development of electrical engineering.

CREDITS

The portraits of Gauss (p. 3), Laplace (p. 357), Heaviside (p. 357), Fourier (p. 613), and Michelson (p. 623) have been reprinted courtesy of the Smithsonian Institution Libraries. The likenesses of Cardano (p. 3) and Gibbs (p. 623) have been reprinted courtesy of the Library of Congress. The engraving of Napoleon (p. 613) has been reprinted courtesy of Bettmann/Corbis.

ACKNOWLEDGMENTS

Several individuals have helped me in the preparation of this book. I am grateful for the helpful suggestions of the several reviewers. I am most grateful to Prof. Yannis Tsividis of Columbia University, who provided his comprehensively thorough and insightful feedback for the book. I also appreciate another comprehensive review by Prof. Roger Green. I thank Profs. Joe Anderson of Tennessee Technological University, Kai S. Yeung of the University of Texas at Arlington, and Alexander Poularikis of the University of Alabama at Huntsville for very thoughtful reviews. Thanks for helpful suggestions are also due to Profs. Babajide Familoni of the University of Memphis, Leslie Collins of Duke University, R. Rajgopalan of the University of Arizona, and William Edward Pierson from the U.S. Air Force Research Laboratory. Only those who write a book understand that writing a book such as this is an obsessively time-consuming activity, which causes much hardship for the family members, where the wife suffers the most. So what can I say except to thank my wife, Rajani, for enormous but invisible sacrifices.

B. P. Lathi

MATLAB

MATLAB is a sophisticated language that serves as a powerful tool to better understand a myriad of topics, including control theory, filter design, and, of course, linear systems and signals. MATLAB's flexible programming structure promotes rapid development and analysis. Outstanding visualization capabilities provide unique insight into system behavior and signal character. By exploring concepts with MATLAB, you will substantially increase your comfort with and understanding of course topics.

As with any language, learning MATLAB is incremental and requires practice. This book provides two levels of exposure to MATLAB. First, short computer examples are interspersed throughout the text to reinforce concepts and perform various computations. These examples utilize standard MATLAB functions as well as functions from the control system, signal processing, and symbolic math toolboxes. MATLAB has many more toolboxes available, but these three are commonly available in many engineering departments.

A second and deeper level of exposure to MATLAB is achieved by concluding each chapter with a separate MATLAB session. Taken together, these eleven sessions provide a self-contained introduction to the MATLAB environment that allows even novice users to quickly gain MATLAB proficiency and competence. These sessions provide detailed instruction on how to use MATLAB to solve problems in linear systems and signals. Except for the very last chapter, special care has been taken to avoid the use of toolbox functions in the MATLAB sessions. Rather, readers are shown the process of developing their own code. In this way, those readers without toolbox access are not at a disadvantage.

All computer code is available online (www.mathworks.com/support/books). Code for the computer examples in a given chapter, say Chapter xx, is named CExx.m. Program yy from MATLAB Session xx is named MSxxPyy.m. Additionally, complete code for each individual MATLAB session is named MSxx.m.

Roger Green

CHAPTER B

BACKGROUND

The topics discussed in this chapter are not entirely new to students taking this course. You have already studied many of these topics in earlier courses or are expected to know them from your previous training. Even so, this background material deserves a review because it is so pervasive in the area of signals and systems. Investing a little time in such a review will pay big dividends later. Furthermore, this material is useful not only for this course but also for several courses that follow. It will also be helpful later, as reference material in your professional career.

B.1 COMPLEX NUMBERS

Complex numbers are an extension of ordinary numbers and are an integral part of the modern number system. Complex numbers, particularly *imaginary numbers,* sometimes seem mysterious and unreal. This feeling of unreality derives from their unfamiliarity and novelty rather than their supposed nonexistence! Mathematicians blundered in calling these numbers "imaginary," for the term immediately prejudices perception. Had these numbers been called by some other name, they would have become demystified long ago, just as irrational numbers or negative numbers were. Many futile attempts have been made to ascribe some physical meaning to imaginary numbers. However, this effort is needless. In mathematics we assign symbols and operations any meaning we wish as long as internal consistency is maintained. The history of mathematics is full of entities that were unfamiliar and held in abhorrence until familiarity made them acceptable. This fact will become clear from the following historical note.

B.1-1 A Historical Note

Among early people the number system consisted only of natural numbers (positive integers) needed to express the number of children, cattle, and quivers of arrows. These people had no need for fractions. Whoever heard of two and one-half children or three and one-fourth cows!

However, with the advent of agriculture, people needed to measure continuously varying quantities, such as the length of a field and the weight of a quantity of butter. The number system, therefore, was extended to include fractions. The ancient Egyptians and Babylonians knew how to handle fractions, but *Pythagoras* discovered that some numbers (like the diagonal

1

of a unit square) could not be expressed as a whole number or a fraction. Pythagoras, a number mystic, who regarded numbers as the essence and principle of all things in the universe, was so appalled at his discovery that he swore his followers to secrecy and imposed a death penalty for divulging this secret.[1] These numbers, however, were included in the number system by the time of Descartes, and they are now known as *irrational numbers*.

Until recently, *negative numbers* were not a part of the number system. The concept of negative numbers must have appeared absurd to early man. However, the medieval Hindus had a clear understanding of the significance of positive and negative numbers.[2,3] They were also the first to recognize the existence of absolute negative quantities.[4] The works of *Bhaskar* (1114–1185) on arithmetic (*Līlāvatī*) and algebra (*Bījaganit*) not only use the decimal system but also give rules for dealing with negative quantities. Bhaskar recognized that positive numbers have two square roots.[5] Much later, in Europe, the men who developed the banking system that arose in Florence and Venice during the late Renaissance (fifteenth century) are credited with introducing a crude form of negative numbers. The seemingly absurd subtraction of 7 from 5 seemed reasonable when bankers began to allow their clients to draw seven gold ducats while their deposit stood at five. All that was necessary for this purpose was to write the difference, 2, on the debit side of a ledger.[6]

Thus the number system was once again broadened (generalized) to include negative numbers. The acceptance of negative numbers made it possible to solve equations such as $x + 5 = 0$, which had no solution before. Yet for equations such as $x^2 + 1 = 0$, leading to $x^2 = -1$, the solution could not be found in the real number system. It was therefore necessary to define a completely new kind of number with its square equal to -1. During the time of Descartes and Newton, imaginary (or complex) numbers came to be accepted as part of the number system, but they were still regarded as algebraic fiction. The Swiss mathematician *Leonhard Euler* introduced the notation i (for *imaginary*) around 1777 to represent $\sqrt{-1}$. Electrical engineers use the notation j instead of i to avoid confusion with the notation i often used for electrical current. Thus

$$j^2 = -1$$

and

$$\sqrt{-1} = \pm j$$

This notation allows us to determine the square root of any negative number. For example,

$$\sqrt{-4} = \sqrt{4} \times \sqrt{-1} = \pm 2j$$

When imaginary numbers are included in the number system, the resulting numbers are called *complex numbers*.

ORIGINS OF COMPLEX NUMBERS

Ironically (and contrary to popular belief), it was not the solution of a quadratic equation, such as $x^2 + 1 = 0$, but a cubic equation with real roots that made imaginary numbers plausible and acceptable to early mathematicians. They could dismiss $\sqrt{-1}$ as pure nonsense when it appeared as a solution to $x^2 + 1 = 0$ because this equation has no real solution. But in 1545, *Gerolamo Cardano* of Milan published *Ars Magna* (*The Great Art*), the most important algebraic work of the Renaissance. In this book he gave a method of solving a general cubic equation in which

Gerolamo Cardano

Karl Friedrich Gauss

a root of a negative number appeared in an intermediate step. According to his method, the solution to a third-order equation[†]

$$x^3 + ax + b = 0$$

is given by

$$x = \sqrt[3]{-\frac{b}{2} + \sqrt{\frac{b^2}{4} + \frac{a^3}{27}}} + \sqrt[3]{-\frac{b}{2} - \sqrt{\frac{b^2}{4} + \frac{a^3}{27}}}$$

For example, to find a solution of $x^3 + 6x - 20 = 0$, we substitute $a = 6, b = -20$ in the foregoing equation to obtain

$$x = \sqrt[3]{10 + \sqrt{108}} + \sqrt[3]{10 - \sqrt{108}} = \sqrt[3]{20.392} - \sqrt[3]{0.392} = 2$$

We can readily verify that 2 is indeed a solution of $x^3 + 6x - 20 = 0$. But when Cardano tried to solve the equation $x^3 - 15x - 4 = 0$ by this formula, his solution was

$$x = \sqrt[3]{2 + \sqrt{-121}} + \sqrt[3]{2 - \sqrt{-121}}$$

[†]This equation is known as the *depressed cubic* equation. A general cubic equation

$$y^3 + py^2 + qy + r = 0$$

can always be reduced to a depressed cubic form by substituting $y = x - (p/3)$. Therefore any general cubic equation can be solved if we know the solution to the depressed cubic. The depressed cubic was independently solved, first by *Scipione del Ferro* (1465–1526) and then by *Niccolo Fontana* (1499–1557). The latter is better known in the history of mathematics as *Tartaglia* ("Stammerer"). Cardano learned the secret of the depressed cubic solution from Tartaglia. He then showed that by using the substitution $y = x - (p/3)$, a general cubic is reduced to a depressed cubic.

What was Cardano to make of this equation in the year 1545? In those days negative numbers were themselves suspect, and a square root of a negative number was doubly preposterous! Today we know that

$$(2 \pm j)^3 = 2 \pm j11 = 2 \pm \sqrt{-121}$$

Therefore, Cardano's formula gives

$$x = (2 + j) + (2 - j) = 4$$

We can readily verify that $x = 4$ is indeed a solution of $x^3 - 15x - 4 = 0$. Cardano tried to explain halfheartedly the presence of $\sqrt{-121}$ but ultimately dismissed the whole enterprize as being "as subtle as it is useless." A generation later, however, *Raphael Bombelli* (1526–1573), after examining Cardano's results, proposed acceptance of imaginary numbers as a necessary vehicle that would transport the mathematician from the *real* cubic equation to its *real* solution. In other words, although we begin and end with real numbers, we seem compelled to move into an unfamiliar world of imaginaries to complete our journey. To mathematicians of the day, this proposal seemed incredibly strange.[7] Yet they could not dismiss the idea of imaginary numbers so easily because this concept yielded the real solution of an equation. It took two more centuries for the full importance of complex numbers to become evident in the works of Euler, Gauss, and Cauchy. Still, Bombelli deserves credit for recognizing that such numbers have a role to play in algebra.[7]

In 1799 the German mathematician *Karl Friedrich Gauss,* at the ripe age of 22, proved the fundamental theorem of algebra, namely that every algebraic equation in one unknown has a root in the form of a complex number. He showed that every equation of the nth order has exactly n solutions (roots), no more and no less. Gauss was also one of the first to give a coherent account of complex numbers and to interpret them as points in a complex plane. It is he who introduced the term "*complex numbers*" and paved the way for their general and systematic use. The number system was once again broadened or generalized to include imaginary numbers. Ordinary (or real) numbers became a special case of generalized (or complex) numbers.

The utility of complex numbers can be understood readily by an analogy with two neighboring countries X and Y, as illustrated in Fig. B.1. If we want to travel from City a to City b

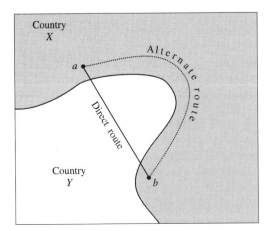

Figure B.1 Use of complex numbers can reduce the work.

(both in Country X), the shortest route is through Country Y, although the journey begins and ends in Country X. We may, if we desire, perform this journey by an alternate route that lies exclusively in X, but this alternate route is longer. In mathematics we have a similar situation with real numbers (Country X) and complex numbers (Country Y). All real-world problems must start with real numbers, and all the final results must also be in real numbers. But the derivation of results is considerably simplified by using complex numbers as an intermediary. It is also possible to solve any real-world problem by an alternate method, using real numbers exclusively, but such procedures would increase the work needlessly.

B.1-2 Algebra of Complex Numbers

A complex number (a, b) or $a + jb$ can be represented graphically by a point whose Cartesian coordinates are (a, b) in a complex plane (Fig. B.2). Let us denote this complex number by z so that

$$z = a + jb \tag{B.1}$$

The numbers a and b (the abscissa and the ordinate) of z are the *real part* and the *imaginary part*, respectively, of z. They are also expressed as

$$\text{Re } z = a$$

$$\text{Im } z = b$$

Note that in this plane all real numbers lie on the horizontal axis, and all imaginary numbers lie on the vertical axis.

Complex numbers may also be expressed in terms of polar coordinates. If (r, θ) are the polar coordinates of a point $z = a + jb$ (see Fig. B.2), then

$$a = r \cos \theta$$

$$b = r \sin \theta$$

and

$$z = a + jb = r \cos \theta + jr \sin \theta$$
$$= r(\cos \theta + j \sin \theta) \tag{B.2}$$

The *Euler formula* states that

$$e^{j\theta} = \cos \theta + j \sin \theta$$

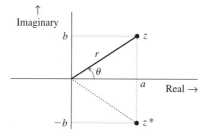

Figure B.2 Representation of a number in the complex plane.

To prove the Euler formula, we use a Maclaurin series to expand $e^{j\theta}$, $\cos\theta$, and $\sin\theta$:

$$e^{j\theta} = 1 + j\theta + \frac{(j\theta)^2}{2!} + \frac{(j\theta)^3}{3!} + \frac{(j\theta)^4}{4!} + \frac{(j\theta)^5}{5!} + \frac{(j\theta)^6}{6!} + \cdots$$

$$= 1 + j\theta - \frac{\theta^2}{2!} - j\frac{\theta^3}{3!} + \frac{\theta^4}{4!} + j\frac{\theta^5}{5!} - \frac{\theta^6}{6!} - \cdots$$

$$\cos\theta = 1 - \frac{\theta^2}{2!} + \frac{\theta^4}{4!} - \frac{\theta^6}{6!} + \frac{\theta^8}{8!} \cdots$$

$$\sin\theta = \theta - \frac{\theta^3}{3!} + \frac{\theta^5}{5!} - \frac{\theta^7}{7!} + \cdots$$

Hence, it follows that[†]

$$e^{j\theta} = \cos\theta + j\sin\theta \tag{B.3}$$

Using Eq. (B.3) in (B.2) yields

$$z = a + jb$$
$$= re^{j\theta} \tag{B.4}$$

Thus, a complex number can be expressed in Cartesian form $a + jb$ or polar form $re^{j\theta}$ with

$$a = r\cos\theta, \qquad b = r\sin\theta \tag{B.5}$$

and

$$r = \sqrt{a^2 + b^2}, \qquad \theta = \tan^{-1}\left(\frac{b}{a}\right) \tag{B.6}$$

Observe that r is the distance of the point z from the origin. For this reason, r is also called the *magnitude* (or *absolute value*) of z and is denoted by $|z|$. Similarly, θ is called the angle of z and is denoted by $\angle z$. Therefore

$$|z| = r, \qquad \angle z = \theta$$

and

$$z = |z|e^{j\angle z} \tag{B.7}$$

Also

$$\frac{1}{z} = \frac{1}{re^{j\theta}} = \frac{1}{r}e^{-j\theta} = \frac{1}{|z|}e^{-j\angle z} \tag{B.8}$$

[†]It can be shown that when we impose the following three desirable properties on an exponential e^z where $z = x + jy$, we come to conclusion that $e^{jy} = \cos y + j\sin y$ (Euler equation). These properties are

1. e^z is a single-valued and analytic function of z.
2. $de^z/dz = e^z$
3. e^z reduces to e^x if $y = 0$.

CONJUGATE OF A COMPLEX NUMBER

We define z^*, the *conjugate* of $z = a + jb$, as

$$z^* = a - jb = re^{-j\theta} \tag{B.9a}$$

$$= |z|e^{-j\angle z} \tag{B.9b}$$

The graphical representation of a number z and its conjugate z^* is depicted in Fig. B.2. Observe that z^* is a mirror image of z about the horizontal axis. *To find the conjugate of any number, we need only replace j by $-j$ in that number* (which is the same as changing the sign of its angle).

The sum of a complex number and its conjugate is a real number equal to twice the real part of the number:

$$z + z^* = (a + jb) + (a - jb) = 2a = 2\,\mathrm{Re}\,z \tag{B.10a}$$

The product of a complex number z and its conjugate is a real number $|z|^2$, the square of the magnitude of the number:

$$zz^* = (a + jb)(a - jb) = a^2 + b^2 = |z|^2 \tag{B.10b}$$

UNDERSTANDING SOME USEFUL IDENTITIES

In a complex plane, $re^{j\theta}$ represents a point at a distance r from the origin and at an angle θ with the horizontal axis, as shown in Fig. B.3a. For example, the number -1 is at a unit distance from the origin and has an angle π or $-\pi$ (in fact, any odd multiple of $\pm\pi$), as seen from Fig. B.3b. Therefore,

$$1e^{\pm j\pi} = -1$$

In fact,

$$e^{\pm jn\pi} = -1 \qquad n \text{ odd integer} \tag{B.11}$$

The number 1, on the other hand, is also at a unit distance from the origin, but has an angle 2π (in fact, $\pm 2n\pi$ for any integer value of n). Therefore,

$$e^{\pm j2n\pi} = 1 \qquad n \text{ integer} \tag{B.12}$$

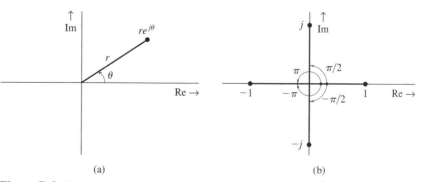

(a)	(b)

Figure B.3 Understanding some useful identities in terms of $re^{j\theta}$.

The number j is at unit distance from the origin and its angle is $\pi/2$ (see Fig. B.3b). Therefore,

$$e^{j\pi/2} = j$$

Similarly,

$$e^{-j\pi/2} = -j$$

Thus

$$e^{\pm j\pi/2} = \pm j \qquad \text{(B.13a)}$$

In fact,

$$e^{\pm jn\pi/2} = \begin{cases} \pm j & n = 1, 5, 9, 13, \ldots \\ \mp j & n = 3, 7, 11, 15, \ldots \end{cases} \qquad \text{(B.13b)}$$

This discussion shows the usefulness of the graphic picture of $re^{j\theta}$. This picture is also helpful in several other applications. For example, to determine the limit of $e^{(\alpha+j\omega)t}$ as $t \to \infty$, we note that

$$e^{(\alpha+j\omega)t} = e^{\alpha t} e^{j\omega t}$$

Now the magnitude of $e^{j\omega t}$ is unity regardless of the value of ω or t because $e^{j\omega t} = re^{j\theta}$ with $r = 1$. Therefore, $e^{\alpha t}$ determines the behavior of $e^{(\alpha+j\omega)t}$ as $t \to \infty$ and

$$\lim_{t\to\infty} e^{(\alpha+j\omega)t} = \lim_{t\to\infty} e^{\alpha t} e^{j\omega t} = \begin{cases} 0 & \alpha < 0 \\ \infty & \alpha > 0 \end{cases} \qquad \text{(B.14)}$$

In future discussions you will find it very useful to remember $re^{j\theta}$ as a number at a distance r from the origin and at an angle θ with the horizontal axis of the complex plane.

A WARNING ABOUT USING ELECTRONIC CALCULATORS IN COMPUTING ANGLES

From the Cartesian form $a + jb$ we can readily compute the polar form $re^{j\theta}$ [see Eq. (B.6)]. Electronic calculators provide ready conversion of rectangular into polar and vice versa. However, if a calculator computes an angle of a complex number by using an inverse trigonometric function $\theta = \tan^{-1}(b/a)$, proper attention must be paid to the quadrant in which the number is located. For instance, θ corresponding to the number $-2 - j3$ is $\tan^{-1}(-3/-2)$. This result is not the same as $\tan^{-1}(3/2)$. The former is $-123.7°$, whereas the latter is $56.3°$. An electronic calculator cannot make this distinction and can give a correct answer only for angles in the first and fourth quadrants. It will read $\tan^{-1}(-3/-2)$ as $\tan^{-1}(3/2)$, which is clearly wrong. When you are computing inverse trigonometric functions, if the angle appears in the second or third quadrant, the answer of the calculator is off by $180°$. The correct answer is obtained by adding or subtracting $180°$ to the value found with the calculator (either adding or subtracting yields the correct answer). For this reason it is advisable to draw the point in the complex plane and determine the quadrant in which the point lies. This issue will be clarified by the following examples.

EXAMPLE B.1

Express the following numbers in polar form:

    (a) $2 + j3$

    (b) $-2 + j1$

    (c) $-2 - j3$

    (d) $1 - j3$

    (a)

$$|z| = \sqrt{2^2 + 3^2} = \sqrt{13} \qquad \angle z = \tan^{-1}\left(\tfrac{3}{2}\right) = 56.3°$$

In this case the number is in the first quadrant, and a calculator will give the correct value of 56.3°. Therefore (see Fig. B.4a), we can write

$$2 + j3 = \sqrt{13}\, e^{j56.3°}$$

    (b)

$$|z| = \sqrt{(-2)^2 + 1^2} = \sqrt{5} \qquad \angle z = \tan^{-1}\left(\tfrac{1}{-2}\right) = 153.4°$$

In this case the angle is in the second quadrant (see Fig. B.4b), and therefore the answer given by the calculator, $\tan^{-1}(1/-2) = -26.6°$, is off by 180°. The correct answer is $(-26.6 \pm 180)° = 153.4°$ or $-206.6°$. Both values are correct because they represent the same angle.

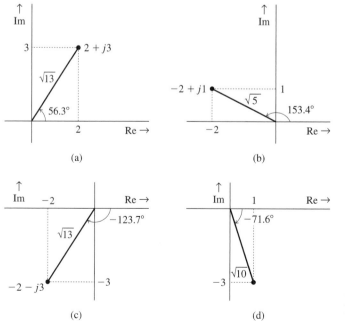

    (a)   (b)

    (c)   (d)

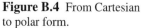

Figure B.4 From Cartesian to polar form.

It is a common practice to choose an angle whose numerical value is less than $180°$. Such a value is called the *principal value* of the angle, which in this case is $153.4°$. Therefore,

$$-2 + j1 = \sqrt{5}e^{j153.4°}$$

(c)

$$|z| = \sqrt{(-2)^2 + (-3)^2} = \sqrt{13} \qquad \angle z = \tan^{-1}\left(\frac{-3}{-2}\right) = -123.7°$$

In this case the angle appears in the third quadrant (see Fig. B.4c), and therefore the answer obtained by the calculator ($\tan^{-1}(-3/-2) = 56.3°$) is off by $180°$. The correct answer is $(56.3 \pm 180)° = 236.3°$ or $-123.7°$. We choose the principal value $-123.7°$ so that (see Fig. B.4c).

$$-2 - j3 = \sqrt{13}e^{-j123.7°}$$

(d)

$$|z| = \sqrt{1^2 + (-3)^2} = \sqrt{10} \qquad \angle z = \tan^{-1}\left(\frac{-3}{1}\right) = -71.6°$$

In this case the angle appears in the fourth quadrant (see Fig. B.4d), and therefore the answer given by the calculator, $\tan^{-1}(-3/1) = -71.6°$, is correct (see Fig. B.4d):

$$1 - j3 = \sqrt{10}e^{-j71.6°}$$

COMPUTER EXAMPLE CB.1

Using the MATLAB function `cart2pol`, convert the following numbers from Cartesian form to polar form:

    (a) $z = 2 + j3$

    (b) $z = -2 + j1$

(a)

```
>> [z_rad,z_mag] = cart2pol(2,3);
>> z_deg = z_rad*(180/pi);
>> disp(['(a) z_mag = ',num2str(z_mag),'; z_rad = ',num2str(z_rad),...
>>      '; z_deg = ',num2str(z_deg)]);
(a) z_mag = 3.6056; z_rad = 0.98279; z_deg = 56.3099
```

Therefore, $z = 2 + j3 = 3.6056e^{j0.98279} = 3.6056e^{j56.3099°}$.

(b)

```
>> [z_rad,z_mag] = cart2pol(-2,1);
>> z_deg = z_rad*(180/pi);
>> disp(['(b) z_mag = ',num2str(z_mag),'; z_rad = ',num2str(z_rad),...
>>      '; z_deg = ',num2str(z_deg)]);
(b) z_mag = 2.2361; z_rad = 2.6779; z_deg = 153.4349
```

Therefore, $z = -2 + j1 = 2.2361e^{j2.6779} = 2.2361e^{j153.4349°}$.

EXAMPLE B.2

Represent the following numbers in the complex plane and express them in Cartesian form:

(a) $2e^{j\pi/3}$

(b) $4e^{-j3\pi/4}$

(c) $2e^{j\pi/2}$

(d) $3e^{-j3\pi}$

(e) $2e^{j4\pi}$

(f) $2e^{-j4\pi}$

(a) $2e^{j\pi/3} = 2(\cos \pi/3 + j \sin \pi/3) = 1 + j\sqrt{3}$ (see Fig. B.5a)

(b) $4e^{-j3\pi/4} = 4(\cos 3\pi/4 - j \sin 3\pi/4) = -2\sqrt{2} - j2\sqrt{2}$ (see Fig. B.5b)

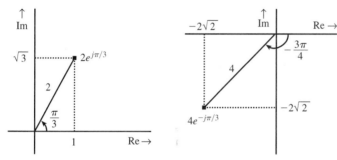

(a)

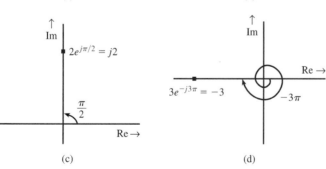

(b)   (c)   (d)

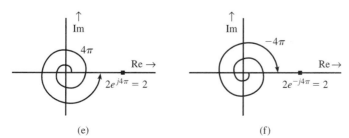

(e)   (f)

Figure B.5 From polar to Cartesian form.

(c) $2e^{j\pi/2} = 2(\cos \pi/2 + j \sin \pi/2) = 2(0 + j1) = j2$ (see Fig. B.5c)

(d) $3e^{-j3\pi} = 3(\cos 3\pi - j \sin 3\pi) = 3(-1 + j0) = -3$ (see Fig. B.5d)

(e) $2e^{j4\pi} = 2(\cos 4\pi + j \sin 4\pi) = 2(1 + j0) = 2$ (see Fig. B.5e)

(f) $2e^{-j4\pi} = 2(\cos 4\pi - j \sin 4\pi) = 2(1 - j0) = 2$ (see Fig. B.5f)

COMPUTER EXAMPLE CB.2

Using the MATLAB function `pol2cart`, convert the number $z = 4e^{-j(3\pi/4)}$ from polar form to Cartesian form.

```
>> [z_real,z_imag] = pol2cart(-3*pi/4,4);
>> disp(['z_real = ',num2str(z_real),'; z_imag = ',num2str(z_imag)]);
z_real = -2.8284; z_imag = -2.8284
```

Therefore, $z = 4e^{-j(3\pi/4)} = -2.8284 - j2.8284$.

ARITHMETICAL OPERATIONS, POWERS, AND ROOTS OF COMPLEX NUMBERS

To perform addition and subtraction, complex numbers should be expressed in Cartesian form. Thus, if

$$z_1 = 3 + j4 = 5e^{j53.1°}$$

and

$$z_2 = 2 + j3 = \sqrt{13}e^{j56.3°}$$

then

$$z_1 + z_2 = (3 + j4) + (2 + j3) = 5 + j7$$

If z_1 and z_2 are given in polar form, we would need to convert them into Cartesian form for the purpose of adding (or subtracting). Multiplication and division, however, can be carried out in either Cartesian or polar form, although the latter proves to be much more convenient. This is because if z_1 and z_2 are expressed in polar form as

$$z_1 = r_1 e^{j\theta_1} \qquad \text{and} \qquad z_2 = r_2 e^{j\theta_2}$$

then

$$z_1 z_2 = (r_1 e^{j\theta_1})(r_2 e^{j\theta_2}) = r_1 r_2 e^{j(\theta_1 + \theta_2)} \tag{B.15a}$$

and

$$\frac{z_1}{z_2} = \frac{r_1 e^{j\theta_1}}{r_2 e^{j\theta_2}} = \frac{r_1}{r_2} e^{j(\theta_1 - \theta_2)} \tag{B.15b}$$

Moreover,

$$z^n = (re^{j\theta})^n = r^n e^{jn\theta} \tag{B.15c}$$

and

$$z^{1/n} = (re^{j\theta})^{1/n} = r^{1/n}e^{j\theta/n} \tag{B.15d}$$

This shows that the operations of multiplication, division, powers, and roots can be carried out with remarkable ease when the numbers are in polar form.

Strictly speaking, there are n values for $z^{1/n}$ (the nth root of z). To find all the n roots, we reexamine Eq. (B.15d).

$$z^{1/n} = [re^{j\theta}]^{1/n} = \left[re^{j(\theta+2\pi k)}\right]^{1/n} = r^{1/n}e^{j(\theta+2\pi k)/n} \qquad k = 0, 1, 2, \ldots, n-1 \tag{B.15e}$$

The value of $z^{1/n}$ given in Eq. (B.15d) is the *principal value* of $z^{1/n}$, obtained by taking the nth root of the principal value of z, which corresponds to the case $k = 0$ in Eq. (B.15e).

EXAMPLE B.3

Determine $z_1 z_2$ and z_1/z_2 for the numbers

$$z_1 = 3 + j4 = 5e^{j53.1°}$$

$$z_2 = 2 + j3 = \sqrt{13}e^{j56.3°}$$

We shall solve this problem in both polar and Cartesian forms.

MULTIPLICATION: CARTESIAN FORM

$$z_1 z_2 = (3 + j4)(2 + j3) = (6 - 12) + j(8 + 9) = -6 + j17$$

MULTIPLICATION: POLAR FORM

$$z_1 z_2 = (5e^{j53.1°})(\sqrt{13}e^{j56.3°}) = 5\sqrt{13}e^{j109.4°}$$

DIVISION: CARTESIAN FORM

$$\frac{z_1}{z_2} = \frac{3 + j4}{2 + j3}$$

To eliminate the complex number in the denominator, we multiply both the numerator and the denominator of the right-hand side by $2 - j3$, the denominator's conjugate. This yields

$$\frac{z_1}{z_2} = \frac{(3 + j4)(2 - j3)}{(2 + j3)(2 - j3)} = \frac{18 - j1}{2^2 + 3^2} = \frac{18 - j1}{13} = \frac{18}{13} - j\frac{1}{13}$$

DIVISION: POLAR FORM

$$\frac{z_1}{z_2} = \frac{5e^{j53.1°}}{\sqrt{13}e^{j56.3°}} = \frac{5}{\sqrt{13}}e^{j(53.1°-56.3°)} = \frac{5}{\sqrt{13}}e^{-j3.2°}$$

It is clear from this example that multiplication and division are easier to accomplish in polar form than in Cartesian form.

EXAMPLE B.4

For $z_1 = 2e^{j\pi/4}$ and $z_2 = 8e^{j\pi/3}$, find the following:

(a) $2z_1 - z_2$

(b) $1/z_1$

(c) z_1/z_2^2

(d) $\sqrt[3]{z_2}$

(a) Since subtraction cannot be performed directly in polar form, we convert z_1 and z_2 to Cartesian form:

$$z_1 = 2e^{j\pi/4} = 2\left(\cos\frac{\pi}{4} + j\sin\frac{\pi}{4}\right) = \sqrt{2} + j\sqrt{2}$$

$$z_2 = 8e^{j\pi/3} = 8\left(\cos\frac{\pi}{3} + j\sin\frac{\pi}{3}\right) = 4 + j4\sqrt{3}$$

Therefore,

$$2z_1 - z_2 = 2(\sqrt{2} + j\sqrt{2}) - (4 + j4\sqrt{3})$$
$$= (2\sqrt{2} - 4) + j(2\sqrt{2} - 4\sqrt{3})$$
$$= -1.17 - j4.1$$

(b)

$$\frac{1}{z_1} = \frac{1}{2e^{j\pi/4}} = \frac{1}{2}e^{-j\pi/4}$$

(c)

$$\frac{z_1}{z_2^2} = \frac{2e^{j\pi/4}}{(8e^{j\pi/3})^2} = \frac{2e^{j\pi/4}}{64e^{j2\pi/3}} = \frac{1}{32}e^{j(\pi/4 - 2\pi/3)} = \frac{1}{32}e^{-j(5\pi/12)}$$

(d) There are three cube roots of $8e^{j(\pi/3)} = 8e^{j(\pi/3 + 2\pi k)}$, $k = 0, 1, 2$.

$$\sqrt[3]{z_2} = z_2^{1/3} = \left[8e^{j(\pi/3 + 2\pi k)}\right]^{1/3} = 8^{1/3}\left(e^{j[(6\pi k + \pi)/3]}\right)^{1/3} = 2e^{j\pi/9}, 2e^{j7\pi/9}, 2e^{j13\pi/9}$$

The principal value (value corresponding to $k = 0$) is $2e^{j\pi/9}$.

COMPUTER EXAMPLE CB.3

Determine $z_1 z_2$ and z_1/z_2 if $z_1 = 3 + j4$ and $z_2 = 2 + j3$

```
>> z_1 = 3+j*4; z_2 = 2+j*3;
>> z_1z_2 = z_1*z_2;
>> z_1divz_2 = z_1/z_2;
>> disp(['z_1*z_2 = ',num2str(z_1z_2),'; z_1/z_2 = ',num2str(z_1divz_2)]);
z_1*z_2 = -6+17i; z_1/z_2 = 1.3846-0.076923i
```

Therefore, $z_1 z_2 = (3 + j4)(2 + j3) = -6 + j17$ and $z_1/z_2 = (3 + j4)/(2 + j3) = 1.3486 - j0.076923$.

EXAMPLE B.5

Consider $X(\omega)$, a complex function of a real variable ω:

$$X(\omega) = \frac{2 + j\omega}{3 + j4\omega}$$

(a) Express $X(\omega)$ in Cartesian form, and find its real and imaginary parts.

(b) Express $X(\omega)$ in polar form, and find its magnitude $|X(\omega)|$ and angle $\angle X(\omega)$.

(a) To obtain the real and imaginary parts of $X(\omega)$, we must eliminate imaginary terms in the denominator of $X(\omega)$. This is readily done by multiplying both the numerator and the denominator of $X(\omega)$ by $3 - j4\omega$, the conjugate of the denominator $3 + j4\omega$ so that

$$X(\omega) = \frac{(2 + j\omega)(3 - j4\omega)}{(3 + j4\omega)(3 - j4\omega)} = \frac{(6 + 4\omega^2) - j5\omega}{9 + 16\omega^2} = \frac{6 + 4\omega^2}{9 + 16\omega^2} - j\frac{5\omega}{9 + \omega^2}$$

This is the Cartesian form of $X(\omega)$. Clearly the real and imaginary parts $X_r(\omega)$ and $X_i(\omega)$ are given by

$$X_r(\omega) = \frac{6 + 4\omega^2}{9 + 16\omega^2} \quad \text{and} \quad X_i(\omega) = \frac{-5\omega}{9 + 16\omega^2}$$

(b)

$$X(\omega) = \frac{2 + j\omega}{3 + j4\omega} = \frac{\sqrt{4 + \omega^2}\, e^{j\tan^{-1}(\omega/2)}}{\sqrt{9 + 16\omega^2}\, e^{j\tan^{-1}(4\omega/3)}}$$

$$= \sqrt{\frac{4 + \omega^2}{9 + 16\omega^2}}\, e^{j[\tan^{-1}(\omega/2) - \tan^{-1}(4\omega/3)]}$$

This is the polar representation of $X(\omega)$. Observe that

$$|X(\omega)| = \sqrt{\frac{4 + \omega^2}{9 + 16\omega^2}} \qquad \angle X(\omega) = \tan^{-1}\left(\frac{\omega}{2}\right) - \tan^{-1}\left(\frac{4\omega}{3}\right)$$

LOGARITHMS OF COMPLEX NUMBERS

We have

$$\log(z_1 z_2) = \log z_1 + \log z_2 \tag{B.16a}$$

$$\log(z_1/z_2) = \log z_1 - \log z_2 \tag{B.16b}$$

$$a^{(z_1 + z_2)} = a^{z_1} \times a^{z_2} \tag{B.16c}$$

$$z^c = e^{c \ln z} \tag{B.16d}$$

$$a^z = e^{z \ln a} \tag{B.16e}$$

If

$$z = re^{j\theta} = re^{j(\theta \pm 2\pi k)} \qquad k = 0, 1, 2, 3, \dots$$

then

$$\ln z = \ln\left(re^{j(\theta \pm 2\pi k)}\right) = \ln r \pm j(\theta + 2\pi k) \qquad k = 0, 1, 2, 3, \dots \tag{B.16f}$$

The value of $\ln z$ for $k = 0$ is called the *principal value* of $\ln z$ and is denoted by $\operatorname{Ln} z$.

$$\ln 1 = \ln(1e^{\pm j2\pi k}) = \pm j2\pi k \qquad k = 0, 1, 2, 3, \dots \tag{B.17a}$$

$$\ln(-1) = \ln[1e^{\pm j\pi(2k+1)}] = \pm j(2k+1)\pi \qquad k = 0, 1, 2, 3, \dots \tag{B.17b}$$

$$\ln j = \ln\left(e^{j\pi(1 \pm 4k)/2}\right) = j\frac{\pi(1 \pm 4k)}{2} \qquad k = 0, 1, 2, 3, \dots \tag{B.17c}$$

$$j^j = e^{j \ln j} = e^{-\pi(1 \pm 4k)/2} \qquad k = 0, 1, 2, 3, \dots \tag{B.17d}$$

In all these expressions, the case of $k = 0$ is the principal value of that expression.

B.2 SINUSOIDS

Consider the sinusoid

$$x(t) = C \cos(2\pi f_0 t + \theta) \tag{B.18}$$

We know that

$$\cos \varphi = \cos(\varphi + 2n\pi) \qquad n = 0, \pm 1, \pm 2, \pm 3, \dots$$

Therefore, $\cos \varphi$ repeats itself for every change of 2π in the angle φ. For the sinusoid in Eq. (B.18), the angle $2\pi f_0 t + \theta$ changes by 2π when t changes by $1/f_0$. Clearly, this sinusoid repeats every $1/f_0$ seconds. As a result, there are f_0 repetitions per second. This is the *frequency* of the sinusoid, and the repetition interval T_0 given by

$$T_0 = \frac{1}{f_0} \tag{B.19}$$

is the *period*. For the sinusoid in Eq. (B.18), C is the *amplitude*, f_0 is the *frequency* (in *hertz*), and θ is the phase. Let us consider two special cases of this sinusoid when $\theta = 0$ and $\theta = -\pi/2$ as follows:

(a) $x(t) = C \cos 2\pi f_0 t \qquad (\theta = 0)$

(b) $x(t) = C \cos(2\pi f_0 t - \pi/2) = C \sin 2\pi f_0 t \qquad (\theta = -\pi/2)$

The angle or phase can be expressed in units of degrees or radians. Although the radian is the proper unit, in this book we shall often use the degree unit because students generally have a better feel for the relative magnitudes of angles expressed in degrees rather than in radians. For example, we relate better to the angle $24°$ than to 0.419 radian. Remember, however, when in doubt, use the radian unit and, above all, be consistent. In other words, in a given problem or an expression do not mix the two units.

It is convenient to use the variable ω_0 (*radian frequency*) to express $2\pi f_0$:

$$\omega_0 = 2\pi f_0 \tag{B.20}$$

With this notation, the sinusoid in Eq. (B.18) can be expressed as

$$x(t) = C \cos(\omega_0 t + \theta)$$

in which the period T_0 is given by [see Eqs. (B.19) and (B.20)]

$$T_0 = \frac{1}{\omega_0/2\pi} = \frac{2\pi}{\omega_0} \tag{B.21a}$$

and

$$\omega_0 = \frac{2\pi}{T_0} \tag{B.21b}$$

In future discussions, we shall often refer to ω_0 as the frequency of the signal $\cos(\omega_0 t + \theta)$, but it should be clearly understood that the frequency of this sinusoid is f_0 Hz ($f_0 = \omega_0/2\pi$), and ω_0 is actually the *radian frequency*.

The signals $C \cos \omega_0 t$ and $C \sin \omega_0 t$ are illustrated in Fig. B.6a and B.6b, respectively. A general sinusoid $C \cos(\omega_0 t + \theta)$ can be readily sketched by shifting the signal $C \cos \omega_0 t$ in Fig. B.6a by the appropriate amount. Consider, for example,

$$x(t) = C \cos(\omega_0 t - 60°)$$

This signal can be obtained by shifting (delaying) the signal $C \cos \omega_0 t$ (Fig. B.6a) to the right by a phase (angle) of $60°$. We know that a sinusoid undergoes a $360°$ change of phase (or angle) in one cycle. A quarter-cycle segment corresponds to a $90°$ change of angle. We therefore shift (delay) the signal in Fig. B.6a by two-thirds of a quarter-cycle segment to obtain $C \cos(\omega_0 t - 60°)$, as shown in Fig. B.6c.

Observe that if we delay $C \cos \omega_0 t$ in Fig. B.6a by a quarter-cycle (angle of $90°$ or $\pi/2$ radians), we obtain the signal $C \sin \omega_0 t$, depicted in Fig. B.6b. This verifies the well-known trigonometric identity

$$C \cos(\omega_0 t - \pi/2) = C \sin \omega_0 t \tag{B.22a}$$

Alternatively, if we advance $C \sin \omega_0 t$ by a quarter-cycle, we obtain $C \cos \omega_0 t$. Therefore,

$$C \sin(\omega_0 t + \pi/2) = C \cos \omega_0 t \tag{B.22b}$$

This observation means $\sin \omega_0 t$ lags $\cos \omega_0 t$ by $90°(\pi/2$ radians), or $\cos \omega_0 t$ leads $\sin \omega_0 t$ by $90°$.

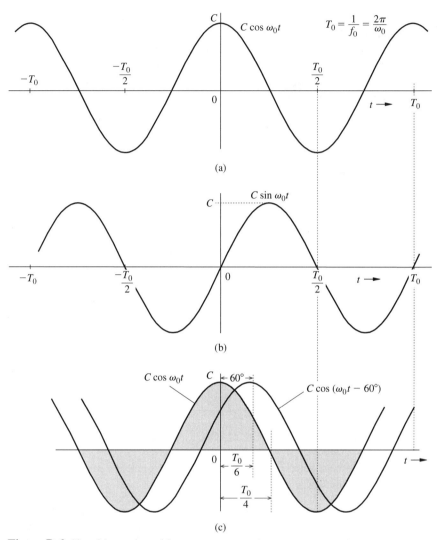

Figure B.6 Sketching a sinusoid.

B.2-1 Addition of Sinusoids

Two sinusoids having the same frequency but different phases add to form a single sinusoid of the same frequency. This fact is readily seen from the well-known trigonometric identity

$$C \cos(\omega_0 t + \theta) = C \cos\theta \cos\omega_0 t - C \sin\theta \sin\omega_0 t$$

$$= a \cos\omega_0 t + b \sin\omega_0 t \tag{B.23a}$$

in which

$$a = C \cos\theta \qquad \text{and} \qquad b = -C \sin\theta$$

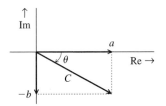

Figure B.7 Phasor addition of sinusoids.

Therefore,

$$C = \sqrt{a^2 + b^2} \tag{B.23b}$$

$$\theta = \tan^{-1}\left(\frac{-b}{a}\right) \tag{B.23c}$$

Equations (B.23b) and (B.23c) show that C and θ are the magnitude and angle, respectively, of a complex number $a - jb$. In other words, $a - jb = Ce^{j\theta}$. Hence, to find C and θ, we convert $a - jb$ to polar form and the magnitude and the angle of the resulting polar number are C and θ, respectively.

To summarize,

$$a \cos \omega_0 t + b \sin \omega_0 t = C \cos (\omega_0 t + \theta) \tag{B.23d}$$

in which C and θ are given by Eqs. (B.23b) and (B.23c), respectively. These happen to be the magnitude and angle, respectively, of $a - jb$.

The process of adding two sinusoids with the same frequency can be clarified by using *phasors* to represent sinusoids. We represent the sinusoid $C \cos (\omega_0 t + \theta)$ by a phasor of length C at an angle θ with the horizontal axis. Clearly, the sinusoid $a \cos \omega_0 t$ is represented by a horizontal phasor of length a ($\theta = 0$), while $b \sin \omega_0 t = b \cos (\omega_0 t - \pi/2)$ is represented by a vertical phasor of length b at an angle $-\pi/2$ with the horizontal (Fig. B.7). Adding these two phasors results in a phasor of length C at an angle θ, as depicted in Fig. B.7. From this figure, we verify the values of C and θ found in Eqs. (B.23b) and (B.23c), respectively.

Proper care should be exercised in computing θ, as explained on page 8 ("A Warning About Using Electronic Calculators in Computing Angles").

EXAMPLE B.6

In the following cases, express $x(t)$ as a single sinusoid:

(a) $x(t) = \cos \omega_0 t - \sqrt{3} \sin \omega_0 t$

(b) $x(t) = -3 \cos \omega_0 t + 4 \sin \omega_0 t$

(a) In this case, $a = 1$, $b = -\sqrt{3}$, and from Eqs. (B.23)

$$C = \sqrt{1^2 + \left(\sqrt{3}\right)^2} = 2$$

$$\theta = \tan^{-1}\left(\frac{\sqrt{3}}{1}\right) = 60°$$

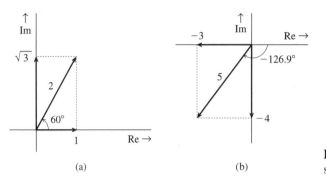

Figure B.8 Phasor addition of sinusoids.

Therefore,

$$x(t) = 2\cos(\omega_0 t + 60°)$$

We can verify this result by drawing phasors corresponding to the two sinusoids. The sinusoid $\cos \omega_0 t$ is represented by a phasor of unit length at a zero angle with the horizontal. The phasor $\sin \omega_0 t$ is represented by a unit phasor at an angle of $-90°$ with the horizontal. Therefore, $-\sqrt{3}\sin \omega_0 t$ is represented by a phasor of length $\sqrt{3}$ at $90°$ with the horizontal, as depicted in Fig. B.8a. The two phasors added yield a phasor of length 2 at $60°$ with the horizontal (also shown in Fig. B.8a).

Alternately, we note that $a - jb = 1 + j\sqrt{3} = 2e^{j\pi/3}$. Hence, $C = 2$ and $\theta = \pi/3$.

Observe that a phase shift of $\pm\pi$ amounts to multiplication by -1. Therefore, $x(t)$ can also be expressed alternatively as

$$x(t) = -2\cos(\omega_0 t + 60° \pm 180°)$$

$$= -2\cos(\omega_0 t - 120°)$$

$$= -2\cos(\omega_0 t + 240°)$$

In practice, the principal value, that is, $-120°$, is preferred.

(b) In this case, $a = -3$, $b = 4$, and from Eqs. (B.23) we have

$$C = \sqrt{(-3)^2 + 4^2} = 5$$

$$\theta = \tan^{-1}\left(\tfrac{-4}{-3}\right) = -126.9°$$

Observe that

$$\tan^{-1}\left(\tfrac{-4}{-3}\right) \neq \tan^{-1}\left(\tfrac{4}{3}\right) = 53.1°$$

Therefore,

$$x(t) = 5\cos(\omega_0 t - 126.9°)$$

This result is readily verified in the phasor diagram in Fig. B.8b. Alternately, $a - jb = -3 - j4 = 5e^{-j126.9°}$. Hence, $C = 5$ and $\theta = -126.9°$.

COMPUTER EXAMPLE CB.4

Express $f(t) = -3\cos(\omega_0 t) + 4\sin(\omega_0 t)$ as a single sinusoid.

Notice that $a\cos(\omega_0 t) + b\sin(\omega_0 t) = C\cos[\omega_0 t + \tan^{-1}(-b/a)]$. Hence, the amplitude C and the angle θ of the resulting sinusoid are the magnitude and angle of a complex number $a - jb$.

```
>> a = -3; b = 4;
>> [theta,C] = cart2pol(a,-b);
>> theta_deg = (180/pi)*theta;
>> disp(['C = ',num2str(C),'; theta = ',num2str(theta),...
>>        '; theta_deg = ',num2str(theta_deg)]);
C = 5; theta = -2.2143; theta_deg = -126.8699
```

Therefore, $f(t) = -3\cos(\omega_0 t) + 4\sin(\omega_0 t) = 5\cos(\omega_0 t - 2.2143) = 5\cos(\omega_0 t - 126.8699°)$.

We can also perform the reverse operation, expressing

$$x(t) = C\cos(\omega_0 t + \theta)$$

in terms of $\cos\omega_0 t$ and $\sin\omega_0 t$ by means of the trigonometric identity

$$C\cos(\omega_0 t + \theta) = C\cos\theta\cos\omega_0 t - C\sin\theta\sin\omega_0 t$$

For example,

$$10\cos(\omega_0 t - 60°) = 5\cos\omega_0 t + 5\sqrt{3}\sin\omega_0 t$$

B.2-2 Sinusoids in Terms of Exponentials: Euler's Formula

Sinusoids can be expressed in terms of exponentials by using Euler's formula [see Eq. (B.3)]

$$\cos\varphi = \frac{1}{2}(e^{j\varphi} + e^{-j\varphi}) \tag{B.24a}$$

$$\sin\varphi = \frac{1}{2j}(e^{j\varphi} - e^{-j\varphi}) \tag{B.24b}$$

Inversion of these equations yields

$$e^{j\varphi} = \cos\varphi + j\sin\varphi \tag{B.25a}$$

$$e^{-j\varphi} = \cos\varphi - j\sin\varphi \tag{B.25b}$$

B.3 Sketching Signals

In this section we discuss the sketching of a few useful signals, starting with exponentials.

B.3-1 Monotonic Exponentials

The signal e^{-at} decays monotonically, and the signal e^{at} grows monotonically with t (assuming $a > 0$) as depicted in Fig. B.9. For the sake of simplicity, we shall consider an exponential e^{-at} starting at $t = 0$, as shown in Fig. B.10a.

The signal e^{-at} has a unit value at $t = 0$. At $t = 1/a$, the value drops to $1/e$ (about 37% of its initial value), as illustrated in Fig. B.10a. This time interval over which the exponential reduces by a factor e (i.e., drops to about 37% of its value) is known as the *time constant* of the exponential. Therefore, the time constant of e^{-at} is $1/a$. Observe that the exponential is reduced to 37% of its initial value over any time interval of duration $1/a$. This can be shown by considering any set of instants t_1 and t_2 separated by one time constant so that

$$t_2 - t_1 = \frac{1}{a}$$

Now the ratio of e^{-at_2} to e^{-at_1} is given by

$$\frac{e^{-at_2}}{e^{-at_1}} = e^{-a(t_2-t_1)} = \frac{1}{e} \approx 0.37$$

We can use this fact to sketch an exponential quickly. For example, consider

$$x(t) = e^{-2t}$$

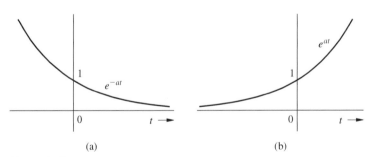

(a)   (b)

Figure B.9 Monotonic exponentials.

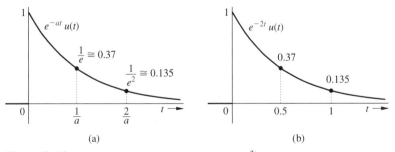

(a)   (b)

Figure B.10 (a) Sketching e^{-at}. (b) Sketching e^{-2t}.

The time constant in this case is 0.5. The value of $x(t)$ at $t = 0$ is 1. At $t = 0.5$ (one time constant) it is $1/e$ (about 0.37). The value of $x(t)$ continues to drop further by the factor $1/e$ (37%) over the next half-second interval (one time constant). Thus $x(t)$ at $t = 1$ is $(1/e)^2$. Continuing in this manner, we see that $x(t) = (1/e)^3$ at $t = 1.5$, and so on. A knowledge of the values of $x(t)$ at $t = 0, 0.5, 1$, and 1.5 allows us to sketch the desired signal as shown in Fig. B.10b.[†]

For a monotonically growing exponential e^{at}, the waveform increases by a factor e over each interval of $1/a$ seconds.

B.3-2 The Exponentially Varying Sinusoid

We now discuss sketching an exponentially varying sinusoid

$$x(t) = Ae^{-at} \cos(\omega_0 t + \theta) \tag{B.26}$$

Let us consider a specific example:

$$x(t) = 4e^{-2t} \cos(6t - 60°) \tag{B.27}$$

We shall sketch $4e^{-2t}$ and $\cos(6t - 60°)$ separately and then multiply them:

(i) **Sketching $4e^{-2t}$.** This monotonically decaying exponential has a time constant of 0.5 second and an initial value of 4 at $t = 0$. Therefore, its values at $t = 0.5, 1, 1.5$, and 2 are $4/e, 4/e^2, 4/e^3$, and $4/e^4$, or about 1.47, 0.54, 0.2, and 0.07, respectively. Using these values as a guide, we sketch $4e^{-2t}$, as illustrated in Fig. B.11a.

(ii) **Sketching $\cos(6t - 60°)$.** The procedure for sketching $\cos(6t - 60°)$ is discussed in Section B.2 (Fig. B.6c). Here the period of the sinusoid is $T_0 = 2\pi/6 \approx 1$, and there is a phase delay of 60°, or two-thirds of a quarter-cycle, which is equivalent to a delay of about $(60/360)(1) \approx 1/6$ second (see Fig. B.11b).

(iii) **Sketching $4e^{-2t} \cos(6t - 60°)$.** We now multiply the waveforms in steps i and ii. This multiplication amounts to forcing the sinusoid $4 \cos(6t - 60°)$ to decrease exponentially with a time constant of 0.5. The initial amplitude (at $t = 0$) is 4, decreasing to $4/e$ ($= 1.47$) at $t = 0.5$, to $1.47/e$ ($= 0.54$) at $t = 1$, and so on. This is depicted in Fig. B.11c. Note that when $\cos(6t - 60°)$ has a value of unity (peak amplitude),

$$4e^{-2t} \cos(6t - 60°) = 4e^{-2t} \tag{B.28}$$

Therefore, $4e^{-2t} \cos(6t - 60°)$ touches $4e^{-2t}$ at the instants at which the sinusoid $\cos(6t - 60°)$ is at its positive peaks. Clearly $4e^{-2t}$ is an envelope for positive amplitudes of $4e^{-2t} \cos(6t - 60°)$. Similar argument shows that $4e^{-2t} \cos(6t - 60°)$ touches $-4e^{-2t}$ at its negative peaks. Therefore, $-4e^{-2t}$ is an envelope for negative amplitudes of $4e^{-2t} \cos(6t - 60°)$. Thus, to sketch $4e^{-2t} \cos(6t - 60°)$, we first draw the envelopes $4e^{-2t}$ and $-4e^{-2t}$ (the mirror image of $4e^{-2t}$ about the horizontal axis), and then sketch the sinusoid $\cos(6t - 60°)$, with these envelopes acting as constraints on the sinusoid's amplitude (see Fig. B.11c).

[†]If we wish to refine the sketch further, we could consider intervals of half the time constant over which the signal decays by a factor $1/\sqrt{e}$. Thus, at $t = 0.25$, $x(t) = 1/\sqrt{e}$, and at $t = 0.75$, $x(t) = 1/e\sqrt{e}$, and so on.

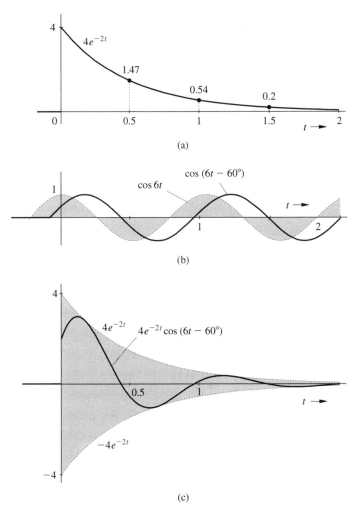

Figure B.11 Sketching an exponentially varying sinusoid.

In general, $Ke^{-at}\cos(\omega_0 t + \theta)$ can be sketched in this manner, with Ke^{-at} and $-Ke^{-at}$ constraining the amplitude of $\cos(\omega_0 t + \theta)$.

B.4 CRAMER'S RULE

Cramer's rule offers a very convenient way to solve simultaneous linear equations. Consider a set of n linear simultaneous equations in n unknowns $x_1, x_2, \ldots, x_n$:

$$
\begin{aligned}
a_{11}x_1 + a_{12}x_2 + \cdots + a_{1n}x_n &= y_1 \\
a_{21}x_1 + a_{22}x_2 + \cdots + a_{2n}x_n &= y_2 \\
&\vdots \\
a_{n1}x_1 + a_{n2}x_2 + \cdots + a_{nn}x_n &= y_n
\end{aligned}
\tag{B.29}
$$

These equations can be expressed in matrix form as

$$
\begin{bmatrix}
a_{11} & a_{12} & \cdots & a_{1n} \\
a_{21} & a_{22} & \cdots & a_{2n} \\
\vdots & \vdots & \cdots & \vdots \\
a_{n1} & a_{n2} & \cdots & a_{nn}
\end{bmatrix}
\begin{bmatrix}
x_1 \\
x_2 \\
\vdots \\
x_n
\end{bmatrix}
=
\begin{bmatrix}
y_1 \\
y_2 \\
\vdots \\
y_n
\end{bmatrix}
\tag{B.30}
$$

We denote the matrix on the left-hand side formed by the elements a_{ij} as $\mathbf{A}$. The determinant of $\mathbf{A}$ is denoted by $|\mathbf{A}|$. If the determinant $|\mathbf{A}|$ is not zero, the set of equations (B.29) has a unique solution given by Cramer's formula

$$
x_k = \frac{|\mathbf{D}_k|}{|\mathbf{A}|} \qquad k = 1, 2, \ldots, n
\tag{B.31}
$$

where $|\mathbf{D}_k|$ is obtained by replacing the kth column of $|\mathbf{A}|$ by the column on the right-hand side of Eq. (B.30) (with elements $y_1, y_2, \ldots, y_n$).

We shall demonstrate the use of this rule with an example.

EXAMPLE B.7

Use Cramer's rule to solve the following simultaneous linear equations in three unknowns:

$$
2x_1 + x_2 + x_3 = 3
$$

$$
x_1 + 3x_2 - x_3 = 7
$$

$$
x_1 + x_2 + x_3 = 1
$$

In matrix form these equations can be expressed as

$$
\begin{bmatrix}
2 & 1 & 1 \\
1 & 3 & -1 \\
1 & 1 & 1
\end{bmatrix}
\begin{bmatrix}
x_1 \\
x_2 \\
x_3
\end{bmatrix}
=
\begin{bmatrix}
3 \\
7 \\
1
\end{bmatrix}
$$

Here,

$$
|\mathbf{A}| =
\begin{vmatrix}
2 & 1 & 1 \\
1 & 3 & -1 \\
1 & 1 & 1
\end{vmatrix}
= 4
$$

Since $|\mathbf{A}| = 4 \neq 0$, a unique solution exists for x_1, x_2, and x_3. This solution is provided by Cramer's rule [Eq. (B.31)] as follows:

$$x_1 = \frac{1}{|\mathbf{A}|} \begin{vmatrix} 3 & 1 & 1 \\ 7 & 3 & -1 \\ 1 & 1 & 1 \end{vmatrix} = \frac{8}{4} = 2$$

$$x_2 = \frac{1}{|\mathbf{A}|} \begin{vmatrix} 2 & 3 & 1 \\ 1 & 7 & -1 \\ 1 & 1 & 1 \end{vmatrix} = \frac{4}{4} = 1$$

$$x_3 = \frac{1}{|\mathbf{A}|} \begin{vmatrix} 2 & 1 & 3 \\ 1 & 3 & 7 \\ 1 & 1 & 1 \end{vmatrix} = \frac{-8}{4} = -2$$

B.5 PARTIAL FRACTION EXPANSION

In the analysis of linear time-invariant systems, we encounter functions that are ratios of two polynomials in a certain variable, say x. Such functions are known as *rational functions*. A rational function $F(x)$ can be expressed as

$$F(x) = \frac{b_m x^m + b_{m-1} x^{m-1} + \cdots + b_1 x + b_0}{x^n + a_{n-1} x^{n-1} + \cdots + a_1 x + a_0} \tag{B.32}$$

$$= \frac{P(x)}{Q(x)} \tag{B.33}$$

The function $F(x)$ is *improper* if $m \geq n$ and *proper* if $m < n$. An improper function can always be separated into the sum of a polynomial in x and a proper function. Consider, for example, the function

$$F(x) = \frac{2x^3 + 9x^2 + 11x + 2}{x^2 + 4x + 3} \tag{B.34a}$$

Because this is an improper function, we divide the numerator by the denominator until the remainder has a lower degree than the denominator.

$$
\begin{array}{r}
2x + 1 \\
x^2 + 4x + 3 \overline{)\, 2x^3 + 9x^2 + 11x + 2} \\
\underline{2x^3 + 8x^2 + 6x} \\
x^2 + 5x + 2 \\
\underline{x^2 + 4x + 3} \\
x - 1
\end{array}
$$

Therefore, $F(x)$ can be expressed as

$$F(x) = \frac{2x^3 + 9x^2 + 11x + 2}{x^2 + 4x + 3} = \underbrace{2x + 1}_{\text{polynomial in } x} + \underbrace{\frac{x - 1}{x^2 + 4x + 3}}_{\text{proper function}} \qquad \text{(B.34b)}$$

A proper function can be further expanded into partial fractions. The remaining discussion in this section is concerned with various ways of doing this.

B.5-1 Method of Clearing Fractions

A rational function can be written as a sum of appropriate partial fractions with unknown coefficients, which are determined by clearing fractions and equating the coefficients of similar powers on the two sides. This procedure is demonstrated by the following example.

EXAMPLE B.8

Expand the following rational function $F(x)$ into partial fractions:

$$F(x) = \frac{x^3 + 3x^2 + 4x + 6}{(x + 1)(x + 2)(x + 3)^2}$$

This function can be expressed as a sum of partial fractions with denominators $(x + 1)$, $(x + 2)$, $(x + 3)$, and $(x + 3)^2$, as follows:

$$F(x) = \frac{x^3 + 3x^2 + 4x + 6}{(x + 1)(x + 2)(x + 3)^2} = \frac{k_1}{x + 1} + \frac{k_2}{x + 2} + \frac{k_3}{x + 3} + \frac{k_4}{(x + 3)^2}$$

To determine the unknowns k_1, k_2, k_3, and k_4 we clear fractions by multiplying both sides by $(x + 1)(x + 2)(x + 3)^2$ to obtain

$$\begin{aligned} x^3 + 3x^2 + 4x + 6 &= k_1(x^3 + 8x^2 + 21x + 18) + k_2(x^3 + 7x^2 + 15x + 9) \\ &\quad + k_3(x^3 + 6x^2 + 11x + 6) + k_4(x^2 + 3x + 2) \\ &= x^3(k_1 + k_2 + k_3) + x^2(8k_1 + 7k_2 + 6k_3 + k_4) \\ &\quad + x(21k_1 + 15k_2 + 11k_3 + 3k_4) + (18k_1 + 9k_2 + 6k_3 + 2k_4) \end{aligned}$$

Equating coefficients of similar powers on both sides yields

$$k_1 + k_2 + k_3 = 1$$

$$8k_1 + 7k_2 + 6k_3 + k_4 = 3$$

$$21k_1 + 15k_2 + 11k_3 + 3k_4 = 4$$

$$18k_1 + 9k_2 + 6k_3 + 2k_4 = 6$$

Solution of these four simultaneous equations yields

$$k_1 = 1, \qquad k_2 = -2, \qquad k_3 = 2, \qquad k_4 = -3$$

Therefore,

$$F(x) = \frac{1}{x+1} - \frac{2}{x+2} + \frac{2}{x+3} - \frac{3}{(x+3)^2}$$

Although this method is straightforward and applicable to all situations, it is not necessarily the most efficient. We now discuss other methods that can reduce numerical work considerably.

B.5-2 The Heaviside "Cover-Up" Method

DISTINCT FACTORS OF $Q(x)$

We shall first consider the partial fraction expansion of $F(x) = P(x)/Q(x)$, in which all the factors of $Q(x)$ are distinct (not repeated). Consider the proper function

$$F(x) = \frac{b_m x^m + b_{m-1} x^{m-1} + \cdots + b_1 x + b_0}{x^n + a_{n-1} x^{n-1} + \cdots + a_1 x + a_0} \qquad m < n$$

$$= \frac{P(x)}{(x - \lambda_1)(x - \lambda_2) \cdots (x - \lambda_n)} \tag{B.35a}$$

We can show that $F(x)$ in Eq. (B.35a) can be expressed as the sum of partial fractions

$$F(x) = \frac{k_1}{x - \lambda_1} + \frac{k_2}{x - \lambda_2} + \cdots + \frac{k_n}{x - \lambda_n} \tag{B.35b}$$

To determine the coefficient k_1, we multiply both sides of Eq. (B.35b) by $x - \lambda_1$ and then let $x = \lambda_1$. This yields

$$(x - \lambda_1)F(x)|_{x=\lambda_1} = k_1 + \frac{k_2(x - \lambda_1)}{(x - \lambda_2)} + \frac{k_3(x - \lambda_1)}{(x - \lambda_3)} + \cdots + \frac{k_n(x - \lambda_1)}{(x - \lambda_n)}\bigg|_{x=\lambda_1}$$

On the right-hand side, all the terms except k_1 vanish. Therefore,

$$k_1 = (x - \lambda_1)F(x)|_{x=\lambda_1} \tag{B.36}$$

Similarly, we can show that

$$k_r = (x - \lambda_r)F(x)|_{x=\lambda_r} \qquad r = 1, 2, \ldots, n \tag{B.37}$$

This procedure also goes under the name *method of residues*.

EXAMPLE B.9

Expand the following rational function $F(x)$ into partial fractions:

$$F(x) = \frac{2x^2 + 9x - 11}{(x + 1)(x - 2)(x + 3)} = \frac{k_1}{x + 1} + \frac{k_2}{x - 2} + \frac{k_3}{x + 3}$$

To determine k_1, we let $x = -1$ in $(x + 1)F(x)$. Note that $(x + 1)F(x)$ is obtained from $F(x)$ by omitting the term $(x + 1)$ from its denominator. Therefore, to compute k_1 corresponding to the factor $(x + 1)$, we cover up the term $(x + 1)$ in the denominator of $F(x)$ and then substitute $x = -1$ in the remaining expression. [Mentally conceal the term $(x + 1)$ in $F(x)$ with a finger and then let $x = -1$ in the remaining expression.] The steps in covering up the function

$$F(x) = \frac{2x^2 + 9x - 11}{(x + 1)(x - 2)(x + 3)}$$

are as follows.

Step 1. Cover up (conceal) the factor $(x + 1)$ from $F(x)$:

$$\frac{2x^2 + 9x - 11}{(x + 1)\,(x - 2)(x + 3)}$$

Step 2. Substitute $x = -1$ in the remaining expression to obtain k_1:

$$k_1 = \frac{2 - 9 - 11}{(-1 - 2)(-1 + 3)} = \frac{-18}{-6} = 3$$

Similarly, to compute k_2, we cover up the factor $(x - 2)$ in $F(x)$ and let $x = 2$ in the remaining function, as follows:

$$k_2 = \left.\frac{2x^2 + 9x - 11}{(x + 1)\,(x - 2)\,(x + 3)}\right|_{x=2} = \frac{8 + 18 - 11}{(2 + 1)(2 + 3)} = \frac{15}{15} = 1$$

and

$$k_3 = \left.\frac{2x^2 + 9x - 11}{(x + 1)(x - 2)\,(x + 3)}\right|_{x=-3} = \frac{18 - 27 - 11}{(-3 + 1)(-3 - 2)} = \frac{-20}{10} = -2$$

Therefore,

$$F(x) = \frac{2x^2 + 9x - 11}{(x + 1)(x - 2)(x + 3)} = \frac{3}{x + 1} + \frac{1}{x - 2} - \frac{2}{x + 3}$$

COMPLEX FACTORS OF $Q(x)$

The procedure just given works regardless of whether the factors of $Q(x)$ are real or complex. Consider, for example,

$$
\begin{aligned}
F(x) &= \frac{4x^2 + 2x + 18}{(x + 1)(x^2 + 4x + 13)} \\
&= \frac{4x^2 + 2x + 18}{(x + 1)(x + 2 - j3)(x + 2 + j3)} \\
&= \frac{k_1}{x + 1} + \frac{k_2}{x + 2 - j3} + \frac{k_3}{x + 2 + j3}
\end{aligned}
\tag{B.38}
$$

where

$$
k_1 = \left[\frac{4x^2 + 2x + 18}{(x + 1)(x^2 + 4x + 13)} \right]_{x=-1} = 2
$$

Similarly,

$$
k_2 = \left[\frac{4x^2 + 2x + 18}{(x + 1)(x + 2 - j3)(x + 2 + j3)} \right]_{x=-2+j3} = 1 + j2 = \sqrt{5}e^{j63.43°}
$$

$$
k_3 = \left[\frac{4x^2 + 2x + 18}{(x + 1)(x + 2 - j3)(x + 2 + j3)} \right]_{x=-2-j3} = 1 - j2 = \sqrt{5}e^{-j63.43°}
$$

Therefore,

$$
F(x) = \frac{2}{x + 1} + \frac{\sqrt{5}e^{j63.43°}}{x + 2 - j3} + \frac{\sqrt{5}e^{-j63.43°}}{x + 2 + j3}
\tag{B.39}
$$

The coefficients k_2 and k_3 corresponding to the complex conjugate factors are also conjugates of each other. This is generally true when the coefficients of a rational function are real. In such a case, we need to compute only one of the coefficients.

QUADRATIC FACTORS

Often we are required to combine the two terms arising from complex conjugate factors into one quadratic factor. For example, $F(x)$ in Eq. (B.38) can be expressed as

$$
F(x) = \frac{4x^2 + 2x + 18}{(x + 1)(x^2 + 4x + 13)} = \frac{k_1}{x + 1} + \frac{c_1 x + c_2}{x^2 + 4x + 13}
$$

The coefficient k_1 is found by the Heaviside method to be 2. Therefore,

$$
\frac{4x^2 + 2x + 18}{(x + 1)(x^2 + 4x + 13)} = \frac{2}{x + 1} + \frac{c_1 x + c_2}{x^2 + 4x + 13}
\tag{B.40}
$$

The values of c_1 and c_2 are determined by clearing fractions and equating the coefficients of similar powers of x on both sides of the resulting equation. Clearing fractions on both sides of

Eq. (B.40) yields

$$4x^2 + 2x + 18 = 2(x^2 + 4x + 13) + (c_1 x + c_2)(x + 1)$$
$$= (2 + c_1)x^2 + (8 + c_1 + c_2)x + (26 + c_2) \tag{B.41}$$

Equating terms of similar powers yields $c_1 = 2$, $c_2 = -8$, and

$$\frac{4x^2 + 2x + 18}{(x + 1)(x^2 + 4x + 13)} = \frac{2}{x + 1} + \frac{2x - 8}{x^2 + 4x + 13} \tag{B.42}$$

SHORTCUTS

The values of c_1 and c_2 in Eq. (B.40) can also be determined by using shortcuts. After computing $k_1 = 2$ by the Heaviside method as before, we let $x = 0$ on both sides of Eq. (B.40) to eliminate c_1. This gives us

$$\frac{18}{13} = 2 + \frac{c_2}{13}$$

Therefore,

$$c_2 = -8$$

To determine c_1, we multiply both sides of Eq. (B.40) by x and then let $x \to \infty$. Remember that when $x \to \infty$, only the terms of the highest power are significant. Therefore,

$$4 = 2 + c_1$$

and

$$c_1 = 2$$

In the procedure discussed here, we let $x = 0$ to determine c_2 and then multiply both sides by x and let $x \to \infty$ to determine c_1. However, nothing is sacred about these values ($x = 0$ or $x = \infty$). We use them because they reduce the number of computations involved. We could just as well use other convenient values for x, such as $x = 1$. Consider the case

$$F(x) = \frac{2x^2 + 4x + 5}{x(x^2 + 2x + 5)}$$
$$= \frac{k}{x} + \frac{c_1 x + c_2}{x^2 + 2x + 5}$$

We find $k = 1$ by the Heaviside method in the usual manner. As a result,

$$\frac{2x^2 + 4x + 5}{x(x^2 + 2x + 5)} = \frac{1}{x} + \frac{c_1 x + c_2}{x^2 + 2x + 5} \tag{B.43}$$

To determine c_1 and c_2, if we try letting $x = 0$ in Eq. (B.43), we obtain ∞ on both sides. So let us choose $x = 1$. This yields

$$\frac{11}{8} = 1 + \frac{c_1 + c_2}{8}$$

or

$$c_1 + c_2 = 3$$

We can now choose some other value for x, such as $x = 2$, to obtain one more relationship to use in determining c_1 and c_2. In this case, however, a simple method is to multiply both sides of Eq. (B.43) by x and then let $x \to \infty$. This yields

$$2 = 1 + c_1$$

so that

$$c_1 = 1 \quad \text{and} \quad c_2 = 2$$

Therefore,

$$F(x) = \frac{1}{x} + \frac{x + 2}{x^2 + 2x + 5}$$

B.5-3 Repeated Factors of $Q(x)$

If a function $F(x)$ has a repeated factor in its denominator, it has the form

$$F(x) = \frac{P(x)}{(x - \lambda)^r (x - \alpha_1)(x - \alpha_2) \cdots (x - \alpha_j)} \tag{B.44}$$

Its partial fraction expansion is given by

$$F(x) = \frac{a_0}{(x - \lambda)^r} + \frac{a_1}{(x - \lambda)^{r-1}} + \cdots + \frac{a_{r-1}}{(x - \lambda)}$$

$$+ \frac{k_1}{x - \alpha_1} + \frac{k_2}{x - \alpha_2} + \cdots + \frac{k_j}{x - \alpha_j} \tag{B.45}$$

The coefficients $k_1, k_2, \ldots, k_j$ corresponding to the unrepeated factors in this equation are determined by the Heaviside method, as before [Eq. (B.37)]. To find the coefficients $a_0, a_1, a_2, \ldots, a_{r-1}$, we multiply both sides of Eq. (B.45) by $(x - \lambda)^r$. This gives us

$$(x - \lambda)^r F(x) = a_0 + a_1(x - \lambda) + a_2(x - \lambda)^2 + \cdots + a_{r-1}(x - \lambda)^{r-1}$$

$$+ k_1 \frac{(x - \lambda)^r}{x - \alpha_1} + k_2 \frac{(x - \lambda)^r}{x - \alpha_2} + \cdots + k_n \frac{(x - \lambda)^r}{x - \alpha_n} \tag{B.46}$$

If we let $x = \lambda$ on both sides of Eq. (B.46), we obtain

$$(x - \lambda)^r F(x)|_{x=\lambda} = a_0 \tag{B.47a}$$

Therefore, a_0 is obtained by concealing the factor $(x - \lambda)^r$ in $F(x)$ and letting $x = \lambda$ in the remaining expression (the Heaviside "cover-up" method). If we take the derivative (with respect to x) of both sides of Eq. (B.46), the right-hand side is $a_1 +$ terms containing a factor $(x - \lambda)$ in their numerators. Letting $x = \lambda$ on both sides of this equation, we obtain

$$\frac{d}{dx}[(x - \lambda)^r F(x)]\bigg|_{x=\lambda} = a_1$$

Thus, a_1 is obtained by concealing the factor $(x - \lambda)^r$ in $F(x)$, taking the derivative of the remaining expression, and then letting $x = \lambda$. Continuing in this manner, we find

$$a_j = \frac{1}{j!} \frac{d^j}{dx^j} [(x - \lambda)^r F(x)] \Big|_{x=\lambda} \qquad (B.47b)$$

Observe that $(x - \lambda)^r F(x)$ is obtained from $F(x)$ by omitting the factor $(x - \lambda)^r$ from its denominator. Therefore, the coefficient a_j is obtained by concealing the factor $(x - \lambda)^r$ in $F(x)$, taking the jth derivative of the remaining expression, and then letting $x = \lambda$ (while dividing by $j!$).

EXAMPLE B.10

Expand $F(x)$ into partial fractions if

$$F(x) = \frac{4x^3 + 16x^2 + 23x + 13}{(x + 1)^3(x + 2)}$$

The partial fractions are

$$F(x) = \frac{a_0}{(x + 1)^3} + \frac{a_1}{(x + 1)^2} + \frac{a_2}{x + 1} + \frac{k}{x + 2}$$

The coefficient k is obtained by concealing the factor $(x + 2)$ in $F(x)$ and then substituting $x = -2$ in the remaining expression:

$$k = \frac{4x^3 + 16x^2 + 23x + 13}{(x + 1)^3 \boxed{(x + 2)}} \Big|_{x=-2} = 1$$

To find a_0, we conceal the factor $(x + 1)^3$ in $F(x)$ and let $x = -1$ in the remaining expression:

$$a_0 = \frac{4x^3 + 16x^2 + 23x + 13}{\boxed{(x + 1)^3} (x + 2)} \Big|_{x=-1} = 2$$

To find a_1, we conceal the factor $(x + 1)^3$ in $F(x)$, take the derivative of the remaining expression, and then let $x = -1$:

$$a_1 = \frac{d}{dx} \left[\frac{4x^3 + 16x^2 + 23x + 13}{\boxed{(x + 1)^3} (x + 2)} \right] \Big|_{x=-1} = 1$$

Similarly,

$$a_2 = \frac{1}{2!} \frac{d^2}{dx^2} \left[\frac{4x^3 + 16x^2 + 23x + 13}{\boxed{(x + 1)^3} (x + 2)} \right] \Big|_{x=-1} = 3$$

Therefore,

$$F(x) = \frac{2}{(x + 1)^3} + \frac{1}{(x + 1)^2} + \frac{3}{x + 1} + \frac{1}{x + 2}$$

B.5-4 Mixture of the Heaviside "Cover-Up" and Clearing Fractions

For multiple roots, especially of higher order, the Heaviside expansion method, which requires repeated differentiation, can become cumbersome. For a function that contains several repeated and unrepeated roots, a hybrid of the two procedures proves to be the best. The simpler coefficients are determined by the Heaviside method, and the remaining coefficients are found by clearing fractions or shortcuts, thus incorporating the best of the two methods. We demonstrate this procedure by solving Example B.10 once again by this method.

In Example B.10, coefficients k and a_0 are relatively simple to determine by the Heaviside expansion method. These values were found to be $k_1 = 1$ and $a_0 = 2$. Therefore,

$$\frac{4x^3 + 16x^2 + 23x + 13}{(x + 1)^3(x + 2)} = \frac{2}{(x + 1)^3} + \frac{a_1}{(x + 1)^2} + \frac{a_2}{x + 1} + \frac{1}{x + 2}$$

We now multiply both sides of this equation by $(x + 1)^3(x + 2)$ to clear the fractions. This yields

$$4x^3 + 16x^2 + 23x + 13$$
$$= 2(x + 2) + a_1(x + 1)(x + 2) + a_2(x + 1)^2(x + 2) + (x + 1)^3$$
$$= (1 + a_2)x^3 + (a_1 + 4a_2 + 3)x^2 + (5 + 3a_1 + 5a_2)x + (4 + 2a_1 + 2a_2 + 1)$$

Equating coefficients of the third and second powers of x on both sides, we obtain

$$\left.\begin{array}{c} 1 + a_2 = 4 \\ a_1 + 4a_2 + 3 = 16 \end{array}\right\} \implies \begin{array}{c} a_1 = 1 \\ a_2 = 3 \end{array}$$

We may stop here if we wish because the two desired coefficients, a_1 and a_2, are now determined. However, equating the coefficients of the two remaining powers of x yields a convenient check on the answer. Equating the coefficients of the x^1 and x^0 terms, we obtain

$$23 = 5 + 3a_1 + 5a_2$$
$$13 = 4 + 2a_1 + 2a_2 + 1$$

These equations are satisfied by the values $a_1 = 1$ and $a_2 = 3$, found earlier, providing an additional check for our answers. Therefore,

$$F(x) = \frac{2}{(x + 1)^3} + \frac{1}{(x + 1)^2} + \frac{3}{x + 1} + \frac{1}{x + 2}$$

which agrees with the earlier result.

A MIXTURE OF THE HEAVISIDE "COVER-UP" AND SHORTCUTS

In Example B.10, after determining the coefficients $a_0 = 2$ and $k = 1$ by the Heaviside method as before, we have

$$\frac{4x^3 + 16x^2 + 23x + 13}{(x + 1)^3(x + 2)} = \frac{2}{(x + 1)^3} + \frac{a_1}{(x + 1)^2} + \frac{a_2}{x + 1} + \frac{1}{x + 2}$$

There are only two unknown coefficients, a_1 and a_2. If we multiply both sides of this equation by x and then let $x \to \infty$, we can eliminate a_1. This yields

$$4 = a_2 + 1 \quad \Longrightarrow \quad a_2 = 3$$

Therefore,

$$\frac{4x^3 + 16x^2 + 23x + 13}{(x+1)^3(x+2)} = \frac{2}{(x+1)^3} + \frac{a_1}{(x+1)^2} + \frac{3}{x+1} + \frac{1}{x+2}$$

There is now only one unknown a_1, which can be readily found by setting x equal to any convenient value, say $x = 0$. This yields

$$\frac{13}{2} = 2 + a_1 + 3 + \frac{1}{2} \quad \Longrightarrow \quad a_1 = 1$$

which agrees with our earlier answer.

There are other possible shortcuts. For example, we can compute a_0 (coefficient of the highest power of the repeated root), subtract this term from both sides, and then repeat the procedure.

B.5-5 Improper $F(x)$ with $m = n$

A general method of handling an improper function is indicated in the beginning of this section. However, for a special case of the numerator and denominator polynomials of $F(x)$ being of the same degree ($m = n$), the procedure is the same as that for a proper function. We can show that for

$$F(x) = \frac{b_n x^n + b_{n-1} x^{n-1} + \cdots + b_1 x + b_0}{x^n + a_{n-1} x^{n-1} + \cdots + a_1 x + a_0}$$

$$= b_n + \frac{k_1}{x - \lambda_1} + \frac{k_2}{x - \lambda_2} + \cdots + \frac{k_n}{x - \lambda_n}$$

the coefficients $k_1, k_2, \ldots, k_n$ are computed as if $F(x)$ were proper. Thus,

$$k_r = (x - \lambda_r) F(x)|_{x = \lambda_r}$$

For quadratic or repeated factors, the appropriate procedures discussed in Sections B.5-2 or B.5-3 should be used as if $F(x)$ were proper. In other words, when $m = n$, the only difference between the proper and improper case is the appearance of an extra constant b_n in the latter. Otherwise the procedure remains the same. The proof is left as an exercise for the reader.

EXAMPLE B.11

Expand $F(x)$ into partial fractions if

$$F(x) = \frac{3x^2 + 9x - 20}{x^2 + x - 6} = \frac{3x^2 + 9x - 20}{(x - 2)(x + 3)}$$

Here $m = n = 2$ with $b_n = b_2 = 3$. Therefore,

$$F(x) = \frac{3x^2 + x - 20}{(x - 2)(x + 3)} = 3 + \frac{k_1}{x - 2} + \frac{k_2}{x + 3}$$

in which

$$k_1 = \frac{3x^2 + 9x - 20}{(x - 2)(x + 3)} \bigg|_{x=2} = \frac{12 + 18 - 20}{(2 + 3)} = \frac{10}{5} = 2$$

and

$$k_2 = \frac{3x^2 + 9x - 20}{(x - 2)(x + 3)} \bigg|_{x=-3} = \frac{27 - 27 - 20}{(-3 - 2)} = \frac{-20}{-5} = 4$$

Therefore,

$$F(x) = \frac{3x^2 + 9x - 20}{(x - 2)(x + 3)} = 3 + \frac{2}{x - 2} + \frac{4}{x + 3}$$

B.5-6 Modified Partial Fractions

In finding the inverse z-transforms (Chapter 5), we require partial fractions of the form $kx/(x - \lambda_i)^r$ rather than $k/(x - \lambda_i)^r$. This can be achieved by expanding $F(x)/x$ into partial fractions. Consider, for example,

$$F(x) = \frac{5x^2 + 20x + 18}{(x + 2)(x + 3)^2}$$

Dividing both sides by x yields

$$\frac{F(x)}{x} = \frac{5x^2 + 20x + 18}{x(x + 2)(x + 3)^2}$$

Expansion of the right-hand side into partial fractions as usual yields

$$\frac{F(x)}{x} = \frac{5x^2 + 20x + 18}{x(x + 2)(x + 3)^2} = \frac{a_1}{x} + \frac{a_2}{x + 2} + \frac{a_3}{(x + 3)} + \frac{a_4}{(x + 3)^2}$$

Using the procedure discussed earlier, we find $a_1 = 1$, $a_2 = 1$, $a_3 = -2$, and $a_4 = 1$. Therefore,

$$\frac{F(x)}{x} = \frac{1}{x} + \frac{1}{x + 2} - \frac{2}{x + 3} + \frac{1}{(x + 3)^2}$$

Now multiplying both sides by x yields

$$F(x) = 1 + \frac{x}{x+2} - \frac{2x}{x+3} + \frac{x}{(x+3)^2}$$

This expresses $F(x)$ as the sum of partial fractions having the form $kx/(x - \lambda_i)^r$.

B.6 VECTORS AND MATRICES

An entity specified by n numbers in a certain order (ordered n-tuple) is an n-dimensional *vector*. Thus, an ordered n-tuple $(x_1, x_2, \ldots, x_n)$ represents an n-dimensional vector $\mathbf{x}$. A vector may be represented as a row (*row vector*):

$$\mathbf{x} = [\,x_1 \quad x_2 \quad \cdots \quad x_n\,]$$

or as a column (*column vector*):

$$\mathbf{x} = \begin{bmatrix} x_1 \\ x_2 \\ \vdots \\ x_n \end{bmatrix}$$

Simultaneous linear equations can be viewed as the transformation of one vector into another. Consider, for example, the n simultaneous linear equations

$$y_1 = a_{11}x_1 + a_{12}x_2 + \cdots + a_{1n}x_n$$

$$y_2 = a_{21}x_1 + a_{22}x_2 + \cdots + a_{2n}x_n$$

$$\vdots \tag{B.48}$$

$$y_m = a_{m1}x_1 + a_{m2}x_2 + \cdots + a_{mn}x_n$$

If we define two column vectors $\mathbf{x}$ and $\mathbf{y}$ as

$$\mathbf{x} = \begin{bmatrix} x_1 \\ x_2 \\ \vdots \\ x_n \end{bmatrix} \quad \text{and} \quad \mathbf{y} = \begin{bmatrix} y_1 \\ y_2 \\ \vdots \\ y_m \end{bmatrix} \tag{B.49}$$

then Eqs. (B.48) may be viewed as the relationship or the function that transforms vector $\mathbf{x}$ into vector $\mathbf{y}$. Such a transformation is called the *linear transformation* of vectors. To perform a linear transformation, we need to define the array of coefficients a_{ij} appearing in Eqs. (B.48). This array is called a *matrix* and is denoted by $\mathbf{A}$ for convenience:

$$\mathbf{A} = \begin{bmatrix} a_{11} & a_{12} & \cdots & a_{1n} \\ a_{21} & a_{22} & \cdots & a_{2n} \\ \vdots & \vdots & \cdots & \vdots \\ a_{m1} & a_{m2} & \cdots & a_{mn} \end{bmatrix} \tag{B.50}$$

A matrix with m rows and n columns is called a matrix of the order (m, n) or an $(m \times n)$ matrix. For the special case of $m = n$, the matrix is called a *square matrix* of order n.

It should be stressed at this point that a matrix is not a number such as a determinant, but an array of numbers arranged in a particular order. It is convenient to abbreviate the representation of matrix $\mathbf{A}$ in Eq. (B.50) with the form $(a_{ij})_{m \times n}$, implying a matrix of order $m \times n$ with a_{ij} as its ijth element. In practice, when the order $m \times n$ is understood or need not be specified, the notation can be abbreviated to (a_{ij}). Note that the first index i of a_{ij} indicates the row and the second index j indicates the column of the element a_{ij} in matrix $\mathbf{A}$.

The simultaneous equations (B.48) may now be expressed in a symbolic form as

$$\mathbf{y} = \mathbf{Ax} \tag{B.51}$$

or

$$\begin{bmatrix} y_1 \\ y_2 \\ \vdots \\ y_m \end{bmatrix} = \begin{bmatrix} a_{11} & a_{12} & \cdots & a_{1n} \\ a_{21} & a_{22} & \cdots & a_{2n} \\ \vdots & \vdots & \cdots & \vdots \\ a_{m1} & a_{m2} & \cdots & a_{mn} \end{bmatrix} \begin{bmatrix} x_1 \\ x_2 \\ \vdots \\ x_n \end{bmatrix} \tag{B.52}$$

Equation (B.51) is the symbolic representation of Eq. (B.48). Yet we have not defined the operation of the multiplication of a matrix by a vector. The quantity $\mathbf{Ax}$ is not meaningful until such an operation has been defined.

B.6-1 Some Definitions and Properties

A square matrix whose elements are zero everywhere except on the main diagonal is a *diagonal matrix*. An example of a diagonal matrix is

$$\begin{bmatrix} 2 & 0 & 0 \\ 0 & 1 & 0 \\ 0 & 0 & 5 \end{bmatrix}$$

A diagonal matrix with unity for all its diagonal elements is called an *identity matrix* or a *unit matrix*, denoted by $\mathbf{I}$. This is a square matrix:

$$\mathbf{I} = \begin{bmatrix} 1 & 0 & 0 & \cdots & 0 \\ 0 & 1 & 0 & \cdots & 0 \\ 0 & 0 & 1 & \cdots & 0 \\ \vdots & \vdots & \vdots & \cdots & \vdots \\ 0 & 0 & 0 & \cdots & 1 \end{bmatrix} \tag{B.53}$$

The order of the unit matrix is sometimes indicated by a subscript. Thus, $\mathbf{I}_n$ represents the $n \times n$ unit matrix (or identity matrix). However, we shall omit the subscript. The order of the unit matrix will be understood from the context.

A matrix having all its elements zero is a *zero matrix*.

A square matrix $\mathbf{A}$ is a *symmetric matrix* if $a_{ij} = a_{ji}$ (symmetry about the main diagonal).

Two matrices of the same order are said to be *equal* if they are equal element by element. Thus, if

$$\mathbf{A} = (a_{ij})_{m \times n} \qquad \text{and} \qquad \mathbf{B} = (b_{ij})_{m \times n}$$

then $\mathbf{A} = \mathbf{B}$ only if $a_{ij} = b_{ij}$ for all i and j.

If the rows and columns of an $m \times n$ matrix $\mathbf{A}$ are interchanged so that the elements in the ith row now become the elements of the ith column (for $i = 1, 2, \ldots, m$), the resulting matrix is called the *transpose* of $\mathbf{A}$ and is denoted by $\mathbf{A}^T$. It is evident that $\mathbf{A}^T$ is an $n \times m$ matrix. For example, if

$$\mathbf{A} = \begin{bmatrix} 2 & 1 \\ 3 & 2 \\ 1 & 3 \end{bmatrix} \quad \text{then} \quad \mathbf{A}^T = \begin{bmatrix} 2 & 3 & 1 \\ 1 & 2 & 3 \end{bmatrix}$$

Thus, if

$$\mathbf{A} = (a_{ij})_{m \times n}$$

then

$$\mathbf{A}^T = (a_{ji})_{n \times m} \tag{B.54}$$

Note that

$$(\mathbf{A}^T)^T = \mathbf{A} \tag{B.55}$$

B.6-2 Matrix Algebra

We shall now define matrix operations, such as addition, subtraction, multiplication, and division of matrices. The definitions should be formulated so that they are useful in the manipulation of matrices.

ADDITION OF MATRICES

For two matrices $\mathbf{A}$ and $\mathbf{B}$, both of the same order ($m \times n$),

$$\mathbf{A} = \begin{bmatrix} a_{11} & a_{12} & \cdots & a_{1n} \\ a_{21} & a_{22} & \cdots & a_{2n} \\ \vdots & \vdots & \cdots & \vdots \\ a_{m1} & a_{m2} & \cdots & a_{mn} \end{bmatrix} \quad \text{and} \quad \mathbf{B} = \begin{bmatrix} b_{11} & b_{12} & \cdots & b_{1n} \\ b_{21} & b_{22} & \cdots & b_{2n} \\ \vdots & \vdots & \cdots & \vdots \\ b_{m1} & b_{m2} & \cdots & b_{mn} \end{bmatrix}$$

we define the sum $\mathbf{A} + \mathbf{B}$ as

$$\mathbf{A} + \mathbf{B} = \begin{bmatrix} (a_{11} + b_{11}) & (a_{12} + b_{12}) & \cdots & (a_{1n} + b_{1n}) \\ (a_{21} + b_{21}) & (a_{22} + b_{22}) & \cdots & (a_{2n} + b_{2n}) \\ \vdots & \vdots & \cdots & \vdots \\ (a_{m1} + b_{m1}) & (a_{m2} + b_{m2}) & \cdots & (a_{mn} + b_{mn}) \end{bmatrix}$$

or

$$\mathbf{A} + \mathbf{B} = (a_{ij} + b_{ij})_{m \times n}$$

Note that two matrices can be added only if they are of the same order.

MULTIPLICATION OF A MATRIX BY A SCALAR

We multiply a matrix $\mathbf{A}$ by a scalar c as follows:

$$cA = c \begin{bmatrix} a_{11} & a_{12} & \cdots & a_{1n} \\ a_{21} & a_{22} & \cdots & a_{2n} \\ \vdots & \vdots & \cdots & \vdots \\ a_{m1} & a_{m2} & \cdots & a_{mn} \end{bmatrix} = \begin{bmatrix} ca_{11} & ca_{12} & \cdots & ca_{1n} \\ ca_{21} & ca_{22} & \cdots & ca_{2n} \\ \vdots & \vdots & \cdots & \vdots \\ ca_{m1} & ca_{m2} & \cdots & ca_{mn} \end{bmatrix} = \mathbf{A}c$$

We also observe that the scalar c and the matrix $\mathbf{A}$ commute, that is,

$$cA = Ac$$

MATRIX MULTIPLICATION

We define the product

$$\mathbf{AB} = \mathbf{C}$$

in which c_{ij}, the element of $\mathbf{C}$ in the ith row and jth column, is found by adding the products of the elements of $\mathbf{A}$ in the ith row with the corresponding elements of $\mathbf{B}$ in the jth column. Thus,

$$c_{ij} = a_{i1}b_{1j} + a_{i2}b_{2j} + \cdots + a_{in}b_{nj}$$

$$= \sum_{k=1}^{n} a_{ik}b_{kj} \tag{B.56}$$

This result is expressed as follows:

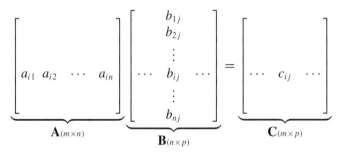

Note carefully that if this procedure is to work, the number of columns of $\mathbf{A}$ must be equal to the number of rows of $\mathbf{B}$. In other words, $\mathbf{AB}$, the product of matrices $\mathbf{A}$ and $\mathbf{B}$, is defined only if the number of columns of $\mathbf{A}$ is equal to the number of rows of $\mathbf{B}$. If this condition is not satisfied, the product $\mathbf{AB}$ is not defined and is meaningless. When the number of columns of $\mathbf{A}$ is equal to the number of rows of $\mathbf{B}$, matrix $\mathbf{A}$ is said to be *conformable* to matrix $\mathbf{B}$ for the product $\mathbf{AB}$. Observe that if $\mathbf{A}$ is an $m \times n$ matrix and $\mathbf{B}$ is an $n \times p$ matrix, $\mathbf{A}$ and $\mathbf{B}$ are conformable for the product, and $\mathbf{C}$ is an $m \times p$ matrix.

We demonstrate the use of the rule in Eq. (B.56) with the following examples.

$$\begin{bmatrix} 2 & 3 \\ 1 & 1 \\ 3 & 1 \end{bmatrix} \begin{bmatrix} 1 & 3 & 1 & 2 \\ 2 & 1 & 1 & 1 \end{bmatrix} = \begin{bmatrix} 8 & 9 & 5 & 7 \\ 3 & 4 & 2 & 3 \\ 5 & 10 & 4 & 7 \end{bmatrix}$$

$$[2 \quad 1 \quad 3] \begin{bmatrix} 2 \\ 1 \\ 1 \end{bmatrix} = 8$$

In both cases, the two matrices are conformable. However, if we interchange the order of the matrices as follows,

$$\begin{bmatrix} 1 & 3 & 1 & 2 \\ 2 & 1 & 1 & 1 \end{bmatrix} \begin{bmatrix} 2 & 3 \\ 1 & 1 \\ 3 & 1 \end{bmatrix}$$

the matrices are no longer conformable for the product. It is evident that in general,

$$\mathbf{AB} \neq \mathbf{BA} \tag{B.57}$$

Indeed, $\mathbf{AB}$ may exist and $\mathbf{BA}$ may not exist, or vice versa, as in our examples. We shall see later that for some special matrices, $\mathbf{AB} = \mathbf{BA}$. When this is true, matrices $\mathbf{A}$ and $\mathbf{B}$ are said to *commute*. We reemphasize that in general, matrices do not commute.

In the matrix product $\mathbf{AB}$, matrix $\mathbf{A}$ is said to be *postmultiplied* by $\mathbf{B}$ or matrix $\mathbf{B}$ is said to be *premultiplied* by $\mathbf{A}$. We may also verify the following relationships:

$$(\mathbf{A} + \mathbf{B})\mathbf{C} = \mathbf{AC} + \mathbf{BC} \tag{B.58}$$

$$\mathbf{C}(\mathbf{A} + \mathbf{B}) = \mathbf{CA} + \mathbf{CB} \tag{B.59}$$

We can verify that any matrix $\mathbf{A}$ premultiplied or postmultiplied by the identity matrix $\mathbf{I}$ remains unchanged:

$$\mathbf{AI} = \mathbf{IA} = \mathbf{A} \tag{B.60}$$

Of course, we must make sure that the order of $\mathbf{I}$ is such that the matrices are conformable for the corresponding product.

We give here, without proof, another important property of matrices:

$$|\mathbf{AB}| = |\mathbf{A}||\mathbf{B}| \tag{B.61}$$

where $|\mathbf{A}|$ and $|\mathbf{B}|$ represent determinants of matrices $\mathbf{A}$ and $\mathbf{B}$.

MULTIPLICATION OF A MATRIX BY A VECTOR

Consider the matrix Eq. (B.52), which represents Eq. (B.48). The right-hand side of Eq. (B.52) is a product of the $m \times n$ matrix $\mathbf{A}$ and a vector $\mathbf{x}$. If, for the time being, we treat the vector $\mathbf{x}$ as if it were an $n \times 1$ matrix, then the product $\mathbf{Ax}$, according to the matrix multiplication rule, yields the right-hand side of Eq. (B.48). Thus, we may multiply a matrix by a vector by treating the vector as if it were an $n \times 1$ matrix. Note that the constraint of conformability still applies. Thus, in this case, $\mathbf{xA}$ is not defined and is meaningless.

MATRIX INVERSION

To define the inverse of a matrix, let us consider the set of equations represented by Eq. (B.52):

$$
\begin{bmatrix} y_1 \\ y_2 \\ \vdots \\ y_n \end{bmatrix} =
\begin{bmatrix}
a_{11} & a_{12} & \cdots & a_{1n} \\
a_{21} & a_{22} & \cdots & a_{2n} \\
\vdots & \vdots & \cdots & \vdots \\
a_{n1} & a_{n2} & \cdots & a_{nn}
\end{bmatrix}
\begin{bmatrix} x_1 \\ x_2 \\ \vdots \\ x_n \end{bmatrix}
\tag{B.62a}
$$

We can solve this set of equations for $x_1, x_2, \ldots, x_n$ in terms of $y_1, y_2, \ldots, y_n$ by using Cramer's rule [see Eq. (B.31)]. This yields

$$
\begin{bmatrix} x_1 \\ x_2 \\ \vdots \\ x_n \end{bmatrix} =
\begin{bmatrix}
\dfrac{|\mathbf{D}_{11}|}{|\mathbf{A}|} & \dfrac{|\mathbf{D}_{21}|}{|\mathbf{A}|} & \cdots & \dfrac{|\mathbf{D}_{n1}|}{|\mathbf{A}|} \\
\dfrac{|\mathbf{D}_{12}|}{|\mathbf{A}|} & \dfrac{|\mathbf{D}_{22}|}{|\mathbf{A}|} & \cdots & \dfrac{|\mathbf{D}_{n2}|}{|\mathbf{A}|} \\
\vdots & \vdots & \cdots & \vdots \\
\dfrac{|\mathbf{D}_{1n}|}{|\mathbf{A}|} & \dfrac{|\mathbf{D}_{2n}|}{|\mathbf{A}|} & \cdots & \dfrac{|\mathbf{D}_{nn}|}{|\mathbf{A}|}
\end{bmatrix}
\begin{bmatrix} y_1 \\ y_2 \\ \vdots \\ y_n \end{bmatrix}
\tag{B.62b}
$$

in which $|\mathbf{A}|$ is the determinant of the matrix $\mathbf{A}$ and $|\mathbf{D}_{ij}|$ is the *cofactor* of element a_{ij} in the matrix $\mathbf{A}$. The cofactor of element a_{ij} is given by $(-1)^{i+j}$ times the determinant of the $(n-1) \times (n-1)$ matrix that is obtained when the ith row and the jth column in matrix $\mathbf{A}$ are deleted.

We can express Eq. (B.62a) in matrix form as

$$
\mathbf{y} = \mathbf{A}\mathbf{x}
\tag{B.63}
$$

We now define $\mathbf{A}^{-1}$, the inverse of a square matrix $\mathbf{A}$, with the property

$$
\mathbf{A}^{-1}\mathbf{A} = \mathbf{I} \qquad \text{(unit matrix)}
\tag{B.64}
$$

Then, premultiplying both sides of Eq. (B.63) by $\mathbf{A}^{-1}$, we obtain

$$
\mathbf{A}^{-1}\mathbf{y} = \mathbf{A}^{-1}\mathbf{A}\mathbf{x} = \mathbf{I}\mathbf{x} = \mathbf{x}
$$

or

$$
\mathbf{x} = \mathbf{A}^{-1}\mathbf{y}
\tag{B.65}
$$

A comparison of Eq. (B.65) with Eq. (B.62b) shows that

$$
\mathbf{A}^{-1} = \frac{1}{|\mathbf{A}|}
\begin{bmatrix}
|\mathbf{D}_{11}| & |\mathbf{D}_{21}| & \cdots & |\mathbf{D}_{n1}| \\
|\mathbf{D}_{12}| & |\mathbf{D}_{22}| & \cdots & |\mathbf{D}_{n2}| \\
\vdots & \vdots & \cdots & \vdots \\
|\mathbf{D}_{1n}| & |\mathbf{D}_{2n}| & \cdots & |\mathbf{D}_{nn}|
\end{bmatrix}
\tag{B.66}
$$

One of the conditions necessary for a unique solution of Eq. (B.62a) is that the number of equations must equal the number of unknowns. This implies that the matrix $\mathbf{A}$ must be a square matrix. In addition, we observe from the solution as given in Eq. (B.62b) that if the solution is to exist, $|\mathbf{A}| \neq 0$.[†] Therefore, the inverse exists only for a square matrix and only under the condition that the determinant of the matrix be nonzero. A matrix whose determinant is nonzero is a *nonsingular* matrix. Thus, an inverse exists only for a nonsingular, square matrix. By definition, we have

$$\mathbf{A}^{-1}\mathbf{A} = \mathbf{I} \tag{B.67a}$$

Postmultiplying this equation by $\mathbf{A}^{-1}$ and then premultiplying by $\mathbf{A}$, we can show that

$$\mathbf{A}\mathbf{A}^{-1} = \mathbf{I} \tag{B.67b}$$

Clearly, the matrices $\mathbf{A}$ and $\mathbf{A}^{-1}$ commute.

The operation of matrix division can be accomplished through matrix inversion.

EXAMPLE B.12

Let us find $\mathbf{A}^{-1}$ if

$$\mathbf{A} = \begin{bmatrix} 2 & 1 & 1 \\ 1 & 2 & 3 \\ 3 & 2 & 1 \end{bmatrix}$$

Here

$$|\mathbf{D}_{11}| = -4, \qquad |\mathbf{D}_{12}| = 8, \qquad |\mathbf{D}_{13}| = -4$$
$$|\mathbf{D}_{21}| = 1, \qquad |\mathbf{D}_{22}| = -1, \qquad |\mathbf{D}_{23}| = -1$$
$$|\mathbf{D}_{31}| = 1, \qquad |\mathbf{D}_{32}| = -5, \qquad |\mathbf{D}_{33}| = 3$$

and $|\mathbf{A}| = -4$. Therefore,

$$\mathbf{A}^{-1} = -\frac{1}{4} \begin{bmatrix} -4 & 1 & 1 \\ 8 & -1 & -5 \\ -4 & -1 & 3 \end{bmatrix}$$

B.6-3 Derivatives and Integrals of a Matrix

Elements of a matrix need not be constants; they may be functions of a variable. For example, if

$$\mathbf{A} = \begin{bmatrix} e^{-2t} & \sin t \\ e^t & e^{-t} + e^{-2t} \end{bmatrix} \tag{B.68}$$

then the matrix elements are functions of t. Here, it is helpful to denote $\mathbf{A}$ by $\mathbf{A}(t)$. Also, it would be helpful to define the derivative and integral of $\mathbf{A}(t)$.

The derivative of a matrix $\mathbf{A}(t)$ (with respect to t) is defined as a matrix whose ijth element is the derivative (with respect to t) of the ijth element of the matrix $\mathbf{A}$. Thus, if

$$\mathbf{A}(t) = [a_{ij}(t)]_{m \times n}$$

[†]These two conditions imply that the number of equations is equal to the number of unknowns and that all the equations are independent.

then

$$\frac{d}{dt}[\mathbf{A}(t)] = \left[\frac{d}{dt}a_{ij}(t)\right]_{m \times n} \tag{B.69a}$$

or

$$\dot{\mathbf{A}}(t) = [\dot{a}_{ij}(t)]_{m \times n} \tag{B.69b}$$

Thus, the derivative of the matrix in Eq. (B.68) is given by

$$\dot{\mathbf{A}}(t) = \begin{bmatrix} -2e^{-2t} & \cos t \\ e^t & -e^{-t} - 2e^{-2t} \end{bmatrix}$$

Similarly, we define the integral of $\mathbf{A}(t)$ (with respect to t) as a matrix whose ijth element is the integral (with respect to t) of the ijth element of the matrix $\mathbf{A}$:

$$\int \mathbf{A}(t)\, dt = \left(\int a_{ij}(t)\, dt \right)_{m \times n} \tag{B.70}$$

Thus, for the matrix $\mathbf{A}$ in Eq. (B.68), we have

$$\int \mathbf{A}(t)\, dt = \begin{bmatrix} \int e^{-2t}\, dt & \int \sin dt \\ \int e^t\, dt & \int (e^{-t} + 2e^{-2t})\, dt \end{bmatrix}$$

We can readily prove the following identities:

$$\frac{d}{dt}(\mathbf{A} + \mathbf{B}) = \frac{d\mathbf{A}}{dt} + \frac{d\mathbf{B}}{dt} \tag{B.71a}$$

$$\frac{d}{dt}(c\mathbf{A}) = c\frac{d\mathbf{A}}{dt} \tag{B.71b}$$

$$\frac{d}{dt}(\mathbf{AB}) = \frac{d\mathbf{A}}{dt}\mathbf{B} + \mathbf{A}\frac{d\mathbf{B}}{dt} = \dot{\mathbf{A}}\mathbf{B} + \mathbf{A}\dot{\mathbf{B}} \tag{B.71c}$$

The proofs of identities (B.71a) and (B.71b) are trivial. We can prove Eq. (B.71c) as follows. Let $\mathbf{A}$ be an $m \times n$ matrix and $\mathbf{B}$ an $n \times p$ matrix. Then, if

$$\mathbf{C} = \mathbf{AB}$$

from Eq. (B.56), we have

$$c_{ik} = \sum_{j=1}^{n} a_{ij}b_{jk}$$

and

$$\dot{c}_{ik} = \underbrace{\sum_{j=1}^{n} \dot{a}_{ij}b_{jk}}_{d_{ik}} + \underbrace{\sum_{j=1}^{n} a_{ij}\dot{b}_{jk}}_{e_{ik}} \tag{B.72}$$

or

$$\dot{c}_{ij} = d_{ij} + e_{ik}$$

Equation (B.72) along with the multiplication rule clearly indicate that d_{ik} is the ikth element of matrix $\dot{\mathbf{A}}\mathbf{B}$ and e_{ik} is the ikth element of matrix $\mathbf{A}\dot{\mathbf{B}}$. Equation (B.71c) then follows.

If we let $\mathbf{B} = \mathbf{A}^{-1}$ in Eq. (B.71c), we obtain

$$\frac{d}{dt}(\mathbf{A}\mathbf{A}^{-1}) = \frac{d\mathbf{A}}{dt}\mathbf{A}^{-1} + \mathbf{A}\frac{d}{dt}\mathbf{A}^{-1}$$

But since

$$\frac{d}{dt}(\mathbf{A}\mathbf{A}^{-1}) = \frac{d}{dt}\mathbf{I} = 0$$

we have

$$\frac{d}{dt}(\mathbf{A}^{-1}) = -\mathbf{A}^{-1}\frac{d\mathbf{A}}{dt}\mathbf{A}^{-1} \tag{B.73}$$

B.6-4 The Characteristic Equation of a Matrix: The Cayley−Hamilton Theorem

For an $(n \times n)$ square matrix $\mathbf{A}$, any vector $\mathbf{x}$ $(\mathbf{x} \neq 0)$ that satisfies the equation

$$\mathbf{A}\mathbf{x} = \lambda\mathbf{x} \tag{B.74}$$

is an *eigenvector* (or *characteristic vector*), and λ is the corresponding *eigenvalue* (or *characteristic value*) of $\mathbf{A}$. Equation (B.74) can be expressed as

$$(\mathbf{A} - \lambda\mathbf{I})\mathbf{x} = 0 \tag{B.75}$$

The solution for this set of homogeneous equations exists if and only if

$$|\mathbf{A} - \lambda\mathbf{I}| = |\lambda\mathbf{I} - \mathbf{A}| = 0 \tag{B.76a}$$

or

$$\begin{vmatrix} a_{11} - \lambda & a_{12} & \cdots & a_{1n} \\ a_{21} & a_{22} - \lambda & \cdots & a_{2n} \\ \vdots & \vdots & \cdots & \vdots \\ a_{n1} & a_{n2} & \cdots & a_{nn} - \lambda \end{vmatrix} = 0 \tag{B.76b}$$

Equation (B.76a) [or (B.76b)] is known as the *characteristic equation* of the matrix $\mathbf{A}$ and can be expressed as

$$Q(\lambda) = |\lambda\mathbf{I} - \mathbf{A}| = \lambda^n + a_{n-1}\lambda^{n-1} + \cdots + a_1\lambda + a_0\lambda^0 = 0 \tag{B.77}$$

$Q(\lambda)$ is called the *characteristic polynomial* of the matrix $\mathbf{A}$. The n zeros of the characteristic polynomial are the eigenvalues of $\mathbf{A}$ and, corresponding to each eigenvalue, there is an eigenvector that satisfies Eq. (B.74).

The *Cayley–Hamilton theorem* states that every $n \times n$ matrix $\mathbf{A}$ satisfies its own characteristic equation. In other words, Eq. (B.77) is valid if λ is replaced by $\mathbf{A}$:

$$Q(\mathbf{A}) = \mathbf{A}^n + a_{n-1}\mathbf{A}^{n-1} + \cdots + a_1\mathbf{A} + a_0\mathbf{A}^0 = 0 \tag{B.78}$$

FUNCTIONS OF A MATRIX

We now demonstrate the use of the Cayley–Hamilton theorem to evaluate functions of a square matrix $\mathbf{A}$.

Consider a function $f(\lambda)$ in the form of an infinite power series:

$$f(\lambda) = \beta_0 + \beta_1\lambda + \beta_2\lambda_2^2 + \cdots + \cdots = \sum_{i=0}^{\infty} \beta_i \lambda^i \qquad \text{(B.79)}$$

Since λ, being an eigenvalue (characteristic root) of $\mathbf{A}$, satisfies the characteristic equation [Eq. (B.77)], we can write

$$\lambda^n = -a_{n-1}\lambda^{n-1} - a_{n-2}\lambda^{n-2} - \cdots - a_1\lambda - a_0 \qquad \text{(B.80)}$$

If we multiply both sides by λ, the left-hand side is λ^{n+1}, and the right-hand side contains the terms λ^n, λ^{n-1}, ..., λ. Using Eq. (B.80), we substitute λ^n in terms of λ^{n-1}, λ^{n-2}, ..., λ so that the highest power on the right-hand side is reduced to $n-1$. Continuing in this way, we see that λ^{n+k} can be expressed in terms of λ^{n-1}, λ^{n-2}, ..., λ for any k. Hence, the infinite series on the right-hand side of Eq. (B.79) can always be expressed in terms of λ^{n-1}, λ^{n-2}, ..., λ and a constant as

$$f(\lambda) = \beta_0 + \beta_1\lambda + \beta_2\lambda^2 + \cdots + \beta_{n-1}\lambda^{n-1} \qquad \text{(B.81)}$$

If we assume that there are n distinct eigenvalues $\lambda_1, \lambda_2, \ldots, \lambda_n$, then Eq. (B.81) holds for these n values of λ. The substitution of these values in Eq. (B.81) yields n simultaneous equations

$$\begin{bmatrix} f(\lambda_1) \\ f(\lambda_2) \\ \vdots \\ f(\lambda_n) \end{bmatrix} = \begin{bmatrix} 1 & \lambda_1 & \lambda_1^2 & \cdots & \lambda_1^{n-1} \\ 1 & \lambda_2 & \lambda_2^2 & \cdots & \lambda_2^{n-1} \\ \vdots & \vdots & \vdots & \cdots & \vdots \\ 1 & \lambda_n & \lambda_n^2 & \cdots & \lambda_n^{n-1} \end{bmatrix} \begin{bmatrix} \beta_0 \\ \beta_1 \\ \vdots \\ \beta_{n-1} \end{bmatrix} \qquad \text{(B.82a)}$$

and

$$\begin{bmatrix} \beta_0 \\ \beta_1 \\ \vdots \\ \beta_{n-1} \end{bmatrix} = \begin{bmatrix} 1 & \lambda_1 & \lambda_1^2 & \cdots & \lambda_1^{n-1} \\ 1 & \lambda_2 & \lambda_2^2 & \cdots & \lambda_2^{n-1} \\ \vdots & \vdots & \vdots & \cdots & \vdots \\ 1 & \lambda_n & \lambda_n^2 & \cdots & \lambda_n^{n-1} \end{bmatrix}^{-1} \begin{bmatrix} f(\lambda_1) \\ f(\lambda_2) \\ \vdots \\ f(\lambda_n) \end{bmatrix} \qquad \text{(B.82b)}$$

Since $\mathbf{A}$ also satisfies Eq. (B.80), we may advance a similar argument to show that if $f(\mathbf{A})$ is a function of a square matrix $\mathbf{A}$ expressed as an infinite power series in $\mathbf{A}$, then

$$f(\mathbf{A}) = \beta_0\mathbf{I} + \beta_1\mathbf{A} + \beta_2\mathbf{A}^2 + \cdots + \cdots = \sum_{i=0}^{\infty} \beta_i \mathbf{A}^i \qquad \text{(B.83a)}$$

and, as argued earlier, the right-hand side can be expressed using terms of power less than or equal to $n-1$,

$$f(\mathbf{A}) = \beta_0\mathbf{I} + \beta_1\mathbf{A} + \beta_2\mathbf{A}^2 + \cdots + \beta_{n-1}\mathbf{A}^{n-1} \qquad \text{(B.83b)}$$

in which the coefficients β_is are found from Eq. (B.82b). If some of the eigenvalues are repeated (multiple roots), the results are somewhat modified.

We shall demonstrate the utility of this result with the following two examples.

B.6-5 Computation of an Exponential and a Power of a Matrix

Let us compute $e^{\mathbf{A}t}$ defined by

$$e^{\mathbf{A}t} = \mathbf{I} + \mathbf{A}t + \frac{\mathbf{A}^2 t^2}{2!} + \cdots + \frac{\mathbf{A}^n t^n}{n!} + \cdots$$

$$= \sum_{k=0}^{\infty} \frac{\mathbf{A}^k t^k}{k!}$$

From Eq. (B.83b), we can express

$$e^{\mathbf{A}t} = \sum_{i=1}^{n-1} \beta_i (\mathbf{A})^i$$

in which the β_is are given by Eq. (B.82b), with $f(\lambda_i) = e^{\lambda_i t}$.

EXAMPLE B.13

Let us consider

$$\mathbf{A} = \begin{bmatrix} 0 & 1 \\ -2 & -3 \end{bmatrix}$$

The characteristic equation is

$$|\lambda \mathbf{I} - \mathbf{A}| = \begin{vmatrix} \lambda & -1 \\ 2 & \lambda + 3 \end{vmatrix} = \lambda^2 + 3\lambda + 2 = (\lambda + 1)(\lambda + 2) = 0$$

Hence, the eigenvalues are $\lambda_1 = -1$, $\lambda_2 = -2$, and

$$e^{\mathbf{A}t} = \beta_0 \mathbf{I} + \beta_1 \mathbf{A}$$

in which

$$\begin{bmatrix} \beta_0 \\ \beta_1 \end{bmatrix} = \begin{bmatrix} 1 & -1 \\ 1 & -2 \end{bmatrix}^{-1} \begin{bmatrix} e^{-t} \\ e^{-2t} \end{bmatrix}$$

$$= \begin{bmatrix} 2 & -1 \\ 1 & -1 \end{bmatrix} \begin{bmatrix} e^{-t} \\ e^{-2t} \end{bmatrix} = \begin{bmatrix} 2e^{-t} - e^{-2t} \\ e^{-t} - e^{-2t} \end{bmatrix}$$

and

$$e^{\mathbf{A}t} = (2e^{-t} - e^{-2t}) \begin{bmatrix} 1 & 0 \\ 0 & 1 \end{bmatrix} + (e^{-t} - e^{-2t}) \begin{bmatrix} 0 & 1 \\ -2 & -3 \end{bmatrix}$$

$$= \begin{bmatrix} 2e^{-t} - e^{-2t} & (e^{-t} - e^{-2t}) \\ -2e^{-t} + 2e^{-2t} & -e^{-t} + 2e^{-2t} \end{bmatrix} \qquad \text{(B.84)}$$

COMPUTATION OF $\mathbf{A}^k$

As Eq. (B.83b) indicates, we can express $\mathbf{A}^k$ as

$$\mathbf{A}^k = \beta_0\mathbf{I} + \beta_1\mathbf{A} + \cdots + \beta_{n-1}\mathbf{A}^{n-1}$$

in which the β_is are given by Eq. (B.82b) with $f(\lambda_i) = \lambda_i^k$. For a completed example of the computation of $\mathbf{A}^k$ by this method, see Example 10.12.

B.7 MISCELLANEOUS

B.7-1 L'Hôpital's Rule

If $\lim f(x)/g(x)$ results in the indeterministic form $0/0$ or ∞/∞, then

$$\lim \frac{f(x)}{g(x)} = \lim \frac{\dot{f}(x)}{\dot{g}(x)}$$

B.7-2 The Taylor and Maclaurin Series

$$f(x) = f(a) + \frac{(x-a)}{1!}\dot{f}(a) + \frac{(x-a)^2}{2!}\ddot{f}(a) + \cdots$$

$$f(x) = f(0) + \frac{x}{1!}\dot{f}(0) + \frac{x^2}{2!}\ddot{f}(0) + \cdots$$

B.7-3 Power Series

$$e^x = 1 + x + \frac{x^2}{2!} + \frac{x^3}{3!} + \cdots + \frac{x^n}{n!} + \cdots$$

$$\sin x = x - \frac{x^3}{3!} + \frac{x^5}{5!} - \frac{x^7}{7!} + \cdots$$

$$\cos x = 1 - \frac{x^2}{2!} + \frac{x^4}{4!} - \frac{x^6}{6!} + \frac{x^8}{8!} - \cdots$$

$$\tan x = x + \frac{x^3}{3} + \frac{2x^5}{15} + \frac{17x^7}{315} + \cdots \qquad x^2 < \pi^2/4$$

$$\tanh x = x - \frac{x^3}{3} + \frac{2x^5}{15} - \frac{17x^7}{315} + \cdots \qquad x^2 < \pi^2/4$$

$$(1+x)^n = 1 + nx + \frac{n(n-1)}{2!}x^2 + \frac{n(n-1)(n-2)}{3!}x^3 + \cdots + \binom{n}{k}x^k + \cdots + x^n$$

$$\approx 1 + nx \qquad |x| \ll 1$$

$$\frac{1}{1-x} = 1 + x + x^2 + x^3 + \cdots \qquad |x| < 1$$

B.7-4 Sums

$$\sum_{k=m}^{n} r^k = \frac{r^{n+1} - r^m}{r - 1} \qquad r \neq 1$$

$$\sum_{k=0}^{n} k = \frac{n(n + 1)}{2}$$

$$\sum_{k=0}^{n} k^2 = \frac{n(n + 1)(2n + 1)}{6}$$

$$\sum_{k=0}^{n} k\, r^k = \frac{r + [n\,(r - 1) - 1]\, r^{n+1}}{(r - 1)^2} \qquad r \neq 1$$

$$\sum_{k=0}^{n} k^2\, r^k = \frac{r[(1 + r)(1 - r^n) - 2n(1 - r)r^n - n^2(1 - r)^2 r^n]}{(1 - r)^3} \qquad r \neq 1$$

B.7-5 Complex Numbers

$$e^{\pm j\pi/2} = \pm j$$

$$e^{\pm jn\pi} = \begin{cases} 1 & n \text{ even} \\ -1 & n \text{ odd} \end{cases}$$

$$e^{\pm j\theta} = \cos \theta \pm j \sin \theta$$

$$a + jb = re^{j\theta} \qquad r = \sqrt{a^2 + b^2}, \theta = \tan^{-1}\left(\frac{b}{a}\right)$$

$$(re^{j\theta})^k = r^k e^{jk\theta}$$

$$(r_1 e^{j\theta_1})(r_2 e^{j\theta_2}) = r_1 r_2 e^{j(\theta_1 + \theta_2)}$$

B.7-6 Trigonometric Identities

$$e^{\pm jx} = \cos x \pm j \sin x$$

$$\cos x = \frac{1}{2}[e^{jx} + e^{-jx}]$$

$$\sin x = \frac{1}{2j}[e^{jx} - e^{-jx}]$$

$$\cos\left(x \pm \frac{\pi}{2}\right) = \mp \sin x$$

$$\sin\left(x \pm \frac{\pi}{2}\right) = \pm \cos x$$

$$2 \sin x \cos x = \sin 2x$$

$$\sin^2 x + \cos^2 x = 1$$

$$\cos^2 x - \sin^2 x = \cos 2x$$

$$\cos^2 x = \frac{1}{2}(1 + \cos 2x)$$

$$\sin^2 x = \frac{1}{2}(1 - \cos 2x)$$

$$\cos^3 x = \frac{1}{4}(3 \cos x + \cos 3x)$$

$$\sin^3 x = \tfrac{1}{4}(3 \sin x - \sin 3x)$$

$$\sin (x \pm y) = \sin x \cos y \pm \cos x \sin y$$

$$\cos (x \pm y) = \cos x \cos y \mp \sin x \sin y$$

$$\tan (x \pm y) = \frac{\tan x \pm \tan y}{1 \mp \tan x \tan y}$$

$$\sin x \sin y = \tfrac{1}{2}[\cos (x - y) - \cos (x + y)]$$

$$\cos x \cos y = \tfrac{1}{2}[\cos (x - y) + \cos (x + y)]$$

$$\sin x \cos y = \tfrac{1}{2}[\sin (x - y) + \sin (x + y)]$$

$$a \cos x + b \sin x = C \cos (x + \theta) \qquad C = \sqrt{a^2 + b^2}, \theta = \tan^{-1}\left(\tfrac{-b}{a}\right)$$

B.7-7 Indefinite Integrals

$$\int u \, dv = uv - \int v \, du$$

$$\int f(x)\dot{g}(x) \, dx = f(x)g(x) - \int \dot{f}(x)g(x) \, dx$$

$$\int \sin ax \, dx = -\frac{1}{a} \cos ax \qquad \int \cos ax \, dx = \frac{1}{a} \sin ax$$

$$\int \sin^2 ax \, dx = \frac{x}{2} - \frac{\sin 2ax}{4a} \qquad \int \cos^2 ax \, dx = \frac{x}{2} + \frac{\sin 2ax}{4a}$$

$$\int x \sin ax \, dx = \frac{1}{a^2}(\sin ax - ax \cos ax)$$

$$\int x \cos ax \, dx = \frac{1}{a^2}(\cos ax + ax \sin ax)$$

$$\int x^2 \sin ax \, dx = \frac{1}{a^3}(2ax \sin ax + 2 \cos ax - a^2 x^2 \cos ax)$$

$$\int x^2 \cos ax \, dx = \frac{1}{a^3}(2ax \cos ax - 2 \sin ax + a^2 x^2 \sin ax)$$

$$\int \sin ax \sin bx \, dx = \frac{\sin (a - b)x}{2(a - b)} - \frac{\sin (a + b)x}{2(a + b)} \qquad a^2 \ne b^2$$

$$\int \sin ax \cos bx \, dx = -\left[\frac{\cos (a - b)x}{2(a - b)} + \frac{\cos (a + b)x}{2(a + b)}\right] \qquad a^2 \ne b^2$$

$$\int \cos ax \cos bx \, dx = \frac{\sin (a - b)x}{2(a - b)} + \frac{\sin (a + b)x}{2(a + b)} \qquad a^2 \ne b^2$$

$$\int e^{ax}\, dx = \frac{1}{a} e^{ax}$$

$$\int x e^{ax}\, dx = \frac{e^{ax}}{a^2}(ax - 1)$$

$$\int x^2 e^{ax}\, dx = \frac{e^{ax}}{a^3}(a^2 x^2 - 2ax + 2)$$

$$\int e^{ax} \sin bx\, dx = \frac{e^{ax}}{a^2 + b^2}(a \sin bx - b \cos bx)$$

$$\int e^{ax} \cos bx\, dx = \frac{e^{ax}}{a^2 + b^2}(a \cos bx + b \sin bx)$$

$$\int \frac{1}{x^2 + a^2}\, dx = \frac{1}{a} \tan^{-1} \frac{x}{a}$$

$$\int \frac{x}{x^2 + a^2}\, dx = \frac{1}{2} \ln(x^2 + a^2)$$

B.7-8 Common Derivative Formulas

$$\frac{d}{dx} f(u) = \frac{d}{du} f(u) \frac{du}{dx}$$

$$\frac{d}{dx}(uv) = u \frac{dv}{dx} + v \frac{du}{dx}$$

$$\frac{d}{dx}\left(\frac{u}{v}\right) = \frac{v \frac{du}{dx} - u \frac{dv}{dx}}{v^2}$$

$$\frac{dx^n}{dx} = nx^{n-1}$$

$$\frac{d}{dx} \ln(ax) = \frac{1}{x}$$

$$\frac{d}{dx} \log(ax) = \frac{\log e}{x}$$

$$\frac{d}{dx} e^{bx} = b e^{bx}$$

$$\frac{d}{dx} a^{bx} = b(\ln a) a^{bx}$$

$$\frac{d}{dx} \sin ax = a \cos ax$$

$$\frac{d}{dx} \cos ax = -a \sin ax$$

$$\frac{d}{dx} \tan ax = \frac{a}{\cos^2 ax}$$

$$\frac{d}{dx}(\sin^{-1} ax) = \frac{a}{\sqrt{1 - a^2 x^2}}$$

$$\frac{d}{dx}(\cos^{-1} ax) = \frac{-a}{\sqrt{1 - a^2 x^2}}$$

$$\frac{d}{dx}(\tan^{-1} ax) = \frac{a}{1 + a^2 x^2}$$

B.7-9 Some Useful Constants

$\pi \approx 3.1415926535$

$e \approx 2.7182818284$

$\dfrac{1}{e} \approx 0.3678794411$

$\log_{10} 2 = 0.30103$

$\log_{10} 3 = 0.47712$

B.7-10 Solution of Quadratic and Cubic Equations

Any *quadratic* equation can be reduced to the form

$$ax^2 + bx + c = 0$$

The solution of this equation is provided by

$$x = \frac{-b \pm \sqrt{b^2 - 4ac}}{2a}$$

A general *cubic* equation

$$y^3 + py^2 + qy + r = 0$$

may be reduced to the *depressed cubic* form

$$x^3 + ax + b = 0$$

by substituting

$$y = x - \frac{p}{3}$$

This yields

$$a = \tfrac{1}{3}(3q - p^2) \qquad b = \tfrac{1}{27}(2p^3 - 9pq + 27r)$$

Now let

$$A = \sqrt[3]{-\frac{b}{2} + \sqrt{\frac{b^2}{4} + \frac{a^3}{27}}} \qquad B = \sqrt[3]{-\frac{b}{2} - \sqrt{\frac{b^2}{4} + \frac{a^3}{27}}}$$

The solution of the depressed cubic is

$$x = A + B, \qquad x = -\frac{A+B}{2} + \frac{A-B}{2}\sqrt{-3}, \qquad x = -\frac{A+B}{2} - \frac{A-B}{2}\sqrt{-3}$$

and

$$y = x - \frac{p}{3}$$

REFERENCES

1. Asimov, Isaac. *Asimov on Numbers.* Bell Publishing, New York, 1982.
2. Calinger, R., ed. *Classics of Mathematics.* Moore Publishing, Oak Park, IL, 1982.
3. Hogben, Lancelot. *Mathematics in the Making.* Doubleday, New York, 1960.
4. Cajori, Florian. *A History of Mathematics,* 4th ed. Chelsea, New York, 1985.
5. Encyclopaedia Britannica. *Micropaedia* IV, 15th ed., vol. 11, p. 1043. Chicago, 1982.
6. Singh, Jagjit. *Great Ideas of Modern Mathematics.* Dover, New York, 1959.
7. Dunham, William. *Journey Through Genius.* Wiley, New York, 1990.

MATLAB SESSION B: ELEMENTARY OPERATIONS

MB.1 MATLAB Overview

Although MATLAB® (a registered trademark of The MathWorks, Inc.) is easy to use, it can be intimidating to new users. Over the years, MATLAB has evolved into a sophisticated computational package with thousands of functions and thousands of pages of documentation. This section provides a brief introduction to the software environment.

When MATLAB is first launched, its command window appears. When MATLAB is ready to accept an instruction or input, a command prompt (>>) is displayed in the command window. Nearly all MATLAB activity is initiated at the command prompt.

Entering instructions at the command prompt generally results in the creation of an object or objects. Many classes of objects are possible, including functions and strings, but usually objects are just data. Objects are placed in what is called the MATLAB workspace. If not visible, the workspace can be viewed in a separate window by typing `workspace` at the command prompt. The workspace provides important information about each object, including the object's name, size, and class.

Another way to view the workspace is the `whos` command. When `whos` is typed at the command prompt, a summary of the workspace is printed in the command window. The `who` command is a short version of `whos` that reports only the names of workspace objects.

Several functions exist to remove unnecessary data and help free system resources. To remove specific variables from the workspace, the `clear` command is typed, followed by the names of the variables to be removed. Just typing `clear` removes all objects from the workspace. Additionally, the `clc` command clears the command window, and the `clf` command clears the current figure window.

Often, important data and objects created in one session need to be saved for future use. The `save` command, followed by the desired filename, saves the entire workspace to a file, which has the .mat extension. It is also possible to selectively save objects by typing `save` followed by the filename and then the names of the objects to be saved. The `load` command followed by the filename is used to load the data and objects contained in a MATLAB data file (.mat file).

Although MATLAB does not automatically save workspace data from one session to the next, lines entered at the command prompt are recorded in the command history. Previous command lines can be viewed, copied, and executed directly from the command history window. From the command window, pressing the up or down arrow key scrolls through previous commands and redisplays them at the command prompt. Typing the first few characters and then pressing the arrow keys scrolls through the previous commands that start with the same characters. The arrow keys allow command sequences to be repeated without retyping.

Perhaps the most important and useful command for new users is `help`. To learn more about a function, simply type `help` followed by the function name. Helpful text is then displayed in the command window. The obvious shortcoming of `help` is that the function name must first be known. This is especially limiting for MATLAB beginners. Fortunately, help screens often conclude by referencing related or similar functions. These references are an excellent way to learn new MATLAB commands. Typing `help help`, for example, displays detailed information on the `help` command itself and also provides reference to relevant functions such as the `lookfor` command. The `lookfor` command helps locate MATLAB functions based on a keyword search. Simply type `lookfor` followed by a single keyword, and MATLAB searches for functions that contain that keyword.

MATLAB also has comprehensive HTML-based help. The HTML help is accessed by using MATLAB's integrated help browser, which also functions as a standard web browser. The HTML help facility includes a function and topic index as well as full text-searching capabilities. Since HTML documents can contain graphics and special characters, HTML help can provide more information than the command-line help. With a little practice, MATLAB makes it very easy to find information.

When MATLAB graphics are created, the `print` command can save figures in a common file format such as postscript, encapsulated postscript, JPEG, or TIFF. The format of displayed data, such as the number of digits displayed, is selected by using the `format` command. MATLAB help provides the necessary details for both these functions. When a MATLAB session is complete, the `exit` command terminates MATLAB.

MB.2 Calculator Operations

MATLAB can function as a simple calculator, working as easily with complex numbers as with real numbers. Scalar addition, subtraction, multiplication, division, and exponentiation are accomplished using the traditional operator symbols +, -, *, /, and ^. Since MATLAB predefines $i = j = \sqrt{-1}$, a complex constant is readily created using cartesian coordinates. For example,

```
>> z = -3-j*4
z = -3.0000 - 4.0000i
```

assigns the complex constant $-3 - j4$ to the variable z.

The real and imaginary components of z are extracted by using the `real` and `imag` operators. In MATLAB, the input to a function is placed parenthetically following the function name.

```
>> z_real = real(z); z_imag = imag(z);
```

When a line is terminated with a semicolon, the statement is evaluated but the results are not displayed to the screen. This feature is useful when one is computing intermediate results, and it allows multiple instructions on a single line. Although not displayed, the results `z_real = -3` and `z_imag = -4` are calculated and available for additional operations such as computing $|z|$.

There are many ways to compute the modulus, or magnitude, of a complex quantity. Trigonometry confirms that $z = -3 - j4$, which corresponds to a 3-4-5 triangle, has modulus $|z| = |-3 - j4| = \sqrt{(-3)^2 + (-4)^2} = 5$. The MATLAB sqrt command provides one way to compute the required square root.

```
>> z_mag = sqrt(z_real^2 + z_imag^2)
z_mag = 5
```

In MATLAB, most commands, include sqrt, accept inputs in a variety of forms including constants, variables, functions, expressions, and combinations thereof.

The same result is also obtained by computing $|z| = \sqrt{zz^*}$. In this case, complex conjugation is performed by using the conj command.

```
>> z_mag = sqrt(z*conj(z))
z_mag = 5
```

More simply, MATLAB computes absolute values directly by using the abs command.

```
>> z_mag = abs(z)
z_mag = 5
```

In addition to magnitude, polar notation requires phase information. The angle command provides the angle of a complex number.

```
>> z_rad = angle(z)
z_rad = -2.2143
```

MATLAB expects and returns angles in a radian measure. Angles expressed in degrees require an appropriate conversion factor.

```
>> z_deg = angle(z)*180/pi
z_deg = -126.8699
```

Notice, MATLAB predefines the variable $pi = \pi$.

It is also possible to obtain the angle of z using a two-argument arc-tangent function, atan2.

```
>> z_rad = atan2(z_imag,z_real)
z_rad = -2.2143
```

Unlike a single-argument arctangent function, the two-argument arctangent function ensures that the angle reflects the proper quadrant. MATLAB supports a full complement of trigonometric functions: standard trigonometric functions cos, sin, tan; reciprocal trigonometric functions sec, csc, cot; inverse trigonometric functions acos, asin, atan, asec, acsc, acot; and hyperbolic variations cosh, sinh, tanh, sech, csch, coth, acosh, asinh, atanh, asech, acsch, and acoth. Of course, MATLAB comfortably supports complex arguments for any trigonometric function. As with the angle command, MATLAB trigonometric functions utilize units of radians.

The concept of trigonometric functions with complex-valued arguments is rather intriguing. The results can contradict what is often taught in introductory mathematics courses. For example, a common claim is that $|\cos(x)| \leq 1$. While this is true for real x, it is not necessarily true for complex x. This is readily verified by example using MATLAB and the cos function.

```
>> cos(j)
ans = 1.5431
```

Problem B.19 investigates these ideas further.

Similarly, the claim that it is impossible to take the logarithm of a negative number is false. For example, the principal value of $\ln(-1)$ is $j\pi$, a fact easily verified by means of Euler's equation. In MATLAB, base-10 and base-e logarithms are computed by using the log10 and log commands, respectively.

```
>> log(-1)
ans = 0 + 3.1416i
```

MB.3 Vector Operations

The power of MATLAB becomes apparent when vector arguments replace scalar arguments. Rather than computing one value at a time, a single expression computes many values. Typically, vectors are classified as row vectors or column vectors. For now, we consider the creation of row vectors with evenly spaced, real elements. To create such a vector, the notation a:b:c is used, where a is the initial value, b designates the step size, and c is the termination value. For example, 0:2:11 creates the length-6 vector of even-valued integers ranging from 0 to 10.

```
>> k = 0:2:11
k =  0    2    4    6    8    10
```

In this case, the termination value does not appear as an element of the vector. Negative and noninteger step sizes are also permissible.

```
>> k = 11:-10/3:0
k = 11.0000    7.6667    4.3333    1.0000
```

If a step size is not specified, a value of one is assumed.

```
>> k = 0:11
k =  0    1    2    3    4    5    6    7    8    9    10    11
```

Vector notation provides the basis for solving a wide variety of problems.

For example, consider finding the three cube roots of minus one, $w^3 = -1 = e^{j(\pi+2\pi k)}$ for integer k. Taking the cube root of each side yields $w = e^{j(\pi/3+2\pi k/3)}$. To find the three unique solutions, use any three consecutive integer values of k and MATLAB's exp function.

```
>> k = 0:2;
>> w = exp(j*(pi/3 + 2*pi*k/3))
w = 0.5000 + 0.8660i   -1.0000 + 0.0000i    0.5000 - 0.8660i
```

The solutions, particularly $w = -1$, are easy to verify.

Finding the 100 unique roots of $w^{100} = -1$ is just as simple.

```
>> k = 0:99;
>> w = exp(j*(pi/100 + 2*pi*k/100));
```

A semicolon concludes the final instruction to suppress the inconvenient display of all 100 solutions. To view a particular solution, the user must specify an index. In MATLAB, ascending positive integer indices specify particular vector elements. For example, the fifth element of w is extracted using an index of 5.

```
>> w(5)
ans = 0.9603 + 0.2790i
```

Notice that this solution corresponds to $k = 4$. The independent variable of a function, in this case k, rarely serves as the index. Since k is also a vector, it can likewise be indexed. In this way, we can verify that the fifth value of k is indeed 4.

```
>> k(5)
ans = 4
```

It is also possible to use a vector index to access multiple values. For example, index vector $98:100$ identifies the last three solutions corresponding to $k = [97, 98, 99]$.

```
>> w(98:100)
ans = 0.9877 - 0.1564i    0.9956 - 0.0941i    0.9995 - 0.0314i
```

Vector representations provide the foundation to rapidly create and explore various signals. Consider the simple 10 Hz sinusoid described by $f(t) = \sin(2\pi 10t + \pi/6)$. Two cycles of this sinusoid are included in the interval $0 \le t < 0.2$. A vector t is used to uniformly represent 500 points over this interval.

```
>> t = 0:0.2/500:0.2-0.2/500;
```

Next, the function $f(t)$ is evaluated at these points.

```
>> f = sin(2*pi*10*t+pi/6);
```

The value of $f(t)$ at $t = 0$ is the first element of the vector and is thus obtained by using an index of one.

```
>> f(1)
ans = 0.5000
```

Unfortunately, MATLAB's indexing syntax conflicts with standard equation notation.[†] That is, the MATLAB indexing command f(1) is not the same as the standard notation $f(1) = f(t)|_{t=1}$. Care must be taken to avoid confusion; remember that the index parameter rarely reflects the independent variable of a function.

MB.4 Simple Plotting

MATLAB's plot command provides a convenient way to visualize data, such as graphing $f(t)$ against the independent variable t.

```
>> plot(t,f);
```

[†] Advanced structures such as MATLAB inline objects are an exception.

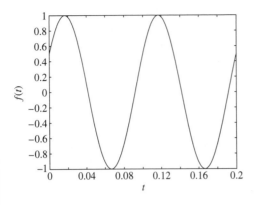

Figure MB.1 $f(t) = \sin(2\pi\, 10t + \pi/6)$.

Axis labels are added using the `xlabel` and `ylabel` commands, where the desired string must be enclosed by single quotation marks. The result is shown in Fig. MB.1.

```
>> xlabel('t'); ylabel('f(t)')
```

The `title` command is used to add a title above the current axis.

By default, MATLAB connects data points with solid lines. Plotting discrete points, such as the 100 unique roots of $w^{100} = -1$, is accommodated by supplying the `plot` command with an additional string argument. For example, the string `'o'` tells MATLAB to mark each data point with a circle rather than connecting points with lines. A full description of the supported plot options is available from MATLAB's help facilities.

```
>> plot(real(w),imag(w),'o');
>> xlabel('Re(w)'); ylabel('Im(w)');
>> axis equal
```

The `axis equal` command ensures that the scale used for the horizontal axis is equal to the scale used for the vertical axis. Without `axis equal`, the plot would appear elliptical rather than circular. Figure MB.2 illustrates that the 100 unique roots of $w^{100} = -1$ lie equally spaced on the unit circle, a fact not easily discerned from the raw numerical data.

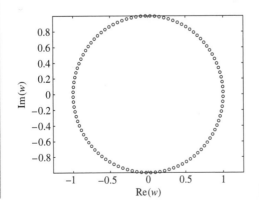

Figure MB.2 Unique roots of $w^{100} = -1$.

MATLAB also includes many specialized plotting functions. For example, MATLAB commands `semilogx`, `semilogy`, and `loglog` operate like the `plot` command but use base-10 logarithmic scales for the horizontal axis, vertical axis, and the horizontal and vertical axes, respectively. Monochrome and color images can be displayed by using the `image` command, and contour plots are easily created with the `contour` command. Furthermore, a variety of three-dimensional plotting routines are available, such as `plot3`, `contour3`, `mesh`, and `surf`. Information about these instructions, including examples and related functions, is available from MATLAB help.

MB.5 Element-by-Element Operations

Suppose a new function $h(t)$ is desired that forces an exponential envelope on the sinusoid $f(t)$, $h(t) = f(t)g(t)$ where $g(t) = e^{-10t}$. First, row vector $g(t)$ is created.

```
>> g = exp(-10*t);
```

Given MATLAB's vector representation of $g(t)$ and $f(t)$, computing $h(t)$ requires some form of vector multiplication. There are three standard ways to multiply vectors: inner product, outer product, and element-by-element product. As a matrix-oriented language, MATLAB defines the standard multiplication operator `*` according to the rules of matrix algebra: the multiplicand must be conformable to the multiplier. A $1 \times N$ row vector times an $N \times 1$ column vector results in the scalar-valued inner product. An $N \times 1$ column vector times a $1 \times M$ row vector results in the outer product, which is an $N \times M$ matrix. Matrix algebra prohibits multiplication of two row vectors or multiplication of two column vectors. Thus, the `*` operator is not used to perform element-by-element multiplication.[†]

Element-by-element operations require vectors to have the same dimensions. An error occurs if element-by-element operations are attempted between row and column vectors. In such cases, one vector must first be transposed to ensure both vector operands have the same dimensions. In MATLAB, most element-by-element operations are preceded by a period. For example, element-by-element multiplication, division, and exponentiation are accomplished using `.*`, `./`, and `.^`, respectively. Vector addition and subtraction are intrinsically element-by-element operations and require no period. Intuitively, we know $h(t)$ should be the same size as both $g(t)$ and $f(t)$. Thus, $h(t)$ is computed using element-by-element multiplication.

```
>> h = f.*g;
```

The `plot` command accommodates multiple curves and also allows modification of line properties. This facilitates side-by-side comparison of different functions, such as $h(t)$ and $f(t)$. Line characteristics are specified by using options that follow each vector pair and are enclosed in single quotes.

```
>> plot(t,f,'-k',t,h,':k');
>> xlabel('t'); ylabel('Amplitude');
>> legend('f(t)','h(t)');
```

[†]While grossly inefficient, element-by-element multiplication can be accomplished by extracting the main diagonal from the outer product of two N-length vectors.

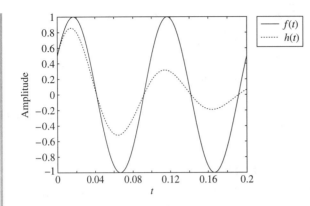

Figure MB.3 Graphical comparison of $f(t)$ and $h(t)$.

Here, `'-k'` instructs MATLAB to plot $f(t)$ using a solid black line, while `':k'` instructs MATLAB to use a dotted black line to plot $h(t)$. A legend and axis labels complete the plot, as shown in Fig. MB.3. It is also possible, although more cumbersome, to use pull-down menus to modify line properties and to add labels and legends directly in the figure window.

MB.6 Matrix Operations

Many applications require more than row vectors with evenly spaced elements; row vectors, column vectors, and matrices with arbitrary elements are typically needed.

MATLAB provides several functions to generate common, useful matrices. Given integers `m`, `n`, and vector `x`, the function `eye(m)` creates the $m \times m$ identity matrix; the function `ones(m,n)` creates the $m \times n$ matrix of all ones; the function `zeros(m,n)` creates the $m \times n$ matrix of all zeros; and the function `diag(x)` uses vector `x` to create a diagonal matrix. The creation of general matrices and vectors, however, requires each individual element to be specified.

Vectors and matrices can be input spreadsheet style by using MATLAB's array editor. This graphical approach is rather cumbersome and is not often used. A more direct method is preferable.

Consider a simple row vector **r**,

$$\mathbf{r} = [1 \quad 0 \quad 0]$$

The MATLAB notation `a:b:c` cannot create this row vector. Rather, square brackets are used to create **r**.

```
>> r = [1 0 0]
r = 1    0    0
```

Square brackets enclose elements of the vector, and spaces or commas are used to separate row elements.

Next, consider the 3×2 matrix **A**,

$$\mathbf{A} = \begin{bmatrix} 2 & 3 \\ 4 & 5 \\ 0 & 6 \end{bmatrix}$$

Matrix **A** can be viewed as a three-high stack of two-element row vectors. With a semicolon to separate rows, square brackets are used to create the matrix.

```
>> A = [2 3;4 5;0 6]
A = 2    3
    4    5
    0    6
```

Each row vector needs to have the same length to create a sensible matrix.

In addition to enclosing string arguments, a single quote performs the complex conjugate transpose operation. In this way, row vectors become column vectors and vice versa. For example, a column vector **c** is easily created by transposing row vector **r**.

```
>> c = r'
c = 1
    0
    0
```

Since vector **r** is real, the complex-conjugate transpose is just the transpose. Had **r** been complex, the simple transpose could have been accomplished by either `r.'` or `(conj(r))'`.

More formally, square brackets are referred to as a concatenation operator. A concatenation combines or connects smaller pieces into a larger whole. Concatenations can involve simple numbers, such as the six-element concatenation used to create the 3×2 matrix **A**. It is also possible to concatenate larger objects, such as vectors and matrices. For example, vector **c** and matrix **A** can be concatenated to form a 3×3 matrix **B**.

```
>> B = [c A]
B = 1    2    3
    0    4    5
    0    0    6
```

Errors will occur if the component dimensions do not sensibly match; a 2×2 matrix would not be concatenated with a 3×3 matrix, for example.

Elements of a matrix are indexed much like vectors, except two indices are typically used to specify row and column.[†] Element $(1, 2)$ of matrix **B**, for example, is 2.

```
>> B(1,2)
ans = 2
```

Indices can likewise be vectors. For example, vector indices allow us to extract the elements common to the first two rows and last two columns of matrix **B**.

```
>> B(1:2,2:3)
ans = 2    3
      4    5
```

[†]Matrix elements can also be accessed by means of a single index, which enumerates along columns. Formally, the element from row m and column n of an $M \times N$ matrix may be obtained with a single index $(n-1)M + m$. For example, element $(1, 2)$ of matrix **B** is accessed by using the index $(2-1)3 + 1 = 4$. That is, `B(4)` yields 2.

One indexing technique is particularly useful and deserves special attention. A colon can be used to specify all elements along a specified dimension. For example, `B(2,:)` selects all column elements along the second row of **B**.

```
>> B(2,:)
ans = 0     4     5
```

Now that we understand basic vector and matrix creation, we turn our attention to using these tools on real problems. Consider solving a set of three linear simultaneous equations in three unknowns.

$$x_1 - 2x_2 + 3x_3 = 1$$
$$-\sqrt{3}x_1 + x_2 - \sqrt{5}x_3 = \pi$$
$$3x_1 - \sqrt{7}x_2 + x_3 = e$$

This system of equations is represented in matrix form according to $\mathbf{Ax} = \mathbf{y}$, where

$$\mathbf{A} = \begin{bmatrix} 1 & -2 & 3 \\ -\sqrt{3} & 1 & -\sqrt{5} \\ 3 & -\sqrt{7} & 1 \end{bmatrix}, \quad \mathbf{x} = \begin{bmatrix} x_1 \\ x_2 \\ x_3 \end{bmatrix}, \text{ and } \quad \mathbf{y} = \begin{bmatrix} 1 \\ \pi \\ e \end{bmatrix}$$

Although Cramer's rule can be used to solve $\mathbf{Ax} = \mathbf{y}$, it is more convenient to solve by multiplying both sides by the matrix inverse of **A**. That is, $\mathbf{x} = \mathbf{A}^{-1}\mathbf{Ax} = \mathbf{A}^{-1}\mathbf{y}$. Solving for **x** by hand or by calculator would be tedious at best, so MATLAB is used. We first create **A** and **y**.

```
>> A = [1 -2 3;-sqrt(3) 1 -sqrt(5);3 -sqrt(7) 1];
>> y = [1;pi;exp(1)];
```

The vector solution is found by using MATLAB's `inv` function.

```
>> x = inv(A)*y
x = -1.9999
    -3.8998
    -1.5999
```

It is also possible to use MATLAB's left divide operator `x = A\y` to find the same solution. The left divide is generally more computationally efficient than matrix inverses. As with matrix multiplication, left division requires that the two arguments be conformable.

Of course, Cramer's rule can be used to compute individual solutions, such as x_1, by using vector indexing, concatenation, and MATLAB's `det` command to compute determinants.

```
>> x1 = det([y,A(:,2:3)])/det(A)
x1 = -1.9999
```

Another nice application of matrices is the simultaneous creation of a family of curves. Consider $h_\alpha(t) = e^{-\alpha t} \sin(2\pi 10t + \pi/6)$ over $0 \le t \le 0.2$. Figure MB.3 shows $h_\alpha(t)$ for $\alpha = 0$ and $\alpha = 10$. Let's investigate the family of curves $h_\alpha(t)$ for $\alpha = [0, 1, \ldots, 10]$.

An inefficient way to solve this problem is create $h_\alpha(t)$ for each α of interest. This requires 11 individual cases. Instead, a matrix approach allows all 11 curves to be computed simultaneously. First, a vector is created that contains the desired values of α.

```
>> alpha = (0:10);
```

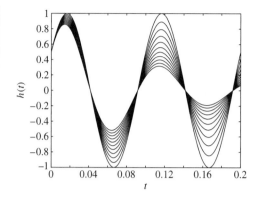

Figure MB.4 $h_\alpha(t)$ for $\alpha = [0, 1, \ldots, 10]$.

By using a sampling interval of one millisecond, $\Delta t = 0.001$, a time vector is also created.

```
>> t = (0:0.001:0.2)';
```

The result is a length-201 column vector. By replicating the time vector for each of the 11 curves required, a time matrix T is created. This replication can be accomplished by using an outer product between t and a 1×11 vector of ones.[†]

```
>> T = t*ones(1,11);
```

The result is a 201×11 matrix that has identical columns. Right multiplying T by a diagonal matrix created from α, columns of T can be individually scaled and the final result is computed.

```
>> H = exp(-T*diag(alpha)).*sin(2*pi*10*T+pi/6);
```

Here, H is a 201×11 matrix, where each column corresponds to a different value of α. That is, $\mathbf{H} = [\mathbf{h}_0, \mathbf{h}_1, \ldots, \mathbf{h}_{10}]$, where $\mathbf{h}_\alpha$ are column vectors. As shown in Fig. MB.4, the 11 desired curves are simultaneously displayed by using MATLAB's plot command, which allows matrix arguments.

```
>> plot(t,H); xlabel('t'); ylabel('h(t)');
```

This example illustrates an important technique called vectorization, which increases execution efficiency for interpretive languages such as MATLAB.[‡] Algorithm vectorization uses matrix and vector operations to avoid manual repetition and loop structures. It takes practice and effort to become proficient at vectorization, but the worthwhile result is efficient, compact code.

MB.7 Partial Fraction Expansions

There are a wide variety of techniques and shortcuts to compute the partial fraction expansion of rational function $F(x) = B(x)/A(x)$, but few are more simple than the MATLAB residue

---

[†]The repmat command provides a more flexible method to replicate or tile objects. Equivalently, T = repmat(t,1,11).
[‡]The benefits of vectorization are less pronounced in recent versions of MATLAB.

command. The basic form of this command is

```
>> [R,P,K] = residue(B,A)
```

The two input vectors B and A specify the polynomial coefficients of the numerator and denominator, respectively. These vectors are ordered in descending powers of the independent variable. Three vectors are output. The vector R contains the coefficients of each partial fraction, and vector P contains the corresponding roots of each partial fraction. For a root repeated r times, the r partial fractions are ordered in ascending powers. When the rational function is not proper, the vector K contains the direct terms, which are ordered in descending powers of the independent variable.

To demonstrate the power of the residue command, consider finding the partial fraction expansion of

$$F(x) = \frac{x^5 + \pi}{(x + \sqrt{2})(x - \sqrt{2})^3} = \frac{x^5 + \pi}{x^4 - \sqrt{8}x^3 + \sqrt{32}x - 4}$$

By hand, the partial fraction expansion of $F(x)$ is difficult to compute. MATLAB, however, makes short work of the expansion.

```
>> [R,P,K] = residue([1 0 0 0 0 pi],[1 -sqrt(8) 0 sqrt(32) -4])
R = 7.8888    5.9713    3.1107    0.1112
P = 1.4142    1.4142    1.4142   -1.4142
K = 1.0000    2.8284
```

Written in standard form, the partial fraction expansion of $F(x)$ is

$$F(x) = x + 2.8284 + \frac{7.8888}{x - \sqrt{2}} + \frac{5.9713}{(x - \sqrt{2})^2} + \frac{3.1107}{(x - \sqrt{2})^3} + \frac{0.1112}{x + \sqrt{2}}$$

The signal processing toolbox function residuez is similar to the residue command and offers more convenient expansion of certain rational functions, such as those commonly encountered in the study of discrete-time systems. Additional information about the residue and residuez commands are available from MATLAB's help facilities.

PROBLEMS

B.1 Given a complex number $w = x + jy$, the complex conjugate of w is defined in rectangular coordinates as $w^* = x - jy$. Use this fact to derive complex conjugation in polar form.

B.2 Express the following numbers in polar form:
(a) $1 + j$
(b) $-4 + j3$
(c) $(1 + j)(-4 + j3)$
(d) $e^{j\pi/4} + 2e^{-j\pi/4}$
(e) $e^j + 1$
(f) $(1 + j)/(-4 + j3)$

B.3 Express the following numbers in Cartesian form:
(a) $3e^{j\pi/4}$
(b) $1/e^j$
(c) $(1 + j)(-4 + j3)$
(d) $e^{j\pi/4} + 2e^{-j\pi/4}$
(e) $e^j + 1$
(f) $1/2^j$

B.4 For complex constant w, prove:
(a) $\text{Re}(w) = (w + w^*)/2$
(b) $\text{Im}(w) = (w - w^*)/2j$

B.5 Given $w = x - jy$, determine:
  (a) $\text{Re}(e^w)$
  (b) $\text{Im}(e^w)$

B.6 For arbitrary complex constants w_1 and w_2, prove or disprove the following:
  (a) $\text{Re}(jw_1) = -\text{Im}(w_1)$
  (b) $\text{Im}(jw_1) = \text{Re}(w_1)$
  (c) $\text{Re}(w_1) + \text{Re}(w_2) = \text{Re}(w_1 + w_2)$
  (d) $\text{Im}(w_1) + \text{Im}(w_2) = \text{Im}(w_1 + w_2)$
  (e) $\text{Re}(w_1)\text{Re}(w_2) = \text{Re}(w_1 w_2)$
  (f) $\text{Im}(w_1)/\text{Im}(w_2) = \text{Im}(w_1/w_2)$

B.7 Given $w_1 = 3 + j4$ and $w_2 = 2e^{j\pi/4}$.
  (a) Express w_1 in standard polar form.
  (b) Express w_2 in standard rectangular form.
  (c) Determine $|w_1|^2$ and $|w_2|^2$.
  (d) Express $w_1 + w_2$ in standard rectangular form.
  (e) Express $w_1 - w_2$ in standard polar form.
  (f) Express $w_1 w_2$ in standard rectangular form.
  (g) Express w_1/w_2 in standard polar form.

B.8 Repeat Prob. B.7 using $w_1 = (3 + j4)^2$ and $w_2 = 2.5 j e^{-j40\pi}$.

B.9 Repeat Prob. B.7 using $w_1 = j + e^{\pi/4}$ and $w_2 = \cos(j)$.

B.10 Use Euler's identity to solve or prove the following:
  (a) Find real, positive constants c and ϕ for all real t such that $2.5\cos(3t) - 1.5\sin(3t + \pi/3) = c\cos(3t + \phi)$
  (b) Prove that $\cos(\theta \pm \phi) = \cos(\theta)\cos(\phi) \mp \sin(\theta)\sin(\phi)$.
  (c) Given real constants a, b, and α, complex constant w, and the fact that

$$\int_a^b e^{wx}\,dx = \frac{1}{w}(e^{wb} - e^{wa})$$

  evaluate the integral

$$\int_a^b e^{wx}\sin(\alpha x)\,dx$$

B.11 In addition to the traditional sine and cosine functions, there are the *hyperbolic* sine and cosine functions, which are defined by $\sinh(w) = (e^w - e^{-w})/2$ and $\cosh(w) = (e^w + e^{-w})/2$. In general, the argument is a complex constant $w = x + jy$.

  (a) Show that $\cosh(w) = \cosh(x)\cos(y) + j\sinh(x)\sin(y)$.
  (b) Determine a similar expression for $\sinh(w)$ in rectangular form that only uses functions of real arguments, such as $\sin(x)$, $\cosh(y)$, and so on.

B.12 Using the complex plane:
  (a) Evaluate and locate the distinct solutions to $(w)^4 = -1$.
  (b) Evaluate and locate the distinct solutions to $(w - (1 + j2))^5 = (32/\sqrt{2})(1 + j)$.
  (c) Sketch the solution to $|w - 2j| = 3$.
  (d) Graph $w(t) = (1 + t)e^{jt}$ for $(-10 \le t \le 10)$.

B.13 The distinct solutions to $(w - w_1)^n = w_2$ lie on a circle in the complex plane, as shown in Fig. PB.13. One solution is located on the real axis at $\sqrt{3} + 1 = 2.732$, and one solution is located on the imaginary axis at $\sqrt{3} - 1 = 0.732$. Determine w_1, w_2, and n.

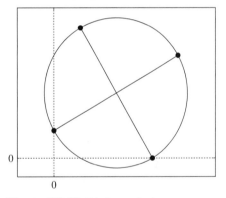

Figure PB.13 Distinct solutions to $(w - w_1)^n = w_2$.

B.14 Find the distinct solutions to $(j - w)^{1.5} = 2 + j2$. Use MATLAB to graph the solution set in the complex plane.

B.15 If $j = \sqrt{-1}$, what is $\sqrt{j}$?

B.16 Find all the values of $\ln(-e)$, expressing your answer in Cartesian form.

B.17 Determine all values of $\log_{10}(-1)$, expressing your answer in Cartesian form. Notice that the logarithm has base 10, not e.

B.18 Express the following in standard rectangular coordinates:
  (a) $\ln(1/(1 + j))$

(b) $\cos(1 + j)$

(c) $(1 - j)^j$

B.19 By constraining w to be purely imaginary, show that the equation $\cos(w) = 2$ can be represented as a standard quadratic equation. Solve this equation for w.

B.20 Determine an expression for an exponentially decaying sinusoid that oscillates three times per second and whose amplitude envelope decreases by 50% every 2 seconds. Use MATLAB to plot the signal over $-2 \le t \le 2$.

B.21 By hand, sketch the following against independent variable t:

(a) $x_1(t) = \text{Re}\left(2e^{(-1+j2\pi)t}\right)$

(b) $x_2(t) = \text{Im}\left(3 - e^{(1-j2\pi)t}\right)$

(c) $x_3(t) = 3 - \text{Im}\left(e^{(1-j2\pi)t}\right)$

B.22 Use MATLAB to produce the plots requested in Prob. B.21.

B.23 Use MATLAB to plot $x(t) = \cos(t) \sin(20t)$ over a suitable range of t.

B.24 Use MATLAB to plot $x(t) = \sum_{k=1}^{10} \cos(2\pi kt)$ over a suitable range of t. The MATLAB command `sum` may prove useful.

B.25 When a bell is struck with a mallet, it produces a ringing sound. Write an equation that approximates the sound produced by a small, light bell. Carefully identify your assumptions. How does your equation change if the bell is large and heavy? You can assess the quality of your models by using the MATLAB `sound` command to listen to your "bell."

B.26 Certain integrals, although expressed in relatively simple form, are quite difficult to solve. For example, $\int e^{-x^2} dx$ cannot be evaluated in terms of elementary functions; most calculators that perform integration cannot handle this indefinite integral. Fortunately, you are smarter than most calculators.

(a) Express e^{-x^2} using a Taylor series expansion.

(b) Using your series expansion for e^{-x^2}, determine $\int e^{-x^2} dx$.

(c) Using a suitably truncated series, evaluate the definite integral $\int_0^1 e^{-x^2} dx$.

B.27 Repeat Prob. B.26 for $\int e^{-x^3} dx$.

B.28 Repeat Prob. B.26 for $\int \cos x^2 dx$.

B.29 For each function, determine a suitable series expansion.

(a) $f_1(x) = (2 - x^2)^{-1}$

(b) $f_2(x) = (0.5)^x$

B.30 You are working on a digital quadrature amplitude modulation (QAM) communication receiver. The QAM receiver requires a pair of quadrature signals: $\cos \Omega n$ and $\sin \Omega n$. These can be simultaneously generated by following a simple procedure: (1) choose a point w on the unit circle, (2) multiply w by itself and store the result, (3) multiply w by the last result and store, and (4) repeat step 3.

(a) Show that this method can generate the desired pair of quadrature sinusoids.

(b) Determine a suitable value of w so that good-quality, periodic, $2\pi \times 100,000$ rad/s signals can be generated. How much time is available for the processing unit to compute each sample?

(c) Simulate this procedure by using MATLAB and report your results.

(d) Identify as many assumptions and limitations to this technique as possible. For example, can your system operate correctly for an indefinite period of time?

B.31 Consider the following system of equations.

$$x_1 + x_2 + x_3 = 1$$
$$x_1 + 2x_2 + 3x_3 = 3$$
$$x_1 - x_2 = -3$$

Use Cramer's rule to determine:

(a) x_1

(b) x_2

(c) x_3

Matrix determinants can be computed by using MATLAB's `det` command.

B.32 A system of equations in terms of unknowns x_1 and x_2 and arbitrary constants $a, b, c, d, e,$ and f is given by

$$ax_1 + bx_2 = c$$
$$dx_1 + ex_2 = f$$

(a) Represent this system of equations in matrix form.

(b) Identify specific constants a, b, c, d, e, and f such that $x_1 = 3$ and $x_2 = -2$. Are the constants you selected unique?

(c) Identify nonzero constants a, b, c, d, e, and f such that no solutions x_1 and x_2 exist.

(d) Identify nonzero constants a, b, c, d, e, and f such that an infinite number of solutions x_1 and x_2 exist.

B.33 Solve the following system of equations:

$$x_1 + x_2 + x_3 + x_4 = 4$$
$$x_1 + x_2 + x_3 - x_4 = 2$$
$$x_1 + x_2 - x_3 - x_4 = 0$$
$$x_1 - x_2 - x_3 - x_4 = -2$$

B.34 Solve the following system of equations:

$$x_1 + x_2 + x_3 + x_4 = 1$$
$$x_1 - 2x_2 + 3x_3 = 2$$
$$x_1 - x_3 + 7x_4 = 3$$
$$-2x_2 + 3x_3 - 4x_4 = 4$$

B.35 Compute by hand the partial fraction expansions of the following rational functions:

(a) $H_1(s) = (s^2 + 5s + 6)/(s^3 + s^2 + s + 1)$, which has denominator poles at $s = \pm j$ and $s = -1$

(b) $H_2(s) = 1/H_1(s) = (s^3 + s^2 + s + 1)/(s^2 + 5s + 6)$

(c) $H_3(s) = 1/((s + 1)^2(s^2 + 1))$

(d) $H_4(s) = (s^2 + 5s + 6)/(3s^2 + 2s + 1)$

B.36 Using MATLAB's `residue` command,

(a) Verify the results of Prob. B.35a.

(b) Verify the results of Prob. B.35b.

(c) Verify the results of Prob. B.35c.

(d) Verify the results of Prob. B.35d.

B.37 Determine the constants a_0, a_1, and a_2 of the partial fraction expansion $F(s) = s/(s + 1)^3 = a_0/(s + 1)^3 + a_1/(s + 1)^2 + a_2/(s + 1)$.

B.38 Let $N = [n_7, n_6, n_5, \ldots, n_2, n_1]$ represent the seven digits of your phone number. Construct a rational function according to

$$H_N(s) = \frac{n_7 s^2 + n_6 s + n_5 + n_4 s^{-1}}{n_3 s^2 + n_2 s + n_1}$$

Use MATLAB's `residue` command to compute the partial fraction expansion of $H_N(s)$.

B.39 When plotted in the complex plane for $-\pi \le \omega \le \pi$, the function $f(\omega) = \cos(\omega) + j0.1 \sin(2\omega)$ results in a so-called Lissajous figure that resembles a two-bladed propeller.

(a) In MATLAB, create two row vectors `fr` and `fi` corresponding to the real and imaginary portions of $f(\omega)$, respectively, over a suitable number N samples of ω. Plot the real portion against the imaginary portion and verify the figure resembles a propeller.

(b) Let complex constant $w = x + jy$ be represented in vector form

$$\mathbf{w} = \begin{bmatrix} x \\ y \end{bmatrix}$$

Consider the 2×2 rotational matrix $\mathbf{R}$:

$$\mathbf{R} = \begin{bmatrix} \cos\theta & -\sin\theta \\ \sin\theta & \cos\theta \end{bmatrix}$$

Show that $\mathbf{Rw}$ rotates vector $\mathbf{w}$ by θ radians.

(c) Create a rotational matrix R corresponding to $10°$ and multiply it by the $2 \times N$ matrix `f = [fr;fi];`. Plot the result to verify that the "propeller" has indeed rotated counterclockwise.

(d) Given the matrix R determined in part c, what is the effect of performing RRf? How about RRRf? Generalize the result.

(e) Investigate the behavior of multiplying $f(\omega)$ by the function $e^{j\theta}$.

SIGNALS AND SYSTEMS

In this chapter we shall discuss certain basic aspects of signals. We shall also introduce important basic concepts and qualitative explanations of the hows and whys of systems theory, thus building a solid foundation for understanding the quantitative analysis in the remainder of the book.

SIGNALS

A *signal* is a set of data or information. Examples include a telephone or a television signal, monthly sales of a corporation, or daily closing prices of a stock market (e.g., the Dow Jones averages). In all these examples, the signals are functions of the independent variable *time*. This is not always the case, however. When an electrical charge is distributed over a body, for instance, the signal is the charge density, a function of *space* rather than time. In this book we deal almost exclusively with signals that are functions of time. The discussion, however, applies equally well to other independent variables.

SYSTEMS

Signals may be processed further by *systems,* which may modify them or extract additional information from them. For example, an antiaircraft gun operator may want to know the future location of a hostile moving target that is being tracked by his radar. Knowing the radar signal, he knows the past location and velocity of the target. By properly processing the radar signal (the input), he can approximately estimate the future location of the target. Thus, a system is an entity that *processes* a set of signals (*inputs*) to yield another set of signals (*outputs*). A system may be made up of physical components, as in electrical, mechanical, or hydraulic systems (hardware realization), or it may be an algorithm that computes an output from an input signal (software realization).

1.1 SIZE OF A SIGNAL

The size of any entity is a number that indicates the largeness or strength of that entity. Generally speaking, the signal amplitude varies with time. How can a signal that exists over a certain time interval with varying amplitude be measured by one number that will indicate the signal size or signal strength? Such a measure must consider not only the signal amplitude, but also its duration. For instance, if we are to devise a single number V as a measure of the size of a human being, we must consider not only his or her width (girth), but also the height. If we make a simplifying assumption that the shape of a person is a cylinder of variable radius r (which varies with the height h) then one possible measure of the size of a person of height H is the person's volume V, given by

$$V = \pi \int_0^H r^2(h)\,dh \tag{1.1}$$

1.1-1 Signal Energy

Arguing in this manner, we may consider the area under a signal $x(t)$ as a possible measure of its size, because it takes account not only of the amplitude but also of the duration. However, this will be a defective measure because even for a large signal $x(t)$, its positive and negative areas could cancel each other, indicating a signal of small size. This difficulty can be corrected by defining the signal size as the area under $x^2(t)$, which is always positive. We call this measure the *signal energy* E_x, defined (for a real signal) as

$$E_x = \int_{-\infty}^{\infty} x^2(t)\,dt \tag{1.2a}$$

This definition can be generalized to a complex valued signal $x(t)$ as

$$E_x = \int_{-\infty}^{\infty} |x(t)|^2\,dt \tag{1.2b}$$

There are also other possible measures of signal size, such as the area under $|x(t)|$. The energy measure, however, is not only more tractable mathematically but is also more meaningful (as shown later) in the sense that it is indicative of the energy that can be extracted from the signal.

1.1-2 Signal Power

The signal energy must be finite for it to be a meaningful measure of the signal size. A necessary condition for the energy to be finite is that the signal amplitude $\to 0$ as $|t| \to \infty$ (Fig. 1.1a). Otherwise the integral in Eq. (1.2a) will not converge.

When the amplitude of $x(t)$ does not $\to 0$ as $|t| \to \infty$ (Fig. 1.1b), the signal energy is infinite. A more meaningful measure of the signal size in such a case would be the time average of the energy, if it exists. This measure is called the *power* of the signal. For a signal $x(t)$, we define its power P_x as

$$P_x = \lim_{T \to \infty} \frac{1}{T} \int_{-T/2}^{T/2} x^2(t)\,dt \tag{1.3a}$$

We can generalize this definition for a complex signal $x(t)$ as

$$P_x = \lim_{T \to \infty} \frac{1}{T} \int_{-T/2}^{T/2} |x(t)|^2\,dt \tag{1.3b}$$

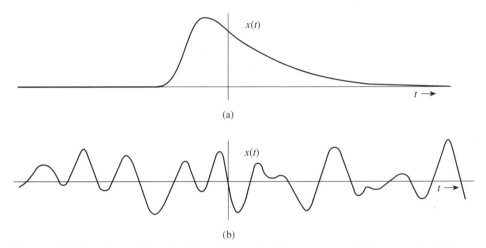

Figure 1.1 Examples of signals: **(a)** a signal with finite energy and **(b)** a signal with finite power.

Observe that the signal power P_x is the time average (mean) of the signal amplitude squared, that is, the *mean-squared* value of $x(t)$. Indeed, the square root of P_x is the familiar *rms* (root-mean-square) value of $x(t)$.

Generally, the mean of an entity averaged over a large time interval approaching infinity exists if the entity either is periodic or has a statistical regularity. If such a condition is not satisfied, the average may not exist. For instance, a ramp signal $x(t) = t$ increases indefinitely as $|t| \to \infty$, and neither the energy nor the power exists for this signal. However, the unit step function, which is not periodic nor has statistical regularity, does have a finite power.

When $x(t)$ is periodic, $|x(t)|^2$ is also periodic. Hence, the power of $x(t)$ can be computed from Eq. (1.3b) by averaging $|x(t)|^2$ over one period.

Comments. The signal energy as defined in Eqs. (1.2) does not indicate the actual energy (in the conventional sense) of the signal because the signal energy depends not only on the signal, but also on the load. It can, however, be interpreted as the energy dissipated in a normalized load of a 1-ohm resistor if a voltage $x(t)$ were to be applied across the 1-ohm resistor (or if a current $x(t)$ were to be passed through the 1-ohm resistor). The measure of "energy" is, therefore indicative of the energy capability of the signal, not the actual energy. For this reason the concepts of conservation of energy should not be applied to this "signal energy." Parallel observation applies to "signal power" defined in Eqs. (1.3). These measures are but convenient indicators of the signal size, which prove useful in many applications. For instance, if we approximate a signal $x(t)$ by another signal $g(t)$, the error in the approximation is $e(t) = x(t) - g(t)$. The energy (or power) of $e(t)$ is a convenient indicator of the goodness of the approximation. It provides us with a quantitative measure of determining the closeness of the approximation. In communication systems, during transmission over a channel, message signals are corrupted by unwanted signals (noise). The quality of the received signal is judged by the relative sizes of the desired signal and the unwanted signal (noise). In this case the ratio of the message signal and noise signal powers (signal to noise power ratio) is a good indication of the received signal quality.

Units of Energy and Power. Equations (1.2) are not correct dimensionally. This is because here we are using the term *energy* not in its conventional sense, but to indicate the signal size. The same observation applies to Eqs. (1.3) for power. The units of energy and power, as defined here, depend on the nature of the signal $x(t)$. If $x(t)$ is a voltage signal, its energy E_x has units of volts squared-seconds (V^2 s), and its power P_x has units of volts squared. If $x(t)$ is a current signal, these units will be amperes squared-seconds (A^2 s) and amperes squared, respectively.

EXAMPLE 1.1

Determine the suitable measures of the signals in Fig. 1.2.

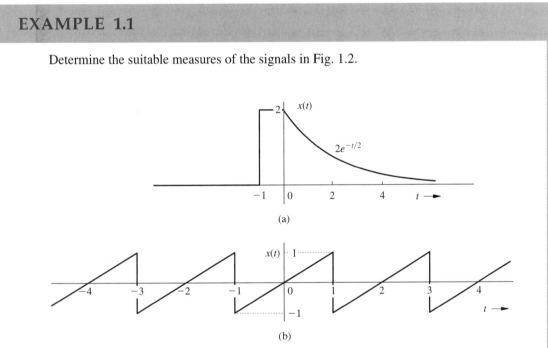

(a)

(b)

Figure 1.2

In Fig. 1.2a, the signal amplitude $\to 0$ as $|t| \to \infty$. Therefore the suitable measure for this signal is its energy E_x given by

$$E_x = \int_{-\infty}^{\infty} x^2(t)\, dt = \int_{-1}^{0} (2)^2\, dt + \int_{0}^{\infty} 4e^{-t}\, dt = 4 + 4 = 8$$

In Fig. 1.2b, the signal amplitude does not $\to 0$ as $|t| \to \infty$. However, it is periodic, and therefore its power exists. We can use Eq. (1.3a) to determine its power. We can simplify the procedure for periodic signals by observing that a periodic signal repeats regularly each period (2 seconds in this case). Therefore, averaging $x^2(t)$ over an infinitely large interval is identical to averaging this quantity over one period (2 seconds in this case). Thus

$$P_x = \frac{1}{2} \int_{-1}^{1} x^2(t)\, dt = \frac{1}{2} \int_{-1}^{1} t^2\, dt = \frac{1}{3}$$

Recall that the signal power is the square of its rms value. Therefore, the rms value of this signal is $1/\sqrt{3}$.

EXAMPLE 1.2

Determine the power and the rms value of

    (a) $x(t) = C \cos{(\omega_0 t + \theta)}$

    (b) $x(t) = C_1 \cos{(\omega_1 t + \theta_1)} + C_2 \cos{(\omega_2 t + \theta_2)}$ $\omega_1 \neq \omega_2$

    (c) $x(t) = De^{j\omega_0 t}$

    (a) This is a periodic signal with period $T_0 = 2\pi/\omega_0$. The suitable measure of this signal is its power. Because it is a periodic signal, we may compute its power by averaging its energy over one period $T_0 = 2\pi/\omega_0$. However, for the sake of demonstration, we shall solve this problem by averaging over an infinitely large time interval using Eq. (1.3a).

$$P_x = \lim_{T \to \infty} \frac{1}{T} \int_{-T/2}^{T/2} C^2 \cos^2{(\omega_0 t + \theta)}\, dt = \lim_{T \to \infty} \frac{C^2}{2T} \int_{-T/2}^{T/2} [1 + \cos{(2\omega_0 t + 2\theta)}]\, dt$$

$$= \lim_{T \to \infty} \frac{C^2}{2T} \int_{-T/2}^{T/2} dt + \lim_{T \to \infty} \frac{C^2}{2T} \int_{-T/2}^{T/2} \cos{(2\omega_0 t + 2\theta)}\, dt$$

The first term on the right-hand side is equal to $C^2/2$. The second term, however, is zero because the integral appearing in this term represents the area under a sinusoid over a very large time interval T with $T \to \infty$. This area is at most equal to the area of half the cycle because of cancellations of the positive and negative areas of a sinusoid. The second term is this area multiplied by $C^2/2T$ with $T \to \infty$. Clearly this term is zero, and

$$P_x = \frac{C^2}{2} \tag{1.4a}$$

This shows that a sinusoid of amplitude C has a power $C^2/2$ regardless of the value of its frequency ω_0 ($\omega_0 \neq 0$) and phase θ. The rms value is $C/\sqrt{2}$. If the signal frequency is zero (dc or a constant signal of amplitude C), the reader can show that the power is C^2.

(b) In Chapter 6, we shall show that a sum of two sinusoids may or may not be periodic, depending on whether the ratio ω_1/ω_2 is a rational number. Therefore, the period of this signal is not known. Hence, its power will be determined by averaging its energy over T seconds with $T \to \infty$. Thus,

$$P_x = \lim_{T \to \infty} \frac{1}{T} \int_{-T/2}^{T/2} [C_1 \cos(\omega_1 t + \theta_1) + C_2 \cos(\omega_2 t + \theta_2)]^2 \, dt$$

$$= \lim_{T \to \infty} \frac{1}{T} \int_{-T/2}^{T/2} C_1^2 \cos^2(\omega_1 t + \theta_1) \, dt + \lim_{T \to \infty} \frac{1}{T} \int_{-T/2}^{T/2} C_2^2 \cos^2(\omega_2 t + \theta_2) \, dt$$

$$+ \lim_{T \to \infty} \frac{2C_1 C_2}{T} \int_{-T/2}^{T/2} \cos(\omega_1 t + \theta_1) \cos(\omega_2 t + \theta_2) \, dt$$

The first and second integrals on the right-hand side are the powers of the two sinusoids, which are $C_1^2/2$ and $C_2^2/2$ as found in part (a). The third term, the product of two sinusoids, can be expressed as a sum of two sinusoids $\cos[(\omega_1 + \omega_2)t + (\theta_1 + \theta_2)]$ and $\cos[(\omega_1 - \omega_2)t + (\theta_1 - \theta_2)]$, respectively. Now, arguing as in part (a), we see that the third term is zero. Hence, we have[†]

$$P_x = \frac{C_1^2}{2} + \frac{C_2^2}{2} \tag{1.4b}$$

and the rms value is $\sqrt{(C_1^2 + C_2^2)/2}$.

We can readily extend this result to a sum of any number of sinusoids with distinct frequencies. Thus, if

$$x(t) = \sum_{n=1}^{\infty} C_n \cos(\omega_n t + \theta_n)$$

assuming that none of the two sinusoids have identical frequencies and $\omega_n \neq 0$, then

$$P_x = \frac{1}{2} \sum_{n=1}^{\infty} C_n^2 \tag{1.4c}$$

If $x(t)$ also has a dc term, as

$$x(t) = C_0 + \sum_{n=1}^{\infty} C_n \cos(\omega_n t + \theta_n)$$

then

$$P_x = C_0^2 + \frac{1}{2} \sum_{n=1}^{\infty} C_n^2 \tag{1.4d}$$

[†]This is true only if $\omega_1 \neq \omega_2$. If $\omega_1 = \omega_2$, the integrand of the third term contains a constant $\cos(\theta_1 - \theta_2)$, and the third term $\to 2C_1 C_2 \cos(\theta_1 - \theta_2)$ as $T \to \infty$.

(c) In this case the signal is complex, and we use Eq. (1.3b) to compute the power.

$$P_x = \lim_{T \to \infty} \frac{1}{T} \int_{-T/2}^{T/2} |De^{j\omega_0 t}|^2 \, dt$$

Recall that $|e^{j\omega_0 t}| = 1$ so that $|De^{j\omega_0 t}|^2 = |D|^2$, and

$$P_x = |D|^2 \tag{1.4e}$$

The rms value is $|D|$.

Comment. In part (b) of Example 1.2, we have shown that the power of the sum of two sinusoids is equal to the sum of the powers of the sinusoids. It may appear that the power of $x_1(t) + x_2(t)$ is $P_{x_1} + P_{x_2}$. Unfortunately, this conclusion is not true in general. It is true only under a certain condition (orthogonality), discussed later (Section 6.5-3).

EXERCISE E1.1

Show that the energies of the signals in Fig. 1.3a, 1.3b, 1.3c, and 1.3d are 4, 1, 4/3, and 4/3, respectively. Observe that doubling a signal quadruples the energy, and time-shifting a signal has no effect on the energy. Show also that the power of the signal in Fig. 1.3e is 0.4323. What is the rms value of signal in Fig. 1.3e?

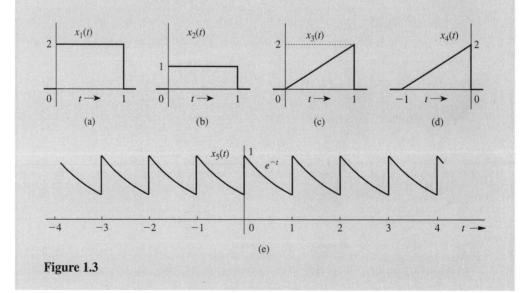

Figure 1.3

EXERCISE E1.2

Redo Example 1.2a to find the power of a sinusoid $C \cos(\omega_0 t + \theta)$ by averaging the signal energy over one period $T_0 = 2\pi/\omega_0$ (rather than averaging over the infinitely large interval). Show also that the power of a dc signal $x(t) = C_0$ is C_0^2, and its rms value is C_0.

EXERCISE E1.3

Show that if $\omega_1 = \omega_2$, the power of $x(t) = C_1 \cos(\omega_1 t + \theta_1) + C_2 \cos(\omega_2 t + \theta_2)$ is $[C_1^2 + C_2^2 + 2C_1 C_2 \cos(\theta_1 - \theta_2)]/2$, which is not equal to $(C_1^2 + C_2^2)/2$.

1.2 SOME USEFUL SIGNAL OPERATIONS

We discuss here three useful signal operations: shifting, scaling, and inversion. Since the independent variable in our signal description is time, these operations are discussed as *time shifting, time scaling,* and *time reversal* (inversion). However, this discussion is valid for functions having independent variables other than time (e.g., frequency or distance).

1.2-1 Time Shifting

Consider a signal $x(t)$ (Fig. 1.4a) and the same signal delayed by T seconds (Fig. 1.4b), which we shall denote by $\phi(t)$. Whatever happens in $x(t)$ (Fig. 1.4a) at some instant t also happens in

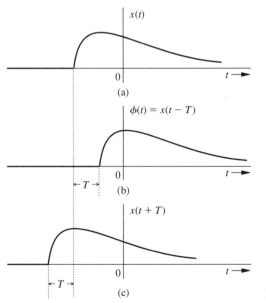

Figure 1.4 Time-shifting a signal.

$\phi(t)$ (Fig. 1.4b) T seconds later at the instant $t + T$. Therefore

$$\phi(t + T) = x(t) \tag{1.5}$$

and

$$\phi(t) = x(t - T) \tag{1.6}$$

Therefore, to time-shift a signal by T, we replace t with $t - T$. Thus $x(t - T)$ represents $x(t)$ time-shifted by T seconds. If T is positive, the shift is to the right (delay), as in Fig. 1.4b. If T is negative, the shift is to the left (advance), as in Fig. 1.4c. Clearly, $x(t - 2)$ is $x(t)$ delayed (right-shifted) by 2 seconds, and $x(t + 2)$ is $x(t)$ advanced (left-shifted) by 2 seconds.

EXAMPLE 1.3

An exponential function $x(t) = e^{-2t}$ shown in Fig. 1.5a is delayed by 1 second. Sketch and mathematically describe the delayed function. Repeat the problem with $x(t)$ advanced by 1 second.

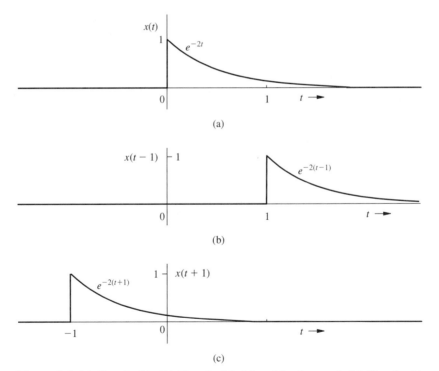

Figure 1.5 (a) Signal $x(t)$. (b) Signal $x(t)$ delayed by 1 second. (c) Signal $x(t)$ advanced by 1 second.

The function $x(t)$ can be described mathematically as

$$x(t) = \begin{cases} e^{-2t} & t \geq 0 \\ 0 & t < 0 \end{cases} \tag{1.7}$$

Let $x_d(t)$ represent the function $x(t)$ delayed (right-shifted) by 1 second as illustrated in Fig. 1.5b. This function is $x(t-1)$; its mathematical description can be obtained from $x(t)$ by replacing t with $t - 1$ in Eq. (1.7). Thus

$$x_d(t) = x(t-1) = \begin{cases} e^{-2(t-1)} & t-1 \geq 0 \quad \text{or} \quad t \geq 1 \\ 0 & t-1 < 0 \quad \text{or} \quad t < 1 \end{cases} \tag{1.8}$$

Let $x_a(t)$ represent the function $x(t)$ advanced (left-shifted) by 1 second as depicted in Fig. 1.5c. This function is $x(t+1)$; its mathematical description can be obtained from $x(t)$ by replacing t with $t + 1$ in Eq. (1.7). Thus

$$x_a(t) = x(t+1) = \begin{cases} e^{-2(t+1)} & t+1 \geq 0 \quad \text{or} \quad t \geq -1 \\ 0 & t+1 < 0 \quad \text{or} \quad t < -1 \end{cases} \tag{1.9}$$

EXERCISE E1.4

Write a mathematical description of the signal $x_3(t)$ in Fig. 1.3c. This signal is delayed by 2 seconds. Sketch the delayed signal. Show that this delayed signal $x_d(t)$ can be described mathematically as $x_d(t) = 2(t-2)$ for $2 \leq t \leq 3$, and equal to 0 otherwise. Now repeat the procedure with the signal advanced (left-shifted) by 1 second. Show that this advanced signal $x_a(t)$ can be described as $x_a(t) = 2(t+1)$ for $-1 \leq t \leq 0$, and equal to 0 otherwise.

1.2-2 Time Scaling

The compression or expansion of a signal in time is known as *time scaling*. Consider the signal $x(t)$ of Fig. 1.6a. The signal $\phi(t)$ in Fig. 1.6b is $x(t)$ compressed in time by a factor of 2. Therefore, whatever happens in $x(t)$ at some instant t also happens to $\phi(t)$ at the instant $t/2$, so that

$$\phi\left(\frac{t}{2}\right) = x(t) \tag{1.10}$$

and

$$\phi(t) = x(2t) \tag{1.11}$$

Observe that because $x(t) = 0$ at $t = T_1$ and T_2, we must have $\phi(t) = 0$ at $t = T_1/2$ and $T_2/2$, as shown in Fig. 1.6b. If $x(t)$ were recorded on a tape and played back at twice the normal

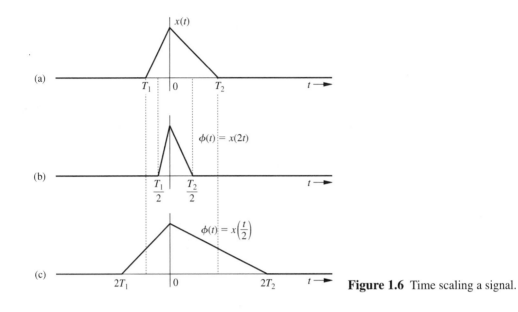

Figure 1.6 Time scaling a signal.

recording speed, we would obtain $x(2t)$. In general, if $x(t)$ is compressed in time by a factor a $(a > 1)$, the resulting signal $\phi(t)$ is given by

$$\phi(t) = x(at) \tag{1.12}$$

Using a similar argument, we can show that $x(t)$ expanded (slowed down) in time by a factor a $(a > 1)$ is given by

$$\phi(t) = x\left(\frac{t}{a}\right) \tag{1.13}$$

Figure 1.6c shows $x(t/2)$, which is $x(t)$ expanded in time by a factor of 2. Observe that in a time-scaling operation, the origin $t = 0$ is the anchor point, which remains unchanged under the scaling operation because at $t = 0$, $x(t) = x(at) = x(0)$.

In summary, to time-scale a signal by a factor a, we replace t with at. If $a > 1$, the scaling results in compression, and if $a < 1$, the scaling results in expansion.

EXAMPLE 1.4

Figure 1.7a shows a signal $x(t)$. Sketch and describe mathematically this signal time-compressed by factor 3. Repeat the problem for the same signal time-expanded by factor 2.

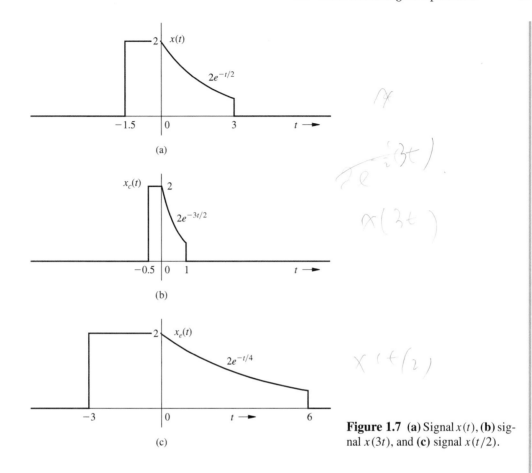

Figure 1.7 (a) Signal $x(t)$, (b) signal $x(3t)$, and (c) signal $x(t/2)$.

The signal $x(t)$ can be described as

$$x(t) = \begin{cases} 2 & -1.5 \le t < 0 \\ 2\,e^{-t/2} & 0 \le t < 3 \\ 0 & \text{otherwise} \end{cases} \tag{1.14}$$

Figure 1.7b shows $x_c(t)$, which is $x(t)$ time-compressed by factor 3; consequently, it can be described mathematically as $x(3t)$, which is obtained by replacing t with $3t$ in the right-hand side of Eq. (1.14). Thus

$$x_c(t) = x(3t) = \begin{cases} 2 & -1.5 \le 3t < 0 \quad \text{or} \quad -0.5 \le t < 0 \\ 2\,e^{-3t/2} & 0 \le 3t < 3 \quad \text{or} \quad 0 \le t < 1 \\ 0 & \text{otherwise} \end{cases} \tag{1.15a}$$

Observe that the instants $t = -1.5$ and 3 in $x(t)$ correspond to the instants $t = -0.5$, and 1 in the compressed signal $x(3t)$.

Figure 1.7c shows $x_e(t)$, which is $x(t)$ time-expanded by factor 2; consequently, it can be described mathematically as $x(t/2)$, which is obtained by replacing t with $t/2$ in $x(t)$. Thus

$$x_e(t) = x\left(\frac{t}{2}\right) = \begin{cases} 2 & -1.5 \le \dfrac{t}{2} < 0 \quad \text{or} \quad -3 \le t < 0 \\ 2e^{-t/4} & 0 \le \dfrac{t}{2} < 3 \quad \text{or} \quad 0 \le t < 6 \\ 0 & \text{otherwise} \end{cases} \qquad (1.15b)$$

Observe that the instants $t = -1.5$ and 3 in $x(t)$ correspond to the instants $t = -3$ and 6 in the expanded signal $x(t/2)$.

EXERCISE E1.5

Show that the time compression by a factor n ($n > 1$) of a sinusoid results in a sinusoid of the same amplitude and phase, but with the frequency increased n-fold. Similarly, the time expansion by a factor n ($n > 1$) of a sinusoid results in a sinusoid of the same amplitude and phase, but with the frequency reduced by a factor n. Verify your conclusion by sketching a sinusoid $\sin 2t$ and the same sinusoid compressed by a factor 3 and expanded by a factor 2.

1.2-3 Time Reversal

Consider the signal $x(t)$ in Fig. 1.8a. We can view $x(t)$ as a rigid wire frame hinged at the vertical axis. To time-reverse $x(t)$, we rotate this frame 180° about the vertical axis. This time reversal [the reflection of $x(t)$ about the vertical axis] gives us the signal $\phi(t)$ (Fig. 1.8b). Observe that

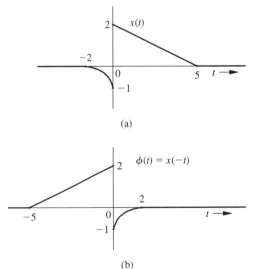

(a)

(b)

Figure 1.8 Time reversal of a signal.

whatever happens in Fig. 1.8a at some instant t also happens in Fig. 1.8b at the instant $-t$, and vice versa. Therefore

$$\phi(t) = x(-t) \qquad (1.16)$$

Thus, to time-reverse a signal we replace t with $-t$, and the time reversal of signal $x(t)$ results in a signal $x(-t)$. We must remember that the reversal is performed about the vertical axis, which acts as an anchor or a hinge. Recall also that the reversal of $x(t)$ about the horizontal axis results in $-x(t)$.

EXAMPLE 1.5

For the signal $x(t)$ illustrated in Fig. 1.9a, sketch $x(-t)$, which is time-reversed $x(t)$.

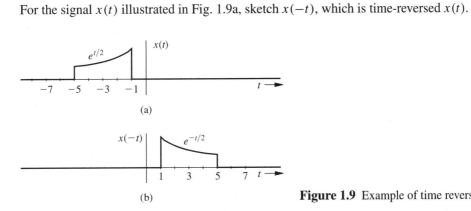

(a)

(b)

Figure 1.9 Example of time reversal.

The instants -1 and -5 in $x(t)$ are mapped into instants 1 and 5 in $x(-t)$. Because $x(t) = e^{t/2}$, we have $x(-t) = e^{-t/2}$. The signal $x(-t)$ is depicted in Fig. 1.9b. We can describe $x(t)$ and $x(-t)$ as

$$x(t) = \begin{cases} e^{t/2} & -1 \geq t > -5 \\ 0 & \text{otherwise} \end{cases}$$

and its time reversed version $x(-t)$ is obtained by replacing t with $-t$ in $x(t)$ as

$$x(-t) = \begin{cases} e^{-t/2} & -1 \geq -t > -5 \quad \text{or} \quad 1 \leq t < 5 \\ 0 & \text{otherwise} \end{cases}$$

1.2-4 Combined Operations

Certain complex operations require simultaneous use of more than one of the operations just described. The most general operation involving all the three operations is $x(at - b)$, which is realized in two possible sequences of operation:

1. Time-shift $x(t)$ by b to obtain $x(t - b)$. Now time-scale the shifted signal $x(t - b)$ by a (i.e., replace t with at) to obtain $x(at - b)$.

2. Time-scale $x(t)$ by a to obtain $x(at)$. Now time-shift $x(at)$ by b/a (i.e., replace t with $t - (b/a)$) to obtain $x[a(t - b/a)] = x(at - b)$. In either case, if a is negative, time scaling involves time reversal.

For example, the signal $x(2t - 6)$ can be obtained in two ways. We can delay $x(t)$ by 6 to obtain $x(t - 6)$, and then time-compress this signal by factor 2 (replace t with $2t$) to obtain $x(2t - 6)$. Alternately, we can first time-compress $x(t)$ by factor 2 to obtain $x(2t)$, then delay this signal by 3 (replace t with $t - 3$) to obtain $x(2t - 6)$.

1.3   CLASSIFICATION OF SIGNALS

There are several classes of signals. Here we shall consider only the following classes, which are suitable for the scope of this book:

1. Continuous-time and discrete-time signals

2. Analog and digital signals

3. Periodic and aperiodic signals

4. Energy and power signals

5. Deterministic and probabilistic signals

1.3-1 Continuous-Time and Discrete-Time Signals

A signal that is specified for a continuum of values of time t (Fig. 1.10a) is a *continuous-time signal,* and a signal that is specified only at discrete values of t (Fig. 1.10b) is a *discrete-time signal*. Telephone and video camera outputs are continuous-time signals, whereas the quarterly gross national product (GNP), monthly sales of a corporation, and stock market daily averages are discrete-time signals.

1.3-2 Analog and Digital Signals

The concept of continuous time is often confused with that of analog. The two are not the same. The same is true of the concepts of discrete time and digital. A signal whose amplitude can take on any value in a continuous range is an *analog signal*. This means that an analog signal amplitude can take on an infinite number of values. A *digital signal,* on the other hand, is one whose amplitude can take on only a finite number of values. Signals associated with a digital computer are digital because they take on only two values (binary signals). A digital signal whose amplitudes can take on M values is an M-ary signal of which binary ($M = 2$) is a special case. The terms *continuous time* and *discrete time* qualify the nature of a signal along the time (horizontal) axis. The terms *analog* and *digital,* on the other hand, qualify the nature of the signal amplitude (vertical axis). Figure 1.11 shows examples of signals of various types. It is clear that analog is not necessarily continuous-time and digital need not be discrete-time. Figure 1.11c

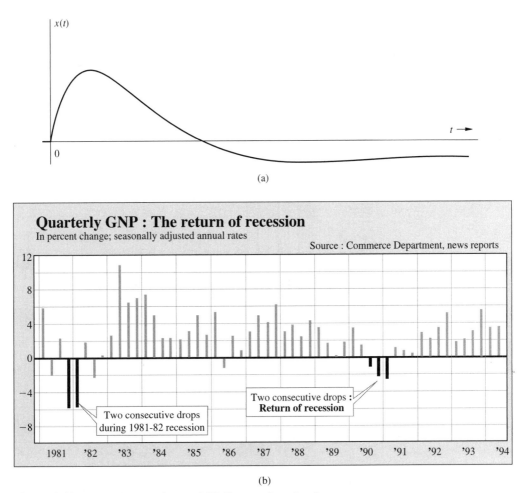

Figure 1.10 (a) Continuous-time and **(b)** discrete-time signals.

shows an example of an analog discrete-time signal. An analog signal can be converted into a digital signal [analog-to-digital (A/D) conversion] through quantization (rounding off), as explained in Section 8.3.

1.3-3 Periodic and Aperiodic Signals

A signal $x(t)$ is said to be *periodic* if for some positive constant T_0

$$x(t) = x(t + T_0) \qquad \text{for all } t \tag{1.17}$$

The *smallest* value of T_0 that satisfies the periodicity condition of Eq. (1.17) is the *fundamental period* of $x(t)$. The signals in Figs. 1.2b and 1.3e are periodic signals with periods 2 and 1, respectively. A signal is *aperiodic* if it is not periodic. Signals in Figs. 1.2a, 1.3a, 1.3b, 1.3c, and 1.3d are all aperiodic.

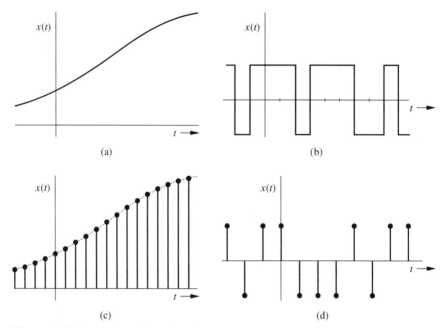

(a)

(b)

(c)

(d)

Figure 1.11 Examples of signals: **(a)** analog, continuous time, **(b)** digital, continuous time, **(c)** analog, discrete time, and **(d)** digital, discrete time.

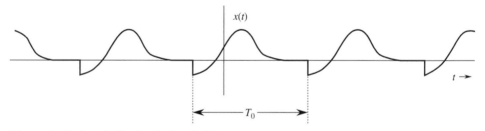

Figure 1.12 A periodic signal of period T_0.

By definition, a periodic signal $x(t)$ remains unchanged when time-shifted by one period. For this reason a periodic signal must start at $t = -\infty$: if it started at some finite instant, say $t = 0$, the time-shifted signal $x(t + T_0)$ would start at $t = -T_0$ and $x(t + T_0)$ would not be the same as $x(t)$. Therefore a *periodic signal, by definition, must start at $t = -\infty$ and continue forever, as illustrated in Fig. 1.12.*

Another important property of a periodic signal $x(t)$ is that $x(t)$ can be generated by *periodic extension* of any segment of $x(t)$ of duration T_0 (the period). As a result we can generate $x(t)$ from any segment of $x(t)$ having a duration of one period by placing this segment and the reproduction thereof end to end ad infinitum on either side. Figure 1.13 shows a periodic signal $x(t)$ of period $T_0 = 6$. The shaded portion of Fig. 1.13a shows a segment of $x(t)$ starting at $t = -1$ and having a duration of one period (6 seconds). This segment, when repeated forever in either direction, results in the periodic signal $x(t)$. Figure 1.13b shows another shaded segment

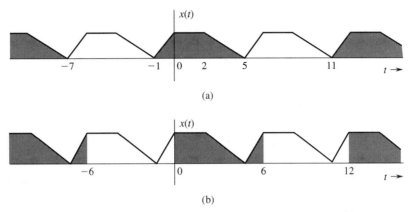

Figure 1.13 Generation of a periodic signal by periodic extension of its segment of one-period duration.

of $x(t)$ of duration T_0 starting at $t = 0$. Again we see that this segment, when repeated forever on either side, results in $x(t)$. The reader can verify that this construction is possible with any segment of $x(t)$ starting at any instant as long as the segment duration is one period.

An additional useful property of a periodic signal $x(t)$ of period T_0 is that the area under $x(t)$ over any interval of duration T_0 is the same; that is, for any real numbers a and b

$$\int_a^{a+T_0} x(t)\, dt = \int_b^{b+T_0} x(t)\, dt \tag{1.18}$$

This result follows from the fact that a periodic signal takes the same values at the intervals of T_0. Hence, the values over any segment of duration T_0 are repeated in any other interval of the same duration. For convenience, the area under $x(t)$ over any interval of duration T_0 will be denoted by

$$\int_{T_0} x(t)\, dt$$

It is helpful to label signals that start at $t = -\infty$ and continue forever as *everlasting* signals. Thus, an everlasting signal exists over the entire interval $-\infty < t < \infty$. The signals in Figs. 1.1b and 1.2b are examples of everlasting signals. Clearly, a periodic signal, by definition, is an everlasting signal.

A signal that does not start before $t = 0$ is a *causal* signal. In other words, $x(t)$ is a causal signal if

$$x(t) = 0 \qquad t < 0 \tag{1.19}$$

Signals in Fig. 1.3a–1.3c are causal signals. A signal that starts before $t = 0$ is a *noncausal* signal. All the signals in Fig. 1.1 and 1.2 are noncausal. Observe that an everlasting signal is always noncausal but a noncausal signal is not necessarily everlasting. The everlasting signal in Fig. 1.2b is noncausal; however, the noncausal signal in Fig. 1.2a is not everlasting. A signal that is zero for all $t \geq 0$ is called an *anticausal* signal.

Comment. A true everlasting signal cannot be generated in practice for obvious reasons. Why should we bother to postulate such a signal? In later chapters we shall see that certain signals (e.g., an impulse and an everlasting sinusoid) that cannot be generated in practice *do* serve a very useful purpose in the study of signals and systems.

1.3-4 Energy and Power Signals

A signal with finite energy is an *energy signal,* and a signal with finite and nonzero power is a *power signal.* Signals in Fig. 1.2a and 1.2b are examples of energy and power signals, respectively. Observe that power is the time average of energy. Since the averaging is over an infinitely large interval, a signal with finite energy has zero power, and a signal with finite power has infinite energy. Therefore, a signal cannot both be an energy signal and a power signal. If it is one, it cannot be the other. On the other hand, there are signals that are neither energy nor power signals. The ramp signal is one such case.

Comments. All practical signals have finite energies and are therefore energy signals. A power signal must necessarily have infinite duration; otherwise its power, which is its energy averaged over an infinitely large interval, will not approach a (nonzero) limit. Clearly, it is impossible to generate a true power signal in practice because such a signal has infinite duration and infinite energy.

Also, because of periodic repetition, periodic signals for which the area under $|x(t)|^2$ over one period is finite are power signals; however, not all power signals are periodic.

EXERCISE E1.6

Show that an everlasting exponential e^{-at} is neither an energy nor a power signal for any real value of a. However, if a is imaginary, it is a power signal with power $P_x = 1$ regardless of the value of a.

1.3-5 Deterministic and Random Signals

A signal whose physical description is known completely, either in a mathematical form or a graphical form, is a *deterministic signal.* A signal whose values cannot be predicted precisely but are known only in terms of probabilistic description, such as mean value or mean-squared value, is a *random signal.* In this book we shall exclusively deal with deterministic signals. Random signals are beyond the scope of this study.

1.4 SOME USEFUL SIGNAL MODELS

In the area of signals and systems, the step, the impulse, and the exponential functions play very important role. Not only do they serve as a basis for representing other signals, but their use can simplify many aspects of the signals and systems.

1.4-1 Unit Step Function $u(t)$

In much of our discussion, the signals begin at $t = 0$ (causal signals). Such signals can be conveniently described in terms of unit step function $u(t)$ shown in Fig. 1.14a. This function is defined by

$$u(t) = \begin{cases} 1 & t \geq 0 \\ 0 & t < 0 \end{cases} \tag{1.20}$$

If we want a signal to start at $t = 0$ (so that it has a value of zero for $t < 0$), we need only multiply the signal by $u(t)$. For instance, the signal e^{-at} represents an everlasting exponential that starts at $t = -\infty$. The causal form of this exponential (Fig. 1.14b) can be described as $e^{-at}u(t)$.

The unit step function also proves very useful in specifying a function with different mathematical descriptions over different intervals. Examples of such functions appear in Fig. 1.7. These functions have different mathematical descriptions over different segments of time as seen from Eqs. (1.14), (1.15a), and (1.15b). Such a description often proves clumsy and inconvenient in mathematical treatment. We can use the unit step function to describe such functions by a single expression that is valid for all t.

Consider, for example, the rectangular pulse depicted in Fig. 1.15a. We can express such a pulse in terms of familiar step functions by observing that the pulse $x(t)$ can be expressed as the sum of the two delayed unit step functions as shown in Fig. 1.15b. The unit step function $u(t)$

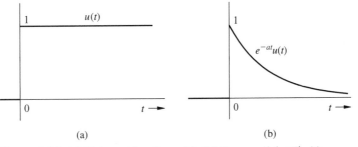

(a)   (b)

Figure 1.14 **(a)** Unit step function $u(t)$. **(b)** Exponential $e^{-at}u(t)$.

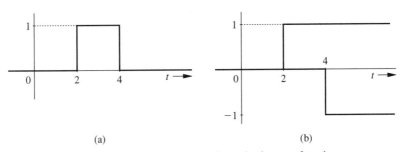

(a)   (b)

Figure 1.15 Representation of a rectangular pulse by step functions.

delayed by T seconds is $u(t - T)$. From Fig. 1.15b, it is clear that

$$x(t) = u(t - 2) - u(t - 4)$$

EXAMPLE 1.6

Describe the signal in Fig. 1.16a.

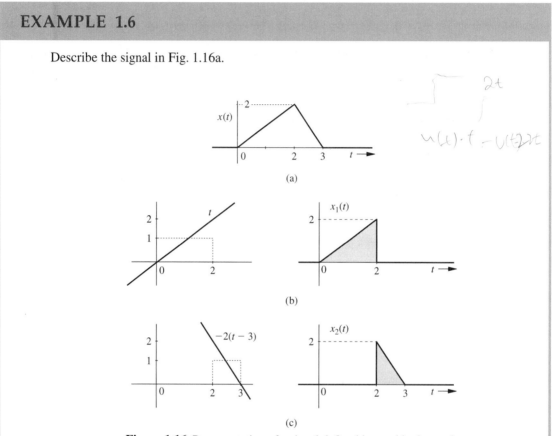

Figure 1.16 Representation of a signal defined interval by interval.

The signal illustrated in Fig. 1.16a can be conveniently handled by breaking it up into the two components $x_1(t)$ and $x_2(t)$, depicted in Fig. 1.16b and 1.16c, respectively. Here, $x_1(t)$ can be obtained by multiplying the ramp t by the gate pulse $u(t) - u(t - 2)$, as shown in Fig. 1.16b. Therefore

$$x_1(t) = t[u(t) - u(t - 2)]$$

The signal $x_2(t)$ can be obtained by multiplying another ramp by the gate pulse illustrated in Fig. 1.16c. This ramp has a slope -2; hence it can be described by $-2t + c$. Now, because the ramp has a zero value at $t = 3$, the constant $c = 6$, and the ramp can be described by $-2(t-3)$. Also, the gate pulse in Fig. 1.16c is $u(t-2) - u(t-3)$. Therefore

$$x_2(t) = -2(t-3)[u(t-2) - u(t-3)]$$

and

$$x(t) = x_1(t) + x_2(t)$$

$$= t\,[u(t) - u(t-2)] - 2(t-3)\,[u(t-2) - u(t-3)]$$

$$= tu(t) - 3(t-2)u(t-2) + 2(t-3)u(t-3)$$

EXAMPLE 1.7

Describe the signal in Fig. 1.7a by a single expression valid for all t.

Over the interval from -1.5 to 0, the signal can be described by a constant 2, and over the interval from 0 to 3, it can be described by $2\,e^{-t/2}$. Therefore

$$x(t) = \underbrace{2[u(t+1.5) - u(t)]}_{x_1(t)} + \underbrace{2e^{-t/2}[u(t) - u(t-3)]}_{x_2(t)}$$

$$= 2u(t+1.5) - 2(1 - e^{-t/2})u(t) - 2e^{-t/2}u(t-3)$$

Compare this expression with the expression for the same function found in Eq. (1.14).

EXERCISE E1.7

Show that the signals depicted in Fig. 1.17a and 1.17b can be described as $u(-t)$ and $e^{-at}u(-t)$, respectively.

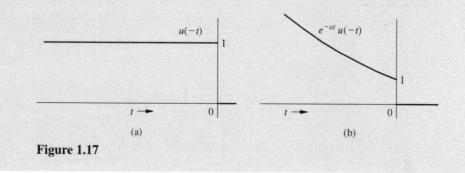

Figure 1.17

EXERCISE E1.8

Show that the signal shown in Fig. 1.18 can be described as

$$x(t) = (t-1)u(t-1) - (t-2)u(t-2) - u(t-4)$$

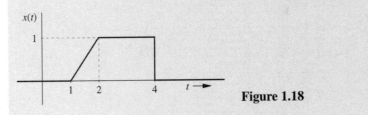

Figure 1.18

1.4-2 The Unit Impulse Function $\delta(t)$

The unit impulse function $\delta(t)$ is one of the most important functions in the study of signals and systems. This function was first defined by P. A. M. Dirac as

$$\delta(t) = 0 \qquad t \neq 0$$

$$\int_{-\infty}^{\infty} \delta(t)\, dt = 1 \tag{1.21}$$

We can visualize an impulse as a tall, narrow, rectangular pulse of unit area, as illustrated in Fig. 1.19b. The width of this rectangular pulse is a very small value $\epsilon \to 0$. Consequently, its height is a very large value $1/\epsilon \to \infty$. The unit impulse therefore can be regarded as a rectangular pulse with a width that has become infinitesimally small, a height that has become infinitely large, and an overall area that has been maintained at unity. Thus $\delta(t) = 0$ everywhere except at $t = 0$, where it is undefined. For this reason a unit impulse is represented by the spearlike symbol in Fig. 1.19a.

Other pulses, such as the exponential, triangular, or Gaussian types, may also be used in impulse approximation. The important feature of the unit impulse function is not its shape but the fact that its effective duration (pulse width) approaches zero while its area remains at unity. For example, the exponential pulse $\alpha e^{-\alpha t} u(t)$ in Fig. 1.20a becomes taller and narrower as α increases. In the limit as $\alpha \to \infty$, the pulse height $\to \infty$, and its width or duration $\to 0$. Yet,

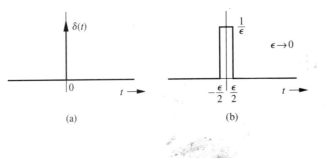

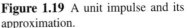

Figure 1.19 A unit impulse and its approximation.

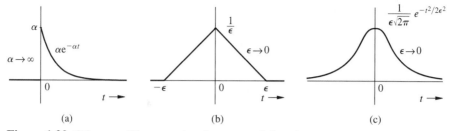

Figure 1.20 Other possible approximations to a unit impulse.

the area under the pulse is unity regardless of the value of α because

$$\int_0^\infty \alpha e^{-\alpha t}\, dt = 1 \tag{1.22}$$

The pulses in Fig. 1.20b and 1.20c behave in a similar fashion. Clearly, the exact impulse function cannot be generated in practice; it can only be approached.

From Eq. (1.21), it follows that the function $k\delta(t) = 0$ for all $t \neq 0$, and its area is k. Thus, $k\delta(t)$ is an impulse function whose area is k (in contrast to the unit impulse function, whose area is 1).

MULTIPLICATION OF A FUNCTION BY AN IMPULSE

Let us now consider what happens when we multiply the unit impulse $\delta(t)$ by a function $\phi(t)$ that is known to be continuous at $t = 0$. Since the impulse has nonzero value only at $t = 0$, and the value of $\phi(t)$ at $t = 0$ is $\phi(0)$, we obtain

$$\phi(t)\delta(t) = \phi(0)\delta(t) \tag{1.23a}$$

Thus, multiplication of a continuous-time function $\phi(t)$ with an unit impulse located at $t = 0$ results in an impulse, which is located at $t = 0$ and has strength $\phi(0)$ (the value of $\phi(t)$ at the location of the impulse). Use of exactly the same argument leads to the generalization of this result, stating that provided $\phi(t)$ is continuous at $t = T$, $\phi(t)$ multiplied by an impulse $\delta(t - T)$ (impulse located at $t = T$) results in an impulse located at $t = T$ and having strength $\phi(T)$ [the value of $\phi(t)$ at the location of the impulse].

$$\phi(t)\delta(t - T) = \phi(T)\delta(t - T) \tag{1.23b}$$

SAMPLING PROPERTY OF THE UNIT IMPULSE FUNCTION

From Eq. (1.23a) it follows that

$$\int_{-\infty}^\infty \phi(t)\delta(t)\, dt = \phi(0) \int_{-\infty}^\infty \delta(t)\, dt$$

$$= \phi(0) \tag{1.24a}$$

provided $\phi(t)$ is continuous at $t = 0$. This result means that *the area under the product of a function with an impulse $\delta(t)$ is equal to the value of that function at the instant at which the unit impulse is located.* This property is very important and useful and is known as the *sampling* or *sifting property* of the unit impulse.

From Eq. (1.23b) it follows that

$$\int_{-\infty}^{\infty} \phi(t)\delta(t - T)\, dt = \phi(T) \tag{1.24b}$$

Equation (1.24b) is just another form of sampling or sifting property. In the case of Eq. (1.24b), the impulse $\delta(t - T)$ is located at $t = T$. Therefore, the area under $\phi(t)\delta(t - T)$ is $\phi(T)$, the value of $\phi(t)$ at the instant at which the impulse is located (at $t = T$). In these derivations we have assumed that the function is continuous at the instant where the impulse is located.

Unit Impulse as a Generalized Function

The definition of the unit impulse function given in Eq. (1.21) is not mathematically rigorous, which leads to serious difficulties. First, the impulse function does not define a unique function: for example, it can be shown that $\delta(t) + \dot{\delta}(t)$ also satisfies Eq. (1.21).[1] Moreover, $\delta(t)$ is not even a true function in the ordinary sense. An ordinary function is specified by its values for all time t. The impulse function is zero everywhere except at $t = 0$, and at this, the only interesting part of its range, it is undefined. These difficulties are resolved by defining the impulse as a generalized function rather than an ordinary function. A *generalized function* is defined by its effect on other functions instead of by its value at every instant of time.

In this approach the impulse function is defined by the sampling property [Eqs. (1.24)]. We say nothing about what the impulse function is or what it looks like. Instead, the impulse function is defined in terms of its effect on a test function $\phi(t)$. We define a unit impulse as a function for which the area under its product with a function $\phi(t)$ is equal to the value of the function $\phi(t)$ at the instant at which the impulse is located. It is assumed that $\phi(t)$ is continuous at the location of the impulse. Therefore, either Eq. (1.24a) or (1.24b) can serve as a definition of the impulse function in this approach. Recall that the sampling property [Eqs. (1.24)] is the consequence of the classical (Dirac) definition of impulse in Eq. (1.21). In contrast, *the sampling property [Eqs. (1.24)] defines the impulse function in the generalized function approach.*

We now present an interesting application of the generalized function definition of an impulse. Because the unit step function $u(t)$ is discontinuous at $t = 0$, its derivative du/dt does not exist at $t = 0$ in the ordinary sense. We now show that this derivative *does* exist in the generalized sense, and it is, in fact, $\delta(t)$. As a proof, let us evaluate the integral of $(du/dt)\phi(t)$, using integration by parts:

$$\int_{-\infty}^{\infty} \frac{du}{dt}\phi(t)\, dt = u(t)\phi(t)\Big|_{-\infty}^{\infty} - \int_{-\infty}^{\infty} u(t)\dot{\phi}(t)\, dt \tag{1.25}$$

$$= \phi(\infty) - 0 - \int_{0}^{\infty} \dot{\phi}(t)\, dt$$

$$= \phi(\infty) - \phi(t)\Big|_{0}^{\infty}$$

$$= \phi(0) \tag{1.26}$$

This result shows that du/dt satisfies the sampling property of $\delta(t)$. Therefore it is an impulse $\delta(t)$ in the generalized sense—that is,

$$\frac{du}{dt} = \delta(t) \tag{1.27}$$

Consequently

$$\int_{-\infty}^{t} \delta(\tau)\, d\tau = u(t) \tag{1.28}$$

These results can also be obtained graphically from Fig. 1.19b. We observe that the area from $-\infty$ to t under the limiting form of $\delta(t)$ in Fig. 1.19b is zero if $t < -\epsilon/2$ and unity if $t \geq \epsilon/2$ with $\epsilon \to 0$. Consequently

$$\int_{-\infty}^{t} \delta(\tau)\, d\tau = \begin{cases} 0 & t < 0 \\ 1 & t \geq 0 \end{cases}$$

$$= u(t) \tag{1.29}$$

This result shows that the unit step function can be obtained by integrating the unit impulse function. Similarly the unit ramp function $x(t) = tu(t)$ can be obtained by integrating the unit step function. We may continue with unit parabolic function $t^2/2$ obtained by integrating the unit ramp, and so on. On the other side, we have derivatives of impulse function, which can be defined as generalized functions (see Prob. 1.4-9). All these functions, derived from the unit impulse function (successive derivatives and integrals) are called *singularity functions*.[†]

EXERCISE E1.9

Show that

(a) $(t^3 + 3)\delta(t) = 3\delta(t)$

(b) $\left[\sin\left(t^2 - \frac{\pi}{2}\right)\right]\delta(t) = -\delta(t)$

(c) $e^{-2t}\delta(t) = \delta(t)$

(d) $\dfrac{\omega^2 + 1}{\omega^2 + 9}\delta(\omega - 1) = \dfrac{1}{5}\delta(\omega - 1)$

[†]Singularity functions were defined by late Prof. S. J. Mason as follows. A singularity is a point at which a function does not possess a derivative. Each of the singularity functions (or if not the function itself, then the function differentiated a finite number of times) has a singular point at the origin and is zero elsewhere.[2]

EXERCISE E1.10

Show that

(a) $\displaystyle\int_{-\infty}^{\infty} \delta(t) e^{-j\omega t}\, dt = 1$

(b) $\displaystyle\int_{-\infty}^{\infty} \delta(t-2) \cos\left(\frac{\pi t}{4}\right) dt = 0$

(c) $\displaystyle\int_{-\infty}^{\infty} e^{-2(x-t)}\delta(2-t)\, dt = e^{-2(x-2)}$

1.4-3 The Exponential Function e^{st}

Another important function in the area of signals and systems is the exponential signal e^{st}, where s is complex in general, given by

$$s = \sigma + j\omega$$

Therefore

$$e^{st} = e^{(\sigma + j\omega)t} = e^{\sigma t}e^{j\omega t} = e^{\sigma t}(\cos \omega t + j \sin \omega t) \qquad (1.30\text{a})$$

Since $s^* = \sigma - j\omega$ (the conjugate of s), then

$$e^{s^* t} = e^{\sigma - j\omega} = e^{\sigma t}e^{-j\omega t} = e^{\sigma t}(\cos \omega t - j \sin \omega t) \qquad (1.30\text{b})$$

and

$$e^{\sigma t} \cos \omega t = \tfrac{1}{2}(e^{st} + e^{s^* t}) \qquad (1.30\text{c})$$

Comparison of this equation with Euler's formula shows that e^{st} is a generalization of the function $e^{j\omega t}$, where the frequency variable $j\omega$ is generalized to a complex variable $s = \sigma + j\omega$. For this reason we designate the variable s as the *complex frequency*. From Eqs. (1.30) it follows that the function e^{st} encompasses a large class of functions. The following functions are either special cases of or can be expressed in terms of e^{st}:

1. A constant $k = ke^{0t}$ $(s = 0)$
2. A monotonic exponential $e^{\sigma t}$ $(\omega = 0,\ s = \sigma)$
3. A sinusoid $\cos \omega t$ $(\sigma = 0,\ s = \pm j\omega)$
4. An exponentially varying sinusoid $e^{\sigma t} \cos \omega t$ $(s = \sigma \pm j\omega)$

These functions are illustrated in Fig. 1.21.

The complex frequency s can be conveniently represented on a *complex frequency plane* (s plane) as depicted in Fig. 1.22. The horizontal axis is the real axis (σ axis), and the vertical axis is the imaginary axis ($j\omega$ axis). The absolute value of the imaginary part of s is $|\omega|$ (the *radian* frequency), which indicates the frequency of oscillation of e^{st}; the real part σ (the *neper* frequency) gives information about the rate of increase or decrease of the amplitude of e^{st}. For signals whose complex frequencies lie on the real axis (σ axis, where $\omega = 0$), the frequency

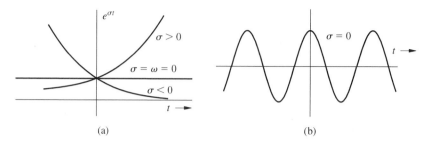

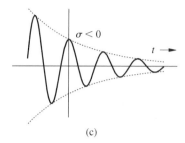

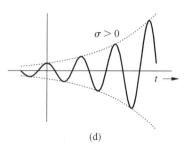

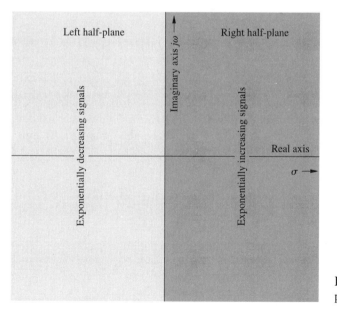

Figure 1.21 Sinusoids of complex frequency $\sigma + j\omega$.

Figure 1.22 Complex frequency plane.

of oscillation is zero. Consequently these signals are monotonically increasing or decreasing exponentials (Fig. 1.21a). For signals whose frequencies lie on the imaginary axis ($j\omega$ axis where $\sigma = 0$), $e^{\sigma t} = 1$. Therefore, these signals are conventional sinusoids with constant amplitude (Fig. 1.21b). The case $s = 0$ ($\sigma = \omega = 0$) corresponds to a constant (dc) signal because $e^{0t} = 1$. For the signals illustrated in Fig. 1.21c and 1.21d, both σ and ω are nonzero; the frequency s is complex and does not lie on either axis. The signal in Fig. 1.21c decays exponentially. Therefore, σ is negative, and s lies to the left of the imaginary axis. In contrast, the signal in Fig. 1.21d *grows* exponentially. Therefore, σ is positive, and s lies to the right of the imaginary axis. Thus the s plane (Fig. 1.21) can be separated into two parts: the *left half-plane* (LHP) corresponding to exponentially decaying signals and the *right half-plane* (RHP) corresponding to exponentially growing signals. The imaginary axis separates the two regions and corresponds to signals of constant amplitude.

An exponentially growing sinusoid $e^{2t} \cos 5t$, for example, can be expressed as a linear combination of exponentials $e^{(2+j5)t}$ and $e^{(2-j5)t}$ with complex frequencies $2 + j5$ and $2 - j5$, respectively, which lie in the RHP. An exponentially decaying sinusoid $e^{-2t} \cos 5t$ can be expressed as a linear combination of exponentials $e^{(-2+j5)t}$ and $e^{(-2-j5)t}$ with complex frequencies $-2 + j5$ and $-2 - j5$, respectively, which lie in the LHP. A constant amplitude sinusoid $\cos 5t$ can be expressed as a linear combination of exponentials e^{j5t} and e^{-j5t} with complex frequencies $\pm j5$, which lie on the imaginary axis. Observe that the monotonic exponentials $e^{\pm 2t}$ are also generalized sinusoids with complex frequencies ± 2.

1.5 EVEN AND ODD FUNCTIONS

A real function $x_e(t)$ is said to be an *even function* of t if[†]

$$x_e(t) = x_e(-t) \tag{1.31}$$

and a real function $x_o(t)$ is said to be an *odd function* of t if

$$x_o(t) = -x_o(-t) \tag{1.32}$$

An even function has the same value at the instants t and $-t$ for all values of t. Clearly, $x_e(t)$ is symmetrical about the vertical axis, as shown in Fig. 1.23a. On the other hand, the value of an odd function at the instant t is the negative of its value at the instant $-t$. Therefore, $x_o(t)$ is antisymmetrical about the vertical axis, as depicted in Fig. 1.23b.

1.5-1 Some Properties of Even and Odd Functions

Even and odd functions have the following properties:

$$\text{even function} \times \text{odd function} = \text{odd function}$$

$$\text{odd function} \times \text{odd function} = \text{even function}$$

$$\text{even function} \times \text{even function} = \text{even function}$$

The proofs are trivial and follow directly from the definition of odd and even functions [Eqs. (1.31) and (1.32)].

[†]A complex signal $x(t)$ is said to be *conjugate symmetrical* if $x(t) = x^*(-t)$. A real conjugate symmetrical signal is an even signal. A signal is *conjugate antisymmetrical* if $x(t) = -x^*(-t)$. A real conjugate antisymmetrical signal is an odd signal.

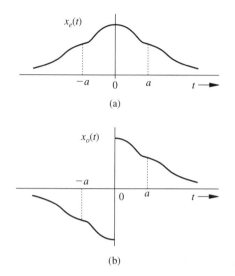

(a)

(b)

Figure 1.23 Functions of t: **(a)** even and **(b)** odd.

AREA

Because $x_e(t)$ is symmetrical about the vertical axis, it follows from Fig. 1.23a that

$$\int_{-a}^{a} x_e(t)\, dt = 2 \int_{0}^{a} x_e(t)\, dt \tag{1.33a}$$

It is also clear from Fig. 1.23b that

$$\int_{-a}^{a} x_o(t)\, dt = 0 \tag{1.33b}$$

These results are valid under the assumption that there is no impulse (or its derivatives) at the origin. The proof of these statements is obvious from the plots of the even and the odd function. Formal proofs, left as an exercise for the reader, can be accomplished by using the definitions in Eqs. (1.31) and (1.32).

Because of their properties, study of odd and even functions proves useful in many applications, as will become evident in later chapters.

1.5-2 Even and Odd Components of a Signal

Every signal $x(t)$ can be expressed as a sum of even and odd components because

$$x(t) = \underbrace{\tfrac{1}{2}[x(t) + x(-t)]}_{\text{even}} + \underbrace{\tfrac{1}{2}[x(t) - x(-t)]}_{\text{odd}} \tag{1.34}$$

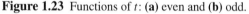

From the definitions in Eqs. (1.31) and (1.32), we can clearly see that the first component on the right-hand side is an even function, while the second component is odd. This is apparent from the fact that replacing t by $-t$ in the first component yields the same function. The same maneuver in the second component yields the negative of that component.

Consider the function

$$x(t) = e^{-at}u(t)$$

Expressing this function as a sum of the even and odd components $x_e(t)$ and $x_o(t)$, we obtain

$$x(t) = x_e(t) + x_o(t)$$

where [from Eq. (1.34)]

$$x_e(t) = \tfrac{1}{2}[e^{-at}u(t) + e^{at}u(-t)] \qquad (1.35\text{a})$$

and

$$x_o(t) = \tfrac{1}{2}[e^{-at}u(t) - e^{at}u(-t)] \qquad (1.35\text{b})$$

The function $e^{-at}u(t)$ and its even and odd components are illustrated in Fig. 1.24.

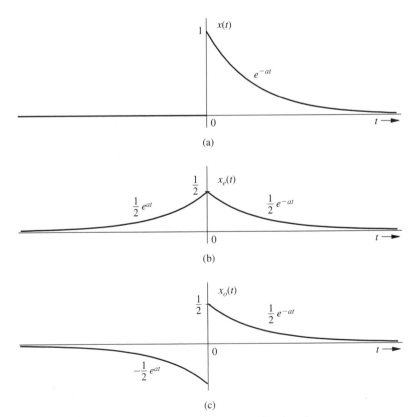

Figure 1.24 Finding even and odd components of a signal.

EXAMPLE 1.8

Find the even and odd components of e^{jt}.

From Eq. (1.34)

$$e^{jt} = x_e(t) + x_o(t)$$

where

$$x_e(t) = \tfrac{1}{2}[e^{jt} + e^{-jt}] = \cos t$$

and

$$x_o(t) = \tfrac{1}{2}[e^{jt} - e^{-jt}] = j \sin t$$

1.6 SYSTEMS

As mentioned in Section 1.1, systems are used to process signals to allow modification or extraction of additional information from the signals. A system may consist of physical components (hardware realization) or of an algorithm that computes the output signal from the input signal (software realization).

Roughly speaking, a physical system consists of interconnected components, which are characterized by their terminal (input–output) relationships. In addition, a system is governed by laws of interconnection. For example, in electrical systems, the terminal relationships are the familiar voltage–current relationships for the resistors, capacitors, inductors, transformers, transistors, and so on, as well as the laws of interconnection (i.e., Kirchhoff's laws). Using these laws, we derive mathematical equations relating the outputs to the inputs. These equations then represent a *mathematical model* of the system.

A system can be conveniently illustrated by a "black box" with one set of accessible terminals where the input variables $x_1(t), x_2(t), \ldots, x_j(t)$ are applied and another set of accessible terminals where the output variables $y_1(t), y_2(t), \ldots, y_k(t)$ are observed (Fig. 1.25).

The study of systems consists of three major areas: mathematical modeling, analysis, and design. Although we shall be dealing with mathematical modeling, our main concern is with analysis and design. The major portion of this book is devoted to the analysis problem—how to determine the system outputs for the given inputs and a given mathematical model of the system

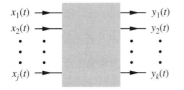

Figure 1.25 Representation of a system.

(or rules governing the system). To a lesser extent, we will also consider the problem of design or synthesis—how to construct a system that will produce a desired set of outputs for the given inputs.

DATA NEEDED TO COMPUTE SYSTEM RESPONSE

To understand what data we need to compute a system response, consider a simple RC circuit with a current source $x(t)$ as its input (Fig. 1.26). The output voltage $y(t)$ is given by

$$y(t) = Rx(t) + \frac{1}{C} \int_{-\infty}^{t} x(\tau)\, d\tau \tag{1.36a}$$

The limits of the integral on the right-hand side are from $-\infty$ to t because this integral represents the capacitor charge due to the current $x(t)$ flowing in the capacitor, and this charge is the result of the current flowing in the capacitor from $-\infty$. Now, Eq. (1.36a) can be expressed as

$$y(t) = Rx(t) + \frac{1}{C} \int_{-\infty}^{0} x(\tau)\, d\tau + \frac{1}{C} \int_{0}^{t} x(\tau)\, d\tau \tag{1.36b}$$

The middle term on the right-hand side is $v_C(0)$, the capacitor voltage at $t = 0$. Therefore

$$y(t) = v_C(0) + Rx(t) + \frac{1}{C} \int_{0}^{t} x(\tau)\, d\tau \qquad t \geq 0 \tag{1.36c}$$

This equation can be readily generalized as

$$y(t) = v_C(t_0) + Rx(t) + \frac{1}{C} \int_{t_0}^{t} x(\tau)\, d\tau \qquad t \geq t_0 \tag{1.36d}$$

From Eq. (1.36a), the output voltage $y(t)$ at an instant t can be computed if we know the input current flowing in the capacitor throughout its entire past $(-\infty$ to $t)$. Alternatively, if we know the input current $x(t)$ from some moment t_0 onward, then, using Eq. (1.36d), we can still calculate $y(t)$ for $t \geq t_0$ from a knowledge of the input current, provided we know $v_C(t_0)$, the initial capacitor voltage (voltage at t_0). Thus $v_C(t_0)$ contains all the relevant information about the circuit's entire past $(-\infty$ to $t_0)$ that we need to compute $y(t)$ for $t \geq t_0$. Therefore, the response of a system at $t \geq t_0$ can be determined from its input(s) during the interval t_0 to t and from certain *initial conditions* at $t = t_0$.

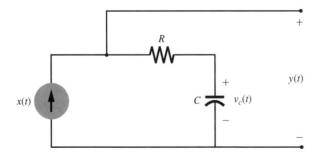

Figure 1.26 Example of a simple electrical system.

In the preceding example, we needed only one initial condition. However, in more complex systems, several initial conditions may be necessary. We know, for example, that in passive *RLC* networks, the initial values of all inductor currents and all capacitor voltages[†] are needed to determine the outputs at any instant $t \geq 0$ if the inputs are given over the interval $[0, t]$.

1.7 CLASSIFICATION OF SYSTEMS

Systems may be classified broadly in the following categories:

1. Linear and nonlinear systems
2. Constant-parameter and time-varying-parameter systems
3. Instantaneous (memoryless) and dynamic (with memory) systems
4. Causal and noncausal systems
5. Continuous-time and discrete-time systems
6. Analog and digital systems
7. Invertible and noninvertible systems
8. Stable and unstable systems

Other classifications, such as deterministic and probabilistic systems, are beyond the scope of this text and are not considered.

1.7-1 Linear and Nonlinear Systems

THE CONCEPT OF LINEARITY

A system whose output is proportional to its input is an *example* of a linear system. But linearity implies more than this; it also implies the *additivity property:* that is, if several inputs are acting on a system, then the total effect on the system due to all these inputs can be determined by considering one input at a time while assuming all the other inputs to be zero. The total effect is then the sum of all the component effects. This property may be expressed as follows: for a linear system, if an input x_1 acting alone has an effect y_1, and if another input x_2, also acting alone, has an effect y_2, then, with both inputs acting on the system, the total effect will be $y_1 + y_2$. Thus, if

$$x_1 \longrightarrow y_1 \quad \text{and} \quad x_2 \longrightarrow y_2 \tag{1.37}$$

then for all x_1 and x_2

$$x_1 + x_2 \longrightarrow y_1 + y_2 \tag{1.38}$$

In addition, a linear system must satisfy the *homogeneity* or scaling property, which states that for arbitrary real or imaginary number k, if an input is increased k-fold, the effect also increases k-fold. Thus, if

$$x \longrightarrow y$$

[†]Strictly speaking, this means independent inductor currents and capacitor voltages.

then for all real or imaginary k

$$kx \longrightarrow ky \tag{1.39}$$

Thus, linearity implies two properties: homogeneity (scaling) and additivity.[†] Both these properties can be combined into one property (*superposition*), which is expressed as follows: If

$$x_1 \longrightarrow y_1 \quad \text{and} \quad x_2 \longrightarrow y_2$$

then for all values of constants k_1 and k_2,

$$k_1 x_1 + k_2 x_2 \longrightarrow k_1 y_1 + k_2 y_2 \tag{1.40}$$

This is true for all x_1 and x_2.

It may appear that additivity implies homogeneity. Unfortunately, homogeneity does not always follow from additivity. Exercise E1.11 demonstrates such a case.

EXERCISE E1.11

Show that a system with the input $x(t)$ and the output $y(t)$ related by $y(t) = \text{Re}\{x(t)\}$ satisfies the additivity property but violates the homogeneity property. Hence, such a system is not linear. [Hint: Show that Eq. (1.39) is not satisfied when k is complex.]

RESPONSE OF A LINEAR SYSTEM

For the sake of simplicity, we discuss only *single-input, single-output (SISO)* systems. But the discussion can be readily extended to *multiple-input, multiple-output (MIMO)* systems.

A system's output for $t \geq 0$ is the result of two independent causes: the initial conditions of the system (or the system state) at $t = 0$ and the input $x(t)$ for $t \geq 0$. If a system is to be linear, the output must be the sum of the two components resulting from these two causes: first, the *zero-input response* component that results only from the initial conditions at $t = 0$ with the input $x(t) = 0$ for $t \geq 0$, and then the *zero-state response* component that results only from the input $x(t)$ for $t \geq 0$ when the initial conditions (at $t = 0$) are assumed to be zero. When all the appropriate initial conditions are zero, the system is said to be in *zero state*. The system output is zero when the input is zero only if the system is in zero state.

In summary, a linear system response can be expressed as the sum of the zero-input and the zero-state component:

$$\textbf{total response} = \textbf{zero-input response} + \textbf{zero-state response} \tag{1.41}$$

This property of linear systems, which permits the separation of an output into components resulting from the initial conditions and from the input, is called the *decomposition property*.

[†]A linear system must also satisfy the additional condition of *smoothness,* where small changes in the system's inputs must result in small changes in its outputs.[3]

For the *RC* circuit of Fig. 1.26, the response $y(t)$ was found to be [see Eq. (1.36c)]

$$y(t) = \underbrace{v_C(0)}_{z-i \text{ component}} + \underbrace{Rx(t) + \frac{1}{C}\int_0^t x(\tau)\,d\tau}_{z-s \text{ component}} \tag{1.42}$$

From Eq. (1.42), it is clear that if the input $x(t) = 0$ for $t \geq 0$, the output $y(t) = v_C(0)$. Hence $v_C(0)$ is the zero-input component of the response $y(t)$. Similarly, if the system state (the voltage v_C in this case) is zero at $t = 0$, the output is given by the second component on the right-hand side of Eq. (1.42). Clearly this is the zero-state component of the response $y(t)$.

In addition to the decomposition property, linearity implies that both the zero-input and zero-state components must obey the principle of superposition with respect to each of their respective causes. For example, if we increase the initial condition k-fold, the zero-input component must also increase k-fold. Similarly, if we increase the input k-fold, the zero-state component must also increase k-fold. These facts can be readily verified from Eq. (1.42) for the *RC* circuit in Fig. 1.26. For instance, if we double the initial condition $v_C(0)$, the zero-input component doubles; if we double the input $x(t)$, the zero-state component doubles.

EXAMPLE 1.9

Show that the system described by the equation

$$\frac{dy}{dt} + 3y(t) = x(t) \tag{1.43}$$

is linear.[†]

Let the system response to the inputs $x_1(t)$ and $x_2(t)$ be $y_1(t)$ and $y_2(t)$, respectively. Then

$$\frac{dy_1}{dt} + 3y_1(t) = x_1(t)$$

and

$$\frac{dy_2}{dt} + 3y_2(t) = x_2(t)$$

[†]Equations such as (1.43) and (1.44) are considered to represent linear systems in the classical definition of linearity. Some authors consider such equations to represent *incrementally linear* systems. According to this definition, *linear system* has only a zero-state component. The zero-input component is absent. Hence, incrementally linear system response can be represented as a response of a linear system (linear in this new definition) plus a zero-input component. We prefer the classical definition to this new definition. It is just a matter of definition and makes no difference in the final results.

Multiplying the first equation by k_1, the second with k_2, and adding them yields

$$\frac{d}{dt}[k_1 y_1(t) + k_2 y_2(t)] + 3[k_1 y_1(t) + k_2 y_2(t)] = k_1 x_1(t) + k_2 x_2(t)$$

But this equation is the system equation [Eq. (1.43)] with

$$x(t) = k_1 x_1(t) + k_2 x_2(t)$$

and

$$y(t) = k_1 y_1(t) + k_2 y_2(t)$$

Therefore, when the input is $k_1 x_1(t) + k_2 x_2(t)$, the system response is $k_1 y_1(t) + k_2 y_2(t)$. Consequently, the system is linear. Using this argument, we can readily generalize the result to show that a system described by a differential equation of the form

$$a_0 \frac{d^N y}{dt^N} + a_1 \frac{d^{N-1} y}{dt^{N-1}} + \cdots + a_N y(t) = b_{N-M} \frac{d^M x}{dt^M} + \cdots + b_{N-1} \frac{dx}{dt} + b_N x(t) \qquad (1.44)$$

is a linear system. The coefficients a_i and b_i in this equation can be constants or functions of time. Although here we proved only zero-state linearity, it can be shown that such systems are also zero-input linear and have the decomposition property.

EXERCISE E1.12

Show that the system described by the following equation is linear:

$$\frac{dy}{dt} + t^2 y(t) = (2t + 3)x(t)$$

EXERCISE E1.13

Show that the system described by the following equation is nonlinear:

$$y(t)\frac{dy}{dt} + 3y(t) = x(t)$$

MORE COMMENTS ON LINEAR SYSTEMS

Almost all systems observed in practice become nonlinear when large enough signals are applied to them. However, it is possible to approximate most of the nonlinear systems by linear systems for small-signal analysis. The analysis of nonlinear systems is generally difficult. Nonlinearities

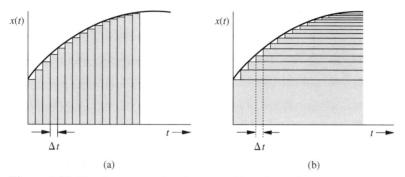

Figure 1.27 Signal representation in terms of impulse and step components.

can arise in so many ways that describing them with a common mathematical form is impossible. Not only is each system a category in itself, but even for a given system, changes in initial conditions or input amplitudes may change the nature of the problem. On the other hand, the superposition property of linear systems is a powerful unifying principle that allows for a general solution. The superposition property (linearity) greatly simplifies the analysis of linear systems. Because of the decomposition property, we can evaluate separately the two components of the output. The zero-input component can be computed by assuming the input to be zero, and the zero-state component can be computed by assuming zero initial conditions. Moreover, if we express an input $x(t)$ as a sum of simpler functions,

$$x(t) = a_1x_1(t) + a_2x_2(t) + \cdots + a_mx_m(t)$$

then, by virtue of linearity, the response $y(t)$ is given by

$$y(t) = a_1y_1(t) + a_2y_2(t) + \cdots + a_my_m(t) \tag{1.45}$$

where $y_k(t)$ is the zero-state response to an input $x_k(t)$. This apparently trivial observation has profound implications. As we shall see repeatedly in later chapters, it proves extremely useful and opens new avenues for analyzing linear systems.

For example, consider an arbitrary input $x(t)$ such as the one shown in Fig. 1.27a. We can approximate $x(t)$ with a sum of rectangular pulses of width Δt and of varying heights. The approximation improves as $\Delta t \to 0$, when the rectangular pulses become impulses spaced Δt seconds apart (with $\Delta t \to 0$).[†] Thus, an arbitrary input can be replaced by a weighted sum of impulses spaced Δt ($\Delta t \to 0$) seconds apart. Therefore, if we know the system response to a unit impulse, we can immediately determine the system response to an arbitrary input $x(t)$ by adding the system response to each impulse component of $x(t)$. A similar situation is depicted in Fig. 1.27b, where $x(t)$ is approximated by a sum of step functions of varying magnitude and spaced Δt seconds apart. The approximation improves as Δt becomes smaller. Therefore, if we know the system response to a unit step input, we can compute the system response to any arbitrary input $x(t)$ with relative ease. Time-domain analysis of linear systems (discussed in Chapter 2) uses this approach.

[†]Here, the discussion of a rectangular pulse approaching an impulse at $\Delta t \to 0$ is somewhat imprecise. It is explained in Section 2.4 with more rigor.

Chapters 4 through 7 employ the same approach but instead use sinusoids or exponentials as the basic signal components. We show that any arbitrary input signal can be expressed as a weighted sum of sinusoids (or exponentials) having various frequencies. Thus a knowledge of the system response to a sinusoid enables us to determine the system response to an arbitrary input $x(t)$.

1.7-2 Time-Invariant and Time-Varying Systems

Systems whose parameters do not change with time are *time-invariant* (also *constant-parameter*) systems. For such a system, if the input is delayed by T seconds, the output is the same as before but delayed by T (assuming initial conditions are also delayed by T). This property is expressed graphically in Fig. 1.28. We can also illustrate this property, as shown in Fig. 1.29. We can delay

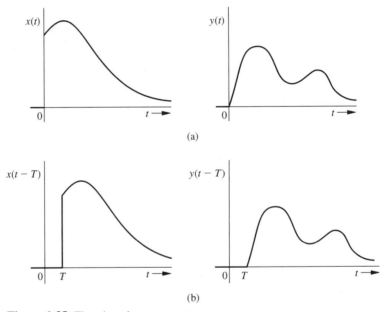

Figure 1.28 Time-invariance property.

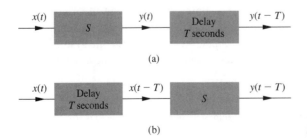

Figure 1.29 Illustration of time-invariance property.

the output $y(t)$ of a system S by applying the output $y(t)$ to a T second delay (Fig. 1.29a). If the system is time invariant, then the delayed output $y(t - T)$ can also be obtained by first delaying the input $x(t)$ before applying it to the system, as shown in Fig. 1.29b. In other words, the system S and the time delay commute if the system S is time invariant. This would not be true for time-varying systems. Consider, for instance, a time-varying system specified by $y(t) = e^{-t}x(t)$. The output for such a system in Fig. 1.29a is $e^{-(t-T)}x(t - T)$. In contrast, the output for the system in Fig. 1.29b is $e^{-t}x(t - T)$.

It is possible to verify that the system in Fig. 1.26 is a time-invariant system. Networks composed of RLC elements and other commonly used active elements such as transistors are time-invariant systems. A system with an input–output relationship described by a linear differential equation of the form given in Example 1.9 [Eq. (1.44)] is a linear time-invariant (LTI) system when the coefficients a_i and b_i of such equation are constants. If these coefficients are functions of time, then the system is a linear *time-varying* system.

The system described in Exercise E1.12 is linear time varying. Another familiar example of a time-varying system is the carbon microphone, in which the resistance R is a function of the mechanical pressure generated by sound waves on the carbon granules of the microphone. The output current from the microphone is thus modulated by the sound waves, as desired.

EXERCISE E1.14

Show that a system described by the following equation is time-varying-parameter system:

$$y(t) = (\sin t)x(t - 2)$$

[Hint: Show that the system fails to satisfy the time-invariance property.]

1.7-3 Instantaneous and Dynamic Systems

As observed earlier, a system's output at any instant t generally depends on the entire past input. However, in a special class of systems, the output at any instant t depends only on its input at that instant. In resistive networks, for example, any output of the network at some instant t depends only on the input at the instant t. In these systems, past history is irrelevant in determining the response. Such systems are said to be *instantaneous* or *memoryless* systems. More precisely, a system is said to be instantaneous (or memoryless) if its output at any instant t depends, at most, on the strength of its input(s) at the same instant t, and not on any past or future values of the input(s). Otherwise, the system is said to be *dynamic* (or a system with memory). A system whose response at t is completely determined by the input signals over the past T seconds [interval from $(t - T)$ to t] is a *finite-memory system* with a memory of T seconds. Networks containing inductive and capacitive elements generally have infinite memory because the response of such networks at any instant t is determined by their inputs over the entire past $(-\infty, t)$. This is true for the RC circuit of Fig. 1.26.

In this book we will generally examine dynamic systems. Instantaneous systems are a special case of dynamic systems.

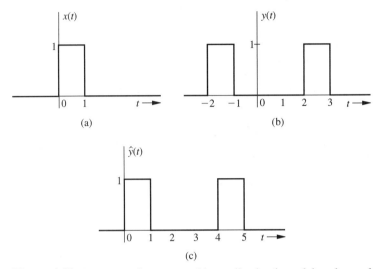

Figure 1.30 A noncausal system and its realization by a delayed causal system.

1.7-4 Causal and Noncausal Systems

A *causal* (also known as a *physical* or *nonanticipative*) system is one for which the output at any instant t_0 depends only on the value of the input $x(t)$ for $t \le t_0$. In other words, the value of the output at the present instant depends only on the past and present values of the input $x(t)$, not on its future values. To put it simply, in a causal system the output cannot start before the input is applied. If the response starts before the input, it means that the system knows the input in the future and acts on this knowledge before the input is applied. A system that violates the condition of causality is called a *noncausal* (or *anticipative*) system.

Any practical system that operates in real time[†] must necessarily be causal. We do not yet know how to build a system that can respond to future inputs (inputs not yet applied). A noncausal system is a prophetic system that knows the future input and acts on it in the present. Thus, if we apply an input starting at $t = 0$ to a noncausal system, the output would begin even before $t = 0$. For example, consider the system specified by

$$y(t) = x(t - 2) + x(t + 2) \tag{1.46}$$

For the input $x(t)$ illustrated in Fig. 1.30a, the output $y(t)$, as computed from Eq. (1.46) (shown in Fig. 1.30b), starts even before the input is applied. Equation (1.46) shows that $y(t)$, the output at t, is given by the sum of the input values 2 seconds before and 2 seconds after t (at $t - 2$ and $t + 2$, respectively). But if we are operating the system in real time at t, we do not know what the value of the input will be 2 seconds later. Thus it is impossible to implement this system in real time. For this reason, noncausal systems are unrealizable in *real time*.

[†]In real-time operations, the response to an input is essentially simultaneous (contemporaneous) with the input itself.

WHY STUDY NONCAUSAL SYSTEMS?

The foregoing discussion may suggest that noncausal systems have no practical purpose. This is not the case; they are valuable in the study of systems for several reasons. First, noncausal systems *are* realizable when the independent variable is other than "time" (e.g., *space*). Consider, for example, an electric charge of density $q(x)$ placed along the x axis for $x \geq 0$. This charge density produces an electric field $E(x)$ that is present at every point on the x axis from $x = -\infty$ to ∞. In this case the input [i.e., the charge density $q(x)$] starts at $x = 0$, but its output [the electric field $E(x)$] begins before $x = 0$. Clearly, this space-charge system is noncausal. This discussion shows that only temporal systems (systems with time as independent variable) must be causal to be realizable. The terms "before" and "after" have a special connection to causality only when the independent variable is time. This connection is lost for variables other than time. Nontemporal systems, such as those occurring in optics, can be noncausal and still realizable.

Moreover, even for temporal systems, such as those used for signal processing, the study of noncausal systems is important. In such systems we may have all input data prerecorded. (This often happens with speech, geophysical, and meteorological signals, and with space probes.) In such cases, the input's future values are available to us. For example, suppose we had a set of input signal records available for the system described by Eq. (1.46). We can then compute $y(t)$ since, for any t, we need only refer to the records to find the input's value 2 seconds before and 2 seconds after t. Thus, noncausal systems can be realized, although not in real time. We may therefore be able to realize a noncausal system, provided we are willing to accept a time delay in the output. Consider a system whose output $\hat{y}(t)$ is the same as $y(t)$ in Eq. (1.46) delayed by 2 seconds (Fig 1.30c), so that

$$\hat{y}(t) = y(t - 2)$$
$$= x(t - 4) + x(t)$$

Here the value of the output $\hat{y}$ at any instant t is the sum of the values of the input x at t and at the instant 4 seconds earlier [at $(t-4)$]. In this case, the output at any instant t does not depend on future values of the input, and the system is causal. The output of this system, which is $\hat{y}(t)$, is identical to that in Eq. (1.46) or Fig. 1.30b except for a delay of 2 seconds. Thus, a noncausal system may be realized or satisfactorily approximated in real time by using a causal system with a delay.

Noncausal systems are realizable with time delay!

A third reason for studying noncausal systems is that they provide an upper bound on the performance of causal systems. For example, if we wish to design a filter for separating a signal from noise, then the optimum filter is invariably a noncausal system. Although unrealizable, this noncausal system's performance acts as the upper limit on what can be achieved and gives us a standard for evaluating the performance of causal filters.

At first glance, noncausal systems may seem to be inscrutable. Actually, there is nothing mysterious about these systems and their approximate realization through physical systems with delay. If we want to know what will happen one year from now, we have two choices: go to a prophet (an unrealizable person) who can give the answers instantly, or go to a wise man and allow him a delay of one year to give us the answer! If the wise man is truly wise, he may even be able, by studying trends, to shrewdly guess the future very closely with a delay of less than a year. Such is the case with noncausal systems—nothing more and nothing less.

EXERCISE E1.15

Show that a system described by the following equation is noncausal:

$$y(t) = \int_{t-5}^{t+5} x(\tau)\,d\tau$$

Show that this system can be realized physically if we accept a delay of 5 seconds in the output.

1.7-5 Continuous-Time and Discrete-Time Systems

Signals defined or specified over a continuous range of time are *continuous-time signals,* denoted by symbols $x(t)$, $y(t)$, and so on. Systems whose inputs and outputs are continuous-time signals are *continuous-time systems.* On the other hand, signals defined only at discrete instants of time $t_0, t_1, t_2, \ldots, t_n, \ldots$ are *discrete-time signals,* denoted by the symbols $x(t_n)$, $y(t_n)$, and so on, where n is some integer. Systems whose inputs and outputs are discrete-time signals are *discrete-time systems.* A digital computer is a familiar example of this type of system. In practice, discrete-time signals can arise from sampling continuous-time signals. For example, when the sampling is uniform, the discrete instants $t_0, t_1, t_2, \ldots$ are uniformly spaced so that

$$t_{k+1} - t_k = T \qquad \text{for all } k$$

In such case, the discrete-time signals represented by the samples of continuous-time signals $x(t)$, $y(t)$, and so on can be expressed as $x(nT)$, $y(nT)$, and so on; for convenience, we further simplify this notation to $x[n]$, $y[n]$, $\ldots$, where it is understood that $x[n] = x(nT)$ and that n is some integer. A typical discrete-time signal is shown in Fig. 1.31. A discrete-time signal may also be viewed as a sequence of numbers $\ldots$, $x[-1]$, $x[0]$, $x[1]$, $x[2]$, $\ldots$. Thus a discrete-time system may be seen as processing a sequence of numbers $x[n]$ and yielding as an output another sequence of numbers $y[n]$.

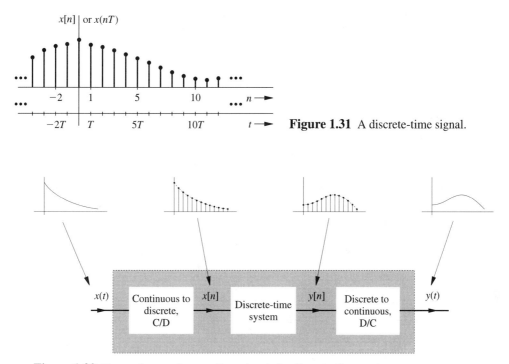

Figure 1.31 A discrete-time signal.

Figure 1.32 Processing continuous-time signals by discrete-time systems.

Discrete-time signals arise naturally in situations that are inherently discrete time, such as population studies, amortization problems, national income models, and radar tracking. They may also arise as a result of sampling continuous-time signals in sampled data systems, digital filtering, and the like. Digital filtering is a particularly interesting application in which continuous-time signals are processed by using discrete-time systems as shown in Fig. 1.32. A continuous-time signal $x(t)$ is first sampled to convert it into a discrete-time signal $x[n]$, which then is processed by the discrete-time system to yield a discrete-time output $y[n]$. A continuous-time signal $y(t)$ is finally constructed from $y[n]$. In this manner we can process a continuous-time signal with appropriate discrete-time system such as a digital computer. Because discrete-time systems have several significant advantages over continuous-time systems, there is an accelerating trend toward processing continuous-time signals with discrete-time systems.

1.7-6 Analog and Digital Systems

Analog and digital signals are discussed in Section 1.3-2. A system whose input and output signals are analog is an *analog system;* a system whose input and output signals are digital is a *digital system.* A digital computer is an example of a digital (binary) system. Observe that a digital computer is digital as well as a discrete-time system.

Figure 1.33 A cascade of a system with its inverse results in an identity system.

1.7-7 Invertible and Noninvertible Systems

A system S performs certain operation(s) on input signal(s). If we can obtain the input $x(t)$ back from the corresponding output $y(t)$ by some operation, the system S is said to be *invertible*. When several different inputs result in the same output (as in a rectifier), it is impossible to obtain the input from the output, and the system is *noninvertible*. Therefore, for an invertible system, it is essential that every input have a unique output so that there is a one-to-one mapping between an input and the corresponding output. The system that achieves the inverse operation [of obtaining $x(t)$ from $y(t)$] is the *inverse system* for S. For instance, if S is an ideal integrator, then its inverse system is an ideal differentiator. Consider a system S connected in tandem with its inverse S_i, as shown in Fig. 1.33. The input $x(t)$ to this tandem system results in signal $y(t)$ at the output of S, and the signal $y(t)$, which now acts as an input to S_i, yields back the signal $x(t)$ at the output of S_i. Thus, S_i undoes the operation of S on $x(t)$, yielding back $x(t)$. A system whose output is equal to the input (for all possible inputs) is an *identity* system. Cascading a system with its inverse system, as shown in Fig. 1.33, results in an identity system.

In contrast, a rectifier, specified by an equation $y(t) = |x(t)|$, is noninvertible because the rectification operation cannot be undone.

Inverse systems are very important in signal processing. In many applications, the signals are distorted during the processing, and it is necessary to undo the distortion. For instance, in transmission of data over a communication channel, the signals are distorted owing to nonideal frequency response and finite bandwidth of a channel. It is necessary to restore the signal as closely as possible to its original shape. Such equalization is also used in audio systems and photographic systems.

1.7-8 Stable and Unstable Systems

Systems can also be classified as *stable* or *unstable* systems. Stability can be *internal* or *external*. If every *bounded input* applied at the input terminal results in a *bounded output*, the system is said to be stable *externally*. The external stability can be ascertained by measurements at the external terminals (input and output) of the system. This type of stability is also known as the stability in the BIBO (bounded-input/bounded-output) sense. The concept of internal stability is postponed to Chapter 2 because it requires some understanding of internal system behavior, introduced in that chapter.

EXERCISE E1.16

Show that a system described by the equation $y(t) = x^2(t)$ is noninvertible but BIBO stable.

1.8 SYSTEM MODEL: INPUT–OUTPUT DESCRIPTION

A system description in terms of the measurements at the input and output terminals is called the *input–output description*. As mentioned earlier, systems theory encompasses a variety of systems, such as electrical, mechanical, hydraulic, acoustic, electromechanical, and chemical, as well as social, political, economic, and biological. The first step in analyzing any system is the construction of a system model, which is a mathematical expression or a rule that satisfactorily approximates the dynamical behavior of the system. In this chapter we shall consider only continuous-time systems. (Modeling of discrete-time systems is discussed in Chapter 3.)

1.8-1 Electrical Systems

To construct a system model, we must study the relationships between different variables in the system. In electrical systems, for example, we must determine a satisfactory model for the voltage–current relationship of each element, such as Ohm's law for a resistor. In addition, we must determine the various constraints on voltages and currents when several electrical elements are interconnected. These are the laws of interconnection—the well-known Kirchhoff laws for voltage and current (KVL and KCL). From all these equations, we eliminate unwanted variables to obtain equation(s) relating the desired output variable(s) to the input(s). The following examples demonstrate the procedure of deriving input-output relationships for some LTI electrical systems.

EXAMPLE 1.10

For the series RLC circuit of Fig. 1.34, find the input–output equation relating the input voltage $x(t)$ to the output current (loop current) $y(t)$.

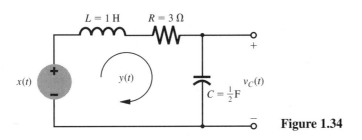

Figure 1.34

Application of Kirchhoff's voltage law around the loop yields

$$v_L(t) + v_R(t) + v_C(t) = x(t) \tag{1.47}$$

By using the voltage–current laws of each element (inductor, resistor, and capacitor), we can express this equation as

$$\frac{dy}{dt} + 3y(t) + 2 \int_{-\infty}^{t} y(\tau)\, d\tau = x(t) \tag{1.48}$$

Differentiating both sides of this equation, we obtain

$$\frac{d^2 y}{dt^2} + 3\frac{dy}{dt} + 2y(t) = \frac{dx}{dt} \tag{1.49}$$

This differential equation is the input–output relationship between the output $y(t)$ and the input $x(t)$.

It proves convenient to use a compact notation D for the differential operator d/dt. Thus

$$\frac{dy}{dt} \equiv Dy(t) \tag{1.50}$$

$$\frac{d^2 y}{dt^2} \equiv D^2 y(t) \tag{1.51}$$

and so on. With this notation, Eq. (1.49) can be expressed as

$$(D^2 + 3D + 2)y(t) = Dx(t) \tag{1.52}$$

The differential operator is the inverse of the integral operator, so we can use the operator $1/D$ to represent integration.[†]

$$\int_{-\infty}^{t} y(\tau)\, d\tau \equiv \frac{1}{D} y(t) \tag{1.53}$$

[†]Use of operator $1/D$ for integration generates some subtle mathematical difficulties because the operators D and $1/D$ do not commute. For instance, we know that $D(1/D) = 1$ because

$$\frac{d}{dt} \left[\int_{-\infty}^{t} y(\tau)\, d\tau \right] = y(t)$$

However, $(1/D)D$ is not necessarily unity. Use of Cramer's rule in solving simultaneous integro-differential equations will always result in cancellation of operators $1/D$ and D. This procedure may yield erroneous results when the factor D occurs in the numerator as well as in the denominator. This happens, for instance, in circuits with all-inductor loops or all-capacitor cut sets. To eliminate this problem, avoid the integral operation in system equations so that the resulting equations are differential rather than integro-differential. In electrical circuits, this can be done by using charge (instead of current) variables in loops containing capacitors and choosing current variables for loops without capacitors. In the literature this problem of commutativity of D and $1/D$ is largely ignored. As mentioned earlier, such procedure gives erroneous results only in special systems, such as the circuits with all-inductor loops or all-capacitor cut sets. Fortunately such systems constitute a very small fraction of the systems we deal with. For further discussion of this topic and a correct method of handling problems involving integrals, see Ref. 4.

Consequently, the loop equation (1.48) can be expressed as

$$\left(D + 3 + \frac{2}{D}\right) y(t) = x(t) \tag{1.54}$$

Multiplying both sides by D, that is, differentiating Eq. (1.54), we obtain

$$(D^2 + 3D + 2) y(t) = Dx(t) \tag{1.55}$$

which is identical to Eq. (1.52).

Recall that Eq. (1.55) is not an algebraic equation, and $D^2 + 3D + 2$ is not an algebraic term that multiplies $y(t)$; it is an operator that operates on $y(t)$. It means that we must perform the following operations on $y(t)$: take the second derivative of $y(t)$ and add to it 3 times the first derivative of $y(t)$ and 2 times $y(t)$. Clearly, a polynomial in D multiplied by $y(t)$ represents a certain differential operation on $y(t)$.

EXAMPLE 1.11

Find the equation relating the input to output for the series RC circuit of Fig. 1.35 if the input is the voltage $x(t)$ and output is

(a) the loop current $i(t)$

(b) the capacitor voltage $y(t)$

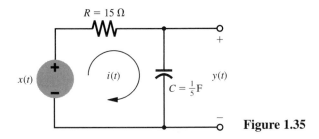

Figure 1.35

(a) The loop equation for the circuit is

$$R\, i(t) + \frac{1}{C} \int_{-\infty}^{t} i(\tau)\, d\tau = x(t) \tag{1.56}$$

or

$$15\, i(t) + 5 \int_{-\infty}^{t} i(\tau)\, d\tau = x(t) \tag{1.57}$$

With operational notation, this equation can be expressed as

$$15\, i(t) + \frac{5}{D} i(t) = x(t) \tag{1.58}$$

(b) Multiplying both sides of Eqs. (1.58) by D (i.e., differentiating the equation), we obtain

$$(15D + 5)\,i(t) = Dx(t) \tag{1.59a}$$

or

$$15\frac{di}{dt} + 5\,i(t) = \frac{dx}{dt} \tag{1.59b}$$

Moreover,

$$i(t) = C\,\frac{dy}{dt}$$

$$= \tfrac{1}{5}Dy(t)$$

Substitution of this result in Eq. (1.59a) yields

$$(3D + 1)y(t) = x(t) \tag{1.60}$$

or

$$3\frac{dy}{dt} + y(t) = x(t) \tag{1.61}$$

EXERCISE E1.17

For the RLC circuit in Fig. 1.34, find the input–output relationship if the output is the inductor voltage $v_L(t)$.

ANSWER
$(D^2 + 3D + 2)v_L(t) = D^2x(t)$

EXERCISE E1.18

For the RLC circuit in Fig. 1.34, find the input–output relationship if the output is the capacitor voltage $v_C(t)$.

ANSWER
$(D^2 + 3D + 2)v_C(t) = 2x(t)$

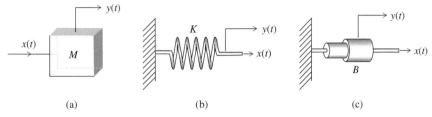

Figure 1.36 Some elements in translational mechanical systems.

1.8-2 Mechanical Systems

Planar motion can be resolved into translational (rectilinear) motion and rotational (torsional) motion. Translational motion will be considered first. We shall restrict ourselves to motions in one dimension.

TRANSLATIONAL SYSTEMS

The basic elements used in modeling translational systems are ideal masses, linear springs, and dashpots providing viscous damping. The laws of various mechanical elements are now discussed.

For a *mass M* (Fig. 1.36a), a force $x(t)$ causes a motion $y(t)$ and acceleration $\ddot{y}(t)$. From Newton's law of motion,

$$x(t) = M\ddot{y}(t) = M\frac{d^2y}{dt^2} = MD^2y(t) \tag{1.62}$$

The force $x(t)$ required to stretch (or compress) a *linear spring* (Fig. 1.36b) by an amount $y(t)$ is given by

$$x(t) = Ky(t) \tag{1.63}$$

where K is the *stiffness* of the spring.

For *a linear dashpot* (Fig. 1.36c), which operates by virtue of viscous friction, the force moving the dashpot is proportional to the relative velocity $\dot{y}(t)$ of one surface with respect to the other. Thus

$$x(t) = B\dot{y}(t) = B\frac{dy}{dt} = BDy(t) \tag{1.64}$$

where B is the *damping coefficient* of the dashpot or the viscous friction.

EXAMPLE 1.12

Find the input–output relationship for the translational mechanical system shown in Fig. 1.37a or its equivalent in Fig. 1.37b. The input is the force $x(t)$, and the output is the mass position $y(t)$.

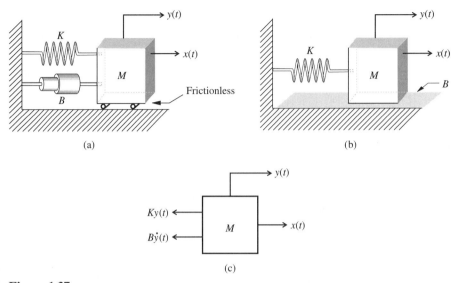

Figure 1.37

In mechanical systems it is helpful to draw a free-body diagram of each junction, which is a point at which two or more elements are connected. In Fig. 1.37, the point representing the mass is a junction. The displacement of the mass is denoted by $y(t)$. The spring is also stretched by the amount $y(t)$, and therefore it exerts a force $-Ky(t)$ on the mass. The dashpot exerts a force $-B\dot{y}(t)$ on the mass as shown in the free-body diagram (Fig. 1.37c). By Newton's second law, the net force must be $M\ddot{y}(t)$. Therefore

$$M\ddot{y}(t) = -B\dot{y}(t) - Ky(t) + x(t)$$

or

$$(MD^2 + BD + K)y(t) = x(t) \tag{1.65}$$

ROTATIONAL SYSTEMS

In rotational systems, the motion of a body may be defined as its motion about a certain axis. The variables used to describe rotational motion are torque (in place of force), angular position (in place of linear position), angular velocity (in place of linear velocity), and angular acceleration (in place of linear acceleration). The system elements are *rotational mass* or *moment of inertia* (in place of mass) and *torsional springs* and *torsional dashpots* (in place of linear springs and dashpots). The terminal equations for these elements are analogous to the corresponding equations for translational elements. If J is the moment of inertia (or rotational mass) of a rotating body about a certain axis, then the external torque required for this motion is equal to J (rotational mass) times the angular acceleration. If θ is the angular position of the body, $\ddot{\theta}$ is its angular acceleration, and

$$\text{torque} = J\ddot{\theta} = J\frac{d^2\theta}{dt^2} = JD^2\theta(t) \tag{1.66}$$

Similarly, if K is the stiffness of a torsional spring (per unit angular twist), and θ is the angular displacement of one terminal of the spring with respect to the other, then

$$\text{torque} = K\theta \tag{1.67}$$

Finally, the torque due to viscous damping of a torsional dashpot with damping coefficient B is

$$\text{torque} = B\dot{\theta}(t) = BD\theta(t) \tag{1.68}$$

EXAMPLE 1.13

The attitude of an aircraft can be controlled by three sets of surfaces (shown shaded in Fig. 1.38): elevators, rudder, and ailerons. By manipulating these surfaces, one can set the aircraft on a desired flight path. The roll angle φ can be controlled by deflecting in the opposite direction the two aileron surfaces as shown in Fig. 1.38. Assuming only rolling motion, find the equation relating the roll angle φ to the input (deflection) θ.

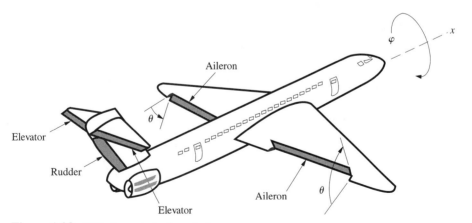

Figure 1.38 Attitude control of an airplane.

The aileron surfaces generate a torque about the roll axis proportional to the aileron deflection angle θ. Let this torque be $c\theta$, where c is the constant of proportionality. Air friction dissipates the torque $B\dot{\varphi}(t)$. The torque available for rolling motion is then $c\theta(t) - B\dot{\varphi}(t)$. If J is the moment of inertia of the plane about the x axis (roll axis), then

$$J\ddot{\varphi}(t) = \text{net torque}$$

$$= c\theta(t) - B\dot{\varphi}(t) \tag{1.69}$$

and

$$J\frac{d^2\varphi}{dt^2} + B\frac{d\varphi}{dt} = c\,\theta(t) \tag{1.70}$$

or

$$(JD^2 + BD)\varphi(t) = c\theta(t) \tag{1.71}$$

This is the desired equation relating the output (roll angle φ) to the input (aileron angle θ).

The roll velocity ω is $\dot{\varphi}(t)$. If the desired output is the roll velocity ω rather than the roll angle φ, then the input–output equation would be

$$J\frac{d\omega}{dt} + B\omega = c\theta \tag{1.72}$$

or

$$(JD + B)\omega(t) = c\theta(t) \tag{1.73}$$

EXERCISE E1.19

Torque $\mathcal{T}(t)$ is applied to the rotational mechanical system shown in Fig. 1.39a. The torsional spring stiffness is K; the rotational mass (the cylinder's moment of inertia about the shaft) is J; the viscous damping coefficient between the cylinder and the ground is B. Find the equation relating the output angle θ to the input torque $\mathcal{T}$. [Hint: A free-body diagram is shown in Fig. 1.39b.]

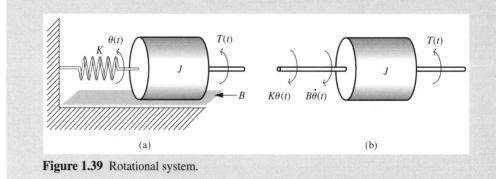

(a)   (b)

Figure 1.39 Rotational system.

ANSWER

$$J\frac{d^2\theta}{dt^2} + B\frac{d\theta}{dt} + K\theta(t) = T(t)$$

or

$$(JD^2 + BD + K)\theta(t) = T(t)$$

1.8-3 Electromechanical Systems

A wide variety of electromechanical systems convert electrical signals into mechanical motion (mechanical energy) and vice versa. Here we consider a rather simple example of an armature-controlled dc motor driven by a current source $x(t)$, as shown in Fig. 1.40a. The torque $T(t)$ generated in the motor is proportional to the armature current $x(t)$. Therefore

$$T(t) = K_T x(t) \tag{1.74}$$

where K_T is a constant of the motor. This torque drives a mechanical load whose free-body diagram is shown in Fig. 1.40b. The viscous damping (with coefficient B) dissipates a torque $B\dot{\theta}(t)$. If J is the moment of inertia of the load (including the rotor of the motor), then the net torque $T(t) - B\dot{\theta}(t)$ must be equal to $J\ddot{\theta}(t)$:

$$J\ddot{\theta}(t) = T(t) - B\dot{\theta}(t) \tag{1.75}$$

Thus

$$(JD^2 + BD)\theta(t) = T(t)$$

$$= K_T x(t) \tag{1.76}$$

which in conventional form can be expressed as

$$J\frac{d^2\theta}{dt^2} + B\frac{d\theta}{dt} = K_T x(t) \tag{1.77}$$

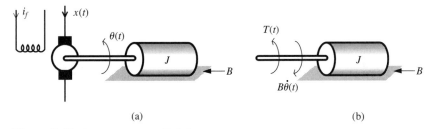

(a)   (b)

Figure 1.40 Armature-controlled dc motor.

1.9 INTERNAL AND EXTERNAL DESCRIPTION OF A SYSTEM

The input–output relationship of a system is an *external description* of that system. We have found an external description (not the *internal description*) of systems in all the examples discussed so far. This may puzzle the reader because in each of these cases, we derived the input–output relationship by analyzing the internal structure of that system. Why is this not an internal description? What makes a description internal? Although it is true that we did find the input–output description by internal analysis of the system, we did so strictly for convenience. We could have obtained the input–output description by making observations at the external (input and output) terminals, for example, by measuring the output for certain inputs such as an impulse or a sinusoid. A description that can be obtained from measurements at the external terminals (even when the rest of the system is sealed inside an inaccessible black box) is an external description. Clearly, the input–output description is an external description. What, then, is an internal description? Internal description is capable of providing the complete information about all possible signals in the system. An external description may not give such complete information. An external description can always be found from an internal description, but the converse is not necessarily true. We shall now give an example to clarify the distinction between an external and an internal description.

Let the circuit in Fig. 1.41a with the input $x(t)$ and the output $y(t)$ be enclosed inside a "black box" with only the input and the output terminals accessible. To determine its external description, let us apply a known voltage $x(t)$ at the input terminals and measure the resulting output voltage $y(t)$.

Let us also assume that there is some initial charge Q_0 present on the capacitor. The output voltage will generally depend on both, the input $x(t)$ and the initial charge Q_0. To compute the output resulting because of the charge Q_0, assume the input $x(t) = 0$ (short across the input).

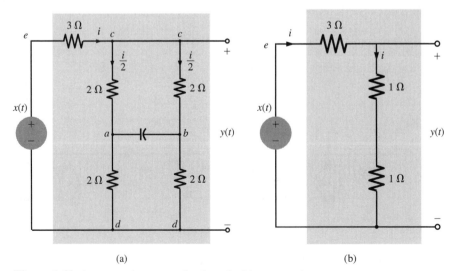

(a)   (b)

Figure 1.41 A system that cannot be described by external measurements.

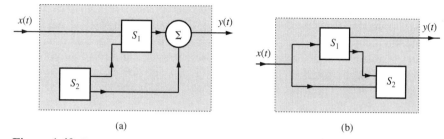

Figure 1.42 Structures of uncontrollable and unobservable systems.

In this case, the currents in the two 2 Ω resistors in the upper and the lower branches at the output terminals are equal and opposite because of the balanced nature of the circuit. Clearly, the capacitor charge results in zero voltage at the output.[†]

Now, to compute the output $y(t)$ resulting from the input voltage $x(t)$, we assume zero initial capacitor charge (short across the capacitor terminals). The current $i(t)$ (Fig. 1.41a), in this case, divides equally between the two parallel branches because the circuit is balanced. Thus, the voltage across the capacitor continues to remain zero. Therefore, for the purpose of computing the current $i(t)$, the capacitor may be removed or replaced by a short. The resulting circuit is equivalent to that shown in Fig. 1.41b, which shows that the input $x(t)$ sees a load of 5 Ω, and

$$i(t) = \tfrac{1}{5}x(t)$$

Also, because $y(t) = 2\,i(t)$,

$$y(t) = \tfrac{2}{5}x(t) \tag{1.78}$$

This is the total response. Clearly, for the external description, the capacitor does not exist. No external measurement or external observation can detect the presence of the capacitor. Furthermore, if the circuit is enclosed inside a "black box" so that only the external terminals are accessible, it is impossible to determine the currents (or voltages) inside the circuit from external measurements or observations. An internal description, however, can provide every possible signal inside the system. In Example 1.15, we shall find the internal description of this system and show that it is capable of determining every possible signal in the system.

For most systems, the external and internal descriptions are equivalent, but there are a few exceptions, as in the present case, where the external description gives an inadequate picture of the systems. This happens when the system is *uncontrollable* and/or *unobservable*.

Figure 1.42 shows structural representations of simple uncontrollable and unobservable systems. In Fig. 1.42a, we note that part of the system (subsystem S_2) inside the box cannot be controlled by the input $x(t)$. In Fig. 1.42b, some of the system outputs (those in subsystem S_2) cannot be observed from the output terminals. If we try to describe either of these systems by applying an external input $x(t)$ and then measuring the output $y(t)$, the measurement will not

---

[†]The output voltage $y(t)$ resulting because of the capacitor charge [assuming $x(t) = 0$] is the zero-input response, which, as argued above, is zero. The output component due to the input $x(t)$ (assuming zero initial capacitor charge) is the zero-state response. Complete analysis of this problem is given later in Example 1.15.

characterize the complete system but only the part of the system (here S_1) that is both controllable and observable (linked to both the input and output). Such systems are undesirable in practice and should be avoided in any system design. The system in Fig. 1.41a can be shown to be neither controllable nor observable. It can be represented structurally as a combination of the systems in Fig. 1.42a and 1.42b.

1.10 INTERNAL DESCRIPTION: THE STATE-SPACE DESCRIPTION

We shall now introduce the *state-space* description of a linear system, which is an internal description of a system. In this approach, we identify certain key variables, called the *state variables,* of the system. These variables have the property that every possible signal in the system can be expressed as a linear combination of these state variables. For example, we can show that every possible signal in a passive *RLC* circuit can be expressed as a linear combination of independent capacitor voltages and inductor currents, which, therefore, are state variables for the circuit.

To illustrate this point, consider the network in Fig. 1.43. We identify two state variables; the capacitor voltage q_1 and the inductor current q_2. If the values of q_1, q_2, and the input $x(t)$ are known at some instant t, we can demonstrate that every possible signal (current or voltage) in the circuit can be determined at t. For example, if $q_1 = 10$, $q_2 = 1$, and the input $x = 20$ at some instant, the remaining voltages and currents at that instant will be

$$i_1 = (x - q_1)/1 = 20 - 10 = 10 \text{ A}$$

$$v_1 = x - q_1 = 20 - 10 = 10 \text{ V}$$

$$v_2 = q_1 = 10 \text{ V}$$

$$i_2 = q_1/2 = 5 \text{ A}$$

$$i_C = i_1 - i_2 - q_2 = 10 - 5 - 1 = 4 \text{ A} \qquad (1.79)$$

$$i_3 = q_2 = 1 \text{ A}$$

$$v_3 = 5q_2 = 5 \text{ V}$$

$$v_L = q_1 - v_3 = 10 - 5 = 5 \text{ V}$$

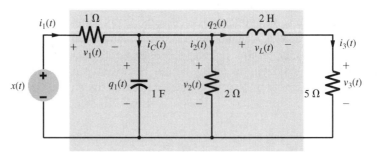

Figure 1.43 Choosing suitable initial conditions in a network.

Thus all signals in this circuit are determined. Clearly, state variables consist of the *key variables* in a system; a knowledge of the state variables allows one to determine every possible output of the system. Note that the *state-variable description is an internal description* of a system because it is capable of describing all possible signals in the system.

EXAMPLE 1.14

This example illustrates how state equations may be natural and easier to determine than other descriptions, such as loop or node equations. Consider again the network in Fig. 1.43 with q_1 and q_2 as the state variables and write the state equations.

This can be done by simple inspection of Fig. 1.43. Since $\dot{q}_1$ is the current through the capacitor,

$$\dot{q}_1 = i_C = i_1 - i_2 - q_2$$
$$= (x - q_1) - 0.5\,q_1 - q_2$$
$$= -1.5\,q_1 - q_2 + x$$

Also $2\dot{q}_2$, the voltage across the inductor, is given by

$$2\dot{q}_2 = q_1 - v_3$$
$$= q_1 - 5q_2$$

or

$$\dot{q}_2 = 0.5\,q_1 - 2.5\,q_2$$

Thus the state equations are

$$\dot{q}_1 = -1.5\,q_1 - q_2 + x$$
$$\dot{q}_2 = 0.5\,q_1 - 2.5\,q_2 \qquad (1.80)$$

This is a set of two simultaneous first-order differential equations. This set of equations is known as the *state equations*. Once these equations have been solved for q_1 and q_2, everything else in the circuit can be determined by using Eqs. (1.79). The set of output equations (1.79) is called the *output equations*. Thus, in this approach, we have two sets of equations, the state equations and the output equations. Once we have solved the state equations, all possible outputs can be obtained from the output equations. In the input–output description, an Nth-order system is described by an Nth-order equation. In the state-variable approach, the same system is described by N simultaneous first-order state equations.[†]

[†]This assumes the system to be controllable and observable. If it is not, the input–output description equation will be of an order lower than the corresponding number of state equations.

EXAMPLE 1.15

In this example, we investigate the nature of state equations and the issue of controllability and observability for the circuit in Fig. 1.41a. This circuit has only one capacitor and no inductors. Hence, there is only one state variable, the capacitor voltage $q(t)$. Since $C = 1$ F, the capacitor current is $\dot{q}$. There are two sources in this circuit: the input $x(t)$ and the capacitor voltage $q(t)$. The response due to $x(t)$, assuming $q(t) = 0$, is the zero-state response, which can be found from Fig. 1.44a, where we have shorted the capacitor [$q(t) = 0$]. The response due to $q(t)$ assuming $x(t) = 0$, is the zero-input response, which can be found from Fig. 1.44b, where we have shorted $x(t)$ to ensure $x(t) = 0$. It is now trivial to find both the components.

Figure 1.44a shows zero-state currents in every branch. It is clear that the input $x(t)$ sees an effective resistance of 5 Ω, and, hence, the current through $x(t)$ is $x/5$ A, which divides in the two parallel branches resulting in the current $x/10$ through each branch.

Examining the circuit in Fig. 1.44b for the zero-input response, we note that the capacitor voltage is q and the current is $\dot{q}$. We also observe that the capacitor sees two loops in parallel, each with resistance 4 Ω and current $\dot{q}/2$. Interestingly, the 3 Ω branch is effectively shorted because the circuit is balanced, and thus the voltage across the terminals cd is zero. The total current in any branch is the sum of the currents in that branch in Fig. 1.44a and 1.44b (principle of superposition).

Branch	Current	Voltage	
ca	$\dfrac{x}{10} + \dfrac{\dot{q}}{2}$	$2\left(\dfrac{x}{10} + \dfrac{\dot{q}}{2}\right)$	
cb	$\dfrac{x}{10} - \dfrac{\dot{q}}{2}$	$2\left(\dfrac{x}{10} - \dfrac{\dot{q}}{2}\right)$	
ad	$\dfrac{x}{10} - \dfrac{\dot{q}}{2}$	$2\left(\dfrac{x}{10} - \dfrac{\dot{q}}{2}\right)$	(1.81)
bd	$\dfrac{x}{10} + \dfrac{\dot{q}}{2}$	$2\left(\dfrac{x}{10} + \dfrac{\dot{q}}{2}\right)$	
ec	$\dfrac{x}{5}$	$3\left(\dfrac{x}{5}\right)$	
ed	$\dfrac{x}{5}$	x	

To find the state equation, we note that the current in branch ca is $(x/10) + \dot{q}/2$ and the current in branch cb is $(x/10) - \dot{q}/2$. Hence, the equation around the loop $acba$ is

$$q = 2\left[-\frac{x}{10} - \frac{\dot{q}}{2}\right] + 2\left[\frac{x}{10} - \frac{\dot{q}}{2}\right] = -2\dot{q}$$

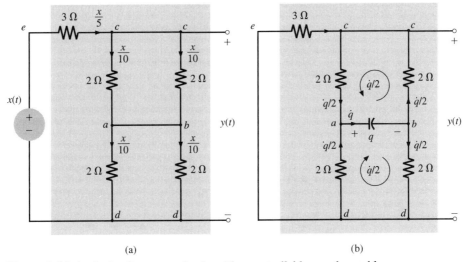

Figure 1.44 Analysis of a system that is neither controllable nor observable.

or

$$\dot{q} = -0.5q \qquad (1.82)$$

This is the desired state equation.

Substitution of $\dot{q} = -0.5q$ in Eqs. (1.81) shows that every possible current and voltage in the circuit can be expressed in terms of the state variable q and the input x, as desired. Hence, the set of Eqs. (1.81) is the output equation for this circuit. Once we have solved the state equation (1.82) for q, we can determine every possible output in the circuit.

The output $y(t)$ is given by

$$y(t) = 2\left[\frac{x}{10} - \frac{\dot{q}}{2}\right] + 2\left[\frac{x}{10} + \frac{\dot{q}}{2}\right]$$

$$= \frac{2}{5}x(t) \qquad (1.83)$$

A little examination of the state and the output equations indicates the nature of this system. The state equation (1.82) shows that the state $q(t)$ is independent of the input $x(t)$, and hence, the system state q cannot be controlled by the input. Moreover, Eq. (1.83) shows that the output $y(t)$ does not depend on the state $q(t)$. Thus, the system state cannot be observed from the output terminals. Hence, the system is neither controllable nor observable. Such is not the case of other systems examined earlier. Consider, for example, the circuit in Fig. 1.43. The state equation (1.80) shows that the states are influenced by the input directly or indirectly. Hence, the system is controllable. Moreover, as the output Eqs. (1.79) show, every possible output is expressed in terms of the state variables and the input. Hence, the states are also observable.

State-space techniques are useful not just because of their ability to provide internal system description, but for several other reasons, including the following.

1. State equations of a system provide a mathematical model of great generality that can describe not just linear systems, but also nonlinear systems; not just time-invariant systems, but also time-varying parameter systems; not just SISO (single-input/single-output) systems, but also multiple-input/multiple-output (MIMO) systems. Indeed, state equations are ideally suited for the analysis, synthesis, and optimization of MIMO systems.

2. Compact matrix notation and the powerful techniques of linear algebra greatly facilitates complex manipulations. Without such features, many important results of the modern system theory would have been difficult to obtain. State equations can yield a great deal of information about a system even when they are not solved explicitly.

3. State equations lend themselves readily to digital computer simulation of complex systems of high order, with or without nonlinearities, and with multiple inputs and outputs.

4. For second-order systems ($N = 2$), a graphical method called *phase-plane analysis* can be used on state equations, whether they are linear or nonlinear.

The real benefits of the state-space approach, however, are realized for highly complex systems of large order. Much of the book is devoted to introduction of the basic concepts of linear systems analysis, which must necessarily begin with simpler systems without using the state-space approach. Chapter 10 deals with the state-space analysis of linear, time invariant, continuous-time, and discrete-time systems.

EXERCISE E1.20

Write the state equations for the series *RLC* circuit shown in Fig. 1.45, using the inductor current $q_1(t)$ and the capacitor voltage $q_2(t)$ as state variables. Express every voltage and current in this circuit as a linear combination of q_1, q_2, and x.

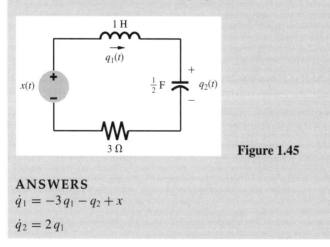

Figure 1.45

ANSWERS

$$\dot{q}_1 = -3\,q_1 - q_2 + x$$

$$\dot{q}_2 = 2\,q_1$$

1.11 SUMMARY

A *signal* is a set of data or information. A *system* processes input signals to modify them or extract additional information from them to produce output signals (response). A system may be made up of physical components (hardware realization), or it may be an algorithm that computes an output signal from an input signal (software realization).

A convenient measure of the size of a signal is its energy, if it is finite. If the signal energy is infinite, the appropriate measure is its power, if it exists. The signal power is the time average of its energy (averaged over the entire time interval from $-\infty$ to ∞). For periodic signals the time averaging need be performed only over one period in view of the periodic repetition of the signal. Signal power is also equal to the mean squared value of the signal (averaged over the entire time interval from $t = -\infty$ to ∞).

Signals can be classified in several ways.

1. A *continuous-time signal* is specified for a continuum of values of the independent variable (such as time t). A *discrete-time signal* is specified only at a finite or a countable set of time instants.

2. An *analog signal* is a signal whose amplitude can take on any value over a continuum. On the other hand, a signal whose amplitudes can take on only a finite number of values is a *digital signal*. The terms *discrete-time* and *continuous-time* qualify the nature of a signal along the time axis (horizontal axis). The terms *analog* and *digital,* on the other hand, qualify the nature of the signal amplitude (vertical axis).

3. A *periodic signal* $x(t)$ is defined by the fact that $x(t) = x(t + T_0)$ for some T_0. The smallest value of T_0 for which this relationship is satisfied is called the *fundamental period*. A periodic signal remains unchanged when shifted by an integer multiple of its period. A periodic signal $x(t)$ can be generated by a periodic extension of any contiguous segment of $x(t)$ of duration T_0. Finally, a periodic signal, by definition, must exist over the entire time interval $-\infty < t < \infty$. A signal is *aperiodic* if it is not periodic.

   An *everlasting signal* starts at $t = -\infty$ and continues forever to $t = \infty$. Hence, periodic signals are everlasting signals. A *causal signal* is a signal that is zero for $t < 0$.

4. A signal with finite energy is an *energy signal*. Similarly a signal with a finite and nonzero power (mean square value) is a *power signal*. A signal can be either an energy signal or a power signal, but not both. However, there are signals that are neither energy nor power signals.

5. A signal whose physical description is known completely in a mathematical or graphical form is a *deterministic signal*. A *random signal* is known only in terms of its probabilistic description such as mean value or mean-square value, rather than by its mathematical or graphical form.

A signal $x(t)$ delayed by T seconds (right-shifted) can be expressed as $x(t - T)$; on the other hand, $x(t)$ advanced by T (left-shifted) is $x(t + T)$. A signal $x(t)$ time-compressed by a factor a $(a > 1)$ is expressed as $x(at)$; on the other hand, the same signal time-expanded by factor a $(a > 1)$ is $x(t/a)$. The signal $x(t)$ when time reversed can be expressed as $x(-t)$.

The unit step function $u(t)$ is very useful in representing causal signals and signals with different mathematical descriptions over different intervals.

In the classical (Dirac) definition, the unit impulse function $\delta(t)$ is characterized by unit area and is concentrated at a single instant $t = 0$. The impulse function has a sampling (or sifting) property, which states that the area under the product of a function with a unit impulse is equal to the value of that function at the instant at which the impulse is located (assuming the function to be continuous at the impulse location). In the modern approach, the impulse function is viewed as a generalized function and is defined by the sampling property.

The exponential function e^{st}, where s is complex, encompasses a large class of signals that includes a constant, a monotonic exponential, a sinusoid, and an exponentially varying sinusoid.

A real signal that is symmetrical about the vertical axis ($t = 0$) is an *even* function of time, and a real signal that is antisymmetrical about the vertical axis is an *odd* function of time. The product of an even function and an odd function is an odd function. However, the product of an even function and an even function or an odd function and an odd function is an even function. The area under an odd function from $t = -a$ to a is always zero regardless of the value of a. On the other hand, the area under an even function from $t = -a$ to a is two times the area under the same function from $t = 0$ to a (or from $t = -a$ to 0). Every signal can be expressed as a sum of odd and even function of time.

A system processes input signals to produce output signals (response). The input is the cause, and the output is its effect. In general, the output is affected by two causes: the internal conditions of the system (such as the initial conditions) and the external input.

Systems can be classified in several ways.

1. Linear systems are characterized by the linearity property, which implies superposition; if several causes (such as various inputs and initial conditions) are acting on a linear system, the total output (response) is the sum of the responses from each cause, assuming that all the remaining causes are absent. A system is nonlinear if superposition does not hold.

2. In time-invariant systems, system parameters do not change with time. The parameters of time-varying-parameter systems change with time.

3. For memoryless (or instantaneous) systems, the system response at any instant t depends only on the value of the input at t. For systems with memory (also known as dynamic systems), the system response at any instant t depends not only on the present value of the input, but also on the past values of the input (values before t).

4. In contrast, if a system response at t also depends on the future values of the input (values of input beyond t), the system is noncausal. In causal systems, the response does not depend on the future values of the input. Because of the dependence of the response on the future values of input, the effect (response) of noncausal systems occurs before the cause. When the independent variable is time (temporal systems), the noncausal systems are prophetic systems, and therefore, unrealizable, although close approximation is possible with some time delay in the response. Noncausal systems with independent variables other than time (e.g., space) are realizable.

5. Systems whose inputs and outputs are continuous-time signals are continuous-time systems; systems whose inputs and outputs are discrete-time signals are discrete-time systems. If a continuous-time signal is sampled, the resulting signal is a discrete-time signal. We can process a continuous-time signal by processing the samples of the signal with a discrete-time system.

6. Systems whose inputs and outputs are analog signals are analog systems; those whose inputs and outputs are digital signals are digital systems.

7. If we can obtain the input $x(t)$ back from the output $y(t)$ of a system S by some operation, the system S is said to be invertible. Otherwise the system is noninvertible.

8. A system is stable if bounded input produces bounded output. This defines the external stability because it can be ascertained from measurements at the external terminals of the system. The external stability is also known as the stability in the BIBO (bounded-input/bounded-output) sense. The internal stability, discussed later in Chapter 2, is measured in terms of the internal behavior of the system.

The system model derived from a knowledge of the internal structure of the system is its internal description. In contrast, an external description is a representation of a system as seen from its input and output terminals; it can be obtained by applying a known input and measuring the resulting output. In the majority of practical systems, an external description of a system so obtained is equivalent to its internal description. At times, however, the external description fails to describe the system adequately. Such is the case with the so-called uncontrollable or unobservable systems.

A system may also be described in terms of certain set of key variables called state variables. In this description, an Nth-order system can be characterized by a set of N simultaneous first-order differential equations in N state variables. State equations of a system represent an internal description of that system.

REFERENCES

1. Papoulis, A. *The Fourier Integral and Its Applications.* McGraw-Hill, New York, 1962.
2. Mason, S. J. *Electronic Circuits, Signals, and Systems.* Wiley, New York, 1960.
3. Kailath, T. *Linear Systems.* Prentice-Hall, Englewood Cliffs, NJ, 1980.
4. Lathi, B. P. *Signals and Systems.* Berkeley-Cambridge Press, Carmichael, CA, 1987.

MATLAB SESSION 1: WORKING WITH FUNCTIONS

Working with functions is fundamental to signals and systems applications. MATLAB provides several methods of defining and evaluating functions. An understanding and proficient use of these methods is therefore necessary and beneficial.

M1.1 Inline Functions

Many simple functions are most conveniently represented by using MATLAB inline objects. An inline object provides a symbolic representation of a function defined in terms of MATLAB operators and functions. For example, consider defining the exponentially damped sinusoid $f(t) = e^{-t} \cos(2\pi t)$.

```
>> f = inline('exp(-t).*cos(2*pi*t)','t')
f =    Inline function:
       f(t) = exp(-t).*cos(2*pi*t)
```

The second argument to the `inline` command identifies the function's input argument as `t`. Input arguments, such as `t`, are local to the inline object and are not related to any workspace variables with the same names.

Once defined, $f(t)$ can be evaluated simply by passing the input values of interest. For example,

```
>> t = 0;
>> f(t)
ans = 1
```

evaluates $f(t)$ at $t = 0$, confirming the expected result of unity. The same result is obtained by passing $t = 0$ directly.

```
>> f(0)
ans = 1
```

Vector inputs allow the evaluation of multiple values simultaneously. Consider the task of plotting $f(t)$ over the interval $(-2 \le t \le 2)$. Gross function behavior is clear: $f(t)$ should oscillate four times with a decaying envelope. Since accurate hand sketches are cumbersome, MATLAB-generated plots are an attractive alternative. As the following example illustrates, care must be taken to ensure reliable results.

Suppose vector `t` is chosen to include only the integers contained in $(-2 \le t \le 2)$, namely $[-2, -1, 0, 1, 2]$.

```
>> t = (-2:2);
```

This vector input is evaluated to form a vector output.

```
>> f(t)
ans = 7.3891    2.7183    1.0000    0.3679    0.1353
```

The `plot` command graphs the result, which is shown in Fig. M1.1.

```
>> plot(t,f(t));
>> xlabel('t'); ylabel('f(t)'); grid;
```

Grid lines, added by using the `grid` command, aid feature identification. Unfortunately, the plot does not illustrate the expected oscillatory behavior. More points are required to adequately represent $f(t)$.

The question, then, is how many points is enough?[†] If too few points are chosen, information is lost. If too many points are chosen, memory and time are wasted. A balance is needed. For oscillatory functions, plotting 20 to 200 points per oscillation is normally adequate. For the present case, `t` is chosen to give 100 points per oscillation.

```
>> t = (-2:0.01:2);
```

[†]Sampling theory, presented later, formally addresses important aspects of this question.

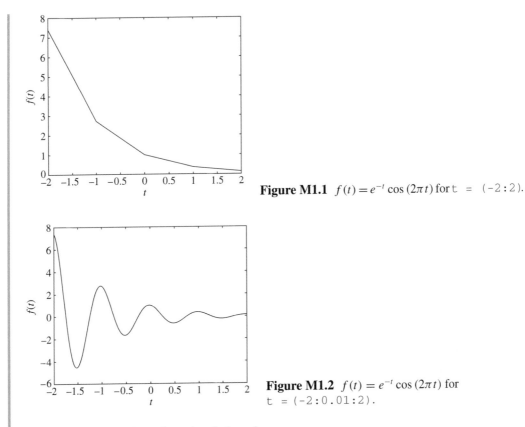

Figure M1.1 $f(t) = e^{-t} \cos(2\pi t)$ for $\mathtt{t} = (-2:2)$.

Figure M1.2 $f(t) = e^{-t} \cos(2\pi t)$ for $\mathtt{t} = (-2:0.01:2)$.

Again, the function is evaluated and plotted.

```
>> plot(t,f(t));
>> xlabel('t'); ylabel('f(t)'); grid;
```

The result, shown in Fig. M1.2, is an accurate depiction of $f(t)$.

M1.2 Relational Operators and the Unit Step Function

The unit step function $u(t)$ arises naturally in many practical situations. For example, a unit step can model the act of turning on a system. With the help of relational operators, inline objects can represent the unit step function.

In MATLAB, a relational operator compares two items. If the comparison is true, a logical true (1) is returned. If the comparison is false, a logical false (0) is returned. Sometimes called indicator functions, relational operators indicates whether a condition is true. Six relational operators are available: <, >, <=, >=, ==, and ~=.

The unit step function is readily defined using the >= relational operator.

```
>> u = inline('(t>=0)','t')
u =   Inline function:
      u(t) = (t>=0)
```

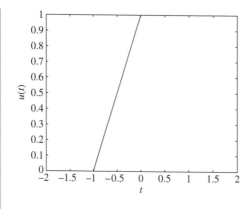

Figure M1.3 $u(t)$ for t = (-2:2).

Any function with a jump discontinuity, such as the unit step, is difficult to plot. Consider plotting $u(t)$ by using t = (-2:2).

```
>> t = (-2:2);
>> plot(t,u(t));
>> xlabel('t'); ylabel('u(t)');
```

Two significant problems are apparent in the resulting plot, shown in Fig. M1.3. First, MATLAB automatically scales plot axes to tightly bound the data. In this case, this normally desirable feature obscures most of the plot. Second, MATLAB connects plot data with lines, making a true jump discontinuity difficult to achieve. The coarse resolution of vector t emphasizes the effect by showing an erroneous sloping line between $t = -1$ and $t = 0$.

The first problem is corrected by vertically enlarging the bounding box with the `axis` command. The second problem is reduced, but not eliminated, by adding points to vector t.

```
>> t = (-2:0.01:2);
>> plot(t,u(t));
>> xlabel('t'); ylabel('u(t)');
>> axis([-2 2 -0.1 1.1]);
```

The four-element vector argument of `axis` specifies x-axis minimum, x-axis maximum, y-axis minimum, and y-axis maximum, respectively. The improved results are shown in Fig. M1.4.

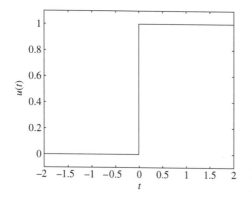

Figure M1.4 $u(t)$ for t = (-2:0.01:2) with axis modification.

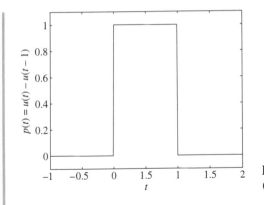

Figure M1.5 $p(t) = u(t) - u(t-1)$ over $(-1 \leq t \leq 2)$.

Relational operators can be combined using logical AND, logical OR, and logical negation: `&`, `|`, and `~`, respectively. For example, `(t>0)&(t<1)` and `~((t<=0)|(t>=1))` both test if $0 < t < 1$. To demonstrate, consider defining and plotting the unit pulse $p(t) = u(t) - u(t-1)$, as shown in Fig. M1.5:

```
>> p = inline('(t>=0)&(t<1)','t');
>> t = (-1:0.01:2); plot(t,p(t));
>> xlabel('t'); ylabel('p(t) = u(t)-u(t-1)');
>> axis([-1 2 -.1 1.1]);
```

For scalar operands, MATLAB also supports two short-circuit logical constructs. A short-circuit logical AND is performed by using `&&`, and a short-circuit logical OR is performed by using `||`. Short-circuit logical operators are often more efficient than traditional logical operators because they test the second portion of the expression only when necessary. That is, when scalar expression A is found false in `(A&&B)`, scalar expression B is not evaluated, since a false result is already guaranteed. Similarly, scalar expression B is not evaluated when scalar expression A is found true in `(A||B)`, since a true result is already guaranteed.

M1.3 Visualizing Operations on the Independent Variable

Two operations on a function's independent variable are commonly encountered: shifting and scaling. Inline objects are well suited to investigate both operations.

Consider $g(t) = f(t)u(t) = e^{-t}\cos(2\pi t)u(t)$, a realizable version of $f(t)$.[†] Unfortunately, MATLAB cannot multiply inline objects. That is, MATLAB reports an error for `g = f*u` when f and u are inline objects. Rather, $g(t)$ needs to be explicitly defined.

```
>> g = inline('exp(-t).*cos(2*pi*t).*(t>=0)','t')
g =    Inline function:
       g(t) = exp(-t).*cos(2*pi*t).*(t>=0)
```

A combined shifting and scaling operation is represented by $g(at + b)$, where a and b are arbitrary real constants. As an example, consider plotting $g(2t + 1)$ over $(-2 \leq t \leq 2)$. With

---

[†]The function $f(t) = e^{-t}\cos(2\pi t)$ can never be realized in practice; it has infinite duration and, as $t \to -\infty$, infinite magnitude.

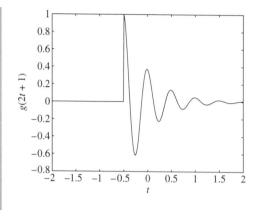

Figure M1.6 $g(2t + 1)$ over $(-2 \leq t \leq 2)$.

$a = 2$, the function is compressed by a factor of 2, resulting in twice the oscillations per unit t. Adding the condition $b > 0$, the waveform shifts to the left. Given inline function g, an accurate plot is nearly trivial to obtain.

```
>>t = (-2:0.01:2);
>> plot(t,g(2*t+1)); xlabel('t'); ylabel('g(2t+1)'); grid;
```

Figure M1.6 confirms the expected waveform compression and left shift. As a final check, realize that function $g(\cdot)$ turns on when the input argument is zero. Therefore, $g(2t + 1)$ should turn on when $2t + 1 = 0$ or at $t = -0.5$, a fact again confirmed by Fig. M1.6.

Next, consider plotting $g(-t + 1)$ over $(-2 \leq t \leq 2)$. Since $a < 0$, the waveform will be reflected. Adding the condition $b > 0$, the final waveform shifts to the right.

```
>> plot(t,g(-t+1)); xlabel('t'); ylabel('g(-t+1)'); grid;
```

Figure M1.7 confirms both the reflection and the right shift.

Up to this point, Figs. M1.6 and M1.7 could be reasonably sketched by hand. Consider plotting the more complicated function $h(t) = g(2t + 1) + g(-t + 1)$ over $(-2 \leq t \leq 2)$ (Fig. M1.8). In this case, an accurate hand sketch is quite difficult. With MATLAB, the work is much less burdensome.

```
>> plot(t,g(2*t+1)+g(-t+1)); xlabel('t'); ylabel('h(t)'); grid;
```

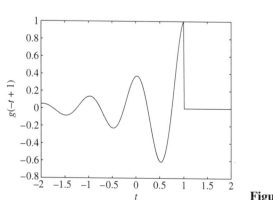

Figure M1.7 $g(-t + 1)$ over $(-2 \leq t \leq 2)$.

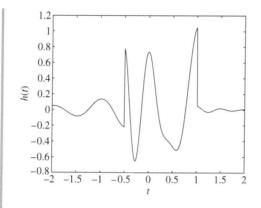

Figure M1.8 $h(t) = g(2t+1) + g(-t+1)$ over $(-2 \leq t \leq 2)$.

M1.4 Numerical Integration and Estimating Signal Energy

Interesting signals often have nontrivial mathematical representations. Computing signal energy, which involves integrating the square of these expressions, can be a daunting task. Fortunately, many difficult integrals can be accurately estimated by means of numerical integration techniques. Even if the integration appears simple, numerical integration provides a good way to verify analytical results.

To start, consider the simple signal $x(t) = e^{-t}(u(t) - u(t - 1))$. The energy of $x(t)$ is expressed as $E_x = \int_{-\infty}^{\infty} |x(t)|^2 \, dt = \int_0^1 e^{-2t} \, dt$. Integrating yields $E_x = 0.5(1 - e^{-2}) \approx 0.4323$. The energy integral can also be evaluated numerically. Figure 1.27 helps illustrate the simple method of rectangular approximation: evaluate the integrand at points uniformly separated by Δt, multiply each by Δt to compute rectangle areas, and then sum over all rectangles. First, we create function $x(t)$.

```
>> x = inline('exp(-t).*((t>=0)&(t<1))','t');
```

Letting $\Delta t = 0.01$, a suitable time vector is created.

```
>> t = (0:0.01:1);
```

The final result is computed by using the sum command.

```
>> E_x = sum(x(t).*x(t)*0.01)
E_x = 0.4367
```

The result is not perfect, but at 1% relative error it is close. By reducing Δt, the approximation is improved. For example, $\Delta t = 0.001$ yields E_x = 0.4328, or 0.1% relative error.

Although simple to visualize, rectangular approximation is not the best numerical integration technique. The MATLAB function quad implements a better numerical integration technique called recursive adaptive Simpson quadrature.[†] To operate, quad requires a function describing

---

[†]A comprehensive treatment of numerical integration is outside the scope of this text. Details of this particular method are not important for the current discussion; it is sufficient to say that it is better than the rectangular approximation.

the integrand, the lower limit of integration, and the upper limit of integration. Notice, that no Δt needs to be specified.

To use quad to estimate E_x, the integrand must first be described.

```
>> x_squared = inline('exp(-2*t).*((t>=0)&(t<1))','t');
```

Estimating E_x immediately follows.

```
>> E_x = quad(x_squared,0,1)
E_x = 0.4323
```

In this case, the relative error is -0.0026%.

The same techniques can be used to estimate the energy of more complex signals. Consider $g(t)$, defined previously. Energy is expressed as $E_g = \int_0^\infty e^{-2t} \cos^2(2\pi t)\,dt$. A closed-form solution exists, but it takes some effort. MATLAB provides an answer more quickly.

```
>> g_squared = inline('exp(-2*t).*(cos(2*pi*t).^2).*(t>=0)','t');
```

Although the upper limit of integration is infinity, the exponentially decaying envelope ensures $g(t)$ is effectively zero well before $t = 100$. Thus, an upper limit of $t = 100$ is used along with $\Delta t = 0.001$.

```
>> t = (0:0.001:100);
>> E_g = sum(g_squared(t)*0.001)
E_g = 0.2567
```

A slightly better approximation is obtained with the quad function.

```
>> E_g = quad(g_squared,0,100)
E_g = 0.2562
```

As an exercise, confirm that the energy of signal $h(t)$, defined previously, is $E_h = 0.3768$.

PROBLEMS

1.1-1 Find the energies of the signals illustrated in Fig. P1.1-1. Comment on the effect on energy of sign change, time shifting, or doubling of the signal. What is the effect on the energy if the signal is multiplied by k?

1.1-2 Repeat Prob. 1.1-1 for the signals in Fig. P1.1-2.

1.1-3 (a) Find the energies of the pair of signals $x(t)$ and $y(t)$ depicted in Fig. P1.1-3a and P1.1-3b. Sketch and find the energies of signals $x(t)+y(t)$ and $x(t)-y(t)$.

Can you make any observation from these results?

(b) Repeat part (a) for the signal pair illustrated in Fig. P1.1-3c. Is your observation in part (a) still valid?

1.1-4 Find the power of the periodic signal $x(t)$ shown in Fig. P1.1-4. Find also the powers and the rms values of:

(a) $-x(t)$
(b) $2x(t)$
(c) $cx(t)$.

Comment.

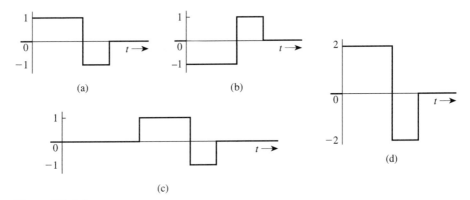

Figure P1.1-1

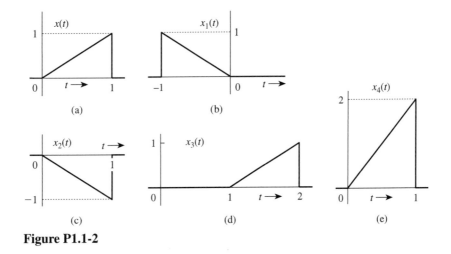

Figure P1.1-2

1.1-5 Determine the power and the rms value for each of the following signals:

(a) $5 + 10\cos(100t + \pi/3)$

(b) $10\cos(100t + \pi/3) + 16\sin(150t + \pi/5)$

(c) $(10 + 2\sin 3t)\cos 10t$

(d) $10\cos 5t \cos 10t$

(e) $10\sin 5t \cos 10t$

(f) $e^{j\alpha t}\cos\omega_0 t$

1.1-6 Figure P1.1-6 shows a periodic 50% duty cycle dc-offset sawtooth wave $x(t)$ with peak amplitude A. Determine the energy and power of $x(t)$.

1.1-7 (a) There are many useful properties related to signal energy. Prove each of the following statements. In each case, let energy signal $x_1(t)$ have energy $E[x_1(t)]$, let energy signal $x_2(t)$ have energy $E[x_2(t)]$, and let T be a nonzero, finite, real-valued constant.

   (i) Prove $E[Tx_1(t)] = T^2 E[x_1(t)]$. That is, amplitude scaling a signal by constant T scales the signal energy by T^2.

   (ii) Prove $E[x_1(t)] = E[x_1(t - T)]$. That is, shifting a signal does not affect its energy.

   (iii) If $(x_1(t) \neq 0) \Rightarrow (x_2(t) = 0)$ and $(x_2(t) \neq 0) \Rightarrow (x_1(t) = 0)$, then

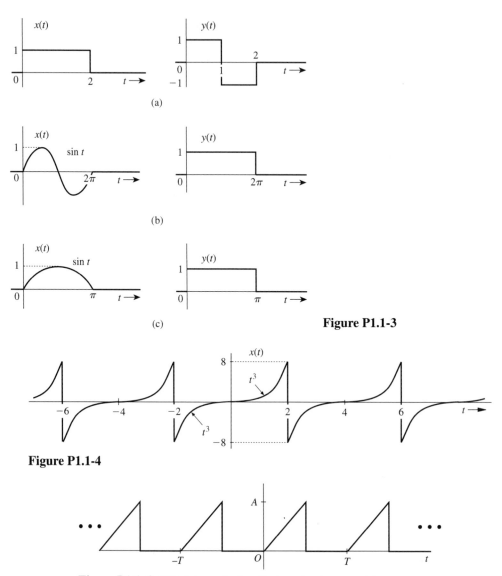

(a)

(b)

(c)

Figure P1.1-3

Figure P1.1-4

Figure P1.1-6 50% duty cycle dc-offset sawtooth wave $x(t)$.

prove $E[x_1(t)+x_2(t)] = E[x_1(t)]+ E[x_2(t)]$. That is, the energy of the sum of two nonoverlapping signals is the sum of the two individual energies.

(iv) Prove $E[x_1(Tt)] = (1/|T|)E[x_1(t)]$. That is, time-scaling a signal by T reciprocally scales the signal energy by $1/|T|$.

(b) Consider the signal $x(t)$ shown in Fig. P1.1-7. Outside the interval shown, $x(t)$ is zero. Determine the signal energy $E[x(t)]$.

1.1-8 (a) Show that the power of a signal

$$x(t) = \sum_{k=m}^{n} D_k e^{j\omega_k t} \quad \text{is} \quad P_x = \sum_{k=m}^{n} |D_k|^2$$

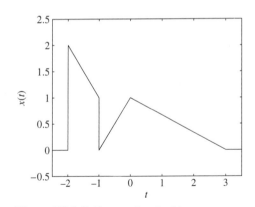

Figure P1.1-7 Energy signal $x(t)$.

assuming all frequencies to be distinct, that is, $\omega_i \neq \omega_k$ for all $i \neq k$.

(b) Using the result in part (a) determine the power of each of the signals in Prob. 1.1-5.

1.1-9 A binary signal $x(t) = 0$ for $t < 0$. For positive time, $x(t)$ toggles between one and zero as follows: one for 1 second, zero for 1 second, one for 1 second, zero for 2 seconds, one for 1 second, zero for 3 seconds, and so forth. That is, the "on" time is always one second but the "off" time successively increases by one second between each toggle. A portion of $x(t)$ is shown in Fig. P1.1-9. Determine the energy and power of $x(t)$.

1.2-1 For the signal $x(t)$ depicted in Fig. P1.2-1, sketch the signals

(a) $x(-t)$

(b) $x(t + 6)$

(c) $x(3t)$

(d) $x(t/2)$

1.2-2 For the signal $x(t)$ illustrated in Fig. P1.2-2, sketch

(a) $x(t - 4)$

(b) $x(t/1.5)$

(c) $x(-t)$

(d) $x(2t - 4)$

(e) $x(2 - t)$

1.2-3 In Fig. P1.2-3, express signals $x_1(t), x_2(t)$, $x_3(t), x_4(t)$, and $x_5(t)$ in terms of signal $x(t)$ and its time-shifted, time-scaled, or time-reversed versions.

1.2-4 For an energy signal $x(t)$ with energy E_x, show that the energy of any one of the signals $-x(t)$, $x(-t)$, and $x(t - T)$ is E_x. Show also that the energy of $x(at)$ as well as $x(at - b)$ is E_x/a, but the energy of $ax(t)$ is $a^2 E_x$. This shows that time inversion and time shifting do not affect signal energy. On the other hand, time compression of a signal ($a > 1$) reduces the energy, and time expansion of a signal ($a < 1$) increases the energy. What is the

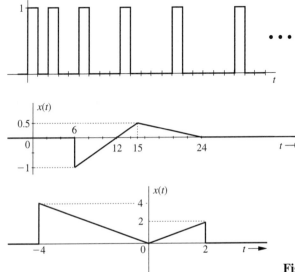

Figure P1.1-9 Binary signal $x(t)$.

Figure P1.2-1

Figure P1.2-2

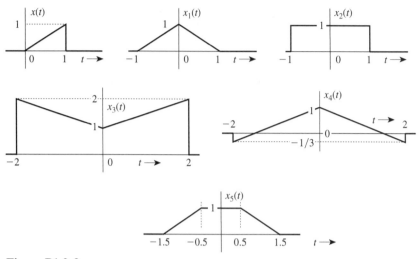

Figure P1.2-3

effect on signal energy if the signal is multiplied by a constant a?

1.2-5 Define $2x(-3t+1)=t(u(-t-1)-u(-t+1))$, where $u(t)$ is the unit step function.

    (a) Plot $2x(-3t+1)$ over a suitable range of t.

    (b) Plot $x(t)$ over a suitable range of t.

1.2-6 Consider the signal $x(t) = 2^{-tu(t)}$, where $u(t)$ is the unit step function.

    (a) Accurately sketch $x(t)$ over $(-1 \le t \le 1)$.

    (b) Accurately sketch $y(t) = 0.5x(1 - 2t)$ over $(-1 \le t \le 1)$.

1.3-1 Determine whether each of the following statements is true or false. If the statement is false, demonstrate this by proof or example.

    (a) Every continuous-time signal is analog signal.

    (b) Every discrete-time signal is digital signal.

    (c) If a signal is not an energy signal, then it must be a power signal and vice versa.

    (d) An energy signal must be of finite duration.

    (e) A power signal cannot be causal.

    (f) A periodic signal cannot be anticausal.

1.3-2 Determine whether each of the following statements is true or false. If the statement is false, demonstrate by proof or example why the statement is false.

    (a) Every bounded periodic signal is a power signal.

    (b) Every bounded power signal is a periodic signal.

    (c) If an energy signal $x(t)$ has energy E, then the energy of $x(at)$ is E/a. Assume a is real and positive.

    (d) If a power signal $x(t)$ has power P, then the power of $x(at)$ is P/a. Assume a is real and positive.

1.3-3 Given $x_1(t) = \cos(t)$, $x_2(t) = \sin(\pi t)$, and $x_3(t) = x_1(t) + x_2(t)$.

    (a) Determine the fundamental periods T_1 and T_2 of signals $x_1(t)$ and $x_2(t)$.

    (b) Show that $x_3(t)$ is not periodic, which requires $T_3 = k_1 T_1 = k_2 T_2$ for some integers k_1 and k_2.

    (c) Determine the powers P_{x_1}, P_{x_2}, and P_{x_3} of signals $x_1(t)$, $x_2(t)$, and $x_3(t)$.

1.3-4 For any constant ω, is the function $f(t) = \sin(\omega t)$ a periodic function of the independent variable t? Justify your answer.

1.3-5 The signal shown in Fig. P1.3-5 is defined as

$$x(t) = \begin{cases} t & 0 \le t < 1 \\ 0.5 + 0.5\cos(2\pi t) & 1 \le t < 2 \\ 3 - t & 2 \le t < 3 \\ 0 & \text{otherwise} \end{cases}$$

The energy of $x(t)$ is $E \approx 1.0417$.

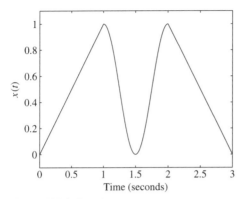

Figure P1.3-5 Energy signal $x(t)$.

(a) What is the energy of $y_1(t) = (1/3)x(2t)$?
(b) A periodic signal $y_2(t)$ is defined as

$$y_2(t) = \begin{cases} x(t) & 0 \le t < 4 \\ y_2(t+4) & \forall t \end{cases}$$

What is the power of $y_2(t)$?
(c) What is the power of $y_3(t) = (1/3)y_2(2t)$?

1.3-6 Let $y_1(t) = y_2(t) = t^2$ over $0 \le t \le 1$. Notice, this statement does not require $y_1(t) = y_2(t)$ for all t.

(a) Define $y_1(t)$ as an even, periodic signal with period $T_1 = 2$. Sketch $y_1(t)$ and determine its power.
(b) Design an odd, periodic signal $y_2(t)$ with period $T_2 = 3$ and power equal to unity. Fully describe $y_2(t)$ and sketch the signal over at least one full period. [Hint: There are an infinite number of possible solutions to this problem—you need to find only one of them!]
(c) We can create a complex-valued function $y_3(t) = y_1(t) + jy_2(t)$. Determine whether this signal is periodic. If yes, determine the period T_3. If no, justify why the signal is not periodic.
(d) Determine the power of $y_3(t)$ defined in part (c). The power of a complex-valued function $z(t)$ is

$$P = \lim_{T \to \infty} \frac{1}{T} \int_{-T/2}^{T/2} z(\tau)z^*(\tau)\,d\tau$$

1.4-1 Sketch the following signal:
(a) $u(t-5) - u(t-7)$

(b) $u(t-5) + u(t-7)$
(c) $t^2[u(t-1) - u(t-2)]$
(d) $(t-4)[u(t-2) - u(t-4)]$

1.4-2 Express each of the signals in Fig. P1.4-2 by a single expression valid for all t.

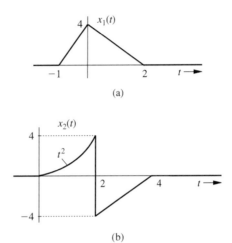

Figure P1.4-2

1.4-3 Simplify the following expressions:

(a) $\left(\dfrac{\sin t}{t^2 + 2} \right) \delta(t)$

(b) $\left(\dfrac{j\omega + 2}{\omega^2 + 9} \right) \delta(\omega)$

(c) $[e^{-t} \cos(3t - 60°)]\delta(t)$

(d) $\left(\dfrac{\sin\left[\frac{\pi}{2}(t-2)\right]}{t^2 + 4} \right) \delta(1 - t)$

(e) $\left(\dfrac{1}{j\omega + 2} \right) \delta(\omega + 3)$

(f) $\left(\dfrac{\sin k\omega}{\omega} \right) \delta(\omega)$

[Hint: Use Eq. (1.23). For part (f) use L'Hôpital's rule.]

1.4-4 Evaluate the following integrals:

(a) $\displaystyle\int_{-\infty}^{\infty} \delta(\tau)x(t - \tau)\,d\tau$

(b) $\displaystyle\int_{-\infty}^{\infty} x(\tau)\delta(t - \tau)\,d\tau$

(c) $\int_{-\infty}^{\infty} \delta(t)e^{-j\omega t}\, dt$

(d) $\int_{-\infty}^{\infty} \delta(2t - 3) \sin \pi t\, dt$

(e) $\int_{-\infty}^{\infty} \delta(t + 3)e^{-t}\, dt$

(f) $\int_{-\infty}^{\infty} (t^3 + 4)\delta(1 - t)\, dt$

(g) $\int_{-\infty}^{\infty} x(2 - t)\delta(3 - t)\, dt$

(h) $\int_{-\infty}^{\infty} e^{(x-1)} \cos\left[\frac{\pi}{2}(x - 5)\right]\delta(x - 3)\, dx$

1.4-5 (a) Find and sketch dx/dt for the signal $x(t)$ shown in Fig. P1.2-2.
      (b) Find and sketch d^2x/dt^2 for the signal $x(t)$ depicted in Fig. P1.4-2a.

1.4-6 Find and sketch $\int_{-\infty}^{t} x(t)\, dt$ for the signal $x(t)$ illustrated in Fig. P1.4-6.

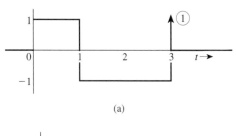

(a)

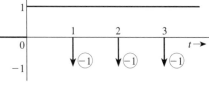

(b)

Figure P1.4-6

1.4-7 Using the generalized function definition of impulse [Eq. (1.24a)], show that $\delta(t)$ is an even function of t.

1.4-8 Using the generalized function definition of impulse [Eq. (1.24a)], show that

$$\delta(at) = \frac{1}{|a|}\delta(t)$$

1.4-9 Show that

$$\int_{-\infty}^{\infty} \dot{\delta}(t)\phi(t)\, dt = -\dot{\phi}(0)$$

where $\phi(t)$ and $\dot{\phi}(t)$ are continuous at $t = 0$, and $\phi(t) \to 0$ as $t \to \pm\infty$. This integral defines $\dot{\delta}(t)$ as a generalized function. [Hint: Use integration by parts.]

1.4-10 A sinusoid $e^{\sigma t} \cos \omega t$ can be expressed as a sum of exponentials e^{st} and e^{-st} [Eq. (1.30c)] with complex frequencies $s = \sigma + j\omega$ and $s = \sigma - j\omega$. Locate in the complex plane the frequencies of the following sinusoids:

(a) $\cos 3t$
(b) $e^{-3t} \cos 3t$
(c) $e^{2t} \cos 3t$
(d) e^{-2t}
(e) e^{2t}
(f) 5

1.5-1 Find and sketch the odd and the even components of the following:

(a) $u(t)$
(b) $tu(t)$
(c) $\sin \omega_0 t$
(d) $\cos \omega_0 t$
(e) $\cos(\omega_0 t + \theta)$
(f) $\sin \omega_0 t\, u(t)$
(g) $\cos \omega_0 t\, u(t)$

1.5-2 (a) Determine even and odd components of the signal $x(t) = e^{-2t}u(t)$.
      (b) Show that the energy of $x(t)$ is the sum of energies of its odd and even components found in part (a).
      (c) Generalize the result in part (b) for any finite energy signal.

1.5-3 (a) If $x_e(t)$ and $x_o(t)$ are even and the odd components of a real signal $x(t)$, then show that

$$\int_{-\infty}^{\infty} x_e(t)x_o(t)\, dt = 0$$

      (b) Show that

$$\int_{-\infty}^{\infty} x(t)\, dt = \int_{-\infty}^{\infty} x_e(t)\, dt$$

1.5-4 An aperiodic signal is defined as $x(t) = \sin(\pi t)u(t)$, where $u(t)$ is the continuous-time step function. Is the odd portion of this signal, $x_o(t)$, periodic? Justify your answer.

1.5-5 An aperiodic signal is defined as $x(t) = \cos(\pi t)u(t)$, where $u(t)$ is the continuous-time step function. Is the even portion of this signal, $x_e(t)$, periodic? Justify your answer.

1.5-6 Consider the signal $x(t)$ shown in Fig. P1.5-6.
   (a) Determine and carefully sketch $v(t) = 3x(-(1/2)(t+1))$.
   (b) Determine the energy and power of $v(t)$.
   (c) Determine and carefully sketch the even portion of $v(t)$, $v_e(t)$.
   (d) Let $a = 2$ and $b = 3$, sketch $v(at+b)$, $v(at)+b$, $av(t+b)$, and $av(t)+b$.
   (e) Let $a = -3$ and $b = -2$, sketch $v(at+b)$, $v(at)+b$, $av(t+b)$, and $av(t)+b$.

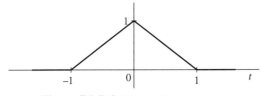

Figure P1.5-6 Input $x(t)$.

1.5-7 Consider the signal $y(t) = (1/5)x(-2t-3)$ shown in Figure P1.5-7.
   (a) Does $y(t)$ have an odd portion, $y_o(t)$? If so, determine and carefully sketch $y_o(t)$. Otherwise, explain why no odd portion exists.
   (b) Determine and carefully sketch the original signal $x(t)$.

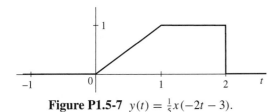

Figure P1.5-7 $y(t) = \frac{1}{5}x(-2t-3)$.

1.5-8 Consider the signal $-(1/2)x(-3t+2)$ shown in Fig. P1.5-8.
   (a) Determine and carefully sketch the original signal $x(t)$.
   (b) Determine and carefully sketch the even portion of the original signal $x(t)$.
   (c) Determine and carefully sketch the odd portion of the original signal $x(t)$.

Figure P1.5-8 $-\frac{1}{2}x(-3t+2)$.

1.5-9 The conjugate symmetric (or Hermitian) portion of a signal is defined as $w_{cs}(t) = (w(t) + w^*(-t))/2$. Show that the real portion of $w_{cs}(t)$ is even and that the imaginary portion of $w_{cs}(t)$ is odd.

1.5-10 The conjugate antisymmetric (or skew-Hermitian) portion of a signal is defined as $w_{ca}(t) = (w(t) - w^*(-t))/2$. Show that the real portion of $w_{ca}(t)$ is odd and that the imaginary portion of $w_{ca}(t)$ is even.

1.5-11 Figure P1.5-11 plots a complex signal $w(t)$ in the complex plane over the time range ($0 \leq t \leq 1$). The time $t = 0$ corresponds with the origin, while the time $t = 1$ corresponds with the point (2,1).

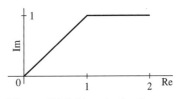

Figure P1.5-11 $w(t)$ for ($0 \leq t \leq 1$).

   (a) In the complex plane, plot $w(t)$ over ($-1 \leq t \leq 1$) if:
      (i) $w(t)$ is an even signal.
      (ii) $w(t)$ is an odd signal.
      (iii) $w(t)$ is a conjugate symmetric signal. [Hint: See Prob. 1.5-9.]
      (iv) $w(t)$ is a conjugate antisymmetric signal. [Hint: See Prob. 1.5-10.]
   (b) In the complex plane, plot as much of $w(3t)$ as possible.

1.5-12 Define complex signal $x(t) = t^2(1+j)$ over interval ($1 \leq t \leq 2$). The remaining portion is defined such that $x(t)$ is a minimum-energy, skew-Hermitian signal.
   (a) Fully describe $x(t)$ for all t.
   (b) Sketch $y(t) = \text{Re}\{x(t)\}$ versus the independent variable t.

(c) Sketch $z(t) = \text{Re}\{jx(-2t+1)\}$ versus the independent variable t.

(d) Determine the energy and power of $x(t)$.

1.6-1 Write the input–output relationship for an ideal integrator. Determine the zero-input and zero-state components of the response.

1.6-2 A force $x(t)$ acts on a ball of mass M (Fig. P1.6-2). Show that the velocity $v(t)$ of the ball at any instant $t > 0$ can be determined if we know the force $x(t)$ over the interval from 0 to t and the ball's initial velocity $v(0)$.

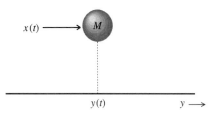

$x(t) \longrightarrow$ M

$y(t)$ $y \longrightarrow$

Figure P1.6-2

1.7-1 For the systems described by the following equations, with the input $x(t)$ and output $y(t)$, determine which of the systems are linear and which are nonlinear.

(a) $\dfrac{dy}{dt} + 2y(t) = x^2(t)$

(b) $\dfrac{dy}{dt} + 3ty(t) = t^2 x(t)$

(c) $3y(t) + 2 = x(t)$

(d) $\dfrac{dy}{dt} + y^2(t) = x(t)$

(e) $\left(\dfrac{dy}{dt}\right)^2 + 2y(t) = x(t)$

(f) $\dfrac{dy}{dt} + (\sin t)y(t) = \dfrac{dx}{dt} + 2x(t)$

(g) $\dfrac{dy}{dt} + 2y(t) = x(t)\dfrac{dx}{dt}$

(h) $y(t) = \displaystyle\int_{-\infty}^{t} x(\tau)\,d\tau$

1.7-2 For the systems described by the following equations, with the input $x(t)$ and output $y(t)$, explain with reasons which of the systems are time-invariant parameter systems and which are time-varying-parameter systems.

(a) $y(t) = x(t-2)$

(b) $y(t) = x(-t)$

(c) $y(t) = x(at)$

(d) $y(t) = t\,x(t-2)$

(e) $y(t) = \displaystyle\int_{-5}^{5} x(\tau)\,d\tau$

(f) $y(t) = \left(\dfrac{dx}{dt}\right)^2$

1.7-3 For a certain LTI system with the input $x(t)$, the output $y(t)$ and the two initial conditions $q_1(0)$ and $q_2(0)$, following observations were made:

$x(t)$	$q_1(0)$	$q_2(0)$	$y(t)$
0	1	−1	$e^{-t}u(t)$
0	2	1	$e^{-t}(3t+2)u(t)$
$u(t)$	−1	−1	$2u(t)$

Determine $y(t)$ when both the initial conditions are zero and the input $x(t)$ is as shown in Fig. P1.7-3. [Hint: There are three causes: the input and each of the two initial conditions. Because of the linearity property, if a cause is increased by a factor k, the response to that cause also increases by the same factor k. Moreover, if causes are added, the corresponding responses add.]

$x(t)$

1

−5   5 $t \rightarrow$

Figure P1.7-3

1.7-4 A system is specified by its input–output relationship as

$$y(t) = \frac{x^2(t)}{dx/dt}$$

Show that the system satisfies the homogeneity property but not the additivity property.

1.7-5 Show that the circuit in Fig. P1.7-5 is zero-state linear but not zero-input linear. Assume all diodes to have identical (matched) characteristics. The output is the current $y(t)$.

1.7-6 The inductor L and the capacitor C in Fig. P1.7-6 are nonlinear, which makes the circuit nonlinear. The remaining three elements are linear. Show that the output $y(t)$ of this nonlinear circuit satisfies the linearity

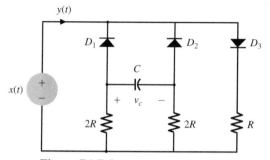

Figure P1.7-5

conditions with respect to the input $x(t)$ and the initial conditions (all the initial inductor currents and capacitor voltages).

1.7-7 For the systems described by the following equations, with the input $x(t)$ and output $y(t)$, determine which are causal and which are non-causal.

(a) $y(t) = x(t-2)$
(b) $y(t) = x(-t)$
(c) $y(t) = x(at)$ $a > 1$
(d) $y(t) = x(at)$ $a < 1$

1.7-8 For the systems described by the following equations, with the input $x(t)$ and output $y(t)$, determine which are invertible and which are noninvertible. For the invertible systems, find the input–output relationship of the inverse system.

(a) $y(t) = \int_{-\infty}^{t} x(\tau)\,d\tau$

(b) $y(t) = x^n(t)$ $x(t)$ real and n integer

(c) $y(t) = \dfrac{dx(t)}{dt}$

(d) $y(t) = x(3t-6)$
(e) $y(t) = \cos[x(t)]$
(f) $y(t) = e^{x(t)}$ $x(t)$ real

1.7-9 Consider a system that multiplies a given input by a ramp function, $r(t) = tu(t)$. That is, $y(t) = x(t)r(t)$.

(a) Is the system linear? Justify your answer.
(b) Is the system memoryless? Justify your answer.
(c) Is the system causal? Justify your answer.
(d) Is the system time invariant? Justify your answer.

1.7-10 A continuous-time system is given by

$$y(t) = 0.5 \int_{-\infty}^{\infty} x(\tau)[\delta(t-\tau) - \delta(t+\tau)]\,d\tau$$

Recall that $\delta(t)$ designates the Dirac delta function.

(a) Explain what this system does.
(b) Is the system BIBO stable? Justify your answer.
(c) Is the system linear? Justify your answer.
(d) Is the system memoryless? Justify your answer.
(e) Is the system causal? Justify your answer.
(f) Is the system time invariant? Justify your answer.

1.7-11 A system is given by

$$y(t) = \frac{d}{dt}x(t-1)$$

(a) Is the system BIBO stable? [Hint: Let system input $x(t)$ be a square wave.]
(b) Is the system linear? Justify your answer.
(c) Is the system memoryless? Justify your answer.
(d) Is the system causal? Justify your answer.
(e) Is the system time invariant? Justify your answer.

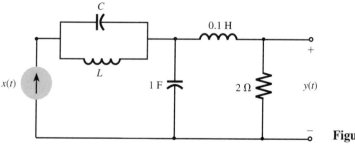

Figure P1.7-6

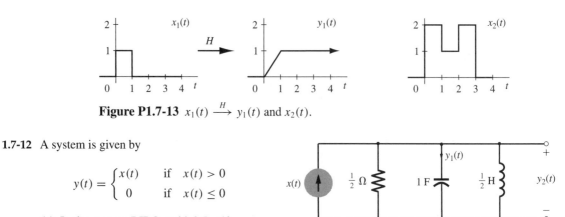

Figure P1.7-13 $x_1(t) \xrightarrow{H} y_1(t)$ and $x_2(t)$.

1.7-12 A system is given by

$$y(t) = \begin{cases} x(t) & \text{if} \quad x(t) > 0 \\ 0 & \text{if} \quad x(t) \le 0 \end{cases}$$

(a) Is the system BIBO stable? Justify your answer.

(b) Is the system linear? Justify your answer.

(c) Is the system memoryless? Justify your answer.

(d) Is the system causal? Justify your answer.

(e) Is the system time invariant? Justify your answer.

1.7-13 Figure P1.7-13 displays an input $x_1(t)$ to a linear time-invariant (LTI) system H, the corresponding output $y_1(t)$, and a second input $x_2(t)$.

(a) Bill suggests that $x_2(t) = 2x_1(3t) - x_1(t - 1)$. Is Bill correct? If yes, prove it. If not, correct his error.

(b) Hoping to impress Bill, Sue wants to know the output $y_2(t)$ in response to the input $x_2(t)$. Provide her with an expression for $y_2(t)$ in terms of $y_1(t)$. Use MATLAB to plot $y_2(t)$.

1.8-1 For the circuit depicted in Fig. P1.8-1, find the differential equations relating outputs $y_1(t)$ and $y_2(t)$ to the input $x(t)$.

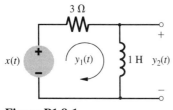

Figure P1.8-1

1.8-2 Repeat Prob. 1.8-1 for the circuit in Fig. P1.8-2.

Figure P1.8-2

1.8-3 A simplified (one-dimensional) model of an automobile suspension system is shown in Fig. P1.8-3. In this case, the input is not a force but a displacement $x(t)$ (the road contour). Find the differential equation relating the output $y(t)$ (auto body displacement) to the input $x(t)$ (the road contour).

1.8-4 A field-controlled dc motor is shown in Fig. P1.8-4. Its armature current i_a is maintained constant. The torque generated by this motor is proportional to the field current i_f (torque $= K_f i_f$). Find the differential equation relating the output position θ to the input voltage $x(t)$. The motor and load together have a moment of inertia J.

1.8-5 Water flows into a tank at a rate of q_i units/s and flows out through the outflow valve at a rate of q_0 units/s (Fig. P1.8-5). Determine the equation relating the outflow q_0 to the input q_i. The outflow rate is proportional to the head h. Thus $q_0 = Rh$, where R is the valve resistance. Determine also the differential equation relating the head h to the input q_i. [Hint: The net inflow of water in time Δt is $(q_i - q_0)\Delta t$. This inflow is also $A\Delta h$ where A is the cross section of the tank.]

1.8-6 Consider the circuit shown in Fig. P1.8-6, with input voltage $x(t)$ and output currents $y_1(t)$, $y_2(t)$, and $y_3(t)$.

(a) What is the order of this system? Explain your answer.

(b) Determine the matrix representation for this system.

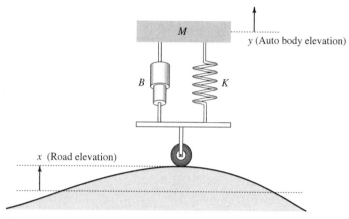

Figure P1.8-3

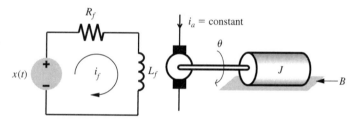

Figure P1.8-4

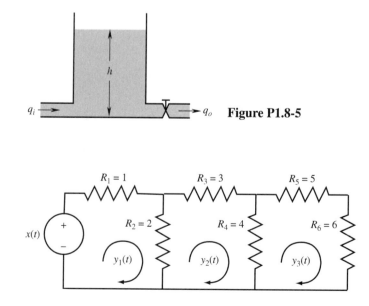

Figure P1.8-5

Figure P1.8-6 Resistor circuit.

(c) Use Cramer's rule to determine the output current $y_3(t)$ for the input voltage $x(t) = (2 - |\cos(t)|)u(t - 1)$.

1.10-1 Write state equations for the parallel RLC circuit in Fig. P1.8-2. Use the capacitor voltage q_1 and the inductor current q_2 as your state variables. Show that every possible current or voltage in the circuit can be expressed in terms of q_1, q_2 and the input $x(t)$.

1.10-2 Write state equations for the third-order circuit shown in Fig. P1.10-2, using the inductor currents q_1, q_2 and the capacitor voltage q_3 as state variables. Show that every possible voltage or current in this circuit can be expressed

as a linear combination of q_1, q_2, q_3, and the input $x(t)$. Also, at some instant t it was found that $q_1 = 5$, $q_2 = 1$, $q_3 = 2$, and $x = 10$. Determine the voltage across and the current through every element in this circuit.

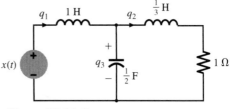

Figure P1.10-2

TIME-DOMAIN ANALYSIS
OF CONTINUOUS-TIME SYSTEMS

In this book we consider two methods of analysis of linear time-invariant (LTI) systems: the time-domain method and the frequency-domain method. In this chapter we discuss the *time-domain analysis* of linear, time-invariant, continuous-time (LTIC) systems.

2.1 INTRODUCTION

For the purpose of analysis, we shall consider *linear differential systems*. This is the class of LTIC systems introduced in Chapter 1, for which the input $x(t)$ and the output $y(t)$ are related by linear differential equations of the form

$$\frac{d^N y}{dt^N} + a_1 \frac{d^{N-1} y}{dt^{N-1}} + \cdots + a_{N-1} \frac{dy}{dt} + a_N y(t)$$

$$= b_{N-M} \frac{d^M x}{dt^M} + b_{N-M+1} \frac{d^{M-1} x}{dt^{M-1}} + \cdots + b_{N-1} \frac{dx}{dt} + b_N x(t) \qquad (2.1a)$$

where all the coefficients a_i and b_i are constants. Using operational notation D to represent d/dt, we can express this equation as

$$(D^N + a_1 D^{N-1} + \cdots + a_{N-1} D + a_N) y(t)$$

$$= (b_{N-M} D^M + b_{N-M+1} D^{M-1} + \cdots + b_{N-1} D + b_N) x(t) \qquad (2.1b)$$

or

$$Q(D) y(t) = P(D) x(t) \qquad (2.1c)$$

where the polynomials $Q(D)$ and $P(D)$ are

$$Q(D) = D^N + a_1 D^{N-1} + \cdots + a_{N-1} D + a_N \qquad (2.2a)$$

$$P(D) = b_{N-M} D^M + b_{N-M+1} D^{M-1} + \cdots + b_{N-1} D + b_N \qquad (2.2b)$$

Theoretically the powers M and N in the foregoing equations can take on any value. However, practical considerations make $M > N$ undesirable for two reasons. In Section 4.3-2, we shall show that an LTIC system specified by Eq. (2.1) acts as an $(M - N)$th-order differentiator.

A differentiator represents an unstable system because a bounded input like the step input results in an unbounded output, $\delta(t)$. Second, the noise is enhanced by a differentiator. Noise is a wideband signal containing components of all frequencies from 0 to a very high frequency approaching ∞.[†] Hence, noise contains a significant amount of rapidly varying components. We know that the derivative of any rapidly varying signal is high. Therefore, any system specified by Eq. (2.1) in which $M > N$ will magnify the high-frequency components of noise through differentiation. It is entirely possible for noise to be magnified so much that it swamps the desired system output even if the noise signal at the system's input is tolerably small. Hence, practical systems generally use $M \leq N$. For the rest of this text we assume implicitly that $M \leq N$. For the sake of generality, we shall assume $M = N$ in Eq. (2.1).

In Chapter 1, we demonstrated that a system described by Eq. (2.1) is linear. Therefore, its response can be expressed as the sum of two components: the zero-input component and the zero-state component (decomposition property).[‡] Therefore,

$$\textbf{total response} = \textbf{zero-input response} + \textbf{zero-state response} \qquad (2.3)$$

The zero-input component is the system response when the input $x(t) = 0$, and thus it is the result of internal system conditions (such as energy storages, initial conditions) alone. It is independent of the external input $x(t)$. In contrast, the zero-state component is the system response to the external input $x(t)$ when the system is in zero state, meaning the absence of all internal energy storages: that is, all initial conditions are zero.

2.2 SYSTEM RESPONSE TO INTERNAL CONDITIONS: THE ZERO-INPUT RESPONSE

The zero-input response $y_0(t)$ is the solution of Eq. (2.1) when the input $x(t) = 0$ so that

$$Q(D)y_0(t) = 0 \qquad (2.4a)$$

or

$$(D^N + a_1 D^{N-1} + \cdots + a_{N-1}D + a_N)y_0(t) = 0 \qquad (2.4b)$$

A solution to this equation can be obtained systematically.[1] However, we will take a shortcut by using heuristic reasoning. Equation (2.4b) shows that a linear combination of $y_0(t)$ and its N

[†]Noise is any undesirable signal, natural or man-made, that interferes with the desired signals in the system. Some of the sources of noise are the electromagnetic radiation from stars, the random motion of electrons in system components, interference from nearby radio and television stations, transients produced by automobile ignition systems, and fluorescent lighting.

[‡]We can verify readily that the system described by Eq. (2.1) has the decomposition property. If $y_0(t)$ is the zero-input response, then, by definition,

$$Q(D)y_0(t) = 0$$

If $y(t)$ is the zero-state response, then $y(t)$ is the solution of

$$Q(D)y(t) = P(D)x(t)$$

subject to zero initial conditions (zero-state). Adding these two equations, we have

$$Q(D)[y_0(t) + y(t)] = P(D)x(t)$$

Clearly, $y_0(t) + y(t)$ is the general solution of Eq. (2.1).

successive derivatives is zero, not at *some* values of t, but for all t. Such a result is possible *if and only if* $y_0(t)$ and all its N successive derivatives are of the same form. Otherwise their sum can never add to zero for all values of t. We know that only an exponential function $e^{\lambda t}$ has this property. So let us assume that

$$y_0(t) = ce^{\lambda t}$$

is a solution to Eq. (2.4b). Then

$$Dy_0(t) = \frac{dy_0}{dt} = c\lambda e^{\lambda t}$$

$$D^2 y_0(t) = \frac{d^2 y_0}{dt^2} = c\lambda^2 e^{\lambda t}$$

$$\vdots$$

$$D^N y_0(t) = \frac{d^N y_0}{dt^N} = c\lambda^N e^{\lambda t}$$

Substituting these results in Eq. (2.4b), we obtain

$$c(\lambda^N + a_1 \lambda^{N-1} + \cdots + a_{N-1}\lambda + a_N)e^{\lambda t} = 0$$

For a nontrivial solution of this equation,

$$\lambda^N + a_1 \lambda^{N-1} + \cdots + a_{N-1}\lambda + a_N = 0 \qquad (2.5a)$$

This result means that $ce^{\lambda t}$ is indeed a solution of Eq. (2.4), provided λ satisfies Eq. (2.5a). Note that the polynomial in Eq. (2.5a) is identical to the polynomial $Q(D)$ in Eq. (2.4), with λ replacing D. Therefore, Eq. (2.5a) can be expressed as

$$Q(\lambda) = 0 \qquad (2.5b)$$

When $Q(\lambda)$ is expressed in factorized form, Eq. (2.5b) can be represented as

$$Q(\lambda) = (\lambda - \lambda_1)(\lambda - \lambda_2)\cdots(\lambda - \lambda_N) = 0 \qquad (2.5c)$$

Clearly, λ has N solutions: $\lambda_1, \lambda_2, \ldots, \lambda_N$, assuming that all λ_i are distinct. Consequently, Eq. (2.4) has N possible solutions: $c_1 e^{\lambda_1 t}, c_2 e^{\lambda_2 t}, \ldots, c_N e^{\lambda_N t}$, with $c_1, c_2, \ldots, c_N$ as arbitrary constants. We can readily show that a general solution is given by the sum of these N solutions,[†]

[†]To prove this assertion, assume that $y_1(t), y_2(t), \ldots, y_N(t)$ are all solutions of Eq. (2.4). Then

$$Q(D)y_1(t) = 0$$
$$Q(D)y_2(t) = 0$$
$$\vdots$$
$$Q(D)y_N(t) = 0$$

Multiplying these equations by $c_1, c_2, \ldots, c_N$, respectively, and adding them together yields

$$Q(D)[c_1 y_1(t) + c_2 y_2(t) + \cdots + c_N y_n(t)] = 0$$

This result shows that $c_1 y_1(t) + c_2 y_2(t) + \cdots + c_N y_n(t)$ is also a solution of the homogeneous equation [Eq. (2.4)].

so that

$$y_0(t) = c_1 e^{\lambda_1 t} + c_2 e^{\lambda_2 t} + \cdots + c_N e^{\lambda_N t} \tag{2.6}$$

where $c_1, c_2, \ldots, c_N$ are arbitrary constants determined by N constraints (the auxiliary conditions) on the solution.

Observe that the polynomial $Q(\lambda)$, which is characteristic of the system, has nothing to do with the input. For this reason the polynomial $Q(\lambda)$ is called the *characteristic polynomial* of the system. The equation

$$Q(\lambda) = 0 \tag{2.7}$$

is called the *characteristic equation* of the system. Equation (2.5c) clearly indicates that λ_1, $\lambda_2, \ldots, \lambda_N$ are the roots of the characteristic equation; consequently, they are called the *characteristic roots* of the system. The terms *characteristic values, eigenvalues,* and *natural frequencies* are also used for characteristic roots.[†] The exponentials $e^{\lambda_i t} (i = 1, 2, \ldots, n)$ in the zero-input response are the *characteristic modes* (also known as *natural modes* or simply as *modes*) of the system. There is a characteristic mode for each characteristic root of the system, and the *zero-input response is a linear combination of the characteristic modes of the system.*

An LTIC system's characteristic modes comprise its single most important attribute. Characteristic modes not only determine the zero-input response but also play an important role in determining the zero-state response. In other words, the entire behavior of a system is dictated primarily by its characteristic modes. In the rest of this chapter we shall see the pervasive presence of characteristic modes in every aspect of system behavior.

REPEATED ROOTS

The solution of Eq. (2.4) as given in Eq. (2.6) assumes that the N characteristic roots $\lambda_1, \lambda_2, \ldots,$ λ_N are distinct. If there are repeated roots (same root occurring more than once), the form of the solution is modified slightly. By direct substitution we can show that the solution of the equation

$$(D - \lambda)^2 y_0(t) = 0$$

is given by

$$y_0(t) = (c_1 + c_2 t) e^{\lambda t}$$

In this case the root λ repeats twice. Observe that the characteristic modes in this case are $e^{\lambda t}$ and $t e^{\lambda t}$. Continuing this pattern, we can show that for the differential equation

$$(D - \lambda)^r y_0(t) = 0 \tag{2.8}$$

the characteristic modes are $e^{\lambda t}, t e^{\lambda t}, t^2 e^{\lambda t}, \ldots, t^{r-1} e^{\lambda t}$, and that the solution is

$$y_0(t) = (c_1 + c_2 t + \cdots + c_r t^{r-1}) e^{\lambda t} \tag{2.9}$$

Consequently, for a system with the characteristic polynomial

$$Q(\lambda) = (\lambda - \lambda_1)^r (\lambda - \lambda_{r+1}) \cdots (\lambda - \lambda_N)$$

[†] *Eigenvalue* is German for "characteristic value."

the characteristic modes are $e^{\lambda_1 t}, te^{\lambda_1 t}, \ldots, t^{r-1}e^{\lambda_1 t}, e^{\lambda_{r+1} t}, \ldots, e^{\lambda_N t}$ and the solution is

$$y_0(t) = (c_1 + c_2 t + \cdots + c_r t^{r-1})e^{\lambda_1 t} + c_{r+1}e^{\lambda_{r+1} t} + \cdots + c_N e^{\lambda_N t}$$

COMPLEX ROOTS

The procedure for handling complex roots is the same as that for real roots. For complex roots the usual procedure leads to complex characteristic modes and the complex form of solution. However, it is possible to avoid the complex form altogether by selecting a real form of solution, as described next.

For a real system, complex roots must occur in pairs of conjugates if the coefficients of the characteristic polynomial $Q(\lambda)$ are to be real. Therefore, if $\alpha + j\beta$ is a characteristic root, $\alpha - j\beta$ must also be a characteristic root. The zero-input response corresponding to this pair of complex conjugate roots is

$$y_0(t) = c_1 e^{(\alpha + j\beta)t} + c_2 e^{(\alpha - j\beta)t} \tag{2.10a}$$

For a real system, the response $y_0(t)$ must also be real. This is possible only if c_1 and c_2 are conjugates. Let

$$c_1 = \frac{c}{2}e^{j\theta} \qquad \text{and} \qquad c_2 = \frac{c}{2}e^{-j\theta}$$

This yields

$$\begin{aligned}
y_0(t) &= \frac{c}{2}e^{j\theta}e^{(\alpha + j\beta)t} + \frac{c}{2}e^{-j\theta}e^{(\alpha - j\beta)t} \\
&= \frac{c}{2}e^{\alpha t}\left[e^{j(\beta t + \theta)} + e^{-j(\beta t + \theta)}\right] \\
&= ce^{\alpha t}\cos(\beta t + \theta) \tag{2.10b}
\end{aligned}$$

Therefore, the zero-input response corresponding to complex conjugate roots $\alpha \pm j\beta$ can be expressed in a complex form (2.10a) or a real form (2.10b).

EXAMPLE 2.1

(a) Find $y_0(t)$, the zero-input component of the response for an LTIC system described by the following differential equation:

$$(D^2 + 3D + 2)y(t) = Dx(t)$$

when the initial conditions are $y_0(0) = 0$, $\dot{y}_0(0) = -5$. Note that $y_0(t)$, being the zero-input component ($x(t) = 0$), is the solution of $(D^2 + 3D + 2)y_0(t) = 0$.

The characteristic polynomial of the system is $\lambda^2 + 3\lambda + 2$. The characteristic equation of the system is therefore $\lambda^2 + 3\lambda + 2 = (\lambda + 1)(\lambda + 2) = 0$. The characteristic roots of the system are $\lambda_1 = -1$ and $\lambda_2 = -2$, and the characteristic modes of the system are e^{-t} and e^{-2t}. Consequently, the zero-input response is

$$y_0(t) = c_1 e^{-t} + c_2 e^{-2t} \qquad (2.11a)$$

To determine the arbitrary constants c_1 and c_2, we differentiate Eq. (2.11a) to obtain

$$\dot{y}_0(t) = -c_1 e^{-t} - 2c_2 e^{-2t} \qquad (2.11b)$$

Setting $t = 0$ in Eqs. (2.11a) and (2.11b), and substituting the initial conditions $y_0(0) = 0$ and $\dot{y}_0(0) = -5$ we obtain

$$0 = c_1 + c_2$$

$$-5 = -c_1 - 2c_2$$

Solving these two simultaneous equations in two unknowns for c_1 and c_2 yields

$$c_1 = -5 \qquad c_2 = 5$$

Therefore

$$y_0(t) = -5e^{-t} + 5e^{-2t} \qquad (2.11c)$$

This is the zero-input component of $y(t)$. Because $y_0(t)$ is present at $t = 0^-$, we are justified in assuming that it exists for $t \geq 0$.[†]

(b) A Similar procedure may be followed for repeated roots. For instance, for a system specified by

$$(D^2 + 6D + 9)y(t) = (3D + 5)x(t)$$

let us determine $y_0(t)$, the zero-input component of the response if the initial conditions are $y_0(0) = 3$ and $\dot{y}_0(0) = -7$.

The characteristic polynomial is $\lambda^2 + 6\lambda + 9 = (\lambda + 3)^2$, and its characteristic roots are $\lambda_1 = -3$, $\lambda_2 = -3$ (repeated roots). Consequently, the characteristic modes of the system are e^{-3t} and te^{-3t}. The zero-input response, being a linear combination of the characteristic modes, is given by

$$y_0(t) = (c_1 + c_2 t)e^{-3t}$$

We can find the arbitrary constants c_1 and c_2 from the initial conditions $y_0(0) = 3$ and $\dot{y}_0(0) = -7$ following the procedure in part (a). The reader can show that $c_1 = 3$ and $c_2 = 2$. Hence,

$$y_0(t) = (3 + 2t)e^{-3t} \qquad t \geq 0$$

[†]$y_0(t)$ may be present even before $t = 0^-$. However, we can be sure of its presence only from $t = 0^-$ onward.

(c) For the case of complex roots, let us find the zero-input response of an LTIC system described by the equation

$$(D^2 + 4D + 40)y(t) = (D + 2)x(t)$$

with initial conditions $y_0(0) = 2$ and $\dot{y}_0(0) = 16.78$.

The characteristic polynomial is $\lambda^2 + 4\lambda + 40 = (\lambda + 2 - j6)(\lambda + 2 + j6)$. The characteristic roots are $-2 \pm j6$.[†] The solution can be written either in the complex form [Eq. (2.10a)] or in the real form [Eq. (2.10b)]. The complex form is $y_0(t) = c_1 e^{\lambda_1 t} + c_2 e^{\lambda_2 t}$, where $\lambda_1 = -2 + j6$ and $\lambda_2 = -2 - j6$. Since $\alpha = -2$ and $\beta = 6$, the real form solution is [see Eq. (2.10b)]

$$y_0(t) = ce^{-2t} \cos(6t + \theta) \tag{2.12a}$$

where c and θ are arbitrary constants to be determined from the initial conditions $y_0(0) = 2$ and $\dot{y}_0(0) = 16.78$. Differentiation of Eq. (2.12a) yields

$$\dot{y}_0(t) = -2ce^{-2t} \cos(6t + \theta) - 6ce^{-2t} \sin(6t + \theta) \tag{2.12b}$$

Setting $t = 0$ in Eqs. (2.12a) and (2.12b), and then substituting initial conditions, we obtain

$$2 = c\cos\theta$$

$$16.78 = -2c\cos\theta - 6c\sin\theta$$

Solution of these two simultaneous equations in two unknowns $c\cos\theta$ and $c\sin\theta$ yields

$$c\cos\theta = 2 \tag{2.13a}$$

$$c\sin\theta = -3.463 \tag{2.13b}$$

Squaring and then adding the two sides of Eqs. (2.13) yields

$$c^2 = (2)^2 + (-3.464)^2 = 16 \Longrightarrow c = 4$$

Next, dividing Eq. (2.13b) by Eq. (2.13a), that is, dividing $c\sin\theta$, by $c\cos\theta$, yields

$$\tan\theta = \frac{-3.463}{2}$$

and

$$\theta = \tan^{-1}\left(\frac{-3.463}{2}\right) = -\frac{\pi}{3}$$

Therefore

$$y_0(t) = 4e^{-2t} \cos\left(6t - \frac{\pi}{3}\right)$$

For the plot of $y_0(t)$, refer again to Fig. B.11c.

[†]The complex conjugate roots of a second-order polynomial can be determined by using the formula in Section B.7-10 or by expressing the polynomial as a sum of two squares. The latter can be accomplished by completing the square with the first two terms, as follows:

$$\lambda^2 + 4\lambda + 40 = (\lambda^2 + 4\lambda + 4) + 36 = (\lambda + 2)^2 + (6)^2 = (\lambda + 2 - j6)(\lambda + 2 + j6)$$

COMPUTER EXAMPLE C2.1

Find the roots λ_1 and λ_2 of the polynomial $\lambda^2 + 4\lambda + k$ for three values of k:

   (a) $k = 3$

   (b) $k = 4$

   (c) $k = 40$

(a)

```
>> r = roots([1 4 3]);
>> disp(['Case (k=3): roots = [',num2str(r.'),']']);
Case (k=3): roots = [-3 -1]
```

For $k = 3$, the polynomial roots are therefore $\lambda_1 = -3$ and $\lambda_2 = -1$.

(b)

```
>> r = roots([1 4 4]);
>> disp(['Case (k=4): roots = [',num2str(r.'),']']);
Case (k=4): roots = [-2 -2]
```

For $k = 4$, the polynomial roots are therefore $\lambda_1 = \lambda_2 = -2$.

(c)

```
>> r = roots([1 4 40]);
>> disp(['Case (k=40): roots = [',num2str(r.',' %0.5g'),']']);
Case (k=40): roots = [-2+6i -2-6i]
```

For $k = 40$, the polynomial roots are therefore $\lambda_1 = -2 + j6$ and $\lambda_2 = -2 - j6$.

COMPUTER EXAMPLE C2.2

Consider an LTIC system specified by the differential equation

$$(D^2 + 4D + k)y(t) = (3D + 5)x(t)$$

Using initial conditions $y_0(0) = 3$ and $\dot{y}_0(0) = -7$, determine the zero-input component of the response for three values of k:

   (a) $k = 3$

   (b) $k = 4$

   (c) $k = 40$

(a)

```
>> y_0 = dsolve('D2y+4*Dy+3*y=0','y(0)=3','Dy(0)=-7','t');
>> disp(['(a) k = 3; y_0 = ',char(y_0)])
(a) k = 3; y_0 = 2*exp(-3*t)+exp(-t)
```

For $k = 3$, the zero-input response is therefore $y_0(t) = 2e^{-3t} + e^{-t}$.

(b)

```
>> y_0 = dsolve('D2y+4*Dy+4*y=0','y(0)=3','Dy(0)=-7','t');
>> disp(['(b) k = 4; y_0 = ',char(y_0)])
(b) k = 4; y_0 = 3*exp(-2*t)-exp(-2*t)*t
```

For $k = 4$, the zero-input response is therefore $y_0(t) = 3e^{-2t} - te^{-2t}$.

(c)

```
>> y_0 = dsolve('D2y+4*Dy+40*y=0','y(0)=3','Dy(0)=-7','t');
>> disp(['(c) k = 40; y_0 = ',char(y_0)])
(c) k = 40; y_0 = -1/6*exp(-2*t)*sin(6*t)+3*exp(-2*t)*cos(6*t)
```

For $k = 40$, the zero-input response is therefore

$$y_0(t) = -\tfrac{1}{6}e^{-2t}\sin(6t) + 3e^{-2t}\cos(6t)$$

EXERCISE E2.1

Find the zero-input response of an LTIC system described by $(D + 5)y(t) = x(t)$ if the initial condition is $y(0) = 5$.

ANSWER
$y_0(t) = 5e^{-5t} \qquad t \geq 0$

EXERCISE E2.2

Solve

$$(D^2 + 2D)y_0(t) = 0$$

if $y_0(0) = 1$ and $\dot{y}_0(0) = 4$.

ANSWER
$y_0(t) = 3 - 2e^{-2t} \qquad t \geq 0$

PRACTICAL INITIAL CONDITIONS
AND THE MEANING OF 0^- AND 0^+

In Example 2.1 the initial conditions $y_0(0)$ and $\dot{y}_0(0)$ were supplied. In practical problems, we must derive such conditions from the physical situation. For instance, in an *RLC* circuit, we may be given the conditions (initial capacitor voltages, initial inductor currents, etc.).

From this information, we need to derive $y_0(0)$, $\dot{y}_0(0)$, ... for the desired variable as demonstrated in the next example.

In much of our discussion, the input is assumed to start at $t = 0$, unless otherwise mentioned. Hence, $t = 0$ is the reference point. The conditions immediately before $t = 0$ (just before the input is applied) are the conditions at $t = 0^-$, and those immediately after $t = 0$ (just after the input is applied) are the conditions at $t = 0^+$ (compare this with the historical time frame B.C. and A.D.). In practice, we are likely to know the initial conditions at $t = 0^-$ rather than at $t = 0^+$. The two sets of conditions are generally different, although in some cases they may be identical.

The total response $y(t)$ consists of two components: the zero-input component $y_0(t)$ [response due to the initial conditions alone with $x(t) = 0$] and the zero-state component resulting from the input alone with all initial conditions zero. At $t = 0^-$, the total response $y(t)$ consists solely of the zero-input component $y_0(t)$ because the input has not started yet. Hence the initial conditions on $y(t)$ are identical to those of $y_0(t)$. Thus, $y(0^-) = y_0(0^-)$, $\dot{y}(0^-) = \dot{y}_0(0^-)$, and so on. Moreover, $y_0(t)$ is the response due to initial conditions alone and does not depend on the input $x(t)$. Hence, application of the input at $t = 0$ does not affect $y_0(t)$. This means the initial conditions on $y_0(t)$ at $t = 0^-$ and 0^+ are identical; that is $y_0(0^-)$, $\dot{y}_0(0^-)$, ... are identical to $y_0(0^+)$, $\dot{y}_0(0^+)$, ..., respectively. It is clear that for $y_0(t)$, there is no distinction between the initial conditions at $t = 0^-$, 0 and 0^+. They are all the same. But this is not the case with the total response $y(t)$, which consists of both the zero-input and the zero-state components. Thus, in general, $y(0^-) \neq y(0^+)$, $\dot{y}(0^-) \neq \dot{y}(0^+)$, and so on.

EXAMPLE 2.2

A voltage $x(t) = 10e^{-3t}u(t)$ is applied at the input of the *RLC* circuit illustrated in Fig. 2.1a. Find the loop current $y(t)$ for $t \geq 0$ if the initial inductor current is zero; that is, $y(0^-) = 0$ and the initial capacitor voltage is 5 volts; that is, $v_C(0^-) = 5$.

The differential (loop) equation relating $y(t)$ to $x(t)$ was derived in Eq. (1.55) as

$$(D^2 + 3D + 2)y(t) = Dx(t)$$

The zero-state component of $y(t)$ resulting from the input $x(t)$, assuming that all initial conditions are zero, that is, $y(0^-) = v_C(0^-) = 0$, will be obtained later in Example 2.6. In this example we shall find the zero-input component $y_0(t)$. For this purpose, we need two initial conditions $y_0(0)$ and $\dot{y}_0(0)$. These conditions can be derived from the given initial conditions, $y(0^-) = 0$ and $v_C(0^-) = 5$, as follows. Recall that $y_0(t)$ is the loop current when the input terminals are shorted so that the input $x(t) = 0$ (zero-input) as depicted in Fig. 2.1b.

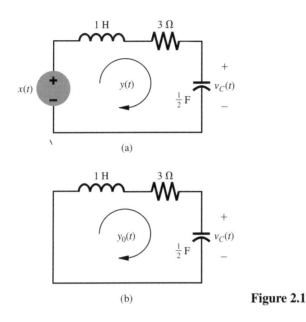

(a)

(b)   **Figure 2.1**

We now compute $y_0(0)$ and $\dot{y}_0(0)$, the values of the loop current and its derivative at $t = 0$, from the initial values of the inductor current and the capacitor voltage. Remember that the inductor current cannot change instantaneously in the absence of an impulsive voltage. Similarly, the capacitor voltage cannot change instantaneously in the absence of an impulsive current. Therefore, when the input terminals are shorted at $t = 0$, the inductor current is still zero and the capacitor voltage is still 5 volts. Thus,

$$y_0(0) = 0$$

To determine $\dot{y}_0(0)$, we use the loop equation for the circuit in Fig. 2.1b. Because the voltage across the inductor is $L(dy_0/dt)$ or $\dot{y}_0(t)$, this equation can be written as follows:

$$\dot{y}_0(t) + 3y_0(t) + v_C(t) = 0$$

Setting $t = 0$, we obtain

$$\dot{y}_0(0) + 3y_0(0) + v_C(0) = 0$$

But $y_0(0) = 0$ and $v_C(0) = 5$. Consequently,

$$\dot{y}_0(0) = -5$$

Therefore, the desired initial conditions are

$$y_0(0) = 0 \quad \text{and} \quad \dot{y}_0(0) = -5$$

Thus, the problem reduces to finding $y_0(t)$, the zero-input component of $y(t)$ of the system specified by the equation $(D^2 + 3D + 2)y(t) = Dx(t)$, when the initial conditions are $y_0(0) = 0$ and $\dot{y}_0(0) = -5$. We have already solved this problem in Example 2.1a, where we found

$$y_0(t) = -5e^{-t} + 5e^{-2t} \qquad t \geq 0 \tag{2.14}$$

This is the zero-input component of the loop current $y(t)$.

It will be interesting to find the initial conditions at $t = 0^-$ and 0^+ for the total response $y(t)$. Let us compare $y(0^-)$ and $\dot{y}(0^-)$ with $y(0^+)$ and $\dot{y}(0^+)$. The two pairs can be compared by writing the loop equation for the circuit in Fig. 2.1a at $t = 0^-$ and $t = 0^+$. The only difference between the two situations is that at $t = 0^-$, the input $x(t) = 0$, whereas at $t = 0^+$, the input $x(t) = 10$ [because $x(t) = 10e^{-3t}$]. Hence, the two loop equations are

$$\dot{y}(0^-) + 3y(0^-) + v_C(0^-) = 0$$

$$\dot{y}(0^+) + 3y(0^+) + v_C(0^+) = 10$$

The loop current $y(0^+) = y(0^-) = 0$ because it cannot change instantaneously in the absence of impulsive voltage. The same is true of the capacitor voltage. Hence, $v_C(0^+) = v_C(0^-) = 5$. Substituting these values in the foregoing equations, we obtain $\dot{y}(0^-) = -5$ and $\dot{y}(0^+) = 5$. Thus

$$y(0^-) = 0, \ \dot{y}(0^-) = -5 \quad \text{and} \quad y(0^+) = 0, \ \dot{y}(0^+) = 5 \tag{2.15}$$

EXERCISE E2.3

In the circuit in Fig. 2.1a, the inductance $L = 0$ and the initial capacitor voltage $v_C(0) = 30$ volts. Show that the zero-input component of the loop current is given by $y_0(t) = -10e^{-2t/3}$ for $t \geq 0$.

INDEPENDENCE OF ZERO-INPUT AND ZERO-STATE RESPONSE

In Example 2.2 we computed the zero-input component without using the input $x(t)$. The zero-state component can be computed from the knowledge of the input $x(t)$ alone; the initial conditions are assumed to be zero (system in zero state). The two components of the system response (the zero-input and zero-state components) are independent of each other. *The two worlds of zero-input response and zero-state response coexist side by side, neither one knowing or caring what the other is doing. For each component, the other is totally irrelevant.*

ROLE OF AUXILIARY CONDITIONS IN SOLUTION OF DIFFERENTIAL EQUATIONS

Solution of a differential equation requires additional pieces of information (the *auxiliary conditions*). Why? We now show heuristically why a differential equation does not, in general, have a unique solution unless some additional constraints (or conditions) on the solution are known.

Differentiation operation is not invertible unless one piece of information about $y(t)$ is given. To get back $y(t)$ from dy/dt, we must know one piece of information, such as $y(0)$. Thus, differentiation is an irreversible (noninvertible) operation during which certain information is lost. To invert this operation, one piece of information about $y(t)$ must be provided to restore the original $y(t)$. Using a similar argument, we can show that, given $d^2 y/dt^2$, we can determine $y(t)$ uniquely only if two additional pieces of information (constraints) about $y(t)$ are given. In general, to determine $y(t)$ uniquely from its Nth derivative, we need N additional pieces of information (constraints) about $y(t)$. These constraints are also called *auxiliary conditions*. When these conditions are given at $t = 0$, they are called *initial conditions*.

2.2-1 Some Insights into the Zero-Input Behavior of a System

By definition, the zero-input response is the system response to its internal conditions, assuming that its input is zero. Understanding this phenomenon provides interesting insight into system behavior. If a system is disturbed momentarily from its rest position and if the disturbance is then removed, the system will not come back to rest instantaneously. In general, it will come back to rest over a period of time and only through a special type of motion that is characteristic of the system.[†] For example, if we press on an automobile fender momentarily and then release it at $t = 0$, there is no external force on the automobile for $t > 0$.[‡] The auto body will eventually come back to its rest (equilibrium) position, but not through any arbitrary motion. It must do so by using only a form of response that is sustainable by the system on its own without any external source, since the input is zero. Only characteristic modes satisfy this condition. *The system uses a proper combination of characteristic modes to come back to the rest position while satisfying appropriate boundary (or initial) conditions.*

If the shock absorbers of the automobile are in good condition (high damping coefficient), the characteristic modes will be monotonically decaying exponentials, and the auto body will come to rest rapidly without oscillation. In contrast, for poor shock absorbers (low damping coefficients), the characteristic modes will be exponentially decaying sinusoids, and the body will come to rest through oscillatory motion. When a series RC circuit with an initial charge on the capacitor is shorted, the capacitor will start to discharge exponentially through the resistor. This response of the RC circuit is caused entirely by its internal conditions and is sustained by this system without the aid of any external input. The exponential current waveform is therefore the characteristic mode of the RC circuit.

Mathematically we know that *any combination of characteristic modes can be sustained by the system alone without requiring an external input*. This fact can be readily verified for the series RL circuit shown in Fig. 2.2. The loop equation for this system is

$$(D + 2)y(t) = x(t)$$

It has a single characteristic root $\lambda = -2$, and the characteristic mode is e^{-2t}. We now verify that a loop current $y(t) = ce^{-2t}$ can be sustained through this circuit without any input voltage.

[†]This assumes that the system will eventually come back to its original rest (or equilibrium) position.
[‡]We ignore the force of gravity, which merely causes a constant displacement of the auto body without affecting the other motion.

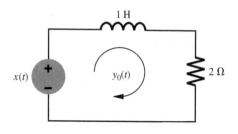

Figure 2.2 Modes always get a free ride.

The input voltage $x(t)$ required to drive a loop current $y(t) = ce^{-2t}$ is given by

$$x(t) = L\frac{dy}{dt} + Ry(t)$$

$$= \frac{d}{dt}(ce^{-2t}) + 2ce^{-2t}$$

$$= -2ce^{-2t} + 2ce^{-2t}$$

$$= 0$$

Clearly, the loop current $y(t) = ce^{-2t}$ is sustained by the RL circuit on its own, without the necessity of an external input.

THE RESONANCE PHENOMENON

We have seen that any signal consisting of a system's characteristic mode is sustained by the system on its own; the system offers no obstacle to such signals. Imagine what would happen if we were to drive the system with an external input that is one of its characteristic modes. This would be like pouring gasoline on a fire in a dry forest or hiring an alcoholic to taste liquor. An alcoholic would gladly do the job without pay. Think what would happen if he were paid by the amount of liquor he tasted! He would work overtime. He would work day and night, until he is burned out. The same thing happens with a system driven by an input of the form of characteristic mode. The system response grows without limit, until it burns out.[†] We call this behavior the *resonance phenomenon*. An intelligent discussion of this important phenomenon requires an understanding of the zero-state response; for this reason we postpone this topic until Section 2.7-7.

2.3 THE UNIT IMPULSE RESPONSE $h(t)$

In Chapter 1 we explained how a system response to an input $x(t)$ may be found by breaking this input into narrow rectangular pulses, as illustrated earlier in Fig. 1.27a, and then summing the system response to all the components. The rectangular pulses become impulses in the limit as their widths approach zero. Therefore, the system response is the sum of its responses to various impulse components. This discussion shows that if we know the system response to an impulse input, we can determine the system response to an arbitrary input $x(t)$. We now discuss

---

[†]In practice, the system in resonance is more likely to go in saturation because of high amplitude levels.

a method of determining $h(t)$, the unit impulse response of an LTIC system described by the Nth-order differential equation [Eq. (2.1a)]

$$Q(D)y(t) = P(D)x(t) \tag{2.16a}$$

where $Q(D)$ and $P(D)$ are the polynomials shown in Eq. (2.2). Recall that noise considerations restrict practical systems to $M \le N$. Under this constraint, the most general case is $M = N$. Therefore, Eq. (2.16a) can be expressed as

$$(D^N + a_1 D^{N-1} + \cdots + a_{N-1}D + a_N)y(t)$$
$$= (b_0 D^N + b_1 D^{N-1} + \cdots + b_{N-1}D + b_N)x(t) \tag{2.16b}$$

Before deriving the general expression for the unit impulse response $h(t)$, it is illuminating to understand qualitatively the nature of $h(t)$. The impulse response $h(t)$ is the system response to an impulse input $\delta(t)$ applied at $t = 0$ with all the initial conditions zero at $t = 0^-$. An impulse input $\delta(t)$ is like lightning, which strikes instantaneously and then vanishes. But in its wake, in that single moment, objects that have been struck are rearranged. Similarly, an impulse input $\delta(t)$ appears momentarily at $t = 0$, and then it is gone forever. But in that moment it generates energy storages; that is, it creates nonzero initial conditions instantaneously within the system at $t = 0^+$. Although the impulse input $\delta(t)$ vanishes for $t > 0$ so that the system has no input after the impulse has been applied, the system will still have a response generated by these newly created initial conditions. The impulse response $h(t)$, therefore, must consist of the system's characteristic modes for $t \ge 0^+$. As a result

$$h(t) = \text{characteristic mode terms} \qquad t \ge 0^+$$

This response is valid for $t > 0$. But what happens at $t = 0$? At a single moment $t = 0$, there can at most be an impulse,[†] so the form of the complete response $h(t)$ is

$$h(t) = A_0 \delta(t) + \text{characteristic mode terms} \qquad t \ge 0 \tag{2.17}$$

because $h(t)$ is the unit impulse response, setting $x(t) = \delta(t)$ and $y(t) = h(t)$ in Eq. (2.16b) yields

$$(D^N + a_1 D^{N-1} + \cdots + a_{N-1}D + a_N)h(t) = (b_0 D^N + b_1 D^{N-1} + \cdots + b_{N-1}D + b_N)\delta(t) \tag{2.18}$$

In this equation we substitute $h(t)$ from Eq. (2.17) and compare the coefficients of similar impulsive terms on both sides. The highest order of the derivative of impulse on both sides is N, with its coefficient value as A_0 on the left-hand side and b_0 on the right-hand side. The two values must be matched. Therefore, $A_0 = b_0$ and

$$h(t) = b_0 \delta(t) + \text{characteristic modes} \tag{2.19}$$

[†]It might be possible for the derivatives of $\delta(t)$ to appear at the origin. However, if $M \le N$, it is impossible for $h(t)$ to have any derivatives of $\delta(t)$. This conclusion follows from Eq. (2.16b) with $x(t) = \delta(t)$ and $y(t) = h(t)$. The coefficients of the impulse and all its derivatives must be matched on both sides of this equation. If $h(t)$ contains $\delta^{(1)}(t)$, the first derivative of $\delta(t)$, the left-hand side of Eq. (2.16b) will contain a term $\delta^{(N+1)}(t)$. But the highest-order derivative term on the right-hand side is $\delta^{(N)}(t)$. Therefore, the two sides cannot match. Similar arguments can be made against the presence of the impulse's higher-order derivatives in $h(t)$.

In Eq. (2.16b), if $M < N$, $b_0 = 0$. Hence, the impulse term $b_0 \delta(t)$ exists only if $M = N$. The unknown coefficients of the N characteristic modes in $h(t)$ in Eq. (2.19) can be determined by using the technique of impulse matching, as explained in the following example.

EXAMPLE 2.3

Find the impulse response $h(t)$ for a system specified by

$$(D^2 + 5D + 6)y(t) = (D + 1)x(t) \qquad (2.20)$$

In this case, $b_0 = 0$. Hence, $h(t)$ consists of only the characteristic modes. The characteristic polynomial is $\lambda^2 + 5\lambda + 6 = (\lambda + 2)(\lambda + 3)$. The roots are -2 and -3. Hence, the impulse response $h(t)$ is

$$h(t) = (c_1 e^{-2t} + c_2 e^{-3t}) u(t) \qquad (2.21)$$

letting $x(t) = \delta(t)$ and $y(t) = h(t)$ in Eq. (2.20), we obtain

$$\ddot{h}(t) + 5\dot{h}(t) + 6h(t) = \dot{\delta}(t) + \delta(t) \qquad (2.22)$$

Recall that initial conditions $h(0^-)$ and $\dot{h}(0^-)$ are both zero. But the application of an impulse at $t = 0$ creates new initial conditions at $t = 0^+$. Let $h(0^+) = K_1$ and $\dot{h}(0^+) = K_2$. These jump discontinuities in $h(t)$ and $\dot{h}(t)$ at $t = 0$ result in impulse terms $\dot{h}(0) = K_1 \delta(t)$ and $\ddot{h}(0) = K_1 \dot{\delta}(t) + K_2 \delta(t)$ on the left-hand side. Matching the coefficients of impulse terms on both sides of Eq. (2.22) yields

$$5K_1 + K_2 = 1, \qquad K_1 = 1 \qquad \Longrightarrow \qquad K_1 = 1, \, K_2 = -4$$

We now use these values $h(0^+) = K_1 = 1$ and $\dot{h}(0^+) = K_2 = -4$ in Eq. (2.21) to find c_1 and c_2. Setting $t = 0^+$ in Eq. (2.21), we obtain $c_1 + c_2 = 1$. Also setting $t = 0^+$ in $\dot{h}(t)$, we obtain $-2c_1 - 3c_1 = -4$. These two simultaneous equations yield $c_1 = -1$ and $c_2 = 2$. Therefore

$$h(t) = (-e^{-2t} + 2e^{-3t})u(t)$$

Although, the method used in this example is relatively simple, we can simplify it still further by using a modified version of impulse matching.

SIMPLIFIED IMPULSE MATCHING METHOD

The alternate technique we present now allows us to reduce the procedure to a simple routine to determine $h(t)$. To avoid the needless distraction, the proof for this procedure is placed in Section 2.8. There, we show that for an LTIC system specified by Eq. (2.16), the unit impulse response $h(t)$ is given by

$$h(t) = b_0 \delta(t) + [P(D)y_n(t)]u(t) \tag{2.23}$$

where $y_n(t)$ is a linear combination of the characteristic modes of the system subject to the following initial conditions:

$$y_n(0) = \dot{y}_n(0) = \ddot{y}_n(0) = \cdots = y_n^{(N-2)}(0) = 0 \quad \text{and} \quad y_n^{(N-1)}(0) = 1 \tag{2.24a}$$

where $y_n^{(k)}(0)$ is the value of the kth derivative of $y_n(t)$ at $t = 0$. We can express this set of conditions for various values of N (the system order) as follows:

$$N = 1 : y_n(0) = 1$$
$$N = 2 : y_n(0) = 0, \dot{y}_n(0) = 1$$
$$N = 3 : y_n(0) = \dot{y}_n(0) = 0, \ddot{y}_n(0) = 1 \tag{2.24b}$$

and so on.

As stated earlier, if the order of $P(D)$ is less than the order of $Q(D)$, that is, if $M < N$, then $b_0 = 0$, and the impulse term $b_0 \delta(t)$ in $h(t)$ is zero.

EXAMPLE 2.4

Determine the unit impulse response $h(t)$ for a system specified by the equation

$$(D^2 + 3D + 2)\, y(t) = Dx(t) \tag{2.25}$$

This is a second-order system ($N = 2$) having the characteristic polynomial

$$(\lambda^2 + 3\lambda + 2) = (\lambda + 1)(\lambda + 2)$$

The characteristic roots of this system are $\lambda = -1$ and $\lambda = -2$. Therefore

$$y_n(t) = c_1 e^{-t} + c_2 e^{-2t} \tag{2.26a}$$

Differentiation of this equation yields

$$\dot{y}_n(t) = -c_1 e^{-t} - 2c_2 e^{-2t} \tag{2.26b}$$

The initial conditions are [see Eq. (2.24b) for $N = 2$]

$$\dot{y}_n(0) = 1 \quad \text{and} \quad y_n(0) = 0$$

Setting $t = 0$ in Eqs. (2.26a) and (2.26b), and substituting the initial conditions just given, we obtain

$$0 = c_1 + c_2$$
$$1 = -c_1 - 2c_2$$

Solution of these two simultaneous equations yields

$$c_1 = 1 \qquad \text{and} \qquad c_2 = -1$$

Therefore

$$y_n(t) = e^{-t} - e^{-2t}$$

Moreover, according to Eq. (2.25), $P(D) = D$, so that

$$P(D)y_n(t) = Dy_n(t) = \dot{y}_n(t) = -e^{-t} + 2e^{-2t}$$

Also in this case, $b_0 = 0$ [the second-order term is absent in $P(D)$]. Therefore

$$h(t) = [P(D)y_n(t)]u(t) = (-e^{-t} + 2e^{-2t})u(t)$$

Comment. In the above discussion, we have assumed $M \leq N$, as specified by Eq. (2.16b). Appendix 2.1 (i.e., Section 2.8) shows that the expression for $h(t)$ applicable to all possible values of M and N is given by

$$h(t) = P(D)[y_n(t)u(t)]$$

where $y_n(t)$ is a linear combination of the characteristic modes of the system subject to initial conditions (2.24). This expression reduces to Eq. (2.23) when $M \leq N$.

Determination of the impulse response $h(t)$ using the procedure in this section is relatively simple. However, in Chapter 4 we shall discuss another, even simpler method using the Laplace transform.

EXERCISE E2.4

Determine the unit impulse response of LTIC systems described by the following equations:

(a) $(D + 2)y(t) = (3D + 5)x(t)$
(b) $D(D + 2)y(t) = (D + 4)x(t)$
(c) $(D^2 + 2D + 1)y(t) = Dx(t)$

ANSWERS

(a) $3\delta(t) - e^{-2t}u(t)$
(b) $(2 - e^{-2t})u(t)$
(c) $(1 - t)e^{-t}u(t)$

COMPUTER EXAMPLE C2.3

Determine the impulse response $h(t)$ for an LTIC system specified by the differential equation

$$(D^2 + 3D + 2)y(t) = Dx(t)$$

This is a second-order system with $b_0 = 0$. First we find the zero-input component for initial conditions $y(0^-) = 0$, and $\dot{y}(0^-) = 1$. Since $P(D) = D$, the zero-input response is differentiated and the impulse response immediately follows.

```
>> y_n = dsolve('D2y+3*Dy+2*y=0','y(0)=0','Dy(0)=1','t');
>> Dy_n = diff(y_n);
>> disp(['h(t) = (',char(Dy_n),')u(t)']);
h(t) = (-exp(-t)+2*exp(-2*t))u(t)
```

Therefore, $h(t) = b_0\delta(t) + [Dy_0(t)]u(t) = (-e^{-t} + 2e^{-2t})u(t)$.

SYSTEM RESPONSE TO DELAYED IMPULSE

If $h(t)$ is the response of an LTIC system to the input $\delta(t)$, then $h(t - T)$ is the response of this same system to the input $\delta(t - T)$. This conclusion follows from the time-invariance property of LTIC systems. Thus, by knowing the unit impulse response $h(t)$, we can determine the system response to a delayed impulse $\delta(t - T)$.

2.4 SYSTEM RESPONSE TO EXTERNAL INPUT: ZERO-STATE RESPONSE

This section is devoted to the determination of the zero-state response of an LTIC system. This is the system response $y(t)$ to an input $x(t)$ when the system is in the zero state, that is, when all initial conditions are zero. *We shall assume that the systems discussed in this section are in the zero state unless mentioned otherwise.* Under these conditions, the zero-state response will be the total response of the system.

We shall use the superposition property for finding the system response to an arbitrary input $x(t)$. Let us define a basic pulse $p(t)$ of unit height and width $\Delta\tau$, starting at $t = 0$ as illustrated in (Fig. 2.3a). Figure 2.3b shows an input $x(t)$ as a sum of narrow rectangular pulses. The pulse starting at $t = n\Delta\tau$ in Fig. 2.3b has a height $x(n\Delta\tau)$, and can be expressed as $x(n\Delta\tau)p(t - n\Delta\tau)$. Now, $x(t)$ is the sum of all such pulses. Hence

$$x(t) = \lim_{\Delta\tau \to 0} \sum_{\tau} x(n\Delta\tau)p(t - n\Delta\tau) = \lim_{\Delta\tau \to 0} \sum_{\tau} \left[\frac{x(n\Delta\tau)}{\Delta\tau}\right] p(t - n\Delta\tau)\Delta\tau \qquad (2.27)$$

The term $[x(n\Delta\tau)/\Delta\tau]p(t - n\Delta\tau)$ represents a pulse $p(t - n\Delta\tau)$ with height $[x(n\Delta\tau)/\Delta\tau]$. As $\Delta\tau \to 0$, the height of this strip $\to \infty$, but its area remains $x(n\Delta\tau)$. Hence, this strip

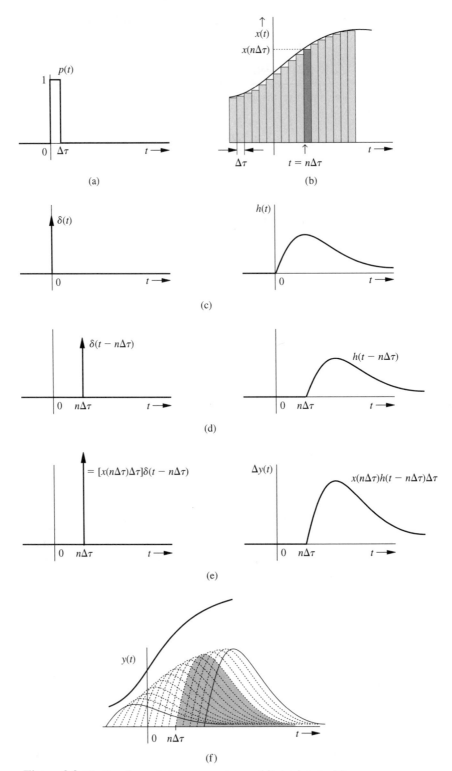

Figure 2.3 Finding the system response to an arbitrary input $x(t)$.

approaches an impulse $x(n\Delta\tau)\delta(t - n\Delta\tau)$ as $\Delta\tau \to 0$ (Fig. 2.3e). Therefore

$$x(t) = \lim_{\Delta\tau \to 0} \sum_{\tau} x(n\Delta\tau)\delta(t - n\Delta\tau)\,\Delta\tau \qquad (2.28)$$

To find the response for this input $x(t)$, we consider the input and the corresponding output pairs, as shown in Fig. 2.3c–2.3f and also shown by directed arrow notation as follows:

$$\text{input} \quad \Longrightarrow \quad \text{output}$$
$$\delta(t) \quad \Longrightarrow \quad h(t)$$
$$\delta(t - n\Delta\tau) \quad \Longrightarrow \quad h(t - n\Delta\tau)$$
$$[x(n\Delta\tau)\Delta\tau]\delta(t - n\Delta\tau) \quad \Longrightarrow \quad [x(n\Delta\tau)\Delta\tau]h(t - n\Delta\tau)$$
$$\underbrace{\lim_{\Delta\tau \to 0} \sum_{\tau} x(n\Delta\tau)\delta(t - n\Delta\tau)\,\Delta\tau}_{x(t) \quad \text{[see Eq. (2.28)]}} \quad \Longrightarrow \quad \underbrace{\lim_{\Delta\tau \to 0} \sum_{\tau} x(n\Delta\tau)h(t - n\Delta\tau)\Delta\tau}_{y(t)}$$

Therefore[†]

$$y(t) = \lim_{\Delta\tau \to 0} \sum_{\tau} x(n\Delta\tau)h(t - n\Delta\tau)\Delta\tau$$
$$= \int_{-\infty}^{\infty} x(\tau)h(t - \tau)\,d\tau \qquad (2.29)$$

This is the result we seek. We have obtained the system response $y(t)$ to an arbitrary input $x(t)$ in terms of the unit impulse response $h(t)$. Knowing $h(t)$, we can determine the response $y(t)$ to any input. *Observe once again the all-pervasive nature of the system's characteristic modes. The system response to any input is determined by the impulse response, which, in turn, is made up of characteristic modes of the system.*

It is important to keep in mind the assumptions used in deriving Eq. (2.29). We assumed a linear, time- invariant (LTI) system. Linearity allowed us to use the principle of superposition, and time invariance made it possible to express the system's response to $\delta(t - n\Delta\tau)$ as $h(t - n\Delta\tau)$.

2.4-1 The Convolution Integral

The zero-state response $y(t)$ obtained in Eq. (2.29) is given by an integral that occurs frequently in the physical sciences, engineering, and mathematics. For this reason this integral is given a special name: the *convolution integral*. The convolution integral of two functions $x_1(t)$ and $x_2(t)$

[†]In deriving this result we have assumed a time-invariant system. If the system is time varying, then the system response to the input $\delta(t - n\Delta\tau)$ cannot be expressed as $h(t - n\Delta\tau)$ but instead has the form $h(t, n\Delta\tau)$. Use of this form modifies Eq. (2.29) to

$$y(t) = \int_{-\infty}^{\infty} x(\tau)h(t, \tau)\,d\tau$$

where $h(t, \tau)$ is the system response at instant t to a unit impulse input located at τ.

is denoted symbolically by $x_1(t) * x_2(t)$ and is defined as

$$x_1(t) * x_2(t) \equiv \int_{-\infty}^{\infty} x_1(\tau) x_2(t - \tau) \, d\tau \tag{2.30}$$

Some important properties of the convolution integral follow.

THE COMMUTATIVE PROPERTY

Convolution operation is commutative; that is, $x_1(t) * x_2(t) = x_2(t) * x_1(t)$. This property can be proved by a change of variable. In Eq. (2.30), if we let $z = t - \tau$ so that $\tau = t - z$ and $d\tau = -dz$, we obtain

$$\begin{aligned}
x_1(t) * x_2(t) &= -\int_{\infty}^{-\infty} x_2(z) x_1(t - z) \, dz \\
&= \int_{-\infty}^{\infty} x_2(z) x_1(t - z) \, dz \\
&= x_2(t) * x_1(t)
\end{aligned} \tag{2.31}$$

THE DISTRIBUTIVE PROPERTY

According to the distributive property,

$$x_1(t) * [x_2(t) + x_3(t)] = x_1(t) * x_2(t) + x_1(t) * x_3(t) \tag{2.32}$$

THE ASSOCIATIVE PROPERTY

According to the associative property,

$$x_1(t) * [x_2(t) * x_3(t)] = [x_1(t) * x_2(t)] * x_3(t) \tag{2.33}$$

The proofs of Eqs. (2.32) and (2.33) follow directly from the definition of the convolution integral. They are left as an exercise for the reader.

THE SHIFT PROPERTY

If

$$x_1(t) * x_2(t) = c(t)$$

then

$$x_1(t) * x_2(t - T) = x_1(t - T) * x_2(t) = c(t - T) \tag{2.34a}$$

and

$$x_1(t - T_1) * x_2(t - T_2) = c(t - T_1 - T_2) \tag{2.34b}$$

Proof. We are given

$$x_1(t) * x_2(t) = \int_{-\infty}^{\infty} x_1(\tau) x_2(t - \tau) \, d\tau = c(t)$$

Therefore

$$x_1(t) * x_2(t - T) = \int_{-\infty}^{\infty} x_1(\tau) x_2(t - T - \tau) \, d\tau$$

$$= c(t - T)$$

Equation (2.34b) follows from Eq. (2.34a).

CONVOLUTION WITH AN IMPULSE

Convolution of a function $x(t)$ with a unit impulse results in the function $x(t)$ itself. By definition of convolution

$$x(t) * \delta(t) = \int_{-\infty}^{\infty} x(\tau) \delta(t - \tau) \, d\tau \tag{2.35}$$

Because $\delta(t - \tau)$ is an impulse located at $\tau = t$, according to the sampling property of the impulse [Eq. (1.24)], the integral in Eq. (2.35) is the value of $x(\tau)$ at $\tau = t$, that is, $x(t)$. Therefore

$$x(t) * \delta(t) = x(t) \tag{2.36}$$

Actually this result was derived earlier [Eq. (2.28)].

THE WIDTH PROPERTY

If the durations (widths) of $x_1(t)$ and $x_2(t)$ are finite, given by T_1 and T_2, respectively, then the duration (width) of $x_1(t) * x_2(t)$ is $T_1 + T_2$ (Fig. 2.4). The proof of this property follows readily from the graphical considerations discussed later in Section 2.4-2.

ZERO-STATE RESPONSE AND CAUSALITY

The (zero-state) response $y(t)$ of an LTIC system is

$$y(t) = x(t) * h(t) = \int_{-\infty}^{\infty} x(\tau) h(t - \tau) \, d\tau \tag{2.37}$$

In deriving Eq. (2.37), we assumed the system to be linear and time invariant. There were no other restrictions either on the system or on the input signal $x(t)$. In practice, most systems are

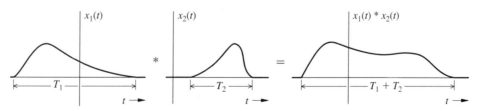

Figure 2.4 Width property of convolution.

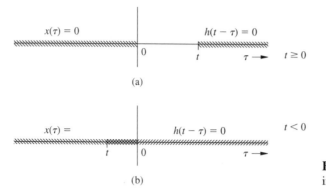

Figure 2.5 Limits of the convolution integral.

causal, so that their response cannot begin before the input. Furthermore, most inputs are also causal, which means they start at $t = 0$.

Causality restriction on both signals and systems further simplifies the limits of integration in Eq. (2.37). By definition, the response of a causal system cannot begin before its input begins. Consequently, the causal system's response to a unit impulse $\delta(t)$ (which is located at $t = 0$) cannot begin before $t = 0$. Therefore, a *causal system's unit impulse response $h(t)$ is a causal signal*.

It is important to remember that the integration in Eq. (2.37) is performed with respect to τ (not t). If the input $x(t)$ is causal, $x(\tau) = 0$ for $\tau < 0$. Therefore, $x(\tau) = 0$ for $\tau < 0$, as illustrated in Fig. 2.5a. Similarly, if $h(t)$ is causal, $h(t - \tau) = 0$ for $t - \tau < 0$; that is, for $\tau > t$, as depicted in Fig. 2.5a. Therefore, the product $x(\tau)h(t - \tau) = 0$ everywhere except over the nonshaded interval $0 \leq \tau \leq t$ shown in Fig. 2.5a (assuming $t \geq 0$). Observe that if t is negative, $x(\tau)h(t - \tau) = 0$ for all τ as shown in Fig. 2.5b. Therefore, Eq. (2.37) reduces to

$$y(t) = x(t) * h(t) = \int_{0^-}^{t} x(\tau)h(t - \tau)\,d\tau \qquad t \geq 0$$
$$= 0 \qquad\qquad\qquad\qquad\qquad\qquad t < 0 \qquad\qquad (2.38a)$$

The lower limit of integration in Eq. (2.38a) is taken as 0^- to avoid the difficulty in integration that can arise if $x(t)$ contains an impulse at the origin. This result shows that if $x(t)$ and $h(t)$ are both causal, the response $y(t)$ is also causal.

Because of the convolution's commutative property [Eq. (2.31)], we can also express Eq. (2.38a) as [assuming causal $x(t)$ and $h(t)$]

$$y(t) = \int_{0^-}^{t} h(\tau)x(t - \tau)\,d\tau \qquad t \geq 0$$
$$= 0 \qquad\qquad\qquad\qquad\qquad\quad t < 0 \qquad\qquad (2.38b)$$

Hereafter, the lower limit of 0^- will be implied even when we write it as 0. As in Eq. (2.38a), this result assumes that both the input and the system are causal.

EXAMPLE 2.5

For an LTIC system with the unit impulse response $h(t) = e^{-2t}u(t)$, determine the response $y(t)$ for the input

$$x(t) = e^{-t}u(t) \qquad (2.39)$$

Here both $x(t)$ and $h(t)$ are causal (Fig. 2.6). Hence, from Eq. (2.38a), we obtain

$$y(t) = \int_0^t x(\tau)h(t - \tau)\,d\tau \qquad t \ge 0$$

Because $x(t) = e^{-t}u(t)$ and $h(t) = e^{-2t}u(t)$

$$x(\tau) = e^{-\tau}u(\tau) \qquad \text{and} \qquad h(t - \tau) = e^{-2(t-\tau)}u(t - \tau)$$

Remember that the integration is performed with respect to τ (not t), and the region of integration is $0 \le \tau \le t$. Hence, $\tau \ge 0$ and $t - \tau \ge 0$. Therefore, $u(\tau) = 1$ and $u(t - \tau) = 1$; consequently

$$y(t) = \int_0^t e^{-\tau}e^{-2(t-\tau)}\,d\tau \qquad t \ge 0$$

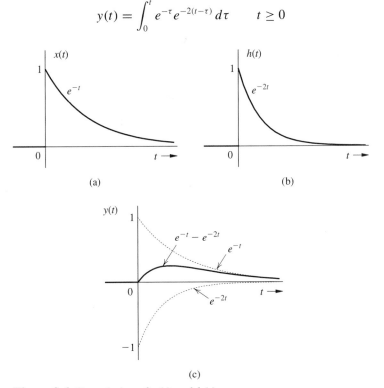

(a)

(b)

(c)

Figure 2.6 Convolution of $x(t)$ and $h(t)$.

Because this integration is with respect to τ, we can pull e^{-2t} outside the integral, giving us

$$y(t) = e^{-2t} \int_0^t e^\tau \, d\tau = e^{-2t}(e^t - 1) = e^{-t} - e^{-2t} \qquad t \geq 0$$

Moreover, $y(t) = 0$ when $t < 0$ [see Eq. (2.38a)]. Therefore

$$y(t) = (e^{-t} - e^{-2t})u(t)$$

The response is depicted in Fig. 2.6c.

EXERCISE E2.5

For an LTIC system with the impulse response $h(t) = 6e^{-t}u(t)$, determine the system response to the input:

(a) $2u(t)$

(b) $3e^{-3t}u(t)$

ANSWERS

(a) $12(1 - e^{-t})u(t)$

(b) $9(e^{-t} - e^{-3t})u(t)$

EXERCISE E2.6

Repeat Exercise E2.5 for the input $x(t) = e^{-t}u(t)$.

ANSWER

$6te^{-t}u(t)$

THE CONVOLUTION TABLE

The task of convolution is considerably simplified by a ready-made convolution table (Table 2.1). This table, which lists several pairs of signals and their convolution, can conveniently determine $y(t)$, a system response to an input $x(t)$, without performing the tedious job of integration. For instance, we could have readily found the convolution in Example 2.5 by using pair 4 (with $\lambda_1 = -1$ and $\lambda_2 = -2$) to be $(e^{-t} - e^{-2t})u(t)$. The following example demonstrates the utility of this table.

TABLE 2.1 Convolution Table

No.	$x_1(t)$	$x_2(t)$	$x_1(t) * x_2(t) = x_2(t) * x_1(t)$
1	$x(t)$	$\delta(t - T)$	$x(t - T)$
2	$e^{\lambda t}u(t)$	$u(t)$	$\dfrac{1 - e^{\lambda t}}{-\lambda}u(t)$
3	$u(t)$	$u(t)$	$tu(t)$
4	$e^{\lambda_1 t}u(t)$	$e^{\lambda_2 t}u(t)$	$\dfrac{e^{\lambda_1 t} - e^{\lambda_2 t}}{\lambda_1 - \lambda_2}u(t) \qquad \lambda_1 \neq \lambda_2$
5	$e^{\lambda t}u(t)$	$e^{\lambda t}u(t)$	$te^{\lambda t}u(t)$
6	$te^{\lambda t}u(t)$	$e^{\lambda t}u(t)$	$\dfrac{1}{2}t^2 e^{\lambda t}u(t)$
7	$t^N u(t)$	$e^{\lambda t}u(t)$	$\dfrac{N! e^{\lambda t}}{\lambda^{N+1}}u(t) - \displaystyle\sum_{k=0}^{N} \dfrac{N! t^{N-k}}{\lambda^{k+1}(N-k)!}u(t)$
8	$t^M u(t)$	$t^N u(t)$	$\dfrac{M! N!}{(M+N+1)!}t^{M+N+1}u(t)$
9	$te^{\lambda_1 t}u(t)$	$e^{\lambda_2 t}u(t)$	$\dfrac{e^{\lambda_2 t} - e^{\lambda_1 t} + (\lambda_1 - \lambda_2)te^{\lambda_1 t}}{(\lambda_1 - \lambda_2)^2}u(t)$
10	$t^M e^{\lambda t}u(t)$	$t^N e^{\lambda t}u(t)$	$\dfrac{M! N!}{(N+M+1)!}t^{M+N+1}e^{\lambda t}u(t)$
11	$t^M e^{\lambda_1 t}u(t)$ $\lambda_1 \neq \lambda_2$	$t^N e^{\lambda_2 t}u(t)$	$\displaystyle\sum_{k=0}^{M} \dfrac{(-1)^k M!(N+k)! t^{M-k}e^{\lambda_1 t}}{k!(M-k)!(\lambda_1 - \lambda_2)^{N+k+1}}u(t)$ $+ \displaystyle\sum_{k=0}^{N} \dfrac{(-1)^k N!(M+k)! t^{N-k}e^{\lambda_2 t}}{k!(N-k)!(\lambda_2 - \lambda_1)^{M+k+1}}u(t)$
12	$e^{-\alpha t}\cos(\beta t + \theta)u(t)$	$e^{\lambda t}u(t)$	$\dfrac{\cos(\theta - \phi)e^{\lambda t} - e^{-\alpha t}\cos(\beta t + \theta - \phi)}{\sqrt{(\alpha + \lambda)^2 + \beta^2}}u(t)$ $\phi = \tan^{-1}[-\beta/(\alpha + \lambda)]$
13	$e^{\lambda_1 t}u(t)$	$e^{\lambda_2 t}u(-t)$	$\dfrac{e^{\lambda_1 t}u(t) + e^{\lambda_2 t}u(-t)}{\lambda_2 - \lambda_1} \qquad \mathrm{Re}\,\lambda_2 > \mathrm{Re}\,\lambda_1$
14	$e^{\lambda_1 t}u(-t)$	$e^{\lambda_2 t}u(-t)$	$\dfrac{e^{\lambda_1 t} - e^{\lambda_2 t}}{\lambda_2 - \lambda_1}u(-t)$

EXAMPLE 2.6

Find the loop current $y(t)$ of the *RLC* circuit in Example 2.2 for the input $x(t) = 10e^{-3t}u(t)$, when all the initial conditions are zero.

The loop equation for this circuit [see Example 1.10 or Eq. (1.55)] is

$$(D^2 + 3D + 2)y(t) = Dx(t)$$

The impulse response $h(t)$ for this system, as obtained in Example 2.4, is

$$h(t) = (2e^{-2t} - e^{-t})u(t)$$

The input is $x(t) = 10e^{-3t}u(t)$, and the response $y(t)$ is

$$y(t) = x(t) * h(t)$$
$$= 10e^{-3t}u(t) * [2e^{-2t} - e^{-t}]u(t)$$

Using the distributive property of the convolution [Eq. (2.32)], we obtain

$$y(t) = 10e^{-3t}u(t) * 2e^{-2t}u(t) - 10e^{-3t}u(t) * e^{-t}u(t)$$
$$= 20[e^{-3t}u(t) * e^{-2t}u(t)] - 10[e^{-3t}u(t) * e^{-t}u(t)]$$

Now the use of pair 4 in Table 2.1 yields

$$y(t) = \frac{20}{-3 - (-2)}[e^{-3t} - e^{-2t}]u(t) - \frac{10}{-3 - (-1)}[e^{-3t} - e^{-t}]u(t)$$
$$= -20(e^{-3t} - e^{-2t})u(t) + 5(e^{-3t} - e^{-t})u(t)$$
$$= (-5e^{-t} + 20e^{-2t} - 15e^{-3t})u(t)$$

EXERCISE E2.7

Rework Exercises E2.5 and E2.6 using the convolution table.

EXERCISE E2.8

Use the convolution table to determine

$$e^{-2t}u(t) * (1 - e^{-t})u(t)$$

ANSWER
$\left(\frac{1}{2} - e^{-t} + \frac{1}{2}e^{-2t}\right)u(t)$

EXERCISE E2.9

For an LTIC system with the unit impulse response $h(t) = e^{-2t}u(t)$, determine the zero-state response $y(t)$ if the input $x(t) = \sin 3t\, u(t)$. [Hint: Use pair 12 from Table 2.1.]

ANSWER

$\frac{1}{13}[3e^{-2t} + \sqrt{13}\cos(3t - 146.32°)]u(t)$

or

$\frac{1}{13}[3e^{-2t} - \sqrt{13}\cos(3t + 33.68°)]u(t)$

RESPONSE TO COMPLEX INPUTS

The LTIC system response discussed so far applies to general input signals, real or complex. However, if the system is real, that is, if $h(t)$ is real, then we shall show that the real part of the input generates the real part of the output, and a similar conclusion applies to the imaginary part.

If the input is $x(t) = x_r(t) + jx_i(t)$, where $x_r(t)$ and $x_i(t)$ are the real and imaginary parts of $x(t)$, then for real $h(t)$

$$y(t) = h(t) * [x_r(t) + jx_i(t)] = h(t) * x_r(t) + jh(t) * x_i(t) = y_r(t) + jy_i(t)$$

where $y_r(t)$ and $y_i(t)$ are the real and the imaginary parts of $y(t)$. Using the right-directed arrow notation to indicate a pair of the input and the corresponding output, the foregoing result can be expressed as follows. If

$$x(t) = x_r(t) + jx_i(t) \implies y(t) = y_r(t) + jy_i(t)$$

then

$$x_r(t) \implies y_r(t)$$
$$x_i(t) \implies y_i(t) \tag{2.40}$$

MULTIPLE INPUTS

Multiple inputs to LTI systems can be treated by applying the superposition principle. Each input is considered separately, with all other inputs assumed to be zero. The sum of all these individual system responses constitutes the total system output when all the inputs are applied simultaneously.

2.4-2 Graphical Understanding of Convolution Operation

Convolution operation can be grasped readily by examining the graphical interpretation of the convolution integral. Such an understanding is helpful in evaluating the convolution integral of more complex signals. In addition, graphical convolution allows us to grasp visually or mentally the convolution integral's result, which can be of great help in sampling, filtering, and many other problems. Finally, many signals have no exact mathematical description, so they can be described only graphically. If two such signals are to be convolved, we have no choice but to perform their convolution graphically.

We shall now explain the convolution operation by convolving the signals $x(t)$ and $g(t)$, illustrated in Fig. 2.7a and 2.7b, respectively. If $c(t)$ is the convolution of $x(t)$ with $g(t)$, then

$$c(t) = \int_{-\infty}^{\infty} x(\tau)g(t-\tau)\,d\tau \tag{2.41}$$

One of the crucial points to remember here is that this integration is performed with respect to τ, so that t is just a parameter (like a constant). This consideration is especially important when we sketch the graphical representations of the functions $x(\tau)$ and $g(t-\tau)$ appearing in the integrand of Eq. (2.41). Both these functions should be sketched as functions of τ, not of t.

The function $x(\tau)$ is identical to $x(t)$, with τ replacing t (Fig. 2.7c). Therefore, $x(t)$ and $x(\tau)$ will have the same graphical representations. Similar remarks apply to $g(t)$ and $g(\tau)$ (Fig. 2.7d).

To appreciate what $g(t-\tau)$ looks like, let us start with the function $g(\tau)$ (Fig. 2.7d). Time reversal of this function (reflection about the vertical axis $\tau = 0$) yields $g(-\tau)$ (Fig. 2.7e). Let us denote this function by $\phi(\tau)$

$$\phi(\tau) = g(-\tau)$$

Now $\phi(\tau)$ shifted by t seconds is $\phi(\tau - t)$, given by

$$\phi(\tau - t) = g[-(\tau - t)] = g(t - \tau)$$

Therefore, we first time-reverse $g(\tau)$ to obtain $g(-\tau)$ and then time-shift $g(-\tau)$ by t to obtain $g(t-\tau)$. For positive t, the shift is to the right (Fig. 2.7f); for negative t, the shift is to the left (Fig. 2.7g, 2.7h).

The preceding discussion gives us a graphical interpretation of the functions $x(\tau)$ and $g(t-\tau)$. The convolution $c(t)$ is the area under the product of these two functions. Thus, to compute $c(t)$ at some positive instant $t = t_1$, we first obtain $g(-\tau)$ by inverting $g(\tau)$ about the vertical axis. Next, we right-shift or delay $g(-\tau)$ by t_1 to obtain $g(t_1 - \tau)$ (Fig. 2.7f), and then we multiply this function by $x(\tau)$, giving us the product $x(\tau)g(t_1 - \tau)$ (shaded portion in Fig. 2.7f). The area A_1 under this product is $c(t_1)$, the value of $c(t)$ at $t = t_1$. We can therefore plot $c(t_1) = A_1$ on a curve describing $c(t)$, as shown in Fig. 2.7i. The area under the product $x(\tau)g(-\tau)$ in Fig. 2.7e is $c(0)$, the value of the convolution for $t = 0$ (at the origin).

A similar procedure is followed in computing the value of $c(t)$ at $t = t_2$, where t_2 is negative (Fig. 2.7g). In this case, the function $g(-\tau)$ is shifted by a negative amount (that is, left-shifted) to obtain $g(t_2 - \tau)$. Multiplication of this function with $x(\tau)$ yields the product $x(\tau)g(t_2 - \tau)$. The area under this product is $c(t_2) = A_2$, giving us another point on the curve $c(t)$ at $t = t_2$ (Fig. 2.7i). This procedure can be repeated for all values of t, from $-\infty$ to ∞. The result will be a curve describing $c(t)$ for all time t. Note that when $t \leq -3$, $x(\tau)$ and $g(t-\tau)$ do not overlap (see Fig. 2.7h); therefore, $c(t) = 0$ for $t \leq -3$.

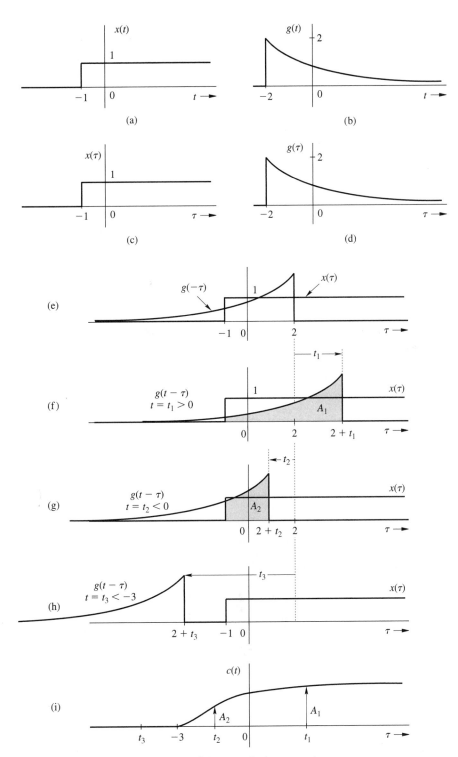

Figure 2.7 Graphical explanation of the convolution operation.

181

SUMMARY OF THE GRAPHICAL PROCEDURE

The procedure for graphical convolution can be summarized as follows

1. Keep the function $x(\tau)$ fixed.

2. Visualize the function $g(\tau)$ as a rigid wire frame, and rotate (or invert) this frame about the vertical axis ($\tau = 0$) to obtain $g(-\tau)$.

3. Shift the inverted frame along the τ axis by t_0 seconds. The shifted frame now represents $g(t_0 - \tau)$.

4. The area under the product of $x(\tau)$ and $g(t_0 - \tau)$ (the shifted frame) is $c(t_0)$, the value of the convolution at $t = t_0$.

5. Repeat this procedure, shifting the frame by different values (positive and negative) to obtain $c(t)$ for all values of t.

The graphical procedure discussed here appears very complicated and discouraging at first reading. Indeed, some people claim that convolution has driven many electrical engineering undergraduates to contemplate theology either for salvation or as an alternative career (*IEEE Spectrum,* March 1991, p. 60).[†] Actually, the bark of convolution is worse than its bite. In graphical convolution, we need to determine the area under the product $x(\tau)g(t - \tau)$ for all values of t from $-\infty$ to ∞. However, a mathematical description of $x(\tau)g(t - \tau)$ is generally valid over a range of t. Therefore, repeating the procedure for every value of t amounts to repeating it only a few times for different ranges of t.

Convolution: its bark is worse than its bite!

[†]Strange that religious establishments are not agitating for compulsory "convolution education" in schools and colleges!

We can also use the commutative property of convolution to our advantage by computing $x(t) * g(t)$ or $g(t) * x(t)$, whichever is simpler. As a rule of thumb, *convolution computations are simplified if we choose to invert (time-reverse) the simpler of the two functions*. For example, if the mathematical description of $g(t)$ is simpler than that of $x(t)$, then $x(t) * g(t)$ will be easier to compute than $g(t) * x(t)$. In contrast, if the mathematical description of $x(t)$ is simpler, the reverse will be true.

We shall demonstrate graphical convolution with the following examples. Let us start by using this graphical method to rework Example 2.5.

EXAMPLE 2.7

Determine graphically $y(t) = x(t) * h(t)$ for $x(t) = e^{-t}u(t)$ and $h(t) = e^{-2t}u(t)$.

In Fig. 2.8a and 2.8b we have $x(t)$ and $h(t)$, respectively; and Fig. 2.8c shows $x(\tau)$ and $h(-\tau)$ as functions of τ. The function $h(t - \tau)$ is now obtained by shifting $h(-\tau)$ by t. If t is positive, the shift is to the right (delay); if t is negative, the shift is to the left (advance). Figure 2.8d shows that for negative t, $h(t - \tau)$ [obtained by left-shifting $h(-\tau)$] does not overlap $x(\tau)$, and the product $x(\tau)h(t - \tau) = 0$, so that

$$y(t) = 0 \qquad t < 0$$

Figure 2.8e shows the situation for $t \geq 0$. Here $x(\tau)$ and $h(t - \tau)$ do overlap, but the product is nonzero only over the interval $0 \leq \tau \leq t$ (shaded interval). Therefore

$$y(t) = \int_0^t x(\tau)h(t - \tau)\, d\tau \qquad t \geq 0$$

All we need to do now is substitute correct expressions for $x(\tau)$ and $h(t - \tau)$ in this integral. From Fig. 2.8a and 2.8b it is clear that the segments of $x(t)$ and $g(t)$ to be used in this convolution (Fig. 2.8e) are described by

$$x(t) = e^{-t} \qquad \text{and} \qquad h(t) = e^{-2t}$$

Therefore

$$x(\tau) = e^{-\tau} \qquad \text{and} \qquad h(t - \tau) = e^{-2(t-\tau)}$$

Consequently

$$y(t) = \int_0^t e^{-\tau} e^{-2(t-\tau)}\, d\tau$$

$$= e^{-2t} \int_0^t e^{\tau}\, d\tau$$

$$= e^{-t} - e^{-2t} \qquad\qquad t \geq 0$$

Moreover, $y(t) = 0$ for $t < 0$, so that

$$y(t) = (e^{-t} - e^{-2t})u(t)$$

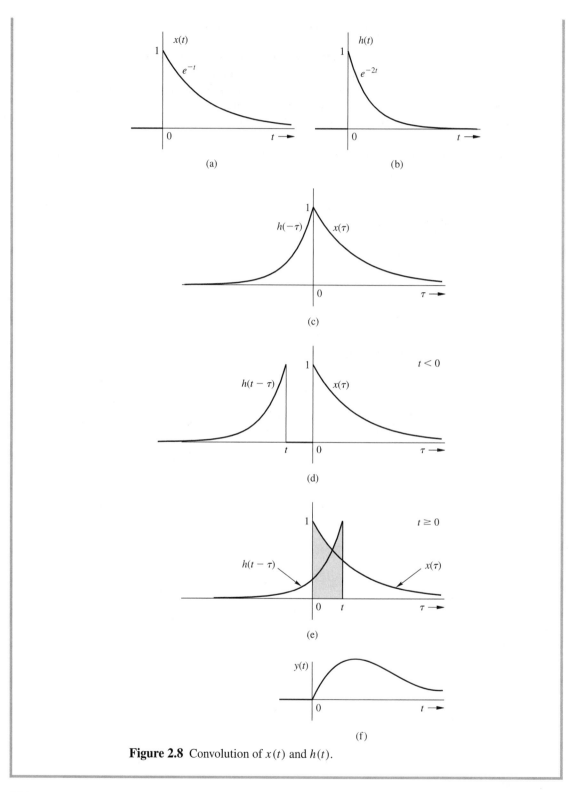

Figure 2.8 Convolution of $x(t)$ and $h(t)$.

EXAMPLE 2.8

Find $c(t) = x(t) * g(t)$ for the signals depicted in Fig. 2.9a and 2.9b.

Since $x(t)$ is simpler than $g(t)$, it is easier to evaluate $g(t) * x(t)$ than $x(t) * g(t)$. However, we shall intentionally take the more difficult route and evaluate $x(t) * g(t)$.

From $x(t)$ and $g(t)$ (Fig. 2.9a and 2.9b, respectively), observe that $g(t)$ is composed of two segments. As a result, it can be described as

$$g(t) = \begin{cases} 2e^{-t} & \text{segment A} \\ -2e^{2t} & \text{segment B} \end{cases}$$

Therefore

$$g(t - \tau) = \begin{cases} 2e^{-(t-\tau)} & \text{segment A} \\ -2e^{2(t-\tau)} & \text{segment B} \end{cases}$$

The segment of $x(t)$ that is used in convolution is $x(t) = 1$, so that $x(\tau) = 1$. Figure 2.9c shows $x(\tau)$ and $g(-\tau)$.

To compute $c(t)$ for $t \geq 0$, we right-shift $g(-\tau)$ to obtain $g(t - \tau)$, as illustrated in Fig. 2.9d. Clearly, $g(t - \tau)$ overlaps with $x(\tau)$ over the shaded interval, that is, over the range $\tau \geq 0$; segment A overlaps with $x(\tau)$ over the interval $(0, t)$, while segment B overlaps with $x(\tau)$ over (t, ∞). Remembering that $x(\tau) = 1$, we have

$$\begin{aligned} c(t) &= \int_0^\infty x(\tau)g(t - \tau)\, d\tau \\ &= \int_0^t 2e^{-(t-\tau)}\, d\tau + \int_t^\infty -2e^{2(t-\tau)}\, d\tau \\ &= 2(1 - e^{-t}) - 1 \\ &= 1 - 2e^{-t} \qquad\qquad t \geq 0 \end{aligned}$$

Figure 2.9e shows the situation for $t < 0$. Here the overlap is over the shaded interval, that is, over the range $\tau \geq 0$, where only the segment B of $g(t)$ is involved. Therefore

$$\begin{aligned} c(t) &= \int_0^\infty x(\tau)g(t - \tau)\, d\tau \\ &= \int_0^\infty g(t - \tau)\, d\tau \\ &= \int_0^\infty -2e^{2(t-\tau)}\, d\tau \\ &= -e^{2t} \qquad\qquad t \leq 0 \end{aligned}$$

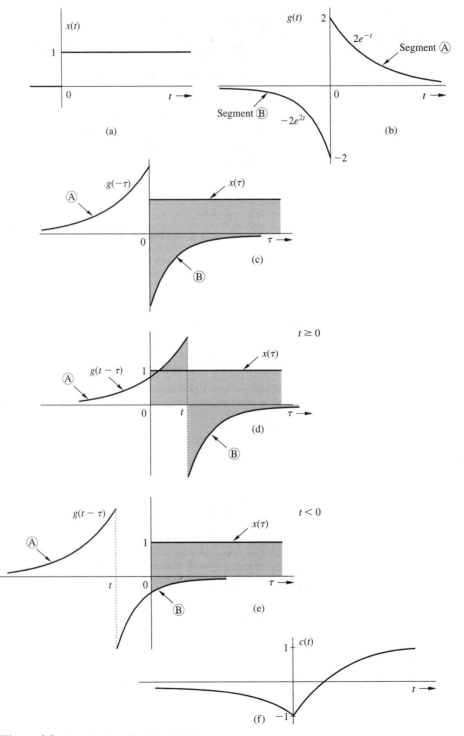

Figure 2.9 Convolution of $x(t)$ and $g(t)$.

Therefore

$$c(t) = \begin{cases} 1 - 2e^{-2t} & t \geq 0 \\ -e^{2t} & t \leq 0 \end{cases}$$

Figure 2.9f shows a plot of $c(t)$.

EXAMPLE 2.9

Find $x(t) * g(t)$ for the functions $x(t)$ and $g(t)$ shown in Fig. 2.10a and 2.10b.

Here, $x(t)$ has a simpler mathematical description than that of $g(t)$, so it is preferable to time-reverse $x(t)$. Hence, we shall determine $g(t) * x(t)$ rather than $x(t) * g(t)$. Thus

$$c(t) = g(t) * x(t)$$

$$= \int_{-\infty}^{\infty} g(\tau)x(t - \tau)\, d\tau$$

First, we determine the expressions for the segments of $x(t)$ and $g(t)$ used in finding $c(t)$. According to Fig. 2.10a and 2.10b, these segments can be expressed as

$$x(t) = 1 \quad \text{and} \quad g(t) = \tfrac{1}{3}t$$

so that

$$x(t - \tau) = 1 \quad \text{and} \quad g(\tau) = \tfrac{1}{3}\tau$$

Figure 2.10c shows $g(\tau)$ and $x(-\tau)$, whereas Fig. 2.10d shows $g(\tau)$ and $x(t - \tau)$, which is $x(-\tau)$ shifted by t. Because the edges of $x(-\tau)$ are at $\tau = -1$ and 1, the edges of $x(t - \tau)$ are at $-1 + t$ and $1 + t$. The two functions overlap over the interval $(0, 1 + t)$ (shaded interval), so that

$$c(t) = \int_{0}^{1+t} g(\tau)x(t - \tau)\, d\tau$$

$$= \int_{0}^{1+t} \tfrac{1}{3}\tau\, d\tau$$

$$= \tfrac{1}{6}(t + 1)^2 \qquad\qquad -1 \leq t \leq 1 \qquad\qquad (2.42a)$$

This situation, depicted in Fig. 2.10d, is valid only for $-1 \leq t \leq 1$. For $t \geq 1$ but ≤ 2, the situation is as illustrated in Fig. 2.10e. The two functions overlap only over the range $-1 + t$ to $1 + t$ (shaded interval). Note that the expressions for $g(\tau)$ and $x(t - \tau)$ do not

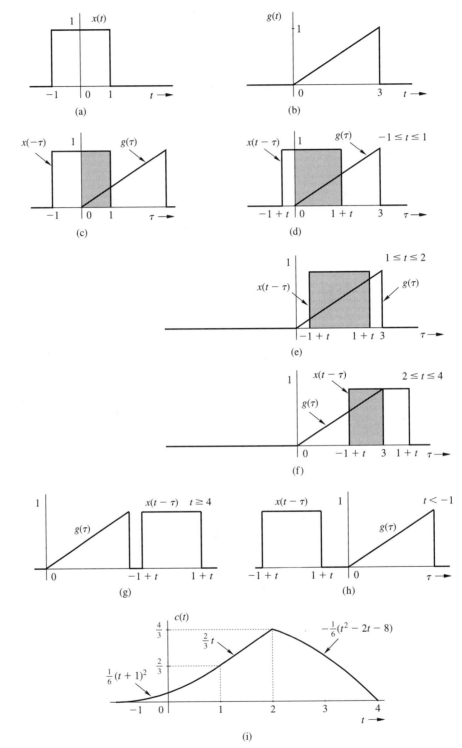

Figure 2.10 Convolution of $x(t)$ and $g(t)$.

change; only the range of integration changes. Therefore

$$c(t) = \int_{-1+t}^{1+t} \tfrac{1}{3}\tau \, d\tau$$

$$= \tfrac{2}{3}t \qquad\qquad 1 \le t \le 2 \qquad\qquad (2.42b)$$

Also note that the expressions in Eqs. (2.42a) and (2.42b) both apply at $t = 1$, the transition point between their respective ranges. We can readily verify that both expressions yield a value of $2/3$ at $t = 1$, so that $c(1) = 2/3$. The continuity of $c(t)$ at transition points indicates a high probability of a right answer. Continuity of $c(t)$ at transition points is assured as long as there are no impulses at the edges of $x(t)$ and $g(t)$.

For $t \ge 2$ but ≤ 4 the situation is as shown in Fig. 2.10f. The functions $g(\tau)$ and $x(t - \tau)$ overlap over the interval from $-1 + t$ to 3 (shaded interval), so that

$$c(t) = \int_{-1+t}^{3} \tfrac{1}{3}\tau \, d\tau$$

$$= -\tfrac{1}{6}(t^2 - 2t - 8) \qquad 2 \le t \le 4 \qquad\qquad (2.42c)$$

Again, both Eqs. (2.42b) and (2.42c) apply at the transition point $t = 2$. We can readily verify that $c(2) = 4/3$ when either of these expressions is used.

For $t \ge 4$, $x(t - \tau)$ has been shifted so far to the right that it no longer overlaps with $g(\tau)$ as depicted in Fig. 2.10g. Consequently

$$c(t) = 0 \qquad t \ge 4 \qquad\qquad (2.42d)$$

We now turn our attention to negative values of t. We have already determined $c(t)$ up to $t = -1$. For $t < -1$ there is no overlap between the two functions, as illustrated in Fig. 2.10h, so that

$$c(t) = 0 \qquad t \le -1 \qquad\qquad (2.42e)$$

Figure 2.10i plots $c(t)$ according to Eqs. (2.42a) through (2.42e).

WIDTH OF THE CONVOLVED FUNCTION

The widths (durations) of $x(t)$, $g(t)$, and $c(t)$ in Example 2.9 (Fig. 2.10) are 2, 3, and 5, respectively. Note that the width of $c(t)$ in this case is the sum of the widths of $x(t)$ and $g(t)$. This observation is not a coincidence. Using the concept of graphical convolution, we can readily see that if $x(t)$ and $g(t)$ have the finite widths of T_1 and T_2 respectively, then the width of $c(t)$ is equal to $T_1 + T_2$. The reason is that the time it takes for a signal of width (duration) T_1 to completely pass another signal of width (duration) T_2 so that they become nonoverlapping is $T_1 + T_2$. When the two signals become nonoverlapping, the convolution goes to zero.

EXERCISE E2.10

Rework Example 2.8 by evaluating $g(t) * x(t)$

EXERCISE E2.11

Use graphical convolution to show that $x(t) * g(t) = g(t) * x(t) = c(t)$ in Fig. 2.11.

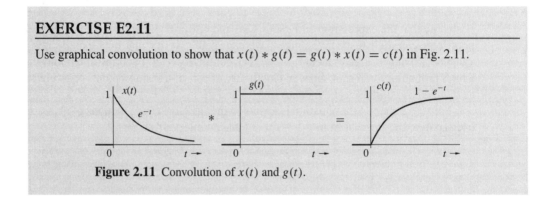

Figure 2.11 Convolution of $x(t)$ and $g(t)$.

EXERCISE E2.12

Repeat Exercise E2.11 for the functions in Fig. 2.12.

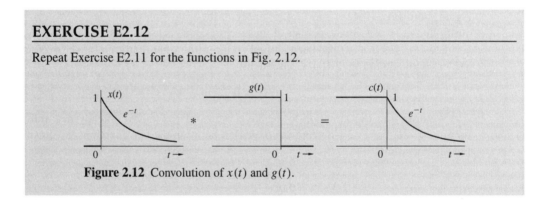

Figure 2.12 Convolution of $x(t)$ and $g(t)$.

EXERCISE E2.13

Repeat Exercise E2.11 for the functions in Fig. 2.13.

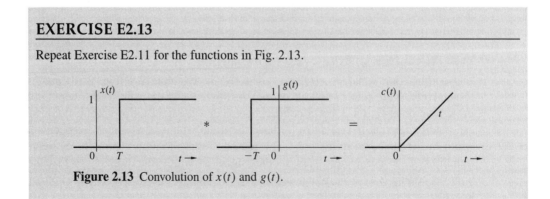

Figure 2.13 Convolution of $x(t)$ and $g(t)$.

THE PHANTOM OF THE SIGNALS
AND SYSTEMS OPERA

In the study of signals and systems we often come across some signals such as an impulse, which cannot be generated in practice and have never been sighted by anyone.[†] One wonders why we even consider such idealized signals. The answer should be clear from our discussion so far in this chapter. Even if the impulse function has no physical existence, we can compute the system response $h(t)$ to this phantom input according to the procedure in Section 2.3, and knowing $h(t)$, we can compute the system response to any arbitrary input. The concept of impulse response, therefore, provides an effective intermediary for computing system response to an arbitrary input. In addition, the impulse response $h(t)$ itself provides a great deal of information and insight about the system behavior. In Section 2.7 we show that the knowledge of impulse response provides much valuable information, such as the response time, pulse dispersion, and filtering properties of the system. Many other useful insights about the system behavior can be obtained by inspection of $h(t)$.

Similarly, in frequency-domain analysis (discussed in later chapters), we use an *everlasting exponential* (or *sinusoid)* to determine system response. An everlasting exponential (or sinusoid) too is a phantom, which no-body has ever seen and which has no physical existence. But it provides another effective intermediary for computing the system response to an arbitrary input. Moreover, the system response to everlasting exponential (or sinusoid) provides valuable information and insight regarding the system's behavior. Clearly, idealized impulse and everlasting sinusoids are friendly and helpful spirits.

Interestingly, the unit impulse and the everlasting exponential (or sinusoid) are the dual of each other in the time-frequency duality, to be studied in Chapter 7. Actually, the time-domain and the frequency-domain methods of analysis are the dual of each other.

WHY CONVOLUTION? AN INTUITIVE
EXPLANATION OF SYSTEM RESPONSE

On the surface, it appears rather strange that the response of linear systems (those gentlest of the gentle systems) should be given by such a tortuous operation of convolution, where one signal is fixed and the other is inverted and shifted. To understand this odd behavior, consider a hypothetical impulse response $h(t)$ that decays linearly with time (Fig. 2.14a). This response is strongest at $t = 0$, the moment the impulse is applied, and it decays linearly at future instants, so that one second later (at $t = 1$ and beyond), it ceases to exist. This means that the closer the impulse input is to an instant t, the stronger is its response at t.

Now consider the input $x(t)$ shown in Fig. 2.14b. To compute the system response, we break the input into rectangular pulses and approximate these pulses with impulses. Generally the response of a causal system at some instant t will be determined by all the impulse components of the input before t. Each of these impulse components will have different weight in determining the response at the instant t, depending on its proximity to t. As seen earlier, the closer the

---

[†]The late Prof. S. J. Mason, the inventor of signal flow graph techniques, used to tell a story of a student frustrated with the impulse function. The student said, "The unit impulse is a thing that is so small you can't see it, except at one place (the origin), where it is so big you can't see it. In other words, you can't see it at all; at least I can't!"[2]

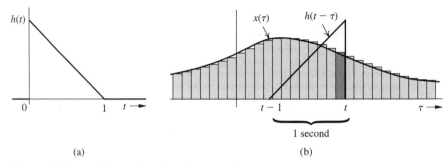

(a)   (b)

Figure 2.14 Intuitive explanation of convolution.

impulse is to t, the stronger is its influence at t. The impulse at t has the greatest weight (unity) in determining the response at t. The weight decreases linearly for all impulses before t until the instant $t - 1$. The input before $t - 1$ has no influence (zero weight). Thus, to determine the system response at t, we must assign a linearly decreasing weight to impulses occurring before t, as shown in Fig. 2.14b. This weighting function is precisely the function $h(t - \tau)$. The system response at t is then determined not by the input $x(\tau)$ but by the weighted input $x(\tau)h(t - \tau)$, and the summation of all these weighted inputs is the convolution integral.

2.4-3 Interconnected Systems

A larger, more complex system can often be viewed as the interconnection of several smaller subsystems, each of which is easier to characterize. Knowing the characterizations of these subsystems, it becomes simpler to analyze such large systems. We shall consider here two basic interconnections, cascade and parallel. Figure 2.15a shows S_1 and S_2, two LTIC subsystems connected in parallel, and Fig. 2.15b shows the same two systems connected in cascade.

In Fig. 2.15a, the device depicted by the symbol Σ inside a circle represents an adder, which adds signals at its inputs. Also the junction from which two (or more) branches radiate out is called the *pickoff node*. Every branch that radiates out from the pickoff node carries the same signal (the signal at the junction). In Fig. 2.15a, for instance, the junction at which the input is applied is a pickoff node from which two branches radiate out, each of which carries the input signal at the node.

Let the impulse response of S_1 and S_2 be $h_1(t)$ and $h_2(t)$, respectively. Further assume that interconnecting these systems, as shown in Fig. 2.15, does not load them. This means that the impulse response of either of these systems remains unchanged whether observed when these systems are unconnected or when they are interconnected.

To find $h_p(t)$, the impulse response of the parallel system S_p in Fig. 2.15a, we apply an impulse at the input of S_p. This results in the signal $\delta(t)$ at the inputs of S_1 and S_2, leading to their outputs $h_1(t)$ and $h_2(t)$, respectively. These signals are added by the adder to yield $h_1(t) + h_2(t)$ as the output of S_p. Consequently

$$h_p(t) = h_1(t) + h_2(t) \tag{2.43a}$$

To find $h_c(t)$, the impulse response of the cascade system S_c in Fig. 2.15b, we apply the input $\delta(t)$ at the input of S_c, which is also the input to S_1. Hence, the output of S_1 is $h_1(t)$, which now

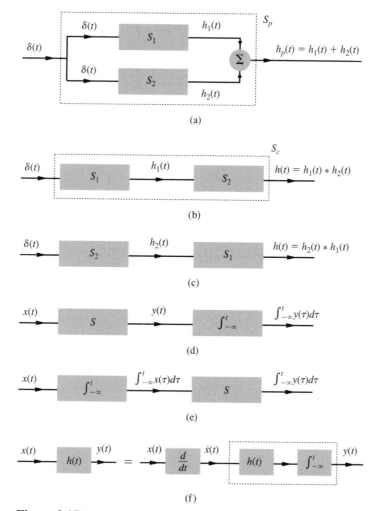

Figure 2.15 Interconnected systems.

acts as the input to S_2. The response of S_2 to input $h_1(t)$ is $h_1(t) * h_2(t)$. Therefore

$$h_c(t) = h_1(t) * h_2(t) \tag{2.43b}$$

Because of commutative property of the convolution, it follows that interchanging the systems S_1 and S_2, as shown in Fig. 2.15c, results in the same impulse response $h_1(t) * h_2(t)$. This means that when several LTIC systems are cascaded, the order of systems does not affect the impulse response of the composite system. In other words, linear operations, performed in cascade, commute. The order in which they are performed is not important, at least theoretically.[†]

[†]Change of order, however, could affect performance because of physical limitations and sensitivities to changes in the subsystems involved.

We shall give here another interesting application of the commutative property of LTIC systems. Figure 2.15d shows a cascade of two LTIC systems: a system S with impulse response $h(t)$, followed by an ideal integrator. Figure 2.15e shows a cascade of the same two systems in reverse order; an ideal integrator followed by S. In Fig. 2.15d, if the input $x(t)$ to S yields the output $y(t)$, then the output of the system 2.15d is the integral of $y(t)$. In Fig. 2.15e, the output of the integrator is the integral of $x(t)$. The output in Fig. 2.15e is identical to the output in Fig. 2.15d. Hence, it follows that if an LTIC system response to input $x(t)$ is $y(t)$, then the response of the same system to integral of $x(t)$ is the integral of $y(t)$. In other words,

$$\text{if} \qquad x(t) \Longrightarrow y(t)$$

$$\text{then} \qquad \int_{-\infty}^{t} x(\tau)\, d\tau \Longrightarrow \int_{-\infty}^{t} y(\tau)\, d\tau \qquad\qquad (2.44a)$$

Replacing the ideal integrator with an ideal differentiator in Fig. 2.15d and 2.15e, and following a similar argument, we conclude that

$$\text{if} \qquad x(t) \Longrightarrow y(t)$$

$$\text{then} \qquad \frac{dx(t)}{dt} \Longrightarrow \frac{dy(t)}{dt} \qquad\qquad (2.44b)$$

If we let $x(t) = \delta(t)$ and $y(t) = h(t)$ in Eq. (2.44a), we find that $g(t)$, the unit step response of an LTIC system with impulse $h(t)$, is given by

$$g(t) = \int_{-\infty}^{t} h(\tau)\, d\tau \qquad\qquad (2.44c)$$

We can also show that the system response to $\dot\delta(t)$ is $dh(t)/dt$. These results can be extended to other singularity functions. For example, the unit ramp response of an LTIC system is the integral of its unit step response, and so on.

INVERSE SYSTEMS

In Fig. 2.15b, if S_1 and S_2 are inverse systems with impulse response $h(t)$ and $h_i(t)$, respectively, then the impulse response of the cascade of these systems is $h(t) * h_i(t)$. But, the cascade of a system with its inverse is an identity system, whose output is the same as the input. In other words, the unit impulse response of the cascade of inverse systems is also an unit impulse $\delta(t)$. Hence

$$h(t) * h_i(t) = \delta(t) \qquad\qquad (2.45)$$

We shall give an interesting application of the commutivity property. As seen from Eq. (2.45), a cascade of inverse systems is an identity system. Moreover, in a cascade of several LTIC subsystems, changing the order of the subsystems in any manner does not affect the impulse response of the cascade system. Using these facts, we observe that the two systems, shown in Fig. 2.15f, are equivalent. We can compute the response of the cascade system on the right-hand side, by computing the response of the system inside the dotted box to the input $\dot x(t)$. The impulse response of the dotted box is $g(t)$, the integral of $h(t)$, as given in Eq. (2.44c). Hence, it follows that

$$y(t) = x(t) * h(t) = \dot x(t) * g(t) \qquad\qquad (2.46)$$

Recall that $g(t)$ is the unit step response of the system. Hence, an LTIC response can also be obtained as a convolution of $\dot{x}(t)$ (the derivative of the input) with the unit step response of the system. This result can be readily extended to higher derivatives of the input. An LTIC system response is the convolution of the nth derivative of the input with the nth integral of the impulse response.

2.4-4 A Very Special Function for LTIC Systems: The Everlasting Exponential e^{st}

There is a very special connection of LTIC systems with the everlasting exponential function e^{st}, where s is a complex variable, in general. We now show that the LTIC system's (zero-state) response to everlasting exponential input e^{st} is also the same everlasting exponential (within a multiplicative constant). Moreover, no other function can make the same claim. Such an input for which the system response is also of the same form is called the *characteristic function* (also *eigenfunction*) of the system. Because a sinusoid is a form of exponential ($s = \pm j\omega$), everlasting sinusoid is also a characteristic function of an LTIC system. Note that we are talking here of an everlasting exponential (or sinusoid), which starts at $t = -\infty$.

If $h(t)$ is the system's unit impulse response, then system response $y(t)$ to an everlasting exponential e^{st} is given by

$$y(t) = h(t) * e^{st}$$

$$= \int_{-\infty}^{\infty} h(\tau)e^{s(t-\tau)}\,d\tau$$

$$= e^{st} \int_{-\infty}^{\infty} h(\tau)e^{-s\tau}\,d\tau$$

The integral on the right-hand side is a function of a complex variable s. Let us denote it by $H(s)$, which is also complex, in general. Thus

$$y(t) = H(s)e^{st} \tag{2.47}$$

where

$$H(s) = \int_{-\infty}^{\infty} h(\tau)e^{-s\tau}\,d\tau \tag{2.48}$$

Equation (2.47) is valid only for the values of s for which $H(s)$ exists, that is, if $\int_{-\infty}^{\infty} h(\tau)e^{-s\tau}\,d\tau$ exists (or converges). The region in the s plane for which this integral converges is called the *region of convergence* for $H(s)$. Further elaboration of the region of convergence is presented in Chapter 4.

Note that $H(s)$ is a constant for a given s. Thus, the input and the output are the same (within a multiplicative constant) for the everlasting exponential signal.

$H(s)$, which is called the *transfer function* of the system, is a function of complex variable s. An alternate definition of the transfer function $H(s)$ of an LTIC system, as seen from Eq. (2.47), is

$$H(s) = \frac{\text{output signal}}{\text{input signal}}\bigg|_{\text{input=everlasting exponential } e^{st}} \tag{2.49}$$

The transfer function is defined for, and is meaningful to, LTIC systems only. It does not exist for nonlinear or time-varying systems in general.

We repeat again that in this discussion we are talking of the everlasting exponential, which starts at $t = -\infty$, not the causal exponential $e^{st}u(t)$, which starts at $t = 0$.

For a system specified by Eq. (2.1), the transfer function is given by

$$H(s) = \frac{P(s)}{Q(s)} \tag{2.50}$$

This follows readily by considering an everlasting input $x(t) = e^{st}$. According to Eq. (2.47), the output is $y(t) = H(s)e^{st}$. Substitution of this $x(t)$ and $y(t)$ in Eq. (2.1) yields

$$H(s)[Q(D)e^{st}] = P(D)e^{st}$$

Moreover

$$D^r e^{st} = \frac{d^r e^{st}}{dt^r} = s^r e^{st}$$

Hence

$$P(D)e^{st} = P(s)e^{st} \qquad \text{and} \qquad Q(D)e^{st} = Q(s)e^{st}$$

Consequently,

$$H(s) = \frac{P(s)}{Q(s)}$$

EXERCISE E2.14

Show that the transfer function of an ideal integrator is $H(s) = 1/s$ and that of an ideal differentiator is $H(s) = s$. Find the answer in two ways: using Eq. (2.48) and Eq. (2.50). [Hint: Find $h(t)$ for the ideal integrator and differentiator. You also may need to use the result in Prob. 1.4-9.]

A FUNDAMENTAL PROPERTY OF LTI SYSTEMS

We can show that Eq. (2.47) is a fundamental property of LTI systems and it follows directly as a consequence of linearity and time invariance. To show this let us assume that the response of an LTI system to an everlasting exponential e^{st} is $y(s, t)$. If we define

$$H(s, t) = \frac{y(s, t)}{e^{st}}$$

then

$$y(s, t) = H(s, t)\, e^{st}$$

Because of the time-invariance property, the system response to input $e^{s(t-T)}$ is $H(s, t-T)\, e^{s(t-T)}$, that is,

$$y(s, t - T) = H(s, t - T)\, e^{s(t-T)} \tag{2.51}$$

The delayed input $e^{s(t-T)}$ represents the input e^{st} multiplied by a constant e^{-sT}. Hence, according to the linearity property, the system response to $e^{s(t-T)}$ must be $y(s,t)\,e^{-sT}$. Hence

$$y(s, t - T) = y(s, t)\,e^{-sT}$$
$$= H(s, t)\,e^{s(t-T)}$$

Comparison of this result with Eq. (2.51) shows that

$$H(s, t) = H(s, t - T) \qquad \text{for all } T$$

This means $H(s, t)$ is independent of t, and we can express $H(s, t) = H(s)$. Hence

$$y(s, t) = H(s)\,e^{st}$$

2.4-5 Total Response

The total response of a linear system can be expressed as the sum of its zero-input and zero-state components:

$$\text{total response} = \underbrace{\sum_{k=1}^{N} c_k e^{\lambda_k t}}_{\text{zero-input component}} + \underbrace{x(t) * h(t)}_{\text{zero-state component}}$$

assuming distinct roots. For repeated roots, the zero-input component should be appropriately modified.

For the series RLC circuit in Example 2.2 with the input $x(t) = 10e^{-3t}u(t)$ and the initial conditions $y(0^-) = 0$, $v_C(0^-) = 5$, we determined the zero-input component in Example 2.1a [Eq. (2.11c)]. We found the zero-state component in Example 2.6. From the results in Examples 2.1a and 2.6, we obtain

$$\text{total current} = \underbrace{(-5e^{-t} + 5e^{-2t})}_{\text{zero-input current}} + \underbrace{(-5e^{-t} + 20e^{-2t} - 15e^{-3t})}_{\text{zero-state current}} \qquad t \geq 0 \qquad (2.52a)$$

Figure 2.16a shows the zero-input, zero-state, and total responses.

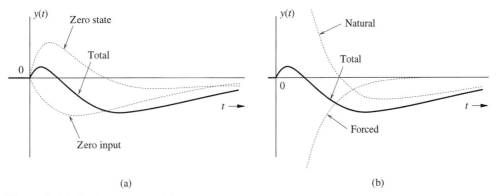

Figure 2.16 Total response and its components.

NATURAL AND FORCED RESPONSE

For the *RLC* circuit in Example 2.2, the characteristic modes were found to be e^{-t} and e^{-2t}. As we expected, the zero-input response is composed exclusively of characteristic modes. Note, however, that even the zero-state response [Eq. (2.52a)] contains characteristic mode terms. This observation is generally true of LTIC systems. We can now lump together all the characteristic mode terms in the total response, giving us a component known as the *natural response* $y_n(t)$. The remainder, consisting entirely of noncharacteristic mode terms, is known as the *forced response* $y_\phi(t)$. The total response of the *RLC* circuit in Example 2.2 can be expressed in terms of natural and forced components by regrouping the terms in Eq. (2.52a) as

$$\text{total current} = \underbrace{(-10e^{-t} + 25e^{-2t})}_{\text{natural response } y_n(t)} + \underbrace{(-15e^{-3t})}_{\text{forced response } y_\phi(t)} \qquad t \geq 0 \qquad (2.52b)$$

Figure 2.16b shows the natural, forced, and total responses.

2.5 CLASSICAL SOLUTION OF DIFFERENTIAL EQUATIONS

In the classical method we solve differential equation to find the natural and forced components rather than the zero-input and zero-state components of the response. Although this method is relatively simple compared with the method discussed so far, as we shall see, it also has several glaring drawbacks.

As Section 2.4-5 showed, when all the characteristic mode terms of the total system response are lumped together, they form the system's *natural response* $y_n(t)$ (also known as the *homogeneous solution* or *complementary solution*). The remaining portion of the response consists entirely of noncharacteristic mode terms and is called the system's *forced response* $y_\phi(t)$ (also known as the *particular solution*). Equation (2.52b) showed these two components for the loop current in the *RLC* circuit of Fig. 2.1a.

The total system response is $y(t) = y_n(t) + y_\phi(t)$. Since $y(t)$ must satisfy the system equation [Eq. (2.1)],

$$Q(D)[y_n(t) + y_\phi(t)] = P(D)x(t)$$

or

$$Q(D)y_n(t) + Q(D)y_\phi(t) = P(D)x(t)$$

But $y_n(t)$ is composed entirely of characteristic modes. Therefore

$$Q(D)y_n(t) = 0$$

so that

$$Q(D)y_\phi(t) = P(D)x(t) \qquad (2.53)$$

The natural response, being a linear combination of the system's characteristic modes, has the same form as that of the zero-input response; only its arbitrary constants are different. These constants are determined from auxiliary conditions, as explained later. We shall now discuss a method of determining the forced response.

2.5-1 Forced Response: The Method of Undetermined Coefficients

It is a relatively simple task to determine $y_\phi(t)$, the forced response of an LTIC system, when the input $x(t)$ is such that it yields only a finite number of independent derivatives. Inputs having the form $e^{\zeta t}$ or t^r fall into this category. For example, the repeated differentiation of $e^{\zeta t}$ yields the same form as this input; that is, $e^{\zeta t}$. Similarly, the repeated differentiation of t^r yields only r independent derivatives. The forced response to such an input can be expressed as a linear combination of the input and its independent derivatives. Consider, for example, the input $at^2 + bt + c$. The successive derivatives of this input are $2at + b$ and $2a$. In this case, the input has only two independent derivatives. Therefore, the forced response can be assumed to be a linear combination of $x(t)$ and its two derivatives. The suitable form for $y_\phi(t)$ in this case is, therefore

$$y_\phi(t) = \beta_2 t^2 + \beta_1 t + \beta_0$$

The undetermined coefficients β_0, β_1, and β_2 are determined by substituting this expression for $y_\phi(t)$ in Eq. (2.53)

$$Q(D)y_\phi(t) = P(D)x(t)$$

and then equating coefficients of similar terms on both sides of the resulting expression.

Although this method can be used only for inputs with a finite number of derivatives, this class of inputs includes a wide variety of the most commonly encountered signals in practice. Table 2.2 shows a variety of such inputs and the form of the forced response corresponding to each input. We shall demonstrate this procedure with an example.

TABLE 2.2   Forced Response

No.	Input $x(t)$	Forced Response
1	$e^{\zeta t}$ $\zeta \neq \lambda_i$ $(i = 1, 2, \ldots, N)$	$\beta e^{\zeta t}$
2	$e^{\zeta t}$ $\zeta = \lambda_i$	$\beta t e^{\zeta t}$
3	k (a constant)	β (a constant)
4	$\cos(\omega t + \theta)$	$\beta \cos(\omega t + \phi)$
5	$(t^r + \alpha_{r-1}t^{r-1} + \cdots + \alpha_1 t + \alpha_0)e^{\zeta t}$	$(\beta_r t^r + \beta_{r-1}t^{r-1} + \cdots + \beta_1 t + \beta_0)e^{\zeta t}$

Note: By definition, $y_\phi(t)$ cannot have any characteristic mode terms. If any term appearing in the right-hand column for the forced response is also a characteristic mode of the system, the correct form of the forced response must be modified to $t^i y_\phi(t)$, where i is the smallest possible integer that can be used and still can prevent $t^i y_\phi(t)$ from having a characteristic mode term. For example, when the input is $e^{\zeta t}$, the forced response (right-hand column) has the form $\beta e^{\zeta t}$. But if $e^{\zeta t}$ happens to be a characteristic mode of the system, the correct form of the forced response is $\beta t e^{\zeta t}$ (see pair 2). If $t e^{\zeta t}$ also happens to be a characteristic mode of the system, the correct form of the forced response is $\beta t^2 e^{\zeta t}$, and so on.

EXAMPLE 2.10

Solve the differential equation

$$(D^2 + 3D + 2)y(t) = Dx(t)$$

if the input

$$x(t) = t^2 + 5t + 3$$

and the initial conditions are $y(0^+) = 2$ and $\dot{y}(0^+) = 3$.

The characteristic polynomial of the system is

$$\lambda^2 + 3\lambda + 2 = (\lambda + 1)(\lambda + 2)$$

Therefore, the characteristic modes are e^{-t} and e^{-2t}. The natural response is then a linear combination of these modes, so that

$$y_n(t) = K_1 e^{-t} + K_2 e^{-2t} \qquad t \geq 0$$

Here the arbitrary constants K_1 and K_2 must be determined from the system's initial conditions.

The forced response to the input $t^2 + 5t + 3$, according to Table 2.2 (pair 5 with $\zeta = 0$), is

$$y_\phi(t) = \beta_2 t^2 + \beta_1 t + \beta_0$$

Moreover, $y_\phi(t)$ satisfies the system equation [Eq. (2.53)]; that is,

$$(D^2 + 3D + 2)y_\phi(t) = Dx(t) \tag{2.54}$$

Now

$$Dy_\phi(t) = \frac{d}{dt}(\beta_2 t^2 + \beta_1 t + \beta_0) = 2\beta_2 t + \beta_1$$

$$D^2 y_\phi(t) = \frac{d^2}{dt^2}(\beta_2 t^2 + \beta_1 t + \beta_0) = 2\beta_2$$

and

$$Dx(t) = \frac{d}{dt}[t^2 + 5t + 3] = 2t + 5$$

Substituting these results in Eq. (2.54) yields

$$2\beta_2 + 3(2\beta_2 t + \beta_1) + 2(\beta_2 t^2 + \beta_1 t + \beta_0) = 2t + 5$$

or

$$2\beta_2 t^2 + (2\beta_1 + 6\beta_2)t + (2\beta_0 + 3\beta_1 + 2\beta_2) = 2t + 5$$

Equating coefficients of similar powers on both sides of this expression yields

$$2\beta_2 = 0$$

$$2\beta_1 + 6\beta_2 = 2$$

$$2\beta_0 + 3\beta_1 + 2\beta_2 = 5$$

Solution of these three simultaneous equations yields $\beta_0 = 1$, $\beta_1 = 1$, and $\beta_2 = 0$. Therefore

$$y_\phi(t) = t + 1 \qquad t \geq 0$$

The total system response $y(t)$ is the sum of the natural and forced solutions. Therefore

$$y(t) = y_n(t) + y_\phi(t)$$
$$= K_1 e^{-t} + K_2 e^{-2t} + t + 1 \qquad t \geq 0$$

so that

$$\dot{y}(t) = -K_1 e^{-t} - 2K_2 e^{-2t} + 1$$

Setting $t = 0$ and substituting $y(0) = 2$ and $\dot{y}(0) = 3$ in these equations, we have

$$2 = K_1 + K_2 + 1$$
$$3 = -K_1 - 2K_2 + 1$$

The solution of these two simultaneous equations is $K_1 = 4$ and $K_2 = -3$. Therefore

$$y(t) = 4e^{-t} - 3e^{-2t} + t + 1 \qquad t \geq 0$$

COMMENTS ON INITIAL CONDITIONS

In the classical method, the initial conditions are required at $t = 0^+$. The reason is that because at $t = 0^-$, only the zero-input component exists, and the initial conditions at $t = 0^-$ can be applied to the zero-input component only. In the classical method, the zero-input and zero-state components cannot be separated. Consequently, the initial conditions must be applied to the total response, which begins at $t = 0^+$.

EXERCISE E2.15

An LTIC system is specified by the equation

$$(D^2 + 5D + 6)y(t) = (D + 1)x(t)$$

The input is $x(t) = 6t^2$. Find the following:

(a) The forced response $y_\phi(t)$

(b) The total response $y(t)$ if the initial conditions are $y(0^+) = 25/18$ and $\dot{y}(0^+) = -2/3$

ANSWERS
(a) $y_\phi(t) = t^2 + \frac{1}{3}t - \frac{11}{18}$

(b) $y(t) = 5e^{-2t} - 3e^{-3t} + \left(t^2 + \frac{1}{3}t - \frac{11}{18}\right)$

The Exponential Input $e^{\zeta t}$

The exponential signal is the most important signal in the study of LTIC systems. Interestingly, the forced response for an exponential input signal turns out to be very simple. From Table 2.2 we see that the forced response for the input $e^{\zeta t}$ has the form $\beta e^{\zeta t}$. We now show that $\beta = Q(\zeta)/P(\zeta)$.[†] To determine the constant β, we substitute $y_\phi(t) = \beta e^{\zeta t}$ in the system equation [Eq. (2.53)] to obtain

$$Q(D)[\beta e^{\zeta t}] = P(D)e^{\zeta t}$$

Now observe that

$$De^{\zeta t} = \frac{d}{dt}(e^{\zeta t}) = \zeta e^{\zeta t}$$

$$D^2 e^{\zeta t} = \frac{d^2}{dt^2}(e^{\zeta t}) = \zeta^2 e^{\zeta t}$$

$$\vdots$$

$$D^r e^{\zeta t} = \zeta^r e^{\zeta t}$$

Consequently

$$Q(D)e^{\zeta t} = Q(\zeta)e^{\zeta t} \qquad \text{and} \qquad P(D)e^{\zeta t} = P(\zeta)e^{\zeta t}$$

Therefore, Eq. (2.53) becomes

$$\beta Q(\zeta)e^{\zeta t} = P(\zeta)e^{\zeta t}$$

and

$$\beta = \frac{P(\zeta)}{Q(\zeta)}$$

Thus, for the input $x(t) = e^{\zeta t}u(t)$, the forced response is given by

$$y_\phi(t) = H(\zeta)e^{\zeta t} \qquad t \geq 0 \tag{2.55}$$

where

$$H(\zeta) = \frac{P(\zeta)}{Q(\zeta)} \tag{2.56}$$

This is an interesting and significant result. It states that for an exponential input $e^{\zeta t}$ the forced response $y_\phi(t)$ is the same exponential multiplied by $H(\zeta) = P(\zeta)/Q(\zeta)$. The total system response $y(t)$ to an exponential input $e^{\zeta t}$ is then given by[‡]

$$y(t) = \sum_{j=1}^{N} K_j e^{\lambda_j t} + H(\zeta)e^{\zeta t} \tag{2.57}$$

[†]This result is valid only if ζ is not a characteristic root of the system.
[‡]Observe the closeness of Eqs. (2.57) and (2.47). Why is there a difference between the two equations? Equation (2.47) is the response to an exponential that started at $-\infty$, while Eq. (2.57) is the response to an exponential that starts at $t = 0$. As $t \to \infty$, Eq. (2.57) approaches Eq. (2.47). In Eq. (2.47), the term $y_n(t)$, which starts at $t = -\infty$, has already decayed at $t = 0$, and hence, is missing.

where the arbitrary constants $K_1, K_2, \ldots, K_N$ are determined from auxiliary conditions. The form of Eq. (2.57) assumes N distinct roots. If the roots are not distinct, proper form of modes should be used.

Recall that the exponential signal includes a large variety of signals, such as a constant ($\zeta = 0$), a sinusoid ($\zeta = \pm j\omega$), and an exponentially growing or decaying sinusoid ($\zeta = \sigma \pm j\omega$). Let us consider the forced response for some of these cases.

THE CONSTANT INPUT $x(t) = C$

Because $C = Ce^{0t}$, the constant input is a special case of the exponential input $Ce^{\zeta t}$ with $\zeta = 0$. The forced response to this input is then given by

$$y_\phi(t) = CH(\zeta)e^{\zeta t} \qquad \text{with } \zeta = 0$$
$$= CH(0) \qquad\qquad\qquad\qquad (2.58)$$

THE EXPONENTIAL INPUT $e^{j\omega t}$

Here $\zeta = j\omega$ and

$$y_\phi(t) = H(j\omega)e^{j\omega t} \qquad\qquad\qquad (2.59)$$

THE SINUSOIDAL INPUT $x(t) = \cos \omega t$

We know that the forced response for the input $e^{\pm j\omega t}$ is $H(\pm j\omega)e^{\pm j\omega t}$. Since $\cos \omega t = (e^{j\omega t} + e^{-j\omega t})/2$, the forced response to $\cos \omega t$ is

$$y_\phi(t) = \tfrac{1}{2}[H(j\omega)e^{j\omega t} + H(-j\omega)e^{-j\omega t}]$$

Because the two terms on the right-hand side are conjugates,

$$y_\phi(t) = \text{Re}[H(j\omega)e^{j\omega t}]$$

But

$$H(j\omega) = |H(j\omega)|e^{j\angle H(j\omega)}$$

so that

$$y_\phi(t) = \text{Re}\{|H(j\omega)|e^{j[\omega t + \angle H(j\omega)]}\}$$
$$= |H(j\omega)| \cos [\omega t + \angle H(j\omega)] \qquad\qquad (2.60)$$

This result can be generalized for the input $x(t) = \cos (\omega t + \theta)$. The forced response in this case is

$$y_\phi(t) = |H(j\omega)| \cos [\omega t + \theta + \angle H(j\omega)] \qquad\qquad (2.61)$$

EXAMPLE 2.11

Solve the differential equation

$$(D^2 + 3D + 2)y(t) = Dx(t)$$

if the initial conditions are $y(0^+) = 2$ and $\dot{y}(0^+) = 3$ and the input is

(a) $10e^{-3t}$

(b) 5

(c) e^{-2t}

(d) $10\cos(3t + 30°)$

According to Example 2.10, the natural response for this case is

$$y_n(t) = K_1 e^{-t} + K_2 e^{-2t}$$

For this case

$$H(\zeta) = \frac{P(\zeta)}{Q(\zeta)} = \frac{\zeta}{\zeta^2 + 3\zeta + 2}$$

(a) For input $x(t) = 10e^{-3t}$, $\zeta = -3$, and

$$y_\phi(t) = 10H(-3)e^{-3t} = 10\left[\frac{-3}{(-3)^2 + 3(-3) + 2}\right]e^{-3t} = -15e^{-3t} \qquad t \geq 0$$

The total solution (the sum of the forced and the natural response) is

$$y(t) = K_1 e^{-t} + K_2 e^{-2t} - 15e^{-3t} \qquad t \geq 0$$

and

$$\dot{y}(t) = -K_1 e^{-t} - 2K_2 e^{-2t} + 45e^{-3t} \qquad t \geq 0$$

The initial conditions are $y(0^+) = 2$ and $\dot{y}(0^+) = 3$. Setting $t = 0$ in the foregoing equations and then substituting the initial conditions yields

$$K_1 + K_2 - 15 = 2 \qquad \text{and} \qquad -K_1 - 2K_2 + 45 = 3$$

Solution of these equations yields $K_1 = -8$ and $K_2 = 25$. Therefore

$$y(t) = -8e^{-t} + 25e^{-2t} - 15e^{-3t} \qquad t \geq 0$$

(b) For input $x(t) = 5 = 5e^{0t}$, $\zeta = 0$, and

$$y_\phi(t) = 5H(0) = 0 \qquad t \geq 0$$

The complete solution is $K_1 e^{-t} + K_2 e^{-2t}$. Using the initial conditions, we determine K_1 and K_2 as in part (a).

(c) Here $\zeta = -2$, which is also a characteristic root of the system. Hence (see pair 2, Table 2.2, or the note at the bottom of the table),

$$y_\phi(t) = \beta t e^{-2t}$$

To find β, we substitute $y_\phi(t)$ in the system equation to obtain

$$(D^2 + 3D + 2)y_\phi(t) = Dx(t)$$

or

$$(D^2 + 3D + 2)[\beta t e^{-2t}] = De^{-2t}$$

But

$$D[\beta t e^{-2t}] = \beta(1 - 2t)e^{-2t}$$
$$D^2[\beta t e^{-2t}] = 4\beta(t - 1)e^{-2t}$$
$$De^{-2t} = -2e^{-2t}$$

Consequently

$$\beta(4t - 4 + 3 - 6t + 2t)e^{-2t} = -2e^{-2t}$$

or

$$-\beta e^{-2t} = -2e^{-2t}$$

Therefore, $\beta = 2$ so that

$$y_\phi(t) = 2te^{-2t}$$

The complete solution is $K_1 e^{-t} + K_2 e^{-2t} + 2te^{-2t}$. Using the initial conditions, we determine K_1 and K_2 as in part (a).

(d) For the input $x(t) = 10\cos(3t + 30°)$, the forced response [see Eq. (2.61)] is

$$y_\phi(t) = 10|H(j3)|\cos[3t + 30° + \angle H(j3)]$$

where

$$H(j3) = \frac{P(j3)}{Q(j3)} = \frac{j3}{(j3)^2 + 3(j3) + 2} = \frac{j3}{-7 + j9} = \frac{27 - j21}{130} = 0.263e^{-j37.9°}$$

Therefore

$$|H(j3)| = 0.263$$
$$\angle H(j3) = -37.9°$$

and

$$y_\phi(t) = 10(0.263)\cos(3t + 30° - 37.9°) = 2.63\cos(3t - 7.9°)$$

The complete solution is $K_1 e^{-t} + K_2 e^{-2t} + 2.63\cos(3t - 7.9°)$. We then use the initial conditions to determine K_1 and K_2 as in part (a).

EXAMPLE 2.12

Use the classical method to find the loop current $y(t)$ in the RLC circuit of Example 2.2 (Fig. 2.1) if the input voltage $x(t) = 10e^{-3t}$ and the initial conditions are $y(0^-) = 0$ and $v_C(0^-) = 5$.

The zero-input and zero-state responses for this problem are found in Examples 2.2 and 2.6, respectively. The natural and forced responses appear in Eq. (2.52b). Here we shall solve this problem by the classical method, which requires the initial conditions at $t = 0^+$. These conditions, already found in Eq. (2.15), are

$$y(0^+) = 0 \quad \text{and} \quad \dot{y}(0^+) = 5$$

The loop equation for this system [see Example 2.2 or Eq. (1.55)] is

$$(D^2 + 3D + 2)y(t) = Dx(t)$$

The characteristic polynomial is $\lambda^2 + 3\lambda + 2 = (\lambda + 1)(\lambda + 2)$. Therefore, the natural response is

$$y_n(t) = K_1 e^{-t} + K_2 e^{-2t}$$

The forced response, already found in part **(a)** of Example 2.11, is

$$y_\phi(t) = -15e^{-3t}$$

The total response is

$$y(t) = K_1 e^{-t} + K_2 e^{-2t} - 15e^{-3t}$$

Differentiation of this equation yields

$$\dot{y}(t) = -K_1 e^{-t} - 2K_2 e^{-2t} + 45e^{-3t}$$

Setting $t = 0^+$ and substituting $y(0^+) = 0$, $\dot{y}(0^+) = 5$ in these equations yields

$$\left.\begin{aligned} 0 &= K_1 + K_2 - 15 \\ 5 &= -K_1 - 2K_2 + 45 \end{aligned}\right\} \implies \begin{aligned} K_1 &= -10 \\ K_2 &= 25 \end{aligned}$$

Therefore

$$y(t) = -10e^{-t} + 25e^{-2t} - 15e^{-3t}$$

which agrees with the solution found earlier in Eq. (2.52b).

COMPUTER EXAMPLE C2.4

Solve the differential equation

$$(D^2 + 3D + 2)y(t) = x(t)$$

using the input $x(t) = 5t + 3$ and initial conditions $y_0(0) = 2$ and $\dot{y}_0(0) = 3$.

```
>> y = dsolve('D2y+3*Dy+2*y=5*t+3','y(0)=2','Dy(0)=3','t');
>> disp(['y(t) = (',char(y),')u(t)']);
y(t) = (-9/4+5/2*t+9*exp(-t)-19/4*exp(-2*t))u(t)
```

Therefore,

$$y(t) = \left(-\tfrac{9}{4} + \tfrac{5}{2}t + 9e^{-t} - \tfrac{19}{4}e^{-2t}\right)u(t)$$

ASSESSMENT OF THE CLASSICAL METHOD

The development in this section shows that the classical method is relatively simple compared with the method of finding the response as a sum of the zero-input and zero-state components. Unfortunately, the classical method has a serious drawback because it yields the total response, which cannot be separated into components arising from the internal conditions and the external input. In the study of systems it is important to be able to express the system response to an input $x(t)$ as an explicit function of $x(t)$. This is not possible in the classical method. Moreover, the classical method is restricted to a certain class of inputs; it cannot be applied to any input. Another minor problem is that because the classical method yields total response, the auxiliary conditions must be on the total response which exists only for $t \geq 0^+$. In practice we are most likely to know the conditions at $t = 0^-$ (before the input is applied). Therefore, we need to derive a new set of auxiliary conditions at $t = 0^+$ from the known conditions at $t = 0^-$.

   If we must solve a particular linear differential equation or find a response of a particular LTIC system, the classical method may be the best. In the theoretical study of linear systems, however, the classical method is not so valuable.

Caution. We have shown in Eq. (2.52a) that the total response of an LTI system can be expressed as a sum of the zero-input and the zero-state components. In Eq. (2.52b), we showed that the same response can also be expressed as a sum of the natural and the forced components. We have also seen that generally the zero-input response is not the same as the natural response (although both are made of natural modes). Similarly, the zero-state response is not the same as the forced response. Unfortunately, such erroneous claims are often encountered in the literature.

2.6 SYSTEM STABILITY

To understand the intuitive basis for the BIBO (bounded-input/bounded-output) stability of a system introduced in Section 1.7, let us examine the stability concept as applied to a right circular cone. Such a cone can be made to stand forever on its circular base, on its apex, or on its side. For this reason, these three states of the cone are said to be *equilibrium states*. Qualitatively, however, the three states show very different behavior. If the cone, standing on its circular base, were to be disturbed slightly and then left to itself, it would eventually return to its original equilibrium position. In such a case, the cone is said to be in *stable equilibrium*. In contrast, if the cone stands on its apex, then the slightest disturbance will cause the cone to move farther and farther away from its equilibrium state. The cone in this case is said to be in an *unstable equilibrium*. The cone lying on its side, if disturbed, will neither go back to the original state nor continue to move farther away from the original state. Thus it is said to be in a *neutral equilibrium*. Clearly, when a system is in stable equilibrium, application of a small disturbance (input) produces a small response. In contrast, when the system is in unstable equilibrium, even a minuscule disturbance (input) produces an unbounded response. BIBO stability definition can be understood in the light of this concept. If every bounded input produces bounded output, the system is (BIBO) stable.[†] In contrast, if even one bounded input results in unbounded response, the system is (BIBO) unstable.

For an LTIC system

$$y(t) = h(t) * x(t)$$

$$= \int_{-\infty}^{\infty} h(\tau)x(t - \tau)\,d\tau \tag{2.62}$$

Therefore

$$|y(t)| \leq \int_{-\infty}^{\infty} |h(\tau)||x(t - \tau)|\,d\tau$$

Moreover, if $x(t)$ is bounded, then $|x(t - \tau)| < K_1 < \infty$, and

$$|y(t)| \leq K_1 \int_{-\infty}^{\infty} |h(\tau)|\,d\tau \tag{2.63}$$

Hence for BIBO stability

$$\int_{-\infty}^{\infty} |h(\tau)|\,d\tau < \infty \tag{2.64}$$

This is a sufficient condition for BIBO stability. We can show that this is also a necessary condition (see Prob. 2.6-4). Therefore, for an LTIC system, if its impulse response $h(t)$ is absolutely integrable, the system is (BIBO) stable. Otherwise it is (BIBO) unstable. In addition, we shall show in Chapter 4 that a necessary (but not sufficient) condition for an LTIC system described by Eq. (2.1) to be BIBO stable is $M \leq N$. If $M > N$, the system is unstable. This is one of the reasons to avoid systems with $M > N$.

Because the BIBO stability of a system can be ascertained by measurements at the external terminals (input and output), this is an external stability criterion. It is no coincidence that the BIBO criterion in (2.64) is in terms of the impulse response, which is an external description of the system.

[†]The system is assumed to be in zero state.

As observed in Section 1.9, the internal behavior of a system is not always ascertainable from the external terminals. Therefore, the external (BIBO) stability may not be correct indication of the internal stability. Indeed, some systems that appear stable by the BIBO criterion may be internally unstable. This is like a room inside a house on fire: no trace of fire is visible from outside, but the entire house will be burned to ashes.

The BIBO stability is meaningful only for systems in which the internal and the external description are equivalent (controllable and observable systems). Fortunately, most practical systems fall into this category, and whenever we apply this criterion, we implicitly assume that the system in fact belongs to this category. The internal stability is all-inclusive, and the external stability can always be determined from the internal stability. For this reason, we now investigate the internal stability criterion.

2.6-1 Internal (Asymptotic) Stability

Because of the great variety of possible system behaviors, there are several definitions of internal stability in the literature. Here we shall consider a definition that is suitable for causal, linear, time-invariant (LTI) systems.

If, in the absence of an external input, a system remains in a particular state (or condition) indefinitely, then that state is said to be an *equilibrium state* of the system. For an LTI system, zero state, in which all initial conditions are zero, is an equilibrium state. Now suppose an LTI system is in zero state and we change this state by creating small nonzero initial conditions (small disturbance). These initial conditions will generate signals consisting of characteristic modes in the system. By analogy with the cone, if the system is stable it should eventually return to zero state. In other words, when left to itself, every mode in a stable system arising as a result of nonzero initial conditions should approach 0 as $t \to \infty$. However, if even one of the modes grows with time, the system will never return to zero state, and the system would be identified as unstable. In the borderline case, some modes neither decay to zero nor grow indefinitely, while all the remaining modes decay to zero. This case is like the neutral equilibrium in the cone. Such a system is said to be *marginally* stable. The internal stability is also called *asymptotic* stability or stability in the sense of *Lyapunov*.

For a system characterized by Eq. (2.1), we can restate the internal stability criterion in terms of the location of the N characteristic roots $\lambda_1, \lambda_2, \ldots, \lambda_N$ of the system in a complex plane. The characteristic modes are of the form $e^{\lambda_k t}$ or $t^r e^{\lambda_k t}$. Location of roots in the complex plane and the corresponding modes are shown in Fig. 2.17. These modes $\to 0$ as $t \to \infty$ if Re $\lambda_k < 0$. In contrast, the modes $\to \infty$ as $t \to \infty$ if Re $\lambda_k > 0$.[†] Hence, a system is (asymptotically) stable if all its characteristic roots lie in the LHP, that is, if Re $\lambda_k < 0$ for all k. If even a single characteristic root lies in the RHP, the system is (asymptotically) unstable. Modes due to roots on

---

[†]This may be seen from the fact that if α and β are the real and the imaginary parts of a root λ, then

$$\lim_{t \to \infty} e^{\lambda t} = \lim_{t \to \infty} e^{(\alpha + j\beta)t} = \lim_{t \to \infty} e^{\alpha t} \, e^{j\beta t} = \begin{cases} 0 & \alpha < 0 \\ \infty & \alpha > 0 \end{cases}$$

This conclusion is also valid for the terms of the form $t^r e^{\lambda t}$.

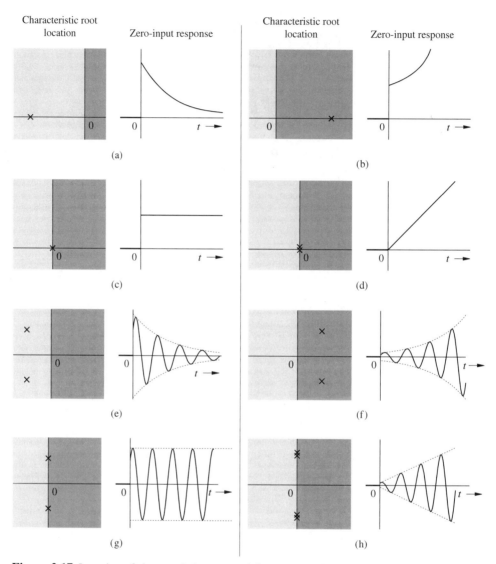

Figure 2.17 Location of characteristic roots and the corresponding characteristic modes.

the imaginary axis ($\lambda = \pm j\omega_0$) are of the form $e^{\pm j\omega_0 t}$. Hence, if some roots are on the imaginary axis, and all the remaining roots are in the LHP, the system is marginally stable (assuming that the roots on the imaginary axis are not repeated). If the imaginary axis roots are repeated, the characteristic modes are of the form $t^r e^{\pm j\omega_k t}$, which *do* grow with time indefinitely. Hence, the system is unstable. Figure 2.18 shows stability regions in the complex plane.

To summarize:

1. An LTIC system is asymptotically stable if, and only if, all the characteristic roots are in the LHP. The roots may be simple (unrepeated) or repeated.

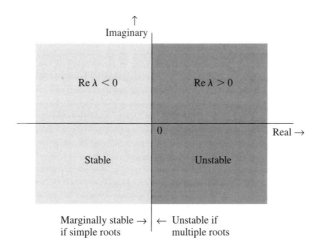

Figure 2.18 Characteristic roots location and system stability.

Marginally stable → | ← Unstable if
if simple roots | multiple roots

2. An LTIC system is unstable if, and only if, one or both of the following conditions exist: (i) at least one root is in the RHP; (ii) there are repeated roots on the imaginary axis.

3. An LTIC system is marginally stable if, and only if, there are no roots in the RHP, and there are some unrepeated roots on the imaginary axis.

2.6-2 Relationship Between BIBO and Asymptotic Stability

The external stability is determined by applying the external input with zero initial conditions, while the internal stability is determined by applying the nonzero initial conditions and no external input. This is why these stabilities are also called the *zero-state stability* and the *zero-input stability*, respectively.

Recall that $h(t)$, the impulse response of an LTIC system, is a linear combination of the system characteristic modes. For an LTIC system, specified by Eq. (2.1), we can readily show that when a characteristic root λ_k is in the LHP, the corresponding mode $e^{\lambda_k t}$ is absolutely integrable. In contrast, if λ_k is in the RHP or on the imaginary axis, $e^{\lambda_k t}$ is not absolutely integrable.[†] This means that an asymptotically stable system is BIBO stable. Moreover, a marginally stable or asymptotically unstable system is BIBO unstable. The converse is not necessarily true; that is, BIBO stability does not necessarily inform us about the internal stability of the system. For instance, if a system is uncontrollable and/or unobservable, some modes of the system are invisible and/or uncontrollable from the external terminals.[3] Hence, the stability picture portrayed by the external description is of questionable value. BIBO (external) stability cannot assure internal (asymptotic) stability, as the following example shows.

[†]Consider a mode of the form $e^{\lambda t}$, where $\lambda = \alpha + j\beta$. Hence, $e^{\lambda t} = e^{\alpha t} e^{j\beta t}$ and $|e^{\lambda t}| = e^{\alpha t}$. Therefore

$$\int_{-\infty}^{\infty} |e^{\lambda \tau} u(\tau)| \, d\tau = \int_{0}^{\infty} e^{\alpha \tau} \, d\tau = \begin{cases} -1/\alpha & \alpha < 0 \\ \infty & \alpha \geq 0 \end{cases}$$

This conclusion is also valid when the integrand is of the form $|t^k e^{\lambda t} u(t)|$.

EXAMPLE 2.13

An LTID systems consists of two subsystems S_1 and S_2 in cascade (Fig. 2.19). The impulse response of these systems are $h_1(t)$ and $h_2(t)$, respectively, given by

$$h_1(t) = \delta(t) - 2e^{-t}u(t) \qquad \text{and} \qquad h_2(t) = e^t u(t)$$

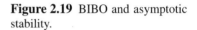

Figure 2.19 BIBO and asymptotic stability.

The composite system impulse response $h(t)$ is given by

$$h(t) = h_1(t) * h_2(t) = h_2(t) * h_1(t) = e^t u(t) * [\delta(t) - 2e^{-t}u(t)]$$

$$= e^t u(t) - 2\left[\frac{e^t - e^{-t}}{2}\right]u(t)$$

$$= e^{-t}u(t)$$

If the composite cascade system were to be enclosed in a black box with only the input and the output terminals accessible, any measurement from these external terminals would show that the impulse response of the system is $e^{-t}u(t)$, without any hint of the dangerously unstable system the system is harboring within.

The composite system is BIBO stable because its impulse response, $e^{-t}u(t)$, is absolutely integrable. Observe, however, the subsystem S_2 has a characteristic root 1, which lies in the RHP. Hence, S_2 is asymptotically unstable. Eventually, S_2 will burn out (or saturate) because of the unbounded characteristic response generated by intended or unintended initial conditions, no matter how small. We shall show in Example 10.11 that this composite system is observable, but not controllable. If the positions of S_1 and S_2 were interchanged (S_2 followed by S_1), the system is still BIBO stable, but asymptotically unstable. In this case, the analysis in Example 10.11 shows that the composite system is controllable, but not observable.

This example shows that BIBO stability does not always imply asymptotic stability. However, asymptotic stability always implies BIBO stability.

Fortunately, uncontrollable and/or unobservable systems are not commonly observed in practice. Henceforth, in determining system stability, we shall assume that unless otherwise mentioned, the internal and the external descriptions of a system are equivalent, implying that the system is controllable and observable.

EXAMPLE 2.14

Investigate the asymptotic and the BIBO stability of LTIC system described by the following equations, assuming that the equations are internal system description.

    (a) $(D+1)(D^2+4D+8)y(t) = (D-3)x(t)$

    (b) $(D-1)(D^2+4D+8)y(t) = (D+2)x(t)$

    (c) $(D+2)(D^2+4)y(t) = (D^2+D+1)x(t)$

    (d) $(D+1)(D^2+4)^2 y(t) = (D^2+2D+8)x(t)$

The characteristic polynomials of these systems are

    (a) $(\lambda+1)(\lambda^2+4\lambda+8) = (\lambda+1)(\lambda+2-j2)(\lambda+2+j2)$

    (b) $(\lambda-1)(\lambda^2+4\lambda+8) = (\lambda-1)(\lambda+2-j2)(\lambda+2+j2)$

    (c) $(\lambda+2)(\lambda^2+4) = (\lambda+2)(\lambda-j2)(\lambda+j2)$

    (d) $(\lambda+1)(\lambda^2+4)^2 = (\lambda+2)(\lambda-j2)^2(\lambda+j2)^2$

Consequently, the characteristic roots of the systems are (see Fig. 2.20):

    (a) $-1, -2 \pm j2$

    (b) $1, -2 \pm j2$

    (c) $-2, \pm j2$

    (d) $-1, \pm j2, \pm j2$

    System (a) is asymptotically stable (all roots in LHP), system (b) is unstable (one root in RHP), system (c) is marginally stable (unrepeated roots on imaginary axis) and no roots in RHP, and system (d) is unstable (repeated roots on the imaginary axis). BIBO stability is readily determined from the asymptotic stability. System (a) is BIBO stable, system (b) is BIBO unstable, system (c) is BIBO unstable, and system (d) is BIBO unstable. We have assumed that these systems are controllable and observable.

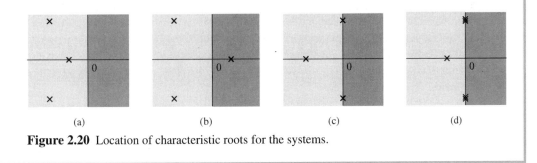

Figure 2.20 Location of characteristic roots for the systems.

EXERCISE E2.16

For each of the systems specified by the equations that follow, plot its characteristic roots in the complex plane and determine whether it is asymptotically stable, marginally stable, or unstable, assuming that the equations describe its internal system. Also, determine BIBO stability for each system.

    (a) $D(D+2)y(t) = 3x(t)$

    (b) $D^2(D+3)y(t) = (D+5)x(t)$

    (c) $(D+1)(D+2)y(t) = (2D+3)x(t)$

    (d) $(D^2+1)(D^2+9)y(t) = (D^2+2D+4)x(t)$

    (e) $(D+1)(D^2-4D+9)y(t) = (D+7)x(t)$

ANSWERS
(a) Marginally stable, but BIBO unstable

(b) Unstable in both senses

(c) Stable in both senses

(d) Marginally stable, but BIBO unstable

(e) Unstable in both senses.

IMPLICATIONS OF STABILITY

All practical signal processing systems must be asymptotically stable. Unstable systems are useless from the viewpoint of signal processing because any set of intended or unintended initial conditions leads to an unbounded response that either destroys the system or (more likely) leads it to some saturation conditions that change the nature of the system. Even if the discernible initial conditions are zero, stray voltages or thermal noise signals generated within the system will act as initial conditions. Because of exponential growth of a mode or modes in unstable systems, a stray signal, no matter how small, will eventually cause an unbounded output.

    Marginally stable systems, though BIBO unstable, do have one important application in the oscillator, which is a system that generates a signal on its own without the application of an external input. Consequently, the oscillator output is a zero-input response. If such a response is to be a sinusoid of frequency ω_0, the system should be marginally stable with characteristic roots at $\pm j\omega_0$. Thus, to design an oscillator of frequency ω_0, we should pick a system with the characteristic polynomial $(\lambda - j\omega_0)(\lambda + j\omega_0) = \lambda^2 + \omega_0^2$. A system described by the differential equation

$$\left(D^2 + \omega_0^2\right)y(t) = x(t) \tag{2.65}$$

will do the job. However, practical oscillators are invariably realized using nonlinear systems.

2.7 Intuitive Insights into System Behavior

This section attempts to provide an understanding of what determines system behavior. Because of its intuitive nature, the discussion is more or less qualitative. We shall now show that the most important attributes of a system are its characteristic roots or characteristic modes because they determine not only the zero-input response but also the entire behavior of the system.

2.7-1 Dependence of System Behavior on Characteristic Modes

Recall that the zero-input response of a system consists of the system's characteristic modes. For a stable system, these characteristic modes decay exponentially and eventually vanish. This behavior may give the impression that these modes do not substantially affect system behavior in general and system response in particular. This impression is totally wrong! We shall now see that the system's characteristic modes leave their imprint on every aspect of the system behavior. *We may compare the system's characteristic modes (or roots) to a seed that eventually dissolves in the ground; however, the plant that springs from it is totally determined by the seed. The imprint of the seed exists on every cell of the plant.*

To understand this interesting phenomenon, recall that the characteristic modes of a system are very special to that system because it can sustain these signals without the application of an external input. In other words, the system offers a free ride and ready access to these signals. Now imagine what would happen if we actually drove the system with an input having the form of a characteristic mode! We would expect the system to respond strongly (this is, in fact, the resonance phenomenon discussed later in this section). If the input is not exactly a characteristic mode but is close to such a mode, we would still expect the system response to be strong. However, if the input is very different from any of the characteristic modes, we would expect the system to respond poorly. We shall now show that these intuitive deductions are indeed true.

Intuition can cut the math jungle instantly!

Although we have devised a measure of similarity of signals later, in Chapter 6, we shall take a simpler approach here. Let us restrict the system's inputs to exponentials of the form $e^{\zeta t}$, where ζ is generally a complex number. The similarity of two exponential signals $e^{\zeta t}$ and $e^{\lambda t}$ will then be measured by the closeness of ζ and λ. If the difference $\zeta - \lambda$ is small, the signals are similar; if $\zeta - \lambda$ is large, the signals are dissimilar.

Now consider a first-order system with a single characteristic mode $e^{\lambda t}$ and the input $e^{\zeta t}$. The impulse response of this system is then given by $Ae^{\lambda t}$, where the exact value of A is not important for this qualitative discussion. The system response $y(t)$ is given by

$$y(t) = h(t) * x(t)$$
$$= Ae^{\lambda t}u(t) * e^{\zeta t}u(t)$$

From the convolution table (Table 2.1), we obtain

$$y(t) = \frac{A}{\zeta - \lambda}[e^{\zeta t} - e^{\lambda t}]u(t) \tag{2.66}$$

Clearly, if the input $e^{\zeta t}$ is similar to $e^{\lambda t}$, $\zeta - \lambda$ is small and the system response is large. *The closer the input $x(t)$ to the characteristic mode, the stronger the system response.* In contrast, if the input is very different from the natural mode, $\zeta - \lambda$ is large and the system responds poorly. This is precisely what we set out to prove.

We have proved the foregoing assertion for a single-mode (first-order) system. It can be generalized to an Nth-order system, which has N characteristic modes. The impulse response $h(t)$ of such a system is a linear combination of its N modes. Therefore, if $x(t)$ is similar to any one of the modes, the corresponding response will be high; if it is similar to none of the modes, the response will be small. Clearly, the characteristic modes are very influential in determining system response to a given input.

It would be tempting to conclude on the basis of Eq. (2.66) that if the input is identical to the characteristic mode, so that $\zeta = \lambda$, then the response goes to infinity. Remember, however, that if $\zeta = \lambda$, the numerator on the right-hand side of Eq. (2.66) also goes to zero. We shall study this complex behavior (resonance phenomenon) later in this section.

We now show that *mere inspection of the impulse response $h(t)$ (which is composed of characteristic modes) reveals a great deal about the system behavior.*

2.7-2 Response Time of a System: The System Time Constant

Like human beings, systems have a certain response time. In other words, when an input (stimulus) is applied to a system, a certain amount of time elapses before the system fully responds to that input. This time lag or response time is called the system *time constant*. As we shall see, a system's time constant is equal to the width of its impulse response $h(t)$.

An input $\delta(t)$ to a system is instantaneous (zero duration), but its response $h(t)$ has a duration T_h. Therefore, the system requires a time T_h to respond fully to this input, and we are justified in viewing T_h as the system's response time or time constant. We arrive at the same conclusion via another argument. The output is a convolution of the input with $h(t)$. If an input is a pulse of width T_x, then the output pulse width is $T_x + T_h$ according to the width property of convolution.

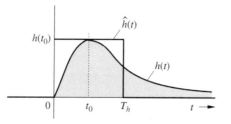

Figure 2.21 Effective duration of an impulse response.

This conclusion shows that the system requires T_h seconds to respond fully to any input. *The system time constant indicates how fast the system is. A system with a smaller time constant is a faster system that responds quickly to an input. A system with a relatively large time constant is a sluggish system that cannot respond well to rapidly varying signals.*

Strictly speaking, the duration of the impulse response $h(t)$ is ∞ because the characteristic modes approach zero asymptotically as $t \to \infty$. However, beyond some value of t, $h(t)$ becomes negligible. It is therefore necessary to use some suitable measure of the impulse response's effective width.

There is no single satisfactory definition of effective signal duration (or width) applicable to every situation. For the situation depicted in Fig. 2.21, a reasonable definition of the duration $h(t)$ would be T_h, the width of the rectangular pulse $\hat{h}(t)$. This rectangular pulse $\hat{h}(t)$ has an area identical to that of $h(t)$ and a height identical to that of $h(t)$ at some suitable instant $t = t_0$. In Fig. 2.21, t_0 is chosen as the instant at which $h(t)$ is maximum. According to this definition,[†]

$$T_h h(t_0) = \int_{-\infty}^{\infty} h(t)\, dt$$

or

$$T_h = \frac{\int_{-\infty}^{\infty} h(t)\, dt}{h(t_0)} \tag{2.67}$$

Now if a system has a single mode

$$h(t) = Ae^{\lambda t} u(t)$$

with λ negative and real, then $h(t)$ is maximum at $t = 0$ with value $h(0) = A$. Therefore, according to Eq. (2.67),

$$T_h = \frac{1}{A} \int_{0}^{\infty} Ae^{\lambda t}\, dt = -\frac{1}{\lambda} \tag{2.68}$$

Thus, the time constant in this case is simply the (negative of the) reciprocal of the system's characteristic root. For the multimode case, $h(t)$ is a weighted sum of the system's characteristic modes, and T_h is a weighted average of the time constants associated with the N modes of the system.

[†]This definition is satisfactory when $h(t)$ is a single, mostly positive (or mostly negative) pulse. Such systems are lowpass systems. This definition should not be applied indiscriminately to all systems.

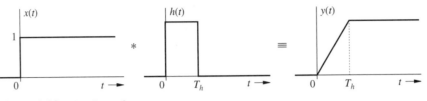

Figure 2.22 Rise time of a system.

2.7-3 Time Constant and Rise Time of a System

Rise time of a system, defined as the time required for the unit step response to rise from 10% to 90% of its steady-state value, is an indication of the speed of response.[†] The system time constant may also be viewed from a perspective of rise time. The unit step response $y(t)$ of a system is the convolution of $u(t)$ with $h(t)$. Let the impulse response $h(t)$ be a rectangular pulse of width T_h, as shown in Fig. 2.22. This assumption simplifies the discussion, yet gives satisfactory results for qualitative discussion. The result of this convolution is illustrated in Fig. 2.22. Note that the output does not rise from zero to a final value instantaneously as the input rises; instead, the output takes T_h seconds to accomplish this. Hence, the rise time T_r of the system is equal to the system time constant

$$T_r = T_h \qquad (2.69)$$

This result and Fig. 2.22 show clearly that a system generally does not respond to an input instantaneously. Instead, it takes time T_h for the system to respond fully.

2.7-4 Time Constant and Filtering

A larger time constant implies a sluggish system because the system takes longer to respond fully to an input. Such a system cannot respond effectively to rapid variations in the input. In contrast, a smaller time constant indicates that a system is capable of responding to rapid variations in the input. Thus, there is a direct connection between a system's time constant and its filtering properties.

A high-frequency sinusoid varies rapidly with time. A system with a large time constant will not be able to respond well to this input. Therefore, such a system will suppress rapidly varying (high-frequency) sinusoids and other high-frequency signals, thereby acting as a lowpass filter (a filter allowing the transmission of low-frequency signals only). We shall now show that a system with a time constant T_h acts as a lowpass filter having a cutoff frequency of $f_c = 1/T_h$ hertz, so that sinusoids with frequencies below f_c hertz are transmitted reasonably well, while those with frequencies above f_c Hz are suppressed.

To demonstrate this fact, let us determine the system response to a sinusoidal input $x(t)$ by convolving this input with the effective impulse response $h(t)$ in Fig. 2.23a. From Fig. 2.23b and 2.23c we see the process of convolution of $h(t)$ with the sinusoidal inputs of two different frequencies. The sinusoid in Fig. 2.23b has a relatively high frequency, while the frequency of

---

[†]Because of varying definitions of rise time, the reader may find different results in the literature. The qualitative and intuitive nature of this discussion should always be kept in mind.

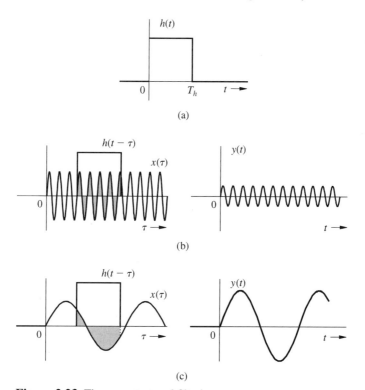

Figure 2.23 Time constant and filtering.

the sinusoid in Fig. 2.23c is low. Recall that the convolution of $x(t)$ and $h(t)$ is equal to the area under the product $x(\tau)h(t - \tau)$. This area is shown shaded in Fig. 2.23b and 2.23c for the two cases. For the high-frequency sinusoid, it is clear from Fig. 2.23b that the area under $x(\tau)h(t - \tau)$ is very small because its positive and negative areas nearly cancel each other out. In this case the output $y(t)$ remains periodic but has a rather small amplitude. This happens when the period of the sinusoid is much smaller than the system time constant T_h. In contrast, for the low-frequency sinusoid, the period of the sinusoid is larger than T_h, rendering the partial cancellation of area under $x(\tau)h(t - \tau)$ less effective. Consequently, the output $y(t)$ is much larger, as depicted in Fig. 2.23c.

Between these two possible extremes in system behavior, a transition point occurs when the period of the sinusoid is equal to the system time constant T_h. The frequency at which this transition occurs is known as the *cutoff frequency* f_c of the system. Because T_h is the period of cutoff frequency f_c,

$$f_c = \frac{1}{T_h} \tag{2.70}$$

The frequency f_c is also known as the bandwidth of the system because the system transmits or passes sinusoidal components with frequencies below f_c while attenuating components with frequencies above f_c. Of course, the transition in system behavior is gradual. There is no dramatic change in system behavior at $f_c = 1/T_h$. Moreover, these results are based on an idealized (rectangular pulse) impulse response; in practice these results will vary somewhat, depending on the exact shape of $h(t)$. Remember that the "feel" of general system behavior is more important than exact system response for this qualitative discussion.

Since the system time constant is equal to its rise time, we have

$$T_r = \frac{1}{f_c} \quad \text{or} \quad f_c = \frac{1}{T_r} \tag{2.71a}$$

Thus, a system's bandwidth is inversely proportional to its rise time. Although Eq. (2.71a) was derived for an idealized (rectangular) impulse response, its implications are valid for lowpass LTIC systems in general. For a general case, we can show that[1]

$$f_c = \frac{k}{T_r} \tag{2.71b}$$

where the exact value of k depends on the nature of $h(t)$. An experienced engineer often can estimate quickly the bandwidth of an unknown system by simply observing the system response to a step input on an oscilloscope.

2.7-5 Time Constant and Pulse Dispersion (Spreading)

In general, the transmission of a pulse through a system causes pulse dispersion (or spreading). Therefore, the output pulse is generally wider than the input pulse. This system behavior can have serious consequences in communication systems in which information is transmitted by pulse amplitudes. Dispersion (or spreading) causes interference or overlap with neighboring pulses, thereby distorting pulse amplitudes and introducing errors in the received information.

Earlier we saw that if an input $x(t)$ is a pulse of width T_x, then T_y, the width of the output $y(t)$, is

$$T_y = T_x + T_h \tag{2.72}$$

This result shows that an input pulse spreads out (disperses) as it passes through a system. Since T_h is also the system's time constant or rise time, the amount of spread in the pulse is equal to the time constant (or rise time) of the system.

2.7-6 Time Constant and Rate of Information Transmission

In pulse communications systems, which convey information through pulse amplitudes, the rate of information transmission is proportional to the rate of pulse transmission. We shall demonstrate that to avoid the destruction of information caused by dispersion of pulses during their transmission through the channel (transmission medium), the rate of information transmission should not exceed the bandwidth of the communications channel.

Since an input pulse spreads out by T_h seconds, the consecutive pulses should be spaced T_h seconds apart to avoid interference between pulses. Thus, the rate of pulse transmission should not exceed $1/T_h$ pulses/second. But $1/T_h = f_c$, the channel's bandwidth, so that we can transmit pulses through a communications channel at a rate of f_c pulses per second and still avoid significant interference between the pulses. The rate of information transmission is therefore proportional to the channel's bandwidth (or to the reciprocal of its time constant).[†]

This discussion (Sections 2.7-2–2.7-6) shows that the system time constant determines much of a system's behavior—its filtering characteristics, rise time, pulse dispersion, and so on. In turn, the time constant is determined by the system's characteristic roots. Clearly the characteristic roots and their relative amounts in the impulse response $h(t)$ determine the behavior of a system.

2.7-7 The Resonance Phenomenon

Finally, we come to the fascinating phenomenon of resonance. As we have already mentioned several times, this phenomenon is observed when the input signal is identical or is very close to a characteristic mode of the system. For the sake of simplicity and clarity, we consider a first-order system having only a single mode, $e^{\lambda t}$. Let the impulse response of this system be[‡]

$$h(t) = Ae^{\lambda t} \tag{2.73}$$

and let the input be

$$x(t) = e^{(\lambda - \epsilon)t}$$

The system response $y(t)$ is then given by

$$y(t) = Ae^{\lambda t} * e^{(\lambda - \epsilon)t}$$

From the convolution table we obtain

$$y(t) = \frac{A}{\epsilon}\left[e^{\lambda t} - e^{(\lambda - \epsilon)t}\right]$$

$$= Ae^{\lambda t}\left(\frac{1 - e^{-\epsilon t}}{\epsilon}\right) \tag{2.74}$$

Now, as $\epsilon \to 0$, both the numerator and the denominator of the term in the parentheses approach zero. Applying L'Hôpital's rule to this term yields

$$\lim_{\epsilon \to 0} y(t) = Ate^{\lambda t} \tag{2.75}$$

Clearly, the response does not go to infinity as $\epsilon \to 0$, but it acquires a factor t, which approaches ∞ as $t \to \infty$. If λ has a negative real part (so that it lies in the LHP), $e^{\lambda t}$ decays faster than t and $y(t) \to 0$ as $t \to \infty$. The resonance phenomenon in this case is present, but its manifestation is aborted by the signal's own exponential decay.

This discussion shows that *resonance is a cumulative phenomenon,* not instantaneous. It builds up linearly[§] with t. When the mode decays exponentially, the signal decays too fast

---

[†]Theoretically, a channel of bandwidth f_c can transmit correctly up to $2f_c$ pulse amplitudes per second.[4] Our derivation here, being very simple and qualitative, yields only half the theoretical limit. In practice it is not easy to attain the upper theoretical limit.

[‡]For convenience we omit multiplying $x(t)$ and $h(t)$ by $u(t)$. Throughout this discussion, we assume that they are causal.

[§]If the characteristic root in question repeats r times, resonance effect increases as t^{r-1}. However, $t^{r-1}e^{\lambda t} \to 0$ as $t \to \infty$ for any value of r, provided Re $\lambda < 0$ (λ in the LHP).

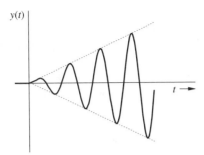

Figure 2.24 Buildup of system response in resonance.

for resonance to counteract the decay; as a result, the signal vanishes before resonance has a chance to build it up. However, if the mode were to decay at a rate less than $1/t$, we should see the resonance phenomenon clearly. This specific condition would be possible if Re $\lambda \geq 0$. For instance, when Re $\lambda = 0$, so that λ lies on the imaginary axis of the complex plane, and therefore

$$\lambda = j\omega$$

and Eq. (2.75) becomes

$$y(t) = Ate^{j\omega t} \tag{2.76}$$

Here, the response does go to infinity linearly with t.

For a real system, if $\lambda = j\omega$ is a root, $\lambda^* = -j\omega$ must also be a root; the impulse response is of the form $Ae^{j\omega t} + Ae^{-j\omega t} = 2A\cos\omega t$. The response of this system to input $A\cos\omega t$ is $2A\cos\omega t * \cos\omega t$. The reader can show that this convolution contains a term of the form $At\cos\omega t$. The resonance phenomenon is clearly visible. The system response to its characteristic mode increases linearly with time, eventually reaching ∞, as indicated in Fig. 2.24.

Recall that when $\lambda = j\omega$, the system is marginally stable. As we have indicated, the full effect of resonance cannot be seen for an asymptotically stable system; only in a marginally stable system does the resonance phenomenon boost the system's response to infinity when the system's input is a characteristic mode. But even in an asymptotically stable system, we see a manifestation of resonance if its characteristic roots are close to the imaginary axis, so that Re λ is a small, negative value. We can show that when the characteristic roots of a system are $\sigma \pm j\omega_0$, then, the system response to the input $e^{j\omega_0 t}$ or the sinusoid $\cos\omega_0 t$ is very large for small σ.[†] The system response drops off rapidly as the input signal frequency moves away from ω_0. This frequency-selective behavior can be studied more profitably after an understanding of frequency-domain analysis has been acquired. For this reason we postpone full discussion of this subject until Chapter 4.

IMPORTANCE OF THE RESONANCE PHENOMENON

The resonance phenomenon is very important because it allows us to design frequency-selective systems by choosing their characteristic roots properly. Lowpass, bandpass, highpass, and

---

[†]This follows directly from Eq. (2.74) with $\lambda = \sigma + j\omega_0$ and $\epsilon = \sigma$.

bandstop filters are all examples of frequency-selective networks. In mechanical systems, the inadvertent presence of resonance can cause signals of such tremendous magnitude that the system may fall apart. A musical note (periodic vibrations) of proper frequency can shatter glass if the frequency is matched to the characteristic root of the glass, which acts as a mechanical system. Similarly, a company of soldiers marching in step across a bridge amounts to applying a periodic force to the bridge. If the frequency of this input force happens to be nearer to a characteristic root of the bridge, the bridge may respond (vibrate) violently and collapse, even though it would have been strong enough to carry many soldiers marching out of step. A case in point is the Tacoma Narrows Bridge failure of 1940. This bridge was opened to traffic in July 1940. Within four months of opening (on November 7, 1940), it collapsed in a mild gale, not because of the wind's brute force but because the frequency of wind-generated vortices, which matched the natural frequency (characteristic roots) of the bridge, causing resonance.

Because of the great damage that may occur, mechanical resonance is generally to be avoided, especially in structures or vibrating mechanisms. If an engine with periodic force (such as piston motion) is mounted on a platform, the platform with its mass and springs should be designed so that their characteristic roots are not close to the engine's frequency of vibration. Proper design of this platform can not only avoid resonance, but also attenuate vibrations if the system roots are placed far away from the frequency of vibration.

2.8 APPENDIX 2.1: DETERMINING THE IMPULSE RESPONSE

In Eq. (2.19), we showed that for an LTIC system S specified by Eq. (2.16), the unit impulse response $h(t)$ can be expressed as

$$h(t) = b_0 \delta(t) + \text{characteristic modes} \tag{2.77}$$

To determine the characteristic mode terms in Eq. (2.77), let us consider a system S_0 whose input $x(t)$ and the corresponding output $w(t)$ are related by

$$Q(D)w(t) = x(t) \tag{2.78}$$

Observe that both the systems S and S_0 have the same characteristic polynomial; namely, $Q(\lambda)$, and, consequently, the same characteristic modes. Moreover, S_0 is the same as S with $P(D) = 1$, that is, $b_0 = 0$. Therefore, according to Eq. (2.77), the impulse response of S_0 consists of characteristic mode terms only without an impulse at $t = 0$. Let us denote this impulse response of S_0 by $y_n(t)$. Observe that $y_n(t)$ consists of characteristic modes of S and therefore may be viewed as a zero-input response of S. Now $y_n(t)$ is the response of S_0 to input $\delta(t)$. Therefore, according to Eq. (2.78)

$$Q(D)y_n(t) = \delta(t) \tag{2.79a}$$

or

$$(D^N + a_1 D^{N-1} + \cdots + a_{N1}D + a_N)y_n(t) = \delta(t) \tag{2.79b}$$

or

$$y_n^{(N)}(t) + a_1 y_n^{(N-1)}(t) + \cdots + a_{N-1} y_n^{(1)}(t) + a_N y_n(t) = \delta(t) \qquad (2.79c)$$

where $y_n^{(k)}(t)$ represents the kth derivative of $y_n(t)$. The right-hand side contains a single impulse term, $\delta(t)$. This is possible only if $y_n^{(N-1)}(t)$ has a unit jump discontinuity at $t = 0$, so that $y_n^{(N)}(t) = \delta(t)$. Moreover, the lower-order terms cannot have any jump discontinuity because this would mean the presence of the derivatives of $\delta(t)$. Therefore $y_n(0) = y_n^{(1)}(0) = \cdots = y_n^{(N-2)}(0) = 0$ (no discontinuity at $t = 0$), and the N initial conditions on $y_n(t)$ are

$$y_n^{(N-1)}(0) = 1$$
$$y_n(0) = y_n^{(1)}(0) = \cdots = y_n^{(N-2)}(0) = 0 \qquad (2.80)$$

This discussion means that $y_n(t)$ is the zero-input response of the system S subject to initial conditions (2.80).

We now show that for the same input $x(t)$ to both systems, S and S_0, their respective outputs $y(t)$ and $w(t)$ are related by

$$y(t) = P(D)w(t) \qquad (2.81)$$

To prove this result, we operate on both sides of Eq. (2.78) by $P(D)$ to obtain

$$Q(D)P(D)w(t) = P(D)x(t)$$

Comparison of this equation with Eq. (2.1c) leads immediately to Eq. (2.81).

Now if the input $x(t) = \delta(t)$, the output of S_0 is $y_n(t)$, and the output of S, according to Eq. (2.81), is $P(D)y_n(t)$. This output is $h(t)$, the unit impulse response of S. Note, however, that because it is an impulse response of a causal system S_0, the function $y_n(t)$ is causal. To incorporate this fact we must represent this function as $y_n(t)u(t)$. Now it follows that $h(t)$, the unit impulse response of the system S, is given by

$$h(t) = P(D)[y_n(t)u(t)] \qquad (2.82)$$

where $y_n(t)$ is a linear combination of the characteristic modes of the system subject to initial conditions (2.80).

The right-hand side of Eq. (2.82) is a linear combination of the derivatives of $y_n(t)u(t)$. Evaluating these derivatives is clumsy and inconvenient because of the presence of $u(t)$. The derivatives will generate an impulse and its derivatives at the origin. Fortunately when $M \leq N$ [Eq. (2.16)], we can avoid this difficulty by using the observation in Eq. (2.77), which asserts that at $t = 0$ (the origin), $h(t) = b_0 \delta(t)$. Therefore, we need not bother to find $h(t)$ at the origin. This simplification means that instead of deriving $P(D)[y_n(t)u(t)]$, we can derive $P(D)y_n(t)$ and add to it the term $b_0 \delta(t)$, so that

$$h(t) = b_0 \delta(t) + P(D)y_n(t) \qquad t \geq 0$$
$$= b_0 \delta(t) + [P(D)y_n(t)]u(t) \qquad (2.83)$$

This expression is valid when $M \leq N$ [the form given in Eq. (2.16b)]. When $M > N$, Eq. (2.82) should be used.

2.9 SUMMARY

This chapter discusses time-domain analysis of LTIC systems. The total response of a linear system is a sum of the zero-input response and zero-state response. The zero-input response is the system response generated only by the internal conditions (initial conditions) of the system, assuming that the external input is zero; hence the adjective "zero-input." The zero-state response is the system response generated by the external input, assuming that all initial conditions are zero, that is, when the system is in zero state.

Every system can sustain certain forms of response on its own with no external input (zero input). These forms are intrinsic characteristics of the system; that is, they do not depend on any external input. For this reason they are called characteristic modes of the system. Needless to say, the zero-input response is made up of characteristic modes chosen in a combination required to satisfy the initial conditions of the system. For an Nth-order system, there are N distinct modes.

The unit impulse function is an idealized mathematical model of a signal that cannot be generated in practice.[†] Nevertheless, introduction of such a signal as an intermediary is very helpful in analysis of signals and systems. The unit impulse response of a system is a combination of the characteristic modes of the system[‡] because the impulse $\delta(t) = 0$ for $t > 0$. Therefore, the system response for $t > 0$ must necessarily be a zero-input response, which, as seen earlier, is a combination of characteristic modes.

The zero-state response (response due to external input) of a linear system can be obtained by breaking the input into simpler components and then adding the responses to all the components. In this chapter we represent an arbitrary input $x(t)$ as a sum of narrow rectangular pulses [staircase approximation of $x(t)$]. In the limit as the pulse width $\rightarrow 0$, the rectangular pulse components approach impulses. Knowing the impulse response of the system, we can find the system response to all the impulse components and add them to yield the system response to the input $x(t)$. The sum of the responses to the impulse components is in the form of an integral, known as the convolution integral. The system response is obtained as the convolution of the input $x(t)$ with the system's impulse response $h(t)$. Therefore, the knowledge of the system's impulse response allows us to determine the system response to any arbitrary input.

LTIC systems have a very special relationship to the everlasting exponential signal e^{st} because the response of an LTIC system to such an input signal is the same signal within a multiplicative constant. The response of an LTIC system to the everlasting exponential input e^{st} is $H(s)e^{st}$, where $H(s)$ is the transfer function of the system.

Differential equations of LTIC systems can also be solved by the classical method, in which the response is obtained as a sum of natural and forced response. These are not the same as the zero-input and zero-state components, although they satisfy the same equations respectively. Although simple, this method is applicable to a restricted class of input signals, and the system response cannot be expressed as an explicit function of the input. These limitations make it useless in the theoretical study of systems.

[†]However, it can be closely approximated by a narrow pulse of unit area and having a width that is much smaller than the time constant of an LTIC system in which it is used.

[‡]There is the possibility of an impulse in addition to the characteristic modes.

If every bounded input results in a bounded output, the system is stable in the bounded-input/bounded-output (BIBO) sense. An LTIC system is BIBO stable if and only if its impulse response is absolutely integrable. Otherwise, it is BIBO unstable. BIBO stability is a stability seen from external terminals of the system. Hence, it is also called external stability or zero-state stability.

In contrast, the internal stability (or the zero-input stability) examines the system stability from inside. When some initial conditions are applied to a system in zero state, then, if the system eventually returns to zero state, the system is said to be stable in the asymptotic or Lyapunov sense. If the system's response increases without bound, it is unstable. If the system does not go to zero state and the response does not increase indefinitely, the system is marginally stable. The internal stability criterion, in terms of the location of a system's characteristic roots, can be summarized as follows.

1. An LTIC system is asymptotically stable if, and only if, all the characteristic roots are in the LHP. The roots may be repeated or unrepeated.

2. An LTIC system is unstable if, and only if, either one or both of the following conditions exist: (i) at least one root is in the RHP; (ii) there are repeated roots on the imaginary axis.

3. An LTIC system is marginally stable if, and only if, there are no roots in the RHP, and there are some unrepeated roots on the imaginary axis.

It is possible for a system to be externally (BIBO) stable but internally unstable. When a system is controllable and observable, its external and internal descriptions are equivalent. Hence, external (BIBO) and internal (asymptotic) stabilities are equivalent and provide the same information. Such a BIBO-stable system is also asymptotically stable, and vice versa. Similarly, a BIBO-unstable system is either marginally stable or asymptotically unstable system.

The characteristic behavior of a system is extremely important because it determines not only the system response to internal conditions (zero-input behavior), but also the system response to external inputs (zero-state behavior) and the system stability. The system response to external inputs is determined by the impulse response, which itself is made up of characteristic modes. The width of the impulse response is called the time constant of the system, which indicates how fast the system can respond to an input. The time constant plays an important role in determining such diverse system behavior as the response time and filtering properties of the system, dispersion of pulses, and the rate of pulse transmission through the system.

REFERENCES

1. Lathi, B. P. *Signals and Systems*. Berkeley-Cambridge Press, Carmichael, CA, 1987.
2. Mason, S. J. *Electronic Circuits, Signals, and Systems*. Wiley, New York, 1960.
3. Kailath, T. *Linear System*. Prentice-Hall, Englewood Cliffs, NJ, 1980.
4. Lathi, B. P. *Modern Digital and Analog Communication Systems,* 3rd ed. Oxford University Press, New York, 1998.

MATLAB SESSION 2: M-FILES

M-files are stored sequences of MATLAB commands and help simplify complicated tasks. There are two types of M-file: script and function. Both types are simple text files and require a .m filename extension.

Although M-files can be created by using any text editor, MATLAB's built-in editor is the preferable choice because of its special features. As with any program, comments improve the readability of an M-file. Comments begin with the % character and continue through the end of the line.

An M-file is executed by simply typing the filename (without the .m extension). To execute, M-files need to be located in the current directory or any other directory in the MATLAB path. New directories are easily added to the MATLAB path by using the addpath command.

M2.1 Script M-Files

Script files, the simplest type of M-file, consist of a series of MATLAB commands. Script files record and automate a series of steps, and they are easy to modify. To demonstrate the utility of a script file, consider the operational amplifier circuit shown in Fig. M2.1.

The system's characteristic modes define the circuit's behavior and provide insight regarding system behavior. Using ideal, infinite gain difference amplifier characteristics, we first derive the differential equation that relates output $y(t)$ to input $x(t)$. Kirchhoff's current law (KCL) at the node shared by R_1 and R_3 provides

$$\frac{x(t) - v(t)}{R_3} + \frac{y(t) - v(t)}{R_2} + \frac{0 - v(t)}{R_1} - C_2 \dot{v}(t) = 0$$

KCL at the inverting input of the op amp gives

$$\frac{v(t)}{R_1} + C_1 \dot{y}(t) = 0$$

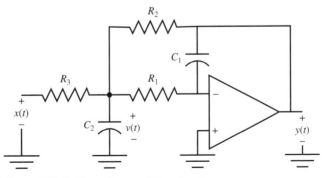

Figure M2.1 Operation amplifier circuit.

Combining and simplifying the KCL equations yields

$$\ddot{y}(t) + \frac{1}{C_2}\left\{\frac{1}{R_1} + \frac{1}{R_2} + \frac{1}{R_3}\right\}\dot{y}(t) + \frac{1}{R_1 R_2 C_1 C_2}y(t) = -\frac{1}{R_1 R_3 C_1 C_2}x(t)$$

which is the desired constant coefficient differential equation. Thus, the characteristic equation is given by

$$\lambda^2 + \frac{1}{C_2}\left\{\frac{1}{R_1} + \frac{1}{R_2} + \frac{1}{R_3}\right\}\lambda + \frac{1}{R_1 R_2 C_1 C_2} = (a_0\lambda^2 + a_1\lambda + a_2) = 0 \qquad \text{(M2.1)}$$

The roots λ_1 and λ_2 of Eq. (M2.1) establish the nature of the characteristic modes $e^{\lambda_1 t}$ and $e^{\lambda_2 t}$.

As a first case, assign nominal component values of $R_1 = R_2 = R_3 = 10\,\text{k}\Omega$ and $C_1 = C_2 = 1\,\mu\text{F}$. A series of MATLAB commands allows convenient computation of the roots $\lambda = [\lambda_1; \lambda_2]$. Although λ can be determined using the quadratic equation, MATLAB's `roots` command is more convenient. The `roots` command requires an input vector that contains the polynomial coefficients in descending order. Even if a coefficient is zero, it must still be included in the vector.

```
% MS2P1.m : MATLAB Session 2, Program 1
% Script M-file determines characteristic roots of op-amp circuit.

% Set component values:
R = [1e4, 1e4, 1e4];
C = [1e-6, 1e-6];
% Determine coefficients for characteristic equation:
a0 = 1;
a1 = (1/R(1)+1/R(2)+1/R(3))/C(2);
a2 = 1/(R(1)*R(2)*C(1)*C(2));
A = [a0 a1 a2];
% Determine characteristic roots:
lambda = roots(A);
```

A script file is created by placing these commands in a text file, which in this case is named MS2P1.m. While comment lines improve program clarity, their removal does not affect program functionality. The program is executed by typing:

```
>> MS2P1
```

After execution, all the resulting variables are available in the workspace. For example, to view the characteristic roots, type:

```
>> lambda
lambda = -261.8034
          -38.1966
```

Thus, the characteristic modes are simple decaying exponentials: $e^{-261.8034t}$ and $e^{-38.1966t}$.

Script files permit simple or incremental changes, thereby saving significant effort. Consider what happens when capacitor C_1 is changed from $1.0 \, \mu$F to 1.0 nF. Changing `MS2P1.m` so that `C = [1e-9, 1e-6]`, allows computation of the new characteristic roots:

```
>> MS2P1
>> lambda
lambda = 1.0e+003 *
            -0.1500 + 3.1587i
            -0.1500 - 3.1587i
```

Perhaps surprisingly, the characteristic modes are now complex exponentials capable of supporting oscillations. The imaginary portion of λ dictates an oscillation rate of 3158.7 rad/s or about 503 Hz. The real portion dictates the rate of decay. The time expected to reduce the amplitude to 25% is approximately $t = \ln 0.25 / \mathrm{Re}(\lambda) \approx 0.01$ second.

M2.2 Function M-Files

It is inconvenient to modify and save a script file each time a change of parameters is desired. Function M-files provide a sensible alternative. Unlike script M-files, function M-files can accept input arguments as well as return outputs. Functions truly extend the MATLAB language in ways that script files cannot.

Syntactically, a function M-file is identical to a script M-file except for the first line. The general form of the first line is:

`function` [*output1, output2, ..., outputN*] = *filename(input1, input2, ..., inputM)*

For example, consider modification of `MS2P1.m` to make function `MS2P2.m`. Component values are passed to the function as two separate inputs: a length-3 vector of resistor values and a length-2 vector of capacitor values. The characteristic roots are returned as a 2×1 complex vector.

```
function [lambda] = MS2P2(R,C)
% MS2P2.m : MATLAB Session 2, Program 2
% Function M-file determines characteristic roots of
% op-amp circuit.
% INPUTS:   R = length-3 vector of resistances
%           C = length-2 vector of capacitances
% OUTPUTS:  lambda = characteristic roots

% Determine coefficients for characteristic equation:
a0 = 1;
a1 = (1/R(1)+1/R(2)+1/R(3))/C(2);
a2 = 1/(R(1)*R(2)*C(1)*C(2));
A = [a0 a1 a2];
% Determine characteristic roots:
lambda = roots(A);
```

As with script M-files, function M-files execute by typing the name at the command prompt. However, inputs must also be included. For example, MS2P2 easily confirms the oscillatory modes of the preceding example.

```
>> lambda = MS2P2([1e4, 1e4, 1e4],[1e-9, 1e-6])
lambda = 1.0e+003 *
          -0.1500 + 3.1587i
          -0.1500 - 3.1587i
```

Although scripts and functions have similarities, they also have distinct differences that are worth pointing out. Scripts operate on workspace data; functions must either be supplied data through inputs or create their own data. Unless passed as an output, variables and data created by functions remain local to the function; variables or data generated by scripts are global and are added to the workspace. To emphasize this point, consider polynomial coefficient vector A, which is created and used in both MS2P1.m and MS2P2.m. Following execution of function MS2P2, the variable A is not added to the workspace. Following execution of script MS2P1, however, A is available in the workspace. Recall, the workspace is easily viewed by typing either who or whos.

M2.3 For-Loops

Real resistors and capacitors never exactly equal their nominal values. Suppose that the circuit components are measured as $R_1 = 10.322 \text{ k}\Omega$, $R_2 = 9.952 \text{ k}\Omega$, $R_3 = 10.115 \text{ k}\Omega$, $C_1 = 1.120 \text{ nF}$, and $C_2 = 1.320 \ \mu\text{F}$. These values are consistent with the 10 and 25% tolerance resistor and capacitor values commonly and readily available. MS2P2.m uses these component values to calculate the new values of λ.

```
>> lambda = MS2P2([10322,9592,10115],[1.12e-9, 1.32e-6])
lambda = 1.0e+003 *
          -0.1136 + 2.6113i
          -0.1136 - 2.6113i
```

Now the natural modes oscillate at 2611.3 rad/s or about 416 Hz. Decay to 25% amplitude is expected in $t = \ln 0.25/(-113.6) \approx 0.012$ second. These values, which differ significantly from the nominal values of 503 Hz and $t \approx 0.01$ second, warrant a more formal investigation of the effect of component variations on the locations of the characteristic roots.

It is sensible to look at three values for each component: the nominal value, a low value, and a high value. Low and high values are based on component tolerances. For example, a 10% 1 kΩ resistor could have an expected low value of $1000(1 - 0.1) = 900 \ \Omega$ and an expected high value of $1000(1 + 0.1) = 1100 \ \Omega$. For the five passive components in the design, $3^5 = 243$ permutations are possible.

Using either MS2P1.m or MS2P2.m to solve each of the 243 cases would be very tedious and boring. For-loops help automate repetitive tasks such as this. In MATLAB, the general structure of a for statement is:

for *variable* = *expression, statement, ..., statement,* end

Five nested for-loops, one for each passive component, are required for the present example.

```
% MS2P3.m : MATLAB Session 2, Program 3
% Script M-file determines characteristic roots over
% a range of component values.

% Pre-allocate memory for all computed roots:
lambda = zeros(2,243);
% Initialize index to identify each permutation:
p=0;
for R1 = 1e4*[0.9,1.0,1.1],
    for R2 = 1e4*[0.9,1.0,1.1],
        for R3 = 1e4*[0.9,1.0,1.1],
            for C1 = 1e-9*[0.75,1.0,1.25],
                for C2 = 1e-6*[0.75,1.0,1.25],
                    p = p+1;
                    lambda(:,p) = MS2P2([R1 R2 R3],[C1 C2]);
                end
            end
        end
    end
end

plot(real(lambda(:)),imag(lambda(:)),'kx',...
    real(lambda(:,1)),imag(lambda(:,1)),'kv',...
    real(lambda(:,end)),imag(lambda(:,end)),'k^')
xlabel('Real'),ylabel('Imaginary')
legend('Characteristic Roots','Min-Component Values Roots',...
    'Max-Component Values Roots',0);
```

The command $lambda = zeros(2,243)$ preallocates a 2×243 array to store the computed roots. When necessary, MATLAB performs dynamic memory allocation, so this command is not strictly necessary. However, preallocation significantly improves script execution speed. Notice also that it would be nearly useless to call script MS2P1 from within the nested loop; script file parameters cannot be changed during execution.

The plot instruction is quite long. Long commands can be broken across several lines by terminating intermediate lines with three dots (...). The three dots tell MATLAB to continue the present command to the next line. Black x's locate roots of each permutation. The command lambda(:) vectorizes the 2×243 matrix lambda into a 486×1 vector. This is necessary in this case to ensure that a proper legend is generated. Because of loop order, permutation $p = 1$ corresponds to the case of all components at the smallest values and permutation $p = 243$ corresponds to the case of all components at the largest values. This information is used to separately highlight the minimum and maximum cases using down-triangles ($\triangledown$) and up-triangles ($\triangle$), respectively. In addition to terminating each for loop, end is used to indicate the final index along a particular dimension, which eliminates the need to remember the particular size of a variable. An overloaded function, such as end, serves multiple uses and is typically interpreted based on context.

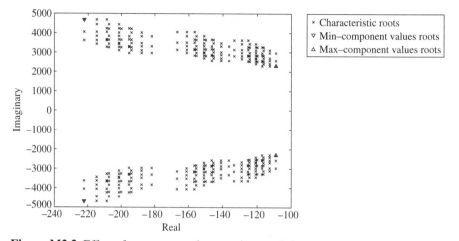

Figure M2.2 Effect of component values on characteristic root locations.

The graphical results provided by MS2P3 are shown in Fig. M2.2. Between extremes, root oscillations vary from 365 to 745 Hz and decay times to 25% amplitude vary from 6.2 to 12.7 ms. Clearly, this circuit's behavior is quite sensitive to ordinary component variations.

M2.4 Graphical Understanding of Convolution

MATLAB graphics effectively illustrate the convolution process. Consider the case of $y(t) = x(t) * h(t)$, where $x(t) = \sin(\pi t)(u(t) - u(t - 1))$ and $h(t) = 1.5(u(t) - u(t - 1.5)) - u(t - 2) + u(t - 2.5)$. Program MS2P4 steps through the convolution over the time interval $(-0.25 \le t \le 3.75)$.

```
% MS2P4.m : MATLAB Session 2, Program 4
% Script M-file graphically demonstrates the convolution process.

figure(1) % Create figure window and make visible on screen
x = inline('1.5*sin(pi*t).*(t>=0&t<1)');
h = inline('1.5*(t>=0&t<1.5)-(t>=2&t<2.5)');
dtau = 0.005; tau = -1:dtau:4;
ti = 0; tvec = -.25:.1:3.75;
y = NaN*zeros(1,length(tvec)); % Pre-allocate memory
for t = tvec,
    ti = ti+1; % Time index
    xh = x(t-tau).*h(tau); lxh = length(xh);
    y(ti) = sum(xh.*dtau); % Trapezoidal approximation of integral
    subplot(2,1,1),plot(tau,h(tau),'k-',tau,x(t-tau),'k--',t,0,'ok');
    axis([tau(1) tau(end) -2.0 2.5]);
    patch([tau(1:end-1);tau(1:end-1);tau(2:end);tau(2:end)],...
        [zeros(1,lxh-1);xh(1:end-1);xh(2:end);zeros(1,lxh-1)],...
        [.8 .8 .8],'edgecolor','none');
```

```
    xlabel('\tau'); legend('h(\tau)','x(t-\tau)','t','h(\tau)x(t-\tau)',3);
    c = get(gca,'children'); set(gca,'children',[c(2);c(3);c(4);c(1)]);
    subplot(2,1,2),plot(tvec,y,'k',tvec(ti),y(ti),'ok');
    xlabel('t'); ylabel('y(t) = \int h(\tau)x(t-\tau) d\tau');
    axis([tau(1) tau(end) -1.0 2.0]); grid;
    drawnow;
end
```

At each step, the program plots $h(\tau)$, $x(t - \tau)$, and shades the area $h(\tau)x(t - \tau)$ gray. This gray area, which reflects the integral of $h(\tau)x(t - \tau)$, is also the desired result, $y(t)$. Figures M2.3, M2.4, and M2.5 display the convolution process at times t of 0.75, 2.25, and 2.85 seconds,

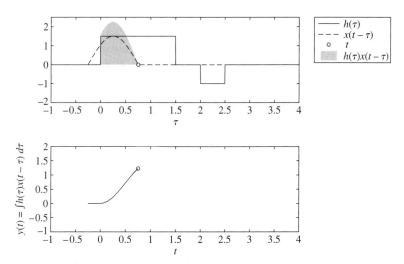

Figure M2.3 Graphical convolution at step $t = 0.75$ second.

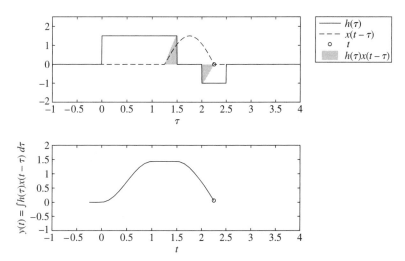

Figure M2.4 Graphical convolution at step $t = 2.25$ seconds.

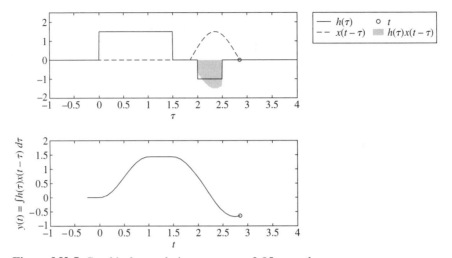

Figure M2.5 Graphical convolution at step $t = 2.85$ seconds.

respectively. These figures help illustrate how the regions of integration change with time. Figure M2.3 has limits of integration from 0 to ($t = 0.75$). Figure M2.4 has two regions of integration, with limits ($t - 1 = 1.25$) to 1.5 and 2.0 to ($t = 2.25$). The last plot, Fig. M2.5, has limits from 2.0 to 2.5.

Several comments regarding MS2P4 are in order. The command `figure(1)` opens the first figure window and, more important, makes sure it is visible. Inline objects are used to represent the functions $x(t)$ and $h(t)$. NaN, standing for not-a-number, usually results from operations such as 0/0 or $\infty - \infty$. MATLAB refuses to plot NaN values, so preallocating $y(t)$ with NaNs ensures that MATLAB displays only values of $y(t)$ that have been computed. As its name suggests, `length` returns the length of the input vector. The `subplot(a,b,c)` command partitions the current figure window into an a-by-b matrix of axes and selects axes c for use. Subplots facilitate graphical comparison by allowing multiple axes in a single figure window. The `patch` command is used to create the gray-shaded area for $h(\tau)x(t - \tau)$. In MS2P4, the `get` and `set` commands are used to reorder `plot` objects so that the gray area does not obscure other lines. Details of the `patch`, `get`, and `set` commands, as used in MS2P4, are somewhat advanced and are not pursued here.[†] MATLAB also prints most Greek letters if the Greek name is preceded by a backslash (\) character. For example, `\tau` in the `xlabel` command produces the symbol τ in the plot's axis label. Similarly, an integral sign is produced by `\int`. Finally, the `drawnow` command forces MATLAB to update the graphics window for each loop iteration. Although slow, this creates an animation-like effect. Replacing `drawnow` with the `pause` command allows users to manually step through the convolution process. The `pause` command still forces the graphics window to update, but the program will not continue until a key is pressed.

[†]Interested students should consult the MATLAB help facilities for further information. Actually, the `get` and `set` commands are extremely powerful and can help modify plots in almost any conceivable way.

PROBLEMS

2.2-1 An LTIC system is specified by the equation

$$(D^2 + 5D + 6)y(t) = (D + 1)x(t)$$

(a) Find the characteristic polynomial, characteristic equation, characteristic roots, and characteristic modes of this system.

(b) Find $y_0(t)$, the zero-input component of the response $y(t)$ for $t \geq 0$, if the initial conditions are $y_0(0^-) = 2$ and $\dot{y}_0(0^-) = -1$.

2.2-2 Repeat Prob. 2.2-1 for

$$(D^2 + 4D + 4)y(t) = Dx(t)$$

and $y_0(0^-) = 3$, $\dot{y}_0(0^-) = -4$.

2.2-3 Repeat Prob. 2.2-1 for

$$D(D + 1)y(t) = (D + 2)x(t)$$

and $y_0(0^-) = \dot{y}_0(0^-) = 1$.

2.2-4 Repeat Prob. 2.2-1 for

$$(D^2 + 9)y(t) = (3D + 2)x(t)$$

and $y_0(0^-) = 0$, $\dot{y}_0(0^-) = 6$.

2.2-5 Repeat Prob. 2.2-1 for

$$(D^2 + 4D + 13)y(t) = 4(D + 2)x(t)$$

with $y_0(0^-) = 5$, $\dot{y}_0(0^-) = 15.98$.

2.2-6 Repeat Prob. 2.2-1 for

$$D^2(D + 1)y(t) = (D^2 + 2)x(t)$$

with $y_0(0^-) = 4$, $\dot{y}_0(0^-) = 3$ and $\ddot{y}_0(0^-) = -1$.

2.2-7 Repeat Prob. 2.2-1 for

$$(D + 1)(D^2 + 5D + 6)y(t) = Dx(t)$$

with $y_0(0^-) = 2$, $\dot{y}_0(0^-) = -1$ and $\ddot{y}_0(0^-) = 5$.

2.2-8 A system is described by a constant-coefficient linear differential equation and has zero-input response given by $y_0(t) = 2e^{-t} + 3$.

(a) Is it possible for the system's characteristic equation to be $\lambda + 1 = 0$? Justify your answer.

(b) Is it possible for the system's characteristic equation to be $\sqrt{3}(\lambda^2 + \lambda) = 0$? Justify your answer.

(c) Is it possible for the system's characteristic equation to be $\lambda(\lambda + 1)^2 = 0$? Justify your answer.

2.3-1 Find the unit impulse response of a system specified by the equation

$$(D^2 + 4D + 3)y(t) = (D + 5)x(t)$$

2.3-2 Repeat Prob. 2.3-1 for

$$(D^2 + 5D + 6)y(t) = (D^2 + 7D + 11)x(t)$$

2.3-3 Repeat Prob. 2.3-1 for the first-order allpass filter specified by the equation

$$(D + 1)y(t) = -(D - 1)x(t)$$

2.3-4 Find the unit impulse response of an LTIC system specified by the equation

$$(D^2 + 6D + 9)y(t) = (2D + 9)x(t)$$

2.4-1 If $c(t) = x(t) * g(t)$, then show that $A_c = A_x A_g$, where A_x, A_g, and A_c are the areas under $x(t)$, $g(t)$, and $c(t)$, respectively. Verify this *area property* of convolution in Examples 2.7 and 2.9.

2.4-2 If $x(t) * g(t) = c(t)$, then show that $x(at) * g(at) = |1/a|c(at)$. This *time-scaling property* of convolution states that if both $x(t)$ and $g(t)$ are time-scaled by a, their convolution is also time-scaled by a (and multiplied by $|1/a|$).

2.4-3 Show that the convolution of an odd and an even function is an odd function and the convolution of two odd or two even functions is an even function. [Hint: Use time-scaling property of convolution in Prob. 2.4-2.]

2.4-4 Using direct integration, find $e^{-at}u(t) * e^{-bt}u(t)$.

2.4-5 Using direct integration, find $u(t) * u(t)$, $e^{-at}u(t) * e^{-at}u(t)$, and $tu(t) * u(t)$.

2.4-6 Using direct integration, find $\sin t\, u(t) * u(t)$ and $\cos t\, u(t) * u(t)$.

2.4-7   The unit impulse response of an LTIC system is

$$h(t) = e^{-t}u(t)$$

Find this system's (zero-state) response $y(t)$ if the input $x(t)$ is:

(a) $u(t)$
(b) $e^{-t}u(t)$
(c) $e^{-2t}u(t)$
(d) $\sin 3t\, u(t)$

Use the convolution table (Table 2.1) to find your answers.

2.4-8   Repeat Prob. 2.4-7 for

$$h(t) = [2e^{-3t} - e^{-2t}]u(t)$$

and if the input $x(t)$ is:

(a) $u(t)$
(b) $e^{-t}u(t)$
(c) $e^{-2t}u(t)$

2.4-9   Repeat Prob. 2.4-7 for

$$h(t) = (1 - 2t)e^{-2t}u(t)$$

and input $x(t) = u(t)$.

2.4-10   Repeat Prob. 2.4-7 for

$$h(t) = 4e^{-2t}\cos 3t\, u(t)$$

and each of the following inputs $x(t)$:

(a) $u(t)$
(b) $e^{-t}u(t)$

2.4-11   Repeat Prob. 2.4-7 for

$$h(t) = e^{-t}u(t)$$

and each of the following inputs $x(t)$:

(a) $e^{-2t}u(t)$
(b) $e^{-2(t-3)}u(t)$
(c) $e^{-2t}u(t-3)$
(d) The gate pulse depicted in Fig. P2.4-11—and provide a sketch of $y(t)$.

2.4-12   A first-order allpass filter impulse response is given by

$$h(t) = -\delta(t) + 2e^{-t}u(t)$$

Figure P2.4-11

(a) Find the zero-state response of this filter for the input $e^t u(-t)$.
(b) Sketch the input and the corresponding zero-state response.

2.4-13   Figure P2.4-13 shows the input $x(t)$ and the impulse response $h(t)$ for an LTIC system. Let the output be $y(t)$.

(a) By inspection of $x(t)$ and $h(t)$, find $y(-1)$, $y(0)$, $y(1)$, $y(2)$, $y(3)$, $y(4)$, $y(5)$, and $y(6)$. Thus, by merely examining $x(t)$ and $h(t)$, you are required to see what the result of convolution yields at $t = -1, 0, 1, 2, 3, 4, 5$, and 6.
(b) Find the system response to the input $x(t)$.

2.4-14   The zero-state response of an LTIC system to an input $x(t) = 2e^{-2t}u(t)$ is $y(t) = [4e^{-2t} + 6e^{-3t}]u(t)$. Find the impulse response of the system. [Hint: We have not yet developed a method of finding $h(t)$ from the knowledge of the input and the corresponding output. Knowing the form of $x(t)$ and $y(t)$, you will have to make the best guess of the general form of $h(t)$.]

2.4-15   Sketch the functions $x(t) = 1/(t^2 + 1)$ and $u(t)$. Now find $x(t) * u(t)$ and sketch the result.

2.4-16   Figure P2.4-16 shows $x(t)$ and $g(t)$. Find and sketch $c(t) = x(t) * g(t)$.

2.4-17   Find and sketch $c(t) = x(t) * g(t)$ for the functions depicted in Fig. P2.4-17.

2.4-18   Find and sketch $c(t) = x_1(t) * x_2(t)$ for the pairs of functions illustrated in Fig. P2.4-18.

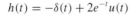

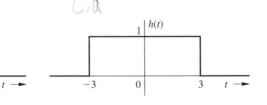

Figure P2.4-13

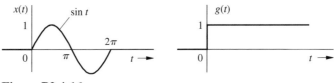

Figure P2.4-16

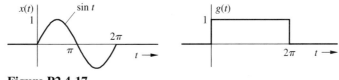

Figure P2.4-17

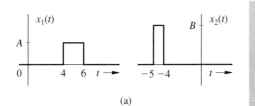

(a)

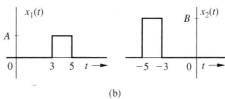

(b)

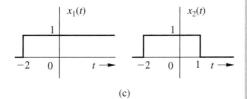

(c)

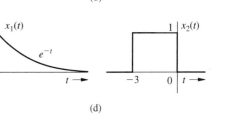

(d)

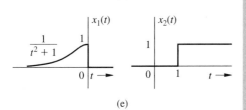

(e)

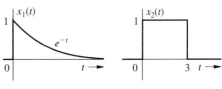

(f)

(g)

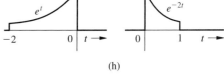

(h)

Figure P2.4-18

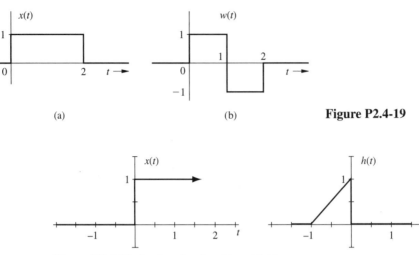

(a)   (b)   **Figure P2.4-19**

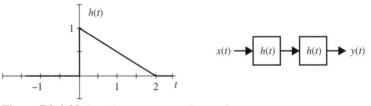

Figure P2.4-21 Analog signals $x(t)$ and $h(t)$.

Figure P2.4-22 Impulse response and cascade system.

2.4-19 Use Eq. (2.46) to find the convolution of $x(t)$ and $w(t)$, shown in Fig. P2.4-19.

2.4-20 Determine $H(s)$, the transfer function of an ideal time delay of T seconds. Find your answer by two methods: using Eq. (2.48) and using Eq. (2.49).

2.4-21 Determine $y(t) = x(t) * h(t)$ for the signals depicted in Fig. P2.4-21.

2.4-22 Two linear time-invariant systems, each with impulse response $h(t)$, are connected in cascade. Refer to Fig. P2.4-22. Given input $x(t) = u(t)$, determine $y(1)$. That is, determine the step response at time $t = 1$ for the cascaded system shown.

2.4-23 Consider the electric circuit shown in Fig. P2.4-23.

(a) Determine the differential equation that relates the input $x(t)$ to output $y(t)$. Recall

that

$$i_C(t) = C\frac{dv_C}{dt} \quad \text{and} \quad v_L(t) = L\frac{di_L}{dt}$$

(b) Find the characteristic equation for this circuit, and express the root(s) of the characteristic equation in terms of L and C.

(c) Determine the zero-input response given an initial capacitor voltage of one volt and an initial inductor current of zero amps. That is, find $y_0(t)$ given $v_C(0) = 1$ V and $i_L(0) = 0$ A. [Hint: The coefficient(s) in $y_0(t)$ are independent of L and C.]

(d) Plot $y_0(t)$ for $t \geq 0$. Does the zero-input response, which is caused solely by initial conditions, ever "die" out?

(e) Determine the total response $y(t)$ to the input $x(t) = e^{-t}u(t)$. Assume an initial inductor current of $i_L(0^-) = 0$ A, an

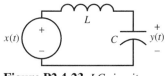

Figure P2.4-23 *LC* circuit.

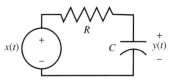
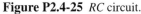

Figure P2.4-25 *RC* circuit.

initial capacitor voltage of $v_C(0^-) = 1$ V, $L = 1$ H, and $C = 1$ F.

2.4-24 Two LTIC systems have impulse response functions given by $h_1(t) = (1 - t)[u(t) - u(t - 1)]$ and $h_2(t) = t[u(t + 2) - u(t - 2)]$.

(a) Carefully sketch the functions $h_1(t)$ and $h_2(t)$.

(b) Assume that the two systems are connected in parallel as shown in Fig. P2.4-24a. Carefully plot the equivalent impulse response function, $h_p(t)$.

(c) Assume that the two systems are connected in cascade as shown in Fig. P2.4-24b. Carefully plot the equivalent impulse response function, $h_s(t)$.

2.4-25 Consider the circuit shown in Fig. P2.4-25.

(a) Find the output $y(t)$ given an initial capacitor voltage of $y(0) = 2$ V and an input $x(t) = u(t)$.

(b) Given an input $x(t) = u(t - 1)$, determine the initial capacitor voltage $y(0)$ so that the output $y(t)$ is 0.5 volt at $t = 2$ seconds.

2.4-26 An analog signal is given by $x(t) = t[u(t) - u(t - 1)]$, as shown in Fig. P2.4-26. Determine and plot $y(t) = x(t) * x(2t)$

2.4-27 Consider the electric circuit shown in Fig. P2.4-27.

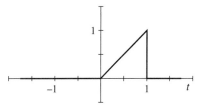

Figure P2.4-26 Short-duration ramp signal $x(t)$.

(a) Determine the differential equation that relates the input current $x(t)$ to output current $y(t)$. Recall that

$$v_L = L\frac{di_L}{dt}$$

(b) Find the characteristic equation for this circuit, and express the root(s) of the characteristic equation in terms of L_1, L_2, and R.

(c) Determine the zero-input response given initial inductor currents of one ampere each. That is, find $y_0(t)$ given $i_{L_1}(0) = i_{L_2}(0) = 1$ A.

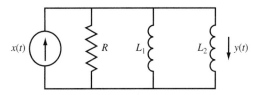

Figure P2.4-27 *RLL* circuit.

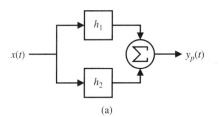

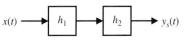

(a)   (b)

Figure P2.4-24 **(a)** Parallel and **(b)** series connections.

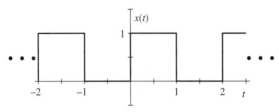

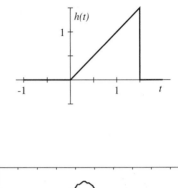

Figure P2.4-29 Periodic input $x(t)$.

2.4-28 An LTI system has step response given by $g(t) = e^{-t}u(t) - e^{-2t}u(t)$. Determine the output of this system $y(t)$ given an input $x(t) = \delta(t - \pi) - \cos(\sqrt{3})u(t)$.

2.4-29 The periodic signal $x(t)$ shown in Fig. P2.4-29 is input to a system with impulse response function $h(t) = t[u(t) - u(t - 1.5)]$, also shown in Fig. P2.4-29. Use convolution to determine the output $y(t)$ of this system. Plot $y(t)$ over $(-3 \le t \le 3)$.

2.4-30 Consider the electric circuit shown in Fig. P2.4-30.
  (a) Determine the differential equation relating input $x(t)$ to output $y(t)$.
  (b) Determine the output $y(t)$ in response to the input $x(t) = 4te^{-3t/2}u(t)$. Assume component values of $R = 1 \ \Omega$, $C_1 = 1$ F, and $C_2 = 2$ F, and initial capacitor voltages of $V_{C_1} = 2$ V and $V_{C_2} = 1$ V.

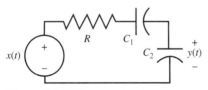

Figure P2.4-30 *RCC* circuit.

2.4-31 A cardiovascular researcher is attempting to model the human heart. He has recorded ventricular pressure, which he believes corresponds to the heart's impulse response function $h(t)$, as shown in Fig. P2.4-31. Comment on the function $h(t)$ shown in Fig. P2.4-31. Can you establish any system properties, such as causality or stability? Do the data suggest any reason to suspect that the measurement is not a true impulse response?

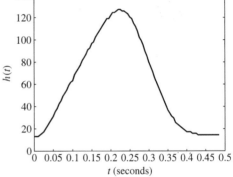

Figure P2.4-31 Measured impulse response function.

2.4-32 The autocorrelation of a function $x(t)$ is given by $r_{xx}(t) = \int_{-\infty}^{\infty} x(\tau)x(\tau - t) \, d\tau$. This equation is computed in a manner nearly identical to convolution.
  (a) Show $r_{xx}(t) = x(t) * x(-t)$
  (b) Determine and plot $r_{xx}(t)$ for the signal $x(t)$ depicted in Fig. P2.4-32. [Hint: $r_{xx}(t) = r_{xx}(-t)$.]

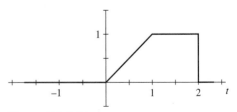

Figure P2.4-32 Analog signal $x(t)$.

2.4-33 Consider the circuit shown in Fig. P2.4-33. This circuit functions as an integrator. Assume ideal op-amp behavior and recall that

$$i_C = C\frac{dV_C}{dt}$$

(a) Determine the differential equation that relates the input $x(t)$ to the output $y(t)$.

(b) This circuit does not behave well at dc. Demonstrate this by computing the zero-state response $y(t)$ for a unit step input $x(t) = u(t)$.

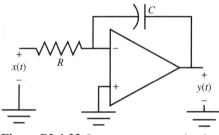

Figure P2.4-33 Integrator op-amp circuit.

2.4-34 Derive the result in Eq. (2.46) in another way. As mentioned in Chapter 1 (Fig. 1.27b), it is possible to express an input in terms of its step components, as shown in Fig. P2.4-34. Find the system response as a sum of the responses to the step components of the input.

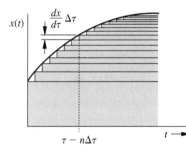

Figure P2.4-34

2.4-35 Show that an LTIC system response to an everlasting sinusoid $\cos \omega_0 t$ is given by

$$y(t) = |H(j\omega_0)| \cos[\omega_0 t + \angle H(j\omega_0)]$$

where

$$H(j\omega) = \int_{-\infty}^{\infty} h(t)e^{-j\omega t}\, dt$$

assuming the integral on the right-hand side exists.

2.4-36 A line charge is located along the x axis with a charge density $Q(x)$ coulombs per meter. Show that the electric field $E(x)$ produced by this line charge at a point x is given by

$$E(x) = Q(x) * h(x)$$

where $h(x) = 1/4\pi\epsilon x^2$. [*Hint:* The charge over an interval $\Delta\tau$ located at $\tau = n\Delta\tau$ is $Q(n\Delta\tau)\Delta\tau$. Also by Coulomb's law, the electric field $E(r)$ at a distance r from a charge q coulombs is given by $E(r) = q/4\pi\epsilon r^2$.]

2.4-37 Consider the circuit shown in Fig. P2.4-37. Assume ideal op-amp behavior and recall that

$$i_C = C\frac{dV_C}{dt}$$

Without a feedback resistor R_f, the circuit functions as an integrator and is unstable, particularly at dc. A feedback resistor R_f corrects this problem and results in a stable circuit that functions as a "lossy" integrator.

(a) Determine the differential equation that relates the input $x(t)$ to the output $y(t)$. What is the corresponding characteristic equation?

(b) To demonstrate that this "lossy" integrator is well behaved at dc, determine the zero-state response $y(t)$ given a unit step input $x(t) = u(t)$.

(c) Investigate the effect of 10% resistor and 25% capacitor tolerances on the system's characteristic root(s).

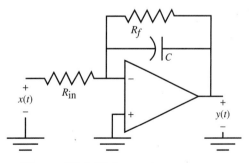

Figure P2.4-37 Lossy integrator op-amp circuit.

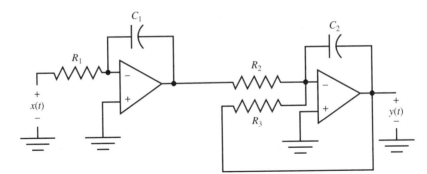

Figure P2.4-38 Op-amp circuit.

2.4-38 Consider the electric circuit shown in Fig. P2.4-38. Let $C_1 = C_2 = 10 \ \mu F$, $R_1 = R_2 = 100 \ k\Omega$, and $R_3 = 50 \ k\Omega$.

(a) Determine the corresponding differential equation describing this circuit. Is the circuit BIBO stable?

(b) Determine the zero-input response $y_0(t)$ if the output of each op amp initially reads one volt.

(c) Determine the zero-state response $y(t)$ to a step input $x(t) = u(t)$.

(d) Investigate the effect of 10% resistor and 25% capacitor tolerances on the system's characteristic roots.

2.4-39 A system is called complex if a real-valued input can produce a complex-valued output. Suppose a linear, time-invariant complex system has impulse response $h(t) = j[u(-t + 2) - u(-t)]$.

(a) Is this system causal? Explain.

(b) Use convolution to determine the zero-state response $y_1(t)$ of this system in response to the unit-duration pulse $x_1(t) = u(t) - u(t - 1)$.

(c) Using the result from part b, determine the zero-state response $y_2(t)$ in response to $x_2(t) = 2u(t - 1) - u(t - 2) - u(t - 3)$.

2.5-1 Use the classical method to solve

$$(D^2 + 7D + 12)y(t) = (D + 2)x(t)$$

for initial conditions of $y(0^+) = 0$, $\dot{y}(0^+) = 1$,

and input $x(t)$ of

(a) $u(t)$
(b) $e^{-t}u(t)$
(c) $e^{-2t}u(t)$

2.5-2 Using the classical method, solve

$$(D^2 + 6D + 25)y(t) = (D + 3)x(t)$$

for the initial conditions of $y(0^+) = 0$, $\dot{y}(0^+) = 2$ and input $x(t) = u(t)$.

2.5-3 Using the classical method, solve

$$(D^2 + 4D + 4)y(t) = (D + 1)x(t)$$

for initial conditions of $y(0^+) = 9/4$, $\dot{y}(0^+) = 5$ and input $x(t)$ of

(a) $e^{-3t}u(t)$
(b) $e^{-t}u(t)$

2.5-4 Using the classical method, solve

$$(D^2 + 2D)y(t) = (D + 1)x(t)$$

for initial conditions of $y(0^+) = 2$, $\dot{y}(0^+) = 1$ and input of $x(t) = u(t)$.

2.5-5 Repeat Prob. 2.5-1 for the input

$$x(t) = e^{-3t}u(t)$$

2.6-1 Explain, with reasons, whether the LTIC systems described by the following equations are (i) stable or unstable in the BIBO sense;

(ii) asymptotically stable, unstable, or marginally stable. Assume that the systems are controllable and observable.

(a) $(D^2 + 8D + 12)y(t) = (D - 1)x(t)$
(b) $D(D^2 + 3D + 2)y(t) = (D + 5)x(t)$
(c) $D^2(D^2 + 2)y(t) = x(t)$
(d) $(D+1)(D^2-6D+5)y(t) = (3D+1)x(t)$

2.6-2 Repeat Prob. 2.6-1 for the following.

(a) $(D + 1)(D^2 + 2D + 5)^2 y(t) = x(t)$
(b) $(D + 1)(D^2 + 9)y(t) = (2D + 9)x(t)$
(c) $(D + 1)(D^2 + 9)^2 y(t) = (2D + 9)x(t)$
(d) $(D^2 + 1)(D^2 + 4)(D^2 + 9)y(t) = 3Dx(t)$

2.6-3 For a certain LTIC system, the impulse response $h(t) = u(t)$.

(a) Determine the characteristic root(s) of this system.
(b) Is this system asymptotically or marginally stable, or is it unstable?
(c) Is this system BIBO stable?
(d) What can this system be used for?

2.6-4 In Section 2.6 we demonstrated that for an LTIC system, condition (2.64) is sufficient for BIBO stability. Show that this is also a necessary condition for BIBO stability in such systems. In other words, show that if condition (2.64) is not satisfied, then there exists a bounded input that produces an unbounded output. [Hint: Assume that a system exists for which $h(t)$ violates condition (2.64) and yet produces an output that is bounded for every bounded input. Establish the contradiction in this statement by considering an input $x(t)$ defined by $x(t_1 - \tau) = 1$ when $h(\tau) \geq 0$ and $x(t_1 - \tau) = -1$ when $h(\tau) < 0$, where t_1 is some fixed instant.]

2.6-5 An analog LTIC system with impulse response function $h(t) = u(t+2) - u(t-2)$ is presented with an input $x(t) = t(u(t) - u(t - 2))$.

(a) Determine and plot the system output $y(t) = x(t) * h(t)$.
(b) Is this system stable? Is this system causal? Justify your answers.

2.6-6 A system has an impulse response function shaped like a rectangular pulse, $h(t) = u(t) - u(t - 1)$. Is the system stable? Is the system causal?

2.6-7 A continuous-time LTI system has impulse response function $h(t) = \sum_{i=0}^{\infty} (0.5)^i \delta(t - i)$.

(a) Is the system causal? Prove your answer.
(b) Is the system stable? Prove your answer.

2.7-1 Data at a rate of 1 million pulses per second are to be transmitted over a certain communications channel. The unit step response $g(t)$ for this channel is shown in Fig. P2.7-1.

(a) Can this channel transmit data at the required rate? Explain your answer.
(b) Can an audio signal consisting of components with frequencies up to 15 kHz be transmitted over this channel with reasonable fidelity?

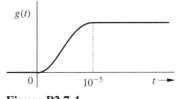

Figure P2.7-1

2.7-2 A certain communication channel has a bandwidth of 10 kHz. A pulse of 0.5 ms duration is transmitted over this channel.

(a) Determine the width (duration) of the received pulse.
(b) Find the maximum rate at which these pulses can be transmitted over this channel without interference between the successive pulses.

2.7-3 A first-order LTIC system has a characteristic root $\lambda = -10^4$.

(a) Determine T_r, the rise time of its unit step input response.
(b) Determine the bandwidth of this system.
(c) Determine the rate at which the information pulses can be transmitted through this system.

2.7-4 Consider a linear, time-invariant system with impulse response $h(t)$ shown in Fig. P2.7-4. Outside the interval shown, $h(t) = 0$.

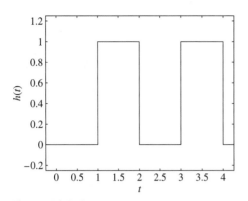

Figure P2.7-4 Impulse response $h(t)$.

(a) What is the rise time T_r of this system? Remember, rise time is the time between the application of a unit step and the moment at which the system has "fully" responded.

(b) Suppose $h(t)$ represents the response of a communication channel. What conditions might cause the channel to have such an impulse response? What is the maximum average number of pulses per unit time that can be transmitted without causing interference? Justify your answer.

(c) Determine the system output $y(t) = x(t) * h(t)$ for $x(t) = [u(t-2) - u(t)]$. Accurately sketch $y(t)$ over $(0 \le t \le 10)$.

TIME-DOMAIN ANALYSIS
OF DISCRETE-TIME SYSTEMS

In this chapter we introduce the basic concepts of discrete-time signals and systems. We shall study convolution method of analysis of linear, time-invariant, discrete-time (LTID) systems. Classical methods of analysis of these systems will also be examined.

3.1 INTRODUCTION

Discrete-time signal is basically a sequence of numbers. Such signals arise naturally in inherently discrete-time situations such as population studies, amortization problems, national income models, and radar tracking. They may also arise as a result of sampling continuous-time signals in sampled data systems and digital filtering. Such signals can be denoted by $x[n]$, $y[n]$, and so on, where the variable n takes integer values, and $x[n]$ denotes the nth number in the sequence labeled x. In this notation, the discrete-time variable n is enclosed in square brackets instead of parentheses, which we have reserved for enclosing continuous-time variable, such as t.

Systems whose inputs and outputs are discrete-time signals are called *discrete-time systems*. A digital computer is a familiar example of this type of system. A discrete-time signal is a sequence of numbers, and a discrete-time system processes a sequence of numbers $x[n]$ to yield another sequence $y[n]$ as the output.[†]

A discrete-time signal, when obtained by uniform sampling of a continuous-time signal $x(t)$, can also be expressed as $x(nT)$, where T is the sampling interval and n, the discrete variable taking on integer values. Thus, $x(nT)$ denotes the value of the signal $x(t)$ at $t = nT$. The signal $x(nT)$ is a sequence of numbers (sample values), and hence, by definition, is a discrete-time signal. Such a signal can also be denoted by the customary discrete-time notation $x[n]$, where $x[n] = x(nT)$. A typical discrete-time signal is depicted in Fig. 3.1, which shows both forms of notation. By way of an example, a continuous-time exponential $x(t) = e^{-t}$, when sampled every $T = 0.1$ second, results in a discrete-time signal $x(nT)$ given by

$$x(nT) = e^{-nT} = e^{-0.1n}$$

[†]There may be more than one input and more than one output.

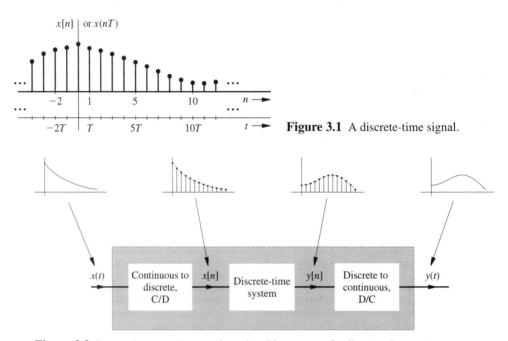

Figure 3.1 A discrete-time signal.

Figure 3.2 Processing a continuous-time signal by means of a discrete-time system.

Clearly, this signal is a function of n and may be expressed as $x[n]$. Such representation is more convenient and will be followed throughout this book even for signals resulting from sampling continuous-time signals.

Digital filters, can process continuous-time signals by discrete-time systems, using appropriate interfaces at the input and the output, as illustrated in Fig. 3.2. A continuous-time signal $x(t)$ is first sampled to convert it into a discrete-time signal $x[n]$, which is then processed by a discrete-time system to yield the output $y[n]$. A continuous-time signal $y(t)$ is finally constructed from $y[n]$. We shall use the notations C/D and D/C for conversion from continuous to discrete-time and from discrete to continuous time. By using the interfaces in this manner, we can use an appropriate discrete-time system to process a continuous-time signal. As we shall see later in our discussion, discrete-time systems have several advantages over continuous-time systems. For this reason, there is an accelerating trend toward processing continuous-time signals with discrete-time systems.

3.1-1 Size of a Discrete-Time Signal

Arguing along the lines similar to those used for continuous-time signals, the size of a discrete-time signal $x[n]$ will be measured by its energy E_x, defined by

$$E_x = \sum_{n=-\infty}^{\infty} |x[n]|^2 \tag{3.1}$$

This definition is valid for real or complex $x[n]$. For this measure to be meaningful, the energy of a signal must be finite. A necessary condition for the energy to be finite is that the signal amplitude must $\to 0$ as $|n| \to \infty$. Otherwise the sum in Eq. (3.1) will not converge. If E_x is finite, the signal is called an *energy signal*.

In some cases, for instance, when the amplitude of $x[n]$ does not $\to 0$ as $|n| \to \infty$, then the signal energy is infinite, and a more meaningful measure of the signal in such a case would be the time average of the energy (if it exists), which is the signal power P_x, defined by

$$P_x = \lim_{N \to \infty} \frac{1}{2N+1} \sum_{-N}^{N} |x[n]|^2 \tag{3.2}$$

In this equation, the sum is divided by $2N + 1$ because there are $2N + 1$ samples in the interval from $-N$ to N. For periodic signals, the time averaging need be performed over only one period in view of the periodic repetition of the signal. If P_x is finite and nonzero, the signal is called a *power signal*. As in the continuous-time case, a discrete-time signal can either be an energy signal or a power signal, but cannot be both at the same time. Some signals are neither energy nor power signals.

EXAMPLE 3.1

Find the energy of the signal $x[n] = n$, shown in Fig. 3.3a and the power for the periodic signal $y[n]$ in Fig. 3.3b.

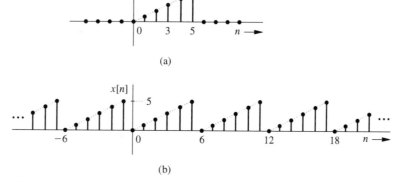

(a)

(b)

Figure 3.3 (a) Energy and (b) power computation for a signal.

By definition

$$E_x = \sum_{n=0}^{5} n^2 = 55$$

A periodic signal $x[n]$ with period N_0 is characterized by the fact that

$$x[n] = x[n + N_0]$$

The smallest value of N_0 for which the preceding equation holds is the *fundamental period*. Such a signal is called N_0 *periodic*. Figure 3.3b shows an example of a periodic signal $y[n]$ of period $N_0 = 6$ because each period contains 6 samples. Note that if the first sample is taken at $n = 0$, the last sample is at $n = N_0 - 1 = 5$, not at $n = N_0 = 6$. Because the signal $y[n]$ is periodic, its power P_y can be found by averaging its energy over one period. Averaging the energy over one period, we obtain

$$P_y = \frac{1}{6} \sum_{n=0}^{5} n^2 = \frac{55}{6}$$

EXERCISE E3.1

Show that the signal $x[n] = a^n u[n]$ is an energy signal of energy $E_x = 1/(1 - |a|^2)$ if $|a| < 1$. It is a power signal of power $P_x = 0.5$ if $|a| = 1$. It is neither an energy signal nor a power signal if $|a| > 1$.

3.2 USEFUL SIGNAL OPERATIONS

Signal operations for *shifting,* and *scaling,* as discussed for continuous-time signals also apply to discrete-time signals with some modifications.

SHIFTING

Consider a signal $x[n]$ (Fig. 3.4a) and the same signal delayed (right-shifted) by 5 units (Fig. 3.4b), which we shall denote by $x_s[n]$.[†] Using the argument employed for a similar operation in continuous-time signals (Section 1.2), we obtain

$$x_s[n] = x[n - 5]$$

Therefore, to shift a sequence by M units (M integer), we replace n with $n - M$. Thus $x[n - M]$ represents $x[n]$ shifted by M units. If M is positive, the shift is to the right (delay). If M is negative, the shift is to the left (advance). Accordingly, $x[n - 5]$ is $x[n]$ delayed (right-shifted) by 5 units, and $x[n + 5]$ is $x[n]$ advanced (left-shifted) by 5 units.

TIME REVERSAL

To time-reverse $x[n]$ in Fig. 3.4a, we rotate $x[n]$ about the vertical axis to obtain the time reversed signal $x_r[n]$ shown in Fig. 3.4c. Using the argument employed for a similar operation in continuous-time signals (Section 1.2), we obtain

$$x_r[n] = x[-n]$$

[†]The terms "delay" and "advance" are meaningful only when the independent variable is time. For other independent variables, such as frequency or distance, it is more appropriate to refer to the "right shift" and "left shift" of a sequence.

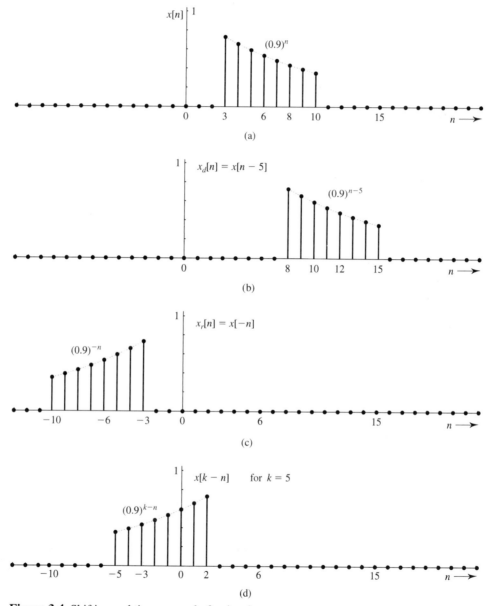

Figure 3.4 Shifting and time reversal of a signal.

Therefore, to time-reverse a signal, we replace n with $-n$ so that $x[-n]$ is the time reversed $x[n]$. For example, if $x[n] = (0.9)^n$ for $3 \le n \le 10$, then $x_r[n] = (0.9)^{-n}$ for $3 \le -n \le 10$; that is, $-3 \ge n \ge -10$, as shown in Fig. 3.4c.

The origin $n = 0$ is the anchor point, which remains unchanged under time-reversal operation because at $n = 0$, $x[n] = x[-n] = x[0]$. Note that while the reversal of $x[n]$ about the vertical axis is $x[-n]$, the reversal of $x[n]$ about the horizontal axis is $-x[n]$.

EXAMPLE 3.2

In the convolution operation, discussed later, we need to find the function $x[k-n]$ from $x[n]$.

This can be done in two steps: (i) time-reverse the signal $x[n]$ to obtain $x[-n]$; (ii) now, right-shift $x[-n]$ by k. Recall that right-shifting is accomplished by replacing n with $n-k$. Hence, right-shifting $x[-n]$ by k units is $x[-(n-k)] = x[k-n]$. Figure 3.4d shows $x[5-n]$, obtained this way. We first time-reverse $x[n]$ to obtain $x[-n]$ in Fig. 3.4c. Next, we shift $x[-n]$ by $k=5$ to obtain $x[k-n] = x[5-n]$, as shown in Fig. 3.4d.

In this particular example, the order of the two operations employed is interchangeable. We can first left-shift $x[k]$ to obtain $x[n+5]$. Next, we time-reverse $x[n+5]$ to obtain $x[-n+5] = x[5-n]$. The reader is encouraged to verify that this procedure yields the same result, as in Fig. 3.4d.

SAMPLING RATE ALTERATION: DECIMATION AND INTERPOLATION

Alteration of the sampling rate is somewhat similar to time-scaling in continuous-time signals. Consider a signal $x[n]$ compressed by factor M. Compressing a signal $x[n]$ by factor M yields $x_d[n]$ given by

$$x_d[n] = x[Mn] \tag{3.3}$$

Because of the restriction that discrete-time signals are defined only for integer values of the argument, we must restrict M to integer values. The values of $x[Mn]$ at $n = 0, 1, 2, 3, \ldots$ are $x[0]$, $x[M]$, $x[2M]$, $x[3M]$, $\ldots$. This means $x[Mn]$ selects every Mth sample of $x[n]$ and deletes all the samples in between. For this reason, this operation is called *decimation*. It reduces the number of samples by factor M. If $x[n]$ is obtained by sampling a continuous-time signal, this operation implies reducing the sampling rate by factor M. For this reason, decimation is also called *downsampling*. Figure 3.5a shows a signals $x[n]$ and Fig. 3.5b shows the signal $x[2n]$, which is obtained by deleting odd numbered samples of $x[n]$.[†]

In the continuous-time case, time compression merely speeds up the signal without loss of any data. In contrast, decimation of $x[n]$ generally causes loss of data. Under certain conditions— for example, if $x[n]$ is the result of oversampling some continuous-time signal—then $x_d[n]$ may still retain the complete information about $x[n]$.

An *interpolated* signal is generated in two steps; first, we expand $x[n]$ by an integer factor L to obtain the expanded signal $x_e[n]$, as

$$x_e[n] = \begin{cases} x[n/L] & n = 0, \pm L \pm 2L, \ldots, \\ 0 & \text{otherwise} \end{cases} \tag{3.4}$$

To understand this expression, consider a simple case of expanding $x[n]$ by a factor 2 ($L = 2$). The expanded signal $x_e[n] = x[n/2]$. When n is odd, $n/2$ is noninteger. But $x[n]$ is defined only for integer values of n and is zero otherwise. Therefore, $x_e[n] = 0$, for odd n, that is,

---

[†]Odd-numbered samples of $x[n]$ can be retained (and even numbered samples deleted) by using the transformation $x_d[n] = x[2n + 1]$.

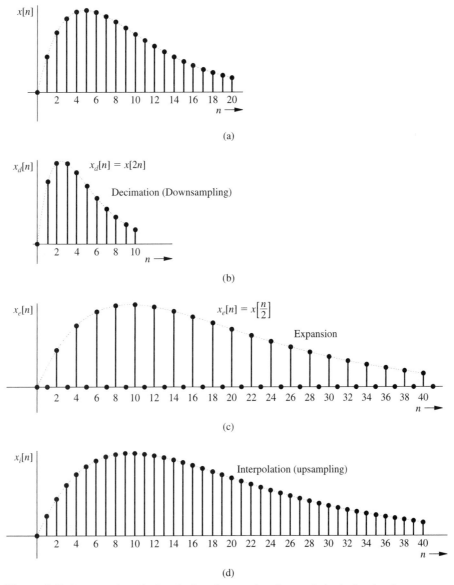

Figure 3.5 Compression (decimation) and expansion (interpolation) of a signal.

$x_e[1] = x_e[3] = x_e[5], \ldots$ are all zero, as depicted in Fig. 3.5c. Moreover, $n/2$ is integer for even n, and the values of $x_e[n] = x[n/2]$ for $n = 0, 2, 4, 6, \ldots$, are $x[0], x[1], x[2], x[3], \ldots$, as shown in Fig. 3.5c. In general, for $n = 0, 1, 2, \ldots$, $x_e[n]$ is given by the sequence

$$x[0], \underbrace{0, 0, \ldots, 0, 0}_{L-1 \text{ zeros}}, x[1], \underbrace{0, 0, \ldots, 0, 0}_{L-1 \text{ zeros}}, x[2], \underbrace{0, 0, \ldots, 0, 0}_{L-1 \text{ zeros}}, \ldots$$

Thus, the sampling rate of $x_e[n]$ is L times that of $x[n]$. Hence, this operation is called *expanding*. The expanded signal $x_e[n]$ contains all the data of $x[n]$, although in an expanded form.

In the expanded signal in Fig. 3.5c, the missing (zero-valued) odd-numbered samples can be reconstructed from the non-zero-valued samples by using some suitable interpolation formula. Figure 3.5d shows such an interpolated signal $x_i[n]$, where the missing samples are constructed by using an interpolating filter. The optimum interpolating filter is usually an ideal lowpass filter, which is realizable only approximately. In practice, we may use an interpolation that is nonoptimum but realizable. Further discussion of interpolation is beyond our scope. This process of filtering to interpolate the zero-valued samples is called *interpolation*. Since the interpolated data are computed from the existing data, interpolation does not result in gain of information.

EXERCISE E3.2

Show that $x[n]$ in Fig. 3.4a left-shifted by 3 units can be expressed as $0.729(0.9)^n$ for $0 \leq n \leq 7$, and zero otherwise. Sketch the shifted signal.

EXERCISE E3.3

Sketch the signal $x[n] = e^{-0.5n}$ for $-3 \leq n \leq 2$, and zero otherwise. Sketch the corresponding time-reversed signal and show that it can be expressed as $x_r[n] = e^{0.5n}$ for $-2 \leq n \leq 3$.

EXERCISE E3.4

Show that $x[-k - n]$ can be obtained from $x[n]$ by first right-shifting $x[n]$ by k units and then time-reversing this shifted signal.

EXERCISE E3.5

A signal $x[n]$ is expanded by factor 2 to obtain signal $x[n/2]$. The odd-numbered samples (n odd) in this signal have zero value. Show that the linearly interpolated odd-numbered samples are given by $x_i[n] = (1/2)\{x[n - 1] + x[n + 1]\}$.

3.3 SOME USEFUL DISCRETE-TIME SIGNAL MODELS

We now discuss some important discrete-time signal models that are encountered frequently in the study of discrete-time signals and systems.

3.3-1 Discrete-Time Impulse Function $\delta[n]$

The discrete-time counterpart of the continuous-time impulse function $\delta(t)$ is $\delta[n]$, a Kronecker delta function, defined by

$$\delta[n] = \begin{cases} 1 & n = 0 \\ 0 & n \neq 0 \end{cases} \tag{3.5}$$

This function, also called the unit impulse sequence, is shown in Fig. 3.6a. The shifted impulse sequence $\delta[n-m]$ is depicted in Fig. 3.6b. Unlike its continuous-time counterpart $\delta(t)$ (the Dirac delta), the Kronecker delta is a very simple function, requiring no special esoteric knowledge of distribution theory.

3.3-2 Discrete-Time Unit Step Function $u[n]$

The discrete-time counterpart of the unit step function $u(t)$ is $u[n]$ (Fig. 3.7a), defined by

$$u[n] = \begin{cases} 1 & \text{for } n \geq 0 \\ 0 & \text{for } n < 0 \end{cases} \tag{3.6}$$

If we want a signal to start at $n=0$ (so that it has a zero value for all $n < 0$), we need only multiply the signal by $u[n]$.

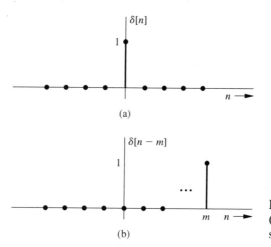

(a)

(b)

Figure 3.6 Discrete-time impulse function: **(a)** unit impulse sequence and **(b)** shifted impulse sequence.

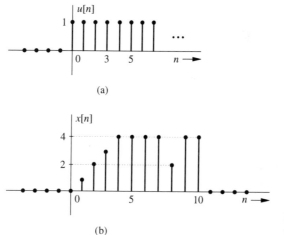

(a)

(b)

Figure 3.7 **(a)** A discrete-time unit step function $u[n]$ and **(b)** its application.

EXAMPLE 3.3

Describe the signal $x[n]$ shown in Fig. 3.7b by a single expression valid for all n.

There are many different ways of viewing $x[n]$. Although each way of viewing yields a different expression, they are all equivalent. We shall consider here just one possible expression.

The signal $x[n]$ can be broken into three components: (1) a ramp component $x_1[n]$ from $n = 0$ to 4, (2) a step component $x_2[n]$ from $n = 5$ to 10, and (3) an impulse component $x_3[n]$ represented by the negative spike at $n = 8$. Let us consider each one separately.

We express $x_1[n] = n(u[n] - u[n-5])$ to account for the signal from $n = 0$ to 4. Assuming that the spike at $n = 8$ does not exist, we can express $x_2[n] = u[n-5] - u[n-11]$ to account for the signal from $n = 5$ to 10. Once these two components have been added, the only part that is unaccounted for is a spike of amplitude -2 at $n = 8$, which can be represented by $x_3[n] = -2\delta[n-8]$. Hence

$$x[n] = x_1[n] + x_2[n] + x_3[n]$$
$$= n(u[n] - u[n-5]) + u[n-5] - u[n-11] - 2\delta[n-8] \qquad \text{for all } n$$

We stress again that the expression is valid for all values of n. The reader can find several other equivalent expressions for $x[n]$. For example, one may consider the step function from $n = 0$ to 10, subtract a ramp over the range $n = 0$–3, and subtract the spike. You can also play with breaking n into different ranges for your expression.

3.3-3 Discrete-Time Exponential γ^n

A continuous-time exponential $e^{\lambda t}$ can be expressed in an alternate form as

$$e^{\lambda t} = \gamma^t \qquad (\gamma = e^\lambda \text{ or } \lambda = \ln \gamma)$$

For example, $e^{-0.3t} = (0.7408)^t$ because $e^{-0.3} = 0.7408$. Conversely, $4^t = e^{1.386t}$ because $e^{1.386} = 4$, that is, $\ln 4 = 1.386$. In the study of continuous-time signals and systems we prefer the form $e^{\lambda t}$ rather than γ^t. In contrast, the exponential form γ^n is preferable in the study of discrete-time signals and systems, as will become apparent later. The discrete-time exponential γ^n can also be expressed by using a natural base, as

$$e^{\lambda n} = \gamma^n \qquad (\gamma = e^\lambda \text{ or } \lambda = \ln \gamma)$$

Because of unfamiliarity with exponentials with bases other than e, exponentials of the form γ^n may seem inconvenient and confusing at first. The reader is urged to plot some exponentials to acquire a sense of these functions.

Nature of γ^n. The signal $e^{\lambda n}$ grows exponentially with n if $\text{Re } \lambda > 0$ (λ in the RHP), and decays exponentially if $\text{Re } \lambda < 0$ (λ in the LHP). It is constant or oscillates with constant amplitude if $\text{Re } \lambda = 0$ (λ on the imaginary axis). Clearly, the location of λ in the complex plane indicates whether the signal $e^{\lambda n}$ will grow exponentially, decay exponentially, or oscillate with constant amplitude (Fig. 3.8a). A constant signal ($\lambda = 0$) is also an oscillation with zero frequency. We now find a similar criterion for determining the nature of γ^n from the location of γ in the complex plane.

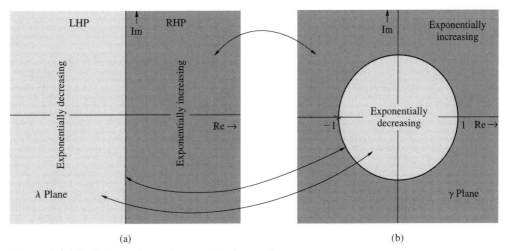

Figure 3.8 The λ plane, the γ plane, and their mapping.

Figure 3.8a shows a complex plane (λ plane). Consider a signal $e^{j\Omega n}$. In this case, $\lambda = j\Omega$ lies on the imaginary axis (Fig. 3.8a), and therefore is a constant-amplitude oscillating signal. This signal $e^{j\Omega n}$ can be expressed as γ^n, where $\gamma = e^{j\Omega}$. Because the magnitude of $e^{j\Omega}$ is unity, $|\gamma| = 1$. Hence, when λ lies on the imaginary axis, the corresponding γ lies on a circle of unit radius, centered at the origin (the *unit circle* illustrated in Fig. 3.8b). Therefore, a signal γ^n oscillates with constant amplitude if γ lies on the unit circle. Thus, the imaginary axis in the λ plane maps into the unit circle in the γ plane.

Next consider the signal $e^{\lambda n}$, where λ lies in the left half-plane in Fig. 3.8a. This means $\lambda = a + jb$, where a is negative ($a < 0$). In this case, the signal decays exponentially. This signal can be expressed as γ^n, where

$$\gamma = e^\lambda = e^{a+jb} = e^a\, e^{jb}$$

and

$$|\gamma| = |e^a|\,|e^{jb}| = e^a \qquad \text{because } |e^{jb}| = 1$$

Also, a is negative ($a < 0$). Hence, $|\gamma| = e^a < 1$. This result means that the corresponding γ lies inside the unit circle. Therefore, a signal γ^n decays exponentially if γ lies within the unit circle (Fig. 3.8b). If, in the preceding case we select a to be positive, (λ in the right half-plane), then $|\gamma| > 1$, and γ lies outside the unit circle. Therefore, a signal γ^n grows exponentially if γ lies outside the unit circle (Fig. 3.8b).

To summarize, the imaginary axis in the λ plane maps into the unit circle in the γ plane. The left half-plane in the λ plane maps into the inside of the unit circle and the right half of the λ plane maps into the outside of the unit circle in the γ plane, as depicted in Fig. 3.8. Observe that

$$\gamma^{-n} = \left(\frac{1}{\gamma}\right)^n$$

Plots of $(0.8)^n$ and $(-0.8)^n$ appear in Fig. 3.9a and 3.9b, respectively. Plots of $(0.5)^n$, and $(1.1)^n$, appear in Fig. 3.9c and 3.9d, respectively. These plots verify our earlier conclusions about the location of γ and the nature of signal growth. Observe that a signal $(-\gamma)^n$ alternates sign successively (is positive for even values of n and negative for odd values of n, as depicted

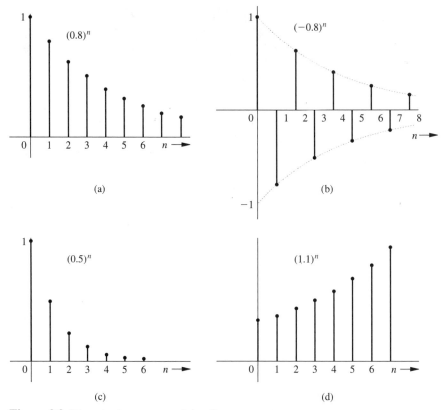

Figure 3.9 Discrete-time exponentials γ^n.

in Fig. 3.9b). Also, the exponential $(0.5)^n$ decays faster than $(0.8)^n$ because 0.5 is closer to the origin than 0.8. The exponential $(0.5)^n$ can also be expressed as 2^{-n} because $(0.5)^{-1} = 2$.

EXERCISE E3.6

Sketch the following signals

    (a) $(1)^n$

    (b) $(-1)^n$

    (c) $(0.5)^n$

    (d) $(-0.5)^n$

    (e) $(0.5)^{-n}$

    (f) 2^{-n}

    (g) $(-2)^n$

Express these exponentials as γ^n, and plot γ in the complex plane for each case. Verify that γ^n decays exponentially with n if γ lies inside the unit circle and that γ^n grows with n if γ is outside the unit circle. If γ is on the unit circle, γ^n is constant or oscillates with a constant amplitude.

EXERCISE E3.7

(a) Show that (i) $(0.25)^{-n} = 4^n$, (ii) $4^{-n} = (0.25)^n$, (iii) $e^{2t} = (7.389)^t$, (iv) $e^{-2t} = (0.1353)^t = (7.389)^{-t}$, (v) $e^{3n} = (20.086)^n$, and (vi) $e^{-1.5n} = (0.2231)^n = (4.4817)^{-n}$.

(b) Show that (i) $2^n = e^{0.693n}$, (ii) $(0.5)^n = e^{-0.693n}$, and (iii) $(0.8)^{-n} = e^{0.2231n}$.

COMPUTER EXAMPLE C3.1

Sketch the following discrete-time signals:

(a) $x_a[n] = (-0.5)^n$

(b) $x_b[n] = (2)^{-n}$

(c) $x_c[n] = (-2)^n$

```
>> n = (0:5);
>> x_a = (-0.5).^n; x_b = 2.^(-n); x_c = (-2).^n;
>> subplot(3,1,1); stem(n,x_a,'k'); ylabel('x_a[n]')
>> subplot(3,1,2); stem(n,x_b,'k'); ylabel('x_b[n]')
>> subplot(3,1,3); stem(n,x_c,'k'); ylabel('x_c[n]'); xlabel('n');
```

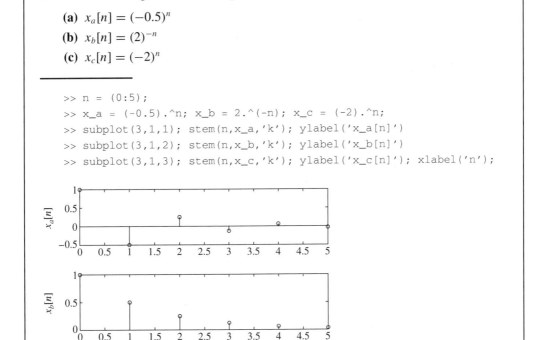

Figure C3.1

3.3-4 Discrete-Time Sinusoid $\cos(\Omega n + \theta)$

A general discrete-time sinusoid can be expressed as $C\cos(\Omega n + \theta)$, where C is the *amplitude*, and θ is the *phase* in radians. Also, Ωn is an angle in radians. Hence, the dimensions of the

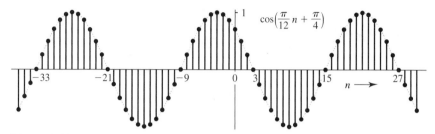

Figure 3.10 A discrete-time sinusoid $\cos\left(\frac{\pi}{12}n + \frac{\pi}{4}\right)$.

frequency Ω are *radians per sample*. This sinusoid may also be expressed as

$$C \cos\left(\Omega n + \theta\right) = C \cos\left(2\pi \mathcal{F}n + \theta\right)$$

where $\mathcal{F} = \Omega/2\pi$. Therefore, the dimensions of the discrete-time frequency $\mathcal{F}$ are (radians/2π) per sample, which is equal to *cycles per sample*. This means if N_0 is the period (samples/cycle) of the sinusoid, then the frequency of the sinusoid $\mathcal{F} = 1/N_0$ (samples/cycle).

Figure 3.10 shows a discrete-time sinusoid $\cos\left(\frac{\pi}{12}n + \frac{\pi}{4}\right)$. For this case, the frequency is $\Omega = \pi/12$ radians/sample. Alternately, the frequency is $\mathcal{F} = 1/24$ cycles/sample. In other words, there are 24 samples in one cycle of the sinusoid.

Because $\cos\left(-x\right) = \cos\left(x\right)$

$$\cos\left(-\Omega n + \theta\right) = \cos\left(\Omega n - \theta\right) \tag{3.7}$$

This shows that both $\cos\left(\Omega n + \theta\right)$ and $\cos\left(-\Omega n + \theta\right)$ have the same frequency (Ω). Therefore, *the frequency of* $\cos\left(\Omega n + \theta\right)$ *is* $|\Omega|$.

SAMPLED CONTINUOUS-TIME SINUSOID YIELDS A DISCRETE-TIME SINUSOID

A continuous-time sinusoid $\cos \omega t$ sampled every T seconds yields a discrete-time sequence whose nth element (at $t = nT$) is $\cos \omega nT$. Thus, the sampled signal $x[n]$ is given by

$$x[n] = \cos \omega nT$$

$$= \cos \Omega n \qquad \text{where } \Omega = \omega T \tag{3.8}$$

Thus, a continuous-time sinusoid $\cos \omega t$ sampled every T seconds yields a discrete-time sinusoid $\cos \Omega n$, where $\Omega = \omega T$.[†]

[†]Superficially, it may appear that a discrete-time sinusoid is a continuous-time sinusoid's cousin in a striped suit. However, some of the properties of discrete-time sinusoids are very different from those of continuous-time sinusoids. For instance, not every discrete-time sinusoid is periodic. A sinusoid $\cos \Omega n$ is periodic only if Ω is a rational multiple of 2π. Also, discrete-time sinusoids are bandlimited to $\Omega = \pi$. Any sinusoid with $\Omega \geq \pi$ can always be expressed as a sinusoid of some frequency $\Omega \leq \pi$. These peculiar properties are the direct consequence of the fact that the period of a discrete-time sinusoid must be an integer. These topics are discussed in Chapters 5 and 9.

3.3-5 Discrete-Time Complex Exponential $e^{j\Omega n}$

Using Euler's formula, we can express an exponential $e^{j\Omega n}$ in terms of sinusoids of the form $\cos(\Omega n + \theta)$, and vice versa

$$e^{j\Omega n} = (\cos \Omega n + j \sin \Omega n)$$

$$e^{-j\Omega n} = (\cos \Omega n - j \sin \Omega n)$$

These equations show that *the frequency of both $e^{j\Omega n}$ and $e^{-j\Omega n}$ is Ω* (radians/sample). Therefore, the frequency of $e^{j\Omega n}$ is $|\Omega|$.

   Observe that for $r = 1$ and $\theta = n\Omega$,

$$e^{j\Omega n} = re^{j\theta}$$

This equation shows that the magnitude and angle of $e^{j\Omega n}$ are 1 and $n\Omega$, respectively. In the complex plane, $e^{j\Omega n}$ is a point on a unit circle at an angle $n\Omega$.

COMPUTER EXAMPLE C3.2

Sketch the discrete-time sinusoid

$$x[n] = \cos\left(\frac{\pi}{12}n + \frac{\pi}{4}\right)$$

```
>> n = (-30:30); x = cos(n*pi/12+pi/4);
>> clf; stem(n,x,'k'); xlabel('n');
```

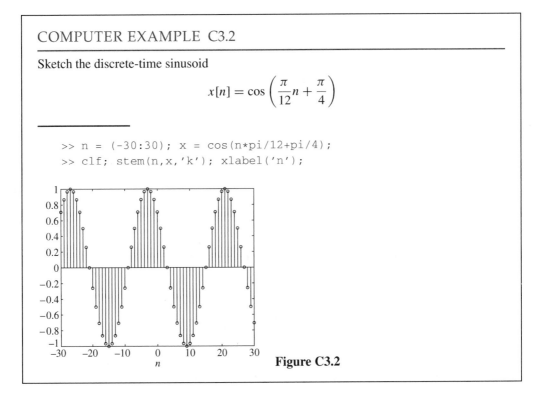

Figure C3.2

3.4 EXAMPLES OF DISCRETE-TIME SYSTEMS

We shall give here four examples of discrete-time systems. In the first two examples, the signals are inherently of the discrete-time variety. In the third and fourth examples, a continuous-time signal is processed by a discrete-time system, as illustrated in Fig. 3.2, by discretizing the signal through sampling.

EXAMPLE 3.4 (Savings Account)

A person makes a deposit (the input) in a bank regularly at an interval of T (say, 1 month). The bank pays a certain interest on the account balance during the period T and mails out a periodic statement of the account balance (the output) to the depositor. Find the equation relating the output $y[n]$ (the balance) to the input $x[n]$ (the deposit).

In this case, the signals are inherently discrete time. Let

$$x[n] = \text{deposit made at the } n\text{th discrete instant}$$

$$y[n] = \text{account balance at the } n\text{th instant computed}$$
$$\text{immediately after receipt of the } n\text{th deposit } x[n]$$

$$r = \text{interest per dollar per period } T$$

The balance $y[n]$ is the sum of (i) the previous balance $y[n-1]$, (ii) the interest on $y[n-1]$ during the period T, and (iii) the deposit $x[n]$

$$y[n] = y[n-1] + ry[n-1] + x[n]$$
$$= (1+r)y[n-1] + x[n]$$

or

$$y[n] - ay[n-1] = x[n] \qquad a = 1+r \tag{3.9a}$$

In this example the deposit $x[n]$ is the input (cause) and the balance $y[n]$ is the output (effect).

A withdrawal from the account is a negative deposit. Therefore, this formulation can handle deposits as well as withdrawals. It also applies to a loan payment problem with the initial value $y[0] = -M$, where M is the amount of the loan. A loan is an initial deposit with a negative value. Alternately, we may treat a loan of M dollars taken at $n = 0$ as an input of $-M$ at $n = 0$ (see Prob. 3.8-16).

We can express Eq. (3.9a) in an alternate form. The choice of index n in Eq. (3.9a) is completely arbitrary, so we can substitute $n + 1$ for n to obtain

$$y[n+1] - ay[n] = x[n+1] \tag{3.9b}$$

We also could have obtained Eq. (3.9b) directly by realizing that $y[n+1]$, the balance at the $(n+1)$st instant, is the sum of $y[n]$ plus $ry[n]$ (the interest on $y[n]$) plus the deposit (input) $x[n+1]$ at the $(n+1)$st instant.

The difference equation in (3.9a) uses delay operation, whereas the form in Eq. (3.9b) uses advance operation. We shall call the form (3.9a) the *delay operator* form and the form (3.9b) the *advance operator* form. The delay operator form is more natural because operation of delay is a causal, and hence realizable. In contrast, advance operation, being noncausal,

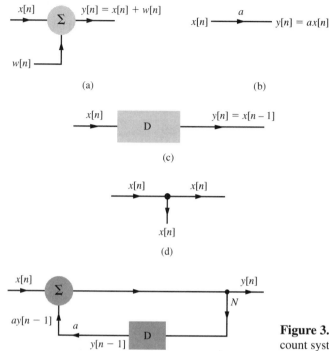

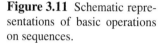

Figure 3.11 Schematic representations of basic operations on sequences.

Figure 3.12 Realization of the savings account system.

is unrealizable. We use the advance operator form primarily for its mathematical convenience over the delay form.[†]

We shall now represent this system in a block diagram form, which is basically a road map to hardware (or software) realization of the system. For this purpose, the causal (realizable) delay operator form in Eq. (3.9a) will be used. There are three basic operations in this equation: *addition, scalar multiplication,* and *delay*. Figure 3.11 shows their schematic representation. In addition, we also have a *pickoff* node (Fig. 3.11d), which is used to provide multiple copies of a signal at its input.

Equation (3.9a) can be rewritten as

$$y[n] = ay[n-1] + x[n] \qquad a = 1 + r \qquad (3.9c)$$

Figure 3.12 shows in block diagram form a system represented by Eq. (3.9c). To understand this realization, assume that the output $y[n]$ is available at the pickoff node N. Unit delay of $y[n]$ results in $y[n-1]$, which is multiplied by a scalar of value a to yield $ay[n-1]$. Next, we generate $y[n]$ by adding the input $x[n]$ and $ay[n-1]$ according to Eq. (3.9c).[‡] Observe that the node N is a pickoff node, from which two copies of the output signal flow out; one as the feedback signal, and the other as the output signal.

[†]Use of the advance operator form results in discrete-time system equations that are identical in form to those for continuous-time systems. This will become apparent later. In the transform analysis, use of the advance operator permits the use of more convenient variable z instead of the clumsy z^{-1} needed in the delay operator form.

[‡]A unit delay represents one unit of time delay. In this example one unit of delay in the output corresponds to period T for the actual output.

EXAMPLE 3.5 (Sales Estimate)

In an nth semester, $x[n]$ students enroll in a course requiring a certain textbook. The publisher sells $y[n]$ new copies of the book in the nth semester. On the average, one-quarter of students with books in salable condition resell the texts at the end of the semester, and the book life is three semesters. Write the equation relating $y[n]$, the new books sold by the publisher, to $x[n]$, the number of students enrolled in the nth semester, assuming that every student buys a book.

In the nth semester, the total books $x[n]$ sold to students must be equal to $y[n]$ (new books from the publisher) plus used books from students enrolled in the preceding two semesters (because the book life is only three semesters). There are $y[n-1]$ new books sold in the $(n-1)$st semester, and one-quarter of these books, that is, $(1/4)y[n-1]$, will be resold in the nth semester. Also, $y[n-2]$ new books are sold in the $(n-2)$nd semester, and one-quarter of these, that is, $(1/4)y[n-2]$, will be resold in the $(n-1)$st semester. Again a quarter of these, that is, $(1/16)y[n-2]$, will be resold in the nth semester. Therefore, $x[n]$ must be equal to the sum of $y[n]$, $(1/4)y[n-1]$, and $(1/16)y[n-2]$.

$$y[n] + \tfrac{1}{4}y[n-1] + \tfrac{1}{16}y[n-2] = x[n] \tag{3.10a}$$

Equation (3.10a) can also be expressed in an alternative form by realizing that this equation is valid for any value of n. Therefore, replacing n by $n+2$, we obtain

$$y[n+2] + \tfrac{1}{4}y[n+1] + \tfrac{1}{16}y[n] = x[n+2] \tag{3.10b}$$

This is the alternative form of Eq. (3.10a).

For a realization of a system with this input–output equation, we rewrite the delay form Eq. (3.10a) as

$$y[n] = -\tfrac{1}{4}y[n-1] - \tfrac{1}{16}y[n-2] + x[n] \tag{3.10c}$$

Figure 3.13 shows a hardware realization of Eq. (3.10c) using two unit delays in cascade.[†]

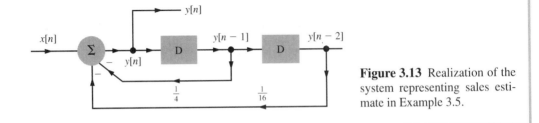

Figure 3.13 Realization of the system representing sales estimate in Example 3.5.

EXAMPLE 3.6 (Digital Differentiator)

Design a discrete-time system, like the one in Fig. 3.2, to differentiate continuous-time signals. This differentiator is used in an audio system having an input signal bandwidth below 20 kHz.

In this case, the output $y(t)$ is required to be the derivative of the input $x(t)$. The discrete-time processor (system) G processes the samples of $x(t)$ to produce the discrete-time output $y[n]$. Let $x[n]$ and $y[n]$ represent the samples T seconds apart of the signals $x(t)$ and $y(t)$, respectively, that is,

$$x[n] = x(nT) \qquad \text{and} \qquad y[n] = y(nT) \tag{3.11}$$

The signals $x[n]$ and $y[n]$ are the input and the output for the discrete-time system G. Now, we require that

$$y(t) = \frac{dx}{dt}$$

Therefore, at $t = nT$ (see Fig. 3.14a)

$$y(nT) = \frac{dx}{dt}\bigg|_{t=nT}$$

$$= \lim_{T \to 0} \frac{1}{T}[x(nT) - x[(n-1)T]]$$

By using the notation in Eq. (3.11), the foregoing equation can be expressed as

$$y[n] = \lim_{T \to 0} \frac{1}{T}\{x[n] - x[n-1]\}$$

This is the input–output relationship for G required to achieve our objective. In practice, the sampling interval T cannot be zero. Assuming T to be sufficiently small, the equation just given can be expressed as

$$y[n] = \frac{1}{T}\{x[n] - x[n-1]\} \tag{3.12}$$

The approximation improves as T approaches 0. A discrete-time processor G to realize Eq. (3.12) is shown inside the shaded box in Fig. 3.14b. The system in Fig. 3.14b acts as a differentiator. This example shows how a continuous-time signal can be processed by a discrete-time system. The considerations for determining the sampling interval T are discussed in Chapters 5 and 8, where it is shown that to process frequencies below 20 kHz, the proper choice is

$$T \leq \frac{1}{2 \times \text{highest frequency}} = \frac{1}{40{,}000} = 25 \ \mu s$$

To see how well this method of signal processing works, let us consider the differentiator in Fig. 3.14b with a ramp input $x(t) = t$, depicted in Fig. 3.14c. If the system were to act as

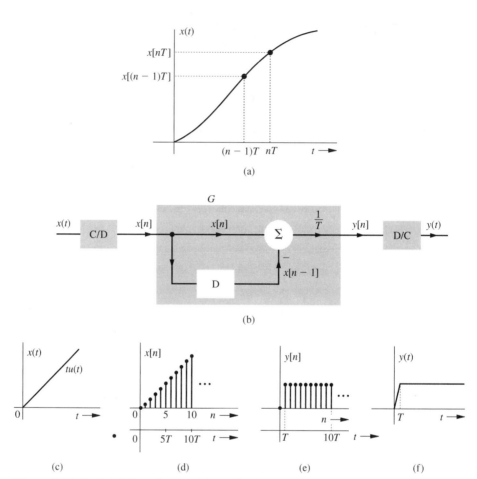

Figure 3.14 Digital differentiator and its realization.

a differentiator, then the output $y(t)$ of the system should be the unit step function $u(t)$. Let us investigate how the system performs this particular operation and how well the system achieves the objective.

The samples of the input $x(t) = t$ at the interval of T seconds act as the input to the discrete-time system G. These samples, denoted by a compact notation $x[n]$, are, therefore,

$$x[n] = x(t)|_{t=nT} = t|_{t=nT} \qquad t \geq 0$$

$$= nT \qquad n \geq 0$$

Figure 3.14d shows the sampled signal $x[n]$. This signal acts as an input to the discrete-time system G. Figure 3.14b shows that the operation of G consists of subtracting a sample from the preceding (delayed) sample and then multiplying the difference with $1/T$. From Fig. 3.14d, it is clear that the difference between the successive samples is a constant $nT - (n-1)T = T$

for all samples, except for the sample at $n = 0$ (because there is no preceding sample at $n = 0$). The output of G is $1/T$ times the difference T, which is unity for all values of n, except $n = 0$, where it is zero. Therefore, the output $y[n]$ of G consists of samples of unit values for $n \geq 1$, as illustrated in Fig. 3.14e. The D/C (discrete-time to continuous-time) converter converts these samples into a continuous-time signal $y(t)$, as shown in Fig. 3.14f. Ideally, the output should have been $y(t) = u(t)$. This deviation from the ideal is due to our use of a nonzero sampling interval T. As T approaches zero, the output $y(t)$ approaches the desired output $u(t)$.

The digital differentiator in Eq. (3.12) is an example of what is known as the *backward difference* system. The reason for calling it so is obvious from Fig. 3.14a. To compute the derivative of $y(t)$, we are using the difference between the present sample value and the preceding (backward) sample value. If we use the difference between the next (forward) sample at $t = (n + 1)T$ and the present sample at $t = nT$, we obtain the forward difference form of differentiator as

$$y[n] = \frac{1}{T}\{x[n + 1] - x[n]\} \tag{3.13}$$

EXAMPLE 3.7 (Digital Integrator)

Design a digital integrator along the same lines as the digital differentiator in Example 3.6.

For an integrator, the input $x(t)$ and the output $y(t)$ are related by

$$y(t) = \int_{-\infty}^{t} x(\tau)\, d\tau$$

Therefore, at $t = nT$ (see Fig. 3.14a)

$$y(nT) = \lim_{T \to 0} \sum_{k=-\infty}^{n} x(kT)T$$

Using the usual notation $x(kT) = x[k]$, $y(nT) = y[n]$, and so on, this equation can be expressed

$$y[n] = \lim_{T \to 0} T \sum_{k=-\infty}^{n} x[k]$$

Assuming that T is small enough to justify the assumption $T \to 0$, we have

$$y[n] = T \sum_{k=-\infty}^{n} x[k] \tag{3.14a}$$

This equation represents an example of *accumulator* system. This digital integrator equation can be expressed in an alternate form. From Eq. (3.14a), it follows that

$$y[n] - y[n-1] = Tx[n] \tag{3.14b}$$

This is an alternate description for the digital integrator. Equations (3.14a) and (3.14b) are equivalent; the one can be derived from the other. Observe that the form of Eq. (3.14b) is similar to that of Eq. (3.9a). Hence, the block diagram representation of the digital differentiator in the form (3.14b) is identical to that in Fig. 3.12 with $a = 1$ and the input multiplied by T.

RECURSIVE AND NONRECURSIVE FORMS OF DIFFERENCE EQUATION

If Eq. (3.14b) expresses Eq. (3.14a) in another form, what is the difference between these two forms? Which form is preferable? To answer these questions, let us examine how the output is computed by each of these forms. In Eq. (3.14a), the output $y[n]$ at any instant n is computed by adding all the past input values till n. This can mean a large number of additions. In contrast, Eq. (3.14b) can be expressed as $y[n] = y[n-1] + Tx[n]$. Hence, computation of $y[n]$ involves addition of only two values; the preceding output value $y[n-1]$ and the present input value $x[n]$. The computations are done recursively by using the preceding output values. For example, if the input starts at $n = 0$, we first compute $y[0]$. Then we use the computed value $y[0]$ to compute $y[1]$. Knowing $y[1]$, we compute $y[2]$, and so on. The computations are recursive. This is why the form (3.14b) is called *recursive* form and the form (3.14a) is called *nonrecursive* form. Clearly, "recursive" and "nonrecursive" describe two different ways of presenting the same information. Equations (3.9), (3.10), and (3.14b) are examples of recursive form, and Eqs. (3.12) and (3.14a) are examples of nonrecursive form.

KINSHIP OF DIFFERENCE EQUATIONS TO DIFFERENTIAL EQUATIONS

We now show that a digitized version of a differential equation results in a difference equation. Let us consider a simple first-order differential equation

$$\frac{dy}{dt} + cy(t) = x(t) \tag{3.15a}$$

Consider uniform samples of $x(t)$ at intervals of T seconds. As usual, we use the notation $x[n]$ to denote $x(nT)$, the nth sample of $x(t)$. Similarly, $y[n]$ denotes $y[nT]$, the nth sample of $y(t)$. From the basic definition of a derivative, we can express Eq. (3.15a) at $t = nT$ as

$$\lim_{T \to 0} \frac{y[n] - y[n-1]}{T} + cy[n] = x[n]$$

Clearing the fractions and rearranging the terms yields (assuming nonzero, but very small T)

$$y[n] + \alpha y[n-1] = \beta x[n] \tag{3.15b}$$

where

$$\alpha = \frac{-1}{1+cT} \quad \text{and} \quad \beta = \frac{T}{1+cT}$$

We can also express Eq. (3.15b) in advance operator form as

$$y[n+1] + \alpha y[n] = \beta x[n+1] \tag{3.15c}$$

It is clear that a differential equation can be approximated by a difference equation of the same order. In this way, we can approximate an nth-order differential equation by a difference equation of nth order. Indeed, a digital computer solves differential equations by using an equivalent difference equation, which can be solved by means of simple operations of addition, multiplication, and shifting. Recall that a computer can perform only these simple operations. It must necessarily approximate complex operation like differentiation and integration in terms of such simple operations. The approximation can be made as close to the exact answer as possible by choosing sufficiently small value for T.

At this stage, we have not developed tools required to choose a suitable value of the sampling interval T. This subject is discussed in Chapter 5 and also in Chapter 8. In Section 5.7, we shall discuss a systematic procedure (impulse invariance method) for finding a discrete-time system with which to realize an Nth-order LTIC system.

ORDER OF A DIFFERENCE EQUATION

Equations (3.9), (3.10), (3.13), (3.14b), and (3.15) are examples of difference equations. The highest-order difference of the output signal or the input signal, whichever is higher represents the *order* of the difference equation. Hence, Eqs. (3.9), (3.13), (3.14b), and (3.15) are first-order difference equations, whereas Eq. (3.10) is of the second order.

EXERCISE E3.8

Design a digital integrator in Example 3.7 using the fact that for an integrator, the output $y(t)$ and the input $x(t)$ are related by $dy/dt = x(t)$. Approximation (similar to that in Example 3.6) of this equation at $t = nT$ yields the recursive form in Eq. (3.14b).

ANALOG, DIGITAL, CONTINUOUS-TIME, AND DISCRETE-TIME SYSTEMS

The basic difference between continuous-time systems and analog systems, as also between discrete-time and digital systems, is fully explained in Sections 1.7-5 and 1.7-6.[†] Historically, discrete-time systems have been realized with digital computers, where continuous-time signals are processed through digitized samples rather than unquantized samples. Therefore, the terms *digital filters* and *discrete-time systems* are used synonymously in the literature. This distinction is

---

[†]The terms *discrete-time* and *continuous-time* qualify the nature of a signal along the time axis (horizontal axis). The terms *analog* and *digital,* in contrast, qualify the nature of the signal amplitude (vertical axis).

irrelevant in the analysis of discrete-time systems. For this reason, we follow this loose convention in this book, where the term *digital filter* implies a *discrete-time system,* and *analog filter* means *continuous-time system.* Moreover, the terms C/D (continuous-to-discrete-time) and D/C will occasionally be used interchangeably with terms A/D (analog-to-digital) and D/A, respectively.

ADVANTAGES OF DIGITAL
SIGNAL PROCESSING

1. Digital systems operation can tolerate considerable variation in signal values, and, hence, are less sensitive to changes in the component parameter values caused by temperature variation, aging and other factors. This results in greater degree of precision and stability. Since they are binary circuits, their accuracy can be increased by using more complex circuitry to increase word length, subject to cost limitations.

2. Digital systems do not require any factory adjustment and can be easily duplicated in volume without having to worry about precise component values. They can be fully integrated, and even highly complex systems can be placed on a single chip by using *VLSI* (very-large-scale integrated) circuits.

3. Digital filters are more flexible. Their characteristics can be easily altered simply by changing the program. Digital hardware implementation permits the use of micropro-cessors, miniprocessors, digital switching, and large-scale integrated circuits.

4. A greater variety of filters can be realized by digital systems.

5. Digital signals can be stored easily and inexpensively on magnetic tapes or disks without deterioration of signal quality. It is also possible to search and select information from distant electronic storehouses.

6. Digital signals can be coded to yield extremely low error rates and high fidelity, as well as privacy. Also, more sophisticated signal processing algorithms can be used to process digital signals.

7. Digital filters can be easily time-shared and therefore can serve a number of inputs simultaneously. Moreover, it is easier and more efficient to multiplex several digital signals on the same channel.

8. Reproduction with digital messages is extremely reliable without deterioration. Analog messages such as photocopies and films, for example, lose quality at each successive stage of reproduction and have to be transported physically from one distant place to another, often at relatively high cost.

   One must weigh these advantages against such disadvantages, as increased system complexity due to use of A/D and D/A interfaces, limited range of frequencies available in practice (about tens of megahertz), and use of more power than is needed for the passive analog circuits. Digital systems use power-consuming active devices.

3.4-1 Classification of Discrete-Time Systems

Before examining the nature of discrete-time system equations, let us consider the concepts of linearity, time invariance (or shift invariance), and causality, which apply to discrete-time systems also.

LINEARITY AND TIME INVARIANCE

For discrete-time systems, the definition of *linearity* is identical to that for continuous-time systems, as given in Eq. (1.40). We can show that the systems in Examples 3.4, 3.5, 3.6, and 3.7 are all linear.

Time invariance (or *shift invariance*) for discrete-time systems is also defined in a way similar to that for continuous-time systems. Systems whose parameters do not change with time (with n) are *time-invariant* or shift-invariant (also *constant-parameter*) systems. For such a system, if the input is delayed by k units or samples, the output is the same as before but delayed by k samples (assuming the initial conditions also are delayed by k). Systems in Examples 3.4, 3.5, 3.6, and 3.7 are time invariant because the coefficients in the system equations are constants (independent of n). If these coefficients were functions of n (time), then the systems would be linear *time-varying* systems. Consider, for example, a system described by

$$y[n] = e^{-n}x[n]$$

For this system, let a signal $x_1[n]$ yield the output $y_1[n]$, and another input $x_2[n]$ yield the output $y_2[n]$. Then

$$y_1[n] = e^{-n}x_1[n] \quad \text{and} \quad y_2[n] = e^{-n}x_2[n]$$

If we let $x_2[n] = x_1[n - N_0]$, then

$$y_2[n] = e^{-n}x_2[n] = e^{-n}x_1[n - N_0] \neq y_1[n - N_0]$$

Clearly, this is a time-varying parameter system.

CAUSAL AND NONCAUSAL SYSTEMS

A *causal* (also known as a *physical* or *nonanticipative*) system is one for which the output at any instant $n = k$ depends only on the value of the input $x[n]$ for $n \leq k$. In other words, the value of the output at the present instant depends only on the past and present values of the input $x[n]$, not on its future values. As we shall see, the systems in Examples 3.4, 3.5, 3.6, and 3.7 are all causal.

INVERTIBLE AND NONINVERTIBLE SYSTEMS

A discrete-time system S is invertible if an inverse system S_i exists such that the cascade of S and S_i results in an *identity* system. An identity system is defined as one whose output is identical to the input. In other words, for an invertible system, the input can be uniquely determined from the corresponding output. For every input there is a unique output. When a signal is processed through such a system, its input can be reconstructed from the corresponding output. There is no loss of information when a signal is processed through an invertible system.

A cascade of a unit delay with a unit advance results in an identity system because the output of such a cascaded system is identical to the input. Clearly, the inverse of an ideal unit delay is ideal unit advance, which is a noncausal (and unrealizable) system. In contrast, a compressor $y[n] = x[Mn]$ is not invertible because this operation loses all but every Mth sample of the input, and, generally, the input cannot be reconstructed. Similarly, an operations, such as, $y[n] = \cos x[n]$ or $y[n] = |x[n]|$ are not invertible.

EXERCISE E3.9

Show that a system specified by equation $y[n] = ax[n] + b$ is invertible, but the system $y[n] = |x[n]|^2$ is noninvertible.

STABLE AND UNSTABLE SYSTEMS

The concept of stability is similar to that in continuous-time systems. Stability can be *internal* or *external*. If every *bounded input* applied at the input terminal results in a *bounded output,* the system is said to be stable *externally*. The external stability can be ascertained by measurements at the external terminals of the system. This type of stability is also known as the stability in the BIBO (bounded-input/bounded-output) sense. Both, internal and external stability are discussed in greater detail in Section 3.10.

MEMORYLESS SYSTEMS AND SYSTEMS WITH MEMORY

The concepts of memoryless (or instantaneous) systems and those without memory (or dynamic) are identical to the corresponding concepts of the continuous-time case. A system is memoryless if its response at any instant n depends at most on the input at the same instant n. The output at any instant of a system with memory generally depends on the past, present, and future values of the input. For example, $y[n] = \sin x[n]$ is an example of instantaneous system, and $y[n] - y[n-1] = x[n]$ is an example of a dynamic system or a system with memory.

3.5 DISCRETE-TIME SYSTEM EQUATIONS

In this section we discuss time-domain analysis of LTID (linear time-invariant, discrete-time systems). With minor differences, the procedure is parallel to that for continuous-time systems.

DIFFERENCE EQUATIONS

Equations (3.9), (3.10), (3.12), and (3.15) are examples of difference equations. Equations (3.9), (3.12), and (3.15) are first-order difference equations, and Eq. (3.10) is a second-order difference equation. All these equations are linear, with constant (not time-varying) coefficients.[†] Before giving a general form of an Nth-order linear difference equation, we recall that a difference equation can be written in two forms; the first form uses delay terms such as $y[n-1]$, $y[n-2]$, $x[n-1]$, $x[n-2]$, and so on, and the alternate form uses advance terms such as $y[n+1]$, $y[n+2]$, and so on. Although the delay form is more natural, we shall often prefer the advance form, not just for the general notational convenience, but also for resulting notational uniformity with the

---

[†]Equations such as (3.9), (3.10), (3.12), and (3.15) are considered to be linear according to the classical definition of linearity. Some authors label such equations as *incrementally linear*. We prefer the classical definition. It is just a matter of individual choice and makes no difference in the final results.

operational form for differential equations. This facilitates the commonality of the solutions and concepts for continuous-time and discrete-time systems.

We start here with a general difference equation, using the advance operator form

$$y[n + N] + a_1 y[n + N - 1] + \cdots + a_{N-1} y[n + 1] + a_N y[n] = b_{N-M} x[n + M]$$
$$+ b_{N-M+1} x[n + M - 1] + \cdots + b_{N-1} x[n + 1] + b_N x[n] \tag{3.16}$$

This is a linear difference equation whose order is $\text{Max}(N, M)$. We have assumed the coefficient of $y[n+N]$ to be unity ($a_0 = 1$) without loss of generality. If $a_0 \neq 1$, we can divide the equation throughout by a_0 to normalize the equation to have $a_0 = 1$.

CAUSALITY CONDITION

For a causal system the output cannot depend on future input values. This means that when the system equation is in the advance operator form (3.16), the causality requires $M \leq N$. If M were to be greater than N, then $y[n + N]$, the output at $n + N$ would depend on $x[n + M]$, which is the input at the later instant $n + M$. For a general causal case, $M = N$, and Eq. (3.16) can be expressed as

$$y[n + N] + a_1 y[n + N - 1] + \cdots + a_{N-1} y[n + 1] + a_N y[n] = b_0 x[n + N]$$
$$+ b_1 x[n + N - 1] + \cdots + b_{N-1} x[n + 1] + b_N x[n] \tag{3.17a}$$

where some of the coefficients on either side can be zero. In this Nth-order equation, a_0, the coefficient of $y[n + N]$, is normalized to unity. Equation (3.17a) is valid for all values of n. Therefore, it is still valid if we replace n by $n - N$ throughout the equation [see Eqs. (3.9a) and (3.9b)]. Such replacement yields the alternative form (the delay operator form) of Eq. (3.17a):

$$y[n] + a_1 y[n - 1] + \cdots + a_{N-1} y[n - N + 1] + a_N y[n - N] = b_0 x[n]$$
$$+ b_1 x[n - 1] + \cdots + b_{N-1} x[n - N + 1] + b_N x[n - N] \tag{3.17b}$$

3.5-1 Recursive (Iterative) Solution of Difference Equation

Equation (3.17b) can be expressed as

$$y[n] = -a_1 y[n - 1] - a_2 y[n - 2] - \cdots - a_N y[n - N]$$
$$+ b_0 x[n] + b_1 x[n - 1] + \cdots + b_N x[n - N] \tag{3.17c}$$

In Eq. (3.17c), $y[n]$ is computed from $2N + 1$ pieces of information; the preceding N values of the output: $y[n - 1]$, $y[n - 2]$, $\ldots$, $y[n - N]$, and the preceding N values of the input: $x[n - 1]$, $x[n - 2]$, $\ldots$, $x[n - N]$, and the present value of the input $x[n]$. Initially, to compute $y[0]$, the N initial conditions $y[-1]$, $y[-2]$, $\ldots$, $y[-N]$ serve as the preceding N output values. Hence, knowing the N initial conditions and the input, we can determine the entire output $y[0]$, $y[1]$, $y[2]$, $y[3]$, $\ldots$ recursively, one value at a time. For instance, to find $y[0]$ we set $n = 0$ in Eq. (3.17c). The left-hand side is $y[0]$, and the right-hand side is expressed in terms of N initial conditions $y[-1]$, $y[-2]$, $\ldots$, $y[-N]$ and the input $x[0]$ if $x[n]$ is causal (because of causality, other input terms $x[-n] = 0$). Similarly, knowing $y[0]$ and the input, we can compute $y[1]$ by setting $n = 1$ in Eq. (3.17c). Knowing $y[0]$ and $y[1]$, we find $y[2]$, and so on. Thus, we can use this recursive procedure to find the complete response $y[0]$, $y[1]$, $y[2]$, ... For this reason, this equation is classed as a recursive form. This method basically reflects the manner in which a computer would solve a recursive difference equation, given the input and initial conditions. Equation (3.17) is nonrecursive if all the $N - 1$ coefficients $a_i = 0$ $(i = 1, 2, \ldots, N - 1)$. In this case, it can be seen that $y[n]$ is computed only from the input values and without using any previous outputs. Generally speaking, the recursive procedure applies only to equations in the recursive form. The recursive (iterative) procedure is demonstrated by the following examples.

EXAMPLE 3.8

Solve iteratively

$$y[n] - 0.5y[n-1] = x[n] \tag{3.18a}$$

with initial condition $y[-1] = 16$ and causal input $x[n] = n^2$ (starting at $n = 0$). This equation can be expressed as

$$y[n] = 0.5y[n-1] + x[n] \tag{3.18b}$$

If we set $n = 0$ in this equation, we obtain

$$y[0] = 0.5y[-1] + x[0]$$
$$= 0.5(16) + 0 = 8$$

Now, setting $n = 1$ in Eq. (3.18b) and using the value $y[0] = 8$ (computed in the first step) and $x[1] = (1)^2 = 1$, we obtain

$$y[1] = 0.5(8) + (1)^2 = 5$$

Next, setting $n = 2$ in Eq. (3.18b) and using the value $y[1] = 5$ (computed in the previous step) and $x[2] = (2)^2$, we obtain

$$y[2] = 0.5(5) + (2)^2 = 6.5$$

Continuing in this way iteratively, we obtain

$$y[3] = 0.5(6.5) + (3)^2 = 12.25$$
$$y[4] = 0.5(12.25) + (4)^2 = 22.125$$

$$\vdots$$

The output $y[n]$ is depicted in Fig. 3.15.

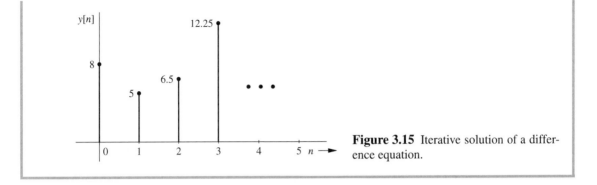

Figure 3.15 Iterative solution of a difference equation.

We now present one more example of iterative solution—this time for a second-order equation. The iterative method can be applied to a difference equation in delay or advance operator form. In Example 3.8 we considered the former. Let us now apply the iterative method to the advance operator form.

EXAMPLE 3.9

Solve iteratively

$$y[n + 2] - y[n + 1] + 0.24y[n] = x[n + 2] - 2x[n + 1] \qquad (3.19)$$

with initial conditions $y[-1] = 2$, $y[-2] = 1$ and a causal input $x[n] = n$ (starting at $n = 0$). The system equation can be expressed as

$$y[n + 2] = y[n + 1] - 0.24y[n] + x[n + 2] - 2x[n + 1] \qquad (3.20)$$

Setting $n = -2$ and then substituting $y[-1] = 2$, $y[-2] = 1$, $x[0] = x[-1] = 0$ (recall that $x[n] = n$ starting at $n = 0$), we obtain

$$y[0] = 2 - 0.24(1) + 0 - 0 = 1.76$$

Setting $n = -1$ in Eq. (3.20) and then substituting $y[0] = 1.76$, $y[-1] = 2$, $x[1] = 1$, $x[0] = 0$, we obtain

$$y[1] = 1.76 - 0.24(2) + 1 - 0 = 2.28$$

Setting $n = 0$ in Eq. (3.20) and then substituting $y[0] = 1.76$, $y[1] = 2.28$, $x[2] = 2$ and $x[1] = 1$ yields

$$y[2] = 2.28 - 0.24(1.76) + 2 - 2(1) = 1.8576$$

and so on.

Note carefully the recursive nature of the computations. From the N initial conditions (and the input) we obtained $y[0]$ first. Then, using this value of $y[0]$ and the preceding $N-1$ initial conditions (along with the input), we find $y[1]$. Next, using $y[0]$, $y[1]$ along with the past $N-2$ initial conditions and input, we obtained $y[2]$, and so on. This method is general and can be applied to a recursive difference equation of any order. It is interesting that the hardware realization of Eq. (3.18a) depicted in Fig. 3.12 (with $a = 0.5$) generates the solution precisely in this (iterative) fashion.

EXERCISE E3.10

Using the iterative method, find the first three terms of $y[n]$ for

$$y[n+1] - 2y[n] = x[n]$$

The initial condition is $y[-1] = 10$ and the input $x[n] = 2$ starting at $n = 0$.

ANSWER
$y[0] = 20$ $y[1] = 42$ $y[2] = 86$

COMPUTER EXAMPLE C3.3

Use MATLAB to solve Example 3.9.

```
>> n = (-2:10)'; y=[1;2;zeros(length(n)-2,1)]; x=[0;0;n(3:end)];
>> for k = 1:length(n)-2,
>> y(k+2) = y(k+1) - 0.24*y(k) + x(k+2) - 2*x(k+1);
>> end;
>> clf; stem(n,y,'k'); xlabel('n'); ylabel('y[n]');
>> disp(' n            y'); disp([num2str([n,y])]);
  n           y
 -2           1
 -1           2
  0         1.76
  1         2.28
  2        1.8576
  3        0.3104
  4       -2.13542
  5       -5.20992
  6       -8.69742
  7        -12.447
  8       -16.3597
  9       -20.3724
 10       -24.4461
```

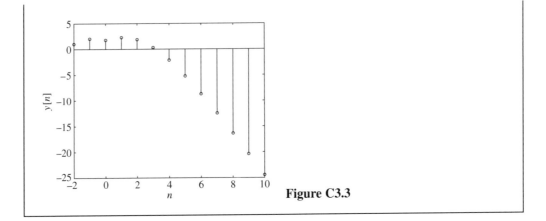

Figure C3.3

We shall see in the future that the solution of a difference equation obtained in this direct (iterative) way is useful in many situations. Despite the many uses of this method, a closed-form solution of a difference equation is far more useful in study of the system behavior and its dependence on the input and the various system parameters. For this reason we shall develop a systematic procedure to analyze discrete-time systems along lines similar to those used for continuous-time systems.

OPERATIONAL NOTATION

In difference equations it is convenient to use operational notation similar to that used in differential equations for the sake of compactness. In continuous-time systems we used the operator D to denote the operation of differentiation. For discrete-time systems we shall use the operator E to denote the operation for advancing a sequence by one time unit. Thus

$$Ex[n] \equiv x[n+1]$$
$$E^2x[n] \equiv x[n+2]$$
$$\vdots$$
$$E^Nx[n] \equiv x[n+N] \tag{3.21}$$

The first-order difference equation of the savings account problem was found to be [see Eq. (3.9b)]

$$y[n+1] - ay[n] = x[n+1] \tag{3.22}$$

Using the operational notation, we can express this equation as

$$Ey[n] - ay[n] = Ex[n]$$

or

$$(E-a)y[n] = Ex[n] \tag{3.23}$$

The second-order difference equation (3.10b)

$$y[n+2] + \tfrac{1}{4}y[n+1] + \tfrac{1}{16}y[n] = x[n+2]$$

can be expressed in operational notation as

$$\left(E^2 + \tfrac{1}{4}E + \tfrac{1}{16}\right)y[n] = E^2 x[n]$$

A general Nth-order difference Eq. (3.17a) can be expressed as

$$(E^N + a_1 E^{N-1} + \cdots + a_{N-1}E + a_N)y[n]$$
$$= (b_0 E^N + b_1 E^{N-1} + \cdots + b_{N-1}E + b_N)x[n] \qquad (3.24a)$$

or

$$Q[E]y[n] = P[E]x[n] \qquad (3.24b)$$

where $Q[E]$ and $P[E]$ are Nth-order polynomial operators

$$Q[E] = E^N + a_1 E^{N-1} + \cdots + a_{N-1}E + a_N \qquad (3.25)$$

$$P[E] = b_0 E^N + b_1 E^{N-1} + \cdots + b_{N-1}E + b_N \qquad (3.26)$$

RESPONSE OF LINEAR DISCRETE-TIME SYSTEMS

Following the procedure used for continuous-time systems, we can show that Eq. (3.24) is a linear equation (with constant coefficients). A system described by such an equation is a linear time-invariant, discrete-time (LTID) system. We can verify, as in the case of LTIC systems (see the footnote on page 152), that the general solution of Eq. (3.24) consists of zero-input and zero-state components.

3.6 SYSTEM RESPONSE TO INTERNAL CONDITIONS: THE ZERO-INPUT RESPONSE

The zero-input response $y_0[n]$ is the solution of Eq. (3.24) with $x[n] = 0$; that is

$$Q[E]y_0[n] = 0 \qquad (3.27a)$$

or

$$(E^N + a_1 E^{N-1} + \cdots + a_{N-1}E + a_N)y_0[n] = 0 \qquad (3.27b)$$

or

$$y_0[n+N] + a_1 y_0[n+N-1] + \cdots + a_{N-1}y_0[n+1] + a_N y_0[n] = 0 \qquad (3.27c)$$

We can solve this equation systematically. But even a cursory examination of this equation points to its solution. This equation states that a linear combination of $y_0[n]$ and advanced $y_0[n]$ is zero, *not for some values of n, but for all n*. Such situation is possible *if and only if* $y_0[n]$ and advanced

$$\gamma^x = (e^\lambda)^\mu$$

$y_0[n]$ have the same form. Only an exponential function γ^n has this property, as the following equation indicates.

$$E^k\{\gamma^n\} = \gamma^{n+k} = \gamma^k \gamma^n \tag{3.28}$$

Equation (3.28) shows that γ^n advanced by k units is a constant (γ^k) times γ^n. Therefore, the solution of Eq. (3.27) must be of the form[†]

$$y_0[n] = c\gamma^n \tag{3.29}$$

To determine c and γ, we substitute this solution in Eq. (3.27b). Equation (3.29) yields

$$E^k y_0[n] = y_0[n+k] = c\gamma^{n+k} \tag{3.30}$$

Substitution of this result in Eq. (3.27b) yields

$$c(\gamma^N + a_1\gamma^{N-1} + \cdots + a_{N-1}\gamma + a_N)\,\gamma^n = 0 \tag{3.31}$$

For a nontrivial solution of this equation

$$\gamma^N + a_1\gamma^{n-1} + \cdots + a_{N-1}\gamma + a_N = 0 \tag{3.32a}$$

or

$$Q[\gamma] = 0 \tag{3.32b}$$

Our solution $c\gamma^n$ [Eq. (3.29)] is correct, provided γ satisfies Eq. (3.32). Now, $Q[\gamma]$ is an Nth-order polynomial and can be expressed in the factored form (assuming all distinct roots):

$$(\gamma - \gamma_1)(\gamma - \gamma_2)\cdots(\gamma - \gamma_N) = 0 \tag{3.32c}$$

Clearly, γ has N solutions $\gamma_1, \gamma_2, \ldots, \gamma_N$ and, therefore, Eq. (3.27) also has N solutions $c_1\gamma_1^n, c_2\gamma_2^n, \ldots, c_n\gamma_N^n$. In such a case, we have shown that the general solution is a linear combination of the N solutions (see the footnote on page 153). Thus

$$y_0[n] = c_1\gamma_1^n + c_2\gamma_2^n + \cdots + c_n\gamma_N^n \tag{3.33}$$

where $\gamma_1, \gamma_2, \ldots, \gamma_n$ are the roots of Eq. (3.32) and $c_1, c_2, \ldots, c_n$ are arbitrary constants determined from N auxiliary conditions, generally given in the form of initial conditions. The polynomial $Q[\gamma]$ is called the *characteristic polynomial* of the system, and

$$Q[\gamma] = 0 \tag{3.34}$$

is the *characteristic equation* of the system. Moreover, $\gamma_1, \gamma_2, \ldots, \gamma_N$, the roots of the characteristic equation, are called *characteristic roots* or *characteristic values* (also *eigenvalues*) of the system. The exponentials γ_i^n $(i = 1, 2, \ldots, N)$ are the *characteristic modes* or *natural*

[†] A signal of the form $n^m\gamma^n$ also satisfies this requirement under certain conditions (repeated roots), discussed later.

modes of the system. A characteristic mode corresponds to each characteristic root of the system, and the *zero-input response is a linear combination of the characteristic modes of the system.*

REPEATED ROOTS

So far we have assumed the system to have N distinct characteristic roots $\gamma_1, \gamma_2, \ldots, \gamma_N$ with corresponding characteristic modes $\gamma_1^n, \gamma_2^n, \ldots, \gamma_N^n$. If two or more roots coincide (repeated roots), the form of characteristic modes is modified. Direct substitution shows that if a root γ repeats r times (root of multiplicity r), the corresponding characteristic modes for this root are $\gamma^n, n\gamma^n, n^2\gamma^n, \ldots, n^{r-1}\gamma^n$. Thus, if the characteristic equation of a system is

$$Q[\gamma] = (\gamma - \gamma_1)^r (\gamma - \gamma_{r+1})(\gamma - \gamma_{r+2}) \cdots (\gamma - \gamma_N) \tag{3.35}$$

the zero-input response of the system is

$$y_0[n] = (c_1 + c_2 n + c_3 n^2 + \cdots + c_r n^{r-1})\gamma_1^n + c_{r+1}\gamma_{r+1}^n + c_{r+2}\gamma_{r+2}^n + \cdots + c_n\gamma_N^n \tag{3.36}$$

COMPLEX ROOTS

As in the case of continuous-time systems, the complex roots of a discrete-time system will occur in pairs of conjugates if the system equation coefficients are real. Complex roots can be treated exactly as we would treat real roots. However, just as in the case of continuous-time systems, we can also use the real form of solution as an alternative.

First we express the complex conjugate roots γ and γ^* in polar form. If $|\gamma|$ is the magnitude and β is the angle of γ, then

$$\gamma = |\gamma|e^{j\beta} \qquad \text{and} \qquad \gamma^* = |\gamma|e^{-j\beta}$$

The zero-input response is given by

$$y_0[n] = c_1\gamma^n + c_2(\gamma^*)^n$$
$$= c_1|\gamma|^n e^{j\beta n} + c_2|\gamma|^n e^{-j\beta n}$$

For a real system, c_1 and c_2 must be conjugates so that $y_0[n]$ is a real function of n. Let

$$c_1 = \frac{c}{2}e^{j\theta} \qquad \text{and} \qquad c_2 = \frac{c}{2}e^{-j\theta} \tag{3.37a}$$

Then

$$y_0[n] = \frac{c}{2}|\gamma|^n \left[e^{j(\beta n + \theta)} + e^{-j(\beta n + \theta)}\right]$$
$$= c|\gamma|^n \cos(\beta n + \theta) \tag{3.37b}$$

where c and θ are arbitrary constants determined from the auxiliary conditions. This is the solution in real form, which avoids dealing with complex numbers.

EXAMPLE 3.10

(a) For an LTID system described by the difference equation

$$y[n + 2] - 0.6y[n + 1] - 0.16y[n] = 5x[n + 2] \tag{3.38a}$$

find the total response if the initial conditions are $y[-1] = 0$ and $y[-2] = 25/4$, and if the input $x[n] = 4^{-n}u[n]$. In this example we shall determine the zero-input component $y_0[n]$ only. The zero-state component is determined later, in Example 3.14.

The system equation in operational notation is

$$(E^2 - 0.6E - 0.16)y[n] = 5E^2x[n] \tag{3.38b}$$

The characteristic polynomial is

$$\gamma^2 - 0.6\gamma - 0.16 = (\gamma + 0.2)(\gamma - 0.8)$$

The characteristic equation is

$$(\gamma + 0.2)(\gamma - 0.8) = 0 \tag{3.39}$$

The characteristic roots are $\gamma_1 = -0.2$ and $\gamma_2 = 0.8$. The zero-input response is

$$y_0[n] = c_1(-0.2)^n + c_2(0.8)^n \tag{3.40}$$

To determine arbitrary constants c_1 and c_2, we set $n = -1$ and -2 in Eq. (3.40), then substitute $y_0[-1] = 0$ and $y_0[-2] = 25/4$ to obtain[†]

$$\left. \begin{array}{l} 0 = -5c_1 + \frac{5}{4}c_2 \\ \frac{25}{4} = 25c_1 + \frac{25}{16}c_2 \end{array} \right\} \quad \Longrightarrow \quad \begin{array}{l} c_1 = \frac{1}{5} \\ c_2 = \frac{4}{5} \end{array}$$

Therefore

$$y_0[n] = \frac{1}{5}(-0.2)^n + \frac{4}{5}(0.8)^n \qquad n \geq 0 \tag{3.41}$$

The reader can verify this solution by computing the first few terms using the iterative method (see Examples 3.8 and 3.9).

(b) A similar procedure may be followed for repeated roots. For instance, for a system specified by the equation

$$(E^2 + 6E + 9)y[n] = (2E^2 + 6E)x[n]$$

Let us determine $y_0[n]$, the zero-input component of the response if the initial conditions are $y_0[-1] = -1/3$ and $y_0[-2] = -2/9$.

[†]The initial conditions $y[-1]$ and $y[-2]$ are the conditions given on the total response. But because the input does not start until $n = 0$, the zero-state response is zero for $n < 0$. Hence, at $n = -1$ and -2 the total response consists of only the zero-input component, so that $y[-1] = y_0[-1]$ and $y[-2] = y_0[-2]$.

The characteristic polynomial is $\gamma^2 + 6\gamma + 9 = (\gamma + 3)^2$, and we have a repeated character-istic root at $\gamma = -3$. The characteristic modes are $(-3)^n$ and $n(-3)^n$. Hence, the zero-input response is

$$y_0[n] = (c_1 + c_2 n)(-3)^n$$

We can determine the arbitrary constants c_1 and c_2 from the initial conditions following the procedure in part (a). It is left as an exercise for the reader to show that $c_1 = 4$ and $c_2 = 3$ so that

$$y_0[n] = (4 + 3n)(-3)^n$$

(c) For the case of complex roots, let us find the zero-input response of an LTID system described by the equation

$$(E^2 - 1.56E + 0.81)y[n] = (E + 3)x[n]$$

when the initial conditions are $y_0[-1] = 2$ and $y_0[-2] = 1$.

The characteristic polynomial is $(\gamma^2 - 1.56\gamma + 0.81) = (\gamma - 0.78 - j0.45)(\gamma - 0.78 + j0.45)$. The characteristic roots are $0.78 \pm j0.45$; that is, $0.9e^{\pm j(\pi/6)}$. We could immediately write the solution as

$$y_0[n] = c(0.9)^n e^{j\pi n/6} + c^*(0.9)^n e^{-j\pi n/6}$$

Setting $n = -1$ and -2 and using the initial conditions $y_0[-1] = 2$ and $y_0[-2] = 1$, we find $c = 2.34\,e^{-j0.17}$ and $c^* = 2.34\,e^{j0.17}$.

Alternately, we could also find the solution by using the real form of solution, as given in Eq. (3.37b). In the present case, the roots are $0.9e^{\pm j(\pi/6)}$. Hence, $|\gamma| = 0.9$ and $\beta = \pi/6$, and the zero-input response, according to Eq. (3.37b), is given by

$$y_0[n] = c(0.9)^n \cos\left(\frac{\pi}{6}n + \theta\right)$$

To determine the arbitrary constants c and θ, we set $n = -1$ and -2 in this equation and substitute the initial conditions $y_0[-1] = 2$, $y_0[-2] = 1$ to obtain

$$2 = \frac{c}{0.9}\cos\left(-\frac{\pi}{6} + \theta\right) = \frac{c}{0.9}\left[\frac{\sqrt{3}}{2}\cos\theta + \frac{1}{2}\sin\theta\right]$$

$$1 = \frac{c}{(0.9)^2}\cos\left(-\frac{\pi}{3} + \theta\right) = \frac{c}{0.81}\left[\frac{1}{2}\cos\theta + \frac{\sqrt{3}}{2}\sin\theta\right]$$

or

$$\frac{\sqrt{3}}{1.8}c\cos\theta + \frac{1}{1.8}c\sin\theta = 2$$

$$\frac{1}{1.62}c\cos\theta + \frac{\sqrt{3}}{1.62}c\sin\theta = 1$$

These are two simultaneous equations in two unknowns $c \cos \theta$ and $c \sin \theta$. Solution of these equations yields

$$c \cos \theta = 2.308$$

$$c \sin \theta = -0.397$$

Dividing $c \sin \theta$ by $c \cos \theta$ yields

$$\tan \theta = \frac{-0.397}{2.308} = \frac{-0.172}{1}$$

$$\theta = \tan^{-1}(-0.172) = -0.17 \text{ rad}$$

Substituting $\theta = -0.17$ radian in $c \cos \theta = 2.308$ yields $c = 2.34$ and

$$y_0[n] = 2.34(0.9)^n \cos\left(\frac{\pi}{6}n - 0.17\right) n \geq 0$$

Observe that here we have used radian unit for both β and θ. We also could have used the degree unit, although this practice is not recommended. The important consideration is to be consistent and to use the same units for both β and θ.

EXERCISE E3.11

Find and sketch the zero-input response for the systems described by the following equations:

(a) $y[n+1] - 0.8y[n] = 3x[n+1]$
(b) $y[n+1] + 0.8y[n] = 3x[n+1]$

In each case the initial condition is $y[-1] = 10$. Verify the solutions by computing the first three terms using the iterative method.

ANSWERS
(a) $8(0.8)^n$
(b) $-8(-0.8)^n$

EXERCISE E3.12

Find the zero-input response of a system described by the equation

$$y[n] + 0.3y[n-1] - 0.1y[n-2] = x[n] + 2x[n-1]$$

The initial conditions are $y_0[-1] = 1$ and $y_0[-2] = 33$. Verify the solution by computing the first three terms iteratively.

ANSWER
$y_0[n] = (0.2)^n + 2(-0.5)^n$

EXERCISE E3.13

Find the zero-input response of a system described by the equation

$$y[n] + 4y[n-2] = 2x[n]$$

The initial conditions are $y_0[-1] = -1/(2\sqrt{2})$ and $y_0[-2] = 1/(4\sqrt{2})$. Verify the solution by computing the first three terms iteratively.

ANSWER

$$y_0[n] = (2)^n \cos\left(\frac{\pi}{2}n - \frac{3\pi}{4}\right)$$

COMPUTER EXAMPLE C3.4

Using the initial conditions $y[-1] = 2$ and $y[-2] = 1$, find and sketch the zero-input response for the system described by $(E^2 - 1.56E + 0.81)y[n] = (E + 3)x[n]$.

```
>> n = (-2:20)'; y = [1;2;zeros(length(n)-2,1)];
>> for k = 1:length(n)-2,
>>     y(k+2) = 1.56*y(k+1)-0.81*y(k);
>> end;
>> clf; stem(n,y,'k'); xlabel('n'); ylabel('y[n]');
```

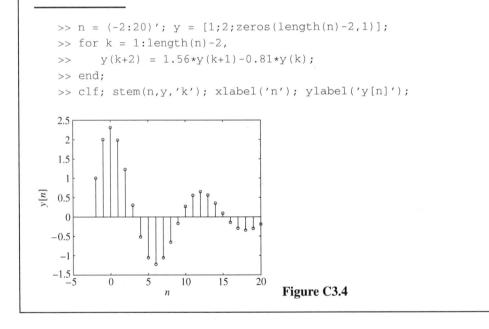

Figure C3.4

3.7 THE UNIT IMPULSE RESPONSE $h[n]$

Consider an nth-order system specified by the equation

$$(E^N + a_1 E^{N-1} + \cdots + a_{N-1}E + a_N)y[n]$$

$$= (b_0 E^N + b_1 E^{N-1} + \cdots + b_{N-1}E + b_N)x[n] \tag{3.42a}$$

or

$$Q[E]y[n] = P[E]x[n] \qquad (3.42b)$$

The unit impulse response $h[n]$ is the solution of this equation for the input $\delta[n]$ with all the initial conditions zero; that is,

$$Q[E]h[n] = P[E]\delta[n] \qquad (3.43)$$

subject to initial conditions

$$h[-1] = h[-2] = \cdots = h[-N] = 0 \qquad (3.44)$$

Equation (3.43) can be solved to determine $h[n]$ iteratively or in a closed form. The following example demonstrates the iterative solution.

EXAMPLE 3.11 (Iterative Determination of $h[n]$)

Find $h[n]$, the unit impulse response of a system described by the equation

$$y[n] - 0.6y[n-1] - 0.16y[n-2] = 5x[n] \qquad (3.45)$$

To determine the unit impulse response, we let the input $x[n] = \delta[n]$ and the output $y[n] = h[n]$ in Eq. (3.45) to obtain

$$h[n] - 0.6h[n-1] - 0.16h[n-2] = 5\delta[n] \qquad (3.46)$$

subject to zero initial state; that is, $h[-1] = h[-2] = 0$.

Setting $n = 0$ in this equation yields

$$h[0] - 0.6(0) - 0.16(0) = 5(1) \quad \Longrightarrow \quad h[0] = 5$$

Next, setting $n = 1$ in Eq. (3.46) and using $h[0] = 5$, we obtain

$$h[1] - 0.6(5) - 0.16(0) = 5(0) \quad \Longrightarrow \quad h[1] = 3$$

Continuing this way, we can determine any number of terms of $h[n]$. Unfortunately, such a solution does not yield a closed-form expression for $h[n]$. Nevertheless, determining a few values of $h[n]$ can be useful in determining the closed-form solution, as the following development shows.

THE CLOSED-FORM SOLUTION OF $h[n]$

Recall that $h[n]$ is the system response to input $\delta[n]$, which is zero for $n > 0$. We know that when the input is zero, only the characteristic modes can be sustained by the system. Therefore, $h[n]$ must be made up of characteristic modes for $n > 0$. At $n = 0$, it may have some nonzero value A_0, so that a general form of $h[n]$ can be expressed as[†]

$$h[n] = A_0\delta[n] + y_c[n]u[n] \tag{3.47}$$

where $y_c[n]$ is a linear combination of the characteristic modes. We now substitute Eq. (3.47) in Eq. (3.43) to obtain $Q[E](A_0\delta[n] + y_c[n]u[n]) = P[E]\delta[n]$. Because $y_c[n]$ is made up of characteristic modes, $Q[E]y_c[n]u[n] = 0$, and we obtain $A_0 Q[E]\delta[n] = P[E]\delta[n]$, that is,

$$A_0\,(\delta[n+N] + a_1\delta[n+N-1] + \cdots + a_N\delta[n]) = b_0\delta[n+N] + \cdots + b_N\delta[n]$$

Setting $n = 0$ in this equation and using the fact that $\delta[m] = 0$ for all $m \neq 0$, and $\delta[0] = 1$, we obtain

$$A_0 a_N = b_N \quad \Longrightarrow \quad A_0 = \frac{b_N}{a_N} \tag{3.48}$$

Hence[‡]

$$h[n] = \frac{b_N}{a_N}\,\delta[n] + y_c[n]u[n] \tag{3.49}$$

The N unknown coefficients in $y_c[n]$ (on the right-hand side) can be determined from a knowledge of N values of $h[n]$. Fortunately, it is a straightforward task to determine values of $h[n]$ iteratively, as demonstrated in Example 3.11. We compute N values $h[0], h[1], h[2], \ldots,$ $h[N-1]$ iteratively. Now, setting $n = 0, 1, 2, \ldots, N - 1$ in Eq. (3.49), we can determine the N unknowns in $y_c[n]$. This point will become clear in the following example.

EXAMPLE 3.12

Determine the unit impulse response $h[n]$ for a system in Example 3.11 specified by the equation

$$y[n] - 0.6y[n-1] - 0.16y[n-2] = 5x[n]$$

This equation can be expressed in the advance operator form as

$$y[n+2] - 0.6y[n+1] - 0.16y[n] = 5x[n+2] \tag{3.50}$$

or

$$(E^2 - 0.6E - 0.16)y[n] = 5E^2 x[n] \tag{3.51}$$

The characteristic polynomial is

$$\gamma^2 - 0.6\gamma - 0.16 = (\gamma + 0.2)(\gamma - 0.8)$$

[†]We assumed that the term $y_c[n]$ consists of characteristic modes for $n > 0$ only. To reflect this behavior, the characteristic terms should be expressed in the form $\gamma_j^n u[n-1]$. But because $u[n-1] = u[n] - \delta[n]$, $c_j\gamma_j^n u[n-1] = c_j\gamma_j^n u[n] - c_j\delta[n]$, and $y_c[n]$ can be expressed in terms of exponentials $\gamma_j^n u[n]$ (which start at $n = 0$), plus an impulse at $n = 0$.

[‡]If $a_N = 0$, then A_0 cannot be determined by Eq. (3.48). In such a case, we show in Section 3.12 that $h[n]$ is of the form $A_0\delta[n] + A_1\delta[n-1] + y_c[n]u[n]$. We have here $N + 2$ unknowns, which can be determined from $N + 2$ values $h[0], h[1], \ldots, h[N+1]$ found iteratively.

The characteristic modes are $(-0.2)^n$ and $(0.8)^n$. Therefore

$$y_c[n] = c_1(-0.2)^n + c_2(0.8)^n \tag{3.52}$$

Also, from Eq. (3.51), we have $a_N = -0.16$ and $b_N = 0$. Therefore, according to Eq. (3.49)

$$h[n] = [c_1(-0.2)^n + c_2(0.8)^n]u[n] \tag{3.53}$$

To determine c_1 and c_2, we need to find two values of $h[n]$ iteratively. This step is already taken in Example 3.11, where we determined that $h[0] = 5$ and $h[1] = 3$. Now, setting $n = 0$ and 1 in Eq. (3.53) and using the fact that $h[0] = 5$ and $h[1] = 3$, we obtain

$$\left.\begin{array}{l} 5 = c_1 + c_2 \\ 3 = -0.2c_1 + 0.8c_2 \end{array}\right\} \quad \Longrightarrow \quad \begin{array}{l} c_1 = 1 \\ c_2 = 4 \end{array}$$

Therefore

$$h[n] = [(-0.2)^n + 4(0.8)^n]\, u[n] \tag{3.54}$$

COMPUTER EXAMPLE C3.5

Use MATLAB to solve Example 3.12.

There are several ways to find the impulse response using MATLAB. In this method, we first specify the input as a unit impulse function. Vectors a and b are created to specify the system. The `filter` command is then used to determine the impulse response. In fact, this method can be used to determine the zero-state response for any input.

```
>> n = (0:19); x = inline('n==0');
>> a = [1 -0.6 -0.16]; b = [5 0 0];
>> h = filter(b,a,x(n));
>> clf; stem(n,h,'k'); xlabel('n'); ylabel('h[n]');
```

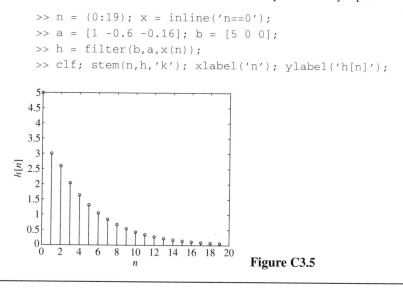

Figure C3.5

Comment. Although it is relatively simple to determine the impulse response $h[n]$ by using the procedure in this section, in Chapter 5 we shall discuss the much simpler method of the z-transform.

EXERCISE E3.14

Find $h[n]$, the unit impulse response of the LTID systems specified by the following equations:

    (a) $y[n+1] - y[n] = x[n]$

    (b) $y[n] - 5y[n-1] + 6y[n-2] = 8x[n-1] - 19x[n-2]$

    (c) $y[n+2] - 4y[n+1] + 4y[n] = 2x[n+2] - 2x[n+1]$

    (d) $y[n] = 2x[n] - 2x[n-1]$

ANSWERS

(a) $h[n] = u[n-1]$

(b) $h[n] = -\frac{19}{6}\delta[n] + \left[\frac{3}{2}(2)^n + \frac{5}{3}(3)^n\right]u[n]$

(c) $h[n] = (2+n)2^n u[n]$

(d) $h[n] = 2\delta[n] - 2\delta[n-1]$

3.8 SYSTEM RESPONSE TO EXTERNAL INPUT: THE ZERO-STATE RESPONSE

The zero-state response $y[n]$ is the system response to an input $x[n]$ when the system is in the zero state. In this section we shall assume that systems are in the zero state unless mentioned otherwise, so that the zero-state response will be the total response of the system. Here we follow the procedure parallel to that used in the continuous-time case by expressing an arbitrary input $x[n]$ as a sum of impulse components. A signal $x[n]$ in Fig. 3.16a can be expressed as a sum of impulse components such as those depicted in Fig. 3.16b–3.16f. The component of $x[n]$ at $n = m$ is $x[m]\delta[n-m]$, and $x[n]$ is the sum of all these components summed from $m = -\infty$ to ∞. Therefore

$$x[n] = x[0]\delta[n] + x[1]\delta[n-1] + x[2]\delta[n-2] + \cdots$$

$$+ x[-1]\delta[n+1] + x[-2]\delta[n+2] + \cdots$$

$$= \sum_{m=-\infty}^{\infty} x[m]\delta[n-m] \tag{3.55}$$

For a linear system, knowing the system response to impulse $\delta[n]$, the system response to any arbitrary input could be obtained by summing the system response to various impulse components. Let $h[n]$ be the system response to impulse input $\delta[n]$. We shall use the

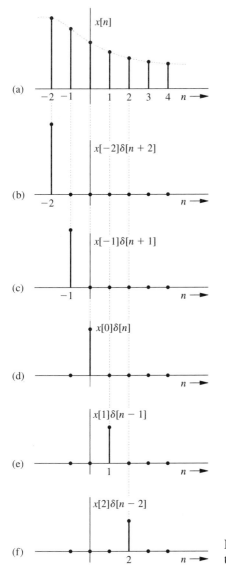

(a)

(b)

(c)

(d)

(e)

(f)

Figure 3.16 Representation of an arbitrary signal $x[n]$ in terms of impulse components.

notation

$$x[n] \Longrightarrow y[n]$$

to indicate the input and the corresponding response of the system. Thus, if

$$\delta[n] \Longrightarrow h[n]$$

then because of time invariance

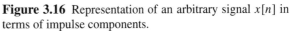

$$\delta[n - m] \Longrightarrow h[n - m]$$

and because of linearity

$$x[m]\delta[n - m] \Longrightarrow x[m]h[n - m]$$

and again because of linearity

$$\sum_{m=-\infty}^{\infty} x[m]\delta[n - m] \underbrace{\qquad}_{x[n]} \quad \Longrightarrow \quad \sum_{m=-\infty}^{\infty} x[m]h[n - m] \underbrace{\qquad}_{y[n]}$$

The left-hand side is $x[n]$ [see Eq. (3.55)], and the right-hand side is the system response $y[n]$ to input $x[n]$. Therefore[†]

$$y[n] = \sum_{m=-\infty}^{\infty} x[m]h[n - m] \tag{3.56}$$

The summation on the right-hand side is known as the *convolution sum* of $x[n]$ and $h[n]$, and is represented symbolically by $x[n] * h[n]$

$$x[n] * h[n] = \sum_{m=-\infty}^{\infty} x[m]h[n - m] \tag{3.57}$$

PROPERTIES OF THE CONVOLUTION SUM

The structure of the convolution sum is similar to that of the convolution integral. Moreover, the properties of the convolution sum are similar to those of the convolution integral. We shall enumerate these properties here without proof. The proofs are similar to those for the convolution integral and may be derived by the reader.

The Commutative Property.

$$x_1[n] * x_2[n] = x_2[n] * x_1[n] \tag{3.58}$$

The Distributive Property.

$$x_1[n] * (x_2[n] + x_3[n]) = x_1[n] * x_2[n] + x_1[n] * x_3[n] \tag{3.59}$$

The Associative Property.

$$x_1[n] * (x_2[n] * x_3[n]) = (x_1[n] * x_2[n]) * x_3[n] \tag{3.60}$$

The Shifting Property. If

$$x_1[n] * x_2[n] = c[n]$$

[†]In deriving this result, we have assumed a time-invariant system. The system response to input $\delta[n - m]$ for a time-varying system cannot be expressed as $h[n - m]$, but instead has the form $h[n, m]$. Using this form, Eq. (3.56) is modified as follows:

$$y[n] = \sum_{m=-\infty}^{\infty} x[m]h[n, m]$$

then

$$x_1[n - m] * x_2[n - p] = c[n - m - p] \tag{3.61}$$

The Convolution with an Impulse.

$$x[n] * \delta[n] = x[n] \tag{3.62}$$

The Width Property. If $x_1[n]$ and $x_2[n]$ have finite widths of W_1 and W_2, respectively, then the width of $x_1[n] * x_2[n]$ is $W_1 + W_2$. The width of a signal is one less than the number of its elements (length). For instance, the signal in Fig. 3.17h has six elements (length of 6) but a width of only 5. Alternately, the property may be stated in terms of lengths as follows: if $x_1[n]$ and $x_2[n]$ have finite lengths of L_1 and L_2 elements, respectively, then the length of $x_1[n] * x_2[n]$ is $L_1 + L_2 - 1$ elements.

CAUSALITY AND ZERO-STATE RESPONSE

In deriving Eq. (3.56), we assumed the system to be linear and time invariant. There were no other restrictions on either the input signal or the system. In our applications, almost all the input signals are causal, and a majority of the systems are also causal. These restrictions further simplify the limits of the sum in Eq. (3.56). If the input $x[n]$ is causal, $x[m] = 0$ for $m < 0$. Similarly, if the system is causal (i.e., if $h[n]$ is causal), then $h[x] = 0$ for negative x, so that $h[n - m] = 0$ when $m > n$. Therefore, if $x[n]$ and $h[n]$ are both causal, the product $x[m]h[n - m] = 0$ for $m < 0$ and for $m > n$, and it is nonzero only for the range $0 \le m \le n$. Therefore, Eq. (3.56) in this case reduces to

$$y[n] = \sum_{m=0}^{n} x[m]h[n - m] \tag{3.63}$$

We shall evaluate the convolution sum first by an analytical method and later with graphical aid.

EXAMPLE 3.13

Determine $c[n] = x[n] * g[n]$ for

$$x[n] = (0.8)^n u[n] \qquad \text{and} \qquad g[n] = (0.3)^n u[n]$$

We have

$$c[n] = \sum_{m=-\infty}^{\infty} x[m]g[n - m]$$

Note that

$$x[m] = (0.8)^m u[m] \qquad \text{and} \qquad g[n - m] = (0.3)^{n-m} u[n - m]$$

Both $x[n]$ and $g[n]$ are causal. Therefore, [see Eq. (3.63)]

$$c[n] = \sum_{m=0}^{n} x[m]g[n - m]$$

$$= \sum_{m=0}^{n} (0.8)^m u[m] (0.3)^{n-m} u[n - m] \tag{3.64}$$

In this summation, m lies between 0 and n ($0 \le m \le n$). Therefore, if $n \ge 0$, then both m and $n - m \ge 0$, so that $u[m] = u[n - m] = 1$. If $n < 0$, m is negative because m lies between 0 and n, and $u[m] = 0$. Therefore, Eq. (3.64) becomes

$$c[n] = \sum_{m=0}^{n} (0.8)^m (0.3)^{n-m} \qquad n \ge 0$$

$$= 0 \qquad n < 0$$

and

$$c[n] = (0.3)^n \sum_{m=0}^{n} \left(\frac{0.8}{0.3} \right)^m u[n]$$

This is a geometric progression with common ratio $(0.8/0.3)$. From Section B.7-4 we have

$$c[n] = (0.3)^n \frac{(0.8)^{n+1} - (0.3)^{n+1}}{(0.3)^n (0.8 - 0.3)} u[n]$$

$$= 2[(0.8)^{n+1} - (0.3)^{n+1}]u[n] \tag{3.65}$$

EXERCISE E3.15

Show that $(0.8)^n u[n] * u[n] = 5[1 - (0.8)^{n+1}]u[n]$.

CONVOLUTION SUM FROM A TABLE

Just as in the continuous-time case, we have prepared a table (Table 3.1) from which convolution sums may be determined directly for a variety of signal pairs. For example, the convolution in Example 3.13 can be read directly from this table (pair 4) as

$$(0.8)^n u[n] * (0.3)^n u[n] = \frac{(0.8)^{n+1} - (0.3)^{n+1}}{0.8 - 0.3} u[n] = 2[(0.8)^{n+1} - (0.3)^{n+1}]u[n]$$

We shall demonstrate the use of the convolution table in the following example.

TABLE 3.1   Convolution Sums

No.	$x_1[n]$	$x_2[n]$	$x_1[n] * x_2[n] = x_2[n] * x_1[n]$
1	$\delta[n-k]$	$x[n]$	$x[n-k]$
2	$\gamma^n u[n]$	$u[n]$	$\left[\dfrac{1-\gamma^{n+1}}{1-\gamma}\right] u[n]$
3	$u[n]$	$u[n]$	$(n+1)u[n]$
4	$\gamma_1^n u[n]$	$\gamma_2^n u[n]$	$\left[\dfrac{\gamma_1^{n+1} - \gamma_2^{n+1}}{\gamma_1 - \gamma_2}\right] u[n] \qquad \gamma_1 \neq \gamma_2$
5	$u[n]$	$nu[n]$	$\dfrac{n(n+1)}{2} u[n]$
6	$\gamma^n u[n]$	$nu[n]$	$\left[\dfrac{\gamma(\gamma^n - 1) + n(1-\gamma)}{(1-\gamma)^2}\right] u[n]$
7	$nu[n]$	$nu[n]$	$\frac{1}{6}n(n-1)(n+1)u[n]$
8	$\gamma^n u[n]$	$\gamma^n u[n]$	$(n+1)\gamma^n u[n]$
9	$n\gamma_1^n u[n]$	$\gamma_2^n u[n]$	$\dfrac{\gamma_1 \gamma_2}{(\gamma_1 - \gamma_2)^2}\left[\gamma_2^n - \gamma_1^n + \dfrac{\gamma_1 - \gamma_2}{\gamma_2}n\gamma_1^n\right] u[n] \qquad \gamma_1 \neq \gamma_2$
10	$\|\gamma_1\|^n \cos(\beta n + \theta)u[n]$	$\gamma_2^n u[n]$	$\dfrac{1}{R}[\|\gamma_1\|^{n+1}\cos[\beta(n+1) + \theta - \phi] - \gamma_2^{n+1}\cos(\theta - \phi)]u[n] \qquad \gamma_2 \text{ real}$ $R = \left[\|\gamma_1\|^2 + \gamma_2^2 - 2\|\gamma_1\|\gamma_2 \cos\beta\right]^{1/2}$ $\phi = \tan^{-1}\left[\dfrac{(\|\gamma_1\|\sin\beta)}{(\|\gamma_1\|\cos\beta - \gamma_2)}\right]$
11	$\gamma_1^n u[n]$	$\gamma_2^n u[-(n+1)]$	$\dfrac{\gamma_1}{\gamma_2 - \gamma_1}\gamma_1^n u[n] + \dfrac{\gamma_2}{\gamma_2 - \gamma_1}\gamma_2^n u[-(n+1)] \qquad \|\gamma_2\| > \|\gamma_1\|$

EXAMPLE 3.14

Find the (zero-state) response $y[n]$ of an LTID system described by the equation

$$y[n+2] - 0.6y[n+1] - 0.16y[n] = 5x[n+2]$$

if the input $x[n] = 4^{-n}u[n]$.

The input can be expressed as $x[n] = 4^{-n}u[n] = (1/4)^n u[n] = (0.25)^n u[n]$. The unit impulse response of this system was obtained in Example 3.12.

$$h[n] = [(-0.2)^n + 4(0.8)^n]u[n]$$

Therefore

$$y[n] = x[n] * h[n]$$

$$= (0.25)^n u[n] * [(-0.2)^n u[n] + 4(0.8)^n u[n]]$$

$$= (0.25)^n u[n] * (-0.2)^n u[n] + (0.25)^n u[n] * 4(0.8)^n u[n]$$

We use pair 4 (Table 3.1) to find the foregoing convolution sums.

$$y[n] = \left[\frac{(0.25)^{n+1} - (-0.2)^{n+1}}{0.25 - (-0.2)} + 4 \frac{(0.25)^{n+1} - (0.8)^{n+1}}{0.25 - 0.8} \right] u[n]$$

$$= (2.22[(0.25)^{n+1} - (-0.2)^{n+1}] - 7.27[(0.25)^{n+1} - (0.8)^{n+1}]) u[n]$$

$$= [-5.05(0.25)^{n+1} - 2.22(-0.2)^{n+1} + 7.27(0.8)^{n+1}] u[n]$$

Recognizing that

$$\gamma^{n+1} = \gamma(\gamma)^n$$

We can express $y[n]$ as

$$y[n] = [-1.26(0.25)^n + 0.444(-0.2)^n + 5.81(0.8)^n] u[n]$$

$$= [-1.26(4)^{-n} + 0.444(-0.2)^n + 5.81(0.8)^n] u[n]$$

EXERCISE E3.16

Show that

$$(0.8)^{n+1} u[n] * u[n] = 4[1 - 0.8(0.8)^n] u[n]$$

EXERCISE E3.17

Show that

$$n\, 3^{-n} u[n] * (0.2)^n u[n] = \tfrac{15}{4} \left[(0.2)^n - \left(1 - \tfrac{2}{3}n\right) 3^{-n} \right] u[n]$$

EXERCISE E3.18

Show that

$$e^{-n} u[n] * 2^{-n} u[n] = \frac{2}{2 - e} \left[e^{-n} - \frac{e}{2} 2^{-n} \right] u[n]$$

COMPUTER EXAMPLE C3.6

Find and sketch the zero-state response for the system described by $(E^2 + 6E + 9)y[n] = (2E^2 + 6E)x[n]$ for the input $x[n] = 4^{-n}u[n]$.

Although the input is bounded and quickly decays to zero, the system itself is unstable and an unbounded output results.

```
>> n = (0:11); x = inline('(4.^(-n)).*(n>=0)');
>> a = [1 6 9]; b = [2 6 0];
>> y = filter(b,a,x(n));
>> clf; stem(n,y,'k'); xlabel('n'); ylabel('y[n]');
```

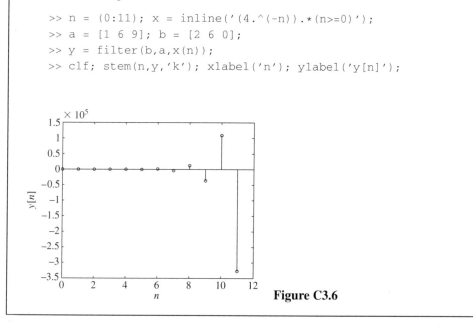

Figure C3.6

RESPONSE TO COMPLEX INPUTS

As in the case of real continuous-time systems, we can show that for an LTID system with real $h[n]$, if the input and the output are expressed in terms of their real and imaginary parts, then the real part of the input generates the real part of the response and the imaginary part of the input generates the imaginary part. Thus, if

$$x[n] = x_r[n] + jx_i[n] \qquad \text{and} \qquad y[n] = y_r[n] + jy_i[n] \qquad (3.66a)$$

using the right-directed arrow to indicate the input–output pair, we can show that

$$x_r[n] \Longrightarrow y_r[n] \qquad \text{and} \qquad x_i[n] \Longrightarrow y_i[n] \qquad (3.66b)$$

The proof is similar to that used to derive Eq. (2.40) for LTIC systems.

MULTIPLE INPUTS

Multiple inputs to LTI systems can be treated by applying the superposition principle. Each input is considered separately, with all other inputs assumed to be zero. The sum of all these individual system responses constitutes the total system output when all the inputs are applied simultaneously.

3.8-1 Graphical Procedure for the Convolution Sum

The steps in evaluating the convolution sum are parallel to those followed in evaluating the convolution integral. The convolution sum of causal signals $x[n]$ and $g[n]$ is given by

$$c[n] = \sum_{m=0}^{n} x[m]g[n-m]$$

We first plot $x[m]$ and $g[n-m]$ as functions of m (not n), because the summation is over m. Functions $x[m]$ and $g[m]$ are the same as $x[n]$ and $g[n]$, plotted respectively as functions of m (see Fig. 3.17). The convolution operation can be performed as follows:

1. Invert $g[m]$ about the vertical axis ($m = 0$) to obtain $g[-m]$ (Fig. 3.17d). Figure 3.17e shows both $x[m]$ and $g[-m]$.

2. Shift $g[-m]$ by n units to obtain $g[n-m]$. For $n > 0$, the shift is to the right (delay); for $n < 0$, the shift is to the left (advance). Figure 3.17f shows $g[n-m]$ for $n > 0$; for $n < 0$, see Fig. 3.17g.

3. Next we multiply $x[m]$ and $g[n-m]$ and add all the products to obtain $c[n]$. The procedure is repeated for each value of n over the range $-\infty$ to ∞.

We shall demonstrate by an example the graphical procedure for finding the convolution sum. Although both the functions in this example are causal, this procedure is applicable to general case.

EXAMPLE 3.15

Find

$$c[n] = x[n] * g[n]$$

where $x[n]$ and $g[n]$ are depicted in Fig. 3.17a and 3.17b, respectively.

We are given

$$x[n] = (0.8)^n \qquad \text{and} \qquad g[n] = (0.3)^n$$

Therefore

$$x[m] = (0.8)^m \qquad \text{and} \qquad g[n-m] = (0.3)^{n-m}$$

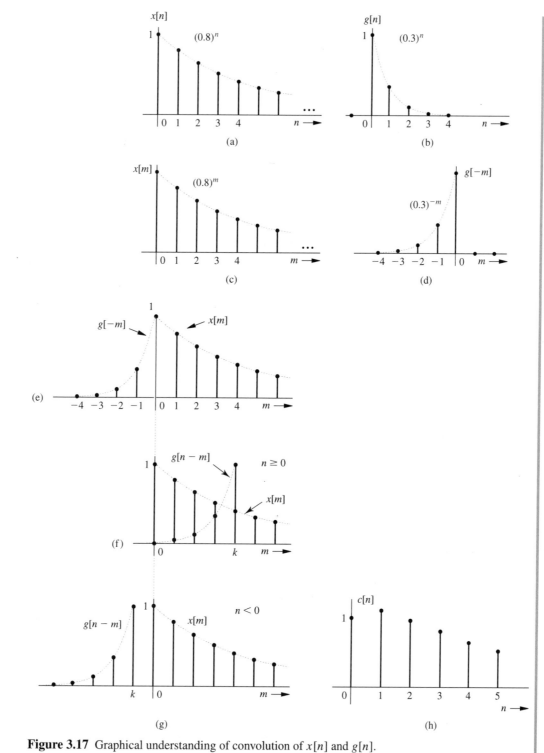

Figure 3.17 Graphical understanding of convolution of $x[n]$ and $g[n]$.

Figure 3.17f shows the general situation for $n \geq 0$. The two functions $x[m]$ and $g[n-m]$ overlap over the interval $0 \leq m \leq n$. Therefore

$$c[n] = \sum_{m=0}^{n} x[m]g[n-m]$$

$$= \sum_{m=0}^{n} (0.8)^m (0.3)^{n-m}$$

$$= (0.3)^n \sum_{m=0}^{n} \left(\frac{0.8}{0.3} \right)^m$$

$$= 2[(0.8)^{n+1} - (0.3)^{n+1}] \qquad n \geq 0 \qquad \text{(see Section B.7-4)}$$

For $n < 0$, there is no overlap between $x[m]$ and $g[n-m]$, as shown in Fig. 3.17g so that

$$c[n] = 0 \qquad n < 0$$

and

$$c[n] = 2[(0.8)^{n+1} - (0.3)^{n+1}]u[n]$$

which agrees with the earlier result in Eq. (3.65).

EXERCISE E3.19

Find $(0.8)^n u[n] * u[n]$ graphically and sketch the result.

ANSWER
$5(1 - (0.8)^{n+1})u[n]$

AN ALTERNATIVE FORM OF GRAPHICAL PROCEDURE: THE SLIDING-TAPE METHOD

This algorithm is convenient when the sequences $x[n]$ and $g[n]$ are short or when they are available only in graphical form. The algorithm is basically the same as the graphical procedure in Fig. 3.17. The only difference is that instead of presenting the data as graphical plots, we display it as a sequence of numbers on tapes. Otherwise the procedure is the same, as will become clear in the following example.

EXAMPLE 3.16

Use the sliding-tape method to convolve the two sequences $x[n]$ and $g[n]$ depicted in Fig. 3.18a and 3.18b, respectively.

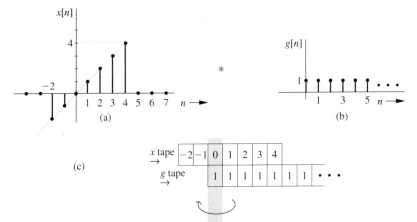

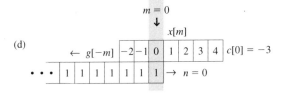

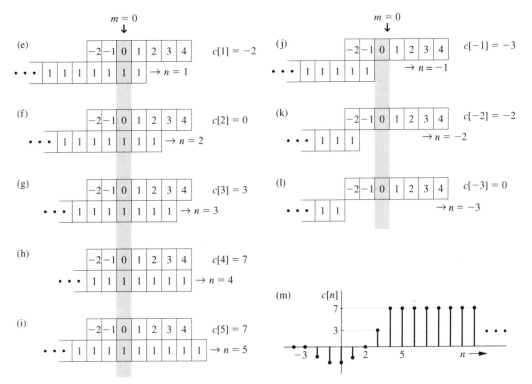

Figure 3.18 Sliding-tape algorithm for discrete-time convolution.

In this procedure we write the sequences $x[n]$ and $g[n]$ in the slots of two tapes: x tape and g tape (Fig. 3.18c). Now leave the x tape stationary (to correspond to $x[m]$). The $g[-m]$ tape is obtained by inverting the $g[m]$ tape about the origin ($m = 0$) so that the slots corresponding to $x[0]$ and $g[0]$ remain aligned (Fig. 3.18d). We now shift the inverted tape by n slots, multiply values on two tapes in adjacent slots, and add all the products to find $c[n]$. Figure 3.18d–3.18i shows the cases for $n = 0$–5. Figure 3.18j–3.18k, and 3.18l shows the cases for $n = -1, -2,$ and -3, respectively.

For the case of $n = 0$, for example (Fig. 3.18d)

$$c[0] = (-2 \times 1) + (-1 \times 1) + (0 \times 1) = -3$$

For $n = 1$ (Fig. 3.18e)

$$c[1] = (-2 \times 1) + (-1 \times 1) + (0 \times 1) + (1 \times 1) = -2$$

Similarly,

$$c[2] = (-2 \times 1) + (-1 \times 1) + (0 \times 1) + (1 \times 1) + (2 \times 1) = 0$$

$$c[3] = (-2 \times 1) + (-1 \times 1) + (0 \times 1) + (1 \times 1) + (2 \times 1) + (3 \times 1) = 3$$

$$c[4] = (-2 \times 1) + (-1 \times 1) + (0 \times 1) + (1 \times 1) + (2 \times 1) + (3 \times 1) + (4 \times 1) = 7$$

$$c[5] = (-2 \times 1) + (-1 \times 1) + (0 \times 1) + (1 \times 1) + (2 \times 1) + (3 \times 1) + (4 \times 1) = 7$$

Figure 3.18i shows that $c[n] = 7$ for $n \geq 4$.

Similarly, we compute $c[n]$ for negative n by sliding the tape backward, one slot at a time, as shown in the plots corresponding to $n = -1, -2,$ and -3, respectively (Fig. 3.18j, 3.18k, and 3.18l).

$$c[-1] = (-2 \times 1) + (-1 \times 1) = -3$$

$$c[-2] = (-2 \times 1) = -2$$

$$c[-3] = 0$$

Figure 3.18l shows that $c[n] = 0$ for $n \leq 3$. Figure 3.18m shows the plot of $c[n]$.

EXERCISE E3.20

Use the graphical procedure of Example 3.16 (sliding-tape technique) to show that $x[n] * g[n] = c[n]$ in Fig. 3.19. Verify the width property of convolution.

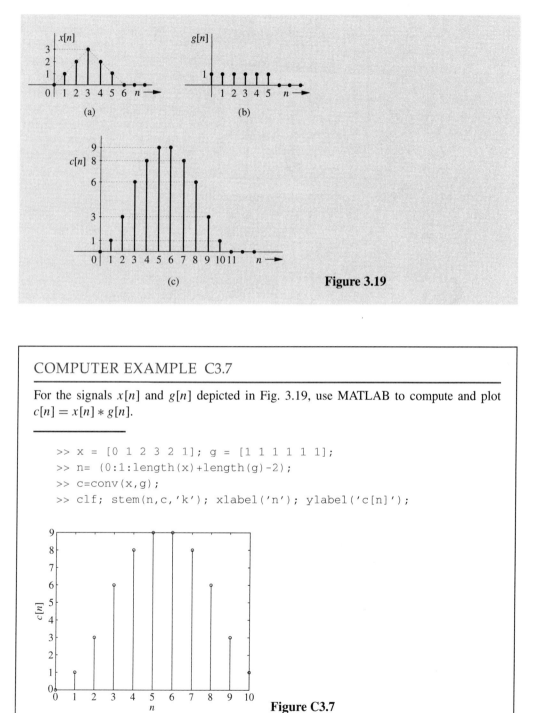

Figure 3.19

COMPUTER EXAMPLE C3.7

For the signals $x[n]$ and $g[n]$ depicted in Fig. 3.19, use MATLAB to compute and plot $c[n] = x[n] * g[n]$.

```
>> x = [0 1 2 3 2 1]; g = [1 1 1 1 1 1];
>> n= (0:1:length(x)+length(g)-2);
>> c=conv(x,g);
>> clf; stem(n,c,'k'); xlabel('n'); ylabel('c[n]');
```

Figure C3.7

3.8-2 Interconnected Systems

As with continuous-time case, we can determine the impulse response of systems connected in parallel (Fig. 3.20a) and cascade (Fig. 3.20b, 3.20c). We can use arguments identical to those used for the continuous-time systems in Section 2.4-3 to show that if two LTID systems $\mathcal{S}_1$ and $\mathcal{S}_2$ with impulse response $h_1[n]$ and $h_2[n]$, respectively, are connected in parallel, the composite parallel system impulse response is $h_1[n] + h_2[n]$. Similarly, if these systems are connected in cascade (in any order), the impulse response of the composite system is $h_1[n] * h_2[n]$. Moreover, because $h_1[n] * h_2[n] = h_2[n] * h_1[n]$, linear systems commute. Their orders can be interchanged without affecting the composite system behavior.

INVERSE SYSTEMS

If the two systems in cascade are inverse of each other, with impulse responses $h[n]$ and $h_i[n]$, respectively, then the impulse response of the cascade of these systems is $h[n] * h_i[n]$. But, the cascade of a system with its inverse is an identity system, whose output is the same as the input.

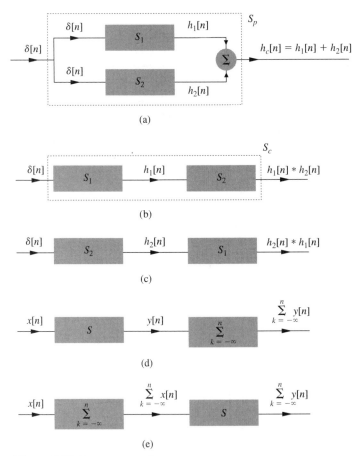

Figure 3.20 Interconnected systems.

Hence, the unit impulse response of an identity system is $\delta[n]$. Consequently

$$h[n] * h_i[n] = \delta[n] \tag{3.67}$$

As an example, we show that an accumulator system and a backward difference system are inverse of each other. An accumulator system is specified by[†]

$$y[n] = \sum_{k=-\infty}^{n} x[k] \tag{3.68a}$$

The backward difference system is specified by

$$y[n] = x[n] - x[n-1] \tag{3.68b}$$

From Eq. (3.68a), we find $h_{acc}[n]$, the impulse response of the accumulator, as

$$h_{acc}[n] = \sum_{k=-\infty}^{n} \delta[k] = u[n] \tag{3.69a}$$

Similarly, from Eq. (3.68b), $h_{bdf}[n]$, the impulse response of the backward difference system is given by

$$h_{bdf}[n] = \delta[n] - \delta[n-1] \tag{3.69b}$$

We can verify that

$$h_{acc} * h_{bdf} = u[n] * \{\delta[n] - \delta[n-1]\} = u[n] - u[n-1] = \delta[n]$$

Roughly speaking, in discrete-time systems, an accumulator is analogous to an integrator in continuous-time systems, and a backward difference system is analogous to a differentiator. We have already encountered examples of these systems in Examples 3.6 and 3.7 (digital differentiator and integrator).

SYSTEM RESPONSE TO $\sum_{k=-\infty}^{n} x[k]$

Figure 3.20d shows a cascade of two LTID systems: a system S with impulse response $h[n]$, followed by an accumulator. Figure 3.20e shows a cascade of the same two systems in reverse order: an accumulator followed by S. In Fig. 3.20d, if the input $x[n]$ to S results in the output $y[n]$, then the output of the system 3.20d is the $\sum y[k]$. In Fig. 3.20e, the output of the accumulator is the sum $\sum x[k]$. Because the output of the system in Fig. 3.20e is identical to that of system Fig. 3.20d, it follows that

$$\text{if} \quad x[n] \Longrightarrow y[n]$$

$$\text{then} \quad \sum_{k=-\infty}^{n} x[k] \Longrightarrow \sum_{k=-\infty}^{n} y[k] \tag{3.70a}$$

[†]Equations (3.68a) and (3.68b) are identical to Eqs. (3.14a) and (3.12), respectively, with $T = 1$.

If we let $x[n] = \delta[n]$ and $y[n] = h[n]$ in Eq. (3.70a), we find that $g[n]$, the unit step response of an LTID system with impulse response $h[n]$, is given by

$$g[n] = \sum_{k=-\infty}^{n} h[k] \qquad (3.70b)$$

The reader can readily prove the inverse relationship

$$h[n] = g[n] - g[n-1] \qquad (3.70c)$$

3.8-3 A Very Special Function for LTID Systems: The Everlasting Exponential z^n

In Section 2.4-4, we showed that there exists one signal for which the response of an LTIC system is the same as the input within a multiplicative constant. The response of an LTIC system to an everlasting exponential input e^{st} is $H(s)e^{st}$, where $H(s)$ is the system transfer function. We now show that for an LTID system, the same role is played by an everlasting exponential z^n. The system response $y[n]$ in this case is given by

$$y[n] = h[n] * z^n$$

$$= \sum_{m=-\infty}^{\infty} h[m]z^{n-m}$$

$$= z^n \sum_{m=-\infty}^{\infty} h[m]z^{-m}$$

For causal $h[n]$, the limits on the sum on the right-hand side would range from 0 to ∞. In any case, this sum is a function of z. Assuming that this sum converges, let us denote it by $H[z]$. Thus,

$$y[n] = H[z]z^n \qquad (3.71a)$$

where

$$H[z] = \sum_{m=-\infty}^{\infty} h[m]z^{-m} \qquad (3.71b)$$

Equation (3.71a) is valid only for values of z for which the sum on the right-hand side of Eq. (3.71b) exists (converges). Note that $H[z]$ is a constant for a given z. Thus, the input and the output are the same (within a multiplicative constant) for the everlasting exponential input z^n.

$H[z]$, which is called the *transfer function* of the system, is a function of the complex variable z. An alternate definition of the transfer function $H[z]$ of an LTID system from Eq. (3.71a) as

$$H[z] = \left. \frac{\text{output signal}}{\text{input signal}} \right|_{\text{input=everlasting exponential } z^n} \qquad (3.72)$$

The transfer function is defined for, and is meaningful to, LTID systems only. It does not exist for nonlinear or time-varying systems in general.

We repeat again that in this discussion we are talking of the everlasting exponential, which starts at $n = -\infty$, not the causal exponential $z^n u[n]$, which starts at $n = 0$.

For a system specified by Eq. (3.24), the transfer function is given by

$$H[z] = \frac{P[z]}{Q[z]} \tag{3.73}$$

This follows readily by considering an everlasting input $x[n] = z^n$. According to Eq. (3.72), the output is $y[n] = H[z]z^n$. Substitution of this $x[n]$ and $y[n]$ in Eq. (3.24b) yields

$$H[z]\{Q[E]z^n\} = P[E]z^n$$

Moreover

$$E^k z^n = z^{n+k} = z^k z^n$$

Hence

$$P[E]z^n = P[z]z^n \quad \text{and} \quad Q[E]z^n = Q[z]z^n$$

Consequently,

$$H[z] = \frac{P[z]}{Q[z]}$$

EXERCISE E3.21

Show that the transfer function of the digital differentiator in Example 3.6 (big shaded block in Fig. 3.14b) is given by $H[z] = (z - 1)/Tz$, and the transfer function of an unit delay, specified by $y[n] = x[n - 1]$, is given by $1/z$.

3.8-4 Total Response

The total response of an LTID system can be expressed as a sum of the zero-input and zero-state components:

$$\text{total response} = \underbrace{\sum_{j=1}^{N} c_j \gamma_j^n}_{\text{zero-input component}} + \underbrace{x[n] * h[n]}_{\text{zero-state component}}$$

In this expression, the zero-input component should be appropriately modified for the case of repeated roots. We have developed procedures to determine these two components. From the system equation, we find the characteristic roots and characteristic modes. The zero-input response is a linear combination of the characteristic modes. From the system equation, we also determine $h[n]$, the impulse response, as discussed in Section 3.7. Knowing $h[n]$ and the input $x[n]$, we find the zero-state response as the convolution of $x[n]$ and $h[n]$. The arbitrary constants $c_1, c_2, \ldots, c_n$ in the zero-input response are determined from the n initial conditions. For the system described by the equation

$$y[n + 2] - 0.6y[n + 1] - 0.16y[n] = 5x[n + 2]$$

with initial conditions $y[-1] = 0$, $y[-2] = 25/4$ and input $x[n] = (4)^{-n}u[n]$, we have determined the two components of the response in Examples 3.10a and 3.14, respectively. From the results in these examples, the total response for $n \geq 0$ is

$$\text{total response} = \underbrace{0.2(-0.2)^n + 0.8(0.8)^n}_{\text{zero-input component}} \underbrace{-1.26(4)^{-n} + 0.444(-0.2)^n + 5.81(0.8)^n}_{\text{zero-state component}} \quad (3.74)$$

NATURAL AND FORCED RESPONSE

The characteristic modes of this system are $(-0.2)^n$ and $(0.8)^n$. The zero-input component is made up of characteristic modes exclusively, as expected, but the characteristic modes also appear in the zero-state response. When all the characteristic mode terms in the total response are lumped together, the resulting component is the *natural response*. The remaining part of the total response that is made up of noncharacteristic modes is the *forced response*. For the present case, Eq. (3.74) yields

$$\text{total response} = \underbrace{0.644(-0.2)^n + 6.61(0.8)^n}_{\text{natural response}} - \underbrace{1.26(4)^{-n}}_{\text{forced response}} \quad n \geq 0 \quad (3.75)$$

3.9 CLASSICAL SOLUTION OF LINEAR DIFFERENCE EQUATIONS

As in the case of LTIC systems, we can use the classical method, in which the response is obtained as a sum of natural and forced components of the response, to analyze LTID systems.

FINDING NATURAL AND FORCED RESPONSE

As explained earlier, the *natural response* of a system consists of all the characteristic mode terms in the response. The remaining noncharacteristic mode terms form the *forced response*. If $y_c[n]$ and $y_\phi[n]$ denote the natural and the forced response respectively, then the total response is given by

$$\text{total response} = \underbrace{y_c[n]}_{\text{modes}} + \underbrace{y_\phi[n]}_{\text{nonmodes}} \quad (3.76)$$

Because the total response $y_c[n] + y_\phi[n]$ is a solution of the system equation (3.24b), we have

$$Q[E](y_c[n] + y_\phi[n]) = P[E]x[n] \quad (3.77)$$

But since $y_c[n]$ is made up of characteristic modes,

$$Q[E]y_c[n] = 0 \quad (3.78)$$

Substitution of this equation in Eq. (3.77) yields

$$Q[E]y_\phi[n] = P[E]x[n] \quad (3.79)$$

The natural response is a linear combination of characteristic modes. The arbitrary constants (multipliers) are determined from suitable auxiliary conditions usually given as $y[0]$, $y[1]$, ..., $y[n-1]$. The reasons for using auxiliary instead of initial conditions are explained later. If we

TABLE 3.2 Forced Response

No.	Input $x[n]$		Forced response $y_\phi[n]$
1	r^n	$r \ne \gamma_i$ $(i = 1, 2, \ldots, N)$	cr^n
2	r^n	$r = \gamma_i$	cnr^n
3	$\cos(\beta n + \theta)$		$c \cos(\beta n + \phi)$
4	$\left(\sum\limits_{i=0}^{m} \alpha_i n^i \right) r^n$		$\left(\sum\limits_{i=0}^{m} c_i n^i \right) r^n$

Note: By definition, $y_\phi[n]$ cannot have any characteristic mode terms. Should any term shown in the right-hand column for the forced response be a characteristic mode of the system, the correct form of the forced response must be modified to $n^i y_\phi[n]$, where i is the smallest integer that will prevent $n^i y_\phi[n]$ from having a characteristic mode term. For example, when the input is r^n, the forced response in the right-hand column is of the form cr^n. But if r^n happens to be a natural mode of the system, the correct form of the forced response is cnr^n (see pair 2).

are given initial conditions $y[-1], y[-2], \ldots, y[-N]$, we can readily use iterative procedure to derive the auxiliary conditions $y[0], y[1], \ldots, y[N-1]$. We now turn our attention to the forced response.

The forced response $y_\phi[n]$ satisfies Eq. (3.79) and, by definition, contains only nonmode terms. To determine the forced response, we shall use the method of undetermined coefficients, the same method used for the continuous-time system. However, rather than retracing all the steps of the continuous-time system, we shall present a table (Table 3.2) listing the inputs and the corresponding forms of forced function with undetermined coefficients. These coefficients can be determined by substituting $y_\phi[n]$ in Eq. (3.79) and equating the coefficients of similar terms.

EXAMPLE 3.17

Solve

$$(E^2 - 5E + 6)y[n] = (E - 5)x[n] \tag{3.80}$$

if the input $x[n] = (3n + 5)u[n]$ and the auxiliary conditions are $y[0] = 4$, $y[1] = 13$.

The characteristic equation is

$$\gamma^2 - 5\gamma + 6 = (\gamma - 2)(\gamma - 3) = 0$$

Therefore, the natural response is

$$y_c[n] = B_1(2)^n + B_2(3)^n$$

To find the form of forced response $y_\phi[n]$, we use Table 3.2, pair 4, with $r = 1, m = 1$. This yields

$$y_\phi[n] = c_1 n + c_0$$

Therefore

$$y_\phi[n + 1] = c_1(n + 1) + c_0 = c_1 n + c_1 + c_0$$
$$y_\phi[n + 2] = c_1(n + 2) + c_0 = c_1 n + 2c_1 + c_0$$

Also

$$x[n] = 3n + 5$$

and

$$x[n + 1] = 3(n + 1) + 5 = 3n + 8$$

Substitution of these results in Eq. (3.79) yields

$$c_1 n + 2c_1 + c_0 - 5(c_1 n + c_1 + c_0) + 6(c_1 n + c_0) = 3n + 8 - 5(3n + 5)$$

or

$$2c_1 n - 3c_1 + 2c_0 = -12n - 17$$

Comparison of similar terms on the two sides yields

$$\left.\begin{array}{r} 2c_1 = -12 \\ -3c_1 + 2c_0 = -17 \end{array}\right\} \implies \begin{array}{l} c_1 = -6 \\ c_2 = -\frac{35}{2} \end{array}$$

Therefore

$$y_\phi[n] = -6n - \frac{35}{2}$$

The total response is

$$y[n] = y_c[n] + y_\phi[n]$$
$$= B_1(2)^n + B_2(3)^n - 6n - \frac{35}{2} \qquad n \geq 0$$

To determine arbitrary constants B_1 and B_2 we set $n = 0$ and 1 and substitute the auxiliary conditions $y[0] = 4$, $y[1] = 13$ to obtain

$$\left.\begin{array}{r} 4 = B_1 + B_2 - \frac{35}{2} \\ 13 = 2B_1 + 3B_2 - \frac{47}{2} \end{array}\right\} \implies \begin{array}{l} B_1 = 28 \\ B_2 = \frac{-13}{2} \end{array}$$

Therefore

$$y_c[n] = 28(2)^n - \frac{13}{2}(3)^n \tag{3.81}$$

and

$$y[n] = \underbrace{28(2)^n - \frac{13}{2}(3)^n}_{y_c[n]} \underbrace{- 6n - \frac{35}{2}}_{y_\phi[n]} \tag{3.82}$$

COMPUTER EXAMPLE C3.8

Use MATLAB to solve Example 3.17.

```
>> n = (0:10)'; y = [4;13;zeros(length(n)-2,1)]; x = (3*n+5).*(n>=0);
>> for k = 1:length(n)-2,
>> y(k+2) = 5*y(k+1) - 6*y(k) + x(k+1) - 5*x(k);
>> end;
>> clf; stem(n,y,'k'); xlabel('n'); ylabel('y[n]');
>> disp(' n        y'); disp([num2str([n,y])]);
 n        y
 0        4
 1       13
 2       24
 3       13
 4     -120
 5     -731
 6    -3000
 7   -10691
 8   -35544
 9  -113675
10  -355224
```

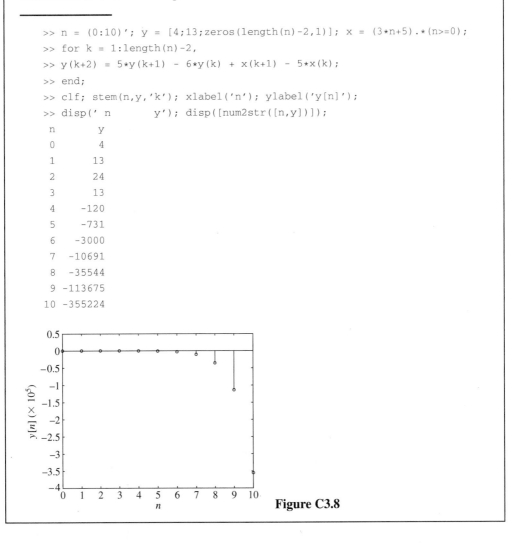

Figure C3.8

EXAMPLE 3.18

Find the sum $y[n]$ if

$$y[n] = \sum_{k=0}^{n} k^2 \qquad (3.83)$$

Such problems can be solved by finding an appropriate difference equation that has $y[n]$ as the response. From Eq. (3.83), we observe that $y[n + 1] = y[n] + (n + 1)^2$. Hence

$$y[n + 1] - y[n] = (n + 1)^2 \qquad (3.84)$$

This is the equation we seek. For this first-order difference equation, we need one auxiliary condition, the value of $y[n]$ at $n = 0$. From Eq. (3.83), it follows that $y[0] = 0$. Thus, we seek the solution of Eq. (3.83) subject to an auxiliary condition $y[0] = 0$.

The characteristic equation of Eq. (3.83) is $\gamma - 1 = 0$, the characteristic root is $\gamma = 1$, and the characteristic mode is $c(1)^n = cu[n]$, where c is an arbitrary constant. Clearly, the natural response is $cu[n]$.

The input, $x[n] = (n + 1)^2 = n^2 + 2n + 1$, is of the form in pair 4 (Table 3.2) with $r = 1$ and $m = 2$. Hence, the desired forced response is

$$y_\phi[n] = \beta_2 n^2 + \beta_1 n + \beta_0$$

Note, however, the term β_0 in $y_\phi[n]$ is of the form of characteristic mode. Hence, the correct form is $y_\phi[n] = \beta_3 n^3 + \beta_1 n^2 + \beta_0 n$. Therefore

$$E y_\phi[n] = y_\phi[n + 1] = \beta_2(n + 1)^3 + \beta_1(n + 1)^2 + \beta_0(n + 1)$$

From Eq. (3.79), we obtain

$$(E - 1)y_\phi[n] = n^2 + 2n + 1$$

or

$$[\beta_2(n + 1)^3 + \beta_1(n + 1)^2 + \beta_0(n + 1)] - [\beta_2 n^3 + \beta_1 n^2 + \beta_0 n] = n^2 + 2n + 1$$

Equating the coefficients of similar powers, we obtain

$$\beta_0 = \tfrac{1}{6} \qquad \beta_1 = \tfrac{1}{2} \qquad \beta_2 = \tfrac{1}{3}$$

Hence

$$y[n] = c + \frac{2n^3 + 3n^2 + n}{6} = c + \frac{n(n + 1)(2n + 1)}{6}$$

Setting $n = 0$ in this equation and using the auxiliary condition $y[0] = 0$, we find $c = 0$ and

$$y[n] = \frac{2n^3 + 3n^2 + n}{6} = \frac{n(n + 1)(2n + 1)}{6}$$

COMMENTS ON AUXILIARY CONDITIONS

This (classical) method requires auxiliary conditions $y[0], y[1], \ldots, y[N - 1]$. This is because at $n = -1, -2, \ldots, -N$, only the zero-input component exists, and these initial conditions can be applied to the zero-input component only. In the classical method, the zero-input and zero-state components cannot be separated. Consequently, the initial conditions must be applied to the total response, which begins at $n = 0$. Hence, we need the auxiliary conditions $y[0], y[1], \ldots, y[N - 1]$. If we are given the initial conditions $y[-1], y[-2], \ldots, y[-N]$, we can use iterative procedure to derive the auxiliary conditions $y[0], y[1], \ldots, y[n - 1]$.

AN EXPONENTIAL INPUT

As in the case of continuous-time systems, we can show that for a system specified by the equation

$$Q[E]y[n] = P[E]x[n] \tag{3.85}$$

the forced response for the exponential input $x[n] = r^n$ is given by

$$y_\phi[n] = H[r]r^n \qquad r \neq \gamma_i \text{ (a characteristic mode)} \tag{3.86}$$

where

$$H[r] = \frac{P[r]}{Q[r]} \tag{3.87}$$

The proof follows from the fact that if the input $x[n] = r^n$, then from Table 3.2 (pair 4), $y_\phi[n] = cr^n$. Therefore

$$E^i x[n] = x[n+i] = r^{n+i} = r^i r^n \qquad \text{and} \qquad P[E]x[n] = P[r]r^n$$

$$E^k y_\phi[n] = y_\phi[n+k] = cr^{n+k} = cr^k r^n \qquad \text{and} \qquad Q[E]y[n] = cQ[r]r^n$$

so that Eq. (3.85) reduces to

$$cQ[r]r^n = P[r]r^n$$

which yields $c = P[r]/Q[r] = H[r]$.

This result is valid only if r is not a characteristic root of the system. If r is a characteristic root, then the forced response is cnr^n where c is determined by substituting $y_\phi[n]$ in Eq. (3.79) and equating coefficients of similar terms on the two sides. Observe that the exponential r^n includes a wide variety of signals such as a constant C, a sinusoid $\cos(\beta n + \theta)$, and an exponentially growing or decaying sinusoid $|\gamma|^n \cos(\beta n + \theta)$.

A Constant Input $x[n] = C$. This is a special case of exponential Cr^n with $r = 1$. Therefore, from Eq. (3.86), we have

$$y_\phi[n] = C\frac{P[1]}{Q[1]} = CH[1] \tag{3.88}$$

A Sinusoidal Input. The input $e^{j\Omega n}$ is an exponential r^n with $r = e^{j\Omega}$. Hence

$$y_\phi[n] = H[e^{j\Omega}]e^{j\Omega n} = \frac{P[e^{j\Omega}]}{Q[e^{j\Omega}]}e^{j\Omega n}$$

Similarly, for the input $e^{-j\Omega n}$

$$y_\phi[n] = H[e^{-j\Omega}]e^{-j\Omega n}$$

Consequently, if the input $x[n] = \cos\Omega n = \frac{1}{2}(e^{j\Omega n} + e^{-j\Omega n})$, then

$$y_\phi[n] = \frac{1}{2}\{H[e^{j\Omega}]e^{j\Omega n} + H[e^{-j\Omega}]e^{-j\Omega n}\}$$

Since the two terms on the right-hand side are conjugates,

$$y_\phi[n] = \text{Re}\{H[e^{j\Omega}]e^{j\Omega n}\}$$

If

$$H[e^{j\Omega}] = |H[e^{j\Omega}]|e^{j\angle H[e^{j\Omega}]}$$

then

$$y_\phi[n] = \text{Re}\{|H[e^{j\Omega}]|e^{j(\Omega n + \angle H[e^{j\Omega}])}\}$$
$$= |H[e^{j\Omega}]|\cos(\Omega n + \angle H[e^{j\Omega}]) \qquad (3.89a)$$

Using a similar argument, we can show that for the input

$$x[n] = \cos(\Omega n + \theta)$$
$$y_\phi[n] = |H[e^{j\Omega}]|\cos(\Omega n + \theta + \angle H[e^{j\Omega}]) \qquad (3.89b)$$

EXAMPLE 3.19

For a system specified by the equation

$$(E^2 - 3E + 2)y[n] = (E + 2)x[n]$$

find the forced response for the input $x[n] = (3)^n u[n]$.

In this case

$$H[r] = \frac{P[r]}{Q[r]} = \frac{r + 2}{r^2 - 3r + 2}$$

and the forced response to input $(3)^n u[n]$ is $H[3](3)^n u[n]$; that is,

$$y_\phi[n] = \frac{3 + 2}{(3)^2 - 3(3) + 2}(3)^n u[n] = \frac{5}{2}(3)^n u[n] \qquad n \geq 0$$

EXAMPLE 3.20

For an LTID system described by the equation

$$(E^2 - E + 0.16)y[n] = (E + 0.32)x[n]$$

determine the forced response $y_\phi[n]$ if the input is

$$x[n] = \cos\left(2n + \frac{\pi}{3}\right)u[n]$$

Here

$$H[r] = \frac{P[r]}{Q[r]} = \frac{r + 0.32}{r^2 - r + 0.16}$$

For the input $\cos(2n + (\pi/3))u[n]$, the forced response is

$$y_\phi[n] = |H[e^{j2}]| \cos\left(2n + \frac{\pi}{3} + \angle H[e^{j2}]\right) u[n]$$

where

$$H[e^{j2}] = \frac{e^{j2} + 0.32}{(e^{j2})^2 - e^{j2} + 0.16} = \frac{(-0.416 + j0.909) + 0.32}{(-0.654 - j0.757) - (-0.416 + j0.909) + 0.16}$$

$$= 0.548e^{j3.294}$$

Therefore

$$|H[e^{j2}]| = 0.548 \qquad \text{and} \qquad \angle H[e^{j2}] = 3.294$$

so that

$$y_\phi[n] = 0.548 \cos\left(2n + \frac{\pi}{3} + 3.294\right) u[n]$$

$$= 0.548 \cos(2n + 4.34)u[n]$$

Assessment of the Classical Method

The remarks in Chapter 2 concerning the classical method for solving differential equations also apply to difference equations.

3.10 System Stability: The External (BIBO) Stability Criterion

Concepts and criteria for the BIBO (external) stability and the internal (asymptotic) stability for discrete-time systems are identical to those corresponding to continuous-time systems. The comments in Section 2.6 for LTIC systems concerning the distinction between the external and the internal stability are also valid for LTID systems.

Recall that

$$y[n] = h[n] * x[n]$$

$$= \sum_{m=-\infty}^{\infty} h[m]x[n-m]$$

and

$$|y[n]| = \left| \sum_{m=-\infty}^{\infty} h[m]x[n-m] \right|$$

$$\leq \sum_{m=-\infty}^{\infty} |h[m]|\,|x[n-m]|$$

If $x[n]$ is bounded, then $|x[n-m]| < K_1 < \infty$, and

$$|y[n]| \leq K_1 \sum_{m=-\infty}^{\infty} |h[m]|$$

Clearly the output is bounded if the summation on the right-hand side is bounded; that is, if

$$\sum_{n=-\infty}^{\infty} |h[n]| < K_2 < \infty \tag{3.90}$$

This is a sufficient condition for BIBO stability. We can show that this is also a necessary condition (see Prob. 3.10-1). Therefore, for an LTID system, if its impulse response $h[n]$ is absolutely summable, the system is (BIBO) stable. Otherwise it is unstable.

All the comments about the nature of the external and the internal stability in Chapter 2 apply to discrete-time case. We shall not elaborate them further.

3.10-1 Internal (Asymptotic) Stability

For LTID systems, as in the case of LTIC systems, the internal stability, called the asymptotical stability or the stability in the sense of Lyapunov (also the zero-input stability) is defined in terms of the zero-input response of a system.

For an LTID system specified by a difference equation of the form (3.17) [or (3.24)], the zero-input response consists of the characteristic modes of the system. The mode corresponding to a characteristic root γ is γ^n. To be more general, let γ be complex so that

$$\gamma = |\gamma| e^{j\beta} \qquad \text{and} \qquad \gamma^n = |\gamma|^n e^{j\beta n}$$

Since the magnitude of $e^{j\beta n}$ is always unity regardless of the value of n, the magnitude of γ^n is $|\gamma|^n$. Therefore

$$
\begin{array}{llll}
\text{if} & |\gamma| < 1, & \gamma^n \to 0 & \text{as } n \to \infty \\
\text{if} & |\gamma| > 1, & \gamma^n \to \infty & \text{as } n \to \infty \\
\text{and if} & |\gamma| = 1, & |\gamma|^n = 1 & \text{for all } n
\end{array}
$$

Figure 3.21 shows the characteristic modes corresponding to characteristic roots at various locations in the complex plane.

These results can be grasped more effectively in terms of the location of characteristic roots in the complex plane. Figure 3.22 shows a circle of unit radius, centered at the origin in a complex plane. Our discussion shows that if all characteristic roots of the system lie inside the *unit circle*, $|\gamma_i| < 1$ for all i and the system is asymptotically stable. On the other hand, even if one characteristic root lies outside the unit circle, the system is unstable. If none of the characteristic roots lie outside the unit circle, but some simple (unrepeated) roots lie on the circle itself, the system is marginally stable. If two or more characteristic roots coincide on the unit circle (repeated roots), the system is unstable. The reason is that for repeated roots, the zero-input response is of the form $n^{r-1}\gamma^n$, and if $|\gamma| = 1$, then $|n^{r-1}\gamma^n| = n^{r-1} \to \infty$ as $n \to \infty$.[†]

[†]If the development of discrete-time systems is parallel to that of continuous-time systems, we wonder why the parallel breaks down here. Why, for instance, are LHP and RHP not the regions demarcating stability and instability? The reason lies in the form of the characteristic modes. In continuous-time systems, we chose the form of characteristic mode as $e^{\lambda_i t}$. In discrete-time systems, for computational convenience, we choose the form to be γ_i^n. Had we chosen this form to be $e^{\lambda_i n}$ where $\gamma_i = e^{\lambda_i}$, then the LHP and RHP (for the location of λ_i) again would demarcate stability and instability. The reason is that if $\gamma = e^\lambda$, $|\gamma| = 1$ implies $|e^\lambda| = 1$, and therefore $\lambda = j\omega$. This shows that the unit circle in γ plane maps into the imaginary axis in the λ plane.

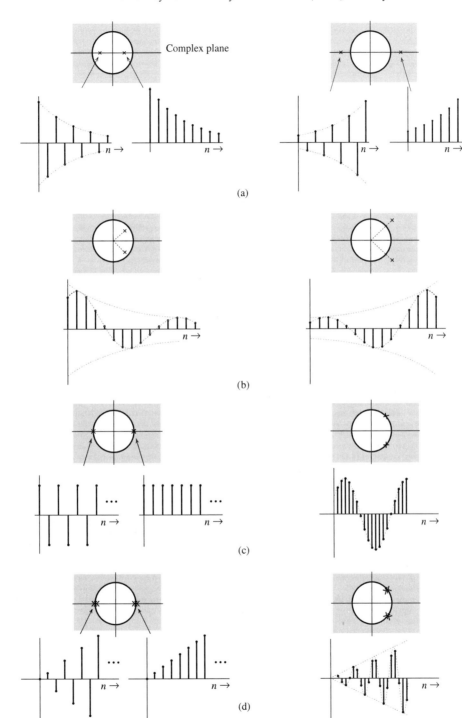

Figure 3.21 Characteristic roots location and the corresponding characteristic modes.

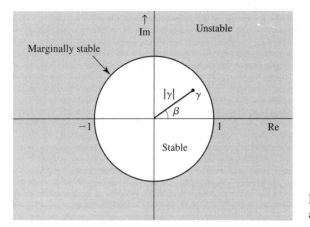

Figure 3.22 Characteristic root locations and system stability.

Note, however, that repeated roots inside the unit circle do not cause instability. To summarize:

1. An LTID system is asymptotically stable if and only if all the characteristic roots are inside the unit circle. The roots may be simple or repeated.

2. An LTID system is unstable if and only if either one or both of the following conditions exist: (i) at least one root is outside the unit circle; (ii) there are repeated roots on the unit circle.

3. An LTID system is marginally stable if and only if there are no roots outside the unit circle and there are some unrepeated roots on the unit circle.

3.10-2 Relationship Between BIBO and Asymptotic Stability

For LTID systems, the relation between the two types of stability is similar to those in LTIC systems. For a system specified by Eq. (3.17), we can readily show that if a characteristic root γ_k is inside the unit circle, the corresponding mode γ_k^n is absolutely summable. In contrast, if γ_k lies outside the unit circle, or on the unit circle, γ_k^n is not absolutely summable.[†]

This means that an asymptotically stable system is BIBO stable. Moreover, a marginally stable or asymptotically unstable system is BIBO unstable. The converse is not necessarily true. The stability picture portrayed by the external description is of questionable value. BIBO (external) stability cannot ensure internal (asymptotic) stability, as the following example shows.

[†]This conclusion follows from the fact that (see Section B.7-4)

$$\sum_{n=-\infty}^{\infty} \left|\gamma_k^n |u[n]\right| = \sum_{n=0}^{\infty} |\gamma_k|^n = \frac{1}{1 - |\gamma_k|} \qquad |\gamma_k| < 1$$

Moreover, if $|\gamma| \geq 1$, the sum diverges and goes to ∞. These conclusions are valid also for the modes of the form $n^r \gamma_k^n$.

EXAMPLE 3.21

An LTID systems consists of two subsystems S_1 and S_2 in cascade (Fig. 3.23). The impulse response of these systems are $h_1[n]$ and $h_2[n]$, respectively, given by

$$h_1[n] = 4\delta[n] - 3(0.5)^n u[n] \qquad \text{and} \qquad h_2[n] = 2^n u[n]$$

Figure 3.23 BIBO and asymptotic stability.

The composite system impulse response $h[n]$ is given by

$$h[n] = h_1[n] * h_2[n] = h_2[n] * h_1[n] = 2^n u[n] * (4\delta[n] - 3(0.5)^n u[n])$$

$$= 4(2)^n u[n] - 3 \left[\frac{2^{n+1} - (0.5)^{n+1}}{2 - 0.5} \right] u[n]$$

$$= (0.5)^n u[n]$$

If the composite cascade system were to be enclosed in a black box with only the input and the output terminals accessible, any measurement from these external terminals would show that the impulse response of the system is $(0.5)^n u[n]$, without any hint of the unstable system sheltered inside the composite system.

The composite system is BIBO stable because its impulse response $(0.5)^n u[n]$ is absolutely summable. However, the system S_2 is asymptotically unstable because its characteristic root, 2, lies outside the unit circle. This system will eventually burn out (or saturate) because of the unbounded characteristic response generated by intended or unintended initial conditions, no matter how small.

The system is asymptotically unstable, though BIBO stable. This example shows that BIBO stability does not necessarily ensure asymptotic stability when a system is either uncontrollable, or unobservable, or both. The internal and the external descriptions of a system are equivalent only when the system is controllable and observable. In such a case, BIBO stability means the system is asymptotically stable, and vice versa.

Fortunately, uncontrollable or unobservable systems are not common in practice. Henceforth, in determining system stability, we shall assume that unless otherwise mentioned, the internal and the external descriptions of the system are equivalent, implying that the system is controllable and observable.

EXAMPLE 3.22

Determine the internal and external stability of systems specified by the following equations. In each case plot the characteristic roots in the complex plane.

(a) $y[n+2] + 2.5y[n+1] + y[n] = x[n+1] - 2x[n]$

(b) $y[n] - y[n-1] + 0.21y[n-2] = 2x[n-1] + 3x[n-2]$

(c) $y[n+3] + 2y[n+2] + \frac{3}{2}y[n+1] + \frac{1}{2}y[n] = x[n+1]$

(d) $(E^2 - E + 1)^2 y[n] = (3E + 1)x[n]$

(a) The characteristic polynomial is

$$\gamma^2 + 2.5\gamma + 1 = (\gamma + 0.5)(\gamma + 2)$$

The characteristic roots are -0.5 and -2. Because $|-2| > 1$ (-2 lies outside the unit circle), the system is BIBO unstable and also asymptotically unstable (Fig. 3.24a).

(b) The characteristic polynomial is

$$\gamma^2 - \gamma + 0.21 = (\gamma - 0.3)(\gamma - 0.7)$$

The characteristic roots are 0.3 and 0.7, both of which lie inside the unit circle. The system is BIBO stable and asymptotically stable (Fig. 3.24b).

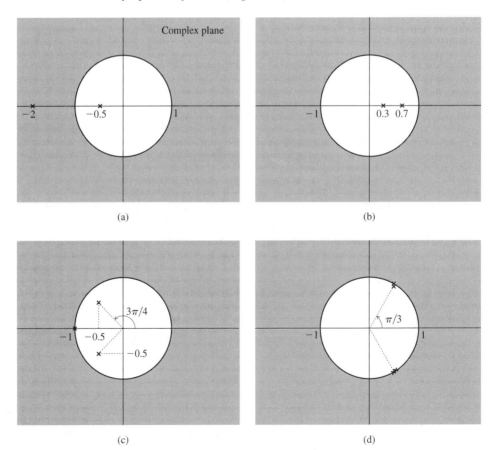

Figure 3.24 Location of characteristic roots for the systems.

(c) The characteristic polynomial is

$$\gamma^3 + 2\gamma^2 + \tfrac{3}{2}\gamma + \tfrac{1}{2} = (\gamma + 1)\left(\gamma^2 + \gamma + \tfrac{1}{2}\right) = (\gamma + 1)(\gamma + 0.5 - j0.5)(\gamma + 0.5 + j0.5)$$

The characteristic roots are -1, $-0.5 \pm j0.5$ (Fig. 3.24c). One of the characteristic roots is on the unit circle and the remaining two roots are inside the unit circle. The system is BIBO unstable but marginally stable.

(d) The characteristic polynomial is

$$(\gamma^2 - \gamma + 1)^2 = \left(\gamma - \tfrac{1}{2} - j\tfrac{\sqrt{3}}{2}\right)^2\left(\gamma - \tfrac{1}{2} + j\tfrac{\sqrt{3}}{2}\right)^2$$

The characteristic roots are $(1/2) \pm j(\sqrt{3}/2) = 1e^{\pm j(\pi/3)}$ repeated twice, and they lie on the unit circle (Fig. 3.24d). The system is BIBO unstable and asymptotically unstable.

EXERCISE E3.22

Find and sketch the location in the complex plane of the characteristic roots of the system specified by the following equation:

$$(E + 1)(E^2 + 6E + 25)y[n] = 3Ex[n]$$

Determine the external and the internal stability of the system.

ANSWER
BIBO and asymptotically unstable

EXERCISE E3.23

Repeat Exercise E3.22 for

$$(E - 1)^2(E + 0.5)y[n] = (E^2 + 2E + 3)x[n]$$

ANSWER
BIBO and asymptotically unstable

3.11 INTUITIVE INSIGHTS INTO SYSTEM BEHAVIOR

The intuitive insights into the behavior of continuous-time systems and their qualitative proofs, discussed in Section 2.7, also apply to discrete-time systems. For this reason, we shall merely mention here without discussion some of the insights presented in Section 2.7.

The system's entire (zero-input and zero-state) behavior is strongly influenced by the characteristic roots (or modes) of the system. The system responds strongly to input signals similar to its characteristic modes and poorly to inputs very different from its characteristic modes. In fact, when the input is a characteristic mode of the system, the response goes to infinity, provided the mode is a nondecaying signal. This is the resonance phenomenon. The width of an impulse response $h[n]$ indicates the response time (time required to respond fully to an input) of the system. It is the time constant of the system.[†] Discrete-time pulses are generally dispersed when passed through a discrete-time system. The amount of dispersion (or spreading out) is equal to the system time constant (or width of $h[n]$). The system time constant also determines the rate at which the system can transmit information. Smaller time constant corresponds to higher rate of information transmission, and vice versa.

3.12 APPENDIX 3.1: IMPULSE RESPONSE FOR A SPECIAL CASE

When $a_N = 0$, $A_0 = b_N/a_N$ becomes indeterminate, and the procedure needs to be modified slightly. When $a_N = 0$, $Q[E]$ can be expressed as $E\hat{Q}[E]$, and Eq. (3.43) can be expressed as

$$E\hat{Q}[E]h[n] = P[E]\delta[n] = P[E]\{E\delta[n-1]\} = EP[E]\delta[n-1]$$

Hence

$$\hat{Q}[E]h[n] = P[E]\delta[n-1]$$

In this case the input vanishes not for $n \geq 1$, but for $n \geq 2$. Therefore, the response consists not only of the zero-input term and an impulse $A_0\delta[n]$ (at $n = 0$), but also of an impulse $A_1\delta[n-1]$ (at $n = 1$). Therefore

$$h[n] = A_0\delta[n] + A_1\delta[n-1] + y_c[n]u[n]$$

We can determine the unknowns A_0, A_1, and the $N-1$ coefficients in $y_c[n]$ from the $N+1$ number of initial values $h[0], h[1], \ldots, h[N]$, determined as usual from the iterative solution of the equation $Q[E]h[n] = P[E]\delta[n]$.[‡] Similarly, if $a_N = a_{N-1} = 0$, we need to use the form $h[n] = A_0\delta[n] + A_1\delta[n-1] + A_2\delta[n-2] + y_c[n]u[n]$. The $N+1$ unknown constants are determined from the $N+1$ values $h[0], h[1], \ldots, h[N]$, determined iteratively, and so on.

3.13 SUMMARY

This chapter discusses time-domain analysis of LTID (linear, time-invariant, discrete-time) systems. The analysis is parallel to that of LTIC systems, with some minor differences. Discrete-time

---

[†]This part of the discussion applies to systems with impulse response $h[n]$ that is a mostly positive (or mostly negative) pulse.

[‡]$\hat{Q}[\gamma]$ is now an $(N-1)$-order polynomial. Hence there are only $N-1$ unknowns in $y_c[n]$.

systems are described by difference equations. For an Nth-order system, N auxiliary conditions must be specified for a unique solution. Characteristic modes are discrete-time exponentials of the form γ^n corresponding to an unrepeated root γ, and the modes are of the form $n^i\gamma^n$ corresponding to a repeated root γ.

The unit impulse function $\delta[n]$ is a sequence of a single number of unit value at $n = 0$. The unit impulse response $h[n]$ of a discrete-time system is a linear combination of its characteristic modes.[†]

The zero-state response (response due to external input) of a linear system is obtained by breaking the input into impulse components and then adding the system responses to all the impulse components. The sum of the system responses to the impulse components is in the form of a sum, known as the convolution sum, whose structure and properties are similar to the convolution integral. The system response is obtained as the convolution sum of the input $x[n]$ with the system's impulse response $h[n]$. Therefore, the knowledge of the system's impulse response allows us to determine the system response to any arbitrary input.

LTID systems have a very special relationship to the everlasting exponential signal z^n because the response of an LTID system to such an input signal is the same signal within a multiplicative constant. The response of an LTID system to the everlasting exponential input z^n is $H[z]z^n$, where $H[z]$ is the transfer function of the system.

Difference equations of LTID systems can also be solved by the classical method, in which the response is obtained as a sum of the natural and the forced components. These are not the same as the zero-input and zero-state components, although they satisfy the same equations, respectively. Although simple, this method is applicable to a restricted class of input signals, and the system response cannot be expressed as an explicit function of the input. These limitations diminish its value considerably in the theoretical study of systems.

The external stability criterion, the bounded-input/bounded-output (BIBO) stability criterion, states that a system is stable if and only if every bounded input produces a bounded output. Otherwise the system is unstable.

The internal stability criterion can be stated in terms of the location of characteristic roots of the system as follows:

1. An LTID system is asymptotically stable if and only if all the characteristic roots are inside the unit circle. The roots may be repeated or unrepeated.

2. An LTID system is unstable if and only if either one or both of the following conditions exist: (i) at least one root is outside the unit circle; (ii) there are repeated roots on the unit circle.

3. An LTID system is marginally stable if and only if there are no roots outside the unit circle and some unrepeated roots on the unit circle.

An asymptotically stable system is always BIBO stable. The converse is not necessarily true.

[†]There is a possibility of an impulse $\delta[n]$ in addition to characteristic modes.

MATLAB SESSION 3: DISCRETE-TIME SIGNALS AND SYSTEMS

MATLAB is naturally and ideally suited to discrete-time signals and systems. Many special functions are available for discrete-time data operations, including the `stem`, `filter`, and `conv` commands. In this session, we investigate and apply these and other commands.

M3.1 Discrete-Time Functions and Stem Plots

Consider the discrete-time function $f[n] = e^{-n/5} \cos(\pi n/5) u[n]$. In MATLAB, there are many ways to represent $f[n]$ including M-files or, for particular n, explicit command line evaluation. In this example, however, we use an inline object

```
>> f = inline('exp(-n/5).*cos(pi*n/5).*(n>=0)','n');
```

A true discrete-time function is undefined (or zero) for noninteger n. Although inline object `f` is intended as a discrete-time function, its present construction does not restrict n to be integer, and it can therefore be misused. For example, MATLAB dutifully returns 0.8606 to `f(0.5)` when a NaN (not-a-number) or zero is more appropriate. The user is responsible for appropriate function use.

Next, consider plotting the discrete-time function $f[n]$ over $(-10 \le n \le 10)$. The `stem` command simplifies this task.

```
>> n = (-10:10)';
>> stem(n,f(n),'k');
>> xlabel('n'); ylabel('f[n]');
```

Here, `stem` operates much like the `plot` command: dependent variable `f(n)` is plotted against independent variable `n` with black lines. The `stem` command emphasizes the discrete-time nature of the data, as Fig. M3.1 illustrates.

For discrete-time functions, the operations of shifting, inversion, and scaling can have surprising results. Compare $f[-2n]$ with $f[-2n + 1]$. Contrary to the continuous case, the second is not a shifted version of first. We can use separate subplots, each over $(-10 \le n \le 10)$, to help illustrate this fact. Notice that unlike the `plot` command, the `stem` command cannot

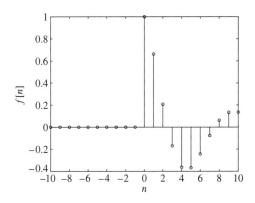

Figure M3.1 $f[n]$ over $(-10 \le n \le 10)$.

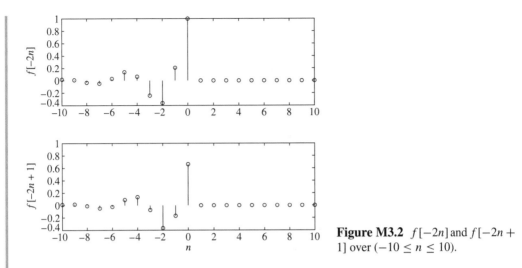

Figure M3.2 $f[-2n]$ and $f[-2n + 1]$ over $(-10 \leq n \leq 10)$.

simultaneously plot multiple functions on a single axis; overlapping stem lines would make such plots difficult to read anyway.

```
>> subplot(2,1,1); stem(n,f(-2*n),'k'); ylabel('f[-2n]');
>> subplot(2,1,2); stem(n,f(-2*n+1),'k'); ylabel('f[-2n+1]'); xlabel('n');
```

The results are shown in Fig. M3.2. Interestingly, the original function $f[n]$ can be recovered by interleaving samples of $f[-2n]$ and $f[-2n + 1]$ and then time-reflecting the result.

Care must always be taken to ensure that MATLAB performs the desired computations. Our inline function f is a case in point: although it correctly downsamples, it does not properly upsample (see Prob. 3.M-1). MATLAB does what it is told, but it is not always told how to do everything correctly!

M3.2 System Responses Through Filtering

MATLAB's `filter` command provides an efficient way to evaluate the system response of a constant coefficient linear difference equation represented in delay form as

$$\sum_{k=0}^{N} a_k y[n - k] = \sum_{k=0}^{N} b_k x[n - k] \qquad (M3.1)$$

In the simplest form, `filter` requires three input arguments: a length-$(N+1)$ vector of feedforward coefficients $[b_0, b_1, \ldots, b_N]$, a length-$(N + 1)$ vector of feedback coefficients $[a_0, a_1, \ldots, a_N]$, and an input vector.[†] Since no initial conditions are specified, the output corresponds to the system's zero-state response.

[†]It is important to pay close attention to the inevitable notational differences found throughout engineering documents. In MATLAB help documents, coefficient subscripts begin at 1 rather than 0 to better conform with MATLAB indexing conventions. That is, MATLAB labels a_0 as `a(1)`, b_0 as `b(1)`, and so forth.

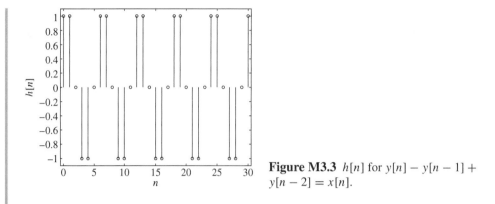

Figure M3.3 $h[n]$ for $y[n] - y[n-1] + y[n-2] = x[n]$.

To serve as an example, consider a system described by $y[n] - y[n-1] + y[n-2] = x[n]$. When $x[n] = \delta[n]$, the zero-state response is equal to the impulse response $h[n]$, which we compute over $(0 \le n \le 30)$.

```
>> b = [1 0 0]; a = [1 -1 1];
>> n = (0:30)'; delta = inline('n==0','n');
>> h = filter(b,a,delta(n));
>> stem(n,h,'k'); axis([-.5 30.5 -1.1 1.1]);
>> xlabel('n'); ylabel('h[n]');
```

As shown in Fig. M3.3, $h[n]$ appears to be ($N_0 = 6$)-periodic for $n \ge 0$. Since periodic signals are not absolutely summable, $\sum_{n=-\infty}^{\infty} |h[n]|$ is not finite and the system is not BIBO stable. Furthermore, the sinusoidal input $x[n] = \cos(2\pi n/6)u[n]$, which is ($N_0 = 6$)-periodic for $n \ge 0$, should generate a resonant zero-state response.

```
>> x = inline('cos(2*pi*n/6).*(n>=0)','n');
>> y = filter(b,a,x(n));
>> stem(n,y,'k'); xlabel('n'); ylabel('y[n]');
```

The response's linear envelope, shown in Fig. M3.4, confirms a resonant response. The characteristic equation of the system is $\gamma^2 - \gamma + 1$, which has roots $\gamma = e^{\pm j\pi/3}$. Since the input $x[n] = \cos(2\pi n/6)u[n] = (1/2)(e^{j\pi n/3} + e^{-j\pi n/3})u[n]$ coincides with the characteristic roots, a resonant response is guaranteed.

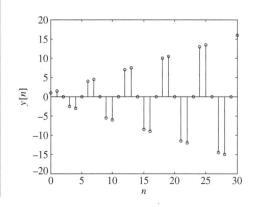

Figure M3.4 Resonant zero-state response $y[n]$ for $x[n] = \cos(2\pi n/6)u[n]$.

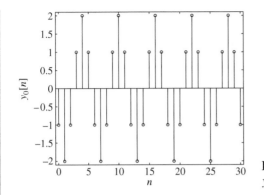

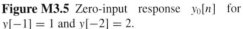

Figure M3.5 Zero-input response $y_0[n]$ for $y[-1] = 1$ and $y[-2] = 2$.

By adding initial conditions, the `filter` command can also compute a system's zero-input response and total response. Continuing the preceding example, consider finding the zero-input response for $y[-1] = 1$ and $y[-2] = 2$ over $(0 \le n \le 30)$.

```
>> z_i = filtic(b,a,[1 2]);
>> y_0 = filter(b,a,zeros(size(n)),z_i);
>> stem(n,y_0,'k'); xlabel('n'); ylabel('y_{0} [n]');
>> axis([-0.5 30.5 -2.1 2.1]);
```

There are many physical ways to implement a particular equation. MATLAB implements Eq. (M3.1) by using the popular direct form II transposed structure.[†] Consequently, initial conditions must be compatible with this implementation structure. The signal processing toolbox function `filtic` converts the traditional $y[-1], y[-2], \ldots, y[-N]$ initial conditions for use with the `filter` command. An input of zero is created with the `zeros` command. The dimensions of this zero input are made to match the vector `n` by using the `size` command. Finally, `_{ }` forces subscript text in the graphics window, and `^{ }` forces superscript text. The results are shown in Fig. M3.5.

Given $y[-1] = 1$ and $y[-2] = 2$ and an input $x[n] = \cos(2\pi n/6)u[n]$, the total response is easy to obtain with the `filter` command.

```
>> y_total = filter(b,a,x(n),z_i);
```

Summing the zero-state and zero-input response gives the same result. Computing the total absolute error provides a check.

```
>> sum(abs(y_total-(y + y_0)))
ans = 1.8430e-014
```

Within computer round-off, both methods return the same sequence.

M3.3 A Custom Filter Function

The `filtic` command is only available if the signal processing toolbox is installed. To accommodate installations without the signal processing toolbox and to help develop your MATLAB skills, consider writing a function similar in syntax to `filter` that directly uses the

---

[†]Implementation structures, such as direct form II transposed, are discussed in Chapter 4.

ICs $y[-1], y[-2], \ldots, y[-N]$. Normalizing $a_0 = 1$ and solving Eq. (M3.1) for $y[n]$ yields

$$y[n] = \sum_{k=0}^{N} b_k x[n-k] - \sum_{k=1}^{N} a_k y[n-k]$$

This recursive form provides a good basis for our custom filter function.

```
function [y] = MS3P1(b,a,x,yi);
% MS3P1.m : MATLAB Session 3, Program 1
% Function M-file filters data x to create y
% INPUTS:    b = vector of feedforward coefficients
%            a = vector of feedback coefficients
%            x = input data vector
%            yi = vector of initial conditions [y[-1], y[-2], ...]
% OUTPUTS:   y = vector of filtered output data

yi = flipud(yi(:)); % Properly format IC's.
y = [yi;zeros(length(x),1)]; % Preinitialize y, beginning with IC's.
x = [zeros(length(yi),1);x(:)]; % Append x with zeros to match size of y.
b = b/a(1);a = a/a(1); % Normalize coefficients.
for n = length(yi)+1:length(y),
    for nb = 0:length(b)-1,
        y(n) = y(n) + b(nb+1)*x(n-nb); % Feedforward terms.
    end
    for na = 1:length(a)-1,
        y(n) = y(n) - a(na+1)*y(n-na); % Feedback terms.
    end
end
y = y(length(yi)+1:end); % Strip off IC's for final output.
```

Most instructions in MS3P1 have been discussed; now we turn to the flipud instruction. The flip up–down command flipud reverses the order of elements in a column vector. Although not used here, the flip left–right command fliplr reverses the order of elements in a row vector. Note that typing help *filename* displays the first contiguous set of comment lines in an M-file. Thus, it is good programming practice to document M-files, as in MS3P1, with an initial block of clear comment lines.

As an exercise, the reader should verify that MS3P1 correctly computes the impulse response $h[n]$, the zero-state response $y[n]$, the zero-input response $y_0[n]$, and the total response $y[n] + y_0[n]$.

M3.4 Discrete-Time Convolution

Convolution of two finite-duration discrete-time signals is accomplished by using the conv command. For example, the discrete-time convolution of two length-4 rectangular pulses, $g[n] = (u[n] - u[n-4]) * (u[n] - u[n-4])$, is a length-$(4 + 4 - 1 = 7)$ triangle. Representing $u[n] - u[n-4]$ by the vector [1, 1, 1, 1], the convolution is computed by:

```
>> conv([1 1 1 1],[1 1 1 1])
ans = 1    2    3    4    3    2    1
```

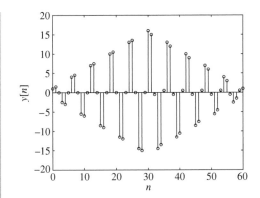

Figure M3.6 $y[n]$ for $x[n] = \cos{(2\pi n/6)}u[n]$ computed with `conv`.

Notice that $(u[n+4] - u[n]) * (u[n] - u[n-4])$ is also computed by `conv([1 1 1 1], [1 1 1 1])` and obviously yields the same result. The difference between these two cases is the regions of support: $(0 \le n \le 6)$ for the first and $(-4 \le n \le 2)$ for the second. Although the `conv` command does not compute the region of support, it is relatively easy to obtain. If vector w begins at $n = n_w$ and vector v begins at $n = n_v$, then `conv(w,v)` begins at $n = n_w + n_v$.

In general, the `conv` command cannot properly convolve infinite-duration signals. This is not too surprising, since computers themselves cannot store an infinite-duration signal. For special cases, however, `conv` can correctly compute a portion of such convolution problems. Consider the common case of convolving two causal signals. By passing the first N samples of each, `conv` returns a length-$(2N - 1)$ sequence. The first N samples of this sequence are valid; the remaining $N - 1$ samples are not.

To illustrate this point, reconsider the zero-state response $y[n]$ over $(0 \le n \le 30)$ for system $y[n] - y[n-1] + y[n-2] = x[n]$ given input $x[n] = \cos{(2\pi n/6)}u[n]$. The results obtained by using a filtering approach are shown in Fig. M3.4.

The response can also be computed using convolution according to $y[n] = h[n] * x[n]$. The impulse response of this system is[†]

$$h[n] = \left\{ \cos{(\pi n/3)} + \frac{1}{\sqrt{3}} \sin{(\pi n/3)} \right\} u[n]$$

Both $h[n]$ and $x[n]$ are causal and have infinite duration, so `conv` can be used to obtain a portion of the convolution.

```
>> h = inline('(cos(pi*n/3)+sin(pi*n/3)/sqrt(3)).*(n>=0)','n');
>> y = conv(h(n),x(n));
>> stem([0:60],y,'k'); xlabel('n'); ylabel('y[n]');
```

The `conv` output is fully displayed in Fig. M3.6. As expected, the results are correct over $(0 \le n \le 30)$. The remaining values are clearly incorrect; the output envelope should continue to grow, not decay. Normally, these incorrect values are not displayed.

```
>> stem(n,y(1:31),'k'); xlabel('n'); ylabel('y[n]');
```

The resulting plot is identical to Fig. M3.4.

[†]Techniques to analytically determine $h[n]$ are presented in Chapter 5.

PROBLEMS

3.1-1 Find the energy of the signals depicted in Figs. P3.1-1.

3.1-2 Find the power of the signals illustrated in Figs. P3.1-2.

3.1-3 Show that the power of a signal $\mathcal{D}e^{j(2\pi/N_0)n}$ is $|\mathcal{D}|^2$. Hence, show that the power of a signal

$$x[n] = \sum_{r=0}^{N_0-1} \mathcal{D}_r e^{jr(2\pi/N_0)n} \text{ is } P_x = \sum_{r=0}^{N_0-1} |\mathcal{D}_r|^2$$

Use the fact that

$$\sum_{k=0}^{N_0-1} e^{j(r-m)2\pi k/N_0} = \begin{cases} N_0 & r = m \\ 0 & \text{otherwise} \end{cases}$$

3.1-4 (a) Determine even and odd components of the signal $x[n] = (0.8)^n u[n]$.

(b) Show that the energy of $x[n]$ is the sum of energies of its odd and even components found in part (a).

(c) Generalize the result in part (b) for any finite energy signal.

3.1-5 (a) If $x_e[n]$ and $x_o[n]$ are the even and the odd components of a real signal $x[n]$, then show that $E_{x_e} = E_{x_0} = 0.5E_x$.

(b) Show that the cross-energy of x_e and x_o is zero, that is,

$$\sum_{n=-\infty}^{\infty} x_e[n]x_o[n] = 0$$

3.2-1 If the energy of a signal $x[n]$ is E_x, then find the energy of the following:

(a) $x[-n]$

(b) $x[n-m]$

(c) $x[m-n]$

(d) $Kx[n]$ (m integer and K constant)

3.2-2 If the power of a periodic signal $x[n]$ is P_x, find and comment on the powers and the rms values of the following:

(a) $-x[n]$

(b) $x[-n]$

(c) $x[n-m]$ (m integer)

(d) $cx[n]$

(e) $x[m-n]$ (m integer)

3.2-3 For the signal shown in Fig. P3.1-1b, sketch the following signals:

(a) $x[-n]$

(b) $x[n+6]$

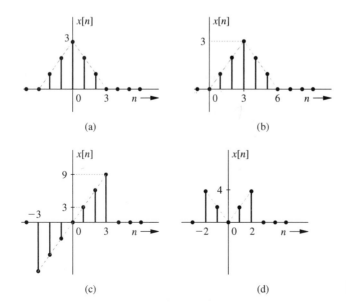

(a)

(b)

(c)

(d)

Figure P3.1-1

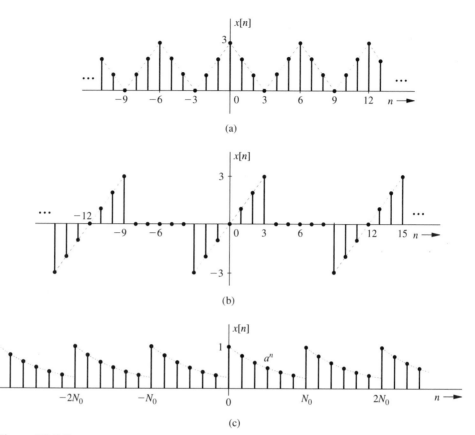

Figure P3.1-2

(c) $x[n-6]$

(d) $x[3n]$

(e) $x\left[\dfrac{n}{3}\right]$

(f) $x[3-n]$

3.2-4 Repeat Prob. 3.2-3 for the signal depicted in Fig. P3.1-1c.

3.3-1 Sketch, and find the power of, the following signals:

(a) $(1)^n$

(b) $(-1)^n$

(c) $u[n]$

(d) $(-1)^n u[n]$

(e) $\cos\left[\dfrac{\pi}{3}n + \dfrac{\pi}{6}\right]$.

3.3-2 Show that

(a) $\delta[n] + \delta[n-1] = u[n] - u[n-2]$

(b) $2^{n-1}\sin\left(\dfrac{\pi n}{3}\right)u[n] = \dfrac{1}{2}2^n \sin\left(\dfrac{\pi n}{3}\right)u[n-1]$

(c) $n(n-1)\gamma^n u[n] = n(n-1)\gamma^n u[n-2]$

(d) $(u[n] + (-1)^n u[n])\sin\left(\dfrac{\pi n}{2}\right) = 0$
for all n

(e) $(u[n]+(-1)^{n+1}u[n])\cos\left(\dfrac{\pi n}{2}\right) = 0$
for all n

3.3-3 Sketch the following signals:

(a) $u[n-2] - u[n-6]$

(b) $n\{u[n] - u[n-7]\}$

(c) $(n-2)\{u[n-2] - u[n-6]\}$

(d) $(-n+8)\{u[n-6] - u[n-9]\}$

(e) $(n-2)\{u[n-2] - u[n-6]\}$
$+ (-n+8)\{u[n-6] - u[n-9]\}$

3.3-4 Describe each of the signals in Fig. P3.1-1 by a single expression valid for all n.

3.3-5 The following signals are in the form $e^{\lambda n}$. Express them in the form γ^n:

(a) $e^{-0.5n}$
(b) $e^{0.5n}$
(c) $e^{-j\pi n}$
(d) $e^{j\pi n}$

In each case show the locations of λ and γ in the complex plane. Verify that an exponential is growing if γ lies outside the unit circle (or if λ lies in the RHP), is decaying if γ lies within the unit circle (or if λ lies in the LHP), and has a constant amplitude if γ lies on the unit circle (or if λ lies on the imaginary axis).

3.3-6 Express the following signals, which are in the form $e^{\lambda n}$, in the form γ^n:

(a) $e^{-(1+j\pi)n}$
(b) $e^{-(1-j\pi)n}$
(c) $e^{(1+j\pi)n}$
(d) $e^{(1-j\pi)n}$
(e) $e^{-[1+j(\pi/3)]n}$
(f) $e^{[1-j(\pi/3)]n}$

3.3-7 The concepts of even and odd functions for discrete-time signals are identical to those of the continuous-time signals discussed in Section 1.5. Using these concepts, find and sketch the odd and the even components of the following:

(a) $u[n]$
(b) $nu[n]$
(c) $\sin\left(\dfrac{\pi n}{4}\right)$
(d) $\cos\left(\dfrac{\pi n}{4}\right)$

3.4-1 A cash register output $y[n]$ represents the total cost of n items rung up by a cashier. The input $x[n]$ is the cost of the nth item.

(a) Write the difference equation relating $y[n]$ to $x[n]$.
(b) Realize this system using a time-delay element.

3.4-2 Let $p[n]$ be the population of a certain country at the beginning of the nth year. The birth and death rates of the population during any year are 3.3 and 1.3%, respectively. If $i[n]$ is the total number of immigrants entering the country during the nth year, write the difference equation relating $p[n+1]$, $p[n]$, and $i[n]$. Assume that the immigrants enter the country throughout the year at a uniform rate.

3.4-3 A moving average is used to detect a trend of a rapidly fluctuating variable such as the stock market average. A variable may fluctuate (up and down) daily, masking its long-term (secular) trend. We can discern the long-term trend by smoothing or averaging the past N values of the variable. For the stock market average, we may consider a 5-day moving average $y[n]$ to be the mean of the past 5 days' market closing values $x[n], x[n-1], \ldots, x[n-4]$.

(a) Write the difference equation relating $y[n]$ to the input $x[n]$.
(b) Use time-delay elements to realize the 5-day moving-average filter.

3.4-4 The digital integrator in Example 3.7 is specified by

$$y[n] - y[n-1] = Tx[n]$$

If an input $u[n]$ is applied to such an integrator, show that the output is $(n+1)Tu[n]$, which approaches the desired ramp $nTu[n]$ as $T \to 0$.

3.4-5 Approximate the following second-order differential equation with a difference equation.

$$\frac{d^2 y}{dt^2} + a_1 \frac{dy}{dt} + a_0 y(t) = x(t)$$

3.4-6 The voltage at the nth node of a resistive ladder in Fig. P3.4-6 is $v[n]$ ($n = 0, 1, 2, \ldots, N$). Show that $v[n]$ satisfies the second-order difference equation

$$v[n+2] - Av[n+1] + v[n] = 0 \qquad A = 2 + \frac{1}{a}$$

[Hint: Consider the node equation at the nth node with voltage $v[n]$.]

3.4-7 Determine whether each of the following statements is true or false. If the statement is false, demonstrate by proof or example why the statement is false. If the statement is true, explain why.

(a) A discrete-time signal with finite power cannot be an energy signal.
(b) A discrete-time signal with infinite energy must be a power signal.
(c) The system described by $y[n] = (n+1)x[n]$ is causal.
(d) The system described by $y[n-1] = x[n]$ is causal.

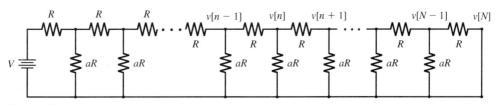

Figure P3.4-6

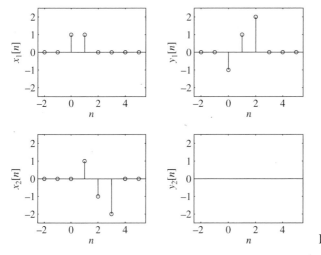

Figure P3.4-8 Input–output plots.

(e) If an energy signal $x[n]$ has energy E, then the energy of $x[an]$ is $E/|a|$.

3.4-8 A linear, time-invariant system produces output $y_1[n]$ in response to input $x_1[n]$, as shown in Fig. P3.4-8. Determine and sketch the output $y_2[n]$ that results when input $x_2[n]$ is applied to the same system.

3.4-9 A system is described by

$$y[n] = \frac{1}{2} \sum_{k=-\infty}^{\infty} x[k](\delta[n-k] + \delta[n+k])$$

(a) Explain what this system does.
(b) Is the system BIBO stable? Justify your answer.
(c) Is the system linear? Justify your answer.
(d) Is the system memoryless? Justify your answer.
(e) Is the system causal? Justify your answer.
(f) Is the system time invariant? Justify your answer.

3.4-10 A discrete-time system is given by

$$y[n+1] = \frac{x[n]}{x[n+1]}$$

(a) Is the system BIBO stable? Justify your answer.
(b) Is the system memoryless? Justify your answer.
(c) Is the system causal? Justify your answer.

3.4-11 Explain why the continuous-time system $y(t) = x(2t)$ is always invertible and yet the corresponding discrete-time system $y[n] = x[2n]$ is not invertible.

3.4-12 Consider the input–output relationships of two similar discrete-time systems:

$$y_1[n] = \sin\left(\frac{\pi}{2}n + 1\right) x[n]$$

and

$$y_2[n] = \sin\left(\frac{\pi}{2}(n+1)\right) x[n]$$

Explain why $x[n]$ can be recovered from $y_1[n]$ yet $x[n]$ cannot be recovered from $y_2[n]$.

3.4-13 Consider a system that multiplies a given input by a ramp function, $r[n]$. That is, $y[n] = x[n]r[n]$.

   (a) Is the system BIBO stable? Justify your answer.

   (b) Is the system linear? Justify your answer.

   (c) Is the system memoryless? Justify your answer.

   (d) Is the system causal? Justify your answer.

   (e) Is the system time invariant? Justify your answer.

3.4-14 A jet-powered car is filmed using a camera operating at 60 frames per second. Let variable n designate the film frame, where $n = 0$ corresponds to engine ignition (film before ignition is discarded). By analyzing each frame of the film, it is possible to determine the car position $x[n]$, measured in meters, from the original starting position $x[0] = 0$.

From physics, we know that velocity is the time derivative of position,

$$v(t) = \frac{d}{dt}x(t)$$

Furthermore, we know that acceleration is the time derivative of velocity,

$$a(t) = \frac{d}{dt}v(t)$$

We can estimate the car velocity from the film data by using a simple difference equation $v[n] = k(x[n] - x[n-1])$

   (a) Determine the appropriate constant k to ensure $v[n]$ has units of meters per second.

   (b) Determine a standard-form constant coefficient difference equation that outputs an estimate of acceleration, $a[n]$, using an input of position, $x[n]$. Identify the advantages and shortcomings of estimating acceleration $a(t)$ with $a[n]$. What is the impulse response $h[n]$ for this system?

3.4-15 Do part (a) of Prob. 3.M-2.

3.5-1 Solve recursively (first three terms only):

   (a) $y[n+1] - 0.5y[n] = 0$, with $y[-1] = 10$

   (b) $y[n+1] + 2y[n] = x[n+1]$, with $x[n] = e^{-n}u[n]$ and $y[-1] = 0$

3.5-2 Solve the following equation recursively (first three terms only):

$$y[n] - 0.6y[n-1] - 0.16y[n-2] = 0$$

with

$$y[-1] = -25,\ y[-2] = 0.$$

3.5-3 Solve recursively the second-order difference Eq. (3.10b) for sales estimate (first three terms only), assuming $y[-1] = y[-2] = 0$ and $x[n] = 100u[n]$.

3.5-4 Solve the following equation recursively (first three terms only):

$$y[n+2] + 3y[n+1] + 2y[n]$$
$$= x[n+2] + 3x[n+1] + 3x[n]$$

with $x[n] = (3)^n u[n]$, $y[-1] = 3$, and $y[-2] = 2$

3.5-5 Repeat Prob. 3.5-4 for

$$y[n] + 2y[n-1] + y[n-2] = 2x[n] - x[n-1]$$

with $x[n] = (3)^{-n}u[n]$, $y[-1] = 2$, and $y[-2] = 3$.

3.6-1 Solve

$$y[n+2] + 3y[n+1] + 2y[n] = 0$$

if $y[-1] = 0$, $y[-2] = 1$.

3.6-2 Solve

$$y[n+2] + 2y[n+1] + y[n] = 0$$

if $y[-1] = 1$, $y[-2] = 1$.

3.6-3 Solve

$$y[n+2] - 2y[n+1] + 2y[n] = 0$$

if $y[-1] = 1$, $y[-2] = 0$.

3.6-4 For the general Nth-order difference Eq. (3.17b), letting

$$a_1 = a_2 = \cdots = a_{N-1} = 0$$

results in a general causal Nth-order LTI *nonrecursive* difference equation

$$y[n] = b_0 x[n] + b_1 x[n-1] + \cdots$$
$$+ b_{N-1}x[n-N+1] + b_N x[n-N]$$

Show that the characteristic roots for this system are zero, and hence, the zero-input response is zero. Consequently, the total response consists of the zero-state component only.

3.6-5 Leonardo Pisano Fibonacci, a famous thirteenth-century mathematician, generated the sequence of integers $\{0, 1, 1, 2, 3, 5, 8, 13, 21, 34, \ldots\}$ while addressing, oddly enough, a problem involving rabbit reproduction. An element of the Fibonacci sequence is the sum of the previous two.

(a) Find the constant-coefficient difference equation whose zero-input response $f[n]$ with auxiliary conditions $f[1] = 0$ and $f[2] = 1$ is a Fibonacci sequence. Given $f[n]$ is the system output, what is the system input?

(b) What are the characteristic roots of this system? Is the system stable?

(c) Designating 0 and 1 as the first and second Fibonacci numbers, determine the fiftieth Fibonacci number. Determine the thousandth Fibonacci number.

3.6-6 Find $v[n]$, the voltage at the nth node of the resistive ladder depicted in Fig. P3.4-6, if $V = 100$ volts and $a = 2$. [Hint 1: Consider the node equation at the nth node with voltage $v[n]$. Hint 2: See Prob. 3.4-6 for the equation for $v[n]$. The auxiliary conditions are $v[0] = 100$ and $v[N] = 0$.]

3.6-7 Consider the discrete-time system $y[n] + y[n-1] + 0.25y[n-2] = \sqrt{3}x[n-8]$. Find the zero input response, $y_0[n]$, if $y_0[-1] = 1$ and $y_0[1] = 1$.

3.7-1 Find the unit impulse response $h[n]$ of systems specified by the following equations:

(a) $y[n+1] + 2y[n] = x[n]$

(b) $y[n] + 2y[n-1] = x[n]$

3.7-2 Repeat Prob. 3.7-1 for

$$(E^2 - 6E + 9)y[n] = Ex[n]$$

3.7-3 Repeat Prob. 3.7-1 for

$$y[n] - 6y[n-1] + 25y[n-2] =$$
$$2x[n] - 4x[n-1]$$

3.7-4 (a) For the general Nth-order difference Eq. (3.17), letting

$$a_0 = a_1 = a_2 = \cdots = a_{N-1} = 0$$

results in a general causal Nth-order LTI *nonrecursive* difference equation

$$y[n] = b_0 x[n] + b_1 x[n-1] + \cdots$$
$$+ b_{N-1} x[n - N + 1] + b_N x[n - N]$$

Find the impulse response $h[n]$ for this system. [Hint: The characteristic equation for this case is $\gamma^n = 0$. Hence, all the characteristic roots are zero. In this case, $y_c[n] = 0$, and the approach in Section 3.7 does not work. Use a direct method to find $h[n]$ by realizing that $h[n]$ is the response to unit impulse input.]

(b) Find the impulse response of a nonrecursive LTID system described by the equation

$$y[n] = 3x[n] - 5x[n-1] - 2x[n-3]$$

Observe that the impulse response has only a finite (N) number of nonzero elements. For this reason, such systems are called *finite-impulse response* (FIR) systems. For a general recursive case [Eq. (3.24)], the impulse response has an infinite number of nonzero elements, and such systems are called *infinite-impulse response* (IIR) systems.

3.8-1 Find the (zero-state) response $y[n]$ of an LTID system whose unit impulse response is

$$h[n] = (-2)^n u[n-1]$$

and the input is $x[n] = e^{-n} u[n+1]$. Find your answer by computing the convolution sum and also by using the convolution table (Table 3.1).

3.8-2 Find the (zero-state) response $y[n]$ of an LTID system if the input is $x[n] = 3^{n-1} u[n+2]$ and

$$h[n] = \tfrac{1}{2}[\delta[n-2] - (-2)^{n+1}]u[n-3]$$

3.8-3 Find the (zero-state) response $y[n]$ of an LTID system if the input $x[n] = (3)^{n+2} u[n+1]$, and

$$h[n] = [(2)^{n-2} + 3(-5)^{n+2}]u[n-1]$$

3.8-4 Find the (zero-state) response $y[n]$ of an LTID system if the input is $x[n] = (3)^{-n+2} u[n+3]$, and

$$h[n] = 3(n-2)(2)^{n-3} u[n-4]$$

3.8-5 Find the (zero-state) response $y[n]$ of an LTID system if its input $x[n] = (2)^n u[n - 1]$, and

$$h[n] = (3)^n \cos\left(\frac{\pi}{3}n - 0.5\right) u[n]$$

Find your answer using only Table 3.1, the convolution table.

3.8-6 Derive the results in entries 1, 2 and 3 in Table 3.1. [Hint: You may need to use the information in Section B.7-4.]

3.8-7 Derive the results in entries 4, 5, and 6 in Table 3.1.

3.8-8 Derive the results in entries 7 and 8 in Table 3.1. [Hint: You may need to use the information in Section B.7-4.]

3.8-9 Derive the results in entries 9 and 11 in Table 3.1. [Hint: You may need to use the information in Section B.7-4.]

3.8-10 Find the total response of a system specified by the equation

$$y[n + 1] + 2y[n] = x[n + 1]$$

if $y[-1] = 10$, and the input $x[n] = e^{-n}u[n]$.

3.8-11 Find an LTID system (zero-state) response if its impulse response $h[n] = (0.5)^n u[n]$, and the input $x[n]$ is

(a) $2^n u[n]$

(b) $2^{n-3} u[n]$

(c) $2^n u[n - 2]$

[Hint: You may need to use the shift property (3.61) of the convolution.]

3.8-12 For a system specified by equation

$$y[n] = x[n] - 2x[n - 1]$$

Find the system response to input $x[n] = u[n]$. What is the order of the system? What type of system (recursive or nonrecursive) is this? Is the knowledge of initial condition(s) necessary to find the system response? Explain.

3.8-13 (a) A discrete-time LTI system is shown in Fig. P3.8-13. Express the overall impulse response of the system, $h[n]$, in terms of $h_1[n], h_2[n], h_3[n], h_4[n]$, and $h_5[n]$.

(b) Two LTID systems in cascade have impulse response $h_1[n]$ and $h_2[n]$, respectively. Show that if $h_1[n] = (0.9)^n u[n] - 0.5(0.9)^{n-1}u[n-1]$ and $h_2[n] = (0.5)^n u[n] - 0.9(0.5)^{n-1}u[n-1]$, the cascade system is an identity system.

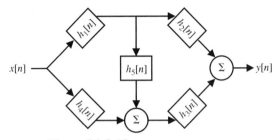

Figure P3.8-13

3.8-14 (a) Show that for a causal system, Eq. (3.70b) can also be expressed as

$$g[n] = \sum_{k=0}^{n} h[n - k]$$

(b) How would the expressions in part (a) change if the system is not causal?

3.8-15 In the savings account problem described in Example 3.4, a person deposits $500 at the beginning of every month, starting at $n = 0$ with the exception at $n = 4$, when instead of depositing $500, she withdraws $1000. Find $y[n]$ if the interest rate is 1% per month ($r = 0.01$).

3.8-16 To pay off a loan of M dollars in N number of payments using a fixed monthly payment of P dollars, show that

$$P = \frac{rM}{1 - (1 + r)^{-N}}$$

where r is the interest rate per dollar per month. [Hint: This problem can be modeled by Eq. (3.9a) with the payments of P dollars starting at $n = 1$. The problem can be approached in two ways (1) Consider the loan as the initial condition $y_0[0] = -M$, and the input $x[n] = Pu[n - 1]$. The loan balance is the sum of the zero-input component (due to the initial condition) and the zero-state component $h[n] * x[n]$. (2) Consider the loan as an input $-M$ at $n = 0$ along with the input due to payments. The loan balance is now exclusively a zero-state component $h[n] * x[n]$. Because the loan is paid off in N payments, set $y[N] = 0$.]

3.8-17 A person receives an automobile loan of $10,000 from a bank at the interest rate of 1.5% per month. His monthly payment is $500, with the first payment due one month after he receives the loan. Compute the number of payments required to pay off the loan. Note that the last payment may not be exactly $500. [Hint: Follow the procedure in Prob. 3.8-16 to determine the balance $y[n]$. To determine N, the number of payments, set $y[N] = 0$. In general, N will not be an integer. The number of payments K is the largest integer $\leq N$. The residual payment is $|y[K]|$.]

3.8-18 Using the sliding-tape algorithm, show that

(a) $u[n] * u[n] = (n + 1)u[n]$

(b) $(u[n] - u[n - m]) * u[n] = (n + 1)u[n] - (n - m + 1)u[n - m]$

3.8-19 Using the sliding-tape algorithm, find $x[n] * g[n]$ for the signals shown in Fig. P3.8-19.

3.8-20 Repeat Prob. 3.8-19 for the signals shown in Fig. P3.8-20.

3.8-21 Repeat Prob. 3.8-19 for the signals depicted in Fig. P3.8-21.

3.8-22 The convolution sum in Eq. (3.63) can be expressed in a matrix form as

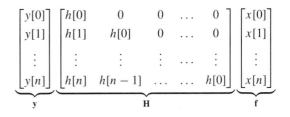

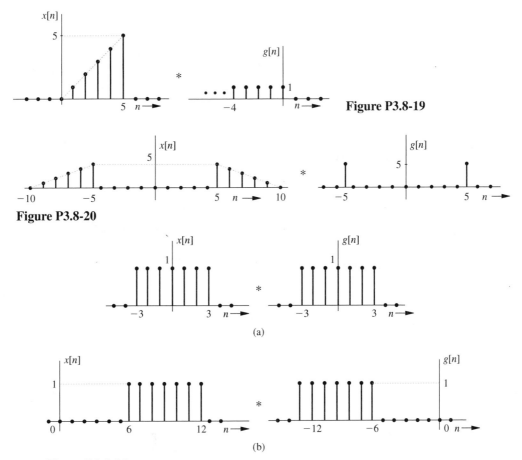

Figure P3.8-19

Figure P3.8-20

(a)

(b)

Figure P3.8-21

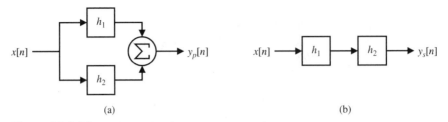

Figure P3.8-24 Parallel and series system connections.

or

$$y = Hf$$

and

$$f = H^{-1}y$$

Knowing $h[n]$ and the output $y[n]$, we can determine the input $x[n]$. This operation is the reverse of the convolution and is known as the *deconvolution*. Moreover, knowing $x[n]$ and $y[n]$, we can determine $h[n]$. This can be done by expressing the foregoing matrix equation as $n+1$ simultaneous equations in terms of $n+1$ unknowns $h[0], h[1], \ldots, h[n]$. These equations can readily be solved iteratively. Thus, we can synthesize a system that yields a certain output $y[n]$ for a given input $x[n]$.

(a) Design a system (i.e., determine $h[n]$) that will yield the output sequence (8, 12, 14, 15, 15.5, 15.75, ...) for the input sequence (1, 1, 1, 1, 1, 1, ...).

(b) For a system with the impulse response sequence (1, 2, 4, ...), the output sequence was (1, 7/3, 43/9, ...). Determine the input sequence.

3.8-23 A second-order LTID system has zero-input response

$$y_0[n] = \left[3, 2\tfrac{1}{3}, 2\tfrac{1}{9}, 2\tfrac{1}{27}, \ldots\right]$$

$$= \sum_{k=0}^{\infty} \left\{2 + \left(\tfrac{1}{3}\right)^k\right\} \delta[n-k]$$

(a) Determine the characteristic equation of this system, $a_0\gamma^2 + a_1\gamma + a_2 = 0$.

(b) Find a bounded, causal input with infinite duration that would cause a strong response from this system. Justify your choice.

(c) Find a bounded, causal input with infinite duration that would cause a weak response from this system. Justify your choice.

3.8-24 An LTID filter has an impulse response function given by $h_1[n] = \delta[n+2] - \delta[n-2]$. A second LTID system has an impulse response function given by $h_2[n] = n(u[n+4] - u[n-4])$.

(a) Carefully sketch the functions $h_1[n]$ and $h_2[n]$ over $(-10 \le n \le 10)$.

(b) Assume that the two systems are connected in parallel as shown in Fig. P3.8-24. Determine the impulse response $h_p[n]$ for the parallel system in terms of $h_1[n]$ and $h_2[n]$. Sketch $h_p[n]$ over $(-10 \le n \le 10)$.

(c) Assume that the two systems are connected in cascade as shown in Fig. P3.8-24. Determine the impulse response $h_s[n]$ for the cascade system in terms of $h_1[n]$ and $h_2[n]$. Sketch $h_s[n]$ over $(-10 \le n \le 10)$.

3.8-25 This problem investigates an interesting application of discrete-time convolution: the expansion of certain polynomial expressions.

(a) By hand, expand $(z^3 + z^2 + z + 1)^2$. Compare the coefficients to [1, 1, 1, 1] * [1, 1, 1, 1].

(b) Formulate a relationship between discrete-time convolution and the expansion of constant-coefficient polynomial expressions.

(c) Use convolution to expand $(z^{-4} - 2z^{-3} + 3z^{-2})^4$.

(d) Use convolution to expand $(z^5 + 2z^4 + 3z^2 + 5)^2(z^{-4} - 5z^{-2} + 13)$.

3.8-26 Joe likes coffee, and he drinks his coffee according to a very particular routine. He begins by adding two teaspoons of sugar to his mug, which he then fills to the brim with hot coffee.

He drinks 2/3 of the mug's contents, adds another two teaspoons of sugar, and tops the mug off with steaming hot coffee. This refill procedure continues, sometimes for many, many cups of coffee. Joe has noted that his coffee tends to taste sweeter with the number of refills.

Let independent variable n designate the coffee refill number. In this way, $n = 0$ indicates the first cup of coffee, $n = 1$ is the first refill, and so forth. Let $x[n]$ represent the sugar (measured in teaspoons) added into the system (a coffee mug) on refill n. Let $y[n]$ designate the amount of sugar (again, teaspoons) contained in the mug on refill n.

(a) The sugar (teaspoons) in Joe's coffee can be represented using a standard second-order constant coefficient difference equation $y[n] + a_1 y[n-1] + a_2 y[n-2] = b_0 x[n] + b_1 x[n-1] + b_2 x[n-2]$. Determine the constants a_1, a_2, b_0, b_1, and b_2.

(b) Determine $x[n]$, the driving function to this system.

(c) Solve the difference equation for $y[n]$. This requires finding the total solution. Joe always starts with a clean mug from the dishwasher, so $y[-1]$ (the sugar content before the first cup) is zero.

(d) Determine the steady-state value of $y[n]$. That is, what is $y[n]$ as $n \rightarrow \infty$? If possible, suggest a way of modifying $x[n]$ so that the sugar content of Joe's coffee remains a constant for all nonnegative n.

3.8-27 A system is called complex if a real-valued input can produce a complex-valued output. Consider a causal complex system described by a first-order constant coefficient linear difference equation:

$$(jE + 0.5)y[n] = (-5E)x[n]$$

(a) Determine the impulse response function $h[n]$ for this system.

(b) Given input $x[n] = u[n-5]$ and initial condition $y_0[-1] = j$, determine the system's total output $y[n]$ for $n \geq 0$.

3.8-28 A discrete-time LTI system has impulse response function $h[n] = n(u[n-2] - u[n+2])$.

(a) Carefully sketch the function $h[n]$ over $(-5 \leq n \leq 5)$.

(b) Determine the difference equation representation of this system, using $y[n]$ to designate the output and $x[n]$ to designate the input.

3.8-29 Do part (a) of Prob. 3.M-3.

3.8-30 Consider three discrete-time signals: $x[n]$, $y[n]$, and $z[n]$. Denoting convolution as $*$, identify the expression(s) that is(are) equivalent to $x[n](y[n] * z[n])$:

(a) $(x[n] * y[n])z[n]$

(b) $(x[n]y[n]) * (x[n]z[n])$

(c) $(x[n]y[n]) * z[n]$

(d) none of the above

Justify your answer!

3.9-1 Use the classical method to solve

$$y[n+1] + 2y[n] = x[n+1]$$

with the input $x[n] = e^{-n}u[n]$, and the auxiliary condition $y[0] = 1$.

3.9-2 Use the classical method to solve

$$y[n] + 2y[n-1] = x[n-1]$$

with the input $x[n] = e^{-n}u[n]$ and the auxiliary condition $y[-1] = 0$. [Hint: You will have to determine the auxiliary condition $y[0]$ by using the iterative method.]

3.9-3 (a) Use the classical method to solve

$$y[n+2] + 3y[n+1] + 2y[n] =$$
$$x[n+2] + 3x[n+1] + 3x[n]$$

with the input $x[n] = (3)^n$ and the auxiliary conditions $y[0] = 1$, $y[1] = 3$.

(b) Repeat part (a) for auxiliary conditions $y[-1] = y[-2] = 1$. [Hint: Use the iterative method to determine to $y[0]$ and $y[1]$.]

3.9-4 Use the classical method to solve

$$y[n] + 2y[n-1] + y[n-2] =$$
$$2x[n] - x[n-1]$$

with the input $x[n] = 3^{-n}u[n]$ and the auxiliary conditions $y[0] = 2$ and $y[1] = -13/3$.

3.9-5 Use the classical method to find the following sums:

(a) $\sum_{k=0}^{n} k$

(b) $\displaystyle\sum_{k=0}^{n} k^3$

3.9-6 Repeat Prob. 3.9-5 to find $\sum_{k=0}^{n} kr^k$.

3.9-7 Use the classical method to solve

$$(E^2 - E + 0.16)y[n] = Ex[n]$$

with the input $x[n] = (0.2)^n u[n]$ and the auxiliary conditions $y[0] = 1$, $y[1] = 2$. [Hint: The input is a natural mode of the system.]

3.9-8 Use the classical method to solve

$$(E^2 - E + 0.16)y[n] = Ex[n]$$

with the input

$$x[n] = \cos\left(\frac{\pi}{2}n + \frac{\pi}{3}\right)u[n]$$

and the initial conditions $y[-1] = y[-2] = 0$. [Hint: Find $y[0]$ and $y[1]$ iteratively.]

3.10-1 In Section 3.10 we showed that for BIBO stability in an LTID system, it is sufficient for its impulse response $h[n]$ to satisfy Eq. (3.90). Show that this is also a necessary condition for the system to be BIBO stable. In other words, show that if Eq. (3.90) is not satisfied, there exists a bounded input that produces unbounded output. [Hint: Assume that a system exists for which $h[n]$ violates Eq. (3.90), yet its output is bounded for every bounded input. Establish the contradiction in

this statement by considering an input $x[n]$ defined by $x[n_1 - m] = 1$ when $h[m] > 0$ and $x[n_1 - m] = -1$ when $h[m] < 0$, where n_1 is some fixed integer.]

3.10-2 Each of the following equations specifies an LTID system. Determine whether each of these systems is BIBO stable or unstable. Determine also whether each is asymptotically stable, unstable, or marginally stable.

(a) $y[n + 2] + 0.6y[n + 1] - 0.16y[n] = x[n + 1] - 2x[n]$

(b) $y[n] + 3y[n - 1] + 2y[n - 2] = x[n - 1] + 2x[n - 2]$

(c) $(E - 1)^2\left(E + \frac{1}{2}\right)y[n] = x[n]$

(d) $y[n] + 2y[n - 1] + 0.96y[n - 2] = x[n]$

(e) $y[n] + y[n - 1] - 2y[n - 2] = x[n] + 2x[n - 1]$

(f) $(E^2 - 1)(E^2 + 1)y[n] = x[n]$

3.10-3 Consider two LTIC systems in cascade, as illustrated in Fig. 3.23. The impulse response of the system S_1 is $h_1[n] = 2^n u[n]$ and the impulse response of the system S_2 is $h_2[n] = \delta[n] - 2\delta[n - 1]$. Is the cascaded system asymptotically stable or unstable? Determine the BIBO stability of the composite system.

3.10-4 Figure P3.10-4 locates the characteristic roots of nine causal, LTID systems, labeled A through I. Each system has only two roots and is described using operator notation as

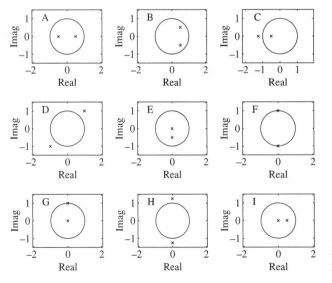

Figure P3.10-4 Characteristic roots for systems A through I.

$Q(E)y[n] = P(E)x[n]$. All plots are drawn to scale, with the unit circle shown for reference. For each of the following parts, identify all the answers that are correct.

(a) Identify all systems that are unstable.
(b) Assuming all systems have $P(E) = E^2$, identify all systems that are real. Recall that a real system always generates a real-valued response to a real-valued input.
(c) Identify all systems that support oscillatory natural modes.
(d) Identify all systems that have at least one mode whose envelop decays at a rate of 2^{-n}.
(e) Identify all systems that have only one mode.

3.10-5 A discrete-time LTI system has impulse response given by

$$h[n] = \delta[n] + \left(\tfrac{1}{3}\right)^n u[n-1]$$

(a) Is the system stable? Is the system causal? Justify your answers.
(b) Plot the signal $x[n] = u[n-3] - u[n+3]$.
(c) Determine the system's zero-state response $y[n]$ to the input $x[n] = u[n-3] - u[n+3]$. Plot $y[n]$ over $(-10 \le n \le 10)$.

3.10-6 An LTID system has impulse response given by

$$h[n] = \left(\tfrac{1}{2}\right)^{|n|}$$

(a) Is the system causal? Justify your answer.
(b) Compute $\sum_{n=-\infty}^{\infty} |h[n]|$. Is this system BIBO stable?
(c) Compute the energy and power of input signal $x[n] = 3u[n-5]$.
(d) Using input $x[n] = 3u[n-5]$, determine the zero-state response of this system at time $n = 10$. That is, determine $y_{zs}[10]$.

3.M-1 Consider the discrete-time function $f[n] = e^{-n/5} \cos(\pi n/5)u[n]$. MATLAB Session 3 uses an inline object in describing this function.

```
>> f = inline('exp(-n/5)
  .*cos(pi*n/5).*(n>=0)','n');
```

While this inline object operates correctly for a downsampling operation such as f[2n], it does not operate correctly for an upsampling operation such as f[n/2]. Modify the inline object f so that it also correctly

accommodates upsampling operations. Test your code by computing and plotting f[n/2] over $(-10 \le n \le 10)$.

3.M-2 An indecisive student contemplates whether he should stay home or take his final exam, which is being held 2 miles away. Starting at home, the student travels half the distance to the exam location before changing his mind. The student turns around and travels half the distance between his current location and his home before changing his mind again. This process of changing direction and traveling half the remaining distance continues until the student either reaches a destination or dies from exhaustion.

(a) Determine a suitable difference equation description of this system.
(b) Use MATLAB to simulate the difference equation in part (a). Where does the student end up as $n \to \infty$? How do your answer change if the student goes two-thirds the way each time, rather than halfway?
(c) Determine a closed-form solution to the equation in part (a). Use this solution to verify the results in part (b).

3.M-3 The cross-correlation function between $x[n]$ and $y[n]$ is given as

$$r_{xy}[k] = \sum_{n=-\infty}^{\infty} x[n]y[n-k]$$

Notice that $r_{xy}[k]$ is quite similar to the convolution sum. The independent variable k corresponds to the relative *shift* between the two inputs.

(a) Express $r_{xy}[k]$ in terms of convolution. Is $r_{xy}[k] = r_{yx}[k]$?
(b) Cross-correlation is said to indicate similarity between two signals. Do you agree? Why or why not?
(c) If $x[n]$ and $y[n]$ are both finite duration, MATLAB's conv command is well suited to compute $r_{xy}[k]$.

   (i) Write a MATLAB function that computes the cross-correlation function using the conv command. Four vectors are passed to the function (x, y, nx, and ny) corresponding to the inputs $x[n]$, $y[n]$, and their respective time vectors. Notice that, x and y are not necessarily the same length. Two outputs should be created (rxy and k) corresponding to $r_{xy}[k]$ and its shift vector.

(ii) Test your code from part c(i), using $x[n] = u[n - 5] - u[n - 10]$ over $(0 \leq n = n_x \leq 20)$ and $y[n] = u[-n - 15] - u[-n - 10] + \delta[n - 2]$ over $(-20 \leq n = n_y \leq 10)$. Plot the result rxy as a function of the shift vector k. What shift k gives the largest magnitude of $r_{xy}[k]$? Does this make sense?

3.M-4 A causal N-point max filter assigns $y[n]$ to the maximum of $\{x[n], \ldots, x[n - (N - 1)]\}$.

(a) Write a MATLAB function that performs N-point max filtering on a length-M input vector x. The two function inputs are vector x and scalar N. To create the length-M output vector y, initially pad the input vector with $N - 1$ zeros. The MATLAB command max may be helpful.

(b) Test your filter and MATLAB code by filtering a length-45 input defined as $x[n] = \cos(\pi n/5) + \delta[n - 30] - \delta[n - 35]$. Separately plot the results for $N = 4$, $N = 8$, and $N = 12$. Comment on the filter behavior.

3.M-5 A causal N-point min filter assigns $y[n]$ to the minimum of $\{x[n], \ldots, x[n - (N - 1)]\}$.

(a) Write a MATLAB function that performs N-point min filtering on a length-M input vector x. The two function inputs are vector x and scalar N. To create the length-M output vector y, initially pad the input vector with $N - 1$ zeros. The MATLAB command min may be helpful.

(b) Test your filter and MATLAB code by filtering a length-45 input defined as $x[n] = \cos(\pi n/5) + \delta[n - 30] - \delta[n - 35]$. Separately plot the results for $N = 4$, $N = 8$, and $N = 12$. Comment on the filter behavior.

3.M-6 A causal N-point median filter assigns $y[n]$ to the median of $\{x[n], \ldots, x[n - (N - 1)]\}$. The median is found by sorting sequence

$\{x[n], \ldots, x[n - (N - 1)]\}$ and choosing the middle value (odd N) or the average of the two middle values (even N).

(a) Write a MATLAB function that performs N-point median filtering on a length-M input vector x. The two function inputs are vector x and scalar N. To create the length-M output vector y, initially pad the input vector with $N - 1$ zeros. The MATLAB command sort or median may be helpful.

(b) Test your filter and MATLAB code by filtering a length-45 input defined as $x[n] = \cos(\pi n/5) + \delta[n - 30] - \delta[n - 35]$. Separately plot the results for $N = 4$, $N = 8$, and $N = 12$. Comment on the filter behavior.

3.M-7 Recall that $y[n] = x[n/N]$ represents an upsample by N operation. An interpolation filter replaces the inserted zeros with more realistic values. A linear interpolation filter has impulse response

$$h[n] = \sum_{k=-(N-1)}^{N-1} \left(1 - \left|\frac{k}{N}\right|\right) \delta(n - k)$$

(a) Determine a constant coefficient difference equation that has impulse response $h[n]$.

(b) The impulse response $h[n]$ is noncausal. What is the smallest time shift necessary to make the filter causal? What is the effect of this shift on the behavior of the filter?

(c) Write a MATLAB function that will compute the parameters necessary to implement an interpolation filter using MATLAB's filter command. That is, your function should output filter vectors b and a given an input scalar N.

(d) Test your filter and MATLAB code. To do this, create $x[n] = \cos(n)$ for $(0 \leq n \leq 9)$. Upsample $x[n]$ by $N = 10$ to create a new signal $x_{up}[n]$. Design the corresponding $N = 10$ linear interpolation filter, filter $x_{up}[n]$ to produce $y[n]$, and plot the results.

3.M-8 A causal N-point moving-average filter has impulse response $h[n] = (u[n] - u[n - N])/N$.

(a) Determine a constant coefficient difference equation that has impulse response $h[n]$.

(b) Write a MATLAB function that will compute the parameters necessary to implement an N-point moving-average filter using MATLAB's `filter` command. That is, your function should output filter vectors b and a given a scalar input N.

(c) Test your filter and MATLAB code by filtering a length-45 input defined as $x[n] = \cos(\pi n/5) + \delta[n - 30] - \delta[n - 35]$. Separately plot the results for $N = 4$, $N = 8$, and $N = 12$. Comment on the filter behavior.

(d) Problem 3.M-7 introduces linear interpolation filters, for use following an upsample by N operation. Within a scale factor, show that a cascade of two N-point moving-average filters is equivalent to the linear interpolation filter. What is the scale factor difference? Test this idea with MATLAB. Create $x[n] = \cos(n)$ for $(0 \le n \le 9)$. Upsample $x[n]$ by $N = 10$ to create a new signal $x_{up}[n]$. Design an $N = 10$ moving-average filter. Filter $x_{up}[n]$ twice and scale to produce $y[n]$. Plot the results. Does the output from the cascaded pair of moving-average filters linearly interpolate the upsampled data?

4

CONTINUOUS-TIME SYSTEM ANALYSIS USING THE LAPLACE TRANSFORM

Because of the linearity (superposition) property of linear time-invariant systems, we can find the response of these systems by breaking the input $x(t)$ into several components and then summing the system response to all the components of $x(t)$. We have already used this procedure in time-domain analysis, in which the input $x(t)$ is broken into impulsive components. In the *frequency-domain analysis* developed in this chapter, we break up the input $x(t)$ into exponentials of the form e^{st}, where the parameter s is the complex frequency of the signal e^{st}, as explained in Section 1.4-3. This method offers an insight into the system behavior complementary to that seen in the time-domain analysis. In fact, the time-domain and the frequency-domain methods are duals of each other.

The tool that makes it possible to represent arbitrary input $x(t)$ in terms of exponential components is the *Laplace transform*, which is discussed in the following section.

4.1 THE LAPLACE TRANSFORM

For a signal $x(t)$, its Laplace transform $X(s)$ is defined by

$$X(s) = \int_{-\infty}^{\infty} x(t)e^{-st}\, dt \tag{4.1}$$

The signal $x(t)$ is said to be the *inverse Laplace transform* of $X(s)$. It can be shown that[1]

$$x(t) = \frac{1}{2\pi j} \int_{c-j\infty}^{c+j\infty} X(s)e^{st}\, ds \tag{4.2}$$

where c is a constant chosen to ensure the convergence of the integral in Eq. (4.1), as explained later.

This pair of equations is known as the *bilateral Laplace transform pair*, where $X(s)$ is the direct Laplace transform of $x(t)$ and $x(t)$ is the inverse Laplace transform of $X(s)$. Symbolically,

$$X(s) = \mathcal{L}[x(t)] \quad \text{and} \quad x(t) = \mathcal{L}^{-1}[X(s)] \tag{4.3}$$

Note that

$$\mathcal{L}^{-1}\{\mathcal{L}[x(t)]\} = x(t) \quad \text{and} \quad \mathcal{L}\{\mathcal{L}^{-1}[X(s)]\} = X(s)$$

It is also common practice to use a bidirectional arrow to indicate a Laplace transform pair, as follows:

$$x(t) \iff X(s)$$

The Laplace transform, defined in this way, can handle signals existing over the entire time interval from $-\infty$ to ∞ (causal and noncausal signals). For this reason it is called the *bilateral* (or *two-sided*) Laplace transform. Later we shall consider a special case—the *unilateral* or *one-sided* Laplace transform—which can handle only causal signals.

LINEARITY OF THE LAPLACE TRANSFORM

We now prove that the Laplace transform is a linear operator by showing that the principle of superposition holds, implying that if

$$x_1(t) \iff X_1(s) \qquad \text{and} \qquad x_2(t) \iff X_2(s)$$

then

$$a_1 x_1(t) + a_2 x_2(t) \iff a_1 X_1(s) + a_2 X_2(s)$$

The proof is simple. By definition

$$\mathcal{L}[a_1 x_1(t) + a_2 x_2(t)] = \int_{-\infty}^{\infty} [a_1 x_1(t) + a_2 x_2(t)] e^{-st} \, dt$$

$$= a_1 \int_{-\infty}^{\infty} x_1(t) e^{-st} \, dt + a_2 \int_{-\infty}^{\infty} x_2(t) e^{-st} \, dt$$

$$= a_1 X_1(s) + a_2 X_2(s) \tag{4.4}$$

This result can be extended to any finite sum.

THE REGION OF CONVERGENCE (ROC)

The *region of convergence* (ROC), also called the region of existence, for the Laplace transform $X(s)$, is the set of values of s (the region in the complex plane) for which the integral in Eq. (4.1) converges. This concept will become clear in the following example.

EXAMPLE 4.1

For a signal $x(t) = e^{-at}u(t)$, find the Laplace transform $X(s)$ and its ROC.

By definition

$$X(s) = \int_{-\infty}^{\infty} e^{-at} u(t) e^{-st} \, dt$$

Because $u(t) = 0$ for $t < 0$ and $u(t) = 1$ for $t \geq 0$,

$$X(s) = \int_{0}^{\infty} e^{-at} e^{-st} \, dt = \int_{0}^{\infty} e^{-(s+a)t} \, dt = -\frac{1}{s+a} e^{-(s+a)t} \Big|_{0}^{\infty} \tag{4.5}$$

Note that s is complex and as $t \to \infty$, the term $e^{-(s+a)t}$ does not necessarily vanish. Here we recall that for a complex number $z = \alpha + j\beta$,

$$e^{-zt} = e^{-(\alpha + j\beta)t} = e^{-\alpha t} e^{-j\beta t}$$

Now $|e^{-j\beta t}| = 1$ regardless of the value of βt. Therefore, as $t \to \infty$, $e^{-zt} \to 0$ only if $\alpha > 0$, and $e^{-zt} \to \infty$ if $\alpha < 0$. Thus

$$\lim_{t \to \infty} e^{-zt} = \begin{cases} 0 & \text{Re } z > 0 \\ \infty & \text{Re } z < 0 \end{cases} \tag{4.6}$$

Clearly

$$\lim_{t \to \infty} e^{-(s+a)t} = \begin{cases} 0 & \text{Re}(s + a) > 0 \\ \infty & \text{Re}(s + a) < 0 \end{cases}$$

Use of this result in Eq. (4.5) yields

$$X(s) = \frac{1}{s + a} \qquad \text{Re}(s + a) > 0 \tag{4.7a}$$

or

$$e^{-at} u(t) \Longleftrightarrow \frac{1}{s + a} \qquad \text{Re } s > -a \tag{4.7b}$$

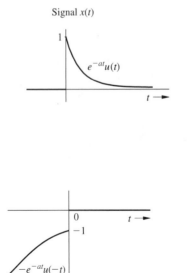

(a)

(b)

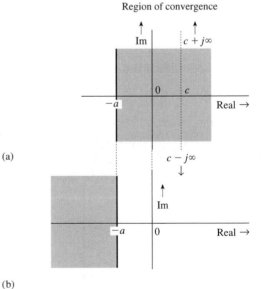

Figure 4.1 Signals **(a)** $e^{-at} u(t)$ and **(b)** $-e^{at} u(-t)$ have the same Laplace transform but different regions of convergence.

The ROC of $X(s)$ is Re $s > -a$, as shown in the shaded area in Fig. 4.1a. This fact means that the integral defining $X(s)$ in Eq. (4.5) exists only for the values of s in the shaded region in Fig. 4.1a. For other values of s, the integral in Eq. (4.5) does not converge. For this reason the shaded region is called the *ROC* (or the *region of existence*) for $X(s)$.

REGION OF CONVERGENCE FOR FINITE-DURATION SIGNALS

A finite-duration signal $x_f(t)$ is a signal that is nonzero only for $t_1 \leq t \leq t_2$, where both t_1 and t_2 are finite numbers and $t_2 > t_1$. For a finite-duration, absolutely integrable signal, the ROC is the entire s plane. This is clear from the fact that if $x_f(t)$ is absolutely integrable and a finite-duration signal, then $x(t)e^{-\sigma t}$ is also absolutely integrable for any value of σ because the integration is over only finite range of t. Hence, the Laplace transform of such a signal converges for every value of s. This means that the ROC of a general signal $x(t)$ remains unaffected by addition of any absolutely integrable, finite-duration signal $x_f(t)$ to $x(t)$. In other words, if $\mathcal{R}$ represents the ROC of a signal $x(t)$, then the ROC of a signal $x(t) + x_f(t)$ is also $\mathcal{R}$.

ROLE OF THE REGION OF CONVERGENCE

The ROC is required for evaluating the inverse Laplace transform $x(t)$ from $X(s)$, as defined by Eq. (4.2). The operation of finding the inverse transform requires an integration in the complex plane, which needs some explanation. The path of integration is along $c + j\omega$, with ω varying from $-\infty$ to ∞.[†] Moreover, the path of integration must lie in the ROC (or existence) for $X(s)$. For the signal $e^{-at}u(t)$, this is possible if $c > -a$. One possible path of integration is shown (dotted) in Fig. 4.1a. Thus, to obtain $x(t)$ from $X(s)$, the integration in Eq. (4.2) is performed along this path. When we integrate $[1/(s + a)]e^{st}$ along this path, the result is $e^{-at}u(t)$. Such integration in the complex plane requires a background in the theory of functions of complex variables. We can avoid this integration by compiling a table of Laplace transforms (Table 4.1), where the Laplace transform pairs are tabulated for a variety of signals. To find the inverse Laplace transform of, say, $1/(s + a)$, instead of using the complex integral of Eq. (4.2), we look up the table and find the inverse Laplace transform to be $e^{-at}u(t)$ (assuming that the ROC is Re $s > -a$). Although the table given here is rather short, it comprises the functions of most practical interest. A more comprehensive table appears in Doetsch.[2]

THE UNILATERAL LAPLACE TRANSFORM

To understand the need for defining unilateral transform, let us find the Laplace transform of signal $x(t)$ illustrated in Fig. 4.1b:

$$x(t) = -e^{-at}u(-t)$$

[†]The discussion about the path of convergence is rather complicated, requiring the concepts of contour integration and understanding of the theory of complex variables. For this reason, the discussion here is somewhat simplified.

TABLE 4.1   A Short Table of (Unilateral) Laplace Transforms

No.	$x(t)$	$X(s)$
1	$\delta(t)$	1
2	$u(t)$	$\dfrac{1}{s}$
3	$tu(t)$	$\dfrac{1}{s^2}$
4	$t^n u(t)$	$\dfrac{n!}{s^{n+1}}$
5	$e^{\lambda t} u(t)$	$\dfrac{1}{s - \lambda}$
6	$te^{\lambda t} u(t)$	$\dfrac{1}{(s - \lambda)^2}$
7	$t^n e^{\lambda t} u(t)$	$\dfrac{n!}{(s - \lambda)^{n+1}}$
8a	$\cos bt \, u(t)$	$\dfrac{s}{s^2 + b^2}$
8b	$\sin bt \, u(t)$	$\dfrac{b}{s^2 + b^2}$
9a	$e^{-at} \cos bt \, u(t)$	$\dfrac{s + a}{(s + a)^2 + b^2}$
9b	$e^{-at} \sin bt \, u(t)$	$\dfrac{b}{(s + a)^2 + b^2}$
10a	$re^{-at} \cos(bt + \theta) \, u(t)$	$\dfrac{(r \cos \theta)s + (ar \cos \theta - br \sin \theta)}{s^2 + 2as + (a^2 + b^2)}$
10b	$re^{-at} \cos(bt + \theta) \, u(t)$	$\dfrac{0.5re^{j\theta}}{s + a - jb} + \dfrac{0.5re^{-j\theta}}{s + a + jb}$
10c	$re^{-at} \cos(bt + \theta) \, u(t)$	$\dfrac{As + B}{s^2 + 2as + c}$

$$r = \sqrt{\frac{A^2 c + B^2 - 2ABa}{c - a^2}}$$

$$\theta = \tan^{-1}\left(\frac{Aa - B}{A\sqrt{c - a^2}}\right)$$

$$b = \sqrt{c - a^2}$$

| 10d | $e^{-at}\left[A \cos bt + \dfrac{B - Aa}{b} \sin bt\right] u(t)$ | $\dfrac{As + B}{s^2 + 2as + c}$ |

$$b = \sqrt{c - a^2}$$

The Laplace transform of this signal is

$$X(s) = \int_{-\infty}^{\infty} -e^{-at}u(-t)e^{-st}\,dt$$

Because $u(-t) = 1$ for $t < 0$ and $u(-t) = 0$ for $t > 0$,

$$X(s) = \int_{-\infty}^{0} -e^{-at}e^{-st}\,dt = -\int_{-\infty}^{0} e^{-(s+a)t}\,dt = \frac{1}{s+a}e^{-(s+a)t}\Big|_{-\infty}^{0}$$

Equation (4.6) shows that

$$\lim_{t \to -\infty} e^{-(s+a)t} = 0 \qquad \text{Re } (s+a) < 0$$

Hence

$$X(s) = \frac{1}{s+a} \qquad \text{Re } s < -a$$

The signal $-e^{-at}u(-t)$ and its ROC (Re $s < -a$) are depicted in Fig. 4.1b. Note that the Laplace transforms for the signals $e^{-at}u(t)$ and $-e^{-at}u(-t)$ are identical except for their regions of convergence. Therefore, for a given $X(s)$, there may be more than one inverse transform, depending on the ROC. In other words, unless the ROC is specified, there is no one-to-one correspondence between $X(s)$ and $x(t)$. This fact increases the complexity in using the Laplace transform. The complexity is the result of trying to handle causal as well as noncausal signals. If we restrict all our signals to the causal type, such an ambiguity does not arise. There is only one inverse transform of $X(s) = 1/(s+a)$, namely, $e^{-at}u(t)$. To find $x(t)$ from $X(s)$, we need not even specify the ROC. In summary, if all signals are restricted to the causal type, then, for a given $X(s)$, there is only one inverse transform $x(t)$.[†]

The unilateral Laplace transform is a special case of the bilateral Laplace transform in which all signals are restricted to being causal; consequently the limits of integration for the integral in Eq. (4.1) can be taken from 0 to ∞. Therefore, the unilateral Laplace transform $X(s)$ of a signal $x(t)$ is defined as

$$X(s) = \int_{0^-}^{\infty} x(t)e^{-st}\,dt \tag{4.8}$$

We choose 0^- (rather than 0^+ used in some texts) as the lower limit of integration. This convention not only ensures inclusion of an impulse function at $t = 0$, but also allows us to use initial conditions at 0^- (rather than at 0^+) in the solution of differential equations via the Laplace transform. In practice, we are likely to know the initial conditions before the input is applied (at 0^-), not after the input is applied (at 0^+). Indeed, the very meaning of the term "initial conditions" implies conditions at $t = 0^-$ (conditions before the input is applied). Detailed analysis of desirability of using $t = 0^-$ appears in Section 4.3.

[†] Actually, $X(s)$ specifies $x(t)$ within a null function $n(t)$, which has the property that the area under $|n(t)|^2$ is zero over any finite interval 0 to t ($t > 0$) (Lerch's theorem). For example, if two functions are identical everywhere except at finite number of points, they differ by a null function.

The unilateral Laplace transform simplifies the system analysis problem considerably because of its *uniqueness property*, which says that for a given $X(s)$, there is a unique inverse transform. But there is a price for this simplification: we cannot analyze noncausal systems or use noncausal inputs. However, in most practical problems, this restriction is of little consequence. For this reason, we shall first consider the unilateral Laplace transform and its application to system analysis. (The bilateral Laplace transform is discussed later, in Section 4.11.)

Basically there is no difference between the unilateral and the bilateral Laplace transform. The unilateral transform is the bilateral transform that deals with a subclass of signals starting at $t = 0$ (causal signals). Therefore, the expression [Eq. (4.2)] for the inverse Laplace transform remains unchanged. In practice, the term *Laplace transform* means *the unilateral Laplace transform*.

EXISTENCE OF THE LAPLACE TRANSFORM

The variable s in the Laplace transform is complex in general, and it can be expressed as $s = \sigma + j\omega$. By definition

$$X(s) = \int_{0^-}^{\infty} x(t)e^{-st}\, dt$$

$$= \int_{0^-}^{\infty} [x(t)e^{-\sigma t}]e^{-j\omega t}\, dt$$

Because $|e^{j\omega t}| = 1$, the integral on the right-hand side of this equation converges if

$$\int_{0^-}^{\infty} \left| x(t)e^{-\sigma t} \right|\, dt < \infty \qquad (4.9)$$

Hence the existence of the Laplace transform is guaranteed if the integral in expression (4.9) is finite for some value of σ. Any signal that grows no faster than an exponential signal $Me^{\sigma_0 t}$ for some M and σ_0 satisfies the condition (4.9). Thus, if for some M and σ_0,

$$|x(t)| \le Me^{\sigma_0 t} \qquad (4.10)$$

we can choose $\sigma > \sigma_0$ to satisfy (4.9).[†] The signal e^{t^2}, in contrast, grows at a rate faster than $e^{\sigma_0 t}$, and consequently not Laplace transformable.[‡] Fortunately such signals (which are not Laplace transformable) are of little consequence from either a practical or a theoretical viewpoint. If σ_0 is the smallest value of σ for which the integral in (4.9) is finite, σ_0 is called the *abscissa of convergence* and the ROC of $X(s)$ is Re $s > \sigma_0$. The abscissa of convergence for $e^{-at}u(t)$ is $-a$ (the ROC is Re $s > -a$).

[†]Condition (4.10) is sufficient but not necessary for the existence of the Laplace transform. For example $x(t) = 1/\sqrt{t}$ is infinite at $t = 0$ and, (4.10) cannot be satisfied, but the transform of $1/\sqrt{t}$ exists and is given by $\sqrt{\pi/s}$.

[‡]However, if we consider a truncated (finite duration) signal e^{t^2}, the Laplace transform exists.

EXAMPLE 4.2

Determine the Laplace transform of the following:

(a) $\delta(t)$

(b) $u(t)$

(c) $\cos \omega_0 t \, u(t)$

(a)

$$\mathcal{L}[\delta(t)] = \int_{0^-}^{\infty} \delta(t)e^{-st} \, dt$$

Using the sampling property [Eq. (1.24a)], we obtain

$$\mathcal{L}[\delta(t)] = 1 \qquad \text{for all } s$$

that is,

$$\delta(t) \Longleftrightarrow 1 \qquad \text{for all } s \tag{4.11}$$

(b) To find the Laplace transform of $u(t)$, recall that $u(t) = 1$ for $t \geq 0$. Therefore

$$\mathcal{L}[u(t)] = \int_{0^-}^{\infty} u(t)e^{-st} \, dt = \int_{0^-}^{\infty} e^{-st} \, dt = -\frac{1}{s}e^{-st} \Big|_{0^-}^{\infty}$$

$$= \frac{1}{s} \qquad \text{Re } s > 0 \tag{4.12}$$

We also could have obtained this result from Eq. (4.7b) by letting $a = 0$.

(c) Because

$$\cos \omega_0 t \, u(t) = \tfrac{1}{2}[e^{j\omega_0 t} + e^{-j\omega_0 t}]u(t)$$

$$\mathcal{L}[\cos \omega_0 t \, u(t)] = \tfrac{1}{2}\mathcal{L}[e^{j\omega_0 t}u(t) + e^{-j\omega_0 t}u(t)] \tag{4.13}$$

From Eq. (4.7), it follows that

$$\mathcal{L}[\cos \omega_0 t \, u(t)] = \frac{1}{2}\left[\frac{1}{s - j\omega_0} + \frac{1}{s + j\omega_0}\right] \qquad \text{Re } (s \pm j\omega) = \text{Re } s > 0$$

$$= \frac{s}{s^2 + \omega_0{}^2} \qquad \text{Re } s > 0 \tag{4.14}$$

For the unilateral Laplace transform, there is a unique inverse transform of $X(s)$; consequently, there is no need to specify the ROC explicitly. For this reason, we shall generally ignore any mention of the ROC for unilateral transforms. Recall, also, that in the unilateral Laplace transform it is understood that every signal $x(t)$ is zero for $t < 0$, and it is appropriate to indicate this fact by multiplying the signal by $u(t)$.

EXERCISE E4.1

By direct integration, find the Laplace transform $X(s)$ and the region of convergence of $X(s)$ for the signals shown in Fig. 4.2.

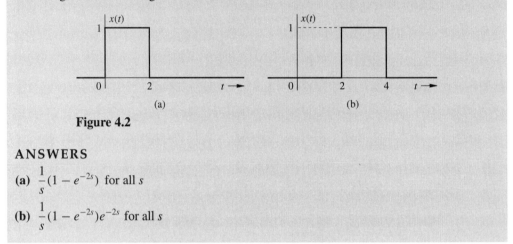

Figure 4.2

ANSWERS

(a) $\dfrac{1}{s}(1 - e^{-2s})$ for all s

(b) $\dfrac{1}{s}(1 - e^{-2s})e^{-2s}$ for all s

4.1-1 Finding the Inverse Transform

Finding the inverse Laplace transform by using the Eq. (4.2) requires integration in the complex plane, a subject beyond the scope of this book (but see, e.g., Ref. 3). For our purpose, we can find the inverse transforms from the transform table (Table 4.1). All we need is to express $X(s)$ as a sum of simpler functions of the forms listed in the table. Most of the transforms $X(s)$ of practical interest are *rational functions*, that is, ratios of polynomials in s. Such functions can be expressed as a sum of simpler functions by using partial fraction expansion (see Section B.5).

Values of s for which $X(s) = 0$ are called the *zeros* of $X(s)$; the values of s for which $X(s) \to \infty$ are called the *poles* of $X(s)$. If $X(s)$ is a rational function of the form $P(s)/Q(s)$, the roots of $P(s)$ are the zeros and the roots of $Q(s)$ are the poles of $X(s)$.

EXAMPLE 4.3

Find the inverse Laplace transforms of

(a) $\dfrac{7s - 6}{s^2 - s - 6}$

(b) $\dfrac{2s^2 + 5}{s^2 + 3s + 2}$

(c) $\dfrac{6(s + 34)}{s(s^2 + 10s + 34)}$

(d) $\dfrac{8s + 10}{(s + 1)(s + 2)^3}$

In no case is the inverse transform of these functions directly available in Table 4.1. Rather, we need to expand these functions into partial fractions, as discussed in Section B.5-1. In this computer age, it is very easy to find partial fractions on a computers. However, just as the availability of a handheld computer does not obviate the need for learning the mechanics of arithmetical operations (addition, multiplication, etc.), the widespread availability of computers does not eliminate the need to learn the mechanics of partial fraction expansion.

(a)

$$X(s) = \frac{7s - 6}{(s + 2)(s - 3)}$$

$$= \frac{k_1}{s + 2} + \frac{k_2}{s - 3}$$

To determine k_1, corresponding to the term $(s + 2)$, we cover up (conceal) the term $(s + 2)$ in $X(s)$ and substitute $s = -2$ (the value of s that makes $s + 2 = 0$) in the remaining expression (see Section B.5-2):

$$k_1 = \frac{7s - 6}{(s + 2)\,(s - 3)}\bigg|_{s=-2} = \frac{-14 - 6}{-2 - 3} = 4$$

Similarly, to determine k_2 corresponding to the term $(s - 3)$, we cover up the term $(s - 3)$ in $X(s)$ and substitute $s = 3$ in the remaining expression

$$k_2 = \frac{7s - 6}{(s + 2)\,(s - 3)}\bigg|_{s=3} = \frac{21 - 6}{3 + 2} = 3$$

Therefore

$$X(s) = \frac{7s - 6}{(s + 2)(s - 3)} = \frac{4}{s + 2} + \frac{3}{s - 3} \tag{4.15a}$$

Checking the Answer

It is easy to make a mistake in partial fraction computations. Fortunately it is simple to check the answer by recognizing that $X(s)$ and its partial fractions must be equal for every value of s if the partial fractions are correct. Let us verify this assertion in Eq. (4.15a) for some convenient value, say $s = 0$. Substitution of $s = 0$ in Eq. (4.15a) yields[†]

$$1 = 2 - 1 = 1$$

[†]Because $X(s) = \infty$ at its poles, we should avoid the pole values (-2 and 3 in the present case) for checking. The answers may check even if partial fractions are wrong. This situation can occur when two or more errors cancel their effects. But the chances of this problem arising for randomly selected values of s are extremely small.

We can now be sure of our answer with a high margin of confidence. Using pair 5 of Table 4.1 in Eq. (4.15a), we obtain

$$x(t) = \mathcal{L}^{-1}\left(\frac{4}{s+2} + \frac{3}{s-3}\right) = (4e^{-2t} + 3e^{3t})u(t) \qquad (4.15b)$$

(b)

$$X(s) = \frac{2s^2 + 5}{s^2 + 3s + 2} = \frac{2s^2 + 5}{(s+1)(s+2)}$$

Observe that $X(s)$ is an improper function with $M = N$. In such a case we can express $X(s)$ as a sum of the coefficient of the highest power in the numerator plus partial fractions corresponding to the poles of $X(s)$ (see Section B.5-5). In the present case the coefficient of the highest power in the numerator is 2. Therefore

$$X(s) = 2 + \frac{k_1}{s+1} + \frac{k_2}{s+2}$$

where

$$k_1 = \frac{2s^2 + 5}{(s+1)(s+2)}\bigg|_{s=-1} = \frac{2+5}{-1+2} = 7$$

and

$$k_2 = \frac{2s^2 + 5}{(s+1)(s+2)}\bigg|_{s=-2} = \frac{8+5}{-2+1} = -13$$

Therefore

$$X(s) = 2 + \frac{7}{s+1} - \frac{13}{s+2}$$

From Table 4.1, pairs 1 and 5, we obtain

$$x(t) = 2\delta(t) + (7e^{-t} - 13e^{-2t})u(t) \qquad (4.16)$$

(c)

$$X(s) = \frac{6(s+34)}{s(s^2 + 10s + 34)}$$

$$= \frac{6(s+34)}{s(s+5-j3)(s+5+j3)}$$

$$= \frac{k_1}{s} + \frac{k_2}{s+5-j3} + \frac{k_2^*}{s+5+j3}$$

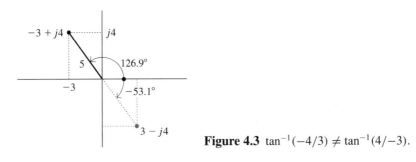

Figure 4.3 $\tan^{-1}(-4/3) \neq \tan^{-1}(4/-3)$.

Note that the coefficients (k_2 and k_2^*) of the conjugate terms must also be conjugate (see Section B.5). Now

$$k_1 = \left. \frac{6(s + 34)}{s \, (s^2 + 10s + 34)} \right|_{s=0} = \frac{6 \times 34}{34} = 6$$

$$k_2 = \left. \frac{6(s + 34)}{s \, (s + 5 - j3) \, (s + 5 + j3)} \right|_{s=-5+j3} = \frac{29 + j3}{-3 - j5} = -3 + j4$$

Therefore

$$k_2^* = -3 - j4$$

To use pair 10b of Table 4.1, we need to express k_2 and k_2^* in polar form.

$$-3 + j4 = \left(\sqrt{3^2 + 4^2} \right) e^{j \tan^{-1}(4/-3)} = 5e^{j \tan^{-1}(4/-3)}$$

Observe that $\tan^{-1}(4/-3) \neq \tan^{-1}(-4/3)$. This fact is evident in Fig. 4.3. For further discussion of this topic, see Example B.1.

From Fig. 4.3, we observe that

$$k_2 = -3 + j4 = 5e^{j126.9°}$$

so that

$$k_2^* = 5e^{-j126.9°}$$

Therefore

$$X(s) = \frac{6}{s} + \frac{5e^{j126.9°}}{s + 5 - j3} + \frac{5e^{-j126.9°}}{s + 5 + j3}$$

From Table 4.1 (pairs 2 and 10b), we obtain

$$x(t) = [6 + 10e^{-5t} \cos(3t + 126.9°)]u(t) \qquad (4.17)$$

ALTERNATIVE METHOD USING QUADRATIC FACTORS

The foregoing procedure involves considerable manipulation of complex numbers. Pair 10c (Table 4.1) indicates that the inverse transform of quadratic terms (with complex conjugate poles) can be found directly without having to find first-order partial fractions. We discussed such a procedure in Section B.5-2. For this purpose, we shall express $X(s)$ as

$$X(s) = \frac{6(s+34)}{s(s^2+10s+34)} = \frac{k_1}{s} + \frac{As+B}{s^2+10s+34}$$

We have already determined that $k_1 = 6$ by the (Heaviside) "cover-up" method. Therefore

$$\frac{6(s+34)}{s(s^2+10s+34)} = \frac{6}{s} + \frac{As+B}{s^2+10s+34}$$

Clearing the fractions by multiplying both sides by $s(s^2+10s+34)$ yields

$$6(s+34) = (6+A)s^2 + (60+B)s + 204$$

Now, equating the coefficients of s^2 and s on both sides yields

$$A = -6 \qquad \text{and} \qquad B = -54$$

and

$$X(s) = \frac{6}{s} + \frac{-6s - 54}{s^2+10s+34}$$

We now use pairs 2 and 10c to find the inverse Laplace transform. The parameters for pair 10c are $A = -6$, $B = -54$, $a = 5$, $c = 34$, $b = \sqrt{c - a^2} = 3$, and

$$r = \sqrt{\frac{A^2 c + B^2 - 2ABa}{c - a^2}} = 10 \qquad \theta = \tan^{-1}\frac{Aa - B}{A\sqrt{c - a^2}} = 126.9°$$

Therefore

$$x(t) = [6 + 10e^{-5t}\cos(3t + 126.9°)]u(t)$$

which agrees with the earlier result.

SHORTCUTS

The partial fractions with quadratic terms also can be obtained by using shortcuts. We have

$$X(s) = \frac{6(s+34)}{s(s^2+10s+34)} = \frac{6}{s} + \frac{As+B}{s^2+10s+34}$$

We can determine A by eliminating B on the right-hand side. This step can be accomplished by multiplying both sides of the equation for $X(s)$ by s and then letting $s \to \infty$. This procedure yields

$$0 = 6 + A \qquad \Longrightarrow \qquad A = -6$$

Therefore

$$\frac{6(s+34)}{s(s^2+10s+34)} = \frac{6}{s} + \frac{-6s+B}{s^2+10s+34}$$

To find B, we let s take on any convenient value, say $s = 1$, in this equation to obtain

$$\frac{210}{45} = 6 + \frac{B - 6}{45} \quad\Longrightarrow\quad B = -54$$

a result that agrees with the answer found earlier.

(d)

$$X(s) = \frac{8s + 10}{(s + 1)(s + 2)^3}$$

$$= \frac{k_1}{s + 1} + \frac{a_0}{(s + 2)^3} + \frac{a_1}{(s + 2)^2} + \frac{a_2}{s + 2}$$

where

$$k_1 = \frac{8s + 10}{(s + 1)(s + 2)^3}\bigg|_{s=-1} = 2$$

$$a_0 = \frac{8s + 10}{(s + 1)(s + 2)^3}\bigg|_{s=-2} = 6$$

$$a_1 = \left\{\frac{d}{ds}\left[\frac{8s + 10}{(s + 1)(s + 2)^3}\right]\right\}_{s=-2} = -2$$

$$a_2 = \frac{1}{2}\left\{\frac{d^2}{ds^2}\left[\frac{8s + 10}{(s + 1)(s + 2)^3}\right]\right\}_{s=-2} = -2$$

Therefore

$$X(s) = \frac{2}{s + 1} + \frac{6}{(s + 2)^3} - \frac{2}{(s + 2)^2} - \frac{2}{s + 2}$$

and

$$x(t) = [2e^{-t} + (3t^2 - 2t - 2)e^{-2t}]u(t) \tag{4.18}$$

ALTERNATIVE METHOD: A HYBRID OF HEAVISIDE AND CLEARING FRACTIONS

In this method, the simpler coefficients k_1 and a_0 are determined by the Heaviside "cover-up" procedure, as discussed earlier. To determine the remaining coefficients, we use the clearing-fraction method. Using the values $k_1 = 2$ and $a_0 = 6$ obtained earlier by the Heaviside "cover-up" method, we have

$$\frac{8s + 10}{(s + 1)(s + 2)^3} = \frac{2}{s + 1} + \frac{6}{(s + 2)^3} + \frac{a_1}{(s + 2)^2} + \frac{a_2}{s + 2}$$

We now clear fractions by multiplying both sides of the equation by $(s + 1)(s + 2)^3$. This procedure yields[†]

$$8s + 10 = 2(s + 2)^3 + 6(s + 1) + a_1(s + 1)(s + 2) + a_2(s + 1)(s + 2)^2$$

$$= (2 + a_2)s^3 + (12 + a_1 + 5a_2)s^2 + (30 + 3a_1 + 8a_2)s + (22 + 2a_1 + 4a_2)$$

Equating coefficients of s^3 and s^2 on both sides, we obtain

$$0 = (2 + a_2) \quad \Longrightarrow \quad a_2 = -2$$

$$0 = 12 + a_1 + 5a_2 = 2 + a_1 \quad \Longrightarrow \quad a_1 = -2$$

We can stop here if we wish, since the two desired coefficients a_1 and a_2 have already been found. However, equating the coefficients of s^1 and s^0 serves as a check on our answers. This step yields

$$8 = 30 + 3a_1 + 8a_2$$

$$10 = 22 + 2a_1 + 4a_2$$

Substitution of $a_1 = a_2 = -2$, obtained earlier, satisfies these equations. This step assures the correctness of our answers.

ANOTHER ALTERNATIVE: A HYBRID OF HEAVISIDE AND SHORTCUTS

In this method, the simpler coefficients k_1 and a_0 are determined by the Heaviside "cover-up" procedure, as discussed earlier. The usual shortcuts are then used to determine the remaining coefficients. Using the values $k_1 = 2$ and $a_0 = 6$, determined earlier by the Heaviside method, we have

$$\frac{8s + 10}{(s + 1)(s + 2)^3} = \frac{2}{s + 1} + \frac{6}{(s + 2)^3} + \frac{a_1}{(s + 2)^2} + \frac{a_2}{s + 2}$$

There are two unknowns, a_1 and a_2. If we multiply both sides by s and then let $s \to \infty$, we eliminate a_1. This procedure yields

$$0 = 2 + a_2 \quad \Longrightarrow \quad a_2 = -2$$

Therefore

$$\frac{8s + 10}{(s + 1)(s + 2)^3} = \frac{2}{s + 1} + \frac{6}{(s + 2)^3} + \frac{a_1}{(s + 2)^2} - \frac{2}{s + 2}$$

There is now only one unknown, a_1. This value can be determined readily by setting s equal to any convenient value, say $s = 0$. This step yields

$$\frac{10}{8} = 2 + \frac{3}{4} + \frac{a_1}{4} - 1 \quad \Longrightarrow \quad a_1 = -2$$

[†]We could have cleared fractions without finding k_1 and a_0. This alternative, however, proves more laborious because it increases the number of unknowns to 4. By predetermining k_1 and a_0, we reduce the unknowns to 2. Moreover, this method provides a convenient check on the solution. This hybrid procedure achieves the best of both methods.

COMPUTER EXAMPLE C4.1

Using the MATLAB `residue` command, determine the inverse Laplace transform of each of the following functions:

(a) $X_a(s) = \dfrac{2s^2 + 5}{s^2 + 3s + 2}$

(b) $X_b(s) = \dfrac{2s^2 + 7s + 4}{(s+1)(s+2)^2}$

(c) $X_c(s) = \dfrac{8s^2 + 21s + 19}{(s+2)(s^2 + s + 7)}$

(a)
```
>> num = [2 0 5]; den = [1 3 2];
>> [r, p, k] = residue(num,den);
>> disp(['(a) r = [',num2str(r.',' %0.5g'),']']);...
>> disp(['    p = [',num2str(p.',' %0.5g'),']']);...
>> disp(['    k = [',num2str(k.',' %0.5g'),']']);
(a) r = [-13  7]
    p = [-2  -1]
    k = [2]
```

Therefore, $X_a(s) = -13/(s+2) + 7/(s+1) + 2$ and $x_a(t) = (-13e^{-2t} + 7e^{-t})u(t) + 2\delta(t)$.

(b)
```
>> num = [2 7 4]; den = [conv([1 1],conv([1 2],[1 2]))];
>> [r, p, k] = residue(num,den);
>> disp(['(b) r = [',num2str(r.',' %0.5g'),']']);...
>> disp(['    p = [',num2str(p.',' %0.5g'),']']);...
>> disp(['    k = [',num2str(k.',' %0.5g'),']']);
(b) r = [3  2  -1]
    p = [-2  -2  -1]
    k = []
```

Therefore, $X_b(s) = 3/(s+2) + 2/(s+2)^2 - 1/(s+1)$ and $x_b(t) = (3e^{-2t} + 2te^{-2t} - e^{-t})u(t)$.

(c)
```
>> num = [8 21 19]; den = [conv([1 2],[1 1 7])];
>> [r, p, k]= residue(num,den);
>> [rad,mag]=cart2pol(real(r),imag(r));
>> disp(['(c) r = [',num2str(r.',' %0.5g'),']']);...
>> disp(['    p = [',num2str(p.',' %0.5g'),']']);...
>> disp(['    k = [',num2str(k.',' %0.5g'),']']);...
>> disp(['    rad = [',num2str(rad.',' %0.5g'),']']);...
>> disp(['    mag = [',num2str(mag.',' %0.5g'),']']);
```

```
(c) r = [3.5-0.48113i   3.5+0.48113i   1+0i]
    p = [-0.5+2.5981i   -0.5-2.5981i   -2+0i]
    k = []
  rad = [-0.13661   0.13661   0]
  mag = [3.5329   3.5329   1]
```

Thus,

$$X_c(s) = \frac{1}{s+2} + \frac{3.5329e^{-j0.13661}}{s+0.5-j2.5981} + \frac{3.5329e^{j0.13661}}{s+0.5+j2.5981}$$

and

$$x_c(t) = [e^{-2t} + 1.7665e^{-0.5t}\cos(2.5981t - 0.1366)]u(t).$$

COMPUTER EXAMPLE C4.2

Using MATLAB's symbolic math toolbox, determine the following:

    (a) the direct Laplace transform of $x_a(t) = \sin(at) + \cos(bt)$

    (b) the inverse Laplace transform of $X_b(s) = as^2/(s^2 + b^2)$

(a)
```
>> x_a = sym('sin(a*t)+cos(b*t)');

>> X_a = laplace(x_a);
>> disp(['(a) X_a(s) = ',char(X_a),' = ',char(simplify(X_a))]);
(a) X_a(s) = a/(s^2+a^2)+s/(s^2+b^2) = (a*s^2+a*b^2+s^3+s*a^2)/
                                      (s^2+a^2)/(s^2+b^2)
```

Therefore, $X_a = (s^3 + as^2 + a^2s + b^2a)/(s^2 + a^2)(s^2 + b^2)$

(b)
```
>> X_b = sym('(a*s^2)/(s^2+b^2)');
>> x_b = ilaplace(X_b);
>> disp(['(b) x_b(t) = ',char(x_b),' = ',char(simplify(x_b))]);
(b) x_b(t) = a*(Dirac(t)-b*sin(b*t)) = a*Dirac(t)-a*b*sin(b*t)
```

Therefore, $x_b(t) = a\delta(t) - ab\sin(bt)u(t)$.

EXERCISE E4.2

(i) Show that the Laplace transform of $10e^{-3t}\cos(4t + 53.13°)$ is $(6s - 14)/(s^2 + 6s + 25)$. Use pair 10a from Table 4.1.

(ii) Find the inverse Laplace transform of the following:

(a) $\dfrac{s + 17}{s^2 + 4s - 5}$

(b) $\dfrac{3s - 5}{(s + 1)(s^2 + 2s + 5)}$

(c) $\dfrac{16s + 43}{(s - 2)(s + 3)^2}$

ANSWERS

(a) $(3e^t - 2e^{-5t})u(t)$

(b) $\left[-2e^{-t} + \frac{5}{2}e^{-t}\cos(2t - 36.87°)\right]u(t)$

(c) $[3e^{2t} + (t - 3)e^{-3t}]u(t)$

A HISTORICAL NOTE: MARQUIS PIERRE-SIMON DE LAPLACE (1749–1827)

The Laplace transform is named after the great French mathematician and astronomer Laplace, who first presented the transform and its applications to differential equations in a paper published in 1779.

Pierre-Simon de Laplace

Oliver Heaviside

Laplace developed the foundations of potential theory and made important contributions to special functions, probability theory, astronomy, and celestial mechanics. In his *Exposition du système du monde* (1796), Laplace formulated a nebular hypothesis of cosmic origin and tried to explain the universe as a pure mechanism. In his *Traité de mécanique céleste* (*celestial mechanics*), which completed the work of Newton, Laplace used mathematics and physics to subject the solar system and all heavenly bodies to the laws of motion and the principle of gravitation. Newton had been unable to explain the irregularities of some heavenly bodies; in desperation, he concluded that God himself must intervene now and then to prevent such catastrophes as Jupiter eventually falling into the sun (and the moon into the earth) as predicted by Newton's calculations. Laplace proposed to show that these irregularities would correct themselves periodically and that a little patience—in Jupiter's case, 929 years—would see everything returning automatically to order; thus there was no reason why the solar and the stellar systems could not continue to operate by the laws of Newton and Laplace to the end of time.[4]

Laplace presented a copy of *Mécanique céleste* to Napoleon, who, after reading the book, took Laplace to task for not including God in his scheme: "You have written this huge book on the system of the world without once mentioning the author of the universe." "Sire," Laplace retorted, "I had no need of that hypothesis." Napoleon was not amused, and when he reported this reply to another great mathematician-astronomer, Louis de Lagrange, the latter remarked, "Ah, but that is a fine hypothesis. It explains so many things."[5]

Napoleon, following his policy of honoring and promoting scientists, made Laplace the minister of the interior. To Napoleon's dismay, however, the new appointee attempted to bring "the spirit of infinitesimals" into administration, and so Laplace was transferred hastily to the senate.

OLIVER HEAVISIDE (1850–1925)

Although Laplace published his transform method to solve differential equations in 1779, the method did not catch on until a century later. It was rediscovered independently in a rather awkward form by an eccentric British engineer, Oliver Heaviside (1850–1925), one of the tragic figures in the history of science and engineering. Despite his prolific contributions to electrical engineering, he was severely criticized during his lifetime and was neglected later to the point that hardly a textbook today mentions his name or credits him with contributions. Nevertheless, his studies had a major impact on many aspects of modern electrical engineering. It was Heaviside who made transatlantic communication possible by inventing cable loading, but no one ever mentions him as a pioneer or an innovator in telephony. It was Heaviside who suggested the use of inductive cable loading, but the credit is given to M. Pupin, who was not even responsible for building the first loading coil.[†] In addition, Heaviside was[6]

- The first to find a solution to the distortionless transmission line.
- The innovator of lowpass filters.
- The first to write Maxwell's equations in modern form.
- The codiscoverer of rate energy transfer by an electromagnetic field.

[†]Heaviside developed the theory for cable loading, George Campbell built the first loading coil, and the telephone circuits using Campbell's coils were in operation before Pupin published his paper. In the legal fight over the patent, however, Pupin won the battle because of his shrewd self-promotion and the poor legal support for Campbell.

- An early champion of the now-common phasor analysis.

- An important contributor to the development of vector analysis. In fact, he essentially created the subject independently of Gibbs.[7]

- An originator of the use of operational mathematics used to solve linear integro-differential equations, which eventually led to rediscovery of the ignored Laplace transform.

- The first to theorize (along with Kennelly of Harvard) that a conducting layer (the Kennelly–Heaviside layer) of atmosphere exists, which allows radio waves to follow earth's curvature instead of traveling off into space in a straight line.

- The first to posit that an electrical charge would increase in mass as its velocity increases, an anticipation of an aspect of Einstein's special theory of relativity.[8] He also forecast the possibility of superconductivity.

Heaviside was a self-made, self-educated man. Although his formal education ended with elementary school, he eventually became a pragmatically successful mathematical physicist. He began his career as a telegrapher, but increasing deafness forced him to retire at the age of 24. He then devoted himself to the study of electricity. His creative work was disdained by many professional mathematicians because of his lack of formal education and his unorthodox methods.

Heaviside had the misfortune to be criticized both by mathematicians, who faulted him for lack of rigor, and by men of practice, who faulted him for using too much mathematics and thereby confusing students. Many mathematicians, trying to find solutions to the distortion-less transmission line, failed because no rigorous tools were available at the time. Heaviside succeeded because he used mathematics not with rigor, but with insight and intuition. Using his much maligned operational method, Heaviside successfully attacked problems that the rigid mathematicians could not solve, problems such as the flow of heat in a body of spatially varying conductivity. Heaviside brilliantly used this method in 1895 to demonstrate a fatal flaw in Lord Kelvin's determination of the geological age of the earth by secular cooling; he used the same flow of heat theory as for his cable analysis. Yet the mathematicians of the Royal Society remained unmoved and were not the least impressed by the fact that Heaviside had found the answer to problems no one else could solve. Many mathematicians who examined his work dismissed it with contempt, asserting that his methods were either complete nonsense or a rehash of known ideas.[6]

Sir William Preece, the chief engineer of the British Post Office, a savage critic of Heaviside, ridiculed Heaviside's work as too theoretical and, therefore, leading to faulty conclusions. Heaviside's work on transmission lines and loading was dismissed by the British Post Office and might have remained hidden, had not Lord Kelvin himself publicly expressed admiration for it.[6]

Heaviside's operational calculus may be formally inaccurate, but in fact it anticipated the operational methods developed in more recent years.[9] Although his method was not fully understood, it provided correct results. When Heaviside was attacked for the vague meaning of his operational calculus, his pragmatic reply was, "Shall I refuse my dinner because I do not fully understand the process of digestion?"

Heaviside lived as a bachelor hermit, often in near-squalid conditions, and died largely unnoticed, in poverty. His life demonstrates the persistent arrogance and snobbishness of the intellectual establishment, which does not respect creativity unless it is presented in the strict language of the establishment.

4.2 SOME PROPERTIES OF THE LAPLACE TRANSFORM

Properties of the Laplace transform are useful not only in the derivation of the Laplace transform of functions but also in the solutions of linear integro-differential equations. A glance at Eqs. (4.2) and (4.1) shows that there is a certain measure of symmetry in going from $x(t)$ to $X(s)$, and vice versa. This symmetry or duality is also carried over to the properties of the Laplace transform. This fact will be evident in the following development.

We are already familiar with two properties; linearity [Eq. (4.4)] and the uniqueness property of the Laplace transform discussed earlier.

4.2-1 Time Shifting

The time-shifting property states that if

$$x(t) \Longleftrightarrow X(s)$$

then for $t_0 \geq 0$

$$x(t - t_0) \Longleftrightarrow X(s)e^{-st_0} \qquad (4.19a)$$

Observe that $x(t)$ starts at $t = 0$, and, therefore, $x(t - t_0)$ starts at $t = t_0$. This fact is implicit, but is not explicitly indicated in Eq. (4.19a). This often leads to inadvertent errors. To avoid such a pitfall, we should restate the property as follows. If

$$x(t)u(t) \Longleftrightarrow X(s)$$

then

$$x(t - t_0)u(t - t_0) \Longleftrightarrow X(s)e^{-st_0} \qquad t_0 \geq 0 \qquad (4.19b)$$

Proof.

$$\mathcal{L}\left[x(t - t_0)u(t - t_0)\right] = \int_0^\infty x(t - t_0)u(t - t_0)e^{-st}\, dt$$

Setting $t - t_0 = \tau$, we obtain

$$\mathcal{L}\left[x(t - t_0)u(t - t_0)\right] = \int_{-t_0}^\infty x(\tau)u(\tau)e^{-s(\tau + t_0)}\, d\tau$$

Because $u(\tau) = 0$ for $\tau < 0$ and $u(\tau) = 1$ for $\tau \geq 0$, the limits of integration can be taken from 0 to ∞. Thus

$$\mathcal{L}\left[x(t - t_0)u(t - t_0)\right] = \int_0^\infty x(\tau)e^{-s(\tau + t_0)}\, d\tau$$

$$= e^{-st_0} \int_0^\infty x(\tau)e^{-s\tau}\, d\tau$$

$$= X(s)e^{-st_0}$$

Note that $x(t - t_0)u(t - t_0)$ is the signal $x(t)u(t)$ delayed by t_0 seconds. The time-shifting property states that *delaying a signal by t_0 seconds amounts to multiplying its transform e^{-st_0}*.

This property of the unilateral Laplace transform holds only for positive t_0 because if t_0 were negative, the signal $x(t - t_0)u(t - t_0)$ may not be causal.

We can readily verify this property in Exercise E4.1. If the signal in Fig. 4.2a is $x(t)u(t)$, then the signal in Fig. 4.2b is $x(t - 2)u(t - 2)$. The Laplace transform for the pulse in Fig. 4.2a is $(1/s)(1 - e^{-2s})$. Therefore, the Laplace transform for the pulse in Fig. 4.2b is $(1/s)(1 - e^{-2s})e^{-2s}$.

The time-shifting property proves very convenient in finding the Laplace transform of functions with different descriptions over different intervals, as the following example demonstrates.

EXAMPLE 4.4

Find the Laplace transform of $x(t)$ depicted in Fig. 4.4a.

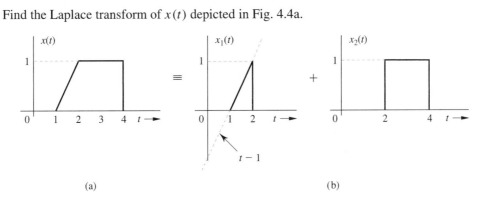

(a) (b)

Figure 4.4 Finding a mathematical description of a function $x(t)$.

Describing mathematically a function such as the one in Fig. 4.4a is discussed in Section 1.4. The function $x(t)$ in Fig. 4.4a can be described as a sum of two components shown in Fig. 4.4b. The equation for the first component is $t - 1$ over $1 \le t \le 2$, so that this component can be described by $(t - 1)[u(t - 1) - u(t - 2)]$. The second component can be described by $u(t - 2) - u(t - 4)$. Therefore

$$x(t) = (t - 1)[u(t - 1) - u(t - 2)] + [u(t - 2) - u(t - 4)]$$
$$= (t - 1)u(t - 1) - (t - 1)u(t - 2) + u(t - 2) - u(t - 4) \qquad (4.20a)$$

The first term on the right-hand side is the signal $tu(t)$ delayed by 1 second. Also, the third and fourth terms are the signal $u(t)$ delayed by 2 and 4 seconds, respectively. The second term, however, cannot be interpreted as a delayed version of any entry in Table 4.1. For this reason, we rearrange it as

$$(t - 1)u(t - 2) = (t - 2 + 1)u(t - 2) = (t - 2)u(t - 2) + u(t - 2)$$

We have now expressed the second term in the desired form as $tu(t)$ delayed by 2 seconds plus $u(t)$ delayed by 2 seconds. With this result, Eq. (4.20a) can be expressed as

$$x(t) = (t - 1)u(t - 1) - (t - 2)u(t - 2) - u(t - 4) \qquad (4.20b)$$

Application of the time-shifting property to $tu(t) \Longleftrightarrow 1/s^2$ yields

$$(t-1)u(t-1) \Longleftrightarrow \frac{1}{s^2}e^{-s} \quad \text{and} \quad (t-2)u(t-2) \Longleftrightarrow \frac{1}{s^2}e^{-2s}$$

Also

$$u(t) \Longleftrightarrow \frac{1}{s} \quad \text{and} \quad u(t-4) \Longleftrightarrow \frac{1}{s}e^{-4s} \tag{4.21}$$

Therefore

$$X(s) = \frac{1}{s^2}e^{-s} - \frac{1}{s^2}e^{-2s} - \frac{1}{s}e^{-4s} \tag{4.22}$$

EXAMPLE 4.5

Find the inverse Laplace transform of

$$X(s) = \frac{s+3+5e^{-2s}}{(s+1)(s+2)}$$

Observe the exponential term e^{-2s} in the numerator of $X(s)$, indicating time delay. In such a case we should separate $X(s)$ into terms with and without delay factor, as

$$X(s) = \underbrace{\frac{s+3}{(s+1)(s+2)}}_{X_1(s)} + \underbrace{\frac{5e^{-2s}}{(s+1)(s+2)}}_{X_2(s)e^{-2s}}$$

where

$$X_1(s) = \frac{s+3}{(s+1)(s+2)} = \frac{2}{s+1} - \frac{1}{s+2}$$

$$X_2(s) = \frac{5}{(s+1)(s+2)} = \frac{5}{s+1} - \frac{5}{s+2}$$

Therefore

$$x_1(t) = (2e^{-t} - e^{-2t})u(t)$$

$$x_2(t) = 5(e^{-t} - e^{-2t})u(t)$$

Also, because

$$X(s) = X_1(s) + X_2(s)e^{-2s}$$

We can write

$$x(t) = x_1(t) + x_2(t-2)$$
$$= (2e^{-t} - e^{-2t})u(t) + 5\left[e^{-(t-2)} - e^{-2(t-2)}\right]u(t-2)$$

EXERCISE E4.3

Find the Laplace transform of the signal illustrated in Fig. 4.5.

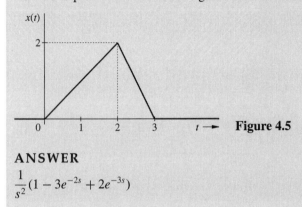

Figure 4.5

ANSWER

$$\frac{1}{s^2}(1 - 3e^{-2s} + 2e^{-3s})$$

EXERCISE E4.4

Find the inverse Laplace transform of

$$X(s) = \frac{3e^{-2s}}{(s-1)(s+2)}$$

ANSWER

$$\left[e^{t-2} - e^{-2(t-2)}\right]u(t-2)$$

4.2-2 Frequency Shifting

The frequency-shifting property states that if

$$x(t) \Longleftrightarrow X(s)$$

then

$$x(t)e^{s_0 t} \Longleftrightarrow X(s - s_0) \qquad\qquad (4.23)$$

Observe the symmetry (or duality) between this property and the time-shifting property (4.19a).

Proof.

$$\mathcal{L}[x(t)e^{s_0 t}] = \int_{0^-}^{\infty} x(t)e^{s_0 t} e^{-st}\, dt = \int_{0^-}^{\infty} x(t)e^{-(s-s_0)t}\, dt = X(s - s_0)$$

EXAMPLE 4.6

Derive pair 9a in Table 4.1 from pair 8a and the frequency-shifting property.

Pair 8a is

$$\cos bt\, u(t) \Longleftrightarrow \frac{s}{s^2 + b^2}$$

From the frequency-shifting property [Eq. (4.23)] with $s_0 = -a$ we obtain

$$e^{-at} \cos bt\, u(t) \Longleftrightarrow \frac{s + a}{(s + a)^2 + b^2}$$

EXERCISE E4.5

Derive pair 6 in Table 4.1 from pair 3 and the frequency-shifting property.

We are now ready to consider the two of the most important properties of the Laplace transform: time differentiation and time integration.

4.2-3 The Time-Differentiation Property[†]

The time-differentiation property states that if

$$x(t) \Longleftrightarrow X(s)$$

then

$$\frac{dx}{dt} \Longleftrightarrow sX(s) - x(0^-) \tag{4.24a}$$

Repeated application of this property yields

$$\frac{d^2x}{dt^2} \Longleftrightarrow s^2 X(s) - sx(0^-) - \dot{x}(0^-) \tag{4.24b}$$

$$\frac{d^n x}{dt^n} \Longleftrightarrow s^n X(s) - s^{n-1}x(0^-) - s^{n-2}\dot{x}(0^-) - \cdots - x^{(n-1)}(0^-)$$

$$= s^n X(s) - \sum_{k=1}^{n} s^{n-k} x^{(k-1)}(0^-) \tag{4.24c}$$

where $x^{(r)}(0^-)$ is $d^r x/dt^r$ at $t = 0^-$.

[†]The dual of the time-differentiation property is the frequency-differentiation property, which states that

$$tx(t) \Longleftrightarrow -\frac{d}{ds} X(s)$$

Proof.

$$\mathcal{L}\left[\frac{dx}{dt}\right] = \int_{0^-}^{\infty} \frac{dx}{dt} e^{-st} dt$$

Integrating by parts, we obtain

$$\mathcal{L}\left[\frac{dx}{dt}\right] = x(t)e^{-st}\Big|_{0^-}^{\infty} + s \int_{0^-}^{\infty} x(t)e^{-st} dt$$

For the Laplace integral to converge [i.e., for $X(s)$ to exist], it is necessary that $x(t)e^{-st} \to 0$ as $t \to \infty$ for the values of s in the ROC for $X(s)$. Thus,

$$\mathcal{L}\left[\frac{dx}{dt}\right] = -x(0^-) + sX(s)$$

Repeated application of this procedure yields Eq. (4.24c).

EXAMPLE 4.7

Find the Laplace transform of the signal $x(t)$ in Fig. 4.6a by using Table 4.1 and the time-differentiation and time-shifting properties of the Laplace transform.

Figures 4.6b and 4.6c show the first two derivatives of $x(t)$. Recall that the derivative at a point of jump discontinuity is an impulse of strength equal to the amount of jump [see Eq. (1.27)]. Therefore

$$\frac{d^2x}{dt^2} = \delta(t) - 3\delta(t-2) + 2\delta(t-3)$$

The Laplace transform of this equation yields

$$\mathcal{L}\left(\frac{d^2x}{dt^2}\right) = \mathcal{L}[\delta(t) - 3\delta(t-2) + 2\delta(t-3)]$$

Using the time-differentiation property Eq. (4.24b), the time-shifting property (4.19a), and the facts that $x(0^-) = \dot{x}(0^-) = 0$, and $\delta(t) \Longleftrightarrow 1$, we obtain

$$s^2 X(s) - 0 - 0 = 1 - 3e^{-2s} + 2e^{-3s}$$

Therefore

$$X(s) = \frac{1}{s^2}(1 - 3e^{-2s} + 2e^{-3s})$$

which confirms the earlier result in Exercise E4.3.

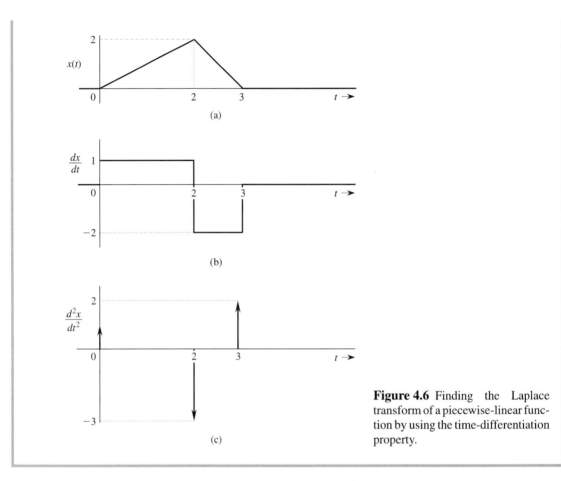

Figure 4.6 Finding the Laplace transform of a piecewise-linear function by using the time-differentiation property.

4.2-4 The Time-Integration Property

The time-integration property states that if

$$x(t) \Longleftrightarrow X(s)$$

then[†]

$$\int_{0^-}^{t} x(\tau)\, d\tau \Longleftrightarrow \frac{X(s)}{s} \tag{4.25}$$

and

$$\int_{-\infty}^{t} x(\tau)\, d\tau \Longleftrightarrow \frac{X(s)}{s} + \frac{\int_{-\infty}^{0^-} x(\tau)\, d\tau}{s} \tag{4.26}$$

[†]The dual of the time-integration property is the frequency-integration property, which states that

$$\frac{x(t)}{t} \Longleftrightarrow \int_{s}^{\infty} X(z)\, dz$$

Proof. We define

$$g(t) = \int_{0^-}^{t} x(\tau)\, d\tau$$

so that

$$\frac{d}{dt} g(t) = x(t) \qquad \text{and} \qquad g(0^-) = 0$$

Now, if

$$g(t) \Longleftrightarrow G(s)$$

then

$$X(s) = \mathcal{L}\left[\frac{d}{dt} g(t)\right] = sG(s) - g(0^-) = sG(s)$$

Therefore

$$G(s) = \frac{X(s)}{s}$$

or

$$\int_{0^-}^{t} x(\tau)\, d\tau \Longleftrightarrow \frac{X(s)}{s}$$

To prove Eq. (4.26), observe that

$$\int_{-\infty}^{t} x(\tau)\, d\tau = \int_{-\infty}^{0^-} x(\tau)\, d\tau + \int_{0^-}^{t} x(\tau)\, d\tau$$

Note that the first term on the right-hand side is a constant for $t \geq 0$. Taking the Laplace transform of the foregoing equation and using Eq. (4.25), we obtain

$$\int_{-\infty}^{t} x(\tau)\, d\tau \Longleftrightarrow \frac{\int_{-\infty}^{0^-} x(\tau)\, d\tau}{s} + \frac{X(s)}{s}$$

SCALING

The scaling property states that if

$$x(t) \Longleftrightarrow X(s)$$

then for $a > 0$

$$x(at) \Longleftrightarrow \frac{1}{a} F\left(\frac{s}{a}\right) \tag{4.27}$$

The proof is given in Chapter 7. Note that a is restricted to positive values because if $x(t)$ is causal, then $x(at)$ is anticausal (is zero for $t \geq 0$) for negative a, and anticausal signals are not permitted in the (unilateral) Laplace transform.

Recall that $x(at)$ is the signal $x(t)$ time-compressed by the factor a, and $X(\frac{s}{a})$ is $X(s)$ expanded along the s-scale by the same factor a (see Section 1.2-2). The scaling property states that *time compression of a signal by a factor a causes expansion of its Laplace transform in the s-scale by the same factor. Similarly, time expansion x(t) causes compression of X(s) in the s-scale by the same factor.*

4.2-5 Time Convolution and Frequency Convolution

Another pair of properties states that if

$$x_1(t) \Longleftrightarrow X_1(s) \qquad \text{and} \qquad x_2(t) \Longleftrightarrow X_2(s)$$

then (*time-convolution property*)

$$x_1(t) * x_2(t) \Longleftrightarrow X_1(s)X_2(s) \tag{4.28}$$

and (*frequency-convolution property*)

$$x_1(t)x_2(t) \Longleftrightarrow \frac{1}{2\pi j}[X_1(s) * X_2(s)] \tag{4.29}$$

Observe the symmetry (or duality) between the two properties. Proofs of these properties are postponed to Chapter 7.

Equation (2.48) indicates that $H(s)$, the transfer function of an LTIC system, is the Laplace transform of the system's impulse response $h(t)$; that is,

$$h(t) \Longleftrightarrow H(s) \tag{4.30}$$

If the system is causal, $h(t)$ is causal, and, according to Eq. (2.48), $H(s)$ is the unilateral Laplace transform of $h(t)$. Similarly, if the system is noncausal, $h(t)$ is noncausal, and $H(s)$ is the bilateral transform of $h(t)$.

We can apply the time-convolution property to the LTIC input–output relationship $y(t) = x(t) * h(t)$ to obtain

$$Y(s) = X(s)H(s) \tag{4.31}$$

The response $y(t)$ is the zero-state response of the LTIC system to the input $x(t)$. From Eq. (4.31), it follows that

$$H(s) = \frac{Y(s)}{X(s)} = \frac{\mathcal{L}[\text{zero-state response}]}{\mathcal{L}[\text{input}]} \tag{4.32}$$

This may be considered to be an alternate definition of the LTIC system transfer function $H(s)$. It is the *ratio of the transform of zero-state response to the transform of the input.*

EXAMPLE 4.8

Using the time-convolution property of the Laplace transform to determine $c(t) = e^{at}u(t) * e^{bt}u(t)$.

From Eq. (4.28), it follows that

$$C(s) = \frac{1}{(s-a)(s-b)} = \frac{1}{a-b}\left[\frac{1}{s-a} - \frac{1}{s-b}\right]$$

The inverse transform of this equation yields

$$c(t) = \frac{1}{a-b}(e^{at} - e^{bt})u(t)$$

TABLE 4.2 The Laplace Transform Properties

Operation	$x(t)$	$X(s)$
Addition	$x_1(t) + x_2(t)$	$X_1(s) + X_2(s)$
Scalar multiplication	$kx(t)$	$kX(s)$
Time differentiation	$\dfrac{dx}{dt}$	$sX(s) - x(0^-)$
	$\dfrac{d^2x}{dt^2}$	$s^2X(s) - sx(0^-) - \dot{x}(0^-)$
	$\dfrac{d^3x}{dt^3}$	$s^3X(s) - s^2x(0^-) - s\dot{x}(0^-) - \ddot{x}(0^-)$
	$\dfrac{d^nx}{dt^n}$	$s^nX(s) - \displaystyle\sum_{k=1}^{n} s^{n-k}x^{(k-1)}(0^-)$
Time integration	$\displaystyle\int_{0^-}^{t} x(\tau)\,d\tau$	$\dfrac{1}{s}X(s)$
	$\displaystyle\int_{-\infty}^{t} x(\tau)\,d\tau$	$\dfrac{1}{s}X(s) + \dfrac{1}{s}\displaystyle\int_{-\infty}^{0^-} x(t)\,dt$
Time shifting	$x(t - t_0)u(t - t_0)$	$X(s)e^{-st_0} \qquad t_0 \geq 0$
Frequency shifting	$x(t)e^{s_0 t}$	$X(s - s_0)$
Frequency differentiation	$-tx(t)$	$\dfrac{dX(s)}{ds}$
Frequency integration	$\dfrac{x(t)}{t}$	$\displaystyle\int_{s}^{\infty} X(z)\,dz$
Scaling	$x(at), a \geq 0$	$\dfrac{1}{a}X\left(\dfrac{s}{a}\right)$
Time convolution	$x_1(t) * x_2(t)$	$X_1(s)X_2(s)$
Frequency convolution	$x_1(t)x_2(t)$	$\dfrac{1}{2\pi j}X_1(s) * X_2(s)$
Initial value	$x(0^+)$	$\displaystyle\lim_{s\to\infty} sX(s) \qquad (n > m)$
Final value	$x(\infty)$	$\displaystyle\lim_{s\to 0} sX(s) \qquad$ [poles of $sX(s)$ in LHP]

INITIAL AND FINAL VALUES

In certain applications, it is desirable to know the values of $x(t)$ as $t \to 0$ and $t \to \infty$ [initial and final values of $x(t)$] from the knowledge of its Laplace transform $X(s)$. Initial and final value theorems provide such information.

   The initial value theorem states that if $x(t)$ and its derivative dx/dt are both Laplace transformable, then

$$x(0^+) = \lim_{s\to\infty} sX(s) \tag{4.33}$$

provided the limit on the right-hand side of Eq. (4.33) exists.

The final value theorem states that if both $x(t)$ and dx/dt are Laplace transformable, then

$$\lim_{t \to \infty} x(t) = \lim_{s \to 0} s X(s) \qquad (4.34)$$

provided $s X(s)$ has no poles in the RHP or on the imaginary axis. To prove these theorems, we use Eq. (4.24a)

$$s X(s) - x(0^-) = \int_{0^-}^{\infty} \frac{dx}{dt} e^{-st}\, dt$$

$$= \int_{0^-}^{0^+} \frac{dx}{dt} e^{-st}\, dt + \int_{0^+}^{\infty} \frac{dx}{dt} e^{-st}\, dt$$

$$= x(t) \Big|_{0^-}^{0^+} + \int_{0^+}^{\infty} \frac{dx}{dt} e^{-st}\, dt$$

$$= x(0^+) - x(0^-) + \int_{0^+}^{\infty} \frac{dx}{dt} e^{-st}\, dt$$

Therefore

$$s X(s) = x(0^+) + \int_{0^+}^{\infty} \frac{dx}{dt} e^{-st}\, dt$$

and

$$\lim_{s \to \infty} s X(s) = x(0^+) + \lim_{s \to \infty} \int_{0^+}^{\infty} \frac{dx}{dt} e^{-st}\, dt$$

$$= x(0^+) + \int_{0^+}^{\infty} \frac{dx}{dt} \left(\lim_{s \to \infty} e^{-st} \right) dt$$

$$= x(0^+)$$

Comment. The initial value theorem applies only if $X(s)$ is strictly proper ($M < N$), because for $M \geq N$, $\lim_{s \to \infty} s X(s)$ does not exist, and the theorem does not apply. In such a case, we can still find the answer by using long division to express $X(s)$ as a polynomial in s plus a strictly proper fraction, where $M < N$. For example, by using long division, we can express

$$\frac{s^3 + 3s^2 + s + 1}{s^2 + 2s + 1} = (s + 1) - \frac{2s}{s^2 + 2s + 1}$$

The inverse transform of the polynomial in s is in terms of $\delta(t)$, and its derivatives, which are zero at $t = 0^+$. In the foregoing case, the inverse transform of $s + 1$ is $\dot{\delta}(t) + \delta(t)$. Hence, the desired $x(0^+)$ is the value of the remainder (strictly proper) fraction, for which the initial value theorem applies. In the present case

$$x(0^+) = \lim_{s \to \infty} \frac{-2s^2}{s^2 + 2s + 1} = -2$$

To prove the final value theorem, we let $s \to 0$ in Eq. (4.24a) to obtain

$$\lim_{s \to 0} [s X(s) - x(0^-)] = \lim_{s \to 0} \int_{0^-}^{\infty} \frac{dx}{dt} e^{-st}\, dt = \int_{0^-}^{\infty} \frac{dx}{dt}\, dt$$

$$= x(t) \Big|_{0^-}^{\infty} = \lim_{t \to \infty} x(t) - x(0^-)$$

a deduction that leads to the desired result, Eq. (4.34).

Comment. The final value theorem applies only if the poles of $X(s)$ are in the LHP (including $s = 0$). If $X(s)$ has a pole in the RHP, $x(t)$ contains an exponentially growing term and $x(\infty)$ does not exist. If there is a pole on the imaginary axis, then $x(t)$ contains an oscillating term and $x(\infty)$ does not exist. However, if there is a pole at the origin, then $x(t)$ contains a constant term, and hence, $x(\infty)$ exists and is a constant.

EXAMPLE 4.9

Determine the initial and final values of $y(t)$ if its Laplace transform $Y(s)$ is given by

$$Y(s) = \frac{10(2s + 3)}{s(s^2 + 2s + 5)}$$

Equations (4.33) and (4.34) yield

$$y(0^+) = \lim_{s \to \infty} sY(s) = \lim_{s \to \infty} \frac{10(2s + 3)}{(s^2 + 2s + 5)} = 0$$

$$y(\infty) = \lim_{s \to 0} sY(s) = \lim_{s \to 0} \frac{10(2s + 3)}{(s^2 + 2s + 5)} = 6$$

4.3 SOLUTION OF DIFFERENTIAL AND INTEGRO-DIFFERENTIAL EQUATIONS

The time-differentiation property of the Laplace transform has set the stage for solving linear differential (or integro-differential) equations with constant coefficients. Because $d^k y/dt^k \Longleftrightarrow s^k Y(s)$, the Laplace transform of a differential equation is an algebraic equation that can be readily solved for $Y(s)$. Next we take the inverse Laplace transform of $Y(s)$ to find the desired solution $y(t)$. The following examples demonstrate the Laplace transform procedure for solving linear differential equations with constant coefficients.

EXAMPLE 4.10

Solve the second-order linear differential equation

$$(D^2 + 5D + 6)y(t) = (D + 1)x(t) \qquad (4.35a)$$

for the initial conditions $y(0^-) = 2$ and $\dot{y}(0^-) = 1$ and the input $x(t) = e^{-4t}u(t)$.

The equation is

$$\frac{d^2y}{dt^2} + 5\frac{dy}{dt} + 6y(t) = \frac{dx}{dt} + x(t) \qquad (4.35b)$$

Let

$$y(t) \Longleftrightarrow Y(s)$$

Then from Eqs. (4.24)

$$\frac{dy}{dt} \Longleftrightarrow sY(s) - y(0^-) = sY(s) - 2$$

and

$$\frac{d^2y}{dt^2} \Longleftrightarrow s^2Y(s) - sy(0^-) - \dot{y}(0^-) = s^2Y(s) - 2s - 1$$

Moreover, for $x(t) = e^{-4t}u(t)$,

$$X(s) = \frac{1}{s+4} \quad \text{and} \quad \frac{dx}{dt} \Longleftrightarrow sX(s) - x(0^-) = \frac{s}{s+4} - 0 = \frac{s}{s+4}$$

Taking the Laplace transform of Eq. (4.35b), we obtain

$$[s^2Y(s) - 2s - 1] + 5[sY(s) - 2] + 6Y(s) = \frac{s}{s+4} + \frac{1}{s+4} \qquad (4.36a)$$

Collecting all the terms of $Y(s)$ and the remaining terms separately on the left-hand side, we obtain

$$(s^2 + 5s + 6)Y(s) - (2s + 11) = \frac{s+1}{s+4} \qquad (4.36b)$$

Therefore

$$(s^2 + 5s + 6)Y(s) = (2s + 11) + \frac{s+1}{s+4} = \frac{2s^2 + 20s + 45}{s+4}$$

and

$$Y(s) = \frac{2s^2 + 20s + 45}{(s^2 + 5s + 6)(s+4)}$$

$$= \frac{2s^2 + 20s + 45}{(s+2)(s+3)(s+4)}$$

Expanding the right-hand side into partial fractions yields

$$Y(s) = \frac{13/2}{s+2} - \frac{3}{s+3} - \frac{3/2}{s+4}$$

The inverse Laplace transform of this equation yields

$$y(t) = \left(\tfrac{13}{2}e^{-2t} - 3e^{-3t} - \tfrac{3}{2}e^{-4t}\right)u(t) \qquad (4.37)$$

Example 4.10 demonstrates the ease with which the Laplace transform can solve linear differential equations with constant coefficients. The method is general and can solve a linear differential equation with constant coefficients of any order.

ZERO-INPUT AND ZERO-STATE COMPONENTS OF RESPONSE

The Laplace transform method gives the total response, which includes zero-input and zero-state components. It is possible to separate the two components if we so desire. The initial condition terms in the response give rise to the zero-input response. For instance, in Example 4.10, the terms attributable to initial conditions $y(0^-) = 2$ and $\dot{y}(0^-) = 1$ in Eq. (4.36a) generate the zero-input response. These initial condition terms are $-(2s + 11)$, as seen in Eq. (4.36b). The terms on the right-hand side are exclusively due to the input. Equation (4.36b) is reproduced below with the proper labeling of the terms.

$$\left(s^2 + 5s + 6\right) Y(s) - (2s + 11) = \frac{s + 1}{s + 4}$$

so that

$$\left(s^2 + 5s + 6\right) Y(s) = \underbrace{(2s + 11)}_{\text{initial condition terms}} + \underbrace{\frac{s + 1}{s + 4}}_{\text{input terms}}$$

Therefore

$$Y(s) = \underbrace{\frac{2s + 11}{s^2 + 5s + 6}}_{\text{zero-input component}} + \underbrace{\frac{s + 1}{(s + 4)(s^2 + 5s + 6)}}_{\text{zero-state component}}$$

$$= \left[\frac{7}{s + 2} - \frac{5}{s + 3}\right] + \left[\frac{-1/2}{s + 2} + \frac{2}{s + 3} - \frac{3/2}{s + 4}\right]$$

Taking the inverse transform of this equation yields

$$y(t) = \underbrace{\left(7e^{-2t} - 5e^{-3t}\right) u(t)}_{\text{zero-input response}} + \underbrace{\left(-\tfrac{1}{2}e^{-2t} + 2e^{-3t} - \tfrac{3}{2}e^{-4t}\right) u(t)}_{\text{zero-state response}}$$

COMMENTS ON INITIAL CONDITIONS AT 0^- AND AT 0^+

The initial conditions in Example 4.10 are $y(0^-) = 2$ and $\dot{y}(0^-) = 1$. If we let $t = 0$ in the total response in Eq. (4.37), we find $y(0) = 2$ and $\dot{y}(0) = 2$, which is at odds with the given initial conditions. Why? Because the initial conditions are given at $t = 0^-$ (just before the input is applied), when only the zero-input response is present. The zero-state response is the result of the input $x(t)$ applied at $t = 0$. Hence, this component does not exist at $t = 0^-$. Consequently, the initial conditions at $t = 0^-$ are satisfied by the zero-input response, not by the total response. We can readily verify in this example that the zero-input response does indeed satisfy the given initial conditions at $t = 0^-$. It is the total response that satisfies the initial conditions at $t = 0^+$, which are generally different from the initial conditions at 0^-.

There also exists a $\mathcal{L}_+$ version of the Laplace transform, which uses the initial conditions at $t = 0^+$ rather than at 0^- (as in our present $\mathcal{L}_-$ version). The $\mathcal{L}_+$ version, which was in vogue till the early 1960s, is identical to the $\mathcal{L}_-$ version except the limits of Laplace integral [Eq. (4.8)] are from 0^+ to ∞. Hence, by definition, the origin $t = 0$ is excluded from the domain. This version, still used in some math books, has some serious difficulties. For instance, the Laplace transform of $\delta(t)$ is zero because $\delta(t) = 0$ for $t \geq 0^+$. Moreover, this approach is rather clumsy in the theoretical study of linear systems because the response obtained cannot be separated into zero-input and zero-state components. As we know, the zero-state component represents the system response as an explicit function of the input, and without knowing this component, it is not possible to assess the effect of the input on the system response in a general way. The $\mathcal{L}_+$ version can separate the response in terms of the natural and the forced components, which are not as interesting as the zero-input and the zero-state components. Note that we can always determine the natural and the forced components from the zero-input and the zero-state components [see Eqs. (2.52)], but the converse is not true. Because of these and some other problems, electrical engineers (wisely) started discarding the $\mathcal{L}_+$ version in the early sixties.

It is interesting to note the time-domain duals of these two Laplace versions. The classical method is the dual of the $\mathcal{L}_+$ method, and the convolution (zero-input/zero-state) method is the dual of the $\mathcal{L}_-$ method. The first pair uses the initial conditions at 0^+, and the second pair uses those at $t = 0^-$. The first pair (the classical method and the $\mathcal{L}_+$ version) is awkward in the theoretical study of linear system analysis. It was no coincidence that the $\mathcal{L}_-$ version was adopted immediately after the introduction to the electrical engineering community of state-space analysis (which uses zero-input/zero-state separation of the output).

EXERCISE E4.6

Solve

$$\frac{d^2y}{dt^2} + 4\frac{dy}{dt} + 3y(t) = 2\frac{dx}{dt} + x(t)$$

for the input $x(t) = u(t)$. The initial conditions are $y(0^-) = 1$ and $\dot{y}(0^-) = 2$.

ANSWER

$y(t) = \frac{1}{3}(1 + 9e^{-t} - 7e^{-3t})u(t)$

EXAMPLE 4.11

In the circuit of Fig. 4.7a, the switch is in the closed position for a long time before $t = 0$, when it is opened instantaneously. Find the inductor current $y(t)$ for $t \geq 0$.

When the switch is in the closed position (for a long time), the inductor current is 2 amperes and the capacitor voltage is 10 volts. When the switch is opened, the circuit is equivalent to that depicted in Fig. 4.7b, with the initial inductor current $y(0^-) = 2$ and the initial capacitor voltage $v_C(0^-) = 10$. The input voltage is 10 volts, starting at $t = 0$, and, therefore, can be represented by $10u(t)$.

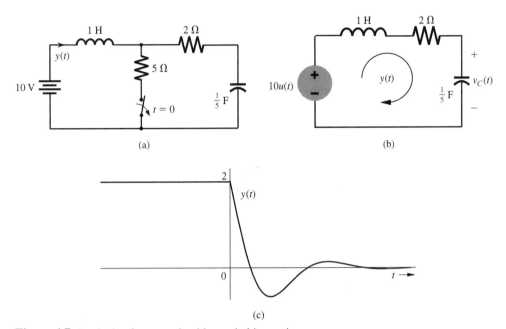

Figure 4.7 Analysis of a network with a switching action.

The loop equation of the circuit in Fig. 4.7b is

$$\frac{dy}{dt} + 2y(t) + 5 \int_{-\infty}^{t} y(\tau)\, d\tau = 10u(t) \tag{4.38}$$

If

$$y(t) \Longleftrightarrow Y(s) \tag{4.39a}$$

then

$$\frac{dy}{dt} \Longleftrightarrow sY(s) - y(0^-) = sY(s) - 2 \tag{4.39b}$$

and [see Eq. (4.26)]

$$\int_{-\infty}^{t} y(\tau)\, d\tau \Longleftrightarrow \frac{Y(s)}{s} + \frac{\int_{-\infty}^{0^-} y(\tau)\, d\tau}{s} \tag{4.39c}$$

Because $y(t)$ is the capacitor current, the integral $\int_{-\infty}^{0^-} y(\tau)\, d\tau$ is $q_C(0^-)$, the capacitor charge at $t = 0^-$, which is given by C times the capacitor voltage at $t = 0^-$. Therefore

$$\int_{-\infty}^{0^-} y(\tau)\, d\tau = q_C(0^-) = Cv_C(0^-) = \tfrac{1}{5}(10) = 2$$

From Eq. (4.39c) it follows that

$$\int_{-\infty}^{t} y(\tau)\, d\tau \Longleftrightarrow \frac{Y(s)}{s} + \frac{2}{s} \tag{4.40}$$

Taking the Laplace transform of Eq. (4.38) and using Eqs. (4.39a), (4.39b), and (4.40), we obtain

$$sY(s) - 2 + 2Y(s) + \frac{5Y(s)}{s} + \frac{10}{s} = \frac{10}{s}$$

or

$$\left[s + 2 + \frac{5}{s} \right] Y(s) = 2$$

and

$$Y(s) = \frac{2s}{s^2 + 2s + 5}$$

To find the inverse Laplace transform of $Y(s)$, we use pair 10c (Table 4.1) with values $A = 2$, $B = 0$, $a = 1$, and $c = 5$. This yields

$$r = \sqrt{\tfrac{20}{4}} = \sqrt{5}, \quad b = \sqrt{c - a^2} = 2 \quad \text{and} \quad \theta = \tan^{-1}\left(\tfrac{2}{4}\right) = 26.6°$$

Therefore

$$y(t) = \sqrt{5}e^{-t} \cos\left(2t + 26.6°\right)u(t)$$

This response is shown in Fig. 4.7c.

Comment. In our discussion so far, we have multiplied input signals by $u(t)$, implying that the signals are zero prior to $t = 0$. This is needlessly restrictive. These signals can have any arbitrary value prior to $t = 0$. As long as the initial conditions at $t = 0$ are specified, we need only the knowledge of the input for $t \geq 0$ to compute the response for $t \geq 0$. Some authors use the notation $1(t)$ to denote a function that is equal to $u(t)$ for $t \geq 0$ and that has arbitrary value for negative t. We have abstained from this usage to avoid needless confusion caused by introduction of a new function, which is very similar to $u(t)$.

4.3-1 Zero-State Response

Consider an Nth-order LTIC system specified by the equation

$$Q(D)y(t) = P(D)x(t)$$

or

$$(D^N + a_1 D^{N-1} + \cdots + a_{N-1}D + a_N)y(t) = (b_0 D^N + b_1 D^{N-1} + \cdots + b_{N-1}D + b_N)x(t) \quad (4.41)$$

We shall now find the general expression for the zero-state response of an LTIC system. Zero-state response $y(t)$, by definition, is the system response to an input when the system is initially relaxed (in zero state). Therefore, $y(t)$ satisfies the system equation (4.41) with zero initial conditions

$$y(0^-) = \dot{y}(0^-) = \ddot{y}(0^-) = \cdots = y^{(N-1)}(0^-) = 0$$

Moreover, the input $x(t)$ is causal, so that

$$x(0^-) = \dot{x}(0^-) = \ddot{x}(0^-) = \cdots = x^{(N-1)}(0^-) = 0$$

Let

$$y(t) \Longleftrightarrow Y(s) \qquad \text{and} \qquad x(t) \Longleftrightarrow X(s)$$

Because of zero initial conditions

$$D^r y(t) = \frac{d^r}{dt^r} y(t) \Longleftrightarrow s^r Y(s)$$

$$D^k x(t) = \frac{d^k}{dt^k} x(t) \Longleftrightarrow s^k X(s)$$

Therefore, the Laplace transform of Eq. (4.41) yields

$$(s^N + a_1 s^{N-1} + \cdots + a_{N-1} s + a_N) Y(s) = (b_0 s^N + b_1 s^{N-1} + \cdots + b_{N-1} s + b_N) X(s)$$

or

$$Y(s) = \frac{b_0 s^N + b_1 s^{N-1} + \cdots + b_{N-1} s + b_N}{s^N + a_1 s^{N-1} + \cdots + a_{N-1} s + a_N} X(s) \tag{4.42a}$$

$$= \frac{P(s)}{Q(s)} X(s) \tag{4.42b}$$

But we have shown in Eq. (4.31) that $Y(s) = H(s)X(s)$. Consequently,

$$H(s) = \frac{P(s)}{Q(s)} \tag{4.43}$$

This is the transfer function of a linear differential system specified in Eq. (4.41). The same result has been derived earlier in Eq. (2.50) using an alternate (time-domain) approach.

We have shown that $Y(s)$, the Laplace transform of the zero-state response $y(t)$, is the product of $X(s)$ and $H(s)$, where $X(s)$ is the Laplace transform of the input $x(t)$ and $H(s)$ is the system transfer function [relating the particular output $y(t)$ to the input $x(t)$].

INTUITIVE INTERPRETATION OF THE LAPLACE TRANSFORM

So far we have treated the Laplace transform as a machine, which converts linear integro-differential equations into algebraic equations. There is no physical understanding of how this is accomplished or what it means. We now discuss a more intuitive interpretation and meaning of the Laplace transform.

In Chapter 2, Eq. (2.47), we showed that LTI system response to an everlasting exponential e^{st} is $H(s)e^{st}$. If we could express every signal as a linear combination of everlasting exponentials of the form e^{st}, we could readily obtain the system response to any input. For example, if

$$x(t) = \sum_{k=1}^{K} X(s_i) e^{s_i t}$$

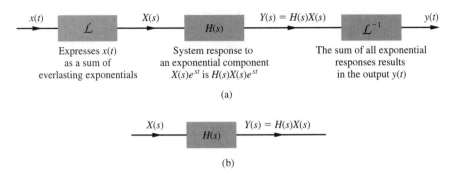

Figure 4.8 Alternate interpretation of the Laplace transform.

The response of an LTIC system to such input $x(t)$ is given by

$$y(t) = \sum_{k=1}^{K} X(s_i)H(s_i)e^{s_i t}$$

Unfortunately, a very small class of signals can be expressed in this form. However, we can express almost all signals of practical utility as a sum of everlasting exponentials over a continuum of frequencies. This is precisely what the Laplace transform in Eq. (4.2) does.

$$x(t) = \frac{1}{2\pi j} \int_{c-j\infty}^{c+j\infty} X(s)e^{st}\, ds \qquad (4.44)$$

Invoking the linearity property of the Laplace transform, we can find the system response $y(t)$ to input $x(t)$ in Eq. (4.44) as[†]

$$y(t) = \frac{1}{2\pi j} \int_{c'-j\infty}^{c'+j\infty} X(s)H(s)e^{st}\, ds$$

$$= \mathcal{L}^{-1}X(s)H(s) \qquad (4.45)$$

Clearly

$$Y(s) = X(s)H(s)$$

We can now represent the transformed version of the system, as depicted in Fig. 4.8a. The input $X(s)$ is the Laplace transform of $x(t)$, and the output $Y(s)$ is the Laplace transform of (the zero-input response) $y(t)$. The system is described by the transfer function $H(s)$. The output $Y(s)$ is the product $X(s)H(s)$.

Recall that s is the complex frequency of e^{st}. This explains why the Laplace transform method is also called the *frequency-domain* method. Note that $X(s)$, $Y(s)$, and $H(s)$ are the frequency-domain representations of $x(t)$, $y(t)$, and $h(t)$, respectively. We may view the boxes marked $\mathcal{L}$ and $\mathcal{L}^{-1}$ in Fig. 4.8a as the interfaces that convert the time-domain entities into the

---

[†]Recall that $H(s)$ has its own region of validity. Hence, the limits of integration for the integral in Eq. (4.44) are modified in Eq. (4.45) to accommodate the region of existence (validity) of $X(s)$ as well as $H(s)$.

corresponding frequency-domain entities, and vice versa. All real-life signals begin in the time domain, and the final answers must also be in the time domain. First, we convert the time-domain input(s) into the frequency-domain counterparts. The problem itself is solved in the frequency domain, resulting in the answer $Y(s)$, also in the frequency domain. Finally, we convert $Y(s)$ to $y(t)$. Solving the problem is relatively simpler in the frequency domain than in the time domain. Henceforth, we shall omit the explicit representation of the interface boxes $\mathcal{L}$ and $\mathcal{L}^{-1}$, representing signals and systems in the frequency domain, as shown in Fig. 4.8b.

THE DOMINANCE CONDITION

In this intuitive interpretation of the Laplace transform, one problem should have puzzled the reader. In Section 2.5 (classical solution of differential equations), we showed in Eq. (2.57) that an LTI system response to input e^{st} is $H(s)\,e^{st}$ plus characteristic mode terms. In the intuitive interpretation, an LTI system response is found by adding the system responses to all the infinite exponential components of the input. These components are exponentials of the form e^{st} starting at $t = -\infty$. We showed in Eq. (2.47) that the response to everlasting input e^{st} is also an everlasting exponential $H(s)\,e^{st}$. Does this result not conflict with the result in Eq. (2.57)? Why are there no characteristic mode terms in Eq. (2.47), as predicted by Eq. (2.57)? The answer is that the mode terms *are* also present. The system response to an everlasting input e^{st} is indeed an everlasting exponential $H(s)\,e^{st}$ plus mode terms. All these signals start at $t = -\infty$. Now, if a mode $e^{\lambda_i t}$ is such that it decays faster than (or grows slower than) e^{st}, that is, if Re λ_i < Re s, then after some time interval, e^{st} will be overwhelmingly stronger than $e^{\lambda_i t}$, and hence will completely dominate such a mode term. In such a case, at any finite time (which is long time after the start at $t = -\infty$), we can ignore the mode terms and say that the complete response is $H(s)\,e^{st}$. Hence, we can reconcile Eq. (2.47) to Eq. (2.57) only if the *dominance condition* is satisfied, that is, if Re λ_i < Re s for all i. If the dominance condition is not satisfied, the mode terms dominate e^{st} and Eq. (2.47) does not hold.[10]

Careful examination shows that the dominance condition is implied in Eq. (2.47). This is because of the caviat in Eq. (2.47) that the response of an LTIC system to everlasting e^{st} is $H(s)\,e^{st}$, provided $H(s)$ exists (or converges). We can show that this condition amounts to dominance condition. If a system has characteristic roots $\lambda_1, \lambda_2, \ldots, \lambda_N$, then $h(t)$ consists of exponentials of the form $e^{\lambda_i t}$ ($i = 1, 2, \ldots, N$) and the convergence of $H(s)$ requires that Re s > Re λ_i for $i = 1, 2, \ldots, N$, which precisely is the dominance condition. Clearly, the dominance condition is implied in Eq. (2.47), and also in the entire fabric of the Laplace transform. It is interesting to note that the elegant structure of convergence in Laplace transform is rooted in such a lowly, mundane origin, as Eq. (2.57).

EXAMPLE 4.12

Find the response $y(t)$ of an LTIC system described by the equation

$$\frac{d^2 y}{dt^2} + 5\frac{dy}{dt} + 6y(t) = \frac{dx}{dt} + x(t)$$

if the input $x(t) = 3e^{-5t}u(t)$ and all the initial conditions are zero; that is, the system is in the zero state.

The system equation is

$$\underbrace{(D^2 + 5D + 6)}_{Q(D)} y(t) = \underbrace{(D + 1)}_{P(D)} x(t)$$

Therefore

$$H(s) = \frac{P(s)}{Q(s)} = \frac{s + 1}{s^2 + 5s + 6}$$

Also

$$X(s) = \mathcal{L}[3e^{-5t}u(t)] = \frac{3}{s + 5}$$

and

$$Y(s) = X(s)H(s) = \frac{3(s + 1)}{(s + 5)(s^2 + 5s + 6)}$$

$$= \frac{3(s + 1)}{(s + 5)(s + 2)(s + 3)}$$

$$= \frac{-2}{s + 5} - \frac{1}{s + 2} + \frac{3}{s + 3}$$

The inverse Laplace transform of this equation is

$$y(t) = (-2e^{-5t} - e^{-2t} + 3e^{-3t})u(t)$$

EXAMPLE 4.13

Show that the transfer function of

(a) an ideal delay of T seconds is e^{-sT}

(b) an ideal differentiator is s

(c) an ideal integrator is $1/s$

(a) **Ideal Delay.** For an ideal delay of T seconds, the input $x(t)$ and output $y(t)$ are related by

$$y(t) = x(t - T)$$

or

$$Y(s) = X(s)e^{-sT} \qquad \text{[see Eq. (4.19a)]}$$

Therefore

$$H(s) = \frac{Y(s)}{X(s)} = e^{-sT} \qquad (4.46)$$

(b) Ideal Differentiator. For an ideal differentiator, the input $x(t)$ and the output $y(t)$ are related by

$$y(t) = \frac{dx}{dt}$$

The Laplace transform of this equation yields

$$Y(s) = sX(s) \qquad [x(0^-) = 0 \text{ for a causal signal}]$$

and

$$H(s) = \frac{Y(s)}{X(s)} = s \qquad (4.47)$$

(c) Ideal Integrator. For an ideal integrator with zero initial state, that is, $y(0^-) = 0$,

$$y(t) = \int_0^t x(\tau)\, d\tau$$

and

$$Y(s) = \frac{1}{s} X(s)$$

Therefore

$$H(s) = \frac{1}{s} \qquad (4.48)$$

EXERCISE E4.7

For an LTIC system with transfer function

$$H(s) = \frac{s+5}{s^2 + 4s + 3}$$

(a) Describe the differential equation relating the input $x(t)$ and output $y(t)$.

(b) Find the system response $y(t)$ to the input $x(t) = e^{-2t}u(t)$ if the system is initially in zero state.

ANSWERS

(a) $\dfrac{d^2 y}{dt^2} + 4\dfrac{dy}{dt} + 3y(t) = \dfrac{dx}{dt} + 5x(t)$

(b) $y(t) = (2e^{-t} - 3e^{-2t} + e^{-3t})u(t)$

4.3-2 Stability

Equation (4.43) shows that the denominator of $H(s)$ is $Q(s)$, which is apparently identical to the characteristic polynomial $Q(\lambda)$ defined in Chapter 2. Does this mean that the denominator of $H(s)$ is the characteristic polynomial of the system? This may or may not be the case, since if $P(s)$

Beware of the RHP poles!

and $Q(s)$ in Eq. (4.43) have any common factors, they cancel out, and the effective denominator of $H(s)$ is not necessarily equal to $Q(s)$. Recall also that the system transfer function $H(s)$, like $h(t)$, is defined in terms of measurements at the external terminals. Consequently, $H(s)$ and $h(t)$ are both external descriptions of the system. In contrast, the characteristic polynomial $Q(s)$ is an internal description. Clearly, we can determine only external stability, that is, BIBO stability, from $H(s)$. If all the poles of $H(s)$ are in LHP, all the terms in $h(t)$ are decaying exponentials, and $h(t)$ is absolutely integrable [see Eq. (2.64)].[†] Consequently, the system is BIBO stable. Otherwise the system is BIBO unstable.

So far, we have assumed that $H(s)$ is a proper function, that is, $M \le N$. We now show that if $H(s)$ is improper, that is, if $M > N$, the system is BIBO unstable. In such a case, using long division, we obtain $H(s) = R(s) + H'(s)$, where $R(s)$ is an $(M - N)$th-order polynomial and $H'(s)$ is a proper transfer function. For example,

$$H(s) = \frac{s^3 + 4s^2 + 4s + 5}{s^2 + 3s + 2} = s + \frac{s^2 + 2s + 5}{s^2 + 3s + 2} \tag{4.49}$$

As shown in Eq. (4.47), the term s is the transfer function of an ideal differentiator. If we apply step function (bounded input) to this system, the output will contain an impulse (unbounded output). Clearly, the system is BIBO unstable. Moreover, such a system greatly amplifies noise because differentiation enhances higher frequencies, which generally predominate in a noise signal. These are two good reasons to avoid improper systems ($M > N$). In our future discussion, we shall implicitly assume that the systems are proper, unless stated otherwise.

If $P(s)$ and $Q(s)$ do not have common factors, then the denominator of $H(s)$ is identical to $Q(s)$, the characteristic polynomial of the system. In this case, we can determine internal stability by using the criterion described in Section 2.6. Thus, if $P(s)$ and $Q(s)$ have no common factors,

---

[†]Values of s for which $H(s)$ is ∞ are the *poles* of $H(s)$. Thus poles of $H(s)$ are the values of s for which the denominator of $H(s)$ is zero.

the asymptotic stability criterion in Section 2.6 can be restated in terms of the poles of the transfer function of a system, as follows:

1. An LTIC system is asymptotically stable if and only if all the poles of its transfer function $H(s)$ are in the LHP. The poles may be simple or repeated.

2. An LTIC system is unstable if and only if either one or both of the following conditions exist: (i) at least one pole of $H(s)$ is in the RHP; (ii) there are repeated poles of $H(s)$ on the imaginary axis.

3. An LTIC system is marginally stable if and only if there are no poles of $H(s)$ in the RHP and some unrepeated poles on the imaginary axis.

Location of zeros of $H(s)$ have no role in determining the system stability.

EXAMPLE 4.14

Figure 4.9a shows a cascade connection of two LTIC systems S_1 followed by S_2. The transfer functions of these systems are $H_1(s) = 1/(s - 1)$ and $H_2(s) = (s - 1)/(s + 1)$, respectively. We shall find the BIBO and asymptotic stability of the composite (cascade) system.

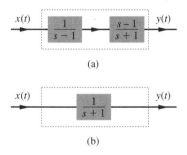

(a)

(b)

Figure 4.9 Distinction between BIBO and asymptotic stability.

If the impulse response of S_1 and S_2 are $h_1(t)$ and $h_2(t)$, respectively, then the impulse response of the cascade system is $h(t) = h_1(t) * h_2(t)$. Hence, $H(s) = H_1(s)H_2(s)$. In the present case,

$$H(s) = \left(\frac{1}{s-1}\right)\left(\frac{s-1}{s+1}\right) = \frac{1}{s+1}$$

The pole of S_1 at $s = 1$ cancels with the zero at $s = 1$ of S_2. This results in a composite system having a single pole at $s = -1$. If the composite cascade system were to be enclosed inside a black box with only the input and the output terminals accessible, any measurement from these external terminals would show that the transfer function of the system is $1/(s+1)$, without any hint of the fact that the system is housing an unstable system (Fig. 4.9b). The impulse response of the cascade system is $h(t) = e^{-t}u(t)$, which is absolutely integrable. Consequently, the system is BIBO stable.

To determine the asymptotic stability, we note that $\mathcal{S}_1$ has one characteristic root at 1, and $\mathcal{S}_2$ also has one root at -1. Recall that the two systems are independent (one does not load the other), and the characteristic modes generated in each subsystem are independent of the other. Clearly, the mode e^t will not be eliminated by the presence of $\mathcal{S}_2$. Hence, the composite system has two characteristic roots, located at ± 1, and the system is asymptotically unstable, though BIBO stable.

Interchanging the positions of $\mathcal{S}_1$ and $\mathcal{S}_2$ makes no difference in this conclusion. This example shows that BIBO stability can be misleading. If a system is asymptotically unstable, it will destroy itself (or, more likely, lead to saturation condition) because of unchecked growth of the response due to intended or unintended stray initial conditions. BIBO stability is not going to save the system. Control systems are often compensated to realize certain desirable characteristics. One should never try to stabilize an unstable system by canceling its RHP pole(s) by RHP zero(s). Such a misguided attempt will fail, not because of the practical impossibility of exact cancellation but for the more fundamental reason, as just explained.

EXERCISE E4.8

Show that an ideal integrator is marginally stable but BIBO unstable.

4.3-3 Inverse Systems

If $H(s)$ is the transfer function of a system $\mathcal{S}$, then $\mathcal{S}_i$, its inverse system has a transfer function $H_i(s)$ given by

$$H_i(s) = \frac{1}{H(s)}$$

This follows from the fact the cascade of $\mathcal{S}$ with its inverse system $\mathcal{S}_i$ is an identity system, with impulse response $\delta(t)$, implying $H(s)H_i(s) = 1$. For example, an ideal integrator and its inverse, an ideal differentiator, have transfer functions $1/s$ and s, respectively, leading to $H(s)H_i(s) = 1$.

4.4 ANALYSIS OF ELECTRICAL NETWORKS: THE TRANSFORMED NETWORK

Example 4.10 shows how electrical networks may be analyzed by writing the integro-differential equation(s) of the system and then solving these equations by the Laplace transform. We now show that it is also possible to analyze electrical networks directly without having to write the integro-differential equations. This procedure is considerably simpler because it permits us to treat an electrical network as if it were a resistive network. For this purpose, we need to represent a network in the "frequency domain" where all the voltages and currents are represented by their Laplace transforms.

For the sake of simplicity, let us first discuss the case with zero initial conditions. If $v(t)$ and $i(t)$ are the voltage across and the current through an inductor of L henries, then

$$v(t) = L\frac{di}{dt}$$

The Laplace transform of this equation (assuming zero initial current) is

$$V(s) = LsI(s)$$

Similarly, for a capacitor of C farads, the voltage–current relationship is $i(t) = C(dv/dt)$ and its Laplace transform, assuming zero initial capacitor voltage, yields $I(s) = CsV(s)$; that is,

$$V(s) = \frac{1}{Cs}I(s)$$

For a resistor of R ohms, the voltage–current relationship is $v(t) = Ri(t)$, and its Laplace transform is

$$V(s) = RI(s)$$

Thus, in the "frequency domain," the voltage–current relationships of an inductor and a capacitor are algebraic; these elements behave like resistors of "resistance" Ls and $1/Cs$, respectively. The generalized "resistance" of an element is called its *impedance* and is given by the ratio $V(s)/I(s)$ for the element (under zero initial conditions). The impedances of a resistor of R ohms, an inductor of L henries, and a capacitance of C farads are R, Ls, and $1/Cs$, respectively.

Also, the interconnection constraints (Kirchhoff's laws) remain valid for voltages and currents in the frequency domain. To demonstrate this point, let $v_j(t)$ $(j = 1, 2, \ldots, k)$ be the voltages across k elements in a loop and let $i_j(t)(j = 1, 2, \ldots, m)$ be the j currents entering a node. Then

$$\sum_{j=1}^{k} v_j(t) = 0 \qquad \text{and} \qquad \sum_{j=1}^{m} i_j(t) = 0$$

Now if

$$v_j(t) \Longleftrightarrow V_j(s) \qquad \text{and} \qquad i_j(t) \Longleftrightarrow I_j(s)$$

then

$$\sum_{j=1}^{k} V_j(s) = 0 \qquad \text{and} \qquad \sum_{j=1}^{m} I_j(s) = 0 \qquad (4.50)$$

This result shows that if we represent all the voltages and currents in an electrical network by their Laplace transforms, we can treat the network as if it consisted of the "resistances" R, Ls, and $1/Cs$ corresponding to a resistor R, an inductor L, and a capacitor C, respectively. The system equations (loop or node) are now algebraic. Moreover, the simplification techniques that have been developed for resistive circuits—equivalent series and parallel impedances, voltage and current divider rules, Thévenin and Norton theorems—can be applied to general electrical networks. The following examples demonstrate these concepts.

EXAMPLE 4.15

Find the loop current $i(t)$ in the circuit shown in Fig. 4.10a if all the initial conditions are zero.

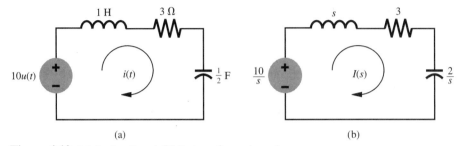

(a)   (b)

Figure 4.10 (a) A circuit and (b) its transformed version.

In the first step, we represent the circuit in the frequency domain, as illustrated in Fig. 4.10b. All the voltages and currents are represented by their Laplace transforms. The voltage $10u(t)$ is represented by $10/s$ and the (unknown) current $i(t)$ is represented by its Laplace transform $I(s)$. All the circuit elements are represented by their respective impedances. The inductor of 1 henry is represented by s, the capacitor of $1/2$ farad is represented by $2/s$, and the resistor of 3 ohms is represented by 3. We now consider the frequency-domain representation of voltages and currents. The voltage across any element is $I(s)$ times its impedance. Therefore, the total voltage drop in the loop is $I(s)$ times the total loop impedance, and it must be equal to $V(s)$, (transform of) the input voltage. The total impedance in the loop is

$$Z(s) = s + 3 + \frac{2}{s} = \frac{s^2 + 3s + 2}{s}$$

The input "voltage" is $V(s) = 10/s$. Therefore, the "loop current" $I(s)$ is

$$I(s) = \frac{V(s)}{Z(s)} = \frac{10/s}{(s^2 + 3s + 2)/s} = \frac{10}{s^2 + 3s + 2} = \frac{10}{(s+1)(s+2)} = \frac{10}{s+1} - \frac{10}{s+2}$$

The inverse transform of this equation yields the desired result:

$$i(t) = 10(e^{-t} - e^{-2t})u(t)$$

INITIAL CONDITION GENERATORS

The discussion in which we assumed zero initial conditions can be readily extended to the case of nonzero initial conditions because the initial condition in a capacitor or an inductor can be represented by an equivalent source. We now show that a capacitor C with an initial voltage $v(0^-)$ (Fig. 4.11a) can be represented in the frequency domain by an uncharged capacitor of impedance $1/Cs$ in series with a voltage source of value $v(0^-)/s$ (Fig. 4.11b) or as the same uncharged capacitor in parallel with a current source of value $Cv(0^-)$ (Fig. 4.11c). Similarly, an inductor L with an initial current $i(0^-)$ (Fig. 4.11d) can be represented in the frequency domain

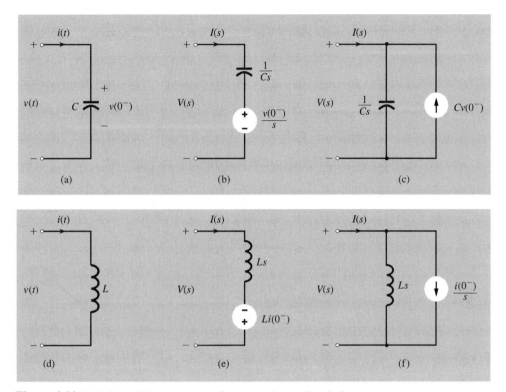

Figure 4.11 Initial condition generators for a capacitor and an inductor.

by an inductor of impedance Ls in series with a voltage source of value $Li(0^-)$ (Fig. 4.11e) or by the same inductor in parallel with a current source of value $i(0^-)/s$ (Fig. 4.11f). To prove this point, consider the terminal relationship of the capacitor in Fig. 4.11a

$$i(t) = C\frac{dv}{dt}$$

The Laplace transform of this equation yields

$$I(s) = C[sV(s) - v(0^-)]$$

This equation can be rearranged as

$$V(s) = \frac{1}{Cs}I(s) + \frac{v(0^-)}{s} \tag{4.51a}$$

Observe that $V(s)$ is the voltage (in the frequency domain) across the charged capacitor and $I(s)/Cs$ is the voltage across the same capacitor without any charge. Therefore, the charged capacitor can be represented by the uncharged capacitor in series with a voltage source of value $v(0^-)/s$, as depicted in Fig. 4.11b. Equation (4.51a) can also be rearranged as

$$V(s) = \frac{1}{Cs}[I(s) + Cv(0^-)] \tag{4.51b}$$

This equation shows that the charged capacitor voltage $V(s)$ is equal to the uncharged capacitor

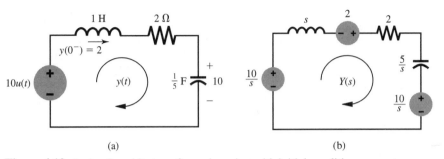

(a)   (b)

Figure 4.12 A circuit and its transformed version with initial condition generators.

voltage caused by a current $I(s) + Cv(0^-)$. This result is reflected precisely in Fig. 4.11c, where the current through the uncharged capacitor is $I(s) + Cv(0^-)$.[†]

For the inductor in Fig. 4.11d, the terminal equation is

$$v(t) = L\frac{di}{dt}$$

and

$$V(s) = L[sI(s) - i(0^-)]$$

$$= LsI(s) - Li(0^-) \tag{4.52a}$$

$$= Ls\left[I(s) - \frac{i(0^-)}{s}\right] \tag{4.52b}$$

We can verify that Fig. 4.11e satisfies Eq. (4.52a) and that Fig. 4.11f satisfies Eq. (4.52b).

Let us rework Example 4.11 using these concepts. Figure 4.12a shows the circuit in Fig. 4.7b with the initial conditions $y(0^-) = 2$ and $v_C(0^-) = 10$. Figure 4.12b shows the frequency-domain representation (transformed circuit) of the circuit in Fig. 4.12a. The resistor is represented by its impedance 2; the inductor with initial current of 2 amperes is represented according to the arrangement in Fig. 4.11e with a series voltage source $Ly(0^-) = 2$. The capacitor with initial voltage of 10 volts is represented according to the arrangement in Fig. 4.11b with a series voltage source $v(0^-)/s = 10/s$. Note that the impedance of the inductor is s and that of the capacitor is $5/s$. The input of $10u(t)$ is represented by its Laplace transform $10/s$.

The total voltage in the loop is $(10/s) + 2 - (10/s) = 2$, and the loop impedance is $(s + 2 + (5/s))$. Therefore

$$Y(s) = \frac{2}{s + 2 + 5/s}$$

$$= \frac{2s}{s^2 + 2s + 5}$$

which confirms our earlier result in Example 4.11.

[†]In the time domain, a charged capacitor C with initial voltage $v(0^-)$ can be represented as the same capacitor uncharged in series with a voltage source $v(0^-)u(t)$, or in parallel with a current source $Cv(0^-)\delta(t)$. Similarly, an inductor L with initial current $i(0^-)$ can be represented by the same inductor with zero initial current in series with a voltage source $Li(0^-)\delta(t)$ or with a parallel current source $i(0^-)u(t)$.

EXAMPLE 4.16

The switch in the circuit of Fig. 4.13a is in the closed position for a long time before $t = 0$, when it is opened instantaneously. Find the currents $y_1(t)$ and $y_2(t)$ for $t \geq 0$.

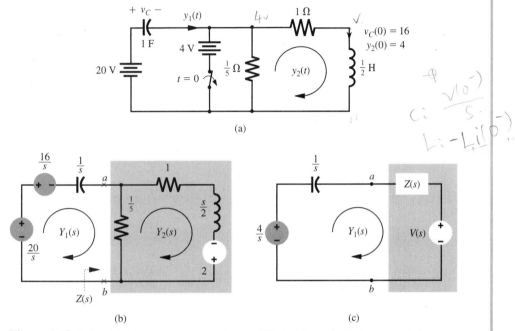

(a)

(b)   (c)

Figure 4.13 Using initial condition generators and Thévenin equivalent representation.

Inspection of this circuit shows that when the switch is closed and the steady-state conditions are reached, the capacitor voltage $v_C = 16$ volts, and the inductor current $y_2 = 4$ amperes. Therefore, when the switch is opened (at $t = 0$), the initial conditions are $v_C(0^-) = 16$ and $y_2(0^-) = 4$. Figure 4.13b shows the transformed version of the circuit in Fig. 4.13a. We have used equivalent sources to account for the initial conditions. The initial capacitor voltage of 16 volts is represented by a series voltage of $16/s$ and the initial inductor current of 4 amperes is represented by a source of value $Ly_2(0^-) = 2$.

From Fig. 4.13b, the loop equations can be written directly in the frequency domain as

$$\frac{Y_1(s)}{s} + \frac{1}{5}[Y_1(s) - Y_2(s)] = \frac{4}{s}$$

$$-\frac{1}{5}Y_1(s) + \frac{6}{5}Y_2(s) + \frac{s}{2}Y_2(s) = 2$$

$$\begin{bmatrix} \frac{1}{s} + \frac{1}{5} & -\frac{1}{5} \\ -\frac{1}{5} & \frac{6}{5} + \frac{s}{2} \end{bmatrix} \begin{bmatrix} Y_1(s) \\ Y_2(s) \end{bmatrix} = \begin{bmatrix} \frac{4}{s} \\ 2 \end{bmatrix}$$

Application of Cramer's rule to this equation yields

$$Y_1(s) = \frac{24(s+2)}{s^2 + 7s + 12}$$

$$= \frac{24(s+2)}{(s+3)(s+4)}$$

$$= \frac{-24}{s+3} + \frac{48}{s+4}$$

and

$$y_1(t) = (-24e^{-3t} + 48e^{-4t})u(t)$$

Similarly, we obtain

$$Y_2(s) = \frac{4(s+7)}{s^2 + 7s + 12}$$

$$= \frac{16}{s+3} - \frac{12}{s+4}$$

and

$$y_2(t) = (16e^{-3t} - 12e^{-4t})u(t)$$

We also could have used Thévenin's theorem to compute $Y_1(s)$ and $Y_2(s)$ by replacing the circuit to the right of the capacitor (right of terminals ab) with its Thévenin equivalent, as shown in Fig. 4.13c. Figure 4.13b shows that the Thévenin impedance $Z(s)$ and the Thévenin source $V(s)$ are

$$Z(s) = \frac{\frac{1}{5}\left(\frac{s}{2}+1\right)}{\frac{1}{5}+\frac{s}{2}+1} = \frac{s+2}{5s+12}$$

$$V(s) = \frac{-\frac{1}{5}}{\frac{1}{5}+\frac{s}{2}+1}\cdot 2 = \frac{-4}{5s+12}$$

According to Fig. 4.13c, the current $Y_1(s)$ is given by

$$Y_1(s) = \frac{\frac{4}{s} - V(s)}{\frac{1}{s} + Z(s)} = \frac{24(s+2)}{s^2 + 7s + 12}$$

which confirms the earlier result. We may determine $Y_2(s)$ in a similar manner.

EXAMPLE 4.17

The switch in the circuit in Fig. 4.14a is at position a for a long time before $t = 0$, when it is moved instantaneously to position b. Determine the current $y_1(t)$ and the output voltage $v_0(t)$ for $t \geq 0$.

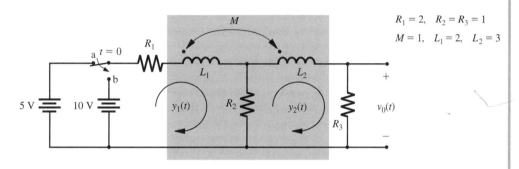

$R_1 = 2, \quad R_2 = R_3 = 1$
$M = 1, \quad L_1 = 2, \quad L_2 = 3$

(a)

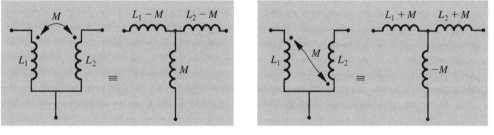

(b)   (c)

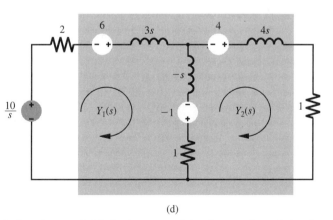

(d)

Figure 4.14 Solution of a coupled inductive network by the transformed circuit method.

Just before switching, the values of the loop currents are 2 and 1, respectively, that is, $y_1(0^-) = 2$ and $y_2(0^-) = 1$.

The equivalent circuits for two types of inductive coupling are illustrated in Fig. 4.14b and 4.14c. For our situation, the circuit in Fig. 4.14c applies. Figure 4.14d shows the transformed version of the circuit in Fig. 4.14a after switching. Note that the inductors $L_1 + M$, $L_2 + M$, and $-M$ are 3, 4, and -1 henries with impedances $3s$, $4s$, and $-s$ respectively. The initial condition voltages in the three branches are $(L_1 + M)y_1(0^-) = 6$, $(L_2 + M)y_2(0^-) = 4$, and $-M[y_1(0^-) - y_2(0^-)] = -1$, respectively. The two loop equations of the circuit are[†]

$$(2s + 3)Y_1(s) + (s - 1)Y_2(s) = \frac{10}{s} + 5$$

$$(s - 1)Y_1(s) + (3s + 2)Y_2(s) = 5$$

(4.53)

or

$$\begin{bmatrix} 2s + 3 & s - 1 \\ s - 1 & 3s + 2 \end{bmatrix} \begin{bmatrix} Y_1(s) \\ Y_2(s) \end{bmatrix} \begin{bmatrix} \frac{5s+10}{s} \\ 5 \end{bmatrix}$$

and

$$Y_1(s) = \frac{2s^2 + 9s + 4}{s(s^2 + 3s + 1)}$$

$$= \frac{4}{s} - \frac{1}{s + 0.382} - \frac{1}{s + 2.618}$$

Therefore

$$y_1(t) = (4 - e^{-0.382t} - e^{-2.618t})u(t)$$

Similarly

$$Y_2(s) = \frac{s^2 + 2s + 2}{s(s^2 + 3s + 1)}$$

$$= \frac{2}{s} - \frac{1.618}{s + 0.382} + \frac{0.618}{s + 2.618}$$

and

$$y_2(t) = (2 - 1.618e^{-0.382t} + 0.618e^{-2.618t})u(t)$$

The output voltage

$$v_0(t) = y_2(t) = (2 - 1.618e^{-0.382t} + 0.618e^{-2.618t})u(t)$$

[†]The time domain equations (loop equations) are

$$L_1\frac{dy_1}{dt} + (R_1 + R_2)y_1(t) - R_2y_2(t) + M\frac{dy_2}{dt} = 10u(t)$$

$$M\frac{dy_1}{dt} - R_2y_1(t) + L_2\frac{dy_2}{dt} + (R_2 + R_3)y_2(t) = 0$$

The Laplace transform of these equations yields Eq. (4.53).

EXERCISE E4.9

For the *RLC* circuit in Fig. 4.15, the input is switched on at $t = 0$. The initial conditions are $y(0^-) = 2$ amperes and $v_C(0^-) = 50$ volts. Find the loop current $y(t)$ and the capacitor voltage $v_C(t)$ for $t \geq 0$.

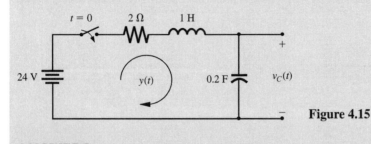

Figure 4.15

ANSWERS

$$y(t) = 10\sqrt{2}e^{-t}\cos{(2t + 81.8°)}u(t)$$

$$v_C(t) = [24 + 31.62e^{-t}\cos{(2t - 34.7°)}]u(t)$$

4.4-1 Analysis of Active Circuits

Although we have considered examples of only passive networks so far, the circuit analysis procedure using the Laplace transform is also applicable to active circuits. All that is needed is to replace the active elements with their mathematical models (or equivalent circuits) and proceed as before.

The operational amplifier (depicted by the triangular symbol in Fig. 4.16a) is a well-known element in modern electronic circuits. The terminals with the positive and the negative signs correspond to noninverting and inverting terminals, respectively. This means that the polarity of the output voltage v_2 is the same as that of the input voltage at the terminal marked by the positive sign (noninverting). The opposite is true for the inverting terminal, marked by the negative sign.

Figure 4.16b shows the model (equivalent circuit) of the operational amplifier (op amp) in Fig. 4.16a. A typical op amp has a very large gain. The output voltage $v_2 = -Av_1$, where A is typically 10^5 to 10^6. The input impedance is very high, of the order of 10^{12} Ω and the output impedance is very low (50–100 Ω). For most applications, we are justified in assuming the gain A and the input impedance to be infinite and the output impedance to be zero. For this reason we see an ideal voltage source at the output.

Consider now the operational amplifier with resistors R_a and R_b connected, as shown in Fig. 4.16c. This configuration is known as the *noninverting amplifier*. Observe that the input polarities in this configuration are inverted in comparison to those in Fig. 4.16a. We now show that the output voltage v_2 and the input voltage v_1 in this case are related by

$$v_2 = Kv_1 \qquad K = 1 + \frac{R_b}{R_a} \tag{4.54}$$

First, we recognize that because the input impedance and the gain of the operational amplifier approach infinity, the input current i_x and the input voltage v_x in Fig. 4.16c are infinitesimal and

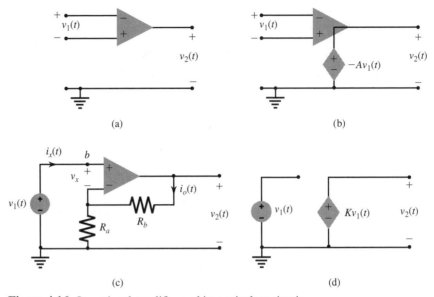

Figure 4.16 Operational amplifier and its equivalent circuit.

may be taken as zero. The dependent source in this case is Av_x instead of $-Av_x$ because of the input polarity inversion. The dependent source Av_x (see Fig. 4.16b) at the output will generate current i_o, as illustrated in Fig. 4.16c. Now

$$v_2 = (R_b + R_a)i_o$$

also

$$v_1 = v_x + R_a i_o$$

$$= R_a i_o$$

Therefore

$$\frac{v_2}{v_1} = \frac{R_b + R_a}{R_a} = 1 + \frac{R_b}{R_a} = K$$

or

$$v_2(t) = K v_1(t)$$

The equivalent circuit of the noninverting amplifier is depicted in Fig. 4.16d.

EXAMPLE 4.18

The circuit in Fig. 4.17a is called the *Sallen–Key* circuit, which is frequently used in filter design. Find the transfer function $H(s)$ relating the output voltage $v_o(t)$ to the input voltage $v_i(t)$.

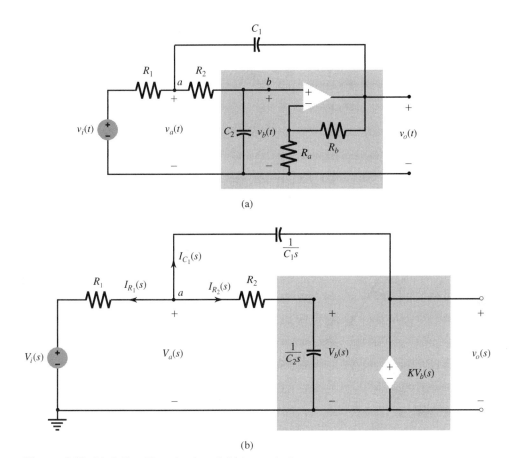

(a)

(b)

Figure 4.17 (a) Sallen–Key circuit and (b) its equivalent.

We are required to find

$$H(s) = \frac{V_o(s)}{V_i(s)}$$

assuming all initial conditions to be zero.

Figure 4.17b shows the transformed version of the circuit in Fig. 4.17a. The noninverting amplifier is replaced by its equivalent circuit. All the voltages are replaced by their Laplace transforms and all the circuit elements are shown by their impedances. All the initial conditions are assumed to be zero, as required for determining $H(s)$.

We shall use node analysis to derive the result. There are two unknown node voltages, $V_a(s)$ and $V_b(s)$, requiring two node equations.

At node a, $I_{R_1}(s)$, the current in R_1 (leaving the node a), is $[V_a(s) - V_i(s)]/R_1$. Similarly, $I_{R_2}(s)$, the current in R_2 (leaving the node a), is $[V_a(s) - V_b(s)]/R_2$, and $I_{C_1}(s)$, the current in capacitor C_1 (leaving the node a), is $[V_a(s) - V_o(s)]C_1s = [V_a(s) - K V_b(s)]C_1s$.

The sum of all the three currents is zero. Therefore

$$\frac{V_a(s) - V_i(s)}{R_1} + \frac{V_a(s) - V_b(s)}{R_2} + [V_a(s) - KV_b(s)]C_1 s = 0$$

or

$$\left(\frac{1}{R_1} + \frac{1}{R_2} + C_1 s\right) V_a(s) - \left(\frac{1}{R_2} + KC_1 s\right) V_b(s) = \frac{1}{R_1} V_i(s) \qquad (4.55a)$$

Similarly, the node equation at node b yields

$$\frac{V_b(s) - V_a(s)}{R_2} + C_2 s V_b(s) = 0$$

or

$$-\frac{1}{R_2} V_a(s) + \left(\frac{1}{R_2} + C_2 s\right) V_b(s) = 0 \qquad (4.55b)$$

The two node equations (4.55a) and (4.55b) in two unknown node voltages $V_a(s)$ and $V_b(s)$ can be expressed in matrix form as

$$\begin{bmatrix} G_1 + G_2 + C_1 s & -(G_2 + KC_1 s) \\ -G_2 & (G_2 + C_2 s) \end{bmatrix} \begin{bmatrix} V_a(s) \\ V_b(s) \end{bmatrix} = \begin{bmatrix} G_1 V_i(s) \\ 0 \end{bmatrix} \qquad (4.56)$$

where

$$G_1 = \frac{1}{R_1} \qquad \text{and} \qquad G_2 = \frac{1}{R_2}$$

Application of Cramer's rule to Eq. (4.56) yields

$$\frac{V_b(s)}{V_i(s)} = \frac{G_1 G_2}{C_1 C_2 s^2 + [G_1 C_2 + G_2 C_2 + G_2 C_1 (1 - K)]s + G_1 G_2}$$

$$= \frac{\omega_0{}^2}{s^2 + 2\alpha s + \omega_0{}^2}$$

where

$$K = 1 + \frac{R_b}{R_a} \qquad \text{and} \qquad \omega_0{}^2 = \frac{G_1 G_2}{C_1 C_2} = \frac{1}{R_1 R_2 C_1 C_2} \qquad (4.57a)$$

$$2\alpha = \frac{G_1 C_2 + G_2 C_2 + G_2 C_1 (1 - K)}{C_1 C_2} = \frac{1}{R_1 C_1} + \frac{1}{R_2 C_1} + \frac{1}{R_2 C_2}(1 - K) \qquad (4.57b)$$

Now

$$V_o(s) = K V_b(s)$$

Therefore

$$H(s) = \frac{V_o(s)}{V_i(s)} = K \frac{V_b(s)}{V_i(s)} = \frac{K \omega_0{}^2}{s^2 + 2\alpha s + \omega_0{}^2} \qquad (4.58)$$

4.5 BLOCK DIAGRAMS

Large systems may consist of an enormous number of components or elements. As anyone who has seen the circuit diagram of a radio or a television receiver can appreciate, analyzing such systems all at once could be next to impossible. In such cases, it is convenient to represent a system by suitably interconnected subsystems, each of which can be readily analyzed. Each subsystem can be characterized in terms of its input–output relationships. A linear system can be characterized by its transfer function $H(s)$. Figure 4.18a shows a block diagram of a system with a transfer function $H(s)$ and its input and output $X(s)$ and $Y(s)$, respectively.

Subsystems may be interconnected by using cascade, parallel, and feedback interconnections (Fig. 4.18b, 4.18c, 4.18d), the three elementary types. When transfer functions appear in cascade, as depicted in Fig. 4.18b, then, as shown earlier, the transfer function of the overall system is the product of the two transfer functions. This result can also be proved by observing that in Fig. 4.18b

$$\frac{Y(s)}{X(s)} = \frac{W(s)}{X(s)} \frac{Y(s)}{W(s)} = H_1(s)H_2(s)$$

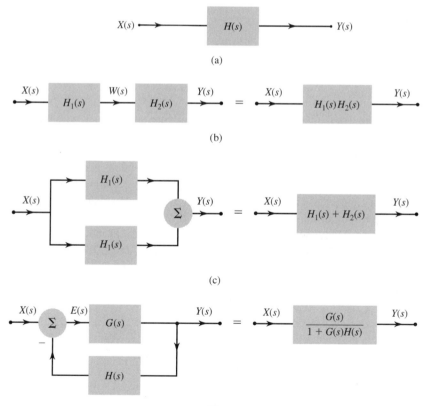

Figure 4.18 Elementary connections of blocks and their equivalents.

We can extend this result to any number of transfer functions in cascade. It follows from this discussion that the subsystems in cascade can be interchanged without affecting the overall transfer function. This commutation property of LTI systems follows directly from the commutative (and associative) property of convolution. We have already proved this property in Section 2.4-3. Every possible ordering of the subsystems yields the same overall transfer function. However, there may be practical consequences (such as sensitivity to parameter variation) affecting the behavior of different ordering.

Similarly, when two transfer functions, $H_1(s)$ and $H_2(s)$, appear in parallel, as illustrated in Fig. 4.18c, the overall transfer function is given by $H_1(s) + H_2(s)$, the sum of the two transfer functions. The proof is trivial. This result can be extended to any number of systems in parallel.

When the output is fed back to the input, as shown in Fig. 4.18d, the overall transfer function $Y(s)/X(s)$ can be computed as follows. The inputs to the adder are $X(s)$ and $-H(s)Y(s)$. Therefore, $E(s)$, the output of the adder, is

$$E(s) = X(s) - H(s)Y(s)$$

But

$$Y(s) = G(s)E(s)$$

$$= G(s)[X(s) - H(s)Y(s)]$$

Therefore

$$Y(s)[1 + G(s)H(s)] = G(s)X(s)$$

so that

$$\frac{Y(s)}{X(s)} = \frac{G(s)}{1 + G(s)H(s)} \tag{4.59}$$

Therefore, the feedback loop can be replaced by a single block with the transfer function shown in Eq. (4.59) (see Fig. 4.18d).

In deriving these equations, we implicitly assume that when the output of one subsystem is connected to the input of another subsystem, the latter does not load the former. For example, the transfer function $H_1(s)$ in Fig. 4.18b is computed by assuming that the second subsystem $H_2(s)$ was not connected. This is the same as assuming that $H_2(s)$ does not load $H_1(s)$. In other words, the input–output relationship of $H_1(s)$ will remain unchanged regardless of whether $H_2(s)$ is connected. Many modern circuits use op amps with high input impedances, so this assumption is justified. When such an assumption is not valid, $H_1(s)$ must be computed under operating conditions [i.e., with $H_2(s)$ connected].

COMPUTER EXAMPLE C4.3

Using the feedback system of Fig. 4.18d with $G(s) = K/(s(s+8))$ and $H(s) = 1$, determine the transfer function for each of the following cases:

    (a) $K = 7$

    (b) $K = 16$

    (c) $K = 80$

(a)

```
>> H = tf(1,1); K = 7; G = tf([0 0 K],[1 8 0]);
>> disp(['(a) K = ',num2str(K)]); TFa = feedback(G,H)
(a) K = 7

Transfer function:
     7
   -------------
s^2 + 8 s + 7
```

Thus, $H_a(s) = 7/(s^2 + 8s + 7)$.

(b)

```
>> H = tf(1,1); K = 16; G = tf([0 0 K],[1 8 0]);
>> disp(['(b) K = ',num2str(K)]); TFb = feedback(G,H)
(b) K = 16

Transfer function:
     16
   -------------
s^2 + 8 s + 16
```

Thus, $H_b(s) = 16/(s^2 + 8s + 16)$.

(c)

```
>> H = tf(1,1); K = 80; G = tf([0 0 K],[1 8 0]);
>> disp(['(c) K = ',num2str(K)]); TFc = feedback(G,H)
(c) K = 80

Transfer function:
     80
   -------------
s^2 + 8 s + 80
```

Thus, $H_c(s) = 80/(s^2 + 8s + 80)$.

4.6 SYSTEM REALIZATION

We now develop a systematic method for realization (or implementation) of an arbitrary Nth-order transfer function. The most general transfer function with $M = N$ is given by

$$H(s) = \frac{b_0 s^N + b_1 s^{N-1} + \cdots + b_{N-1}s + b_N}{s^N + a_1 s^{N-1} + \cdots + a_{N-1}s + a_N} \tag{4.60}$$

Since realization is basically a synthesis problem, there is no unique way of realizing a system. A given transfer function can be realized in many different ways. A transfer function $H(s)$

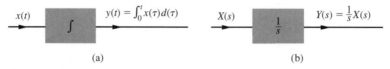

(a)   (b)

Figure 4.19 (a) Time-domain and (b) frequency-domain representations of an integrator.

can be realized by using integrators or differentiators along with adders and multipliers. We avoid use of differentiators for practical reasons discussed in Sections 2.1 and 4.3-2. Hence, in our implementation, we shall use integrators along with scalar multipliers and adders. We are already familiar with representation of all these elements except the integrator. The integrator can be represented by a box with integral sign (time-domain representation, Fig. 4.19a) or by a box with transfer function $1/s$ (frequency-domain representation, 4.19b).

4.6-1 Direct Form I Realization

Rather than realize the general Nth-order system described by Eq. (4.60), we begin with a specific case of the following third-order system and then extend the results to the Nth-order case

$$H(s) = \frac{b_0 s^3 + b_1 s^2 + b_2 s + b_3}{s^3 + a_1 s^2 + a_2 s + a_3} = \frac{b_0 + \dfrac{b_1}{s} + \dfrac{b_2}{s^2} + \dfrac{b_3}{s^3}}{1 + \dfrac{a_1}{s} + \dfrac{a_2}{s^2} + \dfrac{a_3}{s^3}} \tag{4.61}$$

We can express $H(s)$ as

$$H(s) = \underbrace{\left(b_0 + \frac{b_1}{s} + \frac{b_2}{s^2} + \frac{b_3}{s^3} \right)}_{H_1(s)} \underbrace{\left(\frac{1}{1 + \dfrac{a_1}{s} + \dfrac{a_2}{s^2} + \dfrac{a_3}{s^3}} \right)}_{H_2(s)} \tag{4.62}$$

We can realize $H(s)$ as a cascade of transfer function $H_1(s)$ followed by $H_2(s)$, as depicted in Fig. 4.20a, where the output of $H_1(s)$ is denoted by $W(s)$. Because of the commutative property of LTI system transfer functions in cascade, we can also realize $H(s)$ as a cascade of $H_2(s)$ followed by $H_1(s)$, as illustrated in Fig. 4.20b, where the (intermediate) output of $H_2(s)$ is denoted by $V(s)$.

The output of $H_1(s)$ in Fig. 4.20a is given by $W(s) = H_1(s)X(s)$. Hence

$$W(s) = \left(b_0 + \frac{b_1}{s} + \frac{b_2}{s^2} + \frac{b_3}{s^3} \right) X(s) \tag{4.63}$$

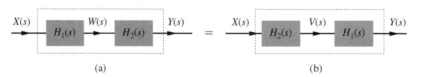

(a)   (b)

Figure 4.20 Realization of a transfer function in two steps.

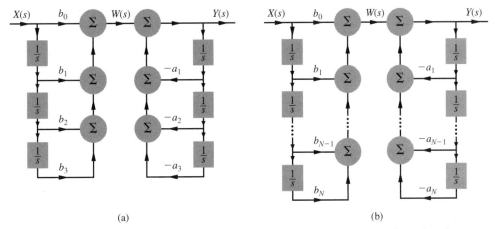

(a)   (b)

Figure 4.21 Direct form I realization of an LTIC system: **(a)** third-order and **(b)** Nth-order.

Also, the output $Y(s)$ and the input $W(s)$ of $H_2(s)$ in Fig. 4.20a are related by $Y(s) = H_2(s)W(s)$. Hence

$$W(s) = \left(1 + \frac{a_1}{s} + \frac{a_2}{s^2} + \frac{a_3}{s^3}\right) Y(s) \tag{4.64}$$

We shall first realize $H_1(s)$. Equation (4.63) shows that the output $W(s)$ can be synthesized by adding the input $b_0 X(s)$ to $b_1(X(s)/s)$, $b_2(X(s)/s^2)$, and $b_3(X(s)/s^3)$. Because the transfer function of an integrator is $1/s$, the signals $X(s)/s$, $X(s)/s^2$, and $X(s)/s^3$ can be obtained by successive integration of the input $x(t)$. The left-half section of Fig. 4.21a shows how $W(s)$ can be synthesized from $X(s)$, according to Eq. (4.63). Hence, this section represents a realization of $H_1(s)$.

To complete the picture, we shall realize $H_2(s)$, which is specified by Eq. (4.64). We can rearrange Eq. (4.64) as

$$Y(s) = W(s) - \left(\frac{a_1}{s} + \frac{a_2}{s^2} + \frac{a_3}{s^3}\right) Y(s) \tag{4.65}$$

Hence, to obtain $Y(s)$, we subtract $a_1 Y(s)/s$, $a_2 Y(s)/s^2$, and $a_3 Y(s)/s^3$ from $W(s)$. We have already obtained $W(s)$ from the first step [output of $H_1(s)$]. To obtain signals $Y(s)/s$, $Y(s)/s^2$, and $Y(s)/s^3$, we assume that we already have the desired output $Y(s)$. Successive integration of $Y(s)$ yields the needed signals $Y(s)/s$, $Y(s)/s^2$, and $Y(s)/s^3$. We now synthesize the final output $Y(s)$ according to Eq. (4.65), as seen in the right-half section of Fig. 4.21a.[†] The left-half section in Fig. 4.21a represents $H_1(s)$ and the right-half is $H_2(s)$. We can generalize this procedure, known as the *direct form I* (DFI) realization, for any value of N. This procedure requires $2N$ integrators to realize an Nth-order transfer function, as shown in Fig. 4.21b.

[†]It may seem odd that we first assumed the existence of $Y(s)$, integrated it successively, and then in turn generated $Y(s)$ from $W(s)$ and the three successive integrals of signal $Y(s)$. This procedure poses a dilemma similar to "Which came first, the chicken or the egg?" The problem here is satisfactorily resolved by writing the expression for $Y(s)$ at the output of the right-hand adder (at the top) in Fig. 4.21a and verifying that this expression is indeed the same as Eq. (4.64).

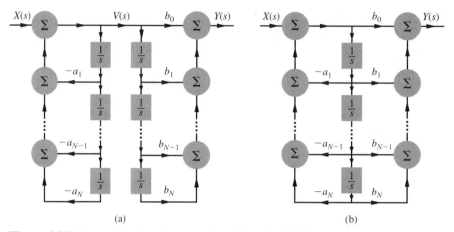

Figure 4.22 Direct form II realization of an Nth-order LTIC system.

4.6-2 Direct Form II Realization

In the direct form I, we realize $H(s)$ by implementing $H_1(s)$ followed by $H_2(s)$, as shown in Fig. 4.20a. We can also realize $H(s)$, as shown in Fig. 4.20b, where $H_2(s)$ is followed by $H_1(s)$. This procedure is known as the *direct form II* realization. Figure 4.22a shows direct form II realization, where we have interchanged sections representing $H_1(s)$ and $H_2(s)$ in Fig. 4.21b. The output of $H_2(s)$ in this case is denoted by $V(s)$.[†]

An interesting observation in Fig. 4.22a is that the input signal to both the chains of integrators is $V(s)$. Clearly, the outputs of integrators in the left-side chain are identical to the corresponding outputs of the right-side integrator chain, thus making the right-side chain redundant. We can eliminate this chain and obtain the required signals from the left-side chain, as shown in Fig. 4.22b. This implementation halves the number of integrators to N, and, thus, is more efficient in hardware utilization than either Fig. 4.21b or 4.22a. This is the *direct form II* (DFII) realization.

An Nth-order differential equation with $N = M$ has a property that its implementation requires a minimum of N integrators. A realization is *canonic* if the number of integrators used in the realization is equal to the order of the transfer function realized. Thus, canonic realization has no redundant integrators. The DFII form in Fig. 4.22b is a canonic realization, and is also called the *direct canonic* form. Note that the DFI is noncanonic.

The direct form I realization (Fig. 4.21b) implements zeros first [the left-half section represented by $H_1(s)$] followed by realization of poles [the right-half section represented by $H_2(s)$] of $H(s)$. In contrast, canonic direct implements poles first followed by zeros. Although both

---

[†]The reader can show that the equations relating $X(s)$, $V(s)$, and $Y(s)$ in Fig. 4.22a are

$$V(s) = X(s) - \left(\frac{a_1}{s} + \frac{a_2}{s^2} + \cdots + \frac{a_N}{s^N} \right) V(s)$$

and

$$Y(s) = \left(b_0 + \frac{b_1}{s} + \frac{b_2}{s^2} + \cdots + \frac{b_N}{s^N} \right) V(s)$$

these realizations result in the same transfer function, they generally behave differently from the viewpoint of sensitivity to parameter variations.

EXAMPLE 4.19

Find the canonic direct form realization of the following transfer functions:

(a) $\dfrac{5}{s+7}$

(b) $\dfrac{s}{s+7}$

(c) $\dfrac{s+5}{s+7}$

(d) $\dfrac{4s+28}{s^2+6s+5}$

All four of these transfer functions are special cases of $H(s)$ in Eq. (4.60).

(a) The transfer function $5/(s+7)$ is of the first order ($N=1$); therefore, we need only one integrator for its realization. The feedback and feedforward coefficients are

$$a_1 = 7 \qquad \text{and} \qquad b_0 = 0, \quad b_1 = 5$$

The realization is depicted in Fig. 4.23a. Because $N=1$, there is a single feedback connection from the output of the integrator to the input adder with coefficient $a_1 = 7$. For $N=1$, generally, there are $N+1 = 2$ feedforward connections. However, in this case, $b_0 = 0$, and there is only one feedforward connection with coefficient $b_1 = 5$ from the output of the integrator to the output adder. Because there is only one input signal to the output adder, we can do away with the adder, as shown in Fig. 4.23a.

(b)

$$H(s) = \frac{s}{s+7}$$

In this first-order transfer function, $b_1 = 0$. The realization is shown in Fig. 4.23b. Because there is only one signal to be added at the output adder, we can discard the adder.

(c)

$$H(s) = \frac{s+5}{s+7}$$

The realization appears in Fig. 4.23c. Here $H(s)$ is a first-order transfer function with $a_1 = 7$ and $b_0 = 1$, $b_1 = 5$. There is a single feedback connection (with coefficient 7) from the integrator output to the input adder. There are two feedforward connections (Fig. 4.23c).[†]

[†]When $M = N$ (as in this case), $H(s)$ can also be realized in another way by recognizing that

$$H(s) = 1 - \frac{2}{s+7}$$

We now realize $H(s)$ as a parallel combination of two transfer functions, as indicated by this equation.

(d)

$$H(s) = \frac{4s + 28}{s^2 + 6s + 5}$$

This is a second-order system with $b_0 = 0$, $b_1 = 4$, $b_2 = 28$, $a_1 = 6$, and $a_2 = 5$. Figure 4.23d shows a realization with two feedback connections and two feedforward connections.

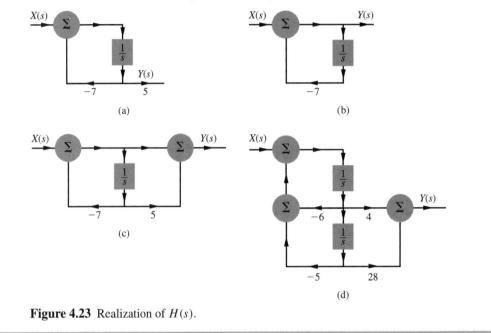

Figure 4.23 Realization of $H(s)$.

EXERCISE E4.10

Give canonic direct realization of

$$H(s) = \frac{2s}{s^2 + 6s + 25}$$

4.6-3 Cascade and Parallel Realizations

An Nth-order transfer function $H(s)$ can be expressed as a product or a sum of N first-order transfer functions. Accordingly, we can also realize $H(s)$ as a cascade (series) or parallel form of these N first-order transfer functions. Consider, for instance, the transfer function in part (d) of the Example 4.19.

$$H(s) = \frac{4s + 28}{s^2 + 6s + 5}$$

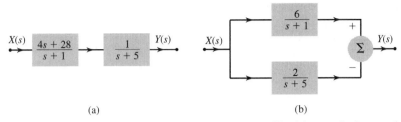

(a)   (b)

Figure 4.24 Realization of $(4s + 28)/((s + 1)(s + 5))$: **(a)** cascade form and **(b)** parallel form.

We can express $H(s)$ as

$$H(s) = \frac{4s + 28}{(s + 1)(s + 5)} = \underbrace{\left(\frac{4s + 28}{s + 1}\right)}_{H_1(s)} \underbrace{\left(\frac{1}{s + 5}\right)}_{H_2(s)} \tag{4.66a}$$

We can also express $H(s)$ as a sum of partial fractions as

$$H(s) = \frac{4s + 28}{(s + 1)(s + 5)} = \underbrace{\frac{6}{s + 1}}_{H_3(s)} - \underbrace{\frac{2}{s + 5}}_{H_4(s)} \tag{4.66b}$$

Equations (4.66) give us the option of realizing $H(s)$ as a cascade of $H_1(s)$ and $H_2(s)$, as shown in Fig. 4.24a, or a parallel of $H_3(s)$ and $H_4(s)$, as depicted in Fig. 4.24b. Each of the first-order transfer functions in Fig. 4.24 can be implemented by using canonic direct realizations, discussed earlier.

This discussion by no means exhausts all the possibilities. In the cascade form alone, there are different ways of grouping the factors in the numerator and the denominator of $H(s)$, and each grouping can be realized in DFI or canonic direct form. Accordingly, several cascade forms are possible. In Section 4.6-4, we shall discuss yet another form that essentially doubles the numbers of realizations discussed so far.

From a practical viewpoint, parallel and cascade forms are preferable because parallel and certain cascade forms are numerically less sensitive than canonic direct form to small parameter variations in the system. Qualitatively, this difference can be explained by the fact that in a canonic realization all the coefficients interact with each other, and a change in any coefficient will be magnified through its repeated influence from feedback and feedforward connections. In a parallel realization, in contrast, the change in a coefficient will affect only a localized segment; the case with a cascade realization is similar.

In the examples of cascade and parallel realization, we have separated $H(s)$ into first-order factors. For $H(s)$ of higher orders, we could group $H(s)$ into factors, not all of which are necessarily of the first order. For example, if $H(s)$ is a third-order transfer function, we could realize this function as a cascade (or a parallel) combination of a first-order and a second-order factor.

REALIZATION OF COMPLEX
CONJUGATE POLES

The complex poles in $H(s)$ should be realized as a second-order (quadratic) factor because we cannot implement multiplication by complex numbers. Consider, for example,

$$H(s) = \frac{10s + 50}{(s + 3)(s^2 + 4s + 13)}$$

$$= \frac{10s + 50}{(s + 3)(s + 2 - j3)(s + 2 + j3)}$$

$$= \frac{2}{s + 3} - \frac{1 + j2}{s + 2 - j3} - \frac{1 - j2}{s + 2 + j3}$$

We cannot realize first-order transfer functions individually with the poles $-2 \pm j3$ because they require multiplication by complex numbers in the feedback and the feedforward paths. Therefore, we need to combine the conjugate poles and realize them as a second-order transfer function.[†] In the present case, we can express $H(s)$ as

$$H(s) = \left(\frac{10}{s + 3}\right)\left(\frac{s + 5}{s^2 + 4s + 13}\right) \tag{4.67a}$$

$$= \frac{2}{s + 3} - \frac{2s - 8}{s^2 + 4s + 13} \tag{4.67b}$$

Now we can realize $H(s)$ in cascade form by using Eq. (4.67a) or in parallel form by using Eq. (4.67b).

REALIZATION OF REPEATED POLES

When repeated poles occur, the procedure for canonic and cascade realization is exactly the same as before. In parallel realization, however, the procedure requires special handling, as explained in Example 4.20.

EXAMPLE 4.20

Determine the parallel realization of

$$H(s) = \frac{7s^2 + 37s + 51}{(s + 2)(s + 3)^2}$$

$$= \frac{5}{s + 2} + \frac{2}{s + 3} - \frac{3}{(s + 3)^2}$$

[†]It is possible to realize complex, conjugate poles indirectly by using a cascade of two first-order transfer functions and feedback. A transfer function with poles $-a \pm jb$ can be realized by using a cascade of two identical first-order transfer functions, each having a pole at $-a$. (See Prob. 4.6-13.)

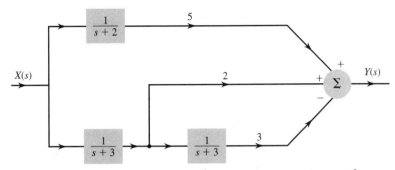

Figure 4.25 Parallel realization of $(7s^2 + 37s + 51)/((s + 2)(s + 3)^2)$.

This third-order transfer function should require no more than three integrators. But if we try to realize each of the three partial fractions separately, we require four integrators because of the one second-order term. This difficulty can be avoided by observing that the terms $1/(s+3)$ and $1/(s+3)^2$ can be realized with a cascade of two subsystems, each having a transfer function $1/(s + 3)$, as shown in Fig. 4.25. Each of the three first-order transfer functions in Fig. 4.25 may now be realized as in Fig. 4.23.

EXERCISE E4.11

Find the canonic, cascade, and parallel realization of

$$H(s) = \frac{s + 3}{s^2 + 7s + 10} = \left(\frac{s + 3}{s + 2}\right)\left(\frac{1}{s + 5}\right)$$

4.6-4 Transposed Realization

Two realizations are said to be *equivalent* if they have the same transfer function. A simple way to generate an equivalent realization from a given realization is to use its *transpose*. To generate a transpose of any realization, we change the given realization as follows:

1. Reverse all the arrow directions without changing the scalar multiplier values.

2. Replace pickoff nodes by adders and vice versa.

3. Replace the input $X(s)$ with the output $Y(s)$ and vice versa.

Figure 4.26a shows the transposed version of the canonic direct form realization in Fig. 4.22b found according to the rules just listed. Figure 4.26b is Fig. 4.26a reoriented in the conventional form so that the input $X(s)$ appears at the left and the output $Y(s)$ appears at the right. Observe that this realization is also canonic.

Rather than prove the theorem on equivalence of the transposed realizations, we shall verify that the transfer function of the realization in Fig. 4.26b is identical to that in Eq. (4.60).

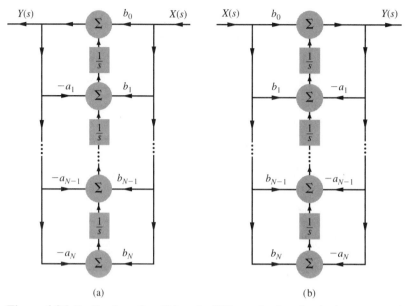

Figure 4.26 Realization of an Nth-order LTI transfer function in the transposed form.

Figure 4.26b shows that $Y(s)$ is being fed back through N paths. The fed-back signal appearing at the input of the top adder is

$$\left(\frac{-a_1}{s} + \frac{-a_2}{s^2} + \cdots + \frac{-a_{N-1}}{s^{N-1}} + \frac{-a_N}{s^N}\right) Y(s)$$

The signal $X(s)$, fed to the top adder through $N+1$ forward paths, contributes

$$\left(b_0 + \frac{b_1}{s} + \cdots + \frac{b_{N-1}}{s^{N-1}} + \frac{b_N}{s^N}\right) X(s)$$

The output $Y(s)$ is equal to the sum of these two signals (feed forward and feed back). Hence

$$Y(s) = \left(\frac{-a_1}{s} + \frac{-a_2}{s^2} + \cdots + \frac{-a_{N-1}}{s^{N-1}} + \frac{-a_N}{s^N}\right) Y(s)$$

$$+ \left(b_0 + \frac{b_1}{s} + \cdots + \frac{b_{N-1}}{s^{N-1}} + \frac{b_N}{s^N}\right) X(s)$$

Transporting all the $Y(s)$ terms to the left side and multiplying throughout by s^N, we obtain

$$(s^N + a_1 s^{N-1} + \cdots + a_{N-1}s + a_N)Y(s) = (b_0 s^N + b_1 s^{N-1} + \cdots + b_{N-1}s + b_N)X(s)$$

Consequently

$$H(s) = \frac{Y(s)}{X(s)} = \frac{b_0 s^N + b_1 s^{N-1} + \cdots + b_{N-1}s + b_N}{s^N + a_1 s^{N-1} + \cdots + a_{N-1}s + a_N}$$

Hence, the transfer function $H(s)$ is identical to that in Eq. (4.60).

We have essentially doubled the number of possible realizations. Every realization that was found earlier has a transpose. Note that the transpose of a transpose results in the same realization.

EXAMPLE 4.21

Find the transpose of the canonic direct realization found in parts (a) and (d) of Example 4.19 (Fig. 4.23c and 4.23d).

The transfer functions are

(a) $\dfrac{s+5}{s+7}$

(b) $\dfrac{4s+28}{s^2+6s+5}$

Both these realizations are special cases of the one in Fig. 4.26b.

(a) In this case, $N = 1$ with $a_1 = 7$, $b_0 = 1$, $b_1 = 5$. The desired realization can be obtained by transposing Fig. 4.23c. However, we already have the general model of the transposed realization in Fig. 4.26b. The desired solution is a special case of Fig. 4.26b with $N = 1$ and $a_1 = 7$, $b_0 = 1$, $b_1 = 5$, as shown in Fig. 4.27a.

(b) In this case $N = 2$ with $b_0 = 0$, $b_1 = 4$, $b_2 = 28$, $a_1 = 6$, $a_2 = 5$. Using the model of Fig. 4.26b, we obtain the desired realization, as shown in Fig. 4.27b.

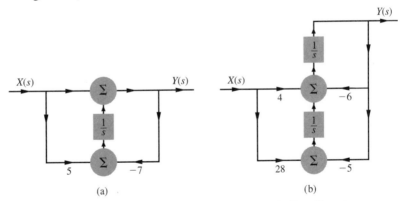

(a)   (b)

Figure 4.27 Transposed form realization of **(a)** $(s+5)/(s+7)$ and **(b)** $(4s+28)/(s^2+6s+5)$.

EXERCISE E4.12

Find the realization that is the transposed version of (a) DFI realization and (b) canonic direct realization of $H(s)$ in Exercise E4.10.

4.6-5 Using Operational Amplifiers for System Realization

In this section, we discuss practical implementation of the realizations described in Section 4.6-4. Earlier we saw that the basic elements required for the synthesis of an LTIC system (or a given transfer function) are (scalar) multipliers, integrators, and adders. All these elements can be realized by operational amplifier (op-amp) circuits.

OPERATIONAL AMPLIFIER CIRCUITS

Figure 4.28 shows an op-amp circuit in the frequency domain (the transformed circuit). Because the input impedance of the op amp is infinite (very high), all the current $I(s)$ flows in the feedback path, as illustrated. Moreover $V_x(s)$, the voltage at the input of the op amp, is zero (very small) because of the infinite (very large) gain of the op amp. Therefore, for all practical purposes,

$$Y(s) = -I(s)Z_f(s)$$

Moreover, because $v_x \approx 0$,

$$I(s) = \frac{X(s)}{Z(s)}$$

Substitution of the second equation in the first yields

$$Y(s) = -\frac{Z_f(s)}{Z(s)}X(s)$$

Therefore, the op-amp circuit in Fig. 4.28 has the transfer function

$$H(s) = -\frac{Z_f(s)}{Z(s)} \tag{4.68}$$

By properly choosing $Z(s)$ and $Z_f(s)$, we can obtain a variety of transfer functions, as the following development shows.

THE SCALAR MULTIPLIER

If we use a resistor R_f in the feedback and a resistor R at the input (Fig. 4.29a), then $Z_f(s) = R_f$, $Z(s) = R$, and

$$H(s) = -\frac{R_f}{R} \tag{4.69a}$$

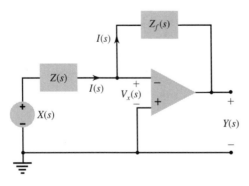

Figure 4.28 A basic inverting configuration op-amp circuit.

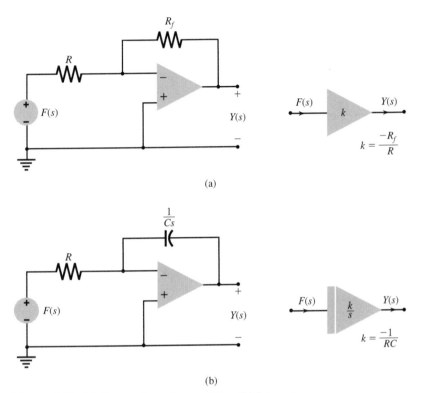

Figure 4.29 (a) Op-amp inverting amplifier. (b) Integrator.

The system acts as a scalar multiplier (or an amplifier) with a negative gain R_f/R. A positive gain can be obtained by using two such multipliers in cascade or by using a single noninverting amplifier, as depicted in Fig. 4.16c. Figure 4.29a also shows the compact symbol used in circuit diagrams for a scalar multiplier.

THE INTEGRATOR

If we use a capacitor C in the feedback and a resistor R at the input (Fig. 4.29b), then $Z_f(s) = 1/Cs$, $Z(s) = R$, and

$$H(s) = \left(-\frac{1}{RC}\right)\frac{1}{s} \tag{4.69b}$$

The system acts as an ideal integrator with a gain $-1/RC$. Figure 4.29b also shows the compact symbol used in circuit diagrams for an integrator.

THE ADDER

Consider now the circuit in Fig. 4.30a with r inputs $X_1(s), X_2(s), \ldots, X_r(s)$. As usual, the input voltage $V_x(s) \simeq 0$ because the gain of op amp $\rightarrow \infty$. Moreover, the current going into the op amp is very small ($\simeq 0$) because the input impedance $\rightarrow \infty$. Therefore, the total current in the feedback resistor R_f is $I_1(s) + I_2(s) + \cdots + I_r(s)$. Moreover, because $V_x(s) = 0$,

$$I_j(s) = \frac{X_j(s)}{R_j} \qquad j = 1, 2, \ldots, r$$

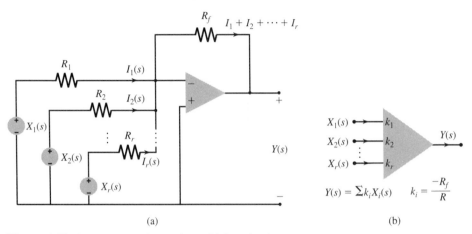

Figure 4.30 Op-amp summing and amplifying circuit.

Also

$$Y(s) = -R_f[I_1(s) + I_2(s) + \cdots + I_r(s)]$$

$$= -\left[\frac{R_f}{R_1}X_1(s) + \frac{R_f}{R_2}X_2(s) + \cdots + \frac{R_f}{R_r}X_r(s)\right]$$

$$= k_1 X_1(s) + k_2 X_2(s) + \cdots + k_r X_r(s) \tag{4.70}$$

where

$$k_i = \frac{-R_f}{R_i}$$

Clearly, the circuit in Fig. 4.30 serves a adder and an amplifier with any desired gain for each of the input signals. Figure 4.30b shows the compact symbol used in circuit diagrams for a adder with r inputs.

EXAMPLE 4.22

Use op-amp circuits to realize the canonic direct form of the transfer function

$$H(s) = \frac{2s + 5}{s^2 + 4s + 10}$$

The basic canonic realization is shown in Fig. 4.31a. The same realization with horizontal reorientation is shown in Fig. 4.31b. Signals at various points are also indicated in the realization. For convenience, we denote the output of the last integrator by $W(s)$. Consequently, the signals at the inputs of the two integrators are $sW(s)$ and $s^2 W(s)$, as shown in Fig. 4.31a and 4.31b. Op-amp elements (multipliers, integrators, and adders) change the polarity of

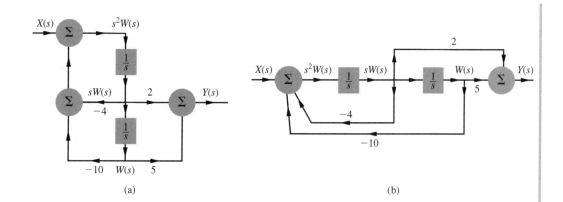

(a)

(b)

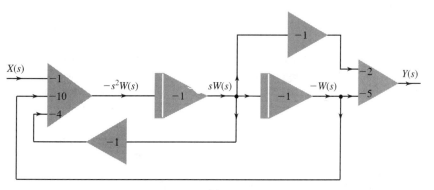

(c)

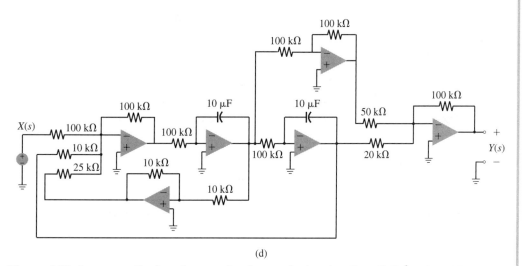

(d)

Figure 4.31 Op-amp realization of a second-order transfer function $(2s + 5)/(s^2 + 4s + 10)$.

the output signals. To incorporate this fact, we modify the canonic realization in Fig. 4.31b to that depicted in Fig. 4.31c. In Fig. 4.31b, the successive outputs of the adder and the integrators are $s^2W(s)$, $sW(s)$, and $W(s)$ respectively. Because of polarity reversals in op-amp circuits, these outputs are $-s^2W(s)$, $sW(s)$, and $-W(s)$, respectively, in Fig. 4.31c. This polarity reversal requires corresponding modifications in the signs of feedback and feedforward gains. According to Fig. 4.31b

$$s^2W(s) = X(s) - 4sW(s) - 10W(s)$$

Therefore

$$-s^2W(s) = -X(s) + 4sW(s) + 10W(s)$$

Because the adder gains are always negative (see Fig. 4.30b), we rewrite the foregoing equation as

$$-s^2W(s) = -1[X(s)] - 4[-sW(s)] - 10[-W(s)]$$

Figure 4.31c shows the implementation of this equation. The hardware realization appears in Fig. 4.31d. Both integrators have a unity gain which requires $RC = 1$. We have used $R = 100$ kΩ and $C = 10\,\mu$F. The gain of 10 in the outer feedback path is obtained in the adder by choosing the feedback resistor of the adder to be 100 kΩ and an input resistor of 10 kΩ. Similarly, the gain of 4 in the inner feedback path is obtained by using the corresponding input resistor of 25 kΩ. The gains of 2 and 5, required in the feedforward connections, are obtained by using a feedback resistor of 100 kΩ and input resistors of 50 and 20 kΩ, respectively.[†]

The op-amp realization in Fig. 4.31 is not necessarily the one that uses the fewest op amps. This example is given just to illustrate a systematic procedure for designing an op-amp circuit of an arbitrary transfer function. There are more efficient circuits (such as Sallen–Key or biquad) that use fewer op amp to realize a second-order transfer function.

EXERCISE E4.13

Show that the transfer functions of the op-amp circuits in Fig. 4.32a and 4.32b are $H_1(s)$ and $H_2(s)$, respectively, where

$$H_1(s) = \frac{-R_f}{R}\left(\frac{a}{s+a}\right) \qquad a = \frac{1}{R_fC_f}$$

$$H_2(s) = -\frac{C}{C_f}\left(\frac{s+b}{s+a}\right) \qquad a = \frac{1}{R_fC_f} \qquad b = \frac{1}{RC}$$

[†] It is possible to avoid the two inverting op amps (with gain -1) in Fig. 4.31d by adding signal $sW(s)$ to the input and output adders directly, using the noninverting amplifier configuration in Fig. 4.16d.

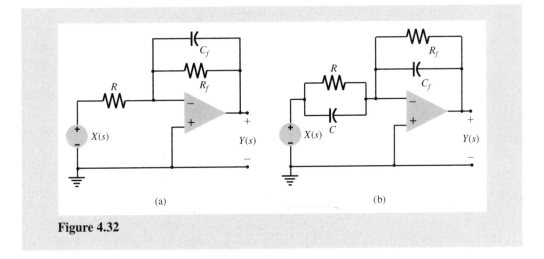

Figure 4.32

4.7 APPLICATION TO FEEDBACK AND CONTROLS

Generally, systems are designed to produce a desired output $y(t)$ for a given input $x(t)$. Using the given performance criteria, we can design a system, as shown in Fig. 4.33a. Ideally, such an open-loop system should yield the desired output. In practice, however, the system characteristics change with time, as a result of aging or replacement of some components or because of changes in the environment in which the system is operating. Such variations cause changes in the output for the same input. Clearly, this is undesirable in precision systems.

A possible solution to this problem is to add a signal component to the input that is not a predetermined function of time but will change to counteract the effects of changing system characteristics and the environment. In short, we must provide a correction at the system input to account for the undesired changes just mentioned. Yet since these changes are generally unpredictable, it is not clear how to preprogram appropriate corrections to the input. However, the difference between the actual output and the desired output gives an indication of the suitable correction to be applied to the system input. It may be possible to counteract the variations by feeding the output (or some function of output) back to the input.

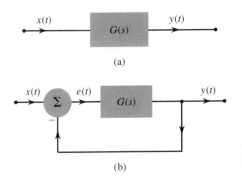

Figure 4.33 **(a)** Open-loop and **(b)** closed-loop (feedback) systems.

We unconsciously apply this principle in daily life. Consider an example of marketing a certain product. The optimum price of the product is the value that maximizes the profit of a merchant. The output in this case is the profit, and the input is the price of the item. The output (profit) can be controlled (within limits) by varying the input (price). The merchant may price the product too high initially, in which case, he will sell too few items, reducing the profit. Using feedback of the profit (output), he adjusts the price (input), to maximize his profit. If there is a sudden or unexpected change in the business environment, such as a strike-imposed shutdown of a large factory in town, the demand for the item goes down, thus reducing his output (profit). He adjusts his input (reduces price) using the feedback of the output (profit) in a way that will optimize his profit in the changed circumstances. If the town suddenly becomes more prosperous because of a new factory opens, he will increase the price to maximize the profit. Thus, by continuous feedback of the output to the input, he realizes his goal of maximum profit (optimum output) in any given circumstances. We observe thousands of examples of feedback systems around us in everyday life. Most social, economical, educational, and political processes are, in fact, feedback processes. A block diagram of such a system, called the *feedback* or *closed-loop* system is shown in Fig. 4.33b.

A feedback system can address the problems arising because of unwanted disturbances such as random-noise signals in electronic systems, a gust of wind affecting a tracking antenna, a meteorite hitting a spacecraft, and the rolling motion of antiaircraft gun platforms mounted on ships or moving tanks. Feedback may also be used to reduce nonlinearities in a system or to control its rise time (or bandwidth). Feedback is used to achieve, with a given system, the desired objective within a given tolerance, despite partial ignorance of the system and the environment. A feedback system, thus, has an ability for supervision and self-correction in the face of changes in the system parameters and external disturbances (change in the environment).

Consider the feedback amplifier in Fig. 4.34. Let the forward amplifier gain $G = 10,000$. One-hundredth of the output is fed back to the input ($H = 0.01$). The gain T of the feedback amplifier is obtained by [see Eq. (4.59)]

$$T = \frac{G}{1 + GH} = \frac{10,000}{1 + 100} = 99.01$$

Suppose that because of aging or replacement of some transistors, the gain G of the forward amplifier changes from 10,000 to 20,000. The new gain of the feedback amplifier is given by

$$T = \frac{G}{1 + GH} = \frac{20,000}{1 + 200} = 99.5$$

Surprisingly, 100% variation in the forward gain G causes only 0.5% variation in the feedback amplifier gain T. Such reduced sensitivity to parameter variations is a must in precision amplifiers. In this example, we reduced the sensitivity of gain to parameter variations at the cost of forward gain, which is reduced from 10,000 to 99. There is no dearth of forward gain (obtained by cascading stages). But low sensitivity is extremely precious in precision systems.

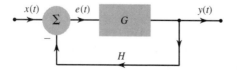

Figure 4.34 Effects of negative and positive feedback.

Now, consider what happens when we add (instead of subtract) the signal fed back to the input. Such addition means the sign on the feedback connection is $+$ instead of $-$ (which is same as changing the sign of H in Fig. 4.34). Consequently

$$T = \frac{G}{1 - GH}$$

If we let $G = 10,000$ as before and $H = 0.9 \times 10^{-4}$, then

$$T = \frac{10,000}{1 - 0.9(10^4)(10^{-4})} = 100,000$$

Suppose that because of aging or replacement of some transistors, the gain of the forward amplifier changes to 11,000. The new gain of the feedback amplifier is

$$T = \frac{11,000}{1 - 0.9(11,000)(10^{-4})} = 1,100,000$$

Observe that in this case, a mere 10% increase in the forward gain G caused 1000% increase in the gain T (from 100,000 to 1,100,000). Clearly, the amplifier is very sensitive to parameter variations. This behavior is exactly opposite of what was observed earlier, when the signal fed back was subtracted from the input.

What is the difference between the two situations? Crudely speaking, the former case is called the *negative feedback* and the latter is the *positive feedback*. The positive feedback increases system gain but tends to make the system more sensitive to parameter variations. It can also lead to instability. In our example, if G were to be 111,111, then $GH = 1$, $T = \infty$, and the system would become unstable because the signal fed back was exactly equal to the input signal itself, since $GH = 1$. Hence, once a signal has been applied, no matter how small and how short in duration, it comes back to reinforce the input undiminished, which further passes to the output, and is fed back again and again and again. In essence, the signal perpetuates itself forever. This perpetuation, even when the input ceases to exist, is precisely the symptom of instability.

Generally speaking, a feedback system cannot be described in black and white terms, such as positive or negative. Usually H is a frequency-dependent component, more accurately represented by $H(s)$, varies with frequency. Consequently, what was negative feedback at lower frequencies can turn into positive feedback at higher frequencies and may give rise to instability. This is one of the serious aspects of feedback systems, which warrants careful attention of a designer.

4.7-1 Analysis of a Simple Control System

Figure 4.35a represents an automatic position control system, which can be used to control the angular position of a heavy object (e.g., a tracking antenna, an antiaircraft gun mount, or the position of a ship). The input θ_i is the desired angular position of the object, which can be set at any given value. The actual angular position θ_o of the object (the output) is measured by a potentiometer whose wiper is mounted on the output shaft. The difference between the input θ_i (set at the desired output position) and the output θ_o (actual position) is amplified; the amplified output, which is proportional to $\theta_i - \theta_o$, is applied to the motor input. If $\theta_i - \theta_o = 0$ (the output being equal to the desired angle), there is no input to the motor, and the motor stops. But if $\theta_o \neq \theta_i$, there will be a nonzero input to the motor, which will turn the shaft until $\theta_o = \theta_i$. It

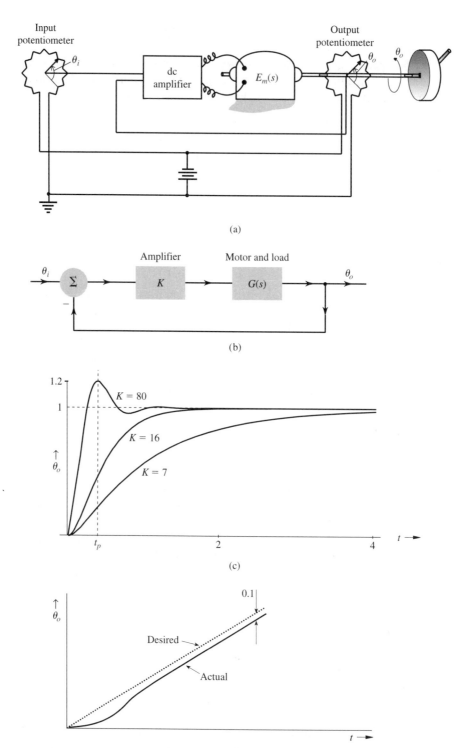

Figure 4.35 (a) An automatic position control system. (b) Its block diagram. (c) The unit step response. (d) The unit ramp response.

is evident that by setting the input potentiometer at a desired position in this system, we can control the angular position of a heavy remote object.

The block diagram of this system is shown in Fig. 4.35b. The amplifier gain is K, where K is adjustable. Let the motor (with load) transfer function that relates the output angle θ_o to the motor input voltage be $G(s)$ [see Eq. (1.77)]. This feedback arrangement is identical to that in Fig. 4.18d with $H(s) = 1$. Hence, $T(s)$, the (closed-loop) system transfer function relating the output θ_o to the input θ_i, is

$$\frac{\Theta_o(s)}{\Theta_i(s)} = T(s) = \frac{KG(s)}{1 + KG(s)}$$

From this equation, we shall investigate the behavior of the automatic position control system in Fig. 4.35a for a step and a ramp input.

STEP INPUT

If we desire to change the angular position of the object instantaneously, we need to apply a step input. We may then want to know how long the system takes to position itself at the desired angle, whether it reaches the desired angle, and whether it reaches the desired position smoothly (monotonically) or oscillates about the final position. If the system oscillates, we may want to know how long it takes for the oscillations to settle down. All these questions can be readily answered by finding the output $\theta_o(t)$ when the input $\theta_i(t) = u(t)$. A step input implies instantaneous change in the angle. This input would be one of the most difficult to follow; if the system can perform well for this input, it is likely to give a good account of itself under most other expected situations. This is why we test control systems for a step input.

For the step input $\theta_i(t) = u(t)$, $\Theta_i(s) = 1/s$ and

$$\Theta_o(s) = \frac{1}{s}T(s) = \frac{KG(s)}{s[1 + KG(s)]}$$

Let the motor (with load) transfer function relating the load angle $\theta_o(t)$ to the motor input voltage be $G(s) = 1/(s(s + 8))$ [see Eq. (1.77)]. This yields

$$\Theta_o(s) = \frac{\dfrac{K}{s(s + 8)}}{s\left[1 + \dfrac{K}{s(s + 8)}\right]} = \frac{K}{s(s^2 + 8s + K)}$$

Let us investigate the system behavior for three different values of gain K.

For $K = 7$,

$$\Theta_o(s) = \frac{7}{s(s^2 + 8s + 7)} = \frac{7}{s(s + 1)(s + 7)} = \frac{1}{s} - \frac{\frac{7}{6}}{s + 1} + \frac{\frac{1}{6}}{s + 7}$$

and

$$\theta_o(t) = \left(1 - \tfrac{7}{6}e^{-t} + \tfrac{1}{6}e^{-7t}\right)u(t)$$

This response, illustrated in Fig. 4.35c, shows that the system reaches the desired angle, but at a rather leisurely pace. To speed up the response let us increase the gain to, say, 80.

For $K = 80$,

$$\Theta_o(s) = \frac{80}{s(s^2 + 8s + 80)} = \frac{80}{s(s + 4 - j8)(s + 4 + j8)}$$

$$= \frac{1}{s} + \frac{\frac{\sqrt{5}}{4}e^{j153^\circ}}{s + 4 - j8} + \frac{\frac{\sqrt{5}}{4}e^{-j153^\circ}}{s + 4 + j8}$$

and

$$\theta_o(t) = \left[1 + \frac{\sqrt{5}}{2}e^{-4t}\cos(8t + 153^\circ)\right]u(t)$$

This response, also depicted in Fig. 4.35c, achieves the goal of reaching the final position at a faster rate than that in the earlier case ($K = 7$). Unfortunately the improvement is achieved at the cost of ringing (oscillations) with high overshoot. In the present case the *percent overshoot* (PO) is 21%. The response reaches its peak value at *peak time* $t_p = 0.393$ second. The *rise time*, defined as the time required for the response to rise from 10% to 90% of its steady-state value, indicates the speed of response.[†] In the present case $t_r = 0.175$ second. The steady-state value of the response is unity so that the *steady-state error* is zero. Theoretically it takes infinite time for the response to reach the desired value of unity. In practice, however, we may consider the response to have settled to the final value if it closely approaches the final value. A widely accepted measure of closeness is within 2% of the final value. The time required for the response to reach and stay within 2% of the final value is called the *settling time* t_s.[‡] In Fig. 4.35c, we find $t_s \approx 1$ second (when $K = 80$). A good system has a small overshoot, small t_r and t_s and a small steady-state error.

A large overshoot, as in the present case, may be unacceptable in many applications. Let us try to determine K (the gain) that yields the fastest response without oscillations. Complex characteristic roots lead to oscillations; to avoid oscillations, the characteristic roots should be real. In the present case the characteristic polynomial is $s^2 + 8s + K$. For $K > 16$, the characteristic roots are complex; for $K < 16$, the roots are real. The fastest response without oscillations is obtained by choosing $K = 16$. We now consider this case.

For $K = 16$,

$$\Theta_o(s) = \frac{16}{s(s^2 + 8s + 16)} = \frac{16}{s(s + 4)^2}$$

$$= \frac{1}{s} - \frac{1}{s + 4} - \frac{4}{(s + 4)^2}$$

and

$$\theta_o(t) = [1 - (4t + 1)e^{-4t}]u(t)$$

This response also appears in Fig. 4.35c. The system with $K > 16$ is said to be *underdamped* (oscillatory response), whereas the system with $K < 16$ is said to be *overdamped*. For $K = 16$, the system is said to be *critically damped*.

[†]*Delay time t_d*, defined as the time required for the response to reach 50% of its steady-state value, is another indication of speed. For the present case, $t_d = 0.141$ second.

[‡]Typical percentage values used are 2 to 5% for t_s.

There is a trade-off between undesirable overshoot and rise time. Reducing overshoots leads to higher rise time (sluggish system). In practice, a small overshoot, which is still faster than the critical damping, may be acceptable. Note that percent overshoot PO and peak time t_p are meaningless for the overdamped or critically damped cases. In addition to adjusting gain K, we may need to augment the system with some type of compensator if the specifications on overshoot and the speed of response are too stringent.

RAMP INPUT

If the antiaircraft gun in Fig. 4.35a is tracking an enemy plane moving with a uniform velocity, the gun-position angle must increase linearly with t. Hence, the input in this case is a ramp; that is, $\theta_i(t) = tu(t)$. Let us find the response of the system to this input when $K = 80$. In this case, $\Theta_i(s) = 1/s^2$, and

$$\Theta_o(s) = \frac{80}{s^2(s^2 + 8s + 80)} = -\frac{0.1}{s} + \frac{1}{s^2} + \frac{0.1(s-2)}{s^2 + 8s + 80}$$

Use of Table 4.1 yields

$$\theta_o(t) = \left[-0.1 + t + \tfrac{1}{8}e^{-8t}\cos(8t + 36.87°)\right] u(t)$$

This response, sketched in Fig. 4.35d, shows that there is a steady-state error $e_r = 0.1$ radian. In many cases such a small steady-state error may be tolerable. If, however, a zero steady-state error to a ramp input is required, this system in its present form is unsatisfactory. We must add some form of compensator to the system.

COMPUTER EXAMPLE C4.4

Using the feedback system of Fig. 4.18d with $G(s) = K/(s(s+8))$ and $H(s) = 1$, determine the step response for each of the following cases:

(a) $K = 7$

(b) $K = 16$

(c) $K = 80$

Additionally,

(d) Find the unit ramp response when $K = 80$.

Computer Example C4.3 computes the transfer functions of these feedback systems in a simple way. In this example, the `conv` command is used to demonstrate polynomial multiplication of the two denominator factors of $G(s)$. Step responses are computed by using the `step` command.

(a–c)

```
>> H = tf(1,1); K = 7; G = tf([0 0 K],conv([0 1 0],[0 1 8]));
>> TFa = feedback(G,H);
>> H = tf(1,1); K = 16; G = tf([0 0 K],conv([0 1 0],[0 1 8]));
>> TFb = feedback(G,H);
>> H = tf(1,1); K = 80; G = tf([0 0 K],conv([0 1 0],[0 1 8]));
>> TFc = feedback(G,H);
>> figure(1); clf; step(TFa,'k-',TFb,'k--',TFc,'k-.');
>> legend('(a) K = 7','(b) K = 16','(c) K = 80',0);
```

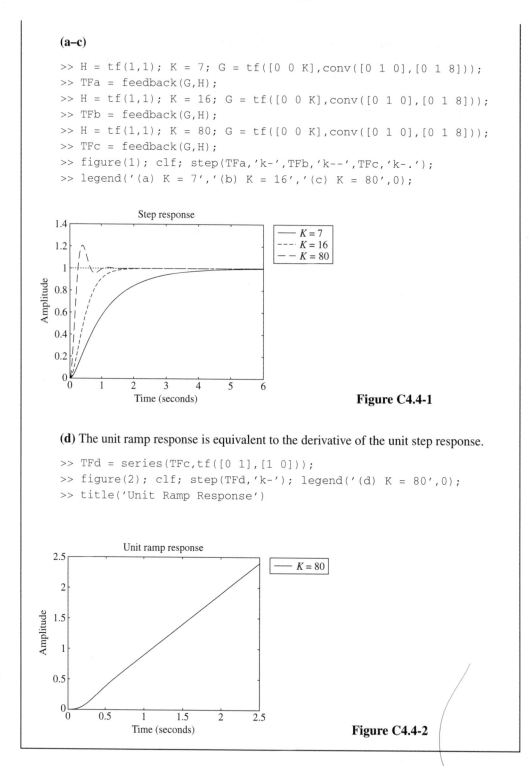

Figure C4.4-1

(d) The unit ramp response is equivalent to the derivative of the unit step response.

```
>> TFd = series(TFc,tf([0 1],[1 0]));
>> figure(2); clf; step(TFd,'k-'); legend('(d) K = 80',0);
>> title('Unit Ramp Response')
```

Figure C4.4-2

DESIGN SPECIFICATIONS

Now the reader has some idea of the various specifications a control system might require. Generally, a control system is designed to meet given transient specifications, steady-state error specifications, and sensitivity specifications. Transient specifications include overshoot, rise time, and settling time of the response to step input. The steady-state error is the difference between the desired response and the actual response to a test input in steady-state. The system should also satisfy a specified sensitivity specifications to some system parameter variations, or to certain disturbances. Above all, the system must remain stable under operating conditions. Discussion of design procedures used to realize given specifications is beyond the scope of this book.

4.8 FREQUENCY RESPONSE OF AN LTIC SYSTEM

Filtering is an important area of signal processing. Filtering characteristics of a system are indicated by its response to sinusoids of various frequencies varying from 0 to ∞. Such characteristics are called the frequency response of the system. In this section, we shall find the frequency response of LTIC systems.

In Section 2.4-4 we showed that an LTIC system response to an everlasting exponential input $x(t) = e^{st}$ is also an everlasting exponential $H(s)e^{st}$. As before, we use an arrow directed from the input to the output to represent an input–output pair:

$$e^{st} \Longrightarrow H(s)e^{st} \qquad (4.71)$$

Setting $s = j\omega$ in this relationship yields

$$e^{j\omega t} \Longrightarrow H(j\omega)e^{j\omega t} \qquad (4.72)$$

Noting that $\cos \omega t$ is the real part of $e^{j\omega t}$, use of Eq. (2.40) yields

$$\cos \omega t \Longrightarrow \mathrm{Re}[H(j\omega)e^{j\omega t}] \qquad (4.73)$$

We can express $H(j\omega)$ in the polar form as

$$H(j\omega) = |H(j\omega)|e^{j\angle H(j\omega)} \qquad (4.74)$$

With this result, the relationship (4.73) becomes

$$\cos \omega t \Longrightarrow |H(j\omega)| \cos [\omega t + \angle H(j\omega)]$$

In other words, the system response $y(t)$ to a sinusoidal input $\cos \omega t$ is given by

$$y(t) = |H(j\omega)| \cos [\omega t + \angle H(j\omega)] \qquad (4.75a)$$

Using a similar argument, we can show that the system response to a sinusoid $\cos (\omega t + \theta)$ is

$$y(t) = |H(j\omega)| \cos [\omega t + \theta + \angle H(j\omega)] \qquad (4.75b)$$

This result is valid only for BIBO-stable systems. The frequency response is meaningless for BIBO-unstable systems. This follows from the fact that the frequency response in Eq. (4.72) is obtained by setting $s = j\omega$ in Eq. (4.71). But, as shown in Section 2.4-4 [Eqs. (2.47) and (2.48)], the relationship (4.71) applies only for the values of s for which $H(s)$ exists. For BIBO-unstable systems, the ROC for $H(s)$ does not include the $j\omega$ axis where $s = j\omega$ [see Eq. (4.14)]. This means that $H(s)$ when $s = j\omega$, is meaningless for BIBO-unstable systems.[†]

Equation (4.75b) shows that for a sinusoidal input of radian frequency ω, the system response is also a sinusoid of the same frequency ω. *The amplitude of the output sinusoid is $|H(j\omega)|$ times the input amplitude, and the phase of the output sinusoid is shifted by $\angle H(j\omega)$ with respect to the input phase* (see later Fig. 4.36 in Example 4.23). For instance, a certain system with $|H(j10)| = 3$ and $\angle H(j10) = -30°$ amplifies a sinusoid of frequency $\omega = 10$ by a factor of 3 and delays its phase by $30°$. The system response to an input $5 \cos(10t + 50°)$ is $3 \times 5 \cos(10t + 50° - 30°) = 15 \cos(10t + 20°)$.

Clearly $|H(j\omega)|$ is the amplitude *gain* of the system, and a plot of $|H(j\omega)|$ versus ω shows the amplitude gain as a function of frequency ω. We shall call $|H(j\omega)|$ the *amplitude response*. It also goes under the name *magnitude response* in the literature.[‡] Similarly, $\angle H(j\omega)$ is the *phase response* and a plot of $\angle H(j\omega)$ versus ω shows how the system modifies or changes the phase of the input sinusoid. Observe that $H(j\omega)$ has the information of $|H(j\omega)|$ and $\angle H(j\omega)$. For this reason, $H(j\omega)$ is also called the *frequency response of the system*. The frequency response plots $|H(j\omega)|$ and $\angle H(j\omega)$ show at a glance how a system responds to sinusoids of various frequencies. Thus, the frequency response of a system represents its filtering characteristic.

EXAMPLE 4.23

Find the frequency response (amplitude and phase response) of a system whose transfer function is

$$H(s) = \frac{s + 0.1}{s + 5}$$

Also, find the system response $y(t)$ if the input $x(t)$ is

(a) $\cos 2t$

(b) $\cos(10t - 50°)$

In this case

$$H(j\omega) = \frac{j\omega + 0.1}{j\omega + 5}$$

[†]This may also be argued as follows. For BIBO-unstable systems, the zero-input response contains nondecaying natural mode terms of the form $\cos \omega_0 t$ or $e^{at} \cos \omega_0 t$ ($a > 0$). Hence, the response of such a system to a sinusoid $\cos \omega t$ will contain not just the sinusoid of frequency ω, but also nondecaying natural modes, rendering the concept of frequency response meaningless. Alternately, we can argue that when $s = j\omega$, a BIBO-unstable system violates the dominance condition Re $\lambda_i < $ Re $j\omega$ for all i, where λ_i represents ith characteristic root of the system (see Section 4.3-1).

[‡]Strictly speaking, $|H(\omega)|$ is magnitude response. There is a fine distinction between amplitude and magnitude. Amplitude A can be positive and negative. In contrast, the magnitude $|A|$ is always nonnegative. We refrain from relying on this useful distinction between amplitude and magnitude in the interest of avoiding proliferation of essentially similar entities. This is also why we shall use the "amplitude" (instead of "magnitude") spectrum for $|H(\omega)|$.

Therefore

$$|H(j\omega)| = \frac{\sqrt{\omega^2 + 0.01}}{\sqrt{\omega^2 + 25}} \qquad \text{and} \qquad \angle H(j\omega) = \tan^{-1}\left(\frac{\omega}{0.1}\right) - \tan^{-1}\left(\frac{\omega}{5}\right)$$

Both the amplitude and the phase response are depicted in Fig. 4.36a as functions of ω. These plots furnish the complete information about the frequency response of the system to sinusoidal inputs.

(a) For the input $x(t) = \cos 2t$, $\omega = 2$, and

$$|H(j2)| = \frac{\sqrt{(2)^2 + 0.01}}{\sqrt{(2)^2 + 25}} = 0.372$$

$$\angle H(j2) = \tan^{-1}\left(\frac{2}{0.1}\right) - \tan^{-1}\left(\frac{2}{5}\right) = 87.1° - 21.8° = 65.3°$$

We also could have read these values directly from the frequency response plots in Fig. 4.36a corresponding to $\omega = 2$. This result means that for a sinusoidal input with frequency $\omega = 2$, the amplitude gain of the system is 0.372, and the phase shift is 65.3°. In other words, the

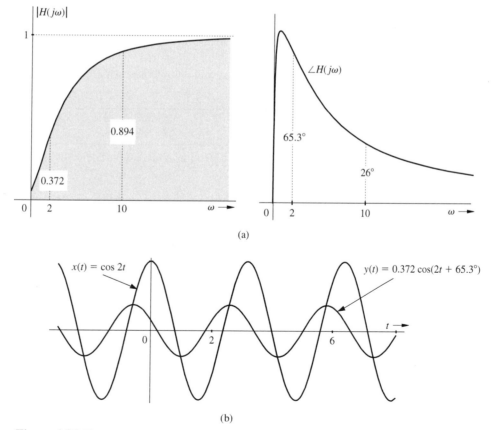

(a)

(b)

Figure 4.36 Frequency responses of the LTIC system.

output amplitude is 0.372 times the input amplitude, and the phase of the output is shifted with respect to that of the input by 65.3°. Therefore, the system response to an input $\cos 2t$ is

$$y(t) = 0.372 \cos (2t + 65.3°)$$

The input $\cos 2t$ and the corresponding system response $0.372 \cos (2t + 65.3°)$ are illustrated in Fig. 4.36b.

(b) For the input $\cos (10t - 50°)$, instead of computing the values $|H(j\omega)|$ and $\angle H(j\omega)$ as in part (a), we shall read them directly from the frequency response plots in Fig. 4.36a corresponding to $\omega = 10$. These are:

$$|H(j10)| = 0.894 \qquad \text{and} \qquad \angle H(j10) = 26°$$

Therefore, for a sinusoidal input of frequency $\omega = 10$, the output sinusoid amplitude is 0.894 times the input amplitude, and the output sinusoid is shifted with respect to the input sinusoid by 26°. Therefore, $y(t)$, the system response to an input $\cos (10t - 50°)$, is

$$y(t) = 0.894 \cos (10t - 50° + 26°) = 0.894 \cos (10t - 24°)$$

If the input were $\sin (10t - 50°)$, the response would be $0.894 \sin (10t - 50° + 26°) = 0.894 \sin (10t - 24°)$.

The frequency response plots in Fig. 4.36a show that the system has highpass filtering characteristics; it responds well to sinusoids of higher frequencies (ω well above 5), and suppresses sinusoids of lower frequencies (ω well below 5).

COMPUTER EXAMPLE C4.5

Plot the frequency response of the transfer function $H(s) = (s + 5)/(s^2 + 3s + 2)$.

```
>> H = tf([1 5],[1 3 2]);
>> bode(H,'k-',{0.1 100});
```

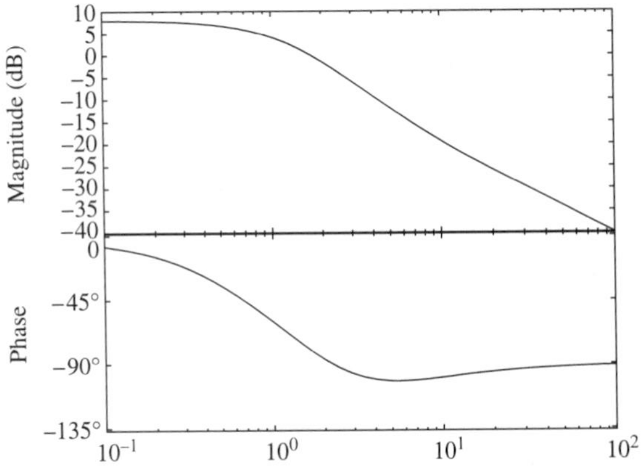

Figure C4.5

EXAMPLE 4.24

Find and sketch the frequency response (amplitude and phase response) for the following.

    (a) an ideal delay of T seconds

    (b) an ideal differentiator

    (c) an ideal integrator

    (a) Ideal delay of T seconds. The transfer function of an ideal delay is [see Eq. (4.46)]

$$H(s) = e^{-sT}$$

Therefore

$$H(j\omega) = e^{-j\omega T}$$

Consequently

$$|H(j\omega)| = 1 \qquad \text{and} \qquad \angle H(j\omega) = -\omega T \tag{4.76}$$

These amplitude and phase responses are shown in Fig. 4.37a. The amplitude response is constant (unity) for all frequencies. The phase shift increases linearly with frequency with a slope of $-T$. This result can be explained physically by recognizing that if a sinusoid $\cos \omega t$ is passed through an ideal delay of T seconds, the output is $\cos \omega (t - T)$. The output sinusoid amplitude is the same as that of the input for all values of ω. Therefore, the amplitude response (gain) is unity for all frequencies. Moreover, the output $\cos \omega (t - T) = \cos (\omega t - \omega T)$ has a phase shift $-\omega T$ with respect to the input $\cos \omega t$. Therefore, the phase response is linearly proportional to the frequency ω with a slope $-T$.

    (b) An ideal differentiator. The transfer function of an ideal differentiator is [see Eq. (4.47)]

$$H(s) = s$$

Therefore

$$H(j\omega) = j\omega = \omega e^{j\pi/2}$$

Consequently

$$|H(j\omega)| = \omega \qquad \text{and} \qquad \angle H(j\omega) = \frac{\pi}{2} \tag{4.77}$$

This amplitude and phase response are depicted in Fig. 4.37b. The amplitude response increases linearly with frequency, and phase response is constant ($\pi/2$) for all frequencies. This result can be explained physically by recognizing that if a sinusoid $\cos \omega t$ is passed through an ideal differentiator, the output is $-\omega \sin \omega t = \omega \cos [\omega t + (\pi/2)]$. Therefore, the output sinusoid amplitude is ω times the input amplitude; that is, the amplitude response (gain) increases linearly with frequency ω. Moreover, the output sinusoid undergoes a phase shift $\pi/2$ with respect to the input $\cos \omega t$. Therefore, the phase response is constant ($\pi/2$) with frequency.

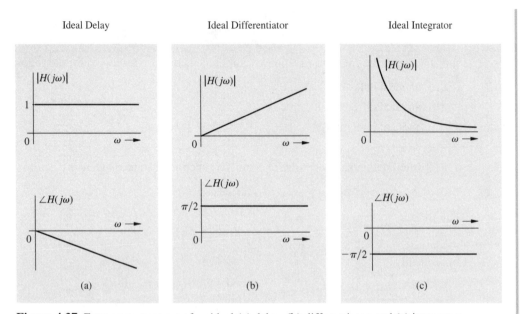

Figure 4.37 Frequency response of an ideal (a) delay, (b) differentiator, and (c) integrator.

In an ideal differentiator, the amplitude response (gain) is proportional to frequency $[|H(j\omega)| = \omega]$, so that the higher-frequency components are enhanced (see Fig. 4.37b). All practical signals are contaminated with noise, which, by its nature, is a broadband (rapidly varying) signal containing components of very high frequencies. A differentiator can increase the noise disproportionately to the point of drowning out the desired signal. This is why ideal differentiators are avoided in practice.

(c) An ideal integrator. The transfer function of an ideal integrator is [see Eq. (4.48)]

$$H(s) = \frac{1}{s}$$

Therefore

$$H(j\omega) = \frac{1}{j\omega} = \frac{-j}{\omega} = \frac{1}{\omega}e^{-j\pi/2}$$

Consequently

$$|H(j\omega)| = \frac{1}{\omega} \qquad \text{and} \qquad \angle H(j\omega) = -\frac{\pi}{2} \tag{4.78}$$

These amplitude and the phase responses are illustrated in Fig. 4.37c. The amplitude response is inversely proportional to frequency, and the phase shift is constant $(-\pi/2)$ with frequency. This result can be explained physically by recognizing that if a sinusoid $\cos \omega t$ is passed through an ideal integrator, the output is $(1/\omega) \sin \omega t = (1/\omega) \cos [\omega t - (\pi/2)]$. Therefore, the amplitude response is inversely proportional to ω, and the phase response is constant

$(-\pi/2)$ with frequency.[†] Because its gain is $1/\omega$, the ideal integrator suppresses higher-frequency components but enhances lower-frequency components with $\omega < 1$. Consequently, noise signals (if they do not contain an appreciable amount of very-low-frequency components) are suppressed (smoothed out) by an integrator.

EXERCISE E4.14

Find the response of an LTIC system specified by

$$\frac{d^2y}{dt^2} + 3\frac{dy}{dt} + 2y(t) = \frac{dx}{dt} + 5x(t)$$

if the input is a sinusoid $20 \sin (3t + 35°)$.

ANSWER
$10.23 \sin (3t - 61.91°)$

4.8-1 Steady-State Response to Causal Sinusoidal Inputs

So far we have discussed the LTIC system response to everlasting sinusoidal inputs (starting at $t = -\infty$). In practice, we are more interested in causal sinusoidal inputs (sinusoids starting at $t = 0$). Consider the input $e^{j\omega t}u(t)$, which starts at $t = 0$ rather than at $t = -\infty$. In this case $X(s) = 1/(s + j\omega)$. Moreover, according to Eq. (4.43), $H(s) = P(s)/Q(s)$, where $Q(s)$ is the characteristic polynomial given by $Q(s) = (s - \lambda_1)(s - \lambda_2) \cdots (s - \lambda_N)$.[‡] Hence,

$$Y(s) = X(s)H(s) = \frac{P(s)}{(s - \lambda_1)(s - \lambda_2) \cdots (s - \lambda_N)(s - j\omega)}$$

[†]A puzzling aspect of this result is that in deriving the transfer function of the integrator in Eq. (4.48), we have assumed that the input starts at $t = 0$. In contrast, in deriving its frequency response, we assume that the everlasting exponential input $e^{j\omega t}$ starts at $t = -\infty$. There appears to be a fundamental contradiction between the everlasting input, which starts at $t = -\infty$, and the integrator, which opens its gates only at $t = 0$. What use is everlasting input, since the integrator starts integrating at $t = 0$? The answer is that the integrator gates are always open, and integration begins whenever the input starts. We restricted the input to start at $t = 0$ in deriving Eq. (4.48) because we were finding the transfer function using the unilateral transform, where the inputs begin at $t = 0$. So the integrator starting to integrate at $t = 0$ is restricted because of the limitations of the unilateral transform method, not because of the limitations of the integrator itself. If we were to find the integrator transfer function using Eq. (2.49), where there is no such restriction on the input, we would still find the transfer function of an integrator as $1/s$. Similarly, even if we were to use the bilateral Laplace transform, where t starts at $-\infty$, we would find the transfer function of an integrator to be $1/s$. The transfer function of a system is the property of the system and does not depend on the method used to find it.

[‡]For simplicity, we have assumed nonrepeating characteristic roots. The procedure is readily modified for repeated roots, and the same conclusion results.

In the partial fraction expansion of the right-hand side, let the coefficients corresponding to the N terms $(s - \lambda_1)$, $(s - \lambda_2)$, ..., $(s - \lambda_N)$ be $k_1, k_2, \ldots, k_N$. The coefficient corresponding to the last term $(s - j\omega)$ is $P(s)/Q(s)|_{s=j\omega} = H(j\omega)$. Hence,

$$Y(s) = \sum_{i=1}^{n} \frac{k_i}{s - \lambda_i} + \frac{H(j\omega)}{s - j\omega}$$

and

$$y(t) = \underbrace{\sum_{i=1}^{n} k_i e^{\lambda_i t} u(t)}_{\text{transient component } y_{tr}(t)} + \underbrace{H(j\omega)e^{j\omega t} u(t)}_{\text{steady-state component } y_{ss}(t)} \qquad (4.79)$$

For an asymptotically stable system, the characteristic mode terms $e^{\lambda_i t}$ decay with time, and, therefore, constitute the so-called *transient* component of the response. The last term $H(j\omega)e^{j\omega t}$ persists forever, and is the *steady-state* component of the response given by

$$y_{ss}(t) = H(j\omega)e^{j\omega t} u(t)$$

This result also explains why an everlasting exponential input $e^{j\omega t}$ results in the total response $H(j\omega)e^{j\omega t}$ for BIBO systems. Because the input started at $t = -\infty$, at any finite time the decaying transient component has long vanished, leaving only the steady-state component. Hence, the total response appears to be $H(j\omega)e^{j\omega t}$.

From the argument that led to Eq. (4.75a), it follows that for a causal sinusoidal input $\cos \omega t$, the steady-state response $y_{ss}(t)$ is given by

$$y_{ss}(t) = |H(j\omega)|\cos[\omega t + \angle H(j\omega)]u(t) \qquad (4.80)$$

In summary, $|H(j\omega)|\cos[\omega t + \angle H(j\omega)]$ is the total response to everlasting sinusoid $\cos \omega t$. In contrast, it is the steady-state response to the same input applied at $t = 0$.

4.9 BODE PLOTS

Sketching frequency response plots ($|H(j\omega)|$ and $\angle H(j\omega)$ versus ω) is considerably facilitated by the use of logarithmic scales. The amplitude and phase response plots as a function of ω on a logarithmic scale are known as *Bode plots*. By using the asymptotic behavior of the amplitude and the phase responses, we can sketch these plots with remarkable ease, even for higher-order transfer functions.

Let us consider a system with the transfer function

$$H(s) = \frac{K(s + a_1)(s + a_2)}{s(s + b_1)(s^2 + b_2 s + b_3)} \qquad (4.81a)$$

where the second-order factor $(s^2 + b_2 s + b_3)$ is assumed to have complex conjugate roots.[†] We

---

[†]Coefficients a_1, a_2 and b_1, b_2, b_3 used in this section are not to be confused with those used in the representation of Nth-order LTIC system equations given earlier [Eqs. (2.1) or (4.41)].

shall rearrange Eq. (4.81a) in the form

$$H(s) = \frac{Ka_1a_2}{b_1b_3} \frac{\left(\dfrac{s}{a_1}+1\right)\left(\dfrac{s}{a_2}+1\right)}{s\left(\dfrac{s}{b_1}+1\right)\left(\dfrac{s^2}{b_3}+\dfrac{b_2}{b_3}s+1\right)} \tag{4.81b}$$

and

$$H(j\omega) = \frac{Ka_1a_2}{b_1b_3} \frac{\left(1+\dfrac{j\omega}{a_1}\right)\left(1+\dfrac{j\omega}{a_2}\right)}{j\omega\left(1+\dfrac{j\omega}{b_1}\right)\left[1+j\dfrac{b_2\omega}{b_3}+\dfrac{(j\omega)^2}{b_3}\right]} \tag{4.81c}$$

This equation shows that $H(j\omega)$ is a complex function of ω. The amplitude response $|H(j\omega)|$ and the phase response $\angle H(j\omega)$ are given by

$$|H(j\omega)| = \frac{Ka_1a_2}{b_1b_3} \frac{\left|1+\dfrac{j\omega}{a_1}\right|\left|1+\dfrac{j\omega}{a_2}\right|}{|j\omega|\left|1+\dfrac{j\omega}{b_1}\right|\left|1+j\dfrac{b_2\omega}{b_3}+\dfrac{(j\omega)^2}{b_3}\right|} \tag{4.82a}$$

and

$$\angle H(j\omega) = \angle\left(1+\frac{j\omega}{a_1}\right) + \angle\left(1+\frac{j\omega}{a_2}\right) - \angle j\omega$$

$$- \angle\left(1+\frac{j\omega}{b_1}\right) - \angle\left[1+\frac{jb_2\omega}{b_3}+\frac{(j\omega)^2}{b_3}\right] \tag{4.82b}$$

From Eq. (4.82b) we see that the phase function consists of the addition of only three kinds of term: (i) the phase of $j\omega$, which is $90°$ for all values of ω, (ii) the phase for the first-order term of the form $1 + j\omega/a$, and (iii) the phase of the second-order term

$$\left[1+\frac{jb_2\omega}{b_3}+\frac{(j\omega)^2}{b_3}\right]$$

We can plot these three basic phase functions for ω in the range 0 to ∞ and then, using these plots, we can construct the phase function of any transfer function by properly adding these basic responses. Note that if a particular term is in the numerator, its phase is added, but if the term is in the denominator, its phase is subtracted. This makes it easy to plot the phase function $\angle H(j\omega)$ as a function of ω. Computation of $|H(j\omega)|$, unlike that of the phase function, however, involves the multiplication and division of various terms. This is a formidable task, especially when we have to plot this function for the entire range of ω (0 to ∞).

We know that a log operation converts multiplication and division to addition and subtraction. So, instead of plotting $|H(j\omega)|$, why not plot log $|H(j\omega)|$ to simplify our task? We can take advantage of the fact that logarithmic units are desirable in several applications, where the variables considered have a very large range of variation. This is particularly true in frequency response plots, where we may have to plot frequency response over a range from a very low frequency, near 0, to a very high frequency, in the range of 10^{10} or higher. A plot on a linear scale of frequencies for such a large range will bury much of the useful information at lower frequencies. Also, the amplitude response may have a very large dynamic range from a low of 10^{-6} to a high of 10^6. A linear plot would be unsuitable for such a situation. Therefore, logarithmic plots not only simplify our task of plotting, but, fortunately, they are also desirable in this situation.

There is another important reason for using logarithmic scale. The Weber–Fechner law (first observed by Weber in 1834) states that human senses (sight, touch, hearing, etc.) generally respond in logarithmic way. For instance, when we hear sound at two different power levels, we judge one sound twice as loud when the ratio of the two sound powers is 10. Human senses respond to equal ratios of power, not equal increments in power.[11] This is clearly a logarithmic response.[†]

The logarithmic unit is the *decibel* and is equal to 20 times the logarithm of the quantity (log to the base 10). Therefore, $20 \log_{10} |H(j\omega)|$ is simply the log amplitude in decibels (dB).[‡] Thus, instead of plotting $|H(j\omega)|$, we shall plot $20 \log_{10} |H(j\omega)|$ as a function of ω. These plots (log amplitude and phase) are called *Bode plots*. For the transfer function in Eq. (4.82a), the *log amplitude* is

$$20 \log |H(j\omega)| = 20 \log \frac{K a_1 a_2}{b_1 b_3} + 20 \log \left| 1 + \frac{j\omega}{a_1} \right| + 20 \log \left| 1 + \frac{j\omega}{a_2} \right| - 20 \log |j\omega|$$

$$- 20 \log \left| 1 + \frac{j\omega}{b_1} \right| - 20 \log \left| 1 + \frac{j b_2 \omega}{b_3} + \frac{(j\omega)^2}{b_3} \right| \tag{4.83}$$

The term $20 \log(K a_1 a_2 / b_1 b_3)$ is a constant. We observe that the log amplitude is a sum of four basic terms corresponding to a constant, a pole or zero at the origin ($20 \log |j\omega|$), a first-order pole or zero ($20 \log |1 + j\omega/a|$), and complex-conjugate poles or zeros ($20 \log |1 + j\omega b_2 / b_3 + (j\omega)^2 / b_3|$). We can sketch these four basic terms as functions of ω and use them to construct the log–amplitude plot of any desired transfer function. Let us discuss each of the terms.

4.9-1 Constant $k a_1 a_2 / b_1 b_3$

The log amplitude of the constant $k a_1 a_2 / b_1 b_2$ term is also a constant, $20 \log(K a_1 a_2 / b_1 b_3)$. The phase contribution from this term is zero for positive value and π for negative value of the constant.

[†]Observe that that the frequencies of musical notes are spaced logarithmically (not linearly). The octave is a ratio of 2. The frequencies of the same note in the successive octaves have a ratio of 2. On the Western musical scale, there are 12 distinct notes in each octave. The frequency of each note is about 6% higher than the frequency of the preceding note. Thus, the successive notes are separated not by some constant frequency, but by constant ratio of 1.06.

[‡]Originally, the unit *bel* (after the inventor of telephone, Alexander Graham Bell) was introduced to represent power ratio as $\log_{10} P_2/P_1$ bels. A tenth of this unit is a decibel, as in $10 \log_{10} P_2/P_1$ decibels. Since, the power ratio of two signals is proportional to the amplitude ratio squared, or $|H(j\omega)|^2$, we have $10 \log_{10} P_2/P_1 = 10 \log_{10} |H(j\omega)|^2 = 20 \log_{10} |H(j\omega)|$ dB.

4.9-2 Pole (or Zero) at the Origin

LOG MAGNITUDE

A pole at the origin gives rise to the term $-20 \log |j\omega|$, which can be expressed as

$$-20 \log |j\omega| = -20 \log \omega$$

This function can be plotted as a function of ω. However, we can effect further simplification by using the logarithmic scale for the variable ω itself. Let us define a new variable u such that

$$u = \log \omega$$

Hence

$$-20 \log \omega = -20u$$

The log–amplitude function $-20u$ is plotted as a function of u in Fig. 4.38a. This is a straight line with a slope of -20. It crosses the u axis at $u = 0$. The ω-scale ($u = \log \omega$) also appears in Fig. 4.38a. Semilog graphs can be conveniently used for plotting, and we can directly plot ω on semilog paper. A ratio of 10 is a *decade*, and a ratio of 2 is known as an *octave*. Furthermore, a decade along the ω-scale is equivalent to 1 unit along the u-scale. We can also show that a ratio of 2 (an octave) along the ω-scale equals to 0.3010 (which is $\log_{10} 2$) along the u-scale.[†]

Note that equal increments in u are equivalent to equal ratios on the ω-scale. Thus, one unit along the u-scale is the same as one decade along the ω-scale. This means that the amplitude plot has a slope of -20 dB/decade or $-20(0.3010) = -6.02$ dB/octave (commonly stated as -6 dB/octave). Moreover, the amplitude plot crosses the ω axis at $\omega = 1$, since $u = \log_{10} \omega = 0$ when $\omega = 1$.

For the case of a zero at the origin, the log–amplitude term is $20 \log \omega$. This is a straight line passing through $\omega = 1$ and having a slope of 20 dB/decade (or 6 dB/octave). This plot is a mirror image about the ω axis of the plot for a pole at the origin and is shown dashed in Fig. 4.38a.

PHASE

The phase function corresponding to the pole at the origin is $-\angle j\omega$ [see Eq. (4.82b)]. Thus

$$\angle H(j\omega) = -\angle j\omega = -90°$$

[†]This point can be shown as follows. Let ω_1 and ω_2 along the ω-scale correspond to u_1 and u_2 along the u-scale so that $\log \omega_1 = u_1$, and $\log \omega_2 = u_2$. Then

$$u_2 - u_1 = \log_{10} \omega_2 - \log_{10} \omega_1 = \log_{10} (\omega_2/\omega_1)$$

Thus, if $(\omega_2/\omega_1) = 10$ (which is a decade)

then $u_2 - u_1 = \log_{10} 10 = 1$

and if $(\omega_2/\omega_1) = 2$ (which is an octave)

then $u_2 - u_1 = \log_{10} 2 = 0.3010$

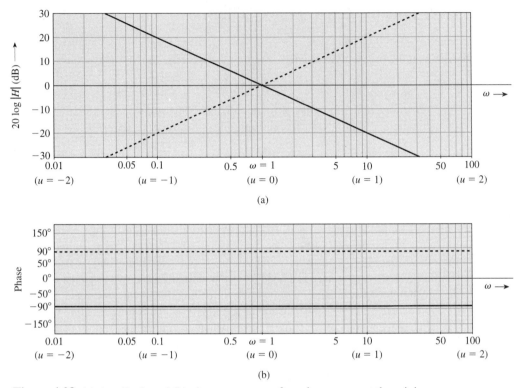

Figure 4.38 (a) Amplitude and (b) phase responses of a pole or a zero at the origin.

The phase is constant $(-90°)$ for all values of ω, as depicted in Fig. 4.38b. For a zero at the origin, the phase is $\angle j\omega = 90°$. This is a mirror image of the phase plot for a pole at the origin and is shown dotted in Fig. 4.38b.

4.9-3 First-Order Pole (or Zero)

THE LOG MAGNITUDE

The log amplitude of a first-order pole at $-a$ is $-20 \log |1 + j\omega/a|$. Let us investigate the asymptotic behavior of this function for extreme values of ω ($\omega \ll a$ and $\omega \gg a$).

(a) For $\omega \ll a$,

$$-20 \log \left| 1 + \frac{j\omega}{a} \right| \approx -20 \log 1 = 0$$

Hence, the log–amplitude function $\rightarrow 0$ asymptotically for $\omega \ll a$ (Fig. 4.39a).

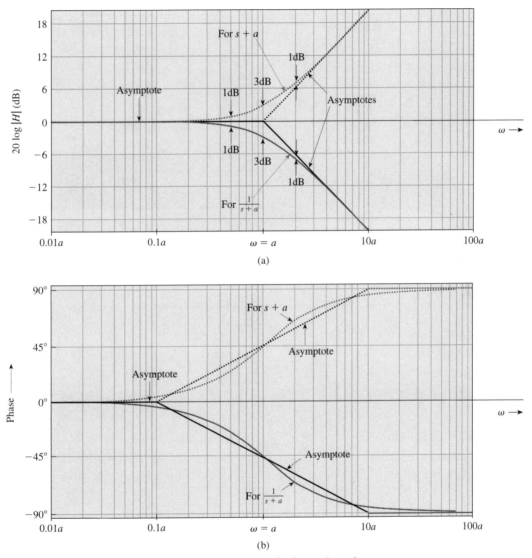

Figure 4.39 (a) Amplitude and (b) phase responses of a first-order pole or zero at $s = -a$.

(b) For the other extreme case, where $\omega \gg a$,

$$-20\log\left|1 + \frac{j\omega}{a}\right| \approx -20\log\left(\frac{\omega}{a}\right) \tag{4.84a}$$

$$= -20\log\omega + 20\log a \tag{4.84b}$$

$$= -20u + 20\log a$$

This represents a straight line (when plotted as a function of u, the log of ω) with a slope of -20 dB/decade (or -6 dB/octave). When $\omega = a$, the log amplitude is zero [Eq. (4.84b)].

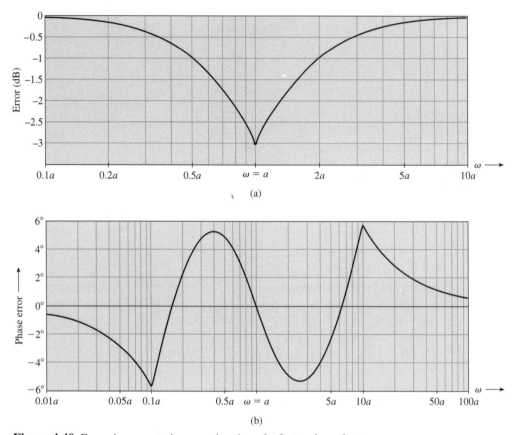

Figure 4.40 Errors in asymptotic approximation of a first-order pole at $s = -a$.

Hence, this line crosses the ω axis at $\omega = a$, as illustrated in Fig. 4.39a. Note that the asymptotes in (a) and (b) meet at $\omega = a$.

The exact log amplitude for this pole is

$$-20\log\left|1 + \frac{j\omega}{a}\right| = -20\log\left(1 + \frac{\omega^2}{a^2}\right)^{1/2}$$

$$= -10\log\left(1 + \frac{\omega^2}{a^2}\right)$$

This exact log magnitude function also appears in Fig. 4.39a. Observe that the actual and the asymptotic plots are very close. A maximum error of 3 dB occurs at $\omega = a$. This frequency is known as the *corner frequency* or *break frequency*. The error everywhere else is less than 3 dB. A plot of the error as a function of ω is shown in Fig. 4.40a. This figure shows that the error at one octave above or below the corner frequency is 1 dB and the error at two octaves above or below the corner frequency is 0.3 dB. The actual plot can be obtained by adding the error to the asymptotic plot.

The amplitude response for a zero at $-a$ (shown dotted in Fig. 4.39a) is identical to that of the pole at $-a$ with a sign change and therefore is the mirror image (about the 0 dB line) of the amplitude plot for a pole at $-a$.

PHASE

The phase for the first-order pole at $-a$ is

$$\angle H(j\omega) = -\angle\left(1 + \frac{j\omega}{a}\right) = -\tan^{-1}\left(\frac{\omega}{a}\right)$$

Let us investigate the asymptotic behavior of this function. For $\omega \ll a$,

$$-\tan^{-1}\left(\frac{\omega}{a}\right) \approx 0$$

and, for $\omega \gg a$,

$$-\tan^{-1}\left(\frac{\omega}{a}\right) \approx -90°$$

The actual plot along with the asymptotes is depicted in Fig. 4.39b. In this case, we use a three-line segment asymptotic plot for greater accuracy. The asymptotes are a phase angle of $0°$ for $\omega \le a/10$, a phase angle of $-90°$ for $\omega \ge 10a$, and a straight line with a slope $-45°$/decade connecting these two asymptotes (from $\omega = a/10$ to $10a$) crossing the ω axis at $\omega = a/10$. It can be seen from Fig. 4.39b that the asymptotes are very close to the curve and the maximum error is $5.7°$. Figure 4.40b plots the error as a function of ω; the actual plot can be obtained by adding the error to the asymptotic plot.

The phase for a zero at $-a$ (shown dotted in Fig. 4.39b) is identical to that of the pole at $-a$ with a sign change, and therefore is the mirror image (about the $0°$ line) of the phase plot for a pole at $-a$.

4.9-4 Second-Order Pole (or Zero)

Let us consider the second-order pole in Eq. (4.81a). The denominator term is $s^2 + b_2 s + b_3$. We shall introduce the often-used standard form $s^2 + 2\zeta\omega_n s + \omega_n^2$ instead of $s^2 + b_2 s + b_3$. With this form, the log amplitude function for the second-order term in Eq. (4.83) becomes

$$-20\log\left|1 + 2j\zeta\frac{\omega}{\omega_n} + \left(\frac{j\omega}{\omega_n}\right)^2\right| \tag{4.85a}$$

and the phase function is

$$-\angle\left[1 + 2j\zeta\frac{\omega}{\omega_n} + \left(\frac{j\omega}{\omega_n}\right)^2\right] \tag{4.85b}$$

THE LOG MAGNITUDE

The log amplitude is given by

$$\text{log amplitude} = -20\log\left|1 + 2j\zeta\left(\frac{\omega}{\omega_n}\right) + \left(\frac{j\omega}{\omega_n}\right)^2\right| \tag{4.86}$$

For $\omega \ll \omega_n$, the log amplitude becomes

$$\text{log amplitude} \approx -20 \log 1 = 0$$

For $\omega \gg \omega_n$, the log amplitude is

$$\text{log amplitude} \approx -20 \log \left| \left(-\frac{\omega}{\omega_n} \right)^2 \right| = -40 \log \left(\frac{\omega}{\omega_n} \right) \tag{4.87a}$$

$$= -40 \log \omega - 40 \log \omega_n \tag{4.87b}$$

$$= -40u - 40 \log \omega_n \tag{4.87c}$$

The two asymptotes are zero for $\omega < \omega_n$ and $-40u - 40 \log \omega_n$ for $\omega > \omega_n$. The second asymptote is a straight line with a slope of -40 dB/decade (or -12 dB/octave) when plotted against the log ω scale. It begins at $\omega = \omega_n$ [see Eq. (4.87b)]. The asymptotes are depicted in Fig. 4.41a. The exact log amplitude is given by [see Eq. (4.86)]

$$\text{log amplitude} = -20 \log \left\{ \left[1 - \left(\frac{\omega}{\omega_n} \right)^2 \right]^2 + 4\zeta^2 \left(\frac{\omega}{\omega_n} \right)^2 \right\}^{1/2} \tag{4.88}$$

The log amplitude in this case involves a parameter ζ, resulting in a different plot for each value of ζ. For complex-conjugate poles,[†] $\zeta < 1$. Hence, we must sketch a family of curves for a number of values of ζ in the range 0 to 1. This is illustrated in Fig. 4.41a. The error between the actual plot and the asymptotes is shown in Fig. 4.42. The actual plot can be obtained by adding the error to the asymptotic plot.

For second-order zeros (complex-conjugate zeros), the plots are mirror images (about the 0 dB line) of the plots depicted in Fig. 4.41a. Note the resonance phenomenon of the complex-conjugate poles. This phenomenon is barely noticeable for $\zeta > 0.707$ but becomes pronounced as $\zeta \to 0$.

PHASE

The phase function for second-order poles, as apparent in Eq. (4.85b), is

$$\angle H(j\omega) = -\tan^{-1} \left[\frac{2\zeta \left(\dfrac{\omega}{\omega_n} \right)}{1 - \left(\dfrac{\omega}{\omega_n} \right)^2} \right] \tag{4.89}$$

For $\omega \ll \omega_n$,

$$\angle H(j\omega) \approx 0$$

[†]For $\zeta \geq 1$, the two poles in the second-order factor are no longer complex but real, and each of these two real poles can be dealt with as a separate first-order factor.

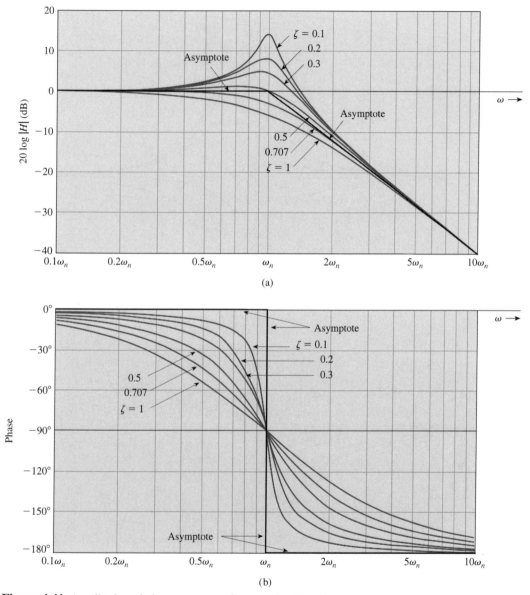

Figure 4.41 Amplitude and phase response of a second-order pole.

For $\omega \gg \omega_n$,

$$\angle H(j\omega) \simeq -180°$$

Hence, the phase $\to -180°$ as $\omega \to \infty$. As in the case of amplitude, we also have a family of phase plots for various values of ζ, as illustrated in Fig. 4.41b. A convenient asymptote for

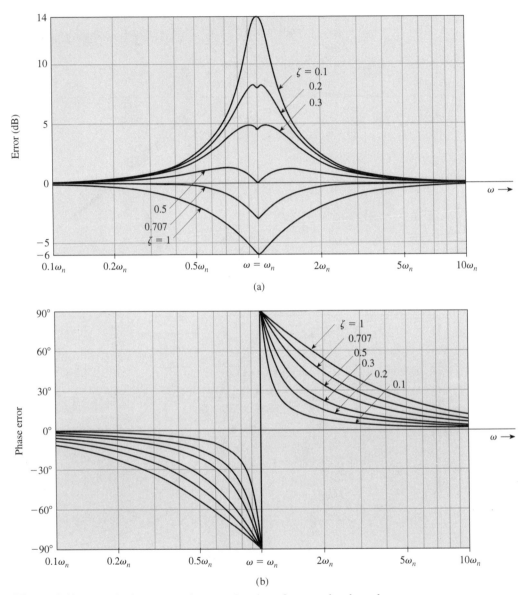

Figure 4.42 Errors in the asymptotic approximation of a second-order pole.

the phase of complex-conjugate poles is a step function that is $0°$ for $\omega < \omega_n$ and $-180°$ for $\omega > \omega_n$. An error plot for such an asymptote is shown in Fig. 4.42 for various values of ζ. The exact phase is the asymptotic value plus the error.

For complex-conjugate zeros, the amplitude and phase plots are mirror images of those for complex conjugate-poles.

We shall demonstrate the application of these techniques with two examples.

EXAMPLE 4.25

Sketch Bode plots for the transfer function

$$H(s) = \frac{20s(s+100)}{(s+2)(s+10)}$$

First, we write the transfer function in normalized form

$$H(s) = \frac{20 \times 100}{2 \times 10} \frac{s\left(1 + \dfrac{s}{100}\right)}{\left(1 + \dfrac{s}{2}\right)\left(1 + \dfrac{s}{10}\right)} = 100 \frac{s\left(1 + \dfrac{s}{100}\right)}{\left(1 + \dfrac{s}{2}\right)\left(1 + \dfrac{s}{10}\right)}$$

Here, the constant term is 100; that is, 40 dB (20 log 100 = 40). This term can be added to the plot by simply relabeling the horizontal axis (from which the asymptotes begin) as the 40 dB line (see Fig. 4.43a). Such a step implies shifting the horizontal axis upward by 40 dB. This is precisely what is desired.

In addition, we have two first-order poles at -2 and -10, one zero at the origin, and one zero at -100.

Step 1. For each of these terms, we draw an asymptotic plot as follows (shown in Fig. 4.43a by dotted lines):

   (i) For the zero at the origin, draw a straight line with a slope of 20 dB/decade passing through $\omega = 1$.

   (ii) For the pole at -2, draw a straight line with a slope of -20 dB/decade (for $\omega > 2$) beginning at the corner frequency $\omega = 2$.

   (iii) For the pole at -10, draw a straight line with a slope of -20 dB/decade beginning at the corner frequency $\omega = 10$.

   (iv) For the zero at -100, draw a straight line with a slope of 20 dB/decade beginning at the corner frequency $\omega = 100$.

Step 2. Add all the asymptotes, as depicted in Fig. 4.43a by solid line segments.

Step 3. Apply the following corrections (see Fig. 4.40a):

   (i) The correction at $\omega = 1$ because of the corner frequency at $\omega = 2$ is -1 dB. The correction at $\omega = 1$ because of the corner frequencies at $\omega = 10$ and $\omega = 100$ are quite small (see Fig. 4.40a) and may be ignored. Hence, the net correction at $\omega = 1$ is -1 dB.

   (ii) The correction at $\omega = 2$ because of the corner frequency at $\omega = 2$ is -3 dB, and the correction because of the corner frequency at $\omega = 10$ is -0.17 dB. The correction because of the corner frequency $\omega = 100$ can be safely ignored. Hence the net correction at $\omega = 2$ is -3.17 dB.

   (iii) The correction at $\omega = 10$ because of the corner frequency at $\omega = 10$ is -3 dB, and the correction because of the corner frequency at $\omega = 2$ is -0.17 dB. The correction because of $\omega = 100$ can be ignored. Hence the net correction at $\omega = 10$ is -3.17 dB.

(iv) The correction at $\omega = 100$ because of the corner frequency at $\omega = 100$ is 3 dB, and the corrections because of the other corner frequencies may be ignored.

(v) In addition to the corrections at corner frequencies, we may consider corrections at intermediate points for more accurate plots. For instance, the corrections at $\omega = 4$ because of corner frequencies at $\omega = 2$ and 10 are -1 and about -0.65, totaling -1.65 dB. In the same way the correction at $\omega = 5$ because of corner frequencies at $\omega = 2$ and 10 are -0.65 and -1, totaling -1.65 dB.

With these corrections, the resulting amplitude plot is illustrated in Fig. 4.43a.

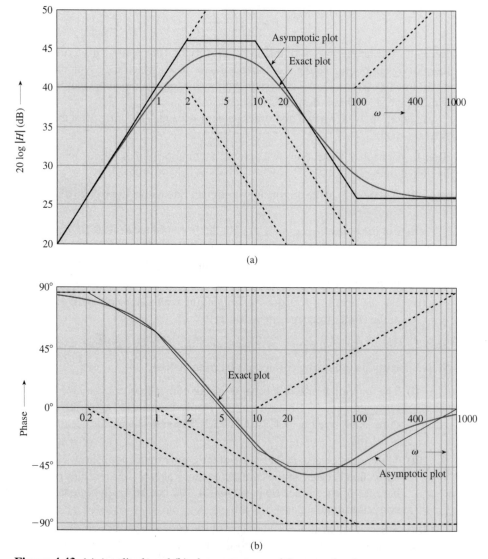

(a)

(b)

Figure 4.43 (a) Amplitude and (b) phase responses of the second-order system.

PHASE PLOTS

We draw the asymptotes corresponding to each of the four factors:

(i) The zero at the origin causes a 90° phase shift.

(ii) The pole at $s = -2$ has an asymptote with a zero value for $-\infty < \omega < 0.2$ and a slope of $-45°/$decade beginning at $\omega = 0.2$ and going up to $\omega = 20$. The asymptotic value for $\omega > 20$ is $-90°$.

(iii) The pole at $s = -10$ has an asymptote with a zero value for $-\infty < \omega < 1$ and a slope of $-45°/$decade beginning at $\omega = 1$ and going up to $\omega = 100$. The asymptotic value for $\omega > 100$ is $-90°$.

(iv) The zero at $s = -100$ has an asymptote with a zero value for $-\infty < \omega < 10$ and a slope of $-45°/$decade beginning at $\omega = 10$ and going up to $\omega = 1000$. The asymptotic value for $\omega > 1000$ is $90°$. All the asymptotes are added, as shown in Fig. 4.43b. The appropriate corrections are applied from Fig. 4.40b, and the exact phase plot is depicted in Fig. 4.43b.

EXAMPLE 4.26

Sketch the amplitude and phase response (Bode plots) for the transfer function

$$H(s) = \frac{10(s + 100)}{s^2 + 2s + 100} = 10\frac{1 + \dfrac{s}{100}}{1 + \dfrac{s}{50} + \dfrac{s^2}{100}}$$

Here, the constant term is 10: that is, 20 dB(20 log 10 = 20). To add this term, we simply label the horizontal axis (from which the asymptotes begin) as the 20 dB line, as before (see Fig. 4.44a).

In addition, we have a real zero at $s = -100$ and a pair of complex conjugate poles. When we express the second-order factor in standard form,

$$s^2 + 2s + 100 = s^2 + 2\zeta\omega_n s + \omega_n^2$$

we have

$$\omega_n = 10 \quad \text{and} \quad \zeta = 0.1$$

Step 1. Draw an asymptote of -40 dB/decade (-12 dB/octave) starting at $\omega = 10$ for the complex conjugate poles, and draw another asymptote of 20 dB/decade, starting at $\omega = 100$ for the (real) zero.

Step 2. Add both asymptotes.

Step 3. Apply the correction at $\omega = 100$, where the correction because of the corner frequency $\omega = 100$ is 3 dB. The correction because of the corner frequency $\omega = 10$, as seen from Fig. 4.42a for $\zeta = 0.1$, can be safely ignored. Next, the correction at $\omega = 10$, because of the corner frequency $\omega = 10$ is 13.90 dB (see Fig. 4.42a for $\zeta = 0.1$). The correction because of the real zero at -100 can be safely ignored at $\omega = 10$. We may find corrections at a few more points. The resulting plot is illustrated in Fig. 4.44a.

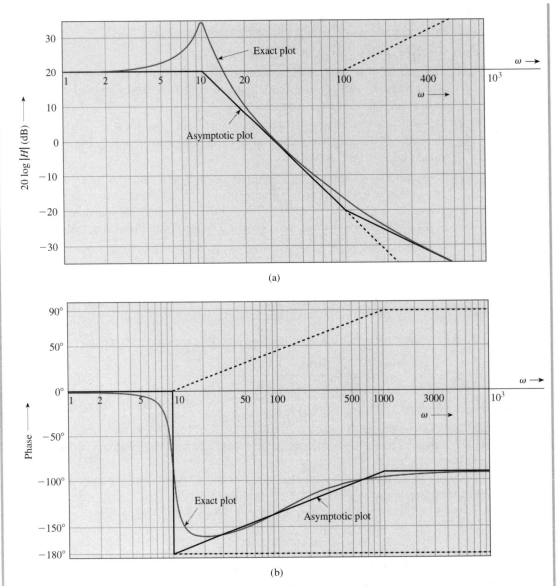

(a)

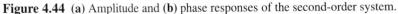

(b)

Figure 4.44 **(a)** Amplitude and **(b)** phase responses of the second-order system.

PHASE PLOT

The asymptote for the complex conjugate poles is a step function with a jump of $-90°$ at $\omega = 10$. The asymptote for the zero at $s = -100$ is zero for $\omega \le 10$ and is a straight line with a slope of $45°$/decade, starting at $\omega = 10$ and going to $\omega = 1000$. For $\omega \ge 1000$, the asymptote is $90°$. The two asymptotes add to give the sawtooth shown in Fig. 4.44b. We now apply the corrections from Figs. 4.42b and 4.40b to obtain the exact plot.

Comment. These two examples demonstrate that actual frequency response plots are very close to asymptotic plots, which are so easy to construct. Thus, by mere inspection of $H(s)$ and its poles and zeros, one can rapidly construct a mental image of the frequency response of a system. This is the principal virtue of Bode plots.

COMPUTER EXAMPLE C4.6

Use MATLAB function bode to solve Examples 4.25 and 4.26.

```
>> bode(tf(conv([20 0],[1 100]),conv([1 2],[1 10])),'k-',...
>>      tf([10 1000],[1 2 100]),'k:');
>> legend('Ex. 4.25','Ex. 4.26',0)
```

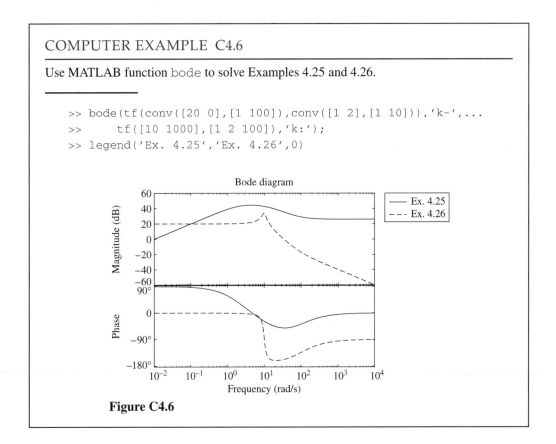

Figure C4.6

POLES AND ZEROS IN THE
RIGHT-HAND PLANE

In our discussion so far, we have assumed the poles and zeros of the transfer function to be in the left-hand plane. What if some of the poles and/or zeros of $H(s)$ lie in the RHP? If there is a pole in the RHP, the system is unstable. Such systems are useless for any signal processing application. For this reason, we shall consider only the case of the RHP zero. The term corresponding to RHP zero at $s = a$ is $(s/a) - 1$, and the corresponding frequency response is $(j\omega/a) - 1$. The amplitude response is

$$\left| \frac{j\omega}{a} - 1 \right| = \left(\frac{\omega^2}{a^2} + 1 \right)^{1/2}$$

This shows that the amplitude response of an RHP zero at $s = a$ is identical to that of an LHP zero or $s = -a$. Therefore, the log amplitude plots remain unchanged whether the zeros are in the LHP or the RHP. However, the phase corresponding to the RHP zero at $s = a$ is

$$\angle\left(\frac{j\omega}{a} - 1\right) = \angle - \left(1 - \frac{j\omega}{a}\right) = \pi + \tan^{-1}\left(\frac{-\omega}{a}\right) = \pi - \tan^{-1}\left(\frac{\omega}{a}\right)$$

whereas the phase corresponding to the LHP zero at $s = -a$ is $\tan^{-1}(\omega/a)$.

The complex-conjugate zeros in the RHP give rise to a term $s^2 - 2\zeta\omega_n s + \omega_n^2$, which is identical to the term $s^2 + 2\zeta\omega_n s + \omega_n^2$ with a sign change in ζ. Hence, from Eqs. (4.88) and (4.89) it follows that the amplitudes are identical, but the phases are of opposite signs for the two terms.

Systems whose poles and zeros are restricted to the LHP are classified as *minimum phase* systems.

4.9-5 The Transfer Function from the Frequency Response

In the preceding section we were given the transfer function of a system. From a knowledge of the transfer function, we developed techniques for determining the system response to sinusoidal inputs. We can also reverse the procedure to determine the transfer function of a minimum phase system from the system's response to sinusoids. This application has significant practical utility. If we are given a system in a black box with only the input and output terminals available, the transfer function has to be determined by experimental measurements at the input and output terminals. The frequency response to sinusoidal inputs is one of the possibilities that is very attractive because the measurements involved are so simple. One only needs to apply a sinusoidal signal at the input and observe the output. We find the amplitude gain $|H(j\omega)|$ and the output phase shift $\angle H(j\omega)$ (with respect to the input sinusoid) for various values of ω over the entire range from 0 to ∞. This information yields the frequency response plots (Bode plots) when plotted against $\log\omega$. From these plots we determine the appropriate asymptotes by taking advantage of the fact that the slopes of all asymptotes must be multiples of ± 20 dB/decade if the transfer function is a rational function (function that is a ratio of two polynomials in s). From the asymptotes, the corner frequencies are obtained. Corner frequencies determine the poles and zeros of the transfer function. Because of the ambiguity about the location of zeros since LHP and RHP zeros (zeros at $s = \pm a$) have identical magnitudes, this procedure works only for minimum phase systems.

4.10 FILTER DESIGN BY PLACEMENT OF POLES AND ZEROS OF $H(s)$

In this section we explore the strong dependence of frequency response on the location of poles and zeros of $H(s)$. This dependence points to a simple intuitive procedure to filter design.

4.10-1 Dependence of Frequency Response on Poles and Zeros of $H(s)$

Frequency response of a system is basically the information about the filtering capability of the system. A system transfer function can be expressed as

$$H(s) = \frac{P(s)}{Q(s)} = b_0 \frac{(s - z_1)(s - z_2) \cdots (s - z_N)}{(s - \lambda_1)(s - \lambda_2) \cdots (s - \lambda_N)} \tag{4.90a}$$

where $z_1, z_2, \ldots, z_N$ are $\lambda_1, \lambda_2, \ldots, \lambda_N$ are the poles of $H(s)$. Now the value of the transfer function $H(s)$ at some frequency $s = p$ is

$$H(s)|_{s=p} = b_0 \frac{(p - z_1)(p - z_2) \cdots (p - z_N)}{(p - \lambda_1)(p - \lambda_2) \cdots (p - \lambda_N)} \tag{4.90b}$$

This equation consists of factors of the form $p - z_i$ and $p - \lambda_i$. The factor $p - z_i$ is a complex number represented by a vector drawn from point z to the point p in the complex plane, as illustrated in Fig. 4.45a. The length of this line segment is $|p - z_i|$, the magnitude of $p - z_i$. The angle of this directed line segment (with the horizontal axis) is $\angle(p - z_i)$. To compute $H(s)$ at $s = p$, we draw line segments from all poles and zeros of $H(s)$ to the point p, as shown in Fig. 4.45b. The vector connecting a zero z_i to the point p is $p - z_i$. Let the length of this vector be r_i, and let its angle with the horizontal axis be ϕ_i. Then $p - z_i = r_i e^{j\phi_i}$. Similarly, the vector connecting a pole λ_i to the point p is $p - \lambda_i = d_i e^{j\theta_i}$, where d_i and θ_i are the length and the angle (with the horizontal axis), respectively, of the vector $p - \lambda_i$. Now from Eq. (4.90b) it follows that

$$H(s)|_{s=p} = b_0 \frac{(r_1 e^{j\phi_1})(r_2 e^{j\phi_2}) \cdots (r_N e^{j\phi_N})}{(d_1 e^{j\theta_1})(d_2 e^{j\theta_2}) \cdots (d_N e^{j\theta_N})}$$

$$= b_0 \frac{r_1 r_2 \cdots r_N}{d_1 d_2 \cdots d_N} e^{j[(\phi_1 + \phi_2 + \cdots + \phi_N) - (\theta_1 + \theta_2 + \cdots + \theta_N)]}$$

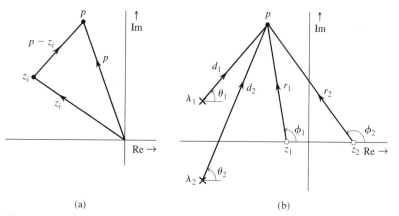

(a)   (b)

Figure 4.45 Vector representations of **(a)** complex numbers and **(b)** factors of $H(s)$.

Therefore

$$|H(s)|_{s=p} = b_0 \frac{r_1 r_2 \cdots r_N}{d_1 d_2 \cdots d_N}$$

$$= b_0 \frac{\text{product of the distances of zeros to } p}{\text{product of the distances of poles to } p} \tag{4.91a}$$

and

$$\angle H(s)|_{s=p} = (\phi_1 + \phi_2 + \cdots + \phi_N) - (\theta_1 + \theta_2 + \cdots + \theta_N)$$

$$= \text{sum of zero angles to } p - \text{sum of pole angles to } p \tag{4.91b}$$

Here, we have assumed positive b_0. If b_0 is negative, there is an additional phase π. Using this procedure, we can determine $H(s)$ for any value of s. To compute the frequency response $H(j\omega)$, we use $s = j\omega$ (a point on the imaginary axis), connect all poles and zeros to the point $j\omega$, and determine $|H(j\omega)|$ and $\angle H(j\omega)$ from Eqs. (4.91). We repeat this procedure for all values of ω from 0 to ∞ to obtain the frequency response.

GAIN ENHANCEMENT BY A POLE

To understand the effect of poles and zeros on the frequency response, consider a hypothetical case of a single pole $-\alpha + j\omega_0$, as depicted in Fig. 4.46a. To find the amplitude response $|H(j\omega)|$ for a certain value of ω, we connect the pole to the point $j\omega$ (Fig. 4.46a). If the length of this line is d, then $|H(j\omega)|$ is proportional to $1/d$,

$$|H(j\omega)| = \frac{K}{d} \tag{4.92}$$

where the exact value of constant K is not important at this point. As ω increases from zero, d decreases progressively until ω reaches the value ω_0. As ω increases beyond ω_0, d increases progressively. Therefore, according to Eq. (4.92), the amplitude response $|H(j\omega)|$ increases from $\omega = 0$ until $\omega = \omega_0$, and it decreases continuously as ω increases beyond ω_0, as illustrated in Fig. 4.46b. Therefore, a pole at $-\alpha + j\omega_0$ results in a frequency-selective behavior that enhances the gain at the frequency ω_0 (resonance). Moreover, as the pole moves closer to the imaginary axis (as α is reduced), this enhancement (resonance) becomes more pronounced. This is because α, the distance between the pole and $j\omega_0$ (d corresponding to $j\omega_0$), becomes smaller, which increases the gain K/d. In the extreme case, when $\alpha = 0$ (pole on the imaginary axis), the gain at ω_0 goes to infinity. Repeated poles further enhance the frequency-selective effect. To summarize, we can enhance a gain at a frequency ω_0 by placing a pole opposite the point $j\omega_0$. The closer the pole is to $j\omega_0$, the higher is the gain at ω_0, and the gain variation is more rapid (more frequency selective) in the vicinity of frequency ω_0. Note that a pole must be placed in the LHP for stability.

Here we have considered the effect of a single complex pole on the system gain. For a real system, a complex pole $-\alpha + j\omega_0$ must accompany its conjugate $-\alpha - j\omega_0$. We can readily show that the presence of the conjugate pole does not appreciably change the frequency-selective behavior in the vicinity of ω_0. This is because the gain in this case is K/dd', where d' is the distance of a point $j\omega$ from the conjugate pole $-\alpha - j\omega_0$. Because the conjugate pole is far from $j\omega_0$, there is no dramatic change in the length d' as ω varies in the vicinity of ω_0. There is a

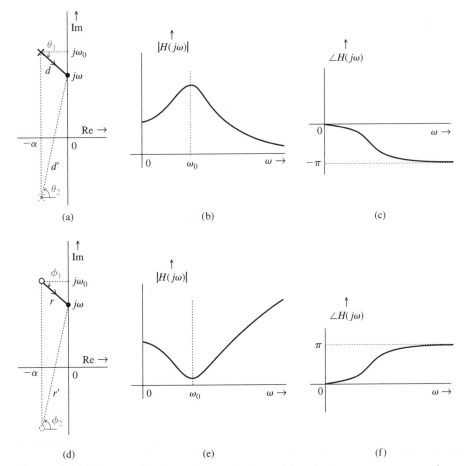

Figure 4.46 The role of poles and zeros in determining the frequency response of an LTIC system.

gradual increase in the value of d' as ω increases, which leaves the frequency-selective behavior as it was originally, with only minor changes.

GAIN SUPPRESSION BY A ZERO

Using the same argument, we observe that zeros at $-\alpha \pm j\omega_0$ (Fig. 4.46d) will have exactly the opposite effect of suppressing the gain in the vicinity of ω_0, as shown in Fig. 4.46e). A zero on the imaginary axis at $j\omega_0$ will totally suppress the gain (zero gain) at frequency ω_0. Repeated zeros will further enhance the effect. Also, a closely placed pair of a pole and a zero (dipole) tend to cancel out each other's influence on the frequency response. Clearly, a proper placement of poles and zeros can yield a variety of frequency-selective behavior. We can use these observations to design lowpass, highpass, bandpass, and bandstop (or notch) filters.

Phase response can also be computed graphically. In Fig. 4.46a, angles formed by the complex conjugate poles $-\alpha \pm j\omega_0$ at $\omega = 0$ (the origin) are equal and opposite. As ω increases from 0 up, the angle θ_1 (due to the pole $-\alpha + j\omega_0$), which has a negative value at $\omega = 0$, is reduced in magnitude; the angle θ_2 because of the pole $-\alpha - j\omega_0$, which has a positive value at $\omega = 0$, increases in magnitude. As a result, $\theta_1 + \theta_2$, the sum of the two angles, increases continuously, approaching a value π as $\omega \to \infty$. The resulting phase response $\angle H(j\omega) = -(\theta_1 + \theta_2)$ is illustrated in Fig. 4.46c. Similar arguments apply to zeros at $-\alpha \pm j\omega_0$. The resulting phase response $\angle H(j\omega) = (\phi_1 + \phi_2)$ is depicted in Fig. 4.46f.

We now focus on simple filters, using the intuitive insights gained in this discussion. The discussion is essentially qualitative.

4.10-2 Lowpass Filters

A typical lowpass filter has a maximum gain at $\omega = 0$. Because a pole enhances the gain at frequencies in its vicinity, we need to place a pole (or poles) on the real axis opposite the origin ($j\omega = 0$), as shown in Fig. 4.47a. The transfer function of this system is

$$H(s) = \frac{\omega_c}{s + \omega_c}$$

We have chosen the numerator of $H(s)$ to be ω_c to normalize the dc gain $H(0)$ to unity. If d is the distance from the pole $-\omega_c$ to a point $j\omega$ (Fig. 4.47a), then

$$|H(j\omega)| = \frac{\omega_c}{d}$$

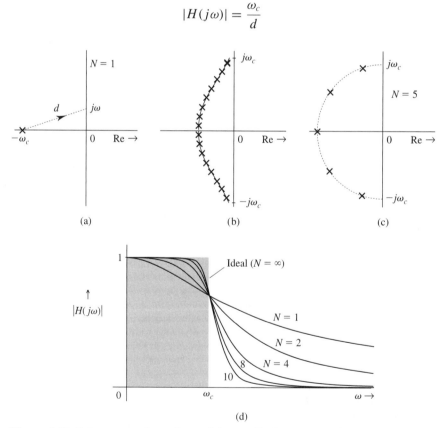

Figure 4.47 Pole–zero configuration and the amplitude response of a lowpass (Butterworth) filter.

with $H(0) = 1$. As ω increases, d increases and $|H(j\omega)|$ decreases monotonically with ω, as illustrated in Fig. 4.47d with label $N = 1$. This is clearly a lowpass filter with gain enhanced in the vicinity of $\omega = 0$.

WALL OF POLES

An ideal lowpass filter characteristic (shaded in Fig. 4.47d) has a constant gain of unity up to frequency ω_c. Then the gain drops suddenly to 0 for $\omega > \omega_c$. To achieve the ideal lowpass characteristic, we need enhanced gain over the entire frequency band from 0 to ω_c. We know that to enhance a gain at any frequency ω, we need to place a pole opposite ω. To achieve an enhanced gain for all frequencies over the band (0 to ω_c), we need to place a pole opposite every frequency in this band. In other words, we need a *continuous wall of poles* facing the imaginary axis opposite the frequency band 0 to ω_c (and from 0 to $-\omega_c$ for conjugate poles), as depicted in Fig. 4.47b. At this point, the optimum shape of this wall is not obvious because our arguments are qualitative and intuitive. Yet, it is certain that to have enhanced gain (constant gain) at every frequency over this range, we need an infinite number of poles on this wall. We can show that for a maximally flat[†] response over the frequency range (0 to ω_c), the wall is a semicircle with an infinite number of poles uniformly distributed along the wall.[12] In practice, we compromise by using a finite number (N) of poles with less-than-ideal characteristics. Figure 4.47c shows the pole configuration for a fifth-order ($N = 5$) filter. The amplitude response for various values of N are illustrated in Fig. 4.47d. As $N \to \infty$, the filter response approaches the ideal. This family of filters is known as the *Butterworth* filters. There are also other families. In *Chebyshev* filters, the wall shape is a semiellipse rather than a semicircle. The characteristics of a Chebyshev filter are inferior to those of Butterworth over the passband $(0, \omega_c)$, where the characteristics show a rippling effect instead of the maximally flat response of Butterworth. But in the stopband $(\omega > \omega_c)$, Chebyshev behavior is superior in the sense that Chebyshev filter gain drops faster than that of the Butterworth.

4.10-3 Bandpass Filters

The shaded characteristic in Fig. 4.48b shows the ideal bandpass filter gain. In the bandpass filter, the gain is enhanced over the entire passband. Our earlier discussion indicates that this can be realized by a wall of poles opposite the imaginary axis in front of the passband centered at ω_0. (There is also a wall of conjugate poles opposite $-\omega_0$.) Ideally, an infinite number of poles is required. In practice, we compromise by using a finite number of poles and accepting less-than-ideal characteristics (Fig. 4.48).

4.10-4 Notch (Bandstop) Filters

An ideal notch filter amplitude response (shaded in Fig. 4.49b) is a complement of the amplitude response of an ideal bandpass filter. Its gain is zero over a small band centered at some frequency ω_0 and is unity over the remaining frequencies. Realization of such a characteristic requires an infinite number of poles and zeros. Let us consider a practical second-order notch filter to obtain zero gain at a frequency $\omega = \omega_0$. For this purpose we must have zeros at $\pm j\omega_0$. The requirement

---

[†]Maximally flat amplitude response means the first $2N - 1$ derivatives of $|H(j\omega)|$ with respect to ω are zero at $\omega = 0$.

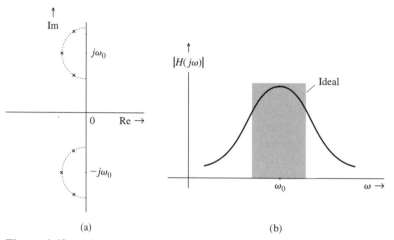

(a)

(b)

Figure 4.48 (a) Pole–zero configuration and (b) the amplitude response of a bandpass filter.

of unity gain at $\omega = \infty$ requires the number of poles to be equal to the number of zeros ($M = N$). This ensures that for very large values of ω, the product of the distances of poles from ω will be equal to the product of the distances of zeros from ω. Moreover, unity gain at $\omega = 0$ requires a pole and the corresponding zero to be equidistant from the origin. For example, if we use two (complex-conjugate) zeros, we must have two poles; the distance from the origin of the poles and of the zeros should be the same. This requirement can be met by placing the two conjugate poles on the semicircle of radius ω_0, as depicted in Fig. 4.49a. The poles can be anywhere on the semicircle to satisfy the equidistance condition. Let the two conjugate poles be at angles $\pm\theta$ with respect to the negative real axis. Recall that a pole and a zero in the vicinity tend to cancel out each other's influences. Therefore, placing poles closer to zeros (selecting θ closer to $\pi/2$) results in a rapid recovery of the gain from value 0 to 1 as we move away from ω_0 in either direction. Figure 4.49b shows the gain $|H(j\omega)|$ for three different values of θ.

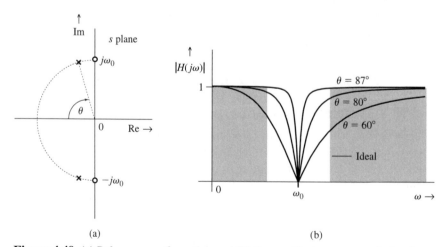

(a)

(b)

Figure 4.49 (a) Pole–zero configuration and (b) the amplitude response of a bandstop (notch) filter.

EXAMPLE 4.27

Design a second-order notch filter to suppress 60 Hz hum in a radio receiver.

___

We use the poles and zeros in Fig. 4.49a with $\omega_0 = 120\pi$. The zeros are at $s = \pm j\omega_0$. The two poles are at $-\omega_0 \cos \theta \pm j\omega_0 \sin \theta$. The filter transfer function is (with $\omega_0 = 120\pi$)

$$H(s) = \frac{(s - j\omega_0)(s + j\omega_0)}{(s + \omega_0 \cos \theta + j\omega_0 \sin \theta)(s + \omega_0 \cos \theta - j\omega_0 \sin \theta)}$$

$$= \frac{s^2 + \omega_0^2}{s^2 + (2\omega_0 \cos \theta)s + \omega_0^2} = \frac{s^2 + 142122.3}{s^2 + (753.98 \cos \theta)s + 142122.3}$$

and

$$|H(j\omega)| = \frac{-\omega^2 + 142122.3}{\sqrt{(-\omega^2 + 142122.3)^2 + (753.98\omega \cos \theta)^2}}$$

The closer the poles are to the zeros (the closer θ is to $\pi/2$), the faster the gain recovery from 0 to 1 on either side of $\omega_0 = 120\pi$. Figure 4.49b shows the amplitude response for three different values of θ. This example is a case of very simple design. To achieve zero gain over a band, we need an infinite number of poles as well as of zeros.

___

COMPUTER EXAMPLE C4.7

The transfer function of a second-order notch filter is

$$H(s) = \frac{s^2 + \omega_0^2}{s^2 + (2\omega_0 \cos(\theta))s + \omega_0^2}$$

Using $\omega_0 = 2\pi 60$, plot the magnitude response for the following cases:

(a) $\theta_a = 60°$

(b) $\theta_b = 80°$

(c) $\theta_c = 87°$

___

```
>> omega_0 = 2*pi*60; theta = [60 80 87]*(pi/180);
>> omega = (0:.5:1000)'; mag = zeros(3,length(omega));
>> for m=1:length(theta)
>>     H = tf([1 0 omega_0^2],[1 2*omega_0*cos(theta(m)) omega_0^2]);
>>     [mag(m,:),phase] = bode(H,omega);
>> end
>> f = omega/(2*pi); plot(f,mag(1,:),'k-',f,mag(2,:),'k--',f,mag(3,:),'k-.');
>> xlabel('f [Hz]'); ylabel('|H(j2\pi f)|');
>> legend('\theta = 60^\circ','\theta = 80^\circ','\theta = 87^\circ',0)
```

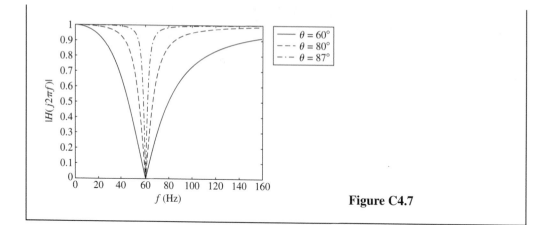

Figure C4.7

EXERCISE E4.15

Use the qualitative method of sketching the frequency response to show that the system with the pole–zero configuration in Fig. 4.50a is a highpass filter and the configuration in Fig. 4.50b is a bandpass filter.

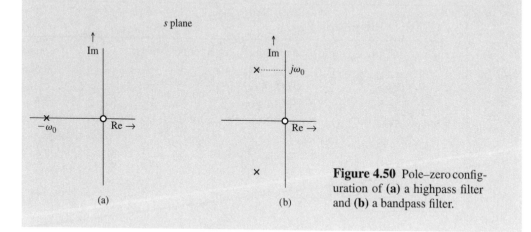

Figure 4.50 Pole–zero configuration of **(a)** a highpass filter and **(b)** a bandpass filter.

4.10-5 Practical Filters and Their Specifications

For ideal filters, everything is black and white; the gains are either zero or unity over certain bands. As we saw earlier, real life does not permit such a worldview. Things have to be gray or shades of gray. In practice, we can realize a variety of filter characteristics that can only approach ideal characteristics.

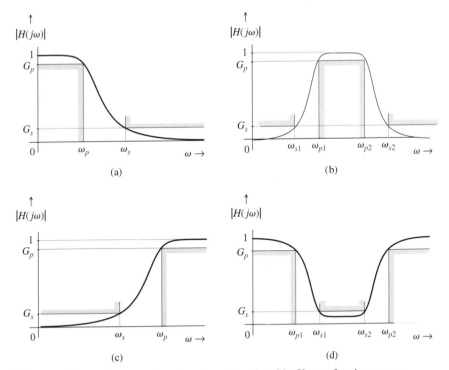

Figure 4.51 Passband, stopband, and transition band in filters of various types.

An ideal filter has a passband (unity gain) and a stopband (zero gain) with a sudden transition from the passband to the stopband. There is no transition band. For practical (or realizable) filters, on the other hand, the transition from the passband to the stopband (or vice versa) is gradual and takes place over a finite band of frequencies. Moreover, for realizable filters, the gain cannot be zero over a finite band (Paley–Wiener condition). As a result, there can no true stopband for practical filters. We therefore define a *stopband* to be a band over which the gain is below some small number G_s, as illustrated in Fig. 4.51. Similarly, we define a *passband* to be a band over which the gain is between 1 and some number G_p $(G_p < 1)$, as shown in Fig. 4.51. We have selected the passband gain of unity for convenience. It could be any constant. Usually the gains are specified in terms of decibels. This is simply 20 times the log (to base 10) of the gain. Thus

$$\hat{G}(\text{dB}) = 20 \log_{10} G$$

A gain of unity is 0 dB and a gain of $\sqrt{2}$ is 3.01 dB, usually approximated by 3 dB. Sometimes the specification may be in terms of attenuation, which is the negative of the gain in dB. Thus a gain of $1/\sqrt{2}$; that is, 0.707, is −3 dB, but is an attenuation of 3 dB.

In a typical design procedure, G_p (*minimum passband gain*) and G_s (*maximum stopband gain*) are specified. Figure 4.51 shows the passband, the stopband, and the transition band for typical lowpass, bandpass, highpass, and bandstop filters. Fortunately, the highpass, bandpass, and bandstop filters can be obtained from a basic lowpass filter by simple frequency transformations. For example, replacing s with ω_c/s in the lowpass filter transfer function results in a highpass filter. Similarly, other frequency transformations yield the bandpass and bandstop filters. Hence, it is necessary to develop a design procedure only for a basic lowpass filter. Then, by using appropriate transformations, we can design filters of other types. The design procedures are beyond our scope here and will not be discussed. Interested reader is referred to Ref. 1.

4.11 THE BILATERAL LAPLACE TRANSFORM

Situations involving noncausal signals and/or systems cannot be handled by the (unilateral) Laplace transform discussed so far. These cases can be analyzed by the *bilateral* (or *two-sided*) Laplace transform defined by

$$X(s) = \int_{-\infty}^{\infty} x(t)e^{-st}\,dt$$

and $x(t)$ can be obtained from $X(s)$ by the inverse transformation

$$x(t) = \frac{1}{2\pi j} \int_{c-j\infty}^{c+j\infty} X(s)e^{st}\,ds$$

Observe that the unilateral Laplace transform discussed so far is a special case of the bilateral Laplace transform, where the signals are restricted to the causal type. Basically, the two transforms are the same. For this reason we use the same notation for the bilateral Laplace transform.

Earlier we showed that the Laplace transforms of $e^{-at}u(t)$ and of $-e^{at}u(-t)$ are identical. The only difference is in their regions of convergence (ROC). The ROC for the former is $\operatorname{Re} s > -a$; that for the latter is $\operatorname{Re} s < -a$, as illustrated in Fig. 4.1. Clearly, the inverse Laplace transform of $X(s)$ is not unique unless the ROC is specified. If we restrict all our signals to the causal type, however, this ambiguity does not arise. The inverse transform of $1/(s+a)$ is $e^{-at}u(t)$. Thus, in the unilateral Laplace transform, we can ignore the ROC in determining the inverse transform of $X(s)$.

We now show that any bilateral transform can be expressed in terms of two unilateral transforms. It is, therefore, possible to evaluate bilateral transforms from a table of unilateral transforms.

Consider the function $x(t)$ appearing in Fig. 4.52a. We separate $x(t)$ into two components, $x_1(t)$ and $x_2(t)$, representing the positive time (*causal*) component and the negative time (*anticausal*) component of $x(t)$, respectively (Fig. 4.52b and 4.52c):

$$x_1(t) = x(t)u(t)$$

$$x_2(t) = x(t)u(-t)$$

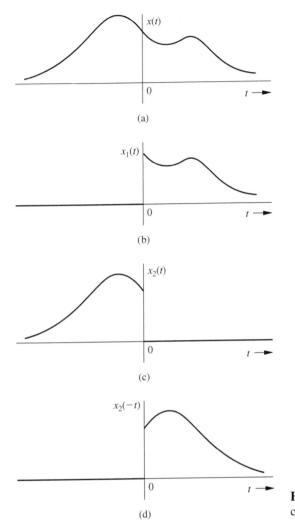

(a)

(b)

(c)

(d)

Figure 4.52 Expressing a signal as a sum of causal and anticausal components.

The bilateral Laplace transform of $x(t)$ is given by

$$X(s) = \int_{-\infty}^{\infty} x(t)e^{-st}\, dt$$

$$= \int_{-\infty}^{0^-} x_2(t)e^{-st}\, dt + \int_{0^-}^{\infty} x_1(t)e^{-st}\, dt$$

$$= X_2(s) + X_1(s) \tag{4.93}$$

where $X_1(s)$ is the Laplace transform of the causal component $x_1(t)$ and $X_2(s)$ is the Laplace transform of the anticausal component $x_2(t)$. Consider $X_2(s)$, given by

$$X_2(s) = \int_{-\infty}^{0^-} x_2(t)e^{-st}\, dt$$

$$= \int_{0^+}^{\infty} x_2(-t)e^{st}\, dt$$

Therefore

$$X_2(-s) = \int_{0^+}^{\infty} x_2(-t)e^{-st}\, dt$$

If $x(t)$ has any impulse or its derivative(s) at the origin, they are included in $x_1(t)$. Consequently, $x_2(t) = 0$ at the origin; that is, $x_2(0) = 0$. Hence, the lower limit on the integration in the preceding equation can be taken as 0^- instead of 0^+. Therefore

$$X_2(-s) = \int_{0^-}^{\infty} x_2(-t)e^{-st}\, dt$$

Because $x_2(-t)$ is causal (Fig. 4.52d), $X_2(-s)$ can be found from the unilateral transform table. Changing the sign of s in $X_2(-s)$ yields $X_2(s)$.

To summarize, the bilateral transform $X(s)$ in Eq. (4.93) can be computed from the unilateral transforms in two steps:

1. Split $x(t)$ into its causal and anticausal components, $x_1(t)$ and $x_2(t)$, respectively.

2. The signals $x_1(t)$ and $x_2(-t)$ are both causal. Take the (unilateral) Laplace transform of $x_1(t)$ and add to it the (unilateral) Laplace transform of $x_2(-t)$, with s replaced by $-s$. This procedure gives the (bilateral) Laplace transform of $x(t)$.

Since $x_1(t)$ and $x_2(-t)$ are both causal, $X_1(s)$ and $X_2(-s)$ are both unilateral Laplace transforms. Let σ_{c1} and σ_{c2} be the abscissas of convergence of $X_1(s)$ and $X_2(-s)$, respectively. This statement implies that $X_1(s)$ exists for all s with Re $s > \sigma_{c1}$, and $X_2(-s)$ exists for all s with Re $s > \sigma_{c2}$. Therefore, $X_2(s)$ exists for all s with Re $s < -\sigma_{c2}$.[†] Therefore, $X(s) = X_1(s) + X_2(s)$ exists for all s such that

$$\sigma_{c1} < \text{Re } s < -\sigma_{c2}$$

The regions of convergence (or existence) of $X_1(s)$, $X_2(s)$, and $X(s)$ are shown in Fig. 4.53. Because $X(s)$ is finite for all values of s lying in the strip of convergence ($\sigma_{c1} < \text{Re } s < -\sigma_{c2}$), poles of $X(s)$ must lie outside this strip. The poles of $X(s)$ arising from the causal component $x_1(t)$ lie to the left of the *strip* (region) *of convergence*, and those arising from its anticausal component $x_2(t)$ lie to its right (see Fig. 4.53). This fact is of crucial importance in finding the inverse bilateral transform.

This result can be generalized to left-sided and right-sided signals. We define a signal $x(t)$ as a *right-sided* signal if $x(t) = 0$ for $t < T_1$ for some finite positive or negative number T_1. A causal signal is always a right-sided signal, but the converse is not necessarily true. A signal is said to *left sided* if it is zero for $t > T_2$ for some finite, positive, or negative number T_2.

[†]For instance, if $x(t)$ exists for all $t > 10$, then $x(-t)$, its time-inverted form, exists for $t < -10$.

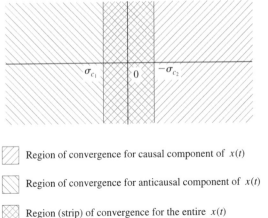

☐ Region of convergence for causal component of $x(t)$

☐ Region of convergence for anticausal component of $x(t)$

☐ Region (strip) of convergence for the entire $x(t)$ **Figure 4.53**

An anticausal signal is always a left-sided signal, but the converse is not necessarily true. A *two-sided* signal is of infinite duration on both positive and negative sides of t and is neither right sided nor left sided.

We can show that the conclusions for ROC for causal signals also hold for right-sided signals, and those for anticausal signals hold for left-sided signals. In other words, if $x(t)$ is causal or right-sided, the poles of $X(s)$ lie to the left of the ROC, and if $x(t)$ is anticausal or left-sided, the poles of $X(s)$ lie to the right of the ROC.

To prove this generalization, we observe that a right-sided signal can be expressed as $x(t) + x_f(t)$, where $x(t)$ is a causal signal and $x_f(t)$ is some finite-duration signal. ROC of any finite-duration signal is the entire s-plane (no finite poles). Hence, the ROC of the right-sided signal $x(t) + x_f(t)$ is the region common to the ROCs of $x(t)$ and $x_f(t)$, which is same as the ROC for $x(t)$. This proves the generalization for right-sided signals. We can use a similar argument to generalize the result for left-sided signals.

Let us find the bilateral Laplace transform of

$$x(t) = e^{bt}u(-t) + e^{at}u(t) \tag{4.94}$$

We already know the Laplace transform of the causal component

$$e^{at}u(t) \Longleftrightarrow \frac{1}{s-a} \qquad \mathrm{Re}\, s > a \tag{4.95}$$

For the anticausal component, $x_2(t) = e^{bt}u(-t)$, we have

$$x_2(-t) = e^{-bt}u(t) \Longleftrightarrow \frac{1}{s+b} \qquad \mathrm{Re}\, s > -b$$

so that

$$X_2(s) = \frac{1}{-s+b} = \frac{-1}{s-b} \qquad \mathrm{Re}\, s < b$$

Therefore

$$e^{bt}u(-t) \Longleftrightarrow \frac{-1}{s-b} \qquad \mathrm{Re}\, s < b \tag{4.96}$$

and the Laplace transform of $x(t)$ in Eq. (4.94) is

$$X(s) = -\frac{1}{s-b} + \frac{1}{s-a} \qquad \mathrm{Re}\, s > a \quad \text{and} \quad \mathrm{Re}\, s < b$$

$$= \frac{a-b}{(s-b)(s-a)} \qquad a < \mathrm{Re}\, s < b \tag{4.97}$$

Figure 4.54 shows $x(t)$ and the ROC of $X(s)$ for various values of a and b. Equation (4.97) indicates that the ROC of $X(s)$ does not exist if $a > b$, which is precisely the case in Fig. 4.54g.

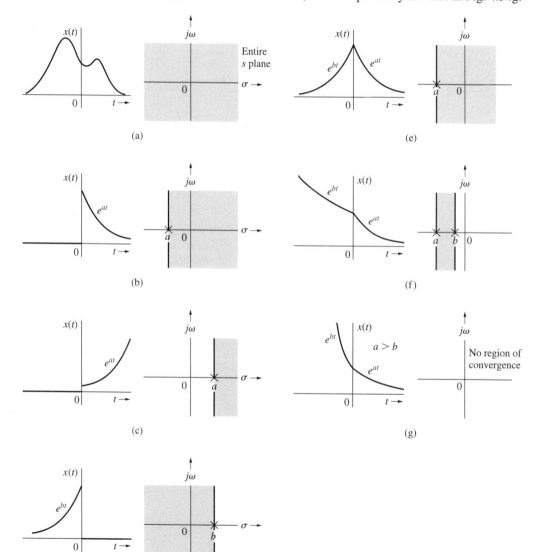

Figure 4.54

Observe that the poles of $X(s)$ are outside (on the edges) of the ROC. The poles of $X(s)$ because of the anticausal component of $x(t)$ lie to the right of the ROC, and those due to the causal component of $x(t)$ lie to its left.

When $X(s)$ is expressed as a sum of several terms, the ROC for $X(s)$ is the intersection of (region common to) the ROCs of all the terms. In general, if $x(t) = \sum_{i=1}^{k} x_i(t)$, then the ROC for $X(s)$ is the intersection of the ROCs (region common to all ROCs) for the transforms $X_1(s), X_2(s), \ldots, X_k(s)$.

EXAMPLE 4.28

Find the inverse Laplace transform of

$$X(s) = \frac{-3}{(s+2)(s-1)}$$

if the ROC is

(a) $-2 < \mathrm{Re}\, s < 1$

(b) $\mathrm{Re}\, s > 1$

(c) $\mathrm{Re}\, s < -2$

_____

(a)

$$X(s) = \frac{1}{s+2} - \frac{1}{s-1}$$

Now, $X(s)$ has poles at -2 and 1. The strip of convergence is $-2 < \mathrm{Re}\, s < 1$. The pole at -2, being to the left of the strip of convergence, corresponds to the causal signal. The pole at 1, being to the right of the strip of convergence, corresponds to the anticausal signal. Equations (4.95) and (4.96) yield

$$x(t) = e^{-2t}u(t) + e^{t}u(-t)$$

(b) Both poles lie to the left of the ROC, so both poles correspond to causal signals. Therefore

$$x(t) = (e^{-2t} - e^{t})u(t)$$

(c) Both poles lie to the right of the region of convergence, so both poles correspond to anticausal signals, and

$$x(t) = (-e^{-2t} + e^{t})u(-t)$$

Figure 4.55 shows the three inverse transforms corresponding to the same $X(s)$ but with different regions of convergence.

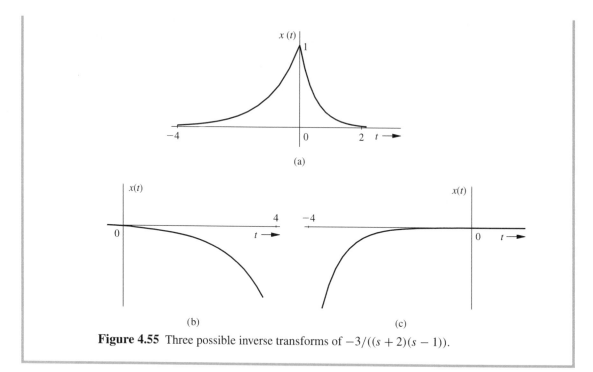

Figure 4.55 Three possible inverse transforms of $-3/((s+2)(s-1))$.

4.11-1 Properties of Bilateral Laplace Transform

Properties of the bilateral Laplace transform are similar to those of the unilateral transform. We shall merely state the properties here without proofs. Let the ROC of $X(s)$ be $a < \mathrm{Re}\ s < b$. Similarly, let the ROC of $X_i(s)$ is $a_i < \mathrm{Re}\ s < b_i$ for $(i = 1, 2)$.

LINEARITY

$$a_1 x_1(t) + a_2 x_2(t) \Longleftrightarrow a_1 X_1(s) + a_2 X_2(s)$$

The ROC for $a_1 X_1(s) + a_2 X_2(s)$ is the region common to (intersection of) the ROCs for $X_1(s)$ and $X_2(s)$.

TIME SHIFT

$$x(t - T) \Longleftrightarrow X(s)e^{-sT}$$

The ROC for $X(s)e^{-sT}$ is identical to the ROC for $X(s)$.

FREQUENCY SHIFT

$$x(t)e^{s_0 t} \Longleftrightarrow X(s - s_0)$$

The ROC for $X(s - s_0)$ is $a + c < \mathrm{Re}\ s < b + c$, where $c = \mathrm{Re}\ s_0$.

TIME DIFFERENTIATION

$$\frac{dx(t)}{dt} \Longleftrightarrow sX(s)$$

The ROC for $sX(s)$ contains the ROC for $X(s)$ and may be larger than that of $X(s)$ under certain conditions [e.g., if $X(s)$ has a first-order pole at $s = 0$ that is canceled by the factor s in $sX(s)$].

TIME INTEGRATION

$$\int_{-\infty}^{t} x(\tau)\, d\tau \Longleftrightarrow X(s)/s$$

The ROC for $sX(s)$ is max $(a, 0) < \text{Re } s < b$.

TIME SCALING

$$x(\beta t) \Longleftrightarrow \frac{1}{|\beta|} X\left(\frac{s}{\beta}\right)$$

The ROC for $X(s/\beta)$ is $\beta a < \text{Re } s < \beta b$. For $\beta > 1$, $x(\beta t)$ represents time compression and the corresponding ROC expands by factor β. For $0 > \beta > 1$, $x(\beta t)$ represents time expansion and the corresponding ROC is compressed by factor β.

TIME CONVOLUTION

$$x_1(t) * x_2(t) \Longleftrightarrow X_1(s)X_2(s)$$

The ROC for $X_1(s)X_2(s)$ is the region common to (intersection of) the ROCs for $X_1(s)$ and $X_2(s)$.

FREQUENCY CONVOLUTION

$$x_1(t)x_2(t) \Longleftrightarrow \frac{1}{2\pi j} \int_{c-j\infty}^{c+j\infty} X_1(w)X_2(s - w)\, dw$$

The ROC for $X_1(s) * X_2(s)$ is $a_1 + a_2 < \text{Re } s < b_1 + b_2$.

TIME REVERSAL

$$x(-t) \Longleftrightarrow X(-s)$$

The ROC for $X(-s)$ is $-b < \text{Re } s < -a$.

4.11-2 Using the Bilateral Transform for Linear System Analysis

Since the bilateral Laplace transform can handle noncausal signals, we can analyze noncausal LTIC systems using the bilateral Laplace transform. We have shown that the (zero-state) output $y(t)$ is given by

$$y(t) = \mathcal{L}^{-1}[X(s)H(s)] \tag{4.98}$$

This expression is valid only if $X(s)H(s)$ exists. The ROC of $X(s)H(s)$ is the region in which both $X(s)$ and $H(s)$ exist. In other words, the ROC of $X(s)H(s)$ is the region common to the regions of convergence of both $X(s)$ and $H(s)$. These ideas are clarified in the following examples.

EXAMPLE 4.29

Find the current $y(t)$ for the RC circuit in Fig. 4.56a if the voltage $x(t)$ is

$$x(t) = e^t u(t) + e^{2t} u(-t)$$

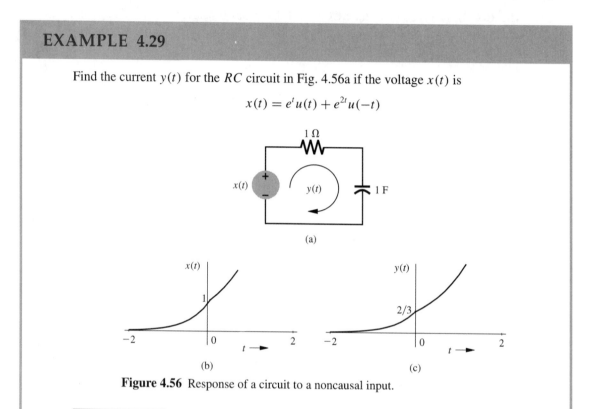

Figure 4.56 Response of a circuit to a noncausal input.

The transfer function $H(s)$ of the circuit is given by

$$H(s) = \frac{s}{s+1} \qquad \text{Re } s > -1$$

Because $h(t)$ is a causal function, the ROC of $H(s)$ is Re $s > -1$. Next, the bilateral Laplace transform of $x(t)$ is given by

$$X(s) = \frac{1}{s-1} - \frac{1}{s-2} = \frac{-1}{(s-1)(s-2)} \qquad 1 < \text{Re } s < 2$$

The response $y(t)$ is the inverse transform of $X(s)H(s)$

$$y(t) = \mathcal{L}^{-1}\left[\frac{-s}{(s+1)(s-1)(s-2)}\right]$$

$$= \mathcal{L}^{-1}\left[\frac{1}{6}\frac{1}{s+1} + \frac{1}{2}\frac{1}{s-1} - \frac{2}{3}\frac{1}{s-2}\right]$$

The ROC of $X(s)H(s)$ is that ROC common to both $X(s)$ and $H(s)$. This is $1 < \text{Re } s < 2$. The poles $s = \pm 1$ lie to the left of the ROC and, therefore, correspond to causal signals; the pole $s = 2$ lies to the right of the ROC and thus represents an anticausal signal. Hence

$$y(t) = \tfrac{1}{6}e^{-t}u(t) + \tfrac{1}{2}e^{t}u(t) + \tfrac{2}{3}e^{2t}u(-t)$$

Figure 4.56c shows $y(t)$. Note that in this example, if

$$x(t) = e^{-4t}u(t) + e^{-2t}u(-t)$$

then the ROC of $X(s)$ is $-4 < \text{Re } s < -2$. Here no region of convergence exists for $X(s)H(s)$. Such a situation means that the dominance condition cannot be satisfied by any permissible value of s in the range $(1 < \text{Re } s < 2)$. Hence, the response $y(t)$ goes to infinity.

EXAMPLE 4.30

Find the response $y(t)$ of a noncausal system with the transfer function

$$H(s) = \frac{-1}{s-1} \qquad \text{Re } s < 1$$

to the input $x(t) = e^{-2t}u(t)$.

We have

$$X(s) = \frac{1}{s+2} \qquad \text{Re } s > -2$$

and

$$Y(s) = X(s)H(s) = \frac{-1}{(s-1)(s+2)}$$

The ROC of $X(s)H(s)$ is the region $-2 < \text{Re } s < 1$. By partial fraction expansion

$$Y(s) = \frac{-1/3}{s-1} + \frac{1/3}{s+2} \qquad -2 < \text{Re } s < 1$$

and

$$y(t) = \tfrac{1}{3}[e^{t}u(-t) + e^{-2t}u(t)]$$

Note that the pole of $H(s)$ lies in the RHP at 1. Yet the system is not unstable. The pole(s) in the RHP may indicate instability or noncausality, depending on its location with respect to the region of convergence of $H(s)$. For example, if $H(s) = -1/(s-1)$ with $\text{Re } s > 1$, the system is causal and unstable, with $h(t) = -e^{t}u(t)$. In contrast, if $H(s) = -1/(s-1)$ with $\text{Re } s < 1$, the system is noncausal and stable, with $h(t) = e^{t}u(-t)$.

EXAMPLE 4.31

Find the response $y(t)$ of a system with the transfer function

$$H(s) = \frac{1}{s+5} \qquad \text{Re } s > -5$$

and the input

$$x(t) = e^{-t}u(t) + e^{-2t}u(-t)$$

The input $x(t)$ is of the type depicted in Fig. 4.54g, and the region of convergence for $X(s)$ does not exist. In this case, we must determine separately the system response to each of the two input components, $x_1(t) = e^{-t}u(t)$ and $x_2(t) = e^{-2t}u(-t)$.

$$X_1(s) = \frac{1}{s+1} \qquad \text{Re } s > -1$$

$$X_2(s) = \frac{-1}{s+2} \qquad \text{Re } s < -2$$

If $y_1(t)$ and $y_2(t)$ are the system responses to $x_1(t)$ and $x_2(t)$, respectively, then

$$Y_1(s) = \frac{1}{(s+1)(s+5)} \qquad \text{Re } s > -1$$

$$= \frac{1/4}{s+1} - \frac{1/4}{s+5}$$

so that

$$y_1(t) = \tfrac{1}{4}(e^{-t} - e^{-5t})u(t)$$

and

$$Y_2(s) = \frac{-1}{(s+2)(s+5)} \qquad -5 < \text{Re } s < -2$$

$$= \frac{-1/3}{s+2} + \frac{1/3}{s+5}$$

so that

$$y_2(t) = \tfrac{1}{3}[e^{-2t}u(-t) + e^{-5t}u(t)]$$

Therefore

$$y(t) = y_1(t) + y_2(t)$$

$$= \tfrac{1}{3}e^{-2t}u(-t) + \left(\tfrac{1}{4}e^{-t} + \tfrac{1}{12}e^{-5t}\right)u(t)$$

4.12 SUMMARY

This chapter discusses analysis of LTIC (linear, time-invariant, continuous-time) systems by the Laplace transform, which transforms integro-differential equations of such systems into algebraic equations. Therefore solving these integro-differential equations reduces to solving algebraic equations. The Laplace transform method cannot be used for time-varying-parameter systems or for nonlinear systems in general.

The transfer function $H(s)$ of an LTIC system is the Laplace transform of its impulse response. It may also be defined as a ratio of the Laplace transform of the output to the Laplace transform of the input when all initial conditions are zero (system in zero state). If $X(s)$ is the Laplace transform of the input $x(t)$ and $Y(s)$ is the Laplace transform of the corresponding output $y(t)$ (when all initial conditions are zero), then $Y(s) = X(s)H(s)$. For an LTIC system described by an Nth-order differential equation $Q(D)y(t) = P(D)x(t)$, the transfer function $H(s) = P(s)/Q(s)$. Like the impulse response $h(t)$, the transfer function $H(s)$ is also an external description of the system.

Electrical circuit analysis can also be carried out by using a transformed circuit method, in which all signals (voltages and currents) are represented by their Laplace transforms, all elements by their impedances (or admittances), and initial conditions by their equivalent sources (initial condition generators). In this method, a network can be analyzed as if it were a resistive circuit.

Large systems can be depicted by suitably interconnected subsystems represented by blocks. Each subsystem, being a smaller system, can be readily analyzed and represented by its input–output relationship, such as its transfer function. Analysis of large systems can be carried out with the knowledge of input–output relationships of its subsystems and the nature of interconnection of various subsystems.

LTIC systems can be realized by scalar multipliers, adders, and integrators. A given transfer function can be synthesized in many different ways, such as canonic, cascade, and parallel. Moreover, every realization has a transpose, which also has the same transfer function. In practice, all the building blocks (scalar multipliers, adders, and integrators) can be obtained from operational amplifiers.

The system response to an everlasting exponential e^{st} is also an everlasting exponential $H(s)e^{st}$. Consequently, the system response to an everlasting exponential $e^{j\omega t}$ is $H(j\omega)e^{j\omega t}$. Hence, $H(j\omega)$ is the frequency response of the system. For a sinusoidal input of unit amplitude and having frequency ω, the system response is also a sinusoid of the same frequency (ω) with amplitude $|H(j\omega)|$, and its phase is shifted by $\angle H(j\omega)$ with respect to the input sinusoid. For this reason $|H(j\omega)|$ is called the amplitude response (gain) and $\angle H(j\omega)$ is called the phase response of the system. Amplitude and phase response of a system indicate the filtering characteristics of the system. The general nature of the filtering characteristics of a system can be quickly determined from a knowledge of the location of poles and zeros of the system transfer function.

Most of the input signals and practical systems are causal. Consequently we are required most of the time to deal with causal signals. When all signals must be causal, the Laplace transform analysis is greatly simplified; the region of convergence of a signal becomes irrelevant to the analysis process. This special case of the Laplace transform (which is restricted to causal signals) is called the unilateral Laplace transform. Much of the chapter deals with this variety of Laplace transform. Section 4.11 discusses the general Laplace transform (the bilateral Laplace transform), which can handle causal and noncausal signals and systems. In the bilateral transform, the inverse transform of $X(s)$ is not unique but depends on the region of convergence of $X(s)$. Thus the region of convergence plays a very crucial role in the bilateral Laplace transform.

REFERENCES

1. Lathi, B. P. *Signal Processing and Linear Systems*, 1st ed. Oxford University Press, New York, 1998.
2. Doetsch, G. *Introduction to the Theory and Applications of the Laplace Transformation with a Table of Laplace Transformations.* Springer-Verlag, New York, 1974.
3. LePage, W. R. *Complex Variables and the Laplace Transforms for Engineers.* McGraw-Hill, New York, 1961.
4. Durant, Will, and Ariel Durant. *The Age of Napoleon*, Part XI in *The Story of Civilization Series.* Simon & Schuster, New York, 1975.
5. Bell, E. T. *Men of Mathematics.* Simon & Schuster, New York, 1937.
6. Nahin, P. J. "Oliver Heaviside: Genius and Curmudgeon." *IEEE Spectrum*, vol. 20, pp. 63–69, July 1983.
7. Berkey, D. *Calculus*, 2nd ed. Saunders, Philadelphia, 1988.
8. Encyclopaedia Britannica. *Micropaedia IV*, 15th ed., p. 981, Chicago, 1982.
9. Churchill, R. V. *Operational Mathematics*, 2nd ed. McGraw-Hill, New York, 1958.
10. Lathi, B. P. *Signals, Systems, and Communication.* Wiley, New York, 1965.
11. Truxal, J. G. *The Age of Electronic Messages.* McGraw-Hill, New York, 1990.
12. Van Valkenberg, M. *Analog Filter Design.* Oxford University Press, New York, 1982.

MATLAB SESSION 4: CONTINUOUS-TIME FILTERS

Continuous-time filters are essential to many if not most engineering systems, and MATLAB is an excellent assistant for filter design and analysis. Although a comprehensive treatment of continuous-time filter techniques is outside the scope of this book, quality filters can be designed and realized with minimal additional theory.

A simple yet practical example demonstrates basic filtering concepts. Telephone voice signals are often lowpass-filtered to eliminate frequencies above a cutoff of 3 kHz, or $\omega_c = 3000(2\pi) \approx 18,850$ rad/s. Filtering maintains satisfactory speech quality and reduces signal bandwidth, thereby increasing the phone company's call capacity. How, then, do we design and realize an acceptable 3 kHz lowpass filter?

M4.1 Frequency Response and Polynomial Evaluation

Magnitude response plots help assess a filter's performance and quality. The magnitude response of an ideal filter is a brick-wall function with unity passband gain and perfect stopband attenuation. For a lowpass filter with cutoff frequency ω_c, the ideal magnitude response is

$$|H_{\text{ideal}}(j\omega)| = \begin{cases} 1 & |\omega| \leq \omega_c \\ 0 & |\omega| > \omega_c \end{cases}$$

Unfortunately, ideal filters cannot be implemented in practice. Realizable filters require compromises, although good designs will closely approximate the desired brick-wall response.

A realizable LTIC system often has a rational transfer function that is represented in the s-domain as

$$H(s) = \frac{Y(s)}{X(s)} = \frac{B(s)}{A(s)} = \frac{\sum\limits_{k=0}^{M} b_{k+N-M} s^{M-k}}{\sum\limits_{k=0}^{N} a_k s^{N-k}}$$

Frequency response $H(j\omega)$ is obtained by letting $s = j\omega$, where frequency ω is in radians per second.

MATLAB is ideally suited to evaluate frequency response functions. Defining a length-$(N+1)$ coefficient vector $\mathbf{A} = [a_0, a_1, \ldots, a_N]$ and a length-$(M+1)$ coefficient vector $\mathbf{B} = [b_{N-M}, b_{N-M+1}, \ldots, b_N]$, program MS4P1 computes $H(j\omega)$ for each frequency in the input vector ω.

```
function [H] = MS4P1(B,A,omega);
% MS4P1.m : MATLAB Session 4, Program 1
% Function M-file computes frequency response for LTIC system
% INPUTS:    B = vector of feedforward coefficients
%            A = vector of feedback coefficients
%            omega = vector of frequencies [rad/s].
% OUTPUTS:   H =  frequency response

H = polyval(B,j*omega)./polyval(A,j*omega);
```

The function polyval efficiently evaluates simple polynomials and makes the program nearly trivial. For example, when A is the vector of coefficients $[a_0, a_1, \ldots, a_N]$, polyval (A,j*omega) computes

$$\sum_{k=0}^{N} a_k (j\omega)^{N-k}$$

for each value of the frequency vector omega. It is also possible to compute frequency responses by using the signal processing toolbox function freqs.

M4.2 Design and Evaluation of a Simple RC Filter

One of the simplest lowpass filters is realized by using the RC circuit shown in Fig. M4.1. This one-pole system has transfer function $H_{RC}(s) = (RCs + 1)^{-1}$ and magnitude response $|H_{RC}(j\omega)| = |(j\omega RC + 1)^{-1}| = 1/\sqrt{1 + (RC\omega)^2}$. Independent of component values R and C, this magnitude function possesses many desirable characteristics such as a gain that is unity at $\omega = 0$ and that monotonically decreases to zero as $\omega \to \infty$.

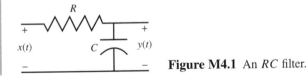

Figure M4.1 An RC filter.

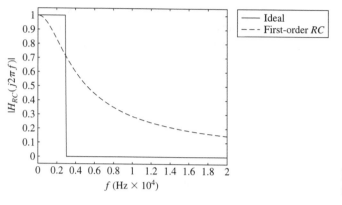

Figure M4.2 First-order RC magnitude response $|H_{RC}(j2\pi f)|$.

Components R and C are chosen to set the desired 3 kHz cutoff frequency. For many filter types, the cutoff frequency corresponds to the half-power point, or $|H_{RC}(j\omega_c)| = 1/\sqrt{2}$. Assigning C a realistic capacitance of 1 nF, the required resistance is computed by $R = 1/\sqrt{C^2\omega_c^2} = 1/\sqrt{(10^{-9})^2(2\pi\,3000)^2}$.

```
>> omega_c = 2*pi*3000;
>> C = 1e-9;
>> R = 1/sqrt(C^2*omega_c^2)
R = 5.3052e+004
```

The root of this first-order RC filter is directly related to the cutoff frequency, $\lambda = -1/RC = -18,850 = -\omega_c$.

To evaluate the RC filter performance, the magnitude response is plotted over the mostly audible frequency range ($0 \le f \le 20\,\text{kHz}$).

```
>> f = linspace(0,20000,200);
>> B = 1; A = [R*C 1]; Hmag_RC = abs(MS4P1(B,A,f*2*pi));
>> plot(f,abs(f*2*pi)<=omega_c,'k-',f,Hmag_RC,'k--');
>> xlabel('f [Hz]'); ylabel('|H_{RC}(j2\pi f)|');
>> axis([0 20000 -0.05 1.05]); legend('Ideal','First-Order RC');
```

The `linspace(X1,X2,N)` command generates an N-length vector of linearly spaced points between X1 and X2.

As shown in Fig. M4.2, the first-order RC response is indeed lowpass with a half-power cutoff frequency equal to 3 kHz. It rather poorly approximates the desired brick-wall response: the passband is not very flat, and stopband attenuation increases very slowly to less than 20 dB at 20 kHz.

M4.3 A Cascaded RC Filter and Polynomial Expansion

A first-order RC filter is destined for poor performance; one pole is simply insufficient to obtain good results. A cascade of RC circuits increases the number of poles and improves the filter response. To simplify the analysis and prevent loading between stages, op-amp followers buffer the output of each section as shown in Fig. M4.3. A cascade of N stages results in an Nth-order

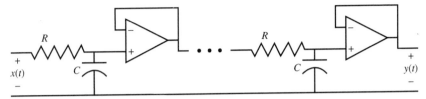

Figure M4.3 A cascaded RC filter.

filter with transfer function given by

$$H_{\text{cascade}}(s) = (H_{RC}(s))^N = (RCs + 1)^{-N}$$

Upon choosing a cascade of 10 sections and $C = 1\,\text{nF}$, a 3 kHz cutoff frequency is obtained by setting $R = \sqrt{2^{1/10} - 1}/(C\omega_c) = \sqrt{2^{1/10} - 1}/(6\pi(10)^{-6})$.

```
>> R = sqrt(2^(1/10)-1)/(C*omega_c)
R = 1.4213e+004
```

This cascaded filter has a 10th-order pole at $\lambda = -1/RC$ and no finite zeros. To compute the magnitude response, polynomial coefficient vectors **A** and **B** are needed. Setting **B** = [1] ensures there are no finite zeros or, equivalently, that all zeros are at infinity. The `poly` command, which expands a vector of roots into a corresponding vector of polynomial coefficients, is used to obtain **A**.

```
>> B = 1; A = poly(-1/(R*C)*ones(10,1));A = A/A(end);
>> Hmag_cascade = abs(MS4P1(B,A,f*2*pi));
```

Notice that scaling a polynomial by a constant does not change its roots. Conversely, the roots of a polynomial specify a polynomial within a scale factor. The command `A = A/A(end)` properly scales the denominator polynomial to ensure unity gain at $\omega = 0$.

The magnitude response plot of the cascaded RC filter is included in Fig. M4.4. The passband remains relatively unchanged, but stopband attenuation is greatly improved to over 60 dB at 20 kHz.

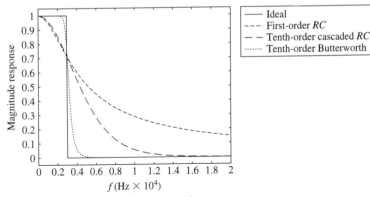

Figure M4.4 Lowpass filter comparison.

M4.4 Butterworth Filters and the `Find` Command

The pole location of a first-order lowpass filter is necessarily fixed by the cutoff frequency. There is little reason, however, to place all the poles of a 10th-order filter at one location. Better pole placement will improve our filter's magnitude response. One strategy, discussed in Section 4.10, is to place a wall of poles opposite the passband frequencies. A semicircular wall of poles leads to the Butterworth family of filters, and a semielliptical shape leads to the Chebyshev family of filters. Butterworth filters are considered first.

To begin, notice that a transfer function $H(s)$ with real coefficients has a squared magnitude response given by $|H(j\omega)|^2 = H(j\omega)H^*(j\omega) = H(j\omega)H(-j\omega) = H(s)H(-s)|_{s=j\omega}$. Thus, half the poles of $|H(j\omega)|^2$ correspond to the filter $H(s)$ and the other half correspond to $H(-s)$. Filters that are both stable and causal require $H(s)$ to include only left-half-plane poles.

The squared magnitude response of a Butterworth filter is

$$|H_{\mathrm{B}}(j\omega)|^2 = \frac{1}{1 + (j\omega/j\omega_c)^{2N}}$$

This function has the same appealing characteristics as the first-order RC filter: a gain that is unity at $\omega = 0$ and monotonically decreases to zero as $\omega \to \infty$. By construction, the half-power gain occurs at ω_c. Perhaps most importantly, however, the first $2N - 1$ derivatives of $|H_{\mathrm{B}}(j\omega)|$ with respect to ω are zero at $\omega = 0$. Put another way, the passband is constrained to be very flat for low frequencies. For this reason, Butterworth filters are sometimes called maximally flat filters.

As discussed in MATLAB Session B, the roots of minus one must lie equally spaced on a circle centered at the origin. Thus, the $2N$ poles of $|H_{\mathrm{B}}(j\omega)|^2$ naturally lie equally spaced on a circle of radius ω_c centered at the origin. Figure M4.5 displays the 20 poles corresponding to the case $N = 10$ and $\omega_c = 3000(2\pi)$ rad/s. An Nth-order Butterworth filter that is both causal and stable uses the N left-half-plane poles of $|H_{\mathrm{B}}(j\omega)|^2$.

To design a 10th-order Butterworth filter, we first compute the 20 poles of $|H_{\mathrm{B}}(j\omega)|^2$:

```
>> N=10;
>> poles = roots([(j*omega_c)^(-2*N),zeros(1,2*N-1),1]);
```

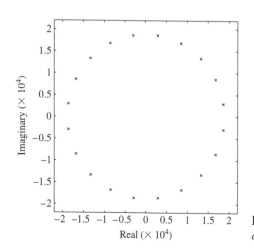

Figure M4.5 Roots of $|H_{\mathrm{B}}(j\omega)|^2$ for $N = 10$ and $\omega_c = 3000(2\pi)$.

The `find` command is an extremely powerful function that returns the indices of a vector's nonzero elements. Combined with relational operators, the `find` command allows us to extract the 10 left-half-plane roots that correspond to the poles of our Butterworth filter.

```
>> B_poles = poles(find(real(poles)<0));
```

To compute the magnitude response, these roots are converted to coefficient vector **A**.

```
>> A = poly(B_poles); A = A/A(end);
>> Hmag_B = abs(MS4P1(B,A,f*2*pi));
```

The magnitude response plot of the Butterworth filter is included in Fig. M4.4. The Butterworth response closely approximates the brick-wall function and provides excellent filter characteristics: flat passband, rapid transition to the stopband, and excellent stopband attenuation (>40 dB at 5 kHz).

M4.5 Using Cascaded Second-Order Sections for Butterworth Filter Realization

For our RC filters, realization preceded design. For our Butterworth filter, however, design has preceded realization. For our Butterworth filter to be useful, we must be able to implement it.

Since the transfer function $H_B(s)$ is known, the differential equation is also known. Therefore, it is possible to try to implement the design by using op-amp integrators, summers, and scalar multipliers. Unfortunately, this approach will not work well. To understand why, consider the denominator coefficients $a_0 = 1.766 \times 10^{-43}$ and $a_{10} = 1$. The smallest coefficient is 43 orders of magnitude smaller than the largest coefficient! It is practically impossible to accurately realize such a broad range in scale values. To understand this, skeptics should try to find realistic resistors such that $R_f/R = 1.766 \times 10^{-43}$. Additionally, small component variations will cause large changes in actual pole location.

A better approach is to cascade five second-order sections, where each section implements one complex conjugate pair of poles. By pairing poles in complex conjugate pairs, each of the resulting second-order sections have real coefficients. With this approach, the smallest coefficients are only about nine orders of magnitude smaller than the largest coefficients. Furthermore, pole placement is typically less sensitive to component variations for cascaded structures.

The Sallen–Key circuit shown in Fig. M4.6 provides a good way to realize a pair of complex-conjugate poles.[†] The transfer function of this circuit is

$$H_{SK}(s) = \frac{\dfrac{1}{R_1 R_2 C_1 C_2}}{s^2 + \left(\dfrac{1}{R_1 C_1} + \dfrac{1}{R_2 C_1}\right)s + \dfrac{1}{R_1 R_2 C_1 C_2}} = \frac{\omega_0^2}{s^2 + \left(\dfrac{\omega_0}{Q}\right)s + \omega_0^2}$$

[†]A more general version of the Sallen–Key circuit has a resistor R_a from the negative terminal to ground and a resistor R_b between the negative terminal and the output. In Fig. M4.6, $R_a = \infty$ and $R_b = 0$.

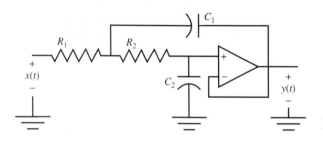

Figure M4.6 Sallen–Key filter stage.

Geometrically, ω_0 is the distance from the origin to the poles and $Q = 1/2 \cos \psi$, where ψ is the angle between the negative real axis and the pole. Termed the "quality factor" of a circuit, Q provides a measure of the peakedness of the response. High-Q filters have poles close to the ω axis, which boost the magnitude response near those frequencies.

Although many ways exist to determine suitable component values, a simple method is to assign R_1 a realistic value and then let $R_2 = R_1$, $C_1 = 2Q/\omega_0 R_1$, and $C_2 = 1/2Q\omega_0 R_2$. Butterworth poles are a distance ω_c from the origin, so $\omega_0 = \omega_c$. For our 10th-order Butterworth filter, the angles ψ are regularly spaced at 9, 27, 45, 63, and 81 degrees. MATLAB program MS4P2 automates the task of computing component values and magnitude responses for each stage.

```
% MS4P2.m : MATLAB Session 4, Program 2
% Script M-file computes Sallen-Key component values and magnitude
% responses for each of the five cascaded second-order filter sections.

omega_0 = 3000*2*pi; % Filter cut-off frequency
psi = [9 27 45 63 81]*pi/180; % Butterworth pole angles
f = linspace(0,6000,200); % Frequency range for magnitude response calculations
Hmag_SK = zeros(5,200); % Pre-allocate array for magnitude responses
for stage = 1:5,
    Q = 1/(2*cos(psi(stage))); % Compute Q for current stage
    % Compute and display filter components to the screen:
    disp(['Stage ',num2str(stage),...
         ' (Q = ',num2str(Q),...
         '): R1 = R2 = ',num2str(56000),...
         ', C1 = ',num2str(2*Q/(omega_0*56000)),...
         ', C2 = ',num2str(1/(2*Q*omega_0*56000))]);
    B = omega_0^2; A = [1 omega_0/Q omega_0^2]; % Compute filter coefficients
    Hmag_SK(stage,:) = abs(MS4P1(B,A,2*pi*f)); % Compute magnitude response
end
plot(f,Hmag_SK,'k',f,prod(Hmag_SK),'k:')
xlabel('f [Hz]'); ylabel('Magnitude Response')
```

The disp command displays a character string to the screen. Character strings must be enclosed in single quotation marks. The num2str command converts numbers to character strings and facilitates the formatted display of information. The prod command multiplies along the columns of a matrix; it computes the total magnitude response as the product of the magnitude responses of the five stages.

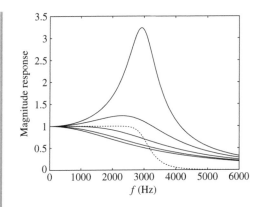

Figure M4.7 Magnitude responses for Sallen–Key filter stages.

Executing the program produces the following output:

```
>> MS4P2
Stage 1 (Q = 0.50623):  R1 = R2 = 56000, C1 = 9.5916e-010, C2 = 9.3569e-010
Stage 2 (Q = 0.56116):  R1 = R2 = 56000, C1 = 1.0632e-009, C2 = 8.441e-010
Stage 3 (Q = 0.70711):  R1 = R2 = 56000, C1 = 1.3398e-009, C2 = 6.6988e-010
Stage 4 (Q = 1.1013):   R1 = R2 = 56000, C1 = 2.0867e-009, C2 = 4.3009e-010
Stage 5 (Q = 3.1962):   R1 = R2 = 56000, C1 = 6.0559e-009, C2 = 1.482e-010
```

Since all the component values are practical, this filter is possible to implement. Figure M4.7 displays the magnitude responses for all five stages (solid lines). The total response (dotted line) confirms a 10th-order Butterworth response. Stage 5, which has the largest Q and implements the pair of conjugate poles nearest the ω axis, is the most peaked response. Stage 1, which has the smallest Q and implements the pair of conjugate poles furthest from the ω axis, is the least peaked response. In practice, it is best to order high-Q stages last; this reduces the risk of the high gains saturating the filter hardware.

M4.6 Chebyshev Filters

Like an order-N Butterworth lowpass filter (LPF), an order-N Chebyshev LPF is an all-pole filter that possesses many desirable characteristics. Compared with an equal-order Butterworth filter, the Chebyshev filter achieves better stopband attenuation and reduced transition band width by allowing an adjustable amount of ripple within the passband.

The squared magnitude response of a Chebyshev filter is

$$|H(j\omega)|^2 = \frac{1}{1 + \epsilon^2 C_N^2(\omega/\omega_c)}$$

where ϵ controls the passband ripple, $C_N(\omega/\omega_c)$ is a degree-N Chebyshev polynomial, and ω_c is the radian cutoff frequency. Several characteristics of Chebyshev LPFs are noteworthy:

- An order-N Chebyshev LPF is equiripple in the passband ($|\omega| \leq \omega_c$), has a total of N maxima and minima over ($0 \leq \omega \leq \omega_c$), and is monotonic decreasing in the stopband ($|\omega| > \omega_c$).

- In the passband, the maximum gain is 1 and the minimum gain is $1/\sqrt{1+\epsilon^2}$. For odd-valued N, $|H(j0)| = 1$. For even-valued N, $|H(j0)| = 1/\sqrt{1+\epsilon^2}$.
- Ripple is controlled by setting $\epsilon = \sqrt{10^{R/10} - 1}$, where R is the allowable passband ripple expressed in decibels. Reducing ϵ adversely affects filter performance (see Prob. 4.M-8).
- Unlike Butterworth filters, the cutoff frequency ω_c rarely specifies the 3 dB point. For $\epsilon \neq 1$, $|H(j\omega_c)|^2 = 1/(1+\epsilon^2) \neq 0.5$. The cutoff frequency ω_c simply indicates the frequency after which $|H(j\omega)| < 1/\sqrt{1+\epsilon^2}$.

The Chebyshev polynomial $C_N(x)$ is defined as

$$C_N(x) = \cos[N \cos^{-1}(x)] = \cosh[N \cosh^{-1}(x)]$$

In this form, it is difficult to verify that $C_N(x)$ is a degree-N polynomial in x. A recursive form of $C_N(x)$ makes this fact more clear (see Prob. 4.M-11).

$$C_N(x) = 2xC_{N-1}(x) - C_{N-2}(x)$$

With $C_0(x) = 1$ and $C_1(x) = x$, the recursive form shows that any C_N is a linear combination of degree-N polynomials and is therefore a degree-N polynomial itself. For $N \geq 2$, MATLAB program MS4P3 generates the $(N + 1)$ coefficients of Chebyshev polynomial $C_N(x)$.

```
function [C_N] = MS4P3(N);
% MS4P3.m : MATLAB Session 4, Program 3
% Function M-file computes Chebyshev polynomial coefficients
% using the recursion relation C_N(x) = 2xC_{N-1}(x) - C_{N-2}(x)
% INPUTS:    N = degree of Chebyshev polynomial
% OUTPUTS:   C_N =  vector of Chebyshev polynomial coefficients

C_Nm2 = 1; C_Nm1 = [1 0];       % Initial polynomial coefficients:
for t = 2:N;
    C_N = 2*conv([1 0],C_Nm1)-[zeros(1,length(C_Nm1)-length(C_Nm2)+1),C_Nm2];
    C_Nm2 = C_Nm1; C_Nm1 = C_N;
end
```

As examples, consider $C_2(x) = 2xC_1(x) - C_0(x) = 2x(x) - 1 = 2x^2 - 1$ and $C_3(x) = 2xC_2(x) - C_1(x) = 2x(2x^2 - 1) - x = 4x^3 - 3x$. MS4P3 easily confirms these cases.

```
>> MS4P3(2)
ans =   2       0     -1
>> MS4P3(3)
ans =   4       0     -3       0
```

Since $C_N(\omega/\omega_c)$ is a degree-N polynomial, $|H(j\omega)|^2$ is an all-pole rational function with $2N$ finite poles. Similar to the Butterworth case, the N poles specifying a causal and stable Chebyshev filter can be found by selecting the N left-half plane roots of $1 + \epsilon^2 C_N^2[s/(j\omega_c)]$.

Root locations and dc gain are sufficient to specify a Chebyshev filter for a given N and ϵ. To demonstrate, consider the design of an order-8 Chebyshev filter with cutoff frequency $f_c = 1$ kHz and allowable passband ripple $R = 1$ dB. First, filter parameters are specified.

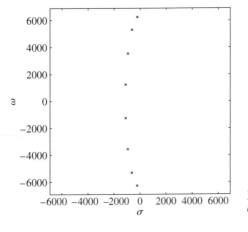

Figure M4.8 Pole–zero plot for an order-8 Chebyshev LPF with $f_c = 1$ kHz and $r = 1$ dB.

```
>> omega_c = 2*pi*1000; R = 1; N = 8;
>> epsilon = sqrt(10^(R/10)-1);
```

The coefficients of $C_N(s/(j\omega_c))$ are obtained with the help of MS4P3, and then the coefficients of $(1 + \epsilon^2 C_N^2(s/(j\omega_c)))$ are computed by using convolution to perform polynomial multiplication.

```
>> CN  = MS4P3(N).*((1/(j*omega_c)).^[N:-1:0]);
>> CP = epsilon^2*conv(CN,CN); CP(end) = CP(end)+1;
```

Next, the polynomial roots are found, and the left-half-plane poles are retained and plotted.

```
>> poles = roots(CP); i = find(real(poles)<0); C_poles = poles(i);
>> plot(real(C_poles),imag(C_poles),'kx'); axis equal;
>> axis(omega_c*[-1.1 1.1 -1.1 1.1]);
>> xlabel('\sigma'); ylabel('\omega');
```

As shown in Fig. M4.8, the roots of a Chebyshev filter lie on an ellipse[†] (see Prob. 4.M-12).

To compute the filter's magnitude response, the poles are expanded into a polynomial, the dc gain is set based on the even value of N, and MS4P1 is used.

```
>> A = poly(C_poles);
>> B = A(end)/sqrt(1+epsilon^2);
>> omega = linspace(0,2*pi*2000,2001);
>> H = MS4P1(B,A,omega);
>> plot(omega/2/pi,abs(H),'k'); grid
>> xlabel('f [Hz]'); ylabel('|H(j2\pi f)|');
>> axis([0 2000 0 1.1]);
```

As seen in Fig. M4.9, the magnitude response exhibits correct Chebyshev filter characteristics: passband ripples are equal in height and never exceed $R = 1$ dB; there are a total of $N = 8$ maxima and minima in the passband; and the gain rapidly and monotonically decreases after the cutoff frequency of $f_c = 1$ kHz.

[†]E. A. Guillemin demonstrates a wonderful relationship between the Chebyshev ellipse and the Butterworth circle in his book, *Synthesis of Passive Networks* (Wiley, New York, 1957).

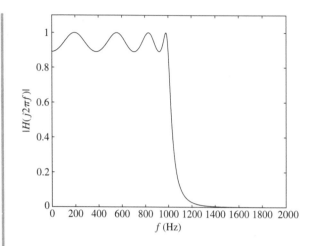

Figure M4.9 Magnitude responses for an order-8 Chebyshev LPF with $f_c = 1\,\text{kHz}$ and $r = 1\,\text{dB}$.

For higher-order filters, polynomial rooting may not provide reliable results. Fortunately, Chebyshev roots can also be determined analytically. For

$$\phi_k = \frac{2k+1}{2N}\pi \qquad \text{and} \qquad \xi = \frac{1}{N}\sinh^{-1}\left(\frac{1}{\epsilon}\right)$$

the Chebyshev poles are

$$p_k = \omega_c \sinh(\xi)\sin(\phi_k) + j\omega_c \cosh(\xi)\cos(\phi_k)$$

Continuing the same example, the poles are recomputed and again plotted. The result is identical to Fig. M4.8.

```
>> k = [1:N]; xi = 1/N*asinh(1/epsilon); phi = (k*2-1)/(2*N)*pi;
>> C_poles = omega_c*(-sinh(xi)*sin(phi)+j*cosh(xi)*cos(phi));
>> plot(real(C_poles),imag(C_poles),'kx'); axis equal;
>> axis(omega_c*[-1.1 1.1 -1.1 1.1]);
>> xlabel('\sigma'); ylabel('\omega');
```

As in the case of high-order Butterworth filters, a cascade of second-order filter sections facilitates practical implementation of Chebyshev filters. Problems 4.M-3 and 4.M-6 use second-order Sallen–Key circuit stages to investigate such implementations.

PROBLEMS

4.1-1 By direct integration [Eq. (4.1)] find the Laplace transforms and the region of convergence of the following functions:

(a) $u(t) - u(t-1)$
(b) $te^{-t}u(t)$
(c) $t\cos\omega_0 t\, u(t)$

(d) $(e^{2t} - 2e^{-t})u(t)$
(e) $\cos\omega_1 t\,\cos\omega_2 t\, u(t)$
(f) $\cosh(at)\,u(t)$
(g) $\sinh(at)\,u(t)$
(h) $e^{-2t}\cos(5t + \theta)\,u(t)$

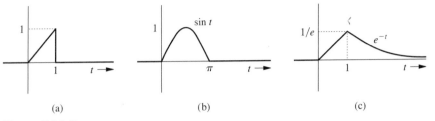

(a)   (b)   (c)

Figure P4.1-2

4.1-2 By direct integration find the Laplace transforms of the signals shown in Fig. P4.1-2.

4.1-3 Find the inverse (unilateral) Laplace transforms of the following functions:

(a) $\dfrac{2s + 5}{s^2 + 5s + 6}$

(b) $\dfrac{3s + 5}{s^2 + 4s + 13}$

(c) $\dfrac{(s + 1)^2}{s^2 - s - 6}$

(d) $\dfrac{5}{s^2(s + 2)}$

(e) $\dfrac{2s + 1}{(s + 1)(s^2 + 2s + 2)}$

(f) $\dfrac{s + 2}{s(s + 1)^2}$

(g) $\dfrac{1}{(s + 1)(s + 2)^4}$

(h) $\dfrac{s + 1}{s(s + 2)^2(s^2 + 4s + 5)}$

(i) $\dfrac{s^3}{(s + 1)^2(s^2 + 2s + 5)}$

4.2-1 Find the Laplace transforms of the following functions using only Table 4.1 and the time-shifting property (if needed) of the unilateral Laplace transform:

(a) $u(t) - u(t - 1)$

(b) $e^{-(t-\tau)}u(t - \tau)$

(c) $e^{-(t-\tau)}u(t)$

(d) $e^{-t}u(t - \tau)$

(e) $te^{-t}u(t - \tau)$

(f) $\sin[\omega_0(t - \tau)]u(t - \tau)$

(g) $\sin[\omega_0(t - \tau)]u(t)$

(h) $\sin \omega_0 t\, u(t - \tau)$

4.2-2 Using only Table 4.1 and the time-shifting property, determine the Laplace transform of the signals in Fig. P4.1-2. [Hint: See Section 1.4 for discussion of expressing such signals analytically.]

4.2-3 Find the inverse Laplace transforms of the following functions:

(a) $\dfrac{(2s + 5)e^{-2s}}{s^2 + 5s + 6}$

(b) $\dfrac{se^{-3s} + 2}{s^2 + 2s + 2}$

(c) $\dfrac{e^{-(s-1)} + 3}{s^2 - 2s + 5}$

(d) $\dfrac{e^{-s} + e^{-2s} + 1}{s^2 + 3s + 2}$

4.2-4 The Laplace transform of a causal periodic signal can be determined from the knowledge of the Laplace transform of its first cycle (period).

(a) If the Laplace transform of $x(t)$ in Fig. P4.2-4a is $X(s)$, then show that $G(s)$, the Laplace transform of $g(t)$ (Fig. P4.2-4b), is

$$G(s) = \dfrac{X(s)}{1 - e^{-sT_0}} \qquad \text{Re } s > 0$$

(b) Use this result to find the Laplace transform of the signal $p(t)$ illustrated in Fig. P4.2-4c.

4.2-5 Starting only with the fact that $\delta(t) \iff 1$, build pairs 2 through 10b in Table 4.1, using various properties of the Laplace transform.

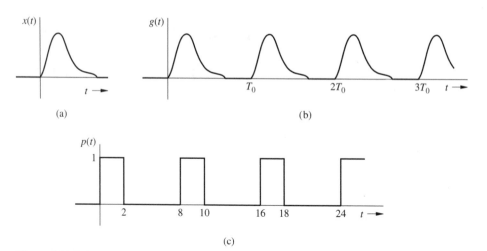

Figure P4.2-4

4.2-6 (a) Find the Laplace transform of the pulses in Fig. 4.2 in the text by using only the time-differentiation property, the time-shifting property, and the fact that $\delta(t) \Longleftrightarrow 1$.

(b) In Example 4.7, the Laplace transform of $x(t)$ is found by finding the Laplace transform of d^2x/dt^2. Find the Laplace transform of $x(t)$ in that example by finding the Laplace transform of dx/dt and using Table 4.1, if necessary.

4.2-7 Determine the inverse unilateral Laplace transform of

$$X(s) = \frac{1}{e^{s+3}} \frac{s^2}{(s+1)(s+2)}$$

4.2-8 Since 13 is such a lucky number, determine the inverse Laplace transform of $X(s) = 1/(s+1)^{13}$ given region of convergence $\sigma > -1$. [Hint: What is the nth derivative of $1/(s+a)$?]

4.2-9 It is difficult to compute the Laplace transform $X(s)$ of signal

$$x(t) = \frac{1}{t}u(t)$$

by using direct integration. Instead, properties provide a simpler method.

(a) Use Laplace transform properties to express the Laplace transform of $tx(t)$ in terms of the unknown quantity $X(s)$.

(b) Use the definition to determine the Laplace transform of $y(t) = tx(t)$.

(c) Solve for $X(s)$ by using the two pieces from a and b. Simplify your answer.

4.3-1 Using the Laplace transform, solve the following differential equations:

(a) $(D^2 + 3D + 2)y(t) = Dx(t)$ if $y(0^-) = \dot{y}(0^-) = 0$ and $x(t) = u(t)$

(b) $(D^2 + 4D + 4)y(t) = (D + 1)x(t)$ if $y(0^-) = 2, \dot{y}(0^-) = 1$ and $x(t) = e^{-t}u(t)$

(c) $(D^2 + 6D + 25)y(t) = (D + 2)x(t)$ if $y(0^-) = \dot{y}(0^-) = 1$ and $x(t) = 25u(t)$

4.3-2 Solve the differential equations in Prob. 4.3-1 using the Laplace transform. In each case determine the zero-input and zero-state components of the solution.

4.3-3 Solve the following simultaneous differential equations using the Laplace transform, assuming all initial conditions to be zero and the input $x(t) = u(t)$:

(a) $(D + 3)y_1(t) - 2y_2(t) = x(t)$
   $- 2y_1(t) + (2D + 4)y_2(t) = 0$

(b) $(D + 2)y_1(t) - (D + 1)y_2(t) = 0$
   $- (D + 1)y_1(t) + (2D + 1)y_2(t) = x(t)$

Determine the transfer functions relating outputs $y_1(t)$ and $y_2(t)$ to the input $x(t)$.

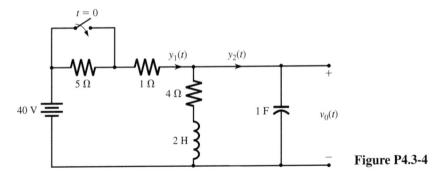

Figure P4.3-4

4.3-4 For the circuit in Fig. P4.3-4, the switch is in open position for a long time before $t = 0$, when it is closed instantaneously.
   (a) Write loop equations (in time domain) for $t \geq 0$.
   (b) Solve for $y_1(t)$ and $y_2(t)$ by taking the Laplace transform of loop equations found in part (a).

4.3-5 For each of the systems described by the following differential equations, find the system transfer function:

(a) $\dfrac{d^2 y}{dt^2} + 11\dfrac{dy}{dt} + 24y(t) = 5\dfrac{dx}{dt} + 3x(t)$

(b) $\dfrac{d^3 y}{dt^3} + 6\dfrac{d^2 y}{dt^2} - 11\dfrac{dy}{dt} + 6y(t)$

$\qquad = 3\dfrac{d^2 x}{dt^2} + 7\dfrac{dx}{dt} + 5x(t)$

(c) $\dfrac{d^4 y}{dt^4} + 4\dfrac{dy}{dt} = 3\dfrac{dx}{dt} + 2x(t)$

(d) $\dfrac{d^2 y}{dt^2} - y(t) = \dfrac{dx}{dt} - x(t)$

4.3-6 For each of the systems specified by the following transfer functions, find the differential equation relating the output $y(t)$ to the input $x(t)$ assuming that the systems are controllable and observable:

(a) $H(s) = \dfrac{s + 5}{s^2 + 3s + 8}$

(b) $H(s) = \dfrac{s^2 + 3s + 5}{s^3 + 8s^2 + 5s + 7}$

(c) $H(s) = \dfrac{5s^2 + 7s + 2}{s^2 - 2s + 5}$

4.3-7 For a system with transfer function

$$H(s) = \dfrac{2s + 3}{s^2 + 2s + 5}$$

(a) Find the (zero-state) response for input $x(t)$ of (i) $10u(t)$ and (ii) $u(t - 5)$.
(b) For this system write the differential equation relating the output $y(t)$ to the input $x(t)$ assuming that the systems are controllable and observable.

4.3-8 For a system with transfer function

$$H(s) = \dfrac{s}{s^2 + 9}$$

(a) Find the (zero-state) response if the input $x(t) = (1 - e^{-t})u(t)$
(b) For this system write the differential equation relating the output $y(t)$ to the input $x(t)$ assuming that the systems are controllable and observable.

4.3-9 For a system with transfer function

$$H(s) = \dfrac{s + 5}{s^2 + 5s + 6}$$

(a) Find the (zero-state) response for the following values of input $x(t)$:
   (i) $e^{-3t}u(t)$
   (ii) $e^{-4t}u(t)$
   (iii) $e^{-4(t-5)}u(t - 5)$
   (iv) $e^{-4(t-5)}u(t)$
   (v) $e^{-4t}u(t - 5)$
(b) For this system write the differential equation relating the output $y(t)$ to the input $x(t)$ assuming that the systems are controllable and observable.

4.3-10 An LTI system has a step response given by $s(t) = e^{-t}u(t) - e^{-2t}u(t)$. Determine the output of this system $y(t)$ given an input $x(t) = \delta(t - \pi) - \cos(\sqrt{3})u(t)$.

4.3-11 For an LTIC system with zero initial conditions (system initially in zero state), if an input $x(t)$ produces an output $y(t)$, then using the Laplace transform show the following.

(a) The input dx/dt produces an output dy/dt.

(b) The input $\int_0^t x(\tau)\,d\tau$ produces an output $\int_0^t y(\tau)\,d\tau$. Hence, show that the unit step response of a system is an integral of the impulse response; that is, $\int_0^t h(\tau)\,d\tau$.

4.3-12 (a) Discuss asymptotic and BIBO stabilities for the systems described by the following transfer functions assuming that the systems are controllable and observable:

(i) $\dfrac{(s+5)}{s^2 + 3s + 2}$

(ii) $\dfrac{s+5}{s^2(s+2)}$

(iii) $\dfrac{s(s+2)}{s+5}$

(iv) $\dfrac{s+5}{s(s+2)}$

(v) $\dfrac{s+5}{s^2 - 2s + 3}$

(b) Repeat part (a) for systems described by the following differential equations. Systems may be uncontrollable and/or unobservable.

(i) $(D^2 + 3D + 2)\,y(t) = (D + 3)\,x(t)$

(ii) $(D^2 + 3D + 2)\,y(t) = (D + 1)\,x(t)$

(iii) $(D^2 + D - 2)\,y(t) = (D - 1)\,x(t)$

(iv) $(D^2 - 3D + 2)\,y(t) = (D - 1)\,x(t)$

4.4-1 Find the zero-state response $y(t)$ of the network in Fig. P4.4-1 if the input voltage $x(t) =$ $te^{-t}u(t)$. Find the transfer function relating the output $y(t)$ to the input $x(t)$. From the transfer function, write the differential equation relating $y(t)$ to $x(t)$.

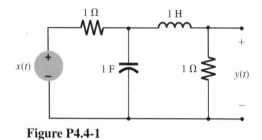

Figure P4.4-1

4.4-2 The switch in the circuit of Fig. P4.4-2 is closed for a long time and then opened instantaneously at $t = 0$. Find and sketch the current $y(t)$.

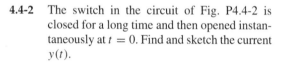

Figure P4.4-2

4.4-3 Find the current $y(t)$ for the parallel resonant circuit in Fig. P4.4-3 if the input is

(a) $x(t) = A \cos \omega_0 t\, u(t)$

(b) $x(t) = A \sin \omega_0 t\, u(t)$

Assume all initial conditions to be zero and, in both cases, $\omega_0^2 = 1/LC$.

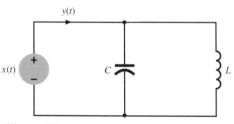

Figure P4.4-3

4.4-4 Find the loop currents $y_1(t)$ and $y_2(t)$ for $t \geq 0$ in the circuit of Fig. P4.4-4a for the input $x(t)$ in Fig. P4.4-4b.

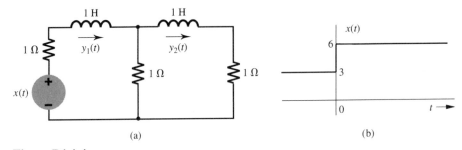

Figure P4.4-4

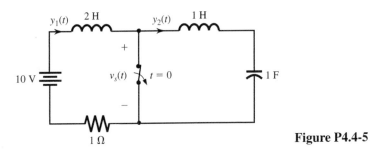

Figure P4.4-5

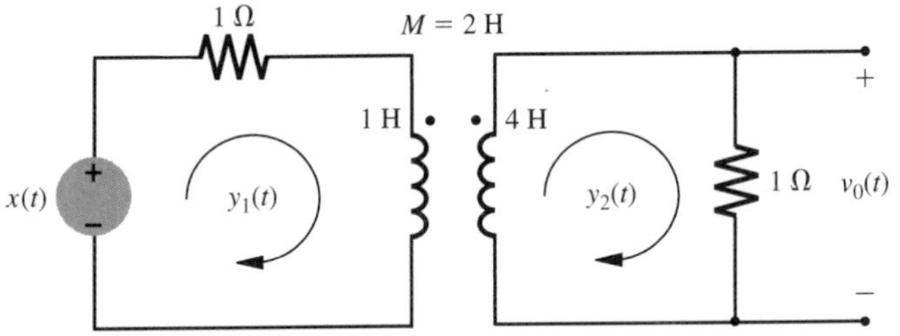

Figure P4.4-6

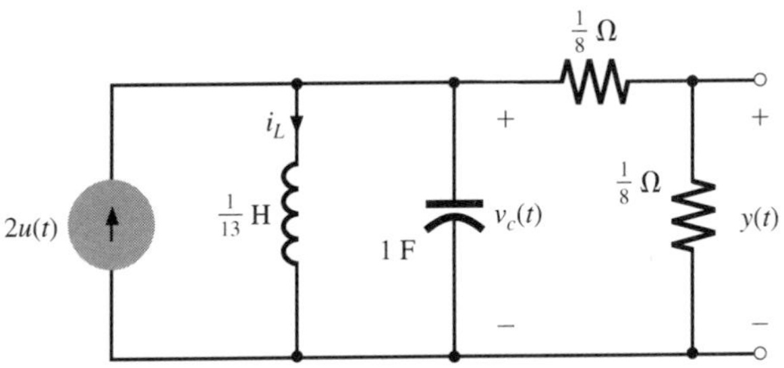

Figure P4.4-7

4.4-5 For the network in Fig. P4.4-5 the switch is in a closed position for a long time before $t = 0$, when it is opened instantaneously. Find $y_1(t)$ and $v_s(t)$ for $t \geq 0$.

4.4-6 Find the output voltage $v_0(t)$ for $t \geq 0$ for the circuit in Fig. P4.4-6, if the input $x(t) =$

$100u(t)$. The system is in the zero state initially.

4.4-7 Find the output voltage $y(t)$ for the network in Fig. P4.4-7 for the initial conditions $i_L(0) = 1$ A and $v_C(0) = 3$ V.

4.4-8   For the network in Fig. P4.4-8, the switch is in position a for a long time and then is moved to position b instantaneously at $t = 0$. Determine the current $y(t)$ for $t > 0$.

4.4-9   Show that the transfer function that relates the output voltage $y(t)$ to the input voltage $x(t)$ for the op-amp circuit in Fig. P4.4-9a is given by

$$H(s) = \frac{Ka}{s+a} \qquad \text{where}$$

$$K = 1 + \frac{R_b}{R_a} \qquad \text{and} \qquad a = \frac{1}{RC}$$

and that the transfer function for the circuit in Fig. P4.4-9b is given by

$$H(s) = \frac{Ks}{s+a}$$

4.4-10   For the second-order op-amp circuit in Fig. P4.4-10, show that the transfer function $H(s)$ relating the output voltage $y(t)$ to the input voltage $x(t)$ is given by

$$H(s) = \frac{-s}{s^2 + 8s + 12}$$

4.4-11   (a)   Using the initial and final value theorems, find the initial and final value of the zero-state response of a system with the transfer

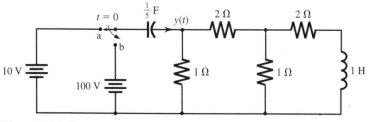

Figure P4.4-8

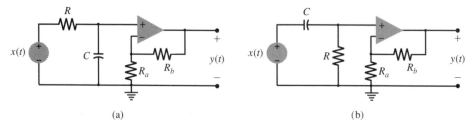

(a)   (b)

Figure P4.4-9

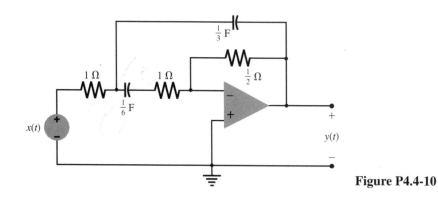

Figure P4.4-10

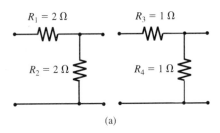

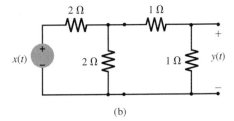

(a)   (b)

Figure P4.5-1

function

$$H(s) = \frac{6s^2 + 3s + 10}{2s^2 + 6s + 5}$$

and the input (i) $u(t)$ and (ii) $e^{-t}u(t)$.

(b) Find $y(0^+)$ and $y(\infty)$ if $Y(s)$ is given by

(i) $\dfrac{s^2 + 5s + 6}{s^2 + 3s + 2}$

(ii) $\dfrac{s^3 + 4s^2 + 10s + 7}{s^2 + 2s + 3}$

4.5-1 Figure P4.5-1a shows two resistive ladder segments. The transfer function of each segment (ratio of output to input voltage) is $1/2$. Figure P4.5-1b shows these two segments connected in cascade.

(a) Is the transfer function (ratio of output to input voltage) of this cascaded network $(1/2)(1/2) = 1/4$?

(b) If your answer is affirmative, verify the answer by direct computation of the transfer function. Does this computation confirm the earlier value $1/4$? If not, why?

(c) Repeat the problem with $R_3 = R_4 = 20 \text{ k}\Omega$. Does this result suggest the answer to the problem in part (b)?

4.5-2 In communication channels, transmitted signal is propagated simultaneously by several paths of varying lengths. This causes the signal to reach the destination with varying time delays and varying gains. Such a system generally distorts the received signal. For error-free communication, it is necessary to undo this distortion as much as possible by using the system that is inverse of the channel model.

For simplicity, let us assume that a signal is propagated by two paths whose time delays differ by τ seconds. The channel over the

intended path has a delay of T seconds and unity gain. The signal over the unintended path has a delay of $T + \tau$ seconds and gain a. Such a channel can be modeled, as shown in Fig. P4.5-2. Find the inverse system transfer function to correct the delay distortion and show that the inverse system can be realized by a feedback system. The inverse system should be causal to be realizable. [Hint: We want to correct only the distortion caused by the relative delay τ seconds. For distortionless transmission, the signal may be delayed. What is important is to maintain the shape of $x(t)$. Thus a received signal of the form $c\,x(t - T)$ is considered to be distortionless.]

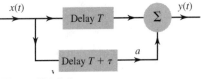

Figure P4.5-2

4.5-3 Discuss BIBO stability of the feedback systems depicted in Fig. P4.5-3. For the case in Fig. P4.5-3b, consider three cases:

(i) $K = 10$

(ii) $K = 50$

(iii) $K = 48$

4.6-1 Realize

$$H(s) = \frac{s(s + 2)}{(s + 1)(s + 3)(s + 4)}$$

by canonic direct, series, and parallel forms.

4.6-2 Realize the transfer function in Prob. 4.6-1 by using the transposed form of the realizations found in Prob. 4.6-1.

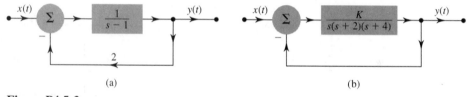

(a)   (b)

Figure P4.5-3

4.6-3   Repeat Prob. 4.6-1 for

(a) $H(s) = \dfrac{3s(s+2)}{(s+1)(s^2+2s+2)}$

(b) $H(s) = \dfrac{2s-4}{(s+2)(s^2+4)}$

4.6-4   Realize the transfer functions in Prob. 4.6-3 by using the transposed form of the realizations found in Prob. 4.6-3.

4.6-5   Repeat Prob. 4.6-1 for

$$H(s) = \frac{2s+3}{5s(s+2)^2(s+3)}$$

4.6-6   Realize the transfer function in Prob. 4.6-5 by using the transposed form of the realizations found in Prob. 4.6-5.

4.6-7   Repeat Prob. 4.6-1 for

$$H(s) = \frac{s(s+1)(s+2)}{(s+5)(s+6)(s+8)}$$

4.6-8   Realize the transfer function in Prob. 4.6-7 by using the transposed form of the realizations found in Prob. 4.6-7.

4.6-9   Repeat Prob. 4.6-1 for

$$H(s) = \frac{s^3}{(s+1)^2(s+2)(s+3)}$$

4.6-10 Realize the transfer function in Prob. 4.6-9 by using the transposed form of the realizations found in Prob. 4.6-9.

4.6-11 Repeat Prob. 4.6-1 for

$$H(s) = \frac{s^3}{(s+1)(s^2+4s+13)}$$

4.6-12 Realize the transfer function in Prob. 4.6-11 by using the transposed form of the realizations found in Prob. 4.6-11.

4.6-13 In this problem we show how a pair of complex conjugate poles may be realized by using

a cascade of two first-order transfer functions and feedback. Show that the transfer functions of the block diagrams in Fig. P4.6-13a and P4.6-13b are

(a) $H(s) = \dfrac{1}{(s+a)^2+b^2}$

$$= \frac{1}{s^2+2as+(a^2+b^2)}$$

(b) $H(s) = \dfrac{s+a}{(s+a)^2+b^2}$

$$= \frac{s+a}{s^2+2as+(a^2+b^2)}$$

Hence, show that the transfer function of the block diagram in Fig. P4.6-13c is

(c) $H(s) = \dfrac{As+B}{(s+a)^2+b^2}$

$$= \frac{As+B}{s^2+2as+(a^2+b^2)}$$

4.6-14 Show op-amp realizations of the following transfer functions:

(i) $\dfrac{-10}{s+5}$

(ii) $\dfrac{10}{s+5}$

(iii) $\dfrac{s+2}{s+5}$

4.6-15 Show two different op-amp circuit realizations of the transfer function

$$H(s) = \frac{s+2}{s+5} = 1 - \frac{3}{s+5}$$

4.6-16 Show an op-amp canonic direct realization of the transfer function

$$H(s) = \frac{3s+7}{s^2+4s+10}$$

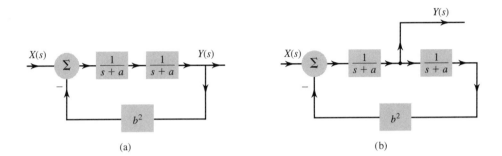

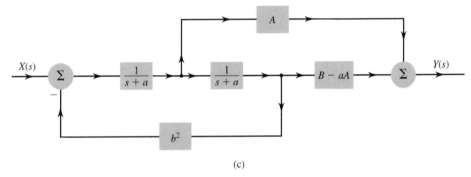

Figure P4.6-13

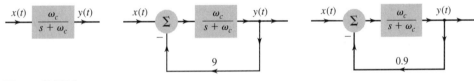

Figure P4.7-1

4.6-17 Show an op-amp canonic direct realization of the transfer function

$$H(s) = \frac{s^2 + 5s + 2}{s^2 + 4s + 13}$$

4.7-1 Feedback can be used to increase (or decrease) the system bandwidth. Consider the system in Fig. P4.7-1a with transfer function $G(s) = \omega_c/(s + \omega_c)$.

(a) Show that the 3 dB bandwidth of this system is ω_c and the dc gain is unity, that is, $|H(j0)| = 1$.

(b) To increase the bandwidth of this system, we use negative feedback with $H(s) = 9$, as depicted in Fig. P4.7-1b. Show that the 3 dB bandwidth of this system is $10\omega_c$. What is the dc gain?

(c) To decrease the bandwidth of this system, we use positive feedback with $H(s) = -0.9$, as illustrated in Fig. P4.7-1c. Show that the 3 dB bandwidth of this system is $\omega_c/10$. What is the dc gain?

(d) The system gain at dc times its 3 dB bandwidth is the *gain–bandwidth product* of a system. Show that this product is the same for all the three systems in Fig. P4.7-1. This result shows that if we increase the bandwidth, the gain decreases and vice versa.

4.8-1 For an LTIC system described by the transfer function

$$H(s) = \frac{s + 2}{s^2 + 5s + 4}$$

find the response to the following everlasting sinusoidal inputs:

(a) $5\cos(2t + 30°)$

(b) $10\sin(2t + 45°)$

(c) $10\cos(3t + 40°)$

Observe that these are everlasting sinusoids.

4.8-2 For an LTIC system described by the transfer function

$$H(s) = \frac{s+3}{(s+2)^2}$$

find the steady-state system response to the following inputs:

(a) $10u(t)$

(b) $\cos(2t + 60°)u(t)$

(c) $\sin(3t - 45°)u(t)$

(d) $e^{j3t}u(t)$

4.8-3 For an allpass filter specified by the transfer function

$$H(s) = \frac{-(s-10)}{s+10}$$

find the system response to the following (everlasting) inputs:

(a) $e^{j\omega t}$

(b) $\cos(\omega t + \theta)$

(c) $\cos t$

(d) $\sin 2t$

(e) $\cos 10t$

(f) $\cos 100t$

Comment on the filter response.

4.8-4 The pole–zero plot of a second-order system $H(s)$ is shown in Fig. P4.8-4. The dc response of this system is minus one, $H(j0) = -1$.

(a) Letting $H(s) = k(s^2 + b_1s + b_2)/(s^2 + a_1s + a_2)$, determine the constants k, b_1, b_2, a_1, and a_2.

(b) What is the output $y(t)$ of this system in response to the input $x(t) = 4 + \cos(t/2 + \pi/3)$?

4.9-1 Sketch Bode plots for the following transfer functions:

(a) $\dfrac{s(s+100)}{(s+2)(s+20)}$

(b) $\dfrac{(s+10)(s+20)}{s^2(s+100)}$

(c) $\dfrac{(s+10)(s+200)}{(s+20)^2(s+1000)}$

4.9-2 Repeat Prob. 4.9-1 for

(a) $\dfrac{s^2}{(s+1)(s^2+4s+16)}$

(b) $\dfrac{s}{(s+1)(s^2+14.14s+100)}$

(c) $\dfrac{(s+10)}{s(s^2+14.14s+100)}$

4.9-3 Using the lowest order possible, determine a system function $H(s)$ with real-valued roots that matches the frequency response in Fig. P4.9-3 Verify your answer with MATLAB.

4.9-4 A graduate student recently implemented an analog phase lock loop (PLL) as part of his thesis. His PLL consists of four basic components: a phase/frequency detector, a charge pump, a loop filter, and a voltage-controlled

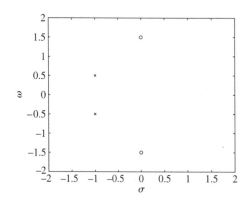

Figure P4.8-4 System pole–zero plot.

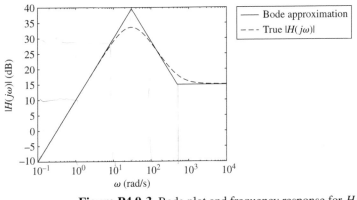

Figure P4.9-3 Bode plot and frequency response for $H(s)$.

oscillator. This problem considers only the loop filter, which is shown in Fig. P4.9-4a. The loop filter input is the current $x(t)$, and the output is the voltage $y(t)$.

(a) Derive the loop filter's transfer function $H(s)$. Express $H(s)$ in standard form.

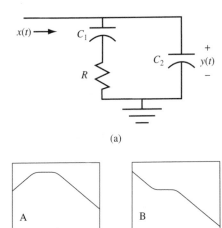

(a)

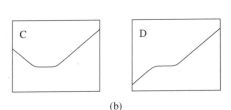

(b)

Figure P4.9-4 (a) Circuit diagram for PLL loop filter. **(b)** Possible magnitude response plots for PLL loop filter.

(b) Figure P4.9-4b provides four possible frequency response plots, labeled A through D. Each log–log plot is drawn to the same scale, and line slopes are either 20 dB/decade, 0 dB/decade, or −20 dB/decade. Clearly identify which plot(s), if any, could represent the loop filter.

(c) Holding the other components constant, what is the general effect of increasing the resistance R on the magnitude response for low-frequency inputs?

(d) Holding the other components constant, what is the general effect of increasing the resistance R on the magnitude-response for high-frequency inputs?

4.10-1 Using the graphical method of Section 4.10-1, draw a rough sketch of the amplitude and phase response of an LTIC system described by the transfer function

$$H(s) = \frac{s^2 - 2s + 50}{s^2 + 2s + 50}$$

$$= \frac{(s - 1 - j7)(s - 1 + j7)}{(s + 1 - j7)(s + 1 + j7)}$$

What kind of filter is this?

4.10-2 Using the graphical method of Section 4.10-1, draw a rough sketch of the amplitude and phase response of LTIC systems whose pole–zero plots are shown in Fig. P4.10-2.

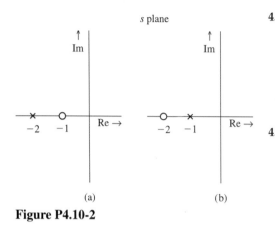

Figure P4.10-2

4.10-3 Design a second-order bandpass filter with center frequency $\omega = 10$. The gain should be zero at $\omega = 0$ and at $\omega = \infty$. Select poles at $-a \pm j10$. Leave your answer in terms of a. Explain the influence of a on the frequency response.

4.10-4 The LTIC system described by $H(s) = (s - 1)/(s + 1)$ has unity magnitude response $|H(j\omega)| = 1$. Positive Pat claims that the output $y(t)$ of this system is equal the input $x(t)$ since the system is allpass. Cynical Cynthia doesn't think so. "This is *signals and systems* class," she complains. "It *has* to be more complicated!" Who is correct, Pat or Cynthia? Justify your answer.

4.10-5 Two students, Amy and Jeff, disagree about an analog system function given by $H_1(s) = s$. Sensible Jeff claims the system has a zero at $s = 0$. Rebellious Amy, however, notes that the system function can be rewritten as $H_1(s) = 1/s^{-1}$ and claims that this implies a system pole at $s = \infty$. Who is correct? Why? What are the poles and zeros of the system $H_2(s) = 1/s$?

4.10-6 A rational transfer function $H(s)$ is often used to represent an analog filter. Why must $H(s)$ be strictly proper for lowpass and bandpass filters? Why must $H(s)$ be proper for highpass and bandstop filters?

4.10-7 For a given filter order N, why is the stopband attenuation rate of an all-pole lowpass filter better than filters with finite zeros?

4.10-8 Is it possible, with real coefficients ($[k, b_1, b_2, a_1, a_2] \in \mathcal{R}$), for a system

$$H(s) = k\frac{s^2 + b_1 s + b_2}{s^2 + a_1 s + a_2}$$

to function as a lowpass filter? Explain your answer.

4.10-9 Nick recently built a simple second-order Butterworth lowpass filter for his home stereo. Although the system performs pretty well, Nick is an overachiever and hopes to improve the system performance. Unfortunately, Nick is pretty lazy and doesn't want to design another filter. Thinking "twice the filtering gives twice the performance," he suggests filtering the audio signal not once but twice with a cascade of two identical filters. His overworked, underpaid signals professor is skeptical and states, "If you are using *identical* filters, it makes no difference whether you filter once or twice!" Who is correct? Why?

4.10-10 An LTIC system impulse response is given by $h(t) = u(t) - u(t - 1)$.

(a) Determine the transfer function $H(s)$. Using $H(s)$, determine and plot the magnitude response $|H(j\omega)|$. Which type of filter most accurately describes the behavior of this system: lowpass, highpass, bandpass, or bandstop?

(b) What are the poles and zeros of $H(s)$? Explain your answer.

(c) Can you determine the impulse response of the inverse system? If so, provide it. If not, suggest a method that could be used to approximate the impulse response of the inverse system.

4.10-11 An ideal lowpass filter $H_{\text{LP}}(s)$ has magnitude response that is unity for low frequencies and zero for high frequencies. An ideal highpass filter $H_{\text{HP}}(s)$ has an opposite magnitude response: zero for low frequencies and unity for high frequencies. A student suggests a possible lowpass-to-highpass filter transformation: $H_{\text{HP}}(s) = 1 - H_{\text{LP}}(s)$. In general, will this transformation work? Explain your answer.

4.10-12 An LTIC system has a rational transfer function $H(s)$. When appropriate, assume that all initial conditions are zero.

(a) Is is possible for this system to output $y(t) = \sin(100\pi t)u(t)$ in response to an input $x(t) = \cos(100\pi t)u(t)$? Explain.

(b) Is is possible for this system to output $y(t) = \sin(100\pi t)u(t)$ in response to an input $x(t) = \sin(50\pi t)u(t)$? Explain.

(c) Is is possible for this system to output $y(t) = \sin(100\pi t)$ in response to an input $x(t) = \cos(100\pi t)$? Explain.

(d) Is is possible for this system to output $y(t) = \sin(100\pi t)$ in response to an input $x(t) = \sin(50\pi t)$? Explain.

4.11-1 Find the ROC, if it exists, of the (bilateral) Laplace transform of the following signals:

(a) $e^{tu(t)}$

(b) $e^{-tu(t)}$

(c) $\dfrac{1}{1 + t^2}$

(d) $\dfrac{1}{1 + e^t}$

(e) e^{-kt^2}

4.11-2 Find the (bilateral) Laplace transform and the corresponding region of convergence for the following signals:

(a) $e^{-|t|}$

(b) $e^{-|t|} \cos t$

(c) $e^t u(t) + e^{2t} u(-t)$

(d) $e^{-tu(t)}$

(e) $e^{tu(-t)}$

(f) $\cos \omega_0 t\, u(t) + e^t\, u(-t)$

4.11-3 Find the inverse (bilateral) Laplace transforms of the following functions:

(a) $\dfrac{2s + 5}{(s + 2)(s + 3)}$ $-3 < \sigma < -2$

(b) $\dfrac{2s - 5}{(s - 2)(s - 3)}$ $2 < \sigma < 3$

(c) $\dfrac{2s + 3}{(s + 1)(s + 2)}$ $\sigma > -1$

(d) $\dfrac{2s + 3}{(s + 1)(s + 2)}$ $\sigma < -2$

(e) $\dfrac{3s^2 - 2s - 17}{(s + 1)(s + 3)(s - 5)}$

$-1 < \sigma < 5$

4.11-4 Find

$$\mathcal{L}^{-1} \left[\frac{2s^2 - 2s - 6}{(s + 1)(s - 1)(s + 2)} \right]$$

if the ROC is

(a) Re $s > 1$

(b) Re $s < -2$

(c) $-1 <$ Re $s < 1$

(d) $-2 <$ Re $s < -1$

4.11-5 For a causal LTIC system having a transfer function $H(s) = 1/(s + 1)$, find the output $y(t)$ if the input $x(t)$ is given by

(a) $e^{-|t|/2}$

(b) $e^t u(t) + e^{2t} u(-t)$

(c) $e^{-t/2} u(t) + e^{-t/4} u(-t)$

(d) $e^{2t} u(t) + e^t u(-t)$

(e) $e^{-t/4} u(t) + e^{-t/2} u(-t)$

(f) $e^{-3t} u(t) + e^{-2t} u(-t)$

4.11-6 The autocorrelation function $r_{xx}(t)$ of a signal $x(t)$ is given by

$$r_{xx}(t) = \int_{-\infty}^{\infty} x(\tau)x(\tau + t)\, d\tau$$

Derive an expression for $R_{xx}(s) = \mathcal{L}(r_{xx}(t))$ in terms of $X(s)$, where $X(s) = \mathcal{L}(x(t))$.

4.11-7 Determine the inverse Laplace transform of

$$X(s) = \frac{2}{s} + \frac{s}{2}$$

given that the region of convergence is $\sigma < 0$.

4.11-8 An absolutely integrable signal $x(t)$ has a pole at $s = \pi$. It is possible that other poles may be present. Recall that an absolutely integrable signal satisfies

$$\int_{-\infty}^{\infty} |x(t)|\, dt < \infty$$

(a) Can $x(t)$ be left sided? Explain.

(b) Can $x(t)$ be right sided? Explain.

(c) Can $x(t)$ be two sided? Explain.

(d) Can $x(t)$ be of finite duration? Explain.

4.11-9 Using the definition, compute the bilateral Laplace transform, including the region of convergence (ROC), of the following complex-valued functions:

(a) $x_1(t) = (j + e^{jt})u(t)$

(b) $x_2(t) = j \cosh(t)u(-t)$

(c) $x_3(t) = e^{j(\frac{\pi}{4})}u(-t+1) + j\delta(t-5)$

(d) $x_4(t) = j'u(-t) + \delta(t - \pi)$

4.11-10 A bounded-amplitude signal $x(t)$ has bilateral Laplace transform $X(s)$ given by

$$X(s) = \frac{s2^s}{(s-1)(s+1)}$$

(a) Determine the corresponding region of convergence.

(b) Determine the time-domain signal $x(t)$.

4.M-1 Express the polynomial $C_{20}(x)$ in standard form. That is, determine the coefficients a_k of $C_{20}(x) = \sum_{k=0}^{20} a_k x^{20-k}$.

4.M-2 Design an order-12 Butterworth lowpass filter with a cutoff frequency of $\omega_c = 2\pi 5000$ by completing the following.

(a) Locate and plot the filter's poles and zeros in the complex plane. Plot the corresponding magnitude response $|H_{\mathrm{LP}}(j\omega)|$ to verify proper design.

(b) Setting all resistor values to 100,000, determine the capacitor values to implement the filter using a cascade of six second-order Sallen–Key circuit sections. The form of a Sallen–Key stage is shown in Fig. P4.M-2. On a single plot, plot the magnitude response of each section as well as the overall magnitude response. Identify the poles that correspond to each section's magnitude response curve. Are the capacitor values realistic?

4.M-3 Rather than a Butterworth filter, repeat Prob. P4.M-2 for a Chebyshev LPF with $R = 3\,\mathrm{dB}$ of passband ripple. Since each Sallen–Key stage is constrained to have unity gain at dc, an overall gain error of $1/\sqrt{1 + \epsilon^2}$ is acceptable.

4.M-4 An analog lowpass filter with cutoff frequency ω_c can be transformed into a highpass filter with cutoff frequency ω_c by using an RC–CR transformation rule: each resistor R_i is replaced by a capacitor $C_i' = 1/R_i\omega_c$ and each capacitor C_i is replaced by a resistor $R_i' = 1/C_i\omega_c$.

Use this rule to design an order-8 Butterworth highpass filter with $\omega_c = 2\pi 4000$ by completing the following.

(a) Design an order-8 Butterworth lowpass filter with $\omega_c = 2\pi 4000$ by using four second-order Sallen–Key circuit stages, the form of which is shown in Fig. P4.M-2. Give resistor and capacitor values for each stage. Choose the resistors so that the RC–CR transformation will result in 1 nF capacitors. At this point, are the component values realistic?

(b) Draw an RC–CR transformed Sallen–Key circuit stage. Determine the transfer function $H(s)$ of the transformed stage in terms of the variables R_1', R_2', C_1', and C_2'.

(c) Transform the LPF designed in part a by using an RC–CR transformation. Give the resistor and capacitor values for each stage. Are the component values realistic?

Using $H(s)$ derived in part b, plot the magnitude response of each section as well as the overall magnitude response. Does the overall response look like a highpass Butterworth filter?

Plot the HPF system poles and zeros in the complex s plane. How do these

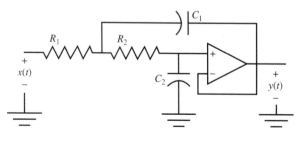

Figure P4.M-2 Sallen–Key filter stage.

locations compare with those of the Butterworth LPF?

4.M-5 Repeat Prob. P4.M-4, using $\omega_c = 2\pi 1500$ and an order-16 filter. That is, eight second-order stages need to be designed.

4.M-6 Rather than a Butterworth filter, repeat Prob. P4.M-4 for a Chebyshev LPF with $R = 3\,\text{dB}$ of passband ripple. Since each transformed Sallen–Key stage is constrained to have unity gain at $\omega = \infty$, an overall gain error of $1/\sqrt{1 + \epsilon^2}$ is acceptable.

4.M-7 The MATLAB signal processing toolbox function `butter` helps design analog Butterworth filters. Use MATLAB help to learn how `butter` works. For each of the following cases, design the filter, plot the filter's poles and zeros in the complex s plane, and plot the decibel magnitude response $20\log_{10}|H(j\omega)|$.

(a) Design a sixth-order analog lowpass filter with $\omega_c = 2\pi 3500$.

(b) Design a sixth-order analog highpass filter with $\omega_c = 2\pi 3500$.

(c) Design a sixth-order analog bandpass filter with a passband between 2 and 4 kHz.

(d) Design a sixth-order analog bandstop filter with a stopband between 2 and 4 kHz.

4.M-8 The MATLAB signal processing toolbox function `cheby1` helps design analog Chebyshev type I filters. A Chebyshev type I filter has a passband ripple and a smooth stopband.

Setting the passband ripple to $R_p = 3\,\text{dB}$, repeat Prob. P4.M-7 using the `cheby1` command. With all other parameters held constant, what is the general effect of reducing R_p, the allowable passband ripple?

4.M-9 The MATLAB signal processing toolbox function `cheby2` helps design analog Chebyshev type II filters. A Chebyshev type II filter has a smooth passband and ripple in the stopband. Setting the stopband ripple $R_s = 20\,\text{dB}$ down, repeat Prob. P4.M-7 using the `cheby2` command. With all other parameters held constant, what is the general effect of increasing R_s, the minimum stopband attenuation?

4.M-10 The MATLAB signal processing toolbox function `ellip` helps design analog elliptic filters. An elliptic filter has ripple in both the passband and the stopband. Setting the passband ripple to $R_p = 3\,\text{dB}$ and the stopband ripple $R_s = 20\,\text{dB}$ down, repeat Prob. P4.M-7 using the `ellip` command.

4.M-11 Using the definition $C_N(x) = \cosh(N\cosh^{-1}(x))$, prove the recursive relation $C_N(x) = 2xC_{N-1}(x) - C_{N-2}(x)$.

4.M-12 Prove that the poles of a Chebyshev filter, which are located at $p_k = \omega_c \sinh(\xi)\sin(\phi_k) + j\omega_c \cosh(\xi)\cos(\phi_k)$, lie on an ellipse. [Hint: The equation of an ellipse in the x–y plane is $(x/a)^2 + (y/b)^2 = 1$, where constants a and b define the major and minor axes of the ellipse.]

5

DISCRETE-TIME SYSTEM ANALYSIS USING THE z-TRANSFORM

The counterpart of the Laplace transform for discrete-time systems is the z-transform. The Laplace transform converts integro-differential equations into algebraic equations. In the same way, the z-transforms changes difference equations into algebraic equations, thereby simplifying the analysis of discrete-time systems. The z-transform method of analysis of discrete-time systems parallels the Laplace transform method of analysis of continuous-time systems, with some minor differences. In fact, we shall see that *the z-transform is the Laplace transform in disguise*.

The behavior of discrete-time systems is similar to that of continuous-time systems (with some differences). The frequency-domain analysis of discrete-time systems is based on the fact (proved in Section 3.8-3) that the response of a linear, time-invariant, discrete-time (LTID) system to an everlasting exponential z^n is the same exponential (within a multiplicative constant) given by $H[z]z^n$. We then express an input $x[n]$ as a sum of (everlasting) exponentials of the form z^n. The system response to $x[n]$ is then found as a sum of the system's responses to all these exponential components. The tool that allows us to represent an arbitrary input $x[n]$ as a sum of (everlasting) exponentials of the form z^n is the z-transform.

5.1 THE z-TRANSFORM

We define $X[z]$, the direct z-transform of $x[n]$, as

$$X[z] = \sum_{n=-\infty}^{\infty} x[n]z^{-n} \tag{5.1}$$

where z is a complex variable. The signal $x[n]$, which is the inverse z-transform of $X[z]$, can be obtained from $X[z]$ by using the following inverse z-transformation:

$$x[n] = \frac{1}{2\pi j} \oint X[z]z^{n-1}\, dz \tag{5.2}$$

The symbol $\oint$ indicates an integration in counterclockwise direction around a closed path in the complex plane (see Fig. 5.1). We derive this z-transform pair later, in Chapter 9, as an extension of the discrete-time Fourier transform pair.

As in the case of the Laplace transform, we need not worry about this integral at this point because inverse z-transforms of many signals of engineering interest can be found in a z-transform table. The direct and inverse z-transforms can be expressed symbolically as

$$X[z] = \mathcal{Z}\{x[n]\} \quad \text{and} \quad x[n] = \mathcal{Z}^{-1}\{X[z]\}$$

or simply as

$$x[n] \Longleftrightarrow X[z]$$

Note that

$$\mathcal{Z}^{-1}[\mathcal{Z}\{x[n]\}] = x[n] \quad \text{and} \quad \mathcal{Z}[\mathcal{Z}^{-1}\{X[z]\}] = X[z]$$

LINEARITY OF THE z-TRANSFORM

Like the Laplace transform, the z-transform is a linear operator. If

$$x_1[n] \Longleftrightarrow X_1[z] \quad \text{and} \quad x_2[n] \Longleftrightarrow X_2[z]$$

then

$$a_1 x_1[n] + a_2 x_2[n] \Longleftrightarrow a_1 X_1[z] + a_2 X_2[z] \tag{5.3}$$

The proof is trivial and follows from the definition of the z-transform. This result can be extended to finite sums.

THE UNILATERAL z-TRANSFORM

For the same reasons discussed in Chapter 4, we find it convenient to consider the unilateral z-transform. As seen for the Laplace case, the bilateral transform has some complications because of nonuniqueness of the inverse transform. In contrast, the unilateral transform has a unique inverse. This fact simplifies the analysis problem considerably, but at a price: the unilateral version can handle only causal signals and systems. Fortunately, most of the practical cases are causal. The more general *bilateral z-transform* is discussed later, in Section 5.9. In practice, the term *z-transform* generally means *the unilateral z-transform*.

In a basic sense, there is no difference between the unilateral and the bilateral z-transform. The unilateral transform is the bilateral transform that deals with a subclass of signals starting at $n = 0$ (causal signals). Hence, the definition of the unilateral transform is the same as that of the bilateral [Eq. (5.1)], except that the limits of the sum are from 0 to ∞

$$X[z] = \sum_{n=0}^{\infty} x[n]z^{-n} \tag{5.4}$$

The expression for the inverse z-transform in Eq. (5.2) remains valid for the unilateral case also.

THE REGION OF CONVERGENCE (ROC) OF $X[z]$

The sum in Eq. (5.1) [or (5.4)] defining the direct z-transform $X[z]$ may not converge (exist) for all values of z. The values of z (the region in the complex plane) for which the sum in Eq. (5.1) converges (or exists) is called the *region of existence,* or more commonly the *region of convergence* (ROC), for $X[z]$. This concept will become clear in the following example.

EXAMPLE 5.1

Find the z-transform and the corresponding ROC for the signal $\gamma^n u[n]$.

By definition

$$X[z] = \sum_{n=0}^{\infty} \gamma^n u[n] z^{-n}$$

Since $u[n] = 1$ for all $n \geq 0$,

$$X[z] = \sum_{n=0}^{\infty} \left(\frac{\gamma}{z}\right)^n$$

$$= 1 + \left(\frac{\gamma}{z}\right) + \left(\frac{\gamma}{z}\right)^2 + \left(\frac{\gamma}{z}\right)^3 + \cdots + \cdots \tag{5.5}$$

It is helpful to remember the following well-known geometric progression and its sum:

$$1 + x + x^2 + x^3 + \cdots = \frac{1}{1-x} \qquad \text{if} \quad |x| < 1 \tag{5.6}$$

Use of Eq. (5.6) in Eq. (5.5) yields

$$X[z] = \frac{1}{1 - \dfrac{\gamma}{z}} \qquad \left|\frac{\gamma}{z}\right| < 1$$

$$= \frac{z}{z - \gamma} \qquad |z| > |\gamma| \tag{5.7}$$

Observe that $X[z]$ exists only for $|z| > |\gamma|$. For $|z| < |\gamma|$, the sum in Eq. (5.5) may not converge; it goes to infinity. Therefore, the ROC of $X[z]$ is the shaded region outside the circle of radius $|\gamma|$, centered at the origin, in the z-plane, as depicted in Fig. 5.1b.

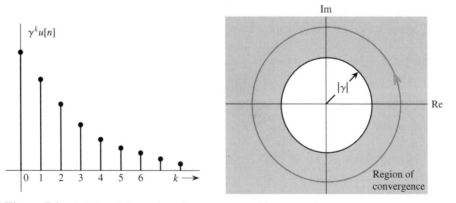

Figure 5.1 $\gamma^n u[n]$ and the region of convergence of its z-transform.

Later in Eq. (5.85), we show that the z-transform of another signal $-\gamma^n u[-(n+1)]$ is also $z/(z-\gamma)$. However, the ROC in this case is $|z| < |\gamma|$. Clearly, the inverse z-transform of $z/(z-\gamma)$ is not unique. However, if we restrict the inverse transform to be causal, then the inverse transform is unique, namely, $\gamma^n u[n]$.

The ROC is required for evaluating $x[n]$ from $X[z]$, according to Eq. (5.2). The integral in Eq. (5.2) is a contour integral, implying integration in a counterclockwise direction along a closed path centered at the origin and satisfying the condition $|z| > |\gamma|$. Thus, any circular path centered at the origin and with a radius greater than $|\gamma|$ (Fig. 5.1b) will suffice. We can show that the integral in Eq. (5.2) along any such path (with a radius greater than $|\gamma|$) yields the same result, namely $x[n]$.[†] Such integration in the complex plane requires a background in the theory of functions of complex variables. We can avoid this integration by compiling a table of z-transforms (Table 5.1), where z-transform pairs are tabulated for a variety of signals. To find the inverse z-transform of say, $z/(z-\gamma)$, instead of using the complex integration in Eq. (5.2), we consult the table and find the inverse z-transform of $z/(z-\gamma)$ as $\gamma^n u[n]$. Because of uniqueness property of the unilateral z-transform, there is only one inverse for each $X[z]$. Although the table given here is rather short, it comprises the functions of most practical interest.

The situation of the z-transform regarding the uniqueness of the inverse transform is parallel to that of the Laplace transform. For the bilateral case, the inverse z-transform is not unique unless the ROC is specified. For the unilateral case, the inverse transform is unique; the region of convergence need not be specified to determine the inverse z-transform. For this reason, we shall ignore the ROC in the unilateral z-transform Table 5.1.

EXISTENCE OF THE z-TRANSFORM

By definition

$$X[z] = \sum_{n=0}^{\infty} x[n]z^{-n} = \sum_{n=0}^{\infty} \frac{x[n]}{z^n}$$

The existence of the z-transform is guaranteed if

$$|X[z]| \leq \sum_{n=0}^{\infty} \frac{|x[n]|}{|z|^n} < \infty$$

for some $|z|$. Any signal $x[n]$ that grows no faster than an exponential signal r_0^n, for some r_0, satisfies this condition. Thus, if

$$|x[n]| \leq r_0^n \qquad \text{for some } r_0 \tag{5.8}$$

then

$$|X[z]| \leq \sum_{n=0}^{\infty} \left(\frac{r_0}{|z|} \right)^n = \frac{1}{1 - \dfrac{r_0}{|z|}} \qquad |z| > r_0$$

[†]Indeed, the path need not even be circular. It can have any odd shape, as long as it encloses the pole(s) of $X[z]$ and the path of integration is counterclockwise.

TABLE 5.1   (Unilateral) z-Transform Pairs

No.	$x[n]$	$X[z]$
1	$\delta[n - k]$	z^{-k}
2	$u[n]$	$\dfrac{z}{z - 1}$
3	$nu[n]$	$\dfrac{z}{(z - 1)^2}$
4	$n^2 u[n]$	$\dfrac{z(z + 1)}{(z - 1)^3}$
5	$n^3 u[n]$	$\dfrac{z(z^2 + 4z + 1)}{(z - 1)^4}$
6	$\gamma^n u[n]$	$\dfrac{z}{z - \gamma}$
7	$\gamma^{n-1} u[n - 1]$	$\dfrac{1}{z - \gamma}$
8	$n\gamma^n u[n]$	$\dfrac{\gamma z}{(z - \gamma)^2}$
9	$n^2 \gamma^n u[n]$	$\dfrac{\gamma z(z + \gamma)}{(z - \gamma)^3}$
10	$\dfrac{n(n - 1)(n - 2) \cdots (n - m + 1)}{\gamma^m m!} \gamma^n u[n]$	$\dfrac{z}{(z - \gamma)^{m+1}}$
11a	$|\gamma|^n \cos \beta n \, u[n]$	$\dfrac{z(z - |\gamma| \cos \beta)}{z^2 - (2|\gamma| \cos \beta)z + |\gamma|^2}$
11b	$|\gamma|^n \sin \beta n \, u[n]$	$\dfrac{z|\gamma| \sin \beta}{z^2 - (2|\gamma| \cos \beta)z + |\gamma|^2}$
12a	$r|\gamma|^n \cos (\beta n + \theta) u[n]$	$\dfrac{rz[z \cos \theta - |\gamma| \cos (\beta - \theta)]}{z^2 - (2|\gamma| \cos \beta)z + |\gamma|^2}$
12b	$r|\gamma|^n \cos (\beta n + \theta) u[n] \qquad \gamma = |\gamma| e^{j\beta}$	$\dfrac{(0.5 re^{j\theta})z}{z - \gamma} + \dfrac{(0.5 re^{-j\theta})z}{z - \gamma^*}$
12c	$r|\gamma|^n \cos (\beta n + \theta) u[n]$	$\dfrac{z(Az + B)}{z^2 + 2az + |\gamma|^2}$

$$r = \sqrt{\frac{A^2 |\gamma|^2 + B^2 - 2AaB}{|\gamma|^2 - a^2}}$$

$$\beta = \cos^{-1} \frac{-a}{|\gamma|}$$

$$\theta = \tan^{-1} \frac{Aa - B}{A\sqrt{|\gamma|^2 - a^2}}$$

Therefore, $X[z]$ exists for $|z| > r_0$. Almost all practical signals satisfy condition (5.8) and are therefore z-transformable. Some signal models (e.g., γ^{n^2}) grow faster than the exponential signal r_0^n (for any r_0) and do not satisfy Eq. (5.8) and therefore are not z-transformable. Fortunately, such signals are of little practical or theoretical interest. Even such signals over a finite interval are z-transformable.

EXAMPLE 5.2

Find the z-transforms of

(a) $\delta[n]$

(b) $u[n]$

(c) $\cos \beta n\, u[n]$

(d) The signal shown in Fig. 5.2

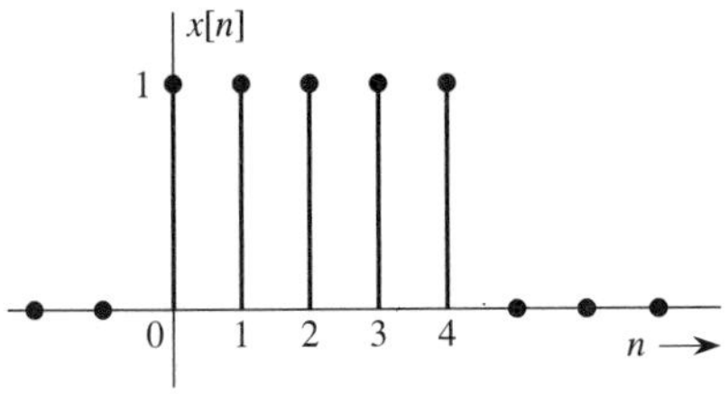

Figure 5.2

Recall that by definition

$$X[z] = \sum_{n=0}^{\infty} x[n]z^{-n}$$

$$= x[0] + \frac{x[1]}{z} + \frac{x[2]}{z^2} + \frac{x[3]}{z^3} + \cdots \tag{5.9}$$

(a) For $x[n] = \delta[n]$, $x[0] = 1$ and $x[2] = x[3] = x[4] = \cdots = 0$. Therefore

$$\delta[n] \Longleftrightarrow 1 \qquad \text{for all } z \tag{5.10}$$

(b) For $x[n] = u[n]$, $x[0] = x[1] = x[3] = \cdots = 1$. Therefore

$$X[z] = 1 + \frac{1}{z} + \frac{1}{z^2} + \frac{1}{z^3} + \cdots$$

From Eq. (5.6) it follows that

$$X[z] = \frac{1}{1 - \dfrac{1}{z}} \qquad \left|\frac{1}{z}\right| < 1$$

$$= \frac{z}{z - 1} \qquad |z| > 1$$

Therefore

$$u[n] \Longleftrightarrow \frac{z}{z-1} \qquad |z| > 1 \tag{5.11}$$

(c) Recall that $\cos \beta n = (e^{j\beta n} + e^{-j\beta n})/2$. Moreover, according to Eq. (5.7),

$$e^{\pm j\beta n} u[n] \Longleftrightarrow \frac{z}{z - e^{\pm j\beta}} \qquad |z| > |e^{\pm j\beta}| = 1$$

Therefore

$$X[z] = \frac{1}{2} \left[\frac{z}{z - e^{j\beta}} + \frac{z}{z - e^{-j\beta}} \right]$$

$$= \frac{z(z - \cos \beta)}{z^2 - 2z \cos \beta + 1} \qquad |z| > 1$$

(d) Here $x[0] = x[1] = x[2] = x[3] = x[4] = 1$ and $x[5] = x[6] = \cdots = 0$. Therefore, according to Eq. (5.9)

$$X[z] = 1 + \frac{1}{z} + \frac{1}{z^2} + \frac{1}{z^3} + \frac{1}{z^4}$$

$$= \frac{z^4 + z^3 + z^2 + z + 1}{z^4} \qquad \text{for all } z \neq 0$$

We can also express this result in a more compact form by summing the geometric progression on the right-hand side of the foregoing equation. From the result in Section B.7-4 with $r = 1/z$, $m = 0$, and $n = 4$, we obtain

$$X[z] = \frac{\left(\dfrac{1}{z}\right)^5 - \left(\dfrac{1}{z}\right)^0}{\dfrac{1}{z} - 1} = \frac{z}{z-1}(1 - z^{-5})$$

EXERCISE E5.1

(a) Find the z-transform of a signal shown in Fig. 5.3.

(b) Use pair 12a (Table 5.1) to find the z-transform of $x[n] = 20.65(\sqrt{2})^n \cos((\pi/4)n - 1.415)u[n]$.

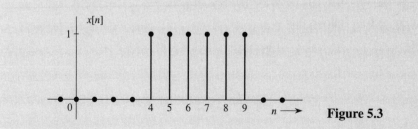

Figure 5.3

ANSWERS

(a) $X[z] = \dfrac{z^5 + z^4 + +z^3 + z^2 + z + 1}{z^9}$ or $\dfrac{z}{z-1}(z^{-4} - z^{-10})$

(b) $\dfrac{z(3.2z + 17.2)}{z^2 - 2z + 2}$

5.1-1 Finding the Inverse Transform

As in the Laplace transform, we shall avoid the integration in the complex plane required to find the inverse z-transform [Eq. (5.2)] by using the (unilateral) transform table (Table 5.1). Many of the transforms $X[z]$ of practical interest are rational functions (ratio of polynomials in z), which can be expressed as a sum of partial fractions, whose inverse transforms can be readily found in a table of transform. The partial fraction method works because for every transformable $x[n]$ defined for $n \geq 0$, there is a corresponding unique $X[z]$ defined for $|z| > r_0$ (where r_0 is some constant), and vice versa.

EXAMPLE 5.3

Find the inverse z-transform of

(a) $\dfrac{8z - 19}{(z - 2)(z - 3)}$

(b) $\dfrac{z(2z^2 - 11z + 12)}{(z - 1)(z - 2)^3}$

(c) $\dfrac{2z(3z + 17)}{(z - 1)(z^2 - 6z + 25)}$

(a) Expanding $X[z]$ into partial fractions yields

$$X[z] = \frac{8z - 19}{(z - 2)(z - 3)} = \frac{3}{z - 2} + \frac{5}{z - 3}$$

From Table 5.1, pair 7, we obtain

$$x[n] = [3(2)^{n-1} + 5(3)^{n-1}]u[n - 1] \qquad (5.12a)$$

If we expand rational $X[z]$ into partial fractions directly, we shall always obtain an answer that is multiplied by $u[n - 1]$ because of the nature of pair 7 in Table 5.1. This form is rather awkward as well as inconvenient. We prefer the form that contains $u[n]$ rather than $u[n - 1]$. A glance at Table 5.1 shows that the z-transform of every signal that is multiplied by $u[n]$ has a factor z in the numerator. This observation suggests that we expand $X[z]$ into *modified partial fractions*, where each term has a factor z in the numerator. This goal can be accomplished by expanding $X[z]/z$ into partial fractions and then multiplying both sides by z. We shall demonstrate this procedure by reworking part **(a)**. For this case

$$\frac{X[z]}{z} = \frac{8z - 19}{z(z - 2)(z - 3)}$$

$$= \frac{(-19/6)}{z} + \frac{(3/2)}{z - 2} + \frac{(5/3)}{z - 3}$$

Multiplying both sides by z yields

$$X[z] = -\frac{19}{6} + \frac{3}{2}\left(\frac{z}{z - 2}\right) + \frac{5}{3}\left(\frac{z}{z - 3}\right)$$

From pairs 1 and 6 in Table 5.1, it follows that

$$x[n] = -\frac{19}{6}\delta[n] + \left[\frac{3}{2}(2)^n + \frac{5}{3}(3)^n\right]u[n] \tag{5.12b}$$

The reader can verify that this answer is equivalent to that in Eq. (5.12a) by computing $x[n]$ in both cases for $n = 0, 1, 2, 3, \ldots$, and comparing the results. The form in Eq. (5.12b) is more convenient than that in Eq. (5.12a). For this reason, we shall always expand $X[z]/z$ rather than $X[z]$ into partial fractions and then multiply both sides by z to obtain modified partial fractions of $X[z]$, which have a factor z in the numerator.

(b)

$$X[z] = \frac{z(2z^2 - 11z + 12)}{(z - 1)(z - 2)^3}$$

and

$$\frac{X[z]}{z} = \frac{2z^2 - 11z + 12}{(z - 1)(z - 2)^3}$$

$$= \frac{k}{z - 1} + \frac{a_0}{(z - 2)^3} + \frac{a_1}{(z - 2)^2} + \frac{a_2}{(z - 2)}$$

where

$$k = \frac{2z^2 - 11z + 12}{(z - 1)(z - 2)^3}\bigg|_{z=1} = -3$$

$$a_0 = \frac{2z^2 - 11z + 12}{(z - 1)(z - 2)^3}\bigg|_{z=2} = -2$$

Therefore

$$\frac{X[z]}{z} = \frac{2z^2 - 11z + 12}{(z - 1)(z - 2)^3} = \frac{-3}{z - 1} - \frac{2}{(z - 2)^3} + \frac{a_1}{(z - 2)^2} + \frac{a_2}{(z - 2)} \tag{5.13}$$

We can determine a_1 and a_2 by clearing fractions. Or we may use a shortcut. For example, to determine a_2, we multiply both sides of Eq. (5.13) by z and let $z \to \infty$. This yields

$$0 = -3 - 0 + 0 + a_2 \implies a_2 = 3$$

This result leaves only one unknown, a_1, which is readily determined by letting z take any convenient value, say $z = 0$, on both sides of Eq. (5.13). This step yields

$$\frac{12}{8} = 3 + \frac{1}{4} + \frac{a_1}{4} - \frac{3}{2}$$

which yields $a_1 = -1$. Therefore

$$\frac{X[z]}{z} = \frac{-3}{z-1} - \frac{2}{(z-2)^3} - \frac{1}{(z-2)^2} + \frac{3}{z-2}$$

and

$$X[z] = -3\frac{z}{z-1} - 2\frac{z}{(z-2)^3} - \frac{z}{(z-2)^2} + 3\frac{z}{z-2}$$

Now the use of Table 5.1, pairs 6 and 10, yields

$$x[n] = \left[-3 - 2\frac{n(n-1)}{8}(2)^n - \frac{n}{2}(2)^n + 3(2)^n\right]u[n]$$

$$= -\left[3 + \frac{1}{4}(n^2 + n - 12)2^n\right]u[n]$$

(c) Complex Poles.

$$X[z] = \frac{2z(3z + 17)}{(z-1)(z^2 - 6z + 25)}$$

$$= \frac{2z(3z + 17)}{(z-1)(z-3-j4)(z-3+j4)}$$

Poles of $X[z]$ are 1, $3 + j4$, and $3 - j4$. Whenever there are complex conjugate poles, the problem can be worked out in two ways. In the first method we expand $X[z]$ into (modified) first-order partial fractions. In the second method, rather than obtain one factor corresponding to each complex conjugate pole, we obtain quadratic factors corresponding to each pair of complex conjugate poles. This procedure is explained next.

METHOD OF FIRST-ORDER FACTORS

$$\frac{X[z]}{z} = \frac{2(3z + 17)}{(z-1)(z^2 - 6z + 25)} = \frac{2(3z + 17)}{(z-1)(z-3-j4)(z-3+j4)}$$

We find the partial fraction of $X[z]/z$ using the Heaviside "cover-up" method:

$$\frac{X[z]}{z} = \frac{2}{z-1} + \frac{1.6e^{-j2.246}}{z-3-j4} + \frac{1.6e^{j2.246}}{z-3+j4}$$

and

$$X[z] = 2\frac{z}{z-1} + (1.6e^{-j2.246})\frac{z}{z-3-j4} + (1.6e^{j2.246})\frac{z}{z-3+j4}$$

The inverse transform of the first term on the right-hand side is $2u[n]$. The inverse transform of the remaining two terms (complex conjugate poles) can be obtained from pair 12b (Table 5.1) by identifying $r/2 = 1.6$, $\theta = -2.246$ rad, $\gamma = 3 + j4 = 5e^{j0.927}$, so that $|\gamma| = 5$, $\beta = 0.927$. Therefore

$$x[n] = [2 + 3.2(5)^n \cos (0.927n - 2.246)]u[n]$$

METHOD OF QUADRATIC FACTORS

$$\frac{X[z]}{z} = \frac{2(3z + 17)}{(z - 1)(z^2 - 6z + 25)} = \frac{2}{z - 1} + \frac{Az + B}{z^2 - 6z + 25}$$

Multiplying both sides by z and letting $z \to \infty$, we find

$$0 = 2 + A \Longrightarrow A = -2$$

and

$$\frac{2(3z + 17)}{(z - 1)(z^2 - 6z + 25)} = \frac{2}{z - 1} + \frac{-2z + B}{z^2 - 6z + 25}$$

To find B, we let z take any convenient value, say $z = 0$. This step yields

$$\frac{-34}{25} = -2 + \frac{B}{25} \Longrightarrow B = 16$$

Therefore

$$\frac{X[z]}{z} = \frac{2}{z - 1} + \frac{-2z + 16}{z^2 - 6z + 25}$$

and

$$X[z] = \frac{2z}{z - 1} + \frac{z(-2z + 16)}{z^2 - 6z + 25}$$

We now use pair 12c, where we identify $A = -2$, $B = 16$, $|\gamma| = 5$, and $a = -3$. Therefore

$$r = \sqrt{\frac{100 + 256 - 192}{25 - 9}} = 3.2, \quad \beta = \cos^{-1}\left(\frac{3}{5}\right) = 0.927 \text{ rad}$$

and

$$\theta = \tan^{-1}\left(\frac{-10}{-8}\right) = -2.246 \text{ rad}$$

so that

$$x[n] = [2 + 3.2(5)^n \cos (0.927n - 2.246)]u[n]$$

EXERCISE E5.2

Find the inverse z-transform of the following functions:

(a) $\dfrac{z(2z - 1)}{(z - 1)(z + 0.5)}$

(b) $\dfrac{1}{(z - 1)(z + 0.5)}$

(c) $\dfrac{9}{(z+2)(z-0.5)^2}\left(\dfrac{z}{z}\right)$

(d) $\dfrac{5z(z-1)}{z^2-1.6z+0.8}$

[Hint: $\sqrt{0.8} = 2/\sqrt{5}$.]

ANSWERS

(a) $\left[\frac{2}{3} + \frac{4}{3}(-0.5)^n\right]u[n]$

(b) $-2\delta[n] + \left[\frac{2}{3} + \frac{4}{3}(-0.5)^n\right]u[n]$

(c) $18\delta[n] - [0.72(-2)^n + 17.28(0.5)^n - 14.4n(0.5)^n]u[n]$

(d) $\frac{5\sqrt{5}}{2}\left(\frac{2}{\sqrt{5}}\right)^n \cos(0.464n + 0.464)u[n]$

INVERSE TRANSFORM BY EXPANSION OF $X[z]$ IN POWER SERIES OF z^{-1}

By definition

$$X[z] = \sum_{n=0}^{\infty} x[n]z^{-n}$$

$$= x[0] + \frac{x[1]}{z} + \frac{x[2]}{z^2} + \frac{x[3]}{z^3} + \cdots$$

$$= x[0]z^0 + x[1]z^{-1} + x[2]z^{-2} + x[3]z^{-3} + \cdots$$

This result is a power series in z^{-1}. Therefore, if we can expand $X[z]$ into the power series in z^{-1}, the coefficients of this power series can be identified as $x[0]$, $x[1]$, $x[2]$, $x[3]$, ... A rational $X[z]$ can be expanded into a power series of z^{-1} by dividing its numerator by the denominator. Consider, for example,

$$X[z] = \frac{z^2(7z-2)}{(z-0.2)(z-0.5)(z-1)}$$

$$= \frac{7z^3 - 2z^2}{z^3 - 1.7z^2 + 0.8z - 0.1}$$

To obtain a series expansion in powers of z^{-1}, we divide the numerator by the denominator as follows:

$$
\begin{array}{r}
7 + 9.9z^{-1} + 11.23z^{-2} + 11.87z^{-3} + \cdots \\
z^3 - 1.7z^2 + 0.8z - 0.1 \overline{)\, 7z^3 - 2z^2 } \\
7z^3 - 11.9z^2 + 5.60z - 0.7 \\
\hline
9.9z^2 - 5.60z + 0.7 \\
9.9z^2 - 16.83z + 7.92 - 0.99z^{-1} \\
\hline
11.23z - 7.22 + 0.99z^{-1} \\
11.23z - 19.09 + 8.98z^{-1} \\
\hline
11.87 - 7.99z^{-1}
\end{array}
$$

Thus

$$X[z] = \frac{z^2(7z-2)}{(z-0.2)(z-0.5)(z-1)} = 7 + 9.9z^{-1} + 11.23z^{-2} + 11.87z^{-3} + \cdots$$

Therefore

$$x[0] = 7, \ x[1] = 9.9, \ x[2] = 11.23, \ x[3] = 11.87, \ldots$$

Although this procedure yields $x[n]$ directly, it does not provide a closed-form solution. For this reason, it is not very useful unless we want to know only the first few terms of the sequence $x[n]$.

EXERCISE E5.3

Using long division to find the power series in z^{-1}, show that the inverse z-transform of $z/(z-0.5)$ is $(0.5)^n u[n]$ or $(2)^{-n} u[n]$.

RELATIONSHIP BETWEEN $h[n]$ AND $H[z]$

For an LTID system, if $h[n]$ is its unit impulse response, then from Eq. (3.71b), where we defined $H[z]$, the system transfer function, we write

$$H[z] = \sum_{n=-\infty}^{\infty} h[n]z^{-n} \tag{5.14a}$$

For causal systems, the limits on the sum are from $n = 0$ to ∞. This equation shows that the transfer function $H[z]$ is the z-transform of the impulse response $h[n]$ of an LTID system; that is,

$$h[n] \Longleftrightarrow H[z] \tag{5.14b}$$

This important result relates the time-domain specification $h[n]$ of a system to $H[z]$, the frequency-domain specification of a system. The result is parallel to that for LTIC systems.

EXERCISE E5.4

Redo Exercise E3.14 by taking the inverse z-transform of $H[z]$, as given by Eq. (3.73).

5.2 SOME PROPERTIES OF THE z-TRANSFORM

The z-transform properties are useful in the derivation of z-transforms of many functions and also in the solution of linear difference equations with constant coefficients. Here we consider a few important properties of the z-transform.

In our discussion, the variable n appearing in signals, such as $x[n]$ and $y[n]$, may or may not stand for time. However, in most applications of our interest, n is proportional to time. For this reason, we shall loosely refer to the variable n as time.

In the following discussion of the shift property, we deal with shifted signals $x[n]u[n]$, $x[n-k]u[n-k]$, $x[n-k]u[n]$, and $x[n+k]u[n]$. Unless we physically understand the meaning of such shifts, our understanding of the shift property remains mechanical rather than intuitive or heuristic. For this reason using a hypothetical signal $x[n]$, we have illustrated various shifted signals for $k = 1$ in Fig. 5.4.

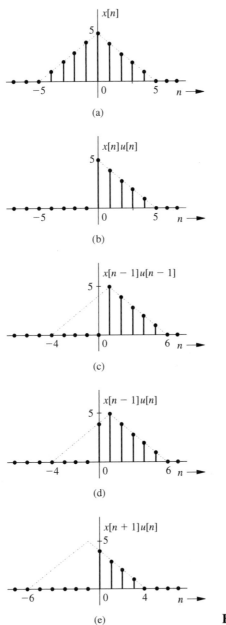

Figure 5.4 A signal $x[n]$ and its shifted versions.

RIGHT SHIFT (DELAY)

If

$$x[n]u[n] \Longleftrightarrow X[z]$$

then

$$x[n-1]u[n-1] \Longleftrightarrow \frac{1}{z}X[z] \tag{5.15a}$$

In general,

$$x[n-m]u[n-m] \Longleftrightarrow \frac{1}{z^m}X[z] \tag{5.15b}$$

Moreover,

$$x[n-1]u[n] \Longleftrightarrow \frac{1}{z}X[z] + x[-1] \tag{5.16a}$$

Repeated application of this property yields

$$x[n-2]u[n] \Longleftrightarrow \frac{1}{z}\left[\frac{1}{z}X[z] + x[-1]\right] + x[-2]$$

$$= \frac{1}{z^2}X[z] + \frac{1}{z}x[-1] + x[-2] \tag{5.16b}$$

In general, for integer value of m

$$x[n-m]u[n] \Longleftrightarrow z^{-m}X[z] + z^{-m}\sum_{n=1}^{m}x[-n]z^n \tag{5.16c}$$

A look at Eqs. (5.15a) and (5.16a) shows that they are identical except for the extra term $x[-1]$ in Eq. (5.16a). We see from Fig. 5.4c and 5.4d that $x[n-1]u[n]$ is the same as $x[n-1]u[n-1]$ plus $x[-1]\delta[n]$. Hence, the difference between their transforms is $x[-1]$.

Proof. For integer value of m

$$\mathcal{Z}\{x[n-m]u[n-m]\} = \sum_{n=0}^{\infty}x[n-m]u[n-m]z^{-n}$$

Recall that $x[n-m]u[n-m] = 0$ for $n < m$, so that the limits on the summation on the right-hand side can be taken from $n = m$ to ∞. Therefore

$$\mathcal{Z}\{x[n-m]u[n-m]\} = \sum_{n=m}^{\infty}x[n-m]z^{-n}$$

$$= \sum_{r=0}^{\infty}x[r]z^{-(r+m)}$$

$$= \frac{1}{z^m}\sum_{r=0}^{\infty}x[r]z^{-r} = \frac{1}{z^m}X[z]$$

To prove Eq. (5.16c), we have

$$\mathcal{Z}\{x[n-m]u[n]\} = \sum_{n=0}^{\infty} x[n-m]z^{-n} = \sum_{r=-m}^{\infty} x[r]z^{-(r+m)}$$

$$= z^{-m}\left[\sum_{r=-m}^{-1} x[r]z^{-r} + \sum_{r=0}^{\infty} x[r]z^{-r}\right]$$

$$= z^{-m}\sum_{n=1}^{m} x[-n]z^{n} + z^{-m}X[z]$$

LEFT SHIFT (ADVANCE)

If

$$x[n]u[n] \Longleftrightarrow X[z]$$

then

$$x[n+1]u[n] \Longleftrightarrow zX[z] - zx[0] \qquad (5.17a)$$

Repeated application of this property yields

$$x[n+2]u[n] \Longleftrightarrow z\{z(X[z] - zx[0]) - x[1]\}$$
$$= z^{2}X[z] - z^{2}x[0] - zx[1] \qquad (5.17b)$$

and for integer value of m

$$x[n+m]u[n] \Longleftrightarrow z^{m}X[z] - z^{m}\sum_{n=0}^{m-1} x[n]z^{-n} \qquad (5.17c)$$

Proof. By definition

$$\mathcal{Z}\{x[n+m]u[n]\} = \sum_{n=0}^{\infty} x[n+m]z^{-n}$$

$$= \sum_{r=m}^{\infty} x[r]z^{-(r-m)}$$

$$= z^{m}\sum_{r=m}^{\infty} x[r]z^{-r}$$

$$= z^{m}\left[\sum_{r=0}^{\infty} x[r]z^{-r} - \sum_{r=0}^{m-1} x[r]z^{-r}\right]$$

$$= z^{m}X[z] - z^{m}\sum_{r=0}^{m-1} x[r]z^{-r}$$

EXAMPLE 5.4

Find the z-transform of the signal $x[n]$ depicted in Fig. 5.5.

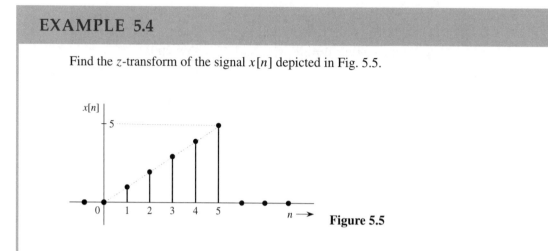

Figure 5.5

The signal $x[n]$ can be expressed as a product of n and a gate pulse $u[n] - u[n-6]$. Therefore

$$x[n] = n\{u[n] - u[n-6]\}$$

$$= nu[n] - nu[n-6]$$

We cannot find the z-transform of $nu[n-6]$ directly by using the right-shift property [Eq. (5.15b)]. So we rearrange it in terms of $(n-6)u[n-6]$ as follows:

$$x[n] = nu[n] - (n-6+6)u[n-6]$$

$$= nu[n] - (n-6)u[n-6] - 6u[n-6]$$

We can now find the z-transform of the bracketed term by using the right-shift property [Eq. (5.15b)]. Because $u[n] \Longleftrightarrow z/(z-1)$

$$u[n-6] \Longleftrightarrow \frac{1}{z^6}\frac{z}{z-1} = \frac{1}{z^5(z-1)}$$

Also, because $nu[n] \Longleftrightarrow z/(z-1)^2$

$$(n-6)u[n-6] \Longleftrightarrow \frac{1}{z^6}\frac{z}{(z-1)^2} = \frac{1}{z^5(z-1)^2}$$

Therefore

$$X[z] = \frac{z}{(z-1)^2} - \frac{1}{z^5(z-1)^2} - \frac{6}{z^5(z-1)}$$

$$= \frac{z^6 - 6z + 5}{z^5(z-1)^2}$$

EXERCISE E5.5

Using only the fact that $u[n] \iff z/(z-1)$ and the right-shift property [Eq. (5.15)], find the z-transforms of the signals in Figs. 5.2 and 5.3.

ANSWERS

See Example 5.2d and Exercise E5.1a.

CONVOLUTION

The time-convolution property states that if[†]

$$x_1[n] \iff X_1[z] \qquad \text{and} \qquad x_2[n] \iff X_2[z],$$

then (*time convolution*)

$$x_1[n] * x_2[n] \iff X_1[z]X_2[z] \tag{5.18}$$

Proof. This property applies to causal as well as noncausal sequences. We shall prove it for the more general case of noncausal sequences, where the convolution sum ranges from $-\infty$ to ∞.
    We have

$$\mathcal{Z}\{x_1[n] * x_2[n]\} = \mathcal{Z}\left[\sum_{m=-\infty}^{\infty} x_1[m]x_2[n-m]\right]$$

$$= \sum_{n=-\infty}^{\infty} z^{-n} \sum_{m=-\infty}^{\infty} x_1[m]x_2[n-m]$$

Interchanging the order of summation, we have

$$\mathcal{Z}[x_1[n] * x_2[n]] = \sum_{m=-\infty}^{\infty} x_1[m] \sum_{n=-\infty}^{\infty} x_2[n-m]z^{-n}$$

$$= \sum_{m=-\infty}^{\infty} x_1[m] \sum_{r=-\infty}^{\infty} x_2[r]z^{-(r+m)}$$

$$= \sum_{m=-\infty}^{\infty} x_1[m]z^{-m} \sum_{r=-\infty}^{\infty} x_2[r]z^{-r}$$

$$= X_1[z]X_2[z]$$

[†]There is also the frequency-convolution property, which states that if

$$x_1[n]x_2[n] \iff \frac{1}{2\pi j} \oint X_1[u]X_2\left[\frac{z}{u}\right]u^{-1}\,du$$

LTID SYSTEM RESPONSE

It is interesting to apply the time-convolution property to the LTID input–output equation $y[n] = x[n] * h[n]$. Since from Eq. (5.14b), we know that $h[n] \iff H[z]$, it follows from Eq. (5.18) that

$$Y[z] = X[z]H[z] \tag{5.19}$$

EXERCISE E5.6

Use the time-convolution property and appropriate pairs in Table 5.1 to show that $u[n] * u[n-1] = nu[n]$.

MULTIPLICATION BY γ^n
(SCALING IN THE z-DOMAIN)

If

$$x[n]u[n] \iff X[z]$$

then

$$\gamma^n x[n]u[n] \iff X\left[\frac{z}{\gamma}\right] \tag{5.20}$$

Proof.

$$\mathcal{Z}\{\gamma^n x[n]u[n]\} = \sum_{n=0}^{\infty} \gamma^n x[n]z^{-n}$$

$$= \sum_{n=0}^{\infty} x[n]\left(\frac{z}{\gamma}\right)^{-n} = X\left[\frac{z}{\gamma}\right]$$

EXERCISE E5.7

Use Eq. (5.20) to derive pairs 6 and 8 in Table 5.1 from pairs 2 and 3, respectively.

MULTIPLICATION BY n

If

$$x[n]u[n] \iff X[z]$$

then

$$nx[n]u[n] \iff -z\frac{d}{dz}X[z] \tag{5.21}$$

Proof.

$$-z \frac{d}{dz} X[z] = -z \frac{d}{dz} \sum_{n=0}^{\infty} x[n] z^{-n}$$

$$= -z \sum_{n=0}^{\infty} -nx[n] z^{-n-1}$$

$$= \sum_{n=0}^{\infty} nx[n] z^{-n} = \mathcal{Z}\{nx[n]u[n]\}$$

EXERCISE E5.8

Use Eq. (5.21) to derive pairs 3 and 4 in Table 5.1 from pair 2. Similarly, derive pairs 8 and 9 from pair 6.

TIME REVERSAL

If

$$x[n] \Longleftrightarrow X[z]$$

then[†]

$$x[-n] \Longleftrightarrow X[1/z] \qquad (5.22)$$

Proof.

$$\mathcal{Z}\{x[-n]\} = \sum_{n=-\infty}^{\infty} x[-n] z^{-n}$$

Changing the sign of the dummy variable n yields

$$\mathcal{Z}\{x[-n]\} = \sum_{n=-\infty}^{\infty} x[n] z^{n}$$

$$= \sum_{n=-\infty}^{\infty} x[n] (1/z)^{-n}$$

$$= X[1/z]$$

The region of convergence is also inverted, that is, if ROC of $x[n]$ is $|z| > |\gamma|$, then, the ROC of $x[-n]$ is $|z| < 1/|\gamma|$.

[†]For complex signal $x[n]$, the time-reversal property is modified as follows:
$$x^*[-n] \Longleftrightarrow X^*[1/z^*]$$

EXERCISE E5.9

Use the time-reversal property and pair 2 in Table 5.1 to show that $u[-n] \Longleftrightarrow -1/(z-1)$ with the ROC $|z| < 1$.

TABLE 5.2   z-Transform Operations

Operation	$x[n]$	$X[z]$
Addition	$x_1[n] + x_2[n]$	$X_1[z] + X_2[z]$
Scalar multiplication	$ax[n]$	$aX[z]$
Right-shifting	$x[n-m]u[n-m]$	$\dfrac{1}{z^m}X[z]$
	$x[n-m]u[n]$	$\dfrac{1}{z^m}X[z] + \dfrac{1}{z^m}\displaystyle\sum_{n=1}^{m} x[-n]z^n$
	$x[n-1]u[n]$	$\dfrac{1}{z}X[z] + x[-1]$
	$x[n-2]u[n]$	$\dfrac{1}{z^2}X[z] + \dfrac{1}{z}x[-1] + x[-2]$
	$x[n-3]u[n]$	$\dfrac{1}{z^3}X[z] + \dfrac{1}{z^2}x[-1] + \dfrac{1}{z}x[-2] + x[-3]$
Left-shifting	$x[n+m]u[n]$	$z^m X[z] - z^m \displaystyle\sum_{n=0}^{m-1} x[n]z^{-n}$
	$x[n+1]u[n]$	$zX[z] - zx[0]$
	$x[n+2]u[n]$	$z^2 X[z] - z^2 x[0] - zx[1]$
	$x[n+3]u[n]$	$z^3 X[z] - z^3 x[0] - z^2 x[1] - zx[2]$
Multiplication by γ^n	$\gamma^n x[n]u[n]$	$X\left[\dfrac{z}{\gamma}\right]$
Multiplication by n	$nx[n]u[n]$	$-z\dfrac{d}{dz}X[z]$
Time convolution	$x_1[n] * x_2[n]$	$X_1[z]X_2[z]$
Time reversal	$x[-n]$	$X[1/z]$
Initial value	$x[0]$	$\displaystyle\lim_{z\to\infty} X[z]$
Final value	$\displaystyle\lim_{N\to\infty} x[N]$	$\displaystyle\lim_{z\to 1}(z-1)X[z]$ poles of
		$(z-1)X[z]$ inside the unit circle

INITIAL AND FINAL VALUES

For a causal $x[n]$,

$$x[0] = \lim_{z \to \infty} X[z] \qquad (5.23a)$$

This result follows immediately from Eq. (5.9).

We can also show that if $(z-1)X(z)$ has no poles outside the unit circle, then[†]

$$\lim_{N \to \infty} x[N] = \lim_{z \to 1}(z-1)X[z] \qquad (5.23b)$$

All these properties of the z-transform are listed in Table 5.2.

5.3 z-TRANSFORM SOLUTION OF LINEAR DIFFERENCE EQUATIONS

The time-shifting (left-shift or right-shift) property has set the stage for solving linear difference equations with constant coefficients. As in the case of the Laplace transform with differential equations, the z-transform converts difference equations into algebraic equations that are readily solved to find the solution in the z domain. Taking the inverse z-transform of the z-domain solution yields the desired time-domain solution. The following examples demonstrate the procedure.

EXAMPLE 5.5

Solve

$$y[n+2] - 5y[n+1] + 6y[n] = 3x[n+1] + 5x[n] \qquad (5.24)$$

if the initial conditions are $y[-1] = 11/6$, $y[-2] = 37/36$, and the input $x[n] = (2)^{-n}u[n]$.

As we shall see, difference equations can be solved by using the right-shift or the left-shift property. Because the difference equation, Eq. (5.24), is in advance operator form, the

---

[†]This can be shown from the fact that

$$x[n] - x[n-1] \iff \left\{1 - \frac{1}{z}\right\} X[z] = \frac{(z-1)X[z]}{z}$$

and

$$\frac{(z-1)X[z]}{z} = \sum_{n=-\infty}^{\infty} \{x[n] - x[n-1]\}z^{-n}$$

and

$$\lim_{z \to 1} \frac{(z-1)X[z]}{z} = \lim_{z \to 1}(z-1)X[z] = \lim_{z \to 1}\lim_{N \to \infty}\sum_{n=-\infty}^{N}\{x[n] - x[n-1]\}z^{-n} = \lim_{N \to \infty} x[N]$$

use of the left-shift property in Eqs. (5.17a) and (5.17b) may seem appropriate for its solution. Unfortunately, as seen from Eqs. (5.17a) and (5.17b), these properties require a knowledge of auxiliary conditions $y[0], y[1], \ldots, y[N-1]$ rather than of the initial conditions $y[-1], y[-2], \ldots, y[-n]$, which are generally given. This difficulty can be overcome by expressing the difference equation (5.24) in delay operator form (obtained by replacing n with $n-2$) and then using the right-shift property.[†] Equation (5.24) in delay operator form is

$$y[n] - 5y[n-1] + 6y[n-2] = 3x[n-1] + 5x[n-2] \qquad (5.25)$$

We now use the right-shift property to take the z-transform of this equation. But before proceeding, we must be clear about the meaning of a term like $y[n-1]$ here. Does it mean $y[n-1]u[n-1]$ or $y[n-1]u[n]$? In any equation, we must have some time reference $n = 0$, and every term is referenced from this instant. Hence, $y[n-k]$ means $y[n-k]u[n]$. Remember also that although we are considering the situation for $n \geq 0$, $y[n]$ is present even before $n = 0$ (in the form of initial conditions). Now

$$y[n]u[n] \Longleftrightarrow Y[z]$$

$$y[n-1]u[n] \Longleftrightarrow \frac{1}{z}Y[z] + y[-1] = \frac{1}{z}Y[z] + \frac{11}{6}$$

$$y[n-2]u[n] \Longleftrightarrow \frac{1}{z^2}Y[z] + \frac{1}{z}y[-1] + y[-2] = \frac{1}{z^2}Y[z] + \frac{11}{6z} + \frac{37}{36}$$

Noting that for causal input $x[n]$,

$$x[-1] = x[-2] = \cdots = x[-n] = 0$$

We obtain

$$x[n] = (2)^{-n}u[n] = (2^{-1})^n u[n] = (0.5)^n u[n] \Longleftrightarrow \frac{z}{z-0.5}$$

$$x[n-1]u[n] \Longleftrightarrow \frac{1}{z}X[z] + x[-1] = \frac{1}{z}\frac{z}{z-0.5} + 0 = \frac{1}{z-0.5}$$

$$x[n-2]u[n] \Longleftrightarrow \frac{1}{z^2}X[z] + \frac{1}{z}x[-1] + x[-2] = \frac{1}{z^2}X[z] + 0 + 0 = \frac{1}{z(z-0.5)}$$

In general

$$x[n-r]u[n] \Longleftrightarrow \frac{1}{z^r}X[z]$$

Taking the z-transform of Eq. (5.25) and substituting the foregoing results, we obtain

$$Y[z] - 5\left[\frac{1}{z}Y[z] + \frac{11}{6}\right] + 6\left[\frac{1}{z^2}Y[z] + \frac{11}{6z} + \frac{37}{36}\right] = \frac{3}{z-0.5} + \frac{5}{z(z-0.5)} \qquad (5.26a)$$

[†]Another approach is to find $y[0], y[1], y[2], \ldots, y[N-1]$ from $y[-1], y[-2], \ldots, y[-n]$ iteratively, as in Section 3.5-1, and then apply the left-shift property to Eq. (5.24).

or

$$\left(1 - \frac{5}{z} + \frac{6}{z^2}\right) Y[z] - \left(3 - \frac{11}{z}\right) = \frac{3}{z - 0.5} + \frac{5}{z(z - 0.5)} \tag{5.26b}$$

from which we obtain

$$(z^2 - 5z + 6) Y[z] = \frac{z(3z^2 - 9.5z + 10.5)}{(z - 0.5)}$$

so that

$$Y[z] = \frac{z(3z^2 - 9.5z + 10.5)}{(z - 0.5)(z^2 - 5z + 6)} \tag{5.27}$$

and

$$\frac{Y[z]}{z} = \frac{3z^2 - 9.5z + 10.5}{(z - 0.5)(z - 2)(z - 3)}$$

$$= \frac{(26/15)}{z - 0.5} - \frac{(7/3)}{z - 2} + \frac{(18/5)}{z - 3}$$

Therefore

$$Y[z] = \frac{26}{15}\left(\frac{z}{z - 0.5}\right) - \frac{7}{3}\left(\frac{z}{z - 2}\right) + \frac{18}{5}\left(\frac{z}{z - 3}\right)$$

and

$$y[n] = \left[\tfrac{26}{15}(0.5)^n - \tfrac{7}{3}(2)^n + \tfrac{18}{5}(3)^n\right] u[n] \tag{5.28}$$

This example demonstrates the ease with which linear difference equations with constant coefficients can be solved by *z*-transform. This method is general; it can be used to solve a single difference equation or a set of simultaneous difference equations of any order as long as the equations are linear with constant coefficients.

Comment. Sometimes, instead of initial conditions $y[-1]$, $y[-2]$, ..., $y[-n]$, auxiliary conditions $y[0]$, $y[1]$, ..., $y[N - 1]$ are given to solve a difference equation. In this case, the equation can be solved by expressing it in the <u>advance operator form</u> and then using the left-shift property (see later: Exercise E5.11).

EXERCISE E5.10

Solve the following equation if the initial conditions $y[-1] = 2$, $y[-2] = 0$, and the input $x[n] = u[n]$:

$$y[n + 2] - \tfrac{5}{6}y[n + 1] + \tfrac{1}{6}y[n] = 5x[n + 1] - x[n]$$

ANSWER

$$y[n] = \left[12 - 15\left(\tfrac{1}{2}\right)^n + \tfrac{14}{3}\left(\tfrac{1}{3}\right)^n\right] u[n]$$

EXERCISE E5.11

Solve the following equation if the auxiliary conditions are $y[0] = 1$, $y[1] = 2$, and the input $x[n] = u[n]$:

$$y[n] + 3y[n-1] + 2y[n-2] = x[n-1] + 3x[n-2]$$

ANSWER

$y[n] = \left[\frac{2}{3} + 2(-1)^n - \frac{5}{3}(-2)^n \right] u[n]$

ZERO-INPUT AND ZERO-STATE COMPONENTS

In Example 5.5 we found the total solution of the difference equation. It is relatively easy to separate the solution into zero-input and zero-state components. All we have to do is to separate the response into terms arising from the input and terms arising from initial conditions. We can separate the response in Eq. (5.26b) as follows:

$$\left(1 - \frac{5}{z} + \frac{6}{z^2} \right) Y[z] - \underbrace{\left(3 - \frac{11}{z} \right)}_{\text{initial condition terms}} = \underbrace{\frac{3}{z - 0.5} + \frac{5}{z(z - 0.5)}}_{\text{terms arising from input}} \tag{5.29}$$

Therefore

$$\left(1 - \frac{5}{z} + \frac{6}{z^2} \right) Y[z] = \underbrace{\left(3 - \frac{11}{z} \right)}_{\text{initial condition terms}} + \underbrace{\frac{(3z + 5)}{z(z - 0.5)}}_{\text{input terms}}$$

Multiplying both sides by z^2 yields

$$(z^2 - 5z + 6)Y[z] = \underbrace{z(3z - 11)}_{\text{initial condition terms}} + \underbrace{\frac{z(3z + 5)}{z - 0.5}}_{\text{input terms}}$$

and

$$Y[z] = \underbrace{\frac{z(3z - 11)}{z^2 - 5z + 6}}_{\text{zero-input response}} + \underbrace{\frac{z(3z + 5)}{(z - 0.5)(z^2 - 5z + 6)}}_{\text{zero-state response}} \tag{5.30}$$

We expand both terms on the right-hand side into modified partial fractions to yield

$$Y[z] = \underbrace{\left[5 \left(\frac{z}{z - 2} \right) - 2 \left(\frac{z}{z - 3} \right) \right]}_{\text{zero input}} + \underbrace{\left[\frac{26}{15} \left(\frac{z}{z - 0.5} \right) - \frac{22}{3} \left(\frac{z}{z - 2} \right) + \frac{28}{5} \left(\frac{z}{z - 3} \right) \right]}_{\text{zero state}}$$

and

$$y[n] = \underbrace{\left[5(2)^n - 2(3)^n \right.}_{\text{zero input}} - \underbrace{\left. \frac{22}{3}(2)^n + \frac{28}{5}(3)^n + \frac{26}{15}(0.5)^n \right]}_{\text{zero state}} u[n]$$

$$= \left[-\frac{7}{3}(2)^n + \frac{18}{5}(3)^n + \frac{26}{15}(0.5)^n \right] u[n]$$

which agrees with the result in Eq. (5.28).

EXERCISE E5.12

Solve

$$y[n + 2] - \tfrac{5}{6}y[n + 1] + \tfrac{1}{6}y[n] = 5x[n + 1] - x[n]$$

if the initial conditions are $y[-1] = 2$, $y[-2] = 0$, and the input $x[n] = u[n]$. Separate the response into zero-input and zero-state components.

ANSWER

$$y[n] = \left(\underbrace{\left[3\left(\tfrac{1}{2}\right)^n - \tfrac{4}{3}\left(\tfrac{1}{3}\right)^n\right]}_{\text{zero input}} + \underbrace{\left[12 - 18\left(\tfrac{1}{2}\right)^n + 6\left(\tfrac{1}{3}\right)^n\right]}_{\text{zero state}} \right) u[n]$$

$$= \left[12 - 15\left(\tfrac{1}{2}\right)^n + \tfrac{14}{3}\left(\tfrac{1}{3}\right)^n\right] u[n]$$

5.3-1 Zero-State Response of LTID Systems: The Transfer Function

Consider an Nth-order LTID system specified by the difference equation

$$Q[E]y[n] = P[E]x[n] \tag{5.31a}$$

or

$$(E^N + a_1 E^{N-1} + \cdots + a_{N-1}E + a_N)y[n]$$
$$= (b_0 E^N + b_1 E^{N-1} + \cdots + b_{N-1}E + b_N)x[n] \tag{5.31b}$$

or

$$y[n + N] + a_1 y[n + N - 1] + \cdots + a_{N-1}y[n + 1] + a_N y[n]$$
$$= b_0 x[n + N] + \cdots + b_{N-1}x[n + 1] + b_N x[n] \tag{5.31c}$$

We now derive the general expression for the zero-state response: that is, the system response to input $x[n]$ when all the initial conditions $y[-1] = y[-2] = \cdots = y[-N] = 0$ (zero state). The input $x[n]$ is assumed to be causal so that $x[-1] = x[-2] = \cdots = x[-N] = 0$.

Equation (5.31c) can be expressed in the delay operator form as

$$y[n] + a_1 y[n - 1] + \cdots + a_N y[n - N]$$
$$= b_0 x[n] + b_1 x[n - 1] + \cdots + b_N x[n - N] \tag{5.31d}$$

Because $y[-r] = x[-r] = 0$ for $r = 1, 2, \ldots, N$

$$y[n - m]u[n] \Longleftrightarrow \frac{1}{z^m}Y[z]$$

$$x[n - m]u[n] \Longleftrightarrow \frac{1}{z^m}X[z] \qquad m = 1, 2, \ldots, N$$

Now the z-transform of Eq. (5.31d) is given by

$$\left(1 + \frac{a_1}{z} + \frac{a_2}{z^2} + \cdots + \frac{a_N}{z^N}\right) Y[z] = \left(b_0 + \frac{b_1}{z} + \frac{b_2}{z^2} + \cdots + \frac{b_N}{z^N}\right) X[z]$$

Multiplication of both sides by z^N yields

$$(z^N + a_1 z^{N-1} + \cdots + a_{N-1}z + a_N)Y[z]$$
$$= (b_0 z^N + b_1 z^{N-1} + \cdots + b_{N-1}z + b_N)X[z]$$

Therefore

$$Y[z] = \left(\frac{b_0 z^N + b_1 z^{N-1} + \cdots + b_{N-1}z + b_N}{z^N + a_1 z^{N-1} + \cdots + a_{N-1}z + a_N}\right) X[z] \tag{5.32}$$

$$= \frac{P[z]}{Q[z]} X[z] \tag{5.33}$$

We have shown in Eq. (5.19) that $Y[z] = X[z]H[z]$. Hence, it follows that

$$H[z] = \frac{P[z]}{Q[z]} = \frac{b_0 z^N + b_1 z^{N-1} + \cdots + b_{N-1}z + b_N}{z^N + a_1 z^{N-1} + \cdots + a_{N-1}z + a_N} \tag{5.34}$$

As in the case of LTIC systems, this result leads to an alternative definition of the LTID system transfer function as the ratio of $Y[z]$ to $X[z]$ (assuming all initial conditions zero).

$$H[z] \equiv \frac{Y[z]}{X[z]} = \frac{\mathcal{Z}[\text{zero-state response}]}{\mathcal{Z}[\text{input}]} \tag{5.35}$$

ALTERNATE INTERPRETATION OF THE z-TRANSFORM

So far we have treated the z-transform as a machine, which converts linear difference equations into algebraic equations. There is no physical understanding of how this is accomplished or what it means. We now discuss more intuitive interpretation and meaning of the z-transform.

In Chapter 3, Eq. (3.71a), we showed that LTID system response to an everlasting exponential z^n is $H[z]z^n$. If we could express every discrete-time signal as a linear combination of everlasting exponentials of the form z^n, we could readily obtain the system response to any input. For example, if

$$x[n] = \sum_{k=1}^{K} X[z_k]z_k^n \tag{5.36a}$$

the response of an LTID system to this input is given by

$$y[n] = \sum_{k=1}^{K} X[z_k]H[z_k]z_k^n \tag{5.36b}$$

Unfortunately, a very small class of signals can be expressed in the form of Eq. (5.36a). However, we can express almost all signals of practical utility as a sum of everlasting exponentials over a

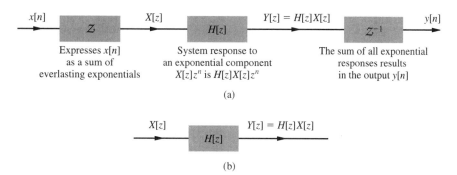

(a)

(b)

Figure 5.6 The transformed representation of an LTID system.

continuum of values of z. This is precisely what the z-transform in Eq. (5.2) does.

$$x[n] = \frac{1}{2\pi j} \oint X[z]z^{n-1}\, dz \tag{5.37}$$

Invoking the linearity property of the z-transform, we can find the system response $y[n]$ to input $x[n]$ in the Eq. (5.37) as[†]

$$y[n] = \frac{1}{2\pi j} \oint X[z]H[z]z^{n-1}\, dz = \mathcal{Z}^{-1}\{X[z]H[z]\}$$

Clearly

$$Y[z] = X[z]H[z]$$

This viewpoint of finding the response of LTID system is illustrated in Fig. 5.6a. Just as in continuous-time systems, we can model discrete-time systems in the transformed manner by representing all signals by their z-transforms and all system components (or elements) by their transfer functions, as shown in Fig. 5.6b.

The result $Y[z] = H[z]X[z]$ greatly facilitates derivation of the system response to a given input. We shall demonstrate this assertion by an example.

EXAMPLE 5.6

for:
zero
State
response

Find the response $y[n]$ of an LTID system described by the difference equation

$$y[n+2] + y[n+1] + 0.16y[n] = x[n+1] + 0.32x[n]$$

or

$$(E^2 + E + 0.16)y[n] = (E + 0.32)x[n]$$

[†]In computing $y[n]$, the contour along which the integration is performed is modified to consider the ROC of $X[z]$ as well as $H[z]$. We ignore this consideration in this intuitive discussion.

for the input $x[n] = (-2)^{-n}u[n]$ and with all the initial conditions zero (system in the zero state).

From the difference equation we find

$$H[z] = \frac{P[z]}{Q[z]} = \frac{z + 0.32}{z^2 + z + 0.16}$$

For the input $x[n] = (-2)^{-n}u[n] = [(-2)^{-1}]^n u(n) = (-0.5)^n u[n]$

$$X[z] = \frac{z}{z + 0.5}$$

and

$$Y[z] = X[z]H[z] = \frac{z(z + 0.32)}{(z^2 + z + 0.16)(z + 0.5)}$$

Therefore

$$\frac{Y[z]}{z} = \frac{(z + 0.32)}{(z^2 + z + 0.16)(z + 0.5)} = \frac{(z + 0.32)}{(z + 0.2)(z + 0.8)(z + 0.5)}$$

$$= \frac{2/3}{z + 0.2} - \frac{8/3}{z + 0.8} + \frac{2}{z + 0.5} \tag{5.38}$$

so that

$$Y[z] = \frac{2}{3}\left(\frac{z}{z + 0.2}\right) - \frac{8}{3}\left(\frac{z}{z + 0.8}\right) + 2\left(\frac{z}{z + 0.5}\right) \tag{5.39}$$

and

$$y[n] = \left[\tfrac{2}{3}(-0.2)^n - \tfrac{8}{3}(-0.8)^n + 2(-0.5)^n\right] u[n]$$

EXAMPLE 5.7 (The Transfer Function of a Unit Delay)

Show that the transfer function of a unit delay is $1/z$.

If the input to the unit delay is $x[n]u[n]$, then its output (Fig. 5.7) is given by

$$y[n] = x[n - 1]u[n - 1]$$

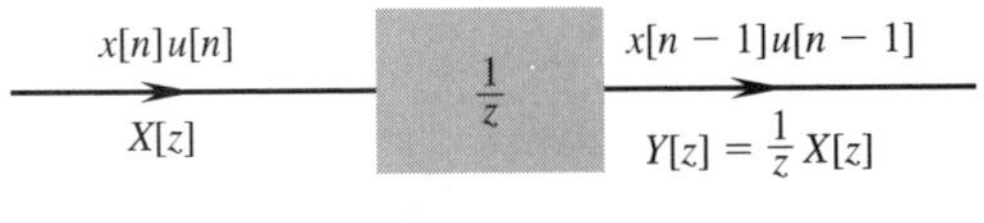

Figure 5.7 Ideal unit delay and its transfer function.

The z-transform of this equation yields [see Eq. (5.15a)]

$$Y[z] = \frac{1}{z}X[z]$$

$$= H[z]X[z]$$

It follows that the transfer function of the unit delay is

$$H[z] = \frac{1}{z} \qquad\qquad (5.40)$$

EXERCISE E5.13

A discrete-time system is described by the following transfer function:

$$H[z] = \frac{z - 0.5}{(z + 0.5)(z - 1)}$$

(a) Find the system response to input $x[n] = 3^{-(n+1)}u[n]$ if all initial conditions are zero.

(b) Write the difference equation relating the output $y[n]$ to input $x[n]$ for this system.

ANSWERS

(a) $y[n] = \frac{1}{3}\left[\frac{1}{2} - 0.8(-0.5)^n + 0.3\left(\frac{1}{3}\right)^n\right]u[n]$

(b) $y[n+2] - 0.5y[n+1] - 0.5y[n] = x[n+1] - 0.5x[n]$

5.3-2 Stability

Equation (5.34) shows that the denominator of $H[z]$ is $Q[z]$, which is apparently identical to the characteristic polynomial $Q[\gamma]$ defined in Chapter 3. Does this mean that the denominator of $H[z]$ is the characteristic polynomial of the system? This may or may not be the case: if $P[z]$ and $Q[z]$ in Eq. (5.34) have any common factors, they cancel out, and the effective denominator of $H[z]$ is not necessarily equal to $Q[z]$. Recall also that the system transfer function $H[z]$, like $h[n]$, is defined in terms of measurements at the external terminals. Consequently, $H[z]$ and $h[n]$ are both external descriptions of the system. In contrast, the characteristic polynomial $Q[z]$ is an internal description. Clearly, we can determine only external stability, that is, BIBO stability, from $H[z]$. If all the poles of $H[z]$ are within the unit circle, all the terms in $h[z]$ are decaying exponentials, and as shown in Section 3.10, $h[n]$ is absolutely summable. Consequently, the system is BIBO stable. Otherwise the system is BIBO unstable.

If $P[z]$ and $Q[z]$ do not have common factors, then the denominator of $H[z]$ is identical to $Q[z]$.[†] The poles of $H[z]$ are the characteristic roots of the system. We can now determine

---

[†]There is no way of determining whether there were common factors in $P[z]$ and $Q[z]$ that were canceled out because in our derivation of $H[z]$, we generally get the final result after the cancellations are already effected. When we use internal description of the system to derive $Q[z]$, however, we find pure $Q[z]$ unaffected by any common factor in $P[z]$.

internal stability. The internal stability criterion in Section 3.10-1 can be restated in terms of the poles of $H[z]$, as follows.

1. An LTID system is asymptotically stable if and only if all the poles of its transfer function $H[z]$ are within the unit circle. The poles may be repeated or simple.

2. An LTID system is unstable if and only if either one or both of the following conditions exist: (i) at least one pole of $H[z]$ is outside the unit circle; (ii) there are repeated poles of $H[z]$ on the unit circle.

3. An LTID system is marginally stable if and only if there are no poles of $H[z]$ outside the unit circle, and there are some simple poles on the unit circle.

EXERCISE E5.14

Show that an *accumulator* whose impulse response is $h[n] = u[n]$ is marginally stable but BIBO unstable.

5.3-3 Inverse Systems

If $H[z]$ is the transfer function of a system $\mathcal{S}$, then $\mathcal{S}_i$, its inverse system has a transfer function $H_i[z]$ given by

$$H_i[z] = \frac{1}{H[z]}$$

This follows from the fact the inverse system $\mathcal{S}_i$ undoes the operation of $\mathcal{S}$. Hence, if $H[z]$ is placed in cascade with $H_i[z]$, the transfer function of the composite system (identity system) is unity. For example, an *accumulator*, whose transfer function is $H[z] = z/(z-1)$ and a *backward difference system* whose transfer function is $H_i[z] = (z-1)/z$ are inverse of each other. Similarly if

$$H[z] = \frac{z - 0.4}{z - 0.7}$$

its inverse system transfer function is

$$H_i[z] = \frac{z - 0.7}{z - 0.4}$$

as required by the property $H[z]H_i[z] = 1$. Hence, it follows that

$$h[n] * h_i[n] = \delta[n]$$

EXERCISE E5.15

Find the impulse response of an accumulator and the backward difference system. Show that the convolution of the two impulse responses yields $\delta[n]$.

5.4 SYSTEM REALIZATION

Because of the similarity between LTIC and LTID systems, conventions for block diagrams and rules of interconnection for LTID are identical to those for continuous-time (LTIC) systems. It is not necessary to rederive these relationships. We shall merely restate them to refresh the reader's memory.

The block diagram representation of the basic operations such as an adder a scalar multiplier, unit delay, and pick off points were shown in Fig. 3.11. In our development, the unit delay, which was represented by a box marked D in Fig. 3.11, will be represented by its transfer function $1/z$. All the signals will also be represented in terms of their z-transforms. Thus, the input and the output will be labeled $X[z]$ and $Y[z]$, respectively.

When two systems with transfer functions $H_1[z]$ and $H_2[z]$ are connected in cascade (as in Fig. 4.18b), the transfer function of the composite system is $H_1[z]H_2[z]$. If the same two systems are connected in parallel (as in Fig. 4.18c), the transfer function of the composite system is $H_1[z]H_2[z]$. For a feedback system (as in Fig. 4.18d), the transfer function is $G[z]/(1 + G[z]H[z])$.

We now consider a systematic method for realization (or simulation) of an arbitrary Nth-order LTID transfer function. Since realization is basically a synthesis problem, there is no unique way of realizing a system. A given transfer function can be realized in many different ways. We present here the two forms of *direct realization*. Each of these forms can be executed in several other ways, such as cascade and parallel. Furthermore, a system can be realized by the transposed version of any known realization of that system. This artifice doubles the number of system realizations. A transfer function $H[z]$ can be realized by using time delays along with adders and multipliers.

We shall consider a realization of a general Nth-order causal LTID system, whose transfer function is given by

$$H[z] = \frac{b_0 z^N + b_1 z^{N-1} + \cdots + b_{N-1}z + b_N}{z^N + a_1 z^{N-1} + \cdots + a_{N-1}z + a_N} \tag{5.41}$$

This equation is identical to the transfer function of a general Nth-order proper LTIC system given in Eq. (4.60). The only difference is that the variable z in the former is replaced by the variable s in the latter. Hence, the procedure for realizing an LTID transfer function is identical to that for the LTIC transfer function with the basic element $1/s$ (integrator) replaced by the element $1/z$ (unit delay). The reader is encouraged to follow the steps in Section 4.6 and rederive the results for the LTID transfer function in Eq. (5.41). Here we shall merely reproduce the realizations from Section 4.6 with integrators $(1/s)$ replaced by unit delays $(1/z)$. The direct form I (DFI) is shown in Fig. 5.8a, the canonic direct form (DFII) is shown in Fig. 5.8b and the transpose of canonic direct is shown in Fig. 5.8c. The DFII and its transpose are canonic because they require N delays, which is the minimum number needed to implement an Nth-order LTID transfer function in Eq. (5.41). In contrast, the form DFI is a noncanonic because it generally requires $2N$ delays. The DFII realization in Fig. 5.8b is also called a *canonic direct* form.

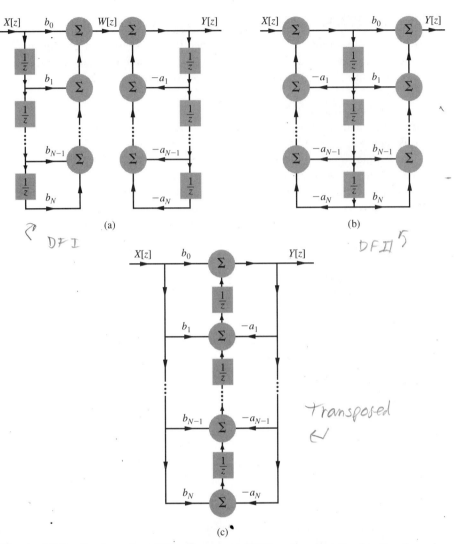

Figure 5.8 Realization of an Nth-order causal LTID system transfer function by using **(a)** DFI, **(b)** canonic direct (DFII), and **(c)** the transpose form of DFII.

EXAMPLE 5.8

Find the canonic direct and the transposed canonic direct realizations of the following transfer functions.

(i) $\dfrac{2}{z+5}$

(ii) $\dfrac{4z+28}{z+1}$

(iii) $\dfrac{z}{z+7}$

(iv) $\dfrac{4z+28}{z^2+6z+5}$

All four of these transfer functions are special cases of $H[z]$ in Eq. (5.41).

(i)

$$H[z] = \frac{2}{z+5}$$

For this case, the transfer function is of the first order ($N = 1$); therefore, we need only one delay for its realization. The feedback and feedforward coefficients are

$$a_1 = 5 \quad \text{and} \quad b_0 = 0, \quad b_1 = 2$$

We use Fig. 5.8 as our model and reduce it to the case of $N = 1$. Figure 5.9a shows the canonic direct (DFII) form, and Fig. 5.9b its transpose. The two realizations are almost the same. The minor difference is that in the DFII form, the gain 2 is provided at the output, and in the transpose, the same gain is provided at the input.

In a similar way, we realize the remaining transfer functions.

(ii)

$$H[z] = \frac{4z+28}{z+1}$$

In this case also, the transfer function is of the first order ($N = 1$); therefore, we need only one delay for its realization. The feedback and feedforward coefficients are

$$a_1 = 1 \quad \text{and} \quad b_0 = 4, \quad b_1 = 28$$

Figure 5.10 illustrates the canonic direct and its transpose for this case.[†]

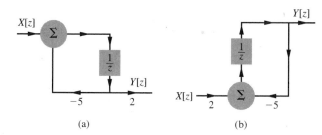

(a)   (b)

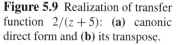

Figure 5.9 Realization of transfer function $2/(z+5)$: **(a)** canonic direct form and **(b)** its transpose.

[†]Transfer functions with $N = M$ may also be expressed as a sum of a constant and a strictly proper transfer function. For example,

$$H[z] = \frac{4z+28}{z+1} = 4 + \frac{24}{z+1}$$

Hence, this transfer function can also be realized as two transfer functions in parallel.

(iii)

$$H[z] = \frac{z}{z+7}$$

Here $N = 1$ and $b_0 = 1$, $b_1 = 0$ and $a_1 = 7$. Figure 5.11 shows the direct and the transposed realizations. Observe that the realizations are almost alike.

(iv)

$$H[z] = \frac{4z + 28}{z^2 + 6z + 5}$$

This is a second-order system ($N = 2$) with $b_0 = 0$, $b_1 = 4$, $b_2 = 28$, $a_1 = 6$, $a_2 = 5$. Figure 5.12 shows the canonic direct and transposed canonic direct realizations.

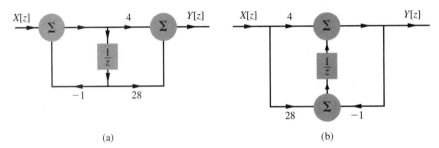

(a)   (b)

Figure 5.10 Realization of $(4z + 28)/(z + 1)$: **(a)** canonic direct form and **(b)** its transpose.

(a)   (b)

Figure 5.11 Realization of $z/(z + 7)$: **(a)** canonic direct form and **(b)** its transpose.

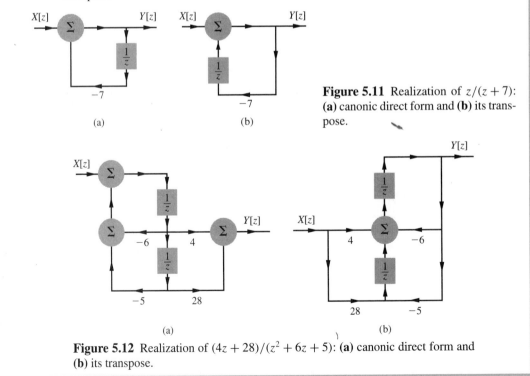

(a)   (b)

Figure 5.12 Realization of $(4z + 28)/(z^2 + 6z + 5)$: **(a)** canonic direct form and **(b)** its transpose.

EXERCISE E5.16

Realize the transfer function

$$H[z] = \frac{2z}{z^2 + 6z + 25}$$

CASCADE AND PARALLEL REALIZATIONS, COMPLEX AND REPEATED POLES

The considerations and observations for cascade and parallel realizations as well as complex and multiple poles are identical to those discussed for LTIC systems in Section 4.6-3.

EXERCISE E5.17

Find canonic direct realizations of the following transfer function by using the cascade and parallel forms. The specific cascade decomposition is as follows:

$$H[z] = \frac{z + 3}{z^2 + 7z + 10} = \left(\frac{z + 3}{z + 2}\right)\left(\frac{1}{z + 5}\right)$$

REALIZATION OF FINITE IMPULSE RESPONSE (FIR) FILTERS

So far we have been quite general in our development of realization techniques. They can be applied to infinite impulse response (IIR) or FIR filters. For FIR filters, the coefficients $a_i = 0$ for all $i \neq 0$.[†] Hence, FIR filters can be readily implemented by means of the schemes developed so far by eliminating all branches with a_i coefficients. The condition $a_i = 0$ implies that all the poles of a FIR filter are at $z = 0$.

EXAMPLE 5.9

Realize $H[z] = (z^3 + 4z^2 + 5z + 2)/z^3$ using canonic direct and transposed forms.

We can express $H[z]$ as

$$H[z] = \frac{z^3 + 4z^2 + 5z + 2}{z^3}$$

[†]This statement is true for all $i \neq 0$ because a_0 is assumed to be unity.

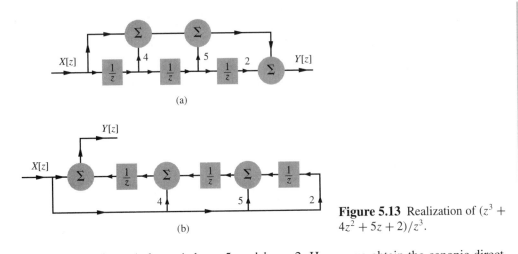

Figure 5.13 Realization of $(z^3 + 4z^2 + 5z + 2)/z^3$.

For $H[z]$, $b_0 = 1$, $b_1 = 4$, $b_2 = 5$, and $b_3 = 2$. Hence, we obtain the canonic direct realization, shown in Fig. 5.13a. We have shown the horizontal orientation because it is easier to see that this filter is basically a tapped delay line. That is why this structure is also known as a *tapped delay line* or *transversal filter*. Figure 5.13b shows the corresponding transposed implementation.

DO ALL REALIZATIONS LEAD TO THE SAME PERFORMANCE?

For a given transfer function, we have presented here several possible different realizations (DFI, canonic form DFII, and its transpose). There are also cascade and parallel versions, and there are many possible grouping of the factors in the numerator and the denominator of $H[z]$, leading to different realizations. We can also use various combinations of these forms in implementing different subsections of a system. Moreover, the transpose of each version doubles the number. However, this discussion by no means exhausts all the possibilities. Transforming variables affords limitless potential realizations of the same transfer function.

Theoretically, all these realizations are equivalent; that is, they lead to the same transfer function. This, however, is true only when we implement them with infinite precision. In practice, finite wordlength restriction causes each realization to behave differently in terms of sensitivity to parameter variation, stability, frequency response distortion error, and so on. These effects are serious for higher-order transfer functions, which require correspondingly higher numbers of delay elements. The finite wordlength errors that plague these implementations are coefficient quantization, overflow errors, and round-off errors. From a practical viewpoint, parallel and cascade forms using low-order filters minimize the effects of finite wordlength. Parallel and certain cascade forms are numerically less sensitive than the canonic direct form to small parameter variations in the system. In the canonic direct form structure with large N, a small change in a filter coefficient due to parameter quantization results in a large change in the location of the poles and the zeros of the system. Qualitatively, this difference can be explained by the fact that in a direct form (or its transpose), all the coefficients interact with each other, and a change in any coefficient will be magnified through its repeated influence from feedback and feedforward connections. In a parallel realization, in contrast, a change in a coefficient will affect only a localized segment; the case of a cascade realization is similar. For this reason, the most popular technique for minimizing finite wordlength effects is to design filters by using cascade or parallel forms employing low-order filters. In practice, high-order filters are realized by using multiple second-order sections in cascade, because second-order filters are not only easier to design but are less susceptible to coefficient quantization and round-off errors and their implementations allow easier data word scaling to reduce the potential overflow effects of data word size growth. A cascaded system using second-order building blocks usually requires fewer multiplications for a given filter frequency response.[1]

There are several ways to pair the poles and zeros of an Nth-order $H[z]$ into a cascade of second-order sections, and several ways to order the resulting sections. Quantizing error will be different for each combination. Although several papers published provide guidelines in predicting and minimizing finite wordlength errors, it is advisable to resort to computer simulation of the filter design. This way, one can vary filter hardware characteristic, such as coefficient wordlengths, accumulator register sizes, sequencing of cascaded sections, and input signal sets. Such an approach is both reliable and economical.[1]

5.5 FREQUENCY RESPONSE OF DISCRETE-TIME SYSTEMS

For (asymptotically or BIBO-stable) continuous-time systems, we showed that the system response to an input $e^{j\omega t}$ is $H(j\omega)e^{j\omega t}$ and that the response to an input cos ωt is $|H(j\omega)| \cos[\omega t + \angle H(j\omega)]$. Similar results hold for discrete-time systems. We now show that for an (asymptotically or BIBO-stable) LTID system, the system response to an input $e^{j\Omega n}$ is $H[e^{j\Omega}]e^{j\Omega n}$ and the response to an input cos Ωn is $|H[e^{j\Omega}]| \cos(\Omega n + \angle H[e^{j\Omega}])$.

The proof is similar to the one used for continuous-time systems. In Section 3.8-3, we showed that an LTID system response to an (everlasting) exponential z^n is also an (everlasting) exponential $H[z]z^n$. This result is valid only for values of z for which $H[z]$, as defined in Eq. (5.14a), exists (converges). As usual, we represent this input–output relationship by a directed arrow notation as

$$z^n \implies H[z]z^n \tag{5.42}$$

Setting $z = e^{j\Omega}$ in this relationship yields

$$e^{j\Omega n} \implies H[e^{j\Omega}]e^{j\Omega n} \tag{5.43}$$

Noting that cos Ωn is the real part of $e^{j\Omega n}$, use of Eq. (3.66b) yields

$$\cos \Omega n \implies \text{Re}\,\{H[e^{j\Omega}]e^{j\Omega n}\} \tag{5.44}$$

Expressing $H[e^{j\Omega}]$ in the polar form

$$H[e^{j\Omega}] = |H[e^{j\Omega}]|e^{j\angle H[e^{j\Omega}]} \tag{5.45}$$

Eq. (5.44) can be expressed as

$$\cos \Omega n \implies |H[e^{j\Omega}]| \cos (\Omega n + \angle H[e^{j\Omega}])$$

In other words, the system response $y[n]$ to a sinusoidal input $\cos \Omega n$ is given by

$$y[n] = |H[e^{j\Omega}]| \cos (\Omega n + \angle H[e^{j\Omega}]) \tag{5.46a}$$

Following the same argument, the system response to a sinusoid $\cos (\Omega n + \theta)$ is

$$y[n] = |H[e^{j\Omega}]| \cos (\Omega n + \theta + \angle H[e^{j\Omega}]) \tag{5.46b}$$

This result is valid only for BIBO-stable or asymptotically stable systems. The frequency response is meaningless for BIBO-unstable systems (which include marginally stable and asymptotically unstable systems). This follows from the fact that the frequency response in Eq. (5.43) is obtained by setting $z = e^{j\Omega}$ in Eq. (5.42). But, as shown in Section 3.8-3 [Eqs. (3.71)], the relationship (5.42) applies only for values of z for which $H[z]$ exists. For BIBO unstable systems, the ROC for $H[z]$ does not include the unit circle where $z = e^{j\Omega}$. This means, for BIBO-unstable systems, that $H[z]$ is meaningless when $z = e^{j\Omega}$.[†]

This important result shows that the response of an asymptotically or BIBO-stable LTID system to a discrete-time sinusoidal input of frequency Ω is also a discrete-time sinusoid of the same frequency. *The amplitude of the output sinusoid is $|H[e^{j\Omega}]|$ times the input amplitude, and the phase of the output sinusoid is shifted by $\angle H[e^{j\Omega}]$ with respect to the input phase.* Clearly $|H[e^{j\Omega}]|$ is the amplitude gain, and a plot of $|H[e^{j\Omega}]|$ versus Ω is the amplitude response of the discrete-time system. Similarly, $\angle H[e^{j\Omega}]$ is the phase response of the system, and a plot of $\angle H[e^{j\Omega}]$ versus Ω shows how the system modifies or shifts the phase of the input sinusoid. Note that $H[e^{j\Omega}]$ incorporates the information of both amplitude and phase responses and therefore is called the *frequency responses* of the system.

STEADY-STATE RESPONSE TO CAUSAL SINUSOIDAL INPUT

As in the case of continuous-time systems, we can show that the response of an LTID system to a causal sinusoidal input $\cos \Omega n \, u[n]$ is $y[n]$ in Eq. (5.46a), plus a natural component consisting of the characteristic modes (see Prob. 5.5-6). For a stable system, all the modes decay exponentially, and only the sinusoidal component in Eq. (5.46a) persists. For this reason, this component is called the sinusoidal *steady-state* response of the system. Thus, $y_{ss}[n]$, the steady-state response of a system to a causal sinusoidal input $\cos \Omega n \, u[n]$, is

$$y_{ss}[n] = |H[e^{j\Omega}]| \cos (\Omega n + \angle H[e^{j\Omega}])u[n]$$

[†]This may also be argued as follows. For BIBO-unstable systems, the zero-input response contains nondecaying natural mode terms of the form $\cos \Omega_0 n$ or $\gamma^n \cos \Omega_0 n$ ($\gamma > 1$). Hence, the response of such a system to a sinusoid $\cos \Omega n$ will contain not just the sinusoid of frequency Ω but also nondecaying natural modes, rendering the concept of frequency response meaningless. Alternately, we can argue that when $z = e^{j\Omega}$, a BIBO-unstable system violates the dominance condition $|\gamma_i| < |e^{j\Omega}|$ for all i, where γ_i represents ith characteristic root of the system (see Section 4.3).

SYSTEM RESPONSE TO SAMPLED CONTINUOUS-TIME SINUSOIDS

So far we have considered the response of a discrete-time system to a discrete-time sinusoid $\cos \Omega n$ (or exponential $e^{j\Omega n}$). In practice, the input may be a sampled continuous-time sinusoid $\cos \omega t$ (or an exponential $e^{j\omega t}$). When a sinusoid $\cos \omega t$ is sampled with sampling interval T, the resulting signal is a discrete-time sinusoid $\cos \omega n T$, obtained by setting $t = nT$ in $\cos \omega t$. Therefore, all the results developed in this section apply if we substitute ωT for Ω:

$$\Omega = \omega T \tag{5.47}$$

EXAMPLE 5.10

For a system specified by the equation

$$y[n + 1] - 0.8y[n] = x[n + 1]$$

find the system response to the input

(a) $1^n = 1$

(b) $\cos \left[\dfrac{\pi}{6} n - 0.2 \right]$

(c) a sampled sinusoid $\cos 1500t$ with sampling interval $T = 0.001$

The system equation can be expressed as

$$(E - 0.8)y[n] = Ex[n]$$

Therefore, the transfer function of the system is

$$H[z] = \frac{z}{z - 0.8} = \frac{1}{1 - 0.8z^{-1}}$$

The frequency response is

$$H[e^{j\Omega}] = \frac{1}{1 - 0.8e^{-j\Omega}}$$

$$= \frac{1}{1 - 0.8(\cos \Omega - j \sin \Omega)}$$

$$= \frac{1}{(1 - 0.8 \cos \Omega) + j0.8 \sin \Omega} \tag{5.48}$$

Therefore

$$|H[e^{j\Omega}]| = \frac{1}{\sqrt{(1 - 0.8 \cos \Omega)^2 + (0.8 \sin \Omega)^2}}$$

$$= \frac{1}{\sqrt{1.64 - 1.6 \cos \Omega}} \tag{5.49a}$$

and

$$\angle H[e^{j\Omega}] = -\tan^{-1}\left[\frac{0.8\sin\Omega}{1 - 0.8\cos\Omega}\right] \tag{5.49b}$$

The amplitude response $|H[e^{j\Omega}]|$ can also be obtained by observing that $|H|^2 = HH^*$. Therefore

$$|H[e^{j\Omega}]|^2 = H[e^{j\Omega}]H^*[e^{j\Omega}]$$
$$= H[e^{j\Omega}]H[e^{-j\Omega}] \tag{5.50}$$

From Eq. (5.48) it follows that

$$|H[e^{j\Omega}]|^2 = \left(\frac{1}{1 - 0.8e^{-j\Omega}}\right)\left(\frac{1}{1 - 0.8e^{j\Omega}}\right) = \frac{1}{1.64 - 1.6\cos\Omega}$$

which yields the result found earlier in Eq. (5.49a).

Figure 5.14 shows plots of amplitude and phase response as functions of Ω. We now compute the amplitude and the phase response for the various inputs.

(a) $x[n] = 1^n = 1$

Since $1^n = (e^{j\Omega})^n$ with $\Omega = 0$, the amplitude response is $H[e^{j0}]$. From Eq. (5.49a) we obtain

$$H[e^{j0}] = \frac{1}{\sqrt{1.64 - 1.6\cos(0)}} = \frac{1}{\sqrt{0.04}} = 5 = 5\angle 0$$

Therefore

$$|H[e^{j0}]| = 5 \quad \text{and} \quad \angle H[e^{j0}] = 0$$

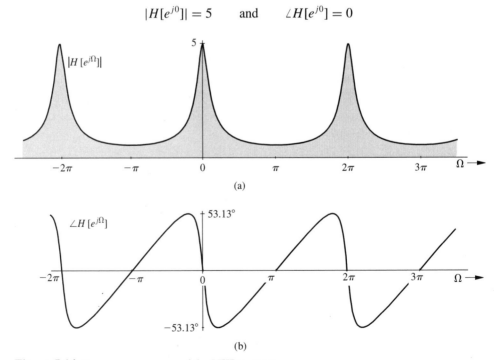

Figure 5.14 Frequency response of the LTID system.

These values also can be read directly from Fig. 5.14a and 5.14b, respectively, corresponding to $\Omega = 0$. Therefore, the system response to input 1 is

$$y[n] = 5(1^n) = 5 \qquad \text{for all } n \tag{5.51}$$

(b) $x[n] = \cos[(\pi/6)n - 0.2]$
Here $\Omega = \pi/6$. According to Eqs. (5.49)

$$|H[e^{j\pi/6}]| = \frac{1}{\sqrt{1.64 - 1.6\cos\dfrac{\pi}{6}}} = 1.983$$

$$\angle H[e^{j\pi/6}] = -\tan^{-1}\left[\frac{0.8\sin\dfrac{\pi}{6}}{1 - 0.8\cos\dfrac{\pi}{6}}\right] = -0.916 \text{ rad}$$

These values also can be read directly from Fig. 5.14a and 5.14b, respectively, corresponding to $\Omega = \pi/6$. Therefore

$$y[n] = 1.983\cos\left(\frac{\pi}{6}n - 0.2 - 0.916\right)$$

$$= 1.983\cos\left(\frac{\pi}{6}n - 1.116\right) \tag{5.52}$$

Figure 5.15 shows the input $x[n]$ and the corresponding system response.
   (c) A sinusoid $\cos 1500t$ sampled every T seconds ($t = nT$) results in a discrete-time sinusoid

$$x[n] = \cos 1500nT \tag{5.53}$$

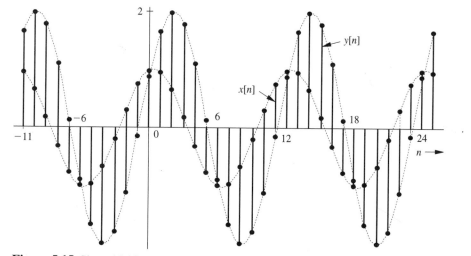

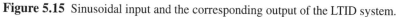

Figure 5.15 Sinusoidal input and the corresponding output of the LTID system.

For $T = 0.001$, the input is

$$x[n] = \cos(1.5n)$$

In this case, $\Omega = 1.5$. According to Eqs. (5.49a) and (5.49b),

$$|H[e^{j1.5}]| = \frac{1}{\sqrt{1.64 - 1.6\cos(1.5)}} = 0.809 \tag{5.54}$$

$$\angle H[e^{j1.5}] = -\tan^{-1}\left[\frac{0.8\sin(1.5)}{1 - 0.8\cos(1.5)}\right] = -0.702 \text{ rad} \tag{5.55}$$

These values also could be read directly from Fig. 5.14 corresponding to $\Omega = 1.5$. Therefore

$$y[n] = 0.809\cos(1.5n - 0.702)$$

COMPUTER EXAMPLE C5.1

Using MATLAB, find the frequency response of the system in Example 5.10.

```
>> Omega = linspace(-pi,pi,400);
>> H = tf([1 0],[1 -0.8],-1);
>> H_Omega = squeeze(freqresp(H,Omega));
>> subplot(2,1,1); plot(Omega,abs(H_Omega),'k'); axis tight;
>> xlabel('\Omega'); ylabel('|H[e^{j \Omega}]|');
>> subplot(2,1,2); plot(Omega,angle(H_Omega)*180/pi,'k'); axis tight;
>> xlabel('\Omega'); ylabel('\angle H[e^{j \Omega}] [deg]');
```

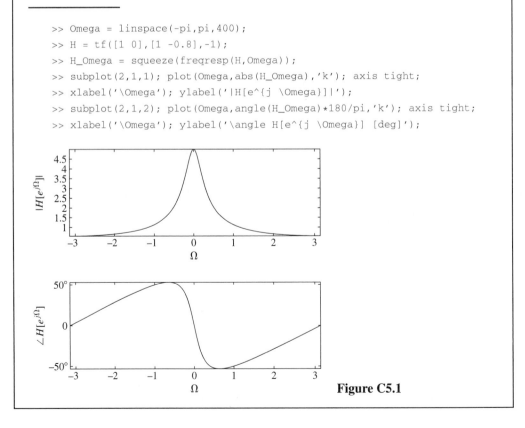

Figure C5.1

Comment. Figure 5.14 shows plots of amplitude and phase response as functions of Ω. These plots as well as Eqs. (5.49) indicate that the frequency response of a discrete-time system is a continuous (rather than discrete) function of frequency Ω. There is no contradiction here. This behavior is merely an indication of the fact that the frequency variable Ω is continuous (takes on all possible values) and therefore the system response exists at every value of Ω.

EXERCISE E5.18

For a system specified by the equation

$$y[n + 1] - 0.5y[n] = x[n]$$

find the amplitude and the phase response. Find the system response to sinusoidal input $\cos(1000t - (\pi/3))$ sampled every $T = 0.5$ ms.

ANSWER

$$|H[e^{j\Omega}]| = \frac{1}{\sqrt{1.25 - \cos \Omega}}$$

$$\angle H[e^{j\Omega}] = -\tan^{-1}\left[\frac{\sin \Omega}{\cos \Omega - 0.5}\right]$$

$$y[n] = 1.639 \cos\left(0.5n - \frac{\pi}{3} - 0.904\right) = 1.639 \cos(0.5n - 1.951)$$

EXERCISE E5.19

Show that for an ideal delay ($H[z] = 1/z$), the amplitude response $|H[e^{j\Omega}]| = 1$, and the phase response $\angle H[e^{j\Omega}] = -\Omega$. Thus, a pure time delay does not affect the amplitude gain of sinusoidal input, but it causes a phase shift (delay) of Ω radians in a discrete sinusoid of frequency Ω. Thus, for an ideal delay, the phase shift of the output sinusoid is proportional to the frequency of the input sinusoid (linear phase shift).

5.5-1 The Periodic Nature of the Frequency Response

In Example 5.10 and Fig. 5.14, we saw that the frequency response $H[e^{j\Omega}]$ is a periodic function of Ω. This is not a coincidence. Unlike continuous-time systems, all LTID systems have periodic frequency response. This is seen clearly from the nature of the expression of the frequency response of an LTID system. Because $e^{\pm j2\pi m} = 1$ for all integer values of m [see Eq. (B.12)],

$$H[e^{j\Omega}] = H\left[e^{j(\Omega + 2\pi m)}\right] \qquad m \text{ integer} \tag{5.56}$$

Therefore the frequency response $H[e^{j\Omega}]$ is a periodic function of Ω with a period 2π. This is the mathematical explanation of the periodic behavior. The physical explanation that follows provides a much better insight into the periodic behavior.

Nonuniqueness of Discrete-Time Sinusoid Waveforms

A continuous-time sinusoid cos ωt has a unique waveform for every real value of ω in the range 0 to ∞. Increasing ω results in a sinusoid of ever increasing frequency. Such is not the case for the discrete-time sinusoid cos Ωn because

$$\cos\left[(\Omega \pm 2\pi m)n\right] = \cos \Omega n \qquad m \text{ integer} \tag{5.57a}$$

and

$$e^{j(\Omega \pm 2\pi m)n} = e^{j\Omega n} \qquad m \text{ integer} \tag{5.57b}$$

This shows that the discrete-time sinusoids cos Ωn (and exponentials $e^{j\Omega n}$) separated by values of Ω in integral multiples of 2π are identical. The reason for the periodic nature of the frequency response of an LTID system is now clear. Since the sinusoids (or exponentials) with frequencies separated by interval 2π are identical, the system response to such sinusoids is also identical and, hence, is periodic with period 2π.

This discussion shows that the discrete-time sinusoid cos Ωn has a unique waveform only for the values of Ω in the range $-\pi$ to π. This band is called the *fundamental band*. Every frequency Ω, no matter how large, is identical to some frequency, Ω_a, in the fundamental band $(-\pi \leq \Omega_a < \pi)$, where

$$\Omega_a = \Omega - 2\pi m \qquad -\pi \leq \Omega_a < \pi \quad \text{and} \quad m \text{ integer} \tag{5.58}$$

The integer m can be positive or negative. We use Eq. (5.58) to plot the fundamental band frequency Ω_a versus the frequency Ω of a sinusoid (Fig. 5.16a). The frequency Ω_a is modulo 2π value of Ω.

All these conclusions are also valid for exponential $e^{j\Omega n}$.

All Discrete-Time Signals Are Inherently Bandlimited

This discussion leads to the surprising conclusion that all discrete-time signals are inherently bandlimited, with frequencies lying in the range $-\pi$ to π radians per sample. In terms of frequency $\mathcal{F} = \Omega/2\pi$, where $\mathcal{F}$ is in cycles per sample, all frequencies $\mathcal{F}$ separated by an integer number are identical. For instance, all discrete-time sinusoids of frequencies 0.3, 1.3, 2.3, ... cycles per sample are identical. The fundamental range of frequencies is -0.5 to 0.5 cycles per sample.

Any discrete-time sinusoid of frequency beyond the fundamental band, when plotted, appears and behaves, in every way, like a sinusoid having its frequency in the fundamental band. It is impossible to distinguish between the two signals. Thus, in a basic sense, discrete-time frequencies beyond $|\Omega| = \pi$ or $|\mathcal{F}| = 1/2$ do not exist. Yet, in a "mathematical" sense, we must admit the existence of sinusoids of frequencies beyond $\Omega = \pi$. What does this mean?

A Man Named Robert

To give an analogy, consider a fictitious person Mr. Robert Thompson. His mother calls him Robby; his acquaintances call him Bob, his close friends call him by his nickname, Shorty. Yet, Robert, Robby, Bob, and Shorty are one and the same person. However, we cannot say that only Mr. Robert Thompson exists, or only Robby exists, or only Shorty exists, or only Bob exists. All these four persons exist, although they are one and the same individual. In a same way, we cannot say that the frequency $\pi/2$ exists and frequency $5\pi/2$ does not exist; they are both the same entity, called by different names.

It is in this sense that we have to admit the existence of frequencies beyond the fundamental band. Indeed, mathematical expressions in frequency-domain automatically cater to this need by their built-in periodicity. As seen earlier, the very structure of the frequency response is 2π periodic. We shall also see later, in Chapter 9, that discrete-time signal spectra are also 2π periodic.

Admitting the existence of frequencies beyond π also serves mathematical and computational convenience in digital signal processing applications. Values of frequencies beyond π may also originate naturally in the process of sampling continuous-time sinusoids. Because there is no upper limit on the value of ω, there is no upper limit on the value of the resulting discrete-time frequency $\Omega = \omega T$ either.[†]

The highest possible frequency is π and the lowest frequency is 0 (dc or constant). Clearly, the high frequencies are those in the vicinity of $\Omega = (2m + 1)\pi$ and the low frequencies are those in the vicinity of $\Omega = 2\pi m$ for all positive or negative integer values of m.

FURTHER REDUCTION IN THE FREQUENCY RANGE

Because $\cos(-\Omega n + \theta) = \cos(\Omega n - \theta)$, a frequency in the range $-\pi$ to 0 is identical to the frequency (of the same magnitude) in the range 0 to π (but with a change in phase sign). Consequently the *apparent frequency* for a discrete-time sinusoid of any frequency is equal to some value in the range 0 to π. Thus, $\cos(8.7\pi n + \theta) = \cos(0.7\pi n + \theta)$, and the apparent frequency is 0.7π. Similarly,

$$\cos(9.6\pi n + \theta) = \cos(-0.4\pi n + \theta) = \cos(0.4\pi n - \theta)$$

Hence, the frequency 9.6π is identical (in every respect) to frequency -0.4π, which, in turn, is equal (within the sign of its phase) to frequency 0.4π. In this case, the apparent frequency reduces to $|\Omega_a| = 0.4\pi$. We can generalize the result to say that the apparent frequency of a discrete-time sinusoid Ω is $|\Omega_a|$, as found from Eq. (5.58), and if $\Omega_a < 0$, there is a phase reversal. Figure 5.16b plots Ω versus the apparent frequency $|\Omega_a|$. The shaded bands represent the ranges of Ω for which there is a phase reversal, when represented in terms of $|\Omega_a|$. For example, the apparent frequency for both the sinusoids $\cos(2.4\pi + \theta)$ and $\cos(3.6\pi + \theta)$ is $|\Omega_a| = 0.4\pi$, as seen from Fig. 5.16b. But 2.4π is in a clear band and 3.6π is in a shaded band. Hence, these sinusoids appear as $\cos(0.4\pi + \theta)$ and $\cos(0.4\pi - \theta)$, respectively.

Although every discrete-time sinusoid can be expressed as having frequency in the range from 0 to π, we generally use the frequency range from $-\pi$ to π instead of 0 to π for two reasons. First, exponential representation of sinusoids with frequencies in the range 0 to π requires a frequency range $-\pi$ to π. Second, even when we are using trigonometric representation, we generally need the frequency range $-\pi$ to π to have exact identity (without phase reversal) of a higher-frequency sinusoid.

For certain practical advantages, in place of the range $-\pi$ to π, we often use other contiguous ranges of width 2π. The range 0 to 2π, for instance, is used in many applications. It is left as an exercise for the reader to show that the frequencies in the range from π to 2π are identical to those in the range from $-\pi$ to 0.

[†]However, if Ω goes beyond π, the resulting aliasing reduces the apparent frequency to $\Omega_a < \pi$.

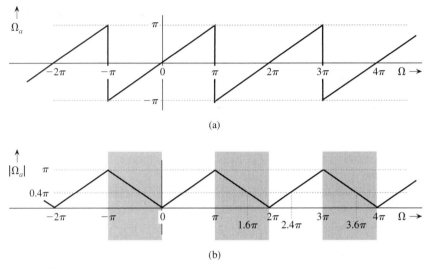

(a)

(b)

Figure 5.16 (a) Actual frequency versus (b) apparent frequency.

EXAMPLE 5.11

Express the following signals in terms of their apparent frequencies.

(a) $\cos{(0.5\pi n + \theta)}$

(b) $\cos{(1.6\pi n + \theta)}$

(c) $\sin{(1.6\pi n + \theta)}$

(d) $\cos{(2.3\pi n + \theta)}$

(e) $\cos{(34.699n + \theta)}$

(a) $\Omega = 0.5\pi$ is in the reduced range already. This is also apparent from Fig. 5.16a or 5.16b. Because $\Omega_a = 0.5\pi$, and there is no phase reversal, and the apparent sinusoid is $\cos{(0.5\pi n + \theta)}$.

(b) We express $1.6\pi = -0.4\pi + 2\pi$ so that $\Omega_a = -0.4\pi$ and $|\Omega_a| = 0.4$. Also, Ω_a is negative, implying sign change for the phase. Hence, the apparent sinusoid is $\cos{(0.4\pi n - \theta)}$. This fact is also apparent from Fig. 5.16b.

(c) We first convert the sine form to cosine form as $\sin{(1.6\pi n + \theta)} = \cos{(1.6\pi n - (\pi/2) + \theta)}$. In part (b) we found $\Omega_a = -0.4\pi$. Hence, the apparent sinusoid is $\cos{(0.4\pi n + (\pi/2) - \theta)} = -\sin{(0.4\pi n - \theta)}$. In this case, both the phase and the amplitude change signs.

(d) $2.3\pi = 0.3\pi + 2\pi$ so that $\Omega_a = 0.3\pi$. Hence, the apparent sinusoid is $\cos{(0.3\pi n + \theta)}$.

(e) We have $34.699 = -3 + 6(2\pi)$. Hence, $\Omega_a = -3$, and the apparent frequency $|\Omega_a| = 3$ rad/sample. Because Ω_a is negative, there is a sign change of the phase. Hence, the apparent sinusoid is $\cos{(3n - \theta)}$.

EXERCISE E5.20

Show that the sinusoids having frequencies Ω of

    (a) 2π

    (b) 3π

    (c) 5π

    (d) 3.2π

    (e) 22.1327

    (f) $\pi + 2$

can be expressed, respectively, as sinusoids of frequencies

    (a) 0

    (b) π

    (c) π

    (d) 0.8π

    (e) 3

    (f) $\pi - 2$

Show that in cases (d), (e), and (f), phase changes sign.

5.5-2 Aliasing and Sampling Rate

The nonuniqueness of discrete-time sinusoids and the periodic repetition of the same waveforms at intervals of 2π may seem innocuous, but in reality it leads to a serious problem for processing continuous-time signals by digital filters. A continuous-time sinusoid $\cos \omega t$ sampled every T seconds ($t = nT$) results in a discrete-time sinusoid $\cos \omega nT$, which is $\cos \Omega n$ with $\Omega = \omega T$. The discrete-time sinusoids $\cos \Omega n$ have unique waveforms only for the values of frequencies in the range $\Omega < \pi$ or $\omega T < \pi$. Therefore samples of continuous-time sinusoids of two (or more) different frequencies can generate the same discrete-time signal, as shown in Fig. 5.17. *This phenomenon is known as aliasing because through sampling, two entirely different analog sinusoids take on the same "discrete-time" identity.*[†]

Aliasing causes ambiguity in digital signal processing, which makes it impossible to determine the true frequency of the sampled signal. Consider, for instance, digitally processing a continuous-time signal that contains two distinct components of frequencies ω_1 and ω_2. The samples of these components appear as discrete-time sinusoids of frequencies $\Omega_1 = \omega_1 T$ and $\Omega_2 = \omega_2 T$. If Ω_1 and Ω_2 happen to differ by an integer multiple of 2π (if $\omega_2 - \omega_1 = 2k\pi / T$), the

---

[†]Figure 5.17 shows samples of two sinusoids $\cos 12\pi t$ and $\cos 2\pi t$ taken every 0.2 second. The corresponding discrete-time frequencies ($\Omega = \omega T = 0.2\omega$) are $\cos 2.4\pi$ and $\cos 0.4\pi$. The apparent frequency of 2.4π is 0.4π, identical to the discrete-time frequency corresponding to the lower sinusoid. This shows that the samples of both these continuous-time sinusoids at 0.2-second intervals are identical, as verified from Fig. 5.17.

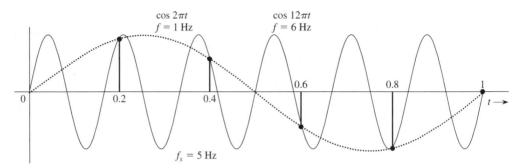

Figure 5.17 Demonstration of aliasing effect.

two frequencies will be read as the same (lower of the two) frequency by the digital processor.[†]
As a result, the higher-frequency component ω_2 not only is lost for good (by losing its identity
to ω_1), but it reincarnates as a component of frequency ω_1, thus distorting the true amplitude of
the original component of frequency ω_1. Hence, the resulting processed signal will be distorted.
Clearly, aliasing is highly undesirable and should be avoided. To avoid aliasing, the frequencies
of the continuous-time sinusoids to be processed should be kept within the fundamental band
$\omega T \le \pi$ or $\omega \le \pi/T$. Under this condition the question of ambiguity or aliasing does not arise
because any continuous-time sinusoid of frequency in this range has a unique waveform when
it is sampled. Therefore if ω_h is the highest frequency to be processed, then, to avoid aliasing,

$$\omega_h < \frac{\pi}{T} \tag{5.59}$$

If f_h is the highest frequency in hertz, $f_h = \omega_h/2\pi$, and, according to Eq. (5.59),

$$f_h < \frac{1}{2T} \tag{5.60a}$$

or

$$T < \frac{1}{2f_h} \tag{5.60b}$$

This shows that discrete-time signal processing places the limit on the highest frequency f_h that
can be processed for a given value of the sampling interval T according to Eq. (5.60a). But
we can process a signal of any frequency (without aliasing) by choosing a suitable value of T
according to Eq. (5.60b). The sampling rate or sampling frequency f_s is the reciprocal of the
sampling interval T, and, according to Eq. (5.60b),

$$f_s = \frac{1}{T} > 2f_h \tag{5.61a}$$

so that

$$f_h < \frac{f_s}{2} \tag{5.61b}$$

[†]In the case shown in Fig. 5.17, $\omega_1 = 12\pi$, $\omega_2 = 2\pi$, and $T = 0.2$. Hence, $\omega_2 - \omega_1 = 10\pi T = 2\pi$, and the
two frequencies are read as the same frequency $\Omega = 0.4\pi$ by the digital processor.

This result is a special case of the well-known *sampling theorem* (to be proved in Chapter 8). It states that to process a continuous-time sinusoid by a discrete-time system, the sampling rate must be greater than twice the frequency (in hertz) of the sinusoid. In short, *a sampled sinusoid must have a minimum of two samples per cycle.*[†] For sampling rate below this minimum value, the output signal will be aliased, which means it will be mistaken for a sinusoid of lower frequency.

ANTIALIASING FILTER

If the sampling rate fails to satisfy condition (5.61), aliasing occurs, causing the frequencies beyond $f_s/2$ Hz to masquerade as lower frequencies to corrupt the spectrum at frequencies below $f_s/2$. To avoid such a corruption, a signal to be sampled is passed through an *antialiasing* filter of bandwidth $f_s/2$ prior to sampling. This operation ensures the condition (5.61). The drawback of such a filter is that we lose the spectral components of the signal beyond frequency $f_s/2$, which is a preferable alternative to the aliasing corruption of the signal at frequencies below $f_s/2$. Chapter 8 presents a detailed analysis of the aliasing problem.

EXAMPLE 5.12

Determine the maximum sampling interval T that can be used in a discrete-time oscillator which generates a sinusoid of 50 kHz.

Here the highest significant frequency $f_h = 50$ kHz. Therefore from Eq. (5.60b)

$$T < \frac{1}{2f_h} = 10\,\mu s$$

The sampling interval must be less than $10\,\mu s$. The sampling frequency is $f_s = 1/T > 100$ kHz.

EXAMPLE 5.13

A discrete-time amplifier uses a sampling interval $T = 25\,\mu s$. What is the highest frequency of a signal that can be processed with this amplifier without aliasing?

From Eq. (5.60a)

$$f_h < \frac{1}{2T} = 20 \text{ kHz}$$

[†] Strictly speaking, we must have more than two samples per cycle.

5.6 FREQUENCY RESPONSE FROM POLE–ZERO LOCATION

The frequency responses (amplitude and phase responses) of a system are determined by pole–zero locations of the transfer function $H[z]$. Just as in continuous-time systems, it is possible to determine quickly the amplitude and the phase response and to obtain physical insight into the filter characteristics of a discrete-time system by using a graphical technique. The general Nth-order transfer function $H[z]$ in Eq. (5.34) can be expressed in factored form as

$$H[z] = b_0 \frac{(z - z_1)(z - z_2) \cdots (z - z_N)}{(z - \gamma_1)(z - \gamma_2) \cdots (z - \gamma_N)} \tag{5.62}$$

We can compute $H[z]$ graphically by using the concepts discussed in Section 4.10. The directed line segment from z_i to z in the complex plane (Fig. 5.18a) represents the complex number $z - z_i$. The length of this segment is $|z - z_i|$ and its angle with the horizontal axis is $\angle(z - z_i)$.

To compute the frequency response $H[e^{j\Omega}]$ we evaluate $H[z]$ at $z = e^{j\Omega}$. But for $z = e^{j\Omega}$, $|z| = 1$ and $\angle z = \Omega$ so that $z = e^{j\Omega}$ represents a point on the unit circle at an angle Ω with the horizontal. We now connect all zeros ($z_1, z_2, \ldots, z_N$) and all poles ($\gamma_1, \gamma_2, \ldots, \gamma_N$) to the point $e^{j\Omega}$, as indicated in Fig. 5.18b. Let $r_1, r_2, \ldots, r_N$ be the lengths and $\phi_1, \phi_2, \ldots, \phi_N$ be the angles, respectively, of the straight lines connecting $z_1, z_2, \ldots, z_N$ to the point $e^{j\Omega}$. Similarly, let $d_1, d_2, \ldots, d_N$ be the lengths and $\theta_1, \theta_2, \ldots, \theta_N$ be the angles, respectively, of the lines connecting

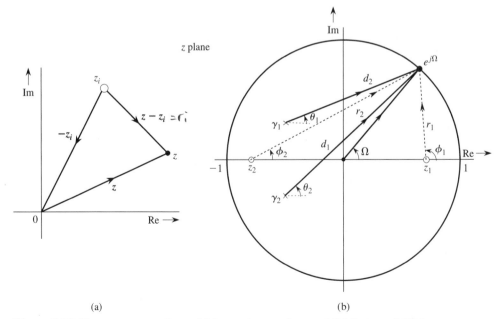

(a)   (b)

Figure 5.18 Vector representations of (a) complex numbers and (b) factors of $H[z]$.

$\gamma_1, \gamma_2, \ldots, \gamma_N$ to $e^{j\Omega}$. Then

$$H[e^{j\Omega}] = H[z]\big|_{z=e^{j\Omega}} = b_0 \frac{(r_1 e^{j\phi_1})(r_2 e^{j\phi_2}) \cdots (r_N e^{j\phi_N})}{(d_1 e^{j\theta_1})(d_2 e^{j\theta_2}) \cdots (d_N e^{j\theta_N})} \tag{5.63}$$

$$= b_0 \frac{r_1 r_2 \cdots r_N}{d_1 d_2 \cdots d_N} e^{j[(\phi_1 + \phi_2 + \cdots + \phi_N) - (\theta_1 + \theta_2 + \cdots + \theta_N)]} \tag{5.64}$$

Therefore (assuming $b_0 > 0$)

$$|H[e^{j\Omega}]| = b_0 \frac{r_1 r_2 \cdots r_N}{d_1 d_2 \cdots d_N}$$

$$= b_0 \frac{\text{product of the distances of zeros to } e^{j\Omega}}{\text{product of distances of poles to } e^{j\Omega}} \tag{5.65a}$$

and

$$\angle H[e^{j\Omega}] = (\phi_1 + \phi_2 + \cdots + \phi_N) - (\theta_1 + \theta_2 + \cdots + \theta_N)$$

$$= \text{sum of zero angles to } e^{j\Omega} - \text{sum of pole angles to } e^{j\Omega} \tag{5.65b}$$

In this manner, we can compute the frequency response $H[e^{j\Omega}]$ for any value of Ω by selecting the point on the unit circle at an angle Ω. This point is $e^{j\Omega}$. To compute the frequency response $H[e^{j\Omega}]$, we connect all poles and zeros to this point and use the foregoing equations to determine $|H[e^{j\Omega}]|$ and $\angle H[e^{j\Omega}]$. We repeat this procedure for all values of Ω from 0 to π to obtain the frequency response.

CONTROLLING GAIN BY PLACEMENT OF POLES AND ZEROS

The nature of the influence of pole and zero locations on the frequency response is similar to that observed in continuous-time systems, with minor differences. In place of the imaginary axis of the continuous-time systems, we have the unit circle in the discrete-time case. The nearer the pole (or zero) is to a point $e^{j\Omega}$ (on the unit circle) representing some frequency Ω, the more influence that pole (or zero) wields on the amplitude response at that frequency because the length of the vector joining that pole (or zero) to the point $e^{j\Omega}$ is small. The proximity of a pole (or a zero) has similar effect on the phase response. From Eq. (5.65a), it is clear that to enhance the amplitude response at a frequency Ω, we should place a pole as close as possible to the point $e^{j\Omega}$ (which is on the unit circle).[†] Similarly, to suppress the amplitude response at a frequency Ω, we should place a zero as close as possible to the point $e^{j\Omega}$ on the unit circle. Placing repeated poles or zeros will further enhance their influence.

Total suppression of signal transmission at any frequency can be achieved by placing a zero on the unit circle at a point corresponding to that frequency. This observation is used in the notch (bandstop) filter design.

Placing a pole or a zero at the origin does not influence the amplitude response because the length of the vector connecting the origin to any point on the unit circle is unity. However,

---

[†]The closest we can place a pole is on the unit circle at the point representing Ω. This choice would lead to infinite gain, but should be avoided because it will render the system marginally stable (BIBO unstable). The closer the point to the unit circle, the more sensitive the system gain to parameter variations.

a pole (or a zero) at the origin adds angle $-\Omega$ (or Ω) to $\angle H[e^{j\Omega}]$. Hence, the phase spectrum $-\Omega$ (or Ω) is a linear function of frequency and therefore represents a pure time delay (or time advance) of T seconds (see Exercise E5.19). Therefore, a pole (a zero) at the origin causes a time delay (or a time advance) of T seconds in the response. There is no change in the amplitude response.

For a stable system, all the poles must be located inside the unit circle. The zeros may lie anywhere. Also, for a physically realizable system, $H[z]$ must be a proper fraction, that is, $N \geq M$. If, to achieve a certain amplitude response, we require $M > N$, we can still make the system realizable by placing a sufficient number of poles at the origin to make $N = M$. This will not change the amplitude response, but it will increase the time delay of the response.

In general, a pole at a point has the opposite effect of a zero at that point. Placing a zero closer to a pole tends to cancel the effect of that pole on the frequency response.

LOWPASS FILTERS

A lowpass filter generally has a maximum gain at or near $\Omega = 0$, which corresponds to point $e^{j0} = 1$ on the unit circle. Clearly, placing a pole inside the unit circle near the point $z = 1$ (Fig. 5.19a) would result in a lowpass response.[†] The corresponding amplitude and phase response appear in Fig. 5.19a. For smaller values of Ω, the point $e^{j\Omega}$ (a point on the unit circle at an angle Ω) is closer to the pole, and consequently the gain is higher. As Ω increases, the distance of the point $e^{j\Omega}$ from the pole increases. Consequently the gain decreases, resulting in a lowpass characteristic. Placing a zero at the origin does not change the amplitude response but it does modify the phase response, as illustrated in Fig. 5.19b. Placing a zero at $z = -1$, however, changes both the amplitude and the phase response (Fig. 5.19c). The point $z = -1$ corresponds to frequency $\Omega = \pi$ ($z = e^{j\Omega} = e^{j\pi} = -1$). Consequently, the amplitude response now becomes more attenuated at higher frequencies, with a zero gain at $\Omega = \pi$. We can approach ideal lowpass characteristics by using more poles staggered near $z = 1$ (but within the unit circle). Figure 5.19d shows a third-order lowpass filter with three poles near $z = 1$ and a third-order zero at $z = -1$, with corresponding amplitude and phase response. For an ideal lowpass filter we need an enhanced gain at every frequency in the band $(0, \Omega_c)$. This can be achieved by placing a continuous wall of poles (requiring an infinite number of poles) opposite this band.

HIGHPASS FILTERS

A highpass filter has a small gain at lower frequencies and a high gain at higher frequencies. Such a characteristic can be realized by placing a pole or poles near $z = -1$ because we want the gain at $\Omega = \pi$ to be the highest. Placing a zero at $z = 1$ further enhances suppression of gain at lower frequencies. Figure 5.19e shows a possible pole–zero configuration of the third-order highpass filter with corresponding amplitude and phase responses.

[†]Placing the pole at $z = 1$ results in maximum (infinite) gain, but renders the system BIBO unstable and hence should be avoided.

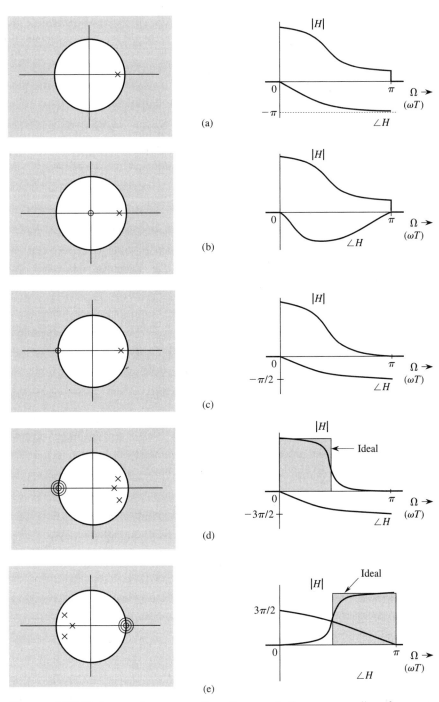

Figure 5.19 Various pole–zero configurations and the corresponding frequency responses.

In the following two examples, we shall realize analog filters by using digital processors and suitable interface devices (C/D and D/C), as shown in Fig. 3.2. At this point we shall examine the design of a digital processor with the transfer function $H[z]$ for the purpose of realizing bandpass and bandstop filters in the following examples.

As Fig. 3.2 showed, the C/D device samples the continuous-time input $x(t)$ to yield a discrete-time signal $x[n]$, which serves as the input to $H[z]$. The output $y[n]$ of $H[z]$ is converted to a continuous-time signal $y(t)$ by a D/C device. We also saw in Eq. (5.47) that a continuous-time sinusoid of frequency ω, when sampled, results in a discrete-time sinusoid $\Omega = \omega T$.

EXAMPLE 5.14 (Bandpass Filter)

By trial and error, design a tuned (bandpass) analog filter with zero transmission at 0 Hz and also at the highest frequency $f_h = 500$ Hz. The resonant frequency is required to be 125 Hz.

Because $f_h = 500$, we require $T < 1/1000$ [see Eq. (5.60b)]. Let us select $T = 10^{-3}.$[†] Recall that the analog frequencies ω corresponds to digital frequencies $\Omega = \omega T$. Hence, analog frequencies $\omega = 0$ and 1000π correspond to $\Omega = 0$ and π, respectively. The gain is required to be zero at these frequencies. Hence, we need to place zeros at $e^{j\Omega}$ corresponding to $\Omega = 0$ and $\Omega = \pi$. For $\Omega = 0$, $z = e^{j\Omega} = 1$; for $\Omega = \pi$, $e^{j\Omega} = -1$. Hence, there must be zeros at $z = \pm 1$. Moreover, we need enhanced response at the resonant frequency $\omega = 250\pi$, which corresponds to $\Omega = \pi/4$, which, in turn, corresponds to $z = e^{j\Omega} = e^{j\pi/4}$. Therefore, to enhance the frequency response at $\omega = 250\pi$, we place a pole in the vicinity of $e^{j\pi/4}$. Because this is a complex pole, we also need its conjugate near $e^{-j\pi/4}$, as indicated in Fig. 5.20a. Let us choose these poles γ_1 and γ_2 as

$$\gamma_1 = |\gamma|e^{j\pi/4} \qquad \text{and} \qquad \gamma_2 = |\gamma|e^{-j\pi/4}$$

where $|\gamma| < 1$ for stability. The closer the γ is to the unit circle, the more sharply peaked is the response around $\omega = 250\pi$. We also have zeros at ± 1. Hence

$$H[z] = K \frac{(z-1)(z+1)}{(z - |\gamma|e^{j\pi/4})(z - |\gamma|e^{-j\pi/4})}$$

$$= K \frac{z^2 - 1}{z^2 - \sqrt{2}|\gamma|z + |\gamma|^2} \qquad (5.66)$$

[†]Strictly speaking, we need $T < 0.001$. However, we shall show in Chapter 8 that if the input does not contain a finite amplitude component of 500 Hz, $T = 0.001$ is adequate. Generally practical signals satisfy this condition.

For convenience we shall choose $K = 1$. The amplitude response is given by

$$|H[e^{j\Omega}]| = \frac{|e^{j2\Omega} - 1|}{|e^{j\Omega} - |\gamma|e^{j\pi/4}||e^{j\Omega} - |\gamma|e^{-j\pi/4}|}$$

Now, by using Eq. (5.50), we obtain

$$|H[e^{j\Omega}]|^2 = \frac{2(1 - \cos 2\Omega)}{\left[1 + |\gamma|^2 - 2|\gamma|\cos\left(\Omega - \frac{\pi}{4}\right)\right]\left[1 + |\gamma|^2 - 2|\gamma|\cos\left(\Omega + \frac{\pi}{4}\right)\right]} \quad (5.67)$$

Figure 5.20b shows the amplitude response as a function of ω, as well as $\Omega = \omega T = 10^{-3}\omega$ for values of $|\gamma| = 0.83$, 0.96, and 1. As expected, the gain is zero at $\omega = 0$ and at 500 Hz ($\omega = 1000\pi$). The gain peaks at about 125 Hz ($\omega = 250\pi$). The resonance (peaking) becomes pronounced as $|\gamma|$ approaches 1. Figure 5.20c shows a canonical realization of this filter [see Eq. (5.66)].

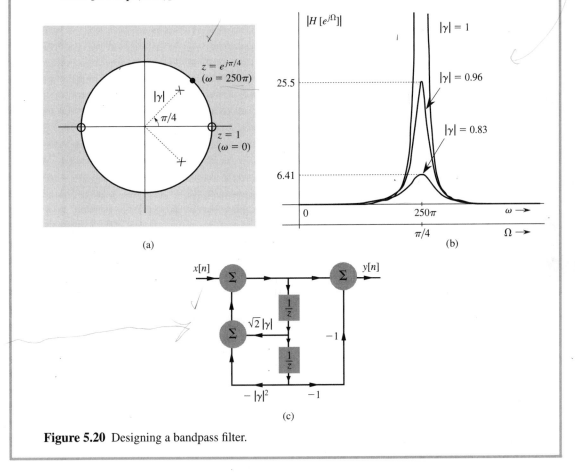

Figure 5.20 Designing a bandpass filter.

COMPUTER EXAMPLE C5.2

Use MATLAB to compute and plot the frequency response of the bandpass filter in Example 5.14 for the following cases:

(a) $|\gamma| = 0.83$

(b) $|\gamma| = 0.96$

(c) $|\gamma| = 0.99$

```
>> Omega = linspace(-pi,pi,4097); g_mag = [0.83 0.96 0.99];
>> H = zeros(length(g_mag),length(Omega));
>> for m = 1:length(g_mag),
>>      H(m,:) = freqz([1 0 -1],[1 -sqrt(2)*g_mag(m) g_mag(m)^2],Omega);
>> end
>> subplot(2,1,1); plot(Omega,abs(H(1,:)),'k-',...
>>      Omega,abs(H(2,:)),'k--',Omega,abs(H(3,:)),'k-.');
>> axis tight; xlabel('\Omega'); ylabel('|H[e^{j \Omega}]|');
>> legend('(a) |\gamma| = 0.83','(b) |\gamma| = 0.96','(c) |\gamma| = 0.99',0)
>> subplot(2,1,2); plot(Omega,angle(H(1,:)),'k-',...
>>      Omega,angle(H(2,:)),'k--',Omega,angle(H(3,:)),'k-.');
>> axis tight; xlabel('\Omega'); ylabel('\angle H[e^{j \Omega}] [rad]');
>> legend('(a) |\gamma| = 0.83','(b) |\gamma| = 0.96','(c) |\gamma| = 0.99',0)
```

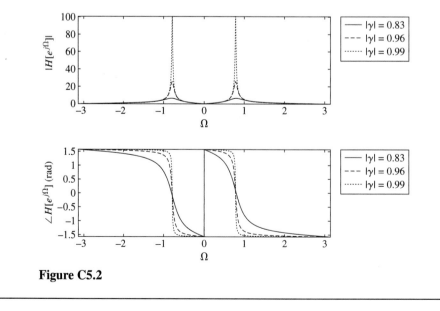

Figure C5.2

EXAMPLE 5.15 [Notch (Bandstop) Filter]

Design a second-order notch filter to have zero transmission at 250 Hz and a sharp recovery of gain to unity on both sides of 250 Hz. The highest significant frequency to be processed is $f_h = 400$ Hz.

In this case, $T < 1/2 f_h = 1.25 \times 10^{-3}$. Let us choose $T = 10^{-3}$. For the frequency 250 Hz, $\Omega = 2\pi(250)T = \pi/2$. Thus, the frequency 250 Hz is represented by a point $e^{j\Omega} = e^{j\pi/2} = j$ on the unit circle, as depicted in Fig. 5.21a. Since we need zero transmission at this frequency, we must place a zero at $z = e^{j\pi/2} = j$ and its conjugate at $z = e^{-j\pi/2} = -j$. We also require a sharp recovery of gain on both sides of frequency 250 Hz. To accomplish this goal, we place two poles close to the two zeros, to cancel out the effect of the two zeros as we move away from the point j (corresponding to frequency 250 Hz). For this reason, let us use poles at $\pm ja$ with $a < 1$ for stability. The closer the poles are to zeros (the closer the a to 1), the faster is the gain recovery on either side of 250 Hz. The resulting transfer function is

$$H[z] = K \frac{(z - j)(z + j)}{(z - ja)(z + ja)}$$

$$= K \frac{z^2 + 1}{z^2 + a^2}$$

The dc gain (gain at $\Omega = 0$, or $z = 1$) of this filter is

$$H[1] = K \frac{2}{1 + a^2}$$

Because we require a dc gain of unity, we must select $K = (1 + a^2)/2$. The transfer function is therefore

$$H[z] = \frac{(1 + a^2)(z^2 + 1)}{2(z^2 + a^2)} \tag{5.68}$$

and according to Eq. (5.50)

$$|H[e^{j\Omega}]|^2 = \frac{(1 + a^2)^2}{4} \frac{(e^{j2\Omega} + 1)(e^{-j2\Omega} + 1)}{(e^{j2\Omega} + a^2)(e^{-j2\Omega} + a^2)}$$

$$= \frac{(1 + a^2)^2(1 + \cos 2\Omega)}{2(1 + a^4 + 2a^2 \cos 2\Omega)}$$

Figure 5.21b shows $|H[e^{j\Omega}]|$ for values of $a = 0.3$, 0.6, and 0.95. Figure 5.21c shows a realization of this filter.

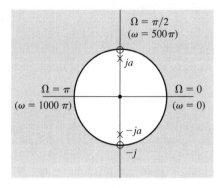

(a)

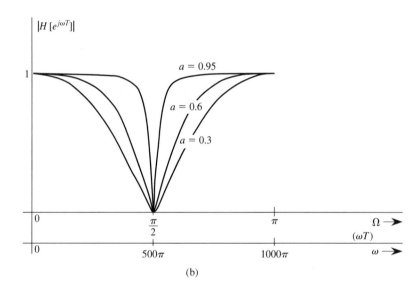

(b)

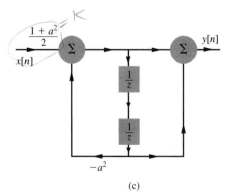

(c)

Figure 5.21 Designing a notch (bandstop) filter.

EXERCISE E5.21

Use the graphical argument to show that a filter with transfer function
$$H[z] = \frac{z - 0.9}{z}$$
acts like a highpass filter. Make a rough sketch of the amplitude response.

5.7 DIGITAL PROCESSING OF ANALOG SIGNALS

An analog (meaning continuous-time) signal can be processed digitally by sampling the analog signal and processing the samples by a digital (meaning discrete-time) processor. The output of the processor is then converted back to analog signal, as shown in Fig. 5.22a. We saw some simple cases of such processing in Examples 3.6, 3.7, 5.14, and 5.15. In this section, we shall derive a criterion for designing such a digital processor for a general LTIC system.

Suppose that we wish to realize an equivalent of an analog system with transfer function $H_a(s)$, shown in Fig. 5.22b. Let the digital processor transfer function in Fig. 5.22a that realizes this desired $H_a(s)$ be $H[z]$. In other words, we wish to make the two systems in Fig. 5.22 equivalent (at least approximately).

By "equivalence" we mean that for a given input $x(t)$, the systems in Fig. 5.22 yield the same output $y(t)$. Therefore $y(nT)$, the samples of the output in Fig. 5.22b, are identical to $y[n]$, the output of $H[z]$ in Fig. 5.22a.

For the sake of generality, we are assuming a noncausal system. The argument and the results are also valid for causal systems. The output $y(t)$ of the system in Fig. 5.22b is

$$y(t) = \int_{-\infty}^{\infty} x(\tau) h_a(t - \tau) \, d\tau$$

$$= \lim_{\Delta\tau \to 0} \sum_{m=-\infty}^{\infty} x(m\Delta\tau) h_a(t - m\Delta\tau) \Delta\tau \qquad (5.69a)$$

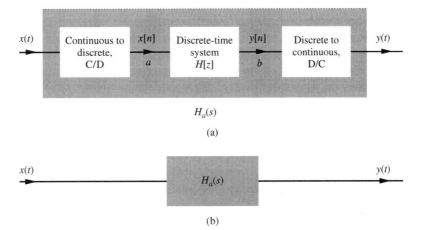

Figure 5.22 Analog filter realization with a digital filter.

For our purpose, it is convenient to use the notation T for $\Delta\tau$ in Eq. (5.69a). Assuming T (the sampling interval) to be small enough, such a change of notation yields

$$y(t) = T \sum_{m=-\infty}^{\infty} x(mT)h_a(t - mT) \qquad (5.69b)$$

The response at the nth sampling instant is $y(nT)$ obtained by setting $t = nT$ in the equation is

$$y(nT) = T \sum_{m=-\infty}^{\infty} x(mT)h_a[(n - m)T] \qquad (5.69c)$$

In Fig. 5.22a, the input to $H[z]$ is $x(nT) = x[n]$. If $h[n]$ is the unit impulse response of $H[z]$, then $y[n]$, the output of $H[z]$, is given by

$$y[n] = \sum_{m=-\infty}^{\infty} x[m]h[n - m] \qquad (5.70)$$

If the two systems are to be equivalent, $y(nT)$ in Eq. (5.69c) must be equal to $y[n]$ in Eq. (5.70). Therefore

$$h[n] = Th_a(nT) \qquad (5.71)$$

This is the time-domain criterion for equivalence of the two systems.[†] According to this criterion, $h[n]$, the unit impulse response of $H[z]$ in Fig. 5.22a, should be T times the samples of $h_a(t)$, the unit impulse response of the system in Fig. 5.22b. This is known as the *impulse invariance criterion* of filter design.

Strictly speaking, this realization guarantees the output equivalence only at the sampling instants, that is, $y(nT) = y[n]$, and that also requires the assumption that $T \to 0$. Clearly, this criterion leads to an approximate realization of $H_a(s)$. However, it can be shown that when the frequency response of $|H_a(j\omega)|$ is bandlimited, the realization is exact,[2] provided the sampling rate is high enough to avoid any aliasing ($T < 1/2f_h$).

REALIZATION OF RATIONAL $H(s)$

If we wish to realize an analog filter with transfer function

$$H_a(s) = \frac{c}{s - \lambda} \qquad (5.72a)$$

The impulse response $h(t)$, given by the inverse Laplace transform of $H_a(s)$, is

$$h_a(t) = ce^{\lambda t}u(t) \qquad (5.72b)$$

The corresponding digital filter unit impulse response $h[n]$, per Eq. (5.71), is

$$h[n] = Th_a(nT) = Tce^{n\lambda T}$$

[†]Because T is a constant, some authors ignore the factor T, which yields a simplified criterion $h[n] = h_a(nT)$. Ignoring T merely scales the amplitude response of the resulting filter.

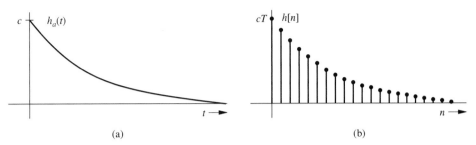

(a)   (b)

Figure 5.23 Impulse response for analog and digital systems in the impulse invariance method of filter design.

Figure 5.23 shows $h_a(t)$ and $h[n]$. The corresponding $H[z]$, the z-transform of $h[n]$, as found from Table 5.1, is

$$H[z] = \frac{Tcz}{z - e^{\lambda T}} \tag{5.73}$$

The procedure of finding $H[z]$ can be systematized for any Nth-order system. First we express an Nth-order analog transfer function $H_a(s)$ as a sum of partial fractions as[†]

$$H_a(s) = \sum_{i=1}^{n} \frac{c_i}{s - \lambda_i} \tag{5.74}$$

Then the corresponding $H[z]$ is given by

$$H[z] = T \sum_{i=1}^{n} \frac{c_i z}{z - e^{\lambda_i T}}$$

This transfer function can be readily realized as explained in Section 5.4. Table 5.3 lists several pairs of $H_a(s)$ and their corresponding $H[z]$. For instance, to realize a digital integrator, we examine its $H_a(s) = 1/s$. From Table 5.3, corresponding to $H_a(s) = 1/s$ (pair 2), we find $H[z] = Tz/(z-1)$. This is exactly the result we obtained in Example 3.7 using another approach.

Note that $H_a(j\omega)$ in Eq. (5.72a) [or (5.74)] is not bandlimited. Consequently, all these realizations are approximate.

CHOOSING THE SAMPLING INTERVAL T

The impulse invariance criterion (5.71) was derived under the assumption that $T \to 0$. Such an assumption is neither practical nor necessary for satisfactory design. Avoiding of aliasing is the most important consideration for the choice of T. In Eq. (5.60a), we showed that for a sampling interval T seconds, the highest frequency that can be sampled without aliasing is $1/2T$ Hz or π/T radians per second. This implies that $H_a(j\omega)$, the frequency response of the analog filter in Fig. 5.22b should not have spectral components beyond frequency π/T radians per second. In other words, to avoid aliasing, the frequency response of the system $H_a(s)$ must be

---

[†]Assuming $H_a(s)$ has simple poles. For repeated poles, the form changes accordingly. Entry 6 in Table 5.3 is suitable for repeated poles.

TABLE 5.3

No.	$H_a(s)$	$h_a(t)$	$h[n]$	$H[z]$
1	K	$K\delta(t)$	$TK\delta[n]$	TK
2	$\dfrac{1}{s}$	$u(t)$	$Tu[n]$	$\dfrac{Tz}{z-1}$
3	$\dfrac{1}{s^2}$	t	nT^2	$\dfrac{T^2z}{(z-1)^2}$
4	$\dfrac{1}{s^3}$	$\dfrac{t^2}{2}$	$\dfrac{k^2T^3}{2}$	$\dfrac{T^3z(z+1)}{2(z-1)^3}$
5	$\dfrac{1}{s-\lambda}$	$e^{\lambda t}$	$Te^{\lambda nT}$	$\dfrac{Tz}{z-e^{\lambda T}}$
6	$\dfrac{1}{(s-\lambda)^2}$	$te^{\lambda t}$	$nT^2e^{\lambda nT}$	$\dfrac{T^2ze^{\lambda T}}{(z-e^{\lambda T})^2}$
7	$\dfrac{As+B}{s^2+2as+c}$	$Tre^{-at}\cos(bt+\theta)$	$Tre^{-anT}\cos(bnT+\theta)$	$\dfrac{Trz[z\cos\theta-e^{-aT}\cos(bT-\theta)]}{z^2-(2e^{-aT}\cos bT)z+e^{-2aT}}$

$$r=\sqrt{\frac{A^2c+B^2-2ABa}{c-a^2}} \qquad b=\sqrt{c-a^2} \qquad \theta=\tan^{-1}\left(\frac{Aa-B}{A\sqrt{c-a^2}}\right)$$

bandlimited to π/T radians per second. We shall see later in Chapter 7 that frequency response of a realizable LTIC system cannot be bandlimited; that is, the response generally exists for all frequencies up to ∞. Therefore, it is impossible to digitally realize an LTIC system exactly without aliasing. The saving grace is that the frequency response of every realizable LTIC system decays with frequency. This allows for a compromise in digitally realizing an LTIC system with an acceptable level of aliasing. The smaller the value of T, the smaller the aliasing, and the better the approximation. Since it is impossible to make $|H_a(j\omega)|$ zero, we are satisfied with making it negligible beyond the frequency π/T. As a rule of thumb,[3] we choose T such that $|H_a(j\omega)|$ at the frequency $\omega=\pi/T$ is less than a certain fraction (often taken as 1%) of the peak value of $|H_a(j\omega)|$. This ensures that aliasing is negligible. The peak $|H_a(j\omega)|$ usually occurs at $\omega=0$ for lowpass filters and at the band center frequency ω_c for bandpass filters.

EXAMPLE 5.16

Design a digital filter to realize a first-order lowpass Butterworth filter with the transfer function

$$H_a(s)=\frac{\omega_c}{s+\omega_c} \qquad \omega_c=10^5 \tag{5.75}$$

For this filter, we find the corresponding $H[z]$ according to Eq. (5.73) (or pair 5 in Table 5.3) as

$$H[z] = \frac{\omega_c T z}{z - e^{-\omega_c T}} \tag{5.76}$$

Next, we select the value of T by means of the criterion according to which the gain at $\omega = \pi/T$ drops to 1% of the maximum filter gain. However, this choice results in such a good design that aliasing is imperceptible. The resulting amplitude response is so close to the desired response that we can hardly notice the aliasing effect in our plot. For the sake of demonstrating the aliasing effect, we shall deliberately select a 10% criterion (instead of 1%). We have

$$|H_a(j\omega)| = \left| \frac{\omega_c}{\sqrt{\omega^2 + \omega_c^2}} \right|$$

In this case $|H_a(j\omega)|_{\max} = 1$, which occurs at $\omega = 0$. Use of 10% criterion leads to $|H_a(\pi/T)| = 0.1$. Observe that

$$|H_a(j\omega)| \approx \frac{\omega_c}{\omega} \qquad \omega \gg \omega_c$$

Hence,

$$|H_a(\pi/T)| \approx \frac{\omega_c}{\pi/T} = 0.1 \quad \Longrightarrow \quad \pi/T = 10\omega_c = 10^6$$

Thus, the 10% criterion yields $T = 10^{-6}\pi$. The 1% criterion would have given $T = 10^{-7}\pi$. Substitution of $T = 10^{-6}\pi$ in Eq. (5.76) yields

$$H[z] = \frac{0.3142z}{z - 0.7304} \tag{5.77}$$

A canonical realization of this filter is shown in Fig. 5.24a. To find the frequency response of this digital filter, we rewrite $H[z]$ as

$$H[z] = \frac{0.3142}{1 - 0.7304z^{-1}}$$

Therefore

$$H[e^{j\omega T}] = \frac{0.3142}{1 - 0.7304e^{-j\omega T}} = \frac{0.3142}{(1 - 0.7304\cos\omega T) + j0.7304\sin\omega T}$$

Consequently

$$|H[e^{j\omega T}]| = \frac{0.3142}{\sqrt{(1 - 0.7304\cos\omega T)^2 + (0.7304\sin\omega T)^2}}$$

$$= \frac{0.3142}{\sqrt{1.533 - 1.4608\cos\omega T}} \tag{5.78a}$$

$$\angle H[e^{j\omega T}] = -\tan^{-1}\left(\frac{0.7304\sin\omega T}{1 - 0.7304\cos\omega T} \right) \tag{5.78b}$$

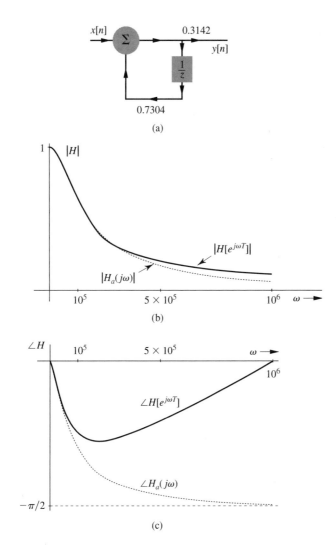

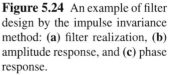

Figure 5.24 An example of filter design by the impulse invariance method: **(a)** filter realization, **(b)** amplitude response, and **(c)** phase response.

This frequency response differs from the desired response $H_a(j\omega)$ because aliasing causes frequencies above π/T to appear as frequencies below π/T. This generally results in increased gain for frequencies below π/T. For instance, the realized filter gain at $\omega = 0$ is $H[e^{j0}] = H[1]$. This value, as obtained from Eq. (5.77), is 1.1654 instead of the desired value 1. We can partly compensate for this distortion by multiplying $H[z]$ or $H[e^{j\omega T}]$ by a normalizing constant $K = H_a(0)/H[1] = 1/1.1654 = 0.858$. This forces the resulting gain of $H[e^{j\omega T}]$ to be equal to 1 at $\omega = 0$. The normalized $H_n[z] = 0.858H[z] = 0.858(0.1\pi z/(z - 0.7304))$. The amplitude response in Eq. (5.78a) is multiplied by $K = 0.858$ and plotted in Fig. 5.24b over the frequency range $0 \leq \omega \leq \pi/T = 10^6$. The multiplying constant K has no effect on the phase response in Eq. (5.78b), which is shown in Fig. 5.24c.

Also, the desired frequency response, according to Eq. (5.75) with $\omega_c = 10^5$, is

$$H_a(j\omega) = \frac{\omega_c}{j\omega + \omega_c} = \frac{10^5}{j\omega + 10^5}$$

Therefore

$$|H_a(j\omega)| = \frac{10^5}{\sqrt{\omega^2 + 10^{10}}} \qquad \text{and} \qquad \angle H_a(j\omega) = -\tan^{-1}\frac{\omega}{10^5}$$

This desired amplitude and phase response are plotted (dotted) in Fig. 5.24b and 5.24c for comparison with realized digital filter response. Observe that the amplitude response behavior of the analog and the digital filter is very close over the range $\omega \le \omega_c = 10^5$. However, for higher frequencies, there is considerable aliasing, especially in the phase spectrum. Had we used the 1% rule, the realized frequency response would have been closer over another decade of the frequency range.

COMPUTER EXAMPLE C5.3

Using the MATLAB `impinvar` command, find the impulse invariance digital filter to realize the first-order analog Butterworth filter presented in Example 5.16.

The analog filter transfer function is $10^5/(s + 10^5)$ and the sampling interval is $T = 10^{-6}\pi$.

```
>> num=[0 10^5]; den=[1 10^5];
>> T = pi/10^6; Fs = 1/T;
>> [b,a] = impinvar(num,den,Fs);
>> tf(b,a,T)


Transfer function:
 0.3142 z
----------
z - 0.7304

Sampling time: 3.1416e-006
```

EXERCISE E5.22

Design a digital filter to realize an analog transfer function

$$H_a(s) = \frac{20}{s + 20}$$

ANSWER

$$H[z] = \frac{20Tz}{z - e^{-20T}} \qquad \text{with} \qquad T = \frac{\pi}{2000}$$

5.8 CONNECTION BETWEEN THE LAPLACE TRANSFORM AND THE z-TRANSFORM

We now show that discrete-time systems also can be analyzed by means of the Laplace transform. In fact, we shall see that *the z-transform is the Laplace transform in disguise* and that discrete-time systems can be analyzed as if they were continuous-time systems.

So far we have considered the discrete-time signal as a sequence of numbers and not as an electrical signal (voltage or current). Similarly, we considered a discrete-time system as a mechanism that processes a sequence of numbers (input) to yield another sequence of numbers (output). The system was built by using delays (along with adders and multipliers) that delay sequences of numbers. A digital computer is a perfect example: every signal is a sequence of numbers, and the processing involves delaying sequences of numbers (along with addition and multiplication).

Now suppose we have a discrete-time system with transfer function $H[z]$ and input $x[n]$. Consider a continuous-time signal $x(t)$ such that its nth sample value is $x[n]$, as shown in Fig. 5.25.[†] Let the sampled signal be $\bar{x}(t)$, consisting of impulses spaced T seconds apart with the nth impulse of strength $x[n]$. Thus

$$\bar{x}(t) = \sum_{n=0}^{\infty} x[n]\delta(t - nT) \tag{5.79}$$

Figure 5.25 shows $x[n]$ and the corresponding $\bar{x}(t)$. The signal $x[n]$ is applied to the input of a discrete-time system with transfer function $H[z]$, which is generally made up of delays, adders, and scalar multipliers. Hence, processing $x[n]$ through $H[z]$ amounts to operating on the sequence $x[n]$ by means of delays, adders, and scalar multipliers. Suppose for $\bar{x}(t)$ samples, we perform operations identical to those performed on the samples of $x[n]$ by $H[z]$. For this purpose, we need a continuous-time system with transfer function $H(s)$ that is identical in structure to the discrete-time system $H[z]$ except that the delays in $H[z]$ are replaced by elements that delay continuous-time signals (such as voltages or currents). There is no other difference between realizations of $H[z]$ and $H(s)$. If a continuous-time impulse $\delta(t)$ is applied to such a delay of T seconds, the output will be $\delta(t - T)$. The continuous-time transfer function of such a delay is e^{-sT} [see Eq. (4.46)]. Hence the delay elements with transfer function $1/z$ in the realization of $H[z]$ will be replaced by the delay elements with transfer function e^{-sT} in the realization of the corresponding $H(s)$. This is same as z being replaced by e^{sT}. Therefore $H(s) = H[e^{sT}]$. Let us now apply $x[n]$ to the input of $H[z]$ and apply $\bar{x}(t)$ at the input of $H[e^{sT}]$. Whatever operations are performed by the discrete-time system $H[z]$ on $x[n]$ (Fig. 5.25a) are also performed by the corresponding continuous-time system $H[e^{sT}]$ on the impulse sequence $\bar{x}(t)$ (Fig. 5.25b). The delaying of a sequence in $H[z]$ would amount to delaying of an impulse train in $H[e^{sT}]$. Adding and multiplying operations are the same in both cases. In other words, one-to-one correspondence of the two systems is preserved in every aspect. Therefore if $y[n]$ is the output of the discrete-time system in Fig. 5.25a, then $\bar{y}(t)$, the output of the continuous-time system in Fig. 5.25b, would

---

[†]We can construct such $x(t)$ from the sample values, as explained in Chapter 8.

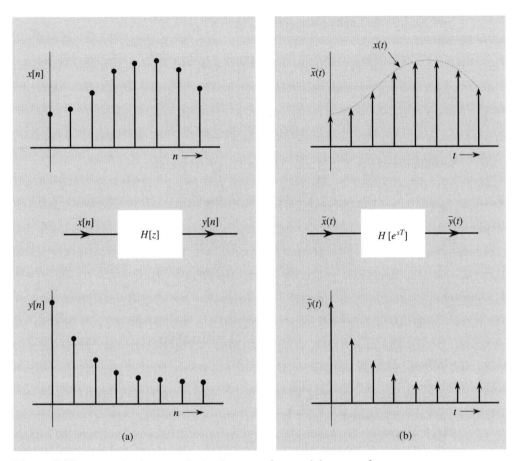

Figure 5.25 Connection between the Laplace transform and the z-transform.

be a sequence of impulse whose nth impulse strength is $y[n]$. Thus

$$\overline{y}(t) = \sum_{n=0}^{\infty} y[n]\delta(t - nT) \tag{5.80}$$

The system in Fig. 5.25b, being a continuous-time system, can be analyzed via the Laplace transform. If

$$\overline{x}(t) \Longleftrightarrow \overline{X}(s) \qquad \text{and} \qquad \overline{y}(t) \Longleftrightarrow \overline{Y}(s)$$

then

$$\overline{Y}(s) = H[e^{sT}]\overline{X}(s) \tag{5.81}$$

Also

$$\overline{X}(s) = \mathcal{L}\left[\sum_{n=0}^{\infty} x[n]\delta(t - nT)\right]$$

Now because the Laplace transform of $\delta(t - nT)$ is e^{-snT}

$$\overline{X}(s) = \sum_{n=0}^{\infty} x[n]e^{-snT} \tag{5.82}$$

and

$$\overline{Y}(s) = \sum_{n=0}^{\infty} y[n]e^{-snT} \tag{5.83}$$

Substitution of Eqs. (5.82) and (5.83) in Eq. (5.81) yields

$$\sum_{n=0}^{\infty} y[n]e^{-snT} = H[e^{sT}]\left[\sum_{n=0}^{\infty} x[n]e^{-snT}\right]$$

By introducing a new variable $z = e^{sT}$, this equation can be expressed as

$$\sum_{n=0}^{\infty} y[n]z^{-n} = H[z]\sum_{n=0}^{\infty} x[n]z^{-n}$$

or

$$Y[z] = H[z]X[z]$$

where

$$X[z] = \sum_{n=0}^{\infty} x[n]z^{-n} \qquad \text{and} \qquad Y[z] = \sum_{n=0}^{\infty} y[n]z^{-n}$$

It is clear from this discussion that the z-transform can be considered to be the Laplace transform with a change of variable $z = e^{sT}$ or $s = (1/T)\ln z$. Note that the transformation $z = e^{sT}$ transforms the imaginary axis in the s plane ($s = j\omega$) into a unit circle in the z plane ($z = e^{sT} = e^{j\omega T}$, or $|z| = 1$). The LHP and RHP in the s plane map into the inside and the outside, respectively, of the unit circle in the z plane.

5.9 THE BILATERAL z-TRANSFORM

Situations involving noncausal signals or systems cannot be handled by the (unilateral) z-transform discussed so far. Such cases can be analyzed by the *bilateral* (or two-sided) z-transform defined in Eq. (5.1), as

$$X[z] = \sum_{n=-\infty}^{\infty} x[n]z^{-n}$$

As in Eq. (5.2), the inverse z-transform is given by

$$x[n] = \frac{1}{2\pi j} \oint X[z]z^{n-1}\,dz$$

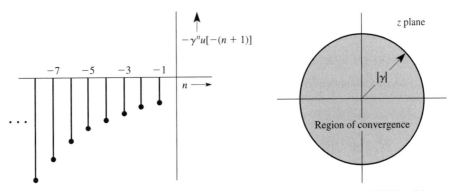

Figure 5.26 (a) $-\gamma^n u[-(n+1)]$ and **(b)** the region of convergence (ROC) of its z-transform.

These equations define the bilateral z-transform. Earlier, we showed that

$$\gamma^n u[n] \Longleftrightarrow \frac{z}{z-\gamma} \qquad |z| > |\gamma| \tag{5.84}$$

In contrast, the z-transform of the signal $-\gamma^n u[-(n+1)]$, illustrated in Fig. 5.26a, is

$$\mathcal{Z}\{-\gamma^n u[-(n+1)]\} = \sum_{-\infty}^{-1} -\gamma^n z^{-n} = \sum_{-\infty}^{-1} -\left(\frac{\gamma}{z}\right)^n$$

$$= -\left[\frac{z}{\gamma} + \left(\frac{z}{\gamma}\right)^2 + \left(\frac{z}{\gamma}\right)^3 + \cdots\right]$$

$$= 1 - \left[1 + \frac{z}{\gamma} + \left(\frac{z}{\gamma}\right)^2 + \left(\frac{z}{\gamma}\right)^3 + \cdots\right]$$

$$= 1 - \frac{1}{1 - \frac{z}{\gamma}} \qquad \left|\frac{z}{\gamma}\right| < 1$$

$$= \frac{z}{z-\gamma} \qquad |z| < |\gamma|$$

Therefore

$$\mathcal{Z}\{-\gamma^n u[-(n+1)]\} = \frac{z}{z-\gamma} \qquad |z| < |\gamma| \tag{5.85}$$

A comparison of Eqs. (5.84) and (5.85) shows that the z-transform of $\gamma^n u[n]$ is identical to that of $-\gamma^n u[-(n+1)]$. The regions of convergence, however, are different. In the former case, $X[z]$ converges for $|z| > |\gamma|$; in the latter, $X[z]$ converges for $|z| < |\gamma|$ (see Fig. 5.26b). Clearly, the inverse transform of $X[z]$ is not unique unless the region of convergence is specified. If we add the restriction that all our signals be causal, however, this ambiguity does not arise. The inverse transform of $z/(z-\gamma)$ is $\gamma^n u[n]$ even without specifying the ROC. Thus, in the unilateral transform, we can ignore the ROC in determining the inverse z-transform of $X[z]$.

As in the case the bilateral Laplace transform, if $x[n] = \sum_{i=1}^k x_i[n]$, then the ROC for $X[z]$ is the intersection of the ROCs (region common to all ROCs) for the transforms $X_1[z]$, $X_2[z], \ldots, X_k[z]$.

The preceding results lead to the conclusion (similar to that for the Laplace transform) that if $z = \beta$ is the largest magnitude pole for a causal sequence, its ROC is $|z| > |\beta|$. If $z = \alpha$ is the smallest magnitude nonzero pole for an anticausal sequence, its ROC is $|z| < |\alpha|$.

REGION OF CONVERGENCE FOR LEFT-SIDED AND RIGHT-SIDED SEQUENCES

Let us first consider a finite duration sequence $x_f[n]$, defined as a sequence that is nonzero for $N_1 \leq n \leq N_2$, where both N_1 and N_2 are finite numbers and $N_2 > N_1$. Also

$$X_f[z] = \sum_{n=N_1}^{N_2} x_f[n]z^{-n}$$

For example, if $N_1 = -2$ and $N_2 = 1$, then

$$X_f[z] = x_f[-2]z^2 + x_f[-1]z + x_f[0] + \frac{x_f[1]}{z}$$

Assuming all the elements in $x_f[n]$ are finite, we observe that $X_f[z]$ has two poles at $z = \infty$ because of terms $x_f[-2]z^2 + x_f[-1]z$ and one pole at $z = 0$ because of term $x_f[1]/z$. Thus, a finite duration sequence could have poles at $z = 0$ and $z = \infty$. Observe that $X_f[z]$ converges for all values of z except possibly $z = 0$ and $z = \infty$.

This means that the ROC of a general signal $x[n] + x_f[n]$ is the same as the ROC of $x[n]$ with the possible exception of $z = 0$ and $z = \infty$.

A *right-sided* sequence is zero for $n < N_2 < \infty$ and a left-sided sequence is zero for $n > N_1 > -\infty$. A causal sequence is always a right-sided sequence, but the converse is not necessarily true. An anticausal sequence is always a left-sided sequence, but the converse is not necessarily true. A *two-sided* sequence is of infinite duration and is neither right sided nor left sided.

A right-sided sequence $x_r[n]$ can be expressed as $x_r[n] = x_c[n] + x_f[n]$, where $x_c[n]$ is a causal signal and $x_f[n]$ is a finite duration signal. Therefore, the ROC for $x_r[n]$ is the same as the ROC for $x_c[n]$ except possibly $z = \infty$. If $z = \beta$ is the largest magnitude pole for a right-sided sequence $x_r[n]$, its ROC is $|\beta| < |z| \leq \infty$. Similarly, a left-sided sequence can be expressed as $x_l[n] = x_a[n] + x_f[n]$, where $x_a[n]$ is an anticausal sequence and $x_f[n]$ is a finite duration signal. Therefore, the ROC for $x_l[n]$ is the same as the ROC for $x_a[n]$ except possibly $z = 0$. Thus, if $z = \alpha$ is the smallest magnitude nonzero pole for a left-sided sequence, its ROC is $0 \leq |z| < |\alpha|$.

EXAMPLE 5.17

Determine the z-transform of

$$x[n] = (0.9)^n u[n] + (1.2)^n u[-(n+1)]$$
$$= x_1[n] + x_2[n]$$

From the results in Eqs. (5.84) and (5.85), we have

$$X_1[z] = \frac{z}{z - 0.9} \qquad |z| > 0.9$$

$$X_2[z] = \frac{-z}{z - 1.2} \qquad |z| < 1.2$$

The common region where both $X_1[z]$ and $X_2[z]$ converge is $0.9 < |z| < 1.2$ (Fig. 5.27b). Hence

$$X[z] = X_1[z] + X_2[z]$$
$$= \frac{z}{z - 0.9} - \frac{z}{z - 1.2}$$
$$= \frac{-0.3z}{(z - 0.9)(z - 1.2)} \qquad 0.9 < |z| < 1.2 \qquad (5.86)$$

The sequence $x[n]$ and the ROC of $X[z]$ are depicted in Fig. 5.27.

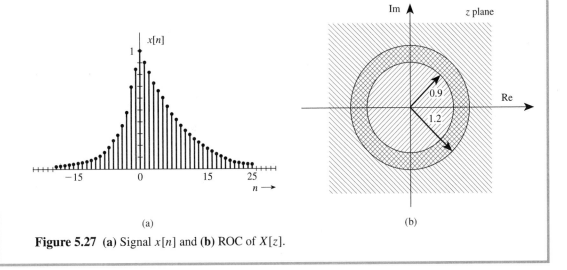

(a) (b)

Figure 5.27 (a) Signal $x[n]$ and (b) ROC of $X[z]$.

EXAMPLE 5.18

Find the inverse z-transform of

$$X[z] = \frac{-z(z + 0.4)}{(z - 0.8)(z - 2)}$$

if the ROC is

(a) $|z| > 2$

(b) $|z| < 0.8$

(c) $0.8 < |z| < 2$

_____

(a)

$$\frac{X[z]}{z} = \frac{-(z + 0.4)}{(z - 0.8)(z - 2)}$$

$$= \frac{1}{z - 0.8} - \frac{2}{z - 2}$$

and

$$X[z] = \frac{z}{z - 0.8} - 2\frac{z}{z - 2}$$

Since the ROC is $|z| > 2$, both terms correspond to causal sequences and

$$x[n] = [(0.8)^n - 2(2)^n]u[n]$$

This sequence appears in Fig. 5.28a.

(b) In this case, $|z| < 0.8$, which is less than the magnitudes of both poles. Hence, both terms correspond to anticausal sequences, and

$$x[n] = [-(0.8)^n + 2(2)^n]u[-(n + 1)]$$

This sequence appears in Fig. 5.28b.

(c) In this case, $0.8 < |z| < 2$; the part of $X[z]$ corresponding to the pole at 0.8 is a causal sequence, and the part corresponding to the pole at 2 is an anticausal sequence:

$$x[n] = (0.8)^n u[n] + 2(2)^n u[-(n + 1)]$$

This sequence appears in Fig. 5.28c.

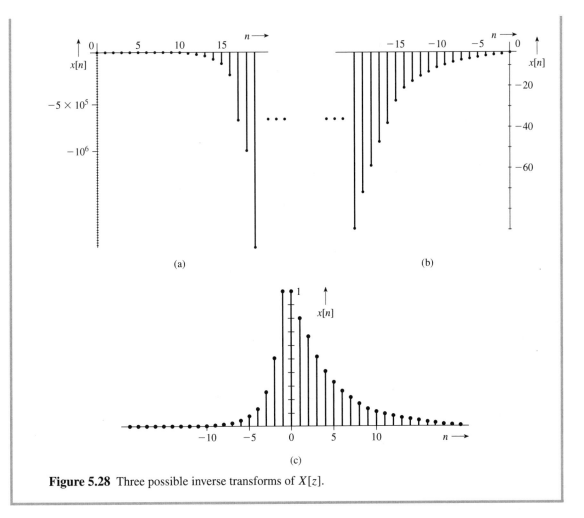

Figure 5.28 Three possible inverse transforms of $X[z]$.

EXERCISE E5.23

Find the inverse z-transform of

$$X[z] = \frac{z}{z^2 + \frac{5}{6}z + \frac{1}{6}} \qquad \tfrac{1}{2} > |z| > \tfrac{1}{3}$$

ANSWER

$6\left(-\frac{1}{3}\right)^n u[n] + 6\left(-\frac{1}{2}\right)^n u[-(n+1)]$

INVERSE TRANSFORM BY EXPANSION OF $X[z]$
IN POWER SERIES OF z

We have

$$X[z] = \sum_n x[n]z^{-n}$$

For an anticausal sequence, which exists only for $n \leq -1$, this equation becomes

$$X[z] = x[-1]z + x[-2]z^2 + x[-3]z^3 + \cdots$$

We can find the inverse z-transform of $X[z]$ by dividing the numerator polynomial by the denominator polynomial, both in ascending powers of z, to obtain a polynomial in ascending powers of z. Thus, to find the inverse transform of $z/(z-0.5)$ (when the ROC is $|z| < 0.5$), we divide z by $-0.5 + z$ to obtain $-2z - 4z^2 - 8z^3 - \cdots$. Hence, $x[-1] = -2$, $x[-2] = -4$, $x[-3] = -8$, and so on.

5.9-1 Properties of the Bilateral z-Transform

Properties of the bilateral z-transform are similar to those of the unilateral transform. We shall merely state the properties here, without proofs, for $x_i[n] \Longleftrightarrow X_i[z]$.

LINEARITY

$$a_1 x_1[n] + a_2 x_2[n] \Longleftrightarrow a_1 X_1[z] + a_2 X_2[z]$$

The ROC for $a_1 X_1[z] + a_2 X_2[z]$ is the region common to (intersection of) the ROCs for $X_1[z]$ and $X_2[z]$.

SHIFT

$$x[n-m] \Longleftrightarrow \frac{1}{z^m} X[z] \qquad m \text{ is positive or negative integer}$$

The ROC for $X[z]/z^m$ is the ROC for $X[z]$ (except for the addition or deletion of $z = 0$ or $z = \infty$ caused by the factor $1/z^m$.

CONVOLUTION

$$x_1[n] * x_2[n] \Longleftrightarrow X_1[z]X_2[z]$$

The ROC for $X_1[z]X_2[z]$ is the region common to (intersection of) the ROCs for $X_1[z]$ and $X_2[z]$.

MULTIPLICATION BY γ^n

$$\gamma^n x[n] \Longleftrightarrow X\left[\frac{z}{\gamma}\right]$$

If the ROC for $X[z]$ is $|\gamma_1| < |z| < |\gamma_2|$, then the ROC for $X[z/\gamma]$ is $|\gamma\gamma_1| < |z| < |\gamma\gamma_2|$, indicating that the ROC is scaled by the factor $|\gamma|$.

Multiplication by n

$$nx[n]u[n] \iff -z\frac{d}{dz}X[z]$$

The ROC for $-z(dX/dz)$ is the same as the ROC for $X[z]$.

Time Reversal[†]

$$x[-n] \iff X[1/z]$$

If the ROC for $X[z]$ is $|\gamma_1| < |z| < |\gamma_2|$ then the ROC for $X[1/z]$ is $1/|\gamma_1| > |z| > |1/\gamma_2|$

5.9-2 Using the Bilateral z-Transform for Analysis of LTID Systems

Because the bilateral z-transform can handle noncausal signals, we can use this transform to analyze noncausal linear systems. The zero-state response $y[n]$ is given by

$$y[n] = \mathcal{Z}^{-1}\{X[z]H[z]\}$$

provided $X[z]H[z]$ exists. The ROC of $X[z]H[z]$ is the region in which both $X[z]$ and $H[z]$ exist, which means that the region is the common part of the ROC of both $X[z]$ and $H[z]$.

EXAMPLE 5.19

For a causal system specified by the transfer function

$$H[z] = \frac{z}{z-0.5}$$

find the zero-state response to input

$$x[n] = (0.8)^n u[n] + 2(2)^n u[-(n+1)]$$

$$X[z] = \frac{z}{z-0.8} - \frac{2z}{z-2} = \frac{-z(z+0.4)}{(z-0.8)(z-2)}$$

The ROC corresponding to the causal term is $|z| > 0.8$, and that corresponding to the anticausal term is $|z| < 2$. Hence, the ROC for $X[z]$ is the common region, given by $0.8 < |z| < 2$. Hence

$$X[z] = \frac{-z(z+0.4)}{(z-0.8)(z-2)} \qquad 0.8 < |z| < 2$$

Therefore

$$Y[z] = X[z]H[z] = \frac{-z^2(z+0.4)}{(z-0.5)(z-0.8)(z-2)}$$

[†]For complex signal $x[n]$, the property is modified, as follows:

$$x^*[-n] \iff X^*[1/z^*]$$

Since the system is causal, the ROC of $H[z]$ is $|z| > 0.5$. The ROC of $X[z]$ is $0.8 < |z| < 2$. The common region of convergence for $X[z]$ and $H[z]$ is $0.8 < |z| < 2$. Therefore

$$Y[z] = \frac{-z^2(z+0.4)}{(z-0.5)(z-0.8)(z-2)} \qquad 0.8 < |z| < 2$$

Expanding $Y[z]$ into modified partial fractions yields

$$Y[z] = -\frac{z}{z-0.5} + \frac{8}{3}\left(\frac{z}{z-0.8}\right) - \frac{8}{3}\left(\frac{z}{z-2}\right) \qquad 0.8 < |z| < 2$$

Since the ROC extends outward from the pole at 0.8, both poles at 0.5 and 0.8 correspond to causal sequence. The ROC extends inward from the pole at 2. Hence, the pole at 2 corresponds to anticausal sequence. Therefore

$$y[n] = \left[-(0.5)^n + \tfrac{8}{3}(0.8)^n\right] u[n] + \tfrac{8}{3}(2)^n u[-(n+1)]$$

EXAMPLE 5.20

For the system in Example 5.19, find the zero-state response to input

$$x[n] = \underbrace{(0.8)^n u[n]}_{x_1[n]} + \underbrace{(0.6)^n u[-(n+1)]}_{x_2[n]}$$

The z-transforms of the causal and anticausal components $x_1[n]$ and $x_2[n]$ of the output are

$$X_1[z] = \frac{z}{z-0.8} \qquad |z| > 0.8$$

$$X_2[z] = \frac{-z}{z-0.6} \qquad |z| < 0.6$$

Observe that a common ROC for $X_1[z]$ and $X_2[z]$ does not exist. Therefore $X[z]$ does not exist. In such a case we take advantage of the superposition principle and find $y_1[n]$ and $y_2[n]$, the system responses to $x_1[n]$ and $x_2[n]$, separately. The desired response $y[n]$ is the sum of $y_1[n]$ and $y_2[n]$. Now

$$H[z] = \frac{z}{z-0.5} \qquad |z| > 0.5$$

$$Y_1[z] = X_1[z]H[z] = \frac{z^2}{(z-0.5)(z-0.8)} \qquad |z| > 0.8$$

$$Y_2[z] = X_2[z]H[z] = \frac{-z^2}{(z-0.5)(z-0.6)} \qquad 0.5 < |z| < 0.6$$

Expanding $Y_1[z]$ and $Y_2[z]$ into modified partial fractions yields

$$Y_1[z] = -\frac{5}{3}\left(\frac{z}{z-0.5}\right) + \frac{8}{3}\left(\frac{z}{z-0.8}\right) \qquad |z| > 0.8$$

$$Y_2[z] = 5\left(\frac{z}{z-0.5}\right) - 6\left(\frac{z}{z-0.6}\right) \qquad 0.5 < |z| < 0.6$$

Therefore

$$y_1[n] = \left[-\tfrac{5}{3}(0.5)^n + \tfrac{8}{3}(0.8)^n\right]u[n]$$

$$y_2[n] = 5(0.5)^n u[n] + 6(0.6)^n u[-(n+1)]$$

and

$$y[n] = y_1[n] + y_2[n]$$

$$= \left[\tfrac{10}{3}(0.5)^n + \tfrac{8}{3}(0.8)^n\right]u[n] + 6(0.6)^n u[-(n+1)]$$

EXERCISE E5.24

For a causal system in Example 5.19, find the zero-state response to input

$$x[n] = \left(\tfrac{1}{4}\right)^n u[n] + 5(3)^n u[-(n+1)]$$

ANSWER

$$\left[-\left(\tfrac{1}{4}\right)^n + 3\left(\tfrac{1}{2}\right)^n\right]u[n] + 6(3)^n u[-(n+1)]$$

5.10 SUMMARY

In this chapter we discussed the analysis of linear, time-invariant, discrete-time (LTID) systems by means of the z-transform. The z-transform changes the difference equations of LTID systems into algebraic equations. Therefore, solving these difference equations reduces to solving algebraic equations.

The transfer function $H[z]$ of an LTID system is equal to the ratio of the z-transform of the output to the z-transform of the input when all initial conditions are zero. Therefore, if $X[z]$ is the z-transform of the input $x[n]$ and $Y[z]$ is the z-transform of the corresponding output $y[n]$ (when all initial conditions are zero), then $Y[z] = H[z]X[z]$. For an LTID system specified by the difference equation $Q[E]y[n] = P[E]x[n]$, the transfer function $H[z] = P[z]/Q[z]$. Moreover, $H[z]$ is the z-transform of the system impulse response $h[n]$. We showed in Chapter 3 that the system response to an everlasting exponential z^n is $H[z]z^n$.

We may also view the z-transform as a tool that expresses a signal $x[n]$ as a sum of exponentials of the form z^n over a continuum of the values of z. Using the fact that an LTID system response to z^n is $H[z]z^n$, we find the system response to $x[n]$ as a sum of the system's responses to all the components of the form z^n over the continuum of values of z.

LTID systems can be realized by scalar multipliers, adders, and time delays. A given transfer function can be synthesized in many different ways. We discussed canonical, transposed canonical, cascade, and parallel forms of realization. The realization procedure is identical to that for continuous-time systems with $1/s$ (integrator) replaced by $1/z$ (unit delay).

In Section 5.8, we showed that discrete-time systems can be analyzed by the Laplace transform as if they were continuous-time systems. In fact, we showed that the z-transform is the Laplace transform with a change in variable.

The majority of the input signals and practical systems are causal. Consequently, we are required to deal with causal signals most of the time. Restricting all signals to the causal type greatly simplifies z-transform analysis; the ROC of a signal becomes irrelevant to the analysis process. This special case of z-transform (which is restricted to causal signals) is called the unilateral z-transform. Much of the chapter deals with this transform. Section 5.9 discusses the general variety of the z-transform (bilateral z-transform), which can handle causal and noncausal signals and systems. In the bilateral transform, the inverse transform of $X[z]$ is not unique, but depends on the ROC of $X[z]$. Thus, the ROC plays a crucial role in the bilateral z-transform.

REFERENCES

1.   Lyons, R. G. *Understanding Digital Signal Processing*. Addison-Wesley, Reading, MA, 1997.

2.   Oppenheim, A. V., and R. W. Schafer. *Discrete-Time Signal Processing*, 2nd ed. Prentice-Hall, Upper Saddle River, NJ, 1999.

3.   Mitra, S. K. *Digital Signal Processing*, 2nd ed. McGraw-Hill, New York, 2001.

MATLAB SESSION 5: DISCRETE-TIME IIR FILTERS

Recent technological advancements have dramatically increased the popularity of discrete-time filters. Unlike their continuous-time counterparts, the performance of discrete-time filters is not affected by component variations, temperature, humidity, or age. Furthermore, digital hardware is easily reprogrammed, which allows convenient change of device function. For example, certain digital hearing aids are individually programmed to match the required response of a user.

Typically, discrete-time filters are categorized as infinite-impulse response (IIR) or finite-impulse response (FIR). A popular method to obtain a discrete-time IIR filter is by transformation of a corresponding continuous-time filter design. MATLAB greatly assists this process. Although discrete-time IIR filter design is the emphasis of this session, methods for discrete-time FIR filter design are considered in MATLAB Session 9.

M5.1 Frequency Response and Pole–Zero Plots

Frequency response and pole–zero plots help characterize filter behavior. Similar to continuous-time systems, rational transfer functions for realizable LTID systems are represented in the z-domain as

$$H[z] = \frac{Y[z]}{X[z]} = \frac{B[z]}{A[z]} = \frac{\sum_{k=0}^{N} b_k z^{-k}}{\sum_{k=0}^{N} a_k z^{-k}} = \frac{\sum_{k=0}^{N} b_k z^{N-k}}{\sum_{k=0}^{N} a_k z^{N-k}} \tag{M5.1}$$

When only the first $(N_1 + 1)$ numerator coefficients are nonzero and only the first $(N_2 + 1)$ denominator coefficients are nonzero, Eq. (M5.1) simplifies to

$$H[z] = \frac{Y[z]}{X[z]} = \frac{B[z]}{A[z]} = \frac{\sum_{k=0}^{N_1} b_k z^{-k}}{\sum_{k=0}^{N_2} a_k z^{-k}} = \frac{\sum_{k=0}^{N_1} b_k z^{N_1-k}}{\sum_{k=0}^{N_2} a_k z^{N_2-k}} z^{N_2-N_1} \qquad \text{(M5.2)}$$

The form of Eq. (M5.2) has many advantages. It can be more efficient than Eq. (M5.1); it still works when $N_1 = N_2 = N$; and it more closely conforms to the notation of built-in MATLAB discrete-time signal processing functions.

The right-hand side of Eq. (M5.2) is a form that is convenient for MATLAB computations. The frequency response $H(e^{j\Omega})$ is obtained by letting $z = e^{j\Omega}$, where Ω has units of radians. Often, $\Omega = \omega T$, where ω is the continuous-time frequency in radians per second and T is the sampling period in seconds. Defining length-$(N_2 + 1)$ coefficient vector $\mathbf{A} = [a_0, a_1, \ldots, a_{N_2}]$ and length-$(N_1 + 1)$ coefficient vector $\mathbf{B} = [b_0, b_1, \ldots, b_{N_1}]$, program MS5P1 computes $H(e^{j\Omega})$ by using Eq. (M5.2) for each frequency in the input vector Ω.

```
function [H] = MS5P1(B,A,Omega);
% MS5P1.m : MATLAB Session 5, Program 1
% Function M-file computes frequency response for LTID systems
% INPUTS:   B = vector of feedforward coefficients
%           A = vector of feedback coefficients
%           Omega = vector of frequencies [rad], typically -pi<=Omega<=pi
% OUTPUTS:  H =  frequency response

N_1 = length(B)-1; N_2 = length(A)-1;
H = polyval(B,exp(j*Omega))./polyval(A,exp(j*Omega)).*exp(j*Omega*(N_2-N_1));
```

Note that owing to MATLAB's indexing scheme, A(k) corresponds to coefficient a_{k-1} and B(k) corresponds to coefficient b_{k-1}. It is also possible to use the signal processing toolbox function freqz to evaluate the frequency response of a system described by Eq. (M5.2). Under special circumstances, the control system toolbox function bode can also be used.

Program MS5P2 computes and plots the poles and zeros of an LTID system described by Eq. (M5.2) again using vectors $\mathbf{B}$ and $\mathbf{A}$.

```
function [p,z] = MS5P2(B,A);
% MS5P2.m : MATLAB Session 5, Program 2
% Function M-file computes and plots poles and zeros for LTID systems
% INPUTS:   B = vector of feedforward coefficients
%           A = vector of feedback coefficients

N_1 = length(B)-1; N_2 = length(A)-1;
p = roots([A,zeros(1,N_1-N_2)]); z = roots([B,zeros(1,N_2-N_1)]);
ucirc = exp(j*linspace(0,2*pi,200)); % Compute unit circle for pole-zero plot
plot(real(p),imag(p),'xk',real(z),imag(z),'ok',real(ucirc),imag(ucirc),'k:');
xlabel('Real'); ylabel('Imaginary');
ax = axis; dx = 0.05*(ax(2)-ax(1)); dy = 0.05*(ax(4)-ax(3));
axis(ax+[-dx,dx,-dy,dy]);
```

The right-hand side of Eq. (M5.2) helps explain how the roots are computed. When $N_1 \neq N_2$, the term $z^{N_2-N_1}$ implies additional roots at the origin. If $N_1 > N_2$, the roots are poles, which are added by concatenating A with `zeros(N_1-N_2,1)`; since $N_2 - N_1 \leq 0$, `zeros(N_2-N_1,1)` produces the empty set and B is unchanged. If $N_2 > N_1$, the roots are zeros, which are added by concatenating B with `zeros(N_2-N_1,1)`; since $N_1 - N_2 \leq 0$, `zeros(N_1-N_2,1)` produces the empty set and A is unchanged. Poles and zeros are indicated with black x's and o's, respectively. For visual reference, the unit circle is also plotted. The last line in MS5P2 expands the plot axis box so that root locations are not obscured.

M5.2 Transformation Basics

Transformation of a continuous-time filter to a discrete-time filter begins with the desired continuous-time transfer function

$$H(s) = \frac{Y(s)}{X(s)} = \frac{B(s)}{A(s)} = \frac{\sum_{k=0}^{M} b_{k+N-M} s^{M-k}}{\sum_{k=0}^{N} a_k s^{N-k}}$$

As a matter of convenience, $H(s)$ is represented in factored form as

$$H(s) = \frac{b_{N-M}}{a_0} \frac{\prod_{k=1}^{M}(s - z_k)}{\prod_{k=1}^{N}(s - p_k)} \tag{M5.3}$$

where z_k and p_k are the system poles and zeros, respectively.

A mapping rule converts the rational function $H(s)$ to a rational function $H[z]$. Requiring that the result be rational ensures that the system realization can proceed with only delay, sum, and multiplier blocks. There are many possible mapping rules. For obvious reasons, good transformations tend to map the ω axis to the unit circle, $\omega = 0$ to $z = 1$, $\omega = \infty$ to $z = -1$, and the left half-plane to the interior of the unit circle. Put another way, sinusoids map to sinusoids, zero frequency maps to zero frequency, high frequency maps to high frequency, and stable systems map to stable systems.

Section 5.8 suggests that the z-transform can be considered to be a Laplace transform with a change of variable $z = e^{sT}$ or $s = (1/T) \ln z$, where T is the sampling interval. It is tempting, therefore, to convert a continuous-time filter to a discrete-time filter by substituting $s = (1/T) \ln z$ into $H(s)$, or $H[z] = H(s)|_{s=(1/T)\ln z}$. Unfortunately, this approach is impractical since the resulting $H[z]$ is not rational and therefore cannot be implemented by using standard blocks. Although not considered here, the so-called matched-z transformation relies on the relationship $z = e^{sT}$ to transform system poles and zeros, so the connection is not completely without merit.

M5.3 Transformation by First-Order Backward Difference

Consider the transfer function $H(s) = Y(s)/X(s) = s$, which corresponds to the first-order continuous-time differentiator

$$y(t) = \frac{d}{dt}x(t)$$

An approximation that resembles the fundamental theorem of calculus is the first-order backward difference

$$y(t) = \frac{x(t) - x(t - T)}{T}$$

For sampling interval T and $t = nT$, the corresponding discrete-time approximation is

$$y[n] = \frac{x[n] - x[n - 1]}{T}$$

which has transfer function

$$H[z] = Y[z]/X[z] = \frac{1 - z^{-1}}{T}$$

This implies a transformation rule that uses the change of variable $s = (1 - z^{-1})/T$ or $z = 1/(1 - sT)$. This transformation rule is appealing since the resulting $H[z]$ is rational and has the same number of poles and zeros as $H(s)$. Section 3.4 discusses this transformation strategy in a different way in describing the kinship of difference equations to differential equations.

After some algebra, substituting $s = (1 - z^{-1})/T$ into Eq. (M5.3) yields

$$H[z] = \left(\frac{b_{N-M} \prod_{k=1}^{M}(1/T - z_k)}{a_0 \prod_{k=1}^{N}(1/T - p_k)} \right) \frac{\prod_{k=1}^{M}\left(1 - \frac{1}{1 - Tz_k}z^{-1}\right)}{\prod_{k=1}^{N}\left(1 - \frac{1}{1 - Tp_k}z^{-1}\right)} \qquad (M5.4)$$

The discrete-time system has M zeros at $1/(1 - Tz_k)$ and N poles at $1/(1 - Tp_k)$. This transformation rule preserves system stability but does not map the ω axis to the unit circle (see Prob. 5.7-9).

MATLAB program MS5P3 uses the first-order backward difference method of Eq. (M5.4) to convert a continuous-time filter described by coefficient vectors $\mathbf{A} = [a_0, a_1, \ldots, a_N]$ and $\mathbf{B} = [b_{N-M}, b_{N-M+1}, \ldots, b_N]$ into a discrete-time filter. The form of the discrete-time filter follows Eq. (M5.2).

```
function [Bd,Ad] = MS5P3(B,A,T);
% MS5P3.m : MATLAB Session 5, Program 3
% Function M-file first-order backward difference transformation
% of a continuous-time filter described by B and A into a discrete-time filter.
% INPUTS:   B = vector of continuous-time filter feedforward coefficients
%           A = vector of continuous-time filter feedback coefficients
%           T = sampling interval
% OUTPUTS:  Bd = vector of discrete-time filter feedforward coefficients
%           Ad = vector of discrete-time filter feedback coefficients

z = roots(B); p = roots(A); % s-domain roots
gain = B(1)/A(1)*prod(1/T-z)/prod(1/T-p);
zd = 1./(1-T*z); pd = 1./(1-T*p); % z-domain roots
Bd = gain*poly(zd); Ad = poly(pd);
```

M5.4 Bilinear Transformation

The bilinear transformation is based on a better approximation than first-order backward differences. Again, consider the continuous-time integrator

$$y(t) = \frac{d}{dt}x(t)$$

Represent signal $x(t)$ as

$$x(t) = \int_{t-T}^{t} \frac{d}{d\tau}x(\tau)\,d\tau + x(t-T)$$

Letting $t = nT$ and replacing the integral with a trapezoidal approximation yields

$$x(nT) = \frac{T}{2}\left[\frac{d}{dt}x(nT) + \frac{d}{dt}x(nT-T)\right] + x(nT-T)$$

Substituting $y(t)$ for $(d/dt)x(t)$, the equivalent discrete-time system is

$$x[n] = \frac{T}{2}(y[n] + y[n-1]) + x[n-1]$$

From z-transforms, the transfer function is

$$H[z] = \frac{Y[z]}{X[z]} = \frac{2(1-z^{-1})}{T(1+z^{-1})}$$

The implied change of variable $s = 2(1-z^{-1})/T(1+z^{-1})$ or $z = (1+sT/2)/(1-sT/2)$ is called the bilinear transformation. Not only does the bilinear transformation result in a rational function $H[z]$, the ω-axis is correctly mapped to the unit circle (see Prob. 5.6-11a).

After some algebra, substituting $s = 2(1-z^{-1})/T(1+z^{-1})$ into Eq. (M5.3) yields

$$H[z] = \left(\frac{b_{N-M}\prod_{k=1}^{M}(2/T - z_k)}{a_0\prod_{k=1}^{N}(2/T - p_k)}\right)\frac{\prod_{k=1}^{M}\left(1 - \frac{1+z_kT/2}{1-z_kT/2}z^{-1}\right)}{\prod_{k=1}^{N}\left(1 - \frac{1+p_kT/2}{1-p_kT/2}z^{-1}\right)}(1+z^{-1})^{N-M} \tag{M5.5}$$

In addition to the M zeros at $(1 + z_kT/2)/(1 - z_kT/2)$ and N poles at $(1 + p_kT/2)/(1 - p_kT/2)$, there are $N - M$ zeros at minus one. Since practical continuous-time filters require $M \leq N$ for stability, the number of added zeros is thankfully always nonnegative.

MATLAB program MS5P4 converts a continuous-time filter described by coefficient vectors $\mathbf{A} = [a_0, a_1, \ldots, a_N]$ and $\mathbf{B} = [b_{N-M}, b_{N-M+1}, \ldots, b_N]$ into a discrete-time filter by using the bilinear transformation of Eq. (M5.5). The form of the discrete-time filter follows Eq. (M5.2). If available, it is also possible use the signal processing toolbox function bilinear to perform the bilinear transformation.

```
function [Bd,Ad] = MS5P4(B,A,T);
% MS5P4.m : MATLAB Session 5, Program 4
% Function M-file bilinear transformation of a continuous-time filter
% described by vectors B and A into a discrete-time filter.
% Length of B must not exceed A.
% INPUTS:    B = vector of continuous-time filter feedforward coefficients
%            A = vector of continuous-time filter feedback coefficients
%            T = sampling interval
% OUTPUTS:   Bd = vector of discrete-time filter feedforward coefficients
%            Ad = vector of discrete-time filter feedback coefficients

if (length(B)>length(A)),
    disp('Numerator order must not exceed denominator order.');
    return
end
z = roots(B); p = roots(A); % s-domain roots
gain = real(B(1)/A(1)*prod(2/T-z)/prod(2/T-p));
zd = (1+z*T/2)./(1-z*T/2); pd = (1+p*T/2)./(1-p*T/2); % z-domain roots
Bd = gain*poly([zd;-ones(length(A)-length(B),1)]); Ad = poly(pd);
```

As with most high-level languages, MATLAB supports general if-structures:

```
if expression,
        statements;
elseif expression,
        statements;
else,
        statements;
end
```

In this program, the `if` statement tests $M > N$. When true, an error message is displayed and the `return` command terminates program execution to prevent errors.

M5.5 Bilinear Transformation with Prewarping

The bilinear transformation maps the entire infinite-length ω axis onto the finite-length unit circle ($z = e^{j\Omega}$) according to $\omega = (2/T)\tan(\Omega/2)$ (see Prob. 5.6-11b). Equivalently, $\Omega = 2\arctan(\omega T/2)$. The nonlinearity of the tangent function causes a frequency compression, commonly called frequency warping, that distorts the transformation.

To illustrate the warping effect, consider the bilinear transformation of a continuous-time lowpass filter with cutoff frequency $\omega_c = 2\pi 3000$ rad/s. If the target digital system uses a sampling rate of 10 kHz, then $T = 1/(10,000)$ and ω_c maps to $\Omega_c = 2\arctan(\omega_c T/2) = 1.5116$. Thus, the transformed cutoff frequency is short of the desired $\Omega_c = \omega_c T = 0.6\pi = 1.8850$.

Cutoff frequencies are important and need to be as accurate as possible. By adjusting the parameter T used in the bilinear transform, one continuous-time frequency can be exactly mapped to one discrete-time frequency; the process is called prewarping. Continuing the last example, adjusting $T = (2/\omega_c) \tan (\Omega_c/2) \approx 1/6848$ achieves the appropriate prewarping to ensure $\omega_c = 2\pi 3000$ maps to $\Omega_c = 0.6\pi$.

M5.6 Example: Butterworth Filter Transformation

To illustrate the transformation techniques, consider a continuous-time 10th-order Butterworth lowpass filter with cutoff frequency $\omega_c = 2\pi 3000$, as designed in MATLAB Session 4. First, we determine continuous-time coefficient vectors **A** and **B**.

```
>> omega_c = 2*pi*3000; N=10;
>> poles = roots([(j*omega_c)^(-2*N),zeros(1,2*N-1),1]);
>> poles = poles(find(poles<0));
>> B = 1; A = poly(poles); A = A/A(end);
```

Programs `MS5P3` and `MS5P4` are used to perform first-order forward difference and bilinear transformations, respectively.

```
>> Omega = linspace(0,pi,200); T = 1/10000; Omega_c = omega_c*T;
>> [B1,A1] = MS5P3(B,A,T); % First-order backward difference transformation
>> [B2,A2] = MS5P4(B,A,T); % Bilinear transformation
>> [B3,A3] = MS5P4(B,A,2/omega_c*tan(Omega_c/2)); % Bilinear with prewarping
```

Magnitude responses are computed by using `MS5P1` and then plotted.

```
>> H1mag = abs(MS5P1(B1,A1,Omega));
>> H2mag = abs(MS5P1(B2,A2,Omega));
>> H3mag = abs(MS5P1(B3,A3,Omega));
>> plot(Omega,(Omega<=Omega_c),'k',Omega,H1mag,'k-.',...
        Omega,H2mag,'k--',Omega,H3mag,'k:');
>> axis([0 pi -.05 1.5]);
>> xlabel('\Omega [rad]'); ylabel('Magnitude Response');
>> legend('Ideal','First-Order Backward Difference',...
        'Bilinear','Bilinear with Prewarping');
```

The result of each transformation method is shown in Fig. M5.1.

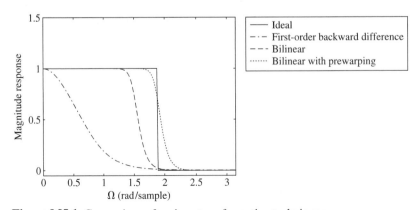

Figure M5.1 Comparison of various transformation techniques.

Although the first-order backward difference results in a lowpass filter, the method causes significant distortion that makes the resulting filter unacceptable with regard to cutoff frequency. The bilinear transformation is better, but, as predicted, the cutoff frequency falls short of the desired value. Bilinear transformation with prewarping properly locates the cutoff frequency and produces a very acceptable filter response.

M5.7 Problems Finding Polynomial Roots

Numerically, it is difficult to accurately determine the roots of a polynomial. Consider, for example, a simple polynomial that has a root at minus one repeated four times, $(s + 1)^4 = s^4 + 4s^3 + 6s^2 + 4s + 1$. The MATLAB `roots` command returns a surprising result:

```
>> roots([1 4 6 4 1])'
ans = -1.0002          -1.0000 - 0.0002i  -1.0000 + 0.0002i  -0.9998
```

Even for this low-degree polynomial, MATLAB does not return the true roots.

The problem worsens as polynomial degree increases. The bilinear transformation of the 10th-order Butterworth filter, for example, should have 10 zeros at minus one. Figure M5.2 shows that the zeros, computed by `MS5P2` with the `roots` command, are not correctly located.

When possible, programs should avoid root computations that may limit accuracy. For example, results from the transformation programs `MS5P3` and `MS5P4` are more accurate if the true transfer function poles and zeros are passed directly as inputs rather than the polynomial coefficient vectors. When roots must be computed, result accuracy should always be verified.

M5.8 Using Cascaded Second-Order Sections to Improve Design

The dynamic range of high-degree polynomial coefficients is often large. Adding the difficulties associated with factoring a high-degree polynomial, it is little surprise that high-order designs are difficult.

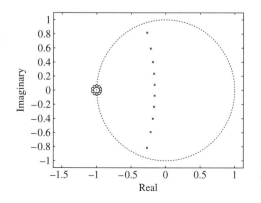

Figure M5.2 Pole–zero plot computed by using `roots`.

As with continuous-time filters, performance is improved by using a cascade of second-order sections to design and realize a discrete-time filter. Cascades of second-order sections are also more robust to the coefficient quantization that occurs when discrete-time filters are implemented on fixed-point digital hardware.

To illustrate the performance possible with a cascade of second-order sections, consider a 180th-order transformed Butterworth discrete-time filter with cutoff frequency $\Omega_c = 0.6\pi \approx 1.8850$. Program MS5P5 completes this design, taking care to initially locate poles and zeros without root computations.

```
% MS5P5.m : MATLAB Session 5, Program 5
% Script M-file designs a 180th-order Butterworth lowpass discrete-time filter
% with cutoff Omega_c = 0.6*pi using 90 cascaded second-order filter sections.

omega_0 = 1; % Use normalized cutoff frequency for analog prototype
psi = [0.5:1:90]*pi/180; % Butterworth pole angles
Omega_c = 0.6*pi; % Discrete-time cutoff frequency
Omega = linspace(0,pi,1000); % Frequency range for magnitude response
Hmag = zeros(90,1000); p = zeros(1,180); z = zeros(1,180); % Pre-allocation
for stage = 1:90,
    Q = 1/(2*cos(psi(stage))); % Compute Q for stage
    B = omega_0^2; A = [1 omega_0/Q omega_0^2]; % Compute stage coefficients
    [B1,A1] = MS5P4(B,A,2/omega_0*tan(0.6*pi/2)); % Transform stage to DT
    p(stage*2-1:stage*2) = roots(A1); % Compute z-domain poles for stage
    z(stage*2-1:stage*2) = roots(B1); % Compute z-domain zeros for stage
    Hmag(stage,:) = abs(MS5P1(B1,A1,Omega)); % Compute stage mag response
end
ucirc = exp(j*linspace(0,2*pi,200)); % Compute unit circle for pole-zero plot
figure;
plot(real(p),imag(p),'kx',real(z),imag(z),'ok',real(ucirc),imag(ucirc),'k:');
axis equal; xlabel('Real'); ylabel('Imaginary');
figure; plot(Omega,Hmag,'k'); axis tight
xlabel('\Omega [rad]'); ylabel('Magnitude Response');
figure; plot(Omega,prod(Hmag),'k'); axis([0 pi -0.05 1.05]);
xlabel('\Omega [rad]'); ylabel('Magnitude Response');
```

The figure command preceding each plot command opens a separate window for each plot.

The filter's pole–zero plot is shown in Fig. M5.3, along with the unit circle, for reference. All 180 zeros of the cascaded design are properly located at minus one. The wall of poles provides an amazing approximation to the desired brick-wall response, as shown by the magnitude response in Fig. M5.4. It is virtually impossible to realize such high-order designs with continuous-time filters, which adds another reason for the popularity of discrete-time filters. Still, the design is not trivial; even functions from the MATLAB signal processing toolbox fail to properly design such a high-order discrete-time Butterworth filter!

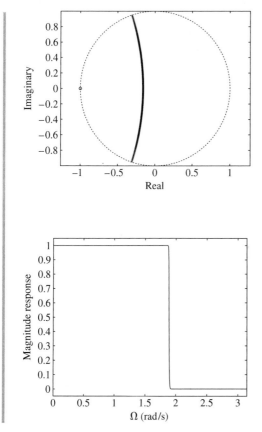

Figure M5.3 Pole–zero plot for 180th-order discrete-time Butterworth filter.

Figure M5.4 Magnitude response for a 180th-order discrete-time Butterworth filter.

PROBLEMS

5.1-1 Using the definition, compute the z-transform of $x[n] = (-1)^n (u[n] - u[n-8])$. Sketch the poles and zeros of $X(z)$ in the z plane. No calculator is needed to do this problem!

5.1-2 Using the definition of the z-transform, find the z-transform and the ROC for each of the following signals.

(a) $u[n-m]$

(b) $\gamma^n \sin \pi n \, u[n]$

(c) $\gamma^n \cos \pi n \, u[n]$

(d) $\gamma^n \sin \dfrac{\pi n}{2} \, u[n]$

(e) $\gamma^n \cos \dfrac{\pi n}{2} \, u[n]$

(f) $\displaystyle\sum_{k=0}^{\infty} 2^{2k} \, \delta[n-2k]$

(g) $\gamma^{n-1} u[n-1]$

(h) $n \gamma^n \, u[n]$

(i) $n \, u[n]$

(j) $\dfrac{\gamma^n}{n!} u[n]$

(k) $[2^{n-1} - (-2)^{n-1}] u[n]$

(l) $\dfrac{(\ln \alpha)^n}{n!} u[n]$

5.1-3 Showing all work, evaluate $\sum_{n=0}^{\infty} n(-3/2)^{-n}$.

5.1-4   Using only the z-transforms of Table 5.1, determine the z-transform of each of the following signals.

(a) $u[n] - u[n-2]$

(b) $\gamma^{n-2}u[n-2]$

(c) $2^{n+1}u[n-1] + e^{n-1}u[n]$

(d) $\left[2^{-n}\cos\left(\dfrac{\pi}{3}n\right)\right]u[n-1]$

(e) $n\gamma^n u[n-1]$

(f) $n(n-1)(n-2)2^{n-3}u[n-m]$
    for $m = 0, 1, 2, 3$

(g) $(-1)^n nu[n]$

(h) $\displaystyle\sum_{k=0}^{\infty} k\delta(n-2k+1)$

5.1-5   Find the inverse z-transform of the following:

(a) $\dfrac{z(z-4)}{z^2 - 5z + 6}$

(b) $\dfrac{z-4}{z^2 - 5z + 6}$

(c) $\dfrac{(e^{-2} - 2)z}{(z - e^{-2})(z - 2)}$

(d) $\dfrac{(z-1)^2}{z^3}$

(e) $\dfrac{z(2z+3)}{(z-1)(z^2 - 5z + 6)}$

(f) $\dfrac{z(-5z+22)}{(z+1)(z-2)^2}$

(g) $\dfrac{z(1.4z + 0.08)}{(z - 0.2)(z - 0.8)^2}$

(h) $\dfrac{z(z-2)}{z^2 - z + 1}$

(i) $\dfrac{2z^2 - 0.3z + 0.25}{z^2 + 0.6z + 0.25}$

(j) $\dfrac{2z(3z - 23)}{(z-1)(z^2 - 6z + 25)}$

(k) $\dfrac{z(3.83z + 11.34)}{(z-2)(z^2 - 5z + 25)}$

(l) $\dfrac{z^2(-2z^2 + 8z - 7)}{(z-1)(z-2)^3}$

5.1-6   (a) Expanding $X[z]$ as a power series in z^{-1}, find the first three terms of $x[n]$ if

$$X[z] = \frac{2z^3 + 13z^2 + z}{z^3 + 7z^2 + 2z + 1}$$

(b) Extend the procedure used in part (a) to find the first four terms of $x[n]$ if

$$X[z] = \frac{2z^4 + 16z^3 + 17z^2 + 3z}{z^3 + 7z^2 + 2z + 1}$$

5.1-7   Find $x[n]$ by expanding

$$X[z] = \frac{\gamma z}{(z-\gamma)^2}$$

as a power series in z^{-1}.

5.1-8   (a) In Table 5.1, if the numerator and the denominator powers of $X[z]$ are M and N, respectively, explain why in some cases $N - M = 0$, while in others $N - M = 1$ or $N - M = m$ (m any positive integer).

(b) Without actually finding the z-transform, state what is $N - M$ for $X[z]$ corresponding to $x[n] = \gamma^n u[n-4]$

5.2-1   For a discrete-time signal shown in Fig. P5.2-1 show that

$$X[z] = \frac{1 - z^{-m}}{1 - z^{-1}}$$

Find your answer by using the definition in Eq. (5.1) and by using Table 5.1 and an appropriate property of the z-transform.

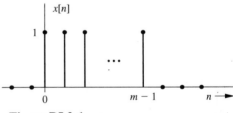

Figure P5.2-1

5.2-2   Find the z-transform of the signal illustrated in Fig. P5.2-2. Solve this problem in two ways, as in Examples 5.2d and 5.4. Verify that the two answers are equivalent.

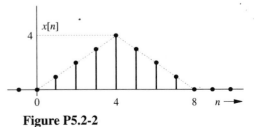

Figure P5.2-2

5.2-3 Using only the fact that $\gamma^n u[n] \Longleftrightarrow z/(z - \gamma)$ and properties of the z-transform, find the z-transform of each of the following:

(a) $n^2 u[n]$

(b) $n^2 \gamma^n u[n]$

(c) $n^3 u[n]$

(d) $a^n [u[n] - u[n - m]]$

(e) $ne^{-2n} u[n - m]$

(f) $(n - 2)(0.5)^{n-3} u[n - 4]$

5.2-4 Using only pair 1 in Table 5.1 and appropriate properties of the z-transform, derive iteratively pairs 2 through 9. In other words, first derive pair 2. Then use pair 2 (and pair 1, if needed) to derive pair 3, and so on.

5.2-5 Find the z-transform of $\cos(\pi n/4) u[n]$ using only pairs 1 and 11b in Table 5.1 and a suitable property of the z-transform.

5.2-6 Apply the time-reversal property to pair 6 of Table 5.1 to show that $\gamma^n u[-(n + 1)] \Longleftrightarrow -z/(z - \gamma)$ and the ROC is given by $|z| < |\gamma|$.

5.2-7 (a) If $x[n] \Longleftrightarrow X[z]$, then show that $(-1)^n x[n] \Longleftrightarrow X[-z]$.

(b) Use this result to show that $(-\gamma)^n u[n] \Longleftrightarrow z/(z + \gamma)$.

(c) Use these results to find the z-transform of

(i) $[2^{n-1} - (-2)^{n-1}] u[n]$

(ii) $\gamma^n \cos \pi n \, u[n]$

5.2-8 (a) If $x[n] \Longleftrightarrow X[z]$, then show that

$$\sum_{k=0}^{n} x[k] \Longleftrightarrow \frac{z X[z]}{z - 1}$$

(b) Use this result to derive pair 2 from pair 1 in Table 5.1.

5.2-9 A number of causal time-domain functions are shown in Fig. P5.2-9.

List the function of time that corresponds to each of the following functions of z. Few or no calculations are necessary! Be careful, the graphs may be scaled differently.

(a) $\dfrac{z^2}{(z - 0.75)^2}$

(b) $\dfrac{z^2 - 0.9z/\sqrt{2}}{z^2 - 0.9\sqrt{2}z + 0.81}$

(c) $\displaystyle\sum_{k=0}^{4} z^{-2k}$

(d) $\dfrac{z^{-5}}{1 - z^{-1}}$

(e) $\dfrac{z^2}{z^4 - 1}$

(f) $\dfrac{0.75z}{(z - 0.75)^2}$

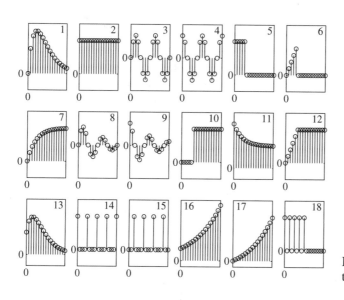

Figure P5.2-9 Various causal time-domain functions.

(g) $\dfrac{z^2 - z/\sqrt{2}}{z^2 - \sqrt{2}z + 1}$

(h) $\dfrac{z^{-1} - 5z^{-5} + 4z^{-6}}{5(1 - z^{-1})^2}$

(i) $\dfrac{z}{z - 1.1}$

(j) $\dfrac{0.25z^{-1}}{(1 - z^{-1})(1 - 0.75z^{-1})}$

5.3-1 Solve Prob. 3.8-16 by the z-transform method.

5.3-2 (a) Solve

$$y[n + 1] + 2y[n] = x[n + 1]$$

when $y[0] = 1$ and $x[n] = e^{-(n-1)}u[n]$
   (b) Find the zero-input and the zero-state components of the response.

5.3-3 (a) Find the output $y[n]$ of an LTID system specified by the equation

$$2y[n + 2] - 3y[n + 1] + y[n]$$
$$= 4x[n + 2] - 3x[n + 1]$$

if the initial conditions are $y[-1]=0$, $y[-2] = 1$, and the input $x[n] = (4)^{-n}u[n]$.
   (b) Find the zero-input and the zero-state components of the response.
   (c) Find the transient and the steady-state components of the response.

5.3-4 Solve Prob. 5.3-3 if instead of initial conditions $y[-1]$, $y[-2]$ you are given the auxiliary conditions $y[0] = 3/2$ and $y[1] = 35/4$.

5.3-5 (a) Solve

$$4y[n + 2] + 4y[n + 1] + y[n] = x[n + 1]$$

with $y[-1]=0$, $y[-2]=1$, and $x[n]=u[n]$.
   (b) Find the zero-input and the zero-state components of the response.
   (c) Find the transient and the steady-state components of the response.

5.3-6 Solve

$$y[n + 2] - 3y[n + 1] + 2y[n] = x[n + 1]$$

if $y[-1]=2$, $y[-2]=3$, and $x[n] = (3)^n u[n]$.

5.3-7 Solve

$$y[n + 2] - 2y[n + 1] + 2y[n] = x[n]$$

with $y[-1] = 1$, $y[-2] = 0$, and $x[n] = u[n]$.

5.3-8 Solve

$$y[n] + 2y[n - 1] + 2y[n - 2]$$
$$= x[n - 1] + 2x[n - 2]$$

with $y[0] = 0$, $y[1] = 1$, and $x[n] = e^n u[n]$.

5.3-9 A system with impulse response $h[n] = 2(1/3)^n u[n - 1]$ produces an output $y[n] = (-2)^n u[n - 1]$. Determine the corresponding input $x[n]$.

5.3-10 Sally deposits $100 into her savings account on the first day of every month except for each December, when she uses her money to buy holiday gifts. Define $b[m]$ as the balance in Sally's account on the first day of month m. Assume Sally opens her account in January ($m = 0$), continues making monthly payments forever (except each December!), and that her monthly interest rate is 1%. Sally's account balance satisfies a simple difference equation $b[m] = (1.01)b[m - 1] + p[m]$, where $p[m]$ designates Sally's monthly deposits. Determine a closed-form expression for $b[m]$ that is only a function of the month m.

5.3-11 For each impulse response, determine the number of system poles, whether the poles are real or complex, and whether the system is BIBO stable.
   (a) $h_1[n] = (-1 + (0.5)^n)u[n]$
   (b) $h_2[n] = (j)^n (u[n] - u[n - 10])$

5.3-12 Find the following sums:

(i) $\displaystyle\sum_{k=0}^{n} k$ (ii) $\displaystyle\sum_{k=0}^{n} k^2$

[Hint: Consider a system whose output $y[n]$ is the desired sum. Examine the relationship between $y[n]$ and $y[n - 1]$. Note also that $y[0] = 0$.]

5.3-13 Find the following sum:

$$\sum_{k=0}^{n} k^3$$

[Hint: See the hint for Prob. 5.3-12.]

5.3-14 Find the following sum:

$$\sum_{k=0}^{n} ka^k \qquad a \neq 1$$

[Hint: See the hint for Prob. 5.3-12.]

5.3-15 Redo Prob. 5.3-12 using the result in Prob. 5.2-8a.

5.3-16 Redo Prob. 5.3-13 using the result in Prob. 5.2-8a.

5.3-17 Redo Prob. 5.3-14 using the result in Prob. 5.2-8a.

5.3-18 (a) Find the zero-state response of an LTID system with transfer function

$$H[z] = \frac{z}{(z + 0.2)(z - 0.8)}$$

and the input $x[n] = e^{(n+1)}u[n]$.

(b) Write the difference equation relating the output $y[n]$ to input $x[n]$.

5.3-19 Repeat Prob. 5.3-18 for $x[n] = u[n]$ and

$$H[z] = \frac{2z + 3}{(z - 2)(z - 3)}$$

5.3-20 Repeat Prob. 5.3-18 for

$$H[z] = \frac{6(5z - 1)}{6z^2 - 5z + 1}$$

and the input $x[n]$ is

(a) $(4)^{-n}u[n]$
(b) $(4)^{-(n-2)}u[n - 2]$
(c) $(4)^{-(n-2)}u[n]$
(d) $(4)^{-n}u[n - 2]$

5.3-21 Repeat Prob. 5.3-18 for $x[n] = u[n]$ and

$$H[z] = \frac{2z - 1}{z^2 - 1.6z + 0.8}$$

5.3-22 Find the transfer functions corresponding to each of the systems specified by difference equations in Probs. 5.3-2, 5.3-3, 5.3-5, and 5.3-8.

5.3-23 Find $h[n]$, the unit impulse response of the systems described by the following equations:

(a) $y[n] + 3y[n - 1] + 2y[n - 2] = x[n] + 3x[n - 1] + 3x[n - 2]$
(b) $y[n + 2] + 2y[n + 1] + y[n] = 2x[n + 2] - x[n + 1]$
(c) $y[n] - y[n - 1] + 0.5y[n - 2] = x[n] + 2x[n - 1]$

5.3-24 Find $h[n]$, the unit impulse response of the systems in Probs. 5.3-18, 5.3-19, and 5.3-21.

5.3-25 A system has impulse response $h[n] = u[n - 3]$.

(a) Determine the impulse response of the inverse system $h^{-1}[n]$.

(b) Is the inverse stable? Is the inverse causal?
(c) Your boss asks you to implement $h^{-1}[n]$ to the best of your ability. Describe your realizable design, taking care to identify any deficiencies.

5.4-1 A system has impulse response given by

$$h[n] = \left[\left(\frac{1 + j}{\sqrt{8}}\right)^n + \left(\frac{1 - j}{\sqrt{8}}\right)^n \right] u[n]$$

This system can be implemented according to Fig. P5.4-1.

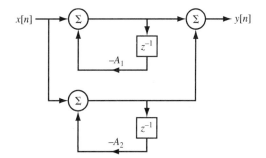

Figure P5.4-1 Structure to implement $h[n]$.

(a) Determine the coefficients A_1 and A_2 to implement $h[n]$ using the structure shown in Fig. P5.4-1.

(b) What is the zero-state response $y_0[n]$ of this system, given a shifted unit step input $x[n] = u[n + 3]$?

5.4-2 (a) Show the canonic direct form, a cascade and, a parallel realization of

$$H[z] = \frac{z(3z - 1.8)}{z^2 - z + 0.16}$$

(b) Find the transpose of the realizations obtained in part (a).

5.4-3 Repeat Prob. 5.4-2 for

$$H[z] = \frac{5z + 2.2}{z^2 + z + 0.16}$$

5.4-4 Repeat Prob. 5.4-2 for

$$H[z] = \frac{3.8z - 1.1}{(z - 0.2)(z^2 - 0.6z + 0.25)}$$

5.4-5 Repeat Prob. 5.4-2 for

$$H[z] = \frac{z(1.6z - 1.8)}{(z - 0.2)(z^2 + z + 0.5)}$$

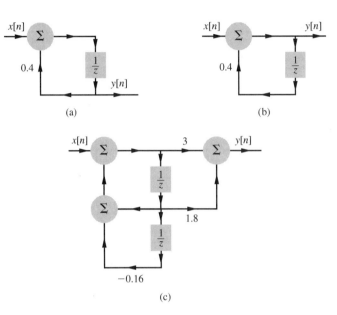

(a)   (b)

(c)

Figure P5.5-1

5.4-6 Repeat Prob. 5.4-2 for

$$H[z] = \frac{z(2z^2 + 1.3z + 0.96)}{(z + 0.5)(z - 0.4)^2}$$

5.4-7 Realize a system whose transfer function is

$$H[z] = \frac{2z^4 + z^3 + 0.8z^2 + 2z + 8}{z^4}$$

5.4-8 Realize a system whose transfer function is given by

$$H[z] = \sum_{n=0}^{6} nz^{-n}$$

5.4-9 This problem demonstrates the enormous number of ways of implementing even a relatively low-order transfer function. A second-order transfer function has two real zeros and two real poles. Discuss various ways of realizing such a transfer function. Consider canonic direct, cascade, parallel, and the corresponding transposed forms. Note also that interchange of cascaded sections yields a different realization.

5.5-1 Find the amplitude and phase response of the digital filters depicted in Fig. P5.5-1.

5.5-2 Find the amplitude and the phase response of the filters shown in Fig. P5.5-2. [Hint: Express $H[e^{j\Omega}]$ as $e^{-j2.5\Omega} H_a[e^{j\Omega}]$.]

5.5-3 Find the frequency response for the moving-average system in Prob. 3.4-3. The input–output equation of this system is given by

$$y[n] = \frac{1}{5} \sum_{k=0}^{4} x[n - k]$$

5.5-4 (a) Input–output relationships of two filters are described by
 (i) $y[n] = -0.9y[n - 1] + x[n]$
 (ii) $y[n] = 0.9y[n - 1] + x[n]$
 For each case, find the transfer function, the amplitude response, and the phase response. Sketch the amplitude response, and state the type (highpass, lowpass, etc.) of each filter.

 (b) Find the response of each of these filters to a sinusoid $x[n] = \cos \Omega n$ for $\Omega = 0.01\pi$ and 0.99π. In general show that the gain (amplitude response) of the filter (i) at frequency Ω_0 is the same as the gain of the filter (ii) at frequency $\pi - \Omega_0$

5.5-5 For an LTID system specified by the equation

$$y[n + 1] - 0.5y[n] = x[n + 1] + 0.8x[n]$$

 (a) Find the amplitude and the phase response.
 (b) Find the system response $y[n]$ for the input $x[n] = \cos (0.5k - (\pi/3))$.

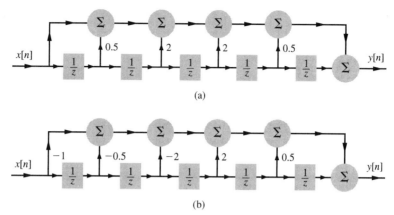

(a)

(b)

Figure P5.5-2

5.5-6 For an asymptotically stable LTID system, show that the steady-state response to input $e^{j\Omega n}u[n]$ is $H[e^{j\Omega}]e^{j\Omega n}u[n]$. The steady-state response is that part of the response which does not decay with time and persists forever.

5.5-7 Express the following signals in terms of apparent frequencies.
(a) $\cos(0.8\pi n + \theta)$
(b) $\sin(1.2\pi n + \theta)$
(c) $\cos(6.9n + \theta)$
(d) $\cos(2.8\pi n + \theta) + 2\sin(3.7\pi n + \theta)$
(e) $\text{sinc}(\pi n/2)$
(f) $\text{sinc}(3\pi n/2)$
(g) $\text{sinc}(2\pi n)$

5.5-8 Show that $\cos(0.6\pi n + (\pi/6)) + \sqrt{3}\cos(1.4\pi n + (\pi/3)) = 2\cos(0.6\pi n - (\pi/6))$.

5.5-9 (a) A digital filter has the sampling interval $T = 50\,\mu s$. Determine the highest frequency that can be processed by this filter without aliasing.

(b) If the highest frequency to be processed is 50 kHz, determine the minimum value of the sampling frequency $\mathcal{F}_s$ and the maximum value of the sampling interval T that can be used.

5.5-10 Consider the discrete-time system represented by

$$y[n] = \sum_{k=0}^{\infty} (0.5)^k x[n-k]$$

(a) Determine and plot the magnitude response $|H(e^{j\Omega})|$ of the system.
(b) Determine and plot the phase response $\angle H(e^{j\Omega})$ of the system.
(c) Find an efficient block representation that implements this system.

5.6-1 Pole–zero configurations of certain filters are shown in Fig. P5.6-1. Sketch roughly the amplitude response of these filters.

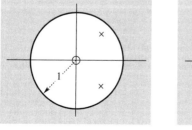

(a)

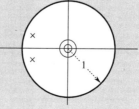

(b)

Figure P5.6-1

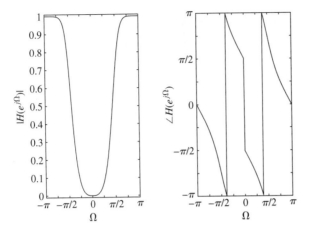

Figure P5.6-3 Frequency response of a real, stable, LTI system.

5.6-2 The system $y[n] - y[n-1] = x[n] - x[n-1]$ is an all-pass system that has zero phase response. Is there any difference between this system and the system $y[n] = x[n]$? Justify your answer.

5.6-3 The magnitude and phase responses of a real, stable, LTI system are shown in Fig. P5.6-3.
  (a) What type of system is this: lowpass, highpass, bandpass, or bandstop?
  (b) What is the output of this system in response to

$$x_1[n] = 2 \sin\left(\frac{\pi}{2}n + \frac{\pi}{4}\right)$$

  (c) What is the output of this system in response to

$$x_2[n] = \cos\left(\frac{7\pi}{4}n\right)$$

5.6-4 Do Prob. 5.M-1 by graphical procedure. Do the sketches approximately, without using MATLAB.

5.6-5 Do Prob. 5.M-4 by graphical procedure. Do the sketches approximately, without using MATLAB.

5.6-6 (a) Realize a digital filter whose transfer function is given by

$$H[z] = K \frac{z+1}{z-a}$$

  (b) Sketch the amplitude response of this filter, assuming $|a| < 1$.

  (c) The amplitude response of this lowpass filter is maximum at $\Omega = 0$. The 3 dB bandwidth is the frequency at which the amplitude response drops to 0.707 (or $1/\sqrt{2}$) times its maximum value. Determine the 3 dB bandwidth of this filter when $a = 0.2$.

5.6-7 Design a digital notch filter to reject frequency 5000 Hz completely and to have a sharp recovery on either side of 5000 Hz to a gain of unity. The highest frequency to be processed is 20 kHz ($\mathcal{F}_h = 20,000$). [Hint: See Example 5.15. The zeros should be at $e^{\pm j\omega T}$ for ω corresponding to 5000 Hz, and the poles are at $ae^{\pm j\omega T}$ with $a < 1$. Leave your answer in terms of a. Realize this filter using the canonical form. Find the amplitude response of the filter.]

5.6-8 Show that a first-order LTID system with a pole at $z = r$ and a zero at $z = 1/r$ ($r \le 1$) is an allpass filter. In other words, show that the amplitude response $|H[e^{j\Omega}]|$ of a system with the transfer function

$$H[z] = \frac{z - \dfrac{1}{r}}{z - r} \qquad r \le 1$$

is constant with frequency. This is a first-order allpass filter. [Hint: Show that the ratio of the distances of any point on the unit circle from the zero (at $z = 1/r$) and the pole (at $z = r$) is a constant $1/r$.]

Generalize this result to show that an LTID system with two poles at $z = re^{\pm j\theta}$ and two zeros at $z = (1/r)e^{\pm j\theta}$ $(r \leq 1)$ is an allpass filter. In other words, show that the amplitude response of a system with the transfer function

$$H[z] = \frac{\left(z - \frac{1}{r}e^{j\theta}\right)\left(z - \frac{1}{r}e^{-j\theta}\right)}{(z - re^{j\theta})(z - re^{-j\theta})}$$

$$= \frac{z^2 - \left(\frac{2}{r}\cos\theta\right)z + \frac{1}{r^2}}{z^2 - (2r\cos\theta)z + r^2} \qquad r \leq 1$$

is constant with frequency.

5.6-9 (a) If $h_1[n]$ and $h_2[n]$, the impulse responses of two LTID systems are related by $h_2[n] = (-1)^n h_1[n]$, then show that

$$H_2[e^{j\Omega}] = H_1[e^{j(\Omega \pm \pi)}]$$

How are the frequency response spectrum $H_2[e^{j\Omega}]$ related to the $H_1[e^{j\Omega}]$.

(b) If $H_1[z]$ represents an ideal lowpass filter with cutoff frequency Ω_c, sketch $H_2[e^{j\Omega}]$. What type of filter is $H_2[e^{j\Omega}]$?

5.6-10 Mappings such as the bilinear transformation are useful in the conversion of continuous-time filters to discrete-time filters. Another useful type of transformation is one that converts a discrete-time filter into a different type of discrete-time filter. Consider a transformation that replaces z with $-z$.

(a) Show that this transformation converts lowpass filters into highpass filters and highpass filters into lowpass filters.

(b) If the original filter is an FIR filter with impulse response $h[n]$, what is the impulse response of the transformed filter?

5.6-11 The bilinear transformation is defined by the rule $s = 2(1 - z^{-1})/T(1 + z^{-1})$.

(a) Show that this transformation maps the ω axis in the s plane to the unit circle $z = e^{j\Omega}$ in the z plane.

(b) Show that this transformation maps Ω to $2\arctan(\omega T/2)$.

5.7-1 In Chapter 3, we used another approximation to find a digital system to realize an analog system. We showed that an analog system specified by Eq. (3.15a) can be realized by using the digital system specified by Eq. (3.15c). Compare that solution with the one resulting from the impulse invariance method. Show that one result is a close approximation of the other and that the approximation improves as $T \to 0$.

5.7-2 (a) Using the impulse invariance criterion, design a digital filter to realize an analog filter with transfer function

$$H_a(s) = \frac{7s + 20}{2(s^2 + 7s + 10)}$$

(b) Show a canonical and a parallel realization of the filter. Use the 1% criterion for the choice of T.

5.7-3 Use the impulse invariance criterion to design a digital filter to realize the second-order analog Butterworth filter with transfer function

$$H_a(s) = \frac{1}{s^2 + \sqrt{2}s + 1}$$

Use the 1% criterion for the choice of T.

5.7-4 Design a digital integrator using the impulse invariance method. Find and give a rough sketch of the amplitude response, and compare it with that of the ideal integrator. If this integrator is used primarily for integrating audio signals (whose bandwidth is 20 kHz), determine a suitable value for T.

5.7-5 An oscillator by definition is a source (no input) that generates a sinusoid of a certain frequency ω_0. Therefore an oscillator is a system whose zero-input response is a sinusoid of the desired frequency. Find the transfer function of a digital oscillator to oscillate at 10 kHz by the methods described in parts a and b. In both methods select T so that there are 10 samples in each cycle of the sinusoid.

(a) Choose $H[z]$ directly so that its zero-input response is a discrete-time sinusoid of frequency $\Omega = \omega T$ corresponding to 10 kHz.

(b) Choose $H_a(s)$ whose zero-input response is an analog sinusoid of 10 kHz. Now use the impulse invariance method to determine $H[z]$.

(c) Show a canonical realization of the oscillator.

5.7-6 A variant of the impulse invariance method is the *step invariance* method of digital filter synthesis. In this method, for a given $H_a(s)$, we design $H[z]$ in Fig. 5.22a such that $y(nT)$ in Fig. 5.22b is identical to $y[n]$ in Fig. 5.22a when $x(t) = u(t)$.
(a) Show that in general

$$H[z] = \frac{z-1}{z} \mathcal{Z}\left[\left(\mathcal{L}^{-1}\frac{H_a(s)}{s}\right)_{t=kT}\right]$$

(b) Use this method to design $H[z]$ for

$$H_a(s) = \frac{\omega_c}{s + \omega_c}$$

(c) Use the step invariant method to synthesize a discrete-time integrator and compare its amplitude response with that of the ideal integrator.

5.7-7 Use the *ramp-invariance* method to synthesize a discrete-time differentiator and integrator. In this method, for a given $H_a(s)$, we design $H[z]$ such that $y(nT)$ in Fig. 5.22b is identical to $y[n]$ in Fig. 5.22a when $x(t) = tu(t)$.

5.7-8 In an impulse invariance design, show that if $H_a(s)$ is a transfer function of a stable system, the corresponding $H[z]$ is also a transfer function of a stable system.

5.7-9 First-order backward differences provide the transformation rule $s = (1 - z^{-1})/T$.
(a) Show that this transformation maps the ω axis in the s plane to a circle of radius $1/2$ centered at $(1/2, 0)$ in the z plane.
(b) Show that this transformation maps the left-half s plane to the interior of the unit circle in the z plane, which ensures that stability is preserved.

5.9-1 Find the z-transform (if it exists) and the corresponding ROC for each of the following signals:
(a) $(0.8)^n u[n] + 2^n u[-(n+1)]$
(b) $2^n u[n] - 3^n u[-(n+1)]$
(c) $(0.8)^n u[n] + (0.9)^n u[-(n+1)]$
(d) $[(0.8)^n + 3(0.4)^n]u[-(n+1)]$
(e) $[(0.8)^n + 3(0.4)^n]u[n]$
(f) $(0.8)^n u[n] + 3(0.4)^n u[-(n+1)]$
(g) $(0.5)^{|n|}$
(h) $n\,u[-(n+1)]$

5.9-2 Find the inverse z-transform of

$$X[z] = \frac{(e^{-2} - 2)z}{(z - e^{-2})(z - 2)}$$

when the ROC is
(a) $|z| > 2$
(b) $e^{-2} < |z| < 2$
(c) $|z| < e^{-2}$

5.9-3 Use partial fraction expansions, z-transform tables, and a region of convergence ($|z| < 1/2$) to determine the inverse z-transform of

$$X(z) = \frac{1}{(2z + 1)(z + 1)\left(z + \frac{1}{2}\right)}$$

5.9-4 Consider the system

$$H(z) = \frac{z\left(z - \frac{1}{2}\right)}{\left(z^3 - \frac{27}{8}\right)}$$

(a) Draw the pole–zero diagram for $H(z)$ and identify all possible regions of convergence.
(b) Draw the pole–zero diagram for $H^{-1}(z)$ and identify all possible regions of convergence.

5.9-5 A discrete-time signal $x[n]$ has a rational z-transform that contains a pole at $z = 0.5$. Given $x_1[n] = (1/3)^n x[n]$ is absolutely summable and $x_2[n] = (1/4)^n x[n]$ is NOT absolutely summable, determine whether $x[n]$ is left sided, right sided, or two sided. Justify your answer!

5.9-6 Let $x[n]$ be an absolutely summable signal with rational z-transform $X(z)$. $X(z)$ is known to have a pole at $z = (0.75 + 0.75j)$, and other poles may be present. Recall that an absolutely summable signal satisfies $\sum_{-\infty}^{\infty} |x[n]| < \infty$.
(a) Can $x[n]$ be left sided? Explain.
(b) Can $x[n]$ be right sided? Explain.
(c) Can $x[n]$ be two sided? Explain.
(d) Can $x[n]$ be of finite duration? Explain.

5.9-7 Consider a causal system that has transfer function

$$H(z) = \frac{z - 0.5}{z + 0.5}$$

When appropriate, assume initial conditions of zero.
(a) Determine the output $y_1[n]$ of this system in response to $x_1[n] = (3/4)^n u[n]$.

(b) Determine the output $y_2[n]$ of this system in response to $x_2[n] = (3/4)^n$.

5.9-8 Let $x[n] = (-1)^n u[n - n_0] + \alpha^n u[-n]$. Determine the constraints on the complex number α and the integer n_0 so that the z-transform $X(z)$ exists with region of convergence $1 < |z| < 2$.

5.9-9 Using the definition, compute the bilateral z-transform, including the region of convergence (ROC), of the following complex-valued functions:
(a) $x_1[n] = (-j)^{-n} u[-n] + \delta[-n]$
(b) $x_2[n] = (j)^n \cos(n + 1) u[n]$
(c) $x_3[n] = j \sinh[n] u[-n + 1]$
(d) $x_4[n] = \sum_{k=-\infty}^{0} (2j)^n \delta[n - 2k]$

5.9-10 Use partial fraction expansions, z-transform tables, and a region of convergence $(0.5 < |z| < 2)$ to determine the inverse z-transform of
(a) $X_1(z) = \dfrac{1}{1 + \frac{13}{6} z^{-1} + \frac{1}{6} z^{-2} - \frac{1}{3} z^{-3}}$

(b) $X_2(z) = \dfrac{1}{z^{-3}(2 - z^{-1})(1 + 2z^{-1})}$

5.9-11 Use partial fraction expansions, z-transform tables, and the fact that the systems are stable to determine the inverse z-transform of
(a) $H_1(z) = \dfrac{z^{-1}}{\left(z - \frac{1}{2}\right)\left(1 + \frac{1}{2} z^{-1}\right)}$

(b) $H_2(z) = \dfrac{z + 1}{z^3 (z - 2)\left(z + \frac{1}{2}\right)}$

5.9-12 By inserting N zeros between every sample of a unit step, we obtain a signal

$$h[n] = \sum_{k=0}^{\infty} \delta[n - Nk]$$

Determine $H(z)$, the bilateral z-transform of $h[n]$. Identify the number and location(s) of the poles of $H(z)$.

5.9-13 Determine the zero-state response of a system having a transfer function

$$H[z] = \frac{z}{(z + 0.2)(z - 0.8)} \qquad |z| > 0.8$$

and an input $x[n]$ given by
(a) $x[n] = e^n u[n]$
(b) $x[n] = 2^n u[-(n + 1)]$
(c) $x[n] = e^n u[n] + 2^n u[-(n + 1)]$

5.9-14 For the system in Prob. 5.9-13, determine the zero-state response for the input

$$x[n] = 2^n u[n] + u[-(n + 1)]$$

5.9-15 For the system in Prob. 5.9-13, determine the zero-state response for the input

$$x[n] = e^{-2n}[-(n + 1)]$$

5.M-1 Consider an LTID system described by the difference equation $4y[n + 2] - y[n] = x[n + 2] + x[n]$.
(a) Plot the pole–zero diagram for this system.
(b) Plot the system's magnitude response $|H(e^{j\Omega})|$ over $-\pi \leq \Omega \leq \pi$.
(c) What type of system is this: lowpass, highpass, bandpass, or bandstop?
(d) Is this system stable? Justify your answer.
(e) Is this system real? Justify your answer.
(f) If the system input is of the form $x[n] = \cos(\Omega n)$, what is the greatest possible amplitude of the output? Justify your answer.
(g) Draw an efficient, causal implementation of this system using only add, scale, and delay blocks.

5.M-2 One interesting and useful application of discrete systems is the implementation of complex (rather than real) systems. A complex system is one in which a real-valued input can produce a complex-valued output. Complex systems that are described by constant coefficient difference equations require at least one complex-valued coefficient, and they are capable of operating on complex-valued inputs.

Consider the complex discrete time system

$$H(z) = \frac{z^2 - j}{z - 0.9 e^{j3\pi/4}}$$

(a) Determine and plot the system zeros and poles.
(b) Sketch the magnitude response $|H(e^{j\omega})|$ of this system over $-2\pi \leq \omega \leq 2\pi$. Comment on the system's behavior.

5.M-3 Consider the complex system

$$H(z) = \frac{z^4 - 1}{2(z^2 + 0.81 j)}$$

Refer to Prob. 5.M-2 for an introduction to complex systems.

(a) Plot the pole–zero diagram for $H(z)$.

(b) Plot the system's magnitude response $|H(e^{j\Omega})|$ over $-\pi \le \Omega \le \pi$.

(c) Explain why $H(z)$ is a noncausal system. Do not give a general definition of causality; specifically identify what makes this system noncausal.

(d) One way to make this system causal is to add two poles to $H(z)$. That is,

$$H_{\text{causal}}(z) = H(z)\frac{1}{(z-a)(z-b)}$$

Find poles a and b such that $|H_{\text{causal}}(e^{j\Omega})| = |H(e^{j\Omega})|$.

(e) Draw an efficient block implementation of $H_{\text{causal}}(z)$.

5.M-4 A discrete-time LTI system is shown in Fig. P5.M-4.

(a) Determine the difference equation that describes this system.

(b) Determine the magnitude response $|H(e^{j\Omega})|$ for this system and simplify your answer. Plot the magnitude response over $-\pi \le \Omega \le \pi$. What type of standard filter (lowpass, highpass, bandpass, or bandstop) best describes this system?

(c) Determine the impulse response $h[n]$ of this system.

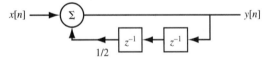

1/2

Figure P5.M-4 Second-order discrete-time system.

5.M-5 Determine the impulse response $h[n]$ for the system shown in Fig. P5.M-5. Is the system stable? Is the system causal?

5.M-6 An LTID filter has an impulse response function given by $h[n] = \delta[n-1] + \delta[n+1]$.

Determine and carefully sketch the magnitude response $|H(e^{j\Omega})|$ over the range $-\pi \le \Omega \le \pi$. For this range of frequencies, is this filter lowpass, highpass, bandpass, or bandstop?

5.M-7 A causal, stable discrete system has the rather strange transfer function $H(z) = \cos(z^{-1})$.

(a) Write MATLAB code that will compute and plot the magnitude response of this system over an appropriate range of digital frequencies Ω. Comment on the system.

(b) Determine the impulse response $h[n]$. Plot $h[n]$ over $(0 \le n \le 10)$.

(c) Determine a difference equation description for an FIR filter that closely approximates the system $H(z) = \cos(z^{-1})$. To verify proper behavior, plot the FIR filter's magnitude response and compare it with the magnitude response computed in Prob. 5.M-7a.

5.M-8 The MATLAB signal processing toolbox function `butter` helps design digital Butterworth filters. Use MATLAB help to learn how `butter` works. For each of the following cases, design the filter, plot the filter's poles and zeros in the complex z plane, and plot the decibel magnitude response $20 \log_{10} |H(e^{j\Omega})|$.

(a) Design an eighth-order digital lowpass filter with $\Omega_c = \pi/3$.

(b) Design an eighth-order digital highpass filter with $\Omega_c = \pi/3$.

(c) Design an eighth-order digital bandpass filter with passband between $5\pi/24$ and $11\pi/24$.

(d) Design an eighth-order digital bandstop filter with stopband between $5\pi/24$ and $11\pi/24$.

5.M-9 The MATLAB signal processing toolbox function `cheby1` helps design digital Chebyshev type I filters. A Chebyshev type I filter has passband ripple and smooth stopband. Setting the passband ripple to $R_p = 3$ dB, repeat

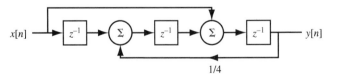

1/4

Figure P5.M-5 Third-order discrete-time system.

Prob. 5.M-8 using the `cheby1` command. With all other parameters held constant, what is the general effect of reducing R_p, the allowable passband ripple?

5.M-10 The MATLAB signal processing toolbox function `cheby2` helps design digital Chebyshev type II filters. A Chebyshev type II filter has smooth passband and ripple in the stopband. Setting the stopband ripple $R_s = 20\,\text{dB}$ down, repeat Prob. 5.M-8 using the `cheby2` command. With all other parameters held constant, what is the general effect of increasing R_s, the minimum stopband attenuation?

5.M-11 The MATLAB signal processing toolbox function `ellip` helps design digital elliptic filters. An elliptic filter has ripple in both the passband and the stopband. Setting the passband ripple to $R_p = 3$ dB and the stopband ripple $R_s = 20$ dB down, repeat Prob. 5.M-8 using the `ellip` command.

CONTINUOUS-TIME SIGNAL ANALYSIS: THE FOURIER SERIES

Electrical engineers instinctively think of signals in terms of their frequency spectra and think of systems in terms of their frequency response. Even teenagers know about audible portion of audio signals having a bandwidth of about 20 kHz and the need for good-quality speakers to respond up to 20 kHz. This is basically thinking in frequency domain. In Chapters 4 and 5 we discussed extensively the frequency-domain representation of systems and their spectral response (system response to signals of various frequencies). In Chapters 6, 7, 8, and 9 we discuss spectral representation of signals, where signals are expressed as a sum of sinusoids or exponentials. Actually, we touched on this topic in Chapters 4 and 5. Recall that the Laplace transform of a continuous-time signal is its spectral representation in terms of exponentials (or sinusoids) of complex frequencies. Similarly the z-transform of a discrete-time signal is its spectral representation in terms of discrete-time exponentials. However, in the earlier chapters we were concerned mainly with system representation, and the spectral representation of signals was incidental to the system analysis. Spectral analysis of signals is an important topic in its own right, and now we turn to this subject.

In this chapter we show that a periodic signal can be represented as a sum of sinusoids (or exponentials) of various frequencies. These results are extended to aperiodic signals in Chapter 7 and to discrete-time signals in Chapter 9. The fascinating subject of sampling of continuous-time signals is discussed in Chapter 8, leading to A/D (analog-to-digital) and D/A conversion. Chapter 8 forms the bridge between the continuous-time and the discrete-time worlds.

6.1 PERIODIC SIGNAL REPRESENTATION BY TRIGONOMETRIC FOURIER SERIES

As seen in Section 1.3-3 [Eq. (1.17)], a periodic signal $x(t)$ with period T_0 (Fig. 6.1) has the property

$$x(t) = x(t + T_0) \qquad \text{for all } t \tag{6.1}$$

The *smallest* value of T_0 that satisfies the periodicity condition (6.1) is the *fundamental period* of $x(t)$. As argued in Section 1.3-3, this equation implies that $x(t)$ starts at $-\infty$ and continues

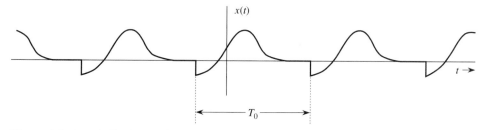

Figure 6.1 A periodic signal of period T_0.

to ∞. Moreover, the area under a periodic signal $x(t)$ over any interval of duration T_0 is the same; that is, for any real numbers a and b

$$\int_a^{a+T_0} x(t)\,dt = \int_b^{b+T_0} x(t)\,dt \tag{6.2}$$

This result follows from the fact that a periodic signal takes the same values at the intervals of T_0. Hence, the values over any segment of duration T_0 are repeated in any other interval of the same duration. For convenience, the area under $x(t)$ over any interval of duration T_0 will be denoted by

$$\int_{T_0} x(t)\,dt$$

The frequency of a sinusoid $\cos 2\pi f_0 t$ or $\sin 2\pi f_0 t$ is f_0 and the period is $T_0 = 1/f_0$. These sinusoids can also be expressed as $\cos \omega_0 t$ or $\sin \omega_0 t$, where $\omega_0 = 2\pi f_0$ is the *radian frequency*, although for brevity, it is often referred to as frequency (see Section B.2). A sinusoid of frequency $n f_0$ is said to be the *nth harmonic* of the sinusoid of frequency f_0.

Let us consider a signal $x(t)$ made up of a sines and cosines of frequency ω_0 and all of its harmonics (including the zeroth harmonic; i.e., dc) with arbitrary amplitudes[†]:

$$x(t) = a_0 + \sum_{n=1}^{\infty} a_n \cos n\omega_0 t + b_n \sin n\omega_0 t \tag{6.3}$$

The frequency ω_0 is called the *fundamental frequency*.

We now prove an extremely important property: $x(t)$ in Eq. (6.3) is a periodic signal with the same period as that of the fundamental, regardless of the values of the amplitudes a_n and b_n. Note that the period T_0 of the fundamental is

$$T_0 = \frac{1}{f_0} = \frac{2\pi}{\omega_0} \tag{6.4}$$

and

$$\omega_0 T_0 = 2\pi \tag{6.5}$$

[†]In Eq. (6.3), the constant term a_0 corresponds to the cosine term for $n = 0$ because $\cos (0 \times \omega_0)t = 1$. However, $\sin (0 \times \omega_0)t = 0$. Hence, the sine term for $n = 0$ is nonexistent.

To prove the periodicity of $x(t)$, all we need is to show that $x(t) = x(t + T_0)$. From Eq. (6.3)

$$x(t + T_0) = a_0 + \sum_{n=1}^{\infty} a_n \cos n\omega_0(t + T_0) + b_n \sin n\omega_0(t + T_0)$$

$$= a_0 + \sum_{n=1}^{\infty} a_n \cos(n\omega_0 t + n\omega_0 T_0) + b_n \sin(n\omega_0 t + n\omega_0 T_0)$$

From Eq. (6.5), we have $n\omega_0 T_0 = 2\pi n$, and

$$x(t + T_0) = a_0 + \sum_{n=1}^{\infty} a_n \cos(n\omega_0 t + 2\pi n) + b_n \sin(n\omega_0 t + 2\pi n)$$

$$= a_0 + \sum_{n=1}^{\infty} a_n \cos n\omega_0 t + b_n \sin n\omega_0 t$$

$$= x(t)$$

We could also infer this result intuitively. In one fundamental period T_0, the nth harmonic executes n complete cycles. Hence, every sinusoid on the right-hand side of Eq. (6.3) executes a complete number of cycles in one fundamental period T_0. Therefore, at $t = T_0$, every sinusoid starts as if it were the origin and repeats the same drama over the next T_0 seconds, and so on ad infinitum. Hence, the sum of such harmonics results in a periodic signal of period T_0.

This result shows that any combination of sinusoids of frequencies $0, f_0, 2f_0, \ldots, kf_0$ is a periodic signal of period $T_0 = 1/f_0$ regardless of the values of amplitudes a_k and b_k of these sinusoids. By changing the values of a_k and b_k in Eq. (6.3), we can construct a variety of periodic signals, all of the same period T_0 ($T_0 = 1/f_0$ or $2\pi/\omega_0$).

The converse of this result is also true. We shall show in Section 6.5-4 that *a periodic signal $x(t)$ with a period T_0 can be expressed as a sum of a sinusoid of frequency f_0 ($f_0 = 1/T_0$) and all its harmonics, as shown in Eq. (6.3).*[†] The infinite series on the right-hand side of Eq. (6.3) is known as the *trigonometric Fourier series* of a periodic signal $x(t)$.

COMPUTING THE COEFFICIENTS
OF A FOURIER SERIES

To determine the coefficients of a Fourier series, consider an integral I defined by

$$I = \int_{T_0} \cos n\omega_0 t \cos m\omega_0 t \, dt \tag{6.6a}$$

[†]Strictly speaking, this statement applies only if a periodic signal $x(t)$ is a continuous function of t. However, Section 6.5-4 shows that it can be applied even for discontinuous signals, if we interpret the equality in Eq. (6.3), not in the ordinary sense, but in the mean-square sense. This means that the power of the difference between the periodic signal $x(t)$ and its Fourier series on the right-hand side of Eq. (6.3) approaches zero as the number of terms in the series approach infinity.

where $\int_{T_0}$ stands for integration over any contiguous interval of T_0 seconds. By using a trigonometric identity (see Section B.7-6), Eq. (6.6a) can be expressed as

$$I = \tfrac{1}{2}\left[\int_{T_0} \cos(n+m)\omega_0 t \, dt + \int_{T_0} \cos(n-m)\omega_0 t \, dt\right] \tag{6.6b}$$

Because $\cos \omega_0 t$ executes one complete cycle during any interval of duration T_0, $\cos(n+m)\omega_0 t$ executes $(n+m)$ complete cycles during any interval of duration T_0. Therefore the first integral in Eq. (6.6b), which represents the area under $n+m$ complete cycles of a sinusoid, equals zero. The same argument shows that the second integral in Eq. (6.6b) is also zero, except when $n = m$. Hence I in Eq. (6.6) is zero for all $n \neq m$. When $n = m$, the first integral in Eq. (6.6b) is still zero, but the second integral yields

$$I = \tfrac{1}{2}\int_{T_0} dt = \frac{T_0}{2}$$

Thus

$$\int_{T_0} \cos n\omega_0 t \cos m\omega_0 t \, dt = \begin{cases} 0 & n \neq m \\ \dfrac{T_0}{2} & m = n \neq 0 \end{cases} \tag{6.7a}$$

Using similar arguments, we can show that

$$\int_{T_0} \sin n\omega_0 t \sin m\omega_0 t \, dt = \begin{cases} 0 & n \neq m \\ \dfrac{T_0}{2} & n = m \neq 0 \end{cases} \tag{6.7b}$$

and

$$\int_{T_0} \sin n\omega_0 t \cos m\omega_0 t \, dt = 0 \qquad \text{for all } n \text{ and } m \tag{6.7c}$$

To determine a_0 in Eq. (6.3) we integrate both sides of Eq. (6.3) over one period T_0 to yield

$$\int_{T_0} x(t)\, dt = a_0 \int_{T_0} dt + \sum_{n=1}^{\infty}\left[a_n \int_{T_0} \cos n\omega_0 t \, dt + b_n \int_{T_0} \sin n\omega_0 t \, dt\right]$$

Recall that T_0 is the period of a sinusoid of frequency ω_0. Therefore functions $\cos n\omega_0 t$ and $\sin n\omega_0 t$ execute n complete cycles over any interval of T_0 seconds, so that the area under these functions over an interval T_0 is zero, and the last two integrals on the right-hand side of the foregoing equation are zero. This yields

$$\int_{T_0} x(t)\, dt = a_0 \int_{T_0} dt = a_0 T_0$$

and

$$a_0 = \frac{1}{T_0}\int_{T_0} x(t)\, dt \tag{6.8a}$$

Next we multiply both sides of Eq. (6.3) by $\cos m\omega_0 t$ and integrate the resulting equation over an interval T_0:

$$\int_{T_0} x(t) \cos m\omega_0 t \, dt = a_0 \int_{T_0} \cos m\omega_0 t \, dt + \sum_{n=1}^{\infty} \left[a_n \int_{T_0} \cos n\omega_0 t \, \cos m\omega_0 t \, dt \right.$$

$$\left. + b_n \int_{T_0} \sin n\omega_0 t \, \cos m\omega_0 t \, dt \right]$$

The first integral on the right-hand side is zero because it is an area under m integral number of cycles of a sinusoid. Also, the last integral on the right-hand side vanishes because of Eq. (6.7c). This leaves only the middle integral, which is also zero for all $n \neq m$ because of Eq. (6.7a). But n takes on all values from 1 to ∞, including m. When $n = m$, this integral is $T_0/2$, according to Eq. (6.7a). Therefore, from the infinite number of terms on the right-hand side, only one term survives to yield $a_n T_0/2 = a_m T_0/2$ (recall that $n = m$). Therefore

$$\int_{T_0} x(t) \cos m\omega_0 t \, dt = \frac{a_m T_0}{2}$$

and

$$a_m = \frac{2}{T_0} \int_{T_0} x(t) \cos m\omega_0 t \, dt \tag{6.8b}$$

Similarly, by multiplying both sides of Eq. (6.3) by $\sin n\omega_0 t$ and then integrating over an interval T_0, we obtain

$$b_m = \frac{2}{T_0} \int_{T_0} x(t) \sin m\omega_0 t \, dt \tag{6.8c}$$

To sum up our discussion, which applies to real or complex $x(t)$, we have shown that a periodic signal $x(t)$ with period T_0 can be expressed as a sum of a sinusoid of period T_0 and its harmonics:

$$x(t) = a_0 + \sum_{n=1}^{\infty} a_n \cos n\omega_0 t + b_n \sin n\omega_0 t \tag{6.9}$$

where

$$\omega_0 = 2\pi f_0 = \frac{2\pi}{T_0} \tag{6.10}$$

$$a_0 = \frac{1}{T_0} \int_{T_0} x(t) \, dt \tag{6.11a}$$

$$a_n = \frac{2}{T_0} \int_{T_0} x(t) \cos n\omega_0 t \, dt \tag{6.11b}$$

$$b_n = \frac{2}{T_0} \int_{T_0} x(t) \sin n\omega_0 t \, dt \tag{6.11c}$$

COMPACT FORM OF FOURIER SERIES

The results derived so far are general and apply whether $x(t)$ is a real or a complex function of t. However, when $x(t)$ is real, coefficients a_n and b_n are real for all n, and the trigonometric Fourier series can be expressed in a *compact form,* using the results in Eqs. (B.23)

$$x(t) = C_0 + \sum_{n=1}^{\infty} C_n \cos(n\omega_0 t + \theta_n) \tag{6.12}$$

where C_n and θ_n are related to a_n and b_n, as [see Eqs. (B.23b) and (B.23c)]

$$C_0 = a_0 \tag{6.13a}$$

$$C_n = \sqrt{a_n{}^2 + b_n{}^2} \tag{6.13b}$$

$$\theta_n = \tan^{-1}\left(\frac{-b_n}{a_n}\right) \tag{6.13c}$$

These results are summarized in Table 6.1.

The compact form in Eq. (6.12) uses the cosine form. We could just as well have used the sine form, with terms $\sin(n\omega_0 t + \theta_n)$ instead of $\cos(n\omega_0 t + \theta_n)$. The literature overwhelmingly favors the cosine form, for no apparent reason except possibly that the cosine phasor is represented by the horizontal axis, which happens to be the reference axis in phasor representation.

Equation (6.11a) shows that a_0 (or C_0) is the average value of $x(t)$ (averaged over one period). This value can often be determined by inspection of $x(t)$.

Because a_n and b_n are real, C_n and θ_n are also real. In the following discussion of trigonometric Fourier series, we shall assume real $x(t)$, unless mentioned otherwise.

TABLE 6.1   Fourier Series Representation of a Periodic Signal of Period T_0 ($\omega_0 = 2\pi/T_0$)

Series Form	Coefficient Computation	Conversion Formulas
Trigonometric	$a_0 = \dfrac{1}{T_0} \displaystyle\int_{T_0} f(t)\, dt$	$a_0 = C_0 = D_0$
$f(t) = a_0 + \displaystyle\sum_{n=1}^{\infty} a_n \cos n\omega_0 t + b_n \sin n\omega_0 t$	$a_n = \dfrac{2}{T_0} \displaystyle\int_{T_0} f(t) \cos n\omega_0 t\, dt$	$a_n - jb_n = C_n e^{j\theta_n} = 2D_n$
	$b_n = \dfrac{2}{T_0} \displaystyle\int_{T_0} f(t) \sin n\omega_0 t\, dt$	$a_n + jb_n = C_n e^{-j\theta_n} = 2D_{-n}$
Compact trigonometric	$C_0 = a_0$	$C_0 = D_0$
$f(t) = C_0 + \displaystyle\sum_{n=1}^{\infty} C_n \cos(n\omega_0 t + \theta_n)$	$C_n = \sqrt{a_n{}^2 + b_n{}^2}$	$C_n = 2|D_n| \qquad n \geq 1$
	$\theta_n = \tan^{-1}\left(\dfrac{-b_n}{a_n}\right)$	$\theta_n = \angle D_n$
Exponential		
$f(t) = \displaystyle\sum_{n=-\infty}^{\infty} D_n e^{jn\omega_0 t}$	$D_n = \dfrac{1}{T_0} \displaystyle\int_{T_0} f(t) e^{-jn\omega_0 t}\, dt$	

6.1-1 The Fourier Spectrum

The compact trigonometric Fourier series in Eq. (6.12) indicates that a periodic signal $x(t)$ can be expressed as a sum of sinusoids of frequencies 0 (dc), $\omega_0, 2\omega_0, \ldots, n\omega_0, \ldots$, whose amplitudes are $C_0, C_1, C_2, \ldots, C_n, \ldots$, and whose phases are $0, \theta_1, \theta_2, \ldots, \theta_n, \ldots$, respectively. We can readily plot amplitude C_n versus n (*the amplitude spectrum*) and θ_n versus n (the *phase spectrum*).[†] Because n is proportional to the frequency $n\omega_0$, these plots are scaled plots of C_n versus ω and θ_n versus ω. The two plots together are the *frequency spectra* of $x(t)$. These spectra show at a glance the frequency contents of the signal $x(t)$ with their amplitudes and phases. Knowing these spectra, we can reconstruct or synthesize the signal $x(t)$ according to Eq. (6.12). Therefore frequency spectra, which are an alternative way of describing a periodic signal $x(t)$, are in every way equivalent to the plot of $x(t)$ as a function of t. The frequency spectra of a signal constitute the *frequency-domain description* of $x(t)$, in contrast to the *time-domain description*, where $x(t)$ is specified as a function of time.

In computing θ_n, the phase of the nth harmonic from Eq. (6.13c), the quadrant in which θ_n lies should be determined from the signs of a_n and b_n. For example, if $a_n = -1$ and $b_n = 1$, θ_n lies in the third quadrant, and

$$\theta_n = \tan^{-1}\left(\tfrac{-1}{-1}\right) = -135°$$

Observe that

$$\tan^{-1}\left(\tfrac{-1}{-1}\right) \neq \tan^{-1}(1) = 45°$$

Although C_n, the amplitude of the nth harmonic as defined in Eq. (6.13b), is positive, we shall find it convenient to allow C_n to take on negative values when $b_n = 0$. This will become clear in later examples.

EXAMPLE 6.1

Find the compact trigonometric Fourier series for the periodic signal $x(t)$ shown in Fig. 6.2a. Sketch the amplitude and phase spectra for $x(t)$.

In this case the period $T_0 = \pi$ and the fundamental frequency $f_0 = 1/T_0 = 1/\pi$ Hz, and

$$\omega_0 = \frac{2\pi}{T_0} = 2 \text{ rad/s}$$

[†]The amplitude C_n, by definition here, is nonnegative. Some authors define amplitude A_n that can take positive or negative values and magnitude $C_n = |A_n|$ that can only be nonnegative. Thus, what we call amplitude spectrum becomes magnitude spectrum. The distinction between amplitude and magnitude, although useful, is avoided in this book in the interest of keeping definitions of essentially similar entities to a minimum.

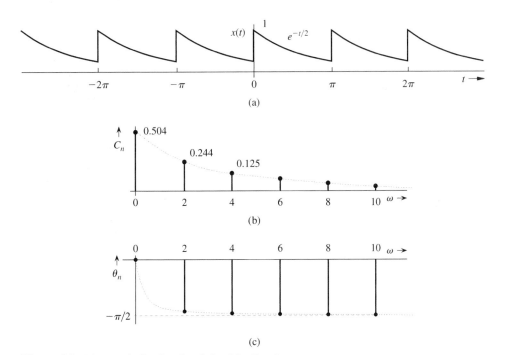

Figure 6.2 (a) A periodic signal and **(b, c)** its Fourier spectra.

Therefore

$$x(t) = a_0 + \sum_{n=1}^{\infty} a_n \cos 2nt + b_n \sin 2nt$$

where

$$a_0 = \frac{1}{\pi} \int_{T_0} x(t)\, dt$$

In this example the obvious choice for the interval of integration is from 0 to π. Hence

$$a_0 = \frac{1}{\pi} \int_0^\pi e^{-t/2}\, dt = 0.504$$

$$a_n = \frac{2}{\pi} \int_0^\pi e^{-t/2} \cos 2nt\, dt = 0.504 \left(\frac{2}{1 + 16n^2} \right)$$

and

$$b_n = \frac{2}{\pi} \int_0^\pi e^{-t/2} \sin 2nt\, dt = 0.504 \left(\frac{8n}{1 + 16n^2} \right)$$

TABLE 6.2

n	C_n	θ_n
0	0.504	0
1	0.244	−75.96
2	0.125	−82.87
3	0.084	−85.24
4	0.063	−86.42
5	0.0504	−87.14
6	0.042	−87.61
7	0.036	−87.95

Therefore

$$x(t) = 0.504 \left[1 + \sum_{n=1}^{\infty} \frac{2}{1 + 16n^2} (\cos 2nt + 4n \sin 2nt) \right] \tag{6.14}$$

Also from Eqs. (6.13)

$$C_0 = a_0 = 0.504$$

$$C_n = \sqrt{a_n^2 + b_n^2} = 0.504 \sqrt{\frac{4}{(1 + 16n^2)^2} + \frac{64n^2}{(1 + 16n^2)^2}} = 0.504 \left(\frac{2}{\sqrt{1 + 16n^2}} \right)$$

$$\theta_n = \tan^{-1} \left(\frac{-b_n}{a_n} \right) = \tan^{-1}(-4n) = -\tan^{-1} 4n$$

Amplitude and phases of the dc and the first seven harmonics are computed from the above equations and displayed in Table 6.2. We can use these numerical values to express $x(t)$ as

$$x(t) = 0.504 + 0.504 \sum_{n=1}^{\infty} \frac{2}{\sqrt{1 + 16n^2}} \cos (2nt - \tan^{-1} 4n) \tag{6.15a}$$

$$x(t) = 0.504 + 0.244 \cos (2t - 75.96°) + 0.125 \cos (4t - 82.87°)$$

$$+ 0.084 \cos (6t - 85.24°) + 0.063 \cos (8t - 86.42°) + \cdots \tag{6.15b}$$

COMPUTER EXAMPLE C6.1

Following Example 6.1, compute and plot the Fourier coefficients for the periodic signal in Fig. 6.2a.

In this example, $T_0 = \pi$ and $\omega_0 = 2$. The expressions for a_0, a_n, b_n, C_n, and θ_n are derived in Example 6.1.

```
>> n = 1:10; a_n(1) = 0.504; a_n(n+1) = 0.504*2./(1+16*n.^2);
>> b_n(1) = 0; b_n(n+1) = 0.504*8*n./(1+16*n.^2);
>> C_n(1) = a_n(1); C_n(n+1) = sqrt(a_n(n+1).^2+b_n(n+1).^2);
>> theta_n(1) = 0; theta_n(n+1) = atan2(-b_n(n+1),a_n(n+1));
>> n = [0,n];
>> clf; subplot(2,2,1); stem(n,a_n,'k'); ylabel('a_n'); xlabel('n');
>> subplot(2,2,2); stem(n,b_n,'k'); ylabel('b_n'); xlabel('n');
>> subplot(2,2,3); stem(n,C_n,'k'); ylabel('C_n'); xlabel('n');
>> subplot(2,2,4); stem(n,theta_n,'k'); ylabel('\theta_n [rad]'); xlabel('n');
```

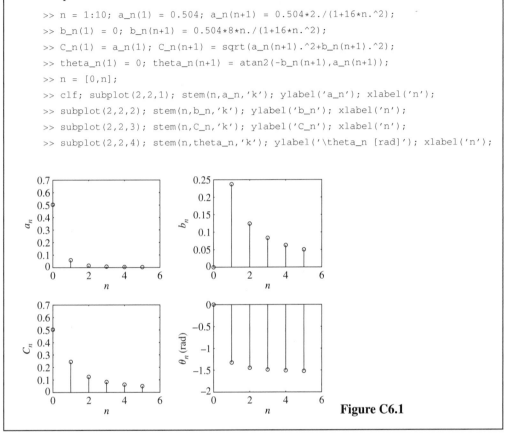

Figure C6.1

The amplitude and phase spectra for $x(t)$, in Fig. 6.2b and 6.2c, tell us at a glance the frequency composition of $x(t)$, that is, the amplitudes and phases of various sinusoidal components of $x(t)$. Knowing the frequency spectra, we can reconstruct $x(t)$, as shown on the right-hand side of Eq. (6.15b). Therefore the frequency spectra (Fig. 6.2b, 6.2c) provide an alternative description—the frequency-domain description of $x(t)$. The time-domain description of $x(t)$ is shown in Fig. 6.2a. *A signal, therefore, has a dual identity: the time-domain identity $x(t)$ and the frequency-domain identity (Fourier spectra). The two identities complement each other; taken together, they provide a better understanding of a signal.*

An interesting aspect of Fourier series is that whenever there is a jump discontinuity in $x(t)$, the series at the point of discontinuity converges to an average of the left-hand and right-hand limits of $x(t)$ at the instant of discontinuity.[†] In the present example, for instance, $x(t)$ is

---

[†]This behavior of the Fourier series is dictated by its convergence in the mean, discussed later in Sections 6.2 and 6.5.

discontinuous at $t = 0$ with $x(0^+) = 1$ and $x(0^-) = x(\pi) = e^{-\pi/2} = 0.208$. The corresponding Fourier series converges to a value $(1 + 0.208)/2 = 0.604$ at $t = 0$. This is easily verified from Eq. (6.15b) by setting $t = 0$.

EXAMPLE 6.2

Find the compact trigonometric Fourier series for the triangular periodic signal $x(t)$ shown in Fig. 6.3a, and sketch the amplitude and phase spectra for $x(t)$.

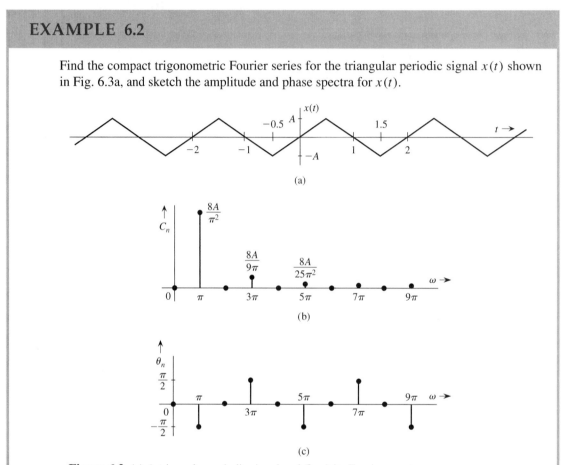

Figure 6.3 (a) A triangular periodic signal and (b, c) its Fourier spectra.

In this case the period $T_0 = 2$. Hence

$$\omega_0 = \frac{2\pi}{2} = \pi$$

and

$$x(t) = a_0 + \sum_{n=1}^{\infty} a_n \cos n\pi t + b_n \sin n\pi t$$

where

$$x(t) = \begin{cases} 2At & |t| < \frac{1}{2} \\ 2A(1-t) & \frac{1}{2} < t < \frac{3}{2} \end{cases}$$

Here it will be advantageous to choose the interval of integration from $-1/2$ to $3/2$ rather than 0 to 2.

A glance at Fig. 6.3a shows that the average value (dc) of $x(t)$ is zero, so that $a_0 = 0$. Also

$$a_n = \frac{2}{2} \int_{-1/2}^{3/2} x(t) \cos n\pi t \, dt$$

$$= \int_{-1/2}^{1/2} 2At \cos n\pi t \, dt + \int_{1/2}^{3/2} 2A(1-t) \cos n\pi t \, dt$$

Detailed evaluation of these integrals shows that both have a value of zero. Therefore

$$a_n = 0$$

$$b_n = \int_{-1/2}^{1/2} 2At \sin n\pi t \, dt + \int_{1/2}^{3/2} 2A(1-t) \sin n\pi t \, dt$$

Detailed evaluation of these integrals yields, in turn,

$$b_n = \frac{8A}{n^2 \pi^2} \sin \left(\frac{n\pi}{2} \right)$$

$$= \begin{cases} 0 & n \text{ even} \\ \dfrac{8A}{n^2 \pi^2} & n = 1, 5, 9, 13, \ldots \\ -\dfrac{8A}{n^2 \pi^2} & n = 3, 7, 11, 15, \ldots \end{cases}$$

Therefore

$$x(t) = \frac{8A}{\pi^2} \left[\sin \pi t - \frac{1}{9} \sin 3\pi t + \frac{1}{25} \sin 5\pi t - \frac{1}{49} \sin 7\pi t + \cdots \right] \qquad (6.16)$$

To plot Fourier spectra, the series must be converted into compact trigonometric form as in Eq. (6.12). In this case this is readily done by converting sine terms into cosine terms with a suitable phase shift. For example,

$$\sin kt = \cos (kt - 90°)$$

$$-\sin kt = \cos (kt + 90°)$$

By using these identities, Eq. (6.16) can be expressed as

$$x(t) = \frac{8A}{\pi^2} \left[\cos (\pi t - 90°) + \frac{1}{9} \cos (3\pi t + 90°) + \frac{1}{25} \cos (5\pi t - 90°) \right.$$

$$\left. + \frac{1}{49} \cos (7\pi t + 90°) + \cdots \right]$$

In this series all the even harmonics are missing. The phases of the odd harmonics alternate from $-90°$ to $90°$. Figure 6.3 shows amplitude and phase spectra for $x(t)$.

EXAMPLE 6.3

A periodic signal $x(t)$ is represented by a trigonometric Fourier series

$$x(t) = 2 + 3\cos 2t + 4\sin 2t + 2\sin(3t + 30°) - \cos(7t + 150°)$$

Express this series as a compact trigonometric Fourier series and sketch amplitude and phase spectra for $x(t)$.

___

In compact trigonometric Fourier series, the sine and cosine terms of the same frequency are combined into a single term and all terms are expressed as cosine terms with positive amplitudes. Using Eqs. (6.12), (6.13b), and (6.13c), we have

$$3\cos 2t + 4\sin 2t = 5\cos(2t - 53.13°)$$

Also

$$\sin(3t + 30°) = \cos(3t + 30° - 90°) = \cos(3t - 60°)$$

and

$$-\cos(7t + 150°) = \cos(7t + 150° - 180°) = \cos(7t - 30°)$$

Therefore

$$x(t) = 2 + 5\cos(2t - 53.13°) + 2\cos(3t - 60°) + \cos(7t - 30°)$$

In this case only four components (including dc) are present. The amplitude of dc is 2. The remaining three components are of frequencies $\omega = 2$, 3, and 7 with amplitudes 5, 2, and 1 and phases $-53.13°$, $-60°$, and $-30°$, respectively. The amplitude and phase spectra for this signal are shown in Fig. 6.4a and 6.4b, respectively.

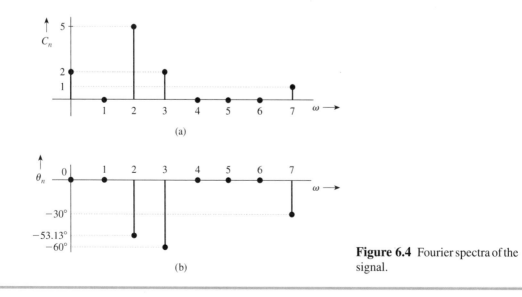

Figure 6.4 Fourier spectra of the signal.

EXAMPLE 6.4

Find the compact trigonometric Fourier series for the square-pulse periodic signal shown in Fig. 6.5a and sketch its amplitude and phase spectra.

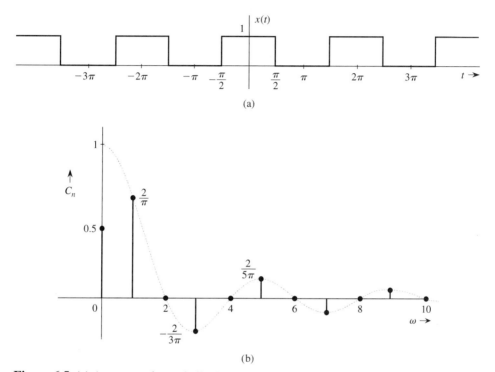

(a)

(b)

Figure 6.5 (a) A square pulse periodic signal and (b) its Fourier spectra.

Here the period is $T_0 = 2\pi$ and $\omega_0 = 2\pi/T_0 = 1$. Therefore

$$x(t) = a_0 + \sum_{n=1}^{\infty} a_n \cos nt + b_n \sin nt$$

where

$$a_0 = \frac{1}{T_0} \int_{T_0} x(t)\, dt$$

From Fig. 6.5a, it is clear that a proper choice of region of integration is from $-\pi$ to π. But since $x(t) = 1$ only over $(-\pi/2, \pi/2)$ and $x(t) = 0$ over the remaining segment,

$$a_0 = \frac{1}{2\pi} \int_{-\pi/2}^{\pi/2} dt = \frac{1}{2}$$

We could have found a_0, the average value of $x(t)$, to be $1/2$ merely by inspection of $x(t)$ in Fig. 6.5a. Also,

$$a_n = \frac{1}{\pi} \int_{-\pi/2}^{\pi/2} \cos nt \, dt = \frac{2}{n\pi} \sin \left(\frac{n\pi}{2} \right)$$

$$= \begin{cases} 0 & n \text{ even} \\ \dfrac{2}{\pi n} & n = 1, 5, 9, 13, \ldots \\ -\dfrac{2}{\pi n} & n = 3, 7, 11, 15, \ldots \end{cases}$$

$$b_n = \frac{1}{\pi} \int_{-\pi/2}^{\pi/2} \sin nt \, dt = 0$$

Therefore

$$x(t) = \frac{1}{2} + \frac{2}{\pi} \left(\cos t - \frac{1}{3} \cos 3t + \frac{1}{5} \cos 5t - \frac{1}{7} \cos 7t + \cdots \right) \tag{6.17}$$

Observe that $b_n = 0$ and all the sine terms are zero. Only the cosine terms appear in the trigonometric series. The series is therefore already in the compact form except that the amplitudes of alternating harmonics are negative. Now by definition, amplitudes C_n are positive [see Eq. (6.13b)]. The negative sign can be accommodated by associating a proper phase, as seen from the trigonometric identity[†]

$$-\cos x = \cos (x - \pi)$$

Using this fact, we can express the series in (6.17) as

$$x(t) = \frac{1}{2} + \frac{2}{\pi} \left[\cos \omega_0 t + \frac{1}{3} \cos (3\omega_0 t - \pi) + \frac{1}{5} \cos 5\omega_0 t \right.$$

$$\left. + \frac{1}{7} \cos (7\omega_0 t - \pi) + \frac{1}{9} \cos 9\omega_0 t + \cdots \right]$$

This is the desired form of the compact trigonometric Fourier series. The amplitudes are

$$C_0 = \frac{1}{2}$$

$$C_n = \begin{cases} 0 & n \text{ even} \\ \dfrac{2}{\pi n} & n \text{ odd} \end{cases}$$

$$\theta_n = \begin{cases} 0 & \text{for all } n \neq 3, 7, 11, 15, \ldots \\ -\pi & n = 3, 7, 11, 15, \ldots \end{cases}$$

[†]Because $\cos (x \pm \pi) = -\cos x$, we could have chosen the phase π or $-\pi$. In fact, $\cos (x \pm N\pi) = -\cos x$ for any odd integral value of N. Therefore the phase can be chosen as $\pm N\pi$, where N is any convenient odd integer.

We might use these values to plot amplitude and phase spectra. However, we can simplify our task in this special case if we allow amplitude C_n to take on negative values. If this is allowed, we do not need a phase of $-\pi$ to account for the sign as seen from Eq. (6.17). This means that phases of all components are zero, and we can discard the phase spectrum and manage with only the amplitude spectrum, as shown in Fig. 6.5b. Observe that there is no loss of information in doing so and that the amplitude spectrum in Fig. 6.5b has the complete information about the Fourier series in Eq. (6.17). *Therefore, whenever all sine terms vanish* $(b_n = 0)$, *it is convenient to allow C_n to take on negative values.* This permits the spectral information to be conveyed by a single spectrum.[†]

Let us investigate the behavior of the series at the points of discontinuities. For the discontinuity at $t = \pi/2$, the values of $x(t)$ on either sides of the discontinuity are $x((\pi/2)^-) = 1$ and $x((\pi/2)^+) = 0$. We can verify by setting $t = \pi/2$ in Eq. (6.17) that $x(\pi/2) = 0.5$, which is a value midway between the values of $x(t)$ on the either sides of the discontinuity at $t = \pi/2$.

6.1-2 The Effect of Symmetry

The Fourier series for the signal $x(t)$ in Fig. 6.2a (Example 6.1) consists of sine and cosine terms, but the series for the signal $x(t)$ in Fig. 6.3a (Example 6.2) consists of sine terms only and the series for the signal $x(t)$ in Fig. 6.5a (Example 6.4) consists of cosine terms only. This is no accident. We can show that the Fourier series of any even periodic function $x(t)$ consists of cosine terms only and the series for any odd periodic function $x(t)$ consists of sine terms only. Moreover, because of symmetry (even or odd), the information of one period of $x(t)$ is implicit in only half the period, as seen in Figs. 6.3a and 6.5a. In these cases, knowing the signal over a half-period and knowing the kind of symmetry (even or odd), we can determine the signal waveform over a complete period. For this reason, the Fourier coefficients in these cases can be computed by integrating over only half the period rather than a complete period. To prove this result, recall that

$$a_0 = \frac{1}{T_0} \int_{-T_0/2}^{T_0/2} x(t)\, dt$$

$$a_n = \frac{2}{T_0} \int_{-T_0/2}^{T_0/2} x(t) \cos n\omega_0 t\, dt$$

$$b_n = \frac{2}{T_0} \int_{-T_0/2}^{T_0/2} x(t) \sin n\omega_0 t\, dt$$

Recall also that $\cos n\omega_0 t$ is an even function and $\sin n\omega_0 t$ is an odd function of t. If $x(t)$ is an even function of t, then $x(t) \cos n\omega_0 t$ is also an even function and $x(t) \sin n\omega_0 t$ is an odd

---

[†]Here, the distinction between amplitude A_n and magnitude $C_n = |A_n|$ would have been useful. But, for the reason mentioned in the footnote on page 600, we refrain from this distinction formally.

function of t (see Section 1.5-1). Therefore, following from Eqs. (1.33a) and (1.33b),

$$a_0 = \frac{2}{T_0} \int_0^{T_0/2} x(t)\, dt \tag{6.18a}$$

$$a_n = \frac{4}{T_0} \int_0^{T_0/2} x(t) \cos n\omega_0 t\, dt \tag{6.18b}$$

$$b_n = 0 \tag{6.18c}$$

Similarly, if $x(t)$ is an odd function of t, then $x(t) \cos n\omega_0 t$ is an odd function of t and $x(t) \sin n\omega_0 t$ is an even function of t. Therefore

$$a_n = 0 \qquad \text{for all } n \tag{6.19a}$$

$$b_n = \frac{4}{T_0} \int_0^{T_0/2} x(t) \sin n\omega_0 t\, dt. \tag{6.19b}$$

Observe that because of symmetry, the integration required to compute the coefficients need be performed over only half the period.

If a periodic signal $x(t)$ shifted by half the period remains unchanged except for a sign—that is, if

$$x\left(t - \frac{T_0}{2}\right) = -x(t) \tag{6.20}$$

the signal is said to have a *half-wave* symmetry. It can be shown that for a signal with a half-wave symmetry, all the even-numbered harmonics vanish (see Prob. 6.1-5). The signal in Fig. 6.3a is an example of such a symmetry. The signal in Fig. 6.5a also has this symmetry, although it is not obvious owing to a dc component. If we subtract the dc component of 0.5 from this signal, the remaining signal has a half-wave symmetry. For this reason this signal has a dc component 0.5 and only the odd harmonics.

EXERCISE E6.1

Find the compact trigonometric Fourier series for periodic signals shown in Fig. 6.6. Sketch their amplitude and phase spectra. Allow C_n to take on negative values if $b_n = 0$ so that the

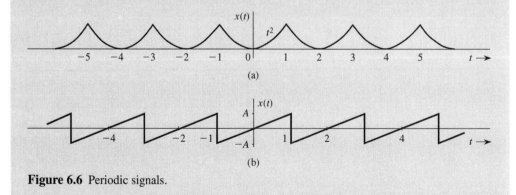

Figure 6.6 Periodic signals.

phase spectrum can be eliminated. [Hint: Use Eqs. (6.18) and (6.19) for appropriate symmetry conditions.]

ANSWERS

(a) $x(t) = \dfrac{1}{3} - \dfrac{4}{\pi^2}\left(\cos \pi t - \dfrac{1}{4}\cos 2\pi t + \dfrac{1}{9}\cos 3\pi t - \dfrac{1}{16}\cos 4\pi t + \cdots\right)$

$= \dfrac{1}{3} + \dfrac{4}{\pi^2}\displaystyle\sum_{n=1}^{\infty}\dfrac{(-1)^n}{n^2}\cos n\pi t$

(b) $x(t) = \dfrac{2A}{\pi}\left[\sin \pi t - \dfrac{1}{2}\sin 2\pi t + \dfrac{1}{3}\sin 3\pi t - \dfrac{1}{4}\sin 4\pi t + \cdots\right]$

$= \dfrac{2A}{\pi}\left[\cos (\pi t - 90°) + \dfrac{1}{2}\cos (2\pi t + 90°) + \dfrac{1}{3}\cos (3\pi t - 90°)\right.$

$\left. + \dfrac{1}{4}\cos (4\pi t + 90°) + \cdots\right]$

6.1-3   Determining the Fundamental Frequency and Period

We have seen that every periodic signal can be expressed as a sum of sinusoids of a fundamental frequency ω_0 and its harmonics. One may ask whether a sum of sinusoids of *any* frequencies represents a periodic signal. If so, how does one determine the period? Consider the following three functions:

$$x_1(t) = 2 + 7\cos \left(\tfrac{1}{2}t + \theta_1\right) + 3\cos \left(\tfrac{2}{3}t + \theta_2\right) + 5\cos \left(\tfrac{7}{6}t + \theta_3\right)$$

$$x_2(t) = 2\cos (2t + \theta_1) + 5\sin (\pi t + \theta_2)$$

$$x_3(t) = 3\sin \left(3\sqrt{2}t + \theta\right) + 7\cos \left(6\sqrt{2}t + \phi\right)$$

Recall that every frequency in a periodic signal is an integer multiple of the fundamental frequency ω_0. Therefore the ratio of any two frequencies is of the form m/n, where m and n are integers. This means that the ratio of any two frequencies is a rational number. When the ratio of two frequencies is a rational number, the frequencies are said to be *harmonically related*.

The largest number of which all the frequencies are integer multiples is the fundamental frequency. In other words, the fundamental frequency is the *greatest common factor* (GCF) of all the frequencies in the series. The frequencies in the spectrum of $x_1(t)$ are 1/2, 2/3, and 7/6 (we do not consider dc). The ratios of the successive frequencies are 3:4 and 4:7, respectively. Because both these numbers are rational, all the three frequencies in the spectrum are harmonically related, and the signal $x_1(t)$ is periodic. The GCF, that is, the greatest number of which $1/2$, $2/3$, and $7/6$ are integer multiples, is $1/6$.[†] Moreover, $3(1/6) = 1/2$, $4(1/6) = 2/3$, and $7(1/6) = 7/6$. Therefore the fundamental frequency is $1/6$, and the three frequencies in the spectrum are the third, fourth, and seventh harmonics. Observe that the fundamental frequency component is absent in this Fourier series.

The signal $x_2(t)$ is not periodic because the ratio of two frequencies in the spectrum is $2/\pi$, which is not a rational number. The signal $x_3(t)$ is periodic because the ratio of frequencies $3\sqrt{2}$ and $6\sqrt{2}$ is $1/2$, a rational number. The greatest common factor of $3\sqrt{2}$ and $6\sqrt{2}$ is $3\sqrt{2}$. Therefore the fundamental frequency $\omega_0 = 3\sqrt{2}$, and the period

$$T_0 = \frac{2\pi}{(3\sqrt{2})} = \frac{\sqrt{2}}{3}\pi \qquad (6.21)$$

EXERCISE E6.2

Determine whether the signal

$$x(t) = \cos\left(\tfrac{2}{3}t + 30°\right) + \sin\left(\tfrac{4}{5}t + 45°\right)$$

is periodic. If it is periodic, find the fundamental frequency and the period. What harmonics are present in $x(t)$?

ANSWER
Periodic with $\omega_0 = 2/15$ and period $T_0 = 15\pi$. The fifth and sixth harmonics.

A HISTORICAL NOTE:
BARON JEAN-BAPTISTE-JOSEPH FOURIER (1768−1830)

The Fourier series and integral comprise a most beautiful and fruitful development, which serves as an indispensable instrument in the treatment of many problems in mathematics, science, and engineering. Maxwell was so taken by the beauty of the Fourier series that he called it a great mathematical poem. In electrical engineering, it is central to the areas of communication, signal processing, and several other fields, including antennas, but its initial reception by the scientific world was not enthusiastic. In fact, Fourier could not get his results published as a paper.

Fourier, a tailor's son, was orphaned at age 8 and educated at a local military college (run by Benedictine monks), where he excelled in mathematics. The Benedictines prevailed upon the young genius to choose the priesthood as his vocation, but revolution broke out before he could take his vows. Fourier joined the people's party. But in its early days, the French Revolution, like most such upheavals, liquidated a large segment of the intelligentsia, including prominent scientists such as Lavoisier. Observing this trend, many intellectuals decided to leave France to save themselves from a rapidly rising tide of barbarism. Fourier, despite his early enthusiasm for

---

[†]The greatest common factor of $a_1/b_1, a_2/b_2, \ldots, a_m/b_m$ is the ratio of the GCF of the numerators set $(a_1, a_2, \ldots, a_m)$ to the LCM (least common multiple) of the denominator set $(b_1, b_2, \ldots, b_m)$. For instance, for the set $(2/3, 6/7, 2)$, the GCF of the numerator set $(2, 6, 2)$ is 2; the LCM of the denominator set $(3, 7, 1)$ is 21. Therefore $2/21$ is the largest number of which $2/3, 6/7$, and 2 are integer multiples.

Jean-Baptiste-Joseph Fourier

Napoleon

the Revolution, narrowly escaped the guillotine twice. It was to the everlasting credit of Napoleon that he stopped the persecution of the intelligentsia and founded new schools to replenish their ranks. The 26-year-old Fourier was appointed chair of mathematics at the newly created École Normale in 1794.[1]

Napoleon was the first modern ruler with a scientific education, and he was one of the rare persons who are equally comfortable with soldiers and scientists. The age of Napoleon was one of the most fruitful in the history of science. Napoleon liked to sign himself aś "member of *Institut de France*" (a fraternity of scientists), and he once expressed to Laplace his regret that "force of circumstances has led me so far from the career of a scientist."[2] Many great figures in science and mathematics, including Fourier and Laplace, were honored and promoted by Napoleon. In 1798 he took a group of scientists, artists, and scholars—Fourier among them—on his Egyptian expedition, with the promise of an exciting and historic union of adventure and research. Fourier proved to be a capable administrator of the newly formed Institut d'Égypte, which, incidentally, was responsible for the discovery of the Rosetta stone. The inscription on this stone in two languages and three scripts (hieroglyphic, demotic, and Greek) enabled Thomas Young and Jean-François Champollion, a protégé of Fourier, to invent a method of translating hieroglyphic writings of ancient Egypt—the only significant result of Napoleon's Egyptian expedition.

Back in France in 1801, Fourier briefly served in his former position as professor of mathematics at the École Polytechnique in Paris. In 1802 Napoleon appointed him the prefect of Isère (with its headquarters in Grenoble), a position in which Fourier served with distinction. Fourier was created Baron of the Empire by Napoleon in 1809. Later, when Napoleon was exiled to Elba, his route was to take him through Grenoble. Fourier had the route changed to avoid meeting

Napoleon, which would have displeased Fourier's new master, King Louis XVIII. Within a year, Napoleon escaped from Elba and returned to France. At Grenoble, Fourier was brought before him in chains. Napoleon scolded Fourier for his ungrateful behavior but reappointed him the prefect of Rhône at Lyons. Within four months Napoleon was defeated at Waterloo and was exiled to St. Helena, where he died in 1821. Fourier once again was in disgrace as a Bonapartist and had to pawn his effects to keep himself alive. But through the intercession of a former student, who was now a prefect of Paris, he was appointed director of the statistical bureau of the Seine, a position that allowed him ample time for scholarly pursuits. Later, in 1827, he was elected to the powerful position of perpetual secretary of the Paris Academy of Science, a section of the Institut de France.[3]

While serving as the prefect of Grenoble, had Fourier carried on his elaborate investigation of the propagation of heat in solid bodies, which led him to the Fourier series and the Fourier integral. On December 21, 1807, he announced these results in a prize paper on the theory of heat. Fourier claimed that an arbitrary function (continuous or with discontinuities) defined in a finite interval by an arbitrarily capricious graph can always be expressed as a sum of sinusoids (Fourier series). The judges, who included the great French mathematicians Laplace, Lagrange, Legendre, Monge, and LaCroix, admitted the novelty and importance of Fourier's work but criticized it for lack of mathematical rigor and generality. Lagrange thought it incredible that a sum of sines and cosines could add up to anything but an infinitely differentiable function. Moreover, one of the properties of an infinitely differentiable function is that if we know its behavior over an arbitrarily small interval, we can determine its behavior over the entire range (the Taylor–Maclaurin series). Such a function is far from an arbitrary or a capriciously drawn graph.[4] Laplace had additional reason to criticize Fourier's work. Laplace and his students had already approached the problem of heat conduction by a different angle, and Laplace was reluctant to accept the superiority of Fourier's method.[5] Fourier thought the criticism unjustified but was unable to prove his claim because the tools required for operations with infinite series were not available at the time. However, posterity has proved Fourier to be closer to the truth than his critics. This is the classic conflict between pure mathematicians and physicists or engineers, as we saw earlier (Chapter 4) in the life of Oliver Heaviside. In 1829 Dirichlet proved Fourier's claim concerning capriciously drawn functions with a few restrictions (Dirichlet conditions).

Although three of the four judges were in favor of publication, Fourier's paper was rejected because of vehement opposition by Lagrange. Fifteen years later, after several attempts and disappointments, Fourier published the results in expanded form as a text, *Théorie analytique de la chaleur*, which is now a classic.

6.2 EXISTENCE AND CONVERGENCE OF THE FOURIER SERIES

For the existence of the Fourier series, coefficients a_0, a_n, and b_n in Eqs. (6.11) must be finite. It follows from Eqs. (6.11a), (6.11b), and (6.11c) that the existence of these coefficients is guaranteed if $x(t)$ is absolutely integrable over one period; that is,

$$\int_{T_0} |x(t)|\, dt < \infty \tag{6.22}$$

However, existence, by itself, does not inform us about the nature and the manner in which the series converges. We shall first discuss the notion of convergence.

6.2-1 Convergence of a Series

The key to many puzzles lies in the nature of the convergence of the Fourier series. Convergence of infinite series is a complex problem. It took mathematicians several decades to understand the convergence aspect of the Fourier series. We shall barely scratch the surface here.

   Nothing annoys a student more than the discussion of convergence. Have we not proved, they ask, that a periodic signal $x(t)$ can be expressed as a Fourier series? Then why spoil the fun by this annoying discussion? All we have shown so far is that a signal represented by a Fourier series in Eq. (6.3) is periodic. We have not proved the converse, that every periodic signal can be expressed as a Fourier series. This issue will be tackled later, in Section 6.5-4, where it will be shown that a periodic signal can be represented by a Fourier series, as in Eq. (6.3), where the equality of the two sides of the equation is not in the ordinary sense, but in the mean-square sense (explained later in this discussion). But the astute reader should have been skeptical of the claims of the Fourier series to represent discontinuous functions in Figs. 6.2a and 6.5a. If $x(t)$ has a jump discontinuity, say at $t = 0$, then $x(0^+)$, $x(0)$, and $x(0^-)$ are generally different. How could a series consisting of the sum of continuous functions of the smoothest type (sinusoids) add to one value at $t = 0^-$ and a different value at $t = 0$ and yet another value at $t = 0^+$? The demand is impossible to satisfy unless the math involved executes some spectacular acrobatics. How does a Fourier series act under such conditions? Precisely for this reason, the great mathematicians Lagrange and Laplace, two of the judges examining Fourier's paper, were skeptical of Fourier's claims and voted against publication of the paper that later became a classic.

   There are also other issues. In any practical application, we can use only a finite number of terms in a series. If, using a fixed number of terms, the series guarantees convergence within an arbitrarily small error at every value of t, such a series is highly desirable, and is called an *uniformly convergent* series. If a series converges at every value of t, but to guarantee convergence within a given error requires different number of terms at different t, then the series is still convergent, but is less desirable. It goes under the name *pointwise convergent* series.

   Finally, we have the case of a series that refuses to converge at some t, no matter how many terms are added. But the series may *converge in the mean,* that is, the energy of the difference between $x(t)$ and the corresponding finite term series approaches zero as the number of terms approaches infinity.[†] To explain this concept, let us consider representation of a function $x(t)$ by an infinite series

$$x(t) = \sum_{n=1}^{\infty} z_n(t) \tag{6.23}$$

Let the partial sum of the first N terms of the series on the right-hand side be denoted by $x_N(t)$,

---

[†]The reason for calling this behavior "convergence in the mean" is that minimizing the error energy over a certain interval is equivalent to minimizing the mean-squared value of the error over the same interval.

that is,

$$x_N(t) = \sum_{n=1}^{N} z_n(t) \tag{6.24}$$

If we approximate $x(t)$ by $x_N(t)$ (the partial sum of the first N terms of the series), the *error* in the approximation is the difference $x(t) - x_N(t)$. The series converges *in the mean* to $x(t)$ in the interval $(0, T_0)$ if

$$\int_0^{T_0} |x(t) - x_N(t)|^2 \, dt \to 0 \qquad \text{as} \quad N \to \infty \tag{6.25}$$

Hence, the energy of the error $x(t) - x_N(t)$ approaches zero as $N \to \infty$. This form of convergence does not require for the series to be equal to $x(t)$ for all t. It just requires the energy of the difference (area under $|x(t) - x_N(t)|^2$) to vanish as $N \to \infty$. Superficially it may appear that if the energy of a signal over an interval is zero, the signal (the error) must be zero everywhere. This is not true. The signal energy can be zero even if there are nonzero values at finite number of isolated points. This is because although the signal is nonzero at a point (and zero everywhere else), the area under its square is still zero. Thus, a series that converges in the mean to $x(t)$ need not converge to $x(t)$ at a finite number of points. This is precisely what happens to the Fourier series when $x(t)$ has jump discontinuities. This is also what makes Fourier series convergence compatible with the Gibbs phenomenon, to be discussed later in this section.

There is a simple criterion for ensuring that a periodic signal $x(t)$ has a Fourier series that converges in the mean. The Fourier series for $x(t)$ converges to $x(t)$ in the mean if $x(t)$ has a finite energy over one period, that is,

$$\int_{T_0} |x(t)|^2 \, dt < \infty \tag{6.26}$$

Thus, the periodic signal $x(t)$, having a finite energy over one period, guarantees the convergence in the mean of its Fourier series. In all the examples discussed so far, condition (6.26) is satisfied, and hence the corresponding Fourier series converges in the mean. Condition (6.26), like condition (6.22), guarantees that the Fourier coefficients are finite.

We shall now discuss an alternate set of criteria, due to Dirichlet, for convergence of the Fourier series.

DIRICHLET CONDITIONS

Dirichlet showed that if $x(t)$ satisfies certain conditions (*Dirichlet conditions*), its Fourier series is guaranteed to converge pointwise at all points where $x(t)$ is continuous. Moreover, at the points of discontinuities, $x(t)$ converges to the value midway between the two values of $x(t)$ on either sides of the discontinuity. These conditions are:

1. The function $x(t)$ must be absolutely integrable; that is, it must satisfy Eq. (6.22).

2. The function $x(t)$ must have only a finite number of finite discontinuities in one period.

3. The function $x(t)$ must contain only a finite number of maxima and minima in one period.

All practical signals, including those in Examples 6.1–6.4, satisfy these conditions.

6.2-2 The Role of Amplitude and Phase Spectra in Waveshaping

The trigonometric Fourier series of a signal $x(t)$ shows explicitly the sinusoidal components of $x(t)$. We can synthesize $x(t)$ by adding the sinusoids in the spectrum of $x(t)$. Let us synthesize the square-pulse periodic signal $x(t)$ of Fig. 6.5a by adding successive harmonics in its spectrum step by step and observing the similarity of the resulting signal to $x(t)$. The Fourier series for this function as found in Eq. (6.17) is

$$x(t) = \frac{1}{2} + \frac{2}{\pi}\left(\cos t - \frac{1}{3}\cos 3t + \frac{1}{5}\cos 5t - \frac{1}{7}\cos 7t + \cdots\right)$$

We start the synthesis with only the first term in the series ($n = 0$), a constant $1/2$ (dc); this is a gross approximation of the square wave, as shown in Fig. 6.7a. In the next step we add the dc ($n = 0$) and the first harmonic (fundamental), which results in a signal shown in Fig. 6.7b. Observe that the synthesized signal somewhat resembles $x(t)$. It is a smoothed-out version of $x(t)$. The sharp corners in $x(t)$ are not reproduced in this signal because sharp corners mean rapid changes, and their reproduction requires rapidly varying (i.e., higher-frequency) components, which are excluded. Figure 6.7c shows the sum of dc, first, and third harmonics (even harmonics are absent). As we increase the number of harmonics progressively, as shown in Fig. 6.7d (sum up to the fifth harmonic) and 6.7e (sum up to the nineteenth harmonic), the edges of the pulses become sharper and the signal resembles $x(t)$ more closely.

ASYMPTOTIC RATE OF AMPLITUDE SPECTRUM DECAY

Figure 6.7 brings out one interesting aspect of the Fourier series. Lower frequencies in the Fourier series affect the large-scale behavior of $x(t)$, whereas the higher frequencies determine the fine structure such as rapid wiggling. Hence, sharp changes in $x(t)$, being a part of fine structure, necessitate higher frequencies in the Fourier series. The sharper the change [the higher the time derivative $\dot{x}(t)$], the higher are the frequencies needed in the series.

The amplitude spectrum indicates the amounts (amplitudes) of various frequency components of $x(t)$. If $x(t)$ is a smooth function, its variations are less rapid. Synthesis of such a function requires predominantly lower-frequency sinusoids and relatively small amounts of rapidly varying (higher-frequency) sinusoids. The amplitude spectrum of such a function would decay swiftly with frequency. To synthesize such a function we require fewer terms in the Fourier series for a good approximation. On the other hand, a signal with sharp changes, such as jump discontinuities, contains rapid variations, and its synthesis requires relatively large amount of high-frequency components. The amplitude spectrum of such a signal would decay slowly with frequency, and to synthesize such a function, we require many terms in its Fourier series for a good approximation. The square wave $x(t)$ is a discontinuous function with jump discontinuities, and therefore its amplitude spectrum decays rather slowly, as $1/n$ [see Eq. (6.17)]. On the other hand, the triangular pulse periodic signal in Fig. 6.3a is smoother because it is a continuous function (no jump discontinuities). Its spectrum decays rapidly with frequency as $1/n^2$ [see Eq. (6.16)].

We can show that[6] that if the first $k - 1$ derivatives of a periodic signal $x(t)$ are continuous and the kth derivative is discontinuous, then its amplitude spectrum C_n decays with frequency

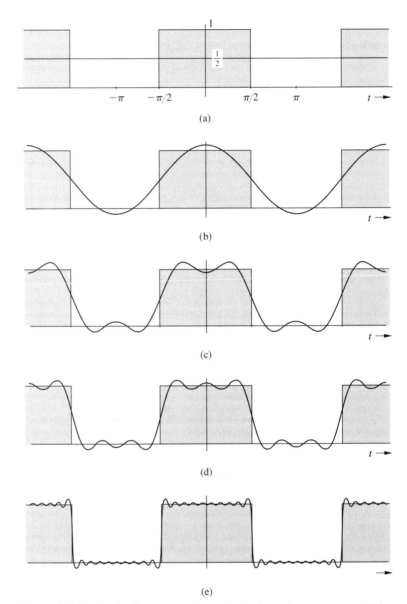

Figure 6.7 Synthesis of a square-pulse periodic signal by successive addition of its harmonics.

at least as rapidly as $1/n^{k+1}$. This result provides a simple and useful means for predicting the asymptotic rate of convergence of the Fourier series. In the case of the square-wave signal (Fig. 6.5a), the zeroth derivative of the signal (the signal itself) is discontinuous, so that $k = 0$. For the triangular periodic signal in Fig. 6.3a, the first derivative is discontinuous; that is, $k = 1$. For this reason the spectra of these signals decay as $1/n$ and $1/n^2$, respectively.

PHASE SPECTRUM: THE WOMAN BEHIND A SUCCESSFUL MAN[†]

The role of the amplitude spectrum in shaping the waveform $x(t)$ is quite clear. However, the role of the phase spectrum in shaping this waveform is less obvious. Yet, the phase spectrum, like the woman behind a successful man, plays an equally important role in waveshaping. We can explain this role by considering a signal $x(t)$ that has rapid changes such as jump discontinuities. To synthesize an instantaneous change at a jump discontinuity, the phases of the various sinusoidal components in its spectrum must be such that all (or most) of the harmonic components will have one sign before the discontinuity and the opposite sign after the discontinuity. This will result in a sharp change in $x(t)$ at the point of discontinuity. We can verify this fact in any waveform with jump discontinuity. Consider, for example, the sawtooth waveform in Fig. 6.6b. This waveform has a discontinuity at $t = 1$. The Fourier series for this waveform, as given in Exercise E6.1b, is

$$x(t) = \frac{2A}{\pi}\left[\cos{(\pi t - 90°)} + \frac{1}{2}\cos{(2\pi t + 90°)} + \frac{1}{3}\cos{(3\pi t - 90°)}\right.$$
$$\left. + \frac{1}{4}\cos{(4\pi t + 90°)} + \cdots\right]$$

Figure 6.8 shows the first three components of this series. The phases of all the (infinite) components are such that all the components are positive just before $t = 1$ and turn negative

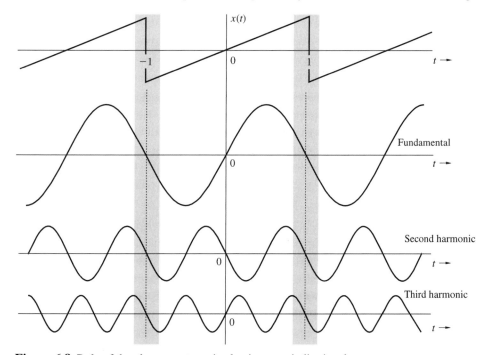

Figure 6.8 Role of the phase spectrum in shaping a periodic signal.

[†]Or, to keep up with times, "the man behind a successful woman."

just after $t = 1$, the point of discontinuity. The same behavior is also observed at $t = -1$, where a similar discontinuity occurs. This sign change in all the harmonics adds up to produce very nearly a jump discontinuity. The role of the phase spectrum is crucial in achieving a sharp change in the waveform. If we ignore the phase spectrum when trying to reconstruct this signal, the result will be a smeared and spread-out waveform. In general, the phase spectrum is just as crucial as the amplitude spectrum in determining the waveform. *The synthesis of any signal $x(t)$ is achieved by using a proper combination of amplitudes and phases of various sinusoids. This unique combination is the Fourier spectrum of $x(t)$.*

FOURIER SYNTHESIS OF DISCONTINUOUS FUNCTIONS: THE GIBBS PHENOMENON

Figure 6.7 showed the square function $x(t)$ and its approximation by a truncated trigonometric Fourier series that includes only the first N harmonics for $N = 1, 3, 5$, and 19. The plot of the truncated series approximates closely the function $x(t)$ as N increases, and we expect that the series will converge exactly to $x(t)$ as $N \to \infty$. Yet the curious fact, as seen from Fig. 6.7, is that even for large N, the truncated series exhibits an oscillatory behavior and an overshoot approaching a value of about 9% in the vicinity of the discontinuity at the nearest peak of oscillation.[†] Regardless of the value of N, the overshoot remains at about 9%. Such strange behavior certainly would undermine anyone's faith in the Fourier series. In fact, this behavior puzzled many scholars at the turn of the century. Josiah Willard Gibbs, an eminent mathematical physicist who was the inventor of vector analysis, gave a mathematical explanation of this behavior (now called the *Gibbs phenomenon*).

We can reconcile the apparent aberration in the behavior of the Fourier series by observing from Fig. 6.7 that the frequency of oscillation of the synthesized signal is $N f_0$, so the width of the spike with 9% overshoot is approximately $1/2N f_0$. As we increase N, the frequency of oscillation increases and the spike width $1/2N f_0$ diminishes. As $N \to \infty$, the error power $\to 0$ because the error consists mostly of the spikes, whose widths $\to 0$. Therefore, as $N \to \infty$, the corresponding Fourier series differs from $x(t)$ by about 9% at the immediate left and right of the points of discontinuity, and yet the error power $\to 0$. The reason for all this confusion is that in this case, the Fourier series converges in the mean. When this happens, all we promise is that the error energy (over one period) $\to 0$ as $N \to \infty$. Thus, the series may differ from $x(t)$ at some points and yet have the error signal power zero, as verified earlier. Note that the series, in this case also converges pointwise at all points except the points of discontinuity. It is precisely at the discontinuities that the series differs from $x(t)$ by 9%.[‡]

When we use only the first N terms in the Fourier series to synthesize a signal, we are abruptly terminating the series, giving a unit weight to the first N harmonics and zero weight to all the remaining harmonics beyond N. This abrupt termination of the series causes the Gibbs phenomenon in synthesis of discontinuous functions. Section 7.8 offers more discussion on the Gibbs phenomenon, its ramifications, and cure.

[†]There is also an undershoot of 9% at the other side [at $t = (\pi/2)^+$] of the discontinuity.

[‡]Actually, at discontinuities, the series converges to a value midway between the values on either side of the discontinuity. The 9% overshoot occurs at $t = (\pi/2)^-$ and 9% undershoot occurs at $t = (\pi/2)^+$.

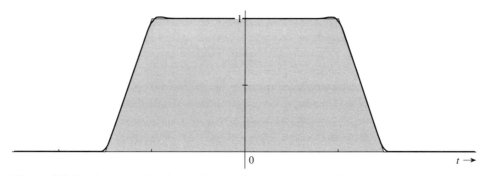

Figure 6.9 Fourier synthesis of a continuous signal using first 19 harmonics.

The Gibbs phenomenon is present only when there is a jump discontinuity in $x(t)$. When a continuous function $x(t)$ is synthesized by using the first N terms of the Fourier series, the synthesized function approaches $x(t)$ for all t as $N \to \infty$. No Gibbs phenomenon appears. This can be seen in Fig. 6.9, which shows one cycle of a continuous periodic signal being synthesized from the first 19 harmonics. Compare the similar situation for a discontinuous signal in Fig. 6.7.

EXERCISE E6.3

By inspection of signals in Figs. 6.2a, 6.6a, and 6.6b, determine the asymptotic rate of decay of their amplitude spectra.

ANSWER
$1/n$, $1/n^2$, and $1/n$, respectively.

COMPUTER EXAMPLE C6.2

Analogous to Fig. 6.7, demonstrate the synthesis of the square wave of Fig. 6.5a by successively adding its Fourier components.

```
>> x = inline('mod(t+pi/2,2*pi)<=pi'); t = linspace(-2*pi,2*pi,1000);
>> sumterms = zeros(16,length(t)); sumterms(1,:) = 1/2;
>> for n = 1:size(sumterms,1)-1;
>>      sumterms(n+1,:) = (2/(pi*n)*sin(pi*n/2))*cos(n*t);
>> end
>> x_N = cumsum(sumterms); figure(1); clf; ind = 0;
>> for N = [0,1:2:size(sumterms,1)-1],
>>      ind = ind+1; subplot(3,3,ind);
>>      plot(t,x_N(N+1,:),'k',t,x(t),'k--'); axis([-2*pi 2*pi -0.2 1.2]);
>>      xlabel('t'); ylabel(['x_{',num2str(N),'}(t)']);
>> end
```

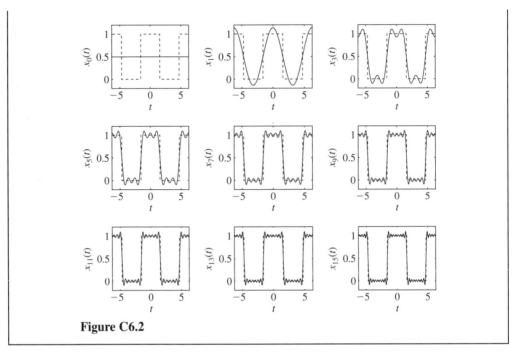

Figure C6.2

A HISTORICAL NOTE ON THE
GIBBS PHENOMENON

Normally speaking, troublesome functions with strange behavior are invented by mathematicians; we rarely see such oddities in practice. In the case of the Gibbs phenomenon, however, the tables were turned. A rather puzzling behavior was observed in a mundane object, a mechanical wave synthesizer, and then well-known mathematicians of the day were dispatched on the scent of it to discover its hideout.

Albert Michelson (of Michelson–Morley fame) was an intense, practical man who developed ingenious physical instruments of extraordinary precision, mostly in the field of optics. His harmonic analyzer, developed in 1898, could compute the first 80 coefficients of the Fourier series of a signal $x(t)$ specified by any graphical description. The instrument could also be used as a harmonic synthesizer, which could plot a function $x(t)$ generated by summing the first 80 harmonics (Fourier components) of arbitrary amplitudes and phases. This analyzer, therefore, had the ability of self-checking its operation by analyzing a signal $x(t)$ and then adding the resulting 80 components to see whether the sum yielded a close approximation of $x(t)$.

Michelson found that the instrument checked very well with most of signals analyzed. However, when he tried a discontinuous function, such as a square wave,[†] a curious behavior was observed. The sum of 80 components showed oscillatory behavior (ringing), with an overshoot of 9% in the vicinity of the points of discontinuity. Moreover, this behavior was a constant

---

[†]Actually it was a periodic sawtooth signal.

Albert Michelson

Josiah Willard Gibbs

feature regardless of the number of terms added. A larger number of terms made the oscillations proportionately faster, but regardless of the number of terms added, the overshoot remained 9%. This puzzling behavior caused Michelson to suspect some mechanical defect in his synthesizer. He wrote about his observation in a letter to *Nature* (December 1898). Josiah Willard Gibbs, who was a professor at Yale, investigated and clarified this behavior for a sawtooth periodic signal in a letter to *Nature*.[7] Later, in 1906, Bôcher generalized the result for any function with discontinuity.[8] It was Bôcher who gave the name *Gibbs phenomenon* to this behavior. Gibbs showed that the peculiar behavior in the synthesis of a square wave was inherent in the behavior of the Fourier series because of nonuniform convergence at the points of discontinuity.

This, however, is not the end of the story. Both Bôcher and Gibbs were under the impression that this property had remained undiscovered until Gibbs's work published in 1899. It is now known that what is called the Gibbs phenomenon had been observed in 1848 by Wilbraham of Trinity College, Cambridge, who clearly saw the behavior of the sum of the Fourier series components in the periodic sawtooth signal later investigated by Gibbs.[9] Apparently, this work was not known to most people, including Gibbs and Bôcher.

6.3   EXPONENTIAL FOURIER SERIES

By using Euler's equality, we can express $\cos n\omega_0 t$ and $\sin n\omega_0 t$ in terms of exponentials $e^{jn\omega_0 t}$ and $e^{-jn\omega_0 t}$. Clearly, we should be able to express the trigonometric Fourier series in Eq. (6.9) in terms of exponentials of the form $e^{jn\omega_0 t}$ with the index n taking on all integer values from $-\infty$ to ∞, including zero. Derivation of the exponential Fourier series from the results already derived for the trigonometric Fourier series is straightforward, involving conversion of sinusoids to exponentials. We shall, however, derive them here independently, without using the prior results of the trigonometric series.

This discussion shows that the *exponential Fourier series* for a periodic signal $x(t)$ can be expressed as

$$x(t) = \sum_{n=-\infty}^{\infty} D_n e^{jn\omega_0 t}$$

To derive the coefficients D_n, we multiply both sides of this equation by $e^{-jm\omega_0 t}$ (m integer) and integrate over one period. This yields

$$\int_{T_0} x(t)\, e^{-jm\omega_0 t}\, dt = \sum_{n=-\infty}^{\infty} D_n \int_{T_0} e^{j(n-m)\omega_0 t}\, dt \qquad (6.27)$$

Here we use the *orthogonality* property of exponentials (proved in the footnote below)[†]

$$\int_{T_0} e^{jn\omega_0 t}\, e^{-jm\omega_0 t}\, dt = \begin{cases} 0 & m \neq n \\ T_0 & m = n \end{cases} \qquad (6.28)$$

$$e^{j(n-m)\omega_0 t}$$

Using this result in Eq. (6.27), we obtain

$$D_m T_0 = \int_{T_0} x(t)\, e^{-jm\omega_0 t}\, dt$$

from which we obtain

$$D_m = \frac{1}{T_0} \int_{T_0} x(t)\, e^{-jm\omega_0 t}\, dt.$$

To summarize, the exponential Fourier series can be expressed as

$$x(t) = \sum_{n=-\infty}^{\infty} D_n e^{jn\omega_0 t} \qquad (6.29a)$$

where

$$D_n = \frac{1}{T_0} \int_{T_0} x(t)\, e^{-jn\omega_0 t}\, dt \qquad (6.29b)$$

Observe the compactness of expressions (6.29a) and (6.29b) and compare them with expressions corresponding to the trigonometric Fourier series. These two equations demonstrate very clearly the principal virtue of the exponential Fourier series. First, the form of the series is most compact. Second, the mathematical expression for deriving the coefficients of the series is also compact. It is much more convenient to handle the exponential series than the trigonometric one. For this reason we shall use exponential (rather than trigonometric) representation of signals in the rest of the book.

[†] We can readily prove this property as follows. For the case of $m = n$, the integrand in Eq. (6.28) is unity and the integral is T_0. When $m \neq n$, the integral on the left-hand side of Eq. (6.28) can be expressed as

$$\int_{T_0} e^{j(n-m)\omega_0 t}\, dt = \int_{T_0} \cos{(n-m)\omega_0 t}\, dt + j \int_{T_0} \sin{(n-m)\omega_0 t}\, dt$$

Both the integrals on the right-hand side represent area under $n - m$ number of cycles. Because $n - m$ is an integer, both the areas are zero. Hence follows Eq. (6.28).

We can now relate D_n to trigonometric series coefficients a_n and b_n. Setting $n = 0$ in Eq. (6.29b), we obtain

$$D_0 = a_0 \tag{6.30a}$$

Moreover, for $n \neq 0$

$$D_n = \frac{1}{T_0} \int_{T_0} x(t) \cos n\omega_0 t \, dt - \frac{j}{T_0} \int_{T_0} x(t) \sin n\omega_0 t \, dt = \frac{1}{2}(a_n - jb_n) \tag{6.30b}$$

and

$$D_{-n} = \frac{1}{T_0} \int_{T_0} x(t) \cos n\omega_0 t \, dt + \frac{j}{T_0} \int_{T_0} x(t) \sin n\omega_0 t \, dt = \frac{1}{2}(a_n + jb_n) \tag{6.30c}$$

These results are valid for general $x(t)$, real or complex. When $x(t)$ is real, a_n and b_n are real, and Eqs. (6.30b) and (6.30c) show that D_n and D_{-n} are conjugates.

$$D_{-n} = D_n^* \tag{6.31}$$

Moreover, from Eqs. (6.13), we observe that

$$a_n - jb_n = \sqrt{a_n^2 + b_n^2} \; e^{j \tan^{-1}\left(\frac{-b_n}{a_n}\right)} = C_n e^{j\theta_n}$$

Hence

$$D_0 = a_0 = C_0 \tag{6.32a}$$

and

$$D_n = \tfrac{1}{2}C_n e^{j\theta_n} \qquad D_{-n} = \tfrac{1}{2}C_n e^{-j\theta_n} \tag{6.32b}$$

Therefore

$$|D_n| = |D_{-n}| = \tfrac{1}{2}C_n \qquad n \neq 0 \tag{6.33a}$$

$$\angle D_n = \theta_n \qquad \angle D_{-n} = -\theta_n \tag{6.33b}$$

Note that $|D_n|$ are the amplitudes and $\angle D_n$ are the angles of various exponential components. From Eqs. (6.33) it follows that when $x(t)$ is real, the amplitude spectrum ($|D_n|$ versus ω) is an even function of ω and the angle spectrum ($\angle D_n$ versus ω) is an odd function of ω. For complex $x(t)$, D_n and D_{-n} are generally not conjugates.

EXAMPLE 6.5

Find the exponential Fourier series for the signal in Fig. 6.2a (Example 6.1).

In this case $T_0 = \pi$, $\omega_0 = 2\pi/T_0 = 2$, and

$$x(t) = \sum_{n=-\infty}^{\infty} D_n e^{j2nt}$$

where

$$D_n = \frac{1}{T_0} \int_{T_0} x(t) e^{-j2nt} \, dt$$

$$= \frac{1}{\pi} \int_0^\pi e^{-t/2} e^{-j2nt} \, dt$$

$$= \frac{1}{\pi} \int_0^\pi e^{-(1/2 + j2n)t} \, dt$$

$$= \frac{-1}{\pi \left(\frac{1}{2} + j2n \right)} e^{-(1/2 + j2n)t} \Big|_0^\pi$$

$$= \frac{0.504}{1 + j4n} \tag{6.34}$$

and

$$x(t) = 0.504 \sum_{n=-\infty}^{\infty} \frac{1}{1 + j4n} e^{j2nt} \tag{6.35a}$$

$$= 0.504 \left[1 + \frac{1}{1 + j4} e^{j2t} + \frac{1}{1 + j8} e^{j4t} + \frac{1}{1 + j12} e^{j6t} + \cdots \right.$$

$$\left. + \frac{1}{1 - j4} e^{-j2t} + \frac{1}{1 - j8} e^{-j4t} + \frac{1}{1 - j12} e^{-j6t} + \cdots \right] \tag{6.35b}$$

Observe that the coefficients D_n are complex. Moreover, D_n and D_{-n} are conjugates, as expected.

COMPUTER EXAMPLE C6.3

Following Example 6.5, compute and plot the exponential Fourier spectra for the periodic signal $x(t)$ shown in Fig. 6.2a.

The expression for D_n is derived in Example 6.5.

```
>> n = (-10:10); D_n = 0.504./(1+j*4*n);
>> clf; subplot(2,1,1); stem(n,abs(D_n),'k');
>> xlabel('n'); ylabel('|D_n|');
>> subplot(2,1,2); stem(n,angle(D_n),'k');
>> xlabel('n'); ylabel('\angle D_n [rad]');
```

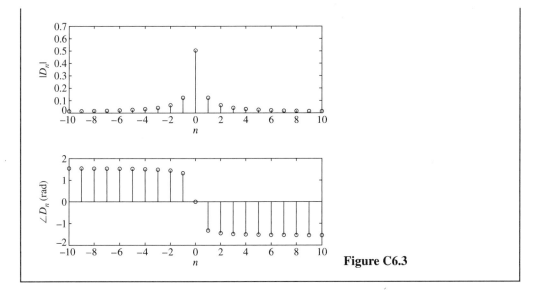

Figure C6.3

6.3-1 Exponential Fourier Spectra

In exponential spectra, we plot coefficients D_n as a function of ω. But since D_n is complex in general, we need both parts of one of two sets of plots: the real and the imaginary parts of D_n, or the magnitude and the angle of D_n. We prefer the latter because of its close connection to the amplitudes and phases of corresponding components of the trigonometric Fourier series. We therefore plot $|D_n|$ versus ω and $\angle D_n$ versus ω. This requires that the coefficients D_n be expressed in polar form as $|D_n|e^{j\angle D_n}$, where $|D_n|$ are the amplitudes and $\angle D_n$ are the angles of various exponential components. Equations (6.33) show that for real $x(t)$, the amplitude spectrum ($|D_n|$ versus ω) is an even function of ω and the angle spectrum ($\angle D_n$ versus ω) is an odd function of ω.

For the series in Example 6.5 [Eq. (6.35b)], for instance,

$$D_0 = 0.504$$

$$D_1 = \frac{0.504}{1+j4} = 0.122e^{-j75.96^\circ} \implies |D_1| = 0.122, \ \angle D_1 = -75.96^\circ$$

$$D_{-1} = \frac{0.504}{1-j4} = 0.122e^{j75.96^\circ} \implies |D_{-1}| = 0.122, \ \angle D_{-1} = 75.96^\circ$$

and

$$D_2 = \frac{0.504}{1+j8} = 0.0625e^{-j82.87^\circ} \implies |D_2| = 0.0625, \ \angle D_2 = -82.87^\circ$$

$$D_{-2} = \frac{0.504}{1-j8} = 0.0625e^{j82.87^\circ} \implies |D_{-2}| = 0.0625, \ \angle D_{-2} = 82.87^\circ$$

and so on. Note that D_n and D_{-n} are conjugates, as expected [see Eqs. (6.33)].

Figure 6.10 shows the frequency spectra (amplitude and angle) of the exponential Fourier series for the periodic signal $x(t)$ in Fig. 6.2a.

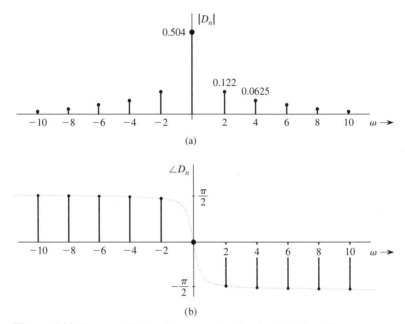

Figure 6.10 Exponential Fourier spectra for the signal in Fig. 6.2a.

We notice some interesting features of these spectra. First, the spectra exist for positive as well as negative values of ω (the frequency). Second, the amplitude spectrum is an even function of ω and the angle spectrum is an odd function of ω.

At times it may appear that the phase spectrum of a real periodic signal fails to satisfy the odd symmetry: for example, when $D_k = D_{-k} = -10$. In this case, $D_k = 10e^{j\pi}$, and therefore, $D_{-k} = 10e^{-j\pi}$. Recall that $e^{\pm j\pi} = -1$. Here, although $D_k = D_{-k}$, their phases should be taken as π and $-\pi$.

WHAT IS A NEGATIVE FREQUENCY?

The existence of the spectrum at negative frequencies is somewhat disturbing because, by definition, the frequency (number of repetitions per second) is a positive quantity. How do we interpret a negative frequency? We can use a trigonometric identity to express a sinusoid of a negative frequency $-\omega_0$ as

$$\cos(-\omega_0 t + \theta) = \cos(\omega_0 t - \theta)$$

This equation clearly shows that the frequency of a sinusoid $\cos(\omega_0 t + \theta)$ is $|\omega_0|$, which is a positive quantity. The same conclusion is reached by observing that

$$e^{\pm j\omega_0 t} = \cos \omega_0 t \pm j \sin \omega_0 t$$

Thus, the frequency of exponentials $e^{\pm j\omega_0 t}$ is indeed $|\omega_0|$. How do we then interpret the spectral plots for negative values of ω? A more satisfying way of looking at the situation is to say that *exponential spectra are a graphical representation of coefficients D_n as a function of ω. Existence of the spectrum at $\omega = -n\omega_0$ is merely an indication that an exponential component*

$e^{-jn\omega_0 t}$ *exists in the series.* We know that a sinusoid of frequency $n\omega_0$ can be expressed in terms of a pair of exponentials $e^{jn\omega_0 t}$ and $e^{-jn\omega_0 t}$.

We see a close connection between the exponential spectra in Fig. 6.10 and the spectra of the corresponding trigonometric Fourier series for $x(t)$ (Fig. 6.2b, 6.2c). Equations (6.33) explain the reason for close connection, for real $x(t)$, between the trigonometric spectra (C_n and θ_n) with exponential spectra ($|D_n|$ and $\angle D_n$). The dc components D_0 and C_0 are identical in both spectra. Moreover, the exponential amplitude spectrum $|D_n|$ is half the trigonometric amplitude spectrum C_n for $n \geq 1$. The exponential angle spectrum $\angle D_n$ is identical to the trigonometric phase spectrum θ_n for $n \geq 0$. We can therefore produce the exponential spectra merely by inspection of trigonometric spectra, and vice versa. The following example demonstrates this feature.

EXAMPLE 6.6

The trigonometric Fourier spectra of a certain periodic signal $x(t)$ are shown in Fig. 6.11a. By inspecting these spectra, sketch the corresponding exponential Fourier spectra and verify your results analytically.

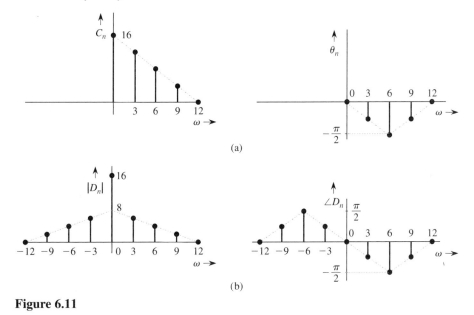

Figure 6.11

The trigonometric spectral components exist at frequencies 0, 3, 6, and 9. The exponential spectral components exist at 0, 3, 6, 9, and −3, −6, −9. Consider first the amplitude spectrum. The dc component remains unchanged: that is, $D_0 = C_0 = 16$. Now $|D_n|$ is an even function of ω and $|D_n| = |D_{-n}| = C_n/2$. Thus, all the remaining spectrum $|D_n|$ for positive n is half the trigonometric amplitude spectrum C_n, and the spectrum $|D_n|$ for negative n is a reflection about the vertical axis of the spectrum for positive n, as shown in Fig. 6.11b.

The angle spectrum $\angle D_n = \theta_n$ for positive n and is $-\theta_n$ for negative n, as depicted in Fig. 6.11b. We shall now verify that both sets of spectra represent the same signal.

Signal $x(t)$, whose trigonometric spectra are shown in Fig. 6.11a, has four spectral components of frequencies 0, 3, 6, and 9. The dc component is 16. The amplitude and the phase of the component of frequency 3 are 12 and $-\pi/4$, respectively. Therefore, this component can be expressed as $12\cos(3t - \pi/4)$. Proceeding in this manner, we can write the Fourier series for $x(t)$ as

$$x(t) = 16 + 12\cos\left(3t - \frac{\pi}{4}\right) + 8\cos\left(6t - \frac{\pi}{2}\right) + 4\cos\left(9t - \frac{\pi}{4}\right)$$

Consider now the exponential spectra in Fig. 6.11b. They contain components of frequencies 0 (dc), ± 3, ± 6, and ± 9. The dc component is $D_0 = 16$. The component e^{j3t} (frequency 3) has magnitude 6 and angle $-\pi/4$. Therefore, this component strength is $6e^{-j\pi/4}$, and it can be expressed as $(6e^{-j\pi/4})e^{j3t}$. Similarly, the component of frequency -3 is $(6e^{j\pi/4})e^{-j3t}$. Proceeding in this manner, $\hat{x}(t)$, the signal corresponding to the spectra in Fig. 6.11b, is

$$\hat{x}(t) = 16 + \left[6e^{-j\pi/4}e^{j3t} + 6e^{j\pi/4}e^{-j3t}\right] + \left[4e^{-j\pi/2}e^{j6t} + 4e^{j\pi/2}e^{-j6t}\right]$$

$$+ \left[2e^{-j\pi/4}e^{j9t} + 2e^{j\pi/4}e^{-j9t}\right]$$

$$= 16 + 6\left[e^{j(3t-\pi/4)} + e^{-j(3t-\pi/4)}\right] + 4\left[e^{j(6t-\pi/2)} + e^{-j(6t-\pi/2)}\right]$$

$$+ 2\left[e^{j(9t-\pi/4)} + e^{-j(9t-\pi/4)}\right]$$

$$= 16 + 12\cos\left(3t - \frac{\pi}{4}\right) + 8\cos\left(6t - \frac{\pi}{2}\right) + 4\cos\left(9t - \frac{\pi}{4}\right)$$

Clearly both sets of spectra represent the same periodic signal.

BANDWIDTH OF A SIGNAL

The difference between the highest and the lowest frequencies of the spectral components of a signal is the *bandwidth* of the signal. The bandwidth of the signal whose exponential spectra are shown in Fig. 6.11b is 9 (in radians). The highest and lowest frequencies are 9 and 0, respectively. Note that the component of frequency 12 has zero amplitude and is nonexistent. Moreover, the lowest frequency is 0, not -9. Recall that the frequencies (in the conventional sense) of the spectral components at $\omega = -3$, -6, and -9 in reality are 3, 6, and 9.[†] The bandwidth can be more readily seen from the trigonometric spectra in Fig. 6.11a.

[†]Some authors *do* define bandwidth as the difference between the highest and the lowest (negative) frequency in the exponential spectrum. The bandwidth according to this definition is twice that defined here. In reality, this phrasing defines not the signal bandwidth but the *spectral width* (width of the exponential spectrum of the signal).

EXAMPLE 6.7

Find the exponential Fourier series and sketch the corresponding spectra for the impulse train $\delta_{T_0}(t)$ depicted in Fig. 6.12a. From this result sketch the trigonometric spectrum and write the trigonometric Fourier series for $\delta_{T_0}(t)$.

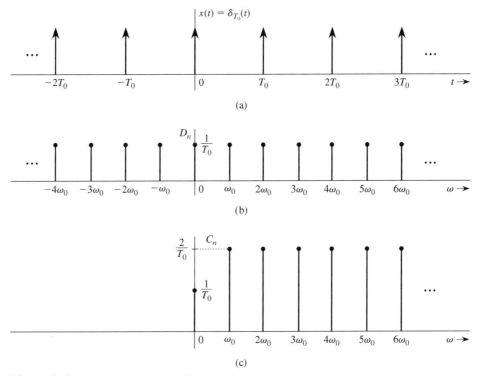

Figure 6.12 (a) Impulse train and (b, c) its Fourier spectra.

The unit impulse train shown in Fig. 6.12a can be expressed as

$$\sum_{n=-\infty}^{\infty} \delta(t - nT_0)$$

Following Papoulis, we shall denote this function as $\delta_{T_0}(t)$ for the sake of notational brevity.

The exponential Fourier series is given by

$$\delta_{T_0}(t) = \sum_{n=-\infty}^{\infty} D_n e^{jn\omega_0 t} \qquad \omega_0 = \frac{2\pi}{T_0} \tag{6.36}$$

where

$$D_n = \frac{1}{T_0} \int_{T_0} \delta_{T_0}(t) e^{-jn\omega_0 t} \, dt$$

Choosing the interval of integration $(-T_0/2, T_0/2)$ and recognizing that over this interval $\delta_{T_0}(t) = \delta(t)$,

$$D_n = \frac{1}{T_0} \int_{-T_0/2}^{T_0/2} \delta(t) e^{-jn\omega_0 t} \, dt$$

In this integral the impulse is located at $t = 0$. From the sampling property (1.24a), the integral on the right-hand side is the value of $e^{-jn\omega_0 t}$ at $t = 0$ (where the impulse is located). Therefore

$$D_n = \frac{1}{T_0} \tag{6.37}$$

Substitution of this value in Eq. (6.36) yields the desired exponential Fourier series

$$\delta_{T_0}(t) = \frac{1}{T_0} \sum_{n=-\infty}^{\infty} e^{jn\omega_0 t} \qquad \omega_0 = \frac{2\pi}{T_0} \tag{6.38}$$

Equation (6.37) shows that the exponential spectrum is uniform ($D_n = 1/T_0$) for all the frequencies, as shown in Fig. 6.12b. The spectrum, being real, requires only the amplitude plot. All phases are zero.

To sketch the trigonometric spectrum, we use Eq. (6.33) to obtain

$$C_0 = D_0 = \frac{1}{T_0}$$

$$C_n = 2|D_n| = \frac{2}{T_0} \qquad n = 1, 2, 3, \ldots$$

$$\theta_n = 0$$

Figure 6.12c shows the trigonometric Fourier spectrum. From this spectrum we can express $\delta_{T_0}(t)$ as

$$\delta_{T_0}(t) = \frac{1}{T_0} [1 + 2(\cos \omega_0 t + \cos 2\omega_0 t + \cos 3\omega_0 t + \cdots)] \qquad \omega_0 = \frac{2\pi}{T_0} \tag{6.39}$$

EFFECT OF SYMMETRY IN EXPONENTIAL FOURIER SERIES

When $x(t)$ has an even symmetry, $b_n = 0$, and from Eq. (6.30b), $D_n = a_n/2$, which is real (positive or negative). Hence, $\angle D_n$ can only be 0 or $\pm\pi$. Moreover, we may compute $D_n = a_n/2$ by using Eq. (6.18b), which requires integration over a half-period only. Similarly, when $x(t)$ has an odd symmetry, $a_n = 0$, and $D_n = -jb_n/2$ is imaginary (positive or negative). Hence, $\angle D_n$ can only be 0 or $\pm\pi/2$. Moreover, we may compute $D_n = -jb_n/2$ by using Eq. (6.19b), which requires integration over a half-period only. Note, however, that in the exponential case, we are using the symmetry property indirectly by finding the trigonometric coefficients. We cannot apply it directly to Eq. (6.29b).

EXERCISE E6.4

The exponential Fourier spectra of a certain periodic signal $x(t)$ are shown in Fig. 6.13. Determine and sketch the trigonometric Fourier spectra of $x(t)$ by inspection of Fig. 6.13. Now write the (compact) trigonometric Fourier series for $x(t)$.

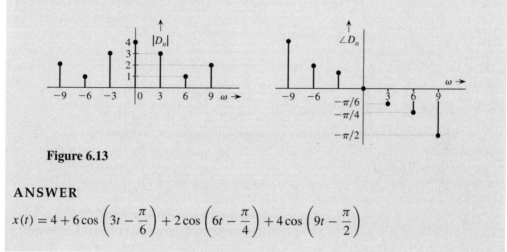

Figure 6.13

ANSWER

$$x(t) = 4 + 6\cos\left(3t - \frac{\pi}{6}\right) + 2\cos\left(6t - \frac{\pi}{4}\right) + 4\cos\left(9t - \frac{\pi}{2}\right)$$

EXERCISE E6.5

Find the exponential Fourier series and sketch the corresponding Fourier spectrum D_n versus ω for the full-wave rectified sine wave depicted in Fig. 6.14.

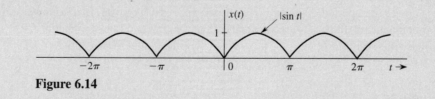

Figure 6.14

ANSWER

$$x(t) = \frac{2}{\pi}\sum_{n=-\infty}^{\infty}\frac{1}{1 - 4n^2}e^{j2nt}$$

EXERCISE E6.6

Find the exponential Fourier series and sketch the corresponding Fourier spectra for the periodic signals shown in Fig. 6.6.

ANSWERS

(a) $x(t) = \dfrac{1}{3} + \dfrac{2}{\pi^2} \displaystyle\sum_{n=-\infty}^{\infty} \dfrac{(-1)^n}{n^2} e^{jn\pi t}$

(b) $x(t) = \dfrac{jA}{\pi} \displaystyle\sum_{n=-\infty\,(n\neq 0)}^{\infty} \dfrac{(-1)^n}{n} e^{jn\pi t}$

6.3-2 Parseval's Theorem

The trigonometric Fourier series of a periodic signal $x(t)$ is given by

$$x(t) = C_0 + \sum_{n=1}^{\infty} C_n \cos(n\omega_0 t + \theta_n)$$

Every term on the right-hand side of this equation is a power signal. As shown in Example 1.2, Eq. (1.4d), the power of $x(t)$ is equal to the sum of the powers of all the sinusoidal components on the right-hand side.

$$P_x = C_0^2 + \frac{1}{2}\sum_{n=1}^{\infty} C_n^2 \tag{6.40}$$

This result is one form of the *Parseval's theorem,* as applied to power signals. It states that the power of a periodic signal is equal to the sum of the powers of its Fourier components.

We can apply the same argument to the exponential Fourier series (see Prob. 1.1-8). The power of a periodic signal $x(t)$ can be expressed as a sum of the powers of its exponential components. In Eq. (1.4e), we showed that the power of an exponential $De^{j\omega_0 t}$ is $|D^2|$. We can use this result to express the power of a periodic signal $x(t)$ in terms of its exponential Fourier series coefficients as

$$P_x = \sum_{n=-\infty}^{\infty} |D_n|^2 \tag{6.41a}$$

For a real $x(t)$, $|D_{-n}| = |D_n|$. Therefore

$$P_x = D_0^2 + 2\sum_{n=1}^{\infty} |D_n|^2 \tag{6.41b}$$

EXAMPLE 6.8

The input signal to an audio amplifier of gain 100 is given by $x(t) = 0.1 \cos \omega_0 t$. Hence, the output is a sinusoid $10 \cos \omega_0 t$. However, the amplifier, being nonlinear at higher amplitude levels, clips all amplitudes beyond ± 8 volts as shown in Fig. 6.15a. We shall determine the harmonic distortion incurred in this operation.

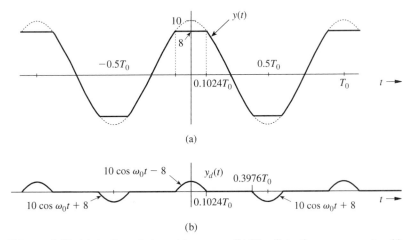

(a)

(b)

Figure 6.15 (a) A clipped sinusoid $\cos \omega_0 t$. (b) The distortion component $x_d(t)$ of the signal in (a).

The output $y(t)$ is the clipped signal in Fig. 6.15a. The distortion signal $y_d(t)$, shown in Fig. 6.15b, is the difference between the undistorted sinusoid $10 \cos \omega_0 t$ and the output signal $y(t)$. The signal $y_d(t)$, whose period is T_0 [the same as that of $y(t)$] can be described over the first cycle as

$$y_d(t) = \begin{cases} 10 \cos \omega_0 t - 8 & |t| \leq 0.1024 T_0 \\[2mm] 10 \cos \omega_0 t + 8 & \dfrac{T_0}{2} - 0.1024 T_0 \leq |t| \leq \dfrac{T_0}{2} + 0.1024 T_0 \\[2mm] 0 & \text{everywhere else} \end{cases}$$

Observe that $y_d(t)$ is an even function of t and its mean value is zero. Hence, $a_0 = C_0 = 0$, and $b_n = 0$. Thus, $C_n = a_n$ and the Fourier series for $y_d(t)$ can be expressed as

$$y_d(t) = \sum_{n=1}^{\infty} C_n \cos n\omega_0 t$$

As usual, we can compute the coefficients C_n (which is equal to a_n) by integrating $y_d(t) \cos n\omega_0 t$ over one cycle (and then dividing by $2/T_0$). Because $y_d(t)$ has even symmetry, we can find a_n by integrating the expression over a half-cycle only using Eq. (6.18b). The straightforward evaluation of the appropriate integral yields[†]

$$C_n = \begin{cases} \dfrac{20}{\pi}\left[\dfrac{\sin[0.6435(n+1)]}{n+1} + \dfrac{\sin[0.6435(n-1)]}{n-1}\right] - \dfrac{32}{\pi}\left[\dfrac{\sin(0.6435n)}{n}\right] & n \text{ odd} \\ 0 & n \text{ even} \end{cases}$$

Computing the coefficients C_1, C_2, C_3, ... from this expression, we can write

$$y_d(t) = 1.04 \cos \omega_0 t + 0.733 \cos 3\omega_0 t + 0.311 \cos 5\omega_0 t + \cdots$$

COMPUTING HARMONIC DISTORTION

We can compute the amount of harmonic distortion in the output signal by computing the power of the distortion component $y_d(t)$. Because $y_d(t)$ is an even function of t and because the energy in the first half-cycle is identical to the energy in the second half-cycle, we can compute the power by averaging the energy over a quarter-cycle. Thus

$$P_{y_d} = \frac{1}{T_0}\int_{-T_0/2}^{T_0/2} y_d{}^2(t)\, dt = \frac{1}{T_0/4}\int_0^{T_0/4} y_d{}^2(t)\, dt$$

$$= \frac{4}{T_0}\int_0^{0.1024T_0} (10 \cos \omega_0 t - 8)^2\, dt = 0.865$$

The power of the desired signal $10 \cos \omega_0 t$ is $(10)^2/2 = 50$. Hence, the total harmonic distortion is

$$D_{\text{tot}} = \frac{0.865}{50} \times 100 = 1.73\%$$

The power of the third harmonic components of $y_d(t)$ is $(0.733)^2/2 = 0.2686$. The third harmonic distortions is[‡]

$$D_3 = \frac{0.2686}{50} \times 100 = 0.5372\%$$

[†]In addition, $y_d(t)$ exhibits half-wave symmetry (see Prob. 6.1-5), where the second half-cycle is the negative of the first. Because of this property, all the even harmonics vanish, and the odd harmonics can be computed by integrating the appropriate expressions over the first half-cycle only (from $-T_0/4$ to $T_0/4$) and doubling the resulting values. Moreover, because of even symmetry, we can integrate the appropriate expressions over 0 to $T_0/4$ (instead of from $-T_0/4$ to $T_0/4$) and double the resulting values. In essence, this allows us to compute C_n by integrating the expression over the quarter-cycle only and then quadrupling the resulting values. Thus

$$C_n = a_n = \frac{8}{T_0}\int_0^{0.1024T_0} [10 \cos \omega_0 t - 8] \cos n\omega_0 t\, dt$$

[‡]In the literature, the harmonic distortion often refers to the rms distortion rather than the power distortion. The rms values are the square-root values of the corresponding powers. Thus, the third harmonic distortion in this sense is $\sqrt{(0.2686/50)} \times 100 = 7.33\%$. Alternately, we may also compute this value directly from the amplitudes of the third harmonic 0.733 and that of the fundamental as 10. The ratio of the rms values is $(0.733/\sqrt{2}) \cdot (10/\sqrt{2}) = 0.0733$ and the percentage distortion is 7.33%.

6.4 LTIC SYSTEM RESPONSE TO PERIODIC INPUTS

A periodic signal can be expressed as a sum of everlasting exponentials (or sinusoids). We also know how to find the response of an LTIC system to an everlasting exponential. From this information we can readily determine the response of an LTIC system to periodic inputs. A periodic signal $x(t)$ with period T_0 can be expressed as an exponential Fourier series

$$x(t) = \sum_{n=-\infty}^{\infty} D_n e^{jn\omega_0 t} \qquad \omega_0 = \frac{2\pi}{T_0}$$

In Section 4.8, we showed that the response of an LTIC system with transfer function $H(s)$ to an everlasting exponential input $e^{j\omega t}$ is an everlasting exponential $H(j\omega)e^{j\omega t}$. This input–output pair can be displayed as[†]

$$\underbrace{e^{j\omega t}}_{\text{input}} \Longrightarrow \underbrace{H(j\omega)e^{j\omega t}}_{\text{output}}$$

Therefore, from the linearity property

$$\underbrace{\sum_{n=-\infty}^{\infty} D_n e^{jn\omega_0 t}}_{\text{input } x(t)} \Longrightarrow \underbrace{\sum_{n=-\infty}^{\infty} D_n H(jn\omega_0)e^{jn\omega_0 t}}_{\text{response } y(t)} \tag{6.42}$$

The response $y(t)$ is obtained in the form of an exponential Fourier series and is therefore a periodic signal of the same period as that of the input.

We shall demonstrate the utility of these results by the following example.

EXAMPLE 6.9

A full-wave rectifier (Fig. 6.16a) is used to obtain a dc signal from a sinusoid $\sin t$. The rectified signal $x(t)$, depicted in Fig. 6.14, is applied to the input of a lowpass RC filter, which suppresses the time-varying component and yields a dc component with some residual ripple. Find the filter output $y(t)$. Find also the dc output and the rms value of the ripple voltage.

[†]This result applies only to the asymptotically stable systems. This is because when $s = j\omega$, the integral on the right-hand side of Eq. (2.48) does not converge for unstable systems. Moreover, for marginally stable systems also, that integral does not converge in the ordinary sense, and $H(j\omega)$ cannot be obtained from $H(s)$ by replacing s with $j\omega$.

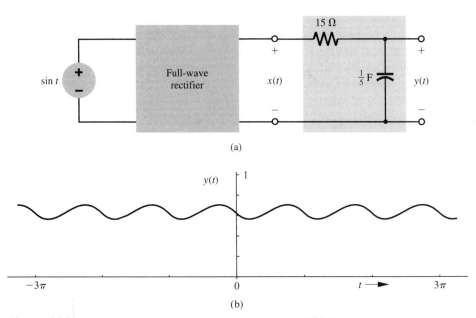

Figure 6.16 (a) Full-wave rectifier with a lowpass filter and (b) its output.

First, we shall find the Fourier series for the rectified signal $x(t)$, whose period is $T_0 = \pi$. Consequently, $\omega_0 = 2$, and

$$x(t) = \sum_{n=-\infty}^{\infty} D_n e^{j2nt}$$

where

$$D_n = \frac{1}{\pi} \int_0^{\pi} \sin t \, e^{-j2nt} \, dt = \frac{2}{\pi(1 - 4n^2)} \tag{6.43}$$

Therefore

$$x(t) = \sum_{n=-\infty}^{\infty} \frac{2}{\pi(1 - 4n^2)} e^{j2nt}$$

Next, we find the transfer function of the RC filter in Fig. 6.16a. This filter is identical to the RC circuit in Example 1.11 (Fig. 1.35) for which the differential equation relating the output (capacitor voltage) to the input $x(t)$ was found to be [Eq. (1.60)]

$$(3D + 1)y(t) = x(t)$$

The transfer function $H(s)$ for this system is found from Eq. (2.50) as

$$H(s) = \frac{1}{3s + 1}$$

and

$$H(j\omega) = \frac{1}{3j\omega + 1} \tag{6.44}$$

From Eq. (6.42), the filter output $y(t)$ can be expressed as (with $\omega_0 = 2$)

$$y(t) = \sum_{n=-\infty}^{\infty} D_n H(jn\omega_0)e^{jn\omega_0 t} = \sum_{n=-\infty}^{\infty} D_n H(j2n)e^{j2nt}$$

Substituting D_n and $H(j2n)$ from Eqs. (6.43) and (6.44) in the foregoing equation, we obtain

$$y(t) = \sum_{n=-\infty}^{\infty} \frac{2}{\pi(1 - 4n^2)(j6n + 1)} e^{j2nt} \tag{6.45}$$

Note that the output $y(t)$ is also a periodic signal given by the exponential Fourier series on the right-hand side. The output is numerically computed from Eq. (6.45) and plotted in Fig. 6.16b.

The output Fourier series coefficient corresponding to $n = 0$ is the dc component of the output, given by $2/\pi$. The remaining terms in the Fourier series constitute the unwanted component called the ripple. We can determine the rms value of the ripple voltage by using Eq. (6.41) to find the power of the ripple component. The power of the ripple is the power of all the components except the dc ($n = 0$). Note that $\hat{D}_n$, the exponential Fourier coefficient for the output $y(t)$ is

$$\hat{D}_n = \frac{2}{\pi(1 - 4n^2)(j6n + 1)}$$

Therefore, from Eq. (6.41b), we have

$$P_{\text{ripple}} = 2\sum_{n=1}^{\infty} |D_n|^2 = 2\sum_{n=1}^{\infty} \left| \frac{2}{\pi(1 - 4n^2)(j6n + 1)} \right|^2 = \frac{8}{\pi^2} \sum_{n=1}^{\infty} \frac{1}{(1 - 4n^2)^2(36n^2 + 1)}$$

Numerical computation of the right-hand side yields $P_{\text{ripple}} = 0.0025$, and the ripple rms value $= \sqrt{P_{\text{ripple}}} = 0.05$. This shows that the rms ripple voltage is 5% of the amplitude of the input sinusoid.

WHY USE EXPONENTIALS?

The exponential Fourier series is just another way of representing trigonometric Fourier series (or vice versa). The two forms carry identical information—no more, no less. The reasons for preferring the exponential form have already been mentioned: this form is more compact, and the expression for deriving the exponential coefficients is also more compact, than those in the trigonometric series. Furthermore, the LTIC system response to exponential signals is also simpler (more compact) than the system response to sinusoids. In addition, the exponential form proves to be much easier than the trigonometric form to manipulate mathematically and otherwise handle in the area of signals as well as systems. Moreover, exponential representation proves much more convenient for analysis of complex $x(t)$. For these reasons, in our future discussion we shall use the exponential form exclusively.

A minor disadvantage of the exponential form is that it cannot be visualized as easily as sinusoids. For intuitive and qualitative understanding, the sinusoids have the edge over exponentials. Fortunately, this difficulty can be overcome readily because of the close connection between exponential and Fourier spectra. For the purpose of mathematical analysis, we shall continue to use exponential signals and spectra; but to understand the physical situation intuitively or qualitatively, we shall speak in terms of sinusoids and trigonometric spectra. Thus, although all mathematical manipulation will be in terms of exponential spectra, we shall now speak of exponential and sinusoids interchangeably when we discuss intuitive and qualitative insights in attempting to arrive at an understanding of physical situations. This is an important point; readers should make an extra effort to familiarize themselves with the two forms of spectra, their relationships, and their convertibility.

DUAL PERSONALITY OF A SIGNAL

The discussion so far shows that a periodic signal has a dual personality—the time domain and the frequency domain. It can be described by its waveform or by its Fourier spectra. The time- and frequency-domain descriptions provide complementary insights into a signal. For in-depth perspective, we need to understand both these identities. It is important to learn to think of a signal from both perspectives. In the next chapter, we shall see that aperiodic signals also have this dual personality. Moreover, we shall show that even LTI systems have this dual personality, which offers complementary insights into the system behavior.

LIMITATIONS OF THE FOURIER SERIES
METHOD OF ANALYSIS

We have developed here a method of representing a periodic signal as a weighted sum of everlasting exponentials whose frequencies lie along the $j\omega$ axis in the s plane. This representation (Fourier series) is valuable in many applications. However, as a tool for analyzing linear systems, it has serious limitations and consequently has limited utility for the following reasons:

1. The Fourier series can be used only for periodic inputs. All practical inputs are aperiodic (remember that a periodic signal starts at $t = -\infty$).
2. The Fourier methods can be applied readily to BIBO-stable (or asymptotically stable) systems. It cannot handle unstable or even marginally stable systems.

The first limitation can be overcome by representing aperiodic signals in terms of everlasting exponentials. This representation can be achieved through the Fourier integral, which may be considered to be an extension of the Fourier series. We shall therefore use the Fourier series as a stepping stone to the Fourier integral developed in the next chapter. The second limitation can be overcome by using exponentials e^{st}, where s is not restricted to the imaginary axis but is free to take on complex values. This generalization leads to the Laplace integral, discussed in Chapter 4 (the Laplace transform).

6.5 GENERALIZED FOURIER SERIES: SIGNALS AS VECTORS[†]

We now consider a very general approach to signal representation with far-reaching consequences. There is a perfect analogy between signals and vectors; the analogy is so strong that the term *analogy* understates the reality. Signals are not just *like* vectors. Signals *are* vectors! A vector can be represented as a sum of its components in a variety of ways, depending on the choice of coordinate system. A signal can also be represented as a sum of its components in a variety of ways. Let us begin with some basic vector concepts and then apply these concepts to signals.

6.5-1 Component of a Vector

A vector is specified by its magnitude and its direction. We shall denote all vectors by boldface. For example, $\mathbf{x}$ is a certain vector with magnitude or length $|\mathbf{x}|$. For the two vectors $\mathbf{x}$ and $\mathbf{y}$ shown in Fig. 6.17, we define their dot (inner or scalar) product as

$$\mathbf{x} \cdot \mathbf{y} = |\mathbf{x}||\mathbf{y}| \cos \theta \tag{6.46}$$

where θ is the angle between these vectors. Using this definition we can express $|\mathbf{x}|$, the length of a vector $\mathbf{x}$, as

$$|\mathbf{x}|^2 = \mathbf{x} \cdot \mathbf{x} \tag{6.47}$$

Let the component of $\mathbf{x}$ along $\mathbf{y}$ be $c\mathbf{y}$ as depicted in Fig. 6.17. Geometrically, the component of $\mathbf{x}$ along $\mathbf{y}$ is the projection of $\mathbf{x}$ on $\mathbf{y}$ and is obtained by drawing a perpendicular from the tip of $\mathbf{x}$ on the vector $\mathbf{y}$, as illustrated in Fig. 6.17. What is the mathematical significance of a component of a vector along another vector? As seen from Fig. 6.17, the vector $\mathbf{x}$ can be expressed in terms of vector $\mathbf{y}$ as

$$\mathbf{x} = c\mathbf{y} + \mathbf{e} \tag{6.48}$$

However, this is not the only way to express $\mathbf{x}$ in terms of $\mathbf{y}$. From Fig. 6.18, which shows two of the infinite other possibilities, we have

$$\mathbf{x} = c_1\mathbf{y} + \mathbf{e}_1 = c_2\mathbf{y} + \mathbf{e}_2 \tag{6.49}$$

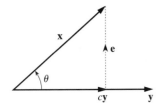

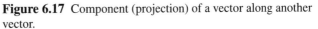

Figure 6.17 Component (projection) of a vector along another vector.

[†]This section closely follows the material from the author's earlier book.[10] Omission of this section will not cause any discontinuity in understanding the rest of the book. Derivation of Fourier series through the signal–vector analogy provides an interesting insight into signal representation and other topics such as signal correlation, data truncation, and signal detection.

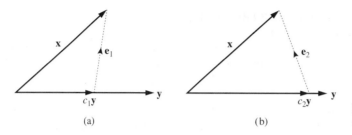

Figure 6.18 Approximation of a vector in terms of another vector.

(a)   (b)

In each of these three representations, $\mathbf{x}$ is represented in terms of $\mathbf{y}$ plus another vector called the *error vector*. If we approximate $\mathbf{x}$ by $c\mathbf{y}$,

$$\mathbf{x} \simeq c\mathbf{y} \tag{6.50}$$

the error in the approximation is the vector $\mathbf{e} = \mathbf{x} - c\mathbf{y}$. Similarly, the errors in approximations in these drawings are $\mathbf{e}_1$ (Fig. 6.18a) and $\mathbf{e}_2$ (Fig. 6.18b). What is unique about the approximation in Fig. 6.17 is that the error vector is the smallest. We can now define mathematically the component of a vector $\mathbf{x}$ along vector $\mathbf{y}$ to be $c\mathbf{y}$ where c is chosen to minimize the length of the error vector $\mathbf{e} = \mathbf{x} - c\mathbf{y}$. Now, the length of the component of $\mathbf{x}$ along $\mathbf{y}$ is $|\mathbf{x}| \cos \theta$. But it is also $c|\mathbf{y}|$ as seen from Fig. 6.17. Therefore

$$c|\mathbf{y}| = |\mathbf{x}| \cos \theta$$

Multiplying both sides by $|\mathbf{y}|$ yields

$$c|\mathbf{y}|^2 = |\mathbf{x}||\mathbf{y}| \cos \theta = \mathbf{x} \cdot \mathbf{y}$$

Therefore

$$c = \frac{\mathbf{x} \cdot \mathbf{y}}{\mathbf{y} \cdot \mathbf{y}} = \frac{1}{|\mathbf{y}|^2} \mathbf{x} \cdot \mathbf{y} \tag{6.51}$$

From Fig. 6.17, it is apparent that when $\mathbf{x}$ and $\mathbf{y}$ are perpendicular, or orthogonal, then $\mathbf{x}$ has a zero component along $\mathbf{y}$; consequently, $c = 0$. Keeping an eye on Eq. (6.51), we therefore define $\mathbf{x}$ and $\mathbf{y}$ to be *orthogonal* if the inner (scalar or dot) product of the two vectors is zero, that is, if

$$\mathbf{x} \cdot \mathbf{y} = 0 \tag{6.52}$$

6.5-2 Signal Comparison and Component of a Signal

The concept of a vector component and orthogonality can be extended to signals. Consider the problem of approximating a real signal $x(t)$ in terms of another real signal $y(t)$ over an interval (t_1, t_2):

$$x(t) \simeq cy(t) \qquad t_1 < t < t_2 \tag{6.53}$$

The error $e(t)$ in this approximation is

$$e(t) = \begin{cases} x(t) - cy(t) & t_1 < t < t_2 \\ 0 & \text{otherwise} \end{cases} \tag{6.54}$$

We now select a criterion for the "best approximation." We know that the signal energy is one possible measure of a signal size. For best approximation, we shall use the criterion that minimizes the size or energy of the error signal $e(t)$ over the interval (t_1, t_2). This energy E_e is given by

$$E_e = \int_{t_1}^{t_2} e^2(t)\, dt$$

$$= \int_{t_1}^{t_2} [x(t) - cy(t)]^2\, dt$$

Note that the right-hand side is a definite integral with t as the dummy variable. Hence, E_e is a function of the parameter c (not t) and E_e is minimum for some choice of c. To minimize E_e, a necessary condition is

$$\frac{dE_e}{dc} = 0 \tag{6.55}$$

or

$$\frac{d}{dc}\left[\int_{t_1}^{t_2} [x(t) - cy(t)]^2\, dt \right] = 0$$

Expanding the squared term inside the integral, we obtain

$$\frac{d}{dc}\left[\int_{t_1}^{t_2} x^2(t)\, dt \right] - \frac{d}{dc}\left[2c \int_{t_1}^{t_2} x(t)y(t)\, dt \right] + \frac{d}{dc}\left[c^2 \int_{t_1}^{t_2} y^2(t)\, dt \right] = 0$$

From which we obtain

$$-2 \int_{t_1}^{t_2} x(t)y(t)\, dt + 2c \int_{t_1}^{t_2} y^2(t)\, dt = 0$$

and

$$c = \frac{\displaystyle\int_{t_1}^{t_2} x(t)y(t)\, dt}{\displaystyle\int_{t_1}^{t_2} y^2(t)\, dt} = \frac{1}{E_y} \int_{t_1}^{t_2} x(t)y(t)\, dt \tag{6.56}$$

We observe a remarkable similarity between the behavior of vectors and signals, as indicated by Eqs. (6.51) and (6.56). It is evident from these two parallel expressions that *the area under the product of two signals corresponds to the inner (scalar or dot) product of two vectors*. In fact, the area under the product of $x(t)$ and $y(t)$ is called the *inner product* of $x(t)$ and $y(t)$, and is denoted by (x, y). The energy of a signal is the inner product of a signal with itself, and corresponds to the vector length square (which is the inner product of the vector with itself).

To summarize our discussion, if a signal $x(t)$ is approximated by another signal $y(t)$ as

$$x(t) \simeq cy(t)$$

then the optimum value of c that minimizes the energy of the error signal in this approximation is given by Eq. (6.56).

Taking our clue from vectors, we say that a signal $x(t)$ contains a component $cy(t)$, where c is given by Eq. (6.56). Note that in vector terminology, $cy(t)$ is the projection of $x(t)$ on $y(t)$. Continuing with the analogy, we say that if the component of a signal $x(t)$ of the form $y(t)$ is zero (i.e., $c = 0$), the signals $x(t)$ and $y(t)$ are orthogonal over the interval (t_1, t_2). Therefore, we define the real signals $x(t)$ and $y(t)$ to be orthogonal over the interval (t_1, t_2) if[†]

$$\int_{t_1}^{t_2} x(t) y(t) \, dt = 0 \tag{6.57}$$

EXAMPLE 6.10

For the square signal $x(t)$ shown in Fig. 6.19 find the component in $x(t)$ of the form $\sin t$. In other words, approximate $x(t)$ in terms of $\sin t$

$$x(t) \simeq c \sin t \qquad 0 < t < 2\pi$$

so that the energy of the error signal is minimum.

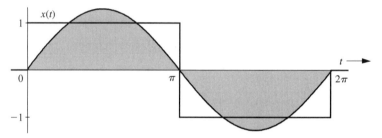

Figure 6.19 Approximation of a square signal in terms of a single sinusoid.

In this case

$$y(t) = \sin t \qquad \text{and} \qquad E_y = \int_0^{2\pi} \sin^2(t) \, dt = \pi$$

From Eq. (6.56), we find

$$c = \frac{1}{\pi} \int_0^{2\pi} x(t) \sin t \, dt = \frac{1}{\pi} \left[\int_0^{\pi} \sin t \, dt + \int_{\pi}^{2\pi} -\sin t \, dt \right] = \frac{4}{\pi} \tag{6.58}$$

Thus

$$x(t) \simeq \frac{4}{\pi} \sin t \tag{6.59}$$

represents the best approximation of $x(t)$ by the function $\sin t$, which will minimize the error energy. This sinusoidal component of $x(t)$ is shown shaded in Fig. 6.19. By analogy with vectors, we say that the square function $x(t)$ depicted in Fig. 6.19 has a component of signal $\sin t$ and that the magnitude of this component is $4/\pi$.

[†]For complex signals the definition is modified as in Eq. (6.65), in Section 6.5–3.

EXERCISE E6.7

Show that over an interval $(-\pi < t < \pi)$, the "best" approximation of the signal $x(t) = t$ in terms of the function $\sin t$ is $2 \sin t$. Verify that the error signal $e(t) = t - 2 \sin t$ is orthogonal to the signal $\sin t$ over the interval $-\pi < t < \pi$. Sketch the signals t and $2 \sin t$ over the interval $-\pi < t < \pi$.

6.5-3 Extension to Complex Signals

So far we have restricted ourselves to real functions of t. To generalize the results to complex functions of t, consider again the problem of approximating a signal $x(t)$ by a signal $y(t)$ over an interval $(t_1 < t < t_2)$:

$$x(t) \simeq cy(t) \tag{6.60}$$

where $x(t)$ and $y(t)$ now can be complex functions of t. Recall that the energy E_y of the complex signal $y(t)$ over an interval (t_1, t_2) is

$$E_y = \int_{t_1}^{t_2} |y(t)|^2 \, dt$$

In this case, both the coefficient c and the error

$$e(t) = x(t) - cy(t) \tag{6.61}$$

are complex (in general). For the "best" approximation, we choose c so that E_e, the energy of the error signal $e(t)$ is minimum. Now,

$$E_e = \int_{t_1}^{t_2} |x(t) - cy(t)|^2 \, dt \tag{6.62}$$

Recall also that

$$|u + v|^2 = (u + v)(u^* + v^*) = |u|^2 + |v|^2 + u^*v + uv^* \tag{6.63}$$

After some manipulation, we can use this result to rearrange Eq. (6.62) as

$$E_e = \int_{t_1}^{t_2} |x(t)|^2 \, dt - \left| \frac{1}{\sqrt{E_y}} \int_a^{t_2} x(t) y^*(t) \, dt \right|^2 + \left| c\sqrt{E_y} - \frac{1}{\sqrt{E_y}} \int_{t_1}^{t_2} x(t) y^*(t) \, dt \right|^2$$

Since the first two terms on the right-hand side are independent of c, it is clear that E_e is minimized by choosing c so that the third term on the right-hand side is zero. This yields

$$c = \frac{1}{E_y} \int_{t_1}^{t_2} x(t) y^*(t) \, dt \tag{6.64}$$

In light of this result, we need to redefine orthogonality for the complex case as follows: two complex functions $x_1(t)$ and $x_2(t)$ are orthogonal over an interval $(t_1 < t < t_2)$ if

$$\int_{t_1}^{t_2} x_1(t)x_2^*(t)\, dt = 0 \qquad \text{or} \qquad \int_{t_1}^{t_2} x_1^*(t)x_2(t)\, dt = 0 \tag{6.65}$$

Either equality suffices. This is a general definition of orthogonality, which reduces to Eq. (6.57) when the functions are real.

EXERCISE E6.8

Show that over an interval $(0 < t < 2\pi)$, the "best" approximation of the square signal $x(t)$ in Fig. 6.19 in terms of the signal e^{jt} is given by $(2/j\pi)\, e^{jt}$. Verify that the error signal $e(t) = x(t) - (2/j\pi)e^{jt}$ is orthogonal to the signal e^{jt}.

ENERGY OF THE SUM OF ORTHOGONAL SIGNALS

We know that the square of the length of a sum of two orthogonal vectors is equal to the sum of the squares of the lengths of the two vectors. Thus, if vectors $\mathbf{x}$ and $\mathbf{y}$ are orthogonal, and if $\mathbf{z} = \mathbf{x} + \mathbf{y}$, then

$$|\mathbf{z}|^2 = |\mathbf{x}|^2 + |\mathbf{y}|^2$$

We have a similar result for signals. The energy of the sum of two orthogonal signals is equal to the sum of the energies of the two signals. Thus, if signals $x(t)$ and $y(t)$ are orthogonal over an interval (t_1, t_2), and if $z(t) = x(t) + y(t)$, then

$$E_z = E_x + E_y \tag{6.66}$$

We now prove this result for complex signals of which real signals are a special case. From Eq. (6.63) it follows that

$$\int_{t_1}^{t_2} |x(t) + y(t)|^2\, dt = \int_{t_1}^{t_2} |x(t)|^2\, dt + \int_{t_1}^{t_2} |y(t)|^2\, dt + \int_{t_1}^{t_2} x(t)y^*(t)\, dt + \int_{t_1}^{t_2} x^*(t)y(t)\, dt$$

$$= \int_{t_1}^{t_2} |x(t)|^2\, dt + \int_{t_1}^{t_2} |y(t)|^2\, dt \tag{6.67}$$

The last result follows from the fact that because of orthogonality, the two integrals of the products $x(t)y^*(t)$ and $x^*(t)y(t)$ are zero [see Eq. (6.65)]. This result can be extended to the sum of any number of mutually orthogonal signals.

6.5-4 Signal Representation by an Orthogonal Signal Set

In this section we show a way of representing a signal as a sum of orthogonal signals. Here again we can benefit from the insight gained from a similar problem in vectors. We know that a vector can be represented as a sum of orthogonal vectors, which form the coordinate system of a vector space. The problem in signals is analogous, and the results for signals are parallel to those for vectors. So, let us review the case of vector representation.

ORTHOGONAL VECTOR SPACE

Let us investigate a three-dimensional Cartesian vector space described by three mutually orthogonal vectors $\mathbf{x}_1$, $\mathbf{x}_2$, and $\mathbf{x}_3$, as illustrated in Fig. 6.20. First, we shall seek to approximate a three-dimensional vector $\mathbf{x}$ in terms of two mutually orthogonal vectors $\mathbf{x}_1$ and $\mathbf{x}_2$:

$$\mathbf{x} \simeq c_1\mathbf{x}_1 + c_2\mathbf{x}_2$$

The error $\mathbf{e}$ in this approximation is

$$\mathbf{e} = \mathbf{x} - (c_1\mathbf{x}_1 + c_2\mathbf{x}_2)$$

or

$$\mathbf{x} = c_1\mathbf{x}_1 + c_2\mathbf{x}_2 + \mathbf{e}$$

As in the earlier geometrical argument, we see from Fig 6.20 that the length of $\mathbf{e}$ is minimum when $\mathbf{e}$ is perpendicular to the $\mathbf{x}_1$–$\mathbf{x}_2$ plane, and $c_1\mathbf{x}_1$ and $c_2\mathbf{x}_2$ are the projections (components) of $\mathbf{x}$ on $\mathbf{x}_1$ and $\mathbf{x}_2$, respectively. Therefore, the constants c_1 and c_2 are given by Eq. (6.51). Observe that the error vector is orthogonal to both the vectors $\mathbf{x}_1$ and $\mathbf{x}_2$.

Now, let us determine the "best" approximation to $\mathbf{x}$ in terms of all three mutually orthogonal vectors $\mathbf{x}_1$, $\mathbf{x}_2$, and $\mathbf{x}_3$:

$$\mathbf{x} \simeq c_1\mathbf{x}_1 + c_2\mathbf{x}_2 + c_3\mathbf{x}_3 \tag{6.68}$$

Figure 6.20 shows that a unique choice of c_1, c_2, and c_3 exists, for which Eq. (6.68) is no longer an approximation but an equality

$$\mathbf{x} = c_1\mathbf{x}_1 + c_2\mathbf{x}_2 + c_3\mathbf{x}_3 \tag{6.69}$$

In this case, $c_1\mathbf{x}_1$, $c_2\mathbf{x}_2$, and $c_3\mathbf{x}_3$ are the projections (components) of $\mathbf{x}$ on $\mathbf{x}_1$, $\mathbf{x}_2$, and $\mathbf{x}_3$, respectively; that is,

$$c_i = \frac{\mathbf{x} \cdot \mathbf{x}_i}{\mathbf{x}_i \cdot \mathbf{x}_i} \tag{6.70a}$$

$$= \frac{1}{|\mathbf{x}_i|^2} \mathbf{x} \cdot \mathbf{x}_i \qquad i = 1, 2, 3 \tag{6.70b}$$

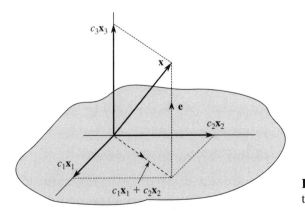

Figure 6.20 Representation of a vector in three-dimensional space.

Note that the error in the approximation is zero when $\mathbf{x}$ is approximated in terms of three mutually orthogonal vectors: $\mathbf{x}_1$, $\mathbf{x}_2$, and $\mathbf{x}_3$. The reason is that $\mathbf{x}$ is a three-dimensional vector, and the vectors $\mathbf{x}_1$, $\mathbf{x}_2$, and $\mathbf{x}_3$ represent a *complete set* of orthogonal vectors in three-dimensional space. Completeness here means that it is impossible to find another vector $\mathbf{x}_4$ in this space, which is orthogonal to all three vectors, $\mathbf{x}_1$, $\mathbf{x}_2$, and $\mathbf{x}_3$. Any vector in this space can then be represented (with zero error) in terms of these three vectors. Such vectors are known as *basis* vectors. If a set of vectors $\{\mathbf{x}_i\}$ is not complete, the error in the approximation will generally not be zero. Thus, in the three-dimensional case discussed earlier, it is generally not possible to represent a vector $\mathbf{x}$ in terms of only two basis vectors without an error.

The choice of basis vectors is not unique. In fact, a set of basis vectors corresponds to a particular choice of coordinate system. Thus, a three-dimensional vector $\mathbf{x}$ may be represented in many different ways, depending on the coordinate system used.

ORTHOGONAL SIGNAL SPACE

We start with real signals first and then extend the discussion to complex signals. We proceed with our signal approximation problem, using clues and insights developed for vector approximation. As before, we define orthogonality of a real signal set $x_1(t), x_2(t), \ldots, x_N(t)$ over interval (t_1, t_2) as

$$\int_{t_1}^{t_2} x_m(t)x_n(t)\,dt = \begin{cases} 0 & m \neq n \\ E_n & m = n \end{cases} \tag{6.71}$$

If the energies $E_n = 1$ for all n, then the set is *normalized* and is called an *orthonormal set*. An orthogonal set can always be normalized by dividing $x_n(t)$ by $\sqrt{E_n}$ for all n.

Now, consider approximating a signal $x(t)$ over the interval (t_1, t_2) by a set of N real, mutually orthogonal signals $x_1(t), x_2(t), \ldots, x_N(t)$ as

$$x(t) \simeq c_1 x_1(t) + c_2 x_2(t) + \cdots + c_N x_N(t) \tag{6.72a}$$

$$\simeq \sum_{n=1}^{N} c_n x_n(t) \tag{6.72b}$$

In the approximation of Eqs. (6.72) the error $e(t)$ is

$$e(t) = x(t) - \sum_{n=1}^{N} c_n x_n(t) \tag{6.73}$$

and E_e, the error signal energy is

$$E_e = \int_{t_1}^{t_2} e^2(t)\,dt$$

$$= \int_{t_1}^{t_2} \left[x(t) - \sum_{n=1}^{N} c_n x_n(t) \right]^2 dt \tag{6.74}$$

According to our criterion for best approximation, we select the values of c_i that minimize E_e.

Hence, the necessary condition is $\partial E_e / \partial c_i = 0$ for $i = 1, 2, \ldots, N$, that is,

$$\frac{\partial}{\partial c_i} \int_{t_1}^{t_2} \left[x(t) - \sum_{n=1}^{N} c_n x_n(t) \right]^2 dt = 0 \tag{6.75}$$

When we expand the integrand, we find that all the cross-multiplication terms arising from the orthogonal signals are zero by virtue of orthogonality: that is, all terms of the form $\int x_m(t) x_n(t) \, dt$ with $m \neq n$ vanish. Similarly, the derivative with respect to c_i of all terms that do not contain c_i is zero. For each i, this leaves only two nonzero terms in Eq. (6.75):

$$\frac{\partial}{\partial c_i} \int_{t_1}^{t_2} \left[-2 c_i x(t) x_i(t) + c_i{}^2 x_i{}^2(t) \right] dt = 0$$

or

$$-2 \int_{t_1}^{t_2} x(t) x_i(t) \, dt + 2 c_i \int_{t_1}^{t_2} x_i{}^2(t) \, dt = 0 \qquad i = 1, 2, \ldots, N$$

Therefore

$$c_i = \frac{\displaystyle\int_{t_1}^{t_2} x(t) x_i(t) \, dt}{\displaystyle\int_{t_1}^{t_2} x_i{}^2(t) \, dt} \tag{6.76a}$$

$$= \frac{1}{E_i} \int_{t_1}^{t_2} x(t) x_i(t) \, dt \qquad i = 1, 2, \ldots, N \tag{6.76b}$$

A comparison of Eqs. (6.76) with Eqs. (6.70) forcefully brings out the analogy of signals with vectors.

Finality Property. Equation (6.76) shows one interesting property of the coefficients of c_1, $c_2, \ldots, c_N$: the optimum value of any coefficient in approximation (6.72a) is independent of the number of terms used in the approximation. For example, if we used only one term ($N = 1$) or two terms ($N = 2$) or any number of terms, the optimum value of the coefficient c_1 would be the same [as given by Eq. (6.76)]. The advantage of this approximation of a signal $x(t)$ by a set of mutually orthogonal signals is that we can continue to add terms to the approximation without disturbing the previous terms. This property of *finality* of the values of the coefficients is very important from a practical point of view.[†]

[†]Contrast this situation with polynomial approximation of $x(t)$. Suppose we wish to find a two-point approximation of $x(t)$ by a polynomial in t; that is, the polynomial is to be equal to $x(t)$ at two points t_1 and t_2. This can be done by choosing a first-order polynomial $a_0 + a_1 t$ with

$$x(t_1) = a_0 + a_1 t_1 \qquad \text{and} \qquad x(t_2) = a_0 + a_1 t_2$$

Solution of these equations yields the desired values of a_0 and a_1. For a three-point approximation, we must choose the polynomial $a_0 + a_1 t + a_2 t^2$ with

$$x(t_i) = a_0 + a_1 t_i + a_2 t_i{}^2 \qquad i = 1, 2, \text{ and } 3$$

The approximation improves with larger number of points (higher-order polynomial), but the coefficients a_0, $a_1, a_2, \ldots$ do not have the finality property. Every time we increase the number of terms in the polynomial, we need to recalculate the coefficients.

ENERGY OF THE ERROR SIGNAL

When the coefficients c_i in the approximation (6.72) are chosen according to Eqs. (6.76), the error signal energy in the approximation (6.72) is minimized. This minimum value of E_e is given by Eq. (6.74):

$$E_e = \int_{t_1}^{t_2} \left[x(t) - \sum_{n=1}^{N} c_n x_n(t) \right]^2 dt$$

$$= \int_{t_1}^{t_2} x^2(t)\, dt + \sum_{n=1}^{N} c_n^2 \int_{t_1}^{t_2} x_n^2(t)\, dt - 2\sum_{n=1}^{N} c_n \int_{t_1}^{t_2} x(t) x_n(t)\, dt$$

Substitution of Eqs. (6.71) and (6.76) in this equation yields

$$E_e = \int_{t_1}^{t_2} x^2(t)\, dt + \sum_{n=1}^{N} c_n^2 E_n - 2\sum_{n=1}^{N} c_n^2 E_n$$

$$= \int_{t_1}^{t_2} x^2(t)\, dt - \sum_{n=1}^{N} c_n^2 E_n \tag{6.77}$$

Observe that because the term $c_k^2 E_k$ is nonnegative, the error energy E_e generally decreases as N, the number of terms, is increased. Hence, it is possible that the error energy $\to 0$ as $N \to \infty$. When this happens, the orthogonal signal set is said to be *complete*. In this case, Eq. (6.72b) is no more an approximation but an equality

$$x(t) = c_1 x_1(t) + c_2 x_2(t) + \cdots + c_n x_n(t) + \cdots$$

$$= \sum_{n=1}^{\infty} c_n x_n(t) \qquad t_1 < t < t_2 \tag{6.78}$$

where the coefficients c_n are given by Eq. (6.76b). Because the error signal energy approaches zero, it follows that the energy of $x(t)$ is now equal to the sum of the energies of its orthogonal components $c_1 x_1(t)$, $c_2 x_2(t)$, $c_3 x_3(t)$, ...

The series on the right-hand side of Eq. (6.78) is called the *generalized Fourier series* of $x(t)$ with respect to the set $\{x_n(t)\}$. When the set $\{x_n(t)\}$ is such that the error energy $E_e \to 0$ as $N \to \infty$ for every member of some particular class, we say that the set $\{x_n(t)\}$ is complete on (t_1, t_2) for that class of $x(t)$, and the set $\{x_n(t)\}$ is called a set of *basis functions* or *basis signals*. Unless otherwise mentioned, in future we shall consider only the class of energy signals.

Thus, when the set $\{x_n(t)\}$ is complete, we have the equality (6.78). One subtle point that must be understood clearly is the meaning of equality in Eq. (6.78). *The equality here is not an equality in the ordinary sense, but in the sense that the error energy, that is, the energy of the difference between the two sides of Eq. (6.78), approaches zero.* If the equality exists in the ordinary sense, the error energy is always zero, but the converse is not necessarily true. The error energy can approach zero even though $e(t)$, the difference between the two sides, is nonzero at some isolated instants. The reason is that even if $e(t)$ is nonzero at such instants, the area under $e^2(t)$ is still zero; thus the Fourier series on the right-hand side of Eq. (6.78) may differ from $x(t)$ at a finite number of points.

In Eq. (6.78), the energy of the left-hand side is E_x, and the energy of the right-hand side is the sum of the energies of all the orthogonal components.[†] Thus

$$\int_{t_1}^{t_2} x^2(t)\, dt = c_1^2 E_1 + c_2^2 E_2 + \cdots$$

$$= \sum_{n=1}^{\infty} c_n^2 E_n \tag{6.79}$$

This is the *Parseval's theorem* applicable to energy signals. In Eqs. (6.40) and (6.41), we have already encountered the form of the Parseval theorem suitable for power signals. Recall that the signal energy (area under the squared value of a signal) is analogous to the square of the length of a vector in the vector–signal analogy. In vector space, we know that the square of the length of a vector is equal to the sum of the squares of the lengths of its orthogonal components. The Parseval theorem in Eq. (6.79) is the statement of this fact as it applies to signals.

GENERALIZATION TO COMPLEX SIGNALS

The foregoing results can be generalized to complex signals as follows: a set of functions $x_1(t)$, $x_2(t), \ldots, x_N(t)$ is mutually orthogonal over the interval (t_1, t_2) if

$$\int_{t_1}^{t_2} x_m(t) x_n^*(t)\, dt = \begin{cases} 0 & m \neq n \\ E_n & m = n \end{cases} \tag{6.80}$$

If this set is complete for a certain class of functions, then a function $x(t)$ in this class can be expressed as

$$x(t) = c_1 x_1(t) + c_2 x_2(t) + \cdots + c_i x_i(t) + \cdots \tag{6.81}$$

where

$$c_n = \frac{1}{E_n} \int_{t_1}^{t_2} x(t) x_n^*(t)\, dt \tag{6.82}$$

EXAMPLE 6.11

As an example, we shall consider again the square signal $x(t)$ in Fig. 6.19. In Example 6.10 this signal was approximated by a single sinusoid $\sin t$. Actually the set $\sin t$, $\sin 2t, \ldots, \sin nt, \ldots$ is orthogonal over the interval over any interval of duration 2π.[‡] The

---

[†]Note that the energy of a signal $cx(t)$ is $c^2 E_x$.

[‡]This sine set, along with the cosine set $\cos 0t$, $\cos t$, $\cos 2t, \ldots, \cos nt, \ldots$, forms a complete set. In this case, however, the coefficients c_i corresponding to the cosine terms are zero. For this reason, we have omitted cosine terms in this example. This composite sine and cosine set is the basis set for the trigonometric Fourier series.

reader can verify this fact by showing that for any real number a

$$\int_a^{a+2\pi} \sin mt \, \sin nt \, dt = \begin{cases} 0 & m \neq n \\ \pi & m = n \end{cases} \tag{6.83}$$

Let us approximate the square signal in Fig. 6.19, using this set, and see how the approximation improves with number of terms

$$x(t) \simeq c_1 \sin t + c_2 \sin 2t + \cdots + c_n \sin Nt$$

where

$$c_n = \frac{\displaystyle\int_0^{2\pi} x(t) \sin nt \, dt}{\displaystyle\int_0^{2\pi} \sin^2 nt \, dt}$$

$$= \frac{1}{\pi} \left[\int_0^\pi \sin nt \, dt + \int_\pi^{2\pi} -\sin nt \, dt \right]$$

$$= \begin{cases} \dfrac{4}{\pi n} & n \text{ odd} \\ 0 & n \text{ even} \end{cases} \tag{6.84}$$

Therefore

$$x(t) \simeq \frac{4}{\pi} \left(\sin t + \frac{1}{3} \sin 3t + \frac{1}{5} \sin 5t + \cdots + \frac{1}{N} \sin Nt \right) \tag{6.85}$$

Note that coefficients of terms $\sin kt$ are zero for even values of k. Figure 6.21 shows how the approximation improves as we increase the number of terms in the series. Let us investigate the error signal energy as $N \to \infty$. From Eq. (6.77)

$$E_e = \int_0^{2\pi} x^2(t) \, dt - \sum_{n=1}^{\infty} c_n^2 E_n$$

Note that

$$\int_0^{2\pi} x^2(t) \, dt = \int_0^\pi 1^2 \, dt + \int_\pi^{2\pi} -1^2 \, dt = 2\pi$$

$$c_n^2 = \begin{cases} \dfrac{16}{n^2 \pi^2} & n \text{ odd} \\ 0 & n \text{ even} \end{cases}$$

and from Eq. (6.83)

$$E_n = \pi$$

Therefore

$$E_e = 2\pi - \sum_{n=1,3,5,\ldots}^{N} \frac{16}{n^2\pi^2}\pi$$

$$= 2\pi - \frac{16}{\pi} \sum_{n=1,3,5,\ldots}^{N} \frac{1}{n^2}$$

For a single-term approximation ($N = 1$),

$$E_e = 2\pi - \frac{16}{\pi} = 1.1938$$

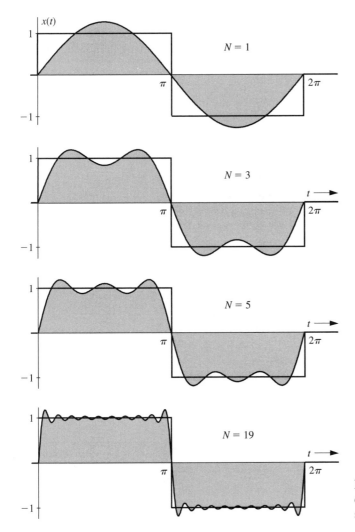

Figure 6.21 Approximation of a square signal by a sum of sinusoids.

TABLE 6.3

N	E_e
1	1.1938
3	0.6243
5	0.4206
7	0.3166
99	0.02545
∞	0

For a two-term approximation ($N = 3$)

$$E_e = 2\pi - \frac{16}{\pi}\left(1 + \frac{1}{9}\right) = 0.6243$$

Table 6.3 shows the error energy E_e for various values of N.

Clearly, $x(t)$ can be represented by the infinite series

$$x(t) = \frac{4}{\pi}\left(\sin t + \frac{1}{3}\sin 3t + \frac{1}{5}\sin 5t + \cdots\right)$$

$$= \frac{4}{\pi}\sum_{n=1,3,5,\ldots}^{\infty}\frac{1}{n}\sin nt \tag{6.86}$$

The equality exists in the sense that the error signal energy $\to 0$ as $N \to \infty$. In this case, the error energy decreases rather slowly with N, indicating that the series converges slowly. This is to be expected because $x(t)$ has jump discontinuities and consequently, according to discussion in Section 6.2-2, the series converges asymptotically as $1/n$.

EXERCISE E6.9

Approximate the signal $x(t) = t - \pi$ (Fig. 6.22) over the interval $(0, 2\pi)$ in terms of the set of sinusoids $\{\sin nt\}$, $n = 0, 1, 2, \ldots$, used in Example 6.11. Find E_e, the error energy. Show that $E_e \to 0$ as $N \to \infty$.

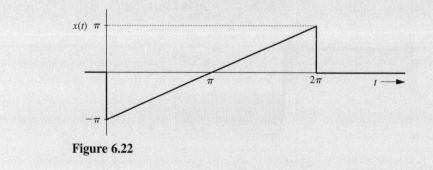

Figure 6.22

ANSWER

$$x(t) \simeq -2 \sum_{n=1}^{N} \frac{1}{N} \sin nt \quad \text{and} \quad E_e = \frac{2}{3}\pi^3 - \sum_{n=1}^{N} \frac{4\pi}{n^2}$$

SOME EXAMPLES OF GENERALIZED FOURIER SERIES

Signals are vectors in every sense. Like a vector, a signal can be represented as a sum of its components in a variety of ways. Just as vector coordinate systems are formed by mutually orthogonal vectors (rectangular, cylindrical, spherical), we also have signal coordinate systems (basis signals) formed by a variety of sets of mutually orthogonal signals. There exist a large number of orthogonal signal sets that can be used as basis signals for generalized Fourier series. Some well-known signal sets are trigonometric (sinusoid) functions, exponential functions, Walsh functions, Bessel functions, Legendre polynomials, Laguerre functions, Jacobi polynomials, Hermite polynomials, and Chebyshev polynomials. The functions that concern us most in this book are the trigonometric and the exponential sets discussed earlier in this chapter.

LEGENDRE FOURIER SERIES

A set of Legendre polynomials $P_n(t)$ ($n = 0, 1, 2, 3, \ldots$) forms a complete set of mutually orthogonal functions over an interval ($-1 < t < 1$). These polynomials can be defined by the Rodrigues formula:

$$P_n(t) = \frac{1}{2^n n!} \frac{d^n}{dt^n}(t^2 - 1)^n \qquad n = 0, 1, 2, \ldots$$

It follows from this equation that

$$P_0(t) = 1$$
$$P_1(t) = t$$
$$P_2(t) = \left(\tfrac{3}{2}t^2 - \tfrac{1}{2}\right)$$
$$P_3(t) = \left(\tfrac{5}{2}t^3 - \tfrac{3}{2}t\right)$$

and so on.

We may verify the orthogonality of these polynomials by showing that

$$\int_{-1}^{1} P_m(t) P_n(t)\, dt = \begin{cases} 0 & m \neq n \\ \dfrac{2}{2m+1} & m = n \end{cases} \tag{6.87}$$

We can express a function $x(t)$ in terms of Legendre polynomials over an interval ($-1 < t < 1$) as

$$x(t) = c_0 P_0(t) + c_1 P_1(t) + \cdots + \cdots \tag{6.88}$$

where

$$c_r = \frac{\displaystyle\int_{-1}^{1} x(t)P_r(t)\,dt}{\displaystyle\int_{-1}^{1} P_r^{2}(t)\,dt}$$

$$= \frac{2r+1}{2} \int_{-1}^{1} x(t)P_r(t)\,dt \tag{6.89}$$

Note that although the series representation is valid over the interval $(-1,\ 1)$, it can be extended to any interval by the appropriate time scaling (see Prob. 6.5-8).

EXAMPLE 6.12

Let us consider the square signal shown in Fig. 6.23. This function can be represented by Legendre Fourier series:

$$x(t) = c_0 P_0(t) + c_1 P_1(t) + \cdots + c_r P_r(t) + \cdots$$

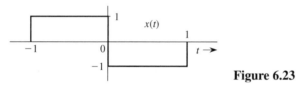

Figure 6.23

The coefficients $c_0, c_1, c_2, \ldots, c_r$ may be found from Eq. (6.89). We have

$$x(t) = \begin{cases} 1 & \cdots -1 < t < 0 \\ -1 & \cdots 0 < t < 1 \end{cases}$$

and

$$c_0 = \frac{1}{2} \int_{-1}^{1} x(t)\,dt = 0$$

$$c_1 = \frac{3}{2} \int_{-1}^{1} tx(t)\,dt = \frac{3}{2}\left(\int_{-1}^{0} t\,dt - \int_{0}^{1} t\,dt\right) = -\frac{3}{2}$$

$$c_2 = \frac{5}{2} \int_{-1}^{1} x(t)\left(\frac{3}{2}t^2 - \frac{1}{2}\right) dt = 0$$

This result follows immediately from the fact that the integrand is an odd function of t. In fact, this is true of all c_r for even values of r, that is,

$$c_0 = c_2 = c_4 = c_6 = \cdots = 0$$

Also,

$$c_3 = \frac{7}{2} \int_{-1}^{1} x(t) \left(\frac{5}{2}t^3 - \frac{3}{2}t\right) dt = \frac{7}{2} \left[\int_{-1}^{0} \left(\frac{5}{2}t^3 - \frac{3}{2}t\right) dt + \int_{0}^{1} - \left(\frac{5}{2}t^3 - \frac{3}{2}t\right) dt \right] = \frac{7}{8}$$

In a similar way, coefficients c_5, c_7, ... can be evaluated. We now have

$$x(t) = -\tfrac{3}{2}t + \tfrac{7}{8}\left(\tfrac{5}{2}t^3 - \tfrac{3}{2}t\right) + \cdots \tag{6.90}$$

TRIGONOMETRIC FOURIER SERIES

We have already proved [see Eqs. (6.7)] that the trigonometric signal set

$$\{1, \ \cos \omega_0 t, \ \cos 2\omega_0 t, \ \ldots, \ \cos n\omega_0 t, \ \ldots; \tag{6.91}$$

$$\sin \omega_0 t, \ \sin 2\omega_0 t, \ \ldots, \ \sin n\omega_0 t, \ \ldots\}$$

is orthogonal over any interval of duration T_0, where $T_0 = 1/f_0$ is the period of the sinusoid of frequency f_0. This is a complete set for a class of signals with finite energies.[11,12] Therefore we can express a signal $x(t)$ by a trigonometric Fourier series over any interval of duration T_0 seconds as

$$x(t) = a_0 + a_1 \cos \omega_0 t + a_2 \cos 2\omega_0 t + \cdots$$

$$+ b_1 \sin \omega_0 t + b_2 \sin 2\omega_0 t + \cdots$$

or

$$x(t) = a_0 + \sum_{n=1}^{\infty} a_n \cos n\omega_0 t + b_n \sin n\omega_0 t \qquad t_1 < t < t_1 + T_0 \tag{6.92a}$$

where

$$\omega_0 = 2\pi f_0 = \frac{2\pi}{T_0} \tag{6.92b}$$

We can use Eq. (6.76) to determine the Fourier coefficients a_0, a_n, and b_n. Thus

$$a_n = \frac{\displaystyle\int_{t_1}^{t_1+T_0} x(t) \cos n\omega_0 t \, dt}{\displaystyle\int_{t_1}^{t_1+T_0} \cos^2 n\omega_0 t \, dt} \tag{6.93}$$

The integral in the denominator of Eq. (6.93) has already been found to be $T_0/2$ when $n \neq 0$ [Eq. (6.7a) with $m = n$]. For $n = 0$, the denominator is T_0. Hence

$$a_0 = \frac{1}{T_0} \int_{t_1}^{t_1+T_0} x(t) \, dt \tag{6.94a}$$

and

$$a_n = \frac{2}{T_0} \int_{t_1}^{t_1+T_0} x(t) \cos n\omega_0 t \, dt \qquad n = 1, 2, 3, \ldots \tag{6.94b}$$

Similarly, we find that

$$b_n = \frac{2}{T_0} \int_{t_1}^{t_1+T_0} x(t) \sin n\omega_0 t \, dt \qquad n = 1, 2, 3, \ldots \tag{6.94c}$$

Note that the Fourier series in Eq. (6.85) of Example 6.11 is indeed the trigonometric Fourier series with $T_0 = 2\pi$ and $\omega_0 = 2\pi/T_0$. In this particular example, it is easy to verify from Eqs. (6.94a) and (6.94b) that $a_n = 0$ for all n, including $n = 0$. Hence the Fourier series in that example consisted only of sine terms.

EXPONENTIAL FOURIER SERIES

As shown in the footnote on page 624, the set of exponentials $e^{jn\omega_0 t}$ ($n = 0, \pm 1, \pm 2, \ldots$) is a set of functions orthogonal over any interval of duration $T_0 = 2\pi/\omega_0$. An arbitrary signal $x(t)$ can now be expressed over an interval $(t_1, t_1 + T_0)$ as

$$x(t) = \sum_{n=-\infty}^{\infty} D_n e^{jn\omega_0 t} \qquad t_1 < t < t_1 + T_0 \tag{6.95a}$$

where [see Eq. (6.82)]

$$D_n = \frac{1}{T_0} \int_{t_1}^{t_1+T_0} x(t) e^{-jn\omega_0 t} \, dt \tag{6.95b}$$

WHY USE THE EXPONENTIAL SET?

If $x(t)$ can be represented in terms of hundreds of different orthogonal sets, why do we exclusively use the exponential (or trigonometric) set for the representation of signals or LTI systems? It so happens that the exponential signal is an eigenfunction of LTI systems. In other words, for an LTI system, only an exponential input e^{st} yields the response that is also an exponential of the same form, given by $H(s)e^{st}$. The same is true of the trigonometric set. This fact makes the use of exponential signals natural for LTI systems in the sense that the system analysis using exponentials as the basis signals is greatly simplified.

6.6 NUMERICAL COMPUTATION OF D_n

We can compute D_n numerically by using the DFT (the discrete Fourier transform discussed in Section 8.5), which uses the samples of a periodic signal $x(t)$ over one period. The sampling interval is T seconds. Hence, there are $N_0 = T_0/T$ number of samples in one period T_0. To find the relationship between D_n and the samples of $x(t)$, consider Eq. (6.29b)

$$\begin{aligned}
D_n &= \frac{1}{T_0} \int_{T_0} x(t) e^{-jn\omega_0 t} \, dt \\
&= \lim_{T \to 0} \frac{1}{N_0 T} \sum_{k=0}^{N_0-1} x(kT) e^{-jn\omega_0 kT} \, T \\
&= \lim_{T \to 0} \frac{1}{N_0} \sum_{k=0}^{N_0-1} x(kT) e^{-jn\Omega_0 k}
\end{aligned} \tag{6.96}$$

where $x(kT)$ is the kth sample of $x(t)$ and

$$N_0 = \frac{T_0}{T} \qquad \Omega_0 = \omega_0 T = \frac{2\pi}{N_0} \tag{6.97}$$

In practice, it is impossible to make $T \to 0$ in computing the right-hand side of Eq. (6.96). We can make T small, but not zero, which will cause the data to increase without limit. Thus, we shall ignore the limit on T in Eq. (6.96) with the implicit understanding that T is reasonably small. Nonzero T will result in some computational error, which is inevitable in any numerical evaluation of an integral. The error resulting from nonzero T is called the *aliasing error*, which is discussed in more detail in Chapter 8. Thus, we can express Eq. (6.96) as

$$D_n = \frac{1}{N_0} \sum_{k=0}^{N_0-1} x(kT) e^{-jn\Omega_0 k} \tag{6.98a}$$

Now, from Eq. (6.97), $\Omega_0 N_0 = 2\pi$. Hence, $e^{jn\Omega_0(k+N_0)} = e^{jn\Omega_0 k}$ and from Eq. (6.98a), it follows that

$$D_{n+N_0} = D_n \tag{6.98b}$$

The periodicity property $D_{n+N_0} = D_n$ means that beyond $n = N_0/2$, the coefficients represent the values for negative n. For instance, when $N_0 = 32$, $D_{17} = D_{-15}$, $D_{18} = D_{-14}, \ldots, D_{31} = D_{-1}$. The cycle repeats again from $n = 32$ on.

We can use the efficient FFT (the *fast Fourier transform* discussed in Section 8.6) to compute the right-hand side of Eq. (6.98b). We shall use MATLAB to implement the FFT algorithm. For this purpose, we need samples of $x(t)$ over one period starting at $t = 0$. In this algorithm, it is also preferable (although not necessary) that N_0 be a power of 2, that is $N_0 = 2^m$, where m is an integer.

COMPUTER EXAMPLE C6.4

Numerically compute and then plot the trigonometric and exponential Fourier spectra for the periodic signal in Fig. 6.2a (Example 6.1).

The samples of $x(t)$ start at $t = 0$ and the last (N_0th) sample is at $t = T_0 - T$. At the points of discontinuity, the sample value is taken as the average of the values of the function on two sides of the discontinuity. Thus, the sample at $t = 0$ is not 1 but $(e^{-\pi/2} + 1)/2 = 0.604$. To determine N_0, we require that D_n for $n \geq N_0/2$ to be negligible. Because $x(t)$ has a jump discontinuity, D_n decays rather slowly as $1/n$. Hence, a choice of $N_0 = 200$ is acceptable because the ($N_0/2$)nd (100th) harmonic is about 1% of the fundamental. However, we also require N_0 to be power of 2. Hence, we shall take $N_0 = 256 = 2^8$.

First, the basic parameters are established.

```
>> T_0 = pi; N_0 = 256; T = T_0/N_0; t = (0:T:T*(N_0-1))'; M = 10;
>> x = exp(-t/2); x(1) = (exp(-pi/2)+1)/2;
```

Next, the DFT, computed by means of the `fft` function, is used to approximate the exponential Fourier spectra over $-M \leq n \leq M$.

```
>> D_n = fft(x)/N_0; n = [-N_0/2:N_0/2-1]';
>> clf; subplot(2,2,1); stem(n,abs(fftshift(D_n)),'k');
>> axis([-M M -.1 .6]); xlabel('n'); ylabel('|D_n|');
>> subplot(2,2,2); stem(n,angle(fftshift(D_n)),'k');
>> axis([-M M -pi pi]); xlabel('n'); ylabel('\angle D_n [rad]');
```

The approximate trigonometric Fourier spectra over $0 \leq n \leq M$ immediately follow.

```
>> n = [0:M]; C_n(1) = abs(D_n(1)); C_n(2:M+1) = 2*abs(D_n(2:M+1));
>> theta_n(1) = angle(D_n(1)); theta_n(2:M+1) = angle(D_n(2:M+1));
>> subplot(2,2,3); stem(n,C_n,'k');
>> xlabel('n'); ylabel('C_n');
>> subplot(2,2,4); stem(n,theta_n,'k');
>> xlabel('n'); ylabel('\theta_n [rad]');
```

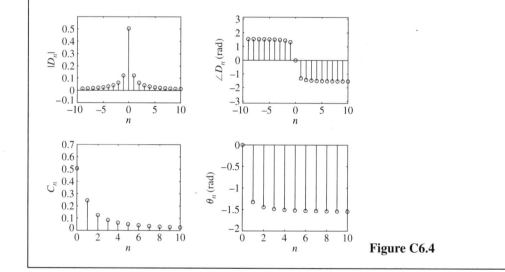

Figure C6.4

6.7 SUMMARY

In this chapter we showed how a periodic signal can be represented as a sum of sinusoids or exponentials. If the frequency of a periodic signal is f_0, then it can be expressed as a weighted sum of a sinusoid of frequency f_0 and its harmonics (the trigonometric Fourier series). We can reconstruct the periodic signal from a knowledge of the amplitudes and phases of these sinusoidal components (amplitude and phase spectra).

If a periodic signal $x(t)$ has an even symmetry, its Fourier series contains only cosine terms (including dc). In contrast, if $x(t)$ has an odd symmetry, its Fourier series contains only sine terms. If $x(t)$ has neither type of symmetry, its Fourier series contains both sine and cosine terms.

At points of discontinuity, the Fourier series for $x(t)$ converges to the mean of the values of $x(t)$ on either sides of the discontinuity. For signals with discontinuities, the Fourier series converges in the mean and exhibits Gibbs phenomenon at the points of discontinuity. The amplitude spectrum of the Fourier series for a periodic signal $x(t)$ with jump discontinuities decays slowly (as $1/n$) with frequency. We need a large number of terms in the Fourier series to approximate $x(t)$ within a given error. In contrast, smoother periodic signal amplitude spectrum decays faster with frequency and we require a fewer number of terms in the series to approximate $x(t)$ within a given error.

A sinusoid can be expressed in terms of exponentials. Therefore the Fourier series of a periodic signal can also be expressed as a sum of exponentials (the exponential Fourier series). The exponential form of the Fourier series and the expressions for the series coefficients are more compact than those of the trigonometric Fourier series. Also, the response of LTIC systems to an exponential input is much simpler than that for a sinusoidal input. Moreover, the exponential form of representation lends itself better to mathematical manipulations than does the trigonometric form. For these reasons, the exponential form of the series is preferred in modern practice in the areas of signals and systems.

The plots of amplitudes and angles of various exponential components of the Fourier series as functions of the frequency are the exponential Fourier spectra (amplitude and angle spectra) of the signal. Because a sinusoid $\cos \omega_0 t$ can be represented as a sum of two exponentials, $e^{j\omega_0 t}$ and $e^{-j\omega_0 t}$, the frequencies in the exponential spectra range from $\omega = -\infty$ to ∞. By definition, frequency of a signal is always a positive quantity. Presence of a spectral component of a negative frequency $-n\omega_0$ merely indicates that the Fourier series contains terms of the form $e^{-jn\omega_0 t}$. The spectra of the trigonometric and exponential Fourier series are closely related, and one can be found by the inspection of the other.

In Section 6.5 we discuss a method of representing signals by the generalized Fourier series, of which the trigonometric and exponential Fourier series are special cases. Signals are vectors in every sense. Just as a vector can be represented as a sum of its components in a variety of ways, depending on the choice of the coordinate system, a signal can be represented as a sum of its components in a variety of ways, of which the trigonometric and exponential Fourier series are only two examples. Just as we have vector coordinate systems formed by mutually orthogonal vectors, we also have signal coordinate systems (basis signals) formed by mutually orthogonal signals. Any signal in this signal space can be represented as a sum of the basis signals. Each set of basis signals yields a particular Fourier series representation of the signal. The signal is equal to its Fourier series, not in the ordinary sense, but in the special sense that the energy of the difference between the signal and its Fourier series approaches zero. This allows for the signal to differ from its Fourier series at some isolated points.

REFERENCES

1. Bell, E. T. *Men of Mathematics.* Simon & Schuster, New York, 1937.

2. Durant, Will, and Ariel Durant. *The Age of Napoleon,* Part XI in *The Story of Civilization Series.* Simon & Schuster, New York, 1975.

3. Calinger, R. *Classics of Mathematics,* 4th ed. Moore Publishing, Oak Park, IL, 1982.

4. Lanczos, C. *Discourse on Fourier Series.* Oliver Boyd, London, 1966.

5. Körner, T. W. *Fourier Analysis.* Cambridge University Press, Cambridge, UK, 1989.

6. Guillemin, E. A. *Theory of Linear Physical Systems.* Wiley, New York, 1963.

7. Gibbs, W. J. *Nature,* vol. 59, p. 606, April 1899.

8. Bôcher, M. *Annals of Mathematics,* vol. 7, no. 2, 1906.

9. Carslaw, H. S. *Bulletin of the American Mathematical Society,* vol. 31, pp. 420–424, October 1925.

10. Lathi, B. P. *Signals, Systems, and Communication.* Wiley, New York, 1965.

11. Walker P. L. *The Theory of Fourier Series and Integrals.* Wiley, New York, 1986.

12. Churchill, R. V., and J. W. Brown. *Fourier Series and Boundary Value Problems,* 3rd ed. McGraw-Hill, New York, 1978.

MATLAB SESSION 6: FOURIER SERIES APPLICATIONS

Computational packages such as MATLAB simplify the Fourier-based analysis, design, and synthesis of periodic signals. MATLAB permits rapid and sophisticated calculations, which promote practical application and intuitive understanding of the Fourier series.

M6.1 Periodic Functions and the Gibbs Phenomenon

It is sufficient to define any T_0-periodic function over the interval $(0 \le t < T_0)$. For example, consider the 2π-periodic function given by

$$x(t) = \begin{cases} \dfrac{1}{A}t & 0 \le t < A \\ 1 & A \le t < \pi \\ 0 & \pi \le t < 2\pi \\ x(t + 2\pi) & \text{otherwise} \end{cases}$$

Although similar to a square wave, $x(t)$ has a linearly rising edge of width A, where $(0 < A < \pi)$. As $A \to 0$, $x(t)$ approaches a square wave; as $A \to \pi$, $x(t)$ approaches a type of sawtooth wave.

In MATLAB, the `mod` command helps represent periodic functions such as $x(t)$.

```
>> x = inline(['mod(t,2*pi)/A.*(mod(t,2*pi)<A)+',...
        '((mod(t,2*pi)>=A)&(mod(t,2*pi)<pi))'],'t','A');
```

Sometimes referred to as the signed remainder after division, `mod(t,2*pi)` returns the value t modulo 2π. Thought of another way, the `mod` operator appropriately shifts t into $[0, T_0)$, where $x(t)$ is conveniently defined. When an inline object is a function of multiple variables, the variables are simply listed in the desired order following the function expression.

The exponential Fourier series coefficients for $x(t)$ (see Prob. 6.3-2) are given by

$$D_n = \begin{cases} \dfrac{2\pi - A}{4\pi} & n = 0 \\ \dfrac{1}{2\pi n}\left(\dfrac{e^{-jnA} - 1}{nA} + je^{-jn\pi}\right) & \text{otherwise} \end{cases} \tag{M6.1}$$

Since $x(t)$ is real, $D_{-n} = D_n^*$. Truncating the Fourier series at $|n| = N$ yields the approximation

$$x(t) \approx x_N(t) = D_0 + \sum_{n=1}^{N} \left(D_n e^{jnt} + D_n^* e^{-jnt} \right) \qquad \text{(M6.2)}$$

Program MS6P1 uses Eq. (M6.2) to compute $x_N(t)$ for $(0 \leq N \leq 100)$, each over $(-\pi/4 \leq t < 2\pi + \pi/4)$.

```
function [x_N,t] = MS6P1(A);
% MS6P1.m : MATLAB Session 6, Program 1
% Function M-file approximates x(t) using Fourier series
% truncated at |n|=N for (0 <= N <= 100).
% INPUTS:   A = width of rising edge
% OUTPUTS:  x_N = output matrix, where x_N(N+1,:) is the |n| = N truncation
%                 t = time vector for x_N

t = linspace(-pi/4,2*pi+pi/4,1000); % Time vector exceeds one period.
sumterms = zeros(101,length(t)); % Pre-allocate memory
sumterms(1,:) = (2*pi-A)/(4*pi); % Compute DC term
for n = 1:100, % Compute N remaining terms
    D_n = 1/(2*pi*n)*((exp(-j*n*A)-1)/(n*A) + j*exp(-j*n*pi));
    sumterms(1+n,:) = real(D_n*exp(j*n*t) + conj(D_n)*exp(-j*n*t));
end
x_N = cumsum(sumterms);
```

Although theoretically not required, the `real` command ensures that small computer round-off errors do not cause a complex-valued result. For a matrix input, the `cumsum` command cumulatively sums along rows: the first output row is identical to the first input row, the second output row is the sum of the first two input rows, the third output row is the sum of the first three input rows, and so forth. Thus, row $(N + 1)$ of x_N corresponds to the truncation of the exponential Fourier series at $|n| = N$.

Setting $A = \pi/2$, Fig. M6.1 compares $x(t)$ and $x_{20}(t)$.

```
>> A = pi/2; [x_N,t] = MS6P1(A);
>> plot(t,x_N(21,:),'k',t,x(t,A),'k:'); axis([-pi/4,2*pi+pi/4,-0.1,1.1]);
>> xlabel('t'); ylabel('Amplitude'); legend('x_20(t)','x(t)',0);
```

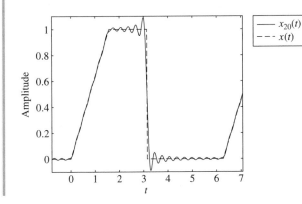

Figure M6.1 Comparison of $x_{20}(t)$ and $x(t)$ using $A = \pi/2$.

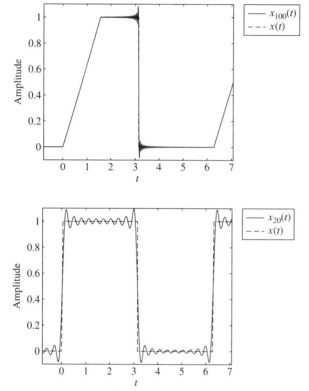

Figure M6.2 Comparison of $x_{100}(t)$ and $x(t)$ using $A = \pi/2$.

Figure M6.3 Comparison of $x_{20}(t)$ and $x(t)$ using $A = \pi/64$.

As expected, the falling edge is accompanied by the overshoot that is characteristic of the Gibbs phenomenon.

Increasing N to 100, as shown in Fig. M6.2, improves the approximation but does not reduce the overshoot.

```
>> plot(t,x_N(101,:),'k',t,x(t,A),'k:'); axis([-pi/4,2*pi+pi/4,-0.1,1.1]);
>> xlabel('t'); ylabel('Amplitude'); legend('x_100(t)','x(t)',0);
```

Reducing A to $\pi/64$ produces a curious result. For $N = 20$, both the rising and falling edges are accompanied by roughly 9% of overshoot, as shown in Fig. M6.3. As the number of terms is increased, overshoot persists only in the vicinity of jump discontinuities. For $x_N(t)$, increasing N decreases the overshoot near the rising edge but not near the falling edge. Remember that it is a true jump discontinuity that causes the Gibbs phenomenon. A continuous signal, no matter how sharply it rises, can always be represented by a Fourier series at every point within any small error by increasing N. This is not the case when a true jump discontinuity is present. Figure M6.4 illustrates this behavior using $N = 100$.

M6.2 Optimization and Phase Spectra

Although magnitude spectra typically receive the most attention, phase spectra are critically important in some applications. Consider the problem of characterizing the frequency response of an unknown system. By applying sinusoids one at a time, the frequency response is empirically

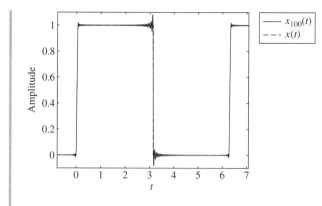

Figure M6.4 Comparison of $x_{100}(t)$ and $x(t)$ using $A = \pi/64$.

measured one point at a time. This process is tedious at best. Applying a superposition of many sinusoids, however, allows simultaneous measurement of many points of the frequency response. Such measurements can be taken by a spectrum analyzer equipped with a transfer function mode or by applying Fourier analysis techniques, which are discussed in later chapters.

A multitone test signal $m(t)$ is constructed using a superposition of N real sinusoids

$$m(t) = \sum_{n=1}^{N} M_n \cos{(\omega_n t + \theta_n)} \tag{M6.3}$$

where M_n and θ_n establish the relative magnitude and phase of each sinusoidal component. It is sensible to constrain all gains to be equal, $M_n = M$ for all n. This ensures equal treatment at each point of the measured frequency response. Although the value M is normally chosen to set the desired signal power, we set $M = 1$ for convenience.

While not required, it is also sensible to space the sinusoidal components uniformly in frequency.

$$m(t) = \sum_{n=1}^{N} \cos{(n\omega_0 t + \theta_n)} \tag{M6.4}$$

Another sensible alternative, which spaces components logarithmically in frequency, is treated in Prob. 6.M-1.

Equation (M6.4) is now a truncated compact-form Fourier series with a flat magnitude spectrum. Frequency resolution and range are set by ω_0 and N, respectively. For example, a 2 kHz range with a resolution of 100 Hz requires $\omega_0 = 2\pi 100$ and $N = 20$. The only remaining unknowns are the θ_n.

While it is tempting to set $\theta_n = 0$ for all n, the results are quite unsatisfactory. MATLAB helps demonstrate the problem by using $\omega_0 = 2\pi 100$ and $N = 20$ sinusoids, each with a peak-to-peak voltage of one volt.

```
>> m = inline('sum(cos(omega*t+theta*ones(size(t))))','theta','t','omega');
>> N = 20; omega = 2*pi*100*[1:N]'; theta = zeros(size(omega));
>> t = linspace(-0.01,0.01,1000);
>> plot(t,m(theta,t,omega),'k'); xlabel('t [sec]'); ylabel('m(t) [volts]');
```

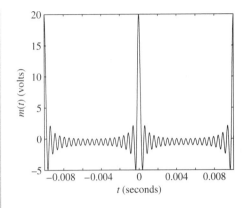

Figure M6.5 Test signal $m(t)$ with $\theta_n = 0$.

As shown in Fig. M6.5, $\theta_n = 0$ causes each sinusoid to constructively add. The resulting 20-volt peak can saturate system components, such as operational amplifiers operating with ± 12-volt rails. To improve signal performance, the maximum amplitude of $m(t)$ over t needs to be reduced.

One way to reduce $\max_t (|m(t)|)$ is to reduce M, the strength of each component. Unfortunately, this approach reduces the system's signal-to-noise ratio and ultimately degrades measurement quality. Therefore, reducing M is not a smart decision. The phases θ_n, however, can be adjusted to reduce $\max_t (|m(t)|)$ while preserving signal power. In fact, since $\theta_n = 0$ maximizes $\max_t (|m(t)|)$, just about any other choice of θ_n will improve the situation. Even a random choice should improve performance.

As with any computer, MATLAB cannot generate truly random numbers. Rather, it generates pseudo-random numbers. Pseudo-random numbers are deterministic sequences that appear to be random. The particular sequence of numbers that is realized depends entirely on the initial state of the pseudo-random-number generator. Setting the generator's initial state to a known value allows a "random" experiment with reproducible results. The command `rand('state',0)` initializes the state of the pseudorandom number generator to a known condition of zero, and the MATLAB command `rand(a,b)` generates an a-by-b matrix of pseudo-random numbers that are uniformly distributed over the interval (0, 1). Radian phases occupy the wider interval $(0, 2\pi)$, so the results from `rand` need to be appropriately scaled.

```
>> rand('state',0); theta_rand0 = 2*pi*rand(N,1);
```

By using the randomly chosen θ_n, $m(t)$ is computed, the maximum magnitude is identified, and the results are plotted as shown in Fig. M6.6.

```
>> m_rand0 = m(theta_rand0,t,omega);
>> [max_mag,max_ind] = max(abs(m_rand0(1:end/2)));
>> plot(t,m_rand0,'k'); axis([-0.01,0.01,-10,10]);
>> xlabel('t [sec]'); ylabel('m(t) [volts]');
>> text(t(max_ind),m_rand0(max_ind),...
        ['\leftarrow max = ',num2str(m_rand0(max_ind))]);
```

For a vector input, the `max` command returns the maximum value and the index corresponding to the first occurrence of the maximum. Similarly, although not used here, the MATLAB command

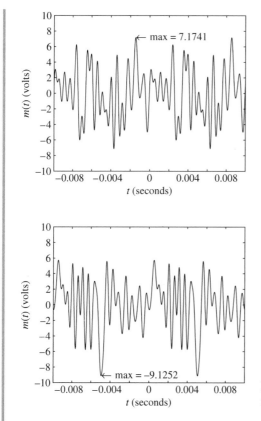

Figure M6.6 Test signal $m(t)$ with random θ_n found by using `rand('state',0)`.

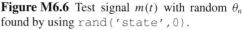

Figure M6.7 Test signal $m(t)$ with random θ_n found by using `rand('state',1)`.

`min` determines and locates minima. The `text(a,b,c)` command annotates the current figure with the string c at the location (a,b). MATLAB's help facilities describe the many properties available to adjust the appearance and format delivered by the `text` command. The `\leftarrow` command produces the symbol ←. Similarly, `\rightarrow`, `\uparrow`, and `\downarrow` produce the symbols →, ↑, and ↓, respectively.

Randomly chosen phases suffer a fatal fault: there is little guarantee of optimal performance. For example, repeating the experiment with `rand('state',1)` produces a maximum magnitude of 9.1 volts, as shown in Fig. M6.7. This value is significantly higher than the previous maximum of 7.2 volts. Clearly, it is better to replace a random solution with an optimal solution.

What constitutes "optimal"? Many choices exist, but desired signal criteria naturally suggest that optimal phases minimize the maximum magnitude of $m(t)$ over all t. To find these optimal phases, MATLAB's `fminsearch` command is useful. First, the function to be minimized, called the objective function, is defined.

```
>> maxmagm = inline('max(abs(sum(cos(omega*t+theta*ones(size(t))))))',...
        'theta','t','omega');
```

The `inline` argument order is important; `fminsearch` uses the first input argument as the variable of minimization. To minimize over θ, as desired, θ must be the first argument of the objective function `maxmagm`.

Next, the time vector is shortened to include only one period of $m(t)$.

```
>> t = linspace(0,0.01,201);
```

A full period ensures that all values of $m(t)$ are considered; the short length of t helps ensure that functions execute quickly. An initial value of θ is randomly chosen to begin the search.

```
>> rand('state',0); theta_init = 2*pi*rand(N,1);
>> theta_opt = fminsearch(maxmagm,theta_init,[],t,omega);
```

Notice that fminsearch finds the minimizer to maxmagm over θ by using an initial value theta_init. Most numerical minimization techniques are capable of finding only local minima, and fminsearch is no exception. As a result, fminsearch does not always produce a unique solution. The empty square brackets indicate no special options are requested, and the remaining ordered arguments are secondary inputs for the objective function. Full format details for fminsearch are available from MATLAB's help facilities.

Figure M6.8 shows the phase-optimized test signal. The maximum magnitude is reduced to a value of 5.4 volts, which is a significant improvement over the original peak of 20 volts.

Although the signals shown in Figs. M6.5 through M6.8 look different, they all possess the same magnitude spectra. They differ only in phase spectra. It is interesting to investigate the similarities and differences of these signals in ways other than graphs and mathematics. For example, is there an audible difference between the signals? For computers equipped with sound capability, the MATLAB sound command can be used to find out.

```
>> Fs = 8000; t = [0:1/Fs:2]; % Two second records at a sampling rate of 8kHz
>> m_0 = m(theta,t,omega);   % m(t) using zero phases
>> sound(m_0/20,Fs);
```

Since the sound command clips magnitudes that exceed one, the input vector is scaled by $1/20$. The remaining signals are created and played in a similar fashion. How well does the human ear discern the differences in phase spectra?

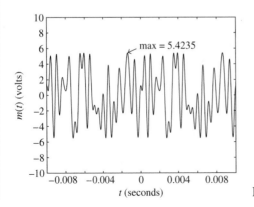

Figure M6.8 Test signal $m(t)$ with optimized phases.

PROBLEMS

6.1-1 For each of the periodic signals shown in Fig. P6.1-1, find the compact trigonometric Fourier series and sketch the amplitude and phase spectra. If either the sine or cosine terms are absent in the Fourier series, explain why.

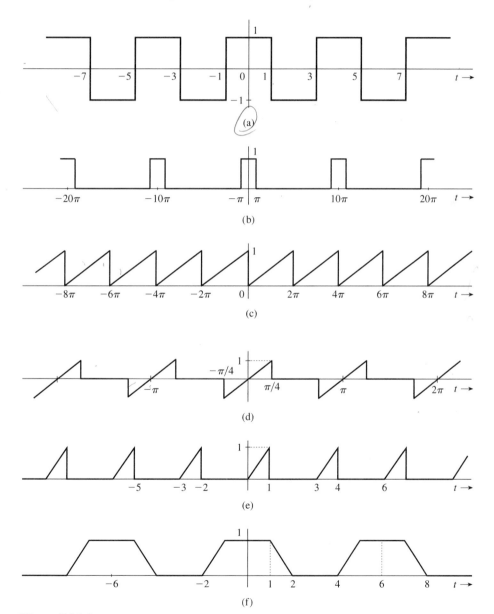

Figure P6.1-1

6.1-2 (a) Find the trigonometric Fourier series for $y(t)$ shown in Fig. P6.1-2.

(b) The signal $y(t)$ can be obtained by time reversal of $x(t)$ shown in Fig. 6.2a. Use this fact to obtain the Fourier series for $y(t)$ from the results in Example 6.1. Verify that the Fourier series thus obtained is identical to that found in part (a).

(c) Show that, in general, time reversal of a periodic signal does not affect the amplitude spectrum, and the phase spectrum is also unchanged except for the change of sign.

6.1-3 (a) Find the trigonometric Fourier series for the periodic signal $y(t)$ depicted in Fig. P6.1-3.

(b) The signal $y(t)$ can be obtained by time compression of $x(t)$ shown in Fig. 6.2a by a factor 2. Use this fact to obtain the Fourier series for $y(t)$ from the results in Example 6.1. Verify that the Fourier series thus obtained is identical to that found in part (a).

(c) Show that, in general, time-compression of a periodic signal by a factor a expands the Fourier spectra along the ω axis by the same factor a. In other words C_0, C_n, and θ_n remain unchanged, but the fundamental frequency is increased by the factor a, thus expanding the spectrum. Similarly time expansion of a periodic signal by a factor a compresses its Fourier spectra along the ω axis by the factor a.

6.1-4 (a) Find the trigonometric Fourier series for the periodic signal $g(t)$ in Fig. P6.1-4. Take advantage of the symmetry.

(b) Observe that $g(t)$ is identical to $x(t)$ in Fig. 6.3a left-shifted by 0.5 second. Use this fact to obtain the Fourier series for $g(t)$ from the results in Example 6.2. Verify that the Fourier series thus obtained is identical to that found in part (a).

(c) Show that, in general, a time shift of T seconds of a periodic signal does not affect the amplitude spectrum. However, the phase of the nth harmonic is increased or decreased $n\omega_0 T$ depending on whether the signal is advanced or delayed by T seconds.

6.1-5 If the two halves of one period of a periodic signal are identical in shape except that one is the negative of the other, the periodic signal is

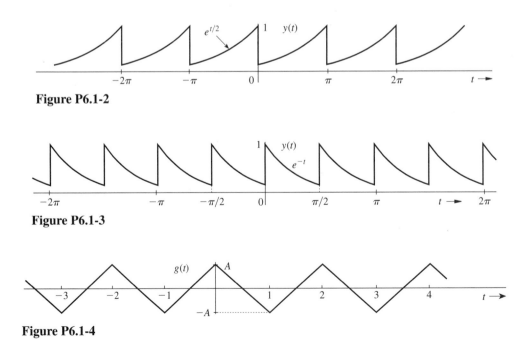

Figure P6.1-2

Figure P6.1-3

Figure P6.1-4

said to have a *half-wave symmetry*. If a periodic signal $x(t)$ with a period T_0 satisfies the half-wave symmetry condition, then

$$x\left(t - \frac{T_0}{2}\right) = -x(t)$$

In this case, show that all the even-numbered harmonics vanish and that the odd-numbered harmonic coefficients are given by

$$a_n = \frac{4}{T_0} \int_0^{T_0/2} x(t) \cos n\omega_0 t \, dt$$

and

$$b_n = \frac{4}{T_0} \int_0^{T_0/2} x(t) \sin n\omega_0 t \, dt$$

Using these results, find the Fourier series for the periodic signals in Fig. P6.1-5.

6.1-6 Over a finite interval, a signal can be represented by more than one trigonometric (or exponential) Fourier series. For instance, if we wish to represent $x(t) = t$ over an interval

$0 < t < 1$ by a Fourier series with fundamental frequency $\omega_0 = 2$, we can draw a pulse $x(t) = t$ over the interval $0 < t < 1$ and repeat the pulse every π seconds so that $T_0 = \pi$ and $\omega_0 = 2$ (Fig. P6.1-6a). If we want the fundamental frequency ω_0 to be 4, we repeat the pulse every $\pi/2$ seconds. If we want the series to contain only cosine terms with $\omega_0 = 2$, we construct a pulse $x(t) = |t|$ over $-1 < t < 1$, and repeat it every π seconds (Fig. P6.1-6b). The resulting signal is an even function with period π. Hence, its Fourier series will have only cosine terms with $\omega_0 = 2$. The resulting Fourier series represents $x(t) = t$ over $0 < t < 1$, as desired. We do not care what it represents outside this interval.

Sketch the periodic signal $x(t)$ such that $x(t) = t$ for $0 < t < 1$ and the Fourier series for $x(t)$ satisfies the following conditions.

(a) $\omega_0 = \pi/2$ and contains all harmonics, but cosine terms only

(b) $\omega_0 = 2$ and contains all harmonics, but sine terms only

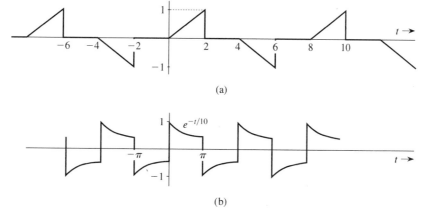

(a)

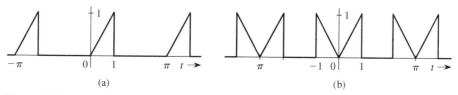

(b)

Figure P6.1-5

(a)   (b)

Figure P6.1-6

(c) $\omega_0 = \pi/2$ and contains all harmonics, which are neither exclusively sine nor cosine

(d) $\omega_0 = 1$ and contains only odd harmonics and cosine terms

(e) $\omega_0 = \pi/2$ and contains only odd harmonics and sine terms

(f) $\omega_0 = 1$ and contains only odd harmonics, which are neither exclusively sine nor cosine.

[Hint: For parts (d), (e), and (f), you need to use half-wave symmetry discussed in Prob. 6.1-5. Cosine terms imply possible dc component.] You are asked only to sketch the periodic signal $x(t)$ satisfying the given conditions. Do not find the values of the Fourier coefficients.

6.1-7   State with reasons whether the following signals are periodic or aperiodic. For periodic signals, find the period and state which of the harmonics are present in the series.

(a) $3 \sin t + 2 \sin 3t$

(b) $2 + 5 \sin 4t + 4 \cos 7t$

(c) $2 \sin 3t + 7 \cos \pi t$

(d) $7 \cos \pi t + 5 \sin 2\pi t$

(e) $3 \cos \sqrt{2}t + 5 \cos 2t$

(f) $\sin \dfrac{5t}{2} + 3 \cos \dfrac{6t}{5} + 3 \sin \left(\dfrac{t}{7} + 30°\right)$

(g) $\sin 3t + \cos \dfrac{15}{4}t$

(h) $(3 \sin 2t + \sin 5t)^2$

(i) $(5 \sin 2t)^3$

6.3-1   For each of the periodic signals in Fig. P6.1-1, find exponential Fourier series and sketch the corresponding spectra.

6.3-2   A 2π-periodic signal $x(t)$ is specified over one period as

$$x(t) = \begin{cases} \dfrac{1}{A}t & 0 \le t < A \\ 1 & A \le t < \pi \\ 0 & \pi \le t < 2\pi \end{cases}$$

Sketch $x(t)$ over two periods from $t = 0$ to 4π. Show that the exponential Fourier series

coefficients D_n for this series are given by

$$D_n = \begin{cases} \dfrac{2\pi - A}{4\pi} & n = 0 \\ \dfrac{1}{2\pi n}\left(\dfrac{e^{-jAn} - 1}{An}\right) & \text{otherwise} \end{cases}$$

6.3-3   A periodic signal $x(t)$ is expressed by the following Fourier series:

$$x(t) = 3 \cos t + \sin\left(5t - \dfrac{\pi}{6}\right) - 2 \cos\left(8t - \dfrac{\pi}{3}\right)$$

(a) Sketch the amplitude and phase spectra for the trigonometric series.

(b) By inspection of spectra in part (a), sketch the exponential Fourier series spectra.

(c) By inspection of spectra in part (b), write the exponential Fourier series for $x(t)$.

(d) Show that the series found in part (c) is equivalent to the trigonometric series for $x(t)$.

6.3-4   The trigonometric Fourier series of a certain periodic signal is given by

$$x(t) = 3 + \sqrt{3}\cos 2t + \sin 2t$$
$$+ \sin 3t - \dfrac{1}{2}\cos\left(5t + \dfrac{\pi}{3}\right)$$

(a) Sketch the trigonometric Fourier spectra.

(b) By inspection of spectra in part (a), sketch the exponential Fourier series spectra.

(c) By inspection of spectra in part (b), write the exponential Fourier series for $x(t)$.

(d) Show that the series found in part (c) is equivalent to the trigonometric series for $x(t)$.

6.3-5   The exponential Fourier series of a certain function is given as

$$x(t) = (2 + j2)e^{-j3t} + j2e^{-jt}$$
$$+ 3 - j2e^{jt} + (2 - j2)e^{j3t}$$

(a) Sketch the exponential Fourier spectra.

(b) By inspection of the spectra in part (a), sketch the trigonometric Fourier spectra for $x(t)$. Find the compact trigonometric Fourier series from these spectra.

(c) Show that the trigonometric series found in part (b) is equivalent to the exponential series for $x(t)$.

(d) Find the signal bandwidth.

6.3-6 Figure P6.3-6 shows the trigonometric Fourier spectra of a periodic signal $x(t)$.

    (a) By inspection of Fig. P6.3-6, find the trigonometric Fourier series representing $x(t)$.

    (b) By inspection of Fig. P6.3-6, sketch the exponential Fourier spectra of $x(t)$.

    (c) By inspection of the exponential Fourier spectra obtained in part (b), find the exponential Fourier series for $x(t)$.

    (d) Show that the series found in parts (a) and (c) are equivalent.

6.3-7 Figure P6.3-7 shows the exponential Fourier spectra of a periodic signal $x(t)$.

    (a) By inspection of Fig. P6.3-7, find the exponential Fourier series representing $x(t)$.

    (b) By inspection of Fig. P6.3-7, sketch the trigonometric Fourier spectra for $x(t)$.

    (c) By inspection of the trigonometric Fourier spectra found in part (b), find the trigonometric Fourier series for $x(t)$.

    (d) Show that the series found in parts (a) and (c) are equivalent.

6.3-8 (a) Find the exponential Fourier series for the signal in Fig. P6.3-8a.

    (b) Using the results in part (a), find the Fourier series for the signal $\hat{x}(t)$ in Fig. P6.3-8b, which is a time-shifted version of the signal $x(t)$.

    (c) Using the results in part (a), find the Fourier series for the signal $\tilde{x}(t)$ in Fig. P6.3-8c, which is a time-scaled version of the signal $x(t)$.

6.3-9 If a periodic signal $x(t)$ is expressed as an exponential Fourier series

$$x(t) = \sum_{n=-\infty}^{\infty} D_n e^{jn\omega_0 t}$$

    (a) Show that the exponential Fourier series for $\hat{x}(t) = x(t - T)$ is given by

$$\hat{x}(t) = \sum_{n=-\infty}^{\infty} \hat{D}_n e^{jn\omega_0 t}$$

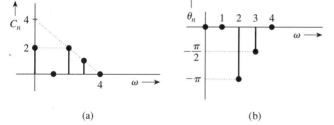

        (a)   (b)   **Figure P6.3-6**

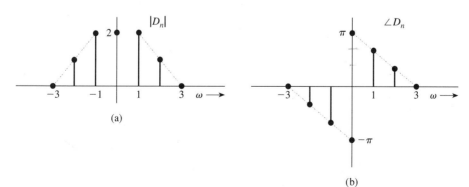

Figure P6.3-7

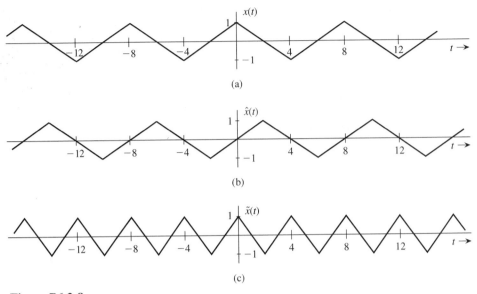

Figure P6.3-8

in which

$$|\hat{D}_n| = |D_n| \quad \text{and} \quad \angle \hat{D}_n = \angle D_n - n\omega_0 T$$

This result shows that time-shifting of a periodic signal by T seconds merely changes the phase spectrum by $n\omega_0 T$. The amplitude spectrum is unchanged.

(b) Show that the exponential Fourier series for $\tilde{x}(t) = x(at)$ is given by

$$\tilde{x}(t) = \sum_{n=-\infty}^{\infty} D_n e^{jn(a\omega_0)t}$$

This result shows that time compression of a periodic signal by a factor a expands its Fourier spectra along the ω axis by the same factor a. Similarly, time expansion of a periodic signal by a factor a compresses its Fourier spectra along the ω axis by the factor a. Can you explain this result intuitively?

6.3-10 (a) The Fourier series for the periodic signal in Fig. 6.6a is given in Exercise E6.1. Verify Parseval's theorem for this series, given

that

$$\sum_{n=1}^{\infty} \frac{1}{n^4} = \frac{\pi^4}{90}$$

(b) If $x(t)$ is approximated by the first N terms in this series, find N so that the power of the error signal is less than 1% of P_x.

6.3-11 (a) The Fourier series for the periodic signal in Fig. 6.6b is given in Exercise E6.1. Verify Parseval's theorem for this series, given that

$$\sum_{n=1}^{\infty} \frac{1}{n^2} = \frac{\pi^2}{6}$$

(b) If $x(t)$ is approximated by the first N terms in this series, find N so that the power of the error signal is less than 10% of P_x.

6.3-12 The signal $x(t)$ in Fig. 6.14 is approximated by the first $2N + 1$ terms (from $n = -N$ to N) in its exponential Fourier series given in Exercise E6.5. Determine the value of N if this $(2N + 1)$-term Fourier series power is to be no less than 99.75% of the power of $x(t)$.

6.4-1 Find the response of an LTIC system with transfer function

$$H(s) = \frac{s}{s^2 + 2s + 3}$$

to the periodic input shown in Fig. 6.2a.

6.4-2 (a) Find the exponential Fourier series for a signal $x(t) = \cos 5t \sin 3t$. You can do this without evaluating any integrals.
  (b) Sketch the Fourier spectra.
  (c) The signal $x(t)$ is applied at the input of an LTIC system with frequency response, as shown in Fig. P6.4-2. Find the output $y(t)$.

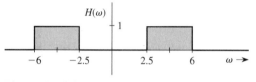

Figure P6.4-2

6.4-3 (a) Find the exponential Fourier series for a periodic signal $x(t)$ shown in Fig. P6.4-3a.
  (b) The signal $x(t)$ is applied at the input of an LTIC system shown in Fig. P6.4-3b. Find the expression for the output $y(t)$.

6.5-1 Derive Eq. (6.51) in an alternate way by observing that $\mathbf{e} = (\mathbf{x} - c\mathbf{y})$ and $|\mathbf{e}|^2 = (\mathbf{x} - c\mathbf{y}) \cdot (\mathbf{x} - c\mathbf{y}) = |\mathbf{x}|^2 + c^2|\mathbf{y}|^2 - 2c\mathbf{x} \cdot \mathbf{y}$.

6.5-2 A signal $x(t)$ is approximated in terms of a signal $y(t)$ over an interval (t_1, t_2):

$$x(t) \simeq cy(t) \qquad t_1 < t < t_2$$

where c is chosen to minimize the error energy.
  (a) Show that $y(t)$ and the error $e(t) = x(t) - cy(t)$ are orthogonal over the interval (t_1, t_2).
  (b) Can you explain the result in terms of the signal–vector analogy?
  (c) Verify this result for the square signal $x(t)$ in Fig. 6.19 and its approximation in terms of signal $\sin t$.

6.5-3 If $x(t)$ and $y(t)$ are orthogonal, then show that the energy of the signal $x(t) + y(t)$ is identical to the energy of the signal $x(t) - y(t)$ and is given by $E_x + E_y$. Explain this result by using the vector analogy. In general, show that for orthogonal signals $x(t)$ and $y(t)$ and for any pair of arbitrary real constants c_1 and c_2, the energies of $c_1x(t) + c_2y(t)$ and $c_1x(t) - c_2y(t)$ are identical, given by $c_1^2 E_x + c_2^2 E_y$.

6.5-4 (a) For the signals $x(t)$ and $y(t)$ depicted in Fig. P6.5-4, find the component of the form $y(t)$ contained in $x(t)$. In other words find the optimum value of c in the

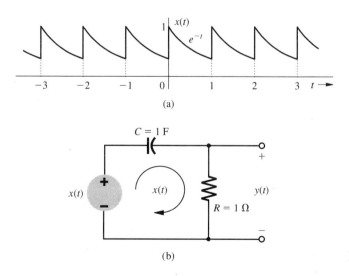

(a)

(b)

Figure P6.4-3

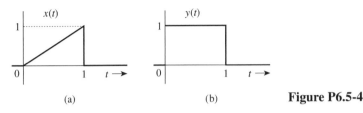

Figure P6.5-4

approximation $x(t) \approx cy(t)$ so that the error signal energy is minimum.

(b) Find the error signal $e(t)$ and its energy E_e. Show that the error signal is orthogonal to $y(t)$, and that $E_x = c^2 E_y + E_e$. Can you explain this result in terms of vectors.

6.5-5 For the signals $x(t)$ and $y(t)$ shown in Fig. P6.5-4, find the component of the form $x(t)$ contained in $y(t)$. In other words, find the optimum value of c in the approximation $y(t) \approx cx(t)$ so that the error signal energy is minimum. What is the error signal energy?

6.5-6 Represent the signal $x(t)$ shown in Fig. P6.5-4a over the interval from 0 to 1 by a trigonometric Fourier series of fundamental frequency $\omega_0 = 2\pi$. Compute the error energy in the representation of $x(t)$ by only the first N terms of this series for $N = 1, 2, 3$, and 4.

6.5-7 Represent $x(t) = t$ over the interval $(0, 1)$ by a trigonometric Fourier series that has

(a) $\omega_0 = 2\pi$ and only sine terms
(b) $\omega_0 = \pi$ and only sine terms
(c) $\omega_0 = \pi$ and only cosine terms

You may use a dc term in these series if necessary.

6.5-8 In Example 6.12, we represented the function in Fig. 6.23 by Legendre polynomials.

(a) Use the results in Example 6.12 to represent the signal $g(t)$ in Fig. P6.5-8 by Legendre polynomials.

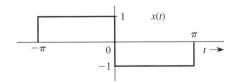

Figure P6.5-8

(b) Compute the error energy for the approximations having one and two- (nonzero) terms.

6.5-9 Walsh functions, which can take on only two amplitude values, form a complete set of orthonormal functions and are of great practical importance in digital applications because they can be easily generated by logic circuitry and because multiplication with these functions can be implemented by simply using a polarity-reversing switch. Figure P6.5-9 shows the first eight functions in this set. Represent $x(t)$ in Fig. P6.5-4a over the interval $(0, 1)$ by using a Walsh Fourier series with these eight basis functions. Compute the energy of $e(t)$, the error in the approximation, using the first N nonzero terms in the series for $N = 1, 2, 3$, and 4. We found the trigonometric Fourier series for $x(t)$ in Prob. 6.5-6. How does the Walsh series compare with the trigonometric series in Prob. 6.5-6 from the viewpoint of the error energy for a given N?

6.M-1 MATLAB Session 6 discusses the construction of a phase-optimized multitone test signal with linearly spaced frequency components. This problem investigates a similar signal with logarithmically spaced frequency components.

A multitone test signal $m(t)$ is constructed by using a superposition of N real sinusoids

$$m(t) = \sum_{n=1}^{N} \cos(\omega_n t + \theta_n)$$

where θ_n establishes the relative phase of each sinusoidal component.

(a) Determine a suitable set of $N = 10$ frequencies ω_n that logarithmically spans $[(2\pi) \leq \omega \leq 100(2\pi)]$ yet still results in a periodic test signal $m(t)$. Determine the period T_0 of your signal. Using $\theta_n = 0$,

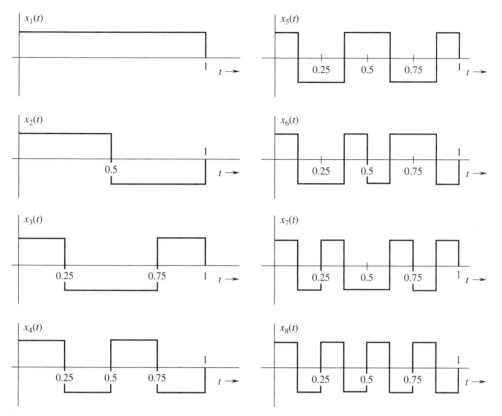

Figure P6.5-9

plot the resulting (T_0)-periodic signal over $-T_0/2 \leq t \leq T_0/2$.

(b) Determine a suitable set of phases θ_n that minimize the maximum magnitude of $m(t)$. Plot the resulting signal and identify the maximum magnitude that results.

(c) Many systems suffer from what is called one-over-f noise. The power of this undesirable noise is proportional to $1/f$. Thus, low-frequency noise is stronger than high-frequency noise. What modifications to $m(t)$ are appropriate for use in environments with $1/f$ noise? Justify your answer.

CHAPTER 7

CONTINUOUS-TIME SIGNAL ANALYSIS: THE FOURIER TRANSFORM

We can analyze linear systems in many different ways by taking advantage of the property of linearity, where by the input is expressed as a sum of simpler components. The system response to any complex input can be found by summing the system's response to these simpler components of the input. In time-domain analysis, we separated the input into impulse components. In the frequency-domain analysis in Chapter 4, we separated the input into exponentials of the form e^{st} (the Laplace transform), where the complex frequency $s = \sigma + j\omega$. The Laplace transform, although very valuable for system analysis, proves somewhat awkward for signal analysis, where we prefer to represent signals in terms of exponentials $e^{j\omega t}$ instead of e^{st}. This is accomplished by the Fourier transform. In a sense, the Fourier transform may be considered to be a special case of the Laplace transform with $s = j\omega$. Although this view is true most of the time, it does not always hold because of the nature of convergence of the Laplace and Fourier integrals.

In Chapter 6, we succeeded in representing periodic signals as a sum of (everlasting) sinusoids or exponentials of the form $e^{j\omega t}$. The Fourier integral developed in this chapter extends this spectral representation to aperiodic signals.

7.1 APERIODIC SIGNAL REPRESENTATION BY FOURIER INTEGRAL

Applying a limiting process, we now show that an aperiodic signal can be expressed as a continuous sum (integral) of everlasting exponentials. To represent an aperiodic signal $x(t)$ such as the one depicted in Fig. 7.1a by everlasting exponentials, let us construct a new periodic signal $x_{T_0}(t)$ formed by repeating the signal $x(t)$ at intervals of T_0 seconds, as illustrated in Fig. 7.1b. The period T_0 is made long enough to avoid overlap between the repeating pulses. The periodic signal $x_{T_0}(t)$ can be represented by an exponential Fourier series. If we let $T_0 \to \infty$, the pulses in the periodic signal repeat after an infinite interval and, therefore

$$\lim_{T_0 \to \infty} x_{T_0}(t) = x(t)$$

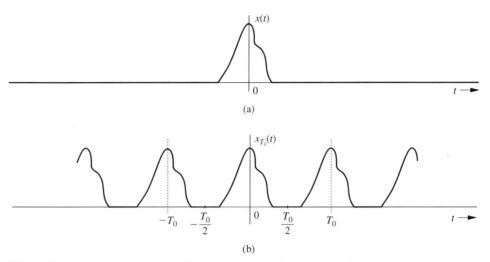

Figure 7.1 Construction of a periodic signal by periodic extension of $x(t)$.

Thus, the Fourier series representing $x_{T_0}(t)$ will also represent $x(t)$ in the limit $T_0 \to \infty$. The exponential Fourier series for $x_{T_0}(t)$ is given by

$$x_{T_0}(t) = \sum_{n=-\infty}^{\infty} D_n e^{jn\omega_0 t} \tag{7.1}$$

where

$$D_n = \frac{1}{T_0} \int_{-T_0/2}^{T_0/2} x_{T_0}(t) e^{-jn\omega_0 t} \, dt \tag{7.2a}$$

and

$$\omega_0 = \frac{2\pi}{T_0} \tag{7.2b}$$

Observe that integrating $x_{T_0}(t)$ over $(-T_0/2, T_0/2)$ is the same as integrating $x(t)$ over $(-\infty, \infty)$. Therefore, Eq. (7.2a) can be expressed as

$$D_n = \frac{1}{T_0} \int_{-\infty}^{\infty} x(t) e^{-jn\omega_0 t} \, dt \tag{7.2c}$$

It is interesting to see how the nature of the spectrum changes as T_0 increases. To understand this fascinating behavior, let us define $X(\omega)$, a continuous function of ω, as

$$X(\omega) = \int_{-\infty}^{\infty} x(t) e^{-j\omega t} \, dt \tag{7.3}$$

A glance at Eqs. (7.2c) and (7.3) shows that

$$D_n = \frac{1}{T_0} X(n\omega_0) \tag{7.4}$$

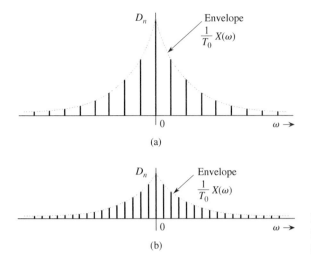

Figure 7.2 Change in the Fourier spectrum when the period T_0 in Fig. 7.1 is doubled.

This means that the Fourier coefficients D_n are $1/T_0$ times the samples of $X(\omega)$ uniformly spaced at intervals of ω_0, as depicted in Fig. 7.2a.[†] Therefore, $(1/T_0)X(\omega)$ is the envelope for the coefficients D_n. We now let $T_0 \to \infty$ by doubling T_0 repeatedly. Doubling T_0 halves the fundamental frequency ω_0 [Eq. (7.2b)], so that there are now twice as many components (samples) in the spectrum. However, by doubling T_0, the envelope $(1/T_0)X(\omega)$ is halved, as shown in Fig. 7.2b. If we continue this process of doubling T_0 repeatedly, the spectrum progressively becomes denser while its magnitude becomes smaller. Note, however, that the relative shape of the envelope remains the same [proportional to $X(\omega)$ in Eq. (7.3)]. In the limit as $T_0 \to \infty$, $\omega_0 \to 0$ and $D_n \to 0$. This result makes for a spectrum so dense that the spectral components are spaced at zero (infinitesimal) intervals. At the same time, the amplitude of each component is zero (infinitesimal). We have *nothing of everything, yet we have something!* This paradox sounds like *Alice in Wonderland,* but as we shall see, these are the classic characteristics of a very familiar phenomenon.[‡]

Substitution of Eq. (7.4) in Eq. (7.1) yields

$$x_{T_0}(t) = \sum_{n=-\infty}^{\infty} \frac{X(n\omega_0)}{T_0} e^{jn\omega_0 t} \tag{7.5}$$

As $T_0 \to \infty$, ω_0 becomes infinitesimal ($\omega_0 \to 0$). Hence, we shall replace ω_0 by a more appropriate notation, $\Delta\omega$. In terms of this new notation, Eq. (7.2b) becomes

$$\Delta\omega = \frac{2\pi}{T_0}$$

[†] For the sake of simplicity, we assume D_n, and therefore $X(\omega)$, in Fig. 7.2, to be real. The argument, however, is also valid for complex D_n [or $X(\omega)$].

[‡] If nothing else, the reader now has irrefutable proof of the proposition that 0% ownership of everything is better than 100% ownership of nothing.

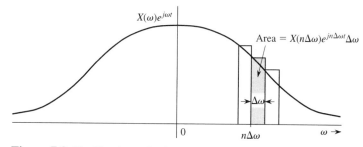

Figure 7.3 The Fourier series becomes the Fourier integral in the limit as $T_0 \to \infty$.

and Eq. (7.5) becomes

$$x_{T_0}(t) = \sum_{n=-\infty}^{\infty} \left[\frac{X(n\Delta\omega)\Delta\omega}{2\pi} \right] e^{(jn\Delta\omega)t} \tag{7.6a}$$

Equation (7.6a) shows that $x_{T_0}(t)$ can be expressed as a sum of everlasting exponentials of frequencies $0, \pm\Delta\omega, \pm2\Delta\omega, \pm3\Delta\omega, \ldots$ (the Fourier series). The amount of the component of frequency $n\Delta\omega$ is $[X(n\Delta\omega)\Delta\omega]/2\pi$. In the limit as $T_0 \to \infty$, $\Delta\omega \to 0$ and $x_{T_0}(t) \to x(t)$. Therefore

$$x(t) = \lim_{T_0 \to \infty} x_{T_0}(t)$$

$$= \lim_{\Delta\omega \to 0} \frac{1}{2\pi} \sum_{n=-\infty}^{\infty} X(n\Delta\omega)e^{(jn\Delta\omega)t} \Delta\omega \tag{7.6b}$$

The sum on the right-hand side of Eq. (7.6b) can be viewed as the area under the function $X(\omega)e^{j\omega t}$, as illustrated in Fig. 7.3. Therefore

$$x(t) = \frac{1}{2\pi} \int_{-\infty}^{\infty} X(\omega)e^{j\omega t} d\omega \tag{7.7}$$

The integral on the right-hand side is called the *Fourier integral*. We have now succeeded in representing an aperiodic signal $x(t)$ by a Fourier integral (rather than a Fourier series).[†] This integral is basically a Fourier series (in the limit) with fundamental frequency $\Delta\omega \to 0$, as seen from Eq. (7.6). The amount of the exponential $e^{jn\Delta\omega t}$ is $X(n\Delta\omega)\Delta\omega/2\pi$. Thus, the function $X(\omega)$ given by Eq. (7.3) acts as a spectral function.

We call $X(\omega)$ the *direct* Fourier transform of $x(t)$, and $x(t)$ the *inverse* Fourier transform of $X(\omega)$. The same information is conveyed by the statement that $x(t)$ and $X(\omega)$ are a Fourier transform pair. Symbolically, this statement is expressed as

$$X(\omega) = \mathcal{F}[x(t)] \qquad \text{and} \qquad x(t) = \mathcal{F}^{-1}[X(\omega)]$$

[†]This derivation should not be considered to be a rigorous proof of Eq. (7.7). The situation is not as simple as we have made it appear.[1]

or

$$x(t) \Longleftrightarrow X(\omega)$$

To recapitulate,

$$X(\omega) = \int_{-\infty}^{\infty} x(t)e^{-j\omega t}\,dt \tag{7.8a}$$

and

$$x(t) = \frac{1}{2\pi}\int_{-\infty}^{\infty} X(\omega)e^{j\omega t}\,d\omega \tag{7.8b}$$

It is helpful to keep in mind that the Fourier integral in Eq. (7.8b) is of the nature of a Fourier series with fundamental frequency $\Delta\omega$ approaching zero [Eq. (7.6b)]. Therefore, most of the discussion and properties of Fourier series apply to the Fourier transform as well. *The transform $X(\omega)$ is the frequency-domain specification of $x(t)$.*

We can plot the spectrum $X(\omega)$ as a function of ω. Since $X(\omega)$ is complex, we have both amplitude and angle (or phase) spectra

$$X(\omega) = |X(\omega)|e^{j\angle X(\omega)} \tag{7.9}$$

in which $|X(\omega)|$ is the amplitude and $\angle X(\omega)$ is the angle (or phase) of $X(\omega)$. According to Eq. (7.8a),

$$X(-\omega) = \int_{-\infty}^{\infty} x(t)e^{j\omega t}\,dt$$

Taking the conjugates of both sides yields

$$x^*(t) \Longleftrightarrow X^*(-\omega) \tag{7.10}$$

This property is known as the *conjugation property*. Now, if $x(t)$ is a real function of t, then $x(t) = x^*(t)$, and from the conjugation property, we find that

$$X(-\omega) = X^*(\omega) \tag{7.11a}$$

This is the *conjugate symmetry* property of the Fourier transform, applicable to real $x(t)$. Therefore, for real $x(t)$

$$|X(-\omega)| = |X(\omega)| \tag{7.11b}$$

$$\angle X(-\omega) = -\angle X(\omega) \tag{7.11c}$$

Thus, for real $x(t)$, the amplitude spectrum $|X(\omega)|$ is an even function, and the phase spectrum $\angle X(\omega)$ is an odd function of ω. These results were derived earlier for the Fourier spectrum of a periodic signal [Eq. (6.33)] and should come as no surprise.

EXAMPLE 7.1

Find the Fourier transform of $e^{-at}u(t)$.

_____

By definition [Eq. (7.8a)],

$$X(\omega) = \int_{-\infty}^{\infty} e^{-at}u(t)e^{-j\omega t}\,dt = \int_{0}^{\infty} e^{-(a+j\omega)t}\,dt = \frac{-1}{a+j\omega}e^{-(a+j\omega)t}\Big|_{0}^{\infty}$$

But $|e^{-j\omega t}| = 1$. Therefore, as $t \to \infty$, $e^{-(a+j\omega)t} = e^{-at}e^{-j\omega t} = \infty$ if $a < 0$, but it is equal to 0 if $a > 0$. Therefore

$$X(\omega) = \frac{1}{a+j\omega} \qquad a > 0$$

Expressing $a + j\omega$ in the polar form as $\sqrt{a^2 + \omega^2}\,e^{j\tan^{-1}(\omega/a)}$, we obtain

$$X(\omega) = \frac{1}{\sqrt{a^2 + \omega^2}}e^{-j\tan^{-1}(\omega/a)}$$

Therefore

$$|X(\omega)| = \frac{1}{\sqrt{a^2 + \omega^2}} \qquad \text{and} \qquad \angle X(\omega) = -\tan^{-1}\left(\frac{\omega}{a}\right) \tag{7.12}$$

The amplitude spectrum $|X(\omega)|$ and the phase spectrum $\angle X(\omega)$ are depicted in Fig. 7.4b. Observe that $|X(\omega)|$ is an even function of ω, and $\angle X(\omega)$ is an odd function of ω, as expected.

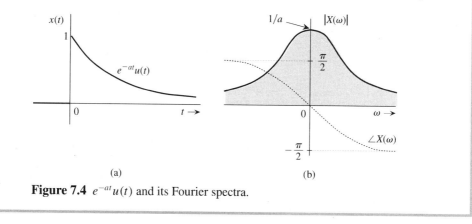

(a)   (b)

Figure 7.4 $e^{-at}u(t)$ and its Fourier spectra.

EXISTENCE OF THE FOURIER TRANSFORM

In Example 7.1 we observed that when $a < 0$, the Fourier integral for $e^{-at}u(t)$ does not converge. Hence, the Fourier transform for $e^{-at}u(t)$ does not exist if $a < 0$ (growing exponential). Clearly, not all signals are Fourier transformable.

Because the Fourier transform is derived here as a limiting case of the Fourier series, it follows that the basic qualifications of the Fourier series, such as *equality in the mean,* and convergence conditions in suitably modified form apply to the Fourier transform as well. It can be shown that if $x(t)$ has a finite energy, that is if

$$\int_{-\infty}^{\infty} |x(t)|^2 \, dt < \infty \tag{7.13}$$

then the Fourier transform $X(\omega)$ is finite and converges to $x(t)$ in the mean. This means, if we let

$$\hat{x}(t) = \lim_{W \to \infty} \frac{1}{2\pi} \int_{-W}^{W} X(\omega) e^{j\omega t} \, d\omega$$

then Eq. (7.8b) implies

$$\int_{-\infty}^{\infty} |x(t) - \hat{x}(t)|^2 \, dt = 0 \tag{7.14}$$

In other words, $x(t)$ and its Fourier integral [the right-hand side of Eq. (7.8b)] can differ at some values of t without contradicting Eq. (7.14). We shall now discuss an alternate set of criteria due to Dirichlet for convergence of the Fourier transform.

As with the Fourier series, if $x(t)$ satisfies certain conditions (*Dirichlet conditions*), its Fourier transform is guaranteed to converge pointwise at all points where $x(t)$ is continuous. Moreover, at the points of discontinuity, $x(t)$ converges to the value midway between the two values of $x(t)$ on either side of the discontinuity. The Dirichlet conditions are as follows:

1. $x(t)$ should be absolutely integrable, that is,

$$\int_{-\infty}^{\infty} |x(t)| \, dt < \infty \tag{7.15}$$

   If this condition is satisfied, we see that the integral on the right-hand side of Eq. (7.8a) is guaranteed to have a finite value.

2. $x(t)$ must have only finite number of finite discontinuities within any finite interval.

3. $x(t)$ must contain only finite number of maxima and minima within any finite interval.

We stress here that although the Dirichlet conditions are sufficient for the existence and pointwise convergence of the Fourier transform, they are not necessary. For example, we saw in Example 7.1 that a growing exponential, which violates Dirichlet's first condition in (7.15) does not have a Fourier transform. But the signal of the form $(\sin at)/t$, which *does* violate this condition, does have a Fourier transform.

Any signal that can be generated in practice satisfies the Dirichlet conditions and therefore has a Fourier transform. Thus, the physical existence of a signal is a sufficient condition for the existence of its transform.

LINEARITY OF THE FOURIER TRANSFORM

The Fourier transform is linear; that is, if

$$x_1(t) \iff X_1(\omega) \qquad \text{and} \qquad x_2(t) \iff X_2(\omega)$$

then

$$a_1 x_1(t) + a_2 x_2(t) \iff a_1 X_1(\omega) + a_2 X_2(\omega) \tag{7.16}$$

The proof is trivial and follows directly from Eq. (7.8a). This result can be extended to any finite number of terms. It can be extended to an infinite number of terms only if the conditions required for interchangeability of the operations of summation and integration are satisfied.

7.1-1 Physical Appreciation of the Fourier Transform

In understanding any aspect of the Fourier transform, we should remember that Fourier representation is a way of expressing a signal in terms of everlasting sinusoids (or exponentials). The Fourier spectrum of a signal indicates the relative amplitudes and phases of sinusoids that are required to synthesize that signal. A periodic signal Fourier spectrum has finite amplitudes and exists at discrete frequencies (ω_0 and its multiples). Such a spectrum is easy to visualize, but the spectrum of an aperiodic signal is not easy to visualize because it has a continuous spectrum. The continuous spectrum concept can be appreciated by considering an analogous, more tangible phenomenon. One familiar example of a continuous distribution is the loading of a beam. Consider a beam loaded with weights $D_1, D_2, D_3, \ldots, D_n$ units at the uniformly spaced points $y_1, y_2, \ldots, y_n$, as shown in Fig. 7.5a. The total load W_T on the beam is given by the sum of these loads at each of the n points:

$$W_T = \sum_{i=1}^{n} D_i$$

Consider now the case of a continuously loaded beam, as depicted in Fig. 7.5b. In this case, although there appears to be a load at every point, the load at any one point is zero. This does not mean that there is no load on the beam. A meaningful measure of load in this situation is not the load at a point, but rather the loading density per unit length at that point. Let $X(y)$ be the loading density per unit length of beam. It then follows that the load over a beam length $\Delta y(\Delta y \to 0)$, at some point y, is $X(y)\Delta y$. To find the total load on the beam, we divide the beam into segments of interval $\Delta y(\Delta y \to 0)$. The load over the nth such segment of length Δy is $X(n\Delta y)\Delta y$. The total load W_T is given by

$$W_T = \lim_{\Delta y \to 0} \sum_{y_1}^{y_n} X(n\Delta y)\,\Delta y$$

$$= \int_{y_1}^{y_n} X(y)\,dy$$

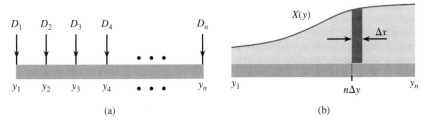

(a)   (b)

Figure 7.5 Weight-loading analogy for the Fourier transform.

The load now exists at every point, and y is now a continuous variable. In the case of discrete loading (Fig. 7.5a), the load exists only at the n discrete points. At other points there is no load. On the other hand, in the continuously loaded case, the load exists at every point, but at any specific point y, the load is zero. The load over a small interval Δy, however, is $[X(n\Delta y)]\Delta y$ (Fig. 7.5b). Thus, even though the load at a point y is zero, the relative load at that point is $X(y)$.

An exactly analogous situation exists in the case of a signal spectrum. When $x(t)$ is periodic, the spectrum is discrete, and $x(t)$ can be expressed as a sum of discrete exponentials with finite amplitudes:

$$x(t) = \sum_n D_n e^{jn\omega_0 t}$$

For an aperiodic signal, the spectrum becomes continuous; that is, the spectrum exists for every value of ω, but the amplitude of each component in the spectrum is zero. The meaningful measure here is not the amplitude of a component of some frequency but the spectral density per unit bandwidth. From Eq. (7.6b) it is clear that $x(t)$ is synthesized by adding exponentials of the form $e^{jn\Delta\omega t}$, in which the contribution by any one exponential component is zero. But the contribution by exponentials in an infinitesimal band $\Delta\omega$ located at $\omega = n\Delta\omega$ is $(1/2\pi)X(n\Delta\omega)\Delta\omega$, and the addition of all these components yields $x(t)$ in the integral form:

$$x(t) = \lim_{\Delta\omega\to 0} \frac{1}{2\pi}\sum_{n=-\infty}^{\infty} X(n\Delta\omega)e^{(jn\Delta\omega)t}\Delta\omega = \frac{1}{2\pi}\int_{-\infty}^{\infty} X(\omega)e^{j\omega t}\,d\omega \qquad (7.17)$$

Thus, $n\Delta\omega$ approaches a continuous variable ω. The spectrum now exists at every ω. The contribution by components within a band $d\omega$ is $(1/2\pi)X(\omega)\,d\omega = X(\omega)\,df$, where df is the bandwidth in hertz. Clearly, $X(\omega)$ is the *spectral density* per unit bandwidth (in hertz).[†] It also follows that even if the amplitude of any one component is infinitesimal, the relative amount of a component of frequency ω is $X(\omega)$. Although $X(\omega)$ is a spectral density, in practice it is customarily called the *spectrum* of $x(t)$ rather than the spectral density of $x(t)$. Deferring to this convention, we shall call $X(\omega)$ the Fourier spectrum (or Fourier transform) of $x(t)$.

A Marvelous Balancing Act

An important point to remember here is that $x(t)$ is represented (or synthesized) by exponentials or sinusoids that are everlasting (not causal). Such conceptualization leads to a rather fascinating picture when we try to visualize the synthesis of a timelimited pulse signal $x(t)$ [Fig. 7.6] by the sinusoidal components in its Fourier spectrum. The signal $x(t)$ exists only over an interval (a, b) and is zero outside this interval. The spectrum of $x(t)$ contains an infinite number of

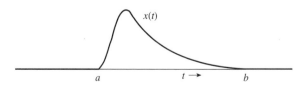

Figure 7.6 The marvel of the Fourier transform.

[†]To stress that the signal spectrum is a *density* function, we shall shade the plot of $|X(\omega)|$ (as in Fig. 7.4b). The representation of $\angle X(\omega)$, however, will be a by a line plot, primarily to avoid visual confusion.

exponentials (or sinusoids), which start at $t = -\infty$ and continue forever. The amplitudes and phases of these components add up exactly to $x(t)$ over the finite interval (a, b) and to zero everywhere outside this interval. Juggling the amplitudes and phases of an infinite number of components to achieve such a perfect and delicate balance boggles the human imagination. Yet the Fourier transform accomplishes it routinely, without much thinking on our part. Indeed, we become so involved in mathematical manipulations that we fail to notice this marvel.

7.2 TRANSFORMS OF SOME USEFUL FUNCTIONS

For convenience, we now introduce a compact notation for the useful gate, triangle, and interpolation functions.

UNIT GATE FUNCTION

We define a unit gate function rect (x) as a gate pulse of unit height and unit width, centered at the origin, as illustrated in Fig. 7.7a:[†]

$$\text{rect}(x) = \begin{cases} 0 & |x| > \frac{1}{2} \\ \frac{1}{2} & |x| = \frac{1}{2} \\ 1 & |x| < \frac{1}{2} \end{cases} \tag{7.18}$$

The gate pulse in Fig. 7.7b is the unit gate pulse rect (x) expanded by a factor τ along the horizontal axis and therefore can be expressed as rect (x/τ) (see Section 1.2-2). Observe that τ, the denominator of the argument of rect (x/τ), indicates the width of the pulse.

UNIT TRIANGLE FUNCTION

We define a unit triangle function $\Delta(x)$ as a triangular pulse of unit height and unit width, centered at the origin, as shown in Fig. 7.8a

$$\Delta(x) = \begin{cases} 0 & |x| \geq \frac{1}{2} \\ 1 - 2|x| & |x| < \frac{1}{2} \end{cases} \tag{7.19}$$

The pulse in Fig. 7.8b is $\Delta(x/\tau)$. Observe that here, as for the gate pulse, the denominator τ of the argument of $\Delta(x/\tau)$ indicates the pulse width.

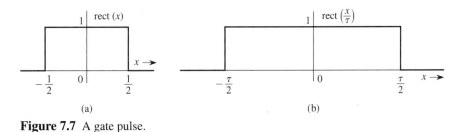

Figure 7.7 A gate pulse.

[†]At $|x| = 0.5$, we require rect $(x) = 0.5$ because the inverse Fourier transform of a discontinuous signal converges to the mean of its two values at the discontinuity.

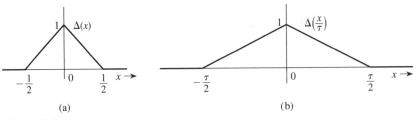

Figure 7.8 A triangle pulse.

INTERPOLATION FUNCTION sinc (x)

The function $\sin x/x$ is the "sine over argument" function denoted by sinc (x).[†] This function plays an important role in signal processing. It is also known as the *filtering or interpolating function*. We define

$$\operatorname{sinc}(x) = \frac{\sin x}{x} \tag{7.20}$$

Inspection of Eq. (7.20) shows the following:

1. sinc (x) is an even function of x.

2. sinc $(x) = 0$ when $\sin x = 0$ except at $x = 0$, where it appears to be indeterminate. This means that sinc $x = 0$ for $x = \pm\pi, \pm 2\pi, \pm 3\pi, \ldots$.

3. Using L'Hôpital's rule, we find sinc $(0) = 1$.

4. sinc (x) is the product of an oscillating signal $\sin x$ (of period 2π) and a monotonically decreasing function $1/x$. Therefore, sinc (x) exhibits damped oscillations of period 2π, with amplitude decreasing continuously as $1/x$.

Figure 7.9a shows sinc (x). Observe that sinc $(x) = 0$ for values of x that are positive and negative integer multiples of π. Figure 7.9b shows sinc $(3\omega/7)$. The argument $3\omega/7 = \pi$ when $\omega = 7\pi/3$. Therefore, the first zero of this function occurs at $\omega = 7\pi/3$.

EXERCISE E7.1

Sketch:

(a) rect $(x/8)$

(b) $\Delta(\omega/10)$

(c) sinc $(3\pi\omega/2)$

(d) sinc (t) rect $(t/4\pi)$

[†]sinc(x) is also denoted by Sa (x) in the literature. Some authors define sinc (x) as

$$\operatorname{sinc}(x) = \frac{\sin \pi x}{\pi x}$$

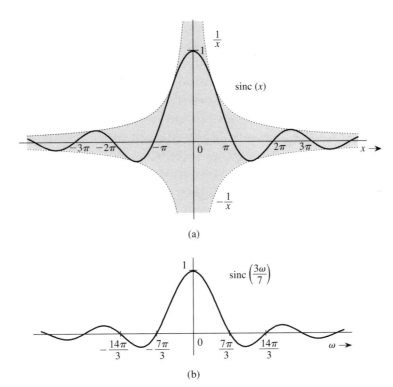

(a)

(b)

Figure 7.9 A sinc pulse.

EXAMPLE 7.2

Find the Fourier transform of $x(t) = \text{rect}\,(t/\tau)$ (Fig. 7.10a).

$$X(\omega) = \int_{-\infty}^{\infty} \text{rect}\left(\frac{t}{\tau}\right) e^{-j\omega t}\, dt$$

Since $\text{rect}\,(t/\tau) = 1$ for $|t| < \tau/2$, and since it is zero for $|t| > \tau/2$,

$$X(\omega) = \int_{-\tau/2}^{\tau/2} e^{-j\omega t}\, dt$$

$$= -\frac{1}{j\omega}(e^{-j\omega\tau/2} - e^{j\omega\tau/2}) = \frac{2\sin\left(\dfrac{\omega\tau}{2}\right)}{\omega}$$

$$= \tau\,\frac{\sin\left(\dfrac{\omega\tau}{2}\right)}{\left(\dfrac{\omega\tau}{2}\right)} = \tau\,\text{sinc}\left(\frac{\omega\tau}{2}\right)$$

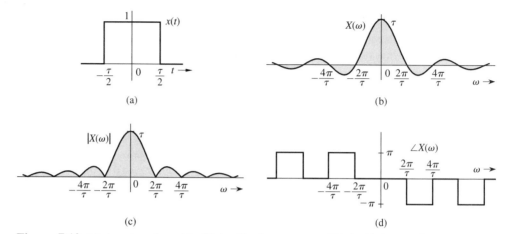

Figure 7.10 (a) A gate pulse $x(t)$, (b) its Fourier spectrum $X(\omega)$, (c) its amplitude spectrum $|X(\omega)|$, and (d) its phase spectrum $\angle X(\omega)$.

Therefore

$$\text{rect}\left(\frac{t}{\tau}\right) \Longleftrightarrow \tau \ \text{sinc}\left(\frac{\omega\tau}{2}\right) \tag{7.21}$$

Recall that sinc $(x) = 0$ when $x = \pm n\pi$. Hence, sinc $(\omega\tau/2) = 0$ when $\omega\tau/2 = \pm n\pi$; that is, when $\omega = \pm 2n\pi/\tau$, $(n = 1, 2, 3, \ldots)$, as depicted in Fig. 7.10b. The Fourier transform $X(\omega)$ shown in Fig. 7.10b exhibits positive and negative values. A negative amplitude can be considered to be a positive amplitude with a phase of $-\pi$ or π. We use this observation to plot the amplitude spectrum $|X(\omega)| = |\text{sinc} \ (\omega\tau/2)|$ (Fig. 7.10c) and the phase spectrum $\angle X(\omega)$ (Fig. 7.10d). The phase spectrum, which is required to be an odd function of ω, may be drawn in several other ways because a negative sign can be accounted for by a phase of $\pm n\pi$, where n is any odd integer. All such representations are equivalent.

BANDWIDTH OF rect (t/τ)

The spectrum $X(\omega)$ in Fig. 7.10 peaks at $\omega = 0$ and decays at higher frequencies. Therefore, rect (t/τ) is a lowpass signal with most of the signal energy in lower-frequency components. Strictly speaking, because the spectrum extends from 0 to ∞, the bandwidth is ∞. However, much of the spectrum is concentrated within the first lobe (from $\omega = 0$ to $\omega = 2\pi/\tau$). Therefore, a rough estimate of the bandwidth of a rectangular pulse of width τ seconds is $2\pi/\tau$ rad/s, or $1/\tau$ Hz.[†] Note the reciprocal relationship of the pulse width with its bandwidth. We shall observe later that this result is true in general.

[†]To compute bandwidth, we must consider the spectrum only for positive values of ω. See discussion in Section 6.3.

EXAMPLE 7.3

Find the Fourier transform of the unit impulse $\delta(t)$.

___

Using the sampling property of the impulse [Eq. (1.24)], we obtain

$$\mathcal{F}[\delta(t)] = \int_{-\infty}^{\infty} \delta(t)e^{-j\omega t}\,dt = 1 \tag{7.22a}$$

or

$$\delta(t) \Longleftrightarrow 1 \tag{7.22b}$$

Figure 7.11 shows $\delta(t)$ and its spectrum.

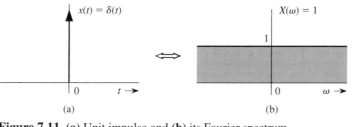

Figure 7.11 (a) Unit impulse and **(b)** its Fourier spectrum.

EXAMPLE 7.4

Find the inverse Fourier transform of $\delta(\omega)$.

___

On the basis of Eq. (7.8b) and the sampling property of the impulse function,

$$\mathcal{F}^{-1}[\delta(\omega)] = \frac{1}{2\pi} \int_{-\infty}^{\infty} \delta(\omega)e^{j\omega t}\,d\omega = \frac{1}{2\pi}$$

Therefore

$$\frac{1}{2\pi} \Longleftrightarrow \delta(\omega) \tag{7.23a}$$

or

$$1 \Longleftrightarrow 2\pi\delta(\omega) \tag{7.23b}$$

This result shows that the spectrum of a constant signal $x(t) = 1$ is an impulse $2\pi\delta(\omega)$, as illustrated in Fig. 7.12.

The result [Eq. (7.23b)] could have been anticipated on qualitative grounds. Recall that the Fourier transform of $x(t)$ is a spectral representation of $x(t)$ in terms of everlasting

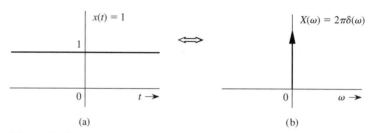

Figure 7.12 (a) A constant (dc) signal and (b) its Fourier spectrum.

exponential components of the form $e^{j\omega t}$. Now, to represent a constant signal $x(t) = 1$, we need a single everlasting exponential $e^{j\omega t}$ with $\omega = 0$.[†] This results in a spectrum at a single frequency $\omega = 0$. Another way of looking at the situation is that $x(t) = 1$ is a dc signal which has a single frequency $\omega = 0$ (dc).

If an impulse at $\omega = 0$ is a spectrum of a dc signal, what does an impulse at $\omega = \omega_0$ represent? We shall answer this question in the next example.

EXAMPLE 7.5

Find the inverse Fourier transform of $\delta(\omega - \omega_0)$.

Using the sampling property of the impulse function, we obtain

$$\mathcal{F}^{-1}[\delta(\omega - \omega_0)] = \frac{1}{2\pi} \int_{-\infty}^{\infty} \delta(\omega - \omega_0)e^{j\omega t}\, d\omega = \frac{1}{2\pi}e^{j\omega_0 t}$$

Therefore

$$\frac{1}{2\pi}e^{j\omega_0 t} \Longleftrightarrow \delta(\omega - \omega_0)$$

or

$$e^{j\omega_0 t} \Longleftrightarrow 2\pi\delta(\omega - \omega_0) \tag{7.24a}$$

This result shows that the spectrum of an everlasting exponential $e^{j\omega_0 t}$ is a single impulse at $\omega = \omega_0$. We reach the same conclusion by qualitative reasoning. To represent the everlasting

---

[†]The constant multiplier 2π in the spectrum $[X(\omega) = 2\pi\delta(\omega)]$ may be a bit puzzling. Since $1 = e^{j\omega t}$ with $\omega = 0$, it appears that the Fourier transform of $x(t) = 1$ should be an impulse of strength unity rather than 2π. Recall, however, that in the Fourier transform $x(t)$ is synthesized by exponentials not of amplitude $X(n\Delta\omega)\Delta\omega$ but of amplitude $1/2\pi$ times $X(n\Delta\omega)\Delta\omega$, as seen from Eq. (7.6b). Had we used variable f (hertz) instead of ω, the spectrum would have been the unit impulse.

exponential $e^{j\omega_0 t}$, we need a single everlasting exponential $e^{j\omega t}$ with $\omega = \omega_0$. Therefore, the spectrum consists of a single component at frequency $\omega = \omega_0$.

From Eq. (7.24a) it follows that

$$e^{-j\omega_0 t} \iff 2\pi\delta(\omega + \omega_0) \tag{7.24b}$$

EXAMPLE 7.6

Find the Fourier transforms of the everlasting sinusoid $\cos \omega_0 t$ (Fig. 7.13a).

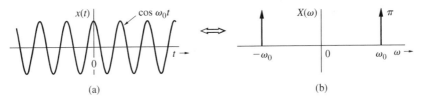

Figure 7.13 **(a)** A cosine signal and **(b)** its Fourier spectrum.

Recall the Euler formula

$$\cos \omega_0 t = \tfrac{1}{2}(e^{j\omega_0 t} + e^{-j\omega_0 t})$$

Adding Eqs. (7.24a) and (7.24b), and using the foregoing result, we obtain

$$\cos \omega_0 t \iff \pi[\delta(\omega + \omega_0) + \delta(\omega - \omega_0)] \tag{7.25}$$

The spectrum of $\cos \omega_0 t$ consists of two impulses at ω_0 and $-\omega_0$, as shown in Fig. 7.13b. The result also follows from qualitative reasoning. An everlasting sinusoid $\cos \omega_0 t$ can be synthesized by two everlasting exponentials, $e^{j\omega_0 t}$ and $e^{-j\omega_0 t}$. Therefore the Fourier spectrum consists of only two components of frequencies ω_0 and $-\omega_0$.

EXAMPLE 7.7 (Fourier Transform of a Periodic Signal)

We can use a Fourier series to express a periodic signal as a sum of exponentials of the form $e^{jn\omega_0 t}$, whose Fourier transform is found in Eq. (7.24a). Hence, we can readily find the Fourier transform of a periodic signal by using the linearity property in Eq. (7.16).

The Fourier series of a periodic signal $x(t)$ with period T_0 is given by

$$x(t) = \sum_{n=-\infty}^{\infty} D_n e^{jn\omega_0 t} \qquad \omega_0 = \frac{2\pi}{T_0}$$

Taking the Fourier transform of both sides, we obtain[†]

$$X(\omega) = 2\pi \sum_{n=-\infty}^{\infty} D_n \delta(\omega - n\omega_0) \qquad (7.26)$$

EXAMPLE 7.8 (Fourier Transform of a Unit Impulse Train)

The Fourier series for the unit impulse train $\delta_{T_0}(t)$, shown in Fig. 7.14a, was found in Example 6.7. The Fourier coefficient D_n for this signal, as seen from Eq. (6.37), is a constant $D_n = 1/T_0$.

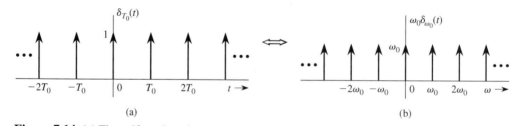

Figure 7.14 (a) The uniform impulse train and (b) its Fourier transform.

From Eq. (7.26), the Fourier transform for the unit impulse train is

$$X(\omega) = \frac{2\pi}{T_0} \sum_{n=-\infty}^{\infty} \delta(\omega - n\omega_0) \qquad \omega_0 = \frac{2\pi}{T_0}$$

$$= \omega_0 \delta_{\omega_0}(\omega) \qquad (7.27)$$

The corresponding spectrum is shown in Fig. 7.14b.

[†]We assume here that the linearity property can be extended to an infinite sum.

EXAMPLE 7.9

Find the Fourier transform of the unit step function $u(t)$.

Trying to find the Fourier transform of $u(t)$ by direct integration leads to an indeterminate result because

$$U(\omega) = \int_{-\infty}^{\infty} u(t) e^{-j\omega t}\, dt = \int_{0}^{\infty} e^{-j\omega t}\, dt = \frac{-1}{j\omega} e^{-j\omega t} \Big|_{0}^{\infty}$$

The upper limit of $e^{-j\omega t}$ as $t \to \infty$ yields an indeterminate answer. So we approach this problem by considering $u(t)$ to be a decaying exponential $e^{-at} u(t)$ in the limit as $a \to 0$ (Fig. 7.15a). Thus

$$u(t) = \lim_{a \to 0} e^{-at} u(t)$$

and

$$U(\omega) = \lim_{a \to 0} \mathcal{F}\{e^{-at} u(t)\} = \lim_{a \to 0} \frac{1}{a + j\omega} \tag{7.28a}$$

Expressing the right-hand side in terms of its real and imaginary parts yields

$$U(\omega) = \lim_{a \to 0} \left[\frac{a}{a^2 + \omega^2} - j \frac{\omega}{a^2 + \omega^2} \right] \tag{7.28b}$$

$$= \lim_{a \to 0} \left[\frac{a}{a^2 + \omega^2} \right] + \frac{1}{j\omega}$$

The function $a/(a^2 + \omega^2)$ has interesting properties. First, the area under this function (Fig. 7.15b) is π regardless of the value of a:

$$\int_{-\infty}^{\infty} \frac{a}{a^2 + \omega^2}\, d\omega = \tan^{-1} \frac{\omega}{a} \Big|_{-\infty}^{\infty} = \pi$$

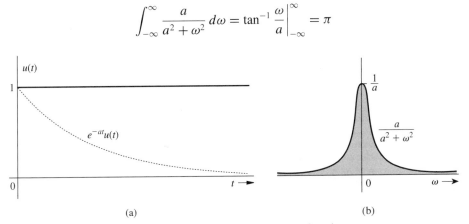

(a)   (b)

Figure 7.15 Derivation of the Fourier transform of the step function.

Second, when $a \rightarrow 0$, this function approaches zero for all $\omega \neq 0$, and all its area (π) is concentrated at a single point $\omega = 0$. Clearly, as $a \rightarrow 0$, this function approaches an impulse of strength π.[†] Thus

$$U(\omega) = \pi \delta(\omega) + \frac{1}{j\omega} \qquad (7.29)$$

Note that $u(t)$ is not a "true" dc signal because it is not constant over the interval $-\infty$ to ∞. To synthesize a "true" dc, we require only one everlasting exponential with $\omega = 0$ (impulse at $\omega = 0$). The signal $u(t)$ has a jump discontinuity at $t = 0$. It is impossible to synthesize such a signal with a single everlasting exponential $e^{j\omega t}$. To synthesize this signal from everlasting exponentials, we need, in addition to an impulse at $\omega = 0$, all the frequency components, as indicated by the term $1/j\omega$ in Eq. (7.29).

EXAMPLE 7.10

Find the Fourier transform of the sign function sgn (t) [pronounced *signum* (t)], depicted in Fig. 7.16.

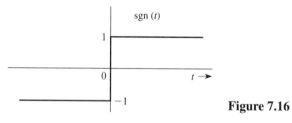

Figure 7.16

Observe that

$$\text{sgn}\,(t) + 1 = 2u(t) \quad \Longrightarrow \quad \text{sgn}\,(t) = 2u(t) - 1$$

Using the results in Eqs. (7.23b), (7.29), and the linearity property, we obtain

$$\text{sgn}\,(t) \Longleftrightarrow \frac{2}{j\omega} \qquad (7.30)$$

[†]The second term on the right-hand side of Eq. (7.28b), being an odd function of ω, has zero area regardless of the value of a. As $a \rightarrow 0$, the second term approaches $1/j\omega$.

EXERCISE E7.2

Show that the inverse Fourier transform of $X(\omega)$ illustrated in Fig. 7.17 is $x(t) = (\omega_0/\pi)$ sinc $(\omega_0 t)$. Sketch $x(t)$.

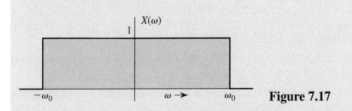

Figure 7.17

EXERCISE E7.3

Show that $\cos(\omega_0 t + \theta) \Longleftrightarrow \pi[\delta(\omega + \omega_0)e^{-j\theta} + \delta(\omega - \omega_0)e^{j\theta}]$.

7.2-1 Connection Between the Fourier and Laplace Transforms

The general (bilateral) Laplace transform of a signal $x(t)$, according to Eq. (4.1) is

$$X(s) = \int_{-\infty}^{\infty} x(t)e^{-st}\, dt \tag{7.31a}$$

Setting $s = j\omega$ in this equation yields

$$X(j\omega) = \int_{-\infty}^{\infty} x(t)e^{-j\omega t}\, dt \tag{7.31b}$$

where $X(j\omega) = X(s)|_{s=j\omega}$. But, the right-hand-side integral defines $X(\omega)$, the Fourier transform of $x(t)$. Does this mean that the Fourier transform can be obtained from the corresponding Laplace transform by setting $s = j\omega$? In other words, is it true that $X(j\omega) = X(\omega)$? Yes and no. Yes, it is true in most cases. For example, when $x(t) = e^{-at}u(t)$, its Laplace transform is $1/(s+a)$, and $X(j\omega) = 1/(j\omega + a)$, which is equal to $X(\omega)$ (assuming $a < 0$). However, for the unit step function $u(t)$, the Laplace transform is

$$u(t) \Longleftrightarrow \frac{1}{s} \qquad \mathrm{Re}\, s > 0$$

The Fourier transform is given by

$$u(t) \Longleftrightarrow \frac{1}{j\omega} + \pi\delta(\omega)$$

Clearly, $X(j\omega) \neq X(\omega)$ in this case.

To understand this puzzle, consider the fact that we obtain $X(j\omega)$ by setting $s = j\omega$ in Eq. (7.31a). This implies that the integral on the right-hand side of Eq. (7.31a) converges for $s = j\omega$, meaning that $s = j\omega$ (the imaginary axis) lies in the ROC for $X(s)$. The general rule is that only when the ROC for $X(s)$ includes the $j\omega$ axis, does setting $s = j\omega$ in $X(s)$ yield the Fourier transform $X(\omega)$, that is, $X(j\omega) = X(\omega)$. This is the case of absolutely integrable $x(t)$. If the ROC of $X(s)$ excludes the $j\omega$ axis, $X(j\omega) \neq X(\omega)$. This is the case for exponentially growing $x(t)$ and also $x(t)$ that is constant or is oscillating with constant amplitude.

The reason for this peculiar behavior has something to do with the nature of convergence of the Laplace and the Fourier integrals when $x(t)$ is not absolutely integrable.[†]

This discussion shows that although the Fourier transform may be considered as a special case of the Laplace transform, we need to circumscribe such a view. This fact can also be seen from the fact that a periodic signal has the Fourier transform, but the Laplace transform does not exist.

7.3 SOME PROPERTIES OF THE FOURIER TRANSFORM

We now study some of the important properties of the Fourier transform and their implications as well as applications. We have already encountered two important properties, linearity [Eq. (7.16)] and the conjugation property [Eq. (7.10)].

Before embarking on this study, we shall explain an important and pervasive aspect of the Fourier transform: the time-frequency duality.

[†]To explain this point, consider the unit step function and its transforms. Both the Laplace and the Fourier transform synthesize $x(t)$, using everlasting exponentials of the form e^{st}. The frequency s can be anywhere in the complex plane for the Laplace transform, but it must be restricted to the $j\omega$ axis in the case of the Fourier transform. The unit step function is readily synthesized in the Laplace transform by a relatively simple spectrum $X(s) \doteq 1/s$, in which the frequencies s are chosen in the RHP [the region of convergence for $u(t)$ is Re $s > 0$]. In the Fourier transform, however, we are restricted to values of s only on the $j\omega$ axis. The function $u(t)$ can still be synthesized by frequencies along the $j\omega$ axis, but the spectrum is more complicated than it is when we are free to choose the frequencies in the RHP. In contrast, when $x(t)$ is absolutely integrable, the region of convergence for the Laplace transform includes the $j\omega$ axis, and we can synthesize $x(t)$ by using frequencies along the $j\omega$ axis in both transforms. This leads to $X(j\omega) = X(\omega)$.

We may explain this concept by an example of two countries, X and Y. Suppose these countries want to construct similar dams in their respective territories. Country X has financial resources but not much manpower. In contrast, Y has considerable manpower but few financial resources. The dams will still be constructed in both countries, although the methods used will be different. Country X will use expensive but efficient equipment to compensate for its lack of manpower, whereas Y will use the cheapest possible equipment in a labor-intensive approach to the project. Similarly, both Fourier and Laplace integrals converge for $u(t)$, but the make-up of the components used to synthesize $u(t)$ will be very different for two cases because of the constraints of the Fourier transform, which are not present for the Laplace transform.

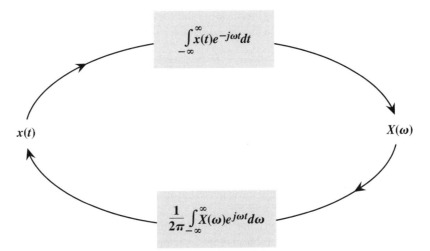

Figure 7.18 A near symmetry between the direct and the inverse Fourier transforms.

TIME-FREQUENCY DUALITY IN THE TRANSFORM OPERATIONS

Equations (7.8) show an interesting fact: the direct and the inverse transform operations are remarkably similar. These operations, required to go from $x(t)$ to $X(\omega)$ and then from $X(\omega)$ to $x(t)$, are depicted graphically in Fig. 7.18. The inverse transform equation can be obtained from the direct transform equation by replacing $x(t)$ with $X(\omega)$, t with ω, and ω with t. In a similar way, we can obtain the direct from the inverse. There are only two minor differences in these operations: the factor 2π appears only in the inverse operator, and the exponential indices in the two operations have opposite signs. Otherwise the two equations are duals of each other.[†] This observation has far-reaching consequences in the study of the Fourier transform. It is the basis of the so-called duality of time and frequency. *The duality principle may be compared with a photograph and its negative. A photograph can be obtained from its negative, and by using an identical procedure, a negative can be obtained from the photograph.* For any result or relationship between $x(t)$ and $X(\omega)$, there exists a dual result or relationship, obtained by interchanging the roles of $x(t)$ and $X(\omega)$ in the original result (along with some

---

[†]Of the two differences, the former can be eliminated by change of variable from ω to f (in hertz). In this case $\omega = 2\pi f$ and $d\omega = 2\pi\, df$.

Therefore, the direct and the inverse transforms are given by

$$X(2\pi f) = \int_{-\infty}^{\infty} x(t)e^{-j2\pi ft}\, dt \qquad \text{and} \qquad x(t) = \int_{-\infty}^{\infty} X(2\pi f)e^{j2\pi ft}\, df$$

This leaves only one significant difference, that of sign change in the exponential index.

minor modifications arising because of the factor 2π and a sign change). For example, the time-shifting property, to be proved later, states that if $x(t) \Longleftrightarrow X(\omega)$, then

$$x(t - t_0) \Longleftrightarrow X(\omega)e^{-j\omega t_0}$$

The dual of this property (the frequency-shifting property) states that

$$x(t)e^{j\omega_0 t} \Longleftrightarrow X(\omega - \omega_0)$$

Observe the role-reversal of time and frequency in these two equations (with the minor difference of the sign change in the exponential index). The value of this principle lies in the fact that *whenever we derive any result, we can be sure that it has a dual.* This possibility can give valuable insights about many unsuspected properties or results in signal processing.

The properties of the Fourier transform are useful not only in deriving the direct and inverse transforms of many functions, but also in obtaining several valuable results in signal processing. The reader should not fail to observe the ever-present duality in this discussion.

LINEARITY

The linearity property [Eq. (7.16)] has already been introduced.

CONJUGATION AND CONJUGATE SYMMETRY

The conjugation property, which has already been introduced, states that if $x(t) \Longleftrightarrow X(\omega)$, then

$$x^*(t) \Longleftrightarrow X^*(-\omega)$$

From this property follows the conjugate symmetry property, also introduced earlier, which states that if $x(t)$ is real, then

$$X(-\omega) = X^*(\omega)$$

DUALITY

The duality property states that if

$$x(t) \Longleftrightarrow X(\omega)$$

then

$$X(t) \Longleftrightarrow 2\pi x(-\omega) \tag{7.32}$$

Proof. From Eq. (7.8b) we can write

$$x(t) = \frac{1}{2\pi} \int_{-\infty}^{\infty} X(u)e^{jut}\, du$$

Hence

$$2\pi x(-t) = \int_{-\infty}^{\infty} X(u)e^{-jut}\, du$$

Changing t to ω yields Eq. (7.32).

EXAMPLE 7.11

In this example we apply the duality property [Eq. (7.32)] to the pair in Fig. 7.19a.

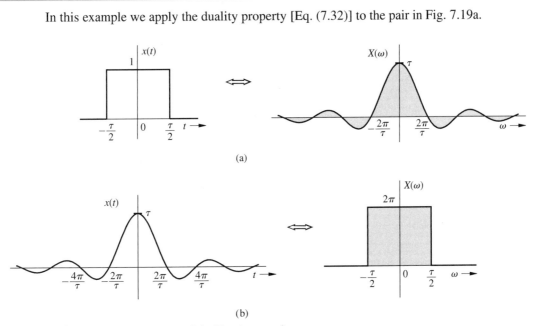

(a)

(b)

Figure 7.19 The duality property of the Fourier transform.

From Eq. (7.21) we have

$$\underbrace{\text{rect}\left(\frac{t}{\tau}\right)}_{x(t)} \Longleftrightarrow \underbrace{\tau\, \text{sinc}\left(\frac{\omega\tau}{2}\right)}_{X(\omega)}$$

Also, $X(t)$ is the same as $X(\omega)$ with ω replaced by t, and $x(-\omega)$ is the same as $x(t)$ with t replaced by $-\omega$. Therefore, the duality property (7.32) yields

$$\underbrace{\tau\, \text{sinc}\left(\frac{\tau t}{2}\right)}_{X(t)} \Longleftrightarrow \underbrace{2\pi\, \text{rect}\left(\frac{-\omega}{\tau}\right)}_{2\pi x(-\omega)} = 2\pi\, \text{rect}\left(\frac{\omega}{\tau}\right) \tag{7.33}$$

In Eq. (7.33) we used the fact that $\text{rect}\,(-x) = \text{rect}\,(x)$ because rect is an even function. Figure 7.19b shows this pair graphically. Observe the interchange of the roles of t and ω (with the minor adjustment of the factor 2π). This result appears as pair 18 in Table 7.1 (with $\tau/2 = W$).

As an interesting exercise, the reader should generate the dual of every pair in Table 7.1 by applying the duality property.

TABLE 7.1   Fourier Transforms

No.	$x(t)$	$X(\omega)$	
1	$e^{-at}u(t)$	$\dfrac{1}{a + j\omega}$	$a > 0$
2	$e^{at}u(-t)$	$\dfrac{1}{a - j\omega}$	$a > 0$
3	$e^{-a\lvert t\rvert}$	$\dfrac{2a}{a^2 + \omega^2}$	$a > 0$
4	$te^{-at}u(t)$	$\dfrac{1}{(a + j\omega)^2}$	$a > 0$
5	$t^n e^{-at}u(t)$	$\dfrac{n!}{(a + j\omega)^{n+1}}$	$a > 0$
6	$\delta(t)$	1	
7	1	$2\pi\delta(\omega)$	
8	$e^{j\omega_0 t}$	$2\pi\delta(\omega - \omega_0)$	
9	$\cos\omega_0 t$	$\pi[\delta(\omega - \omega_0) + \delta(\omega + \omega_0)]$	
10	$\sin\omega_0 t$	$j\pi[\delta(\omega + \omega_0) - \delta(\omega - \omega_0)]$	
11	$u(t)$	$\pi\delta(\omega) + \dfrac{1}{j\omega}$	
12	$\operatorname{sgn} t$	$\dfrac{2}{j\omega}$	
13	$\cos\omega_0 t\, u(t)$	$\dfrac{\pi}{2}[\delta(\omega - \omega_0) + \delta(\omega + \omega_0)] + \dfrac{j\omega}{\omega_0^2 - \omega^2}$	
14	$\sin\omega_0 t\, u(t)$	$\dfrac{\pi}{2j}[\delta(\omega - \omega_0) - \delta(\omega + \omega_0)] + \dfrac{\omega_0}{\omega_0^2 - \omega^2}$	
15	$e^{-at}\sin\omega_0 t\, u(t)$	$\dfrac{\omega_0}{(a + j\omega)^2 + \omega_0^2}$	$a > 0$
16	$e^{-at}\cos\omega_0 t\, u(t)$	$\dfrac{a + j\omega}{(a + j\omega)^2 + \omega_0^2}$	$a > 0$
17	$\operatorname{rect}\left(\dfrac{t}{\tau}\right)$	$\tau\operatorname{sinc}\left(\dfrac{\omega\tau}{2}\right)$	
18	$\dfrac{W}{\pi}\operatorname{sinc}(Wt)$	$\operatorname{rect}\left(\dfrac{\omega}{2W}\right)$	
19	$\Delta\left(\dfrac{t}{\tau}\right)$	$\dfrac{\tau}{2}\operatorname{sinc}^2\left(\dfrac{\omega\tau}{4}\right)$	
20	$\dfrac{W}{2\pi}\operatorname{sinc}^2\left(\dfrac{Wt}{2}\right)$	$\Delta\left(\dfrac{\omega}{2W}\right)$	
21	$\displaystyle\sum_{n=-\infty}^{\infty}\delta(t - nT)$	$\displaystyle\omega_0\sum_{n=-\infty}^{\infty}\delta(\omega - n\omega_0)$	$\omega_0 = \dfrac{2\pi}{T}$
22	$e^{-t^2/2\sigma^2}$	$\sigma\sqrt{2\pi}\,e^{-\sigma^2\omega^2/2}$	

EXERCISE E7.4

Apply the duality property to pairs 1, 3, and 9 (Table 7.1) to show that

(a) $1/(jt + a) \Longleftrightarrow 2\pi e^{a\omega}u(-\omega)$

(b) $2a/(t^2 + a^2) \Longleftrightarrow 2\pi e^{-a|\omega|}$

(c) $\delta(t + t_0) + \delta(t - t_0) \Longleftrightarrow 2\cos t_0\omega$

THE SCALING PROPERTY

If

$$x(t) \Longleftrightarrow X(\omega)$$

then, for any real constant a,

$$x(at) \Longleftrightarrow \frac{1}{|a|}X\left(\frac{\omega}{a}\right) \tag{7.34}$$

Proof. For a positive real constant a,

$$\mathcal{F}[x(at)] = \int_{-\infty}^{\infty} x(at)e^{-j\omega t}\,dt$$

$$= \frac{1}{a}\int_{-\infty}^{\infty} x(u)e^{(-j\omega/a)u}\,du = \frac{1}{a}X\left(\frac{\omega}{a}\right)$$

Similarly, we can demonstrate that if $a < 0$,

$$x(at) \Longleftrightarrow \frac{-1}{a}X\left(\frac{\omega}{a}\right)$$

Hence follows Eq. (7.34).

SIGNIFICANCE OF THE SCALING PROPERTY

The function $x(at)$ represents the function $x(t)$ compressed in time by a factor a (see Section 1.2-2). Similarly, a function $X(\omega/a)$ represents the function $X(\omega)$ expanded in frequency by the same factor a. *The scaling property states that time compression of a signal results in its spectral expansion, and time expansion of the signal results in its spectral compression.* Intuitively, compression in time by factor a means that the signal is varying faster by factor a.[†] To synthesize such a signal, the frequencies of its sinusoidal components must be increased by the factor a, implying that its frequency spectrum is expanded by the factor a. Similarly, a signal expanded in time varies more slowly; hence the frequencies of its components are lowered, implying that its frequency spectrum is compressed. For instance, the signal $\cos 2\omega_0 t$ is the same as the signal $\cos \omega_0 t$ time-compressed by a factor of 2. Clearly, the spectrum of the former (impulse at $\pm 2\omega_0$) is an expanded version of the spectrum of the latter (impulse at $\pm\omega_0$). The effect of this scaling is demonstrated in Fig. 7.20.

[†]We are assuming $a > 1$, although the argument still holds if $a < 1$. In the latter case, compression becomes expansion by factor $1/a$, and vice versa.

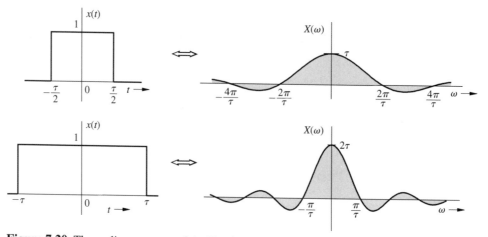

Figure 7.20 The scaling property of the Fourier transform.

RECIPROCITY OF SIGNAL DURATION AND ITS BANDWIDTH

The scaling property implies that if $x(t)$ is wider, its spectrum is narrower, and vice versa. Doubling the signal duration halves its bandwidth, and vice versa. This suggests that the bandwidth of a signal is inversely proportional to the signal duration or width (in seconds).[†] We have already verified this fact for the gate pulse, where we found that the bandwidth of a gate pulse of width τ seconds is $1/\tau$ Hz. More discussion of this interesting topic can be found in the literature.[2]

By letting $a = -1$ in Eq. (7.34), we obtain the *inversion property of time and frequency:*

$$x(-t) \Longleftrightarrow X(-\omega) \tag{7.35}$$

EXAMPLE 7.12

Find the Fourier transforms of $e^{at}u(-t)$ and $e^{-a|t|}$.

Application of Eq. (7.35) to pair 1 (Table 7.1) yields

$$e^{at}u(-t) \Longleftrightarrow \frac{1}{a - j\omega} \qquad a > 0$$

Also

$$e^{-a|t|} = e^{-at}u(t) + e^{at}u(-t)$$

[†]When a signal has infinite duration, we must consider its effective or equivalent duration. There is no unique definition of effective signal duration. One possible definition is given in Eq. (2.67).

Therefore

$$e^{-a|t|} \iff \frac{1}{a+j\omega} + \frac{1}{a-j\omega} = \frac{2a}{a^2+\omega^2} \qquad a > 0 \qquad (7.36)$$

The signal $e^{-a|t|}$ and its spectrum are illustrated in Fig. 7.21.

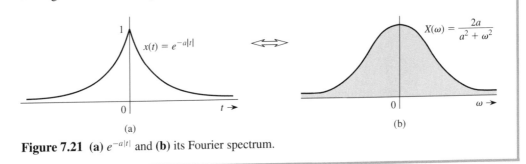

Figure 7.21 (a) $e^{-a|t|}$ and (b) its Fourier spectrum.

THE TIME-SHIFTING PROPERTY

If

$$x(t) \iff X(\omega)$$

then

$$x(t - t_0) \iff X(\omega)e^{-j\omega t_0} \qquad (7.37a)$$

Proof. By definition,

$$\mathcal{F}[x(t - t_0)] = \int_{-\infty}^{\infty} x(t-t_0)e^{-j\omega t}\, dt$$

Letting $t - t_0 = u$, we have

$$\mathcal{F}[x(t - t_0)] = \int_{-\infty}^{\infty} x(u)e^{-j\omega(u+t_0)}\, du$$

$$= e^{-j\omega t_0}\int_{-\infty}^{\infty} x(u)e^{-j\omega u}\, du = X(\omega)e^{-j\omega t_0} \qquad (7.37b)$$

This result shows that *delaying a signal by t_0 seconds does not change its amplitude spectrum. The phase spectrum, however, is changed by $-\omega t_0$.*

PHYSICAL EXPLANATION
OF THE LINEAR PHASE

Time delay in a signal causes a linear phase shift in its spectrum. This result can also be derived by heuristic reasoning. Imagine $x(t)$ being synthesized by its Fourier components, which are sinusoids of certain amplitudes and phases. The delayed signal $x(t - t_0)$ can be synthesized by the same sinusoidal components, each delayed by t_0 seconds. The amplitudes of the components

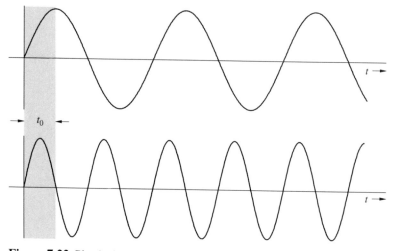

Figure 7.22 Physical explanation of the time-shifting property.

remain unchanged. Therefore, the amplitude spectrum of $x(t - t_0)$ is identical to that of $x(t)$. The time delay of t_0 in each sinusoid, however, does change the phase of each component. Now, a sinusoid $\cos \omega t$ delayed by t_0 is given by

$$\cos \omega (t - t_0) = \cos (\omega t - \omega t_0)$$

Therefore a time delay t_0 in a sinusoid of frequency ω manifests as a phase delay of ωt_0. This is a linear function of ω, meaning that higher-frequency components must undergo proportionately higher phase shifts to achieve the same time delay. This effect is depicted in Fig. 7.22 with two sinusoids, the frequency of the lower sinusoid being twice that of the upper. The same time delay t_0 amounts to a phase shift of $\pi/2$ in the upper sinusoid and a phase shift of π in the lower sinusoid. This verifies the fact that *to achieve the same time delay, higher-frequency sinusoids must undergo proportionately higher phase shifts*. The principle of linear phase shift is very important, and we shall encounter it again in distortionless signal transmission and filtering applications.

EXAMPLE 7.13

Find the Fourier transform of $e^{-a|t-t_0|}$.

This function, shown in Fig. 7.23a, is a time-shifted version of $e^{-a|t|}$ (depicted in Fig. 7.21a). From Eqs. (7.36) and (7.37) we have

$$e^{-a|t-t_0|} \iff \frac{2a}{a^2 + \omega^2} e^{-j\omega t_0} \qquad (7.38)$$

The spectrum of $e^{-a|t-t_0|}$ (Fig. 7.23b) is the same as that of $e^{-a|t|}$ (Fig. 7.21b), except for an added phase shift of $-\omega t_0$.

Observe that the time delay t_0 causes a linear phase spectrum $-\omega t_0$. This example clearly demonstrates the effect of time shift.

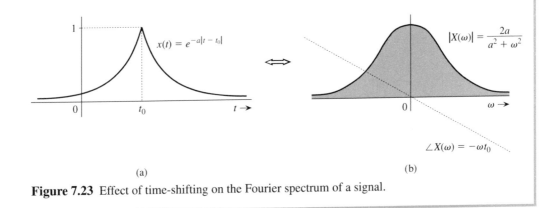

(a)   (b)

Figure 7.23 Effect of time-shifting on the Fourier spectrum of a signal.

EXAMPLE 7.14

Find the Fourier transform of the gate pulse $x(t)$ illustrated in Fig. 7.24a.

The pulse $x(t)$ is the gate pulse rect (t/τ) in Fig. 7.10a delayed by $3\tau/4$ second. Hence, according to Eq. (7.37a), its Fourier transform is the Fourier transform of rect (t/τ) multiplied by $e^{-j\omega(3\tau/4)}$. Therefore

$$X(\omega) = \tau \, \text{sinc}\left(\frac{\omega\tau}{2}\right)e^{-j\omega(3\tau/4)}$$

The amplitude spectrum $|X(\omega)|$ (depicted in Fig. 7.24b) of this pulse is the same as that indicated in Fig. 7.10c. But the phase spectrum has an added linear term $-3\omega\tau/4$. Hence, the phase spectrum of $x(t)$ (Fig. 7.24a) is identical to that in Fig. 7.10d plus a linear term $-3\omega\tau/4$, as shown in Fig. 7.24c.

PHASE SPECTRUM USING PRINCIPAL VALUES

There is an alternate way of spectral representation of $\angle X(\omega)$. The phase angle computed on a calculator or by using a computer subroutine is generally the principal value (modulo 2π value) of the phase angle, which always lies in the range $-\pi$ to π. For instance, the principal value of angle $3\pi/2$ is $-\pi/2$, and so on. The principal value differs from the actual value by $\pm 2\pi$ radians (and its integer multiples) in a way that ensures that the principal value remains within $-\pi$ to π. Thus, the principal value will show jump discontinuities of $\pm 2\pi$ whenever the actual phase crosses $\pm \pi$. The phase plot in Fig. 7.24c is redrawn in Fig. 7.24d using principal value for the phase. This phase pattern, which contains phase discontinuities of magnitudes 2π and π, becomes repetitive at intervals of $\omega = 8\pi/\tau$.

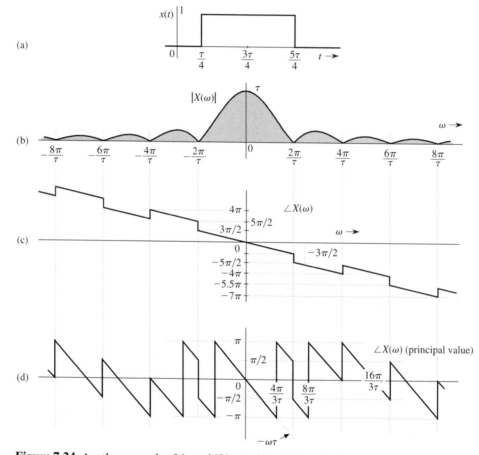

Figure 7.24 Another example of time-shifting and its effect on the Fourier spectrum of a signal.

EXERCISE E7.5

Use pair 18 of Table 7.1 and the time-shifting property to show that the Fourier transform of $\text{sinc}\,[\omega_0(t - T)]$ is $(\pi/\omega_0)\,\text{rect}\,(\omega/2\omega_0)e^{-j\omega T}$. Sketch the amplitude and phase spectra of the Fourier transform.

THE FREQUENCY-SHIFTING PROPERTY

If
$$x(t) \Longleftrightarrow X(\omega)$$

then
$$x(t)e^{j\omega_0 t} \Longleftrightarrow X(\omega - \omega_0) \tag{7.39}$$

Proof. By definition,

$$\mathcal{F}[x(t)e^{j\omega_0 t}] = \int_{-\infty}^{\infty} x(t)e^{j\omega_0 t} e^{-j\omega t}\, dt = \int_{-\infty}^{\infty} x(t)e^{-j(\omega - \omega_0)t}\, dt = X(\omega - \omega_0)$$

According to this property, the multiplication of a signal by a factor $e^{j\omega_0 t}$ shifts the spectrum of that signal by $\omega = \omega_0$. Note the duality between the time-shifting and the frequency-shifting properties.

Changing ω_0 to $-\omega_0$ in Eq. (7.39) yields

$$x(t)e^{-j\omega_0 t} \Longleftrightarrow X(\omega + \omega_0) \tag{7.40}$$

Because $e^{j\omega_0 t}$ is not a real function that can be generated, frequency shifting in practice is achieved by multiplying $x(t)$ by a sinusoid. Observe that

$$x(t)\cos \omega_0 t = \tfrac{1}{2}[x(t)e^{j\omega_0 t} + x(t)e^{-j\omega_0 t}]$$

From Eqs. (7.39) and (7.40), it follows that

$$x(t)\cos \omega_0 t \Longleftrightarrow \tfrac{1}{2}[X(\omega - \omega_0) + X(\omega + \omega_0)] \tag{7.41}$$

This result shows that the multiplication of a signal $x(t)$ by a sinusoid of frequency ω_0 shifts the spectrum $X(\omega)$ by $\pm \omega_0$, as depicted in Fig. 7.25.

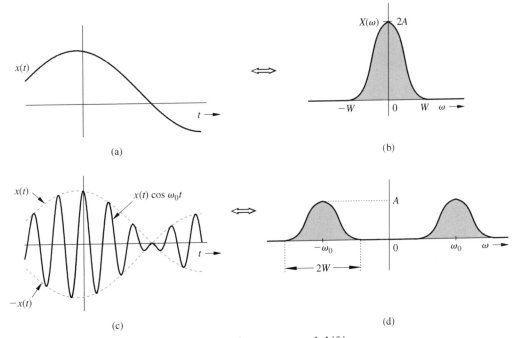

Figure 7.25 Amplitude modulation of a signal causes spectral shifting.

Multiplication of a sinusoid $\cos \omega_0 t$ by $x(t)$ amounts to modulating the sinusoid amplitude. This type of modulation is known as *amplitude modulation*. The sinusoid $\cos \omega_0 t$ is called the *carrier*, the signal $x(t)$ is the *modulating signal*, and the signal $x(t) \cos \omega_0 t$ is the *modulated signal*. Further discussion of modulation and demodulation appears in Section 7.7.

To sketch a signal $x(t) \cos \omega_0 t$, we observe that

$$x(t) \cos \omega_0 t = \begin{cases} x(t) & \text{when } \cos \omega_0 t = 1 \\ -x(t) & \text{when } \cos \omega_0 t = -1 \end{cases}$$

Therefore, $x(t) \cos \omega_0 t$ touches $x(t)$ when the sinusoid $\cos \omega_0 t$ is at its positive peaks and touches $-x(t)$ when $\cos \omega_0 t$ is at its negative peaks. This means that $x(t)$ and $-x(t)$ act as envelopes for the signal $x(t) \cos \omega_0 t$ (see Fig. 7.25). The signal $-x(t)$ is a mirror image of $x(t)$ about the horizontal axis. Figure 7.25 shows the signals $x(t)$ and $x(t) \cos \omega_0 t$ and their spectra.

EXAMPLE 7.15

Find and sketch the Fourier transform of the modulated signal $x(t) \cos 10t$ in which $x(t)$ is a gate pulse $\text{rect}(t/4)$ as illustrated in Figure 7.26a.

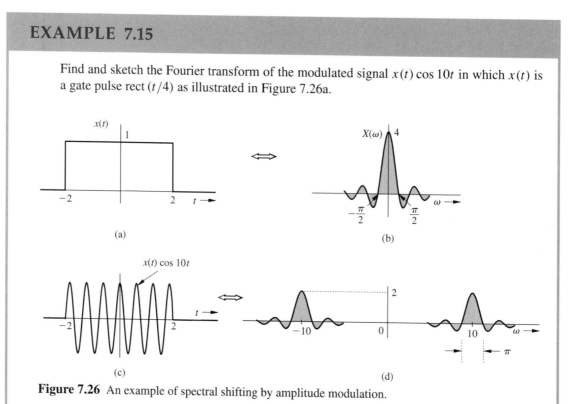

Figure 7.26 An example of spectral shifting by amplitude modulation.

From pair 17 (Table 7.1) we find $\text{rect}(t/4) \iff 4 \text{ sinc}(2\omega)$, which is depicted in Fig. 7.26b. From Eq. (7.41) it follows that

$$x(t) \cos 10t \iff \tfrac{1}{2}[X(\omega + 10) + X(\omega - 10)]$$

In this case, $X(\omega) = 4$ sinc (2ω). Therefore

$$x(t) \cos 10t \Longleftrightarrow 2 \text{ sinc } [2(\omega + 10)] + 2 \text{ sinc } [2(\omega - 10)]$$

The spectrum of $x(t) \cos 10t$ is obtained by shifting $X(\omega)$ in Fig. 7.26b to the left by 10 and also to the right by 10, and then multiplying it by 0.5, as depicted in Fig. 7.26d.

EXERCISE E7.6

Show that

$$x(t) \cos (\omega_0 t + \theta) \Longleftrightarrow \tfrac{1}{2} \left[X(\omega - \omega_0)e^{j\theta} + X(\omega + \omega_0)e^{-j\theta} \right]$$

EXERCISE E7.7

Sketch signal $e^{-|t|} \cos 10t$. Find the Fourier transform of this signal and sketch its spectrum.

ANSWER

$$X(\omega) = \frac{1}{(\omega - 10)^2 + 1} + \frac{1}{(\omega + 10)^2 + 1}$$

See Fig. 7.21b for the spectrum of $e^{-a|t|}$.

APPLICATION OF MODULATION

Modulation is used to shift signal spectra. Some of the situations that call for spectrum shifting are presented next.

1. If several signals, all occupying the same frequency band, are transmitted simultaneously over the same transmission medium, they will all interfere; it will be impossible to separate or retrieve them at a receiver. For example, if all radio stations decide to broadcast audio signals simultaneously, a receiver will not be able to separate them. This problem is solved by using modulation, whereby each radio station is assigned a distinct carrier frequency. Each station transmits a modulated signal. This procedure shifts the signal spectrum to its allocated band, which is not occupied by any other station. A radio receiver can pick up any station by tuning to the band of the desired station. The receiver must now demodulate the received signal (undo the effect of modulation). Demodulation therefore consists of another spectral shift required to restore the signal to its original band. Note that both modulation and demodulation implement spectral shifting; consequently, demodulation operation is similar to modulation (see Section 7.7).

   This method of transmitting several signals simultaneously over a channel by sharing its frequency band is known as *frequency-division multiplexing (FDM)*.

Old is gold, but sometimes it is fool's gold.

2. For effective radiation of power over a radio link, the antenna size must be of the order of the wavelength of the signal to be radiated. Audio signal frequencies are so low (wavelengths are so large) that impracticably large antennas would be required for radiation. Here, shifting the spectrum to a higher frequency (a smaller wavelength) by modulation solves the problem.

CONVOLUTION

The time-convolution property and its dual, the frequency-convolution property, state that if

$$x_1(t) \Longleftrightarrow X_1(\omega) \qquad \text{and} \qquad x_2(t) \Longleftrightarrow X_2(\omega)$$

then

$$x_1(t) * x_2(t) \Longleftrightarrow X_1(\omega)X_2(\omega) \quad \text{(time convolution)} \tag{7.42}$$

and

$$x_1(t)x_2(t) \Longleftrightarrow \frac{1}{2\pi}X_1(\omega) * X_2(\omega) \quad \text{(frequency convolution)} \tag{7.43}$$

Proof. By definition

$$\mathcal{F}|x_1(t) * x_2(t)| = \int_{-\infty}^{\infty} e^{-j\omega t} \left[\int_{-\infty}^{\infty} x_1(\tau)x_2(t-\tau)\,d\tau \right] dt$$

$$= \int_{-\infty}^{\infty} x_1(\tau) \left[\int_{-\infty}^{\infty} e^{-j\omega t} x_2(t-\tau)\,dt \right] d\tau$$

The inner integral is the Fourier transform of $x_2(t - \tau)$, given by [time-shifting property in Eq. (7.37)] $X_2(\omega)e^{-j\omega\tau}$. Hence

$$\mathcal{F}[x_1(t) * x_2(t)] = \int_{-\infty}^{\infty} x_1(\tau)e^{-j\omega\tau} X_2(\omega)\, d\tau = X_2(\omega) \int_{-\infty}^{\infty} x_1(\tau)e^{-j\omega\tau}\, d\tau = X_1(\omega)X_2(\omega)$$

Let $H(\omega)$ be the Fourier transform of the unit impulse response $h(t)$, that is

$$h(t) \Longleftrightarrow H(\omega) \tag{7.44a}$$

Application of the time-convolution property to $y(t) = x(t) * h(t)$ yields (assuming that both $x(t)$ and $h(t)$ are Fourier transformable)

$$Y(\omega) = X(\omega)H(\omega) \tag{7.44b}$$

The frequency-convolution property (7.43) can be proved in exactly the same way by reversing the roles of $x(t)$ and $X(\omega)$.

EXAMPLE 7.16

Use the time-convolution property to show that if

$$x(t) \Longleftrightarrow X(\omega)$$

then

$$\int_{-\infty}^{t} x(\tau)\, d\tau \Longleftrightarrow \frac{X(\omega)}{j\omega} + \pi X(0)\delta(\omega) \tag{7.45}$$

Because

$$u(t - \tau) = \begin{cases} 1 & \tau \le t \\ 0 & \tau > t \end{cases}$$

it follows that

$$x(t) * u(t) = \int_{-\infty}^{\infty} x(\tau)u(t - \tau)\, d\tau = \int_{-\infty}^{t} x(\tau)\, d\tau$$

Now, from the time-convolution property [Eq. (7.42)], it follows that

$$x(t) * u(t) = \int_{-\infty}^{t} x(\tau)\, d\tau \Longleftrightarrow X(\omega) \left[\frac{1}{j\omega} + \pi\delta(\omega) \right]$$

$$= \frac{X(\omega)}{j\omega} + \pi X(0)\delta(\omega)$$

In deriving the last result, we used Eq. (1.23a).

EXERCISE E7.8

Use the time-convolution property to show that $x(t) * \delta(t) = x(t)$.

EXERCISE E7.9

Use the time-convolution property to show that

$$e^{-at}u(t) * e^{-bt}u(t) = \frac{1}{b-a}[e^{-at} - e^{-bt}]u(t)$$

TIME DIFFERENTIATION AND TIME INTEGRATION

If

$$x(t) \Longleftrightarrow X(\omega)$$

then

$$\frac{dx}{dt} \Longleftrightarrow j\omega X(\omega) \quad \text{(time differentiation)}^{\dagger} \tag{7.46}$$

and

$$\int_{-\infty}^{t} x(\tau)\, d\tau \Longleftrightarrow \frac{X(\omega)}{j\omega} + \pi X(0)\delta(\omega) \quad \text{(time integration)} \tag{7.47}$$

Proof. Differentiation of both sides of Eq. (7.8b) yields

$$\frac{dx}{dt} = \frac{1}{2\pi} \int_{-\infty}^{\infty} j\omega X(\omega)e^{j\omega t}\, d\omega$$

This result shows that

$$\frac{dx}{dt} \Longleftrightarrow j\omega X(\omega)$$

Repeated application of this property yields

$$\frac{d^n x}{dt^n} \Longleftrightarrow (j\omega)^n X(\omega) \tag{7.48}$$

---

†Valid only if the transform of dx/dt exists. In other words, dx/dt must satisfy the Dirichlet conditions. The first Dirichlet condition implies

$$\int_{-\infty}^{\infty} \left| \frac{dx}{dt} \right|\, dt < \infty$$

We also require that $x(t) \to 0$ as $t \to \pm\infty$. Otherwise, $x(t)$ has a dc component, which gets lost in differentiation, and there is no one-to-one relationship between $x(t)$ and dx/dt.

TABLE 7.2   Fourier Transform Operations

Operation	$x(t)$	$X(\omega)$
Scalar multiplication	$kx(t)$	$kX(\omega)$
Addition	$x_1(t) + x_2(t)$	$X_1(\omega) + X_2(\omega)$
Conjugation	$x^*(t)$	$X^*(-\omega)$
Duality	$X(t)$	$2\pi x(-\omega)$
Scaling (a real)	$x(at)$	$\dfrac{1}{|a|} X\left(\dfrac{\omega}{a}\right)$
Time shifting	$x(t - t_0)$	$X(\omega)e^{-j\omega t_0}$
Frequency shifting (ω_0 real)	$x(t)e^{j\omega_0 t}$	$X(\omega - \omega_0)$
Time convolution	$x_1(t) * x_2(t)$	$X_1(\omega)X_2(\omega)$
Frequency convolution	$x_1(t)x_2(t)$	$\dfrac{1}{2\pi} X_1(\omega) * X_2(\omega)$
Time differentiation	$\dfrac{d^n x}{dt^n}$	$(j\omega)^n X(\omega)$
Time integration	$\displaystyle\int_{-\infty}^{t} x(u)\, du$	$\dfrac{X(\omega)}{j\omega} + \pi X(0)\delta(\omega)$

The time-integration property [Eq. (7.47)] has already been proved in Example 7.16. The Fourier transform properties are summarized in Table 7.2.

EXAMPLE 7.17

Use the time-differentiation property to find the Fourier transform of the triangle pulse $\Delta(t/\tau)$ illustrated in Fig. 7.27a.

To find the Fourier transform of this pulse we differentiate the pulse successively, as illustrated in Fig. 7.27b and 7.27c. Because dx/dt is constant everywhere, its derivative, d^2x/dt^2, is zero everywhere. But dx/dt has jump discontinuities with a positive jump of $2/\tau$ at $t = \pm\tau/2$, and a negative jump of $4/\tau$ at $t = 0$. Recall that the derivative of a signal at a jump discontinuity is an impulse at that point of strength equal to the amount of jump. Hence, d^2x/dt^2, the derivative of dx/dt, consists of a sequence of impulses, as depicted in Fig. 7.27c; that is,

$$\frac{d^2x}{dt^2} = \frac{2}{\tau}\left[\delta\left(t + \frac{\tau}{2}\right) - 2\delta(t) + \delta\left(t - \frac{\tau}{2}\right)\right] \tag{7.49}$$

From the time-differentiation property [Eq. (7.46)]

$$\frac{d^2x}{dt^2} \iff (j\omega)^2 X(\omega) = -\omega^2 X(\omega) \tag{7.50a}$$

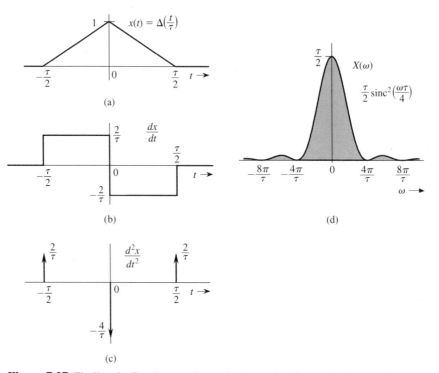

Figure 7.27 Finding the Fourier transform of a piecewise-linear signal using the time-differentiation property.

Also, from the time-shifting property [Eq. (7.37b)]

$$\delta(t - t_0) \Longleftrightarrow e^{-j\omega t_0} \tag{7.50b}$$

Taking the Fourier transform of Eq. (7.49) and using the results in Eqs. (7.50), we obtain

$$-\omega^2 X(\omega) = \frac{2}{\tau}\left[e^{j(\omega\tau/2)} - 2 + e^{-j(\omega\tau/2)}\right] = \frac{4}{\tau}\left(\cos\frac{\omega\tau}{2} - 1\right) = -\frac{8}{\tau}\sin^2\left(\frac{\omega\tau}{4}\right)$$

and

$$X(\omega) = \frac{8}{\omega^2\tau}\sin^2\left(\frac{\omega\tau}{4}\right) = \frac{\tau}{2}\left[\frac{\sin\left(\frac{\omega\tau}{4}\right)}{\frac{\omega\tau}{4}}\right]^2 = \frac{\tau}{2}\operatorname{sinc}^2\left(\frac{\omega\tau}{4}\right) \tag{7.51}$$

The spectrum $X(\omega)$ is depicted in Fig. 7.27d. This procedure of finding the Fourier transform can be applied to any function $x(t)$ made up of straight-line segments with $x(t) \to 0$ as $|t| \to \infty$. The second derivative of such a signal yields a sequence of impulses whose Fourier transform can be found by inspection. This example suggests a numerical method of finding the Fourier transform of an arbitrary signal $x(t)$ by approximating the signal by straight-line segments.

EXERCISE E7.10

Use the time-differentiation property to find the Fourier transform of rect (t/τ).

7.4 SIGNAL TRANSMISSION THROUGH LTIC SYSTEMS

If $x(t)$ and $y(t)$ are the input and output of an LTIC system with impulse response $h(t)$, then, as demonstrated in Eq. (7.44b)

$$Y(\omega) = H(\omega)X(\omega) \tag{7.52}$$

This equation does not apply to (asymptotically) unstable systems because $h(t)$ for such systems is not Fourier transformable. It applies to BIBO stable as well as most of the marginally stable systems.[†] Also the input $x(t)$ itself must be Fourier transformable if we are to use Eq. (7.52).

In Chapter 4, we saw that the Laplace transform is more versatile and capable of analyzing all kinds of LTIC systems whether stable, unstable, or marginally stable. Laplace transform can also handle exponentially growing inputs. In comparison to the Laplace transform, the Fourier transform in system analysis is not just clumsier, but also very restrictive. Hence, the Laplace transform is preferable to the Fourier transform in LTIC system analysis. We shall not belabor the application of the Fourier transform to LTIC system analysis. We consider just one example here.

EXAMPLE 7.18

Find the zero-state response of a stable LTIC system with frequency response[‡]

$$H(s) = \frac{1}{s + 2}$$

and the input is $x(t) = e^{-t}u(t)$.

In this case

$$X(\omega) = \frac{1}{j\omega + 1}$$

Moreover, because the system is stable, the frequency response $H(j\omega) = H(\omega)$. Hence

$$H(\omega) = H(s)|_{s=j\omega} = \frac{1}{j\omega + 2}$$

[†]For marginally stable systems, if the input $x(t)$ contains a finite amplitude sinusoid of the system's natural frequency, which leads to resonance, the output is not Fourier transformable. It does, however, apply to marginally stable systems if the input does not contain a finite amplitude sinusoid of the system's natural frequency.

[‡]Stability implies that the region of convergence of $H(s)$ includes the $j\omega$ axis.

Therefore

$$Y(\omega) = H(\omega)X(\omega)$$

$$= \frac{1}{(j\omega + 2)(j\omega + 1)}$$

Expanding the right-hand side in partial fractions yields

$$Y(\omega) = \frac{1}{j\omega + 1} - \frac{1}{j\omega + 2}$$

and

$$y(t) = (e^{-t} - e^{-2t})u(t)$$

EXERCISE E7.11

For the system in Example 7.18, show that the zero-input response to the input $e^t u(-t)$ is $y(t) = \frac{1}{3}[e^t u(-t) + e^{-2t} u(t)]$ [Hint: Use pair 2 (Table 7.1) to find the Fourier transform of $e^t u(-t)$.]

HEURISTIC UNDERSTANDING OF LINEAR SYSTEM RESPONSE

In finding the linear system response to arbitrary input, the time-domain method uses convolution integral and the frequency-domain method uses the Fourier integral. Despite the apparent dissimilarities of the two methods, their philosophies are amazingly similar. In the time-domain case we express the input $x(t)$ as a sum of its impulse components; in the frequency-domain case, the input is expressed as a sum of everlasting exponentials (or sinusoids). In the former case, the response $y(t)$ obtained by summing the system's responses to impulse components results in the convolution integral; in the latter case, the response obtained by summing the system's response to everlasting exponential components results in the Fourier integral. These ideas can be expressed mathematically as follows:

1. For the time-domain case,

$$\delta(t) \implies h(t) \qquad \text{shows the impulse response of the system is } h(t)$$

$$x(t) = \int_{-\infty}^{\infty} x(\tau)\delta(t - \tau)\,d\tau \qquad \text{expresses } x(t) \text{ as a sum of impulse components}$$

and

$$y(t) = \int_{-\infty}^{\infty} x(\tau)h(t - \tau)\,d\tau \qquad \text{expresses } y(t) \text{ as a sum of responses to impulse components of the input } x(t)$$

2. For the frequency-domain case,

$$e^{j\omega t} \implies H(\omega)e^{j\omega t} \qquad \text{shows the system response}$$
$$\text{to } e^{j\omega t} \text{ is } H(\omega)e^{j\omega t}$$

$$x(t) = \frac{1}{2\pi} \int_{-\infty}^{\infty} X(\omega)e^{j\omega t}\,d\omega \qquad \text{shows } x(t) \text{ as a sum of everlasting}$$
$$\text{exponential components}$$

and

$$y(t) = \frac{1}{2\pi} \int_{-\infty}^{\infty} X(\omega)H(\omega)e^{j\omega t}\,d\omega \qquad \text{expresses } y(t) \text{ as a sum of responses to}$$
$$\text{exponential components of the input } x(t)$$

The frequency-domain view sees a system in terms of its frequency response (system response to various sinusoidal components). It views a signal as a sum of various sinusoidal components. Transmission of an input signal through a (linear) system is viewed as transmission of various sinusoidal components of the input through the system.

It was not by coincidence that we used the impulse function in time-domain analysis and the exponential $e^{j\omega t}$ in studying the frequency domain. The two functions happen to be duals of each other. Thus, the Fourier transform of an impulse $\delta(t - \tau)$ is $e^{-j\omega\tau}$, and the Fourier transform of $e^{j\omega_0 t}$ is an impulse $2\pi\delta(\omega - \omega_0)$. This *time–frequency duality* is a constant theme in the Fourier transform and linear systems.

7.4-1 Signal Distortion During Transmission

For a system with frequency response $H(\omega)$, if $X(\omega)$ and $Y(\omega)$ are the spectra of the input and the output signals, respectively, then

$$Y(\omega) = X(\omega)H(\omega) \tag{7.53}$$

The transmission of the input signal $x(t)$ through the system changes it into the output signal $y(t)$. Equation (7.53) shows the nature of this change or modification. Here $X(\omega)$ and $Y(\omega)$ are the spectra of the input and the output, respectively. Therefore, $H(\omega)$ is the spectral response of the system. The output spectrum is obtained by the input spectrum multiplied by the spectral response of the system. Equation (7.53), which clearly brings out the spectral shaping (or modification) of the signal by the system, can be expressed in polar form as

$$|Y(\omega)|e^{j\angle Y(\omega)} = |X(\omega)||H(\omega)|e^{j[\angle X(\omega)+\angle H(j\omega)]}$$

Therefore

$$|Y(\omega)| = |X(\omega)|\,|H(\omega)| \tag{7.54a}$$

$$\angle Y(\omega) = \angle X(\omega) + \angle H(\omega) \tag{7.54b}$$

During transmission, the input signal amplitude spectrum $|X(\omega)|$ is changed to $|X(\omega)|\,|H(\omega)|$. Similarly, the input signal phase spectrum $\angle X(\omega)$ is changed to $\angle X(\omega) + \angle H(\omega)$. An input signal spectral component of frequency ω is modified in amplitude by a factor $|H(\omega)|$ and is shifted in phase by an angle $\angle H(\omega)$. Clearly, $|H(\omega)|$ is the amplitude response, and $\angle H(\omega)$ is the phase response of the system. The plots of $|H(\omega)|$ and $\angle H(\omega)$ as functions of ω show at a glance how the system modifies the amplitudes and phases of various sinusoidal inputs. This is the reason why $H(\omega)$ is also called the *frequency response* of the system. During transmission through the system, some frequency components may be boosted in amplitude, while others may be attenuated. The relative phases of the various components also change. In general, the output waveform will be different from the input waveform.

DISTORTIONLESS TRANSMISSION

In several applications, such as signal amplification or message signal transmission over a communication channel, we require that the output waveform be a replica of the input waveform. In such cases we need to minimize the distortion caused by the amplifier or the communication channel. It is, therefore, of practical interest to determine the characteristics of a system that allows a signal to pass without distortion (*distortionless transmission*).

Transmission is said to be distortionless if the input and the output have identical waveshapes within a multiplicative constant. A delayed output that retains the input waveform is also considered to be distortionless. Thus, in distortionless transmission, the input $x(t)$ and the output $y(t)$ satisfy the condition

$$y(t) = G_0 x(t - t_d) \tag{7.55}$$

The Fourier transform of this equation yields

$$Y(\omega) = G_0 X(\omega) e^{-j\omega t_d}$$

But

$$Y(\omega) = X(\omega) H(\omega)$$

Therefore

$$H(\omega) = G_0 e^{-j\omega t_d}$$

This is the frequency response required of a system for distortionless transmission. From this equation it follows that

$$|H(\omega)| = G_0 \tag{7.56a}$$

$$\angle H(\omega) = -\omega t_d \tag{7.56b}$$

This result shows that for distortionless transmission, the amplitude response $|H(\omega)|$ must be a constant, and the phase response $\angle H(\omega)$ must be a linear function of ω with slope $-t_d$, where t_d is the delay of the output with respect to input (Fig. 7.28).

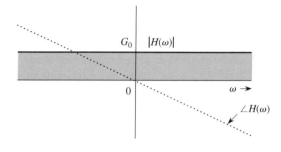

Figure 7.28 LTIC system frequency response for distortionless transmission.

MEASURE OF TIME DELAY VARIATION WITH FREQUENCY

The gain $|H(\omega)| = G_0$ means that every spectral component is multiplied by a constant G_0. Also as seen in connection with Fig. 7.22, a linear phase $\angle H(\omega) = -\omega t_d$ means that every spectral component is delayed by t_d seconds. This results in the output equal to G_0 times the input delayed by t_d seconds. Because each spectral component is attenuated by the same factor (G_0) and delayed by exactly the same amount (t_d), the output signal is an exact replica of the input (except for attenuating factor G_0 and delay t_d).

For distortionless transmission, we require a *linear phase* characteristic. The phase is not only linear function of ω, but it should also pass through the origin $\omega = 0$. In practice, many systems have a phase characteristic that may be only approximately linear. A convenient way of judging phase linearity is to plot the slope of $\angle H(\omega)$ as a function of frequency. This slope, which is constant for an ideal linear phase (ILP) system, is a function of ω in the general case and can be expressed as

$$t_g(\omega) = -\frac{d}{d\omega}\angle H(\omega) \tag{7.57}$$

If $t_g(\omega)$ is constant, all the components are delayed by the same time interval t_g. But if the slope is not constant, the time delay t_g varies with frequency. This variation means that different frequency components undergo different amounts of time delay, and consequently the output waveform will not be a replica of the input waveform. As we shall see, $t_g(\omega)$ plays an important role in bandpass systems and is called the *group delay* or *envelope* delay. Observe that constant t_d [Eq. (7.56b)] implies constant t_g. Note that $\angle H(\omega) = \phi_0 - \omega t_g$ also has a constant t_g. Thus, constant group delay is a more relaxed condition.

It is often thought (erroneously) that flatness of amplitude response $|H(\omega)|$ alone can guarantee signal quality. However, a system that has a flat amplitude response may yet distort a signal beyond recognition if the phase response is not linear (t_d not constant).

THE NATURE OF DISTORTION IN AUDIO AND VIDEO SIGNALS

Generally speaking, the human ear can readily perceive amplitude distortion but is relatively insensitive to phase distortion. For the phase distortion to become noticeable, the variation in delay [variation in the slope of $\angle H(\omega)$] should be comparable to the signal duration (or the physically perceptible duration, in case the signal itself is long). In the case of audio signals, each spoken syllable can be considered to be an individual signal. The average duration of a spoken syllable is of a magnitude of the order of 0.01 to 0.1 second. The audio systems may have nonlinear phases, yet no noticeable signal distortion results because in practical audio systems, maximum variation in the slope of $\angle H(\omega)$ is only a small fraction of a millisecond. This is the real truth underlying the statement that "the human ear is relatively insensitive to phase distortion."[3] As a result, the manufacturers of audio equipment make available only $|H(\omega)|$, the amplitude response characteristic of their systems.

For video signals, in contrast, the situation is exactly the opposite. The human eye is sensitive to phase distortion but is relatively insensitive to amplitude distortion. The amplitude distortion in television signals manifests itself as a partial destruction of the relative half-tone values of the resulting picture, but this effect is not readily apparent to the human eye. The phase distortion (nonlinear phase), on the other hand, causes different time delays in different picture elements. The result is a smeared picture, and this effect is readily perceived by the human eye. Phase distortion is also very important in digital communication systems because the nonlinear phase characteristic of a channel causes pulse dispersion (spreading out), which in turn causes pulses to interfere with neighboring pulses. Such interference between pulses can cause an error in the pulse amplitude at the receiver: a binary **1** may read as **0**, and vice versa.

7.4-2 Bandpass Systems and Group Delay

The distortionless transmission conditions [Eqs. (7.56)] can be relaxed slightly for bandpass systems. For lowpass systems, the phase characteristics not only should be linear over the band of interest but should also pass through the origin. For bandpass systems, the phase characteristics must be linear over the band of interest but need not pass through the origin.

Consider an LTI system with amplitude and phase characteristics as shown in Fig. 7.29, where the amplitude spectrum is a constant G_0 and the phase is $\phi_0 - \omega t_g$ over a band $2W$ centered

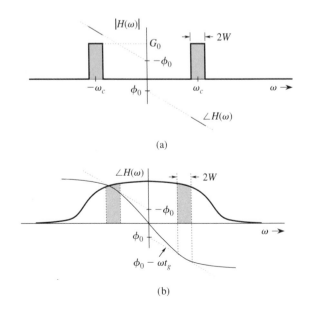

Figure 7.29 Generalized linear phase characteristics.

at frequency ω_c. Over this band, we can describe $H(\omega)$ as[†]

$$H(\omega) = G_0 e^{j(\phi_0 - \omega t_g)} \qquad \omega \geq 0 \tag{7.58}$$

The phase of $H(\omega)$ in Eq. (7.58), shown dotted in Fig. 7.29b, is linear but does not pass through the origin.

Consider a modulated input signal $z(t) = x(t) \cos \omega_c t$. This is a bandpass signal, whose spectrum is centered at $\omega = \omega_c$. The signal $\cos \omega_c t$ is the carrier, and the signal $x(t)$, which is a lowpass signal of bandwidth W (see Fig. 7.25), is the *envelope* of $z(t)$.[‡] We shall now show that the transmission of $z(t)$ through $H(\omega)$ results in distortionless transmission of the envelope $x(t)$. However, the carrier phase changes by ϕ_0. To show this, consider an input $\hat{z}(t) = x(t)e^{j\omega_c t}$ and the corresponding output $\hat{y}(t)$. From Eq. (7.39), $\hat{Z}(\omega) = X(\omega - \omega_c)$, and the corresponding output spectrum $\hat{Y}(\omega)$ is given by

$$\hat{Y}(\omega) = H(\omega)\hat{Z}(\omega) = H(\omega)X(\omega - \omega_c)$$

Recall that the bandwidth of $X(\omega)$ is W, so that the bandwidth of $X(\omega - \omega_c)$ is $2W$, centered at ω_c. Over this range $H(\omega)$ is given by Eq. (7.58). Hence,

$$\hat{Y}(\omega) = G_0 X(\omega - \omega_c)e^{j(\phi_0 - \omega t_g)} = G_0 e^{j\phi_0} X(\omega - \omega_c)e^{-\omega t_g}$$

Use of Eqs. (7.37a) and (7.39) yields $\hat{y}(t)$ as

$$\hat{y}(t) = G_0 e^{j\phi_0} x(t - t_g)e^{j\omega_c(t - t_g)} = G_0 x(t - t_g)e^{j[\omega_c(t - t_g) + \phi_0]}$$

This is the system response to input $\hat{z}(t) = x(t)e^{j\omega_c t}$, which is a complex signal. We are really interested in finding the response to the input $z(t) = x(t) \cos \omega_c t$, which is the real part of $\hat{z}(t) = x(t)e^{j\omega_c t}$. Hence, we use property (2.40) to obtain $y(t)$, the system response to the input $z(t) = x(t) \cos \omega_c t$, as

$$y(t) = G_0 x(t - t_g) \cos [\omega_c(t - t_g) + \phi_0)] \tag{7.59}$$

where t_g, the *group* (or *envelope*) delay, is the negative slope of $\angle H(\omega)$ at ω_c.[§] The output $y(t)$ is basically the delayed input $z(t - t_g)$, except that the output carrier acquires an extra phase ϕ_0. The output envelope $x(t - t_g)$ is the delayed version of the input envelope $x(t)$ and is not affected by extra phase ϕ_0 of the carrier. In a modulated signal, such as, $x(t) \cos \omega_c t$, the information generally resides in the envelope $x(t)$. Hence, the transmission is considered to be distortionless if the envelope $x(t)$ remains undistorted.

[†] Because the phase function is an odd function of ω, if $\angle H(\omega) = \phi_0 - \omega t_g$ for $\omega \geq 0$, over the band $2W$ (centered at ω_c), then $\angle H(\omega) = -\phi_0 - \omega t_g$ for $\omega < 0$ over the band $2W$ (centered at $-\omega_c$), as shown in Fig. 7.29a.

[‡] The envelope of a bandpass signal is well defined only when the bandwidth of the envelope is well below the carrier ω_c ($W \ll \omega_c$).

[§] Equation (7.59) can also be expressed as

$$y(t) = G_o x(t - t_g) \cos \omega_c(t - t_{ph})$$

where t_{ph}, called the *phase delay* at ω_c, is given by $t_{ph}(\omega_c) = (\omega_c t_g - \phi_0)/\omega_c$. Generally, t_{ph} varies with ω, and we can write

$$t_{ph}(\omega) = \frac{\omega t_g - \phi_0}{\omega}$$

Recall also that t_g itself may vary with ω.

Most of the practical systems satisfy conditions (7.58), at least over a very small band. Figure 7.29b shows a typical case in which this condition is satisfied for a small band W centered at frequency ω_c.

A system in Eq. (7.58) is said to have a *generalized linear phase* (GLP), as illustrated in Fig. 7.29. The ideal linear phase (ILP) characteristics is shown in Fig. 7.28. For distortionless transmission of bandpass signals, the system need satisfy Eq. (7.58) only over the bandwidth of the bandpass signal.

Caution. Recall that the phase response associated with the amplitude response may have jump discontinuities when the amplitude response goes negative. Jump discontinuities also arise because of the use of the principal value for phase. Under such conditions, to compute the group delay [Eq. (7.57)], we should ignore the jump discontinuities.

EXAMPLE 7.19

(a) A signal $z(t)$, shown in Fig. 7.30b, is given by

$$z(t) = x(t) \cos \omega_c t$$

where $\omega_c = 2000\pi$. The pulse $x(t)$ (Fig. 7.30a) is a lowpass pulse of duration 0.1 second and has a bandwidth of about 10 Hz. This signal is passed through a filter whose frequency response is shown in Fig. 7.30c (shown only for positive ω). Find and sketch the filter output $y(t)$.

(b) Find the filter response if $\omega_c = 4000\pi$.

(a) The spectrum $Z(\omega)$ is a narrow band of width 20 Hz, centered at frequency $f_0 = 1$ kHz. The gain at the center frequency (1 kHz) is 2. The group delay, which is the negative of the slope of the phase plot, can be found by drawing tangents at ω_c, as shown in Fig. 7.30c. The negative of the slope of the tangent represents t_g, and the intercept along the vertical axis by the tangent represents ϕ_0 at that frequency. From the tangents at ω_c, we find t_g, the group delay, as

$$t_g = \frac{2.4\pi - 0.4\pi}{2000\pi} = 10^{-3}$$

The vertical axis intercept is $\phi_0 = -0.4\pi$. Hence, by using Eq. (7.59) with gain $G_0 = 2$, we obtain

$$y(t) = 2x(t - t_g) \cos [\omega_c(t - t_g) - 0.4\pi] \qquad \omega_c = 2000\pi \quad t_g = 10^{-3}$$

Figure 7.30d shows the output $y(t)$, which consists of the modulated pulse envelope $x(t)$ delayed by 1 ms and the phase of the carrier changed by -0.4π. The output shows no distortion of the envelope $x(t)$, only the delay. The carrier phase change does not affect the shape of envelope. Hence, the transmission is considered distortionless.

(b) Figure 7.30c shows that when $\omega_c = 4000\pi$, the slope of $\angle H(\omega)$ is zero, so that $t_g = 0$. Also the gain $G_0 = 1.5$, and the intercept of the tangent with the vertical axis is $\phi_0 = -3.1\pi$. Hence

$$y(t) = 1.5x(t) \cos (\omega_c t - 3.1\pi)$$

This too is a distortionless transmission for the same reasons as for case (a).

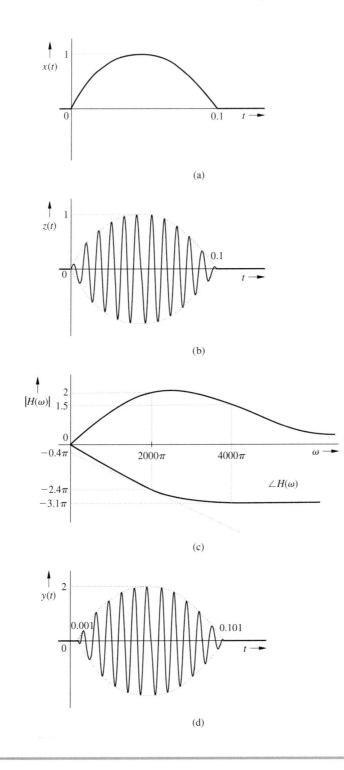

(a)

(b)

(c)

(d)

Figure 7.30

7.5 IDEAL AND PRACTICAL FILTERS

Ideal filters allow distortionless transmission of a certain band of frequencies and completely suppress the remaining frequencies. The ideal lowpass filter (Fig. 7.31), for example, allows all components below $\omega = W$ rad/s to pass without distortion and suppresses all components above $\omega = W$. Figure 7.32 illustrates ideal highpass and bandpass filter characteristics.

The ideal lowpass filter in Fig. 7.31a has a linear phase of slope $-t_d$, which results in a time delay of t_d seconds for all its input components of frequencies below W rad/s. Therefore, if the input is a signal $x(t)$ bandlimited to W rad/s, the output $y(t)$ is $x(t)$ delayed by t_d: that is,

$$y(t) = x(t - t_d)$$

The signal $x(t)$ is transmitted by this system without distortion, but with time delay t_d. For this filter $|H(\omega)| = \text{rect}(\omega/2W)$ and $\angle H(\omega) = e^{-j\omega t_d}$, so that

$$H(\omega) = \text{rect}\left(\frac{\omega}{2W}\right) e^{-j\omega t_d} \tag{7.60a}$$

The unit impulse response $h(t)$ of this filter is obtained from pair 18 (Table 7.1) and the

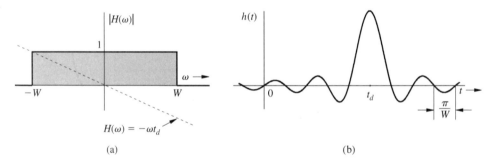

Figure 7.31 Ideal lowpass filter: **(a)** frequency response and **(b)** impulse response.

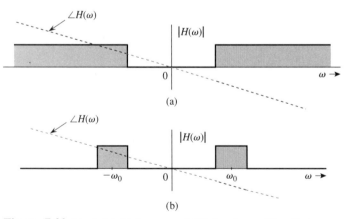

Figure 7.32 Ideal **(a)** highpass and **(b)** bandpass filter frequency responses.

time-shifting property

$$h(t) = \mathcal{F}^{-1}\left[\text{rect}\left(\frac{\omega}{2W}\right)e^{-j\omega t_d}\right]$$

$$= \frac{W}{\pi}\text{sinc}\left[W(t - t_d)\right] \tag{7.60b}$$

Recall that $h(t)$ is the system response to impulse input $\delta(t)$, which is applied at $t = 0$. Figure 7.31b shows a curious fact: the response $h(t)$ begins even before the input is applied (at $t = 0$). Clearly, the filter is noncausal and therefore physically unrealizable. Similarly, one can show that other ideal filters (such as the ideal highpass or ideal bandpass filters depicted in Fig. 7.32) are also physically unrealizable.

For a physically realizable system, $h(t)$ must be causal: that is,

$$h(t) = 0 \qquad \text{for } t < 0$$

In the frequency domain, this condition is equivalent to the well-known *Paley–Wiener criterion*, which states that the necessary and sufficient condition for the amplitude response $|H(\omega)|$ to be realizable is[†]

$$\int_{-\infty}^{\infty} \frac{|\ln|H(\omega)||}{1 + \omega^2}\, d\omega < \infty \tag{7.61}$$

If $H(\omega)$ does not satisfy this condition, it is unrealizable. Note that if $|H(\omega)| = 0$ over any finite band, $|\ln|H(\omega)|| = \infty$ over that band, and condition (7.61) is violated. If, however, $H(\omega) = 0$ at a single frequency (or a set of discrete frequencies), the integral in Eq. (7.61) may still be finite even though the integrand is infinite at those discrete frequencies. Therefore, for a physically realizable system, $H(\omega)$ may be zero at some discrete frequencies, but it cannot be zero over any finite band. In addition, if $|H(\omega)|$ decays exponentially (or at a higher rate) with ω, the integral in (7.61) goes to infinity, and $|H(\omega)|$ cannot be realized. Clearly, $|H(\omega)|$ cannot decay too fast with ω. According to this criterion, ideal filter characteristics (Figs. 7.31 and 7.32) are unrealizable.

The impulse response $h(t)$ in Fig. 7.31 is not realizable. One practical approach to filter design is to cut off the tail of $h(t)$ for $t < 0$. The resulting causal impulse response $\hat{h}(t)$, given by

$$\hat{h}(t) = h(t)u(t)$$

is physically realizable because it is causal (Fig. 7.33). If t_d is sufficiently large, $\hat{h}(t)$ will be a close approximation of $h(t)$, and the resulting filter $\hat{H}(\omega)$ will be a good approximation of an ideal filter. This close realization of the ideal filter is achieved because of the increased value of time delay t_d. This observation means that the price of close realization is higher delay in the output; this situation is common in noncausal systems. Of course, theoretically, a delay $t_d = \infty$ is needed to realize the ideal characteristics. But a glance at Fig. 7.31b shows that a delay t_d of three

---

[†]We are assuming that $|H(\omega)|$ is square integrable, that is,

$$\int_{-\infty}^{\infty} |H(\omega)|^2\, d\omega < \infty$$

Note that the Paley–Wiener criterion is a criterion for the realizability of the amplitude response $|H(\omega)|$.

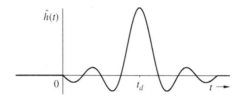

Figure 7.33 Approximate realization of an ideal low-pass filter by truncation of its impulse response.

or four times $\frac{\pi}{W}$ will make $\widehat{h}(t)$ a reasonably close version of $h(t-t_d)$. For instance, an audio filter is required to handle frequencies of up to 20 kHz ($W = 40,000\pi$). In this case, a t_d of about 10^{-4} (0.1 ms) would be a reasonable choice. The truncation operation [cutting the tail of $h(t)$ to make it causal], however, creates some unsuspected problems. We discuss these problems and their cure in Section 7.8.

In practice, we can realize a variety of filter characteristics that approach the ideal. Practical (realizable) filter characteristics are gradual, without jump discontinuities in amplitude response.

EXERCISE E7.12

Show that a filter with Gaussian frequency response $H(\omega) = e^{-\alpha\omega^2}$ is unrealizable. Demonstrate this fact in two ways: first by showing that its impulse response is noncausal, and then by showing that $|H(\omega)|$ violates the Paley–Wiener criterion. [Hint: Use pair 22 in Table 7.1.]

THINKING IN THE TIME AND FREQUENCY DOMAINS: A TWO-DIMENSIONAL VIEW OF SIGNALS AND SYSTEMS

Both signals and systems have dual personalities, the time domain and the frequency domain. For a deeper perspective, we should examine and understand both these identities because they offer complementary insights. An exponential signal, for instance, can be specified by its time-domain description such as $e^{-2t}u(t)$ or by its Fourier transform (its frequency-domain description) $1/(j\omega + 2)$. The time-domain description depicts the waveform of a signal. The frequency-domain description portrays its spectral composition (relative amplitudes of its sinusoidal (or exponential) components and their phases). For the signal e^{-2t}, for instance, the time-domain description portrays the exponentially decaying signal with a time constant 0.5. The frequency-domain description characterizes it as a lowpass signal, which can be synthesized by sinusoids with amplitudes decaying with frequency roughly as $1/\omega$.

An LTIC system can also be described or specified in the time domain by its impulse response $h(t)$ or in the frequency domain by its frequency response $H(\omega)$. In Section 2.7, we studied intuitive insights in the system behavior offered by the impulse response, which consists of characteristic modes of the system. By purely qualitative reasoning, we saw that the system responds well to signals that are similar to the characteristic modes and responds poorly to signals that are very different from those modes. We also saw that the shape of the impulse response $h(t)$ determines the system time constant (speed of response), and pulse dispersion (spreading), which, in turn, determines the rate of pulse transmission.

The frequency response $H(\omega)$ specifies the system response to exponential or sinusoidal input of various frequencies. This is precisely the filtering characteristic of the system.

Experienced electrical engineers instinctively think in both domains (the time and frequency) whenever possible. When they look at a signal, they consider, its waveform, the signal width (duration), and the rate at which the waveform decays. This is basically a time-domain perspective. They also think of the signal in terms of its frequency spectrum, that is, in terms of its sinusoidal components and their relative amplitudes and phases, whether the spectrum is lowpass, bandpass, highpass and so on. This is the frequency-domain perspective. When they think of a system, they think of its impulse response $h(t)$. The width of $h(t)$ indicates the time constant (response time): that is, how fast the system is capable of responding to an input, and how much dispersion (spreading) it will cause. This is the time-domain perspective. From the frequency-domain perspective, these engineers view a system as a filter, which selectively transmits certain frequency components and suppresses the others [frequency response $H(\omega)$]. Knowing the input signal spectrum and the frequency response of the system, they create a mental image of the output signal spectrum. This concept is precisely expressed by $Y(\omega) = X(\omega)H(\omega)$.

We can analyze LTI systems by time-domain techniques or by frequency-domain techniques. Then why learn both? The reason is that the two domains offer complementary insights into system behavior. Some aspects are easily grasped in one domain; other aspects may be easier to see in the other domain. Both the time-domain and frequency-domain methods are as essential for the study of signals and systems as two eyes are essential to a human being for correct visual perception of reality. A person can see with either eye, but for proper perception of three-dimensional reality, both eyes are essential.

It is important to keep the two domains separate, and not to mix the entities in the two domains. If we are using the frequency domain to determine the system response, we must deal with all signals in terms of their spectra (Fourier transforms) and all systems in terms of their frequency responses. For example, to determine the system response $y(t)$ to an input $x(t)$, we must first convert the input signal into its frequency-domain description $X(\omega)$. The system description also must be in the frequency domain, that is, the frequency response $H(\omega)$. The output signal spectrum $Y(\omega) = X(\omega)H(\omega)$. Thus, the result (output) is also in the frequency domain. To determine the final answer $y(t)$, we must take the inverse transform of $Y(\omega)$.

7.6 SIGNAL ENERGY

The signal energy E_x of a signal $x(t)$ was defined in Chapter 1 as

$$E_x = \int_{-\infty}^{\infty} |x(t)|^2 \, dt \tag{7.62}$$

Signal energy can be related to the signal spectrum $X(\omega)$ by substituting Eq. (7.8b) in Eq. (7.62):

$$E_x = \int_{-\infty}^{\infty} x(t)x^*(t) \, dt = \int_{-\infty}^{\infty} x(t) \left[\frac{1}{2\pi} \int_{-\infty}^{\infty} X^*(\omega)e^{-j\omega t} \, d\omega \right] dt$$

Here we used the fact that $x^*(t)$, being the conjugate of $x(t)$, can be expressed as the conjugate of the right-hand side of Eq. (7.8b). Now, interchanging the order of integration yields

$$E_x = \frac{1}{2\pi} \int_{-\infty}^{\infty} X^*(\omega) \left[\int_{-\infty}^{\infty} x(t) e^{-j\omega t}\, dt \right] d\omega$$

$$= \frac{1}{2\pi} \int_{-\infty}^{\infty} X(\omega) X^*(\omega)\, d\omega$$

$$= \frac{1}{2\pi} \int_{-\infty}^{\infty} |X(\omega)|^2\, d\omega \tag{7.63}$$

Consequently,

$$E_x = \int_{-\infty}^{\infty} |x(t)|^2\, dt = \frac{1}{2\pi} \int_{-\infty}^{\infty} |X(\omega)|^2\, d\omega \tag{7.64}$$

This is the statement of the well-known *Parseval's theorem* (for the Fourier transform). A similar result was obtained in Eqs. (6.40) and (6.41) for a periodic signal and its Fourier series. This result allows us to determine the signal energy from either the time-domain specification $x(t)$ or the corresponding frequency-domain specification $X(\omega)$.

Equation (7.63) can be interpreted to mean that the energy of a signal $x(t)$ results from energies contributed by all the spectral components of the signal $x(t)$. The total signal energy is the area under $|X(\omega)^2|$ (divided by 2π). If we consider a small band $\Delta\omega$ ($\Delta\omega \to 0$), as illustrated in Fig. 7.34, the energy ΔE_x of the spectral components in this band is the area of $|X(\omega)|^2$ under this band (divided by 2π):

$$\Delta E_x = \frac{1}{2\pi} |X(\omega)|^2\, \Delta\omega = |X(\omega)|^2\, \Delta f \qquad \frac{\Delta\omega}{2\pi} = \Delta f \text{ Hz} \tag{7.65}$$

Therefore, the energy contributed by the components in this band of Δf (in hertz) is $|X(\omega)|^2 \Delta f$. The total signal energy is the sum of energies of all such bands and is indicated by the area under $|X(\omega)|^2$ as in Eq. (7.63). Therefore, $|X(\omega)|^2$ is the *energy spectral density* (per unit bandwidth in hertz).

For real signals, $X(\omega)$ and $X(-\omega)$ are conjugates, and $|X(\omega)|^2$ is an even function of ω because

$$|X(\omega)|^2 = X(\omega) X^*(\omega) = X(\omega) X(-\omega)$$

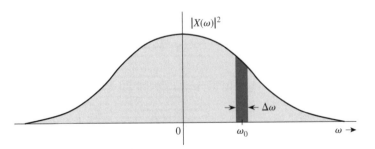

Figure 7.34 Interpretation of energy spectral density of a signal.

Therefore, Eq. (7.63) can be expressed as[†]

$$E_x = \frac{1}{\pi} \int_0^\infty |X(\omega)|^2 \, d\omega \qquad (7.66)$$

The signal energy E_x, which results from contributions from all the frequency components from $\omega = 0$ to ∞, is given by ($1/\pi$ times) the area under $|X(\omega)|^2$ from $\omega = 0$ to ∞. It follows that the energy contributed by spectral components of frequencies between ω_1 and ω_2 is

$$\Delta E_x = \frac{1}{\pi} \int_{\omega_1}^{\omega_2} |X(\omega)|^2 \, d\omega \qquad (7.67)$$

EXAMPLE 7.20

Find the energy of signal $x(t) = e^{-at}u(t)$. Determine the frequency W (rad/s) so that the energy contributed by the spectral components of all the frequencies below W is 95% of the signal energy E_x.

We have

$$E_x = \int_{-\infty}^{\infty} x^2(t) \, dt = \int_0^\infty e^{-2at} \, dt = \frac{1}{2a} \qquad (7.68)$$

We can verify this result by Parseval's theorem. For this signal

$$X(\omega) = \frac{1}{j\omega + a}$$

and

$$E_x = \frac{1}{\pi} \int_0^\infty |X(\omega)|^2 \, d\omega = \frac{1}{\pi} \int_0^\infty \frac{1}{\omega^2 + a^2} \, d\omega = \frac{1}{\pi a} \tan^{-1} \frac{\omega}{a} \Big|_0^\infty = \frac{1}{2a}$$

The band $\omega = 0$ to $\omega = W$ contains 95% of the signal energy, that is, $0.95/2a$. Therefore, from Eq. (7.67) with $\omega_1 = 0$ and $\omega_2 = W$, we obtain

$$\frac{0.95}{2a} = \frac{1}{\pi} \int_0^W \frac{d\omega}{\omega^2 + a^2} = \frac{1}{\pi a} \tan^{-1} \frac{\omega}{a} \Big|_0^W = \frac{1}{\pi a} \tan^{-1} \frac{W}{a}$$

or

$$\frac{0.95\pi}{2} = \tan^{-1} \frac{W}{a} \quad \Longrightarrow \quad W = 12.706a \text{ rad/s} \qquad (7.69)$$

This result indicates that the spectral components of $x(t)$ in the band from 0 (dc) to $12.706a$ rad/s ($2.02a$ Hz) contribute 95% of the total signal energy; all the remaining spectral components (in the band from $12.706a$ rad/s to ∞) contribute only 5% of the signal energy.

[†]In Eq. (7.66) it is assumed that $X(\omega)$ does not contain an impulse at $\omega = 0$. If such an impulse exists, it should be integrated separately with a multiplying factor of $1/2\pi$ rather than $1/\pi$.

EXERCISE E7.13

Use Parseval's theorem to show that the energy of the signal $x(t) = 2a/(t^2 + a^2)$ is $2\pi/a$. [Hint: Find $X(\omega)$ using pair 3 of Table 7.1 and the duality property.]

THE ESSENTIAL BANDWIDTH OF A SIGNAL

The spectra of all practical signals extend to infinity. However, because the energy of any practical signal is finite, the signal spectrum must approach 0 as $\omega \to \infty$. Most of the signal energy is contained within a certain band of B Hz, and the energy contributed by the components beyond B Hz is negligible. We can therefore suppress the signal spectrum beyond B Hz with little effect on the signal shape and energy. The bandwidth B is called the *essential bandwidth* of the signal. The criterion for selecting B depends on the error tolerance in a particular application. We may, for example, select B to be that band which contains 95% of the signal energy.[†] This figure may be higher or lower than 95%, depending on the precision needed. Using such a criterion, we can determine the essential bandwidth of a signal. The essential bandwidth B for the signal $e^{-at}u(t)$, using 95% energy criterion, was determined in Example 7.20 to be $2.02a$ Hz.

Suppression of all the spectral components of $x(t)$ beyond the essential bandwidth results in a signal $\hat{x}(t)$, which is a close approximation of $x(t)$. If we use the 95% criterion for the essential bandwidth, the energy of the error (the difference) $x(t) - \hat{x}(t)$ is 5% of E_x.

7.7 APPLICATION TO COMMUNICATIONS: AMPLITUDE MODULATION

Modulation causes a spectral shift in a signal and is used to gain certain advantages mentioned in our discussion of the frequency-shifting property. Broadly speaking, there are two classes of modulation: amplitude (linear) modulation and angle (nonlinear) modulation. In this section, we shall discuss some practical forms of amplitude modulation.

7.7-1 Double-Sideband, Suppressed-Carrier (DSB-SC) Modulation

In amplitude modulation, the amplitude A of the carrier $A \cos(\omega_c t + \theta_c)$ is varied in some manner with the *baseband* (message)[‡] signal $m(t)$ (known as the *modulating signal*). The frequency ω_c and the phase θ_c are constant. We can assume $\theta_c = 0$ without loss of generality. If the carrier

---

[†]For lowpass signals, the essential bandwidth may also be defined as a frequency at which the value of the amplitude spectrum is a small fraction (about 1%) of its peak value. In Example 7.20, for instance, the peak value, which occurs at $\omega = 0$, is $1/a$.

[‡]The term *baseband* is used to designate the band of frequencies of the signal delivered by the source or the input transducer.

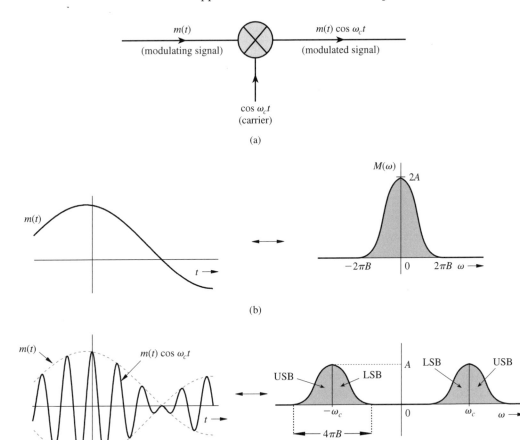

Figure 7.35 DSB-SC modulation.

amplitude A is made directly proportional to the modulating signal $m(t)$, the modulated signal is $m(t) \cos \omega_c t$ (Fig. 7.35). As was indicated earlier [Eq. (7.41)], this type of modulation simply shifts the spectrum of $m(t)$ to the carrier frequency (Fig. 7.35c). Thus, if

$$m(t) \Longleftrightarrow M(\omega)$$

then

$$m(t) \cos \omega_c t \Longleftrightarrow \tfrac{1}{2}\left[M(\omega + \omega_c) + M(\omega - \omega_c)\right] \tag{7.70}$$

Recall that $M(\omega - \omega_c)$ is $M(\omega)$-shifted to the right by ω_c and $M(\omega + \omega_c)$ is $M(\omega)$-shifted to the left by ω_c. Thus, the process of modulation shifts the spectrum of the modulating signal to the left and the right by ω_c. Note also that if the bandwidth of $m(t)$ is B Hz, then, as indicated in Fig. 7.35c, the bandwidth of the modulated signal is $2B$ Hz. We also observe that the modulated signal spectrum centered at ω_c is composed of two parts: a portion that lies above ω_c, known as the *upper sideband (USB)*, and a portion that lies below ω_c, known as the *lower sideband (LSB)*. Similarly, the spectrum centered at $-\omega_c$ has upper and lower sidebands. This form of modulation is called *double sideband (DSB)* modulation for obvious reason.

The relationship of B to ω_c is of interest. Figure 7.35c shows that $\omega_c \geq 2\pi B$ to avoid the overlap of the spectra centered at $\pm\omega_c$. If $\omega_c < 2\pi B$, the spectra overlap and the information of $m(t)$ is lost in the process of modulation, a loss that makes it impossible to get back $m(t)$ from the modulated signal $m(t) \cos \omega_c t$.[†]

EXAMPLE 7.21

For a baseband signal $m(t) = \cos \omega_m t$, find the DSB signal, and sketch its spectrum. Identify the upper and lower sidebands.

We shall work this problem in the frequency domain as well as the time domain to clarify the basic concepts of DSB-SC modulation. In the frequency-domain approach, we work with the signal spectra. The spectrum of the baseband signal $m(t) = \cos \omega_m t$ is given by

$$M(\omega) = \pi [\delta(\omega - \omega_m) + \delta(\omega + \omega_m)]$$

The spectrum consists of two impulses located at $\pm\omega_m$, as depicted in Fig. 7.36a. The DSB-SC (modulated) spectrum, as indicated by Eq. (7.70), is the baseband spectrum in Fig. 7.36a

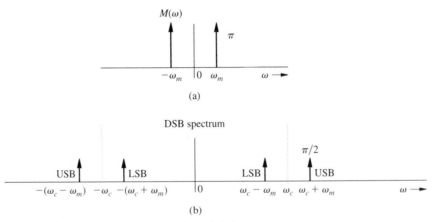

Figure 7.36 An example of DSB-SC modulation.

shifted to the right and the left by ω_c (times 0.5), as depicted in Fig. 7.36b. This spectrum consists of impulses at $\pm(\omega_c - \omega_m)$ and $\pm(\omega_c + \omega_m)$. The spectrum beyond ω_c is the upper sideband (USB), and the one below ω_c is the lower sideband (LSB). Observe that the DSB-SC spectrum does not have as a component the carrier frequency ω_c. This is why the term *double-sideband, suppressed carrier* (DSB-SC) is used for this type of modulation.

In the time-domain approach, we work directly with signals in the time domain. For the baseband signal $m(t) = \cos \omega_m t$, the DSB-SC signal $\varphi_{\text{DSB-SC}}(t)$ is

$$
\begin{aligned}
\varphi_{\text{DSB-SC}}(t) &= m(t) \cos \omega_c t \\
&= \cos \omega_m t \, \cos \omega_c t \\
&= \tfrac{1}{2}[\cos (\omega_c + \omega_m)t + \cos (\omega_c - \omega_m)t]
\end{aligned}
\tag{7.71}
$$

This result shows that when the baseband (message) signal is a single sinusoid of frequency ω_m, the modulated signal consists of two sinusoids: the component of frequency $\omega_c + \omega_m$ (the upper sideband), and the component of frequency $\omega_c - \omega_m$ (the lower sideband). Figure 7.36b illustrates precisely the spectrum of $\varphi_{\text{DSB-SC}}(t)$. Thus, each component of frequency ω_m in the modulating signal results into two components of frequencies $\omega_c + \omega_m$ and $\omega_c - \omega_m$ in the modulated signal. This being a DSB-SC (suppressed carrier) modulation, there is no component of the carrier frequency ω_c on the right-hand side of Eq. (7.71).[†]

DEMODULATION OF DSB-SC SIGNALS

The DSB-SC modulation translates or shifts the frequency spectrum to the left and the right by ω_c (i.e., at $+\omega_c$ and $-\omega_c$), as seen from Eq. (7.70). To recover the original signal $m(t)$ from the modulated signal, we must retranslate the spectrum to its original position. The process of recovering the signal from the modulated signal (retranslating the spectrum to its original position) is referred to as *demodulation*, or *detection*. Observe that if the modulated signal spectrum in Fig. 7.35c is shifted to the left and to the right by ω_c (and halved), we obtain the spectrum illustrated in Fig. 7.37b, which contains the desired baseband spectrum in addition to an unwanted spectrum at $\pm 2\omega_c$. The latter can be suppressed by a lowpass filter. Thus, demodulation, which is almost identical to modulation, consists of multiplication of the incoming modulated signal $m(t) \cos \omega_c t$ by a carrier $\cos \omega_c t$ followed by a lowpass filter, as depicted in Fig. 7.37a. We can verify this conclusion directly in the time domain by observing that the signal $e(t)$ in Fig. 7.37a is

$$
\begin{aligned}
e(t) &= m(t) \cos^2 \omega_c t \\
&= \tfrac{1}{2}[m(t) + m(t) \cos 2\omega_c t]
\end{aligned}
\tag{7.72a}
$$

[†]The term *suppressed carrier* does not necessarily mean absence of the spectrum at the carrier frequency. "Suppressed carrier" merely implies that there is no discrete component of the carrier frequency. Since no discrete component exists, the DSB-SC spectrum does not have impulses at $\pm\omega_c$, which further implies that the modulated signal $m(t) \cos \omega_c t$ does not contain a term of the form $k \cos \omega_c t$ [assuming that $m(t)$ has a zero mean value].

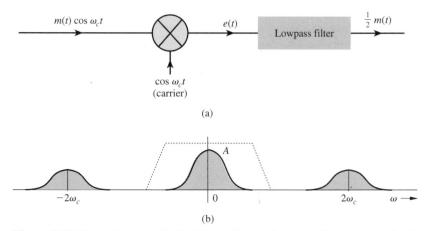

Figure 7.37 Demodulation of DSB-SC: **(a)** demodulator and **(b)** spectrum of $e(t)$.

Therefore, the Fourier transform of the signal $e(t)$ is

$$E(\omega) = \tfrac{1}{2}M(\omega) + \tfrac{1}{4}[M(\omega + 2\omega_c) + M(\omega - 2\omega_c)] \tag{7.72b}$$

Hence, $e(t)$ consists of two components $(1/2)m(t)$ and $(1/2)m(t)\cos 2\omega_c t$, with their spectra, as illustrated in Fig. 7.37b. The spectrum of the second component, being a modulated signal with carrier frequency $2\omega_c$, is centered at $\pm 2\omega_c$. Hence, this component is suppressed by the lowpass filter in Fig. 7.37a. The desired component $(1/2)M(\omega)$, being a lowpass spectrum (centered at $\omega = 0$), passes through the filter unharmed, resulting in the output $(1/2)m(t)$.

A possible form of lowpass filter characteristics is depicted (dotted) in Fig. 7.37b. In this method of recovering the baseband signal, called *synchronous detection,* or *coherent detection,* we use a carrier of exactly the same frequency (and phase) as the carrier used for modulation. Thus, for demodulation, we need to generate a local carrier at the receiver in frequency and phase coherence (synchronism) with the carrier used at the modulator. We shall demonstrate in Example 7.22 that both phase and frequency synchronism are extremely critical.

EXAMPLE 7.22

Discuss the effect of lack of frequency and phase coherence (synchronism) between the carriers at the modulator (transmitter) and the demodulator (receiver) in DSB-SC.

Let the modulator carrier be $\cos \omega_c t$ (Fig. 7.35a). For the demodulator in Fig. 7.37a, we shall consider two cases: with carrier $\cos(\omega_c t + \theta)$ (phase error of θ) and with carrier $\cos(\omega_c + \Delta\omega)t$ (frequency error $\Delta\omega$).

(a) With the demodulator carrier $\cos{(\omega_c t + \theta)}$ (instead of $\cos{\omega_c t}$) in Fig. 7.37a, the multiplier output is $e(t) = m(t)\cos{\omega_c t}\cos{(\omega_c t + \theta)}$ instead of $m(t)\cos^2{\omega_c t}$. From the trigonometric identity, we obtain

$$e(t) = m(t)\cos{\omega_c t}\ \cos{(\omega_c t + \theta)}$$

$$= \tfrac{1}{2}m(t)[\cos{\theta} + \cos{(2\omega_c t + \theta)}]$$

The spectrum of the component $(1/2)m(t)\cos{(2\omega_c t + \theta)}$ is centered at $\pm 2\omega_c$. Consequently, it will be filtered out by the lowpass filter at the output. The component $(1/2)m(t)\cos{\theta}$ is the signal $m(t)$ multiplied by a constant $(1/2)\cos{\theta}$. The spectrum of this component is centered at $\omega = 0$ (lowpass spectrum) and will pass through the lowpass filter at the output, yielding the output $(1/2)m(t)\cos{\theta}$.

If θ is constant, the phase asynchronism merely yields an output that is attenuated (by a factor $\cos{\theta}$). Unfortunately, in practice, θ is often the phase difference between the carriers generated by two distant generators and varies randomly with time. This variation would result in an output whose gain varies randomly with time.

(b) In the case of frequency error, the demodulator carrier is $\cos{(\omega_c + \Delta\omega)t}$. This situation is very similar to the phase error case in part (a) with θ replaced by $(\Delta\omega)t$. Following the analysis in part (a), we can express the demodulator product $e(t)$ as

$$e(t) = m(t)\cos{\omega_c t}\ \cos{(\omega_c + \Delta\omega)t}$$

$$= \tfrac{1}{2}m(t)[\cos{(\Delta\omega)t} + \cos{(2\omega_c + \Delta\omega)t}]$$

The spectrum of the component $(1/2)m(t)\cos{(2\omega_c + \Delta\omega)t}$ is centered at $\pm(2\omega_c + \Delta\omega)$. Consequently, this component will be filtered out by the lowpass filter at the output. The component $(1/2)m(t)\cos{(\Delta\omega)t}$ is the signal $m(t)$ multiplied by a low-frequency carrier of frequency $\Delta\omega$. The spectrum of this component is centered at $\pm\Delta\omega$. In practice, the frequency error $(\Delta\omega)$ is usually very small. Hence, the signal $(1/2)m(t)\cos{(\Delta\omega)t}$ (whose spectrum is centered at $\pm\Delta\omega$) is a lowpass signal and passes through the lowpass filter at the output, resulting in the output $(1/2)m(t)\cos{(\Delta\omega)t}$. The output is the desired signal $m(t)$ multiplied by a very-low-frequency sinusoid $\cos{(\Delta\omega)t}$. The output in this case is not merely an attenuated replica of the desired signal $m(t)$, but represents $m(t)$ multiplied by a time-varying gain $\cos{(\Delta\omega)t}$. If, for instance, the transmitter and the receiver carrier frequencies differ just by 1 Hz, the output will be the desired signal $m(t)$ multiplied by a time-varying signal whose gain goes from the maximum to 0 every half-second. This is like a restless child fiddling with the volume control knob of a receiver, going from maximum volume to zero volume every half-second. This kind of distortion (called the *beat effect*) is beyond repair.

7.7-2 Amplitude Modulation (AM)

For the suppressed-carrier scheme just discussed, a receiver must generate a carrier in frequency and phase synchronism with the carrier at a transmitter that may be located hundreds or thousands of miles away. This situation calls for a sophisticated receiver, which could be quite costly. The other alternative is for the transmitter to transmit a carrier $A \cos \omega_c t$ [along with the modulated signal $m(t) \cos \omega_c t$] so that there is no need to generate a carrier at the receiver. In this case the transmitter needs to transmit much larger power, a rather expensive procedure. In point-to-point communications, where there is one transmitter for each receiver, substantial complexity in the receiver system can be justified, provided there is a large enough saving in expensive high-power transmitting equipment. On the other hand, for a broadcast system with a multitude of receivers for each transmitter, it is more economical to have one expensive high-power transmitter and simpler, less expensive receivers. The second option (transmitting a carrier along with the modulated signal) is the obvious choice in this case. This is amplitude modulation (AM), in which the transmitted signal $\varphi_{AM}(t)$ is given by

$$\varphi_{AM}(t) = A \cos \omega_c t + m(t) \cos \omega_c t \tag{7.73a}$$

$$= [A + m(t)] \cos \omega_c t \tag{7.73b}$$

Recall that the DSB-SC signal is $m(t) \cos \omega_c t$. From Eq. (7.73b) it follows that the AM signal is identical to the DSB-SC signal with $A + m(t)$ as the modulating signal [instead of $m(t)$]. Therefore, to sketch $\varphi_{AM}(t)$, we sketch $A + m(t)$ and $-[A + m(t)]$ as the envelopes and fill in between with the sinusoid of the carrier frequency. Two cases are considered in Fig. 7.38. In the first case, A is large enough so that $A + m(t) \geq 0$ (is nonnegative) for all values of t. In the second case, A is not large enough to satisfy this condition. In the first case, the envelope (Fig. 7.38d) has the same shape as $m(t)$ (although riding on a dc of magnitude A). In the second case the envelope shape is not $m(t)$, for some parts get rectified (Fig. 7.38e). Thus, we can detect the desired signal $m(t)$ by detecting the envelope in the first case. In the second case, such a detection is not possible. We shall see that envelope detection is an extremely simple and inexpensive operation, which does not require generation of a local carrier for the demodulation. But as just noted, the envelope of AM has the information about $m(t)$ only if the AM signal $[A + m(t)] \cos \omega_c t$ satisfies the condition $A + m(t) > 0$ for all t. Thus, the condition for envelope detection of an AM signal is

$$A + m(t) \geq 0 \qquad \text{for all } t \tag{7.74}$$

If m_p is the peak amplitude (positive or negative) of $m(t)$ then condition (7.74) is equivalent to[†]

$$A \geq m_p \tag{7.75}$$

Thus the minimum carrier amplitude required for the viability of envelope detection is m_p. This point is clearly illustrated in Fig. 7.38.

[†] In case the negative and the positive peak amplitudes are not identical, m_p in condition (7.75) is the absolute negative peak amplitude.

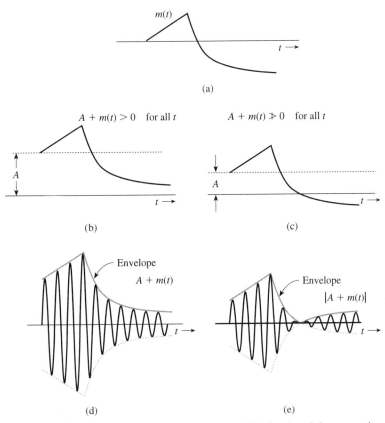

Figure 7.38 An AM signal (**a**) for two values of A (**b, c**) and the respective envelopes (**d, e**).

We define the *modulation index* μ as

$$\mu = \frac{m_p}{A} \tag{7.76}$$

where A is the carrier amplitude. Note that m_p is a constant of the signal $m(t)$. Because $A \geq m_p$ and because there is no upper bound on A, it follows that

$$0 \leq \mu \leq 1 \tag{7.77}$$

as the required condition for the viability of demodulation of AM by an envelope detector.

When $A < m_p$, Eq. (7.76) shows that $\mu > 1$ (overmodulation, shown in Fig. 7.38e). In this case, the option of envelope detection is no longer viable. We then need to use synchronous demodulation. Note that synchronous demodulation can be used for any value of μ (see Prob. 7.7-6). The envelope detector, which is considerably simpler and less expensive than the synchronous detector, can be used only when $\mu \leq 1$.

EXAMPLE 7.23

Sketch $\varphi_{AM}(t)$ for modulation indices of $\mu = 0.5$ (50% modulation) and $\mu = 1$ (100% modulation), when $m(t) = B \cos \omega_m t$. This case is referred to as *tone modulation* because the modulating signal is a pure sinusoid (or tone).

In this case, $m_p = B$ and the modulation index according to Eq. (7.76) is

$$\mu = \frac{B}{A}$$

Hence, $B = \mu A$ and

$$m(t) = B \cos \omega_m t = \mu A \cos \omega_m t$$

Therefore

$$\varphi_{AM}(t) = [A + m(t)] \cos \omega_c t = A[1 + \mu \cos \omega_m t] \cos \omega_c t \qquad (7.78)$$

The modulated signals corresponding to $\mu = 0.5$ and $\mu = 1$ appear in Fig. 7.39a and 7.39b, respectively.

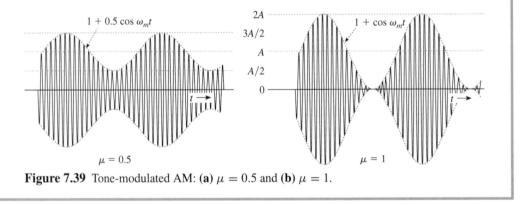

Figure 7.39 Tone-modulated AM: **(a)** $\mu = 0.5$ and **(b)** $\mu = 1$.

DEMODULATION OF AM: THE ENVELOPE DETECTOR

The AM signal can be demodulated coherently by a locally generated carrier (see Prob. 7.7-6). Since, however, coherent, or synchronous, demodulation of AM (with $\mu \leq 1$) will defeat the very purpose of AM, it is rarely used in practice. We shall consider here one of the noncoherent methods of AM demodulation, *envelope detection.*[†]

In an envelope detector, the output of the detector follows the envelope of the (modulated) input signal. The circuit illustrated in Fig. 7.40a functions as an envelope detector. During the

---

[†]There are also other methods of noncoherent detection. The rectifier detector consists of a rectifier followed by a lowpass filter. This method is also simple and almost as inexpensive as the envelope detector.[4] The nonlinear detector, although simple and inexpensive, results in a distorted output.

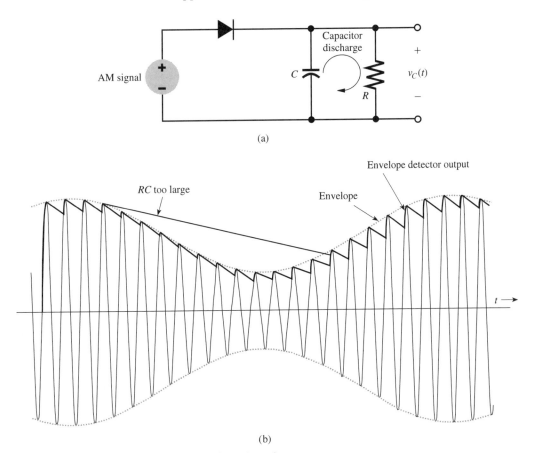

(a)

(b)

Figure 7.40 Demodulation by means of envelope detector.

positive cycle of the input signal, the diode conducts and the capacitor C charges up to the peak voltage of the input signal (Fig. 7.40b). As the input signal falls below this peak value, the diode is cut off, because the capacitor voltage (which is very nearly the peak voltage) is greater than the input signal voltage, a circumstance causing the diode to open. The capacitor now discharges through the resistor R at a slow rate (with a time constant RC). During the next positive cycle, the same drama repeats. When the input signal becomes greater than the capacitor voltage, the diode conducts again. The capacitor again charges to the peak value of this (new) cycle. As the input voltage falls below the new peak value, the diode cuts off again and the capacitor discharges slowly during the cutoff period, a process that changes the capacitor voltage very slightly.

In this manner, during each positive cycle, the capacitor charges up to the peak voltage of the input signal and then decays slowly until the next positive cycle. Thus, the output voltage $v_C(t)$ follows closely the envelope of the input. The capacitor discharge between positive peaks, however, causes a ripple signal of frequency ω_c in the output. This ripple can be reduced by increasing the time constant RC so that the capacitor discharges very little between the positive peaks ($RC \gg 1/\omega_c$). Making RC too large, however, would make it impossible for the capacitor voltage to follow the envelope (see Fig. 7.40b). Thus, RC should be large in comparison to $1/\omega_c$

but small in comparison to $1/2\pi B$, where B is the highest frequency in $m(t)$. Incidentally, these two conditions also require that $\omega_c \gg 2\pi B$, a condition necessary for a well-defined envelope.

The envelope-detector output $v_C(t)$ is $A + m(t)$ plus a ripple of frequency ω_c. The dc term A can be blocked out by a capacitor or a simple RC highpass filter. The ripple is reduced further by another (lowpass) RC filter. In the case of audio signals, the speakers also act as lowpass filters, which further enhances suppression of the high-frequency ripple.

7.7-3 Single-Sideband Modulation (SSB)

Now consider the baseband spectrum $M(\omega)$ (Fig. 7.41a) and the spectrum of the DSB-SC modulated signal $m(t) \cos \omega_c t$ (Fig. 7.41b). The DSB spectrum in Fig. 7.41b has two sidebands: the upper and the lower (USB and LSB), both containing complete information on $M(\omega)$

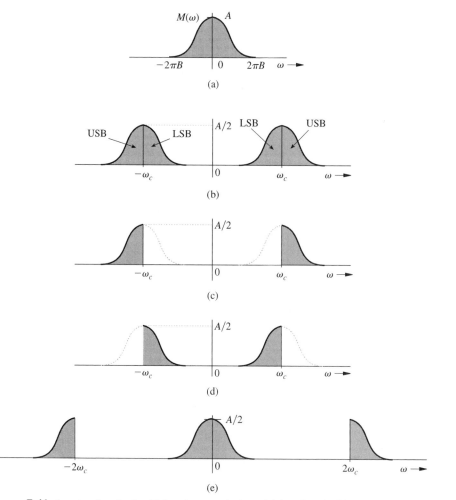

Figure 7.41 Spectra for single-sideband transmission: **(a)** baseband, **(b)** DSB, **(c)** USB, **(d)** LSB, and **(e)** synchronously demodulated signal.

[see Eqs. (7.11)]. Clearly, it is redundant to transmit both sidebands, a process that requires twice the bandwidth of the baseband signal. A scheme in which only one sideband is transmitted is known as *single-sideband (SSB) transmission,* which requires only half the bandwidth of the DSB signal. Thus, we transmit only the upper sidebands (Fig. 7.41c) or only the lower sidebands (Fig. 7.41d).

An SSB signal can be coherently (synchronously) demodulated. For example, multiplication of a USB signal (Fig. 7.41c) by $2 \cos \omega_c t$ shifts its spectrum to the left and to the right by ω_c, yielding the spectrum in Fig. 7.41e. Lowpass filtering of this signal yields the desired baseband signal. The case is similar with an LSB signal. Hence, demodulation of SSB signals is identical to that of DSB-SC signals, and the synchronous demodulator in Fig. 7.37a can demodulate SSB signals. Note that we are talking of SSB signals without an additional carrier. Hence, they are suppressed-carrier signals (SSB-SC).

EXAMPLE 7.24

Find the USB (upper sideband) and LSB (lower sideband) signals when $m(t) = \cos \omega_m t$. Sketch their spectra, and show that these SSB signals can be demodulated using the synchronous demodulator in Fig. 7.37a.

The DSB-SC signal for this case is

$$\varphi_{\text{DSB-SC}}(t) = m(t) \cos \omega_c t$$

$$= \cos \omega_m t \cos \omega_c t$$

$$= \tfrac{1}{2}[\cos (\omega_c - \omega_m)t + \cos (\omega_c + \omega_m)t] \tag{7.79}$$

As pointed out in Example 7.21, the terms $(1/2) \cos (\omega_c + \omega_m)t$ and $(1/2) \cos (\omega_c - \omega_m)t$ represent the upper and lower sidebands, respectively. The spectra of the upper and lower sidebands are given in Fig. 7.42a and 7.42b. Observe that these spectra can be obtained from the DSB-SC spectrum in Fig. 7.36b by using a proper filter to suppress the undesired sidebands. For instance, the USB signal in Fig. 7.42a can be obtained by passing the DSB-SC signal (Fig. 7.36b) through a highpass filter of cutoff frequency ω_c. Similarly, the LSB signal in Fig. 7.42b can be obtained by passing the DSB-SC signal through a lowpass filter of cutoff frequency ω_c.

If we apply the LSB signal $(1/2) \cos (\omega_c - \omega_m)t$ to the synchronous demodulator in Fig. 7.37a, the multiplier output is

$$e(t) = \tfrac{1}{2} \cos (\omega_c - \omega_m)t \cos \omega_c t$$

$$= \tfrac{1}{4}[\cos \omega_m t + \cos (2\omega_c - \omega_m)t]$$

The term $(1/4) \cos (2\omega_c - \omega_m)t$ is suppressed by the lowpass filter, producing the desired output $(1/4) \cos \omega_m t$ [which is $m(t)/4$]. The spectrum of this term is $\pi[\delta(\omega + \omega_m) + \delta(\omega - \omega_m)]/4$, as depicted in Fig. 7.42c. In the same way we can show that the USB signal can be demodulated by the synchronous demodulator.

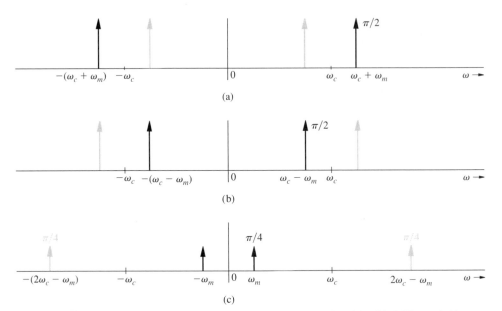

Figure 7.42 Single-sideband spectra for $m(t) = \cos \omega_m t$: **(a)** USB, **(b)** LSB, and **(c)** synchronously demodulated LSB signal.

In the frequency domain, demodulation (multiplication by $\cos \omega_c t$) amounts to shifting the LSB spectrum (Fig. 7.42b) to the left and the right by ω_c (times 0.5) and then suppressing the high frequency, as illustrated in Fig. 7.42c. The resulting spectrum represents the desired signal $(1/4)m(t)$.

GENERATION OF SSB SIGNALS

Two methods are commonly used to generate SSB signals. The *selective-filtering method* uses sharp cutoff filters to eliminate the undesired sideband, and the second method uses phase-shifting networks[4] to achieve the same goal.[†] We shall consider here only the first method.

Selective filtering is the most commonly used method of generating SSB signals. In this method, a DSB-SC signal is passed through a sharp cutoff filter to eliminate the undesired sideband.

To obtain the USB, the filter should pass all components above ω_c unattenuated and completely suppress all components below ω_c. Such an operation requires an ideal filter, which is unrealizable. It can, however, be realized closely if there is some separation between the passband and the stopband. Fortunately, the voice signal provides this condition, because its spectrum shows little power content at the origin (Fig. 7.43). Moreover, articulation tests show that for speech signals, frequency components below 300 Hz are not important. In other words, we may

---

[†]Yet another method, known as Weaver's method, is also used to generate SSB signals.

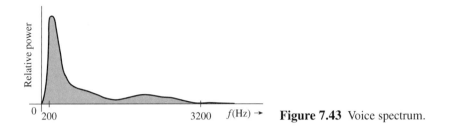

Figure 7.43 Voice spectrum.

suppress all speech components below 300 Hz without appreciably affecting intelligibility.[†] Thus, filtering of the unwanted sideband becomes relatively easy for speech signals because we have a 600 Hz transition region around the cutoff frequency ω_c. For some signals, which have considerable power at low frequencies (around $\omega = 0$), SSB techniques cause considerable distortion. Such is the case with video signals. Consequently, for video signals, instead of SSB, we use another technique, the *vestigial sideband (VSB)*, which is a compromise between SSB and DSB. It inherits the advantages of SSB and DSB but avoids their disadvantages at a cost of slightly increased bandwidth. VSB signals are relatively easy to generate, and their bandwidth is only slightly (typically 25%) greater than that of the SSB signals. In VSB signals, instead of rejecting one sideband completely (as in SSB), we accept a gradual cutoff from one sideband[4].

7.7-4 Frequency-Division Multiplexing

Signal multiplexing allows transmission of several signals on the same channel. Later, in Chapter 8 (Section 8.2-2), we shall discuss time-division multiplexing (TDM), where several signals time-share the same channel, such as a cable or an optical fiber. In frequency-division multiplexing (FDM), the use of modulation, as illustrated in Fig. 7.44, makes several signals share the band of the same channel. Each signal is modulated by a different carrier frequency. The various carriers are adequately separated to avoid overlap (or interference) between the spectra of various modulated signals. These carriers are referred to as *subcarriers*. Each signal may use a different kind of modulation, for example, DSB-SC, AM, SSB-SC, VSB-SC, or even other forms of modulation, not discussed here [such as FM (frequency modulation) or PM (phase modulation)]. The modulated-signal spectra may be separated by a small guard band to avoid interference and to facilitate signal separation at the receiver.

When all the modulated spectra are added, we have a composite signal that may be considered to be a new baseband signal. Sometimes, this composite baseband signal may be used to further modulate a high-frequency (radio frequency, or RF) carrier for the purpose of transmission.

At the receiver, the incoming signal is first demodulated by the RF carrier to retrieve the composite baseband, which is then bandpass-filtered to separate the modulated signals. Then each modulated signal is individually demodulated by an appropriate subcarrier to obtain all the basic baseband signals.

[†]Similarly, suppression of components of a speech signal above 3500 Hz causes no appreciable change in intelligibility.

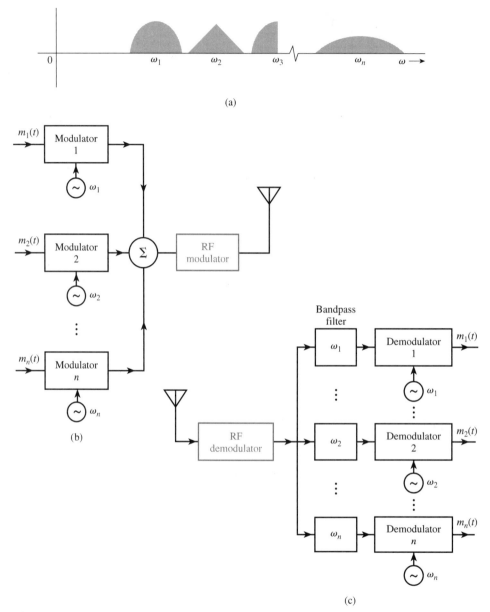

(a)

(b)

(c)

Figure 7.44 Frequency-division multiplexing: **(a)** FDM spectrum **(b)** transmitter, and **(c)** receiver.

7.8 DATA TRUNCATION: WINDOW FUNCTIONS

We often need to truncate data in diverse situations from numerical computations to filter design. For example, if we need to compute numerically the Fourier transform of some signal, say $e^{-t}u(t)$, we will have to truncate the signal $e^{-t}u(t)$ beyond a sufficiently large value of t (typically five time constants and above). The reason is that in numerical computations, we have to deal with data of finite duration. Similarly, the impulse response $h(t)$ of an ideal lowpass filter is noncausal and approaches zero asymptotically as $|t| \to \infty$. For a practical design, we may want to truncate $h(t)$ beyond a sufficiently large value of $|t|$ to make $h(t)$ causal and of finite duration. In signal sampling, to eliminate aliasing, we must use an antialiasing filter to truncate the signal spectrum beyond the half-sampling frequency $\omega_s/2$. Again, we may want to synthesize a periodic signal by adding the first n harmonics and truncating all the higher harmonics. These examples show that data truncation can occur in both time and frequency domains. On the surface, truncation appears to be a simple problem of cutting off the data at a point at which values are deemed to be sufficiently small. Unfortunately, this is not the case. Simple truncation can cause some unsuspected problems.

WINDOW FUNCTIONS

Truncation operation may be regarded as multiplying a signal of a large width by a window function of a smaller (finite) width. Simple truncation amounts to using a *rectangular window* $w_R(t)$ (shown later in Fig. 7.47a) in which we assign unit weight to all the data within the window width ($|t| < T/2$), and assign zero weight to all the data lying outside the window ($|t| > T/2$). It is also possible to use a window in which the weight assigned to the data within the window may not be constant. In a *triangular window* $w_T(t)$, for example, the weight assigned to data decreases linearly over the window width (shown later in Fig. 7.47b).

Consider a signal $x(t)$ and a window function $w(t)$. If $x(t) \Longleftrightarrow X(\omega)$ and $w(t) \Longleftrightarrow W(\omega)$, and if the windowed function $x_w(t) \Longleftrightarrow X_w(\omega)$, then

$$x_w(t) = x(t)w(t) \qquad \text{and} \qquad X_w(\omega) = \frac{1}{2\pi} X(\omega) * W(\omega)$$

According to the width property of convolution, it follows that the width of $X_w(\omega)$ equals the sum of the widths of $X(\omega)$ and $W(\omega)$. Thus, truncation of a signal increases its bandwidth by the amount of bandwidth of $w(t)$. Clearly, the truncation of a signal causes its spectrum to spread (or smear) by the amount of the bandwidth of $w(t)$. Recall that the signal bandwidth is inversely proportional to the signal duration (width). Hence, the wider the window, the smaller its bandwidth, and the smaller the *spectral spreading*. This result is predictable because a wider window means that we are accepting more data (closer approximation), which should cause smaller distortion (smaller spectral spreading). Smaller window width (poorer approximation) causes more spectral spreading (more distortion). In addition, since $W(\omega)$ is really not strictly bandlimited and its spectrum $\to 0$ only asymptotically, the spectrum of $X_w(\omega) \to 0$ asymptotically also at the same rate as that of $W(\omega)$, even if the $X(\omega)$ is in fact strictly bandlimited. Thus, windowing causes the spectrum of $X(\omega)$ to spread into the band where it is supposed to be zero. This effect is called *leakage*. The following example clarifies these twin effects, the spectral spreading and the leakage.

Let us consider $x(t) = \cos \omega_0 t$ and a rectangular window $w_R(t) = \text{rect}(t/T)$, illustrated in Fig. 7.45b. The reason for selecting a sinusoid for $x(t)$ is that its spectrum consists of spectral lines of zero width (Fig. 7.45a). Hence, this choice will make the effect of spectral spreading and leakage easily discernible. The spectrum of the truncated signal $x_w(t)$ is the convolution of the two impulses of $X(\omega)$ with the sinc spectrum of the window function. Because the convolution of any function with an impulse is the function itself (shifted at the location of the impulse), the resulting spectrum of the truncated signal is $1/2\pi$ times the two sinc pulses at $\pm\omega_0$, as depicted in Fig. 7.45c (also see Fig. 7.26). Comparison of spectra $X(\omega)$ and $X_w(\omega)$ reveals the effects of

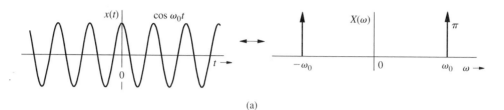

(a)

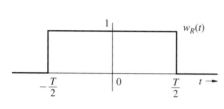

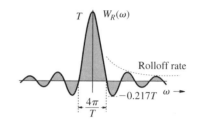

(b)

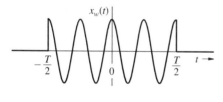

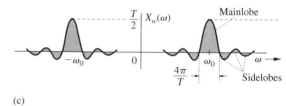

(c)

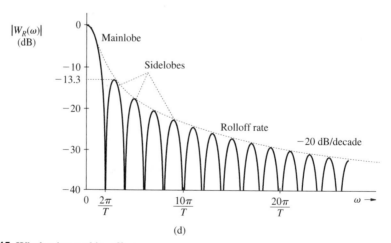

(d)

Figure 7.45 Windowing and its effects.

truncation. These are:

1. The spectral lines of $X(\omega)$ have zero width. But the truncated signal is spread out by $2\pi/T$ about each spectral line. The amount of spread is equal to the width of the mainlobe of the window spectrum. One effect of this *spectral spreading* (or smearing) is that

if $x(t)$ has two spectral components of frequencies differing by less than $4\pi/T$ rad/s ($2/T$ Hz), they will be indistinguishable in the truncated signal. The result is loss of spectral resolution. We would like the spectral spreading [mainlobe width of $W(\omega)$] to be as small as possible.

2. In addition to the mainlobe spreading, the truncated signal has sidelobes, which decay slowly with frequency. The spectrum of $x(t)$ is zero everywhere except at $\pm\omega_0$. On the other hand, the truncated signal spectrum $X_w(\omega)$ is zero nowhere because of the sidelobes. These sidelobes decay asymptotically as $1/\omega$. Thus, the truncation causes spectral *leakage* in the band where the spectrum of the signal $x(t)$ is zero. The peak *sidelobe* magnitude is 0.217 times the mainlobe magnitude (13.3 dB below the peak mainlobe magnitude). Also, the sidelobes decay at a rate $1/\omega$, which is -6 dB/octave (or -20 dB/decade). This is the sidelobes *rolloff rate*. We want smaller sidelobes with a faster rate of decay (high rolloff rate). Figure 7.45d, which plots $|W_R(\omega)|$ as a function of ω, clearly shows the mainlobe and sidelobe features, with the first sidelobe amplitude -13.3 dB below the mainlobe amplitude and the sidelobes decaying at a rate of -6 dB/octave (or -20 dB/decade).

So far, we have discussed the effect on the signal spectrum of signal truncation (truncation in the time domain). Because of the time–frequency duality, the effect of spectral truncation (truncation in frequency domain) on the signal shape is similar.

REMEDIES FOR SIDE EFFECTS OF TRUNCATION

For better results, we must try to minimize the twin side effects of truncations: spectral spreading (mainlobe width) and leakage (sidelobe). Let us consider each of these ills.

1. The spectral spread (mainlobe width) of the truncated signal is equal to the bandwidth of the window function $w(t)$. We know that the signal bandwidth is inversely proportional to the signal width (duration). Hence, to reduce the spectral spread (mainlobe width), we need to increase the window width.

2. To improve the leakage behavior, we must search for the cause of the slow decay of sidelobes. In Chapter 6, we saw that the Fourier spectrum decays as $1/\omega$ for a signal with jump discontinuity, decays as $1/\omega^2$ for a continuous signal whose first derivative is discontinuous, and so on.[†] Smoothness of a signal is measured by the number of continuous derivatives it possesses. The smoother the signal, the faster the decay of its spectrum. Thus, we can achieve a given leakage behavior by selecting a suitably smooth (tapered) window.

3. For a given window width, the remedies for the two effects are incompatible. If we try to improve one, the other deteriorates. For instance, among all the windows of a given width, the rectangular window has the smallest spectral spread (mainlobe width), but its sidelobes have high level and they decay slowly. A tapered (smooth) window of the

---

[†]This result was demonstrated for periodic signals. However, it applies to aperiodic signals also. This is because we showed in the beginning of this chapter that if $x_{T_0}(t)$ is a periodic signal formed by periodic extension of an aperiodic signal $x(t)$, then the spectrum of $x_{T_0}(t)$ is ($1/T_0$ times) the samples of $X(\omega)$. Thus, what is true of the decay rate of the spectrum of $x_{T_0}(t)$ is also true of the rate of decay of $X(\omega)$.

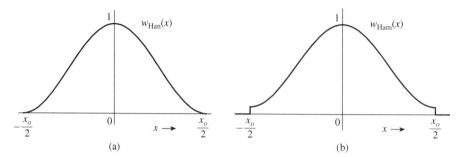

Figure 7.46 **(a)** Hanning and **(b)** Hamming windows.

same width has smaller and faster decaying sidelobes, but it has a wider mainlobe.[†] But we can compensate for the increased mainlobe width by widening the window. Thus, we can remedy both the side effects of truncation by selecting a suitably smooth window of sufficient width.

There are several well-known tapered-window functions, such as Bartlett (triangular), Hanning (von Hann), Hamming, Blackman, and Kaiser, which truncate the data gradually. These windows offer different trade-offs with respect to spectral spread (mainlobe width), the peak sidelobe magnitude, and the leakage rolloff rate as indicated in Table 7.3.[5,6] Observe that all windows are symmetrical about the origin (even functions of t). Because of this feature, $W(\omega)$ is a real function of ω; that is, $\angle W(\omega)$ is either 0 or π. Hence, the phase function of the truncated signal has a minimal amount of distortion.

Figure 7.46 shows two well-known tapered-window functions, the von Hann (or Hanning) window $w_{HAN}(x)$ and the Hamming window $w_{HAM}(x)$. We have intentionally used the independent variable x because windowing can be performed in the time domain as well as in the frequency domain; so x could be t or ω, depending on the application.

There are hundreds of windows, all with different characteristics. But the choice depends on a particular application. The rectangular window has the narrowest mainlobe. The Bartlett (triangle) window (also called the Fejer or Cesaro) is inferior in all respects to the Hanning window. For this reason it is rarely used in practice. Hanning is preferred over Hamming in spectral analysis because it has faster sidelobe decay. For filtering applications, on the other hand, the Hamming window is chosen because it has the smallest sidelobe magnitude for a given mainlobe width. The Hamming window is the most widely used general-purpose window. The Kaiser window, which uses $I_0(\alpha)$, the modified zero-order Bessel function, is more versatile and adjustable. Selecting a proper value of α ($0 \leq \alpha \leq 10$) allows the designer to tailor the window to suit a particular application. The parameter α controls the mainlobe–sidelobe trade-off. When $\alpha = 0$, the Kaiser window is the rectangular window. For $\alpha = 5.4414$, it is the Hamming window, and when $\alpha = 8.885$, it is the Blackman window. As α increases, the mainlobe width increases and the sidelobe level decreases.

[†]A tapered window yields a higher mainlobe width because the effective width of a tapered window is smaller than that of the rectangular window; see Section 2.7-2 [Eq. (2.67)] for the definition of effective width. Therefore, from the reciprocity of the signal width and its bandwidth, it follows that the rectangular window mainlobe is narrower than a tapered window.

TABLE 7.3   Some Window Functions and Their Characteristics

No.	Window $w(t)$	Mainlobe Width	Rolloff Rate (dB/oct)	Peak Sidelobe level (dB)
1	Rectangular: $\text{rect}\left(\dfrac{t}{T}\right)$	$\dfrac{4\pi}{T}$	-6	-13.3
2	Bartlett: $\Delta\left(\dfrac{t}{2T}\right)$	$\dfrac{8\pi}{T}$	-12	-26.5
3	Hanning: $0.5\left[1 + \cos\left(\dfrac{2\pi t}{T}\right)\right]$	$\dfrac{8\pi}{T}$	-18	-31.5
4	Hamming: $0.54 + 0.46\cos\left(\dfrac{2\pi t}{T}\right)$	$\dfrac{8\pi}{T}$	-6	-42.7
5	Blackman: $0.42 + 0.5\cos\left(\dfrac{2\pi t}{T}\right) + 0.08\cos\left(\dfrac{4\pi t}{T}\right)$	$\dfrac{12\pi}{T}$	-18	-58.1
6	Kaiser: $\dfrac{I_0\left[\alpha\sqrt{1 - 4\left(\dfrac{t}{T}\right)^2}\right]}{I_0(\alpha)} \quad 0 \le \alpha \le 10$	$\dfrac{11.2\pi}{T}$	-6	$-59.9\ (\alpha = 8.168)$

7.8-1 Using Windows in Filter Design

We shall design an ideal lowpass filter of bandwidth W rad/s, with frequency response $H(\omega)$, as shown in Fig. 7.47e or 7.47f. For this filter, the impulse response $h(t) = (W/\pi)\text{sinc}\,(Wt)$ (Fig. 7.47c) is noncausal and, therefore, unrealizable. Truncation of $h(t)$ by a suitable window (Fig. 7.47a) makes it realizable, although the resulting filter is now an approximation to the desired ideal filter. [†] We shall use a rectangular window $w_R(t)$ and a triangular (Bartlett) window $w_T(t)$ to truncate $h(t)$, and then examine the resulting filters. The truncated impulse responses $h_R(t) = h(t)w_R(t)$ and $h_T(t) = h(t)w_T(t)$ are depicted in Fig. 7.47d. Hence, the windowed filter frequency response is the convolution of $H(\omega)$ with the Fourier transform of the window, as illustrated in Fig. 7.47e and 7.47f. We make the following observations.

1. The windowed filter spectra show *spectral spreading* at the edges, and instead of a sudden switch there is a gradual transition from the passband to the stopband of the filter. The transition band is smaller ($2\pi/T$ rad/s) for the rectangular case than for the triangular case ($4\pi/T$ rad/s).

2. Although $H(\omega)$ is bandlimited, the windowed filters are not. But the stopband behavior of the triangular case is superior to that of the rectangular case. For the rectangular window, the leakage in the stopband decreases slowly (as $1/\omega$) in comparison to that of the triangular window (as $1/\omega^2$). Moreover, the rectangular case has a higher peak sidelobe amplitude than that of the triangular window.

[†]In addition to truncation, we need to delay the truncated function by $T/2$ to render it causal. However, the time delay only adds a linear phase to the spectrum without changing the amplitude spectrum. Thus, to simplify our discussion, we shall ignore the delay.

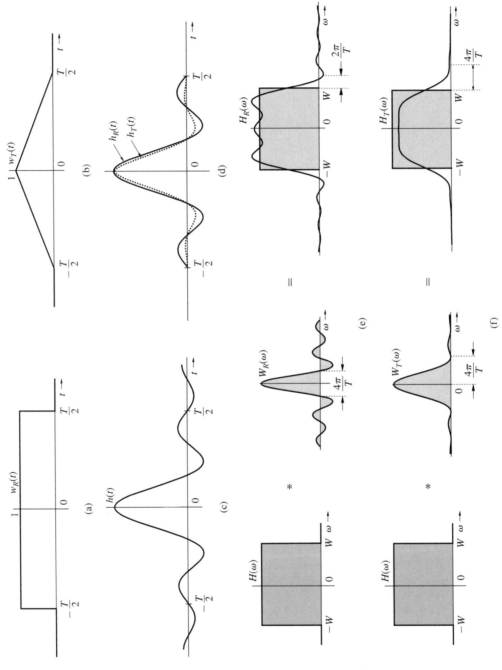

Figure 7.47

7.9 Summary

In Chapter 6, we represented periodic signals as a sum of (everlasting) sinusoids or exponentials (Fourier series). In this chapter we extended this result to aperiodic signals, which are represented by the Fourier integral (instead of the Fourier series). An aperiodic signal $x(t)$ may be regarded as a periodic signal with period $T_0 \to \infty$, so that the Fourier integral is basically a Fourier series with a fundamental frequency approaching zero. Therefore, for aperiodic signals, the Fourier spectra are continuous. This continuity means that a signal is represented as a sum of sinusoids (or exponentials) of all frequencies over a continuous frequency interval. The Fourier transform $X(\omega)$, therefore, is the spectral density (per unit bandwidth in hertz).

An ever present aspect of the Fourier transform is the duality between time and frequency, which also implies duality between the signal $x(t)$ and its transform $X(\omega)$. This duality arises because of near-symmetrical equations for direct and inverse Fourier transforms. The duality principle has far-reaching consequences and yields many valuable insights into signal analysis.

The scaling property of the Fourier transform leads to the conclusion that the signal bandwidth is inversely proportional to signal duration (signal width). Time-shifting of a signal does not change its amplitude spectrum, but it does add a linear phase component to its spectrum. Multiplication of a signal by an exponential $e^{j\omega_0 t}$ shifts the spectrum to the right by ω_0. In practice, spectral shifting is achieved by multiplying a signal by a sinusoid such as $\cos \omega_0 t$ (rather than the exponential $e^{j\omega_0 t}$). This process is known as amplitude modulation. Multiplication of two signals results in convolution of their spectra, whereas convolution of two signals results in multiplication of their spectra.

For an LTIC system with the frequency response $H(\omega)$, the input and output spectra $X(\omega)$ and $Y(\omega)$ are related by the equation $Y(\omega) = X(\omega)H(\omega)$. This is valid only for asymptotically stable systems. It also applies to marginally stable systems if the input does not contain a finite amplitude sinusoid of the natural frequency of the system. For asymptotically unstable systems, the frequency response $H(\omega)$ does not exist. For distortionless transmission of a signal through an LTIC system, the amplitude response $|H(\omega)|$ of the system must be constant, and the phase response $\angle H(\omega)$ should be a linear function of ω over a band of interest. Ideal filters, which allow distortionless transmission of a certain band of frequencies and suppress all the remaining frequencies, are physically unrealizable (noncausal). In fact, it is impossible to build a physical system with zero gain $[H(\omega) = 0]$ over a finite band of frequencies. Such systems (which include ideal filters) can be realized only with infinite time delay in the response.

The energy of a signal $x(t)$ is equal to $1/2\pi$ times the area under $|X(\omega)^2|$ (Parseval's theorem). The energy contributed by spectral components within a band Δf (in hertz) is given by $|X(\omega)|^2 \Delta f$. Therefore, $|X(\omega)|^2$ is the energy spectral density per unit bandwidth (in hertz).

The process of modulation shifts the signal spectrum to different frequencies. Modulation is used for many reasons: to transmit several messages simultaneously over the same channel for the sake of utilizing channel's high bandwidth, to effectively radiate power over a radio link, to shift a signal spectrum at higher frequencies to overcome the difficulties associated with signal processing at lower frequencies, and to effect the exchange of transmission bandwidth and transmission power required to transmit data at a certain rate. Broadly speaking there are two types of modulation, amplitude and angle modulation. Each class has several subclasses.

In practice, we often need to truncate data. Truncating is like viewing data through a window, which permits only certain portions of the data to be seen and hides (suppresses) the remainder.

Abrupt truncation of data amounts to a rectangular window, which assigns a unit weight to data seen from the window and zero weight to the remaining data. Tapered windows, on the other hand, reduce the weight gradually from 1 to 0. Data truncation can cause some unsuspected problems. For example, in computation of the Fourier transform, windowing (data truncation) causes spectral spreading (spectral smearing) that is characteristic of the window function used. A rectangular window results in the least spreading, but it does so at the cost of a high and oscillatory spectral leakage outside the signal band, which decays slowly as $1/\omega$. In comparison to a rectangular window, tapered windows in general have larger spectral spreading (smearing), but the spectral leakage is smaller and decays faster with frequency. If we try to reduce spectral leakage by using a smoother window, the spectral spreading increases. Fortunately, spectral spreading can be reduced by increasing the window width. Therefore, we can achieve a given combination of spectral spread (transition bandwidth) and leakage characteristics by choosing a suitable tapered window function of a sufficiently long width T.

REFERENCES

1. Churchill, R. V., and J. W. Brown. *Fourier Series and Boundary Value Problems,* 3rd ed. McGraw-Hill, New York, 1978.
2. Bracewell, R. N. *Fourier Transform and Its Applications,* rev. 2nd ed. McGraw-Hill, New York, 1986.
3. Guillemin, E. A. *Theory of Linear Physical Systems.* Wiley, New York, 1963.
4. Lathi, B. P. *Modern Digital and Analog Communication Systems,* 3rd ed. Oxford University Press, New York, 1998.
5. Hamming, R. W. *Digital Filters,* 2nd ed. Prentice-Hall, Englewood Cliffs, NJ, 1983.
6. Harris, F. J. On the use of windows for harmonic analysis with the discrete Fourier transform. *Proceedings of the IEEE,* vol. 66, no. 1, pp. 51–83, January 1978.

MATLAB SESSION 7: FOURIER TRANSFORM TOPICS

MATLAB is useful for investigating a variety of Fourier transform topics. In this session, a rectangular pulse is used to investigate the scaling property, Parseval's theorem, essential bandwidth, and spectral sampling. Kaiser window functions are also investigated.

M7.1 The Sinc Function and the Scaling Property

As shown in Example 7.2, the Fourier transform of $x(t) = \text{rect}(t/\tau)$ is $X(\omega) = \tau \, \text{sinc}(\omega\tau/2)$. To represent $X(\omega)$ in MATLAB, a sinc function is first required.[†]

```
function [y] = MS7P1(x);
% MS7P1.m : MATLAB Session 7, Program 1
% Function M-file computes the sinc function, y = sin(x)/x.

y = ones(size(x)); i = find(x~=0);
y(i) = sin(x(i))./x(i);
```

[†]The signal processing toolbox function `sinc(x)`, which computes $(\sin(\pi x))/\pi x$, also works provided that inputs are scaled by $1/\pi$.

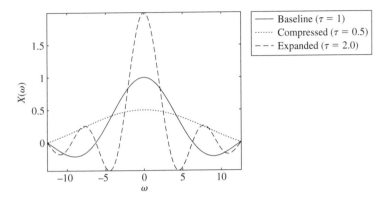

Figure M7.1 Pulse spectra for $\tau = 1.0$, $\tau = 0.5$, and $\tau = 2.0$.

The computational simplicity of $\operatorname{sinc}(x) = \sin(x)/x$ is somewhat deceptive: $\sin(0)/0$ results in a divide-by-zero error. Thus, program MS7P1 assigns $\operatorname{sinc}(0) = 1$ and computes the remaining values according to the definition. Notice that MS7P1 cannot be directly replaced by an inline object. Inline objects prohibit the defining expression from having multiple lines, using other inline objects, and using certain commands such as =, if, or for. M-files, however, can be used to define an inline object. For example, MS7P1 helps represent $X(\omega)$ as an inline object.

```
>> X = inline('tau*MS7P1(omega*tau/2)','omega','tau');
```

Once we have defined $X(\omega)$, it is simple to investigate the effects of scaling the pulse width τ. Consider the three cases $\tau = 1.0$, $\tau = 0.5$, and $\tau = 2.0$.

```
>> omega = linspace(-4*pi,4*pi,200);
>> plot(omega,X(omega,1),'k-',omega,X(omega,0.5),'k:',omega,X(omega,2),'k--');
>> grid; axis tight; xlabel('\omega'); ylabel('X(\omega)');
>> legend('Baseline (\tau = 1)','Compressed (\tau = 0.5)',...
         'Expanded (\tau = 2.0)');
```

Figure M7.1 confirms the reciprocal relationship between signal duration and spectral bandwidth: time compression causes spectral expansion, and time expansion causes spectral compression. Additionally, spectral amplitudes are directly related to signal energy. As a signal is compressed, signal energy and thus spectral magnitude decrease. The opposite effect occurs when the signal is expanded.

M7.2 Parseval's Theorem and Essential Bandwidth

Parseval's theorem concisely relates energy between the time domain and the frequency domain:

$$\int_{-\infty}^{\infty} |x(t)|^2 \, dt = \frac{1}{2\pi} \int_{-\infty}^{\infty} |X(\omega)|^2 \, d\omega$$

This too is easily verified with MATLAB. For example, a unit amplitude pulse $x(t)$ with duration τ has energy $E_x = \tau$. Thus,

$$\int_{-\infty}^{\infty} |X(\omega)|^2 \, d\omega = 2\pi \tau$$

Letting $\tau = 1$, the energy of $X(\omega)$ is computed by using the quad function.

```
>> X_squared = inline('(tau*MS7P1(omega*tau/2)).^2','omega','tau');
>> quad(X_squared,-1e6,1e6,[],[],1)
ans = 6.2817
```

Although not perfect, the result of the numerical integration is consistent with the expected value of $2\pi \approx 6.2832$. For quad, the first argument is the function to be integrated, the next two arguments are the limits of integration, the empty square brackets indicate default values for special options, and the last argument is the secondary input τ for the inline function X_squared. Full format details for quad are available from MATLAB's help facilities.

A more interesting problem involves computing a signal's essential bandwidth. Consider, for example, finding the essential bandwidth W, in radians per second, that contains fraction β of the energy of the square pulse $x(t)$. That is, we want to find W such that

$$\frac{1}{2\pi} \int_{-W}^{W} |X(\omega)|^2 \, d\omega = \beta\tau$$

Program MS7P2 uses a guess-and-check method to find W.

```
function [W,E_W] = MS7P2(tau,beta,tol)
% MS7P2.m : MATLAB Session 7, Program 2
% Function M-file computes essential bandwidth W for square pulse.
% INPUTS:    tau = pulse width
%            beta = fraction of signal energy desired in W
%            tol = tolerance of relative energy error
% OUTPUTS:  W = essential bandwidth [rad/s]
%            E_W = Energy contained in bandwidth W
W = 0; step = 2*pi/tau; % Initial guess and step values
X_squared = inline('(tau*MS7P1(omega*tau/2)).^2','omega','tau');
E = beta*tau; % Desired energy in W
relerr = (E-0)/E; % Initial relative error is 100 percent
while(abs(relerr) > tol),
    if (relerr>0), % W too small
        W=W+step; % Increase W by step
    elseif (relerr<0), % W too large
        step = step/2; W = W-step; % Decrease step size and then W.
    end
    E_W = 1/(2*pi)*quad(X_squared,-W,W,[],[],tau);
    relerr = (E - E_W)/E;
end
```

Although this guess-and-check method is not the most efficient, it is relatively simple to understand: MS7P2 sensibly adjusts W until the relative error is within tolerance. The number of iterations needed to converge to a solution depends on a variety of factors and is not known beforehand. The `while` command is ideal for such situations:

```
while expression,
      statements;
end
```

While the *expression* is true, the *statements* are continually repeated.

To demonstrate MS7P2, consider the 90% essential bandwidth W for a pulse of one second duration. Typing `[W,E_W]=MS7P2(1,0.9,0.001)` returns an essential bandwidth $W = 5.3014$ that contains 89.97% of the energy. Reducing the error tolerance improves the estimate. `MS7P2(1,0.9,0.00005)` returns an essential bandwidth $W = 5.3321$ that contains 90.00% of the energy. These essential bandwidth calculations are consistent with estimates presented after Example 7.2.

M7.3 Spectral Sampling

Consider a signal with finite duration τ. A periodic signal $x_{T_0}(t)$ is constructed by repeating $x(t)$ every T_0 seconds, where $T_0 \geq \tau$. From Eq. (7.4) we can write the Fourier coefficients of $x_{T_0}(t)$ as $D_n = (1/T_0)X(n2\pi/T_0)$. Put another way, the Fourier coefficients are obtained by sampling the spectrum $X(\omega)$.

By using spectral sampling, it is simple to determine the Fourier series coefficients for an arbitrary duty-cycle square-pulse periodic signal. The square pulse $x(t) = \text{rect}(t/\tau)$ has spectrum $X(\omega) = \tau \text{sinc}(\omega\tau/2)$. Thus, the nth Fourier coefficient of the periodic extension $x_{T_0}(t)$ is $D_n = (\tau/T_0)\text{sinc}(n\pi\tau/T_0)$. As in Example 6.4, $\tau = \pi$ and $T_0 = 2\pi$ provide a square-pulse periodic signal. The Fourier coefficients are determined by

```
>> tau = pi; T_0 = 2*pi; n = [0:10];
>> D_n = tau/T_0*MS7P1(n*pi*tau/T_0);
>> stem(n,D_n); xlabel('n'); ylabel('D_n');
>> axis([-0.5 10.5 -0.2 0.55]);
```

The results, shown in Fig. M7.2, agree with Fig. 6.5b. Doubling the period to $T_0 = 4\pi$ effectively doubles the spectral sampling, as shown in Fig. M7.3.

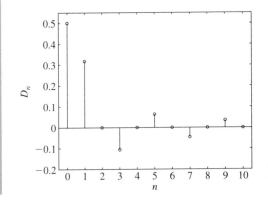

Figure M7.2 Fourier spectra for $\tau = \pi$ and $T_0 = 2\pi$.

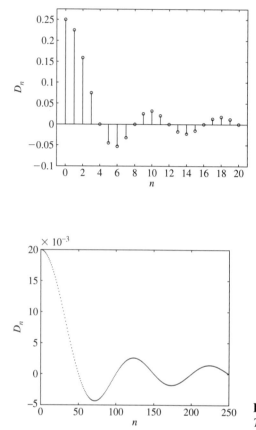

Figure M7.3 Fourier spectra for $\tau = \pi$ and $T_0 = 4\pi$.

Figure M7.4 Fourier spectra for $\tau = \pi$ and $T_0 = 50\pi$.

As T_0 increases, the spectral sampling becomes progressively finer. An evolution of the Fourier series toward the Fourier integral is seen by allowing the period T_0 to become large. Figure M7.4 shows the result for $T_0 = 50\pi$.

If $T_0 = \tau$, the signal x_{T_0} is a constant and the spectrum should concentrate energy at dc. In this case, the sinc function is sampled at the zero crossings and $D_n = 0$ for all n not equal to 0. Only the sample corresponding to $n = 0$ is nonzero, indicating a dc signal, as expected. It is a simple matter to modify the previous code to verify this case.

M7.4 Kaiser Window Functions

A window function is useful only if it can be easily computed and applied to a signal. The Kaiser window, for example, is flexible but appears rather intimidating:

$$
w_K(t) = \begin{cases} \dfrac{I_0\!\left(\alpha\sqrt{1 - 4(t/T)^2}\right)}{I_0(\alpha)} & |t| < T/2 \\[2mm] 0 & \text{otherwise} \end{cases}
$$

Fortunately, the bark of a Kaiser window is worse than its bite! The function $I_0(x)$, a zero-order modified Bessel function of the first kind, can be computed according to

$$I_0(x) = \sum_{k=0}^{\infty} \left(\frac{x^k}{2^k k!} \right)^2$$

or, more simply, by using the MATLAB function `besseli(0,x)`. In fact, MATLAB supports a wide range of Bessel functions, including Bessel functions of the first and second kinds (`besselj` and `bessely`), modified Bessel functions of the first and second kinds (`besseli` and `besselk`), Hankel functions (`besselh`), and Airy functions (`airy`).

Program `MS7P3` computes Kaiser windows at times t by using parameters T and α.

```
function [w_K] = MS7P3(t,T,alpha)
% MS7P3.m : MATLAB Session 7, Program 3
% Function M-file computes a width-T Kaiser window using parameter alpha.
% Alpha can also be a string identifier: 'rectangular', 'Hamming', or
% 'Blackman'.
% INPUTS:   t = independent variable of the window function
%           T = window width
%           alpha = Kaiser parameter or string identifier
% OUTPUTS:  w_K = Kaiser window function
if strncmpi(alpha,'rectangular',1),
    alpha = 0;
elseif strncmpi(alpha,'Hamming',3),
    alpha = 5.4414;
elseif strncmpi(alpha,'Blackman',1),
    alpha = 8.885;
elseif isa(alpha,'char')
    disp('Unrecognized string identifier.'); return
end
w_K = zeros(size(t)); i = find(abs(t)<T/2);
w_K(i) = besseli(0,alpha*sqrt(1-4*t(i).^2/(T^2)))/besseli(0,alpha);
```

Recall that $\alpha = 0$, $\alpha = 5.4414$, and $\alpha = 8.885$ correspond to rectangular, Hamming, and Blackman windows, respectively. `MS7P3` is written to allow these special-case Kaiser windows to be identified by name rather than by α value. While unnecessary, this convenient feature is achieved with the help of the `strncmpi` command.

The `strncmpi(S1,S2,N)` command compares the first `N` characters of strings `S1` and `S2`, ignoring case. More completely, MATLAB has four variants of string comparison: `strcmp`, `strcmpi`, `strncmp`, and `strncmpi`. Comparisons are restricted to the first `N` characters when `n` is present; case is ignored when `i` is present. Thus, `MS7P3` identifies any string `alpha` that starts with the letter `r` or `R` as a rectangular window. To prevent confusion with a Hanning window, the first three characters must match to identify a Hamming window. The `isa(alpha,'char')` command determines whether `alpha` is a character string. MATLAB help documents the many other classes that `isa` can identify. In `MS7P3`, `isa` is used to terminate execution if a string identifier `alpha` has not been recognized as one of the three special cases.

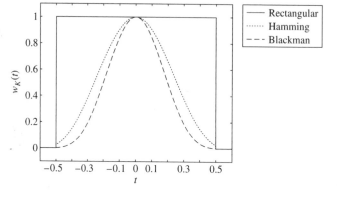

Figure M7.5 Special-case, unit duration Kaiser windows.

Figure M7.5 shows the three special-case, unit duration Kaiser windows generated by

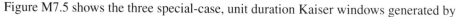

```
>> t = [-0.6:.001:0.6]; T = 1;
>> plot(t,MS7P3(t,T,'r'),'k-',t,MS7P3(t,T,'ham'),'k:',t,MS7P3(t,T,'b'),'k--');
>> axis([-0.6 0.6 -.1 1.1]); xlabel('t'); ylabel('w_K(t)');
>> legend('Rectangular','Hamming','Blackman');
```

PROBLEMS

7.1-1 Show that if $x(t)$ is an even function of t, then

$$X(\omega) = 2 \int_0^\infty x(t) \cos \omega t \, dt$$

and if $x(t)$ is an odd function of t, then

$$X(\omega) = -2j \int_0^\infty x(t) \sin \omega t \, dt$$

Hence, prove that if $x(t)$ is a real and even function of t, then $X(\omega)$ is a real and even function of ω. In addition, if $x(t)$ is a real and odd function of t, then $X(\omega)$ is an imaginary and odd function of ω.

7.1-2 Show that for a real $x(t)$, Eq. (7.8b) can be expressed as

$$x(t) = \frac{1}{\pi} \int_0^\infty |X(\omega)| \cos [\omega t + \angle X(\omega)] \, d\omega$$

This is the trigonometric form of the Fourier integral. Compare this with the compact trigonometric Fourier series.

7.1-3 A signal $x(t)$ can be expressed as the sum of even and odd components (see Section 1.5-2):

$$x(t) = x_e(t) + x_o(t)$$

(a) If $x(t) \Longleftrightarrow X(\omega)$, show that for real $x(t)$,

$$x_e(t) \Longleftrightarrow \text{Re}[X(\omega)]$$

and

$$x_o(t) \Longleftrightarrow j \, \text{Im}[X(\omega)]$$

(b) Verify these results by finding the Fourier transforms of the even and odd components of the following signals: **(i)** $u(t)$ and **(ii)** $e^{-at}u(t)$.

7.1-4 From definition (7.8a), find the Fourier transforms of the signals $x(t)$ in Fig. P7.1-4.

7.1-5 From definition (7.8a), find the Fourier transforms of the signals depicted in Fig. P7.1-5.

7.1-6 Use Eq. (7.8b) to find the inverse Fourier transforms of the spectra in Fig. P7.1-6.

7.1-7 Use Eq. (7.8b) to find the inverse Fourier transforms of the spectra in Fig. P7.1-7.

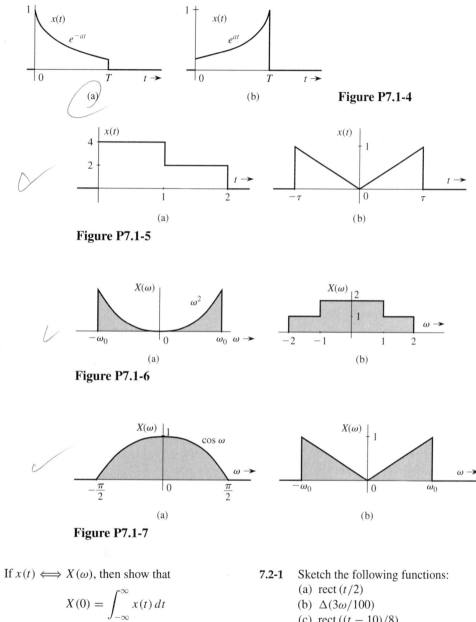

Figure P7.1-4

Figure P7.1-5

Figure P7.1-6

Figure P7.1-7

7.1-8   If $x(t) \iff X(\omega)$, then show that

$$X(0) = \int_{-\infty}^{\infty} x(t)\, dt$$

and

$$x(0) = \frac{1}{2\pi} \int_{-\infty}^{\infty} X(\omega)\, d\omega$$

Also show that

$$\int_{-\infty}^{\infty} \operatorname{sinc}(x)\, dx = \int_{-\infty}^{\infty} \operatorname{sinc}^2(x)\, dx = \pi$$

7.2-1   Sketch the following functions:
(a) $\operatorname{rect}(t/2)$
(b) $\Delta(3\omega/100)$
(c) $\operatorname{rect}((t-10)/8)$
(d) $\operatorname{sinc}(\pi\omega/5)$
(e) $\operatorname{sinc}((\omega/5) - 2\pi)$
(f) $\operatorname{sinc}(t/5)\operatorname{rect}(t/10\pi)$

7.2-2   From definition (7.8a), show that the Fourier transform of $\operatorname{rect}(t-5)$ is $\operatorname{sinc}(\omega/2)e^{-j5\omega}$. Sketch the resulting amplitude and phase spectra.

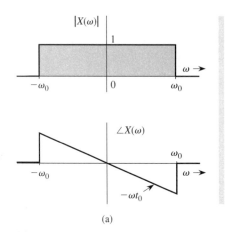

(a)   (b)

Figure P7.2-4

7.2-3 From definition (7.8b), show that the inverse Fourier transform of $\text{rect}\,((\omega - 10)/2\pi)$ is $\text{sinc}\,(\pi t)\,e^{j10t}$.

7.2-4 Find the inverse Fourier transform of $X(\omega)$ for the spectra illustrated in Fig. P7.2-4. [Hint: $X(\omega) = |X(\omega)|e^{j\angle X(\omega)}$. This problem illustrates how different phase spectra (both with the same amplitude spectrum) represent entirely different signals.]

7.2-5 (a) Can you find the Fourier transform of $e^{at}u(t)$ when $a > 1$ by setting $s = j\omega$ in the Laplace transform of $e^{at}u(t)$? Explain.
 (b) Find the Laplace transform of $x(t)$ shown in Fig. P7.2-5. Can you find the Fourier transform of $x(t)$ by setting $s = j\omega$ in its Laplace transform? Explain. Verify your answer by finding the Fourier and the Laplace transforms of $x(t)$.

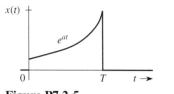

$x(t)$

e^{at}

0 T $t \rightarrow$

Figure P7.2-5

7.3-1 Apply the duality property to the appropriate pair in Table 7.1 to show that
(a) $\frac{1}{2}[\delta(t) + j/\pi t] \Longleftrightarrow u(\omega)$
(b) $\delta(t + T) + \delta(t - T) \Longleftrightarrow 2\cos T\omega$
(c) $\delta(t + T) - \delta(t - T) \Longleftrightarrow 2j\sin T\omega$

7.3-2 The Fourier transform of the triangular pulse $x(t)$ in Fig. P7.3-2 is expressed as

$$X(\omega) = \frac{1}{\omega^2}(e^{j\omega} - j\omega e^{j\omega} - 1)$$

Use this information, and the time-shifting and time-scaling properties, to find the Fourier transforms of the signals $x_i(t)(i = 1, 2, 3, 4, 5)$ shown in Fig. P7.3-2.

7.3-3 Using only the time-shifting property and Table 7.1, find the Fourier transforms of the signals depicted in Fig. P7.3-3.

7.3-4 Use the time-shifting property to show that if $x(t) \Longleftrightarrow X(\omega)$, then

$$x(t + T) + x(t - T) \Longleftrightarrow 2X(\omega)\cos T\omega$$

This is the dual of Eq. (7.41). Use this result and Table 7.1 to find the Fourier transforms of the signals shown in Fig. P7.3-4.

7.3-5 Prove the following results, which are duals of each other:

$$x(t)\sin \omega_0 t \Longleftrightarrow \frac{1}{2j}[X(\omega - \omega_0)$$
$$- X(\omega + \omega_0)]$$

$$\frac{1}{2j}[x(t + T) - x(t - T)] \Longleftrightarrow X(\omega)\sin T\omega$$

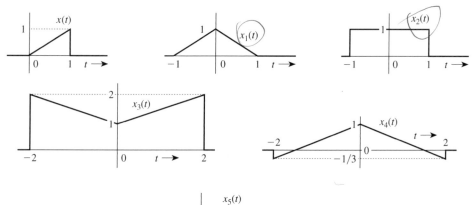

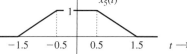

Figure P7.3-2

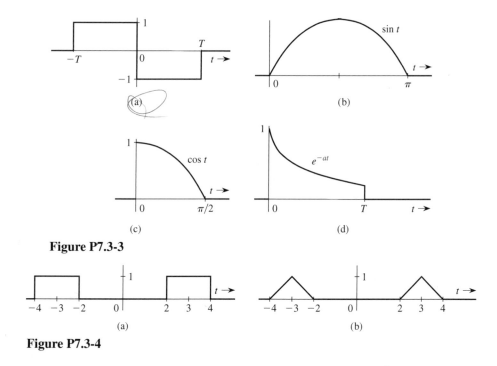

Figure P7.3-3

Figure P7.3-4

Use the latter result and Table 7.1 to find the Fourier transform of the signal in Fig. P7.3-5.

7.3-6 The signals in Fig. P7.3-6 are modulated signals with carrier cos $10t$. Find the Fourier transforms of these signals by using the appropriate properties of the Fourier transform

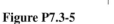

Figure P7.3-5

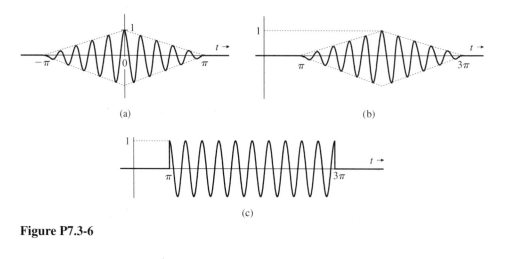

Figure P7.3-6

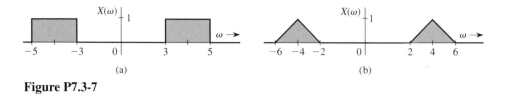

Figure P7.3-7

and Table 7.1. Sketch the amplitude and phase spectra for Fig. P7.3-6a and P7.3-6b.

7.3-7 Use the frequency-shifting property and Table 7.1 to find the inverse Fourier transform of the spectra depicted in Fig. P7.3-7.

7.3-8 Use the time convolution property to prove pairs 2, 4, 13, and 14 in Table 2.1 (assume $\lambda < 0$ in pair 2, λ_1 and $\lambda_2 < 0$ in pair 4, $\lambda_1 < 0$ and $\lambda_2 > 0$ in pair 13, and λ_1 and $\lambda_2 > 0$ in pair 14). These restrictions are placed because of the Fourier transformability issue for the signals concerned. For pair 2, you need to apply the result in Eq. (1.23).

7.3-9 A signal $x(t)$ is bandlimited to B Hz. Show that the signal $x^n(t)$ is bandlimited to nB Hz.

7.3-10 Find the Fourier transform of the signal in Fig. P7.3-3a by three different methods:

(a) By direct integration using the definition (7.8a).

(b) Using only pair 17 (Table 7.1) and the time-shifting property.

(c) Using the time-differentiation and time-shifting properties, along with the fact that $\delta(t) \Longleftrightarrow 1$.

7.3-11 (a) Prove the frequency-differentiation property (dual of the time-differentiation property):

$$-jtx(t) \Longleftrightarrow \frac{d}{d\omega}X(\omega)$$

(b) Use this property and pair 1 (Table 7.1) to determine the Fourier transform of $te^{-at}u(t)$.

7.4-1 For a stable LTIC system with transfer function

$$H(s) = \frac{1}{s+1}$$

find the (zero-state) response if the input $x(t)$ is

(a) $e^{-2t}u(t)$

(b) $e^{-t}u(t)$

(c) $e^{t}u(-t)$

(d) $u(t)$

7.4-2 A stable LTIC system is specified by the frequency response

$$H(\omega) = \frac{-1}{j\omega - 2}$$

Find the impulse response of this system and show that this is a noncausal system. Find the (zero-state) response of this system if the input $x(t)$ is

(a) $e^{-t}u(t)$

(b) $e^t u(-t)$

7.4-3 Signals $x_1(t) = 10^4 \text{rect}\,(10^4 t)$ and $x_2(t) = \delta(t)$ are applied at the inputs of the ideal lowpass filters $H_1(\omega) = \text{rect}\,(\omega/40,000\pi)$ and $H_2(\omega) = \text{rect}\,(\omega/20,000\pi)$ (Fig. P7.4-3). The outputs $y_1(t)$ and $y_2(t)$ of these filters are multiplied to obtain the signal $y(t) = y_1(t)y_2(t)$.

(a) Sketch $X_1(\omega)$ and $X_2(\omega)$.

(b) Sketch $H_1(\omega)$ and $H_2(\omega)$.

(c) Sketch $Y_1(\omega)$ and $Y_2(\omega)$.

(d) Find the bandwidths of $y_1(t)$, $y_2(t)$, and $y(t)$.

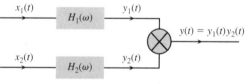

Figure P7.4-3

7.4-4 A lowpass system time constant is often defined as the width of its unit impulse response $h(t)$ (see Section 2.7-2). An input pulse $p(t)$ to this system acts like an impulse of strength equal to the area of $p(t)$ if the width of $p(t)$ is much smaller than the system time constant, and provided $p(t)$ is a lowpass pulse, implying that its spectrum is concentrated at low frequencies. Verify this behavior by considering a system whose unit impulse response is $h(t) = \text{rect}\,(t/10^{-3})$. The input pulse is a triangle pulse $p(t) = \Delta(t/10^{-6})$. Show that the system response to this pulse is very nearly the system response to the input $A\delta(t)$, where A is the area under the pulse $p(t)$.

7.4-5 A lowpass system time constant is often defined as the width of its unit impulse response $h(t)$ (see Section 2.7-2). An input pulse $p(t)$

to this system passes practically without distortion if the width of $p(t)$ is much greater than the system time constant, and provided $p(t)$ is a lowpass pulse, implying that its spectrum is concentrated at low frequencies. Verify this behavior by considering a system whose unit impulse response is $h(t) = \text{rect}\,(t/10^{-3})$. The input pulse is a triangle pulse $p(t) = \Delta(t)$. Show that the system output to this pulse is very nearly $kp(t)$, where k is the system gain to a dc signal, that is, $k = H(0)$.

7.4-6 A causal signal $h(t)$ has a Fourier transform $H(\omega)$. If $R(\omega)$ and $X(\omega)$ are the real and the imaginary parts of $H(\omega)$, that is, $H(\omega) = R(\omega) + jX(\omega)$, then show that

$$R(\omega) = \frac{1}{\pi} \int_{-\infty}^{\infty} \frac{X(\omega)}{\omega - y}\, d\omega$$

and

$$X(\omega) = -\frac{1}{\pi} \int_{-\infty}^{\infty} \frac{R(\omega)}{\omega - y}\, d\omega$$

assuming that $h(t)$ has no impulse at the origin. This pair of integrals defines the *Hilbert transform*. [Hint: Let $h_e(t)$ and $h_o(t)$ be the even and odd components of $h(t)$. Use the results in Prob. 7.1-3. See Fig. 1.24 for the relationship between $h_e(t)$ and $h_o(t)$.]

This problem states one of the important properties of causal systems: that the real and imaginary parts of the frequency response of a causal system are related. If one specifies the real part, the imaginary part cannot be specified independently. The imaginary part is predetermined by the real part, and vice versa. This result also leads to the conclusion that the magnitude and angle of $H(\omega)$ are related provided all the poles and zeros of $H(\omega)$ lie in the LHP.

7.5-1 Consider a filter with the frequency response

$$H(\omega) = e^{-(k\omega^2 + j\omega t_0)}$$

Show that this filter is physically unrealizable by using the time-domain criterion [noncausal $h(t)$] and the frequency–domain (Paley–Wiener) criterion. Can this filter be made approximately realizable by choosing t_0 sufficiently large? Use your own (reasonable) criterion of approximate realizability to determine t_0. [Hint: Use pair 22 in Table 7.1.]

7.5-2 Show that a filter with frequency response

$$H(\omega) = \frac{2(10^5)}{\omega^2 + 10^{10}} e^{-j\omega t_0}$$

is unrealizable. Can this filter be made approximately realizable by choosing a sufficiently large t_0? Use your own (reasonable) criterion of approximate realizability to determine t_0.

7.5-3 Determine whether the filters with the following frequency response $H(\omega)$ are physically realizable. If they are not realizable, can they be realized approximately by allowing a finite time delay in the response?
(a) 10^{-6} sinc $(10^{-6}\omega)$
(b) $10^{-4} \Delta (\omega/40,000\pi)$
(c) $2\pi\, \delta(\omega)$

7.6-1 Show that the energy of a Gaussian pulse

$$x(t) = \frac{1}{\sigma\sqrt{2\pi}} e^{-t^2/2\sigma^2}$$

is $1/(2\sigma\sqrt{\pi})$. Verify this result by using Parseval's theorem to derive the energy E_x from $X(\omega)$. [Hint: See pair 22 in Table 7.1. Use the fact that $\int_{-\infty}^{\infty} e^{-x^2/2}\, dx = \sqrt{2\pi}$.]

7.6-2 Use Parseval's theorem (7.64) to show that

$$\int_{-\infty}^{\infty} \text{sinc}^2 (kx)\, dx = \frac{\pi}{k}$$

7.6-3 A lowpass signal $x(t)$ is applied to a squaring device. The squarer output $x^2(t)$ is applied to a lowpass filter of bandwidth Δf (in hertz) (Fig. P7.6-3). Show that if Δf is very small ($\Delta f \to 0$), then the filter output is a dc signal $y(t) \approx 2E_x\Delta f$. [Hint: If $x^2(t) \iff A(\omega)$, then show that $Y(\omega) \approx [4\pi A(0)\Delta f]\delta(\omega)$ if $\Delta f \to 0$. Now, show that $A(0) = E_x$.]

7.6-4 Generalize Parseval's theorem to show that for real, Fourier-transformable signals $x_1(t)$

and $x_2(t)$

$$\int_{-\infty}^{\infty} x_1(t)x_2(t)\, dt$$

$$= \frac{1}{2\pi} \int_{-\infty}^{\infty} X_1(-\omega)X_2(\omega)\, d\omega$$

$$= \frac{1}{2\pi} \int_{-\infty}^{\infty} X_1(\omega)X_2(-\omega)\, d\omega$$

7.6-5 Show that

$$\int_{-\infty}^{\infty} \text{sinc}\, (Wt - m\pi)\, \text{sinc}\, (Wt - n\pi)\, dt$$

$$= \begin{cases} 0 & m \neq n \\ \dfrac{\pi}{W} & m = n \end{cases}$$

[Hint: Recognize that

$$\text{sinc}\, (Wt - k\pi) = \text{sinc}\left[W\left(t - \frac{k\pi}{W} \right)\right]$$

$$\iff \frac{\pi}{W} \text{rect}\left(\frac{\omega}{2W} \right) e^{-jk\pi\omega/W}$$

Use this fact and the result in Prob. 7.6-4.]

7.6-6 For the signal

$$x(t) = \frac{2a}{t^2 + a^2}$$

determine the essential bandwidth B (in hertz) of $x(t)$ such that the energy contained in the spectral components of $x(t)$ of frequencies below B Hz is 99% of the signal energy E_x.

7.7-1 For each of the following baseband signals (i) $m(t) = \cos 1000t$, (ii) $m(t) = 2\cos 1000t + \cos 2000t$, and (iii) $m(t) = \cos 1000t \cos 3000t$:
(a) Sketch the spectrum of $m(t)$.
(b) Sketch the spectrum of the DSB-SC signal $m(t) \cos 10,000t$.
(c) Identify the upper sideband (USB) and the lower sideband (LSB) spectra.
(d) Identify the frequencies in the baseband, and the corresponding frequencies in

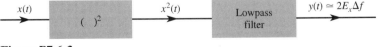

Figure P7.6-3

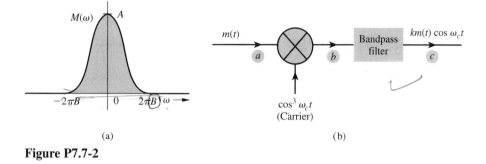

Figure P7.7-2

the DSB-SC, USB, and LSB spectra. Explain the nature of frequency shifting in each case.

7.7-2 You are asked to design a DSB-SC modulator to generate a modulated signal $km(t) \cos \omega_c t$, where $m(t)$ is a signal bandlimited to B Hz (Fig. P7.7-2a). Figure P7.7-2b shows a DSB-SC modulator available in the stockroom. The bandpass filter is tuned to ω_c and has a bandwidth of $2B$ Hz. The carrier generator available generates not $\cos \omega_c t$, but $\cos^3 \omega_c t$.
   (a) Explain whether you would be able to generate the desired signal using only this equipment. If so, what is the value of k?
   (b) Determine the signal spectra at points b and c, and indicate the frequency bands occupied by these spectra.
   (c) What is the minimum usable value of ω_c?
   (d) Would this scheme work if the carrier generator output were $\cos^2 \omega_c t$? Explain.
   (e) Would this scheme work if the carrier generator output were $\cos^n \omega_c t$ for any integer $n \geq 2$?

7.7-3 In practice, the analog multiplication operation is difficult and expensive. For this reason, in amplitude modulators, it is necessary to find some alternative to multiplication of $m(t)$ with $\cos \omega_c t$. Fortunately, for this purpose, we can replace multiplication with switching operation. A similar observation applies to demodulators. In the scheme depicted in Fig. P7.7-3a, the period of the rectangular periodic pulse $x(t)$ shown in Fig. P7.7-3b is $T_0 = 2\pi/\omega_c$. The bandpass filter is centered at $\pm\omega_c$ and has a bandwidth of $2B$ Hz. Note that multiplication by a square periodic pulse $x(t)$ in Fig. P7.7-3b amounts to periodic

on–off switching of $m(t)$, which is bandlimited to B Hz. Such a switching operation is relatively simple and inexpensive.

Show that this scheme can generate an amplitude-modulated signal $k \cos \omega_c t$. Determine the value of k. Show that the same scheme can also be used for demodulation, provided the bandpass filter in Fig. P7.7-3a is replaced by a lowpass (or baseband) filter.

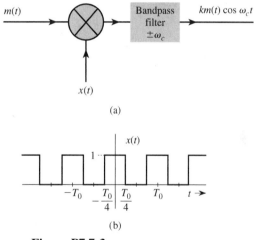

(a)

(b)

Figure P7.7-3

7.7-4 Figure P7.7-4a shows a scheme to transmit two signals $m_1(t)$ and $m_2(t)$ simultaneously on the same channel (without causing spectral interference). Such a scheme, which transmits more than one signal, is known as signal *multiplexing*. In this case, we transmit multiple signals by sharing an available spectral band on the channel, and, hence, this is an example of the *frequency-division* multiplexing. The signal at

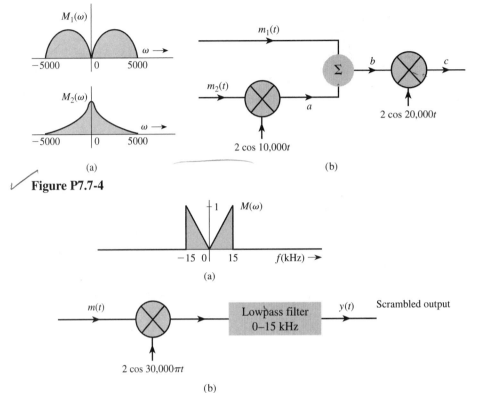

(a)

✓ **Figure P7.7-4**

(b)

Figure P7.7-5

point *b* is the multiplexed signal, which now modulates a carrier of frequency 20,000 rad/s. The modulated signal at point *c* is now transmitted over the channel.

(a) Sketch the spectra at points *a*, *b*, and *c*.

(b) What must be the minimum bandwidth of the channel?

(c) Design a receiver to recover signals $m_1(t)$ and $m_2(t)$ from the modulated signal at point *c*.

7.7-5 The system shown in Fig. P7.7-5 is used for scrambling audio signals. The output $y(t)$ is the scrambled version of the input $m(t)$.

(a) Find the spectrum of the scrambled signal $y(t)$.

(b) Suggest a method of descrambling $y(t)$ to obtain $m(t)$.

A slightly modified version of this scrambler was first used commercially on the 25-mile radio-telephone circuit connecting Los Angeles and Santa Catalina Island.

7.7-6 Figure P7.7-6 presents a scheme for coherent (synchronous) demodulation. Show that this scheme can demodulate the AM signal $[A+m(t)]\cos \omega_c t$ regardless of the value of *A*.

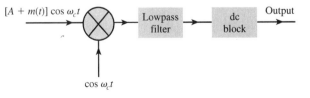

Figure P7.7-6

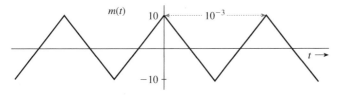

$m(t)$

10

10^{-3}

-10

$t \rightarrow$

Figure P7.7-7

7.7-7 Sketch the AM signal $[A + m(t)] \cos \omega_c t$ for the periodic triangle signal $m(t)$ illustrated in Fig. P7.7-7 corresponding to the following modulation indices:
(a) $\mu = 0.5$
(b) $\mu = 1$
(c) $\mu = 2$
(d) $\mu = \infty$
How do you interpret the case $\mu = \infty$?

7.M-1 Consider the signal $x(t) = e^{-at}u(t)$. Modify MS7P2 to compute the following essential bandwidths.
(a) Setting $a = 1$, determine the essential bandwidth W_1 that contains 95% of the signal energy. Compare this value with the theoretical value presented in Example 7.20.
(b) Setting $a = 2$, determine the essential bandwidth W_2 that contains 90% of the signal energy.
(c) Setting $a = 3$, determine the essential bandwidth W_3 that contains 75% of the signal energy.

7.M-2 A unit amplitude pulse with duration τ is defined as

$$x(t) = \begin{cases} 1 & |t| \leq \tau/2 \\ 0 & \text{otherwise} \end{cases}$$

(a) Determine the duration τ_1 that results in a 95% essential bandwidth of 5 Hz.
(b) Determine the duration τ_2 that results in a 90% essential bandwidth of 10 Hz.
(c) Determine the duration τ_3 that results in a 75% essential bandwidth 20 Hz.

7.M-3 Consider the signal $x(t) = e^{-at}u(t)$.
(a) Determine the decay parameter a_1 that results in a 95% essential bandwidth of 5 Hz.

(b) Determine the decay parameter a_2 that results in a 90% essential bandwidth of 10 Hz.
(c) Determine the decay parameter a_3 that results in a 75% essential bandwidth 20 Hz.

7.M-4 Use MATLAB to determine the 95, 90, and 75% essential bandwidths of a one-second triangle function with a peak amplitude of one. Recall that a triangle function can be constructed by the convolution of two rectangular pulses.

7.M-5 A 1/3 duty-cycle square-pulse T_0-periodic signal $x(t)$ is described as

$$x(t) = \begin{cases} 1 & -T_0/6 \leq t \leq T_0/6 \\ 0 & T_0/t \leq |t| \leq T_0/2 \\ x(t + T_0) & \forall t \end{cases}$$

(a) Use spectral sampling to determine the Fourier series coefficients D_n of $x(t)$ for $T_0 = 2\pi$. Evaluate and plot D_n for ($0 \leq n \leq 10$).
(b) Use spectral sampling to determine the Fourier series coefficients D_n of $x(t)$ for $T_0 = \pi$. Evaluate and plot D_n for ($0 \leq n \leq 10$). How does this result compare with your answer to part a? What can be said about the relation of T_0 to D_n for signal $x(t)$, which has fixed duty cycle of 1/3?

7.M-6 Determine the Fourier transform of a Gaussian pulse defined as $x(t) = e^{-t^2}$. Plot both $x(t)$ and $X(\omega)$. How do the two curves compare? [Hint:

$$\frac{1}{\sqrt{2\pi}} \int_{-\infty}^{\infty} e^{-(t-a)^2/2} dt = 1$$

for any real or imaginary a.]

CHAPTER

8

SAMPLING: THE BRIDGE FROM CONTINUOUS TO DISCRETE

A continuous-time signal can be processed by processing its samples through a discrete-time system. For this purpose, it is important to maintain the signal sampling rate high enough to permit the reconstruction of the original signal from these samples without error (or with an error within a given tolerance). The necessary quantitative framework for this purpose is provided by the sampling theorem derived in Section 8.1.

The sampling theory is the bridge between the continuous-time and the discrete-time worlds. The information inherent in a sampled continuous-time signal is equivalent to that of a discrete-time signal. A sampled continuous-time signal is a sequence of impulses, while a discrete-time signal presents the same information as a sequence of numbers. These are basically two different ways of presenting the same data. Clearly, all the concepts in the analysis of sampled signals apply to discrete-time signals. We should not be surprised to see that the Fourier spectra of the two kinds of signal are also the same (within a multiplicative constant).

8.1 THE SAMPLING THEOREM

We now show that a real signal whose spectrum is bandlimited to B Hz [$X(\omega) = 0$ for $|\omega| > 2\pi B$] can be reconstructed exactly (without any error) from its samples taken uniformly at a rate $f_s > 2B$ samples per second. In other words, the minimum sampling frequency is $f_s = 2B$ Hz.[†]

To prove the sampling theorem, consider a signal $x(t)$ (Fig. 8.1a) whose spectrum is band-limited to B Hz (Fig. 8.1b).[‡] For convenience, spectra are shown as functions of ω as well as of f (hertz). Sampling $x(t)$ at a rate of f_s Hz (f_s samples per second) can be accomplished by multiplying $x(t)$ by an impulse train $\delta_T(t)$(Fig. 8.1c), consisting of unit impulses repeating

---

[†]The theorem stated here (and proved subsequently) applies to lowpass signals. A bandpass signal whose spectrum exists over a frequency band $f_c - (B/2) < |f| < f_c + (B/2)$ has a bandwidth of B Hz. Such a signal is uniquely determined by $2B$ samples per second. In general, the sampling scheme is a bit more complex in this case. It uses two interlaced sampling trains, each at a rate of B samples per second. See, for example, Linden.[1]

[‡]The spectrum $X(\omega)$ in Fig. 8.1b is shown as real, for convenience. However, our arguments are valid for complex $X(\omega)$ as well.

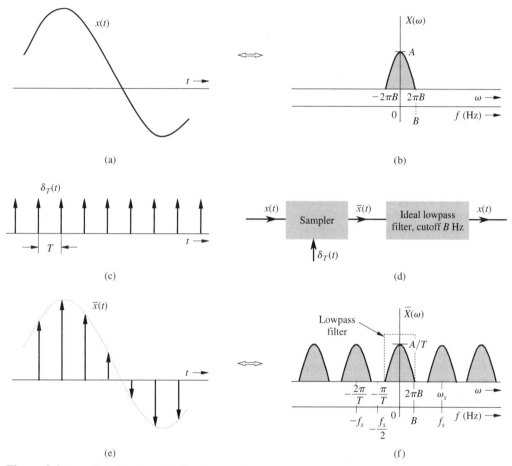

Figure 8.1 Sampled signal and its Fourier spectrum.

periodically every T seconds, where $T = 1/f_s$. The schematic of a sampler is shown in Fig. 8.1d. The resulting sampled signal $\bar{x}(t)$ is shown in Fig. 8.1e. The sampled signal consists of impulses spaced every T seconds (the sampling interval). The nth impulse, located at $t = nT$, has a strength $x(nT)$, the value of $x(t)$ at $t = nT$.

$$\bar{x}(t) = x(t)\delta_T(t) = \sum_n x(nT)\delta(t - nT) \tag{8.1}$$

Because the impulse train $\delta_T(t)$ is a periodic signal of period T, it can be expressed as a trigonometric Fourier series like that already obtained in Example 6.7 [Eq. (6.39)]

$$\delta_T(t) = \frac{1}{T}[1 + 2\cos \omega_s t + 2\cos 2\omega_s t + 2\cos 3\omega_s t + \cdots] \qquad \omega_s = \frac{2\pi}{T} = 2\pi f_s \tag{8.2}$$

Therefore

$$\bar{x}(t) = x(t)\delta_T(t) = \frac{1}{T}[x(t) + 2x(t)\cos \omega_s t + 2x(t)\cos 2\omega_s t + 2x(t)\cos 3\omega_s t + \cdots] \tag{8.3}$$

To find $\overline{X}(\omega)$, the Fourier transform of $\overline{x}(t)$, we take the Fourier transform of the right-hand side of Eq. (8.3), term by term. The transform of the first term in the brackets is $X(\omega)$. The transform of the second term $2x(t)\cos\omega_s t$ is $X(\omega - \omega_s) + X(\omega + \omega_s)$ [see Eq. (7.41)]. This represents spectrum $X(\omega)$ shifted to ω_s and $-\omega_s$. Similarly, the transform of the third term $2x(t)\cos 2\omega_s t$ is $X(\omega - 2\omega_s) + X(\omega + 2\omega_s)$, which represents the spectrum $X(\omega)$ shifted to $2\omega_s$ and $-2\omega_s$, and so on to infinity. This result means that the spectrum $\overline{X}(\omega)$ consists of $X(\omega)$ repeating periodically with period $\omega_s = 2\pi/T$ rad/s, or $f_s = 1/T$ Hz, as depicted in Fig. 8.1f. There is also a constant multiplier $1/T$ in Eq. (8.3). Therefore

$$\overline{X}(\omega) = \frac{1}{T}\sum_{n=-\infty}^{\infty} X(\omega - n\omega_s) \qquad (8.4)$$

If we are to reconstruct $x(t)$ from $\overline{x}(t)$, we should be able to recover $X(\omega)$ from $\overline{X}(\omega)$. This recovery is possible if there is no overlap between successive cycles of $\overline{X}(\omega)$. Figure 8.1f indicates that this requires

$$f_s > 2B \qquad (8.5)$$

Also, the sampling interval $T = 1/f_s$. Therefore

$$T < \frac{1}{2B} \qquad (8.6)$$

Thus, as long as the sampling frequency f_s is greater than twice the signal bandwidth B (in hertz), $\overline{X}(\omega)$ consists of nonoverlapping repetitions of $X(\omega)$. Figure 8.1f shows that the gap between the two adjacent spectral repetitions is $f_s - 2B$ Hz, and $x(t)$ can be recovered from its samples $\overline{x}(t)$ by passing the sampled signal $\overline{x}(t)$ through an ideal lowpass filter having a bandwidth of any value between B and $f_s - B$ Hz. The minimum sampling rate $f_s = 2B$ required to recover $x(t)$ from its samples $\overline{x}(t)$ is called the *Nyquist rate* for $x(t)$, and the corresponding sampling interval $T = 1/2B$ is called the *Nyquist interval* for $x(t)$. Samples of a signal taken at its Nyquist rate are the *Nyquist samples* of that signal.

We are saying that the Nyquist rate $2B$ Hz is the minimum sampling rate required to preserve the information of $x(t)$. This contradicts Eq. (8.5), where we showed that to preserve the information of $x(t)$, the sampling rate f_s needs to be greater than $2B$ Hz. Strictly speaking, Eq. (8.5) is the correct statement. However, if the spectrum $X(\omega)$ contains no impulse or its derivatives at the highest frequency B Hz, then the minimum sampling rate $2B$ Hz is adequate. In practice, it is rare to observe $X(\omega)$ with an impulse or its derivatives at the highest frequency. If the contrary situation were to occur, we should use Eq. (8.5).[†]

[†]An interesting observation is that if the impulse is because of a cosine term, the sampling rate of $2B$ Hz is adequate. However, if the impulse is because of a sine term, then the rate must be greater than $2B$ Hz. This may be seen from the fact that samples of $\sin 2\pi Bt$ using $T = 1/2B$ are all zero because $\sin 2\pi BnT = \sin \pi n = 0$. But, samples of $\cos 2\pi Bt$ are $\cos 2\pi BnT = \cos \pi n = (-1)^n$. We can reconstruct $\cos 2\pi Bt$ from these samples. This peculiar behavior occurs because in the sampled signal spectrum corresponding to the signal $\cos 2\pi Bt$, the impulses, which occur at frequencies $(2n\pm 1)B$ Hz ($n = 0, \pm 1, \pm 2, \ldots$), interact constructively, whereas in the case of $\sin 2\pi Bt$, the impulses, because of their opposite phases ($e^{\pm j\pi/2}$), interact destructively and cancel out in the sampled signal spectrum. Hence, $\sin 2\pi Bt$ cannot be reconstructed from its samples at a rate $2B$ Hz. A similar situation exists for signal $\cos(2\pi Bt + \theta)$, which contains a component of the form $\sin 2\pi Bt$. For this reason, it is advisable to maintain sampling rate above $2B$ Hz if a finite amplitude component of a sinusoid of frequency B Hz is present in the signal.

The sampling theorem proved here uses samples taken at uniform intervals. This condition is not necessary. Samples can be taken arbitrarily at any instants as long as the sampling instants are recorded and there are, on average, $2B$ samples per second.[2] The essence of the sampling theorem was known to mathematicians for a long time in the form of the *interpolation formula* [see later, Eq. (8.11)]. The origin of the sampling theorem was attributed by H. S. Black to Cauchy in 1841. The essential idea of the sampling theorem was rediscovered in the 1920s by Carson, Nyquist, and Hartley.

EXAMPLE 8.1

In this example, we examine the effects of sampling a signal at the Nyquist rate, below the Nyquist rate (undersampling), and above the Nyquist rate (oversampling). Consider a signal $x(t) = \text{sinc}^2(5\pi t)$ (Fig. 8.2a) whose spectrum is $X(\omega) = 0.2\,\Delta(\omega/20\pi)$ (Fig. 8.2b). The bandwidth of this signal is 5 Hz (10π rad/s). Consequently, the Nyquist rate is 10 Hz; that is, we must sample the signal at a rate no less than 10 samples/s. The Nyquist interval is $T = 1/2B = 0.1$ second.

Recall that the sampled signal spectrum consists of $(1/T)X(\omega) = (0.2/T)\,\Delta(\omega/20\pi)$ repeating periodically with a period equal to the sampling frequency f_s Hz. We present this information in Table 8.1 for three sampling rates: $f_s = 5$ Hz (undersampling), 10 Hz (Nyquist rate), and 20 Hz (oversampling).

In the first case (undersampling), the sampling rate is 5 Hz (5 samples/s), and the spectrum $(1/T)X(\omega)$ repeats every 5 Hz (10π rad/s). The successive spectra overlap, as depicted in Fig. 8.2d, and the spectrum $X(\omega)$ is not recoverable from $\overline{X}(\omega)$; that is, $x(t)$ cannot be reconstructed from its samples $\overline{x}(t)$ in Fig. 8.2c. In the second case, we use the Nyquist sampling rate of 10 Hz (Fig. 8.2e). The spectrum $\overline{X}(\omega)$ consists of back-to-back, nonoverlapping repetitions of $(1/T)X(\omega)$ repeating every 10 Hz. Hence, $X(\omega)$ can be recovered from $\overline{X}(\omega)$ using an ideal lowpass filter of bandwidth 5 Hz (Fig. 8.2f). Finally, in the last case of oversampling (sampling rate 20 Hz), the spectrum $\overline{X}(\omega)$ consists of nonoverlapping repetitions of $(1/T)X(\omega)$ (repeating every 20 Hz) with empty bands between successive

TABLE 8.1

Sampling Frequency f_s (Hz)	Sampling Interval T (second)	$\dfrac{1}{T}X(\omega)$	Comments
5	0.2	$\Delta\left(\dfrac{\omega}{20\pi}\right)$	Undersampling
10	0.1	$2\Delta\left(\dfrac{\omega}{20\pi}\right)$	Nyquist rate
20	0.05	$4\Delta\left(\dfrac{\omega}{20\pi}\right)$	Oversampling

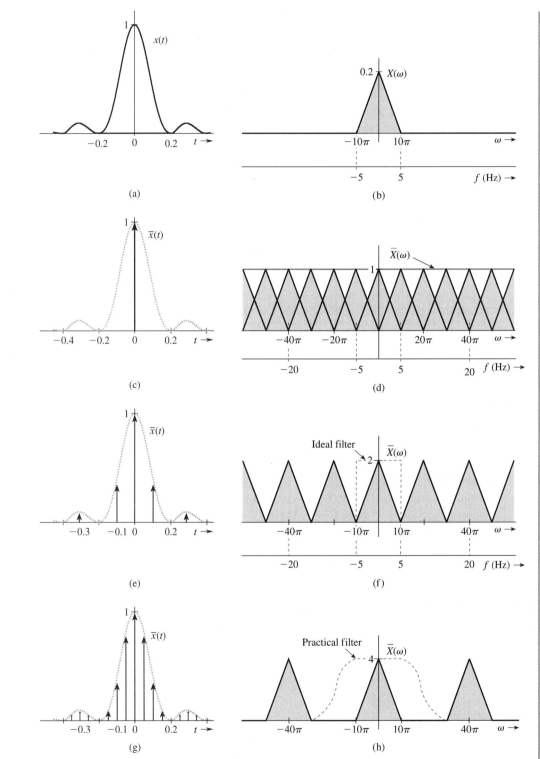

Figure 8.2 Effects of undersampling and oversampling.

cycles (Fig. 8.2h). Hence, $X(\omega)$ can be recovered from $\overline{X}(\omega)$ by using an ideal lowpass filter or even a practical lowpass filter (shown dashed in Fig. 8.2h).[†]

EXERCISE E8.1

Find the Nyquist rate and the Nyquist interval for the signals sinc $(100\pi t)$ and sinc $(100\pi t) +$ sinc $(50\pi t)$.

ANSWER
The Nyquist interval is 0.01 second and the Nyquist sampling rate is 100 Hz for both the signals.

FOR SKEPTICS ONLY

Rare is a reader, who, at first encounter, is not skeptical of the sampling theorem. It seems impossible that Nyquist samples can define the one and the only signal that passes through those sample values. We can easily picture infinite number of signals passing through a given set of samples. However, among all these (infinite number of) signals, only one has the minimum bandwidth $B \leq 1/2T$ Hz, where T is the sampling interval. See Prob. 8.2-9.

To summarize, for a given set of samples taken at a rate f_s Hz, there is only one signal of bandwidth $B \leq f_s/2$ that passes through those samples. All other signals that pass through those samples have bandwidth higher than $f_s/2$, and the samples are sub–Nyquist rate samples for those signals.

8.1-1 Practical Sampling

In proving the sampling theorem, we assumed ideal samples obtained by multiplying a signal $x(t)$ by an impulse train that is physically unrealizable. In practice, we multiply a signal $x(t)$ by a train of pulses of finite width, depicted in Fig. 8.3c. The sampler is shown in Fig. 8.3d. The sampled signal $\overline{x}(t)$ is illustrated in Fig. 8.3e. We wonder whether it is possible to recover or reconstruct $x(t)$ from this $\overline{x}(t)$. Surprisingly, the answer is affirmative, provided the sampling rate is not below the Nyquist rate. The signal $x(t)$ can be recovered by lowpass filtering $\overline{x}(t)$ as if it were sampled by impulse train.

The plausibility of this result becomes apparent when we consider the fact that reconstruction of $x(t)$ requires the knowledge of the Nyquist sample values. This information is available or built in the sampled signal $\overline{x}(t)$ in Fig. 8.3e because the nth sampled pulse strength is $x(nT)$. To prove the result analytically, we observe that the sampling pulse train $p_T(t)$ depicted in Fig. 8.3c,

---

[†]The filter should have a constant gain between 0 and 5 Hz and zero gain beyond 10 Hz. In practice, the gain beyond 10 Hz can be made negligibly small, but not zero.

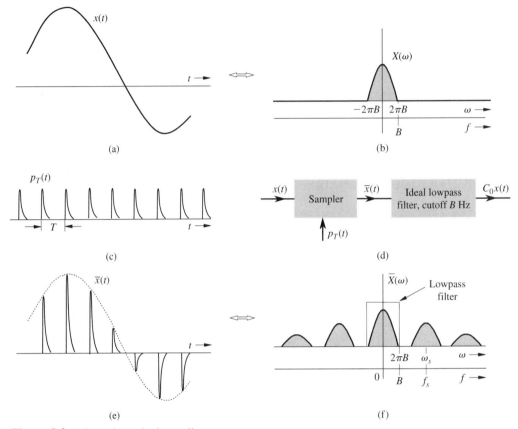

Figure 8.3 Effect of practical sampling.

being a periodic signal, can be expressed as a trigonometric Fourier series

$$p_T(t) = C_0 + \sum_{n=1}^{\infty} C_n \cos(n\omega_s t + \theta_n) \qquad \omega_s = \frac{2\pi}{T}$$

and

$$\bar{x}(t) = x(t) p_T(t) = x(t) \left[C_0 + \sum_{n=1}^{\infty} C_n \cos(n\omega_s t + \theta_n) \right]$$

$$= C_0 x(t) + \sum_{n=1}^{\infty} C_n x(t) \cos(n\omega_s t + \theta_n) \tag{8.7}$$

The sampled signal $\bar{x}(t)$ consists of $C_0 x(t)$, $C_1 x(t) \cos(\omega_s t + \theta_1)$, $C_2 x(t) \cos(2\omega_s t + \theta_2)$, ... Note that the first term $C_0 x(t)$ is the desired signal and all the other terms are modulated signals with spectra centered at $\pm\omega_s$, $\pm 2\omega_s$, $\pm 3\omega_s$, . . . , as illustrated in Fig. 8.3f. Clearly the signal $x(t)$ can be recovered by lowpass filtering of $\bar{x}(t)$, as shown in Fig. 8.3d. As before, it is necessary that $\omega_s > 4\pi B$ (or $f_s > 2B$).

EXAMPLE 8.2

To demonstrate practical sampling, consider a signal $x(t) = \text{sinc}^2(5\pi t)$ sampled by a rectangular pulse sequence $p_T(t)$ illustrated in Fig. 8.4c. The period of $p_T(t)$ is 0.1 second, so that the fundamental frequency (which is the sampling frequency) is 10 Hz. Hence, $\omega_s = 20\pi$. The Fourier series for $p_T(t)$ can be expressed as

$$p_T(t) = C_0 + \sum_{n=1}^{\infty} C_n \cos n\omega_s t$$

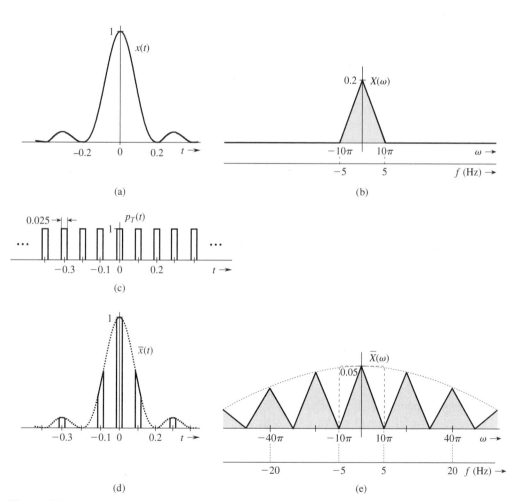

Figure 8.4 An example of practical sampling.

Hence

$$\overline{x}(t) = x(t)p_T(t)$$

$$= C_1 x(t) + C_1 x(t) \cos 20\pi t + C_2 x(t) \cos 40\pi t + C_3 x(t) \cos 60\pi t + \cdots$$

Use of Eqs. (6.18) yields $C_0 = \frac{1}{4}$ and $C_n = \frac{2}{n\pi} \sin\left(\frac{n\pi}{4}\right)$. Consequently, we have

$$\overline{x}(t) = x(t)p_T(t)$$

$$= \tfrac{1}{4}x(t) + C_1 x(t) \cos 20\pi t + C_2 x(t) \cos 40\pi t + C_3 x(t) \cos 60\pi t + \cdots$$

and

$$\overline{X}(\omega) = \frac{1}{4}X(\omega) + \frac{C_1}{2}[X(\omega - 20\pi) + X(\omega + 20\pi)] + \frac{C_2}{2}[X(\omega - 40\pi)$$

$$+ X(\omega + 40\pi)] + \frac{C_3}{2}[X(\omega - 60\pi) + X(\omega + 60\pi)] + \cdots$$

where $C_n = (2/n\pi)\sin(n\pi/4)$. The spectrum consists of $X(\omega)$ repeating periodically at the interval of 20π rad/s (10 Hz). Hence, there is no overlap between cycles, and $X(\omega)$ can be recovered by using an ideal lowpass filter of bandwidth 5 Hz. An ideal lowpass filter of unit gain (and bandwidth 5 Hz) will allow the first term on the righthand side of the foregoing equation to pass fully and suppress all the other terms. Hence, the output $y(t)$ is

$$y(t) = \tfrac{1}{4}x(t)$$

EXERCISE E8.2

Show that the basic pulse $p(t)$ used in the sampling pulse train in Fig. 8.4c cannot have zero area if we wish to reconstruct $x(t)$ by lowpass-filtering the sampled signal.

8.2 SIGNAL RECONSTRUCTION

The process of reconstructing a continuous-time signal $x(t)$ from its samples is also known as *interpolation*. In Section 8.1, we saw that a signal $x(t)$ bandlimited to B Hz can be reconstructed (interpolated) exactly from its samples if the sampling frequency f_s exceeds $2B$ Hz or the sampling interval T is less than $1/2B$. This reconstruction is accomplished by passing the sampled signal through an ideal lowpass filter of gain T and having a bandwidth of any value between B and $f_s - B$ Hz. From a practical viewpoint, a good choice is the middle value $f_s/2 = 1/2T$ Hz or π/T rad/s. This value allows for small deviations in the ideal filter characteristics on either side of the cutoff frequency. With this choice of cutoff frequency and gain T, the ideal lowpass filter required for signal reconstruction (or interpolation) is

$$H(\omega) = T \operatorname{rect}\left(\frac{\omega}{2\pi f_s}\right) = T \operatorname{rect}\left(\frac{\omega T}{2\pi}\right) \tag{8.8}$$

The interpolation process here is expressed in the frequency domain as a filtering operation. Now we shall examine this process from the time-domain viewpoint.

TIME-DOMAIN VIEW:
A SIMPLE INTERPOLATION

Consider the interpolation system shown in Fig. 8.5a. We start with a very simple interpolating filter, whose impulse response is rect (t/T), depicted in Fig. 8.5b. This is a gate pulse centered at the origin, having unit height, and width T (the sampling interval). We shall find the output of this filter when the input is the sampled signal $\overline{x}(t)$ consisting of an impulse train with the nth impulse at $t = nT$ with strength $x(nT)$. Each sample in $\overline{x}(t)$, being an impulse, produces at the output a gate pulse of height equal to the strength of the sample. For instance, the nth sample is an impulse of strength $x(nT)$ located at $t = nT$ and can be expressed as $x(nT)\delta(t - nT)$. When this impulse passes through the filter, it produces at the output a gate pulse of height $x(nT)$, centered at $t = nT$ (shaded in Fig. 8.5c). Each sample in $\overline{x}(t)$ will generate a corresponding gate pulse, resulting in the filter output that is a staircase approximation of $x(t)$, shown dotted in Fig. 8.5c. This filter thus gives a crude form of interpolation.

The frequency response of this filter $H(\omega)$ is the Fourier transform of the impulse response rect (t/T).

$$h(t) = \text{rect}\left(\frac{t}{T}\right) \tag{8.9a}$$

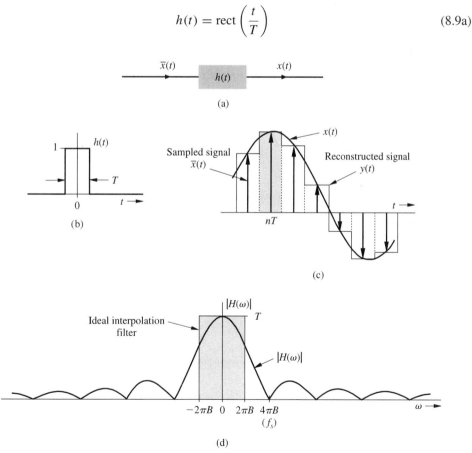

Figure 8.5 Simple interpolation by means of a zero-order hold (ZOH) circuit. (a) ZOH interpolator. (b) Impulse response of a ZOH circuit. (c) Signal reconstruction by ZOH, as viewed in the time domain. (d) Frequency response of a ZOH.

and

$$H(\omega) = T \operatorname{sinc}\left(\frac{\omega T}{2}\right) \tag{8.9b}$$

The amplitude response $|H(\omega)|$ for this filter, illustrated in Fig. 8.5d, explains the reason for the crudeness of this interpolation. This filter, also known as the *zero-order hold* (ZOH) filter, is a poor form of the ideal lowpass filter (shaded in Fig. 8.5d) required for exact interpolation.[†]

We can improve on the ZOH filter by using the *first-order hold* filter, which results in a linear interpolation instead of the staircase interpolation. The linear interpolator, whose impulse response is a triangle pulse $\Delta(t/2T)$, results in an interpolation in which successive sample tops are connected by straight-line segments (see Prob. 8.2-3).

TIME-DOMAIN VIEW:
AN IDEAL INTERPOLATION

The ideal interpolation filter frequency response obtained in Eq. (8.8) is illustrated in Fig. 8.6a. The impulse response of this filter, the inverse Fourier transform of $H(\omega)$ is

$$h(t) = \operatorname{sinc}\left(\frac{\pi t}{T}\right) \tag{8.10a}$$

For the Nyquist sampling rate, $T = 1/2B$, and

$$h(t) = \operatorname{sinc}(2\pi Bt) \tag{8.10b}$$

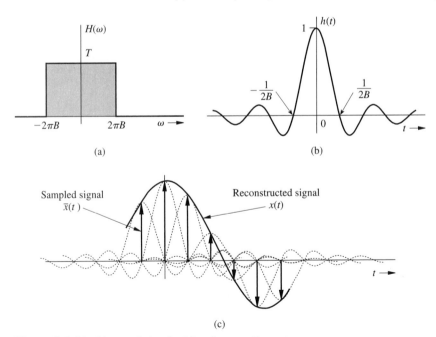

(a)   (b)

(c)

Figure 8.6 Ideal interpolation for Nyquist sampling rate.

[†]Figure 8.5b shows that the impulse response of this filter is noncausal, and this filter is not realizable. In practice, we make it realizable by delaying the impulse response by $T/2$. This merely delays the output of the filter by $T/2$.

This $h(t)$ is depicted in Fig. 8.6b. Observe the interesting fact that $h(t) = 0$ at all Nyquist sampling instants $(t = \pm n/2B)$ except at $t = 0$. When the sampled signal $\bar{x}(t)$ is applied at the input of this filter, the output is $x(t)$. Each sample in $\bar{x}(t)$, being an impulse, generates a sinc pulse of height equal to the strength of the sample, as illustrated in Fig. 8.6c. The process is identical to that depicted in Fig. 8.5c, except that $h(t)$ is a sinc pulse instead of a gate pulse. Addition of the sinc pulses generated by all the samples results in $x(t)$. The nth sample of the input $\bar{x}(t)$ is the impulse $x(nT)\delta(t - nT)$; the filter output of this impulse is $x(nT)h(t - nT)$. Hence, the filter output to $\bar{x}(t)$, which is $x(t)$, can now be expressed as a sum

$$x(t) = \sum_n x(nT)h(t - nT)$$

$$= \sum_n x(nT) \operatorname{sinc}\left[\frac{\pi}{T}(t - nT)\right] \tag{8.11a}$$

For the case of Nyquist sampling rate, $T = 1/2B$, and Eq. (8.11a) simplifies to

$$x(t) = \sum_n x(nT) \operatorname{sinc}(2\pi Bt - n\pi) \tag{8.11b}$$

Equation (8.11b) is the *interpolation formula*, which yields values of $x(t)$ between samples as a weighted sum of all the sample values.

EXAMPLE 8.3

Find a signal $x(t)$ that is bandlimited to B Hz, and whose samples are

$$x(0) = 1 \quad \text{and} \quad x(\pm T) = x(\pm 2T) = x(\pm 3T) = \cdots = 0$$

where the sampling interval T is the Nyquist interval for $x(t)$, that is, $T = 1/2B$.

Because, we are given the Nyquist sample values, we use the interpolation formula (8.11b) to construct $x(t)$ from its samples. Since all but one of the Nyquist samples are zero, only one term (corresponding to $n = 0$) in the summation on the right-hand side of Eq. (8.11b) survives. Thus

$$x(t) = \operatorname{sinc}(2\pi Bt) \tag{8.12}$$

This signal is illustrated in Fig. 8.6b. Observe that this is the only signal that has a bandwidth B Hz and the sample values $x(0) = 1$ and $x(nT) = 0 \ (n \neq 0)$. No other signal satisfies these conditions.

8.2-1 Practical Difficulties in Signal Reconstruction

Consider the signal reconstruction procedure illustrated in Fig. 8.7a. If $x(t)$ is sampled at the Nyquist rate $f_s = 2B$ Hz, the spectrum $\overline{X}(\omega)$ consists of repetitions of $X(\omega)$ without any gap

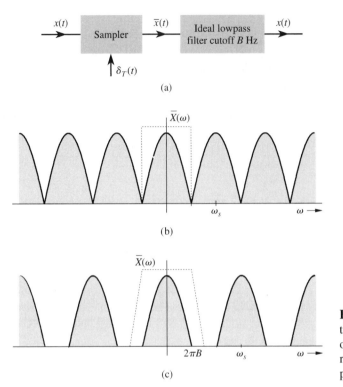

Figure 8.7 (a) Signal reconstruction from its samples. (b) Spectrum of a signal sampled at the Nyquist rate. (c) Spectrum of a signal sampled above the Nyquist rate.

between successive cycles, as depicted in Fig. 8.7b. To recover $x(t)$ from $\overline{x}(t)$, we need to pass the sampled signal $\overline{x}(t)$ through an ideal lowpass filter, shown dotted in Fig. 8.7b. As seen in Section 7.5, such a filter is unrealizable; it can be closely approximated only with infinite time delay in the response. In other words, we can recover the signal $x(t)$ from its samples with infinite time delay. A practical solution to this problem is to sample the signal at a rate higher than the Nyquist rate ($f_s > 2B$ or $\omega_s > 4\pi B$). The result is $\overline{X}(\omega)$, consisting of repetitions of $X(\omega)$ with a finite band gap between successive cycles, as illustrated in Fig. 8.7c. Now, we can recover $X(\omega)$ from $\overline{X}(\omega)$ using a lowpass filter with a gradual cutoff characteristic, shown dotted in Fig. 8.7c. But even in this case, if the unwanted spectrum is to be suppressed, the filter gain must be zero beyond some frequency (see Fig. 8.7c). According to the Paley–Wiener criterion [Eq. (7.61)], it is impossible to realize even this filter. The only advantage in this case is that the required filter can be closely approximated with a smaller time delay. All this means that it is impossible in practice to recover a bandlimited signal $x(t)$ exactly from its samples, even if the sampling rate is higher than the Nyquist rate. However, as the sampling rate increases, the recovered signal approaches the desired signal more closely.

THE TREACHERY OF ALIASING

There is another fundamental practical difficulty in reconstructing a signal from its samples. The sampling theorem was proved on the assumption that the signal $x(t)$ is bandlimited. *All practical signals are timelimited;* that is, they are of finite duration or width. We can demonstrate (see Prob. 8.2-14) that a signal cannot be timelimited and bandlimited simultaneously. If a signal is timelimited, it cannot be bandlimited, and vice versa (but it can be simultaneously nontime-limited and nonbandlimited). Clearly, all practical signals, which are necessarily timelimited, are nonbandlimited, as shown in Fig. 8.8a; they have infinite bandwidth, and the spectrum $\overline{X}(\omega)$ consists of overlapping cycles of $X(\omega)$ repeating every f_s Hz (the sampling frequency), as illustrated in Fig. 8.8b.[†] Because of infinite bandwidth in this case, the spectral overlap is unavoidable, regardless of the sampling rate. Sampling at a higher rate reduces but does not eliminate overlapping between repeating spectral cycles. Because of the overlapping tails, $\overline{X}(\omega)$ no longer has complete information about $X(\omega)$, and it is no longer possible, even theoretically, to recover $x(t)$ exactly from the sampled signal $\overline{x}(t)$. If the sampled signal is passed through an ideal lowpass filter of cutoff frequency $f_s/2$ Hz, the output is not $X(\omega)$ but $X_a(\omega)$ (Fig. 8.8c), which is a version of $X(\omega)$ distorted as a result of two separate causes:

1. The loss of the tail of $X(\omega)$ beyond $|f| > f_s/2$ Hz.

2. The reappearance of this tail inverted or folded onto the spectrum. Note that the spectra cross at frequency $f_s/2 = 1/2T$ Hz. This frequency is called the *folding* frequency. The spectrum may be viewed as if the lost tail is folding back onto itself at the fold-ing frequency. For instance, a component of frequency $(f_s/2) + f_z$ shows up as or "impersonates" as a component of lower frequency $(f_s/2) - f_z$ in the reconstructed signal. Thus, the components of frequencies above $f_s/2$ reappear as components of fre-quencies below $f_s/2$. This tail inversion, known as *spectral folding* or *aliasing,* is shown shaded in Fig. 8.8b and also in Fig. 8.8c. In the process of aliasing, not only are we losing all the components of frequencies above the folding frequency $f_s/2$ Hz, but these very components reappear (aliased) as lower-frequency components, as shown in Fig. 8.8b or 8.8c. Such aliasing destroys the integrity of the frequency components below the folding frequency $f_s/2$, as depicted in Fig. 8.8c.

The aliasing problem is analogous to that of an army with a platoon that has secretly defected to the enemy side. The platoon is, however, ostensibly loyal to the army. The army is in double jeopardy. First, the army has lost this platoon as a fighting force. In addition, during actual fighting, the army will have to contend with the sabotage by the defectors and will have to find another loyal platoon to neutralize the defectors. Thus, the army has lost two platoons in nonproductive activity.

DEFECTORS ELIMINATED:
THE ANTIALIASING FILTER

If you were the commander of the betrayed army, the solution to the problem would be obvious. As soon as the commander got wind of the defection, he would incapacitate, by whatever means, the defecting platoon *before the fighting begins*. This way he loses only one (the defecting)

[†]Figure 8.8b shows that from the infinite number of repeating cycles, only the neighboring spectral cycles over-lap. This is a somewhat simplified picture. In reality, all the cycles overlap and interact with every other cycle because of the infinite width of all practical signal spectra. Fortunately, all practical spectra also must decay at higher frequencies. This results in insignificant amount of interference from cycles other than the immediate neighbors. When such an assumption is not justified, aliasing computations become little more involved.

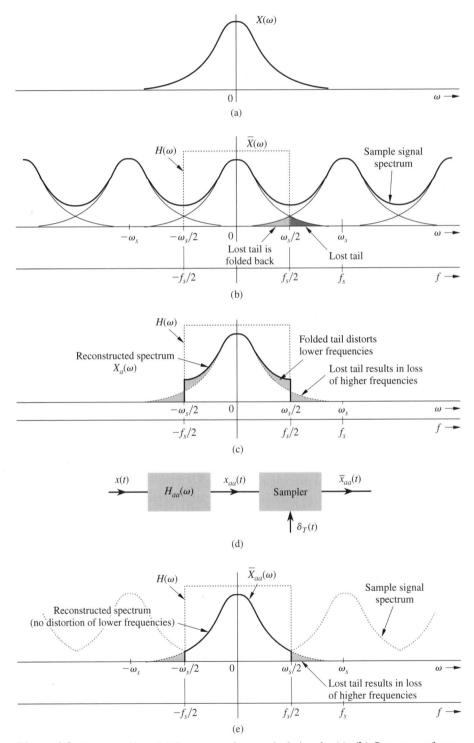

Figure 8.8 Aliasing effect. **(a)** Spectrum of a practical signal $x(t)$. **(b)** Spectrum of sampled $x(t)$. **(c)** Reconstructed signal spectrum. **(d)** Sampling scheme using antialiasing filter. **(e)** Sampled signal spectrum (dotted) and the reconstructed signal spectrum (solid) when antialiasing filter is used.

platoon. This is a partial solution to the double jeopardy of betrayal and sabotage, a solution that partly rectifies the problem and cuts the losses to half.

We follow exactly the same procedure. The potential defectors are all the frequency components beyond the folding frequency $f_s/2 = 1/2T$ Hz. We should eliminate (suppress) these components from $x(t)$ *before sampling* $x(t)$. Such suppression of higher frequencies can be accomplished by an ideal lowpass filter of cutoff $f_s/2$ Hz, as shown in Fig. 8.8d. This is called the *antialiasing filter*. Figure 8.8d also shows that antialiasing filtering is performed before sampling. Figure 8.8e shows the sampled signal spectrum (dotted) and the reconstructed signal $X_{aa}(\omega)$ when antialiasing scheme is used. An antialiasing filter essentially bandlimits the signal $x(t)$ to $f_s/2$ Hz. This way, we lose only the components beyond the folding frequency $f_s/2$ Hz. These suppressed components now cannot reappear to corrupt the components of frequencies below the folding frequency. Clearly, use of an antialiasing filter results in the reconstructed signal spectrum $X_{aa}(\omega) = X(\omega)$ for $|f| < f_s/2$. Thus, although we lost the spectrum beyond $f_s/2$ Hz, the spectrum for all the frequencies below $f_s/2$ remains intact. The effective aliasing distortion is cut in half owing to elimination of folding. We stress again that the antialiasing operation must be performed *before the signal is sampled*.

An antialiasing filter also helps to reduce noise. Noise, generally, has a wideband spectrum, and without antialiasing, the aliasing phenomenon itself will cause the noise lying outside the desired band to appear in the signal band. Antialiasing suppresses the entire noise spectrum beyond frequency $f_s/2$.

The antialiasing filter, being an ideal filter, is unrealizable. In practice, we use a steep cutoff filter, which leaves a sharply attenuated spectrum beyond the folding frequency $f_s/2$.

SAMPLING FORCES NONBANDLIMITED SIGNALS TO APPEAR BANDLIMITED

Figure 8.8b shows the spectrum of a signal $x(t)$ consists of overlapping cycles of $X(\omega)$. This means that $\bar{x}(t)$ are sub-Nyquist samples of $x(t)$. However, we may also view the spectrum in Fig. 8.8b as the spectrum $X_a(\omega)$ (Fig. 8.8c), repeating periodically every f_s Hz without overlap. The spectrum $X_a(\omega)$ is bandlimited to $f_s/2$ Hz. Hence, these (sub-Nyquist) samples of $x(t)$ are actually the Nyquist samples for signal $x_a(t)$. In conclusion, sampling a nonbandlimited signal $x(t)$ at a rate f_s Hz makes the samples appear to be the Nyquist samples of some signal $x_a(t)$, bandlimited to $f_s/2$ Hz. In other words, sampling makes a nonbandlimited signal appear to be a bandlimited signal $x_a(t)$ with bandwidth $f_s/2$ Hz. A similar conclusion applies if $x(t)$ is bandlimited but sampled at a sub-Nyquist rate.

VERIFICATION OF ALIASING IN SINUSOIDS

We showed in Fig. 8.8b how sampling a signal below the Nyquist rate causes aliasing, which makes a signal of higher frequency $(f_s/2) + f_z$ Hz masquerade as a signal of lower frequency $(f_s/2) - f_z$ Hz. Figure 8.8b demonstrates this result in the frequency domain. Let us now verify it in the time domain to gain a deeper appreciation of aliasing.

We can prove our proposition by showing that samples of sinusoids of frequencies $(\omega_s/2) + \omega_z$ and $(\omega_s/2) - \omega_z$ are identical when the sampling frequency is $f_s = \omega_s/2\pi$ Hz.

For a sinusoid $x(t) = \cos \omega t$, sampled at intervals of T seconds, $x(nT)$, its nth sample (at $t = nT$) is

$$x(nT) = \cos \omega nT \qquad n \text{ integer}$$

Hence, samples of sinusoids of frequency $\omega = (\omega_s/2) \pm \omega_z$ are[†]

$$x(nT) = \cos\left(\frac{\omega_s}{2} \pm \omega_z\right)nT \tag{8.13}$$

$$= \cos\left(\frac{\omega_s}{2}\right)nT \cos\omega_z nT \mp \sin\left(\frac{\omega_s}{2}\right)nT \sin\omega_z nT$$

Recognizing that $\omega_s T = 2\pi f_s T = 2\pi$, and $\sin(\omega_s/2)nT = \sin\pi n = 0$ for all integer n, we obtain

$$x(nT) = \cos\left(\frac{\omega_s}{2}\right)nT \cos\omega_z nT$$

Clearly, the samples of a sinusoid of frequency $(f_s/2) + f_z$ are identical to the samples of a sinusoid $(f_s/2) - f_z$.[‡] For instance, when a sinusoid of frequency 100 Hz is sampled at a rate of 120 Hz, the apparent frequency of the sinusoid that results from reconstruction of the samples is 20 Hz. This follows from the fact that here, $100 = (f_s/2) + f_z = 60 + f_z$ so that $f_z = 40$. Hence, $(f_s/2) - f_z = 20$. Such would precisely be the conclusion arrived at from Fig. 8.8b.

This discussion again shows that sampling a sinusoid of frequency f aliasing can be avoided if the sampling rate $f_s > 2f$ Hz.

$$0 \le f < \frac{f_s}{2} \qquad \text{or} \qquad 0 \le \omega < \frac{\pi}{T} \tag{8.14}$$

Violating this condition leads to aliasing, implying that the samples appear to be those of a lower-frequency signal. Because of this loss of identity, it is impossible to reconstruct the signal faithfully from its samples.

GENERAL CONDITION FOR ALIASING
IN SINUSOIDS

We can generalize the foregoing result by showing that samples of a sinusoid of frequency f_0 are identical to those of a sinusoid of frequency $f_0 + mf_s$ Hz (integer m), where f_s is the sampling frequency. The samples of $\cos 2\pi(f_0 + mf_s)t$ are

$$\cos 2\pi(f_0 + mf_s)nT = \cos(2\pi f_0 nT + 2\pi mn) = \cos 2\pi f_0 nT$$

The result follows because mn is an integer and $f_s T = 1$. This result shows that sinusoids of frequencies that differ by integer multiple of f_s result in identical set of samples. In other words,

---

[†]Here we have ignored the phase aspect of the sinusoid. Sampled versions of a sinusoid $x(t) = \cos(\omega t + \theta)$ with two different frequencies $(\omega_s/2) \pm \omega_z$ have identical frequency, but the phase signs may be reversed depending on the value of ω_z.

[‡]The reader is encouraged to verify this result graphically by plotting the spectrum of a sinusoid of frequency $(\omega_s/2) + \omega_z$ [impulses at $\pm[(\omega_s/2) + \omega_z]$] and its periodic repetition at intervals ω_s. Although the result is valid for all values of ω_z, consider the case of $\omega_z < \omega_s/2$ to simplify the graphics.

samples of sinusoids separated by frequency f_s Hz are identical. This implies that samples of sinusoids in any frequency band of f_s Hz are unique; that is, no two sinusoids in that band have the same samples (when sampled at a rate f_s Hz). For instance, frequencies in the band from $-f_s/2$ to $f_s/2$ have unique samples (at the sampling rate f_s). This band is called the *fundamental band*. Recall also that $f_s/2$ is the folding frequency.

From the discussion thus far, we conclude that if a continuous-time sinusoid of frequency f Hz is sampled at a rate of f_s Hz (samples/s), the resulting samples would appear as samples of a continuous-time sinusoid of frequency f_a in the fundamental band, where

$$f_a = f - mf_s \qquad -\frac{f_s}{2} \le f_a < \frac{f_s}{2} \qquad m \text{ an integer} \qquad (8.15a)$$

The frequency f_a lies in the fundamental band from $-f_s/2$ to $f_s/2$. Figure 8.9a shows the plot of f_a versus f, where f is the actual frequency and f_a is the corresponding fundamental band frequency, whose samples are identical to those of the sinusoid of frequency f, when sampling rate is f_s Hz.

Recall, however, that the sign change of a frequency does not alter the actual frequency of the waveform. This is because

$$\cos(-\omega_a t + \theta) = \cos(\omega_a t - \theta)$$

Clearly the *apparent frequency* of a sinusoid of frequency $-f_a$ is also f_a. However, its phase undergoes a sign change. This means the apparent frequency of any sampled sinusoid lies in the range from 0 to $f_s/2$ Hz. To summarize, if a continuous-time sinusoid of frequency f Hz is sampled at a rate of f_s Hz (samples/second), the resulting samples would appear as samples of a continuous-time sinusoid of frequency $|f_a|$ that lies in the band from 0 to $f_s/2$. According to

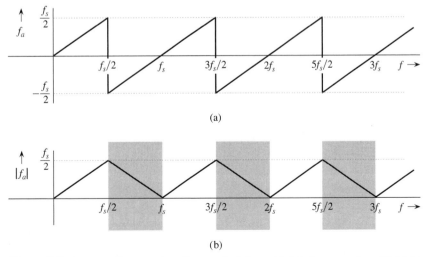

Figure 8.9 Apparent frequencies of a sampled sinusoid: **(a)** f_a versus f and **(b)** $|f_a|$ versus f.

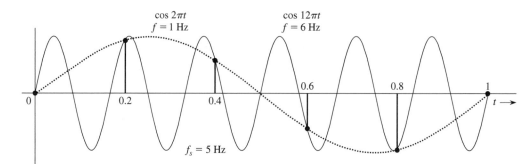

Figure 8.10 Demonstration of aliasing.

Eq. (8.15a)

$$|f_a| = |f - mf_s| \qquad |f_a| \le \frac{f_s}{2} \qquad m \text{ an integer} \qquad (8.15b)$$

The plot of the apparent frequency $|f_a|$ versus f is shown in Fig. 8.9b.[†] As expected, the apparent frequency $|f_a|$ of any sampled sinusoid, regardless of its frequency, is always in the range of 0 to $f_s/2$ Hz. However, when f_a is negative, the phase of the apparent sinusoid undergoes a sign change. The frequency belts in which such phase changes occur are shown shaded in Fig. 8.9b.

Consider, for example, a sinusoid $\cos(2\pi f t + \theta)$ with $f = 8000$ Hz sampled at a rate $f_s = 3000$ Hz. Using Eq. (8.15a), we obtain $f_a = 8000 - 3 \times 3000 = -1000$. Hence, $|f_a| = 1000$. The samples would appear to have come from a sinusoid $\cos(2000\pi t - \theta)$. Observe the sign change of the phase because f_a is negative.[‡]

In the light of the foregoing development, let us consider a sinusoid of frequency $f = (f_s/2) + f_z$, sampled at a rate of f_s Hz. According to Eq. (8.15a),

$$f_a = \frac{f_s}{2} + f_z - (1 \times f_s) = -\frac{f_s}{2} + f_z$$

Hence, the apparent frequency is $|f_a| = (f_s/2) - f_z$, confirming our earlier result. However, the phase of the sinusoid will suffer a sign change because f_a is negative.

Figure 8.10 shows how samples of sinusoids of two different frequencies (sampled at the same rate) generate identical sets of samples. Both the sinusoids are sampled at a rate $f_s = 5$ Hz ($T = 0.2$ second). The frequencies of the two sinusoids, 1 Hz (period 1) and 6 Hz (period 1/6), differ by $f_s = 5$ Hz.

The reason for aliasing can be clearly seen in Fig. 8.10. The root of the problem is the sampling rate, which may be adequate for the lower-frequency sinusoid but is clearly inadequate for the higher-frequency sinusoid. The figure clearly shows that between the successive samples

---

[†]The plots in Figs. 8.9 and Fig. 5.16 are identical. This is because a sampled sinusoid is basically a discrete-time sinusoid.

[‡]For phase sign change, we are assuming the signal of the form $\cos(2\pi f t + \theta)$. If the form is $\sin(2\pi f t + \theta)$, the rule changes slightly. It is left as an exercise for the reader to show that when $f_a < 0$, this sinusoid appears as $-\sin(2\pi |f_a| t - \theta)$. Thus, in addition to phase change, the amplitude also changes sign.

of the higher-frequency sinusoid, there are wiggles, which are bypassed or ignored, and are unrepresented in the samples, indicating a sub-Nyquist rate of sampling. The frequency of the apparent signal $x_a(t)$ is always the lowest possible frequency that lies within the band $|f| \le f_s/2$. Thus, the apparent frequency of the samples in this example is 1 Hz. If these samples are chosen to reconstruct a signal using a lowpass filter of bandwidth $f_s/2$, we shall obtain a sinusoid of frequency 1 Hz.

EXAMPLE 8.4

A continuous-time sinusoid $\cos(2\pi f t + \theta)$ is sampled at a rate $f_s = 1000$ Hz. Determine the apparent (aliased) sinusoid of the resulting samples if the input signal frequency f is:

(a) 400 Hz

(b) 600 Hz

(c) 1000 Hz

(d) 2400 Hz

The folding frequency is $f_s/2 = 500$. Hence, sinusoids below 500 Hz (frequency within the fundamental band) will not be aliased and sinusoids of frequency above 500 Hz will be aliased.

(a) $f = 400$ Hz is less than 500 Hz, and there is no aliasing. The apparent sinusoid is $\cos(2\pi f t + \theta)$ with $f = 400$.

(b) $f = 600$ Hz can be expressed as $600 = -400 + 1000$ so that $f_a = -400$. Hence, the aliased frequency is 400 Hz and the phase changes sign. Hence, the apparent (aliased) sinusoid is $\cos(2\pi f t - \theta)$ with $f = 400$.

(c) $f = 1000$ Hz can be expressed as $1000 = 0 + 1000$ so that $f_a = 0$. Hence, the aliased frequency is 0 Hz (dc), and there is no phase sign change. Hence, the apparent sinusoid is $y(t) = \cos(0\pi t \pm \theta) = \cos(\theta)$. This is a dc signal with constant sample values for all n.

(d) $f = 2400$ Hz can be expressed as $2400 = 400 + (2 \times 1000)$ so that $f_a = 400$. Hence, the aliased frequency is 400 Hz and there is no sign change for the phase. The apparent sinusoid is $\cos(2\pi f t + \theta)$ with $f = 400$.

We could have found these answers directly from Fig. 8.9b. For example, for case b, we read $|f_a| = 400$ corresponding to $f = 600$. Moreover, $f = 600$ lies in the shaded belt. Hence, there is a phase sign change.

EXERCISE E8.3

Show that samples of 90 Hz and 110 Hz sinusoids of the form $\cos \omega t$ are identical when sampled at a rate 200 Hz.

EXERCISE E8.4

A sinusoid of frequency f_0 Hz is sampled at a rate of 100 Hz. Determine the apparent frequency of the samples if f_0 is

   (a) 40 Hz

   (b) 60 Hz

   (c) 140 Hz

   (d) 160

ANSWER
40 Hz for all four cases

8.2-2 Some Applications of the Sampling Theorem

The sampling theorem is very important in signal analysis, processing, and transmission because it allows us to replace a continuous-time signal with a discrete sequence of numbers. Processing a continuous-time signal is therefore equivalent to processing a discrete sequence of numbers. Such processing leads us directly into the area of digital filtering. In the field of communication, the transmission of a continuous-time message reduces to the transmission of a sequence of numbers by means of pulse trains. The continuous-time signal $x(t)$ is sampled, and sample values are used to modify certain parameters of a periodic pulse train. We may vary the amplitudes (Fig. 8.11b), widths (Fig. 8.11c), or positions (Fig. 8.11d) of the pulses in proportion to the sample values of the signal $x(t)$. Accordingly, we may have *pulse-amplitude modulation* (PAM), *pulse-width modulation* (PWM), or *pulse-position modulation* (PPM). The most important form of pulse modulation today is *pulse-code modulation* (PCM), discussed in Section 8.3 in connection with Fig. 8.14b. In all these cases, instead of transmitting $x(t)$, we transmit the corresponding pulse-modulated signal. At the receiver, we read the information of the pulse-modulated signal and reconstruct the analog signal $x(t)$.

One advantage of using pulse modulation is that it permits the simultaneous transmission of several signals on a time-sharing basis—*time-division multiplexing* (TDM). Because a pulse-modulated signal occupies only a part of the channel time, we can transmit several pulse-modulated signals on the same channel by interweaving them. Figure 8.12 shows the TDM of two PAM signals. In this manner, we can multiplex several signals on the same channel by reducing pulse widths.[†]

Digital signals also offer an advantage in the area of communications, where signals must travel over distances. Transmission of digital signals is more rugged than that of analog signals because digital signals can withstand channel noise and distortion much better as long as the noise and the distortion are within limits. An analog signal can be converted to digital binary

----

[†]Another method of transmitting several baseband signals simultaneously is frequency-division multiplexing (FDM) discussed in Section 7.7-4. In FDM various signals are multiplexed by sharing the channel bandwidth. The spectrum of each message is shifted to a specific band not occupied by any other signal. The information of various signals is located in nonoverlapping frequency bands of the channel (Fig. 7.44). In a way, TDM and FDM are duals of each other.

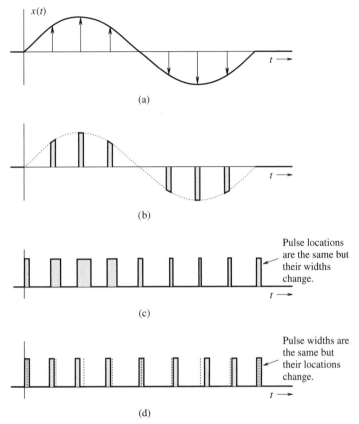

Figure 8.11 Pulse-modulated signals. (a) The signal. (b) The PAM signal. (c) The PWM (PDM) signal. (d) The PAM signal.

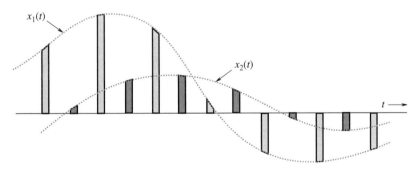

Figure 8.12 Time-division multiplexing of two signals.

form through sampling and quantization (rounding off), as explained in the next section. The digital (binary) message in Fig. 8.13a is distorted by the channel, as illustrated in Fig. 8.13b. Yet if the distortion remains within a limit, we can recover the data without error because we need only make a simple binary decision: Is the received pulse positive or negative? Figure 8.13c shows the same data with channel distortion and noise. Here again the data can be recovered

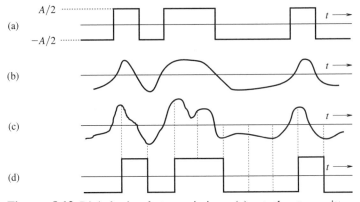

Figure 8.13 Digital signal transmission: **(a)** at the transmitter, **(b)** received distorted signal (without noise), **(c)** received distorted signal (with noise), and **(d)** regenerated signal at the receiver.

correctly as long as the distortion and the noise are within limits. Such is not the case with analog messages. Any distortion or noise, no matter how small, will distort the received signal.

The greatest advantage of digital communication over the analog counterpart, however, is the viability of regenerative repeaters in the former. In an analog transmission system, a message signal grows progressively weaker as it travels along the channel (transmission path), whereas the channel noise and the signal distortion, being cumulative, become progressively stronger. Ultimately the signal, overwhelmed by noise and distortion, is mutilated. Amplification is of little help because it enhances the signal and the noise in the same proportion. Consequently, the distance over which an analog message can be transmitted is limited by the transmitted power. If a transmission path is long enough, the channel distortion and noise will accumulate sufficiently to overwhelm even a digital signal. The trick is to set up repeaters along the transmission path at distances short enough to permit detection of signal pulses before the noise and distortion have a chance to accumulate sufficiently. At each repeater, the pulses are detected, and new, clean pulses are transmitted to the next repeater, which, in turn, duplicates the same process. If the noise and distortion remain within limits (which is possible because of the closely spaced repeaters), pulses can be detected correctly.† This way the digital messages can be transmitted over longer distances with greater reliability. In contrast, analog messages cannot be cleaned up periodically, and their transmission is therefore less reliable. The most significant error in digitized signals comes from quantizing (rounding off). This error, discussed in Section 8.3, can be reduced as much as desired by increasing the number of quantization levels, at the cost of an increased bandwidth of the transmission medium (channel).

8.3 ANALOG-TO-DIGITAL (A/D) CONVERSION

An *analog* signal is characterized by the fact that its amplitude can take on any value over a continuous range. Hence, analog signal amplitude can take on an infinite number of values. In contrast, a *digital* signal amplitude can take on only a finite number of values. An analog

---

†The error in pulse detection can be made negligible.

signal can be converted into a digital signal by means of sampling and *quantizing* (rounding off). Sampling an analog signal alone will not yield a digital signal because a sample of analog signal can still take on any value in a continuous range. It is digitized by rounding off its value to one of the closest permissible numbers (or *quantized levels*), as illustrated in Fig. 8.14a, which represents one possible quantizing scheme. The amplitudes of the analog signal $x(t)$ lie in the range $(-V, V)$. This range is partitioned into L subintervals, each of magnitude $\Delta = 2V/L$. Next, each sample amplitude is approximated by the midpoint value of the subinterval in which the sample falls (see Fig. 8.14a for $L = 16$). It is clear that each sample is approximated to one of the L numbers. Thus, the signal is digitized with quantized samples taking on any one of the L values. This is an L-ary digital signal (see Section 1.3-2). Each sample can now be represented by one of L distinct pulses.

From a practical viewpoint, dealing with a large number of distinct pulses is difficult. We prefer to use the smallest possible number of distinct pulses, the very smallest number being two. A digital signal using only two symbols or values is the binary signal. A binary digital signal (a signal that can take on only two values) is very desirable because of its simplicity, economy, and ease of engineering. We can convert an L-ary signal into a binary signal by using pulse coding. Figure 8.14b shows one such code for the case of $L = 16$. This code, formed by binary representation of the 16 decimal digits from 0 to 15, is known as the *natural binary code (NBC)*. For L quantization levels, we need a minimum of b binary code digits, where $2^b = L$ or $b = \log_2 L$.

Each of the 16 levels is assigned one binary code word of four digits. Thus, each sample in this example is encoded by four binary digits. To transmit or digitally process the binary data, we need to assign a distinct electrical pulse to each of the two binary states. One possible way is to assign a negative pulse to a binary **0** and a positive pulse to a binary **1** so that each sample is now represented by a group of four binary pulses (pulse code), as depicted in Fig. 8.14b. The resulting binary signal is a digital signal obtained from the analog signal $x(t)$ through A/D conversion. In communications jargon, such a signal is known as a pulse-code-modulated (PCM) signal.

The convenient contraction of "*binary digit*" to *bit* has become an industry standard abbreviation.

The audio signal bandwidth is about 15 kHz, but subjective tests show that signal articulation (intelligibility) is not affected if all the components above 3400 Hz are suppressed.[3] Since the objective in telephone communication is intelligibility rather than high fidelity, the components above 3400 Hz are eliminated by a lowpass filter.[†] The resulting signal is then sampled at a rate of 8000 samples/s (8 kHz). This rate is intentionally kept higher than the Nyquist sampling rate of 6.8 kHz to avoid unrealizable filters required for signal reconstruction. Each sample is finally quantized into 256 levels ($L = 256$), which requires a group of eight binary pulses to encode each sample ($2^8 = 256$). Thus, a digitized telephone signal consists of data amounting to $8 \times 8000 = 64,000$ or 64 kbit/s, requiring 64,000 binary pulses per second for its transmission.

The compact disc (CD), a recent high-fidelity application of A/D conversion, requires the audio signal bandwidth of 20 kHz. Although the Nyquist sampling rate is only 40 kHz, an actual sampling rate of 44.1 kHz is used for the reason mentioned earlier. The signal is quantized into a rather large number of levels ($L = 65,536$) to reduce quantizing error. The binary-coded samples are now recorded on the CD.

[†]Components below 300 Hz may also be suppressed without affecting the articulation.

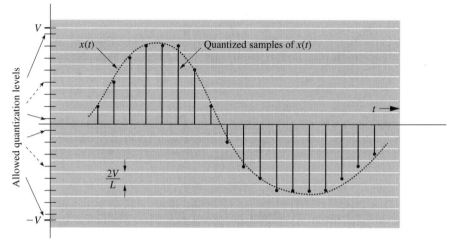

(a)

Digit	Binary equivalent	Pulse code waveform
0	**0000**	
1	**0001**	
2	**0010**	
3	**0011**	
4	**0100**	
5	**0101**	
6	**0110**	
7	**0111**	
8	**1000**	
9	**1001**	
10	**1010**	
11	**1011**	
12	**1100**	
13	**1101**	
14	**1110**	
15	**1111**	

(b)

Figure 8.14 Analog-to-digital (A/D) conversion of a signal: **(a)** quantizing and **(b)** pulse coding.

A HISTORICAL NOTE

The binary system of representing any number by using **1**s and **0**s was invented by Pingala (ca. 200 B.C.) in India. It was again worked out independently in the West by Gottfried Wilhelm Leibniz (1646–1716). He felt a spiritual significance in this discovery, reasoning that **1** representing unity was clearly a symbol for God, while **0** represented the nothingness. He reasoned that if all numbers can be represented merely by the use of **1** and **0**, this surely proves that God created the universe out of nothing!

EXAMPLE 8.5

A signal $x(t)$ bandlimited to 3 kHz is sampled at a rate $33\frac{1}{3}\%$ higher than the Nyquist rate. The maximum acceptable error in the sample amplitude (the maximum error due to quantization) is 0.5% of the peak amplitude V. The quantized samples are binary-coded. Find the required sampling rate, the number of bits required to encode each sample, and the bit rate of the resulting PCM signal.

The Nyquist sampling rate is $f_{\text{Nyq}} = 2 \times 3000 = 6000$ Hz (samples/s). The actual sampling rate is $f_A = 6000 \times (1\frac{1}{3}) = 8000$ Hz.

The quantization step is Δ, and the maximum quantization error is $\pm\Delta/2$, where $\Delta = 2V/L$. The maximum error due to quantization, $\Delta/2$, should be no greater than 0.5% of the signal peak amplitude V. Therefore

$$\frac{\Delta}{2} = \frac{V}{L} = \frac{0.5}{100}V \implies L = 200$$

For binary coding, L must be a power of 2. Hence, the next higher value of L that is a power of 2 is $L = 256$. Because $\log_2 256 = 8$, we need 8 bits to encode each sample. Therefore the bit rate of the PCM signal is

$$8000 \times 8 = 64{,}000 \text{ bits/s}$$

EXERCISE E8.5

The American Standard Code for Information Interchange (ASCII) has 128 characters, which are binary coded. A certain computer generates 100,000 characters per second. Show that

(a) 7 bits (binary digits) are required to encode each character

(b) 700,000 bits/s are required to transmit the computer output.

8.4 DUAL OF TIME SAMPLING: SPECTRAL SAMPLING

As in other cases, the sampling theorem has its dual. In Section 8.1, we discussed the time-sampling theorem and showed that a signal bandlimited to B Hz can be reconstructed from the

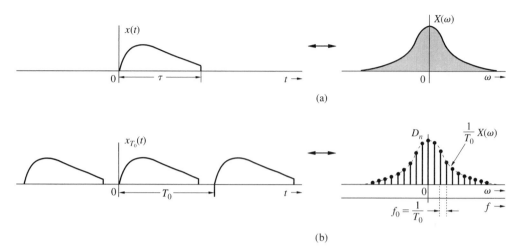

Figure 8.15 Periodic repetition of a signal amounts to sampling its spectrum.

signal samples taken at a rate of $f_s > 2B$ samples/s. Note that the signal spectrum exists over the frequency range (in hertz) of $-B$ to B. Therefore, $2B$ is the spectral width (not the bandwidth, which is B) of the signal. This fact means that a signal $x(t)$ can be reconstructed from samples taken at a rate $f_s >$ the spectral width of $X(\omega)$ in hertz ($f_s > 2B$).

We now prove the dual of the time-sampling theorem. This is the *spectral sampling theorem*, that applies to timelimited signals (the dual of bandlimited signals). A timelimited signal $x(t)$ exists only over a finite interval of τ seconds, as shown in Fig. 8.15a. Generally, a timelimited signal is characterized by $x(t) = 0$ for $t < T_1$ and $t > T_2$ (assuming $T_2 > T_1$). The signal width or duration is $\tau = T_2 - T_1$ seconds.

The spectral sampling theorem states that the spectrum $X(\omega)$ of a signal $x(t)$ timelimited to a duration of τ seconds can be reconstructed from the samples of $X(\omega)$ taken at a rate R samples/Hz, where $R > \tau$ (the signal width or duration) in seconds.

Figure 8.15a shows a timelimited signal $x(t)$ and its Fourier transform $X(\omega)$. Although $X(\omega)$ is complex in general, it is adequate for our line of reasoning to show $X(\omega)$ as a real function.

$$X(\omega) = \int_{-\infty}^{\infty} x(t)e^{-j\omega t}\,dt = \int_{0}^{\tau} x(t)e^{-j\omega t}\,dt \tag{8.16}$$

We now construct $x_{T_0}(t)$, a periodic signal formed by repeating $x(t)$ every T_0 seconds ($T_0 > \tau$), as depicted in Fig. 8.15b. This periodic signal can be expressed by the exponential Fourier series

$$x_{T_0}(t) = \sum_{n=-\infty}^{\infty} D_n e^{jn\omega_0 t} \qquad \omega_0 = \frac{2\pi}{T_0}$$

where (assuming $T_0 > \tau$)

$$D_n = \frac{1}{T_0}\int_{0}^{T_0} x(t)e^{-jn\omega_0 t}\,dt = \frac{1}{T_0}\int_{0}^{\tau} x(t)e^{-jn\omega_0 t}\,dt$$

From Eq. (8.16) it follows that

$$D_n = \frac{1}{T_0}X(n\omega_0)$$

This result indicates that the coefficients of the Fourier series for $x_{T_0}(t)$ are $(1/T_0)$ times the sample values of the spectrum $X(\omega)$ taken at intervals of ω_0. This means that the spectrum of the periodic signal $x_{T_0}(t)$ is the sampled spectrum $X(\omega)$, as illustrated in Fig. 8.15b. Now as long as $T_0 > \tau$, the successive cycles of $x(t)$ appearing in $x_{T_0}(t)$ do not overlap, and $x(t)$ can be recovered from $x_{T_0}(t)$. Such recovery implies indirectly that $X(\omega)$ can be reconstructed from its samples. These samples are separated by the fundamental frequency $f_0 = 1/T_0$ Hz of the periodic signal $x_{T_0}(t)$. Hence, the condition for recovery is $T_0 > \tau$; that is,

$$f_0 < \frac{1}{\tau} \text{ Hz}$$

Therefore, to be able to reconstruct the spectrum $X(\omega)$ from the samples of $X(\omega)$, the samples should be taken at frequency intervals $f_0 < 1/\tau$ Hz. If R is the sampling rate (samples/Hz), then

$$R = \frac{1}{f_0} > \tau \text{ samples/Hz} \tag{8.17}$$

SPECTRAL INTERPOLATION

Consider a signal timelimited to τ seconds and centered at T_c. we now show that the spectrum $X(\omega)$ of $x(t)$ can be reconstructed from the samples of $X(\omega)$. For this case, using the dual of the approach employed to derive the signal interpolation formula in Eq. (8.11b), we obtain the spectral interpolation formula[†]

$$X(\omega) = \sum_{n=-\infty}^{\infty} X(n\omega_0) \text{ sinc}\left(\frac{\omega T_0}{2} - n\pi\right) e^{-j(\omega-n\omega_0)T_c} \qquad \omega_0 = \frac{2\pi}{T_0} \qquad T_0 > \tau \tag{8.18}$$

For the case in Fig. 8.15, $T_c = T_0/2$. If the pulse $x(t)$ were to be centered at the origin, then $T_c = 0$, and the exponential term at the extreme right in Eq. (8.18) would vanish. In such a case Eq. (8.18) would be the exact dual of Eq. (8.11b).

EXAMPLE 8.6

The spectrum $X(\omega)$ of a unit duration signal $x(t)$, centered at the origin, is sampled at the intervals of 1 Hz or 2π rad/s (the Nyquist rate). The samples are:

$$X(0) = 1 \qquad \text{and} \qquad X(\pm 2\pi n) = 0 \qquad n = 1, 2, 3, \ldots$$

Find $x(t)$.

─────────

We use the interpolation formula (8.18) (with $T_c = 0$) to construct $X(\omega)$ from its samples. Since all but one of the Nyquist samples are zero, only one term (corresponding to $n = 0$)

[†]This can be obtained by observing that the Fourier transform of $x_{T_0}(t)$ is $2\pi \sum_n D_n \delta(\omega - n\omega_0)$ [see Eq. (7.26)]. We can recover $x(t)$ from $x_{T_0}(t)$ by multiplying the latter with rect $(t - T_c)/T_0$, whose Fourier transform is T_0 sinc $(\omega T_0/2)e^{-j\omega T_c}$. Hence, $X(\omega)$ is $1/2\pi$ times the convolution of these two Fourier transforms, which yields Eq. (8.18).

in the summation on the right-hand side of Eq. (8.18) survives. Thus, with $X(0) = 1$ and $\tau = T_0 = 1$, we obtain

$$X(\omega) = \text{sinc}\left(\frac{\omega}{2}\right) \qquad \text{and} \qquad x(t) = \text{rect}(t) \tag{8.19}$$

For a signal of unit duration, this is the only spectrum with the sample values $X(0) = 1$ and $X(2\pi n) = 0 \ (n \neq 0)$. No other spectrum satisfies these conditions.

8.5 NUMERICAL COMPUTATION OF THE FOURIER TRANSFORM: THE DISCRETE FOURIER TRANSFORM (DFT)

Numerical computation of the Fourier transform of $x(t)$ requires sample values of $x(t)$ because a digital computer can work only with discrete data (sequence of numbers). Moreover, a computer can compute $X(\omega)$ only at some discrete values of ω [samples of $X(\omega)$]. We therefore need to relate the samples of $X(\omega)$ to samples of $x(t)$. This task can be accomplished by using the results of the two sampling theorems developed in Sections 8.1 and 8.4.

We begin with a timelimited signal $x(t)$ (Fig. 8.16a) and its spectrum $X(\omega)$ (Fig. 8.16b). Since $x(t)$ is timelimited, $X(\omega)$ is nonbandlimited. For convenience, we shall show all spectra as functions of the frequency variable f (in hertz) rather than ω. According to the sampling theorem, the spectrum $\overline{X}(\omega)$ of the sampled signal $\overline{x}(t)$ consists of $X(\omega)$ repeating every f_s Hz, where $f_s = 1/T$, as depicted in Fig. 8.16d.[†] In the next step, the sampled signal in Fig. 8.16c is repeated periodically every T_0 seconds, as illustrated in Fig. 8.16e. According to the spectral sampling theorem, such an operation results in sampling the spectrum at a rate of T_0 samples/Hz. This sampling rate means that the samples are spaced at $f_0 = 1/T_0$ Hz, as depicted in Fig. 8.16f.

The foregoing discussion shows that when a signal $x(t)$ is sampled and then periodically repeated, the corresponding spectrum is also sampled and periodically repeated. Our goal is to relate the samples of $x(t)$ to the samples of $X(\omega)$.

NUMBER OF SAMPLES

One interesting observation from Fig. 8.16e and 8.16f is that N_0, the number of samples of the signal in Fig. 8.16e in one period T_0 is identical to N_0', the number of samples of the spectrum in Fig. 8.16f in one period f_s. The reason is

$$N_0 = \frac{T_0}{T} \qquad \text{and} \qquad N_0' = \frac{f_s}{f_0} \tag{8.20a}$$

[†]There is a multiplying constant $1/T$ for the spectrum in Fig. 8.16d [see Eq. (8.4)], but this is irrelevant to our discussion here.

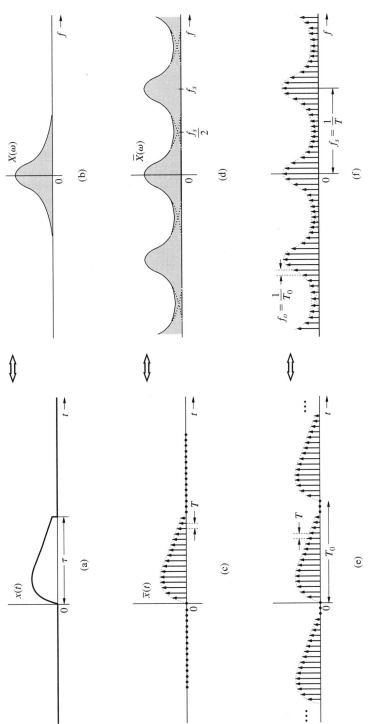

Figure 8.16 Relationship between samples of $x(t)$ and $X(\omega)$.

But, because

$$f_s = \frac{1}{T} \qquad \text{and} \qquad f_0 = \frac{1}{T_0} \tag{8.20b}$$

$$N_0 = \frac{T_0}{T} = \frac{f_s}{f_0} = N_0' \tag{8.20c}$$

ALIASING AND LEAKAGE IN NUMERICAL COMPUTATION

Figure 8.16f shows the presence of aliasing in the samples of the spectrum $X(\omega)$. This aliasing error can be reduced as much as desired by increasing the sampling frequency f_s (decreasing the sampling interval $T = 1/f_s$). The aliasing can never be eliminated for timelimited $x(t)$, however, because its spectrum $X(\omega)$ is nonbandlimited. Had we started out with a signal having a bandlimited spectrum $X(\omega)$, there would be no aliasing in the spectrum in Fig. 8.16f. Unfortunately such a signal is nontimelimited, and its repetition (in Fig. 8.16e) would result in signal overlapping (aliasing in the time domain). In this case we shall have to contend with errors in signal samples. In other words, in computing the direct or inverse Fourier transform numerically, we can reduce the error as much as we wish, but the error can never be eliminated. This is true of numerical computation of direct and inverse Fourier transforms, regardless of the method used. For example, if we determine the Fourier transform by direct integration numerically, by using Eq. (7.8a), there will be an error because the interval of integration Δt can never be made zero. Similar remarks apply to numerical computation of the inverse transform. Therefore, we should always keep in mind the nature of this error in our results. In our discussion (Fig. 8.16), we assumed $x(t)$ to be a timelimited signal. If $x(t)$ were not timelimited, we would need to timelimit it because numerical computations can work only with finite data. Further, this data truncation causes error because of spectral spreading (smearing) and leakage, as discussed in Section 7.8. The leakage also causes aliasing. Leakage can be reduced by using a tapered window for signal truncation. But this choice increases spectral spreading or smearing. Spectral spreading can be reduced by increasing the window width (i.e., more data), which increases T_0, and reduces f_0 (increases *spectral* or *frequency resolution*).

PICKET FENCE EFFECT

The numerical computation method yields only the uniform sample values of $X(\omega)$. The major peaks or valleys of $X(\omega)$ can lie between two samples and may remain hidden, giving a false picture of reality. Viewing samples is like viewing the signal and its spectrum through a "picket fence" with upright posts that are very wide and placed close together. What is hidden behind the pickets is much more than what we can see. Such misleading results can be avoided by using a sufficiently large N_0, the number of samples, to increase resolution. We can also use zero padding (discussed later) or the spectral interpolation formula [Eq. (8.18)] to determine the values of $X(\omega)$ between samples.

POINTS OF DISCONTINUITY

If $x(t)$ or $X(\omega)$ has a jump discontinuity at a sampling point, the sample value should be taken as the average of the values on the two sides of the discontinuity because the Fourier representation at a point of discontinuity converges to the average value.

DERIVATION OF THE DISCRETE FOURIER TRANSFORM (DFT)

If $x(nT)$ and $X(r\omega_0)$ are the nth and rth samples of $x(t)$ and $X(\omega)$, respectively, then we define new variables x_n and X_r as

$$x_n = Tx(nT)$$

$$= \frac{T_0}{N_0}x(nT) \tag{8.21a}$$

and

$$X_r = X(r\omega_0) \tag{8.21b}$$

where

$$\omega_0 = 2\pi f_0 = \frac{2\pi}{T_0} \tag{8.21c}$$

We shall now show that x_n and X_r are related by the following equations[†]:

$$X_r = \sum_{n=0}^{N_0-1} x_n e^{-jr\Omega_0 n} \tag{8.22a}$$

$$x_n = \frac{1}{N_0}\sum_{r=0}^{N_0-1} X_r e^{jr\Omega_0 n} \qquad \Omega_0 = \omega_0 T = \frac{2\pi}{N_0} \tag{8.22b}$$

These equations define the direct and the inverse *discrete Fourier transforms*, with X_r the direct discrete Fourier transform (DFT) of x_n, and x_n the inverse discrete Fourier transform (IDFT) of X_r. The notation

$$x_n \Longleftrightarrow X_r$$

is also used to indicate that x_n and X_r are a DFT pair. Remember that x_n is T_0/N_0 times the nth sample of $x(t)$ and X_r is the rth sample of $X(\omega)$. Knowing the sample values of $x(t)$, we can use the DFT to compute the sample values of $X(\omega)$—and vice versa. Note, however, that x_n is a function of n ($n = 0, 1, 2, \ldots, N_0 - 1$) rather than of t and that X_r is a function of r ($r = 0, 1, 2, \ldots, N_0 - 1$) rather than of ω. Moreover, both x_n and X_r are periodic sequences of period N_0 (Fig. 8.16e, 8.16f). Such sequences are called N_0-*periodic sequences*. The proof of the DFT relationships in Eqs. (8.22) follows directly from the results of the sampling theorem. The sampled signal $\overline{x}(t)$ (Fig. 8.16c) can be expressed as

$$\overline{x}(t) = \sum_{n=0}^{N_0-1} x(nT)\delta(t - nT) \tag{8.23}$$

Since $\delta(t - nT) \Longleftrightarrow e^{-jn\omega T}$, the Fourier transform of Eq. (8.23) yields

$$\overline{X}(\omega) = \sum_{n=0}^{N_0-1} x(nT)e^{-jn\omega T} \tag{8.24}$$

[†]In Eqs. (8.22a) and (8.22b), the summation is performed from 0 to $N_0 - 1$. It is shown in Section 9.1-2 [Eqs. (9.12) and (9.13)] that the summation may be performed over any successive N_0 values of n or r.

But from Fig. 8.1f [or Eq. (8.4)], it is clear that over the interval $|\omega| \leq \omega_s/2$, $\overline{X}(\omega)$, the Fourier transform of $\overline{x}(t)$ is $X(\omega)/T$, assuming negligible aliasing. Hence

$$X(\omega) = T\overline{X}(\omega) = T \sum_{n=0}^{N_0-1} x(nT)e^{-jn\omega T} \qquad |\omega| \leq \frac{\omega_s}{2}$$

and

$$X_r = X(r\omega_0) = T \sum_{n=0}^{N_0-1} x(nT)e^{-nkr\omega_0 T} \tag{8.25}$$

If we let $\omega_0 T = \Omega_0$, then from Eqs. (8.20a) and (8.20b)

$$\Omega_0 = \omega_0 T = 2\pi f_0 T = \frac{2\pi}{N_0} \tag{8.26}$$

Also, from Eq. (8.21a),

$$Tx(nT) = x_n$$

Therefore, Eq. (8.25) becomes

$$X_r = \sum_{n=0}^{N_0-1} x_n e^{-jr\Omega_0 n} \qquad \Omega_0 = \frac{2\pi}{N_0} \tag{8.27}$$

The inverse transform relationship (8.22b) could be derived by using a similar procedure with the roles of t and ω reversed, but here we shall use a more direct proof. To prove the inverse relation in Eq. (8.22b), we multiply both sides of Eq. (8.27) by $e^{jm\Omega_0 r}$ and sum over r as

$$\sum_{r=0}^{N_0-1} X_r e^{jm\Omega_0 r} = \sum_{r=0}^{N_0-1} \left[\sum_{n=0}^{N_0-1} x_n e^{-jr\Omega_0 n} \right] e^{jm\Omega_0 r}$$

By interchanging the order of summation on the right-hand side, we have

$$\sum_{r=0}^{N_0-1} X_r e^{jm\Omega_0 r} = \sum_{n=0}^{N_0-1} x_n \left[\sum_{r=0}^{N_0-1} e^{j(m-n)\Omega_0 r} \right]$$

We can readily show that the inner sum on the right-hand side is zero for $n \neq m$, and that the sum is N_0 when $n = m$. To avoid a break in the flow, the proof is placed in the footnote below.[†]

[†]We show that

$$\sum_{n=0}^{N_0-1} e^{jk\Omega_0 n} = \begin{cases} N_0 & k = 0, \pm N_0, \pm 2N_0, \ldots \\ 0 & \text{otherwise} \end{cases} \tag{8.28}$$

Recall that $\Omega_0 N_0 = 2\pi$. So $e^{jk\omega_0 n} = 1$ when $k = 0, \pm N_0, \pm 2N_0, \ldots$. Hence, the sum on the left-hand side of Eq. (8.28) is N_0. To compute the sum for other values of k, we note that the sum on the left-hand side of Eq. 8.28) is a geometric progression with common ratio $\alpha = e^{jk\Omega_0}$. Therefore, (see Section B.7-4)

$$\sum_{n=0}^{N_0-1} e^{jk\Omega_0 n} = \frac{e^{jk\Omega_0 N_0} - 1}{e^{jk\Omega_0} - 1} = 0 \qquad (e^{jk\Omega_0 N_0} = e^{j2\pi m} = 1)$$

Therefore the outer sum will have only one nonzero term when $n = m$, and it is $N_0 x_n = N_0 x_m$. Therefore

$$x_m = \frac{1}{N_0} \sum_{r=0}^{N_0-1} X_r e^{jm\Omega_0 r} \qquad \Omega_0 = \frac{2\pi}{N_0}$$

Because X_r is N_0 periodic, we need to determine the values of X_r over any one period. It is customary to determine X_r over the range $(0, N_0 - 1)$ rather than over the range $(-N_0/2, (N_0/2) - 1)$.[†]

CHOICE OF T AND T_0

In DFT computation, we first need to select suitable values for N_0 and T or T_0. For this purpose we should first decide on B, the essential bandwidth (in hertz) of the signal. The sampling frequency f_s must be at least $2B$, that is,

$$\frac{f_s}{2} \geq B \tag{8.29a}$$

Moreover, the sampling interval $T = 1/f_s$ [Eq. (8.20b)], and

$$T \leq \frac{1}{2B} \tag{8.29b}$$

Once we pick B, we can choose T according to Eq. (8.29b). Also,

$$f_0 = \frac{1}{T_0} \tag{8.30}$$

where f_0 is the *frequency resolution* [separation between samples of $X(\omega)$]. Hence, if f_0 is given, we can pick T_0 according to Eq. (8.30). Knowing T_0 and T, we determine N_0 from

$$N_0 = \frac{T_0}{T}$$

ZERO PADDING

Recall that observing X_r is like observing the spectrum $X(\omega)$ through a picket fence. If the frequency sampling interval f_0 is not sufficiently small, we could miss out on some significant details and obtain a misleading picture. To obtain a higher number of samples, we need to reduce f_0. Because $f_0 = 1/T_0$, a higher number of samples requires us to increase the value of T_0, the period of repetition for $x(t)$. This option increases N_0, the number of samples of $x(t)$, by adding dummy samples of 0 value. This addition of dummy samples is known as *zero padding*. Thus, zero padding increases the number of samples and may help in getting a better idea of the spectrum $X(\omega)$ from its samples X_r. To continue with our picket fence analogy, zero padding is like using more, and narrower, pickets.

[†]The DFT equations (8.22) represent a transform in their own right, and they are exact. There is no approximation. However, x_n and X_r, thus obtained, are only approximations to the actual samples of a signal $x(t)$ and of its Fourier transform $X(\omega)$.

ZERO PADDING DOES NOT IMPROVE
ACCURACY OR RESOLUTION

Actually, we are not observing $X(\omega)$ through a picket fence. We are observing a distorted version of $X(\omega)$ resulting from the truncation of $x(t)$. Hence, we should keep in mind that even if the fence were transparent, we would see a reality distorted by aliasing. Seeing through the picket fence just gives us an imperfect view of the imperfectly represented reality. Zero padding only allows us to look at more samples of that imperfect reality. It can never reduce the imperfection in what is behind the fence. The imperfection, which is caused by aliasing, can only be lessened by reducing the sampling interval T. Observe that reducing T also increases N_0, the number of samples and is also like increasing the number of pickets while reducing their width. But in this case, the reality behind the fence is also better dressed and we see more of it.

EXAMPLE 8.7

A signal $x(t)$ has a duration of 2 ms and an essential bandwidth of 10 kHz. It is desirable to have a frequency resolution of 100 Hz in the DFT ($f_0 = 100$). Determine N_0.

To have $f_0 = 100$ Hz, the effective signal duration T_0 must be

$$T_0 = \frac{1}{f_0} = \frac{1}{100} = 10 \text{ ms}$$

Since the signal duration is only 2 ms, we need zero padding over 8 ms. Also, $B = 10,000$. Hence, $f_s = 2B = 20,000$ and $T = 1/f_s = 50\,\mu\text{s}$. Further,

$$N_0 = \frac{f_s}{f_0} = \frac{20,000}{100} = 200$$

The *fast Fourier transform* (FFT) algorithm (discussed later: see Section 8.6) is used to compute DFT, where it proves convenient (although not necessary) to select N_0 as a power of 2; that is, $N_0 = 2^n$ (n, integer). Let us choose $N_0 = 256$. Increasing N_0 from 200 to 256 can be used to reduce aliasing error (by reducing T), to improve resolution (by increasing T_0 using zero padding), or a combination of both:

(i) **Reducing Aliasing Error.** We maintain the same T_0 so that $f_0 = 100$. Hence

$$f_s = N_0 f_0 = 256 \times 100 = 25,600 \quad \text{and} \quad T = \frac{1}{f_s} = 39\,\mu\text{s}$$

Thus, increasing N_0 from 200 to 256 permits us to reduce the sampling interval T from 50 μs to 39 μs while maintaining the same frequency resolution ($f_0 = 100$).

(ii) **Improving Resolution.** Here we maintain the same $T = 50\,\mu$s, which yields

$$T_0 = N_0 T = 256(50 \times 10^{-6}) = 12.8 \text{ ms} \quad \text{and} \quad f_0 = \frac{1}{T_0} = 78.125 \text{ Hz}$$

Thus, increasing N_0 from 200 to 256 can improve the frequency resolution from 100 to 78.125 Hz while maintaining the same aliasing error ($T = 50\,\mu s$).

(iii) **Combination of These Two Options.** We could choose $T = 45\,\mu s$ and $T_0 = 11.5$ ms so that $f_0 = 86.96$ Hz.

EXAMPLE 8.8

Use DFT to compute the Fourier transform of $e^{-2t}u(t)$. Plot the resulting Fourier spectra.

We first determine T and T_0. The Fourier transform of $e^{-2t}u(t)$ is $1/(j\omega + 2)$. This lowpass signal is not bandlimited. In Section 7.6, we used the energy criterion to compute the essential bandwidth of a signal. Here, we shall present a simpler, but workable alternative to the energy criterion. The essential bandwidth of a signal will be taken as the frequency at which $|X(\omega)|$ drops to 1% of its peak value (see the footnote on page 732). In this case, the peak value occurs at $\omega = 0$, where $|X(0)| = 0.5$. Observe that

$$|X(\omega)| = \frac{1}{\sqrt{\omega^2 + 4}} \approx \frac{1}{\omega} \qquad \omega \gg 2$$

Also, 1% of the peak value is $0.01 \times 0.5 = 0.005$. Hence, the essential bandwidth B is at $\omega = 2\pi B$, where

$$|X(\omega)| \approx \frac{1}{2\pi B} = 0.005 \quad \Rightarrow \quad B = \frac{100}{\pi}\ \text{Hz}$$

and from Eq. (8.29b),

$$T \le \frac{1}{2B} = \frac{\pi}{200} = 0.015708$$

Had we used 1% energy criterion to determine the essential bandwidth, following the procedure in Example 7.20, we would have obtained $B = 20.26$ Hz, which is somewhat smaller than the value just obtained by using the 1% amplitude criterion.

The second issue is to determine T_0. Because the signal is not timelimited, we have to truncate it at T_0 such that $x(T_0) \ll 1$. A reasonable choice would be $T_0 = 4$ because $x(4) = e^{-8} = 0.000335 \ll 1$. The result is $N_0 = T_0/T = 254.6$, which is not a power of 2. Hence, we choose $T_0 = 4$, and $T = 0.015625 = 1/64$, yielding $N_0 = 256$, which is a power of 2.

Note that there is a great deal of flexibility in determining T and T_0, depending on the accuracy desired and the computational capacity available. We could just as well have chosen $T = 0.03125$, yielding $N_0 = 128$, although this choice would have given a slightly higher aliasing error.

Because the signal has a jump discontinuity at $t = 0$, the first sample (at $t = 0$) is 0.5, the averages of the values on the two sides of the discontinuity. We compute X_r (the DFT) from the samples of $e^{-2t}u(t)$ according to Eq. (8.22a). Note that X_r is the rth sample of $X(\omega)$, and these samples are spaced at $f_0 = 1/T_0 = 0.25$ Hz. ($\omega_0 = \pi/2$ rad/s.)

Because X_r is N_0 periodic, $X_r = X_{(r+256)}$ so that $X_{256} = X_0$. Hence, we need to plot X_r over the range $r = 0$ to 255 (not 256). Moreover, because of this periodicity, $X_{-r} = X_{(-r+256)}$, and the values of X_r over the range $r = -127$ to -1 are identical to those over the range $r = 129$ to 255. Thus, $X_{-127} = X_{129}$, $X_{-126} = X_{130}, \ldots, X_{-1} = X_{255}$. In addition, because of the property of conjugate symmetry of the Fourier transform, $X_{-r} = X_r^*$, it follows that $X_{-1} = X_1^*$, $X_{-2} = X_2^*, \ldots, X_{-128} = X_{128}^*$. Thus, we need X_r only over the range $r = 0$ to $N_0/2$ (128 in this case).

Figure 8.17 shows the computed plots of $|X_r|$ and $\angle X_r$. The exact spectra are depicted by continuous curves for comparison. Note the nearly perfect agreement between the two sets of spectra. We have depicted the plot of only the first 28 points rather than all 128 points, which would have made the figure very crowded, resulting in loss of clarity. The points are at the intervals of $1/T_0 = 1/4$ Hz or $\omega_0 = 1.5708$ rad/s. The 28 samples, therefore, exhibit the plots over the range $\omega = 0$ to $\omega = 28(1.5708) \approx 44$ rad/s or 7 Hz.

In this example, we knew $X(\omega)$ beforehand and hence could make intelligent choices for B (or the sampling frequency f_s). In practice, we generally do not know $X(\omega)$ beforehand. In fact, that is the very thing we are trying to determine. In such a case, we must make an intelligent guess for B or f_s from circumstantial evidence. We should then continue reducing the value of T and recomputing the transform until the result stabilizes within the desired number of significant digits. The MATLAB program, which implements the DFT using the FFT algorithm, is presented in Computer Example C8.1.

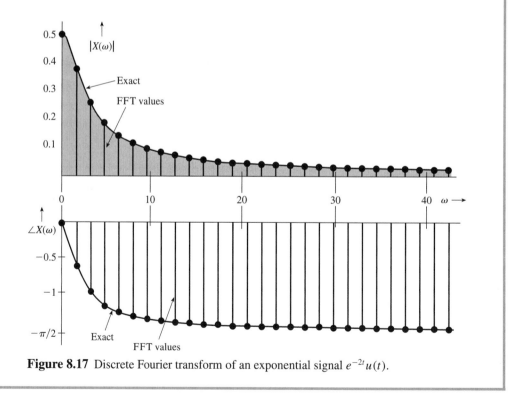

Figure 8.17 Discrete Fourier transform of an exponential signal $e^{-2t}u(t)$.

COMPUTER EXAMPLE C8.1

Using MATLAB, repeat Example 8.8.

First, MATLAB's `fft` command is used to compute the DFT.

```
>> T_0 = 4; N_0 = 256;
>> T = T_0/N_0; t = (0:T:T*(N_0-1))';
>> x = T*exp(-2*t); x(1) = T*(exp(-2*T_0)+1)/2;
>> X_r = fft(x); r = [-N_0/2:N_0/2-1]'; omega_r = r*2*pi/T_0;
```

The true Fourier transform is also computed for comparison.

```
>> omega = linspace(-pi/T,pi/T,4097); X = 1./(j*omega+2);
```

For clarity, we display spectrum over a restricted frequency range.

```
>> subplot(211);
>> plot(omega,abs(X),'k',omega_r,fftshift(abs(X_r)),'ko');
>> xlabel('\omega'); ylabel('|X(\omega)|')
>> axis([-0.01 40 -0.01 0.51]);
>> legend('True FT',['DFT with T_0 = ',num2str(T_0),...
>>            ', N_0 = ',num2str(N_0)],0);
>> subplot(212);
>> plot(omega,angle(X),'k',omega_r,fftshift(angle(X_r)),'ko');
>> xlabel('\omega'); ylabel('\angle X(\omega)')
>> axis([-0.01 40 -pi/2-0.01 0.01]);
>> legend('True FT',['DFT with T_0 = ',num2str(T_0),...
>>            ', N_0 = ',num2str(N_0)],0);
```

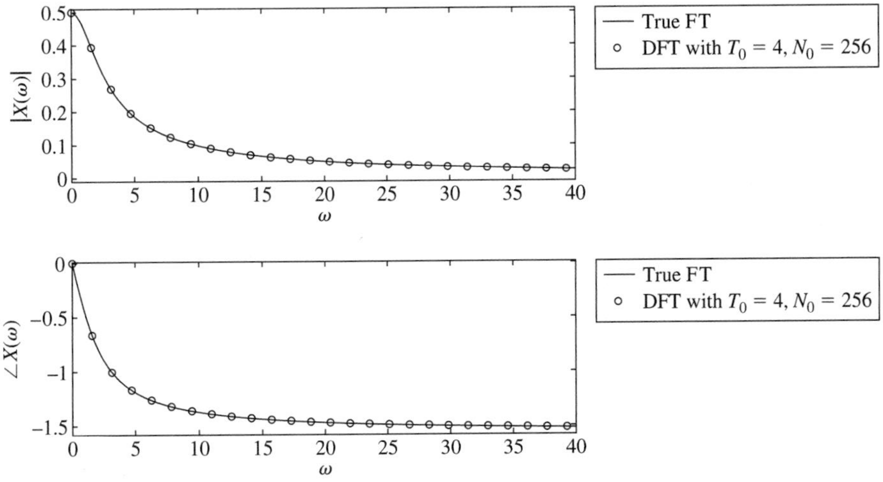

Figure C8.1

EXAMPLE 8.9

Use DFT to compute the Fourier transform of $8 \operatorname{rect}(t)$.

This gate function and its Fourier transform are illustrated in Fig. 8.18a and 8.18b. To determine the value of the sampling interval T, we must first decide on the essential bandwidth B. In Fig. 8.18b, we see that $X(\omega)$ decays rather slowly with ω. Hence, the essential bandwidth B is rather large. For instance, at $B = 15.5$ Hz (97.39 rad/s), $X(\omega) = -0.1643$, which is about 2% of the peak at $X(0)$. Hence, the essential bandwidth is well above 16 Hz if we use the 1% of the peak amplitude criterion for computing the essential bandwidth. However, we shall deliberately take $B = 4$ for two reasons: to show the effect of aliasing and because the use of $B > 4$ would give an enormous number of samples, which could not be conveniently displayed on the page without losing sight of the essentials. Thus, we shall intentionally accept approximation to clarify the concepts of DFT graphically.

The choice of $B = 4$ results in the sampling interval $T = 1/2B = 1/8$. Looking again at the spectrum in Fig. 8.18b, we see that the choice of the frequency resolution $f_0 = 1/4$ Hz is reasonable. Such a choice gives us four samples in each lobe of $X(\omega)$. In this case $T_0 = 1/f_0 = 4$ seconds and $N_0 = T_0/T = 32$. The duration of $x(t)$ is only 1 second. We must repeat it every 4 seconds ($T_0 = 4$), as depicted in Fig. 8.18c, and take samples every $1/8$ second. This choice yields 32 samples ($N_0 = 32$). Also,

$$x_n = Tx(nT)$$

$$= \tfrac{1}{8}x(nT)$$

Since $x(t) = 8 \operatorname{rect}(t)$, the values of x_n are 1, 0, or 0.5 (at the points of discontinuity), as illustrated in Fig. 8.18c, where x_n is depicted as a function of t as well as n, for convenience.

In the derivation of the DFT, we assumed that $x(t)$ begins at $t = 0$ (Fig. 8.16a), and then took N_0 samples over the interval $(0, T_0)$. In the present case, however, $x(t)$ begins at $-1/2$. This difficulty is easily resolved when we realize that the DFT obtained by this procedure is actually the DFT of x_n repeating periodically every T_0 seconds. Figure 8.18c clearly indicates that periodic repeating the segment of x_n over the interval from -2 to 2 seconds yields the same signal as the periodic repeating the segment of x_n over the interval from 0 to 4 seconds. Hence, the DFT of the samples taken from -2 to 2 seconds is the same as that of the samples taken from 0 to 4 seconds. Therefore, regardless of where $x(t)$ starts, we can always take the samples of $x(t)$ and its periodic extension over the interval from 0 to T_0. In the present example, the 32 sample values are

$$x_n = \begin{cases} 1 & 0 \leq n \leq 3 \quad \text{and} \quad 29 \leq n \leq 31 \\ 0 & 5 \leq n \leq 27 \\ 0.5 & n = 4, 28 \end{cases}$$

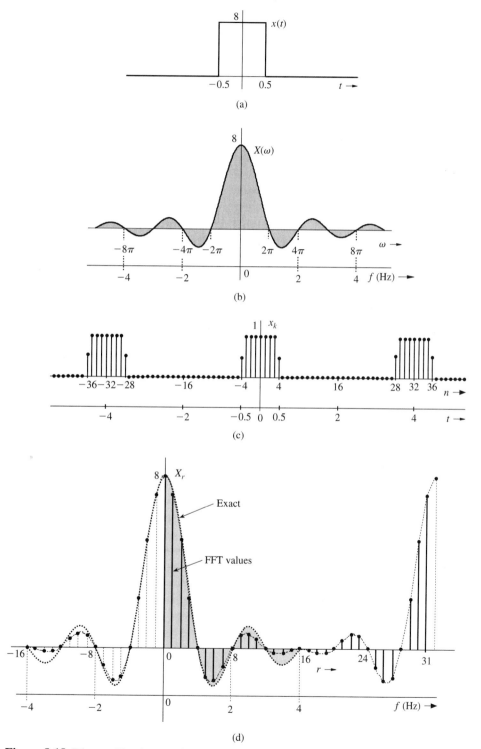

Figure 8.18 Discrete Fourier transform of a gate pulse.

Observe that the last sample is at $t = 31/8$, not at 4, because the signal repetition starts at $t = 4$, and the sample at $t = 4$ is the same as the sample at $t = 0$. Now, $N_0 = 32$ and $\Omega_0 = 2\pi/32 = \pi/16$. Therefore [see Eq. (8.22a)]

$$X_r = \sum_{n=0}^{31} x_n e^{-jr(\pi/16)n}$$

Values of X_r are computed according to this equation and plotted in Fig. 8.18d.

The samples X_r are separated by $f_0 = 1/T_0$ Hz. In this case $T_0 = 4$, so the frequency resolution f_0 is 1/4 Hz, as desired. The folding frequency $f_s/2 = B = 4$ Hz corresponds to $r = N_0/2 = 16$. Because X_r is N_0 periodic ($N_0 = 32$), the values of X_r for $r = -16$ to $n = -1$ are the same as those for $r = 16$ to $n = 31$. For instance, $X_{17} = X_{-15}$, $X_{18} = X_{-14}$, and so on. The DFT gives us the samples of the spectrum $X(\omega)$.

For the sake of comparison, Fig. 8.18d also shows the shaded curve 8 sinc $(\omega/2)$, which is the Fourier transform of 8 rect (t). The values of X_r computed from the DFT equation show aliasing error, which is clearly seen by comparing the two superimposed plots. The error in X_2 is just about 1.3%. However, the aliasing error increases rapidly with r. For instance, the error in X_6 is about 12%, and the error in X_{10} is 33%. The error in X_{14} is a whopping 72%. The percent error increases rapidly near the folding frequency ($r = 16$) because $x(t)$ has a jump discontinuity, which makes $X(\omega)$ decay slowly as $1/\omega$. Hence, near the folding frequency, the inverted tail (due to aliasing) is very nearly equal to $X(\omega)$ itself. Moreover, the final values are the difference between the exact and the folded values (which are very close to the exact values). Hence, the percent error near the folding frequency ($r = 16$ in this case) is very high, although the absolute error is very small. Clearly, for signals with jump discontinuities, the aliasing error near the folding frequency will always be high (in percentage terms), regardless of the choice of N_0. To ensure a negligible aliasing error at any value r, we must make sure that $N_0 \gg r$. This observation is valid for all signals with jump discontinuities.

COMPUTER EXAMPLE C8.2

Using MATLAB, repeat Example 8.9.

First, MATLAB's `fft` command is used to compute the DFT.

```
>> T_0 = 4; N_0 = 32; T = T_0/N_0;
>> x_n = [ones(1,4) 0.5 zeros(1,23) 0.5 ones(1,3)]';
>> X_r = fft(x_n); r = [-N_0/2:N_0/2-1]'; omega_r = r*2*pi/T_0;
```

The true Fourier transform is also computed for comparison.

```
>> omega = linspace(-pi/T,pi/T,4097); X = 8*sinc(omega/(2*pi));
```

Next, the spectrum is displayed.

```
>> figure(1); subplot(2,1,1);
>> plot(omega,abs(X),'k',omega_r,fftshift(abs(X_r)),'ko');
>> xlabel('\omega'); ylabel('|X(\omega)|'); axis tight
>> legend('True FT',['T_0 = ',num2str(T_0),...
>>             ', N_0 = ',num2str(N_0)],0);
>> subplot(2,1,2);
>> plot(omega,angle(X),'k',omega_r,fftshift(angle(X_r)),'ko');
>> xlabel('\omega'); ylabel('\angle X(\omega)'); axis([-25 25 -.5 3.5]);
```

Notice that the DFT approximation does not perfectly follow the true Fourier transform, especially at high frequencies. As in Example 8.9, this is because the parameter B is deliberately set too small.

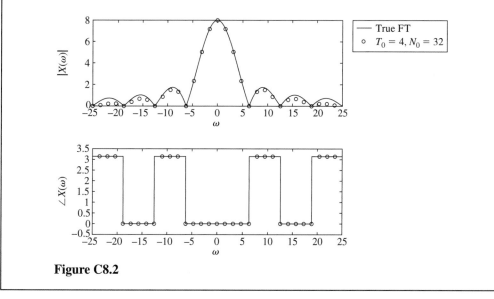

Figure C8.2

8.5-1 Some Properties of the DFT

The discrete Fourier transform is basically the Fourier transform of a sampled signal repeated periodically. Hence, the properties derived earlier for the Fourier transform apply to the DFT as well.

LINEARITY

If $x_n \Longleftrightarrow X_r$ and $g_n \Longleftrightarrow G_r$, then

$$a_1 x_n + a_2 g_n \Longleftrightarrow a_1 X_r + a_2 G_r \qquad (8.31)$$

The proof is trivial.

Conjugate Symmetry

From the conjugation property $x^*(t) \iff X^*(-\omega)$, we have

$$x_n^* \longleftrightarrow X_{-r}^* \tag{8.32a}$$

From this equation and the time-reversal property, we obtain

$$x_{-n}^* \longleftrightarrow X_r^* \tag{8.32b}$$

When $x(t)$ is real, then the conjugate symmetry property states that $X^*(\omega) = X(-\omega)$. Hence, for real x_n

$$X_r^* = X_{-r}$$

Moreover, X_r is N_0 periodic. Hence

$$X_r^* = X_{N_0-r} \tag{8.32c}$$

Because of this property, we need compute only half the DFTs for real x_n. The other half are the conjugates.

Time Shifting (Circular Shifting)[†]

$$x_{n-k} \iff X_r e^{-jr\Omega_0 k} \tag{8.33}$$

Proof. We use Eq. (8.22b) to find the inverse DFT of $X_r e^{-jr\Omega_0 k}$ as

$$\frac{1}{N_0} \sum_{r=0}^{N_0-1} X_r e^{-jr\Omega_0 k} e^{jr\omega_0 n} = \frac{1}{N_0} \sum_{r=0}^{N_0-1} X_r e^{jr\Omega_0(n-k)} = x_{n-k}$$

Frequency Shifting

$$x_n e^{jn\Omega_0 m} \iff X_{r-m} \tag{8.34}$$

Proof. This proof is identical to that of the time-shifting property except that we start with Eq. (8.22a).

Periodic Convolution

$$x_n \ \textcircled{p}\ g_n \iff X_r G_r \tag{8.35a}$$

and

$$x_n g_n \iff \frac{1}{N_0} X_r \ \textcircled{p}\ G_r \tag{8.35b}$$

[†]Also known as *circular shifting* because such a shift can be interpreted as a circular shift of the N_0 samples in the first cycle $0 \le n \le N_0 - 1$.

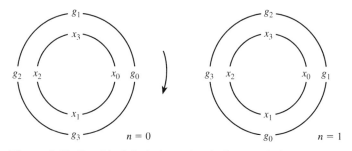

Figure 8.19 Graphical depictions of periodic convolution.

For two N_0-periodic sequences x_n and g_n, the periodic convolution is defined by

$$x_n \circledP g_n = \sum_{k=0}^{N_0-1} x_k g_{n-k} = \sum_{k=0}^{N_0-1} g_k x_{n-k} \tag{8.36}$$

To prove condition (8.35a), we find the DFT of periodic convolution $x_n \circledP g_n$ as

$$\sum_{n=0}^{N_0-1} \left(\sum_{k=0}^{N_0-1} x_k g_{n-k} \right) e^{-jr\omega_0 n} = \sum_{k=0}^{N_0-1} x_k \left(\sum_{n=0}^{N_0-1} g_{n-k} e^{-jr\omega_0 n} \right)$$

$$= \sum_{k=0}^{N_0-1} x_k (G_r e^{-jr\Omega_0 k}) = X_r G_r$$

Equation (8.35b) can be proved in the same way.

For periodic sequences, the convolution can be visualized in terms of two sequences, with one sequence fixed and the other inverted and moved past the fixed sequence, one digit at a time. If the two sequences are N_0 periodic, the same configuration will repeat after N_0 shifts of the sequence. Clearly the convolution $x_n \circledP g_n$ becomes N_0 periodic. Such convolution can be conveniently visualized in terms of N_0 sequences, as illustrated in Fig. 8.19, for the case of $N_0 = 4$. The inner N_0-point sequence x_n is clockwise and fixed. The outer N_0-point sequence g_n is inverted so that it becomes counterclockwise. This sequence is now rotated clockwise one unit at a time. We multiply the overlapping numbers and add. For example, the value of $x_n \circledP g_n$ at $n = 0$ (Fig. 8.19) is

$$x_0 g_0 + x_1 g_3 + x_2 g_2 + x_3 g_1$$

and the value of $x_n \circledP g_n$ at $n = 1$ is (Fig. 8.19)

$$x_0 g_1 + x_1 g_0 + x_2 g_3 + x_3 g_2$$

and so on.

8.5-2 Some Applications of the DFT

The DFT is useful not only in the computation of direct and inverse Fourier transforms, but also in other applications such as convolution, correlation, and filtering. Use of the efficient FFT algorithm, discussed shortly (Section 8.6), makes it particularly appealing.

LINEAR CONVOLUTION

Let $x(t)$ and $g(t)$ be the two signals to be convolved. In general, these signals may have different time durations. To convolve them by using their samples, they must be sampled at the same rate (not below the Nyquist rate of either signal). Let $x_n (0 \leq n \leq N_1 - 1)$ and $g_n (0 \leq n \leq N_2 - 1)$ be the corresponding discrete sequences representing these samples. Now,

$$c(t) = x(t) * g(t)$$

and if we define three sequences as $x_n = Tx(nT)$, $g_n = Tg(nT)$, and $c_n = Tc(nT)$, then[†]

$$c_n = x_n * g_n$$

where we define the linear convolution sum of two discrete sequences x_n and g_n as

$$c_n = x_n * g_n = \sum_{k=-\infty}^{\infty} x_k g_{n-k}$$

Because of the width property of the convolution, c_n exists for $0 \leq n \leq N_1 + N_2 - 1$. To be able to use the DFT periodic convolution technique, we must make sure that the periodic convolution will yield the same result as the linear convolution. In other words, the signal resulting from the periodic convolution must have the same length ($N_1 + N_2 - 1$) as that of the signal resulting from linear convolution. This step can be accomplished by adding $N_2 - 1$ dummy samples of zero value to x_n and $N_1 - 1$ dummy samples of zero value to g_n (zero padding). This procedure changes the length of both x_n and g_n to $N_1 + N_2 - 1$. The periodic convolution now is identical to the linear convolution except that it repeats periodically with period $N_1 + N_2 - 1$. A little reflection will show that in such a case the circular convolution procedure in Fig. 8.19 over one cycle ($0 \leq n \leq N_1 + N_2 - 1$) is identical to the linear convolution of the two sequences x_n and g_n. We can use the DFT to find the convolution $x_n * g_n$ in three steps, as follows:

1. Find the DFTs X_r and G_r corresponding to suitably padded x_n and g_n.
2. Multiply X_r by G_r.
3. Find the IDFT of $X_r G_r$. This procedure of convolution, when implemented by the fast Fourier transform algorithm (discussed later), is known as the *fast convolution*.

FILTERING

We generally think of filtering in terms of some hardware-oriented solution (namely, building a circuit with RLC components and operational amplifiers). However, filtering also has a software-oriented solution [a computer algorithm that yields the filtered output $y(t)$ for a given input $x(t)$]. This goal can be conveniently accomplished by using the DFT. If $x(t)$ is the signal to be filtered, then X_r, the DFT of x_n, is found. The spectrum X_r is then shaped (filtered) as desired by multiplying X_r by H_r, where H_r are the samples of $H(\omega)$ for the filter [$H_r = H(r\omega_0)$]. Finally,

---

[†]We can show that[4] $c_n = \lim_{T \to 0} x_n * g_n$. Because $T \neq 0$ in practice, there will be some error in this equation. This error is inherent in any numerical method used to compute convolution of continuous-time signals.

we take the IDFT of $X_r H_r$ to obtain the filtered output $y_n[y_n = Ty(nT)]$. This procedure is demonstrated in the following example.

EXAMPLE 8.10

The signal $x(t)$ in Fig. 8.20a is passed through an ideal lowpass filter of frequency response $H(\omega)$ depicted in Fig. 8.20b. Use the DFT to find the filter output.

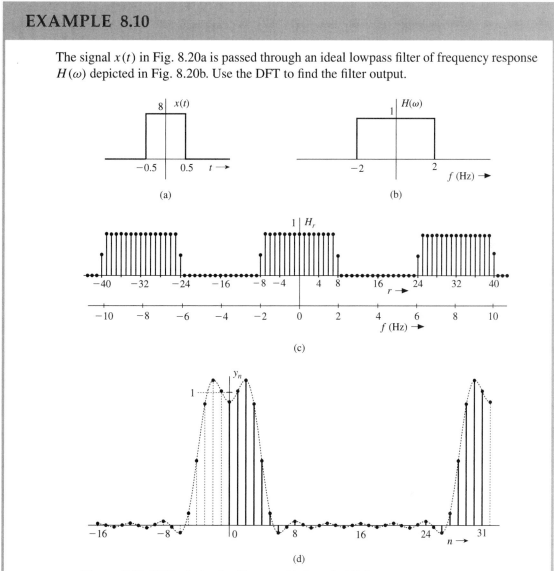

Figure 8.20 DFT solution for filtering $x(t)$ through $H(\omega)$.

We have already found the 32-point DFT of $x(t)$ (see Fig. 8.18d). Next we multiply X_r by H_r. To find H_r, we recall using $f_0 = 1/4$ in computing the 32-point DFT of $x(t)$. Because X_r is 32-periodic, H_r must also be 32-periodic with samples separated by 1/4 Hz. This fact means that H_r must be repeated every 8 Hz or 16π rad/s (see Fig. 8.20c). The resulting 32 samples of H_r over $(0 \leq \omega \leq 16\pi)$ are as follows:

$$H_r = \begin{cases} 1 & 0 \leq r \leq 7 \quad \text{and} \quad 25 \leq r \leq 31 \\ 0 & 9 \leq r \leq 23 \\ 0.5 & r = 8, 24 \end{cases}$$

We multiply X_r with H_r. The desired output signal samples y_n are found by taking the inverse DFT of $X_r H_r$. The resulting output signal is illustrated in Fig. 8.20d. Table 8.2 gives a printout of x_n, X_r, H_r, Y_r, and y_n.

TABLE 8.2

No.	x_k	X_r	H_r	$X_r H_r$	y_k
0	1	8.000	1	8.000	0.9285
1	1	7.179	1	7.179	1.009
2	1	5.027	1	5.027	1.090
3	1	2.331	1	2.331	0.9123
4	0.5	0.000	1	0.000	0.4847
5	0	−1.323	1	−1.323	0.08884
6	0	−1.497	1	−1.497	−0.05698
7	0	−0.8616	1	−0.8616	−0.01383
8	0	0.000	0.5	0.000	0.02933
9	0	0.5803	0	0.000	0.004837
10	0	0.6682	0	0.000	−0.01966
11	0	0.3778	0	0.000	−0.002156
12	0	0.000	0	0.000	0.01534
13	0	−0.2145	0	0.000	0.0009828
14	0	−0.1989	0	0.000	−0.01338
15	0	−0.06964	0	0.000	−0.0002876
16	0	0.000	0	0.000	0.01280
17	0	−0.06964	0	0.000	−0.0002876
18	0	−0.1989	0	0.000	−0.01338
19	0	−0.2145	0	0.000	0.0009828
20	0	0.000	0	0.000	0.01534
21	0	0.3778	0	0.000	−0.002156
22	0	0.6682	0	0.000	−0.01966
23	0	0.5803	0	0.000	0.004837
24	0	0.000	0.5	0.000	0.03933
25	0	−0.8616	1	−0.8616	−0.01383
26	0	−1.497	1	−1.497	−0.05698
27	0	−1.323	1	−1.323	0.08884
28	0.5	0.000	1	0.000	0.4847
29	1	2.331	1	2.331	0.9123
30	1	5.027	1	5.027	1.090
31	1	7.179	1	7.179	1.009

COMPUTER EXAMPLE C8.3

Solve Example 8.10 using MATLAB.

```
>> T_0 = 4; N_0 = 32; T = T_0/N_0; n = (0:N_0-1); r = n;
>> x_n = [ones(1,4) 0.5 zeros(1,23) 0.5 ones(1,3)]'; X_r = fft(x_n);
>> H_r = [ones(1,8) 0.5 zeros(1,15) 0.5 ones(1,7)]';
>> Y_r = H_r.*X_r; y_n = ifft(Y_r);
>> figure(1); subplot(2,2,1); stem(n,x_n,'k');
>> xlabel('n'); ylabel('x_n'); axis([0 31 -.1 1.1]);
>> subplot(2,2,2); stem(r,real(X_r),'k');
>> xlabel('r'); ylabel('X_r'); axis([0 31 -2 8]);
>> subplot(2,2,3); stem(n,real(y_n),'k');
>> xlabel('n'); ylabel('y_n'); axis([0 31 -.1 1.1]);
>> subplot(2,2,4); stem(r,X_r.*H_r,'k');
>> xlabel('r'); ylabel('Y_r = X_RH_r'); axis([0 31 -2 8]);
```

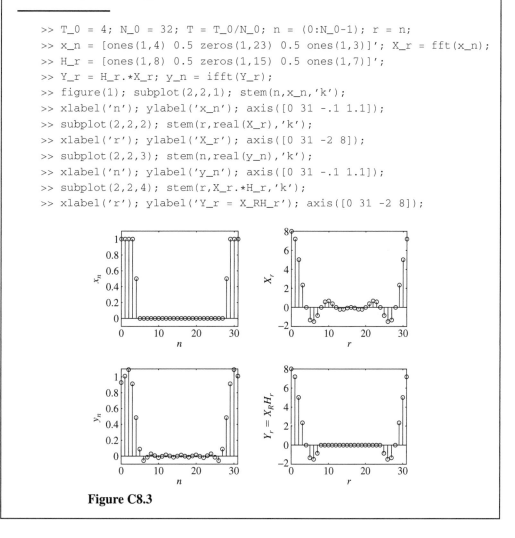

Figure C8.3

8.6 THE FAST FOURIER TRANSFORM (FFT)

The number of computations required in performing the DFT was dramatically reduced by an algorithm developed by Cooley and Tukey in 1965.[5] This algorithm, known as the *fast Fourier transform* (FFT), reduces the number of computations from something on the order of N_0^2 to $N_0 \log N_0$. To compute one sample X_r from Eq. (8.22a), we require N_0 complex multiplications and $N_0 - 1$ complex additions. To compute N_0 such values (X_r for $r = 0, 1, \ldots, N_0 - 1$), we require a total of N_0^2 complex multiplications and $N_0(N_0 - 1)$ complex additions. For a large N_0, these computations can be prohibitively time-consuming, even for a high-speed computer. The FFT algorithm is what made the use of Fourier transform accessible for digital signal processing.

HOW DOES FFT REDUCE NUMBER OF COMPUTATIONS?

It is easy to understand the magic of FFT. The secret is in the linearity of the Fourier transform, and also of the DFT. Because of linearity, we can compute the Fourier transform of a signal $x(t)$ as a sum of the Fourier transforms of segments of $x(t)$ of shorter duration. The same principle applies to computation of DFT. Consider a signal of length $N_0 = 16$ samples. As seen earlier, DFT computation of this sequence requires $N_0^2 = 256$ multiplications and $N_0(N_0 - 1) = 240$ additions. We can split this sequence in two shorter sequences, each of length 8. To compute DFT of each of these segments, we need 64 multiplications and 56 additions. Thus, we need a total of 128 multiplications and 112 additions. Suppose, we split the original sequence in four segments of length 4 each. To compute the DFT of each segment, we require 16 multiplications and 12 additions. Hence, we need a total of 64 multiplications and 48 additions. If we split the sequence in eight segments of length 2 each, we need 4 multiplications and 2 additions for each segment, resulting in a total of 32 multiplications and 8 additions. Thus, we have been able to reduce the number of multiplications from 256 to 32 and the number of additions from 240 to 8. Moreover, some of these multiplications turn out to be multiplications by 1 or -1. All this fantastic economy in number of computations is realized by the FFT without any approximation! The values obtained by the FFT are identical to those obtained by DFT. In this example, we considered a relatively small value of $N_0 = 16$. The reduction in number of computations is much more dramatic for higher values of N_0.

The FFT algorithm is simplified if we choose N_0 to be a power of 2, although such a choice is not essential. For convenience, we define

$$W_{N_0} = e^{-(j2\pi/N_0)} = e^{-j\Omega_0} \tag{8.37}$$

so that

$$X_r = \sum_{n=0}^{N_0-1} x_n W_{N_0}^{nr} \qquad 0 \le r \le N_0 - 1 \tag{8.38a}$$

and

$$x_n = \frac{1}{N_0} \sum_{r=0}^{N_0-1} X_r W_{N_0}^{-nr} \qquad 0 \le n \le N_0 - 1 \tag{8.38b}$$

Although there are many variations of the Tukey–Cooley algorithm, these can be grouped into two basic types: *decimation in time* and *decimation in frequency*.

THE DECIMATION-IN-TIME ALGORITHM

Here we divide the N_0-point data sequence x_n into two $(N_0/2)$-point sequences consisting of even- and odd-numbered samples, respectively, as follows:

$$\underbrace{x_0, x_2, x_4, \ldots, x_{N_0-2}}_{\text{sequence } g_n}, \underbrace{x_1, x_3, x_5, \ldots, x_{N_0-1}}_{\text{sequence } h_n}$$

Then, from Eq. (8.38a),

$$X_r = \sum_{n=0}^{(N_0/2)-1} x_{2n} W_{N_0}^{2nr} + \sum_{n=0}^{(N_0/2)-1} x_{2n+1} W_{N_0}^{(2n+1)r} \tag{8.39}$$

Also, since

$$W_{N_0/2} = W_{N_0}^2 \tag{8.40}$$

we have

$$X_r = \sum_{n=0}^{(N_0/2)-1} x_{2n} W_{N_0/2}^{nr} + W_{N_0}^r \sum_{n=0}^{(N_0/2)-1} x_{2n+1} W_{N_0/2}^{nr}$$

$$= G_r + W_{N_0}^r H_r \qquad 0 \le r \le N_0 - 1 \tag{8.41}$$

where G_r and H_r are the $(N_0/2)$-point DFTs of the even- and odd-numbered sequences, g_n and h_n, respectively. Also, G_r and H_r, being the $(N_0/2)$-point DFTs, are $(N_0/2)$ periodic. Hence

$$G_{r+(N_0/2)} = G_r$$
$$H_{r+(N_0/2)} = H_r \tag{8.42}$$

Moreover,

$$W_{N_0}^{r+(N_0/2)} = W_{N_0}^{N_0/2} W_{N_0}^r = e^{-j\pi} W_{N_0}^r = -W_{N_0}^r \tag{8.43}$$

From Eqs. (8.41), (8.42), and (8.43), we obtain

$$X_{r+(N_0/2)} = G_r - W_{N_0}^r H_r \tag{8.44}$$

This property can be used to reduce the number of computations. We can compute the first $N_0/2$ points ($0 \le n \le (N_0/2) - 1$) of X_r by using Eq. (8.41) and the last $N_0/2$ points by using Eq. (8.44) as

$$X_r = G_r + W_{N_0}^r H_r \qquad 0 \le r \le \frac{N_0}{2} - 1 \tag{8.45a}$$

$$X_{r+(N_0/2)} = G_r - W_{N_0}^r H_r \qquad 0 \le r \le \frac{N_0}{2} - 1 \tag{8.45b}$$

Thus, an N_0-point DFT can be computed by combining the two $(N_0/2)$-point DFTs, as in Eqs. (8.45). These equations can be represented conveniently by the *signal flow* graph depicted in Fig. 8.21. This structure is known as a *butterfly*. Figure 8.22a shows the implementation of Eqs. (8.42) for the case of $N_0 = 8$.

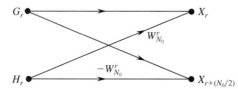

Figure 8.21 Butterfly signal flow graph.

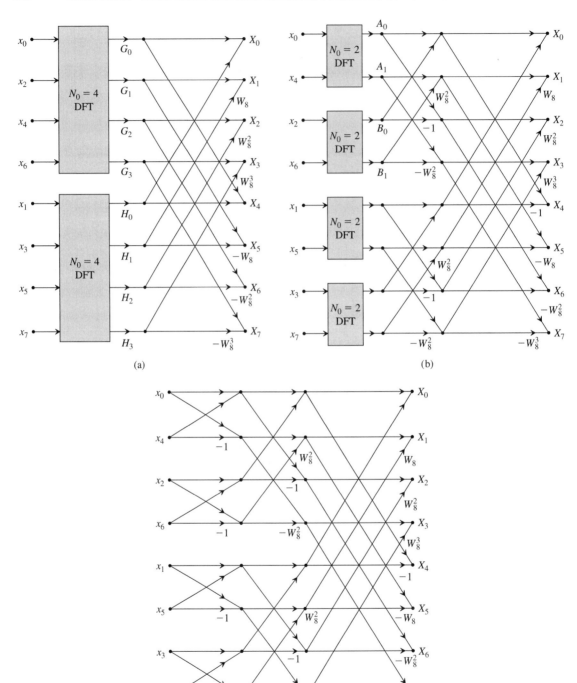

Figure 8.22 Successive steps in an 8-point FFT.

The next step is to compute the $(N_0/2)$-point DFTs G_r and H_r. We repeat the same procedure by dividing g_n and h_n into two $(N_0/4)$-point sequences corresponding to the even- and odd-numbered samples. Then we continue this process until we reach the one-point DFT. These steps for the case of $N_0 = 8$ are shown in Fig. 8.22a, 8.22b, and 8.22c. Figure 8.22c shows that the two-point DFTs require no multiplication.

To count the number of computations required in the first step, assume that G_r and H_r are known. Equations (8.45) clearly show that to compute all the N_0 points of the X_r, we require N_0 complex additions and $N_0/2$ complex multiplications[†] (corresponding to $W_{N_0}^r H_r$).

In the second step, to compute the $(N_0/2)$-point DFT G_r from the $(N_0/4)$-point DFT, we require $N_0/2$ complex additions and $N_0/4$ complex multiplications. We require an equal number of computations for H_r. Hence, in the second step, there are N_0 complex additions and $N_0/2$ complex multiplications. The number of computations required remains the same in each step. Since a total of $\log_2 N_0$ steps is needed to arrive at a one-point DFT, we require, conservatively, a total of $N_0 \log_2 N_0$ complex additions and $(N_0/2) \log_2 N_0$ complex multiplications, to compute the N_0-point DFT. Actually, as Fig. 8.22c shows, many multiplications are multiplications by 1 or -1, which further reduces the number of computations.

The procedure for obtaining IDFT is identical to that used to obtain the DFT except that $W_{N_0} = e^{j(2\pi/N_0)}$ instead of $e^{-j(2\pi/N_0)}$ (in addition to the multiplier $1/N_0$). Another FFT algorithm, the *decimation-in-frequency* algorithm, is similar to the decimation-in-time algorithm. The only difference is that instead of dividing x_n into two sequences of even- and odd-numbered samples, we divide x_n into two sequences formed by the first $N_0/2$ and the last $N_0/2$ digits, proceeding in the same way until a single-point DFT is reached in $\log_2 N_0$ steps. The total number of computations in this algorithm is the same as that in the decimation-in-time algorithm.

8.7 SUMMARY

A signal bandlimited to B Hz can be reconstructed exactly from its samples if the sampling rate $f_s > 2B$ Hz (the sampling theorem). Such a reconstruction, although possible theoretically, poses practical problems such as the need for ideal filters, which are unrealizable or are realizable only with infinite delay. Therefore, in practice, there is always an error in reconstructing a signal from its samples. Moreover, practical signals are not bandlimited, which causes an additional error (aliasing error) in signal reconstruction from its samples. When a signal is sampled at a frequency f_s Hz, samples of a sinusoid of frequency $(f_s/2) + x$ Hz appear as samples of a lower frequency $(f_s/2) - x$ Hz. This phenomenon, in which higher frequencies appear as lower frequencies, is known as aliasing. Aliasing error can be reduced by bandlimiting a signal to $f_s/2$ Hz (half the sampling frequency). Such bandlimiting, done prior to sampling, is accomplished by an antialiasing filter that is an ideal lowpass filter of cutoff frequency $f_s/2$ Hz.

The sampling theorem is very important in signal analysis, processing, and transmission because it allows us to replace a continuous-time signal with a discrete sequence of numbers. Processing a continuous-time signal is therefore equivalent to processing a discrete sequence of numbers. This leads us directly into the area of digital filtering (discrete-time systems). In the field of communication, the transmission of a continuous-time message reduces to the transmission of

---

[†]Actually, $N_0/2$ is a conservative figure because some multiplications corresponding to the cases of $W_{N_0}^r = 1, j$, and so on, are eliminated.

a sequence of numbers. This opens doors to many new techniques of communicating continuous-time signals by pulse trains.

The dual of the sampling theorem states that for a signal timelimited to τ seconds, its spectrum $X(\omega)$ can be reconstructed from the samples of $X(\omega)$ taken at uniform intervals not greater than $1/\tau$ Hz. In other words, the spectrum should be sampled at a rate not less than τ samples/Hz.

To compute the direct or the inverse Fourier transform numerically, we need a relationship between the samples of $x(t)$ and $X(\omega)$. The sampling theorem and its dual provide such a quantitative relationship in the form of a discrete Fourier transform (DFT). The DFT computations are greatly facilitated by a fast Fourier transform (FFT) algorithm, which reduces the number of computations from something on the order of N_0^2 to $N_0 \log N_0$.

REFERENCES

1.   Linden, D. A. A discussion of sampling theorem. *Proceedings of the IRE,* vol. 47, pp. 1219–1226, July 1959.

2.   Siebert, W. M. *Circuits, Signals, and Systems.* MIT–McGraw-Hill, New York, 1986.

3.   Bennett, W. R. *Introduction to Signal Transmission.* McGraw-Hill, New York, 1970.

4.   Lathi, B. P. *Linear Systems and Signals.* Berkeley-Cambridge Press, Carmichael, CA, 1992.

5.   Cooley, J. W., and J. W. Tukey. An algorithm for the machine calculation of complex fourier series. *Mathematics of Computation*, vol. 19, pp. 297–301, April 1965.

MATLAB SESSION 8: THE DISCRETE FOURIER TRANSFORM

As an idea, the discrete Fourier transform (DFT) has been known for hundreds of years. Practical computing devices, however, are responsible for bringing the DFT into common use. MATLAB is capable of DFT computations that would have been impractical not too many years ago.

M8.1 Computing the Discrete Fourier Transform

The MATLAB command `fft(x)` computes the DFT of a vector x that is defined over ($0 \leq n \leq N_0 - 1$) (Problem 8.M-1 considers how to scale the DFT to accommodate signals that do not begin at $n = 0$.) As its name suggests, the function `fft` uses the computationally more efficient fast Fourier transform algorithm when it is appropriate to do so. The inverse DFT is easily computed by using the `ifft` function.

To illustrate MATLAB's DFT capabilities, consider 50 points of a 10 Hz sinusoid sampled at $f_s = 50$ Hz and scaled by $T = 1/f_s$.

```
>> T = 1/50; N_0 = 50; n = (0:N_0-1);
>> x = T*cos(2*pi*10*n*T);
```

In this case, the vector x contains exactly 10 cycles of the sinusoid. The `fft` command computes the DFT.

```
>> X = fft(x);
```

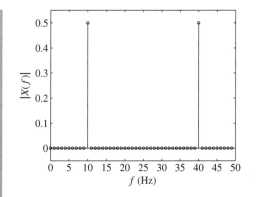

Figure M8.1 $|X(f)|$ computed over $(0 \le f < 50)$ by using `fft`.

Since the DFT is both discrete and periodic, `fft` needs to return only the N_0 discrete values contained in the single period $(0 \le f < f_s)$.

While X_r can be plotted as a function of r, it is more convenient to plot the DFT as a function of frequency f. A frequency vector, in hertz, is created by using N_0 and T.

```
>> f = (0:N_0-1)/(T*N_0);
>> stem(f,abs(X),'k'); axis([0 50 -0.05 0.55]);
>> xlabel('f [Hz]'); ylabel('|X(f)|');
```

As expected, Fig. M8.1 shows content at a frequency of 10 Hz. Since the time-domain signal is real, $X(f)$ is conjugate symmetric. Thus, content at 10 Hz implies equal content at -10 Hz. The content visible at 40 Hz is an alias of the -10 Hz content.

Often, it is preferred to plot a DFT over the principal frequency range $(-f_s/2 \le f < f_s/2)$. The MATLAB function `fftshift` properly rearranges the output of `fft` to accomplish this task.

```
>> stem(f-1/(T*2),fftshift(abs(X)),'k'); axis([-25 25 -0.05 0.55]);
>> xlabel('f [Hz]'); ylabel('|X(f)|');
```

When we use `fftshift`, the conjugate symmetry that accompanies the DFT of a real signal becomes apparent, as shown in Fig. M8.2.

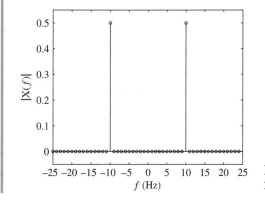

Figure M8.2 $|X(f)|$ displayed over $(-25 \le f < 25)$ by using `fftshift`.

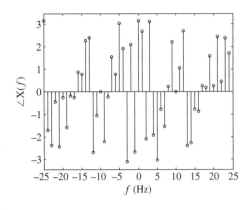

Figure M8.3 $\angle X(f)$ displayed over $(-25 \leq f < 25)$.

Since DFTs are generally complex valued, the magnitude plots of Figs. M8.1 and M8.2 offer only half the picture; the signal's phase spectrum, shown in Fig. M8.3, completes it.

```
>> stem(f-1/(T*2),fftshift(angle(X)),'k'); axis([-25 25 -1.1*pi 1.1*pi]);
>> xlabel('f [Hz]'); ylabel('∠ X(f)');
```

Since the signal is real, the phase spectrum necessarily has odd symmetry. Additionally, the phase at ± 10 Hz is zero, as expected for a zero-phase cosine function. More interesting, however, are the phase values found at the remaining frequencies. Does a simple cosine really have such complicated phase characteristics? The answer, of course, is no. The magnitude plot of Fig. M8.2 helps identify the problem: there is zero content at frequencies other than ± 10 Hz. Phase computations are not reliable at points where the magnitude response is zero. One way to remedy this problem is to assign a phase of zero when the magnitude response is near or at zero.

M8.2 Improving the Picture with Zero Padding

DFT magnitude and phase plots paint a picture of a signal's spectrum. At times, however, the picture can be somewhat misleading. Given a sampling frequency $f_s = 50$ Hz and a sampling interval $T = 1/f_s$, consider the signal

$$y[n] = T e^{j2\pi \left(10\frac{1}{3}\right)nT}$$

This complex-valued, periodic signal contains a single positive frequency at $10\frac{1}{3}$ Hz. Let us compute the signal's DFT using 50 samples.

```
>> y = T*exp(j*2*pi*(10+1/3)*n*T); Y = fft(y);
>> stem(f-25,fftshift(abs(Y)),'k'); axis([-25 25 -0.05 1.05]);
>> xlabel('f [Hz]'); ylabel('|Y(f)|');
```

In this case, the vector y contains a noninteger number of cycles. Figure M8.4 shows the significant frequency leakage that results. Since $y[n]$ is not real, also notice that the DFT is not conjugate symmetric.

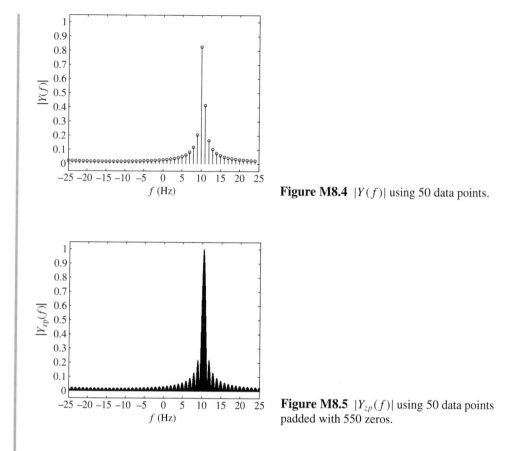

Figure M8.4 $|Y(f)|$ using 50 data points.

Figure M8.5 $|Y_{zp}(f)|$ using 50 data points padded with 550 zeros.

In this example, the discrete DFT frequencies do not include the actual $10\frac{1}{3}$ Hz frequency of the signal. Thus, it is difficult to determine the signal's frequency from Fig. M8.4. To improve the picture, the signal is zero-padded to 12 times its original length.

```
>> y_zp = [y,zeros(1,11*length(y))]; Y_zp = fft(y_zp);
>> f_zp = (0:12*N_0-1)/(T*12*N_0);
>> stem(f_zp-25,fftshift(abs(Y_zp)),'k.'); axis([-25 25 -0.05 1.05]);
>> xlabel('f [Hz]'); ylabel('|Y_{zp}(f)|');
```

Figure M8.5 correctly shows the peak frequency at $10\frac{1}{3}$ Hz and better represents the signal's spectrum.

It is important to keep in mind that zero padding does not increase the resolution or accuracy of the DFT. To return to the picket fence analogy, zero padding increases the number of pickets in our fence but cannot change what is behind the fence. More formally, the characteristics of the sinc function, such as main beam width and sidelobe levels, depend on the fixed width of the pulse, not on the number of zeros that follow. Adding zeros cannot change the characteristics of the sinc function and thus cannot change the resolution or accuracy of the DFT. Adding zeros simply allows the sinc function to be sampled more finely.

M8.3 Quantization

A *b*-bit analog-to-digital converter (ADC) samples an analog signal and quantizes amplitudes by using 2^b discrete levels. This quantization results in signal distortion that is particularly noticeable for small *b*. Typically, quantization is classified as symmetric or asymmetric. As discussed earlier, asymmetric quantization uses zero as a quantization level, which can help suppress low-level noise.

Program `MS8P1` quantizes a signal using either symmetric quantization or asymmetric quantization. If available, the signal processing toolbox functions `uencode` and `udecode` can also be used for symmetric quantization, but not for asymmetric quantization.

```
function [xq] = MS8P1(x,xmax,b,method);
% MS8P1.m : MATLAB Session 8, Program 1
% Function M-file quantizes x over (-xmax,xmax) using 2^b levels.
% Both symmetric and asymmetric quantization are supported.
% INPUTS:   x = input signal
%           xmax = maximum magnitude of signal to be quantized
%           b = number of quantization bits
%           method = default 'sym' for symmetrical, 'asym' for asymmetrical
% OUTPUTS:  xq = quantized signal
if (nargin<3),
    disp('Insufficient number of inputs.'); return
elseif (nargin==3),
    method = 'sym';
elseif (nargin>4),
    disp('Too many inputs.'); return
end

switch lower(method)
    case 'asym'
        offset = 1/2;
    case 'sym'
        offset = 0;
    otherwise
        disp('Unrecognized quantization method.'); return
end
q = floor(2^b*((x+xmax)/(2*xmax))+offset);
i = find(q>2^b-1); q(i) = 2^b-1;
i = find(q<0); q(i) = 0;
xq = (q-(2^(b-1)-(1/2-offset)))*(2*xmax/(2^b));
```

Several MATLAB commands require discussion. First, the `nargin` function returns the number of input arguments. In this program, `nargin` is used to ensure that a correct number of inputs is supplied. If the number of inputs supplied is incorrect, an error message is displayed and the function terminates. If only three input arguments are detected, the quantization type is not explicitly specified and the program assigns the default symmetric method.

As with many high-level languages such as C, MATLAB supports general switch/case structures:[†]

```
switch switch_expr,
case case_expr,
    statements;
...
otherwise,
    statements;
end
```

MS8P1 switches among cases of the string `method`. In this way, method-specific parameters are easily set. The command `lower` is used to convert a string to all lowercase characters. In this way, strings such as SYM, Sym, and sym are all indistinguishable. Similar to `lower`, the MATLAB command `upper` converts a string to all uppercase.

Quantization is accomplished by appropriately scaling and shifting the input and then rounding the result. The `floor(q)` command rounds elements of `q` to the nearest integer toward minus infinity. Mathematically, it computes $\lfloor q \rfloor$. To accommodate different types of rounding, MATLAB supplies three other rounding commands: `ceil`, `round`, and `fix`. The command `ceil(q)` rounds elements of `q` to the nearest integers toward infinity, ($\lceil q \rceil$); the command `round(q)` rounds elements of `q` toward the nearest integer; the command `fix(q)` rounds elements of `q` to the nearest integer toward zero. For example, if `q = [-0.5 0.5];`, `floor(q)` yields `[-1 0]`, `ceil(q)` yields `[0 1]`, `round(q)` yields `[-1 1]`, and `fix(q)` yields `[0 0]`. The final aspect of quantization in MS8P1 is accomplished by means of the `find` command, which is used to identify values outside the maximum allowable range for saturation.

To verify operation, MS8P1 is used to determine the transfer characteristics of a symmetric 3-bit quantizer operating over $(-10, 10)$.

```
>> x = (-10:.0001:10);
>> xsq = MS8P1(x,10,3,'sym');
>> plot(x,xsq,'k'); grid;
>> xlabel('Quantizer input'); ylabel('Quantizer output');
```

Figure M8.6 shows the results. Figure M8.7 shows the transfer characteristics of an asymmetric 3-bit quantizer. The asymmetric quantizer includes zero as a quantization level, but at a price: quantization errors exceed $\Delta/2 = 1.25$ for input values greater than 8.75.

There is no doubt that quantization can change a signal. It follows that the spectrum of a quantized signal can also change. While these changes are difficult to characterize mathematically, they are easy to investigate by using MATLAB. Consider a 1 Hz cosine sampled at $f_s = 50$ Hz over 1 second.

```
>> x = cos(2*pi*n*T); X = fft(x);
```

[†]A functionally equivalent structure can be written by using `if`, `elseif`, and `else` statements.

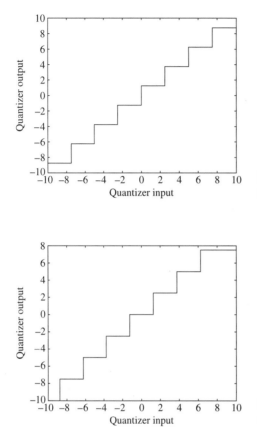

Figure M8.6 Transfer characteristics of a symmetric 3-bit quantizer.

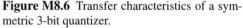

Figure M8.7 Transfer characteristics of an asymmetric 3-bit quantizer.

Upon quantizing by means of a 2-bit asymmetric quantizer, both the signal and spectrum are substantially changed.

```
>> xaq = MS8P1(x,1,2,'asym'); Xaq = fft(xaq);
>> subplot(2,2,1); stem(n,x,'k'); axis([0 49 -1.1 1.1]);
>> xlabel('n');ylabel('x[n]');
>> subplot(2,2,2); stem(f-25,fftshift(abs(X)),'k'); axis([-25,25 -1 26])
>> xlabel('f');ylabel('|X(f)|');
>> subplot(2,2,3); stem(n,xaq,'k');axis([0 49 -1.1 1.1]);
>> xlabel('n');ylabel('x_{aq}[n]');
>> subplot(2,2,4); stem(f-25,fftshift(abs(fft(xaq))),'k'); axis([-25,25 -1 26]);
>> xlabel('f');ylabel('|X_{aq}(f)|');
```

The results are shown in Fig. M8.8. The original signal $x[n]$ appears sinusoidal and has pure spectral content at ± 1 Hz. The asymmetrically quantized signal $x_{aq}[n]$ is significantly distorted. The corresponding magnitude spectrum $|X_{aq}(f)|$ is spread over a broad range of frequencies.

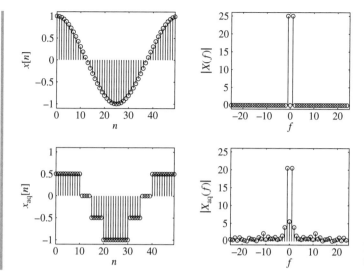

Figure M8.8 Signal and spectrum effects of quantization.

PROBLEMS

[Note: In many problems, the plots of spectra are shown as functions of frequency f Hz for convenience, although we have labeled them as functions of ω as $X(\omega)$, $Y(\omega)$, etc.]

8.1-1 Figure P8.1-1 shows Fourier spectra of signals $x_1(t)$ and $x_2(t)$. Determine the Nyquist sampling rates for signals $x_1(t)$, $x_2(t)$, $x_1^2(t)$, $x_2^3(t)$, and $x_1(t)x_2(t)$.

8.1-2 Determine the Nyquist sampling rate and the Nyquist sampling interval for the signals

   (a) $\operatorname{sinc}^2(100\pi t)$
   (b) $0.01\operatorname{sinc}^2(100\pi t)$
   (c) $\operatorname{sinc}(100\pi t) + 3\operatorname{sinc}^2(60\pi t)$
   (d) $\operatorname{sinc}(50\pi t)\operatorname{sinc}(100\pi t)$

8.1-3 (a) Sketch $|X(\omega)|$, the amplitude spectrum of a signal $x(t) = 3\cos 6\pi t + \sin 18\pi t + 2\cos(28 - \epsilon)\pi t$, where ϵ is a very small number $\to 0$. Determine the minimum sampling rate required to be able to reconstruct $x(t)$ from these samples.

   (b) Sketch the amplitude spectrum of the sampled signal when the sampling rate is 25% above the Nyquist rate (show the spectrum over the frequency range ± 50 Hz only). How would you reconstruct $x(t)$ from these samples?

8.1-4 (a) Derive the sampling theorem by considering the fact that the sampled signal $\bar{x}(t) = x(t)\delta_T(t)$, and using the frequency-convolution property in Eq. (7.43).

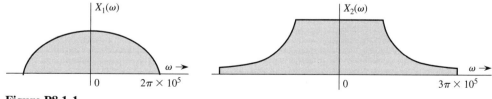

Figure P8.1-1

(b) For a sampling train consisting of shifted unit impulses at instants $nT + \tau$ instead of at nT (for all positive and negative integer values of n), find the spectrum of the sampled signal.

8.1-5   A signal is bandlimited to 12 kHz. The band between 10 and 12 kHz has been so corrupted by excessive noise that the information in this band is nonrecoverable. Determine the minimum sampling rate for this signal so that the uncorrupted portion of the band can be recovered. If we were to filter out the corrupted spectrum prior to sampling, what would be the minimum sampling rate?

8.1-6   A continuous-time signal $x(t) = \Delta((t-1)/2)$ is sampled at three rates: 10, 2, and 1 Hz. Sketch the resulting sampled signals. Because $x(t)$ is timelimited, its bandwidth is infinite. However, most of its energy is concentrated in a small band. Can you determine the reasonable minimum sampling rate that will allow reconstruction of this signal with a small error? The answer is not unique. Make a reasonable assumption of what you define as a "negligible" or "small" error.

8.1-7   (a) A signal $x(t) = 5 \operatorname{sinc}^2(5\pi t) + \cos 20\pi t$ is sampled at a rate of 10 Hz. Find the spectrum of the sampled signal. Can $x(t)$ be reconstructed by lowpass filtering the sampled signal?

(b) Repeat part (a) for a sampling frequency of 20 Hz. Can you reconstruct the signal from this sampled signal? Explain.

(c) If $x(t) = 5 \operatorname{sinc}^2(5\pi t) + \sin 20\pi t$, can you reconstruct $x(t)$ from the samples of $x(t)$ at a rate of 20 Hz? Explain your answer with spectral representation(s).

(d) For $x(t) = 5 \operatorname{sinc}^2(5\pi t) + \sin 20\pi t$, can you reconstruct $x(t)$ from the samples of $x(t)$ at a rate of 21 Hz? Explain your answer with spectral representation(s). Comment on your results.

8.1-8   (a) The highest frequency in the spectrum $X(\omega)$ (Fig. P8.1-8a) of a bandpass signal $x(t)$ is 30 Hz. Hence, the minimum sampling frequency needed to sample $x(t)$ is 60 Hz. Show the spectrum of the signal sampled at a rate of 60 Hz. Can you reconstruct $x(t)$ from these samples? How?

(b) A certain busy student looks at $X(\omega)$, concludes that its bandwidth is really 10 Hz, and decides that the sampling rate 20 Hz is adequate for sampling $x(t)$. Sketch the spectrum of the signal sampled at a rate of 20 Hz. Can she reconstruct $x(t)$ from these samples?

(c) The same student, using the same reasoning, looks at $Y(\omega)$ in Fig. P8.1-8b, the spectrum of another bandpass signal $y(t)$, and concludes that she can use sampling rate of 20 Hz to sample $y(t)$. Sketch the

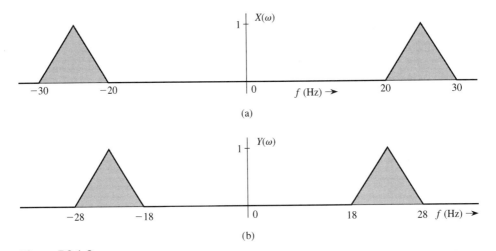

(a)

(b)

Figure P8.1-8

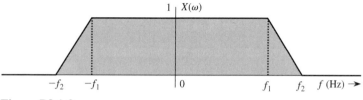

Figure P8.1-9

spectrum of the signal $y(t)$ sampled at a rate of 20 Hz. Can she reconstruct $y(t)$ from these samples?

8.1-9   A signal $x(t)$ whose spectrum $X(\omega)$, as shown in Fig. P8.1-9, is sampled at a frequency $f_s = f_1 + f_2$ Hz. Find all the sample values of $x(t)$ merely by inspection of $X(\omega)$.

8.1-10   In digital data transmission over a communication channel, it is important to know the upper theoretical limit on the rate of digital pulses that can be transmitted over a channel of bandwidth B Hz. In digital transmission, the relative shape of the pulse is not important. We are interested in knowing only the amplitude represented by the pulse. For instance, in binary communication, we are interested in knowing whether the received pulse amplitude is 1 or -1 (positive or negative). Thus, each pulse represents one piece of information. Consider one independent amplitude value (not necessarily binary) as one piece of information. Show that $2B$ independent pieces of information per second can be transmitted correctly (assuming no noise) over a channel of bandwidth B Hz. This important principle in communication theory states that one hertz of bandwidth can transmit two independent pieces of information per second. It represents the upper rate of pulse transmission over a channel without any error in reception in the absence of noise. [Hint: According to the interpolation formula [Eq. (8.11)], a continuous-time signal of bandwidth B Hz can be constructed from $2B$ pieces of information/second.]

8.1-11   This example is one of those interesting situations leading to a curious result in the category of defying gravity. The sinc function can be recovered from its samples taken at extremely low frequencies in apparent defiance of the sampling theorem.

Consider a sinc pulse $x(t) = \text{sinc}\,(4\pi t)$ for which $X(\omega) = (1/4)\,\text{rect}\,(\omega/8\pi)$. The bandwidth of $x(t)$ is $B = 2$ Hz, and its Nyquist rate is 4 Hz.

(a)   Sample $x(t)$ at a rate 4 Hz and sketch the spectrum of the sampled signal.

(b)   To recover $x(t)$ from its samples, we pass the sampled signal through an ideal low-pass filter of bandwidth $B = 2$ Hz and gain $G = T = 1/4$. Sketch this system and show that for this system $H(\omega) = (1/4)\,\text{rect}\,(\omega/8\pi)$. Show also that when the input is the sampled $x(t)$ at a rate 4 Hz, the output of this system is indeed $x(t)$, as expected.

(c)   Now sample $x(t)$ at half the Nyquist rate, at 2 Hz. Apply this sampled signal at the input of the lowpass filter used part (b). Find the output.

(d)   Repeat part (c) for the sampling rate 1 Hz.

(e)   Show that the output of the lowpass filter in part (b) is $x(t)$ to the sampled $x(t)$ if the sampling rate is $4/N$, where N is any positive integer. This means that we can recover $x(t)$ from its samples taken at arbitrarily small rate by letting $N \to \infty$.

(f)   The mystery may be clarified a bit by examining the problem in the time domain. Find the samples of $x(t)$ when the sampling rate is $2/N$ (N integer).

8.2-1   A signal $x(t) = \text{sinc}\,(200\pi t)$ is sampled (multiplied) by a periodic pulse train $p_T(t)$ represented in Fig. P8.2-1. Find and sketch the spectrum of the sampled signal. Explain whether you will be able to reconstruct $x(t)$ from these samples. Find the filter output if the sampled signal is passed through an ideal lowpass filter of bandwidth 100 Hz and unit gain. What is the filter output if its bandwidth B Hz is

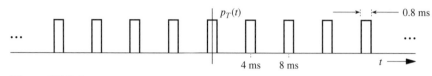

Figure P8.2-1

between 100 and 150 Hz? What happens if the bandwidth exceeds 150 Hz?

8.2-2 Show that the circuit in Fig. P8.2-2 is a realization of the causal ZOH (zero-order hold) circuit. You can do this by showing that the unit impulse response $h(t)$ of this circuit is indeed equal to that in Eq. (8.9a) delayed by $T/2$ seconds to make it causal.

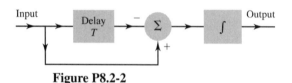

Figure P8.2-2

8.2-3 (a) A first-order hold circuit (FOH) can also be used to reconstruct a signal $x(t)$ from its samples. The impulse response of this circuit is $h(t) = \Delta(t/2T)$, where T is the sampling interval. Consider a typical sampled signal $\bar{x}(t)$ and show that this circuit performs the linear interpolation. In other words, the filter output consists of sample tops connected by straight-line segments. Follow the procedure discussed in Section 8.2 (Fig. 8.5c).

(b) Determine the frequency response of this filter, and its amplitude response, and compare it with
  (i) the ideal filter required for signal reconstruction
  (ii) the ZOH circuit

(c) This filter, being noncausal, is unrealizable. By delaying its impulse response, the filter can be made realizable. What is the minimum delay required to make it realizable? How would this delay affect the reconstructed signal and the filter frequency response?

(d) Show that the causal FOH circuit in part (c) can be realized by the ZOH circuit depicted

in Fig. P8.2-2 followed by an identical filter in cascade.

8.2-4 In the text, for sampling purposes, we used timelimited narrow pulses such as impulses or rectangular pulses of width less than the sampling interval T. Show that it is not necessary to restrict the sampling pulse width. We can use sampling pulses of arbitrarily large duration and still be able to reconstruct the signal $x(t)$ as long as the pulse rate is no less than the Nyquist rate for $x(t)$.

Consider $x(t)$ to be bandlimited to B Hz. The sampling pulse to be used is an exponential $e^{-at}u(t)$. We multiply $x(t)$ by a periodic train of exponential pulses of the form $e^{-at}u(t)$ spaced T seconds apart. Find the spectrum of the sampled signal, and show that $x(t)$ can be reconstructed from this sampled signal provided the sampling rate is no less than $2B$ Hz or $T < 1/2B$. Explain how you would reconstruct $x(t)$ from the sampled signal.

8.2-5 In Example 8.2, the sampling of a signal $x(t)$ was accomplished by multiplying the signal by a pulse train $p_T(t)$, resulting in the sampled signal depicted in Fig. 8.4d. This procedure is known as the *natural sampling*. Figure P8.2-5 shows the so-called *flat-top sampling* of the same signal $x(t) = \text{sinc}^2(5\pi t)$.

(a) Show that the signal $x(t)$ can be recovered from flat-top samples if the sampling rate is no less than the Nyquist rate.

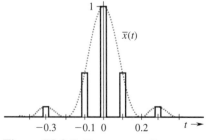

Figure P8.2-5 Flat-top sampling.

(b) Explain how you would recover $x(t)$ from the flat-top samples.

(c) Find the expression for the sampled signal spectrum $\overline{X}(\omega)$ and sketch it roughly.

8.2-6 A sinusoid of frequency f_0 Hz is sampled at a rate $f_s = 20$ Hz. Find the apparent frequency of the sampled signal if f_0 is

(a) 8 Hz
(b) 12 Hz
(c) 20 Hz
(d) 22 Hz
(e) 32 Hz

8.2-7 A sinusoid of unknown frequency f_0 is sampled at a rate 60 Hz. The apparent frequency of the samples is 20 Hz. Determine f_0 if it is known that f_0 lies in the range

(a) 0–30 Hz
(b) 30–60 Hz
(c) 60–90 Hz
(d) 90–120 Hz

8.2-8 A signal $x(t) = 3\cos 6\pi t + \cos 16\pi t + 2\cos 20\pi t$ is sampled at a rate 25% above the Nyquist rate. Sketch the spectrum of the sampled signal. How would you reconstruct $x(t)$ from these samples? If the sampling frequency is 25% below the Nyquist rate, what are the frequencies of the sinusoids present in the output of the filter with cutoff frequency equal to the folding frequency? Do not write the actual output; give just the frequencies of the sinusoids present in the output.

8.2-9 (a) Show that the signal $x(t)$, reconstructed from its samples $x(nT)$, using Eq. (8.11a) has a bandwidth $B \le 1/2T$ Hz.

(b) Show that $x(t)$ is the smallest bandwidth signal that passes through samples $x(nT)$. [Hint: Use the *reductio ad absurdum* method.]

8.2-10 In digital communication systems, the efficient use of channel bandwidth is ensured by transmitting digital data encoded by means of bandlimited pulses. Unfortunately, bandlimited pulses are nontimelimited; that is, they have infinite duration, which causes pulses representing successive digits to interfere and cause errors in the reading of true pulse values. This difficulty can be resolved by shaping a pulse $p(t)$ in such a way that it is bandlimited,

yet causes zero interference at the sampling instants. To transmit R pulses per second, we require a minimum bandwidth $R/2$ Hz (see Prob. 8.1-10). The bandwidth of $p(t)$ should be $R/2$ Hz, and its samples, in order to cause no interference at all other sampling instants, must satisfy the condition

$$p(nT) = \begin{cases} 1 & n = 0 \\ 0 & n \neq 0 \end{cases} \qquad T = \frac{1}{R}$$

Because the pulse rate is R pulses per second, the sampling instants are located at intervals of $1/R$ seconds. Hence, the foregoing condition ensures that any given pulse will not interfere with the amplitude of any other pulse at its center. Find $p(t)$. Is $p(t)$ unique in the sense that no other pulse satisfies the given requirements?

8.2-11 The problem of pulse interference in digital data transmission was outlined in Prob. 8.2-10, where we found a pulse shape $p(t)$ to eliminate the interference. Unfortunately, the pulse found is not only noncausal, (and unrealizable) but also has a serious drawback: because of its slow decay (as $1/t$), it is prone to severe interference due to small parameter deviation. To make the pulse decay rapidly, Nyquist proposed relaxing the bandwidth requirement from $R/2$ Hz to $kR/2$ Hz with $1 \le k \le 2$. The pulse must still have a property of noninterference with other pulses, for example,

$$p(nT) = \begin{cases} 1 & n = 0 \\ 0 & n \neq 0 \end{cases} \qquad T = \frac{1}{R}$$

Show that this condition is satisfied only if the pulse spectrum $P(\omega)$ has an odd symmetry about the set of dotted axes as shown in Fig. P8.2-11. The bandwidth of $P(\omega)$ is $kR/2$ Hz ($1 \le k \le 2$).

8.2-12 The Nyquist samples of a signal $x(t)$ bandlimited to B Hz are

$$x(nT) = \begin{cases} 1 & n = 0, 1 \\ 0 & \text{all } n \neq 0, 1 \end{cases} \qquad T = \frac{1}{2B}$$

Show that

$$x(t) = \frac{\operatorname{sinc}(2\pi Bt)}{1 - 2Bt}$$

This pulse, known as the *duobinary pulse,* is used in digital transmission applications.

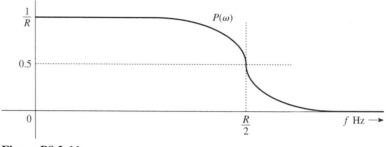

Figure P8.2-11

8.2-13 A signal bandlimited to B Hz is sampled at a rate $f_s = 2B$ Hz. Show that

$$\int_{-\infty}^{\infty} x(t)\, dt = T \sum_{-\infty}^{\infty} x(nT)$$

$$\int_{-\infty}^{\infty} |x(t)|^2\, dt = T \sum_{-\infty}^{\infty} |x(nT)|^2$$

[Hint: Use the orthogonality property of the sinc function in Prob. 7.6-5.]

8.2-14 Prove that a signal cannot be simultaneously timelimited and bandlimited. [Hint: Show that contrary assumption leads to contradiction. Assume a signal to be simultaneously time-limited and bandlimited so that $X(\omega) = 0$ for $|\omega| \geq 2\pi B$. In this case $X(\omega) = X(\omega)$ rect $(\omega/4\pi B')$ for $B' > B$. This fact means that $x(t)$ is equal to $x(t) * 2B'$ sinc $(2\pi B' t)$. The latter cannot be timelimited because the sinc function tail extends to infinity.]

8.3-1 A compact disc (CD) records audio signals digitally by means of a binary code. Assume an audio signal bandwidth of 15 kHz.

(a) What is the Nyquist rate?

(b) If the Nyquist samples are quantized into 65,536 levels ($L = 65,536$) and then binary coded, what number of binary digits is required to encode a sample.

(c) Determine the number of binary digits per second (bits/s) required to encode the audio signal.

(d) For practical reasons discussed in the text, signals are sampled at a rate well above the Nyquist rate. Practical CDs use 44,100 samples/s. If $L = 65,536$, determine the number of pulses per second required to encode the signal.

8.3-2 A TV signal (video and audio) has a bandwidth of 4.5 MHz. This signal is sampled, quantized, and binary coded.

(a) Determine the sampling rate if the signal is to be sampled at a rate 20% above the Nyquist rate.

(b) If the samples are quantized into 1024 levels, what number of binary pulses is required to encode each sample?

(c) Determine the binary pulse rate (bits/s) of the binary–coded signal.

8.3-3 (a) In a certain A/D scheme, there are 16 quantization levels. Give one possible binary code and one possible quaternary (4-ary) code. For the quaternary code use **0, 1, 2,** and **3** as the four symbols. Use the minimum number of digits in your code.

(b) To represent a given number of quantization levels L, we require a minimum of b_M digits for an M-ary code. Show that the ratio of the number of digits in a binary code to the number of digits in a quaternary (4-ary) code is 2, that is, $b_2/b_4 = 2$.

8.3-4 Five telemetry signals, each of bandwidth 1 kHz, are quantized and binary coded. These signals are time-division multiplexed (signal bits interleaved). Choose the number of quantization levels so that the maximum error in sample amplitudes is no greater than 0.2% of the peak signal amplitude. The signals must be sampled at least 20% above the Nyquist rate. Determine the data rate (bits per second) of the multiplexed signal.

8.4-1 The Fourier transform of a signal $x(t)$, bandlimited to B Hz, is $X(\omega)$. The signal $x(t)$ is repeated periodically at intervals T, where $T = 1.25/B$. The resulting signal $y(t)$ is

$$y(t) = \sum_{-\infty}^{\infty} x(t - nT)$$

Show that $y(t)$ can be expressed as

$$\cdot \, y(t) = C_0 + C_1 \cos(1.6\pi Bt + \theta_1)$$

where

$$C_0 = \frac{1}{T} X(0)$$

$$C_1 = \frac{2}{T} \left| X\left(\frac{2\pi}{T}\right) \right|$$

and

$$\theta_1 = \angle X\left(\frac{2\pi}{T}\right)$$

Recall that a bandlimited signal is not time-limited, and hence has infinite duration. The periodic repetitions are all overlapping.

8.5-1 For a signal $x(t)$ that is timelimited to 10 ms and has an essential bandwidth of 10 kHz, determine N_0, the number of signal samples necessary to compute a power-of-2 FFT with a frequency resolution f_0 of at least 50 Hz. Explain whether any zero padding is necessary.

8.5-2 To compute the DFT of signal $x(t)$ in Fig. P8.5-2, write the sequence x_n (for $n = 0$ to $N_0 - 1$) if the frequency resolution f_0 must be at least 0.25 Hz. Assume the essential

bandwidth (the folding frequency) of $x(t)$ to be at least 3 Hz. Do not compute the DFT; just write the appropriate sequence x_n.

8.5-3 Choose appropriate values for N_0 and T and compute the DFT of the signal $e^{-t}u(t)$. Use two different criteria for determining the effective bandwidth of $e^{-t}u(t)$. As the bandwidth, use the frequency at which the amplitude response drops to 1% of its peak value (at $\omega = 0$). Next, use the 99% energy criterion for determining the bandwidth (see Example 7.20).

8.5-4 Repeat Prob. 8.5-3 for the signal

$$x(t) = \frac{2}{t^2 + 1}$$

8.5-5 For the signals $x(t)$ and $g(t)$ represented in Fig. P8.5-5, write the appropriate sequences x_n and g_n necessary for the computation of the convolution of $x(t)$ and $g(t)$ using DFT. Use $T = 1/8$.

8.5-6 For this problem, interpret the N-point DFT as an N-periodic function of r. To stress this fact, we shall change the notation X_r to $X(r)$. Are the following frequency domain signals valid DFTs? Answer yes or no. For each valid DFT, determine the size N of the DFT and whether the time-domain signal is real.

(a) $X(r) = j - \pi$

(b) $X(r) = \sin(r/10)$

(c) $X(r) = \sin(\pi r/10)$

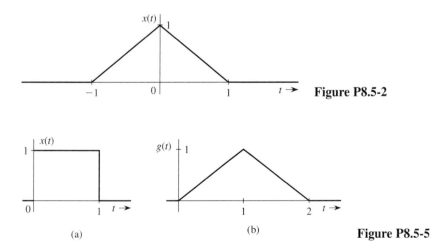

Figure P8.5-2

Figure P8.5-5

(d) $X(r) = \left[(1+j)/\sqrt{2}\right]^r$

(e) $X(r) = \langle r + \pi \rangle_{10}$ where, $\langle \cdot \rangle_{10}$ denotes the modulo-N operation.

8.M-1 MATLAB's `fft` command computes the DFT of a vector x assuming the first sample occurs at time $n = 0$. Given that $X = \mathtt{fft(x)}$ has already been computed, derive a method to correct X to reflect an arbitrary starting time $n = n_0$.

8.M-2 Consider a complex signal composed of two closely spaced complex exponentials: $x_1[n] = e^{j2\pi n 30/100} + e^{j2\pi n 33/100}$. For each of the following cases, plot the length-N DFT magnitude as a function of frequency f_r, where $f_r = r/N$.

(a) Compute and plot the DFT of $x_1[n]$ using 10 samples ($0 \le n \le 9$). From the plot, can both exponentials be identified? Explain.

(b) Zero-pad the signal from part (a) with 490 zeros and then compute and plot the 500-point DFT. Does this improve the picture of the DFT? Explain.

(c) Compute and plot the DFT of $x_1[n]$ using 100 samples ($0 \le n \le 99$). From the plot, can both exponentials be identified? Explain.

(d) Zero-pad the signal from part (c) with 400 zeros and then compute and plot the 500-point DFT. Does this improve the picture of the DFT? Explain.

8.M-3 Repeat Prob. 8.M-2, using the complex signal $x_2[n] = e^{j2\pi n 30/100} + e^{j2\pi n 31.5/100}$.

8.M-4 Consider a complex signal composed of a dc term and two complex exponentials: $y_1[n] = 1 + e^{j2\pi n 30/100} + 0.5 * e^{j2\pi n 43/100}$. For each of the following cases, plot the length-N DFT magnitude as a function of frequency f_r, where $f_r = r/N$.

(a) Use MATLAB to compute and plot the DFT of $y_1[n]$ with 20 samples ($0 \le n \le 19$). From the plot, can the two non-dc exponentials be identified? Given the amplitude relation between the two, the lower-frequency peak should be twice as large as the higher-frequency peak. Is this the case? Explain.

(b) Zero-pad the signal from part (a) to a total length of 500. Does this improve locating the two non-dc exponential components?

Is the lower-frequency peak twice as large as the higher-frequency peak? Explain.

(c) MATLAB's signal processing toolbox function `window` allows window functions to be easily generated. Generate a length-20 Hanning window and apply it to $y_1[n]$. Using this windowed function, repeat parts (a) and (b). Comment on whether the window function helps or hinders the analysis.

8.M-5 Repeat Prob. 8.M-4, using the complex signal $y_2[n] = 1 + e^{j2\pi n 30/100} + 0.5e^{j2\pi n 38/100}$.

8.M-6 This problem investigates the idea of zero padding applied in the frequency domain. When asked, plot the length-N DFT magnitude as a function of frequency f_r, where $f_r = r/N$.

(a) In MATLAB, create a vector x that contains one period of the sinusoid $x[n] = \cos((\pi/2)n)$. Plot the result. How "sinusoidal" does the signal appear?

(b) Use the `fft` command to compute the DFT X of vector x. Plot the magnitude of the DFT coefficients. Do they make sense?

(c) Zero-pad the DFT vector to a total length of 100 by inserting the appropriate number of zeros in the middle of the vector X. Call this zero-padded DFT sequence Y. Why are zeros inserted in the middle rather than the end? Take the inverse DFT of Y and plot the result. What similarities exist between the new signal y and the original signal x? What are the differences between x and y? What is the effect of zero padding in the frequency domain? How is this type of zero padding similar to zero padding in the time domain?

(d) Derive a general modification to the procedure of zero padding in the frequency domain to ensure that the amplitude of the resulting time-domain signal is left unchanged.

(e) Consider one period of a square wave described by the length-8 vector $[1 \quad 1 \quad 1 \quad 1 \quad -1 \quad -1 \quad -1 \quad -1]$. Zero-pad the DFT of this vector to a length of 100, and call the result S. Scale S according to part (d), take the inverse DFT, and plot the result. Does the new time-domain signal $s[n]$ look like a square wave? Explain.

FOURIER ANALYSIS OF DISCRETE-TIME SIGNALS

In Chapters 6 and 7, we studied the ways of representing a continuous-time signal as a sum of sinusoids or exponentials. In this chapter we shall discuss similar development for discrete-time signals. Our approach is parallel to that used for continuous-time signals. We first represent a periodic $x[n]$ as a Fourier series formed by a discrete-time exponential (or sinusoid) and its harmonics. Later we extend this representation to an aperiodic signal $x[n]$ by considering $x[n]$ as a limiting case of a periodic signal with the period approaching infinity.

9.1 DISCRETE-TIME FOURIER SERIES (DTFS)

A continuous-time sinusoid $\cos \omega t$ is a periodic signal regardless of the value of ω. Such is not the case for the discrete-time sinusoid $\cos \Omega n$ (or exponential $e^{j\Omega n}$). A sinusoid $\cos \Omega n$ is periodic only if $\Omega/2\pi$ is a rational number. This can be proved by observing that if this sinusoid is N_0 periodic, then

$$\cos \Omega (n + N_0) = \cos \Omega n$$

This is possible only if

$$\Omega N_0 = 2\pi m \qquad m \text{ integer}$$

Here, both m and N_0 are integers. Hence, $\Omega/2\pi = m/N_0$ is a rational number. Thus, a sinusoid $\cos \Omega n$ (or exponential $e^{j\Omega n}$) is periodic only if

$$\frac{\Omega}{2\pi} = \frac{m}{N_0} \qquad \text{a rational number} \tag{9.1a}$$

When this condition ($\Omega/2\pi$ a rational number) is satisfied, the period N_0 of the sinusoid $\cos \Omega n$ is given by [Eq. (9.1a)]

$$N_0 = m \left(\frac{2\pi}{\Omega} \right) \tag{9.1b}$$

To compute N_0, we must choose the smallest value of m that will make $m(2\pi/\Omega)$ an integer. For example, if $\Omega = 4\pi/17$, then the smallest value of m that will make $m(2\pi/\Omega) = m(17/2)$

an integer is 2. Therefore

$$N_0 = m\left(\frac{2\pi}{\Omega}\right) = 2\left(\frac{17}{2}\right) = 17$$

However, a sinusoid $\cos(0.8n)$ is not a periodic signal because $0.8/2\pi$ is not a rational number.

9.1-1 Periodic Signal Representation by Discrete-Time Fourier Series

A continuous-time periodic signal of period T_0 can be represented as a trigonometric Fourier series consisting of a sinusoid of the fundamental frequency $\omega_0 = 2\pi/T_0$, and all its harmonics. The exponential form of the Fourier series consists of exponentials e^{j0t}, $e^{\pm j\omega_0 t}$, $e^{\pm j2\omega_0 t}$, $e^{\pm j3\omega_0 t}, \dots$.

A discrete-time periodic signal can be represented by a discrete-time Fourier series using a parallel development. Recall that a periodic signal $x[n]$ with period N_0 is characterized by the fact that

$$x[n] = x[n + N_0] \tag{9.2}$$

The smallest value of N_0 for which this equation holds is the *fundamental period*. The *fundamental frequency* is $\Omega_0 = 2\pi/N_0$ rad/sample. An N_0-periodic signal $x[n]$ can be represented by a discrete-time Fourier series made up of sinusoids of fundamental frequency $\Omega_0 = 2\pi/N_0$ and its harmonics. As in the continuous-time case, we may use a trigonometric or an exponential form of the Fourier series. Because of its compactness and ease of mathematical manipulations, the exponential form is preferable to the trigonometric. For this reason we shall bypass the trigonometric form and go directly to the exponential form of the discrete-time Fourier series.

The exponential Fourier series consists of the exponentials e^{j0n}, $e^{\pm j\Omega_0 n}$, $e^{\pm j2\Omega_0 n}, \dots$, $e^{\pm jn\Omega_0 n}, \dots$, and so on. There would be an infinite number of harmonics, except for the property proved in Section 5.5-1, that discrete-time exponentials whose frequencies are separated by 2π (or integer multiples of 2π) are identical because

$$e^{j(\Omega\pm2\pi m)n} = e^{j\Omega n}e^{\pm2\pi mn} = e^{j\Omega n} \qquad m \text{ integer}$$

The consequence of this result is that the rth harmonic is identical to the $(r + N_0)$th harmonic. To demonstrate this, let g_n denote the nth harmonic $e^{jn\Omega_0 n}$. Then

$$g_{r+N_0} = e^{j(r+N_0)\Omega_0 n} = e^{j(r\Omega_0 n + 2\pi n)} = e^{jr\Omega_0 n} = g_r$$

and

$$g_r = g_{r+N_0} = g_{r+2N_0} = \dots = g_{r+mN_0} \qquad m \text{ integer} \tag{9.3}$$

Thus, the first harmonic is identical to the $(N_0 + 1)$st harmonic, the second harmonic is identical to the $(N_0 + 2)$nd harmonic, and so on. In other words, there are only N_0 independent harmonics, and their frequencies range over an interval 2π (because the harmonics are separated by $\Omega_0 = 2\pi/N_0$). This means, unlike the continuous-time counterpart, that the discrete-time Fourier series has only a finite number (N_0) of terms. This result is consistent with our observation in Section 5.5-1 that all discrete-time signals are bandlimited to a band from $-\pi$ to π. Because the harmonics are separated by $\Omega_0 = 2\pi/N_0$, there can only be N_0 harmonics in this band. We also saw that this band can be taken from 0 to 2π or any other contiguous band of width 2π. This means we may choose the N_0 independent harmonics $e^{jr\Omega_0 n}$ over $0 \le r \le N_0 - 1$, or over $-1 \le r \le N_0 - 2$, or over $1 \le r \le N_0$, or over any other suitable choice for that matter. Every one of these sets will have the same harmonics, although in different order.

Let us consider the first choice, which corresponds to exponentials $e^{jr\Omega_0 n}$ for $r = 0, 1, 2, \ldots,$ $N_0 - 1$. The Fourier series for an N_0-periodic signal $x[n]$ consists of only these N_0 harmonics, and can be expressed as

$$x[n] = \sum_{r=0}^{N_0-1} \mathcal{D}_r e^{jr\Omega_0 n} \qquad \Omega_0 = \frac{2\pi}{N_0} \tag{9.4}$$

To compute coefficients $\mathcal{D}_r$ in the Fourier series (9.4), we multiply both sides of (9.4) by $e^{-jm\Omega_0 n}$ and sum over n from $n = 0$ to $(N_0 - 1)$.

$$\sum_{n=0}^{N_0-1} x[n]e^{-jm\Omega_0 n} = \sum_{n=0}^{N_0-1} \sum_{r=0}^{N_0-1} \mathcal{D}_r e^{j(r-m)\Omega_0 n} \tag{9.5}$$

The right-hand sum, after interchanging the order of summation, results in

$$\sum_{r=0}^{N_0-1} \mathcal{D}_r \left[\sum_{n=0}^{N_0-1} e^{j(r-m)\Omega_0 n} \right] \tag{9.6}$$

The inner sum, according to Eq. (8.28) in Section 8.5, is zero for all values of $r \ne m$. It is nonzero with a value N_0 only when $r = m$. This fact means the outside sum has only one term $\mathcal{D}_m N_0$ (corresponding to $r = m$). Therefore, the right-hand side of Eq. (9.5) is equal to $\mathcal{D}_m N_0$, and

$$\sum_{n=0}^{N_0-1} x[n]e^{-jm\Omega_0 n} = \mathcal{D}_m N_0$$

and

$$\mathcal{D}_m = \frac{1}{N_0} \sum_{n=0}^{N_0-1} x[n]e^{-jm\Omega_0 n} \tag{9.7}$$

We now have a discrete-time Fourier series (DTFS) representation of an N_0-periodic signal $x[n]$ as

$$x[n] = \sum_{r=0}^{N_0-1} \mathcal{D}_r e^{jr\Omega_0 n} \tag{9.8}$$

where

$$\mathcal{D}_r = \frac{1}{N_0} \sum_{n=0}^{N_0-1} x[n]e^{-jr\Omega_0 n} \qquad \Omega_0 = \frac{2\pi}{N_0} \tag{9.9}$$

Observe that DTFS equations (9.8) and (9.9) are identical (within a scaling constant) to the DFT equations (8.22b) and (8.22a).[†] Therefore, we can use the efficient FFT algorithm to compute the DTFS coefficients.

[†] If we let $x[n] = N_0 x_k$ and $\mathcal{D}_r = X_r$, Eqs. (9.8) and (9.9) are identical to Eqs. (8.22b) and (8.22a), respectively.

9.1-2 Fourier Spectra of a Periodic Signal $x[n]$

The Fourier series consists of N_0 components

$$\mathcal{D}_0, \mathcal{D}_1 e^{j\Omega_0 n}, \mathcal{D}_2 e^{j2\Omega_0 n}, \ldots, \mathcal{D}_{N_0-1} e^{j(N_0-1)\Omega_0 n}$$

The frequencies of these components are $0, \Omega_0, 2\Omega_0, \ldots, (N_0 - 1)\Omega_0$ where $\Omega_0 = 2\pi/N_0$. The amount of the rth harmonic is $\mathcal{D}_r$. We can plot this amount $\mathcal{D}_r$ (the Fourier coefficient) as a function of index r or frequency Ω. Such a plot, called the *Fourier spectrum* of $x[n]$, gives us, at a glance, the graphical picture of the amounts of various harmonics of $x[n]$.

In general, the Fourier coefficients $\mathcal{D}_r$ are complex, and they can be represented in the polar form as

$$\mathcal{D}_r = |\mathcal{D}_r| e^{j\angle \mathcal{D}_r} \tag{9.10}$$

The plot of $|\mathcal{D}_r|$ versus Ω is called the amplitude spectrum and that of $\angle \mathcal{D}_r$ versus Ω is called the angle (or phase) spectrum. These two plots together are the frequency spectra of $x[n]$. Knowing these spectra, we can reconstruct or synthesize $x[n]$ according to Eq. (9.8). Therefore, the Fourier (or frequency) spectra, which are an alternative way of describing a periodic signal $x[n]$, are in every way equivalent (in terms of the information) to the plot of $x[n]$ as a function of n. The Fourier spectra of a signal constitute the *frequency-domain* description of $x[n]$, in contrast to the time-domain description, where $x[n]$ is specified as a function of index n (representing time).

The results are very similar to the representation of a continuous-time periodic signal by an exponential Fourier series except that, generally, the continuous-time signal spectrum bandwidth is infinite and consists of an infinite number of exponential components (harmonics). The spectrum of the discrete-time periodic signal, in contrast, is bandlimited and has at most N_0 components.

PERIODIC EXTENSION OF FOURIER SPECTRUM

We now show that if $\phi[r]$ is an N_0-periodic function of r, then

$$\sum_{r=0}^{N_0-1} \phi[r] = \sum_{r=\langle N_0 \rangle} \phi[r] \tag{9.11}$$

where $r = \langle N_0 \rangle$ indicates summation over any N_0 consecutive values of r. Because $\phi[r]$ is N_0 periodic, the same values repeat with period N_0. Hence, the sum of any set of N_0 consecutive values of $\phi[r]$ must be the same no matter the value of r at which we start summing. Basically it represents the sum over one cycle.

To apply this result to the DTFS, we observe that $e^{-jr\Omega_0 n}$ is N_0 periodic because

$$e^{-jr\Omega_0(n + N_0)} = e^{-jr\Omega_0 n} e^{-j2\pi r} = e^{-jr\Omega_0 n}$$

Therefore, if $x[n]$ is N_0 periodic, $x[n]e^{-jr\Omega_0 n}$ is also N_0 periodic. Hence, from Eq. (9.9) it follows that $\mathcal{D}_r$ is also N_0 periodic, as is $\mathcal{D}_r e^{jr\Omega_0 n}$. Now, because of the property (9.11), we can express Eqs. (9.8) and (9.9) as

$$x[n] = \sum_{r=\langle N_0 \rangle} \mathcal{D}_r e^{jr\Omega_0 n} \tag{9.12}$$

and

$$\mathcal{D}_r = \frac{1}{N_0} \sum_{n=\langle N_0 \rangle} x[n] e^{-jr\Omega_0 n} \tag{9.13}$$

If we plot $\mathcal{D}_r$ for all values of r (rather than only $0 \le r \le N_0 - 1$), then the spectrum $\mathcal{D}_r$ is N_0 periodic. Moreover, Eq. (9.12) shows that $x[n]$ can be synthesized by not only the N_0 exponentials corresponding to $0 \le r \le N_0 - 1$, but by any successive N_0 exponentials in this spectrum, starting at any value of r (positive or negative). For this reason, it is customary to show the spectrum $\mathcal{D}_r$ for all values of r (not just over the interval $0 \le r \le N_0 - 1$). *Yet we must remember that to synthesize $x[n]$ from this spectrum, we need to add only N_0 consecutive components.* All these observations are consistent with our discussion in Chapter 5, where we showed that a sinusoid of a given frequency is equivalent to multitudes of sinusoids, all separated by integer multiple of 2π in frequency.

Along the Ω scale, $\mathcal{D}_r$ repeats every 2π intervals, and along the r scale, $\mathcal{D}_r$ repeats at intervals of N_0. Equations (9.12) and (9.13) show that both $x[n]$ and its spectrum $\mathcal{D}_r$ are N_0 periodic and both have exactly the same number of components (N_0) over one period.

Equation (9.13) shows that $\mathcal{D}_r$ is complex in general, and $\mathcal{D}_{-r}$ is the conjugate of $\mathcal{D}_r$ if $x[n]$ is real. Thus

$$|\mathcal{D}_r| = |\mathcal{D}_{-r}| \qquad \text{and} \qquad \angle\mathcal{D}_r = -\angle\mathcal{D}_{-r} \tag{9.14}$$

so that the amplitude spectrum $|\mathcal{D}_r|$ is an even function , and $\angle\mathcal{D}_r$ is an odd function of r (or Ω). All these concepts will be clarified by the examples to follow. The first example is rather trivial and serves mainly to familiarize the reader with the basic concepts of DTFS.

EXAMPLE 9.1

Find the discrete-time Fourier series (DTFS) for $x[n] = \sin 0.1\pi n$ (Fig. 9.1a). Sketch the amplitude and phase spectra.

In this case the sinusoid $\sin 0.1\pi n$ is periodic because $\Omega/2\pi = 1/20$ is a rational number and the period N_0 is [see Eq. (9.1b)]

$$N_0 = m\left(\frac{2\pi}{\Omega}\right) = m\left(\frac{2\pi}{0.1\pi}\right) = 20m$$

The smallest value of m that makes $20m$ an integer is $m = 1$. Therefore, the period $N_0 = 20$, so that $\Omega_0 = 2\pi/N_0 = 0.1\pi$, and from Eq. (9.12)

$$x[n] = \sum_{r=\langle 20 \rangle} \mathcal{D}_r e^{j0.1\pi rn}$$

where the sum is performed over any 20 consecutive values of r. We shall select the range $-10 \le r < 10$ (values of r from -10 to 9). This choice corresponds to synthesizing $x[n]$ using the spectral components in the fundamental frequency range $(-\pi \le \Omega < \pi)$. Thus,

$$x[n] = \sum_{r=-10}^{9} \mathcal{D}_r e^{j0.1\pi rn}$$

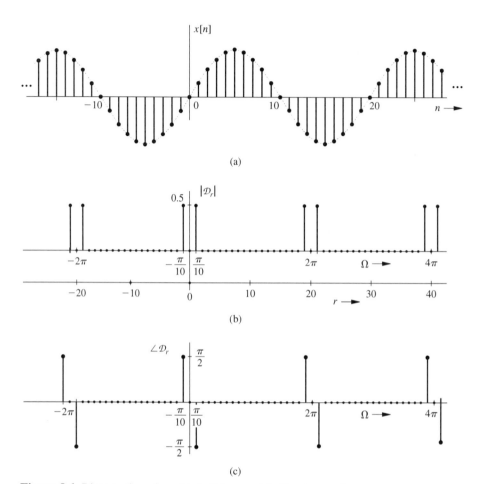

Figure 9.1 Discrete-time sinusoid $\sin 0.1\pi n$ and its Fourier spectra.

where, according to Eq. (9.13),

$$\mathcal{D}_r = \frac{1}{20} \sum_{n=-10}^{9} \sin 0.1\pi n\, e^{-j0.1\pi rn}$$

$$= \frac{1}{20} \sum_{n=-10}^{9} \frac{1}{2j}(e^{j0.1\pi n} - e^{-j0.1\pi n})e^{-j0.1\pi rn}$$

$$= \frac{1}{40j} \left[\sum_{n=-10}^{9} e^{j0.1\pi n(1-r)} - \sum_{n=-10}^{9} e^{-j0.1\pi n(1+r)} \right]$$

In these sums, r takes on all values between -10 and 9. From Eq. (8.28) it follows that the first sum on the right-hand side is zero for all values of r except $r = 1$, when the sum is equal to $N_0 = 20$. Similarly, the second sum is zero for all values of r except $r = -1$, when it is equal to $N_0 = 20$. Therefore

$$\mathcal{D}_1 = \frac{1}{2j} \quad \text{and} \quad \mathcal{D}_{-1} = -\frac{1}{2j}$$

and all other coefficients are zero. The corresponding Fourier series is given by

$$x[n] = \sin 0.1\pi n = \frac{1}{2j}(e^{j0.1\pi n} - e^{-j0.1\pi n}) \tag{9.15}$$

Here the fundamental frequency $\Omega_0 = 0.1\pi$, and there are only two nonzero components:

$$\mathcal{D}_1 = \frac{1}{2j} = \frac{1}{2}e^{-j\pi/2}$$

$$\mathcal{D}_{-1} = -\frac{1}{2j} = \frac{1}{2}e^{j\pi/2}$$

Therefore

$$|\mathcal{D}_1| = |\mathcal{D}_{-1}| = \tfrac{1}{2}$$

$$\angle \mathcal{D}_1 = -\frac{\pi}{2} \quad \text{and} \quad \angle \mathcal{D}_{-1} = \frac{\pi}{2}$$

Sketches of $\mathcal{D}_r$ for the interval $(-10 \leq r < 10)$ appear in Fig. 9.1b and 9.1c. According to Eq. (9.15), there are only two components corresponding to $r = 1$ and -1. The remaining 18 coefficients are zero. The rth component $\mathcal{D}_r$ is the amplitude of the frequency $r\Omega_0 = 0.1r\pi$. Therefore, the frequency interval corresponding to $-10 \leq r < 10$ is $-\pi \leq \Omega < \pi$, as depicted in Fig. 9.1b and 9.1c. This spectrum over the range $-10 \leq r < 10$ (or $-\pi \leq \Omega < \pi$) is sufficient to specify the frequency-domain description (Fourier series), and we can synthesize $x[n]$ by adding these spectral components. Because of the periodicity property discussed in Section 9.1-2, the spectrum $\mathcal{D}_r$ is a periodic function of r with period $N_0 = 20$. For this reason, we repeat the spectrum with period $N_0 = 20$ (or $\Omega = 2\pi$), as illustrated in Fig. 9.1b and 9.1c, which are periodic extensions of the spectrum in the range $-10 \leq r < 10$. Observe that the amplitude spectrum is an even function and the angle or phase spectrum is an odd function of r (or Ω) as expected.

The result (9.15) is a trigonometric identity and could have been obtained immediately without the formality of finding the Fourier coefficients. We have intentionally chosen this trivial example to introduce the reader gently to the new concept of the discrete-time Fourier series and its periodic nature. The Fourier series is a way of expressing a periodic signal $x[n]$ in terms of exponentials of the form $e^{jr\Omega_0 n}$ and its harmonics. The result in Eq. (9.15) is merely a statement of the (obvious) fact that $\sin 0.1\pi n$ can be expressed as a sum of two exponentials $e^{j0.1\pi n}$ and $e^{-j0.1\pi n}$.

Because of the periodicity of the discrete-time exponentials $e^{jr\Omega_0 n}$, the Fourier series components can be selected in any range of length $N_0 = 20$ (or $\Omega = 2\pi$). For example, if we select the frequency range $0 \leq \Omega < 2\pi$ (or $0 \leq r < 20$), we obtain the Fourier series as

$$x[n] = \sin 0.1\pi n = \frac{1}{2j}(e^{j0.1\pi n} - e^{j1.9\pi n}) \tag{9.16}$$

This series is equivalent to that in Eq. (9.15) because the two exponentials $e^{j1.9\pi n}$ and $e^{-j0.1\pi n}$ are equivalent. This follows from the fact that $e^{j1.9\pi n} = e^{j1.9\pi n} \times e^{-j2\pi n} = e^{-j0.1\pi n}$.

We could have selected the spectrum over any other range of width $\Omega = 2\pi$ in Fig. 9.1b and 9.1c as a valid discrete-time Fourier series. The reader may verify this by proving that such a spectrum starting anywhere (and of width $\Omega = 2\pi$) is equivalent to the same two components on the right-hand side of Eq. (9.15).

EXERCISE E9.1

From the spectra in Fig. 9.1 write the Fourier series corresponding to the interval $-10 \geq r > -30$ (or $-\pi \geq \Omega > -3\pi$). Show that this Fourier is equivalent to that in Eq. (9.15).

EXERCISE E9.2

Find the period and the DTFS for

$$x[n] = 4\cos 0.2\pi n + 6\sin 0.5\pi n$$

over the interval $0 \leq r \leq 19$. Use Eq. (9.9) to compute $\mathcal{D}_r$.

ANSWERS
$N_0 = 20$

$x[n] = 2e^{j0.2\pi n} + (3e^{-j\pi/2})e^{j0.5\pi n} + (3e^{j\pi/2})e^{j1.5\pi n} + 2e^{j1.8\pi n}$

EXERCISE E9.3

Find the fundamental periods N_0, if any, for the following sinusoids.

    (a) $\sin(301\pi n/4)$

    (b) $\cos 1.3n$

ANSWERS
(a) $N_0 = 8$

(b) N_0 does not exist because the sinusoid is not periodic.

EXAMPLE 9.2

Find the discrete-time Fourier series for the periodic sampled gate function shown in Fig. 9.2a.

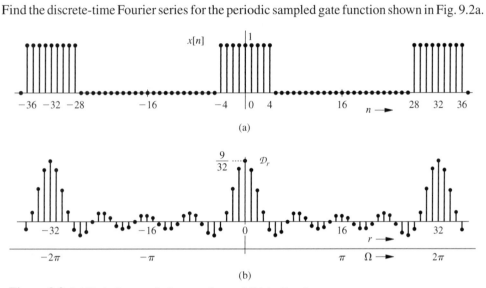

Figure 9.2 (a) Periodic sampled gate pulse and (b) its Fourier spectrum.

In this case $N_0 = 32$ and $\Omega_0 = 2\pi/32 = \pi/16$. Therefore

$$x[n] = \sum_{r=\langle 32 \rangle} \mathcal{D}_r e^{jr(\pi/16)n} \tag{9.17}$$

where

$$\mathcal{D}_r = \frac{1}{32} \sum_{n=\langle 32 \rangle} x[n] e^{-jr(\pi/16)n} \tag{9.18a}$$

For our convenience, we shall choose the interval $-16 \le n \le 15$ for the summation (9.18a), although any other interval of the same width (32 points) would give the same result.[†]

$$\mathcal{D}_r = \frac{1}{32} \sum_{n=-16}^{15} x[n] e^{-jr(\pi/16)n}$$

Now $x[n] = 1$ for $-4 \le n \le 4$ and is zero for all other values of n. Therefore

$$\mathcal{D}_r = \frac{1}{32} \sum_{n=-4}^{4} e^{-jr(\pi/16)n} \tag{9.18b}$$

[†]In this example we have used the same equations as those for DFT in Example 8.9, within a scaling constant. In the present example, the values of $x[n]$ at $n = 4$ and -4 are taken as 1 (full value), whereas in Example 8.9 these values are 0.5 (half the value). This is the reason for the slight difference in spectra in Fig. 9.2b and Fig. 8.18d. Unlike continuous-time signals, discontinuity is a meaningless concept in discrete-time signals.

This is a geometric progression with a common ratio $e^{-j(\pi/16)r}$. Therefore [see Sec. (B.7-4)],

$$\mathcal{D}_r = \frac{1}{32}\left[\frac{e^{-j(5\pi r/16)} - e^{j(4\pi r/16)}}{e^{-j(\pi r/16)} - 1}\right]$$

$$= \left(\frac{1}{32}\right)\frac{e^{-j(0.5\pi r/16)}}{e^{-j(0.5\pi r/16)}}\frac{\left[e^{-j(4.5\pi r/16)} - e^{j(4.5\pi r/16)}\right]}{\left[e^{-j(0.5\pi r/16)} - e^{j(0.5\pi r/16)}\right]}$$

$$= \left(\frac{1}{32}\right)\frac{\sin\left(\dfrac{4.5\pi r}{16}\right)}{\sin\left(\dfrac{0.5\pi r}{16}\right)}$$

$$= \left(\frac{1}{32}\right)\frac{\sin(4.5r\Omega_0)}{\sin(0.5r\Omega_0)} \qquad \Omega_0 = \frac{\pi}{16} \tag{9.19}$$

This spectrum (with its periodic extension) is depicted in Fig. 9.2b.[†]

COMPUTER EXAMPLE C9.1

Repeat Example 9.2 using MATLAB.

```
>> N_0 = 32; n = (0:N_0-1);
>> x_n = [ones(1,5) zeros(1,23) ones(1,4)];
>> for r = 0:31,
>>      X_r(r+1) = sum(x_n.*exp(-j*r*2*pi/N_0*n))/32;
>> end
>> subplot(2,1,1); r = n; stem(r,real(X_r),'k');
>> xlabel('r'); ylabel('X_r'); axis([0 31 -.1 0.3]);
>> legend('DTFS by direct computation',0);
>> X_r = fft(x_n)/N_0;
>> subplot(2,1,2); stem(r,real(X_r),'k');
>> xlabel('r'); ylabel('X_r'); axis([0 31 -.1 0.3]);
>> legend('DTFS by FFT',0);
```

[†]Strictly speaking, geometric progression sum formula applies only if the common ratio $e^{-j(\pi/16)r} \neq 1$. When $r = 0$, this ratio is unity. Hence, Eq. (9.19) is valid for values of $r \neq 0$. For the case $r = 0$, the sum in Eq. (9.18b) is given by

$$\frac{1}{32}\sum_{n=-4}^{4} x[n] = \frac{9}{32}$$

Fortunately, the value of $\mathcal{D}_0$, as computed from Eq. (9.19) also happens to be 9/32. Hence, Eq. (9.19) is valid for all r.

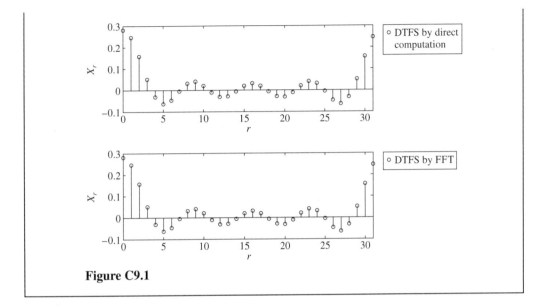

Figure C9.1

9.2 APERIODIC SIGNAL REPRESENTATION BY FOURIER INTEGRAL

In Section 9.1 we succeeded in representing periodic signals as a sum of (everlasting) exponentials. In this section we extend this representation to aperiodic signals. The procedure is identical conceptually to that used in Chapter 7 for continuous-time signals.

Applying a limiting process, we now show that an aperiodic signal $x[n]$ can be expressed as a continuous sum (integral) of everlasting exponentials. To represent an aperiodic signal $x[n]$ such as the one illustrated in Fig. 9.3a by everlasting exponential signals, let us construct a new periodic signal $x_{N_0}[n]$ formed by repeating the signal $x[n]$ every N_0 units, as shown in

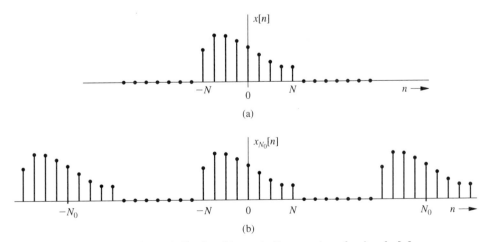

Figure 9.3 Generation of a periodic signal by periodic extension of a signal $x[n]$.

Fig. 9.3b. The period N_0 is made large enough to avoid overlap between the repeating cycles ($N_0 \geq 2N + 1$). The periodic signal $x_{N_0}[n]$ can be represented by an exponential Fourier series. If we let $N_0 \to \infty$, the signal $x[n]$ repeats after an infinite interval, and therefore

$$\lim_{N_0 \to \infty} x_{N_0}[n] = x[n]$$

Thus, the Fourier series representing $x_{N_0}[n]$ will also represent $x[n]$ in the limit $N_0 \to \infty$. The exponential Fourier series for $x_{N_0}[n]$ is given by

$$x_{N_0}[n] = \sum_{r=\langle N_0 \rangle} \mathcal{D}_r e^{jr\Omega_0 n} \qquad \Omega_0 = \frac{2\pi}{N_0} \tag{9.20}$$

where

$$\mathcal{D}_r = \frac{1}{N_0} \sum_{n=-\infty}^{\infty} x[n] e^{-jr\Omega_0 n} \tag{9.21}$$

The limits for the sum on the right-hand side of Eq. (9.21) should be from $-N$ to N. But because $x[n] = 0$ for $|n| > N$, it does not matter if the limits are taken from $-\infty$ to ∞.

It is interesting to see how the nature of the spectrum changes as N_0 increases. To understand this behavior, let us define $X(\Omega)$, a continuous function of Ω, as

$$X(\Omega) = \sum_{n=-\infty}^{\infty} x[n] e^{-j\Omega n} \tag{9.22}$$

From this definition and Eq. (9.21), we have

$$\mathcal{D}_r = \frac{1}{N_0} X(r\Omega_0) \tag{9.23}$$

This result shows that the Fourier coefficients $\mathcal{D}_r$ are $1/N_0$ times the samples of $X(\Omega)$ taken every Ω_0 rad/s.[†] Therefore, $(1/N_0)X(\Omega)$ is the envelope for the coefficients $\mathcal{D}_r$. We now let $N_0 \to \infty$ by doubling N_0 repeatedly. Doubling N_0 halves the fundamental frequency Ω_0, with the result that the spacing between successive spectral components (harmonics) is halved, and there are now twice as many components (samples) in the spectrum. At the same time, by doubling N_0, the envelope of the coefficients $\mathcal{D}_r$ is halved, as seen from Eq. (9.23). If we continue this process of doubling N_0 repeatedly, the number of components doubles in each step; the spectrum progressively becomes denser while its magnitude $\mathcal{D}_r$ becomes smaller. Note, however, that the relative shape of the envelope remains the same [proportional to $X(\Omega)$ in Eq. (9.22)]. In the limit, as $N_0 \to \infty$, the fundamental frequency $\Omega_0 \to 0$, and $\mathcal{D}_r \to 0$. The separation between successive harmonics, which is Ω_0, is approaching zero (infinitesimal), and the spectrum becomes so dense that it appears to be continuous. But as the number of harmonics increases indefinitely, the harmonic amplitudes $\mathcal{D}_r$ become vanishingly small (infinitesimal). We discussed an identical situation in Section 7.1.

We follow the procedure in Section 7.1 and let $N_0 \to \infty$. According to Eq. (9.22)

$$X(r\Omega_0) = \sum_{n=-\infty}^{\infty} x[n] e^{-jr\Omega_0 n} \tag{9.24}$$

[†]For the sake of simplicity we assume $\mathcal{D}_r$ and therefore $X(\Omega)$ to be real. The argument, however, is also valid for complex $\mathcal{D}_r$ [or $X(\Omega)$].

Using Eqs. (9.23), we can express Eq. (9.20) as

$$x_{N_0}[n] = \frac{1}{N_0} \sum_{r=\langle N_0 \rangle} X(r\Omega_0) e^{jr\Omega_0 n} \tag{9.25a}$$

$$= \sum_{r=\langle N_0 \rangle} X(r\Omega_0) e^{jr\Omega_0 n} \left(\frac{\Omega_0}{2\pi} \right) \tag{9.25b}$$

In the limit as $N_0 \to \infty$, $\Omega_0 \to 0$ and $x_{N_0}[n] \to x[n]$. Therefore

$$x[n] = \lim_{\Omega_0 \to 0} \sum_{r=\langle N_0 \rangle} \left[\frac{X(r\Omega_0)\Omega_0}{2\pi} \right] e^{jr\Omega_0 n} \tag{9.26}$$

Because Ω_0 is infinitesimal, it will be appropriate to replace Ω_0 with an infinitesimal notation $\Delta\Omega$:

$$\Delta\Omega = \frac{2\pi}{N_0} \tag{9.27}$$

Equation (9.26) can be expressed as

$$x[n] = \lim_{\Delta\Omega \to 0} \frac{1}{2\pi} \sum_{r=\langle N_0 \rangle} X(r\Delta\Omega) e^{jr\Delta\Omega n} \Delta\Omega \tag{9.28}$$

The range $r = \langle N_0 \rangle$ implies the interval of N_0 number of harmonics, which is $N_0 \Delta\Omega = 2\pi$ according to Eq. (9.27). In the limit, the right-hand side of Eq. (9.28) becomes the integral

$$x[n] = \frac{1}{2\pi} \int_{2\pi} X(\Omega) e^{jn\Omega} \, d\Omega \tag{9.29}$$

where $\int_{2\pi}$ indicates integration over any continuous interval of 2π. The spectrum $X(\Omega)$ is given by [Eq. (9.22)]

$$X(\Omega) = \sum_{n=-\infty}^{\infty} x[n] e^{-j\Omega n} \tag{9.30}$$

The integral on the right-hand side of Eq. (9.29) is called the *Fourier integral*. We have now succeeded in representing an aperiodic signal $x[n]$ by a Fourier integral (rather than a Fourier series). This integral is basically a Fourier series (in the limit) with fundamental frequency $\Delta\Omega \to 0$, as seen in Eq. (9.28). The amount of the exponential $e^{jr\Delta\Omega n}$ is $X(r\Delta\Omega)\Delta\Omega/2\pi$. Thus, the function $X(\Omega)$ given by Eq. (9.30) acts as a spectral function, which indicates the relative amounts of various exponential components of $x[n]$.

We call $X(\Omega)$ the (direct) discrete-time Fourier transform (DTFT) of $x[n]$, and $x[n]$ the inverse discrete-time Fourier transform (IDTFT) of $X(\Omega)$. This nomenclature can be represented as

$$X(\Omega) = \text{DTFT}\{x[n]\} \qquad \text{and} \qquad x[n] = \text{IDTFT}\{X(\Omega)\}$$

The same information is conveyed by the statement that $x[n]$ and $X(\Omega)$ are a (discrete-time) Fourier transform pair. Symbolically, this is expressed as

$$x[n] \Longleftrightarrow X(\Omega)$$

The Fourier transform $X(\Omega)$ is the frequency-domain description of $x[n]$.

9.2-1 Nature of Fourier Spectra

We now discuss several important features of the discrete-time Fourier transform and the spectra associated with it.

THE FOURIER SPECTRA ARE CONTINUOUS FUNCTIONS OF Ω

Although $x[n]$ is a discrete-time signal, $X(\Omega)$, its DTFT is a continuous function of Ω for the simple reason that Ω is a continuous variable, which can take any value over a continuous interval from $-\infty$ to ∞.

THE FOURIER SPECTRA ARE PERIODIC FUNCTIONS OF Ω WITH PERIOD 2π

From Eq. (9.30) it follows that

$$X(\Omega + 2\pi) = \sum_{n=-\infty}^{\infty} x[n]e^{-j(\Omega+2\pi)n} = \sum_{n=-\infty}^{\infty} x[n]e^{-j\Omega n}e^{-j2\pi n} = X(\Omega) \tag{9.31}$$

Clearly, the spectrum $X(\Omega)$ is a continuous, periodic function of Ω with period 2π. We must remember, however, that to synthesize $x[n]$, we need to use the spectrum over a frequency interval of only 2π, starting at any value of Ω [see Eq. (9.29)]. As a matter of convenience, we shall choose this interval to be the fundamental frequency range $(-\pi, \ \pi)$. It is, therefore, not necessary to show discrete-time-signal spectra beyond the fundamental range, although we often do so.

The reason for periodic behavior of $X(\Omega)$ was discussed in Chapter 5, where we showed that, in a basic sense, the discrete-time frequency Ω is bandlimited to $|\Omega| \leq \pi$. However, all discrete-time sinusoids with frequencies separated by an integer multiple of 2π are identical. This is why the spectrum is 2π periodic.

CONJUGATE SYMMETRY OF $X(\Omega)$

From Eq. (9.30), we obtain the DTFT of $x^*[n]$ as

$$\text{DTFT}\{x^*[n]\} = \sum_{n=-\infty}^{\infty} x^*[n]e^{-j\Omega n} = X^*(-\Omega) \tag{9.32a}$$

In other words,

$$x^*[n] \Longleftrightarrow X^*(-\Omega) \tag{9.32b}$$

For real $x[n]$, Eq. (9.32b) reduces to $x[n] \Longleftrightarrow X^*(-\Omega)$, which implies that for real $x[n]$

$$X(\Omega) = X^*(-\Omega)$$

Therefore, for real $x[n]$, $X(\Omega)$ and $X(-\Omega)$ are conjugates. Since $X(\Omega)$ is generally complex, we have both amplitude and angle (or phase) spectra

$$X(\Omega) = |X(\Omega)|e^{j\angle X(\Omega)}$$

Because of conjugate symmetry of $X(\Omega)$, it follows that for real $x[n]$

$$|X(\Omega)| = |X(-\Omega)|$$

$$\angle X(\Omega) = -\angle X(-\Omega)$$

Therefore, the amplitude spectrum $|X(\Omega)|$ is an even function of Ω and the phase spectrum $\angle X(\Omega)$ is an odd function of Ω for real $x[n]$.

Physical Appreciation of the Discrete-Time Fourier Transform

In understanding any aspect of the Fourier transform, we should remember that Fourier representation is a way of expressing a signal $x[n]$ as a sum of everlasting exponentials (or sinusoids). The Fourier spectrum of a signal indicates the relative amplitudes and phases of the exponentials (or sinusoids) required to synthesize $x[n]$.

A detailed explanation of the nature of such sums over a continuum of frequencies is provided in Section 7.1-1.

Existence of the DTFT

Because $|e^{-j\Omega n}| = 1$, from Eq. (9.30), it follows that the existence of $X(\Omega)$ is guaranteed if $x[n]$ is absolutely summable: that is,

$$\sum_{n=-\infty}^{\infty} |x[n]| < \infty \tag{9.33a}$$

This shows that the condition of absolute summability is a sufficient condition for the existence of the DTFT representation. This condition also guarantees its uniform convergence. The inequality

$$\left[\sum_{n=-\infty}^{\infty} |x[n]|\right]^2 \geq \sum_{n=-\infty}^{\infty} |x[n]|^2$$

shows that the energy of an absolutely summable sequence is finite. However, not all finite-energy signals are absolutely summable. Signal $x[n] = \text{sinc}(n)$ is such an example. For such signals, the DTFT converges, not uniformly, but in the mean.[†]

To summarize, $X(\Omega)$ exists under a weaker condition

$$\sum_{n=-\infty}^{\infty} |x[n]|^2 < \infty \tag{9.33b}$$

The DTFT under this condition is guaranteed to converge in the mean. Thus, the DTFT of the exponentially growing signal $\gamma^n u[n]$ does not exist when $|\gamma| > 1$ because the signal violates conditions (9.33a) and (9.33b). But the DTFT exists for the signal $\text{sinc}(n)$, which violates (9.33a) but does satisfy (9.33b) (see later: Example 9.6). In addition, if the use of $\delta(\Omega)$, the continuous-time impulse function, is permitted, we can even find the DTFT of some signals that violate both (9.33a) and (9.33b). Such signals are not absolutely summable, nor do they have finite energy. For example, as seen from pairs 11 and 12 of Table 9.1, the DTFT of $x[n] = 1$ for all n and $x[n] = e^{j\Omega_0 n}$ exist, although they violate conditions (9.33a) and (9.33b).

[†]This means

$$\lim_{M \to \infty} \int_{-\pi}^{\pi} \left| X(\Omega) - \sum_{n=-M}^{M} x[n]e^{-j\Omega n} \right|^2 d\Omega = 0$$

TABLE 9.1   A Short Table of Discrete-Time Fourier Transforms

No.	$x[n]$	$X(\Omega)$	
1	$\delta[n-k]$	$e^{-jk\Omega}$	Integer k
2	$\gamma^n u[n]$	$\dfrac{e^{j\Omega}}{e^{j\Omega}-\gamma}$	$|\gamma|<1$
3	$-\gamma^n u[-(n+1)]$	$\dfrac{e^{j\Omega}}{e^{j\Omega}-\gamma}$	$|\gamma|>1$
4	$\gamma^{|n|}$	$\dfrac{1-\gamma^2}{1-2\gamma\cos\Omega+\gamma^2}$	$|\gamma|<1$
5	$n\gamma^n u[n]$	$\dfrac{\gamma e^{j\Omega}}{(e^{j\Omega}-\gamma)^2}$	$|\gamma|<1$
6	$\gamma^n \cos(\Omega_0 n+\theta)u[n]$	$\dfrac{e^{j\Omega}[e^{j\Omega}\cos\theta-\gamma\cos(\Omega_0-\theta)]}{e^{j2\Omega}-(2\gamma\cos\Omega_0)e^{j\Omega}+\gamma^2}$	$|\gamma|<1$
7	$u[n]-u[n-M]$	$\dfrac{\sin(M\Omega/2)}{\sin(\Omega/2)}\,e^{-j\Omega(M-1)/2}$	
8	$\dfrac{\Omega_c}{\pi}\operatorname{sinc}(\Omega_c n)$	$\displaystyle\sum_{k=-\infty}^{\infty}\operatorname{rect}\left(\dfrac{\Omega-2\pi k}{2\Omega_c}\right)$	$\Omega_c\le\pi$
9	$\dfrac{\Omega_c}{2\pi}\operatorname{sinc}^2\left(\dfrac{\Omega_c n}{2}\right)$	$\displaystyle\sum_{k=-\infty}^{\infty}\Delta\left(\dfrac{\Omega-2\pi k}{2\Omega_c}\right)$	$\Omega_c\le\pi$
10	$u[n]$	$\dfrac{e^{j\Omega}}{e^{j\Omega}-1}+\pi\displaystyle\sum_{k=-\infty}^{\infty}\delta(\Omega-2\pi k)$	
11	$1\quad$ for all n	$2\pi\displaystyle\sum_{k=-\infty}^{\infty}\delta(\Omega-2\pi k)$	
12	$e^{j\Omega_0 n}$	$2\pi\displaystyle\sum_{k=-\infty}^{\infty}\delta(\Omega-\Omega_0-2\pi k)$	
13	$\cos\Omega_0 n$	$\pi\displaystyle\sum_{k=-\infty}^{\infty}\delta(\Omega-\Omega_0-2\pi k)+\delta(\Omega+\Omega_0-2\pi k)$	
14	$\sin\Omega_0 n$	$j\pi\displaystyle\sum_{k=-\infty}^{\infty}\delta(\Omega+\Omega_0-2\pi k)-\delta(\Omega-\Omega_0-2\pi k)$	
15	$(\cos\Omega_0 n)\,u[n]$	$\dfrac{e^{j2\Omega}-e^{j\Omega}\cos\Omega_0}{e^{j2\Omega}-2e^{j\Omega}\cos\Omega_0+1}+\dfrac{\pi}{2}\displaystyle\sum_{k=-\infty}^{\infty}\delta(\Omega-2\pi k-\Omega_0)+\delta(\Omega-2\pi k+\Omega_0)$	
16	$(\sin\Omega_0 n)\,u[n]$	$\dfrac{e^{j\Omega}\sin\Omega_0}{e^{j2\Omega}-2e^{j\Omega}\cos\Omega_0+1}+\dfrac{\pi}{2j}\displaystyle\sum_{k=-\infty}^{\infty}\delta(\Omega-2\pi k-\Omega_0)-\delta(\Omega-2\pi k+\Omega_0)$	

EXAMPLE 9.3

Find the DTFT of $x[n] = \gamma^n u[n]$.

$$X(\Omega) = \sum_{n=0}^{\infty} \gamma^n e^{-j\Omega n}$$

$$= \sum_{n=0}^{\infty} (\gamma e^{-j\Omega})^n$$

This is an infinite geometric series with a common ratio $\gamma e^{-j\Omega}$. Therefore (see Section B.7-4)

$$X(\Omega) = \frac{1}{1 - \gamma e^{-j\Omega}}$$

provided that $|\gamma e^{-j\Omega}| < 1$. But because $|e^{-j\Omega}| = 1$, this condition implies $|\gamma| < 1$. Therefore

$$X(\Omega) = \frac{1}{1 - \gamma e^{-j\Omega}} \qquad |\gamma| < 1 \tag{9.34a}$$

If $|\gamma| > 1$, $X(\Omega)$ does not converge. This result is in conformity with condition (9.33). From Eq. (9.34a)

$$X(\Omega) = \frac{1}{1 - \gamma \cos \Omega + j\gamma \sin \Omega} \tag{9.34b}$$

so that

$$|X(\Omega)| = \frac{1}{\sqrt{(1 - \gamma \cos \Omega)^2 + (\gamma \sin \Omega)^2}} \tag{9.35a}$$

$$= \frac{1}{\sqrt{1 + \gamma^2 - 2\gamma \cos \Omega}}$$

$$\angle X(\Omega) = -\tan^{-1}\left[\frac{\gamma \sin \Omega}{1 - \gamma \cos \Omega}\right] \tag{9.35b}$$

Figure 9.4 shows $x[n] = \gamma^n u[n]$ and its spectra for $\gamma = 0.8$. Observe that the frequency spectra are continuous and periodic functions of Ω with the period 2π. As explained earlier, we need to use the spectrum only over the frequency interval of 2π. We often select this interval to be the fundamental frequency range $(-\pi, \pi)$.

The amplitude spectrum $|X(\Omega)|$ is an even function and the phase spectrum $\angle X(\Omega)$ is an odd function of Ω.

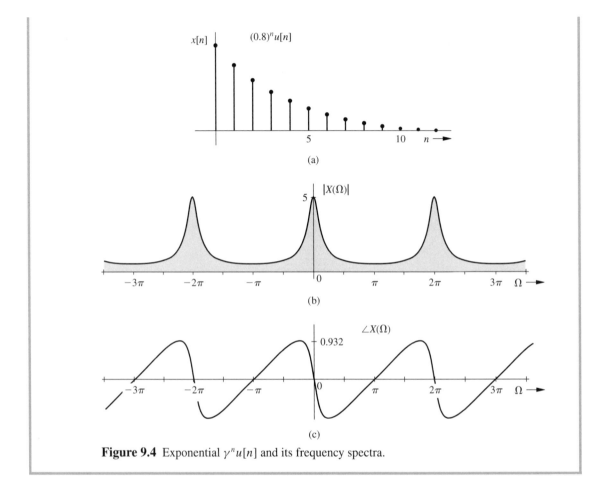

Figure 9.4 Exponential $\gamma^n u[n]$ and its frequency spectra.

EXAMPLE 9.4

Find the DTFT of $\gamma^n u[-(n + 1)]$ depicted in Fig. 9.5.

$$X(\Omega) = \sum_{n=-\infty}^{\infty} \gamma^n u[-(n + 1)]e^{-j\Omega n} = \sum_{n=-1}^{-\infty} (\gamma e^{-j\Omega})^n = \sum_{n=-1}^{-\infty} \left(\frac{1}{\gamma}e^{j\Omega}\right)^{-n}$$

Setting $n = -m$ yields

$$x[n] = \sum_{m=1}^{\infty} \left(\frac{1}{\gamma}e^{j\Omega}\right)^m = \frac{1}{\gamma}e^{j\Omega} + \left(\frac{1}{\gamma}e^{j\Omega}\right)^2 + \left(\frac{1}{\gamma}e^{j\Omega}\right)^3 + \cdots$$

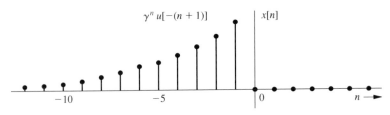

Figure 9.5 Exponential $\gamma^n u[-(n+1)]$.

This is a geometric series with a common ratio $e^{j\Omega}/\gamma$. Therefore, from Section B.7-4,

$$X(\Omega) = \frac{1}{\gamma e^{-j\Omega} - 1} \qquad |\gamma| > 1 \tag{9.36}$$

$$= \frac{1}{(\gamma \cos \Omega - 1) - j\gamma \sin \Omega}$$

Therefore

$$|X(\Omega)| = \frac{1}{\sqrt{1 + \gamma^2 - 2\gamma \cos \Omega}}$$

$$\angle X(\Omega) = \tan^{-1}\left[\frac{\gamma \sin \Omega}{\gamma \cos \Omega - 1}\right] \tag{9.37}$$

Except for the change of sign, this Fourier transform (and the corresponding frequency spectra) is identical to that of $x[n] = \gamma^n u[n]$. Yet there is no ambiguity in determining the IDTFT of $X(\Omega) = 1/(\gamma e^{-j\Omega} - 1)$ because of the restrictions on the value of γ in each case. If $|\gamma| < 1$, then the inverse transform is $x[n] = -\gamma^n u[n]$. If $|\gamma| > 1$, it is $x[n] = \gamma^n[-(n+1)]$.

EXAMPLE 9.5

Find the DTFT of the discrete-time rectangular pulse illustrated in Fig. 9.6a. This pulse is also known as the 9-point rectangular window function.

$$X(\Omega) = \sum_{n=-\infty}^{\infty} x[n]e^{-j\Omega n}$$

$$= \sum_{n=-(M-1)/2}^{(M-1)/2} (e^{-j\Omega})^n \qquad M = 9 \tag{9.38}$$

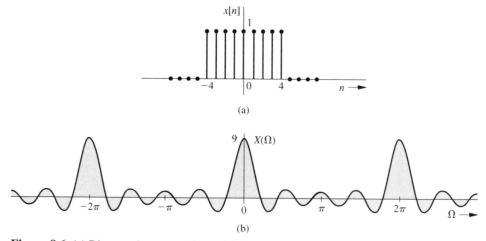

(a)

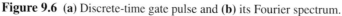

(b)

Figure 9.6 (a) Discrete-time gate pulse and (b) its Fourier spectrum.

This is a geometric progression with a common ratio $e^{-j\Omega}$ and (see Section B.7-4)

$$X(\Omega) = \frac{e^{-j[(M+1)/2]\Omega} - e^{j[(M-1)/2]\Omega}}{e^{-j\Omega} - 1}$$

$$= \frac{e^{-j\Omega/2}\left(e^{-j(M/2)\Omega} - e^{j(M/2)\Omega}\right)}{e^{-j\Omega/2}\left(e^{-j\Omega/2} - e^{j\Omega/2}\right)}$$

$$= \frac{\sin\left(\dfrac{M}{2}\Omega\right)}{\sin(0.5\Omega)} \tag{9.39}$$

$$= \frac{\sin(4.5\Omega)}{\sin(0.5\Omega)} \qquad \text{for } M = 9 \tag{9.40}$$

Figure 9.6b shows the spectrum $X(\Omega)$ for $M = 9$.

COMPUTER EXAMPLE C9.2

Repeat Example 9.5 using MATLAB.

```
>> syms Omega n M
>> X = simplify(symsum(exp(-j*Omega*n),n,-(M-1)/2,(M-1)/2))

X =

1/(-1+exp(i*Omega))*(-exp(-1/2*i*Omega*(M-1))+exp(1/2*i*Omega*(M+1)))
        % Evaluate symbolic expression X(Omega) for M = 9:
>> Omega = linspace(-pi,pi,1000); M = 9*ones(size(Omega)); X = subs(X);
```

```
>> subplot(2,1,1); plot(Omega,abs(X),'k'); axis([-pi pi 0 9]);
>> xlabel('\Omega'); ylabel('X(\Omega)');
>> legend('DTFT by symbolic computation',0);
        % Compute samples of X(Omega) using the FFT:
>> N = 512; M = 9;
>> x = [ones(1,(M+1)/2) zeros(1,N-M) ones(1,(M-1)/2)];
>> X = fft(x); Omega = (-N/2:N/2-1)*2*pi/N;
>> subplot(2,1,2); plot(Omega,fftshift(abs(X)),'k.'); axis([-pi pi 0 9]);
>> xlabel('\Omega'); ylabel('X(\Omega)');
>> legend('DTFT samples by FFT',0);
```

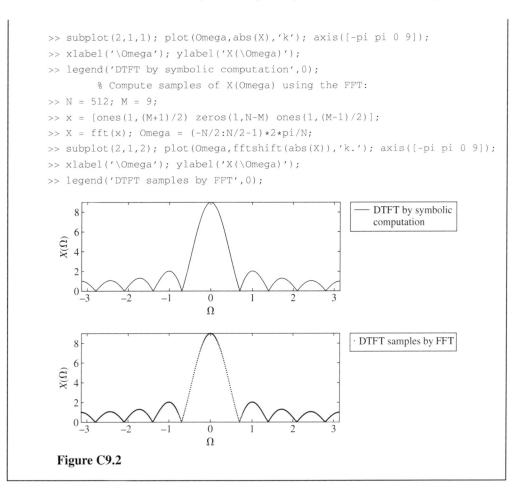

Figure C9.2

EXAMPLE 9.6

Find the inverse DTFT of the rectangular pulse spectrum described over the fundamental band ($|\Omega| \leq \pi$) by $X(\Omega) = \text{rect}(\Omega/2\Omega_c)$ for $\Omega_c \leq \pi$. Because of the periodicity property, $X(\Omega)$ repeats at the intervals of 2π, as shown in Fig. 9.7a.

According to Eq. (9.29)

$$x[n] = \frac{1}{2\pi} \int_{-\pi}^{\pi} X(\Omega)e^{jn\Omega}\,d\Omega = \frac{1}{2\pi}\int_{-\Omega_c}^{\Omega_c} e^{jn\Omega}\,d\Omega$$

$$= \frac{1}{j2\pi n}e^{jn\Omega}\Big|_{-\Omega_c}^{\Omega_c} = \frac{\sin(\Omega_c n)}{\pi n} = \frac{\Omega_c}{\pi}\,\text{sinc}\,(\Omega_c n) \qquad (9.41)$$

The signal $x[n]$ is depicted in Fig. 9.7b (for the case $\Omega_c = \pi/4$).

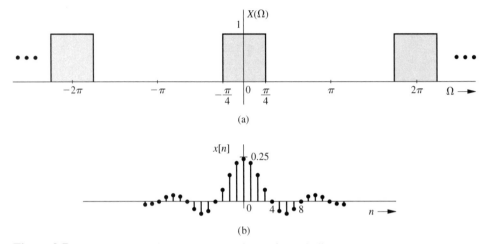

Figure 9.7 Inverse discrete-time Fourier transform of a periodic gate spectrum.

EXERCISE E9.4

Find and sketch the amplitude and phase spectra of the DTFT of the signal $x[n] = \gamma^{|k|}$ with $|\gamma| < 1$.

ANSWER

$$X(\Omega) = \frac{1 - \gamma^2}{1 - 2\gamma \cos \Omega + \gamma^2}$$

EXERCISE E9.5

Find and sketch the amplitude and phase spectra of the DTFT of the signal $x[n] = \delta[n + 1] - \delta[n - 1]$.

ANSWERS

$|X(\Omega)| = 2|\sin \Omega|$

$\angle X(\omega) = (\pi/2)[1 - \text{sgn} (\sin \Omega)]$

9.2-2 Connection Between the DTFT and the z-Transform

The connection between the (bilateral) z-transform and the DTFT is similar to that between the Laplace transform and the Fourier transform. The z-transform of $x[n]$, according to Eq. (5.1) is

$$X[z] = \sum_{n=-\infty}^{\infty} x[n]z^{-n} \tag{9.42a}$$

Setting $z = e^{j\Omega}$ in this equation yields

$$X[e^{j\Omega}] = \sum_{n=-\infty}^{\infty} x[n]e^{-j\Omega n} \tag{9.42b}$$

The right-hand side sum defines $X(\Omega)$, the DTFT of $x[n]$. Does this mean that the DTFT can be obtained from the corresponding z-transform by setting $z = e^{j\Omega}$? In other words, is it true that $X[e^{j\Omega}] = X(\Omega)$? Yes, it is true in most cases. For example, when $x[n] = a^n u[n]$, its z-transform is $z/(z-a)$, and $X[e^{j\Omega}] = e^{j\Omega}/(e^{j\Omega}-a)$, which is equal to $X(\Omega)$ (assuming $|a| < 1$). However, for the unit step function $u[n]$, the z-transform is $z/(z-1)$, and $X[e^{j\Omega}] = e^{j\Omega}/(e^{j\Omega} - 1)$. As seen from Table 9.1, pair 10, this is not equal to $X(\Omega)$ in this case.

We obtained $X[e^{j\Omega}]$ by setting $z = e^{j\Omega}$ in Eq. (9.42a). This implies that the sum on the right-hand side of Eq. (9.42a) converges for $z = e^{j\Omega}$, which means the unit circle (characterized by $z = e^{j\Omega}$) lies in the region of convergence for $X[z]$. Hence, the general rule is that only when the ROC for $X[z]$ includes the unit circle, setting $z = e^{j\Omega}$ in $X[z]$ yields the DTFT $X(\Omega)$. This applies for all $x[n]$ that are absolutely summable. If the ROC of $X[z]$ excludes the unit circle, $X[e^{j\Omega}] \neq X(\Omega)$. This applies to all exponentially growing $x[n]$ and also $x[n]$, which either is constant or oscillates with constant amplitude.

The reason for this peculiar behavior has something to do with the nature of convergence of the z-transform and the DTFT.[†]

This discussion shows that although the DTFT may be considered to be a special case of the z-transform, we need to circumscribe such a view. This cautionary note is supported by the fact that a periodic signal has the DTFT, but its z-transform does not exist.

9.3 PROPERTIES OF THE DTFT

In the next section, we shall see a close connection between the DTFT and the CTFT (continuous-time Fourier transform). For this reason, the properties of the DTFT are very similar to those of the CTFT, as the following discussion shows.

LINEARITY OF THE DTFT

If

$$x_1[n] \Longleftrightarrow X_1(\Omega) \qquad \text{and} \qquad x_2[n] \Longleftrightarrow X_2(\Omega)$$

then

$$a_1 x_1[n] + a_2 x_2[n] \Longleftrightarrow a_1 X_1(\Omega) + a_2 X_2(\Omega) \tag{9.43}$$

The proof is trivial. The result can be extended to any finite sums.

[†]To explain this point, consider the unit step function $u[n]$ and its transforms. Both the z-transform and the DTFT synthesize $x[n]$, using everlasting exponentials of the form z^n. The value of z can be anywhere in the complex z-plane for the z-transform, but it must be restricted to the unit circle ($z = e^{j\Omega}$) in the case of the DTFT. The unit step function is readily synthesized in the z-transform by a relatively simple spectrum $X[z] = z/(z-1)$, by choosing z outside the unit circle (the ROC for $u[n]$ is $|z| > 1$). In the DTFT, however, we are restricted to values of z only on the unit circle ($z = e^{j\Omega}$). The function $u[n]$ can still be synthesized by values of z on the unit circle, but the spectrum is more complicated than when we are free to choose z anywhere including the region outside the unit circle. In contrast, when $x[n]$ is absolutely summable, the region of convergence for the z-transform includes the unit circle, and we can synthesize $x[n]$ by using z along the unit circle in both the transforms. This leads to $X[e^{j\Omega}] = X(\Omega)$.

CONJUGATE SYMMETRY OF $X(\Omega)$

In Eq. (9.32b), we proved the *conjugation property*

$$x^*[n] \Longleftrightarrow X^*(-\Omega) \tag{9.44}$$

We also showed that as a consequence of this, when $x[n]$ is real, $X(\Omega)$ and $X(-\Omega)$ are conjugates, that is,

$$X(-\Omega) = X^*(\Omega) \tag{9.45}$$

This is the *conjugate symmetry* property. Since $X(\Omega)$ is generally complex, we have both amplitude and angle (or phase) spectra

$$X(\Omega) = |X(\Omega)|e^{j\angle X(\Omega)} \tag{9.46}$$

Hence, for real $x[n]$, it follows that

$$|X(\Omega)| = |X(-\Omega)| \tag{9.47a}$$

$$\angle X(\Omega) = -\angle X(-\Omega) \tag{9.47b}$$

Therefore, for real $x[n]$, the amplitude spectrum $|X(\Omega)|$ is an even function of Ω and the phase spectrum $\angle X(\Omega)$ is an odd function of Ω.

TIME AND FREQUENCY REVERSAL

$$x[-n] \Longleftrightarrow X(-\Omega) \tag{9.48}$$

From Eq. (9.30), the DTFT of $x[-n]$ is

$$\text{DTFT}\{x[-n]\} = \sum_{n=-\infty}^{\infty} x[-n]e^{-j\Omega n} = \sum_{m=-\infty}^{\infty} x[m]e^{j\Omega m} = X(-\Omega)$$

EXAMPLE 9.7

Use the time-frequency reversal property (9.48) and pair 2 in Table 9.1 to derive pair 4 in Table 9.1.

Pair 2 states that

$$\gamma^n u[n] = \frac{e^{j\Omega}}{e^{j\Omega} - \gamma} \qquad |\gamma| < 1 \tag{9.49a}$$

Hence, from Eq. (9.48)

$$\gamma^{-n} u[-n] = \frac{e^{-j\Omega}}{e^{-j\Omega} - \gamma} \qquad |\gamma| < 1 \tag{9.49b}$$

Moreover, $\gamma^{|n|}$ could be expressed as a sum of $\gamma^n u[n]$ and $\gamma^{-n} u[-n]$, except that the impulse at $n = 0$ is counted twice (once from each of the two exponentials). Hence

$$\gamma^{|n|} = \gamma^n u[n] + \gamma^{-n} u[-n] - \delta[n]$$

Therefore, using Eqs. (9.49a) and (9.49b), and invoking the linearity property, we can write

$$\text{DTFT}\{\gamma^{|n|}\} = \frac{e^{j\Omega}}{e^{j\Omega} - \gamma} + \frac{e^{-j\Omega}}{e^{-j\Omega} - \gamma} - 1 = \frac{1 - \gamma^2}{1 - 2\gamma \cos \Omega + \gamma^2} \qquad |\gamma| < 1$$

which agrees with pair 4 in Table 9.1.

EXERCISE E9.6

In Table 9.1, derive pair 13 from pair 15 by using the time-reversal property (9.48).

MULTIPLICATION BY n:
FREQUENCY DIFFERENTIATION

$$nx[n] \iff j\frac{dX(\Omega)}{d\Omega} \tag{9.50}$$

The result follows immediately by differentiating both sides of Eq. (9.30) with respect to Ω.

EXAMPLE 9.8

Use the property in Eq. (9.50) [multiplication by n] and pair 2 in Table 9.1 to derive pair 5 in Table 9.1.

Pair 2 states that

$$\gamma^n u[n] = \frac{e^{j\Omega}}{e^{j\Omega} - \gamma} \qquad |\gamma| < 1 \tag{9.51}$$

Hence, from Eq. (9.50)

$$n\gamma^n u[n] = j\frac{d}{d\Omega}\left\{\frac{e^{j\Omega}}{e^{j\Omega} - \gamma}\right\} = \frac{\gamma e^{j\Omega}}{(e^{j\Omega} - \gamma)^2} \qquad |\gamma| < 1$$

which agrees with pair 5 in Table 9.1.

TIME-SHIFTING PROPERTY
If

$$x[n] \Longleftrightarrow X(\Omega)$$

then

$$x[n - k] \Longleftrightarrow X(\Omega)e^{-jk\Omega} \qquad \text{for integer } k \tag{9.52}$$

This property can be proved by direct substitution in the equation defining the direct transform. From Eq. (9.30) we obtain

$$x[n - k] \Longleftrightarrow \sum_{n=-\infty}^{\infty} x[n-k]e^{-j\Omega n} = \sum_{m=-\infty}^{\infty} x[m]e^{-j\Omega[m+k]}$$

$$= e^{-j\Omega k} \sum_{n=-\infty}^{\infty} x[m]e^{-j\Omega m} = e^{-jk\Omega}X(\Omega)$$

This result shows that *delaying a signal by k samples does not change its amplitude spectrum. The phase spectrum, however, is changed by* $-k\Omega$. This added phase is a linear function of Ω with slope $-k$.

PHYSICAL EXPLANATION OF THE LINEAR PHASE

Time delay in a signal causes a linear phase shift in its spectrum. The heuristic explanation of this result is exactly parallel to that for continuous-time signals given in Section 7.3 (see Fig. 7.22).

EXAMPLE 9.9

Find the DTFT of $x[n] = (1/4) \operatorname{sinc}(\pi(n-2)/4)$, shown in Fig. 9.8a.

In Example 9.6, we found that

$$\frac{1}{4} \operatorname{sinc}\left(\frac{\pi n}{4}\right) \Longleftrightarrow \sum_{m=-\infty}^{\infty} \operatorname{rect}\left(\frac{\Omega - 2\pi m}{\pi/2}\right)$$

Use of the time-shifting property [Eq. (9.52)] yields (for integer k)

$$\frac{1}{4} \operatorname{sinc}\left(\frac{\pi(n-2)}{4}\right) \Longleftrightarrow \sum_{m=-\infty}^{\infty} \operatorname{rect}\left(\frac{\Omega - 2\pi m}{\pi/2}\right) e^{-j2\Omega} \tag{9.53}$$

The spectrum of the shifted signal is shown in Fig. 9.8b.

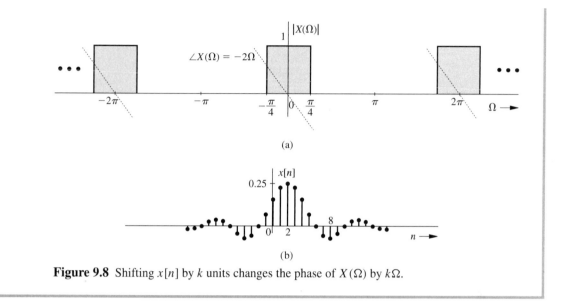

(a)

(b)

Figure 9.8 Shifting $x[n]$ by k units changes the phase of $X(\Omega)$ by $k\Omega$.

EXERCISE E9.7

Verify the result in Eq. (9.40) from pair 7 In Table 9.1, and the time-shifting property of the DTFT.

FREQUENCY-SHIFTING PROPERTY

If

$$x[n] \Longleftrightarrow X(\Omega)$$

then

$$x[n]e^{j\Omega_c n} \Longleftrightarrow X(\Omega - \Omega_c) \tag{9.54}$$

This property is the dual of the time-shifting property. To prove the frequency-shifting property, we have from Eq. (9.30)

$$x[n]e^{j\Omega_c n} \Longleftrightarrow \sum_{n=-\infty}^{\infty} x[n]e^{j\Omega_c n}e^{-j\Omega n} = \sum_{n=-\infty}^{\infty} x[n]e^{-j(\Omega - \Omega_c)n} = X(\Omega - \Omega_c)$$

From this result, it follows that

$$x[n]e^{-j\Omega_c n} \Longleftrightarrow X(\Omega + \Omega_c)$$

Adding this pair to the pair in Eq. (9.54), we obtain

$$x[n]\cos(\Omega_c n) \iff \tfrac{1}{2}\{X(\Omega - \Omega_c) + X(\Omega + \Omega_c)\} \tag{9.55}$$

This is the *modulation property.*
Multiplying both sides of pair (9.54) by $e^{j\theta}$, we obtain

$$x[n]e^{j(\Omega_c n + \theta)} \iff X(\Omega - \Omega_c)e^{j\theta} \tag{9.56}$$

Using this pair, we can generalize the modulation property as

$$x[n]\cos(\Omega_c n + \theta) \iff \tfrac{1}{2}\{X(\Omega - \Omega_c)e^{j\theta} + X(\Omega + \Omega_c)e^{-j\theta}\} \tag{9.57}$$

EXAMPLE 9.10

A signal $x[n] = \text{sinc}(\pi n/4)$ modulates a carrier $\cos \Omega_c n$. Find and sketch the spectrum of the modulated signal $x[n]\cos \Omega_c n$ for

(a) $\Omega_c = \pi/2$
(b) $\Omega_c = 7\pi/8 = 0.875\pi$

(a) For $x[n] = \text{sinc}(\pi n/4)$, we find (Table 9.1, pair 8)

$$X(\Omega) = 4 \sum_{m=-\infty}^{\infty} \text{rect}\left(\frac{\Omega - 2\pi m}{\pi/2}\right)$$

Figure 9.9a shows the DTFT $X(\Omega)$. From the modulation property (9.55), we obtain

$$x[n]\cos(0.5\pi n) \iff 2 \sum_{m=-\infty}^{\infty} \text{rect}\left(\frac{\Omega - 0.5\pi - 2\pi m}{0.5\pi}\right) + \text{rect}\left(\frac{\Omega + 0.5\pi - 2\pi m}{0.5\pi}\right)$$

Figure 9.9b shows half the $X(\Omega)$ shifted by $\pi/2$ and Fig. 9.9c shows half the $X(\Omega)$ shifted by $-\pi/2$. The spectrum of the modulated signal, is obtained by adding these two shifted spectra and multiplying by half, as shown in Fig. 9.9d.

(b) Figure 9.10a shows $X(\Omega)$, which is the same as that in part a. For $\Omega_c = 7\pi/8 = 0.875\pi$, the modulation property (9.55) yields

$$x[n]\cos(0.875\pi n) \iff 2 \sum_{m=-\infty}^{\infty} \text{rect}\left(\frac{\Omega - 0.875\pi - 2\pi m}{0.5\pi}\right) + \text{rect}\left(\frac{\Omega + 0.875\pi - 2\pi m}{0.5\pi}\right)$$

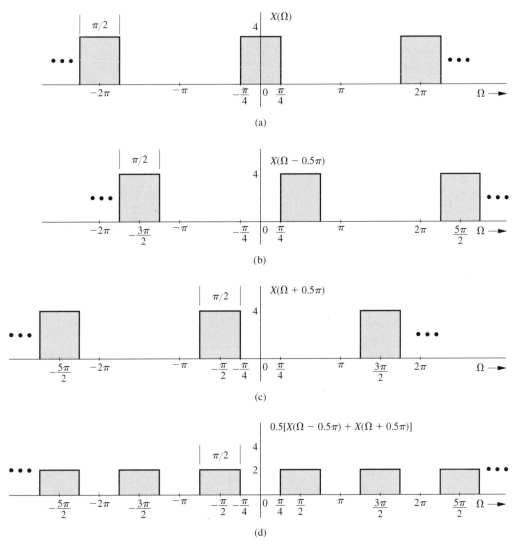

Figure 9.9 Instance of modulation for Example 9.10a.

Figure 9.10b shows $X(\Omega)$ shifted by $7\pi/8$ and Fig. 9.10c shows $X(\Omega)$ shifted by $-7\pi/8$. The spectrum of the modulated signal, is obtained by adding these two shifted spectra and multiplying by half, as shown in Fig. 9.10d. In this case, the two shifted spectra overlap. Since, the operation of modulation thus causes aliasing, it does not achieve the desired effect of spectral shifting. In this example, to realize spectral shifting without aliasing requires $\Omega_c \leq 3\pi/4$.

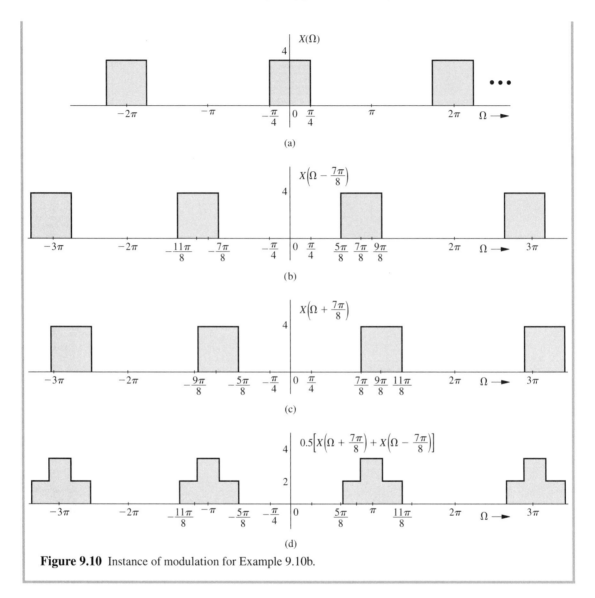

Figure 9.10 Instance of modulation for Example 9.10b.

EXERCISE E9.8

In Table 9.1, derive pairs 12 and 13 from pair 11 and the frequency-shifting/modulation property.

TIME- AND FREQUENCY-CONVOLUTION PROPERTY

If

$$x_1[n] \Longleftrightarrow X_1(\Omega) \qquad \text{and} \qquad x_2[n] \Longleftrightarrow X_2(\Omega)$$

then

$$x_1[n] * x_2[n] \iff X_1(\Omega)X_2(\Omega) \tag{9.58a}$$

and

$$x_1[n]x_2[n] \iff \frac{1}{2\pi} X_1(\Omega) \circledP X_2(\Omega) \tag{9.58b}$$

where

$$x_1[n] * x_2[n] = \sum_{m=-\infty}^{\infty} x_1[m]x_2[n-m]$$

For two continuous, periodic signals, we define the periodic convolution, denoted by symbol $\circledP$ as[†]

$$X_1(\Omega) \circledP X_2(\Omega) = \frac{1}{2\pi} \int_{2\pi} X_1(u)X_2(\Omega - u) \, du$$

The convolution here is not the *linear* convolution used so far. This is a *periodic* (or *circular*) convolution applicable to convolution of two continuous, periodic functions with the same period. The limit of integration in the convolution extends only to one period.

Proof of the time convolution property is identical to that given in Section 5.2 [Eq. (5.18)]. All we have to do is replace z with $e^{j\Omega}$. To prove the frequency-convolution property (9.58b), we have

$$x_1[n]x_2[n] \iff \sum_{n=-\infty}^{\infty} x_1[n]x_2[n]e^{-j\Omega n} = \sum_{n=-\infty}^{\infty} x_2[n]\left[\frac{1}{2\pi}\int_{2\pi} X_1(u)e^{-jnu}\, du\right] e^{-j\Omega n}$$

Interchanging the order of summation and integration, we obtain

$$x_1[n]x_2[n] \iff \frac{1}{2\pi}\int_{2\pi} X_1(u)\left[\sum_{n=-\infty}^{\infty} x_2[n]e^{-j(\Omega-u)n}\right] du = \frac{1}{2\pi}\int_{2\pi} X_1(u)X_2(\Omega - u)\, du$$

EXAMPLE 9.11

If $x[n] \iff X(\Omega)$, then show that

$$\sum_{k=-\infty}^{n} x[k] \iff \pi X(0) \sum_{k=-\infty}^{\infty} \delta(\Omega - 2\pi k) + \frac{e^{j\Omega}}{e^{j\Omega} - 1} X(\Omega) \tag{9.59}$$

Recognize that the sum on the left-hand side of (9.59) is $x[n] * u[n]$ because

$$x[n] * u[n] = \sum_{k=-\infty}^{\infty} x[k]u[n-k] = \sum_{k=-\infty}^{n} x[k]$$

[†]In Eq. (8.36), we defined periodic convolution for two discrete, periodic sequences in a different way. Although, we are using the same symbol $\circledP$ for both discrete and continuous cases, the meaning will be clear from the context.

In deriving this result, we have used the fact that

$$u[n-k] = \begin{cases} 1 & k \le n \\ 0 & k > n \end{cases}$$

Hence, from the time-convolution property (9.58a) and pair 10 in Table 9.1, it follows that

$$\sum_{k=-\infty}^{n} x[k] \Longleftrightarrow X(\Omega)\left(\pi \sum_{k=-\infty}^{\infty} \delta(\Omega - 2\pi k) + \frac{e^{j\Omega}}{e^{j\Omega}-1}\right)$$

Because of 2π periodicity, $X(0) = X(2\pi k)$. Moreover, $X(\Omega)\delta(\Omega - 2\pi k) = X(2\pi k)\delta(\Omega - 2\pi k) = X(0)\delta(\Omega - 2\pi k)$. Hence

$$\sum_{k=-\infty}^{n} x[k] \Longleftrightarrow \pi X(0) \sum_{k=-\infty}^{\infty} \delta(\Omega - 2\pi k) + \frac{e^{j\Omega}}{e^{j\Omega}-1}X(\Omega)$$

EXERCISE E9.9

In Table 9.1, derive pair 9 from pair 8, assuming $\Omega_c \le \pi/2$. Use the frequency-convolution property.

PARSEVAL'S THEOREM

If

$$x[n] \Longleftrightarrow X(\Omega)$$

then E_x, the energy of $x[n]$, is given by

$$E_x = \sum_{n=-\infty}^{\infty} |x[n]|^2 = \frac{1}{2\pi}\int_{2\pi} |X(\Omega)|^2\, d\Omega \qquad (9.60)$$

To prove this property, we have from Eq. (9.44)

$$X^*(\Omega) = \sum_{n=-\infty}^{\infty} x^*[n]e^{j\Omega n} \qquad (9.61)$$

Now

$$\sum_{n=-\infty}^{\infty} |x[n]|^2 = \sum_{n=-\infty}^{\infty} x^*[n]x[n] = \sum_{n=-\infty}^{\infty} x^*[n]\left[\frac{1}{2\pi}\int_{2\pi} X(\Omega)e^{j\Omega n}\, d\Omega\right]$$

$$= \frac{1}{2\pi}\int_{2\pi} X(\Omega)\left[\sum_{n=-\infty}^{\infty} x^*[n]e^{j\Omega n}\right] d\Omega$$

$$= \frac{1}{2\pi}\int_{2\pi} X(\Omega)X^*(\Omega)\, d\Omega = \frac{1}{2\pi}\int_{2\pi} |X(\Omega)|^2\, d\Omega$$

EXAMPLE 9.12

Find the energy of $x[n] = \text{sinc}\,(\Omega_c n)$, assuming $\Omega_c < \pi$.

From pair 8, Table 9.1, the fundamental band spectrum of $x[n]$ is

$$\text{sinc}(\Omega_c n) \Longleftrightarrow \frac{\pi}{\Omega_c} \text{rect}\left(\frac{\Omega}{2\Omega_c}\right) \qquad |\Omega| \le \pi$$

Hence, from the Parseval theorem in Eq. (9.60), we have

$$E_x = \frac{1}{2\pi} \int_{-\pi}^{\pi} \frac{\pi^2}{\Omega_c^2} \left[\text{rect}\left(\frac{\Omega}{2\Omega_c}\right)\right]^2 d\Omega$$

Recognizing that $\text{rect}\,(\Omega/2\Omega_c) = 1$ over $|\Omega| \le \Omega_c$ and is zero otherwise, the preceding integral yields

$$E_x = \frac{1}{2\pi} \left(\frac{\pi^2}{\Omega_c^2}\right)(2\Omega_c) = \frac{\pi}{\Omega_c}$$

TABLE 9.2 Properties of the DTFT

Operation	$x[n]$	$X(\Omega)$
Linearity	$a_1 x_1[n] + a_2 x_2[n]$	$a_1 X_1(\Omega) + a_2 X_2(\Omega)$
Conjugation	$x^*[n]$	$X^*(-\Omega)$
Scalar multiplication	$ax[n]$	$aX(\Omega)$
Multiplication by n	$nx[n]$	$j\dfrac{dX(\Omega)}{d\Omega}$
Time reversal	$x[-n]$	$X(-\Omega)$
Time shifting	$x[n-k]$	$X(\Omega)e^{-jk\Omega} \qquad k$ integer
Frequency shifting	$x[n]\,e^{j\Omega_c n}$	$X(\Omega - \Omega_c)$
Time convolution	$x_1[n] * x_2[n]$	$X_1(\Omega)X_2(\Omega)$
Frequency convolution	$x_1[n]x_2[n]$	$\dfrac{1}{2\pi}\displaystyle\int_{2\pi} X_1[u]X_2[\Omega - u]\,du$
Parseval's theorem	$E_x = \displaystyle\sum_{n=-\infty}^{\infty} |x[n]|^2$	$E_x = \dfrac{1}{2\pi}\displaystyle\int_{2\pi} |X(\Omega)|^2\,d\Omega$

9.4 LTI Discrete-Time System Analysis by DTFT

Consider a linear, time-invariant, discrete-time system with the unit impulse response $h[n]$. We shall find the (zero-state) system response $y[n]$ for the input $x[n]$. Let

$$x[n] \Longleftrightarrow X(\Omega) \qquad y[n] \Longleftrightarrow Y(\Omega) \qquad \text{and} \qquad h[n] \Longleftrightarrow H(\Omega)$$

Because

$$y[n] = x[n] * h[n] \tag{9.62}$$

According to Eq. (9.58a) it follows that

$$Y(\Omega) = X(\Omega)H(\Omega) \tag{9.63}$$

This result is similar to that obtained for continuous-time systems. Let us examine the role of $H(\Omega)$, the DTFT of the unit impulse response $h[n]$.

Equation (9.63) holds only for BIBO-stable systems and also for marginally stable systems if the input does not contain the system's natural mode(s). In other cases, the response grows with n and is not Fourier transformable.[†] Moreover, the input $x[n]$ also has to be DTF transformable. For cases where Eq. (9.63) does not apply, we use the z-transform for system analysis.

Equation (9.63) shows that the output signal frequency spectrum is the product of the input signal frequency spectrum and the frequency response of the system. From this equation, we obtain

$$|Y(\Omega)| = |X(\Omega)|\,|H(\Omega)| \tag{9.64}$$

and

$$\angle Y(\Omega) = \angle X(\Omega) + \angle H(\Omega) \tag{9.65}$$

This result shows that the output amplitude spectrum is the product of the input amplitude spectrum and the amplitude response of the system. The output phase spectrum is the sum of the input phase spectrum and the phase response of the system.

We can also interpret Eq. (9.63) in terms of the frequency-domain viewpoint, which sees a system in terms of its frequency response (system response to various exponential or sinusoidal components). The frequency domain views a signal as a sum of various exponential or sinusoidal components. The transmission of a signal through a (linear) system is viewed as transmission of various exponential or sinusoidal components of the input signal through the system. This concept can be understood by displaying the input–output relationships by a directed arrow as follows:

$$e^{j\Omega n} \implies H(\Omega)e^{j\Omega n}$$

which shows that the system response to $e^{j\Omega n}$ is $H(\Omega)e^{j\Omega n}$, and

$$x[n] = \frac{1}{2\pi} \int_{2\pi} X(\Omega)e^{j\Omega n}\,d\Omega$$

[†]It does not hold for asymptotically unstable systems whose impulse response $h[n]$ does not have DTFT. In case the system is marginally stable and the input does not contain the system's mode term, the response does not grow with n and is, therefore, Fourier transformable.

which shows $x[n]$ as a sum of everlasting exponential components. Invoking the linearity property, we obtain

$$y[n] = \frac{1}{2\pi} \int_{2\pi} X(\Omega) H(\Omega) e^{j\Omega n} \, d\Omega$$

which gives $y[n]$ as a sum of responses to all input components and is equivalent to Eq. (9.63). Thus, $X(\Omega)$ is the input spectrum and $Y(\Omega)$ is the output spectrum, given by $X(\Omega)H(\Omega)$.

EXAMPLE 9.13

An LTID system is specified by the equation

$$y[n] - 0.5y[n-1] = x[n] \tag{9.66}$$

Find $H(\Omega)$, the frequency response of this system. Determine the (zero-state) response $y[n]$ if the input $x[n] = (0.8)^n u[n]$.

Let $x[n] \Longleftrightarrow X(\Omega)$ and $y[n] \Longleftrightarrow Y(\Omega)$. The DTFT of both sides of Eq. (9.66) yields

$$(1 - 0.5e^{-j\Omega})Y(\Omega) = X(\Omega)$$

According to Eq. (9.63)

$$H(\Omega) = \frac{Y(\Omega)}{X(\Omega)} = \frac{1}{1 - e^{-j\Omega}} = \frac{e^{j\Omega}}{e^{j\Omega} - 0.5}$$

Also, $x[n] = (0.8)^n u[n]$. Hence,

$$X(\Omega) = \frac{e^{j\Omega}}{e^{j\Omega} - 0.8}$$

and

$$Y(\Omega) = X(\Omega)H(\Omega) = \frac{2e^{j\Omega}}{(e^{j\Omega} - 0.8)(e^{j\Omega} - 0.5)}$$

We can express the right-hand side as a sum of two first-order terms (modified partial fraction expansion as discussed in Section B.5-6) as follows[†]:

$$\frac{Y(\Omega)}{e^{j\Omega}} = \frac{e^{j\Omega}}{(e^{j\Omega} - 0.5)(e^{j\Omega} - 0.8)}$$

$$= \frac{-\frac{5}{3}}{e^{j\Omega} - 0.5} + \frac{\frac{8}{3}}{e^{j\Omega} - 0.8}$$

[†]Here $Y(\Omega)$ is a function of variable $e^{j\Omega}$. Hence, $x = e^{j\Omega}$ for the purpose of comparison with the expression in Section B.5-6.

Consequently

$$Y(\Omega) = -\left(\frac{5}{3}\right)\frac{e^{j\Omega}}{e^{j\Omega} - 0.5} + \left(\frac{8}{3}\right)\frac{e^{j\Omega}}{e^{j\Omega} - 0.8}$$

$$= -\left(\frac{5}{3}\right)\frac{1}{1 - 0.5e^{-j\Omega}} + \left(\frac{8}{3}\right)\frac{1}{1 - 0.8e^{-j\Omega}}$$

According to Eq. (9.34a), the inverse DTFT of this equation is

$$\dot{y}[n] = \left[-\tfrac{5}{3}(0.5)^n + \tfrac{8}{3}(0.8)^n\right]u[n] \tag{9.67}$$

This example demonstrates the procedure for using DTFT to determine an LTID system response. It is similar to the Fourier transform method in the analysis of LTIC systems. As in the case of the Fourier transform, this method can be used only if the system is asymptotically or BIBO stable and if the input signal is DTF-transformable.[†] We shall not belabor this method further because it is clumsier and more restrictive than the z-transform method discussed in Chapter 5.

9.4-1 Distortionless Transmission

In several applications, digital signals are passed through LTI systems, and we require that the output waveform be a replica of the input waveform. As in the continuous-time case, transmission is said to be distortionless if the input $x[n]$ and the output $y[n]$ satisfy the condition

$$y[n] = G_0\,x[n - n_d] \tag{9.68}$$

Here n_d, the delay (in samples) is assumed to be integer. The Fourier transform of Eq. (9.68) yields

$$Y(\Omega) = G_0\,X(\Omega)\,e^{-j\Omega n_d}$$

But

$$Y(\Omega) = X(\Omega)\,H(\Omega)$$

Therefore

$$H(\Omega) = G_0\,e^{-j\Omega n_d}$$

This is the frequency response required for distortionless transmission. From this equation it follows that

$$|H(\Omega)| = G_0 \tag{9.69a}$$

$$\angle H(\Omega) = -\Omega n_d \tag{9.69b}$$

Thus, for distortionless transmission, the amplitude response $|H(\Omega)|$ must be a constant, and the phase response $\angle H(\Omega)$ must be a linear function of Ω with slope $-n_d$, where n_d is the delay

---

[†]It can also be applied to marginally stable systems if the input does not contain natural mode(s) of the system.

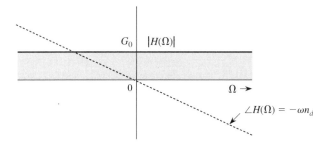

Figure 9.11 LTI system frequency response for distortionless transmission.

in number of samples with respect to input (Fig. 9.11). These are precisely the characteristics of an ideal delay of n_d samples with a gain of G_0 [see Eq. (9.52)].

MEASURE OF DELAY VARIATION

For distortionless transmission, we require a *linear phase* characteristic. In practice, many systems have a phase characteristic that may be only approximately linear. A convenient way of judging phase linearity is to plot the slope of $\angle H(\Omega)$ as a function of frequency. This slope is constant for ideal linear phase (ILP) system, but it may vary with Ω in general case. The slope can be expressed as

$$n_g(\Omega) = -\frac{d}{d\Omega}\angle H(\Omega) \tag{9.70}$$

If $n_g(\Omega)$ is constant, all the components are delayed by n_g samples. But if the slope is not constant, the delay n_g varies with frequency. This variation means that different frequency components undergo different amounts of delay, and consequently the output waveform will not be a replica of the input waveform. As in the case of LTIC systems, $n_g(\Omega)$, as defined in Eq. (9.70), plays an important role in bandpass systems and is called the *group delay* or *envelope* delay. Observe that constant n_d implies constant n_g. Note that $\angle H(\Omega) = \phi_0 - \Omega n_d$ also has a constant n_g. Thus, constant group delay is a more relaxed condition.

DISTORTIONLESS TRANSMISSION
OVER BANDPASS SYSTEMS

As in the case of continuous-time systems, the distortionless transmission conditions can be relaxed for discrete-time bandpass systems. For lowpass systems, the phase characteristic should not only be linear over the band of interest, but should also pass through the origin [condition (9.69b)]. For bandpass systems, the phase characteristic should be linear over the band of interest, but it need not pass through the origin (n_g should be constant). The amplitude response is required to be constant over the passband. Thus, for distortionless transmission over a bandpass system, the frequency response for positive range of Ω is of the form[†]

$$H(\Omega) = G_0 e^{j(\phi_0 - \Omega n_g)} \qquad \Omega \geq 0$$

[†]Because the phase function is an odd function of Ω, if $\angle H(\Omega) = \phi_0 - \Omega n_g$ for $\Omega \geq 0$, over the band $2W$ (centered at Ω_c), then $\angle H(\Omega) = -\phi_0 - \Omega n_g$ for $\Omega < 0$ over the band $2W$ (centered at $-\Omega_c$).

The proof is identical to that for the continuous-time case in Section 7.4-2 and will not be repeated. In using Eq. (9.70) to compute n_g, we should ignore jump discontinuities in the phase function.

9.4-2 Ideal and Practical Filters

Ideal filters allow distortionless transmission of a certain band of frequencies and suppress all the remaining frequencies. The general ideal lowpass filter shown in Fig. 9.12 for $|\Omega| \leq \pi$ allows all components below the cutoff frequency $\Omega = \Omega_c$ to pass without distortion and suppresses all components above Ω_c. Figure 9.13 illustrates ideal highpass and bandpass filter characteristics.

The ideal lowpass filter in Fig. 9.12a has a linear phase of slope $-n_d$, which results in a delay of n_d samples for all its input components of frequencies below Ω_c rad/sample. Therefore, if the input is a signal $x[n]$ bandlimited to Ω_c the output $y[n]$ is $x[n]$ delayed by n_d, that is,

$$y[n] = x[n - n_d]$$

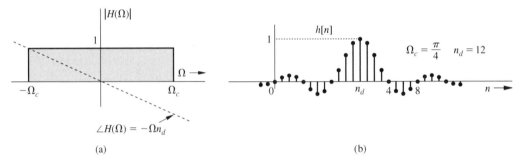

(a)   (b)

Figure 9.12 Ideal lowpass filter: its frequency response and impulse response.

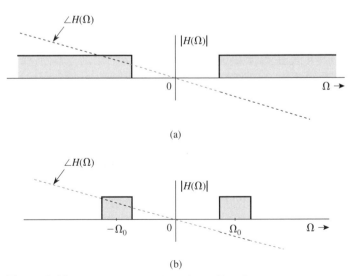

(a)

(b)

Figure 9.13 Ideal highpass and bandpass filter frequency response.

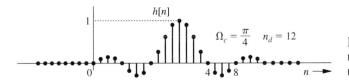

Figure 9.14 Approximate realization of an ideal lowpass filter by truncation of its impulse response.

The signal $x[n]$ is transmitted by this system without distortion, but with delay of n_d samples. For this filter,

$$H(\Omega) = \sum_{m=-\infty}^{\infty} \text{rect}\left(\frac{\Omega - 2\pi m}{2\Omega_c}\right) e^{-j\Omega n_d} \qquad (9.71\text{a})$$

The unit impulse response $h[n]$ of this filter is obtained from pair 8 (Table 9.1) and the time-shifting property

$$h[n] = \frac{\Omega_c}{\pi} \text{sinc}\left[\Omega_c(n - n_d)\right] \qquad (9.71\text{b})$$

Because $h[n]$ is the system response to impulse input $\delta[n]$, which is applied at $n = 0$, it must be causal (i.e., it must not start before $n = 0$) for a realizable system. Figure 9.12b shows $h[n]$ for $\Omega_c = \pi/4$ and $n_d = 12$. This figure also shows that $h[n]$ is noncausal and hence, unrealizable. Similarly, one can show that other ideal filters (such as the ideal highpass or and bandpass filters depicted in Fig. 9.13) are also noncausal and therefore physically unrealizable.

    One practical approach to realize ideal lowpass filter approximately is to truncate both tails (positive and negative) of $h[n]$ so that it has a finite length and then delay sufficiently to make it causal (Fig. 9.14). We now synthesize a system with this truncated (and delayed) impulse response. For closer approximation, the truncating window has to be correspondingly wider. The delay required also increases correspondingly. Thus, the price of closer realization is higher delay in the output; this situation is common in noncausal systems.

9.5 DTFT CONNECTION WITH THE CTFT

Consider a continuous-time signal $x_c(t)$ (Fig. 9.15a) with the Fourier transform $X_c(\omega)$ bandlimited to B Hz (Fig. 9.15b). This signal is sampled with a sampling interval T. The sampling rate is at least equal to the Nyquist rate; that is, $T \leq 1/2B$. The sampled signal $\bar{x}_c(t)$ (Fig. 9.15c) can be expressed as

$$\bar{x}_c(t) = \sum_{n=-\infty}^{\infty} x_c(nT)\,\delta(t - nT)$$

The continuous-time Fourier transform of the foregoing equation yields

$$\bar{X}_c(\omega) = \sum_{n=-\infty}^{\infty} x_c(nT)\,e^{-jnT\omega} \qquad (9.72)$$

In Section 8.1 (Fig. 8.1f), we showed that $\bar{X}_c(\omega)$ is $X_c(\omega)/T$ repeating periodically with a period $\omega_s = 2\pi/T$, as illustrated in Fig. 9.15d. Let us construct a discrete-time signal $x[n]$ such that its nth sample value is equal to the value of the nth sample of $x_c(t)$, as depicted in Fig. 9.15e, that is,

$$x[n] = x_c(nT) \qquad (9.73)$$

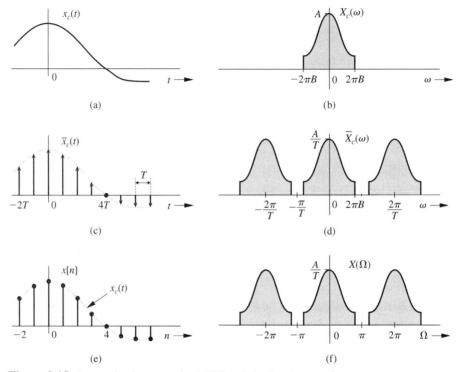

Figure 9.15 Connection between the DTFT and the Fourier transform.

Now, $X(\Omega)$, the DTFT of $x[n]$, is given by

$$X(\Omega) = \sum_{n=-\infty}^{\infty} x[n]\,e^{-jn\Omega} = \sum_{n=-\infty}^{\infty} x_c(nT)\,e^{-jn\Omega} \tag{9.74}$$

Comparison of Eqs. (9.74) and (9.72) shows that letting $\omega T = \Omega$ in $\overline{X}_c(\omega)$ yields $X(\Omega)$, that is,

$$X(\Omega) = \overline{X}_c(\omega)|_{\omega T = \Omega} \tag{9.75}$$

Alternately, $X(\Omega)$ can be obtained from $\overline{X}_c(\omega)$ by replacing ω with Ω/T, that is,

$$X(\Omega) = \overline{X}_c\left(\frac{\Omega}{T}\right) \tag{9.76}$$

Therefore, $X(\Omega)$ is identical to $\overline{X}_c(\omega)$, frequency-scaled by factor T, as shown in Fig. 9.15f. Thus, $\omega = 2\pi/T$ in Fig. 9.15d corresponds to $\Omega = 2\pi$ in Fig. 9.15f.

9.5-1 Use of DFT and FFT for Numerical Computation of DTFT

The discrete Fourier transform (DFT), as discussed in Chapter 8, is a tool for computing the samples of continuous-time Fourier transform (CTFT). Because of the close connection between CTFT and DTFT, as seen in Eq. (9.76), we can also use this same DFT to compute DTFT samples.

In Chapter 8, Eqs. (8.22a) and (8.22b) relate an N_0-point sequence x_n to another N_0-point sequence X_r. Changing the notation x_n to $x[n]$, in these equations, we obtain

$$X_r = \sum_{n=0}^{N_0-1} x[n]e^{-jr\Omega_0 n} \tag{9.77a}$$

$$x[n] = \frac{1}{N_0} \sum_{r=0}^{N_0-1} X_r e^{jr\Omega_0 n} \qquad \Omega_0 = \frac{2\pi}{N_0} \tag{9.77b}$$

Comparing Eq. (9.30) with Eq. (9.77a), we recognize that X_r is the sample of $X(\Omega)$ at $\Omega = r\Omega_0$, that is,

$$X_r = X(r\Omega_0) \qquad \Omega_0 = \frac{2\pi}{N_0}$$

Hence, the pair of DFT equations (9.77) relates an N_0-point sequence $x[n]$ to the N_0-point samples of corresponding $X(\Omega)$. We can now use the efficient algorithm FFT (discussed in Chapter 8) to compute X_r from $x[n]$, and vice versa.

If $x[n]$ is not timelimited, we can still find the approximate values of X_r by suitably windowing $x[n]$. To reduce the error, the window should be tapered and should have sufficient width to satisfy error specifications. In practice, numerical computation of signals, which are generally nontimelimited, is performed in this manner because of the computational economy of the DFT, especially for signals of long duration.

COMPUTATION OF DISCRETE-TIME FOURIER SERIES (DTFS)

The discrete-time Fourier series (DTFS) equations (9.8) and (9.9) are identical to the DFT equations (8.22b) and (8.22a) within a scaling constant N_0. If we let $x[n] = N_0 x_n$ and $\mathcal{D}_r = X_r$ in Eqs. (9.9) and (9.8), we obtain

$$X_r = \sum_{n=0}^{N_0-1} x_n e^{-jr\Omega_0 n}$$

$$x_n = \frac{1}{N_0} \sum_{r=0}^{N_0-1} X_r e^{jr\Omega_0 n} \qquad \Omega_0 = \frac{2\pi}{N_0} \tag{9.78}$$

This is precisely the DFT pair in Eqs. (8.22). For instance, to compute the DTFS for the periodic signal in Fig. 9.2a, we use the values of $x_n = x[n]/N_0$ as

$$x_n = \begin{cases} \frac{1}{32} & 0 \leq n \leq 4 \quad \text{and} \quad 28 \leq n \leq 31 \\ 0 & 5 \leq n \leq 27 \end{cases}$$

Numerical computations in modern digital signal processing are conveniently performed with the discrete Fourier transform, introduced in Section 8.5. The DFT computations can be very efficiently executed by using the fast Fourier transform (FFT) algorithm discussed in Section 8.6. The DFT is indeed the workhorse of modern digital signal processing. The discrete-time Fourier transform (DTFT) and the inverse discrete-time Fourier transform (IDTFT) can be computed by using the DFT. For an N_0-point signal $x[n]$, its DFT yields exactly N_0 samples of $X(\Omega)$ at frequency intervals of $2\pi/N_0$. We can obtain a larger number of samples of $X(\Omega)$ by padding a sufficient number of zero-valued samples to $x[n]$. The N_0-point DFT of $x[n]$ gives exact values of the DTFT samples if $x[n]$ has a finite length N_0. If the length of $x[n]$ is infinite, we need to use the appropriate window function to truncate $x[n]$.

Because of the convolution property, we can use DFT to compute the convolution of two signals $x[n]$ and $h[n]$, as discussed in Section 8.5. This procedure, known as fast convolution, requires padding both signals by a suitable number of zeros, to make the linear convolution of the two signals identical to the circular (or periodic) convolution of the padded signals. Large blocks of data may be processed by sectioning the data into smaller blocks and processing such smaller blocks in sequence. Such a procedure requires smaller memory and reduces the processing time.[1]

9.6 GENERALIZATION OF THE DTFT TO THE z-TRANSFORM

LTID systems can be analyzed by using the DTFT. This method, however, has the following limitations:

1. Existence of the DTFT is guaranteed only for absolutely summable signals. The DTFT does not exist for exponentially or even linearly growing signals. This means that the DTFT method is applicable only for a limited class of inputs.

2. Moreover, this method can be applied only to asymptotically or BIBO stable systems; it cannot be used for unstable or even marginally stable systems.

These are serious limitations in the study of LTID system analysis. Actually it is the first limitation that is also the cause of the second limitation. Because the DTFT is incapable of handling growing signals, it is incapable of handling unstable or marginally stable systems.[†] Our goal is, therefore, to extend the DTFT concept so that it can handle exponentially growing signals.

We may wonder what causes this limitation on DTFT so that it is incapable of handling exponentially growing signals. Recall that in DTFT, we are using sinusoids or exponentials of the form $e^{j\Omega n}$ to synthesize an arbitrary signal $x[n]$. These signals are sinusoids with constant amplitudes. They are incapable of synthesizing exponentially growing signals no matter how many such components we add. Our hope, therefore, lies in trying to synthesize $x[n]$ by using exponentially growing sinusoids or exponentials. This goal can be accomplished by generalizing the frequency variable $j\Omega$ to $\sigma + j\Omega$, that is, by using exponentials of the form $e^{(\sigma+j\Omega)n}$ instead of exponentials $e^{j\Omega n}$. The procedure is almost identical to that used in extending the Fourier transform to the Laplace transform.

Let us define a new variable $\hat{X}(j\Omega) = X(\Omega)$. Hence

$$\hat{X}(j\Omega) = \sum_{n=-\infty}^{\infty} x[n] e^{-j\Omega n} \tag{9.79}$$

[†]Recall that the output of an unstable system grows exponentially. Also, the output of a marginally stable system to characteristic mode input grows with time.

and

$$x[n] = \frac{1}{2\pi} \int_{-\pi}^{\pi} \hat{X}(j\Omega) \, e^{j\Omega n} \, d\Omega \tag{9.80}$$

Consider now the DTFT of $x[n] e^{-\sigma n}$ (σ real)

$$\text{DTFT}\{x[n] e^{-\sigma n}\} = \sum_{n=-\infty}^{\infty} x[n] e^{-\sigma n} e^{-j\Omega n} \tag{9.81}$$

$$= \sum_{n=-\infty}^{\infty} x[n] e^{-(\sigma + j\Omega)n} \tag{9.82}$$

It follows from Eq. (9.79) that the sum in Eq. (9.82) is $\hat{X}(\sigma + j\Omega)$. Thus

$$\text{DTFT}\{x[n] e^{-\sigma n}\} = \sum_{n=-\infty}^{\infty} x[n] e^{-(\sigma + j\Omega)n} = \hat{X}(\sigma + j\Omega) \tag{9.83}$$

Hence, the inverse DTFT of $\hat{X}(\sigma + j\Omega)$ is $x[n] e^{-\sigma n}$. Therefore

$$x[n] e^{-\sigma n} = \frac{1}{2\pi} \int_{-\pi}^{\pi} \hat{X}(\sigma + j\Omega) \, e^{j\Omega n} \, d\Omega \tag{9.84}$$

Multiplying both sides of Eq. (9.84) by $e^{\sigma n}$ yields

$$x[n] = \frac{1}{2\pi} \int_{-\pi}^{\pi} \hat{X}(\sigma + j\Omega) \, e^{(\sigma + j\Omega)n} \, d\Omega \tag{9.85}$$

Let us define a new variable z as

$$z = e^{\sigma + j\Omega} \quad \text{so that} \quad \ln z = \sigma + j\Omega \quad \text{and} \quad \frac{1}{z} \, dz = j \, d\Omega \tag{9.86}$$

Because $z = e^{\sigma + j\Omega}$ is complex, we can express it as $z = re^{j\Omega}$, where $r = e^{\sigma}$. Thus, z lies on a circle of radius r, and as Ω varies from $-\pi$ to π, z circumambulates along this circle, completing exactly one counterclockwise rotation, as illustrated in Fig. 9.16. Changing to variable z in Eq. (9.85) yields

$$x[n] = \frac{1}{2\pi j} \oint \hat{X}(\ln z) \, z^{n-1} dz \tag{9.87a}$$

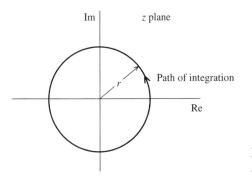

Figure 9.16 Contour of integration for the z-transform.

and from Eq. (9.83) we obtain

$$\hat{X}(\ln z) = \sum_{n=-\infty}^{\infty} x[n]\, z^{-n} \tag{9.87b}$$

where the integral $\oint$ indicates a contour integral around a circle of radius r in the counterclockwise direction.

Equations (9.87a) and (9.87b) are the desired extensions. They are, however, in a clumsy form. For the sake of convenience, we make another notational change by observing that $\hat{X}(\ln z)$ is a function of z. Let us denote it by a simpler notation $X[z]$. Thus, Eqs. (9.87) become

$$x[n] = \frac{1}{2\pi j} \oint X[z]\, z^{n-1} dz \tag{9.88}$$

and

$$X[z] = \sum_{n=-\infty}^{\infty} x[n]\, z^{-n} \tag{9.89}$$

This is the (bilateral) z-transform pair. Equation (9.88) expresses $x[n]$ as a continuous sum of exponentials of the form $z^n = e^{(\sigma+j\Omega)n} = r^n\, e^{j\Omega n}$. Thus, by selecting a proper value for r (or σ), we can make the exponential grow (or decay) at any exponential rate we desire.

If we let $\sigma = 0$, we have $z = e^{j\Omega}$ and

$$X[z]\big|_{z=e^{j\Omega}} = \hat{X}(\ln z)\big|_{z=e^{j\Omega}} = \hat{X}(j\Omega) = X(\Omega) \tag{9.90}$$

Thus, the familiar DTFT is just a special case of the z-transform $X[z]$ obtained by letting $z = e^{j\Omega}$ and assuming that the sum on the right-hand side of Eq. (9.89) converges when $z = e^{j\Omega}$. This also implies that the ROC for $X[z]$ includes the unit circle.

9.7 SUMMARY

This chapter deals with the analysis and processing of discrete-time signals. For analysis, our approach is parallel to that used in continuous-time signals. We first represent a periodic $x[n]$ as a Fourier series formed by a discrete-time exponential and its harmonics. Later we extend this representation to an aperiodic signal $x[n]$ by considering $x[n]$ to be a limiting case of a periodic signal with period approaching infinity.

Periodic signals are represented by discrete-time Fourier series (DTFS); aperiodic signals are represented by the discrete-time Fourier integral. The development, although similar to that of continuous-time signals, also reveals some significant differences. The basic difference in the two cases arises because a continuous-time exponential $e^{j\omega t}$ has a unique waveform for every value of ω in the range $-\infty$ to ∞. In contrast, a discrete-time exponential $e^{j\Omega n}$ has a unique waveform only for values of Ω in a continuous interval of 2π. Therefore, if Ω_0 is the fundamental frequency, then at most $2\pi/\Omega_0$ exponentials in the Fourier series are independent. Consequently, the discrete-time exponential Fourier series has only $N_0 = 2\pi/\Omega_0$ terms.

The discrete-time Fourier transform (DTFT) of an aperiodic signal is a continuous function of Ω and is periodic with period 2π. We can synthesize $x[n]$ from spectral components of $X(\Omega)$ in any band of width 2π. In a basic sense, the DTFT has a finite spectral width of 2π, which makes it bandlimited to π radians.

Linear, time-invariant, discrete-time (LTID) systems can be analyzed by means of the DTFT if the input signals are DTF transformable and if the system is stable. Analysis of unstable (or marginally stable) systems and/or exponentially growing inputs can be handled by the z-transform, which is a generalized DTFT. The relationship of the DTFT to the z-transform is similar to that of the Fourier transform to the Laplace transform. Whereas the z-transform is superior to the DTFT for analysis of LTID systems, the DTFT is preferable in signal analysis.

If $H(\Omega)$ is the DTFT of the system's impulse response $h[n]$, then $|H(\Omega)|$ is the amplitude response, and $\angle H(\Omega)$ is the phase response of the system. Moreover, if $X(\Omega)$ and $Y(\Omega)$ are the DTFTs of the input $x[n]$ and the corresponding output $y[n]$, then $Y(\Omega) = H(\Omega)X(\Omega)$. Therefore the output spectrum is the product of the input spectrum and the system's frequency response.

Because of the similarity between the DFT and DTFT relationships, numerical computations of the DTFT of finite-length signals can be handled by using the DFT and the FFT, introduced in Sections 8.5 and 8.6. For signals of infinite length, we use a window of suitable length to truncate the signal so that the final results are within a given error tolerance.

REFERENCE

1.   Mitra, S. K. *Digital Signal Processing: A Computer-Based Approach,* 2nd ed. McGraw-Hill, New York, 2001.

MATLAB SESSION 9: WORKING WITH THE DTFS AND THE DTFT

This session investigates various methods to compute the discrete-time Fourier series (DTFS). Performance of these methods is assessed by using MATLAB's stopwatch and profiling functions. Additionally, the discrete-time Fourier transform (DTFT) is applied to the important topic of finite impulse response (FIR) filter design.

M9.1 Computing the Discrete-Time Fourier Series

Within a scale factor, the DTFS is identical to the DFT. Thus, methods to compute the DFT can be readily used to compute the DTFS. Specifically, the DTFS is the DFT scaled by $1/N_0$. As an example, consider a 50 Hz sinusoid sampled at 1000 Hz over one-tenth of a second.

```
>> T = 1/1000; N_0 = 100; n = (0:N_0-1)';
>> x = cos(2*pi*50*n*T);
```

The DTFS is obtained by scaling the DFT.

```
>> X = fft(x)/N_0; f = (0:N_0-1)/(T*N_0);
>> stem(f-500,fftshift(abs(X)),'k'); axis([-500 500 -0.1 0.6]);
>> xlabel('f [Hz]'); ylabel('|X(f)|');
```

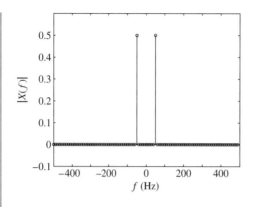

Figure M9.1 DTFS computed by scaling the DFT.

Figure M9.1 shows a peak magnitude of 0.5 at ±50 Hz. This result is consistent with Euler's representation

$$\cos(2\pi 50 nT) = \frac{e^{j2\pi 50 nT} + e^{-j2\pi 50 nT}}{2}$$

Lacking the $1/N_0$ scale factor, the DFT would have a peak amplitude 100 times larger.

The inverse DTFS is obtained by scaling the inverse DFT by N_0.

```
>> x_hat = real(ifft(X)*N_0);
>> stem(n,x_hat,'k'); axis([0 99 -1.1 1.1]);
>> xlabel('n'); ylabel('x_{hat}[n]');
```

Figure M9.2 confirms that the sinusoid $x[n]$ is properly recovered. Although the result is theoretically real, computer round-off errors produce a small imaginary component, which the `real` command removes.

Although the `fft` provides an efficient method to compute the DTFS, other important computational methods exist. A matrix-based approach is one popular way to implement Eq. (9.9). Although not as efficient as an FFT-based algorithm, matrix-based approaches provide insight into the DTFS and serve as an excellent model for solving similarly structured problems.

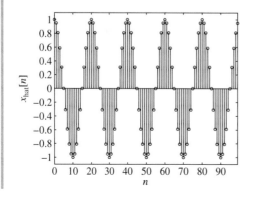

Figure M9.2 Inverse DTFS computed by scaling the inverse DFT.

To begin, define $W_{N_0} = e^{j\Omega_0}$, which is a constant for a given N_0. Substituting W_{N_0} into Eq. (9.9) yields

$$\mathcal{D}_r = \frac{1}{N_0} \sum_{n=0}^{N_0-1} x[n] W_{N_0}^{-nr}$$

An inner product of two vectors computes $\mathcal{D}_r$.

$$\mathcal{D}_r = \frac{1}{N_0} \begin{bmatrix} 1 & W_{N_0}^{-r} & W_{N_0}^{-2r} & \cdots & W_{N_0}^{-(N_0-1)r} \end{bmatrix} \begin{bmatrix} x[0] \\ x[1] \\ x[2] \\ \vdots \\ x[N_0 - 1] \end{bmatrix}$$

Stacking the results for all r yields:

$$\begin{bmatrix} \mathcal{D}_0 \\ \mathcal{D}_1 \\ \mathcal{D}_2 \\ \vdots \\ \mathcal{D}_{N_0-1} \end{bmatrix} = \frac{1}{N_0} \begin{bmatrix} 1 & 1 & 1 & \cdots & 1 \\ 1 & W_{N_0}^{-1} & W_{N_0}^{-2} & \cdots & W_{N_0}^{-(N_0-1)} \\ 1 & W_{N_0}^{-2} & W_{N_0}^{-4} & \cdots & W_{N_0}^{-2(N_0-1)} \\ \vdots & \vdots & \vdots & \cdots & \vdots \\ 1 & W_{N_0}^{-(N_0-1)} & W_{N_0}^{-2(N_0-1)} & \cdots & W_{N_0}^{-(N_0-1)^2} \end{bmatrix} \begin{bmatrix} x[0] \\ x[1] \\ x[2] \\ \vdots \\ x[N_0 - 1] \end{bmatrix} \quad \text{(M9.1)}$$

In matrix notation, Eq. (M9.1) is compactly written as

$$\mathcal{D} = \frac{1}{N_0} \mathbf{W}_{N_0} \mathbf{x}$$

Since it is also used to compute the DFT, matrix $\mathbf{W}_{N_0}$ is often called a DFT matrix.

Program MS9P1 computes the N_0-by-N_0 DFT matrix $\mathbf{W}_{N_0}$. Although not used here, the signal processing toolbox function dftmtx computes the same DFT matrix, although in a less obvious but more efficient fashion.

```
function [W] = MS9P1(N_0);
% MS9P1.m : MATLAB Session 9, Program 1
% Function M-file computes the N_0-by-N_0 DFT matrix W.

W = (exp(-j*2*pi/N_0)).^((0:N_0-1)'*(0:N_0-1));
```

While less efficient than FFT-based methods, the matrix approach correctly computes the DTFS.

```
>> W = MS9P1(100); X = W*x/N_0;
>> stem(f-500,fftshift(abs(X)),'k'); axis([-500 500 -0.1 0.6]);
>> xlabel('f [Hz]'); ylabel('|X(f)|');
```

The resulting plot is indistinguishable from Fig. M9.1. Problem 9.M-1 investigates a matrix-based approach to compute Eq. (9.8), the inverse DTFS.

M9.2 Measuring Code Performance

Writing efficient code is important, particularly if the code is frequently used, requires complicated operations, involves large data sets, or operates in real time. MATLAB provides several tools for assessing code performance. When properly used, the `profile` function provides detailed statistics that help assess code performance. MATLAB help thoroughly describes the use of the sophisticated `profile` command.

A simpler method of assessing code efficiency is to measure execution time and compare it with a reference. The MATLAB command `tic` starts a stopwatch timer. The `toc` command reads the timer. Sandwiching instructions between `tic` and `toc` returns the elapsed time. For example, the execution time of the 100-point matrix-based DTFS computation is:

```
>> tic; W*x/N_0; toc
elapsed_time = 0
```

Different machines operate at different speeds with different operating systems and with different background tasks. Therefore, elapsed time measurements can vary considerably from machine to machine and from execution to execution. In this particular case, however, a result other than zero is reported only by relatively slow machines. Execution times are so brief that MATLAB reports unreliable times or fails to register an elapsed time at all.

To increase the elapsed time and therefore the accuracy of the time measurement, a loop is used to repeat the calculation.

```
>> tic; for i=1:1000, W*x/N_0; end; toc
elapsed_time = 0.4400
```

This elapsed time suggests that each 100-point DTFS calculation takes around one-half millisecond. What exactly does this mean, however? Elapsed time is only meaningful relative to some reference. Consider the time it takes to compute the same DTFS using the FFT-based approach.

```
>> tic; for i=1:1000, fft(x)/N_0; end; toc
elapsed_time = 0.1600
```

With this as a reference, the matrix-based computations appear to be several times slower than the FFT-based computations. This difference is more dramatic as N_0 is increased. Since the two methods provide identical results, there is little incentive to use the slower matrix-based approach, and the FFT-based algorithm is generally preferred. Even so, the FFT can exhibit curious behavior: adding a few data points, even the artificial samples introduced by zero padding, can dramatically increase or decrease execution times. The `tic` and `toc` commands illustrate this strange result. Consider computing the DTFS of 1015 random data points 1000 times.

```
>> y1 = rand(1015,1);
>> tic; for i=1:1000; fft(y1)/1015; end; T1=toc
T1 = 1.9800
```

Next, pad the sequence with four zeros.

```
>> y2 = [y1;zeros(4,1)];
>> tic; for i=1:1000; fft(y2)/1019; end; T2=toc
T2 = 8.3000
```

The ratio of the two elapsed times indicates that adding four points to an already long sequence increases the computation time by a factor of 4. Next, the sequence is zero-padded to a length of $N_0 = 1024$.

```
>> y3 = [y2;zeros(5,1)];
>> tic; for i=1:1000; fft(y3)/1024; end; T3=toc
T3 = 0.9900
```

In this case, the added data decrease the original execution time by a factor of 2 and the second execution time by a factor of 8! These results are particularly surprising when it is realized that the lengths of y1, y2, and y3 differ by less than 1%.

As it turns out, the efficiency of the `fft` command depends on the factorability of N_0. With the `factor` command, $1015 = (5)(7)(29)$, 1019 is prime, and $1024 = (2)^{10}$. The most factorable length, 1024, results in the fastest execution, while the least factorable length, 1019, results in the slowest execution. To ensure the greatest factorability and fastest operation, vector lengths are ideally a power of 2.

M9.3 FIR Filter Design by Frequency Sampling

Finite impulse response (FIR) digital filters are flexible, always stable, and relatively easy to implement. These qualities make FIR filters a popular choice among digital filter designers. The difference equation of a length-N causal FIR filter is conveniently expressed as

$$y[n] = h_0 x[n] + h_1 x[n-1] + \cdots + h_{N-1} x[n-(N-1)] = \sum_{k=0}^{N-1} h_k x[n-k]$$

The filter coefficients, or tap weights as they are sometimes called, are expressed using the variable h to emphasize that the coefficients themselves represent the impulse response of the filter.

The filter's frequency response is

$$H(\Omega) = \frac{Y(\Omega)}{X(\Omega)} = \sum_{k=0}^{N-1} h_k e^{-j\Omega k}$$

Since $H(\Omega)$ is a 2π-periodic function of the continuous variable Ω, it is sufficient to specify $H(\Omega)$ over a single period $(0 \le \Omega < 2\pi)$.

In many filtering applications, the desired magnitude response $|H_d(\Omega)|$ is known but not the filter coefficients $h[n]$. The question, then, is one of determining the filter coefficients from the desired magnitude response.

Consider the design of a lowpass filter with cutoff frequency $\Omega_c = \pi/4$. An inline function represents the desired ideal frequency response.

```
>> H_d = inline('(mod(Omega,2*pi)<pi/4)+(mod(Omega,2*pi)>2*pi-pi/4)');
```

Since the inverse DTFT of $H_d(\Omega)$ is a sampled sinc function, it is impossible to perfectly achieve the desired response with a causal, finite-length FIR filter. A realizable FIR filter is necessarily an approximation, and an infinite number of possible solutions exist. Thought of another way, $H_d(\Omega)$ specifies an infinite number of points, but the FIR filter only has N unknown tap weights. In general, we expect a length-N filter to match only N points of the desired response over $(0 \le \Omega < 2\pi)$. Which frequencies should be chosen?

A simple and sensible method is to select N frequencies uniformly spaced on the interval $(0 \leq \Omega < 2\pi)$, $(0, 2\pi/N, 4\pi/N, 6\pi/N, \ldots, (N-1)2\pi/N)$. By choosing uniformly spaced frequency samples, the N-point inverse DFT can be used to determine the tap weights $h[n]$. Program MS9P2 illustrates this procedure.

```
function [h] = MS9P2(N,H_d);
% MS9P2.m : MATLAB Session 9, Program 2
% Function M-file designs a length-N FIR filter by sampling the desired
% magnitude response H.  Phase of magnitude samples is left as zero.
% INPUTS:   N = desired FIR filter length
%           H_d = inline function that defines the desired magnitude response
% OUTPUTS:  h = impulse response (FIR filter coefficients)

% Create N equally spaced frequency samples:
Omega = linspace(0,2*pi*(1-1/N),N)';
% Sample the desired magnitude response and create h[n]:
H = 1.0*H_d(Omega); h = real(ifft(H));
```

To complete the design, the filter length must be specified. Small values of N reduce the filter's complexity but also reduce the quality of the filter's response. Large values of N improve the approximation of $H_d(\Omega)$ but also increase complexity. A balance is needed. We choose an intermediate value of $N = 21$ and use MS9P2 to design the filter.

```
>> N = 21; h = MS9P2(N,H_d);
```

To assess the filter quality, the frequency response is computed by means of program MS5P1.

```
>> Omega = linspace(0,2*pi,1000); samples = linspace(0,2*pi*(1-1/N),N)';
>> H = MS5P1(h,1,Omega);
>> subplot(2,1,1); stem([0:N-1],h,'k'); xlabel('n'); ylabel('h[n]');
>> subplot(2,1,2);
>> plot(samples,H_d(samples),'ko',Omega,H_d(Omega),'k:',Omega,abs(H),'k');
>> axis([0 2*pi -0.1 1.6]); xlabel('\Omega'); ylabel('|H(\Omega)|');
>> legend('Samples','Desired','Actual',0);
```

As shown in Fig. M9.3, the filter's frequency response intersects the desired response at the sampled values of $H_d(\Omega)$. The overall response, however, has significant ripple between sample points that renders the filter practically useless. Increasing the filter length does not alleviate the ripple problems. Figure M9.4 shows the case $N = 41$.

To understand the poor behavior of filters designed with MS9P2, remember that the impulse response of an ideal lowpass filter is a sinc function with the peak centered at zero. Thought of another way, the peak of the sinc is centered at $n = 0$ because the phase of $H_d(\Omega)$ is zero. Constrained to be causal, the impulse response of the designed filter still has a peak at $n = 0$ but it cannot include values for negative n. As a result, the sinc function is split in an unnatural way with sharp discontinuities on both ends of $h[n]$. Sharp discontinuities in the time domain appear as high-frequency oscillations in the frequency domain, which is why $H(\Omega)$ has significant ripple.

To improve the filter behavior, the peak of the sinc is moved to $n = (N-1)/2$, the center of the length-N filter response. In this way, the peak is not split, no large discontinuities are present, and frequency-response ripple is consequently reduced. With DFT properties, a cyclic

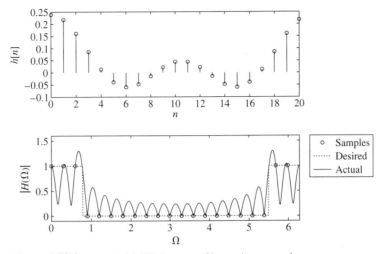

Figure M9.3 Length-21 FIR lowpass filter using zero phase.

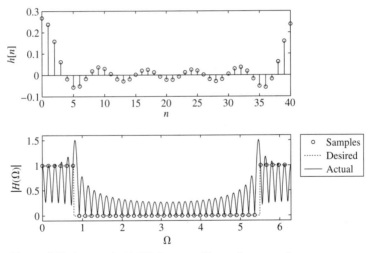

Figure M9.4 Length-41 FIR lowpass filter using zero phase.

shift of $(N - 1)/2$ in the time domain requires a scale factor of $e^{-j\Omega(N-1)/2}$ in the frequency domain.[†] Notice that the scale factor $e^{-j\Omega(N-1)/2}$ affects only phase, not magnitude, and results in a linear phase filter. Program MS9P3 implements the procedure.

[†]Technically, the shift property requires $(N-1)/2$ to be an integer, which occurs only for odd-length filters. The next-to-last line of program MS9P3 implements a correction factor, of sorts, that is required to accommodate the fractional shifts desired for even-length filters. The mathematical derivation of this correction is nontrivial and is not included here. Those hesitant to use this correction factor have an alternative: simply round $(N-1)/2$ to the nearest integer. Although the rounded shift is slightly off center for even-length filters, there is usually little or no appreciable difference in the characteristics of the filter. Even so, true centering is desirable because the resulting impulse response is symmetric, which can reduce by half the number of multiplies required to implement the filter.

```
function [h] = MS9P3(N,H_d);
% MS9P3.m : MATLAB Session 9, Program 3
% Function M-file designs a length-N FIR filter by sampling the desired
% magnitude response H.  Phase is defined to shift h[n] by (N-1)/2.
% INPUTS:   N = desired FIR filter length
%           H_d = inline function that defines the desired magnitude response
% OUTPUTS:  h = impulse response (FIR filter coefficients)

% Create N equally spaced frequency samples:
Omega = linspace(0,2*pi*(1-1/N),N)';
% Sample the desired magnitude response:
H = H_d(Omega);
% Define phase to shift h[n] by (N-1)/2:
H = H.*exp(-j*Omega*((N-1)/2));
H(fix(N/2)+2:N,1) = H(fix(N/2)+2:N,1)*((-1)^(N-1));
h = real(ifft(H));
```

Figure M9.5 shows the results for the $N = 21$ case using MS9P3 to compute $h[n]$. As hoped, the impulse response looks like a sinc function with the peak centered at $n = 10$. Additionally, the frequency-response ripple is greatly reduced. With MS9P3, increasing N improves the quality of the filter, as shown in Fig. M9.6 for the case $N = 41$. While the magnitude response is needed to establish the general shape of the filter response, it is the proper selection of phase that ensures the acceptability of the filter's behavior.

To illustrate the flexibility of the design method, consider a bandpass filter with passband $(\pi/4 < |\Omega| < \pi/2)$.

```
>> H_d = inline(['(mod(Omega,2*pi)>pi/4)&(mod(Omega,2*pi)<pi/2)+',...
        '(mod(Omega,2*pi)>3*pi/2)&(mod(Omega,2*pi)<7*pi/4)']);
```

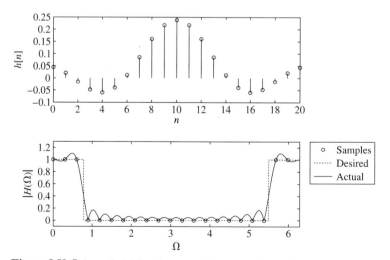

Figure M9.5 Length-21 FIR lowpass filter using linear phase.

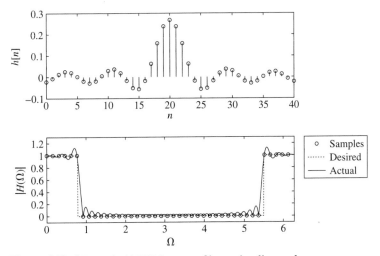

Figure M9.6 Length-41 FIR lowpass filter using linear phase.

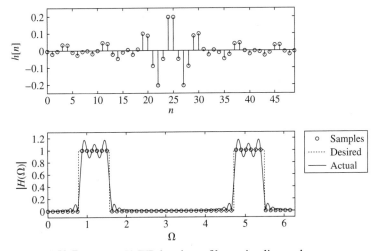

Figure M9.7 Length-50 FIR bandpass filter using linear phase.

Figure M9.7 shows the results for $N = 50$. Notice that this even-length filter uses a fractional shift and is symmetric about $n = 24.5$.

Although FIR filter design by means of frequency sampling is very flexible, it is not always appropriate. Extreme care is needed for filters, such as digital differentiators and Hilbert transformers, that require special phase characteristics for proper operation. Additionally, if frequency samples occur near jump discontinuities of $H_d(\Omega)$, rounding errors may, in rare cases, disrupt the desired symmetry of the sampled magnitude response. Such cases are corrected by slightly adjusting the location of problematic jump discontinuities or by changing the value of N.

PROBLEMS

9.1-1 Find the discrete-time Fourier series (DTFS) and sketch their spectra $|\mathcal{D}_r|$ and $\angle \mathcal{D}_r$ for $0 \leq r \leq N_0 - 1$ for the following periodic signal:
$$x[n] = 4\cos 2.4\pi n + 2\sin 3.2\pi n$$

9.1-2 Repeat Prob. 9.1-1 for $x[n] = \cos 2.2\pi n \cos 3.3\pi n$.

9.1-3 Repeat Prob. 9.1-1 for $x[n] = 2\cos 3.2\pi (n-3)$.

9.1-4 Find the discrete-time Fourier series and the corresponding amplitude and phase spectra for the $x[n]$ shown in Fig. P9.1-4.

9.1-5 Repeat Prob. 9.1-4 for the $x[n]$ depicted in Fig. P9.1-5.

9.1-6 Repeat Prob. 9.1-4 for the $x[n]$ illustrated in Fig. P9.1-6.

9.1-7 An N_0-periodic signal $x[n]$ is represented by its DTFS as in Eq. (9.8). Prove Parseval's theorem (for the DTFS), which states that
$$\frac{1}{N_0} \sum_{n=\langle N_0 \rangle} |x[n]|^2 = \sum_{r=\langle N_0 \rangle} |\mathcal{D}_r|^2$$

In the text [Eq. (9.60)], we obtain the Pareseval's theorem for DTFT. [Hint: If w is complex, then $|w|^2 = ww^*$, and use Eq. (8.28).]

9.1-8 Answer yes or no, and justify your answers with an appropriate example or proof.

(a) Is a sum of aperiodic discrete-time sequences ever periodic?

(b) Is a sum of periodic discrete-time sequences ever aperiodic?

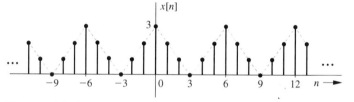

Figure P9.1-4

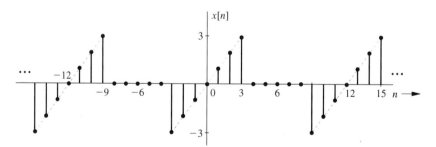

Figure P9.1-5

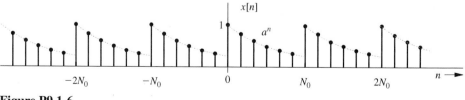

Figure P9.1-6

9.2-1 Show that for a real $x[n]$, Eq. (9.29) can be expressed as

$$x[n] = \frac{1}{\pi} \int_0^\pi |X(\Omega)| \cos{(\Omega n + \angle X(\Omega))}\, d\Omega$$

This is the trigonometric form of the DTFT.

9.2-2 A signal $x[n]$ can be expressed as the sum of even and odd components (Section 1.5-1):

$$x[n] = x_e[n] + x_o[n]$$

(a) If $x[n] \Longleftrightarrow X(\Omega)$, show that for real $x[n]$,

$$x_e[n] \Longleftrightarrow \text{Re}[X(\Omega)]$$

and

$$x_o[n] \Longleftrightarrow j\,\text{Im}[X(\Omega)]$$

(b) Verify these results by finding the DTFT of the even and odd components of the signal $(0.8)^n u[n]$.

9.2-3 For the following signals, find the DTFT directly, using the definition in Eq. (9.30). Assume $|\gamma| < 1$.

(a) $\delta[n]$

(b) $\delta[n - k]$

(c) $\gamma^n u[n - 1]$

(d) $\gamma^n u[n + 1]$

(e) $(-\gamma)^n u[n]$

(f) $\gamma^{|n|}$

9.2-4 Use Eq. (9.29) to find the inverse DTFT for the following spectra, given only over the interval $|\Omega| \le \pi$. Assume Ω_c and $\Omega_0 < \pi$.

(a) $e^{jk\Omega}$ integer k

(b) $\cos k\Omega$ integer k

(c) $\cos^2(\Omega/2)$

(d) $\Delta\left(\dfrac{\Omega}{2\Omega_c}\right)$

(e) $2\pi\delta(\Omega - \Omega_0)$

(f) $\pi[\delta(\Omega - \Omega_0) + \delta(\Omega + \Omega_0)]$

9.2-5 Using Eq. (9.29), show that the inverse DTFT of rect $((\Omega - \pi/4)/\pi)$ is $0.5 \,\text{sinc}\,(\pi n/2)\, e^{j\pi n/4}$.

9.2-6 From definition (9.30), find the DTFT of the signals $x[n]$ in Fig. P9.2-6.

9.2-7 From definition (9.30), find the DTFT of the signals depicted in Fig. P9.2-7.

9.2-8 Use Eq. (9.29) to find the inverse DTFT of the spectra (shown only for $|\Omega| \le \pi$) in Fig. P9.2-8.

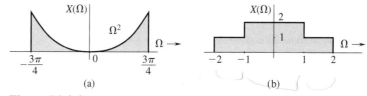

(a)   (b)   **Figure P9.2-6**

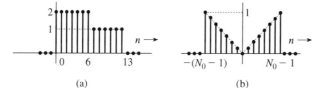

(a)   (b)   **Figure P9.2-7**

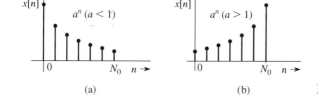

(a)   (b)

Figure P9.2-8

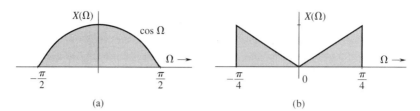

(a)   (b)

Figure P9.2-9

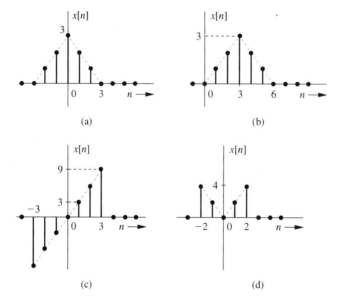

(a)   (b)

(c)   (d)   **Figure P9.2-10**

9.2-9 Use Eq. (9.29) to find the inverse DTFT of the spectra (shown only for $|\Omega| \leq \pi$) in Fig. P9.2-9.

9.2-10 Find the DTFT for the signals shown in Fig. P9.2-10.

9.2-11 Find the inverse DTFT of $X(\Omega)$ (shown only for $|\Omega| \leq \pi$) for the spectra illustrated in Fig. P9.2-11. [Hint: $X(\Omega) = |X(\Omega)|e^{j\angle X(\Omega)}$. This problem illustrates how different phase spectra (both with the same amplitude spectrum) represent entirely different signals.]

9.2-12 (a) Show that time-expanded signal $x_e[n]$ in Eq. (3.4) can also be expressed as

$$x_e[n] = \sum_{k=-\infty}^{\infty} x[k]\delta[n - Lk]$$

(b) Find the DTFT of $x_e[n]$ by finding the DTFT of the right-hand side of the equation in part (a).

(c) Use the result in part (b) and Table 9.1 to find the DTFT of $z[n]$, shown in Fig. P9.2-12.

9.2-13 (a) A glance at Eq. (9.29) shows that the inverse DTFT equation is identical to the inverse (continuous-time) Fourier transform Eq. (7.8b) for a signal $x(t)$ bandlimited to π rad/s. Hence, we should be able to use the continuous-time Fourier transform Table 7.1 to find DTFT pairs that correspond to continuous-time transform pairs for bandlimited signals. Use this fact to derive DTFT pairs 8, 9, 11, 12, 13, and 14 in Table 9.1 by means of the appropriate pairs in Table 7.1.

(b) Can this method be used to derive pairs 2, 3, 4, 5, 6, 7, 10, 15, and 16 in Table 9.1? Justify your answer with specific reason(s).

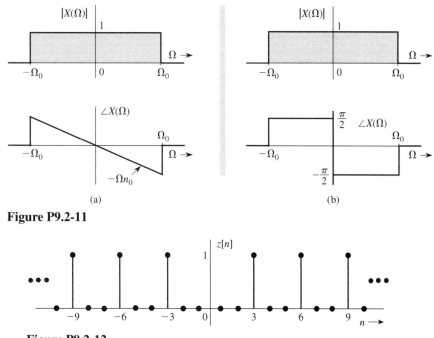

(a)   (b)

Figure P9.2-11

Figure P9.2-12

9.2-14 Are the following frequency domain signals valid DTFT's? Answer yes or no, and justify your answers.

(a) $X(\Omega) = \Omega + \pi$

(b) $X(\Omega) = j + \pi$

(c) $X(\Omega) = \sin(10\Omega)$

(d) $X(\Omega) = \sin(\Omega/10)$

(e) $X(\Omega) = \delta(\Omega)$

9.3-1 Using only pairs 2 and 5 (Table 9.1) and the time-shifting property (9.52), find the DTFT of the following signals, assuming $|a| < 1$

(a) $u[n] - u[n-9]$

(b) $a^{n-m}u[n-m]$

(c) $a^{n-3}(u[n] - u[n-10])$

(d) $a^{n-m}u[n]$

(e) $a^n u[n-m]$

(f) $(n-m)a^{n-m}u[n-m]$

(g) $(n-m)a^n u[n]$

(h) $na^{n-m}u[n-m]$

9.3-2 The triangular pulse $x[n]$ shown in Fig. P9.3-2a is given by

$$X(\Omega) = \frac{4e^{j6\Omega} - 5e^{j5\Omega} + e^{j\Omega}}{(e^{j\Omega} - 1)^2}$$

Use this information and the DTFT properties to find the DTFT of the signals $x_1[n]$, $x_2[n]$, $x_3[n]$, and $x_4[n]$ shown in Fig. P9.3-2b, P9.3-2c, P9.3-2d, and P9.3-2e, respectively.

9.3-3 Show that periodic convolution $X(\Omega)(P)$ $Y(\Omega) = 2\pi X(\Omega)$ if

$$X(\Omega) = \sum_{k=0}^{4} a_k e^{-jk\Omega}$$

and

$$Y(\Omega) = \frac{\sin(5\Omega/2)}{\sin(\Omega/2)} e^{-j2\Omega}$$

where a_k is a set of arbitrary constants.

9.3-4 Using only pair 2 (Table 9.1) and properties of DTFT, find the DTFT of the following signals, assuming $|a| < 1$ and $\Omega_0 < \pi$.

(a) $a^n \cos \Omega_0 n u[n]$

(b) $n^2 a^n u[n]$

(c) $(n-k)a^{2n}u[n-m]$

9.3-5 Use pair 10 in Table 9.1, and some suitable property or properties of the DTFT, to derive pairs 11, 12, 13, 14, 15, and 16.

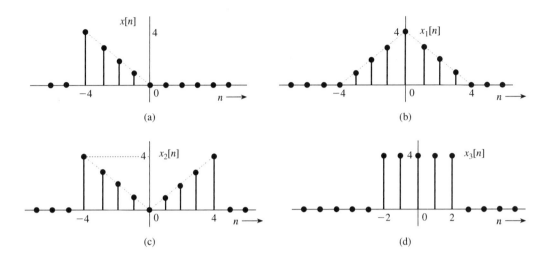

(a)

(b)

(c)

(d)

(e)

Figure P9.3-2

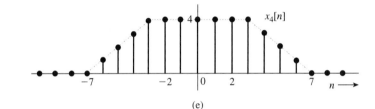

(a)

(b)

Figure P9.3-6

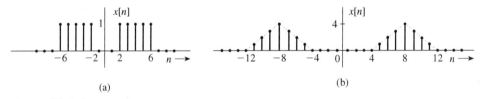

Figure P9.3-7

9.3-6 Use the time-shifting property to show that

$$x[n+k] + x[n-k] \Longleftrightarrow 2X(\Omega) \cos k\Omega$$

Use this result to find the DTFT of the signals shown in Fig. P9.3-6.

9.3-7 Use the time-shifting property to show that

$$x[n+k] - x[n-k] \Longleftrightarrow 2jX(\Omega) \sin k\Omega$$

Use this result to find the DTFT of the signal shown in Fig. P9.3-7.

9.3-8 Using only pair 2 in Table 9.1 and the convolution property, find the inverse DTFT of $X(\Omega) = e^{2j\Omega}/(e^{j\Omega} - \gamma)^2$.

9.3-9 In Table 9.1, you are given pair 1. From this information and using some suitable property or properties of the DTFT, derive pairs 2, 3, 4, 5, 6, and 7 of Table 9.1. For example, starting with pair 1, derive pair 2. From pair 2, use suitable properties of the DTFT to derive pair 3. From pairs 2 and 3, derive pair 4, and so on.

9.3-10 From the pair $e^{j(\Omega_0/2)n} \Longleftrightarrow 2\pi\delta(\Omega - (\Omega_0/2))$ over the fundamental band, and the frequency-convolution property, find the DTFT of $e^{j\Omega_0 n}$. Assume $\Omega_0 < \pi/2$.

9.3-11 From the definition and properties of the DTFT, show that

(a) $\displaystyle\sum_{n=-\infty}^{\infty} \text{sinc}\,(\Omega_c n) = \frac{\pi}{\Omega_c} \quad \Omega_c < \pi$

(b) $\displaystyle\sum_{n=-\infty}^{\infty} (-1)^n \text{sinc}\,(\Omega_c n) = 0 \quad \Omega_c < \pi$

(c) $\displaystyle\sum_{n=-\infty}^{\infty} \text{sinc}^2(\Omega_c n) = \frac{\pi}{\Omega_c} \quad \Omega_c < \pi/2$

(d) $\displaystyle\sum_{n=-\infty}^{\infty} (-1)^n \text{sinc}^2(\Omega_c n) = 0 \quad \Omega_c < \pi/2$

(e) $\displaystyle\int_{-\pi}^{\pi} \frac{\sin(M\Omega/2)}{\sin(\Omega/2)} = 2\pi \quad \text{odd } M$

(f) $\displaystyle\sum_{n=-\infty}^{\infty} |\text{sinc}\,(\Omega_c n)|^4 = 2\pi/3\Omega_c \quad \Omega_c < \pi/2$

9.3-12 Show that the energy of signal $x_c(t)$ specified in Eq. (9.73) is identical to T times the energy of the discrete-time signal $x[n]$, assuming $x_c(t)$ is bandlimited to $B \le 1/2T$ Hz. [Hint: Recall that sinc functions are orthogonal, that is,

$$\int_{-\infty}^{\infty} \text{sinc}\,[\pi(t-m)]\,\text{sinc}\,[\pi(t-n)]\,dt$$

$$= \begin{cases} 0 & m \ne n \\ 1 & m = n \end{cases}$$

9.4-1 Use the DTFT method to find the zero-state response $y[n]$ of a causal system with frequency response

$$H(\Omega) = \frac{e^{j\Omega} + 0.32}{e^{j2\Omega} + e^{j\Omega} + 0.16}$$

and the input

$$x[n] = (-0.5)^n u[n]$$

9.4-2 Repeat Prob. 9.4-1 for

$$H(\Omega) = \frac{e^{j\Omega} + 0.32}{e^{j2\Omega} + e^{j\Omega} + 0.16}$$

and input

$$x[n] = u[n]$$

9.4-3 Repeat Prob. 9.4-1 for

$$H(\Omega) = \frac{e^{j\Omega}}{e^{j\Omega} - 0.5}$$

and

$$x[n] = 0.8^n u[n] + 2(2)^n u[-(n+1)]$$

9.4-4 An accumulator system has the property that an input $x[n]$ results in the output

$$y[n] = \sum_{k=-\infty}^{n} x[k]$$

(a) Find the unit impulse response $h[n]$ and the frequency response $H(\Omega)$ for the accumulator.

(b) Use the results in part a to find the DTFT of $u[n]$.

9.4-5 An LTID system frequency response over $|\Omega| \le \pi$ is

$$H(\Omega) = \text{rect}\left(\frac{\Omega}{\pi}\right)e^{-j2\Omega}$$

Find the output $y[n]$ of this system, if the input $x[n]$ is given by

(a) $\text{sinc}\,(\pi n/2)$
(b) $\text{sinc}\,(\pi n)$
(c) $\text{sinc}^2\,(\pi n/4)$

9.4-6 (a) If $x[n] \Longleftrightarrow X(\Omega)$, then, show that $(-1)^n x[n] \Longleftrightarrow X(\Omega - \pi)$.

(b) Sketch $\gamma^n u[n]$ and $(-\gamma)^n u[n]$ for $\gamma = 0.8$; see the spectra for $\gamma^n u[n]$ in Fig. 9.4b and 9.4c. From these spectra, sketch the spectra for $(-\gamma)^n u[n]$.

(c) An ideal lowpass filter of cutoff frequency Ω_c is specified by the frequency response

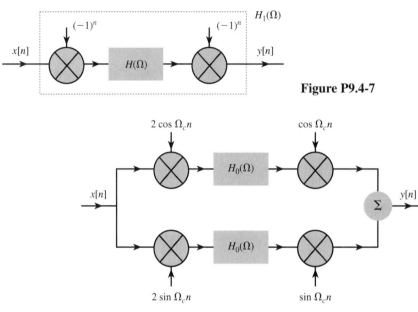

Figure P9.4-7

Figure P9.4-9

$H(\Omega) = \text{rect}(\Omega/2\Omega_c)$. Find its impulse response $h[n]$. Find the frequency response of a filter whose impulse response is $(-1)^n h[n]$. Sketch the frequency response of this filter. What kind of filter is this?

9.4-7 A filter with impulse response $h[n]$ is modified as shown in Fig. P9.4-7. Determine the resulting filter impulse response $h_1[n]$. Find also the resulting filter frequency response $H_1(\Omega)$ in terms of the frequency response $H(\Omega)$. How are $H(\Omega)$ and $H_1(\Omega)$ related?

9.4-8 (a) Consider an LTID system S_1, specified by a difference equation of the form of Eqs. (3.17a) or (3.17b) or (3.24) in Chapter 3. We construct another system S_2 by replacing coefficients a_i ($i = 0, 1, 2, \ldots, N$) by coefficients $(-1)^i a_i$ and replacing all coefficients b_i ($i = 0, 1, 2, \ldots, N$) with coefficients $(-1)^i b_i$. How are the frequency responses of the two systems related?

(b) If S_1 represents a lowpass filter, what kind of filter is specified by S_2?

(c) What type of filter (lowpass, highpass, etc.) is specified by the difference

equation

$$y[n] - 0.8y[n-1] = x[n]$$

What kind of filter is specified by the difference equation

$$y[n] + 0.8y[n-1] = x[n]$$

9.4-9 (a) The system shown in Fig. P9.4-9 contains two identical LTID filters with frequency response $H_0(\Omega)$ and corresponding impulse response $h_0[n]$. It is easy to see that the system is linear. Show that this system is also time invariant. Do this by finding the response of the system to input $\delta[n-k]$ in terms of $h_0[n]$.

(b) If $H_0(\Omega) = \text{rect}(\Omega/2W)$ over the fundamental band, and $\Omega_c + W \leq \pi$, find $H(\Omega)$, the frequency response of this system. What kind of filter is this?

9.M-1 This problem uses a matrix-based approach to investigate the computation of the inverse DTFS.

(a) Implement Eq. (9.8), the inverse DTFS, using a matrix-based approach.

(b) Compare the execution speed of the matrix-based approach to the IFFT-based approach for input vectors of sizes 10, 100, and 1000.

(c) What is the result of multiplying the DFT matrix $\mathbf{W}_{N_0}$ by the inverse DTFS matrix? Discuss your result.

9.M-2 A stable, first-order highpass IIR digital filter has transfer function

$$H(z) = \left(\frac{1+\alpha}{2}\right)\left(\frac{1-z^{-1}}{1-\alpha z^{-1}}\right)$$

(a) Derive an expression relating α to the 3 dB cutoff frequency Ω_c.

(b) Test your expression from part a in the following manner. First, compute α to achieve a 3 dB cutoff frequency of 1 kHz, assuming a sampling rate of $\mathcal{F}_s = 5$ kHz. Determine a difference equation description of the system, and verify that the system is stable. Next, compute and plot the magnitude response of the resulting filter. Verify that the filter is highpass and has the correct cutoff frequency.

(c) Holding α constant, what happens to the cutoff frequency Ω_c as $\mathcal{F}_s$ is increased to 50 kHz? What happens the cutoff frequency f_c as $\mathcal{F}_s$ is increased to 50 kHz?

(d) Is there a well-behaved inverse filter to $H(z)$? Explain.

(e) Determine α for $\Omega_c = \pi/2$. Comment on the resulting filter, particularly $h[n]$.

9.M-3 Figure P9.M-3 provides the desired magnitude response $|H(\Omega)|$ of a real filter. Mathematically,

$$|H(\Omega)| = \begin{cases} 4\Omega/\pi & 0 \le \Omega < \pi/4 \\ 2 - 4\Omega/\pi & \pi/4 \le \Omega < \pi/2 \\ 0 & \pi/2 \le \Omega \le \pi \end{cases}$$

Since the digital filter is real, $|H(\Omega)| = |H(-\Omega)|$ and $|H(\Omega)| = |H(\Omega + 2\pi)|$ for all Ω.

(a) Can a realizable filter have this exact magnitude response? Explain your answer.

(b) Use the frequency sampling method to design an FIR filter with this magnitude response (or a reasonable approximation). Use MATLAB to plot the magnitude response of your filter.

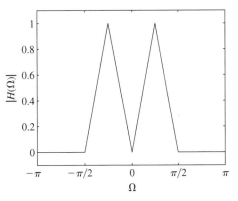

Figure P9.M-3 Desired magnitude response $|H(\Omega)|$.

9.M-4 A real FIR comb filter is needed that has magnitude response $|H(\Omega)| = [0, 3, 0, 3, 0, 3, 0, 3]$ for $\Omega = [0, \pi/4, \pi/2, 3\pi/4, \pi, 5\pi/4, 3\pi/2, 7\pi/4]$, respectively. Provide the impulse response $h[n]$ of a filter that accomplishes these specifications.

9.M-5 A permutation matrix $\mathbf{P}$ has a single one in each row and column with the remaining elements all zero. Permutation matrices are useful for reordering the elements of a vector; the operation $\mathbf{Px}$ reorders the elements of a column vector $\mathbf{x}$ based on the form of $\mathbf{P}$.

(a) Fully describe an $N_0 \times N_0$ permutation matrix named $\mathbf{R}_{N_0}$ that reverses the order of the elements of a column vector $\mathbf{x}$.

(b) Given DFT matrix $\mathbf{W}_{N_0}$, verify that $(\mathbf{W}_{N_0})(\mathbf{W}_{N_0}) = \mathbf{W}_{N_0}^2$ produces a scaled permutation matrix. How does $\mathbf{W}_{N_0}^2 \mathbf{x}$ reorder the elements of $\mathbf{x}$?

(c) What is the result of $(\mathbf{W}_{N_0}^2)(\mathbf{W}_{N_0}^2)\mathbf{x} = \mathbf{W}_{N_0}^4 \mathbf{x}$?

CHAPTER 10

STATE-SPACE ANALYSIS

In Section 1.10, basic notions of *state variables* were introduced. In this chapter, we shall discuss state variables in more depth.

Most of this book deals with external (input–output) description of systems. As noted in Chapter 1, such a description may be inadequate in some cases, and we need a systematic way of finding a system's *internal description*. State-space analysis of systems meets this need. In this method, we first select a set of key variables, called the *state variables,* in the system. Every possible signal or variable in the system at any instant t can be expressed in terms of the state variables and the input(s) at that instant t. If we know all the state variables as a function of t, we can determine every possible signal or variable in the system at any instant with a relatively simple relationship. The system description in this method consists of two parts:

1. A set of equations relating the state variables to the inputs (*the state equation*).

2. A set of equations relating outputs to the state variables and the inputs (*the output equation*).

The analysis procedure, therefore, consists of solving the state equation first, and then solving the output equation. The state-space description is capable of determining every possible system variable (or output) from knowledge of the input and the initial state (conditions) of the system. For this reason, it is an *internal description* of the system.

By its nature, state variable analysis is eminently suited for multiple-input, multiple-output (MIMO) systems. A single-input, single output (SISO) system is a special case of MIMO systems. In addition, the state-space techniques are useful for several other reasons, mentioned in Section 1.10, and repeated here:

1. The state equations of a system provide a mathematical model of great generality that can describe not just linear systems, but also nonlinear systems; not just time-invariant systems, but also time-varying-parameter systems; not just SISO systems, but also MIMO systems. Indeed, state equations are ideally suited for analysis, synthesis, and optimization of MIMO systems.

2. Compact matrix notation along with powerful techniques of linear algebra greatly facilitate complex manipulations. Without such features, many important results of modern

system theory would have been difficult to obtain. State equations can yield a great deal of information about a system even when they are not solved explicitly.

3. State equations lend themselves readily to digital computer simulation of complex systems of high order, with or without nonlinearities, and with multiple inputs and outputs.

4. For second-order systems ($N = 2$), a graphical method called *phase-plane analysis* can be used on state equations, whether they are linear or nonlinear.

This chapter requires some understanding of matrix algebra. Section B.6 is a self-contained treatment of matrix algebra, which should be more than adequate for the purposes of this chapter.

10.1 INTRODUCTION

From the discussion in Chapter 1, we know that to determine a system's response(s) at any instant t, we need to know the system's inputs during its entire past, from $-\infty$ to t. If the inputs are known only for $t > t_0$, we can still determine the system output(s) for any $t > t_0$, provided we know certain initial conditions in the system at $t = t_0$. These initial conditions collectively are called the *initial state* of the system (at $t = t_0$).

The state variables $q_1(t), q_2(t), \ldots, q_N(t)$ are the minimum number of variables of a system such that their initial values at any instant t_0 are sufficient to determine the behavior of the system for all time $t \geq t_0$ when the input(s) to the system is known for $t \geq t_0$. This statement implies that an output of a system at any instant is determined completely from a knowledge of the values of the system state and the input at that instant.

Initial conditions of a system can be specified in many different ways. Consequently, the system state can also be specified in many different ways. This means that state variables are not unique.

This discussion is also valid for multiple-input, multiple-output (MIMO) systems, where every possible system output at any instant t is determined completely from a knowledge of the system state and the input(s) at the instant t. These ideas should become clear from the following example of an *RLC* circuit.

EXAMPLE 10.1

Find a state-space description of the *RLC* circuit shown in Fig. 10.1. Verify that all possible system outputs at some instant t can be determined from knowledge of the system state and the input at that instant t.

It is known that inductor currents and capacitor voltages in an *RLC* circuit can be used as one possible choice of state variables. For this reason, we shall choose q_1 (the capacitor voltage) and q_2 (the inductor current) as our state variables.

The node equation at the intermediate node is

$$i_3 = i_1 - i_2 - q_2$$

but $i_3 = 0.2\dot{q}_1, i_1 = 2(x - q_1), i_2 = 3q_1$. Hence

$$0.2\dot{q}_1 = 2(x - q_1) - 3q_1 - q_2$$

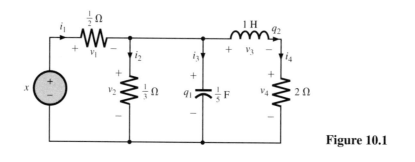

Figure 10.1

or

$$\dot{q}_1 = -25q_1 - 5q_2 + 10x \tag{10.1}$$

This is the first state equation. To obtain the second state equation, we sum the voltages in the extreme right loop formed by C, L, and the $2\,\Omega$ resistor so that they are equal to zero:

$$-q_1 + \dot{q}_2 + 2q_2 = 0$$

or

$$\dot{q}_2 = q_1 - 2q_2 \tag{10.2}$$

Thus, the two state equations are

$$\dot{q}_1 = -25q_1 - 5q_2 + 10x \tag{10.3a}$$

$$\dot{q}_2 = q_1 - 2q_2 \tag{10.3b}$$

Every possible output can now be expressed as a linear combination of q_1, q_2, and x. From Fig. 10.1, we have

$$v_1 = x - q_1$$
$$i_1 = 2(x - q_1)$$
$$v_2 = q_1$$
$$i_2 = 3q_1$$
$$i_3 = i_1 - i_2 - q_2 = 2(x - q_1) - 3q_1 - q_2 = -5q_1 - q_2 + 2x$$
$$i_4 = q_2$$
$$v_4 = 2i_4 = 2q_2$$
$$v_3 = q_1 - v_4 = q_1 - 2q_2 \tag{10.4}$$

This set of equations is known as the *output equation* of the system. It is clear from this set that every possible output at some instant t can be determined from knowledge of $q_1(t)$, $q_2(t)$, and $x(t)$, the system state, and the input at the instant t. Once we have solved the state equations (10.3) to obtain $q_1(t)$ and $q_2(t)$, we can determine every possible output for any given input $x(t)$.

For continuous-time systems, the state equations are N simultaneous first-order differential equations in N state variables $q_1, q_2, \ldots, q_N$ of the form

$$\dot{q}_i = g_i(q_1, q_2, \ldots, q_N, x_1, x_2, \ldots, x_j) \qquad i = 1, 2, \ldots, N \qquad (10.5)$$

where $x_1, x_2, \ldots, x_j$ are the j system inputs. For a linear system, these equations reduce to a simpler linear form

$$\dot{q}_i = a_{i1}q_1 + a_{i2}q_2 + \cdots + a_{iN}q_N + b_{i1}x_1 + b_{i2}x_2 + \cdots + b_{ij}x_j \qquad i = 1, 2, \ldots, N \qquad (10.6a)$$

If there are k outputs $y_1, y_2, \ldots, y_k$, the k output equations are of the form

$$y_m = c_{m1}q_1 + c_{m2}q_2 + \cdots + c_{mN}q_N + d_{m1}x_1 + d_{m2}x_2 + \cdots + d_{mj}x_j \qquad m = 1, 2, \ldots, k \qquad (10.6b)$$

The N simultaneous first-order state equations are also known as the *normal-form* equations.

These equations can be written more conveniently in matrix form:

$$\underbrace{\begin{bmatrix} \dot{q}_1 \\ \dot{q}_2 \\ \vdots \\ \dot{q}_N \end{bmatrix}}_{\mathbf{q}} = \underbrace{\begin{bmatrix} a_{11} & a_{12} & \cdots & a_{1N} \\ a_{21} & a_{22} & \cdots & a_{2N} \\ \vdots & \vdots & \cdots & \vdots \\ a_{N1} & a_{N2} & \cdots & a_{NN} \end{bmatrix}}_{\mathbf{A}} \underbrace{\begin{bmatrix} q_1 \\ q_2 \\ \vdots \\ q_N \end{bmatrix}}_{\mathbf{q}} + \underbrace{\begin{bmatrix} b_{11} & b_{12} & \cdots & b_{1j} \\ b_{21} & b_{22} & \cdots & b_{2j} \\ \vdots & \vdots & \cdots & \vdots \\ b_{N1} & b_{N2} & \cdots & b_{Nj} \end{bmatrix}}_{\mathbf{B}} \underbrace{\begin{bmatrix} x_1 \\ x_2 \\ \vdots \\ x_j \end{bmatrix}}_{\mathbf{x}} \qquad (10.7a)$$

and

$$\underbrace{\begin{bmatrix} y_1 \\ y_2 \\ \vdots \\ y_k \end{bmatrix}}_{\mathbf{y}} = \underbrace{\begin{bmatrix} c_{11} & c_{12} & \cdots & c_{1N} \\ c_{21} & c_{22} & \cdots & c_{2N} \\ \vdots & \vdots & \cdots & \vdots \\ c_{k1} & c_{k2} & \cdots & c_{kN} \end{bmatrix}}_{\mathbf{C}} \underbrace{\begin{bmatrix} q_1 \\ q_2 \\ \vdots \\ q_N \end{bmatrix}}_{\mathbf{q}} + \underbrace{\begin{bmatrix} d_{11} & d_{12} & \cdots & d_{1j} \\ d_{21} & d_{22} & \cdots & d_{2j} \\ \vdots & \vdots & \cdots & \vdots \\ d_{k1} & d_{k2} & \cdots & d_{kj} \end{bmatrix}}_{\mathbf{D}} \underbrace{\begin{bmatrix} x_1 \\ x_2 \\ \vdots \\ x_j \end{bmatrix}}_{\mathbf{x}} \qquad (10.7b)$$

or

$$\dot{\mathbf{q}} = \mathbf{Aq} + \mathbf{Bx} \qquad (10.8a)$$

$$\mathbf{y} = \mathbf{Cq} + \mathbf{Dx} \qquad (10.8b)$$

Equation (10.8a) is the state equation and Eq. (10.8b) is the output equation; $\mathbf{q}$, $\mathbf{y}$, and $\mathbf{x}$ are the state vector, the output vector, and the input vector, respectively.

For discrete-time systems, the state equations are N simultaneous first-order difference equations. Discrete-time systems are discussed in Section 10.6.

10.2 A SYSTEMATIC PROCEDURE FOR DETERMINING STATE EQUATIONS

We shall discuss here a systematic procedure for determining the state-space description of linear time-invariant systems. In particular, we shall consider systems of two types: (1) *RLC* networks and (2) systems specified by block diagrams or Nth-order transfer functions.

10.2-1 Electrical Circuits

The method used in Example 10.1 proves effective in most of the simple cases. The steps are as follows:

1. Choose all independent capacitor voltages and inductor currents to be the state variables.
2. Choose a set of loop currents; express the state variables and their first derivatives in terms of these loop currents.
3. Write loop equations, and eliminate all variables other than state variables (and their first derivatives) from the equations derived in steps 2 and 3.

EXAMPLE 10.2

Write the state equations for the network shown in Fig. 10.2.

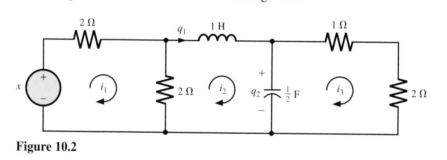

Figure 10.2

Step 1. There is one inductor and one capacitor in the network. Therefore, we shall choose the inductor current q_1 and the capacitor voltage q_2 as the state variables.

Step 2. The relationship between the loop currents and the state variables can be written by inspection:

$$q_1 = i_2 \tag{10.9a}$$

$$\tfrac{1}{2}\dot{q}_2 = i_2 - i_3 \tag{10.9b}$$

Step 3. The loop equations are

$$4i_1 - 2i_2 = x \tag{10.10a}$$

$$2(i_2 - i_1) + \dot{q}_1 + q_2 = 0 \tag{10.10b}$$

$$-q_2 + 3i_3 = 0 \tag{10.10c}$$

Now we eliminate i_1, i_2, and i_3 from Eqs. (10.9) and (10.10) as follows. From Eq. (10.10b), we have

$$\dot{q}_1 = 2(i_1 - i_2) - q_2$$

We can eliminate i_1 and i_2 from this equation by using Eqs. (10.9a) and (10.10a) to obtain

$$\dot{q}_1 = -q_1 - q_2 + \tfrac{1}{2}x$$

The substitution of Eqs. (10.9a) and (10.10c) in Eq. (10.9b) yields

$$\dot{q}_2 = 2q_1 - \tfrac{2}{3}q_2$$

These are the desired state equations. We can express them in matrix form as

$$\begin{bmatrix} \dot{q}_1 \\ \dot{q}_2 \end{bmatrix} = \begin{bmatrix} -1 & -1 \\ 2 & -\tfrac{2}{3} \end{bmatrix} \begin{bmatrix} q_1 \\ q_2 \end{bmatrix} + \begin{bmatrix} \tfrac{1}{2} \\ 0 \end{bmatrix} x \qquad (10.11)$$

The derivation of state equations from loop equations is facilitated considerably by choosing loops in such a way that only one loop current passes through each of the inductors or capacitors.

AN ALTERNATIVE PROCEDURE

We can also determine the state equations by the following procedure:

1. Choose all independent capacitor voltages and inductor currents to be the state variables.
2. Replace each capacitor by a voltage source equal to the capacitor voltage, and replace each inductor by a current source equal to the inductor current. This step will transform the RLC network into a network consisting only of resistors, current sources, and voltage sources.
3. Find the current through each capacitor and equate it to $C\dot{q}_i$, where q_i is the capacitor voltage. Similarly, find the voltage across each inductor and equate it to $L\dot{q}_j$, where q_j is the inductor current.

EXAMPLE 10.3

Use the three-step alternative procedure just outlined to write the state equations for the network in Fig. 10.2.

In the network in Fig. 10.2, we replace the inductor by a current source of current q_1 and the capacitor by a voltage source of voltage q_2, as shown in Fig. 10.3. The resulting network consists of four resistors, two voltage sources, and one current source. We can determine the voltage v_L across the inductor and the current i_c through the capacitor by using the principle of superposition. This step can be accomplished by inspection. For example, v_L has three components arising from three sources. To compute the component due to x, we assume that $q_1 = 0$ (open circuit) and $q_2 = 0$ (short circuit). Under these conditions, the entire network to

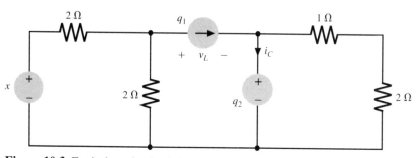

Figure 10.3 Equivalent circuit of the network in Fig. 10.2.

the right of the 2 Ω resistor is opened, and the component of v_L due to x is the voltage across the 2 Ω resistor. This voltage is clearly $(1/2)x$. Similarly, to find the component of v_L due to q_1, we short x and q_2. The source q_1 sees an equivalent resistor of 1 Ω across it, and hence $v_L = -q_1$. Continuing the process, we find that the component of v_L due to q_2 is $-q_2$. Hence

$$v_L = \dot{q}_1 = \tfrac{1}{2}x - q_1 - q_2 \tag{10.12a}$$

Using the same procedure, we find

$$i_c = \tfrac{1}{2}\dot{q}_2 = q_1 - \tfrac{1}{3}q_2 \tag{10.12b}$$

These equations are identical to the state equations (10.11) obtained earlier.[†]

10.2-2 State Equations from a Transfer Function

It is relatively easy to determine the state equations of a system specified by its transfer function.[‡] Consider, for example, a first-order system with the transfer function

$$H(s) = \frac{1}{s+a} \tag{10.13}$$

[†]This procedure requires modification if the system contains all-capacitor and voltage-source tie sets or all-inductor and current-source cut sets. In the case of all-capacitor and voltage-source tie sets, all capacitor voltages cannot be independent. One capacitor voltage can be expressed in terms of the remaining capacitor voltages and the voltage source(s) in that tie set. Consequently, one of the capacitor voltages should not be used as a state variable, and that capacitor should not be replaced by a voltage source. Similarly, in all-inductor current-source tie sets, one inductor should not be replaced by a current source. If there are all-capacitor tie sets or all-inductor cut sets only, no further complications occur. In all-capacitor voltage-source tie sets and/or all-inductor current-source cut sets, we have additional difficulties in that the terms involving derivatives of the input may occur. This problem can be solved by redefining the state variables. The final state variables will not be capacitor voltages and inductor currents.

[‡]We implicitly assume that the system is controllable and observable. This implies that there are no pole–zero cancellations in the transfer function. If such cancellations are present, the state variable description represents only the part of the system that is controllable and observable (the part of the system that is coupled to the input and the output). In other words, the internal description represented by the state equations is no better than the external description represented by the input–output equation.

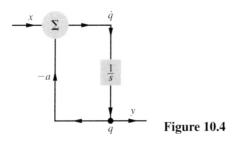

Figure 10.4

The system realization appears in Fig. 10.4. The integrator output q serves as a natural state variable since, in practical realization, initial conditions are placed on the integrator output. The integrator input is naturally $\dot{q}$. From Fig. 10.4, we have

$$\dot{q} = -aq + x \tag{10.14a}$$

$$y = q \tag{10.14b}$$

In Section 4.6 we saw that a given transfer function can be realized in several ways. Consequently, we should be able to obtain different state-space descriptions of the same system by using different realizations. This assertion will be clarified by the following example.

EXAMPLE 10.4

Determine the state-space description of a system specified by the transfer function

$$H(s) = \frac{2s + 10}{s^3 + 8s^2 + 19s + 12} \tag{10.15a}$$

$$= \left(\frac{2}{s+1}\right)\left(\frac{s+5}{s+3}\right)\left(\frac{1}{s+4}\right) \tag{10.15b}$$

$$= \frac{\frac{4}{3}}{s+1} - \frac{2}{s+3} + \frac{\frac{2}{3}}{s+4} \tag{10.15c}$$

We shall use the procedure developed in Section 4.6 to realize $H(s)$ in Eq. (10.15) in four ways: (i) the direct form II (DFII), and (ii) the transpose of DFII [Eq. (10.15a)], (iii) the cascade realization [Eq. (10.15b)], and (iv) the parallel realization [Eq. (10.15c)]. These realizations are depicted in Fig. 10.5. As mentioned earlier, the output of each integrator serves as a natural state variable.

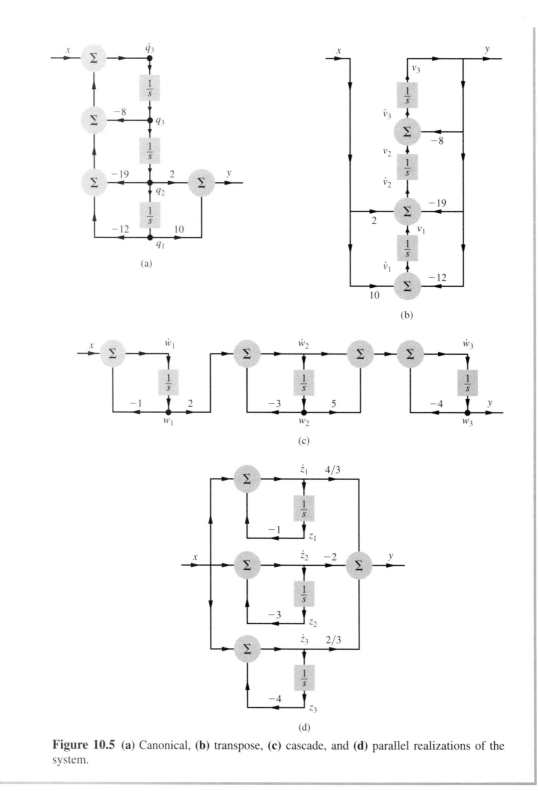

Figure 10.5 (a) Canonical, (b) transpose, (c) cascade, and (d) parallel realizations of the system.

DIRECT FORM II AND ITS TRANSPOSE

Here we shall realize the system using the canonical form (direct form II and its transpose) discussed in Section 4.6. If we choose the state variables to be the three integrator outputs q_1, q_2, and q_3, then, according to Fig. 10.5a,

$$\dot{q}_1 = q_2$$
$$\dot{q}_2 = q_3 \tag{10.16a}$$
$$\dot{q}_3 = -12q_1 - 19q_2 - 8q_3 + x$$

Also, the output y is given by

$$y = 10q_1 + 2q_2 \tag{10.16b}$$

Equations (10.16a) are the state equations, and Eq. (10.16b) is the output equation. In matrix form we have

$$
\begin{bmatrix} \dot{q}_1 \\ \dot{q}_2 \\ \dot{q}_3 \end{bmatrix}
=
\underbrace{\begin{bmatrix} 0 & 1 & 0 \\ 0 & 0 & 1 \\ -12 & -19 & -8 \end{bmatrix}}_{\mathbf{A}}
\begin{bmatrix} q_1 \\ q_2 \\ q_3 \end{bmatrix}
+
\underbrace{\begin{bmatrix} 0 \\ 0 \\ 1 \end{bmatrix}}_{\mathbf{B}} x
\tag{10.17a}
$$

and

$$
\mathbf{y} = \underbrace{[10 \quad 2 \quad 0]}_{\mathbf{C}}
\begin{bmatrix} q_1 \\ q_2 \\ q_3 \end{bmatrix}
\tag{10.17b}
$$

We can also realize $H(s)$ by using the transpose of the DFII form, as shown in Fig. 10.5b. If we label the output of the three integrators as the state variables v_1, v_2, and v_3, then, according to Fig. 10.5b,

$$\dot{v}_1 = -12v_3 + 10x$$
$$\dot{v}_2 = v_1 - 19v_3 + 2x \tag{10.18a}$$
$$\dot{v}_3 = v_2 - 8v_3$$

and the output y is given by

$$y = v_3 \tag{10.18b}$$

Hence

$$
\begin{bmatrix} \dot{v}_1 \\ \dot{v}_2 \\ \dot{v}_3 \end{bmatrix}
=
\underbrace{\begin{bmatrix} 0 & 0 & -12 \\ 1 & 0 & -19 \\ 0 & 1 & -8 \end{bmatrix}}_{\hat{\mathbf{A}}}
\begin{bmatrix} v_1 \\ v_2 \\ v_3 \end{bmatrix}
+
\underbrace{\begin{bmatrix} 10 \\ 2 \\ 0 \end{bmatrix}}_{\hat{\mathbf{B}}} x
\tag{10.19a}
$$

and

$$y = \underbrace{[0 \quad 0 \quad 1]}_{\hat{\mathbf{c}}} \begin{bmatrix} v_1 \\ v_2 \\ v_3 \end{bmatrix} \tag{10.19b}$$

Observe closely the relationship between the state-space description of $H(s)$ by means of the DFII realization [Eqs. (10.17)] and that from the transpose of DFII realization [Eqs. (10.19)]. The $\mathbf{A}$ matrices in these two cases are the transpose of each other; also, the $\mathbf{B}$ of one is the transpose of $\mathbf{C}$ in the other, and vice versa. Hence

$$(\mathbf{A})^T = \hat{\mathbf{A}}$$
$$(\mathbf{B})^T = \hat{\mathbf{C}} \tag{10.20}$$
$$(\mathbf{C})^T = \hat{\mathbf{B}}$$

This is no coincidence. This duality relation is generally true.[1]

CASCADE REALIZATION

The three integrator outputs w_1, w_2, and w_3 in Fig. 10.5c are the state variables. The state equations are

$$\dot{w}_1 = -w_1 + x \tag{10.21a}$$
$$\dot{w}_2 = 2w_1 - 3w_2 \tag{10.21b}$$
$$\dot{w}_3 = 5w_2 + \dot{w}_2 - 4w_3 \tag{10.21c}$$

and the output equation is

$$y = w_3 \tag{10.22}$$

Upon eliminating $\dot{w}_2$ from Eq. (10.21c) by means of Eq. (10.21b), we convert these equations into the desired state form

$$\begin{bmatrix} \dot{w}_1 \\ \dot{w}_2 \\ \dot{w}_3 \end{bmatrix} = \begin{bmatrix} -1 & 0 & 0 \\ 2 & -3 & 0 \\ 2 & 2 & -4 \end{bmatrix} \begin{bmatrix} w_1 \\ w_2 \\ w_3 \end{bmatrix} + \begin{bmatrix} 1 \\ 0 \\ 0 \end{bmatrix} x \tag{10.23a}$$

and

$$y = [0 \quad 0 \quad 1] \begin{bmatrix} w_1 \\ w_2 \\ w_3 \end{bmatrix} \tag{10.23b}$$

PARALLEL REALIZATION (DIAGONAL REPRESENTATION)

The three integrator outputs z_1, z_2, and z_3 in Fig. 10.5d are the state variables. The state equations are

$$\dot{z}_1 = -z_1 + x$$
$$\dot{z}_2 = -3z_2 + x$$
$$\dot{z}_3 = -4z_3 + x \tag{10.24a}$$

and the output equation is

$$y = \tfrac{4}{3}z_1 - 2z_2 + \tfrac{2}{3}z_3 \tag{10.24b}$$

Therefore, the equations in the matrix form are

$$\begin{bmatrix} \dot{z}_1 \\ \dot{z}_2 \\ \dot{z}_3 \end{bmatrix} = \begin{bmatrix} -1 & 0 & 0 \\ 0 & -3 & 0 \\ 0 & 0 & -4 \end{bmatrix} \begin{bmatrix} z_1 \\ z_2 \\ z_3 \end{bmatrix} + \begin{bmatrix} 1 \\ 1 \\ 1 \end{bmatrix} x \tag{10.25a}$$

$$y = \begin{bmatrix} \tfrac{4}{3} & -2 & \tfrac{2}{3} \end{bmatrix} \begin{bmatrix} z_1 \\ z_2 \\ z_3 \end{bmatrix} \tag{10.25b}$$

COMPUTER EXAMPLE C10.1

Use MATLAB to determine the first canonical form (DFI) for the system given in Example 10.4.

[Caution: MATLAB's convention for labeling state variables $q_1, q_2, \ldots, q_n$ in a block diagram, such as shown in Fig. 10.5a, is reversed. That is, MATLAB labels q_1 as q_n, q_2 and q_{n-1}, and so on.]

```
>> num = [2 10]; den = [1 8 19 12];
>> [A,B,C,D] = tf2ss(num,den)

A =     -8   -19   -12
         1     0     0
         0     1     0

B =      1
         0
         0

C =      0     2    10

D =      0
```

It is also possible to determine the transfer function from the state space representation.

```
>> [num,den] = ss2tf(A,B,C,D);
>> tf(num,den)

Transfer function:
4.441e-015 s^2 + 2 s + 10
-------------------------
 s^3 + 8 s^2 + 19 s + 12
```

A GENERAL CASE

It is clear that a system has several state-space descriptions. Notable among these are the variables obtained from the DFII, its transpose, and the diagonalized variables (in the parallel realization). State equations in these forms can be written immediately by inspection of the transfer function. Consider the general Nth-order transfer function

$$H(s) = \frac{b_0 s^N + b_1 s^{N-1} + \cdots + b_{N-1} s + b_N}{s^N + a_1 s^{N-1} + \cdots + a_{N-1} s + a_N} \tag{10.26a}$$

$$= \frac{b_0 s^N + b_1 s^{N-1} + \cdots + b_{N-1} s + b_N}{(s - \lambda_1)(s - \lambda_2) \cdots (s - \lambda_N)}$$

$$= b_0 + \frac{k_1}{s - \lambda_1} + \frac{k_2}{s - \lambda_2} + \cdots + \frac{k_N}{s - \lambda_N} \tag{10.26b}$$

The realizations of $H(s)$ found by using direct form II [Eq. (10.26a)] and the parallel form [Eq. (10.26b)] appear in Fig. 10.6a and 10.6b, respectively.

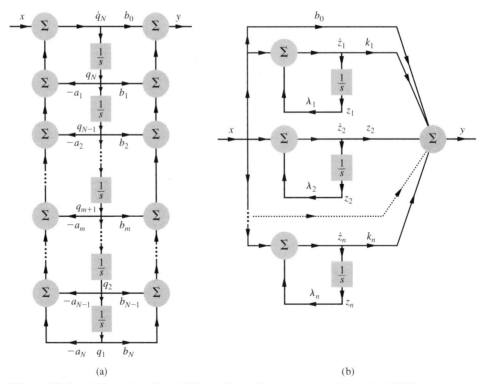

Figure 10.6 (a) Direct form II and (b) parallel realizations for an Nth-order LTIC system.

The N integrator outputs $q_1, q_2, \ldots, q_N$ in Fig. 10.6a are the state variables. By inspection of this figure, we obtain

$$\dot{q}_1 = q_2$$
$$\dot{q}_2 = q_3$$
$$\vdots$$
$$\dot{q}_{N-1} = q_N$$
$$\dot{q}_N = -a_N q_1 - a_{N-1} q_2 - \cdots - a_2 q_{N-1} - a_1 q_N + x$$

(10.27a)

and output y is

$$y = b_N q_1 + b_{N-1} q_2 + \cdots + b_1 q_N + b_0 \dot{q}_N$$

(10.27b)

We can eliminate $\dot{q}_N$ by using the last equation in the set (10.27a) to yield

$$y = (b_N - b_0 a_N) q_1 + (b_{N-1} - b_0 a_{N-1}) q_2 + \cdots + (b_1 - b_0 a_1) q_N + b_0 x$$
$$= \hat{b}_N q_1 + \hat{b}_{N-1} q_2 + \cdots + \hat{b}_1 q_N + b_0 x$$

(10.27c)

where $\hat{b}_i = b_i - b_0 a_i$,

or

$$\begin{bmatrix} \dot{q}_1 \\ \dot{q}_2 \\ \vdots \\ \dot{q}_{N-1} \\ \dot{q}_N \end{bmatrix} = \begin{bmatrix} 0 & 1 & 0 & \cdots & 0 & 0 \\ 0 & 0 & 1 & \cdots & 0 & 0 \\ \vdots & \vdots & \vdots & \cdots & \vdots & \vdots \\ 0 & 0 & 0 & \cdots & 0 & 1 \\ -a_N & -a_{N-1} & -a_{N-2} & \cdots & -a_2 & -a_1 \end{bmatrix} \begin{bmatrix} q_1 \\ q_2 \\ \vdots \\ q_{N-1} \\ q_N \end{bmatrix} + \begin{bmatrix} 0 \\ 0 \\ \vdots \\ 0 \\ 1 \end{bmatrix} x$$

(10.28a)

and

$$y = [\hat{b}_N \quad \hat{b}_{N-1} \quad \cdots \quad \hat{b}_1] \begin{bmatrix} q_1 \\ q_2 \\ \vdots \\ q_N \end{bmatrix} + b_0 x$$

(10.28b)

In Fig. 10.6b, the N integrator outputs $z_1, z_2, \ldots, z_N$ are the state variables. By inspection of this figure, we obtain

$$\dot{z}_1 = \lambda_1 z_1 + x$$
$$\dot{z}_2 = \lambda_2 z_2 + x$$
$$\vdots$$
$$\dot{z}_N = \lambda_N z_N + x$$

(10.29a)

and

$$y = k_1 z_1 + k_2 z_2 + \cdots + k_N z_N + b_0 x$$

(10.29b)

or

$$
\begin{bmatrix} \dot{z}_1 \\ \dot{z}_2 \\ \vdots \\ \dot{z}_{N-1} \\ \dot{z}_N \end{bmatrix} = \begin{bmatrix} \lambda_1 & 0 & \cdots & 0 & 0 \\ 0 & \lambda_2 & \cdots & 0 & 0 \\ \vdots & \vdots & \cdots & \vdots & \vdots \\ 0 & 0 & \cdots & \lambda_{N-1} & 0 \\ 0 & 0 & \cdots & 0 & \lambda_N \end{bmatrix} \begin{bmatrix} z_1 \\ z_2 \\ \vdots \\ z_{N-1} \\ z_N \end{bmatrix} + \begin{bmatrix} 1 \\ 1 \\ \vdots \\ 1 \\ 1 \end{bmatrix} x \tag{10.30a}
$$

and

$$
y = \begin{bmatrix} k_1 & k_2 & \cdots & k_{N-1} & k_N \end{bmatrix} \begin{bmatrix} z_1 \\ z_2 \\ \vdots \\ z_{N-1} \\ z_N \end{bmatrix} + b_0 x \tag{10.30b}
$$

Observe that the diagonalized form of the state matrix [Eq. (10.30a)] has the transfer function poles as its diagonal elements. The presence of repeated poles in $H(s)$ will modify the procedure slightly. The handling of these cases is discussed in Section 4.6.

It is clear from the foregoing discussion that a state-space description is not unique. For any realization of $H(s)$ obtained from integrators, scalar multipliers, and adders, a corresponding state-space description exists. Since there are uncountable possible realizations of $H(s)$, there are uncountable possible state-space descriptions.

The advantages and drawbacks of various types of realization were discussed in Section 4.6.

10.3 SOLUTION OF STATE EQUATIONS

The state equations of a linear system are N simultaneous linear differential equations of the first order. We studied the techniques of solving linear differential equations in Chapters 2 and 4. The same techniques can be applied to state equations without any modification. However, it is more convenient to carry out the solution in the framework of matrix notation.

These equations can be solved in both the time and frequency domains (Laplace transform). The latter is relatively easier to deal with than the time-domain solution. For this reason, we shall first consider the Laplace transform solution.

10.3-1 Laplace Transform Solution of State Equations

The ith state equation [Eq. (10.6a)] is of the form

$$
\dot{q}_i = a_{i1}q_1 + a_{i2}q_2 + \cdots + a_{iN}q_N + b_{i1}x_1 + b_{i2}x_2 + \cdots + b_{ij}x_j \tag{10.31a}
$$

We shall take the Laplace transform of this equation. Let

$$
q_i(t) \Longleftrightarrow Q_i(s)
$$

so that

$$
\dot{q}_i(t) \Longleftrightarrow sQ_i(s) - q_i(0)
$$

Also, let

$$x_i(t) \iff X_i(s)$$

The Laplace transform of Eq. (10.31a) yields

$$sQ_i(s) - q_i(0) = a_{i1}Q_1(s) + a_{i2}Q_2(s) + \cdots + a_{iN}Q_N(s) + b_{i1}X_1(s)$$
$$+ b_{i2}X_2(s) + \cdots + b_{ij}X_j(s) \tag{10.31b}$$

Taking the Laplace transforms of all N state equations, we obtain

$$s\underbrace{\begin{bmatrix} Q_1(s) \\ Q_2(s) \\ \vdots \\ Q_N(s) \end{bmatrix}}_{\mathbf{Q}(s)} - \underbrace{\begin{bmatrix} q_1(0) \\ q_2(0) \\ \vdots \\ q_N(0) \end{bmatrix}}_{\mathbf{q}(0)} = \underbrace{\begin{bmatrix} a_{11} & a_{12} & \cdots & a_{1N} \\ a_{21} & a_{22} & \cdots & a_{2N} \\ \vdots & \vdots & \cdots & \vdots \\ a_{N1} & a_{N2} & \cdots & a_{NN} \end{bmatrix}}_{\mathbf{A}} \underbrace{\begin{bmatrix} Q_1(s) \\ Q_2(s) \\ \vdots \\ Q_N(s) \end{bmatrix}}_{\mathbf{Q}(s)}$$

$$+ \underbrace{\begin{bmatrix} b_{11} & b_{12} & \cdots & b_{1j} \\ b_{21} & b_{22} & \cdots & b_{2j} \\ \vdots & \vdots & \cdots & \vdots \\ b_{N1} & b_{N2} & \cdots & b_{Nj} \end{bmatrix}}_{\mathbf{B}} \underbrace{\begin{bmatrix} X_1(s) \\ X_2(s) \\ \vdots \\ X_j(s) \end{bmatrix}}_{\mathbf{X}(s)} \tag{10.32a}$$

Defining the vectors, as indicated, we have

$$s\mathbf{Q}(s) - \mathbf{q}(0) = \mathbf{A}\mathbf{Q}(s) + \mathbf{B}\mathbf{X}(s)$$

or

$$s\mathbf{Q}(s) - \mathbf{A}\mathbf{Q}(s) = \mathbf{q}(0) + \mathbf{B}\mathbf{X}(s)$$

and

$$(s\mathbf{I} - \mathbf{A})\mathbf{Q}(s) = \mathbf{x}(0) + \mathbf{B}\mathbf{X}(s) \tag{10.32b}$$

where $\mathbf{I}$ is the $N \times N$ identity matrix. From Eq. (10.32b), we have

$$\mathbf{Q}(s) = (s\mathbf{I} - \mathbf{A})^{-1}[\mathbf{q}(0) + \mathbf{B}\mathbf{X}(s)] \tag{10.33a}$$

$$= \mathbf{\Phi}(s)[\mathbf{q}(0) + \mathbf{B}\mathbf{X}(s)] \tag{10.33b}$$

where

$$\mathbf{\Phi}(s) = (s\mathbf{I} - \mathbf{A})^{-1} \tag{10.34}$$

Thus, from Eq. (10.33b),

$$\mathbf{Q}(s) = \mathbf{\Phi}(s)\mathbf{q}(0) + \mathbf{\Phi}(s)\mathbf{B}\mathbf{X}(s) \tag{10.35a}$$

and

$$\mathbf{q}(t) = \underbrace{\mathcal{L}^{-1}[\mathbf{\Phi}(s)]\mathbf{q}(0)}_{\text{zero-input component}} + \underbrace{\mathcal{L}^{-1}[\mathbf{\Phi}(s)\mathbf{B}\mathbf{X}(s)]}_{\text{zero-state component}} \tag{10.35b}$$

Equation (10.35b) gives the desired solution. Observe the two components of the solution. The first component yields $\mathbf{q}(t)$ when the input $x(t) = 0$. Hence the first component is the zero-input component. In a similar manner, we see that the second component is the zero-state component.

EXAMPLE 10.5

Find the state vector $\mathbf{q}(t)$ for the system whose state equation is given by

$$\dot{\mathbf{q}} = \mathbf{A}\mathbf{q} + \mathbf{B}\mathbf{x}$$

where

$$\mathbf{A} = \begin{bmatrix} -12 & \frac{2}{3} \\ -36 & -1 \end{bmatrix} \qquad \mathbf{B} = \begin{bmatrix} \frac{1}{3} \\ 1 \end{bmatrix} \qquad \mathbf{x}(t) = u(t)$$

and the initial conditions are $q_1(0) = 2$, $q_2(0) = 1$.

From Eq. (10.33b), we have

$$\mathbf{Q}(s) = \mathbf{\Phi}(s)[\mathbf{q}(0) + \mathbf{B}X(s)]$$

Let us first find $\mathbf{\Phi}(s)$. We have

$$(s\mathbf{I} - \mathbf{A}) = s \begin{bmatrix} 1 & 0 \\ 0 & 1 \end{bmatrix} - \begin{bmatrix} -12 & \frac{2}{3} \\ -36 & -1 \end{bmatrix} = \begin{bmatrix} s+12 & -\frac{2}{3} \\ 36 & s+1 \end{bmatrix}$$

and

$$\mathbf{\Phi}(s) = (s\mathbf{I} - \mathbf{A})^{-1} = \begin{bmatrix} \frac{s+1}{(s+4)(s+9)} & \frac{2/3}{(s+4)(s+9)} \\ \frac{-36}{(s+4)(s+9)} & \frac{s+12}{(s+4)(s+9)} \end{bmatrix} \qquad (10.36a)$$

Now, $\mathbf{q}(0)$ is given as

$$\mathbf{q}(0) = \begin{bmatrix} 2 \\ 1 \end{bmatrix}$$

Also, $X(s) = 1/s$, and

$$\mathbf{B}X(s) = \begin{bmatrix} \frac{1}{3} \\ 1 \end{bmatrix} \frac{1}{s} = \begin{bmatrix} \frac{1}{3s} \\ \frac{1}{s} \end{bmatrix}$$

Therefore

$$\mathbf{q}(0) + \mathbf{B}X(s) = \begin{bmatrix} 2 + \frac{1}{3s} \\ 1 + \frac{1}{s} \end{bmatrix} = \begin{bmatrix} \frac{6s+1}{3s} \\ \frac{s+1}{s} \end{bmatrix}$$

and

$$\mathbf{Q}(s) = \mathbf{\Phi}(s)[\mathbf{q}(0) + \mathbf{B}\mathbf{X}(s)]$$

$$= \begin{bmatrix} \frac{s+1}{(s+4)(s+9)} & \frac{2/3}{(s+4)(s+9)} \\ \frac{-36}{(s+4)(s+9)} & \frac{s+12}{(s+4)(s+9)} \end{bmatrix} \begin{bmatrix} \frac{6s+1}{3s} \\ \frac{s+1}{s} \end{bmatrix}$$

$$= \begin{bmatrix} \frac{2s^2+3s+1}{s(s+4)(s+9)} \\ \frac{s-59}{(s+4)(s+9)} \end{bmatrix}$$

$$= \begin{bmatrix} \frac{1/36}{s} - \frac{21/20}{s+4} + \frac{136/45}{s+9} \\ \frac{-63/5}{s+4} + \frac{68/5}{s+9} \end{bmatrix}$$

The inverse Laplace transform of this equation yields

$$\begin{bmatrix} q_1(t) \\ q_2(t) \end{bmatrix} = \begin{bmatrix} \left(\frac{1}{36} - \frac{21}{20}e^{-4t} + \frac{136}{45}e^{-9t}\right)u(t) \\ \left(-\frac{63}{5}e^{-4t} + \frac{68}{5}e^{-9t}\right)u(t) \end{bmatrix} \qquad (10.36b)$$

COMPUTER EXAMPLE C10.2

Repeat Example 10.5 using MATLAB. [Caution: See the caution in Computer Example C10.1.]

```
>> syms s
>> A = [-12 2/3;-36 -1]; B = [1/3; 1]; q0 = [2;1]; X = 1/s;
>> q = ilaplace(inv(s*eye(2)-A)*(q0+B*X))
q =

[ 136/45*exp(-9*t)-21/20*exp(-4*t)+1/36]
[         68/5*exp(-9*t)-63/5*exp(-4*t)]
```

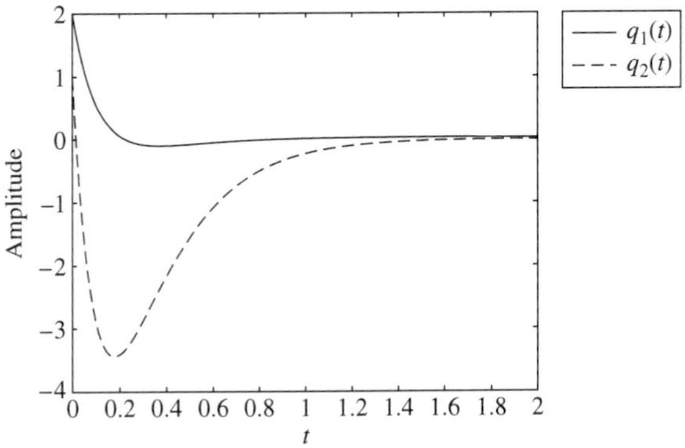

Figure C10.2

Next, a plot is generated of the state vector.

```
>> t = (0:.01:2)'; q = subs(q);
>> q1 = q(1:length(t)); q2 = q(length(t)+1:end);
>> plot(t,q1,'k',t,q2,'k--'); xlabel('t'); ylabel('Amplitude');
>> legend('q_1(t)','q_2(t)');
```

The plot is shown in Fig. C10.2.

THE OUTPUT

The output equation is given by

$$\mathbf{y} = \mathbf{Cq} + \mathbf{Dx}$$

and

$$\mathbf{Y}(s) = \mathbf{CQ}(s) + \mathbf{DX}(s)$$

Upon substituting Eq. (10.33b) into this equation, we have

$$\mathbf{Y}(s) = \mathbf{C}\{\mathbf{\Phi}(s)[\mathbf{q}(0) + \mathbf{Bx\,X}(s)]\} + \mathbf{DX}(s)$$

$$= \underbrace{\mathbf{C\Phi}(s)\mathbf{q}(0)}_{\text{zero-input response}} + \underbrace{[\mathbf{C\Phi}(s)\mathbf{B} + \mathbf{D}]\mathbf{X}(s)}_{\text{zero-state response}} \tag{10.37}$$

The zero-state response [i.e., the response $\mathbf{Y}(s)$ when $\mathbf{q}(0) = \mathbf{0}$] is given by

$$\mathbf{Y}(s) = [\mathbf{C\Phi}(s)\mathbf{B} + \mathbf{D}]\mathbf{X}(s) \tag{10.38a}$$

Note that the transfer function of a system is defined under the zero-state condition [see Eq. (4.32)]. The matrix $\mathbf{C\Phi}(s)\mathbf{B} + \mathbf{D}$ is the *transfer function matrix* $\mathbf{H}(s)$ of the system, which relates the responses $y_1, y_2, \ldots, y_k$ to the inputs $x_1, x_2, \ldots, x_j$:

$$\mathbf{H}(s) = \mathbf{C\Phi}(s)\mathbf{B} + \mathbf{D} \tag{10.38b}$$

and the zero-state response is

$$\mathbf{Y}(s) = \mathbf{H}(s)\mathbf{X}(s) \tag{10.39}$$

The matrix $\mathbf{H}(s)$ is a $k \times j$ matrix (k is the number of outputs and j is the number of inputs). The ijth element $H_{ij}(s)$ of $H(s)$ is the transfer function that relates the output $y_i(t)$ to the input $x_j(t)$.

EXAMPLE 10.6

Let us consider a system with a state equation

$$\begin{bmatrix} \dot{q}_1 \\ \dot{q}_2 \end{bmatrix} = \begin{bmatrix} 0 & 1 \\ -2 & -3 \end{bmatrix} \begin{bmatrix} q_1 \\ q_2 \end{bmatrix} + \begin{bmatrix} 1 & 0 \\ 1 & 1 \end{bmatrix} \begin{bmatrix} x_1 \\ x_2 \end{bmatrix} \tag{10.40a}$$

and an output equation

$$
\begin{bmatrix} y_1 \\ y_2 \\ y_3 \end{bmatrix} = \begin{bmatrix} 1 & 0 \\ 1 & 1 \\ 0 & 2 \end{bmatrix} \begin{bmatrix} q_1 \\ q_2 \end{bmatrix} + \begin{bmatrix} 0 & 0 \\ 1 & 0 \\ 0 & 1 \end{bmatrix} \begin{bmatrix} x_1 \\ x_2 \end{bmatrix} \qquad (10.40b)
$$

In this case,

$$
\mathbf{A} = \begin{bmatrix} 0 & 1 \\ -2 & -3 \end{bmatrix} \qquad \mathbf{B} = \begin{bmatrix} 1 & 0 \\ 1 & 1 \end{bmatrix} \qquad \mathbf{C} = \begin{bmatrix} 1 & 0 \\ 1 & 1 \\ 0 & 2 \end{bmatrix} \qquad \mathbf{D} = \begin{bmatrix} 0 & 0 \\ 1 & 0 \\ 0 & 1 \end{bmatrix} \qquad (10.40c)
$$

and

$$
\mathbf{\Phi}(s) = (s\mathbf{I} - \mathbf{A})^{-1} = \begin{bmatrix} s & -1 \\ 2 & s+3 \end{bmatrix}^{-1} = \begin{bmatrix} \frac{s+3}{(s+1)(s+2)} & \frac{1}{(s+1)(s+2)} \\ \frac{-2}{(s+1)(s+2)} & \frac{s}{(s+1)(s+2)} \end{bmatrix} \qquad (10.41)
$$

Hence, the transfer function matrix $\mathbf{H}(s)$ is given by

$$
\mathbf{H}(s) = \mathbf{C}\mathbf{\Phi}(s)\mathbf{B} + \mathbf{D}
$$

$$
= \begin{bmatrix} 1 & 0 \\ 1 & 1 \\ 0 & 2 \end{bmatrix} \begin{bmatrix} \frac{s+3}{(s+1)(s+2)} & \frac{1}{(s+1)(s+2)} \\ \frac{-2}{(s+1)(s+2)} & \frac{s}{(s+1)(s+2)} \end{bmatrix} \begin{bmatrix} 1 & 0 \\ 1 & 1 \end{bmatrix} + \begin{bmatrix} 0 & 0 \\ 1 & 0 \\ 0 & 1 \end{bmatrix}
$$

$$
= \begin{bmatrix} \frac{s+4}{(s+1)(s+2)} & \frac{1}{(s+1)(s+2)} \\ \frac{s+4}{s+2} & \frac{1}{s+2} \\ \frac{2(s-2)}{(s+1)(s+2)} & \frac{s^2+5s+2}{(s+1)(s+2)} \end{bmatrix} \qquad (10.42)
$$

and the zero-state response is

$$
\mathbf{Y}(s) = \mathbf{H}(s)\mathbf{X}(s)
$$

Remember that the ijth element of the transfer function matrix in Eq. (10.42) represents the transfer function that relates the output $y_i(t)$ to the input $x_j(t)$. For instance, the transfer function that relates the output y_3 to the input x_2 is $H_{32}(s)$, where

$$
H_{32}(s) = \frac{s^2 + 5s + 2}{(s+1)(s+2)}
$$

COMPUTER EXAMPLE C10.3

Repeat Example 10.6 using MATLAB.

```
>> A = [0 1;-2 -3]; B = [1 0;1 1];
>> C = [1 0;1 1;0 2]; D = [0 0;1 0;0 1];
>> syms s; H = simplify(C*inv(s*eye(2)-A)*B+D)

H =

[        (s+4)/(s^2+3*s+2),              1/(s^2+3*s+2)]
[               (s+4)/(s+2),                    1/(s+2)]
[    2*(-2+s)/(s^2+3*s+2), (5*s+s^2+2)/(s^2+3*s+2)]
```

Transfer functions relating particular inputs to particular outputs can be obtained using the
`ss2tf` function.

```
>> disp('Transfer function relating y_3 and x_2:')
>> [num,den] = ss2tf(A,B,C,D,2); tf(num(3,:),den)

Transfer function relating y_3 and x_2:

Transfer function:
s^2 + 5 s + 2
-------------
s^2 + 3 s + 2
```

CHARACTERISTIC ROOTS (EIGENVALUES) OF A MATRIX

It is interesting to observe that the denominator of every transfer function in Eq. (10.42) is
$(s + 1)(s + 2)$ except for $H_{21}(s)$ and $H_{22}(s)$, where the factor $(s + 1)$ is canceled. This is
no coincidence. We see that the denominator of every element of $\mathbf{\Phi}(s)$ is $|s\mathbf{I} - \mathbf{A}|$ because
$\mathbf{\Phi}(s) = (s\mathbf{I} - \mathbf{A})^{-1}$, and the inverse of a matrix has its determinant in the denominator. Since $\mathbf{C}$,
$\mathbf{B}$, and $\mathbf{D}$ are matrices with constant elements, we see from Eq. (10.38b) that the denominator of
$\mathbf{\Phi}(s)$ will also be the denominator of $\mathbf{H}(s)$. Hence, the denominator of every element of $\mathbf{H}(s)$ is
$|s\mathbf{I} - \mathbf{A}|$, except for the possible cancellation of the common factors mentioned earlier. In other
words, the zeros of the polynomial $|s\mathbf{I} - \mathbf{A}|$ are also the poles of all transfer functions of the
system. *Therefore, the zeros of the polynomial $|s\mathbf{I} - \mathbf{A}|$ are the characteristic roots of the system.*
Hence, the characteristic roots of the system are the roots of the equation

$$|s\mathbf{I} - \mathbf{A}| = 0 \tag{10.43a}$$

Since $|s\mathbf{I} - \mathbf{A}|$ is an Nth-order polynomial in s with N zeros $\lambda_1, \lambda_2, \ldots, \lambda_N$, we can write
Eq. (10.43a) as

$$|s\mathbf{I} - \mathbf{A}| = s^N + a_1 s^{N-1} + \cdots + a_{N-1}s + a_N$$

$$= (s - \lambda_1)(s - \lambda_2) \cdots (s - \lambda_N) = 0 \tag{10.43b}$$

For the system in Example 10.6,

$$|s\mathbf{I} - \mathbf{A}| = \begin{vmatrix} s & 0 \\ 0 & s \end{vmatrix} - \begin{vmatrix} 0 & 1 \\ -2 & -3 \end{vmatrix}$$

$$= \begin{vmatrix} s & -1 \\ 2 & s+3 \end{vmatrix}$$

$$= s^2 + 3s + 2 \tag{10.44a}$$

$$= (s+1)(s+2) \tag{10.44b}$$

Hence

$$\lambda_1 = -1 \quad \text{and} \quad \lambda_2 = -2$$

Equation (10.43a) is known as the *characteristic equation of the matrix* $\mathbf{A}$, and $\lambda_1, \lambda_2, \ldots, \lambda_N$ are the characteristic roots of $\mathbf{A}$. The term *eigenvalue*, meaning "characteristic value" in German, is also commonly used in the literature. Thus, we have shown that the characteristic roots of a system are the eigenvalues (characteristic values) of the matrix $\mathbf{A}$.

At this point, the reader will recall that if $\lambda_1, \lambda_2, \ldots, \lambda_N$ are the poles of the transfer function, then the zero-input response is of the form

$$y_0(t) = c_1 e^{\lambda_1 t} + c_2 e^{\lambda_2 t} + \cdots + c_N e^{\lambda_N t} \tag{10.45}$$

This fact is also obvious from Eq. (10.37). The denominator of every element of the zero-input response matrix $\mathbf{C\Phi}(s)\mathbf{q}(0)$ is $|s\mathbf{I} - \mathbf{A}| = (s - \lambda_1)(s - \lambda_2) \cdots (s - \lambda_N)$. Therefore, the partial fraction expansion and the subsequent inverse Laplace transform will yield a zero-input component of the form in Eq. (10.45).

10.3-2 Time-Domain Solution of State Equations

The state equation is

$$\dot{\mathbf{q}} = \mathbf{A}\mathbf{q} + \mathbf{B}\mathbf{x} \tag{10.46}$$

We now show that the solution of the vector differential Equation (10.46) is

$$\mathbf{q}(t) = e^{\mathbf{A}t}\mathbf{q}(0) + \int_0^t e^{\mathbf{A}(t-\tau)}\mathbf{B}\mathbf{x}(\tau)\,d\tau \tag{10.47}$$

Before proceeding further, we must define the exponential of the matrix appearing in Eq. (10.47). An exponential of a matrix is defined by an infinite series identical to that used in defining an exponential of a scalar. We shall define

$$e^{\mathbf{A}t} = \mathbf{I} + \mathbf{A}t + \frac{\mathbf{A}^2 t^2}{2!} + \frac{\mathbf{A}^3 t^3}{3!} + \cdots + \frac{\mathbf{A}^n t^n}{n!} + \cdots \tag{10.48a}$$

$$= \sum_{k=0}^{\infty} \frac{\mathbf{A}^k t^k}{k!} \tag{10.48b}$$

For example, if

$$\mathbf{A} = \begin{bmatrix} 0 & 1 \\ 2 & 1 \end{bmatrix}$$

then

$$\mathbf{A}t = \begin{bmatrix} 0 & 1 \\ 2 & 1 \end{bmatrix} t = \begin{bmatrix} 0 & t \\ 2t & t \end{bmatrix} \tag{10.49}$$

and

$$\frac{\mathbf{A}^2 t^2}{2!} = \begin{bmatrix} 0 & 1 \\ 2 & 1 \end{bmatrix}\begin{bmatrix} 0 & 1 \\ 2 & 1 \end{bmatrix}\frac{t^2}{2} = \begin{bmatrix} 2 & 1 \\ 2 & 3 \end{bmatrix}\frac{t^2}{2} = \begin{bmatrix} t^2 & \frac{t^2}{2} \\ t^2 & \frac{3t^2}{2} \end{bmatrix} \tag{10.50}$$

and so on.

We can show that the infinite series in Eq. (10.48a) is absolutely and uniformly convergent for all values of t. Consequently, it can be differentiated or integrated term by term. Thus, to find $(d/dt)e^{\mathbf{A}t}$, we differentiate the series on the right-hand side of Eq. (10.48a) term by term:

$$\frac{d}{dt}e^{\mathbf{A}t} = \mathbf{A} + \mathbf{A}^2 t + \frac{\mathbf{A}^3 t^2}{2!} + \frac{\mathbf{A}^4 t^3}{3!} + \cdots \tag{10.51a}$$

$$= \mathbf{A}\left[\mathbf{I} + \mathbf{A}t + \frac{\mathbf{A}^2 t^2}{2!} + \frac{\mathbf{A}^3 t^3}{3!} + \cdots\right]$$

$$= \mathbf{A}e^{\mathbf{A}t} \tag{10.51b}$$

The infinite series on the right-hand side of Eq. (10.51a) also may be expressed as

$$\frac{d}{dt}e^{\mathbf{A}t} = \left[\mathbf{I} + \mathbf{A}t + \frac{\mathbf{A}^2 t^2}{2!} + \frac{\mathbf{A}^3 t^3}{3!} + \cdots + \cdots\right]\mathbf{A}$$

$$= e^{\mathbf{A}t}\mathbf{A}$$

Hence

$$\frac{d}{dt}e^{\mathbf{A}t} = \mathbf{A}e^{\mathbf{A}t} = e^{\mathbf{A}t}\mathbf{A} \tag{10.52}$$

Also note that from the definition (10.48a), it follows that

$$e^0 = \mathbf{I} \tag{10.53a}$$

where

$$\mathbf{I} = \begin{bmatrix} 1 & 0 \\ 0 & 1 \end{bmatrix}$$

If we premultiply or postmultiply the infinite series for $e^{\mathbf{A}t}$ [Eq. (10.48a)] by an infinite series for $e^{-\mathbf{A}t}$, we find that

$$(e^{-\mathbf{A}t})(e^{\mathbf{A}t}) = (e^{\mathbf{A}t})(e^{-\mathbf{A}t}) = \mathbf{I} \tag{10.53b}$$

In Section B.6-3, we showed that

$$\frac{d}{dt}(\mathbf{UV}) = \frac{d\mathbf{U}}{dt}\mathbf{V} + \mathbf{U}\frac{d\mathbf{V}}{dt}$$

Using this relationship, we observe that

$$\frac{d}{dt}[e^{-\mathbf{A}t}\mathbf{q}] = \left(\frac{d}{dt}e^{-\mathbf{A}t}\right)\mathbf{q} + e^{-\mathbf{A}t}\dot{\mathbf{q}}$$

$$= -e^{-\mathbf{A}t}\mathbf{A}\mathbf{q} + e^{-\mathbf{A}t}\dot{\mathbf{q}} \qquad (10.54)$$

We now premultiply both sides of Eq. (10.46) by $e^{-\mathbf{A}t}$ to yield

$$e^{-\mathbf{A}t}\dot{\mathbf{q}} = e^{-\mathbf{A}t}\mathbf{A}\mathbf{q} + e^{-\mathbf{A}t}\mathbf{B}\mathbf{x} \qquad (10.55a)$$

or

$$e^{-\mathbf{A}t}\dot{\mathbf{q}} - e^{-\mathbf{A}t}\mathbf{A}\mathbf{q} = e^{-\mathbf{A}t}\mathbf{B}\mathbf{x} \qquad (10.55b)$$

A glance at Eq. (10.54) shows that the left-hand side of Eq. (10.55b) is

$$\frac{d}{dt}[e^{-\mathbf{A}t}]$$

Hence

$$\frac{d}{dt}[e^{-\mathbf{A}t}] = e^{-\mathbf{A}t}\mathbf{B}\mathbf{x}$$

The integration of both sides of this equation from 0 to t yields

$$e^{-\mathbf{A}t}\mathbf{q}\Big|_0^t = \int_0^t e^{-\mathbf{A}\tau}\mathbf{B}\mathbf{x}(\tau)\,d\tau \qquad (10.56a)$$

or

$$e^{-\mathbf{A}t}\mathbf{q}(t) - \mathbf{q}(0) = \int_0^t e^{-\mathbf{A}\tau}\mathbf{B}\mathbf{x}(\tau)\,d\tau \qquad (10.56b)$$

Hence

$$e^{-\mathbf{A}t}\mathbf{q} = \mathbf{q}(0) + \int_0^t e^{-\mathbf{A}\tau}\mathbf{B}\mathbf{x}(\tau)\,d\tau \qquad (10.56c)$$

Premultiplying Eq. (10.56c) by $e^{\mathbf{A}t}$ and using Eq. (10.53b), we have

$$\mathbf{q}(t) = \underbrace{e^{\mathbf{A}t}\mathbf{q}(0)}_{\text{zero-input component}} + \underbrace{\int_0^t e^{\mathbf{A}(t-\tau)}\mathbf{B}\mathbf{x}(\tau)\,d\tau}_{\text{zero-state component}} \qquad (10.57a)$$

This is the desired solution. The first term on the right-hand side represents $q(t)$ when the input $x(t) = 0$. Hence it is the zero-input component. The second term, by a similar argument, is seen to be the zero-state component.

The results of Eq. (10.57a) can be expressed more conveniently in terms of the matrix convolution. We can define the convolution of two matrices in a manner similar to the multiplication of two matrices, except that the multiplication of two elements is replaced by their convolution.

For example,

$$\begin{bmatrix} x_1 & x_2 \\ x_3 & x_4 \end{bmatrix} * \begin{bmatrix} g_1 & g_2 \\ g_3 & g_4 \end{bmatrix} = \begin{bmatrix} (x_1 * g_1 + x_2 * g_3) & (x_1 * g_2 + x_2 * g_4) \\ (x_3 * g_1 + x_4 * g_3) & (x_3 * g_2 + x_4 * g_4) \end{bmatrix}$$

By using this definition of matrix convolution, we can express Eq. (10.57a) as

$$\mathbf{q}(t) = e^{\mathbf{A}t}\mathbf{q}(0) + e^{\mathbf{A}t} * \mathbf{B}\mathbf{x}(t) \tag{10.57b}$$

Note that the limits of the convolution integral [Eq. (10.57a)] are from 0 to t. Hence, all the elements of $e^{\mathbf{A}t}$ in the convolution term of Eq. (10.57b) are implicitly assumed to be multiplied by $u(t)$.

The result of Eqs. (10.57) can be easily generalized for any initial value of t. It is left as an exercise for the reader to show that the solution of the state equation can be expressed as

$$\mathbf{q}(t) = e^{\mathbf{A}(t-t_0)}\mathbf{q}(t_0) + \int_{t_0}^{t} e^{\mathbf{A}(t-\tau)}\mathbf{B}\mathbf{x}(\tau)\,d\tau \tag{10.58}$$

Determining $e^{\mathbf{A}t}$

The exponential $e^{\mathbf{A}t}$ required in Eqs. (10.57) can be computed from the definition in Eq. (10.48a). Unfortunately, this is an infinite series, and its computation can be quite laborious. Moreover, we may not be able to recognize the closed-form expression for the answer. There are several efficient methods of determining $e^{\mathbf{A}t}$ in closed form. It was shown in Section B.6-5 that for an $N \times N$ matrix $\mathbf{A}$,

$$e^{\mathbf{A}t} = \beta_0\mathbf{I} + \beta_1\mathbf{A} + \beta_2\mathbf{A}^2 + \cdots + \beta_{N-1}\mathbf{A}^{N-1} \tag{10.59a}$$

where

$$\begin{bmatrix} \beta_0 \\ \beta_1 \\ \vdots \\ \beta_{N-1} \end{bmatrix} = \begin{bmatrix} 1 & \lambda_1 & \lambda_1^2 & \cdots & \lambda_1^{N-1} \\ 1 & \lambda_2 & \lambda_2^2 & \cdots & \lambda_2^{N-1} \\ \vdots & \vdots & \vdots & \cdots & \vdots \\ 1 & \lambda_N & \lambda_N^2 & \cdots & \lambda_N^{N-1} \end{bmatrix}^{-1} \begin{bmatrix} e^{\lambda_1 t} \\ e^{\lambda_2 t} \\ \vdots \\ e^{\lambda_N t} \end{bmatrix}$$

and $\lambda_1, \lambda_2, \ldots, \lambda_N$ are the N characteristic values (eigenvalues) of $\mathbf{A}$.

We can also determine $e^{\mathbf{A}t}$ by comparing Eqs. (10.57a) and (10.35b). It is clear that

$$e^{\mathbf{A}t} = \mathcal{L}^{-1}[\mathbf{\Phi}(s)] \tag{10.59b}$$

$$= \mathcal{L}^{-1}[(s\mathbf{I} - \mathbf{A})^{-1}] \tag{10.59c}$$

Thus, $e^{\mathbf{A}t}$ and $\mathbf{\Phi}(s)$ are a Laplace transform pair. To be consistent with Laplace transform notation, $e^{\mathbf{A}t}$ is often denoted by $\boldsymbol{\phi}(t)$, *the state transition matrix* (STM):

$$e^{\mathbf{A}t} = \boldsymbol{\phi}(t)$$

EXAMPLE 10.7

Use the time-domain method to find the solution to the problem in Example 10.5.

For this case, the characteristic roots are given by

$$|s\mathbf{I} - \mathbf{A}| = \begin{vmatrix} s + 12 & -\frac{2}{3} \\ 36 & s + 1 \end{vmatrix} = s^2 + 13s + 36 = (s + 4)(s + 9) = 0$$

The roots are $\lambda_1 = -4$ and $\lambda_2 = -9$, so

$$\begin{bmatrix} \beta_0 \\ \beta_1 \end{bmatrix} = \begin{bmatrix} 1 & -4 \\ 1 & -9 \end{bmatrix}^{-1} \begin{bmatrix} e^{-4t} \\ e^{-9t} \end{bmatrix} = \frac{1}{5} \begin{bmatrix} 9e^{-4t} - 4e^{-9t} \\ e^{-4t} - e^{-9t} \end{bmatrix}$$

and

$$e^{\mathbf{A}t} = \beta_0 \mathbf{I} + \beta_1 \mathbf{A}$$

$$= \left(\tfrac{9}{5}e^{-4t} - \tfrac{4}{5}e^{-9t}\right) \begin{bmatrix} 1 & 0 \\ 0 & 1 \end{bmatrix} + \left(\tfrac{1}{5}e^{-4t} - \tfrac{1}{5}e^{-9t}\right) \begin{bmatrix} -12 & \tfrac{2}{3} \\ -36 & -1 \end{bmatrix}$$

$$= \begin{bmatrix} \left(\tfrac{-3}{5}e^{-4t} + \tfrac{8}{5}e^{-9t}\right) & \tfrac{2}{15}(e^{-4t} - e^{-9t}) \\ \tfrac{36}{5}(-e^{-4t} + e^{-9t}) & \left(\tfrac{8}{5}e^{-4t} - \tfrac{3}{5}e^{-9t}\right) \end{bmatrix} \tag{10.60}$$

The zero-input component is given by [see Eq. (10.57a)]

$$e^{\mathbf{A}t}\mathbf{q}(0) = \begin{bmatrix} \left(-\tfrac{3}{5}e^{-4t} + \tfrac{8}{5}e^{-9t}\right) & \tfrac{2}{15}(e^{-4t} - e^{-9t}) \\ \tfrac{36}{5}(-e^{-4t} + e^{-9t}) & \left(\tfrac{8}{5}e^{-4t} - \tfrac{3}{5}e^{-9t}\right) \end{bmatrix} \begin{bmatrix} 2 \\ 1 \end{bmatrix}$$

$$= \begin{bmatrix} \left(\tfrac{-16}{15}e^{-4t} + \tfrac{46}{15}e^{-9t}\right)u(t) \\ \left(\tfrac{-64}{5}e^{-4t} + \tfrac{69}{5}e^{-9t}\right)u(t) \end{bmatrix} \tag{10.61a}$$

Note the presence of $u(t)$ in Eq. (10.61a), indicating that the response begins at $t = 0$. The zero-state component is $e^{\mathbf{A}t} * \mathbf{B}x$ [see Eq. (10.57b)], where

$$\mathbf{B}x = \begin{bmatrix} \tfrac{1}{3} \\ 1 \end{bmatrix} u(t) = \begin{bmatrix} \tfrac{1}{3}u(t) \\ u(t) \end{bmatrix}$$

and

$$e^{\mathbf{A}t} * \mathbf{B}x(t) = \begin{bmatrix} \left(\tfrac{-3}{5}e^{-4t} + \tfrac{8}{5}e^{-9t}\right)u(t) & \tfrac{2}{15}(e^{-4t} - e^{-9t})u(t) \\ \tfrac{36}{5}(-e^{-4t} + e^{-9t}u(t)) & \left(\tfrac{8}{5}e^{-4t} - \tfrac{3}{5}e^{-9t}\right)u(t) \end{bmatrix} * \begin{bmatrix} \tfrac{1}{3}u(t) \\ u(t) \end{bmatrix}$$

Note again the presence of the term $u(t)$ in every element of $e^{\mathbf{A}t}$. This is the case because the limits of the convolution integral run from 0 to t [Eqs. (10.56)]. Thus

$$e^{\mathbf{A}t} * \mathbf{Bx}(t) = \begin{bmatrix} \left(-\frac{3}{5}e^{-4t} + \frac{8}{5}e^{-9t}\right)u(t) * \frac{1}{3}u(t) & \frac{2}{15}(e^{-4t} - e^{-9t})u(t) * u(t) \\ \frac{36}{5}(-e^{-4t} + e^{-9t})u(t) * \frac{1}{3}u(t) & \left(\frac{8}{5}e^{-4t} - \frac{3}{5}e^{-9t}\right)u(t) * u(t) \end{bmatrix}$$

$$= \begin{bmatrix} -\frac{1}{15}e^{-4t}u(t) * u(t) + \frac{2}{5}e^{-9t}u(t) * u(t) \\ -\frac{4}{5}e^{-4t}u(t) * u(t) + \frac{9}{5}e^{-9t}u(t) * u(t) \end{bmatrix}$$

Substitution for the preceding convolution integrals from the convolution table (Table 2.1) yields

$$e^{\mathbf{A}t} * \mathbf{Bx}(t) = \begin{bmatrix} -\frac{1}{60}(1 - e^{-4t})u(t) + \frac{2}{45}(1 - e^{-9t})u(t) \\ -\frac{1}{5}(1 - e^{-4t})u(t) + \frac{1}{5}(1 - e^{-9t})u(t) \end{bmatrix}$$

$$= \begin{bmatrix} \left(\frac{1}{36} + \frac{1}{60}e^{-4t} - \frac{2}{45}e^{-9t}\right)u(t) \\ \frac{1}{5}(e^{-4t} - e^{-9t})u(t) \end{bmatrix} \tag{10.61b}$$

The sum of the two components [Eq. (10.61a) and Eq. (10.61b)] now gives the desired solution for $\mathbf{q}(t)$:

$$\mathbf{q}(t) = \begin{bmatrix} q_1(t) \\ q_2(t) \end{bmatrix} = \begin{bmatrix} \left(\frac{1}{36} - \frac{21}{20}e^{-4t} + \frac{136}{45}e^{-9t}\right)u(t) \\ \left(\frac{-63}{5}e^{-4t} + \frac{68}{5}e^{-9t}\right)u(t) \end{bmatrix} \tag{10.61c}$$

This result confirms the solution obtained by using the frequency-domain method [see Eq. (10.36b)]. Once the state variables q_1 and q_2 have been found for $t \geq 0$, all the remaining variables can be determined from the output equation.

THE OUTPUT

The output equation is given by

$$\mathbf{y}(t) = \mathbf{Cq}(t) + \mathbf{Dx}(t)$$

The substitution of the solution for $\mathbf{q}$ [Eq. (10.57b)] in this equation yields

$$\mathbf{y}(t) = \mathbf{C}[e^{\mathbf{A}t}\mathbf{q}(0) + e^{\mathbf{A}t} * \mathbf{Bx}(t)] + \mathbf{Dx}(t) \tag{10.62a}$$

Since the elements of $\mathbf{B}$ are constants,

$$e^{\mathbf{A}t} * \mathbf{Bx}(t) = e^{\mathbf{A}t}\mathbf{B} * \mathbf{x}(t)$$

With this result, Eq. (10.62a) becomes

$$\mathbf{y}(t) = \mathbf{C}[e^{\mathbf{A}t}\mathbf{q}(0) + e^{\mathbf{A}t}\mathbf{B} * \mathbf{x}(t)] + \mathbf{Dx}(t) \tag{10.62b}$$

Now recall that the convolution of $x(t)$ with the unit impulse $\delta(t)$ yields $x(t)$. Let us define a $j \times j$ diagonal matrix $\boldsymbol{\delta}(t)$ such that all its diagonal terms are unit impulse functions. It is then obvious that

$$\boldsymbol{\delta}(t) * \mathbf{x}(t) = \mathbf{x}(t)$$

and Eq. (10.62b) can be expressed as

$$\mathbf{y}(t) = \mathbf{C}[e^{\mathbf{A}t}\mathbf{q}(0) + e^{\mathbf{A}t}\mathbf{B} * \mathbf{x}(t)] + \mathbf{D}\boldsymbol{\delta}(t) * \mathbf{x}(t) \qquad (10.63a)$$

$$= \mathbf{C}e^{\mathbf{A}t}\mathbf{q}(0) + [\mathbf{C}e^{\mathbf{A}t}\mathbf{B} + \mathbf{D}\boldsymbol{\delta}(t)] * \mathbf{x}(t) \qquad (10.63b)$$

With the notation $\boldsymbol{\phi}(t)$ for $e^{\mathbf{A}t}$, Eq. (10.63b) may be expressed as

$$\mathbf{y}(t) = \underbrace{\mathbf{C}\boldsymbol{\phi}(t)\mathbf{q}(0)}_{\text{zero-input response}} + \underbrace{[\mathbf{C}\boldsymbol{\phi}(t)\mathbf{B} + \mathbf{D}\boldsymbol{\delta}(t)] * \mathbf{x}(t)}_{\text{zero-state response}} \qquad (10.63c)$$

The zero-state response, that is, the response when $\mathbf{q}(0) = \mathbf{0}$, is

$$\mathbf{y}(t) = [\mathbf{C}\boldsymbol{\phi}(t)\mathbf{B} + \mathbf{D}\boldsymbol{\delta}(t)] * \mathbf{x}(t) \qquad (10.64a)$$

$$= \mathbf{h}(t) * \mathbf{x}(t) \qquad (10.64b)$$

where

$$\mathbf{h}(t) = \mathbf{C}\boldsymbol{\phi}(t)\mathbf{B} + \mathbf{D}\boldsymbol{\delta}(t) \qquad (10.65)$$

The matrix $\mathbf{h}(t)$ is a $k \times j$ matrix known as the *impulse response matrix*. The reason for this designation is obvious. The ijth element of $\mathbf{h}(t)$ is $h_{ij}(t)$, which represents the zero-state response y_i when the input $x_j(t) = \delta(t)$ and when all other inputs (and all the initial conditions) are zero. It can also be seen from Eq. (10.39) and (10.64b) that

$$\mathcal{L}[\mathbf{h}(t)] = \mathbf{H}(s)$$

EXAMPLE 10.8

For the system described by Eqs. (10.40a) and (10.40b), use Eq. (10.59b) to determine $e^{\mathbf{A}t}$:

$$\boldsymbol{\phi}(t) = e^{\mathbf{A}t} = \mathcal{L}^{-1}\boldsymbol{\Phi}(s)$$

This problem was solved earlier with frequency-domain techniques. From Eq. (10.41), we have

$$\boldsymbol{\phi}(t) = \mathcal{L}^{-1} \begin{bmatrix} \frac{s+3}{(s+1)(s+2)} & \frac{1}{(s+1)(s+2)} \\ \frac{-2}{(s+1)(s+2)} & \frac{s}{(s+1)(s+2)} \end{bmatrix}$$

$$= \mathcal{L}^{-1} \begin{bmatrix} \frac{2}{s+1} - \frac{1}{s+2} & \frac{1}{s+1} - \frac{1}{s+2} \\ \frac{-2}{s+1} + \frac{2}{s+2} & \frac{-1}{s+1} + \frac{2}{s+2} \end{bmatrix}$$

$$= \begin{bmatrix} 2e^{-t} - e^{-2t} & e^{-t} - e^{-2t} \\ -2e^{-t} + 2e^{-2t} & -e^{-t} + 2e^{-2t} \end{bmatrix}$$

The same result is obtained in Example B.13 (Section B.6-5) by using Eq. (10.59a) [see Eq. (B.84)].

Also, $\boldsymbol{\delta}(t)$ is a diagonal $j \times j$ or 2×2 matrix:

$$\boldsymbol{\delta}(t) = \begin{bmatrix} \delta(t) & 0 \\ 0 & \delta(t) \end{bmatrix}$$

Substituting the matrices $\boldsymbol{\phi}(t)$, $\boldsymbol{\delta}(t)$, $\mathbf{C}$, $\mathbf{D}$, and $\mathbf{B}$ [Eq. (10.40c)] into Eq. (10.65), we have

$$\mathbf{h}(t) = \begin{bmatrix} 1 & 0 \\ 1 & 1 \\ 0 & 2 \end{bmatrix} \begin{bmatrix} 2e^{-t} - e^{-2t} & e^{-t} - e^{-2t} \\ -2e^{-t} + 2e^{-2t} & -e^{-+2} + e^{-2t} \end{bmatrix} \begin{bmatrix} 1 & 0 \\ 1 & 1 \end{bmatrix} + \begin{bmatrix} 0 & 0 \\ 1 & 0 \\ 0 & 1 \end{bmatrix} \begin{bmatrix} \delta(t) & 0 \\ 0 & \delta(t) \end{bmatrix}$$

$$= \begin{bmatrix} 3e^{-t} - 2e^{-2t} & e^{-t} - e^{-2t} \\ \delta(t) + 2e^{-2t} & e^{-2t} \\ -6e^{-t} + 8e^{-2t} & \delta(t) - 2e^{-2t} + 4e^{-2t} \end{bmatrix} \qquad (10.66)$$

The reader can verify that the transfer function matrix $\mathbf{H}(s)$ in Eq. (10.42) is the Laplace transform of the unit impulse response matrix $\mathbf{h}(t)$ in Eq. (10.66).

10.4 LINEAR TRANSFORMATION OF STATE VECTOR

In Section 10.1 we saw that the state of a system can be specified in several ways. The sets of all possible state variables are related—in other words, if we are given one set of state variables, we should be able to relate it to any other set. We are particularly interested in a linear type of relationship. Let $q_1, q_2, \ldots, q_N$ and $w_1, w_2, \ldots, w_N$ be two different sets of state variables specifying the same system. Let these sets be related by linear equations as

$$w_1 = p_{11}q_1 + p_{12}q_2 + \cdots + p_{1N}q_N$$

$$w_2 = p_{21}q_1 + p_{22}q_2 + \cdots + p_{2N}q_N$$

$$\vdots \qquad (10.67a)$$

$$w_N = p_{N1}q_1 + p_{N2}q_2 + \cdots + p_{NN}q_N$$

or

$$\underbrace{\begin{bmatrix} w_1 \\ w_2 \\ \vdots \\ w_N \end{bmatrix}}_{\mathbf{w}} = \underbrace{\begin{bmatrix} p_{11} & p_{12} & \cdots & p_{1N} \\ p_{21} & p_{22} & \cdots & p_{2N} \\ \vdots & \vdots & \cdots & \vdots \\ p_{N1} & p_{N2} & \cdots & p_{NN} \end{bmatrix}}_{\mathbf{P}} \underbrace{\begin{bmatrix} q_1 \\ q_2 \\ \vdots \\ q_N \end{bmatrix}}_{\mathbf{q}} \qquad (10.67b)$$

Defining the vector $\mathbf{w}$ and matrix $\mathbf{P}$ as just shown, we can write Eq. (10.67b) as

$$\mathbf{w} = \mathbf{Pq} \qquad (10.67c)$$

and

$$\mathbf{q} = \mathbf{P}^{-1}\mathbf{w} \tag{10.67d}$$

Thus, the state vector $\mathbf{q}$ is transformed into another state vector $\mathbf{w}$ through the linear transformation in Eq. (10.67c).

If we know $\mathbf{w}$, we can determine $\mathbf{q}$ from Eq. (10.67d), provided $\mathbf{P}^{-1}$ exists. This is equivalent to saying that $\mathbf{P}$ is a nonsingular matrix[†] ($|\mathbf{P}| \neq 0$). Thus, if $\mathbf{P}$ is a nonsingular matrix, the vector $\mathbf{w}$ defined by Eq. (10.67c) is also a state vector. Consider the state equation of a system

$$\dot{\mathbf{q}} = \mathbf{A}\mathbf{q} + \mathbf{B}\mathbf{x} \tag{10.68a}$$

If

$$\mathbf{w} = \mathbf{P}\mathbf{q} \tag{10.68b}$$

then

$$\mathbf{q} = \mathbf{P}^{-1}\mathbf{w}$$

and

$$\dot{\mathbf{q}} = \mathbf{P}^{-1}\dot{\mathbf{w}}$$

Hence the state equation (10.68a) now becomes

$$\mathbf{P}^{-1}\dot{\mathbf{w}} = \mathbf{A}\mathbf{P}^{-1}\mathbf{w} + \mathbf{B}\mathbf{x}$$

or

$$\dot{\mathbf{w}} = \mathbf{P}\mathbf{A}\mathbf{P}^{-1}\mathbf{w} + \mathbf{P}\mathbf{B}\mathbf{x} \tag{10.68c}$$

$$= \hat{\mathbf{A}}\mathbf{w} + \hat{\mathbf{B}}\mathbf{x} \tag{10.68d}$$

where

$$\hat{\mathbf{A}} = \mathbf{P}\mathbf{A}\mathbf{P}^{-1} \tag{10.69a}$$

and

$$\hat{\mathbf{B}} = \mathbf{P}\mathbf{B} \tag{10.69b}$$

Equation (10.68d) is a state equation for the same system, but now it is expressed in terms of the state vector $\mathbf{w}$.

The output equation is also modified. Let the original output equation be

$$\mathbf{y} = \mathbf{C}\mathbf{q} + \mathbf{D}\mathbf{x}$$

In terms of the new state variable $\mathbf{w}$, this equation becomes

$$\mathbf{y} = \mathbf{C}(\mathbf{P}^{-1}\mathbf{w}) + \mathbf{D}\mathbf{x}$$

$$= \hat{\mathbf{C}}\mathbf{w} + \mathbf{D}\mathbf{x}$$

where

$$\hat{\mathbf{C}} = \mathbf{C}\mathbf{P}^{-1} \tag{10.69c}$$

[†]This condition is equivalent to saying that all N equations in Eq. (10.67a) are linearly independent; that is, none of the N equations can be expressed as a linear combination of the remaining equations.

EXAMPLE 10.9

The state equations of a certain system are given by

$$\begin{bmatrix} \dot{q}_1 \\ \dot{q}_2 \end{bmatrix} = \begin{bmatrix} 0 & 1 \\ -2 & -3 \end{bmatrix} \begin{bmatrix} q_1 \\ q_2 \end{bmatrix} + \begin{bmatrix} 1 \\ 2 \end{bmatrix} x(t) \tag{10.70a}$$

Find the state equations for this system when the new state variables w_1 and w_2 are

$$w_1 = q_1 + q_2$$
$$w_2 = q_1 - q_2$$

or

$$\begin{bmatrix} w_1 \\ w_2 \end{bmatrix} = \begin{bmatrix} 1 & 1 \\ 1 & -1 \end{bmatrix} \begin{bmatrix} q_1 \\ q_2 \end{bmatrix} \tag{10.70b}$$

According to Eq. (10.68d), the state equation for the state variable $\mathbf{w}$ is given by

$$\dot{\mathbf{w}} = \hat{\mathbf{A}}\mathbf{w} + \hat{\mathbf{B}}\mathbf{x}$$

where [see Eqs. (10.69)]

$$\hat{\mathbf{A}} = \mathbf{P}\mathbf{A}\mathbf{P}^{-1} = \begin{bmatrix} 1 & 1 \\ 1 & -1 \end{bmatrix} \begin{bmatrix} 0 & 1 \\ -2 & -3 \end{bmatrix} \begin{bmatrix} 1 & 1 \\ 1 & -1 \end{bmatrix}^{-1}$$

$$= \begin{bmatrix} 1 & 1 \\ 1 & -1 \end{bmatrix} \begin{bmatrix} 0 & 1 \\ -2 & -3 \end{bmatrix} \begin{bmatrix} \frac{1}{2} & \frac{1}{2} \\ \frac{1}{2} & -\frac{1}{2} \end{bmatrix}$$

$$= \begin{bmatrix} -2 & 0 \\ 3 & -1 \end{bmatrix}$$

and

$$\hat{\mathbf{B}} = \mathbf{P}\mathbf{B} = \begin{bmatrix} 1 & 1 \\ 1 & -1 \end{bmatrix} \begin{bmatrix} 1 \\ 2 \end{bmatrix} = \begin{bmatrix} 3 \\ -1 \end{bmatrix}$$

Therefore

$$\begin{bmatrix} \dot{w}_1 \\ \dot{w}_2 \end{bmatrix} = \begin{bmatrix} -2 & 0 \\ 3 & -1 \end{bmatrix} \begin{bmatrix} w_1 \\ w_2 \end{bmatrix} + \begin{bmatrix} 3 \\ -1 \end{bmatrix} x(t)$$

This is the desired state equation for the state vector $\mathbf{w}$. The solution of this equation requires a knowledge of the initial state $\mathbf{w}(0)$. This can be obtained from the given initial state $\mathbf{q}(0)$ by using Eq. (10.70b).

COMPUTER EXAMPLE C10.4

Repeat Example 10.9 using MATLAB.

```
>> A = [0 1;-2 -3]; B = [1; 2];
>> P = [1 1;1 -1];
>> Ahat = P*A*inv(P), Bhat = P*B

Ahat =        -2      0
               3     -1

Bhat =         3
              -1
```

Therefore,

$$\begin{bmatrix} \dot{w}_1 \\ \dot{w}_2 \end{bmatrix} = \begin{bmatrix} -2 & 0 \\ 3 & -1 \end{bmatrix} \begin{bmatrix} w_1 \\ w_2 \end{bmatrix} + \begin{bmatrix} 3 \\ -1 \end{bmatrix} x(t)$$

INVARIANCE OF EIGENVALUES

We have seen that the poles of all possible transfer functions of a system are the eigenvalues of the matrix $\mathbf{A}$. If we transform a state vector from $\mathbf{q}$ to $\mathbf{w}$, the variables $w_1, w_2, \ldots, w_N$ are linear combinations of $q_1, q_2, \ldots, q_N$ and therefore may be considered to be outputs. Hence, the poles of the transfer functions relating $w_1, w_2, \ldots, w_N$ to the various inputs must also be the eigenvalues of matrix $\mathbf{A}$. On the other hand, the system is also specified by Eq. (10.68d). This means that the poles of the transfer functions must be the eigenvalues of $\hat{\mathbf{A}}$. Therefore, the eigenvalues of matrix $\mathbf{A}$ remain unchanged for the linear transformation of variables represented by Eq. (10.67), and the eigenvalues of matrix $\mathbf{A}$ and matrix $\hat{\mathbf{A}}(\hat{\mathbf{A}} = \mathbf{PAP}^{-1})$ are identical, implying that the characteristic equations of $\mathbf{A}$ and $\hat{\mathbf{A}}$ are also identical. This result also can be proved alternately as follows.

Consider the matrix $\mathbf{P}(s\mathbf{I} - \mathbf{A})\mathbf{P}^{-1}$. We have

$$\mathbf{P}(s\mathbf{I} - \mathbf{A})\mathbf{P}^{-1} = \mathbf{P}s\mathbf{I}\mathbf{P}^{-1} - \mathbf{PAP}^{-1} = s\mathbf{PIP}^{-1} - \hat{\mathbf{A}} = s\mathbf{I} - \hat{\mathbf{A}}$$

Taking the determinants of both sides, we obtain

$$|\mathbf{P}||s\mathbf{I} - \mathbf{A}||\mathbf{P}^{-1}| = |s\mathbf{I} - \hat{\mathbf{A}}|$$

The determinants $|\mathbf{P}|$ and $|\mathbf{P}^{-1}|$ are reciprocals of each other. Hence

$$|s\mathbf{I} - \mathbf{A}| = |s\mathbf{I} - \hat{\mathbf{A}}| \tag{10.71}$$

This is the desired result. We have shown that the characteristic equations of $\mathbf{A}$ and $\hat{\mathbf{A}}$ are identical. Hence the eigenvalues of $\mathbf{A}$ and $\hat{\mathbf{A}}$ are identical.

In Example 10.9, matrix $\mathbf{A}$ is given as

$$\mathbf{A} = \begin{bmatrix} 0 & 1 \\ -2 & -3 \end{bmatrix}$$

The characteristic equation is

$$|s\mathbf{I} - \mathbf{A}| = \begin{vmatrix} s & -1 \\ 2 & s+3 \end{vmatrix} = s^2 + 3s + 2 = 0$$

Also

$$\hat{\mathbf{A}} = \begin{bmatrix} -2 & 0 \\ 3 & -1 \end{bmatrix}$$

and

$$|s\mathbf{I} - \hat{\mathbf{A}}| = \begin{bmatrix} s+2 & 0 \\ -3 & s+1 \end{bmatrix} = s^2 + 3s + 2 = 0$$

This result verifies that the characteristic equations of $\mathbf{A}$ and $\hat{\mathbf{A}}$ are identical.

10.4-1 Diagonalization of Matrix A

For several reasons, it is desirable to make matrix $\mathbf{A}$ diagonal. If $\mathbf{A}$ is not diagonal, we can transform the state variables such that the resulting matrix $\hat{\mathbf{A}}$ is diagonal.[†] One can show that for any diagonal matrix $\mathbf{A}$, the diagonal elements of this matrix must necessarily be $\lambda_1, \lambda_2, \ldots, \lambda_N$ (the eigenvalues) of the matrix. Consider the diagonal matrix $\mathbf{A}$:

$$\mathbf{A} = \begin{bmatrix} a_1 & 0 & 0 & \cdots & 0 \\ 0 & a_2 & 0 & \cdots & 0 \\ \vdots & \vdots & \vdots & \cdots & \vdots \\ 0 & 0 & 0 & \cdots & a_N \end{bmatrix}$$

The characteristic equation is given by

$$|s\mathbf{I} - \mathbf{A}| = \begin{bmatrix} (s-a_1) & 0 & 0 & \cdots & 0 \\ 0 & (s-a_2) & 0 & \cdots & 0 \\ \vdots & \vdots & \vdots & \cdots & \vdots \\ 0 & 0 & 0 & \cdots & (s-a_N) \end{bmatrix} = 0$$

or

$$(s-a_1)(s-a_2)\cdots(s-a_N) = 0$$

Hence, the eigenvalues of $\mathbf{A}$ are $a_1, a_2, \ldots, a_N$. The nonzero (diagonal) elements of a diagonal matrix are therefore its eigenvalues $\lambda_1, \lambda_2, \ldots, \lambda_N$. We shall denote the diagonal matrix by a

---

[†]In this discussion we assume distinct eigenvalues. If the eigenvalues are not distinct, we can reduce the matrix to a modified diagonalized (Jordan) form.

special symbol, Λ:

$$\Lambda = \begin{bmatrix} \lambda_1 & 0 & 0 & \cdots & 0 \\ 0 & \lambda_2 & 0 & \cdots & 0 \\ \vdots & \vdots & \vdots & \cdots & \vdots \\ 0 & 0 & 0 & \cdots & \lambda_N \end{bmatrix} \tag{10.72}$$

Let us now consider the transformation of the state vector $\mathbf{A}$ such that the resulting matrix $\hat{\mathbf{A}}$ is a diagonal matrix Λ.

Consider the system

$$\dot{\mathbf{q}} = \mathbf{Aq} + \mathbf{Bx}$$

We shall assume that $\lambda_1, \lambda_2, \ldots, \lambda_N$, the eigenvalues of $\mathbf{A}$, are distinct (no repeated roots). Let us transform the state vector $\mathbf{q}$ into the new state vector $\mathbf{z}$, using the transformation

$$\mathbf{z} = \mathbf{Pq} \tag{10.73a}$$

Then, after the development of Eq. (10.68c), we have

$$\dot{\mathbf{z}} = \mathbf{PAP}^{-1}\mathbf{z} + \mathbf{PBx} \tag{10.73b}$$

We desire the transformation to be such that $\mathbf{PAP}^{-1}$ is a diagonal matrix Λ given by Eq. (10.72), or

$$\dot{\mathbf{z}} = \Lambda\mathbf{z} + \hat{\mathbf{B}}\mathbf{x} \tag{10.73c}$$

Hence

$$\Lambda = \mathbf{PAP}^{-1} \tag{10.74a}$$

or

$$\Lambda\mathbf{P} = \mathbf{PA} \tag{10.74b}$$

We know Λ and $\mathbf{A}$. Equation (10.74b) therefore can be solved to determine $\mathbf{P}$.

EXAMPLE 10.10

Find the diagonalized form of the state equation for the system in Example 10.9.

In this case,

$$\mathbf{A} = \begin{bmatrix} 0 & 1 \\ -2 & -3 \end{bmatrix}$$

We found $\lambda_1 = -1$ and $\lambda_2 = -2$. Hence

$$\Lambda = \begin{bmatrix} -1 & 0 \\ 0 & -2 \end{bmatrix}$$

and Eq. (10.74b) becomes

$$\begin{bmatrix} -1 & 0 \\ 0 & -2 \end{bmatrix} \begin{bmatrix} p_{11} & p_{12} \\ p_{21} & p_{22} \end{bmatrix} = \begin{bmatrix} p_{11} & p_{12} \\ p_{21} & p_{22} \end{bmatrix} \begin{bmatrix} 0 & 1 \\ -2 & -3 \end{bmatrix}$$

Equating the four elements on two sides, we obtain

$$-p_{11} = -2p_{12} \tag{10.75a}$$

$$-p_{12} = p_{11} - 3p_{12} \tag{10.75b}$$

$$-2p_{21} = -2p_{22} \tag{10.75c}$$

$$-2p_{22} = p_{21} - 3p_{22} \tag{10.75d}$$

The reader will immediately recognize that Eqs. (10.75a) and (10.75b) are identical. Similarly, Eqs. (10.75c) and (10.75d) are identical. Hence two equations may be discarded, leaving us with only two equations [Eqs. (10.75a) and (10.75c)] and four unknowns. This observation means that there is no unique solution. There is, in fact, an infinite number of solutions. We can assign any value to p_{11} and p_{21} to yield one possible solution.[†] If $p_{11} = k_1$ and $p_{21} = k_2$, then from Eqs. (10.75a) and (10.75c) we have $p_{12} = k_1/2$ and $p_{22} = k_2$:

$$\mathbf{P} = \begin{bmatrix} k_1 & \frac{k_1}{2} \\ k_2 & k_2 \end{bmatrix} \tag{10.75e}$$

We may assign any values to k_1 and k_2. For convenience, let $k_1 = 2$ and $k_2 = 1$. This substitution yields

$$\mathbf{P} = \begin{bmatrix} 2 & 1 \\ 1 & 1 \end{bmatrix} \tag{10.75f}$$

The transformed variables [Eq. (10.73a)] are

$$\begin{bmatrix} z_1 \\ z_2 \end{bmatrix} = \begin{bmatrix} 2 & 1 \\ 1 & 1 \end{bmatrix} \begin{bmatrix} q_1 \\ q_2 \end{bmatrix} = \begin{bmatrix} 2q_1 + q_2 \\ q_1 + q_2 \end{bmatrix} \tag{10.76}$$

Thus, the new state variables z_1 and z_2 are related to q_1 and q_2 by Eq. (10.76). The system equation with $\mathbf{z}$ as the state vector is given by [see Eq. (10.73c)]

$$\dot{\mathbf{z}} = \Lambda\mathbf{z} + \hat{\mathbf{B}}\mathbf{x}$$

where

$$\hat{\mathbf{B}} = \mathbf{PB} = \begin{bmatrix} 2 & 1 \\ 1 & 1 \end{bmatrix} \begin{bmatrix} 1 \\ 2 \end{bmatrix} = \begin{bmatrix} 4 \\ 3 \end{bmatrix}$$

[†]If, however, we want the state equations in diagonalized form, as in Eq. (10.30a), where all the elements of $\hat{\mathbf{B}}$ matrix are unity, there is a unique solution. The reason is that the equation $\hat{\mathbf{B}} = \mathbf{PB}$, where all the elements of $\hat{\mathbf{B}}$ are unity, imposes additional constraints. In the present example, this condition will yield $p_{11} = 1/2$, $p_{12} = 1/4$, $p_{21} = 1/3$, and $p_{22} = 1/3$. The relationship between $\mathbf{z}$ and $\mathbf{q}$ is then

$$z_1 = \tfrac{1}{2}q_1 + \tfrac{1}{4}q_2 \quad \text{and} \quad z_2 = \tfrac{1}{3}q_1 + \tfrac{1}{3}q_2$$

Hence

$$\begin{bmatrix} \dot{z}_1 \\ \dot{z}_2 \end{bmatrix} = \begin{bmatrix} -1 & 0 \\ 0 & -2 \end{bmatrix} \begin{bmatrix} z_1 \\ z_2 \end{bmatrix} + \begin{bmatrix} 4 \\ 3 \end{bmatrix} x \tag{10.77a}$$

or

$$\dot{z}_1 = -z_1 + 4x$$

$$\dot{z}_2 = -2z_2 + 3x \tag{10.77b}$$

Note the distinctive nature of these state equations. Each state equation involves only one variable and therefore can be solved by itself. A general state equation has the derivative of one state variable equal to a linear combination of all state variables. Such is not the case with the diagonalized matrix Λ. Each state variable z_i is chosen so that it is uncoupled from the rest of the variables; hence a system with N eigenvalues is split into N decoupled systems, each with an equation of the form

$$\dot{z}_i = \lambda_i z_i + (\text{input terms})$$

This fact also can be readily seen from Fig. 10.7a, which is a realization of the system represented by Eq. (10.77). In contrast, consider the original state equations [see Eq. 10.70a)]

$$\dot{q}_1 = q_2 + x(t)$$

$$\dot{q}_2 = -2q_1 - 3q_2 + 2x(t)$$

A realization for these equations is shown in Fig. 10.7b. It can be seen from Fig. 10.7a that the states z_1 and z_2 are decoupled, whereas the states q_1 and q_2 (Fig. 10.7b) are coupled. It should be remembered that Fig. 10.7a and 10.7b are realizations of the same system.[†]

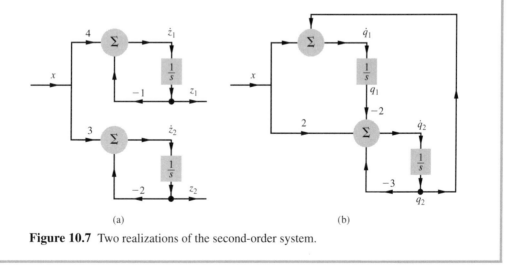

(a)   (b)

Figure 10.7 Two realizations of the second-order system.

[†]Here we only have a simulated state equation; the outputs are not shown. The outputs are linear combinations of state variables (and inputs). Hence, the output equation can be easily incorporated into these diagrams.

COMPUTER EXAMPLE C10.5

Repeat Example 10.10 using MATLAB. [Caution: Neither $\mathbf{P}$ nor $\hat{\mathbf{B}}$ is unique.]

```
>> A = [0 1;-2 -3]; B = [1; 2];
>> [V, Lambda] = eig(A);
>> P = inv(V), Lambda, Bhat = P*B

P =      2.8284     1.4142
         2.2361     2.2361

Lambda =     -1        0
              0       -2

Bhat =      5.6569
            6.7082
```

Therefore,

$$\mathbf{z} = \begin{bmatrix} z_1 \\ z_2 \end{bmatrix} = \begin{bmatrix} 2.8284 & 1.4142 \\ 2.2361 & 2.2361 \end{bmatrix} \begin{bmatrix} q_1 \\ q_2 \end{bmatrix} = \mathbf{Pq}$$

and

$$\dot{\mathbf{z}} = \begin{bmatrix} \dot{z}_1 \\ \dot{z}_2 \end{bmatrix} = \begin{bmatrix} -1 & 0 \\ 0 & -2 \end{bmatrix} \begin{bmatrix} z_1 \\ z_2 \end{bmatrix} + \begin{bmatrix} 5.6569 \\ 6.7082 \end{bmatrix} x(t) = \Lambda \mathbf{z} + \hat{\mathbf{B}}\mathbf{x}$$

10.5 CONTROLLABILITY AND OBSERVABILITY

Consider a diagonalized state-space description of a system

$$\dot{\mathbf{z}} = \Lambda \mathbf{z} + \hat{\mathbf{B}}\mathbf{x} \tag{10.78a}$$

and

$$\mathbf{Y} = \hat{\mathbf{C}}\mathbf{z} + \mathbf{Dx} \tag{10.78b}$$

We shall assume that all N eigenvalues $\lambda_1, \lambda_2, \ldots, \lambda_N$ are distinct. The state equations (10.78a) are of the form

$$\dot{z}_m = \lambda_m z_m + \hat{b}_{m1} x_1 + \hat{b}_{m2} x_2 + \cdots + \hat{b}_{mj} x_j \qquad m = 1, 2, \ldots, N$$

If $\hat{b}_{m1}, \hat{b}_{m2}, \ldots, \hat{b}_{mj}$ (the mth row in matrix $\hat{\mathbf{B}}$) are all zero, then

$$\dot{z}_m = \lambda_m z_m$$

and the variable z_m is uncontrollable because z_m is not coupled to any of the inputs. Moreover, z_m is decoupled from all the remaining $(N-1)$ state variables because of the diagonalized nature of the variables. Hence, there is no direct or indirect coupling of z_m with any of the inputs, and the system is uncontrollable. In contrast, if at least one element in the mth row of $\hat{\mathbf{B}}$ is nonzero, z_m

is coupled to at least one input and is therefore controllable. *Thus, a system with a diagonalized state [Eqs. (10.78)] is completely controllable if and only if the matrix $\hat{\mathbf{B}}$ has no row of zero elements.*

The outputs [Eq. (10.78b)] are of the form

$$y_i = \hat{c}_{i1}z_1 + \hat{c}_{i2}z_2 + \cdots + \hat{c}_{iN}z_N + \sum_{m=1}^{j} d_{im}x_m \qquad i = 1, 2, \ldots, k$$

If $\hat{c}_{im} = 0$, then the state z_m will not appear in the expression for y_i. Since all the states are decoupled because of the diagonalized nature of the equations, the state z_m cannot be observed directly or indirectly (through other states) at the output y_i. Hence the mth mode $e^{\lambda_m t}$ will not be observed at the output y_i. If $\hat{c}_{1m}, \hat{c}_{2m}, \ldots, \hat{c}_{km}$ (the mth column in matrix $\hat{\mathbf{C}}$) are all zero, the state z_m will not be observable at any of the k outputs, and the state z_m is unobservable. In contrast, if at least one element in the mth column of $\hat{\mathbf{C}}$ is nonzero, z_m is observable at least at one output. *Thus, a system with diagonalized equations of the form in Eqs. (10.78) is completely observable if and only if the matrix $\hat{\mathbf{C}}$ has no column of zero elements.* In this discussion, we assumed distinct eigenvalues; for repeated eigenvalues, the modified criteria can be found in the literature.[1,2]

If the state-space description is not in diagonalized form, it may be converted into diagonalized form using the procedure in Example 10.10. It is also possible to test for controllability and observability even if the state-space description is in undiagonalized form.[1,2]

EXAMPLE 10.11

Investigate the controllability and observability of the systems in Fig. 10.8.

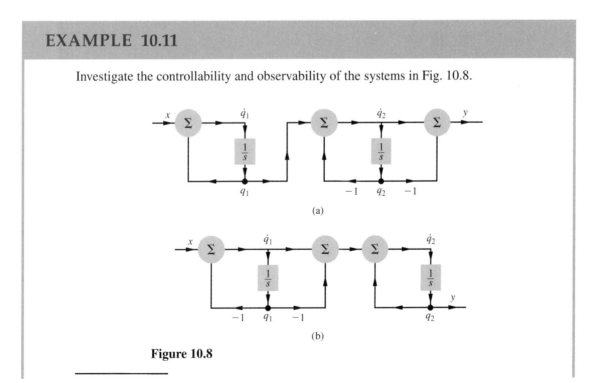

(a)

(b)

Figure 10.8

In both cases, the state variables are identified as the two integrator outputs, q_1 and q_2. The state equations for the system in Fig. 10.8a are

$$\dot{q}_1 = q_1 + x$$
$$\dot{q}_2 = q_1 - q_2 \tag{10.79}$$

and

$$y = \dot{q}_2 - q_2 = q_1 - 2q_2$$

Hence

$$\mathbf{A} = \begin{bmatrix} 1 & 0 \\ 1 & -1 \end{bmatrix} \qquad \mathbf{B} = \begin{bmatrix} 1 \\ 0 \end{bmatrix} \qquad \mathbf{C} = [1 \quad -2] \qquad \mathbf{D} = 0$$

$$|s\mathbf{I} - \mathbf{A}| = \begin{vmatrix} s - 1 & 0 \\ -1 & s + 1 \end{vmatrix} = (s - 1)(s + 1)$$

Therefore

$$\lambda_1 = 1 \qquad \text{and} \qquad \lambda_2 = -1$$

and

$$\Lambda = \begin{bmatrix} 1 & 0 \\ 0 & -1 \end{bmatrix} \tag{10.80}$$

We shall now use the procedure in Section 10.4-1 to diagonalize this system. According to Eq. (10.74b), we have

$$\begin{bmatrix} 1 & 0 \\ 0 & -1 \end{bmatrix} \begin{bmatrix} p_{11} & p_{12} \\ p_{21} & p_{22} \end{bmatrix} = \begin{bmatrix} p_{11} & p_{12} \\ p_{21} & p_{22} \end{bmatrix} \begin{bmatrix} 1 & 0 \\ 1 & -1 \end{bmatrix}$$

The solution of this equation yields

$$p_{12} = 0 \qquad \text{and} \qquad -2p_{21} = p_{22}$$

Choosing $p_{11} = 1$ and $p_{21} = 1$, we have

$$\mathbf{P} = \begin{bmatrix} 1 & 0 \\ 1 & -2 \end{bmatrix}$$

and

$$\hat{\mathbf{B}} = \mathbf{PB} = \begin{bmatrix} 1 & 0 \\ 1 & -2 \end{bmatrix} \begin{bmatrix} 1 \\ 0 \end{bmatrix} = \begin{bmatrix} 1 \\ 1 \end{bmatrix} \tag{10.81a}$$

All the rows of $\hat{\mathbf{B}}$ are nonzero. Hence the system is controllable. Also,

$$\mathbf{Y} = \mathbf{Cq}$$
$$= \mathbf{CP}^{-1}\mathbf{z}$$
$$= \hat{\mathbf{C}}\mathbf{z} \tag{10.81b}$$

and

$$\hat{\mathbf{C}} = \mathbf{CP}^{-1} = [1 \quad -2]\begin{bmatrix} 1 & 0 \\ 1 & -2 \end{bmatrix}^{-1} = [1 \quad -2]\begin{bmatrix} 1 & 0 \\ \frac{1}{2} & -\frac{1}{2} \end{bmatrix} = [0 \quad 1] \qquad (10.81c)$$

The first column of $\hat{\mathbf{C}}$ is zero. Hence the mode z_1 (corresponding to $\lambda_1 = 1$) is unobservable. The system is therefore controllable but not observable. We come to the same conclusion by realizing the system with the diagonalized state variables z_1 and z_2, whose state equations are

$$\dot{\mathbf{z}} = \mathbf{\Lambda z} + \hat{\mathbf{B}} x$$

$$y = \hat{\mathbf{C}} \mathbf{z}$$

According to Eqs. (10.80) and (10.81), we have

$$\dot{z}_1 = z_1 + x$$

$$\dot{z}_2 = -z_2 + x$$

and

$$y = z_2$$

Figure 10.9a shows a realization of these equations. It is clear that each of the two modes is controllable, but the first mode (corresponding to $\lambda = 1$) is not observable at the output.

The state equations for the system in Fig. 10.8b are

$$\dot{q}_1 = -q_1 + x$$

$$\dot{q}_2 = \dot{q}_1 - q_1 + q_2 = -2q_1 + q_2 + x \qquad (10.82)$$

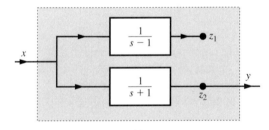

(a)

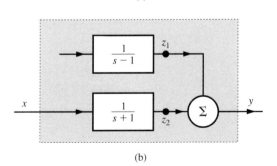

(b)

Figure 10.9 Equivalents of the systems in Fig. 10.8.

and

$$y = q_2$$

Hence

$$\mathbf{A} = \begin{bmatrix} -1 & 0 \\ -2 & 1 \end{bmatrix} \quad \mathbf{B} = \begin{bmatrix} 1 \\ 1 \end{bmatrix} \quad \mathbf{C} = \begin{bmatrix} 0 & 1 \end{bmatrix} \quad \mathbf{D} = 0$$

$$|s\mathbf{I} - \mathbf{A}| = \begin{vmatrix} s+1 & 0 \\ -1 & s-1 \end{vmatrix} = (s+1)(s-1)$$

so that $\lambda_1 = -1$, $\lambda_2 = 1$, and

$$\Lambda = \begin{bmatrix} -1 & 0 \\ 0 & 1 \end{bmatrix} \tag{10.83}$$

Diagonalizing the matrix, we have

$$\begin{bmatrix} 1 & 0 \\ 0 & -1 \end{bmatrix} \begin{bmatrix} p_{11} & p_{12} \\ p_{21} & p_{22} \end{bmatrix} = \begin{bmatrix} p_{11} & p_{12} \\ p_{21} & p_{22} \end{bmatrix} \begin{bmatrix} -1 & 0 \\ -2 & 1 \end{bmatrix}$$

The solution of this equation yields $p_{11} = -p_{12}$ and $p_{22} = 0$. Choosing $p_{11} = -1$ and $p_{21} = 1$, we obtain

$$\mathbf{P} = \begin{bmatrix} -1 & 1 \\ 1 & 0 \end{bmatrix}$$

and

$$\hat{\mathbf{B}} = \mathbf{PB} = \begin{bmatrix} -1 & 1 \\ 1 & 0 \end{bmatrix} \begin{bmatrix} 1 \\ 1 \end{bmatrix} = \begin{bmatrix} 0 \\ 1 \end{bmatrix} \tag{10.84a}$$

$$\hat{\mathbf{C}} = \mathbf{CP}^{-1} = \begin{bmatrix} 0 & 1 \end{bmatrix} \begin{bmatrix} 0 & 1 \\ 1 & 1 \end{bmatrix} = \begin{bmatrix} 1 & 1 \end{bmatrix} \tag{10.84b}$$

The first row of $\hat{\mathbf{B}}$ is zero. Hence the mode corresponding to $\lambda_1 = 1$ is not controllable. However, since none of the columns of $\hat{\mathbf{C}}$ vanish, both modes are observable at the output. Hence the system is observable but not controllable.

We reach the same conclusion by realizing the system with the diagonalized state variables z_1 and z_2. The two state equations are

$$\dot{\mathbf{z}} = \Lambda\mathbf{z} + \hat{\mathbf{B}}x$$

$$y = \hat{\mathbf{C}}\mathbf{z}$$

From Eqs. (10.83) and (10.84), we have

$$\dot{z}_1 = z_1$$

$$\dot{z}_2 = -z_2 + x$$

and thus

$$y = z_1 + z_2 \tag{10.85}$$

Figure 10.9b shows a realization of these equations. Clearly, each of the two modes is observable at the output, but the mode corresponding to $\lambda_1 = 1$ is not controllable.

COMPUTER EXAMPLE C10.6

Repeat Example 10.11 using MATLAB.

(a) System from Fig. 10.8a.

```
>> A = [1 0;1 -1]; B = [1; 0]; C = [1 -2];
>> [V, Lambda] = eig(A); P=inv(V);
>> disp('Part (a):'), Bhat = P*B, Chat = C*inv(P)
Part (a):

Bhat =    -0.5000
           1.1180

Chat =    -2      0
```

Since all the rows of Bhat ($\hat{\mathbf{B}}$) are nonzero, the system is controllable. However, one column of Chat ($\hat{\mathbf{C}}$) is zero, so one mode is unobservable.

(b) System from Fig. 10.8b.

```
>> A = [-1 0;-2 1]; B = [1; 1]; C = [0 1];
>> [V, Lambda] = eig(A); P=inv(V);
>> disp('Part (b):'), Bhat = P*B, Chat = C*inv(P),
Part (b):

Bhat =          0
           1.4142

Chat =     1.0000    0.7071
```

One of the rows of Bhat ($\hat{\mathbf{B}}$) is zero, so one mode is uncontrollable. Since all of the columns of Chat ($\hat{\mathbf{C}}$) are nonzero, the system is observable.

10.5-1 Inadequacy of the Transfer Function Description of a System

Example 10.11 demonstrates the inadequacy of the transfer function to describe an LTI system in general. The systems in Fig. 10.8a and 10.8b both have the same transfer function

$$H(s) = \frac{1}{s + 1}$$

Yet the two systems are very different. Their true nature is revealed in Fig. 10.9a and 10.9b, respectively. Both the systems are unstable, but their transfer function $H(s) = 1/(s+1)$ does not give any hint of it. Moreover, the systems are very different from the viewpoint of controllability and observability. The system in Fig. 10.8a is controllable but not observable, whereas the system in Fig. 10.8b is observable but not controllable.

The transfer function description of a system looks at a system only from the input and output terminals. Consequently, the transfer function description can specify only the part of the system that is coupled to the input and the output terminals. From Fig. 10.9a and 10.9b we see that in both cases only a part of the system that has a transfer function $H(s) = 1/(s + 1)$ is coupled to the input and the output terminals. This is why both systems have the same transfer function $H(s) = 1/(s + 1)$.

The state variable description (Eqs. 10.79 and 10.82), on the other hand, contains all the information about these systems to describe them completely. The reason is that the state variable description is an internal description, not the external description obtained from the system behavior at external terminals.

Apparently, the transfer function fails to describe these systems completely because the transfer functions of these systems have a common factor $s - 1$ in the numerator and denominator; this common factor is canceled out in the systems in Fig. 10.8, with a consequent loss of the information. Such a situation occurs when a system is uncontrollable and/or unobservable. If a system is both controllable and observable (which is the case with most of the practical systems) the transfer function describes the system completely. In such a case the internal and external descriptions are equivalent.

10.6 STATE-SPACE ANALYSIS OF DISCRETE-TIME SYSTEMS

We have shown that an Nth-order differential equation can be expressed in terms of N first-order differential equations. In the following analogous procedure, we show that a general Nth-order difference equation can be expressed in terms of N first-order difference equations.

Consider the z-transfer function

$$H[z] = \frac{b_0 z^N + b_1 z^{N-1} + \cdots + b_{N-1} z + b_N}{z^N + a_1 z^{N-1} + \cdots + a_{N-1} z + a_N} \tag{10.86a}$$

The input $x[n]$ and the output $y[n]$ of this system are related by the difference equation

$$(E^N + a_1 E^{N-1} + \cdots + a_{N-1} E + a_N) y[n]$$
$$= (b_0 E^N + b_1 E^{N-1} + \cdots + b_{N-1} E + b_N) x[n] \tag{10.86b}$$

The DFII realization of this equation is illustrated in Fig. 10.10. Signals appearing at the outputs of N delay elements are denoted by $q_1[n], q_2[n], \ldots, q_N[n]$. The input of the first delay is $q_N[n + 1]$. We can now write N equations, one at the input of each delay:

$$q_1[n + 1] = q_2[n]$$
$$q_2[n + 1] = q_3[n]$$
$$\vdots \tag{10.87}$$
$$q_{N-1}[n + 1] = q_N[n]$$
$$q_N[n + 1] = -a_N q_1[n] - a_{N-1} q_2[n] - \cdots - a_1 q_N[n] + x[n]$$

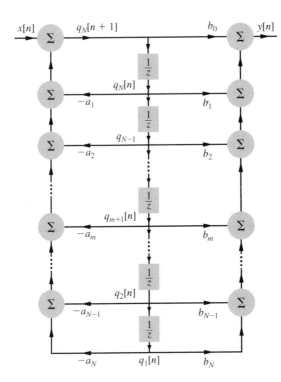

Figure 10.10 Direct form II realization of an Nth-order, discrete-time system.

and

$$y[n] = b_N q_1[n] + b_{N-1} q_2[n] + \cdots + b_1 q_N[n] + b_0 q_{N+1}[n]$$

We can eliminate $q_{N+1}[n]$ from this equation by using the last equation in the set (10.87) to yield

$$y[n] = (b_N - b_0 a_N) q_1[n] + (b_{N-1} - b_0 a_{N-1}) q_2[n] + \cdots + (b_1 - b_0 a_1) q_N[n] + b_0 x[n]$$

$$= \hat{b}_N q_1[n] + \hat{b}_{N-1} q_2[n] + \cdots + \hat{b}_1 q_N[n] + b_0 x[n] \qquad (10.88)$$

where $\hat{b}_i = b_i - b_0 a_i$.

Equations (10.87) are N first-order difference equations in N variables $q_1[n]$, $q_2[n]$, ..., $q_N[n]$. These variables should immediately be recognized as state variables, since the specification of the initial values of these variables in Fig. 10.10 will uniquely determine the response $y[n]$ for a given $x[n]$. Thus, Eqs. (10.87) represent the state equations, and Eq. (10.88) is the output equation. In matrix form we can write these equations as

$$\underbrace{\begin{bmatrix} q_1[n+1] \\ q_2[n+1] \\ \vdots \\ q_{N-1}[n+1] \\ q_N[n+1] \end{bmatrix}}_{\mathbf{q}[n+1]} = \underbrace{\begin{bmatrix} 0 & 1 & 0 & \cdots & 0 & 0 \\ 0 & 0 & 1 & \cdots & 0 & 0 \\ \vdots & \vdots & \vdots & \cdots & \vdots & \vdots \\ 0 & 0 & 0 & \cdots & 0 & 1 \\ -a_N & -a_{N-1} & -a_{N-2} & \cdots & -a_2 & -a_1 \end{bmatrix}}_{\mathbf{A}} \underbrace{\begin{bmatrix} q_1[n] \\ q_2[n] \\ \vdots \\ q_{N-1}[n] \\ q_N[n] \end{bmatrix}}_{\mathbf{q}[n]} + \underbrace{\begin{bmatrix} 0 \\ 0 \\ \vdots \\ 0 \\ 1 \end{bmatrix}}_{\mathbf{B}} x[n]$$

$$(10.89a)$$

and

$$\mathbf{y}[n] = \underbrace{\begin{bmatrix} \hat{b}_N & \hat{b}_{N-1} & \cdots & \hat{b}_1 \end{bmatrix}}_{\mathbf{C}} \begin{bmatrix} q_1[n] \\ q_2[n] \\ \vdots \\ q_N[n] \end{bmatrix} + \underbrace{b_0}_{\mathbf{D}} x[n] \qquad (10.89b)$$

In general,

$$\mathbf{q}[n+1] = \mathbf{Aq}[n] + \mathbf{Bx}[n] \qquad (10.90a)$$

$$\mathbf{y}[n] = \mathbf{Cq}[n] + \mathbf{Dx}[n] \qquad (10.90b)$$

Here we have represented a discrete-time system with state equations for DFII form. There are several other possible representations, as discussed in Section 10.2. We may, for example, use the cascade, parallel, or transpose of DFII forms to realize the system, or we may use some linear transformation of the state vector to realize other forms. In all cases, the output of each delay element qualifies as a state variable. We then write the equation at the input of each delay element. The N equations thus obtained are the N state equations.

10.6-1 Solution in State-Space

Consider the state equation

$$\mathbf{q}[n+1] = \mathbf{Aq}[n] + \mathbf{Bx}[n] \qquad (10.91)$$

From this equation it follows that

$$\mathbf{q}[n] = \mathbf{Aq}[n-1] + \mathbf{Bx}[n-1] \qquad (10.92a)$$

and

$$\mathbf{q}[n-1] = \mathbf{Aq}[n-2] + \mathbf{Bx}[n-2] \qquad (10.92b)$$

$$\mathbf{q}[n-2] = \mathbf{Aq}[n-3] + \mathbf{Bx}[n-3] \qquad (10.92c)$$

$$\vdots$$

$$\mathbf{q}[1] = \mathbf{Aq}[0] + \mathbf{Bx}[0]$$

Substituting Eq. (10.92b) in Eq. (10.92a), we obtain

$$\mathbf{q}[n] = \mathbf{A}^2\mathbf{q}[n-2] + \mathbf{ABx}[n-2] + \mathbf{Bx}[n-1]$$

Substituting Eq. (10.92c) in this equation, we obtain

$$\mathbf{q}[n] = \mathbf{A}^3\mathbf{q}[n-3] + \mathbf{A}^2\mathbf{Bx}[n-3] + \mathbf{ABx}[n-2] + \mathbf{Bx}[n-1]$$

Continuing in this way, we obtain

$$\mathbf{q}[n] = \mathbf{A}^n\mathbf{q}[0] + \mathbf{A}^{n-1}\mathbf{Bx}[0] + \mathbf{A}^{n-2}\mathbf{Bx}[1] + \cdots + \mathbf{Bx}[n-1]$$

$$= \mathbf{A}^n\mathbf{q}[0] + \sum_{m=0}^{n-1} \mathbf{A}^{n-1-m}\mathbf{Bx}[m] \qquad (10.93a)$$

The upper limit on the summation in Eq. (10.93a) is nonnegative. Hence $n \geq 1$, and the summation is recognized as the convolution sum

$$\mathbf{A}^{n-1}u[n-1] * \mathbf{B}\mathbf{x}[n]$$

Consequently

$$\mathbf{q}[n] = \underbrace{\mathbf{A}^n\mathbf{q}[0]}_{\text{zero input}} + \underbrace{\mathbf{A}^{n-1}u[n-1] * \mathbf{B}\mathbf{x}[n]}_{\text{zero state}} \tag{10.93b}$$

and

$$\mathbf{y}[n] = \mathbf{C}\mathbf{q} + \mathbf{D}\mathbf{x}$$

$$= \mathbf{C}\mathbf{A}^n\mathbf{q}[0] + \sum_{m=0}^{n-1} \mathbf{C}\mathbf{A}^{n-1-m}\mathbf{B}\mathbf{x}[m] + \mathbf{D}\mathbf{x} \tag{10.94a}$$

$$= \mathbf{C}\mathbf{A}^n\mathbf{q}[0] + \mathbf{C}\mathbf{A}^{n-1}u[n-1] * \mathbf{B}\mathbf{x}[n] + \mathbf{D}\mathbf{x} \tag{10.94b}$$

In Section B.6-5, we showed that

$$\mathbf{A}^n = \beta_0\mathbf{I} + \beta_1\mathbf{A} + \beta_2\mathbf{A}^2 + \cdots + \beta_{N-1}\mathbf{A}^{N-1} \tag{10.95a}$$

where (assuming N distinct eigenvalues of $\mathbf{A}$)

$$
\begin{bmatrix} \beta_0 \\ \beta_1 \\ \vdots \\ \beta_{N-1} \end{bmatrix}
=
\begin{bmatrix}
1 & \lambda_1 & \lambda_1^2 & \cdots & \lambda_1^{N-1} \\
1 & \lambda_2 & \lambda_2^2 & \cdots & \lambda_2^{N-1} \\
\vdots & \vdots & \vdots & \cdots & \vdots \\
1 & \lambda_N & \lambda_N^2 & \cdots & \lambda_N^{N-1}
\end{bmatrix}^{-1}
\begin{bmatrix} \lambda_1^n \\ \lambda_2^n \\ \vdots \\ \lambda_N^n \end{bmatrix}
\tag{10.95b}
$$

and $\lambda_1, \lambda_2, \ldots, \lambda_N$ are the N eigenvalues of $\mathbf{A}$.

We can also determine $\mathbf{A}^n$ from the z-transform formula, which will be derived later, in Eq. (10.102):

$$\mathbf{A}^n = \mathcal{Z}^{-1}[(\mathbf{I} - z^{-1}\mathbf{A})^{-1}] \tag{10.95c}$$

EXAMPLE 10.12

Give a state-space description of the system in Fig. 10.11. Find the output $y[n]$ if the input $x[n] = u[n]$ and the initial conditions are $q_1[0] = 2$ and $q_2[0] = 3$.

Recognizing that $q_2[n] = q_1[n+1]$, the state equations are [see Eq. (10.89)]

$$
\begin{bmatrix} q_1[n+1] \\ q_2[n+1] \end{bmatrix}
=
\begin{bmatrix} 0 & 1 \\ -\frac{1}{6} & \frac{5}{6} \end{bmatrix}
\begin{bmatrix} q_1[n] \\ q_2[n] \end{bmatrix}
+
\begin{bmatrix} 0 \\ 1 \end{bmatrix} x
\tag{10.96a}
$$

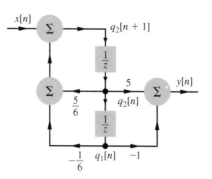

Figure 10.11

and

$$y[n] = \begin{bmatrix} -1 & 5 \end{bmatrix} \begin{bmatrix} q_1[n] \\ q_2[n] \end{bmatrix} \tag{10.96b}$$

To find the solution [Eq. (10.94)], we must first determine $\mathbf{A}^n$. The characteristic equation of $\mathbf{A}$ is

$$|\lambda \mathbf{I} - \mathbf{A}| = \begin{vmatrix} \lambda & -1 \\ \frac{1}{6} & \lambda - \frac{5}{6} \end{vmatrix} = \lambda^2 - \frac{5}{6}\lambda + \frac{1}{6} = \left(\lambda - \frac{1}{3}\right)\left(\lambda - \frac{1}{2}\right) = 0$$

Hence, $\lambda_1 = 1/3$ and $\lambda_2 = 1/2$ are the eigenvalues of $\mathbf{A}$ and [see Eq. (10.95a)]

$$\mathbf{A}^n = \beta_0 \mathbf{I} + \beta_1 \mathbf{A}$$

where [see Eq. (10.95b)]

$$\begin{bmatrix} \beta_0 \\ \beta_1 \end{bmatrix} = \begin{bmatrix} 1 & \frac{1}{3} \\ 1 & \frac{1}{2} \end{bmatrix}^{-1} \begin{bmatrix} \left(\frac{1}{3}\right)^n \\ \left(\frac{1}{2}\right)^n \end{bmatrix} = \begin{bmatrix} 3 & -2 \\ -6 & 6 \end{bmatrix} \begin{bmatrix} (3)^{-n} \\ (2)^{-n} \end{bmatrix} = \begin{bmatrix} 3(3)^{-n} - 2(2)^{-n} \\ -6(3)^{-n} + 6(2)^{-n} \end{bmatrix}$$

and

$$\mathbf{A}^n = [3(3)^{-n} - 2(2)^{-n}] \begin{bmatrix} 1 & 0 \\ 0 & 1 \end{bmatrix} + [-6(3)^{-n} + 6(2)^{-n}] \begin{bmatrix} 0 & 1 \\ -\frac{1}{6} & \frac{5}{6} \end{bmatrix}$$

$$= \begin{bmatrix} 3(3)^{-n} - 2(2)^{-n} & -6(3)^{-n} + 6(2)^{-n} \\ (3)^{-n} - (2)^{-n} & -2(3)^{-n} + 3(2)^{-n} \end{bmatrix} \tag{10.97}$$

We can now determine the state vector $\mathbf{q}[n]$ from Eq. (10.93b). Since we are interested in the output $y[n]$, we shall use Eq. (10.94b) directly. Note that

$$\mathbf{CA}^n = \begin{bmatrix} -1 & 5 \end{bmatrix} \mathbf{A}^n = [2(3)^{-n} - 3(2)^{-n} \quad -4(3)^{-n} + 9(2)^{-n}] \tag{10.98}$$

and the zero-input response is $\mathbf{CA}^n \mathbf{q}[0]$, with

$$\mathbf{q}[0] = \begin{bmatrix} 2 \\ 3 \end{bmatrix}$$

Hence, the zero-input response is

$$\mathbf{CA}^n \mathbf{q}[0] = -8(3)^{-n} + 21(2)^{-n} \tag{10.99a}$$

The zero-state component is given by the convolution sum of $\mathbf{CA}^{n-1}u[n-1]$ and $\mathbf{B}x[n]$. We can use the shifting property of the convolution sum [Eq. (9.46)] to obtain the zero-state component by finding the convolution sum of $\mathbf{CA}^n u[n]$ and $\mathbf{B}x[n]$ and then replacing n with $n-1$ in the result. We use this procedure because the convolution sums are listed in Table 3.1 for functions of the type $x[n]u[n]$ rather than $x[n]u[n-1]$.

$$\mathbf{CA}^n u[n] * \mathbf{B}x[n] = [2(3)^{-n} - 3(2)^{-n} - 4(3)^{-n} + 9(2)^{-n}] * \begin{bmatrix} 0 \\ u[n] \end{bmatrix}$$

$$= -4(3)^{-n} * u[n] + 9(2)^{-n} * u[n]$$

Using Table 3.1 (pair 4), we obtain

$$\mathbf{CA}^n u[n] * \mathbf{B}x[n] = -4\left[\frac{1 - 3^{-(n+1)}}{1 - \frac{1}{3}}\right]u[n] + 9\left[\frac{1 - 2^{-(n+1)}}{1 - \frac{1}{2}}\right]u[n]$$

$$= \left[12 + 6\big(3^{-(n+1)}\big) - 18\big(2^{-(n+1)}\big)\right]u[n]$$

Now the desired (zero-state) response is obtained by replacing n by $n-1$. Hence

$$\mathbf{CA}^n u[n] * \mathbf{B}x[n-1] = [12 + 6(3)^{-n} - 18(2)^{-n}]u[n-1] \tag{10.99b}$$

It follows that

$$y[n] = [-8(3)^{-n} + 21(2)^{-n}u[n] + [12 + 6(3)^{-n} - 18(2)^{-n}]u[n-1] \tag{10.100a}$$

This is the desired answer. We can simplify this answer by observing that $12 + 6(3)^{-n} - 18(2)^{-n} = 0$ for $n = 0$. Hence, $u[n-1]$ may be replaced by $u[n]$ in Eq. (10.99b), and

$$y[n] = [12 - 2(3)^{-n} + 3(2)^{-n}]u[n] \tag{10.100b}$$

COMPUTER EXAMPLE C10.7

Use MATLAB to find a graphical solution for Example 10.12.

```
>> A = [0 1;-1/6 5/6]; B = [0; 1]; C = [-1 5]; D = 0;
>> sys = ss(A,B,C,D,-1); % Discrete-time state space model
>> N = 25; x = ones(1,N+1); n = (0:N); q0 = [2;3];
>> [y,q] = lsim(sys,x,n,q0); % Simulate output and state vector
>> clf; stem(n,y,'k'); xlabel('n'); ylabel('y[n]'); axis([-.5 25.5 11.5 13.5]);
```

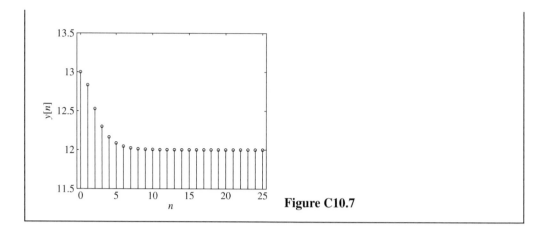

Figure C10.7

COMPUTER EXAMPLE C10.8

Use MATLAB to plot the zero-state response of the system in Example 10.12.

```
>> A = [0 1;-1/6 5/6]; B = [0; 1]; C = [-1 5]; D = 0;
>> N = 25; x = ones(1,N+1); n = (0:N);
>> [num,den] = ss2tf(A,B,C,D);
>> y = filter(num,den,x);
>> clf; stem(n,y,'k'); xlabel('n'); ylabel('y[n] (ZSR)');
>> axis([-.5 25.5 -0.5 12.5]);
```

Similar to Computer Example C10.7, it is also possible to compute the output by using:

```
>> A = [0 1;-1/6 5/6]; B = [0; 1]; C = [-1 5]; D = 0;
>> N = 25; x = ones(1,N+1); n = (0:N);
>> sys = ss(A,B,C,D,-1); y = lsim(sys,x,n);
>> clf; stem(n,y,'k'); xlabel('n'); ylabel('y[n] (ZSR)');
>> axis([-.5 25.5 -0.5 12.5]);
```

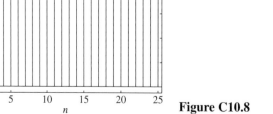

Figure C10.8

10.6-2 The z-Transform Solution

The z-transform of Eq. (10.91) is given by

$$z\mathbf{Q}[z] - z\mathbf{q}[0] = \mathbf{A}\mathbf{Q}[z] + \mathbf{B}\mathbf{X}[z]$$

Therefore

$$(z\mathbf{I} - \mathbf{A})\mathbf{Q}[z] = z\mathbf{q}[0] + \mathbf{B}\mathbf{X}[z]$$

and

$$\mathbf{Q}[z] = (z\mathbf{I} - \mathbf{A})^{-1}z\mathbf{q}[0] + (z\mathbf{I} - \mathbf{A})^{-1}\mathbf{B}\mathbf{X}[z]$$

$$= (\mathbf{I} - z^{-1}\mathbf{A})^{-1}\mathbf{q}[0] + (z\mathbf{I} - \mathbf{A})^{-1}\mathbf{B}\mathbf{X}[z] \qquad (10.101a)$$

Hence

$$\mathbf{q}[n] = \underbrace{\mathcal{Z}^{-1}[(\mathbf{I} - z^{-1}\mathbf{A})^{-1}]\mathbf{q}[0]}_{\text{zero-input component}} + \underbrace{\mathcal{Z}^{-1}[(z\mathbf{I} - \mathbf{A})^{-1}\mathbf{B}\mathbf{X}[z]]}_{\text{zero-state component}} \qquad (10.101b)$$

A comparison of Eq. (10.101b) with Eq. (10.93b) shows that

$$\mathbf{A}^{n} = \mathcal{Z}^{-1}[(\mathbf{I} - z^{-1}\mathbf{A})^{-1}] \qquad (10.102)$$

The output equation is given by

$$\mathbf{Y}[z] = \mathbf{C}\mathbf{Q}[z] + \mathbf{D}\mathbf{X}[z]$$

$$= \mathbf{C}[(\mathbf{I} - z^{-1}\mathbf{A})^{-1}\mathbf{q}[0] + (z\mathbf{I} - \mathbf{A})^{-1}\mathbf{B}\mathbf{X}[z]] + \mathbf{D}\mathbf{X}[z]$$

$$= \mathbf{C}(\mathbf{I} - z^{-1}\mathbf{A})^{-1}\mathbf{q}[0] + [\mathbf{C}(z\mathbf{I} - \mathbf{A})^{-1}\mathbf{B} + \mathbf{D}]\mathbf{X}[z]$$

$$= \underbrace{\mathbf{C}(\mathbf{I} - z^{-1}\mathbf{A})^{-1}\mathbf{q}[0]}_{\text{zero-input response}} + \underbrace{\mathbf{H}[z]\mathbf{X}[z]}_{\text{zero-state response}} \qquad (10.103a)$$

where

$$\mathbf{H}[z] = \mathbf{C}(z\mathbf{I} - \mathbf{A})^{-1}\mathbf{B} + \mathbf{D} \qquad (10.103b)$$

Note that $\mathbf{H}[z]$ is the transfer function matrix of the system, and $H_{ij}[z]$, the ijth element of $\mathbf{H}[z]$, is the transfer function relating the output $y_i[n]$ to the input $x_j[n]$. If we define $\mathbf{h}[n]$ as

$$\mathbf{h}[n] = \mathcal{Z}^{-1}[\mathbf{H}[z]]$$

then $\mathbf{h}[n]$ represents the unit impulse function response matrix of the system. Thus, $h_{ij}[n]$, the ijth element of $\mathbf{h}[n]$, represents the zero-state response $y_i[n]$ when the input $x_j[n] = \delta[n]$ and all other inputs are zero.

EXAMPLE 10.13

Use the z-transform to, find the response $y[n]$ for the system in Example 10.12.

According to Eq. (10.103a)

$$\mathbf{Y}[z] = [-1 \quad 5] \begin{bmatrix} 1 & -\frac{1}{z} \\ \frac{1}{6z} & 1 - \frac{5}{6z} \end{bmatrix}^{-1} \begin{bmatrix} 2 \\ 3 \end{bmatrix} + [-1 \quad 5] \begin{bmatrix} z & -1 \\ \frac{1}{6} & z - \frac{5}{6} \end{bmatrix}^{-1} \begin{bmatrix} 0 \\ \frac{z}{z-1} \end{bmatrix}$$

$$= [-1 \quad 5] \begin{bmatrix} \frac{z(6z-5)}{6z^2 - 5z + 1} & \frac{6z}{6z^2 - 5z + 1} \\ \frac{-z}{6z^2 - 5z + 1} & \frac{6z^2}{6z^2 - 5z + 1} \end{bmatrix} \begin{bmatrix} 2 \\ 3 \end{bmatrix} + [-1 \quad 5] \begin{bmatrix} \frac{z}{(z-1)\left(z^2 - \frac{5}{6}z + \frac{1}{6}\right)} \\ \frac{z^2}{(z-1)\left(z^2 - \frac{5}{6}z + \frac{1}{6}\right)} \end{bmatrix}$$

$$= \frac{13z^2 - 3z}{z^2 - \frac{5}{6}z + \frac{1}{6}} + \frac{(5z-1)z}{(z-1)\left(z^2 - \frac{5}{6}z + \frac{1}{6}\right)}$$

$$= \frac{-8z}{z - \frac{1}{3}} + \frac{21z}{z - \frac{1}{2}} + \frac{12z}{z-1} + \frac{12z}{z-1} + \frac{6z}{z - \frac{1}{3}} - \frac{18z}{z - \frac{1}{2}}$$

Therefore

$$y[n] = [\underbrace{-8(3)^{-n} + 21(2)^{-n}}_{\text{zero-input response}} + \underbrace{12 + 6(3)^{-n} - 18(2)^{-n}}_{\text{zero-state response}}]u[n]$$

LINEAR TRANSFORMATION, CONTROLLABILITY, AND OBSERVABILITY

The procedure for linear transformation is parallel to that in the continuous-time case (Section 10.4). If $\mathbf{w}$ is the transformed-state vector given by

$$\mathbf{w} = \mathbf{Pq}$$

then

$$\mathbf{w}[n+1] = \mathbf{PAP}^{-1}\mathbf{w}[n] + \mathbf{PBx}$$

and

$$\mathbf{y}[n] = (\mathbf{CP}^{-1})\mathbf{w} + \mathbf{Dx}$$

Controllability and observability may be investigated by diagonalizing the matrix, as explained in Section 10.4-1.

10.7 SUMMARY

An Nth-order system can be described in terms of N key variables—the state variables of the system. The state variables are not unique; rather, they can be selected in a variety of ways. Every possible system output can be expressed as a linear combination of the state variables and the inputs. Therefore the state variables describe the entire system, not merely the relationship between certain input(s) and output(s). For this reason, the state variable description is an internal description of the system. Such a description is therefore the most general system description, and it contains the information of the external descriptions, such as the impulse response and the transfer function. The state variable description can also be extended to time-varying-parameter systems and nonlinear systems. An external description of a system may not characterize the system completely.

The state equations of a system can be written directly from knowledge of the system structure, from the system equations, or from the block diagram representation of the system. State equations consist of a set of N first-order differential equations and can be solved by time-domain or frequency-domain (transform) methods. Suitable procedures exist to transform one given set of state variables into another. Because a set of state variables is not unique, we can have an infinite variety of state-space descriptions of the same system. The use of an appropriate transformation allows us to see clearly which of the system states are controllable and which are observable.

REFERENCES

1.   Kailath, Thomas. *Linear Systems.* Prentice-Hall, Englewood Cliffs, NJ, 1980.
2.   Zadeh, L., and C. Desoer. *Linear System Theory.* McGraw-Hill, New York, 1963.

MATLAB SESSION 10: TOOLBOXES AND STATE-SPACE ANALYSIS

The preceding MATLAB sessions provide a comprehensive introduction to the basic MATLAB environment. However, MATLAB also offers a wide range of toolboxes that perform specialized tasks. Once installed, toolbox functions operate no differently from ordinary MATLAB functions. Although toolboxes are purchased at extra cost, they save time and offer the convenience of predefined functions. It would take significant effort to duplicate a toolbox's functionality by using custom user-defined programs.

Three toolboxes are particularly appropriate in the study of signals and systems: the control system toolbox, the signal processing toolbox, and the symbolic math toolbox. Functions from these toolboxes have been utilized throughout the text in the Computer Exercises as well as certain homework problems. This session provides a more formal introduction to a selection of functions, both standard and toolbox, that are appropriate for state-space problems.

M10.1 z-Transform Solutions to Discrete-Time State-Space Systems

As with continuous-time systems, it is often more convenient to solve discrete-time systems in the transform domain rather than in the time domain. As given in Example 10.12, consider the state-space description of the system shown in Fig. 10.11.

$$\begin{bmatrix} q_1[n+1] \\ q_2[n+1] \end{bmatrix} = \begin{bmatrix} 0 & 1 \\ -\frac{1}{6} & \frac{5}{6} \end{bmatrix} \begin{bmatrix} q_1[n] \\ q_2[n] \end{bmatrix} + \begin{bmatrix} 0 \\ 1 \end{bmatrix} x[n]$$

and

$$y[n] = [-1 \quad 5] \begin{bmatrix} q_1[n] \\ q_2[n] \end{bmatrix}$$

We are interested in the output $y[n]$ in response to the input $x[n] = u[n]$ with initial conditions $q_1[0] = 2$ and $q_2[0] = 3$.

To describe this system, the state matrices $\mathbf{A}$, $\mathbf{B}$, $\mathbf{C}$, and $\mathbf{D}$ are first defined.

```
>> A = [0 1;-1/6 5/6]; B = [0; 1]; C = [-1 5]; D = 0;
```

Additionally, the vector of initial conditions is defined.

```
>> q_0 = [2;3];
```

In the transform domain, the solution to the state equation is

$$\mathbf{Q}[z] = (\mathbf{I} - z^{-1}\mathbf{A})^{-1}\mathbf{q}[0] + (z\mathbf{I} - \mathbf{A})^{-1}\mathbf{B}X[z] \qquad \text{(M10.1)}$$

The solution is separated into two parts: the zero-input component and the zero-state component.

MATLAB's symbolic toolbox makes possible a symbolic representation of Eq. (M10.1). First, a symbolic variable z needs to be defined.

```
>> z = sym('z');
```

The `sym` command is used to construct symbolic variables, objects, and numbers. Typing `whos` confirms that z is indeed a symbolic object. The `syms` command is a shorthand command for constructing symbolic objects. For example, `syms z s` is equivalent to the two instructions `z = sym('z');` and `s = sym('s');`.

Next, a symbolic expression for $X[z]$ needs to be constructed for the unit step input, $x[n] = u[n]$. The z-transform is computed by means of the `ztrans` command.

```
>> X = ztrans(sym('1'))
X = z/(z-1)
```

Several comments are in order. First, the `ztrans` command assumes a causal signal. For $n \geq 0$, $u[n]$ has a constant value of one. Second, the argument of `ztrans` needs to be a symbolic expression, even if the expression is a constant. Thus, a symbolic one `sym('1')` is required. Also note that continuous-time systems use Laplace transforms rather than z-transforms. In such cases, the `laplace` command replaces the `ztrans` command.

Construction of $\mathbf{Q}[z]$ is now trivial.

```
>> Q = inv(eye(2)-z^(-1)*A)*q_0 + inv(z*eye(2)-A)*B*X
Q =
[  2*(6*z-5)*z/(6*z^2-5*z+1)+18*z/(6*z^2-5*z+1)+6*z/(z-1)/(6*z^2-5*z+1)]
[     -2*z/(6*z^2-5*z+1)+18/(6*z^2-5*z+1)*z^2+6*z^2/(z-1)/(6*z^2-5*z+1)]
```

Unfortunately, not all MATLAB functions work with symbolic objects. Still, the symbolic toolbox overloads many standard MATLAB functions, such as `inv`, to work with symbolic objects. Recall that overloaded functions have identical names but different behavior; proper function selection is typically determined by context.

The expression Q is somewhat unwieldy. The simplify command uses various algebraic techniques to simplify the result.

```
>> Q = simplify(Q)
Q =
[ 2*z*(6*z^2-2*z-1)/(z-1)/(6*z^2-5*z+1)]
[ 2*z*(-7*z+1+9*z^2)/(z-1)/(6*z^2-5*z+1)]
```

The resulting expression is mathematically equivalent to the original but notationally more compact.

Since $\mathbf{D} = 0$, the output $Y[z]$ is given by $Y[z] = \mathbf{C}Q[z]$.

```
>> Y = simplify(C*Q)
Y = 6*z*(13*z^2-11*z+2)/(z-1)/(6*z^2-5*z+1)
```

The corresponding time-domain expression is obtained using the inverse z-transform command iztrans.

```
>> y = iztrans(Y)
y = 12+3*(1/2)^n-2*(1/3)^n
```

Like ztrans, the iztrans command assumes a causal signal so the result implies multiplication by a unit step. That is, the system output is $y[n] = (12 + 3(1/2)^n - 2(1/3)^n)u[n]$, which is equivalent to Eq. (10.100b) derived in Example 10.12. Continuous-time systems use inverse Laplace transforms rather than inverse z-transforms. In such cases, the ilaplace command therefore replaces the iztrans command.

Following a similar procedure, it is a simple matter to compute the zero-input response $y_0[n]$:

```
>> y_0 = iztrans(simplify(C*inv(eye(2)-z^(-1)*A)*q_0))
y_0 = 21*(1/2)^n-8*(1/3)^n
```

The zero-state response is given by

```
>> y-y_0
ans = 12-18*(1/2)^n+6*(1/3)^n
```

Typing iztrans(simplify(C*inv(z*eye(2)-A)*B*X)) produces the same result.

MATLAB plotting functions, such as plot and stem, do not directly support symbolic expressions. By using the subs command, however, it is easy to replace a symbolic variable with a vector of desired values.

```
>> n = [0:25]; stem(n,subs(y,n),'k'); xlabel('n'); ylabel('y[n]');
```

Figure M10.1 shows the results, which are equivalent to the results obtained using program CE10(7). Although there are plotting commands in the symbolic math toolbox such as ezplot that plot symbolic expression, these plotting routines lack the flexibility needed to satisfactorily plot discrete-time functions.

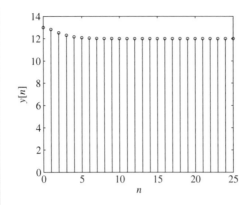

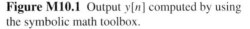

Figure M10.1 Output $y[n]$ computed by using the symbolic math toolbox.

M10.2 Transfer Functions from State-Space Representations

A system's transfer function provides a wealth of useful information. From Eq. (10.103b), the transfer function for the system described in Example 10.12 is:

```
>> H = simplify(C*inv(z*eye(2)-A)*B+D)
H = 6*(-1+5*z)/(6*z^2-5*z+1)
```

It is also possible to determine the numerator and denominator transfer function coefficients from a state-space model by using the signal processing toolbox function ss2tf.

```
>> [num,den] = ss2tf(A,B,C,D)
num = 0     5.0000    -1.0000
den = 1.0000    -0.8333     0.1667
```

The denominator of $H[z]$ provides the characteristic polynomial

$$\gamma^2 - \tfrac{5}{6}\gamma + \tfrac{1}{6}$$

Equivalently, the characteristic polynomial is the determinant of $(z\mathbf{I} - \mathbf{A})$.

```
>> syms gamma; char_poly = subs(det(z*eye(2)-A),z,gamma)
char_poly = gamma^2-5/6*gamma+1/6
```

Here, the subs command replaces the symbolic variable z with the desired symbolic variable gamma.

The roots command does not accommodate symbolic expressions. Thus, the sym2poly command converts the symbolic expression into a polynomial coefficient vector suitable for the roots command.

```
>> roots(sym2poly(char_poly))
ans = 0.5000
       0.3333
```

Taking the inverse z-transform of $H[z]$ yields the impulse response $h[n]$.

```
>> h = iztrans(H)
h = -6*charfcn[0](n)+18*(1/2)^n-12*(1/3)^n
```

As suggested by the characteristic roots, the characteristic modes of the system are $(1/2)^n$ and $(1/3)^n$. Notice that the symbolic math toolbox represents $\delta[n]$ as `charfcn[0](n)`. In general, $\delta[n-a]$ is represented as `charfnc[a](n)`. This notation is frequently encountered. Consider, for example, delaying the input $x[n] = u[n]$ by 2, $x[n-2] = u[n-2]$. In the transform domain, this is equivalent to $z^{-2}X[z]$. Taking the inverse z-transform of $z^{-2}X[z]$ yields:

```
>> iztrans(z^(-2)*X)
ans = -charfcn[1](n)-charfcn[0](n)+1
```

That is, MATLAB represents the delayed unit step $u[n-2]$ as $(-\delta[n-1] - \delta[n-0] + 1)u[n]$. The transfer function also permits convenient calculation of the zero-state response.

```
>> iztrans(H*X)
ans = -18*(1/2)^n+6*(1/3)^n+12
```

The result agrees with previous calculations.

M10.3 Controllability and Observability of Discrete-Time Systems

In their controllability and observability, discrete-time systems one analogous to continuous-time systems. For example, consider the LTID system described by the constant coefficient difference equation

$$y[n] + \tfrac{5}{6}y[n-1] + \tfrac{1}{6}y[n-2] = x[n] + \tfrac{1}{2}x[n-1]$$

Figure M10.2 illustrates the direct form II (DFII) realization of this system. The system input is $x[n]$, the system output is $y[n]$, and the outputs of the delay blocks are designated as state variables $q_1[n]$ and $q_2[n]$.

The corresponding state and output equations (see Prob. 10.M-1) are

$$\mathbf{Q}[n+1] = \begin{bmatrix} q_1[n+1] \\ q_2[n+1] \end{bmatrix} = \begin{bmatrix} 0 & 1 \\ -\tfrac{1}{6} & -\tfrac{5}{6} \end{bmatrix} \begin{bmatrix} q_1[n] \\ q_2[n] \end{bmatrix} + \begin{bmatrix} 0 \\ 1 \end{bmatrix} x[n] = \mathbf{A}\mathbf{Q}[n] + \mathbf{B}x[n]$$

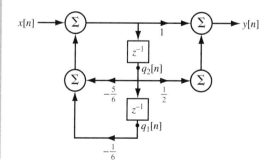

Figure M10.2 Direct form II realization of $y[n] + (5/6)y[n-1] + (1/6)y[n-2] = x[n] + (1/2)x[n-1]$.

and

$$y[n] = \begin{bmatrix} -\frac{1}{6} & -\frac{1}{3} \end{bmatrix} \begin{bmatrix} q_1[n] \\ q_2[n] \end{bmatrix} + 1x[n] = \mathbf{C}\mathbf{Q}[n] + \mathbf{D}x[n]$$

To describe this system in MATLAB, the state matrices $\mathbf{A}$, $\mathbf{B}$, $\mathbf{C}$, and $\mathbf{D}$ are first defined.

```
>> A = [0 1;-1/6 -5/6]; B = [0; 1]; C = [-1/6 -1/3]; D = 1;
```

To assess the controllability and observability of this system, the state matrix $\mathbf{A}$ needs to be diagonalized.[†] As shown in Eq. (10.74b), this requires a transformation matrix $\mathbf{P}$ such that

$$\mathbf{PA} = \mathbf{\Lambda P} \qquad\qquad (\text{M10.2})$$

where $\mathbf{\Lambda}$ is a diagonal matrix containing the unique eigenvalues of $\mathbf{A}$. Recall, the transformation matrix $\mathbf{P}$ is not unique.

To determine a matrix $\mathbf{P}$, it is helpful to review the eigenvalue problem. Mathematically, an eigendecomposition of $\mathbf{A}$ is expressed as

$$\mathbf{AV} = \mathbf{V\Lambda} \qquad\qquad (\text{M10.3})$$

where $\mathbf{V}$ is a matrix of eigenvectors and $\mathbf{\Lambda}$ is a diagonal matrix of eigenvalues. Pre- and post-multiplying both sides of Eq. (M10.3) by $\mathbf{V}^{-1}$ yields

$$\mathbf{V}^{-1}\mathbf{AVV}^{-1} = \mathbf{V}^{-1}\mathbf{V\Lambda V}^{-1}$$

Simplification yields

$$\mathbf{V}^{-1}\mathbf{A} = \mathbf{\Lambda V}^{-1} \qquad\qquad (\text{M10.4})$$

Comparing Eqs. (M10.2) and (M10.4), we see that a suitable transformation matrix $\mathbf{P}$ is given by an inverse eigenvector matrix $\mathbf{V}^{-1}$.

The `eig` command is used to verify that $\mathbf{A}$ has the required distinct eigenvalues as well as compute the needed eigenvector matrix $\mathbf{V}$.

```
>> [V,Lambda] = eig(A)
V =
    0.9487   -0.8944
   -0.3162    0.4472
Lambda =
   -0.3333         0
        0   -0.5000
```

Since the diagonal elements of Lambda are all unique, a transformation matrix $\mathbf{P}$ is given by:

```
>> P = inv(V);
```

[†]This approach requires that the state matrix $\mathbf{A}$ have unique eigenvalues. Systems with repeated roots require that state matrix $\mathbf{A}$ be transformed into a modified diagonal form, also called the Jordan form. The MATLAB function `jordan` is used in these cases.

The transformed state matrices $\hat{\mathbf{A}} = \mathbf{PAP}^{-1}$, $\hat{\mathbf{B}} = \mathbf{PB}$, and $\hat{\mathbf{C}} = \mathbf{CP}^{-1}$ are easily computed by using transformation matrix $\mathbf{P}$. Notice that matrix $\mathbf{D}$ is unaffected by state variable transformations.

```
>> Ahat = P*A*inv(P), Bhat = P*B, Chat = C*inv(P)
Ahat =
    -0.3333    -0.0000
     0.0000    -0.5000
Bhat =

     6.3246
     6.7082
Chat =
    -0.0527    -0.0000
```

The proper operation of $\mathbf{P}$ is verified by the correct diagonalization of $\mathbf{A}$, $\hat{\mathbf{A}} = \mathbf{\Lambda}$. Since no row of $\hat{\mathbf{B}}$ is zero, the system is controllable. Since, however, at least one column of $\hat{\mathbf{C}}$ is zero, the system is not observable. These characteristics are no coincidence. The DFII realization, which is more descriptively called the controller canonical form, is always controllable but not always observable.

As a second example, consider the same system realized using the transposed direct form II structure (TDFII), as shown in Fig. M10.3. The system input is $x[n]$, the system output is $y[n]$, and the outputs of the delay blocks are designated as state variables $v_1[n]$ and $v_2[n]$.

The corresponding state and output equations (see Prob. 10.M-2) are

$$\mathbf{V}[n + 1] = \begin{bmatrix} v_1[n + 1] \\ v_2[n + 1] \end{bmatrix} = \begin{bmatrix} 0 & -\frac{1}{6} \\ 1 & -\frac{5}{6} \end{bmatrix} \begin{bmatrix} v_1[n] \\ v_2[n] \end{bmatrix} + \begin{bmatrix} -\frac{1}{6} \\ -\frac{1}{3} \end{bmatrix} x[n] = \mathbf{AV}[n] + \mathbf{B}x[n]$$

and

$$y[n] = \begin{bmatrix} 0 & 1 \end{bmatrix} \begin{bmatrix} v_1[n] \\ v_2[n] \end{bmatrix} + 1x[n] = \mathbf{CV}[n] + \mathbf{D}x[n]$$

To describe this system in MATLAB, the state matrices $\mathbf{A}$, $\mathbf{B}$, $\mathbf{C}$, and $\mathbf{D}$ are defined.

```
>> A = [0 -1/6;1 -5/6]; B = [-1/6; -1/3]; C = [0 1]; D = 1;
```

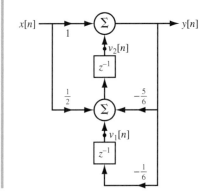

Figure M10.3 Transposed direct form II realization of $y[n] + (5/6)y[n-1] + (1/6)y[n-2] = x[n] + (1/2)x[n-1]$.

To diagonalize **A**, a transformation matrix **P** is created.

```
>> [V,Lambda] = eig(A)
V =
    0.4472    0.3162
    0.8944    0.9487
Lambda =
   -0.3333         0
        0   -0.5000
```

The characteristic modes of a system do not depend on implementation, so the eigenvalues of the DFII and TDFII realizations are the same. However, the eigenvectors of the two realizations are quite different. Since the transformation matrix **P** depends on the eigenvectors, different realizations can possess different observability and controllability characteristics.

Using transformation matrix **P**, the transformed state matrices $\hat{\mathbf{A}} = \mathbf{PAP}^{-1}$, $\hat{\mathbf{B}} = \mathbf{PB}$, and $\hat{\mathbf{C}} = \mathbf{CP}^{-1}$ are computed.

```
>> P = inv(V);
>> Ahat = P*A*inv(P), Bhat = P*B, Chat = C*inv(P)
Ahat =
   -0.3333         0
    0.0000   -0.5000
Bhat =
   -0.3727
   -0.0000
Chat =
    0.8944    0.9487
```

Again, the proper operation of **P** is verified by the correct diagonalization of **A**, $\hat{\mathbf{A}} = \mathbf{\Lambda}$. Since no column of $\hat{\mathbf{C}}$ is zero, the system is observable. However, at least one row of $\hat{\mathbf{B}}$ is zero, and therefore the system is not controllable. The TDFII realization, which is more descriptively called the observer canonical form, is always observable but not always controllable. It is interesting to note that the properties of controllability and observability are influenced by the particular realization of a system.

M10.4 Matrix Exponentiation and the Matrix Exponential

Matrix exponentiation is important to many problems, including the solution of discrete-time state-space equations. Equation (10.93b), for example, shows that the state response requires matrix exponentiation, $\mathbf{A}^n$. For a square **A** and specific n, MATLAB happily returns $\mathbf{A}^n$ by using the $^\wedge$ operator. From the system in Example 10.12 and $n = 3$, we have

```
>> A = [0 1;-1/6 5/6]; n = 3; A^n
ans =
   -0.1389    0.5278
   -0.0880    0.3009
```

The same result is also obtained by typing A*A*A.

Often, it is useful to solve $\mathbf{A}^n$ symbolically. Noting $\mathbf{A}^n = \mathcal{Z}^{-1}[(\mathbf{I} - z^{-1}\mathbf{A})^{-1}]$, the symbolic toolbox can produce a symbolic expression for $\mathbf{A}^n$.

```
>> syms z n; An = simplify(iztrans(inv(eye(2)-z^(-1)*A)))
An =
[ -2^(1-n)+3^(1-n),  6*2^(-n)-6*3^(-n)]
[ -2^(-n)+3^(-n),    3*2^(-n)-2*3^(-n)]
```

Notice that this result is identical to Eq. (10.97), derived earlier. Substituting the case $n = 3$ into An provides a result that is identical to the one elicited by the previous A^n command.

```
>> subs(An,n,3)
ans =
   -0.1389    0.5278
   -0.0880    0.3009
```

For continuous-time systems, the matrix exponential $e^{\mathbf{A}t}$ is commonly encountered. The expm command can compute the matrix exponential symbolically. Using the system from Example 10.7 yields:

```
>> syms t; A = [-12 2/3;-36 -1]; eAt = simplify(expm(A*t))
eAt =
[   -3/5*exp(-4*t)+8/5*exp(-9*t), -2/15*exp(-9*t)+2/15*exp(-4*t)]
[ 36/5*exp(-9*t)-36/5*exp(-4*t),    8/5*exp(-4*t)-3/5*exp(-9*t)]
```

This result is identical to the result given in Eq. (10.60). Similar to the discrete-time case, an identical result is obtained by typing syms s; simplify(ilaplace(inv(s*eye(2)-A))).

For a specific t, the matrix exponential is also easy to compute, either through substitution or direct computation. Consider the case $t = 3$.

```
>> subs(eAt,t,3)
ans = 1.0e-004 *
   -0.0369    0.0082
   -0.4424    0.0983
```

The command expm(A*3) produces the same result.

PROBLEMS

10.1-1 Convert each of the following second-order differential equations into a set of two first-order differential equations (state equations). State which of the sets represent nonlinear equations.

(a) $\ddot{y} + 10\dot{y} + 2y = x$

(b) $\ddot{y} + 2e^y \dot{y} + \log y = x$

(c) $\ddot{y} + \phi_1(y)\dot{y} + \phi_2(y)y = x$

10.2-1 Write the state equations for the RLC network in Fig. P10.2-1.

10.2-2 Write the state and output equations for the network in Fig. P10.2-2.

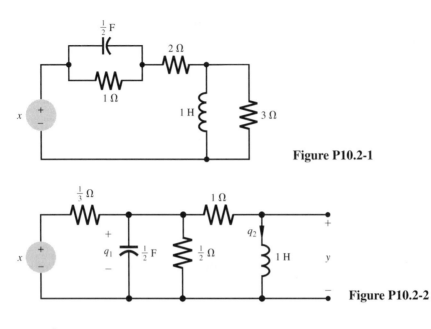

Figure P10.2-1

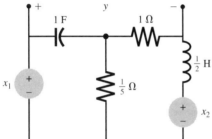

Figure P10.2-2

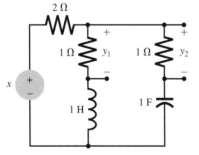

Figure P10.2-3

Figure P10.2-4

10.2-3 Write the state and output equations for the network in Fig. P10.2-3.

10.2-4 Write the state and output equations for the electrical network in Fig. P10.2-4.

10.2-5 Write the state and output equations for the network in Fig. P10.2-5.

10.2-6 Write the state and output equations of the system shown in Fig. P10.2-6.

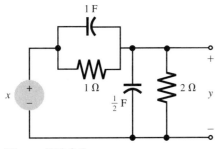

Figure P10.2-5

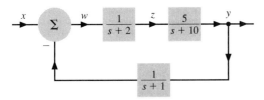

Figure P10.2-6

10.2-7 Write the state and output equations of the system shown in Fig. P10.2-7.

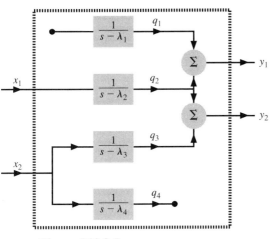

Figure P10.2-7

10.2-8 For a system specified by the transfer function

$$H(s) = \frac{3s + 10}{s^2 + 7s + 12}$$

write sets of state equations for DFII and its transpose, cascade, and parallel forms. Also write the corresponding output equations.

10.2-9 Repeat Prob. 10.2-8 for

(a)

$$H(s) = \frac{4s}{(s + 1)(s + 2)^2}$$

(b)

$$H(s) = \frac{s^3 + 7s^2 + 12s}{(s + 1)^3(s + 2)}$$

10.3-1 Find the state vector $\mathbf{q}(t)$ by using the Laplace transform method if

$$\dot{\mathbf{q}} = \mathbf{A}\mathbf{q} + \mathbf{B}x$$

where

$$\mathbf{A} = \begin{bmatrix} 0 & 2 \\ -1 & -3 \end{bmatrix} \qquad \mathbf{B} = \begin{bmatrix} 0 \\ 1 \end{bmatrix}$$

$$\mathbf{q}(0) = \begin{bmatrix} 2 \\ 1 \end{bmatrix} \qquad x(t) = 0$$

10.3-2 Repeat Prob. 10.3-1 for

$$\mathbf{A} = \begin{bmatrix} -5 & -6 \\ 1 & 0 \end{bmatrix} \qquad \mathbf{B} = \begin{bmatrix} 1 \\ 0 \end{bmatrix}$$

$$\mathbf{q}(0) = \begin{bmatrix} 5 \\ 4 \end{bmatrix} \qquad x(t) = \sin 100t$$

10.3-3 Repeat Prob. 10.3-1 for

$$\mathbf{A} = \begin{bmatrix} -2 & 0 \\ 1 & -1 \end{bmatrix} \qquad \mathbf{B} = \begin{bmatrix} 1 \\ 0 \end{bmatrix}$$

$$\mathbf{q}(0) = \begin{bmatrix} 0 \\ -1 \end{bmatrix} \qquad x(t) = u(t)$$

10.3-4 Repeat Prob. 10.3-1 for

$$\mathbf{A} = \begin{bmatrix} -1 & 1 \\ 0 & -2 \end{bmatrix} \qquad \mathbf{B} = \begin{bmatrix} 1 & 1 \\ 0 & 1 \end{bmatrix}$$

$$\mathbf{q}(0) = \begin{bmatrix} 1 \\ 2 \end{bmatrix} \qquad \mathbf{x} = \begin{bmatrix} u(t) \\ \delta(t) \end{bmatrix}$$

10.3-5 Use the Laplace transform method to find the response y for

$$\dot{\mathbf{q}} = \mathbf{A}\mathbf{q} + \mathbf{B}x(t)$$

$$y = \mathbf{C}\mathbf{q} + \mathbf{D}x(t)$$

where

$$\mathbf{A} = \begin{bmatrix} -3 & 1 \\ -2 & 0 \end{bmatrix} \qquad \mathbf{B} = \begin{bmatrix} 1 \\ 0 \end{bmatrix}$$

$$\mathbf{C} = [0 \quad 1] \qquad \mathbf{D} = 0$$

and

$$x(t) = u(t) \qquad \mathbf{q}(0) = \begin{bmatrix} 2 \\ 0 \end{bmatrix}$$

10.3-6 Repeat Prob. 10.3-5 for

$$\mathbf{A} = \begin{bmatrix} -1 & 1 \\ -1 & -1 \end{bmatrix} \qquad \mathbf{B} = \begin{bmatrix} 0 \\ 1 \end{bmatrix}$$

$$\mathbf{C} = [1 \quad 1] \qquad \mathbf{D} = 1$$

$$x(t) = u(t) \qquad \mathbf{q}(0) = \begin{bmatrix} 2 \\ 1 \end{bmatrix}$$

10.3-7 The transfer function $H(s)$ in Prob. 10.2-8 is realized as a cascade of $H_1(s)$ followed by $H_2(s)$, where

$$H_1(s) = \frac{1}{s+3}$$

$$H_2(s) = \frac{3s+10}{s+4}$$

Let the outputs of these subsystems be state variables q_1 and q_2, respectively. Write the state equations and the output equation for this system and verify that $\mathbf{H}(s) = \mathbf{C}\boldsymbol{\phi}(s)\mathbf{B} + \mathbf{D}$.

10.3-8 Find the transfer function matrix $\mathbf{H}(s)$ for the system in Prob. 10.3-5.

10.3-9 Find the transfer function matrix $\mathbf{H}(s)$ for the system in Prob. 10.3-6.

10.3-10 Find the transfer function matrix $\mathbf{H}(s)$ for the system

$$\dot{\mathbf{q}} = \mathbf{A}\mathbf{q} + \mathbf{B}\mathbf{x}$$

$$y = \mathbf{C}\mathbf{q} + \mathbf{D}\mathbf{x}$$

where

$$\mathbf{A} = \begin{bmatrix} 0 & 1 \\ -1 & -2 \end{bmatrix} \qquad \mathbf{B} = \begin{bmatrix} 0 & 1 \\ 1 & 0 \end{bmatrix} \qquad \mathbf{x} = \begin{bmatrix} x_1(t) \\ x_2(t) \end{bmatrix}$$

$$\mathbf{C} = \begin{bmatrix} 1 & 2 \\ 4 & 1 \\ 1 & 1 \end{bmatrix} \qquad \mathbf{D} = \begin{bmatrix} 0 & 0 \\ 0 & 0 \\ 1 & 0 \end{bmatrix}$$

10.3-11 Repeat Prob. 10.3-1, using the time-domain method.

10.3-12 Repeat Prob. 10.3-2, using the time-domain method.

10.3-13 Repeat Prob. 10.3-3, using the time-domain method.

10.3-14 Repeat Prob. 10.3-4, using the time-domain method.

10.3-15 Repeat Prob. 10.3-5, using the time-domain method.

10.3-16 Repeat Prob. 10.3-6, using the time-domain method.

10.3-17 Find the unit impulse response matrix $\mathbf{h}(t)$ for the system in Prob. 10.3-7, using Eq. (10.65).

10.3-18 Find the unit impulse response matrix $\mathbf{h}(t)$ for the system in Prob. 10.3-6.

10.3-19 Find the unit impulse response matrix $\mathbf{h}(t)$ for the system in Prob. 10.3-10.

10.4-1 The state equations of a certain system are given as

$$\dot{q}_1 = q_2 + 2x$$

$$\dot{q}_2 = -q_1 - q_2 + x$$

Define a new state vector $\mathbf{w}$ such that

$$w_1 = q_2$$

$$w_2 = q_2 - q_1$$

Find the state equations of the system with $\mathbf{w}$ as the state vector. Determine the characteristic roots (eigenvalues) of the matrix $\mathbf{A}$ in the original and the transformed state equations.

10.4-2 The state equations of a certain system are

$$\dot{q}_1 = q_2$$

$$\dot{q}_2 = -2q_1 - 3q_2 + 2x$$

(a) Determine a new state vector $\mathbf{w}$ (in terms of vector $\mathbf{q}$) such that the resulting state equations are in diagonalized form.

(b) For output **y** given by

$$\mathbf{y} = \mathbf{Cq} + \mathbf{Dx}$$

where

$$\mathbf{C} = \begin{bmatrix} 1 & 1 \\ -1 & 2 \end{bmatrix} \qquad \mathbf{D} = 0$$

determine the output **y** in terms of the new state vector **w**.

10.4-3 Given a system

$$\dot{\mathbf{q}} = \begin{bmatrix} 0 & 1 & 0 \\ 0 & 0 & 1 \\ 0 & -2 & -3 \end{bmatrix} \mathbf{q} + \begin{bmatrix} 0 \\ 0 \\ 1 \end{bmatrix} x$$

determine a new state vector **w** such that the state equations are diagonalized.

10.4-4 The state equations of a certain system are given in diagonalized form as

$$\dot{\mathbf{q}} = \begin{bmatrix} -1 & 0 & 0 \\ 0 & -3 & 0 \\ 0 & 0 & -2 \end{bmatrix} \mathbf{q} + \begin{bmatrix} 1 \\ 1 \\ 1 \end{bmatrix} x$$

The output equation is given by

$$y = \begin{bmatrix} 1 & 3 & 1 \end{bmatrix} \mathbf{q}$$

Determine the output y for

$$\mathbf{q}(0) = \begin{bmatrix} 1 \\ 2 \\ 1 \end{bmatrix} \qquad x(t) = u(t)$$

10.5-1 Write the state equations for the systems depicted in Fig. P10.5-1. Determine a new state vector **w** such that the resulting state equations are in diagonalized form. Write the output **y** in terms of **w**. Determine in each case whether the system is controllable and observable.

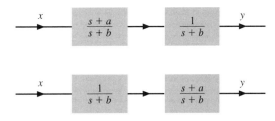

Figure P10.5-1

10.6-1 An LTI discrete-time system is specified by

$$\mathbf{A} = \begin{bmatrix} 2 & 0 \\ 1 & 1 \end{bmatrix} \qquad \mathbf{B} = \begin{bmatrix} 0 \\ 1 \end{bmatrix}$$

$$\mathbf{C} = \begin{bmatrix} 0 & 1 \end{bmatrix} \qquad \mathbf{D} = \begin{bmatrix} 1 \end{bmatrix}$$

and

$$\mathbf{q}(0) = \begin{bmatrix} 2 \\ 1 \end{bmatrix} \qquad x[n] = u[n]$$

(a) Find the output $y[n]$, using the time-domain method.

(b) Find the output $y[n]$, using the frequency-domain method.

10.6-2 An LTI discrete-time system is specified by the difference equation

$$y[n+2] + y[n+1] + 0.16y[n]$$
$$= x[n+1] + 0.32x[n]$$

(a) Show the DFII, its transpose, cascade, and parallel realizations of this system.

(b) Write the state and the output equations from these realizations, using the output of each delay element as a state variable.

10.6-3 Repeat Prob. 10.6-2 for

$$y[n+2] + y[n+1] - 6y[n]$$
$$= 2x[n+2] + x[n+1]$$

10.M-1 Verify the state and output equations for the LTID system shown in Fig. M.10-2.

10.M-2 Verify the state and output equations for the LTID system shown in Fig. M.10-3.

INDEX